# Principles and Applications of Organotransition Metal Chemistry

# Principles and Applications of Organotransition Metal Chemistry

James P. Collman
*Stanford University*

Louis S. Hegedus
*Colorado State University*

University Science Books
Mill Valley, California

Editor: Aidan Kelly
Design and Production: Dick Palmer
Typesetting: Engineering Research Center
Printing and Binding: The Maple-Vail Book Manufacturing Group

University Science Books
20 Edgehill Road
Mill Valley, CA 94941



Library of Congress Catalog Card Number: 79-57228
ISBN 0-935702-03-2

Printed in the United States of America
10 9 8 7 6 5 4 3

# Acknowledgements

This text could not have been written without the advice and help from many friends and associates. We are greatly indebted to them. The following read various chapters of the crude manuscript and in some cases provided unpublished material: Henry Baumgarten, John Bercaw, Bob Bergman, Terry Collins, Rick Finke, Dennis Forster, John Gladysz, Bob Grubbs, Jack Halpern, Roald Hoffmann, Jim Ibers, Rich Larock, Norvell Nelson, Kim Neuberg, Jack Norton, Gary Posner, Chris Reed, Warren Roper, Richard Schrock, Marty Semmelhack, Tom Sorrell, John Stille, Barry Trost, Donald Valentine, and Peter Vollhardt. The following individuals proofread various chapters of the final manuscript: Jim Belmont, Susana Bencosme, Joe Bozell, Eric Evitt, Don Korte, Karen Kosydar, Mike McGuire, Jim McKearin, Dave Olsen, Roger Pettman, Varaprath Sudarsanan, Dave Thompson, and Keith Woo. Jackie Swinehart typed the entire manuscript for copy editing and prepared the structural formulas aided by Burke Blevins. Sandy Page typed the camera-ready copy. John Choi did the drafting. There are undoubtedly many errors in the final copy. For these the authors take responsibility.

# Contents

## GLOSSARY OF ABBREVIATIONS

| | |
|---|---|
| Å | angstrom unit, $10^{-8}$ cm |
| Ac | acetyl, $CH_3CO$ |
| acac | acetylacetonate anion |
| AIBN | azoisobutyronitrile |
| Ar | aryl such as phenyl |
| bipy | 2,2'-dipyridine |
| BTMSA | _bis_-(trimethylsilyl)acetylene |
| _n_-Bu | _n_-butyl |
| _t_-Bu | _t_-butyl |
| Bz | benzyl |
| CG | coordination geometry |
| CIDNP | chemically induced dynamic nuclear polarization |
| $cm^{-1}$ | wave number |
| $Chg_2$ | _bis_-cyclohexylglyoximato |
| COD | 1,5-cyclooctadiene |
| Cp | $\pi$-cyclopentadienyl, $\eta^5$-$C_5H_5$ |
| diars | _o_-phenylenebisdimethylarsine, _o_-$C_6H_4(AsMe_2)_2$ |
| diglyme | diethyleneglycol dimethyl ether, $MeO(CH_2CH_2O)_2Me$ |
| DIOP | a chiral bidentate ligand, see p. 344 |
| diphos | 1,2-_bis_-(diphenylphosphino)ethane, dppe |
| DMF | N,N-dimethylformamide |
| $dmg_2$ | _bis_-dimethylglyoximato, see p. 141 |
| $d^n$ | formal d-electron configuration |
| DMSO | dimethylsulfoxide |
| ee | per cent enantiomorphic excess, see p. 342 |
| en | 1,2-diaminoethane, $(H_2NCH_2CH_2NH_2)$ |
| esr or epr | electron spin (or paramagnetic) resonance |
| Et | ethyl |
| Fp | $FeCp(CO)_2$ |
| glyme | ethylene glycol dimethyl ether, $MeOCH_2CH_2OMe$ |
| HMPA | hexamethylphosphoric triamide, $(Me_2N)_3PO$ |
| HOMO | highest occupied molecular orbital |
| Hz | hertz, $sec^{-1}$ |
| L | any unidentate ligand, but often $Ph_3P$ |
| LUMO | lowest unoccupied molecular orbital |
| M | the central metal in a complex or a metal on a surface |

| | |
|---|---|
| Me | methyl, $CH_3$ |
| Mes | mesityl, 1,3,5-trimethylphenyl |
| MO | molecular orbital |
| NMP | N-methylpyrrolidinone |
| nmr | nuclear magnetic resonance |
| Np | α-naphthyl |
| OAc | acetate anion |
| [O] | oxidizing conditions |
| Pc | phthalocyanine, see p. 141 |
| Ph | phenyl, $C_6H_5$ |
| $Ph_3P$ | triphenylphosphine |
| phen | 1,10-phenanthroline |
| $PPN^+$ | $[(Ph_3P)_2N]^+$ |
| Pr | *n*-propyl |
| i-Pr | isopropyl |
| py | pyridine |
| R | an alkyl group |
| $R_F$ | perfluoroalkyl group |
| salen | *bis*-salicylaldehyde-ethylenediimine, see p. 14 |
| THF | tetrahydrofuran |
| TMED | N,N,N',N'-tetramethylethylenediamine |
| TPP | *meso*-tetraphenylporphyrin, see p. 141 |
| ν | frequency in $cm^{-1}$ |
| μ | micron ($10^{-6}$ M) or magnetic moment in Bohr magnetons |
| X | a halogen |

# 1

# A Perspective

This text derives from a graduate course J. P. Collman has taught for the past twelve years at Stanford, and from courses and annual literature reviews presented by L. S. Hegedus at Colorado State University and elsewhere. During this period the chemistry of organotransition-metal complexes has experienced remarkable growth stimulated by industrial research on homogeneous catalysis, by the inorganic chemist's interest in structural diversity, and by the organic and inorganic chemist's search for new synthetic methods. We intend to introduce organic chemists to the reactions of organotransition-metal compounds. Herein emphasis is given to reactions and to reaction mechanism rather than to structure, bonding, or spectroscopic properties. This text is addressed to well-prepared senior chemistry majors, first-year graduate students, and organic chemists who plan to use transition-metal complexes as reagents or catalysts. This account is not comprehensive, but instructive, dealing with principles and illustrated with practical examples. Indeed, an exhaustive, all-inclusive treatise on organotransition-metal chemistry is currently well beyond the capacity of a single book or the authors.

Modern organic chemistry is usually organized by reaction classes and reaction mechanisms, with emphasis given to the stereochemistry of reactions. This book, with its emphasis on reactions and reaction mechanisms, is based on an analogous approach. It is intended both as an introductory text and as a concise review of catalytic and synthetic applications of organotransition-metal chemistry. To accomplish these dual purposes, in the first part of the text we have incorporated structural and mechanistic principles; and, in the later sections, we have illustrated these principles in the context of catalytic and stoichiometric reactions. Citations of the literature are given in square brackets in the text, and are gathered at the end of each chapter; however, the reader should be warned that these are not complete but are lead references. Whenever possible, we have listed significant reviews. Scientists who have made major contributions to organotransition-metal chemistry are identified in the text, not to herald these individuals, but to facilitate literature searches. The work of many chemists follows well-organized themes; so it is useful for a student to know who is likely to publish work on a particular topic.

Our understanding of reaction mechanism is continually evolving. New developments in homogeneous catalysis and synthetic methodology are often guided by concepts of reaction mechanism. Thus we have devoted a substantial portion of this text to mechanism. Mechanisms can serve as useful guides to plan new research themes. However, the student should be aware, and the experienced chemist reminded, that a mechanism is never proven, but is only a hypothesis which best fits presently known experimental data. In such a rapidly evolving topic as organotransition-metal chemistry, many mechanisms are obscure or at a primitive stage. There are many opportunities in this field for future mechanistic studies. We hope this account will stimulate further mechanisms research, as well as further applications to organic synthesis and homogeneous catalysis. We have learned from writing this text just as we hope students and chemists alike will learn from reading it.

The chemistry of low-valent transition-metal compounds is introduced in Chapter 2, where a rudimentary bonding scheme is given as a set of rules for "electron bookkeeping." This approach is based on the somewhat artificial, but very useful notion of <u>formal oxidation states</u>, which leads simply to the commonly employed <u>d-electron configurations</u> ($d^{\underline{n}}$). <u>Coordination numbers</u> (CN) are then introduced and shown to be simply related to $d^{\underline{n}}$. The concepts of <u>coordinative saturation</u> and <u>coordinative unsaturation</u> follow from this d-electron configuration. Idealized <u>coordination geometries</u> (CG) can also be related to $d^{\underline{n}}$ and coordination numbers. A list of characteristic ligands with assignments of their usual coordination numbers and formal charges is given; this list enables one immediately to classify the vast majority of transition-metal complexes in terms of the oxidation states of the central metal, the coordination number, and the probable coordination geometry. This classification sets the stage for a rational discussion of reaction classes.

Chapter 3 illustrates the vast scope of organotransition-metal compounds and provides examples of the structure-bonding principles developed in Chapter 2. Chapter

3 is organized according to the types of ligands which are found in transition-metal compounds. Both simple and novel ligands are covered, including $\pi$-acid groups. The latter are a distinctive feature of low-valent transition-metal compounds. Structural, chemical, and spectroscopic characteristics of individual ligands are mentioned, but are not discussed in detail. This provides background material which the reader may require to read the literature or to plan research in organometallic chemistry.

The two major categories of organotransition-metal reactions, oxidative-addition/reductive-elimination and insertions, are presented in Chapters 4 and 5, respectively. These reaction classes are meant to imply over-all chemical changes rather than specific reaction mechanisms. Our current understanding of the various mechanisms of these reactions is presented. The dominant path for an individual case depends on many factors: the metal, the substrate, the ligands, and the experimental conditions. We shall find that generalizations about mechanisms are misleading. Nevertheless, the over-all classification of these reaction types is useful in organizing the chemistry. In addition, particular mechanisms are frequently encountered and are signaled by certain experimental criteria. The feature which makes organotransition-metal chemistry different from that of other organometallic compounds is the importance of the oxidative-addition/reductive-elimination reactions discussed in Chapter 3. These require the variable occupancy of the highest occupied electronic levels, which are principally the metal d-levels. This variable occupancy is the characteristic property of the transition elements.

Taken together, oxidative-addition/reductive-elimination and insertion reactions can be shown to account for most of the organic reactions which are catalyzed by soluble organotransition-metal compounds. Thus Chapters 4 and 5 provide a rational basis for the discussion of homogeneous catalysis.

In Chapter 6 these concepts are illustrated by discussion of homogeneous catalytic hydrogenations, which are the most thoroughly studied examples of homogeneously catalyzed reactions. Stoichiometric reactions of transition-metal hydrides are then presented in Chapter 7.

A group of important industrially catalyzed reactions involving carbon monoxide is presented in Chapter 8, and is followed by the stoichiometric reactions of metal carbonyls in Chapter 9.

Chapter 10 treats a rather new topic, metallacycles. The formation and fragmentation of metallacycles are involved in several important catalytic processes, such as olefin metathesis and acetylene cyclo-oligomerizations.

The remaining five chapters illustrate reactions of specific groups: alkyls, olefins, alkynes, $\eta^6$-arenes, and $\eta^3$-allyls. Most of these reactions are stoichiometric. Thus Chapters 11 to 15 illustrate practical applications to organic synthesis of the principles which are developed in the first part of the text.

It is also important for the reader to know those subjects which are not included in this text because of its limited scope. Students may wish to explore elsewhere each of these active, current research topics.

We have confined this text to reactions of low oxidation-state organotransition-metal compounds. Thus catalytic oxidations and oxygenations which usually involve higher oxidation levels have been excluded. The organometallic chemistry of the nontransition metals including lanthanides and actinides is also not discussed. We shall see that distinctive chemical features, such as oxidative-addition and $\pi$-backbonding, usually require the variable occupancy of d-shells and are therefore almost exclusively in the province of transition-metal compounds. Classic coordination chemistry of the transition metals is also not discussed except in bonding theory (Chapter 2), where there is some overlap with organometallic chemistry.

We have not described much of "bioinorganic chemistry" except to mention end-bonded oxygen complexes in Chapter 3. A serious omission from this area concerns the rearrangements of organic substrates which are catalyzed by the enzyme Vitamin $B_{12}$. These reactions are thought to involve homolysis of a carbon-cobalt bond [1]. However, this vast topic is still quite controversial, and is not simply related to other general reaction classes exhibited by organotransition-metal compounds. We have also ignored the vast, complex, and still enigmatic subject of nitrogen fixation [2], except to mention in Chapter 3 the structural features of dinitrogen as a ligand.

Another major omission concerns the structure, reactions, and catalytic properties of transition-metal clusters [3] and of multiple metal-metal bonds [4]. The role of clusters as precatalysts and as hypothetical models for ligands chemisorbed on metal particles is currently very fashionable. However, simple bonding theories do not readily account for cluster structures nor, in some cases, even their stoichiometry. Furthermore, it is very difficult to establish that an intact cluster, as opposed to a fragment or a metastable metallic particle derived from a cluster, is the actual catalyst in a reaction. To date this has been seldom, if ever, established. We have not discussed heterogeneous catalysts in any detail, although these are mentioned in Chapter 6 in the context of catalytic hydrogenation. Multiple metal-metal bonds have so far shown a very limited chemistry, which is of little importance to either organic synthesis or catalysis. Both topics have promise and are being actively investigated.

Photochemistry is not discussed in this account. Although irradiation is occasionally used to form an active catalyst from a precatalyst, it is difficult to discover the nature of the actual catalyst. Furthermore, simply heating the complex or using labile ligands will sometimes afford the same catalyst. The photochemistry of organotransition-metal compounds usually involves dissociation of a CO ligand or homolytic rupture of a metal-metal or a metal-carbon bond. Intermolecular reactions involving photochemically excited states of organotransition-metal complexes are very rare, since the lifetimes of these excited states are too short for even diffusion-controlled intermolecular reactions to occur.

We have not presented, in any detail, the possible role of transition metals in diverting symmetry-controlled organic reactions, although there is a limited discussion of this topic in Chapter 10. The problem is that many such reactions, rather than being concerted, occur in steps, so that Woodward-Hoffmann rules do not apply. Halpern has eloquently argued this point [5], although the subject is still controversial.

There is no separate section on free-radical chemistry [6], although the role of free radicals is mentioned in the context of oxidative addition (Chapter 4), catalytic hydrogenation (Chapter 6), and certain oxo processes (Chapter 8), and in other isolated cases.

Gas-phase chemistry is not discussed herein, nor are the metal-atom synthesis techniques for preparing stable or matrix-isolated metastable compounds [7].

We considered writing a separate section outlining methods and strategies for synthesizing organometallic complexes but decided not to do so. Many individual cases are described and lead references are given, especially in Chapter 3. The general reaction classes discussed in Chapters 4 and 5 suggest strategies.

A separate section on ligand-replacement mechanisms has not been given, although particular cases are described. Three guiding principles apply here. First, coordinatively unsaturated complexes usually exhibit dominant associative ligand-displacement ($S_N2$) mechanisms, whereas coordinatively saturated complexes normally show dissociative ($S_N1$) paths. Second, certain ligands may weaken *trans*-bonds that assist dissociative paths, or they may promote associative displacements of *trans* ligands. These have been distinguished as a *trans-effect* (a kinetic phenomenon) and a *trans-influence* (a thermodynamic, structural bond weakening) [8]. We believe that this distinction is artificial, since the two phenomena are strongly related. *Therefore in this text we use* the term trans-effect to describe both kinetic and structural properties. There are also cis labilizing ligands, such as the acyls which are mentioned in Chapter 5. The third kinetic principle is that the reactivity of isostructural complexes often follows the order: 1st-row > 2nd-row > 3rd-row transition metals. This has an important consequence, that 3rd-row complexes usually react too slowly to be useful as homogeneous catalysts. However, models for intermediate stages of catalytic cycles may be found among 3rd-row transition metal compounds.

Finally, we have not included a separate section concerning spectroscopic methodologies for the characterization of organotransition-metal compounds. Students and chemists with enough experience to use this text should have the necessary background in physical and organic chemistry to understand the limited spectroscopic data presented herein. There is occasional need to know that isotopes of certain metals have non-zero nuclear spins, e.g., $^{103}Rh$, spin ½, 100% natural abundance. It will also be helpful to recognize simple group-theory symbols designating the symmetry of a coordination sphere in a complex. This is now common background for senior chemistry majors [9].

It is instructional to describe briefly the historical development of organotransition-metal chemistry. A perspective on the past may help us predict the prospects of this field. A more authoritative account of this history has been published [10].

A chronological account of major events in organotransition-metal chemistry is outlined in Table 1.1. These landmarks were selected either because they represent firsts or because they stimulated further developments.

The discovery of the first organotransition-metal compound, a platinum-olefin complex, by the Danish pharmacist Zeiss in 1827 [11], was not to be understood or appreciated for more than a century. The first metal carbonyl, a platinum-chloride complex, was described about 40 years later by Schutzenberger [12]. Mond's preparation of $Ni(CO)_4$, the first binary metal carbonyl, had greater significance, since this discovery lead to a commercial process for refining nickel [13]. Imagine, however, the social consequences of handling such an insidiously toxic, volatile substance with the extant primitive chemical-engineering techniques. The work on $Ni(CO)_4$ soon lead to the discovery of $Fe(CO)_5$ [14].

Nearly 30 years passed before Hein discovered the substances we now know to be $\pi$-bonded chromium-arene complexes [15]. These compounds could not be explained by the bonding theories of that day (Lewis' octet rule was just evolving). Some 35 years

*Table 1.1. Historic landmarks in organotransition-metal chemistry.*

| Date | Event | Reference |
|---|---|---|
| 1827 | Zeiss' salt discovered, $K^+[(C_2H_4)PtCl_3]$ | [11] |
| 1868 | Schutzenberger prepared the first carbonyl complex, $[PtCl_2(CO)]_2$. | [12] |
| 1890 | Mond prepares $Ni(CO)_4$. | [13] |
| 1891 | Mond and Berthelot prepare $Fe(CO)_5$. | [14] |
| 1919 | Hein prepares ill-characterized $\eta^6$-arene chromium compounds. | [15] |
| 1925 | The Fischer-Tropsch process is developed. | [17] |
| 1930 | Reihlen prepares 1,3-butadiene-iron tricarbonyl. | [18] |
| 1938 | Roelen discovers the cobalt-catalyzed oxo process. | [19] |
| 1938 | Calvin discovers homogeneous catalytic hydrogenation of quinone by copper acetate. | [21] |
| 1938 | Lucas and Winstein study silver-olefin complexes. | [20] |
| 1939 | Iguchi describes a rhodium-based homogeneous hydrogenation catalyst. | [22] |
| 1938-1945 | Reppe's group develops many homogeneously catalyzed processes. | [23] |
| 1951 | Orgel, Pauling, and Zeiss describe the backbonding in metal carbonyls. | [24] |
| 1951 | Ferrocene is discovered by Kealy and Pauson and by Miller. | [25] |
| 1952 | Wilkinson, Rosenblum, Whiting, and Woodward proposed a sandwich structure for ferrocene. | [26] |
| 1952 | E. O. Fischer describes the cobalticinium cation. | [27] |
| 1955 | Cotton and Wilkinson discover fluxional behavior. | [28] |
| 1955 | Halpern begins to study the mechanism of homogeneous catalytic hydrogenation. | [30] |
| 1955 | Ziegler and Natta discover metal-catalyzed olefin polymerization. | [31a,b] |
| 1956 | Longuet-Higgins and Orgel predict stable cyclobutadiene complexes. | [32] |
| 1958 | The structure of $[CpMo(CO_3]_2$ reveals a covalent metal-metal bond without bridging ligands. | [34] |
| 1958 | Criegee and Hubel prepare stable cyclobutadiene complexes. | [33] |
| 1959 | Shaw and Chatt describe an oxidative-addition reaction. | [35] |
| 1961 | Crowfoot-Hodgkin elucidates structure of the coenzyme Vitamin $B_{12}$. | [37] |
| 1962 | Vaska discovers the "Vaska Complex." | [36] |
| 1964 | Fisher isolates the first carbene complex. | [38] |
| 1964 | Banks reports olefin metathesis. | [39] |
| 1965 | Allen and Senoff discover the first dinitrogen complex. | [40] |
| 1965 | Wilkinson and Caffey independently discover "the Wilkinson Hydrogenation complex." | [41a,b] |

later Onsager proposed structures for these substances [16a], but these were treated with skepticism until Fischer clarified their nature [16b].

The Fischer-Tropsch process, which (as we shall see in Chapter 8) is a topic of great current interest, was discovered in 1925 [17]. The commercialization of this process was soon suppressed by the discovery of cheap petroleum, except for a resurgence under wartime emergency in Germany.

The discovery of a remarkably inert complex between the normally reactive 1,3-butadiene and iron carbonyl was announced by Reihlen in 1930 [18]. However, his work failed to attract attention and did not stimulate further research. In fact, this substance was still obscure 25 years later, when one of us (JPC) began independent research.

In 1938 Roelen discovered the cobalt-catalyzed oxo process while investigating the Fischer-Tropsch reaction [19]. The oxo reaction soon became the first truly viable, commercial, homogeneously catalyzed process and is still practiced today (see Chapter 8). The oxo process greatly stimulated research in homogeneous catalysis and organometallic chemistry by industrial and academic chemists.

Retrospectively, 1938 was a vintage year for organotransition-metal chemistry. In his Ph.D work under Lucas, Winstein studied silver-olefin complexes [20]. However, Winstein, one of the greatest mechanisms chemists, did not again focus his research on organotransition-metal chemistry until just before his death in 1969. In 1938 Calvin reported the first homogeneously catalyzed hydrogenation (of quinone by copper acetate) [21], but he too left this topic to explore photosynthesis. Within a year Iguchi in Japan reported a rhodium-catalyzed hydrogenation of fumaric acid [22], but this work was apparently obscured by the war and limited recognition of Japanese chemistry in the U.S. and Europe.

Note that after an increasing succession of research highlights, little appeared to happen between 1939 and 1951. This appearance resulted from wartime disruption, diversion, and obfuscation of chemical research. We now know that during that period Reppe's group in Germany was developing a vast array of organic reactions catalyzed by soluble organotransition-metal compounds [23].

From 1951 on, research in organotransition-metal chemistry has developed at an ever-accelerating pace. In that year the nature of the metal-carbonyl bond, the prototype of $\pi$-backbonding, was proposed by Orgel, Pauling, and Zeiss [24]. However, the landmark event of 1951 was the independent discovery of ferrocene by two groups: Kealy and Pauson, and Miller, who proposed a $\sigma$-bonded structure [25]. The next year the correct $\pi$-bonded structure of ferrocene was advanced by Wilkinson, Rosenblum, Whiting, and Woodward [26]. In the same year E. O. Fischer described the isoelectronic cationic cobalt(III) compound [27]. The discovery of these novel "sandwich" structures and their chemical reactions (especially of ferrocene) greatly stimulated academic research in this field. Woodward's immense prestige undoubtedly contributed to this interest. One of us (JPC) vividly recalls the then-young German chemist, E. O. Fischer, giving a stimulating lecture on bis-benzene chromium at the University of Illinois in 1956. Fischer and Wilkinson were awarded the Nobel prize in 1973, a belated recognition of these discoveries.

In 1955, G. Wilkinson and his student F. A. Cotton described the fluxional behavior of some organometallic compounds [28]. Such dynamic intramolecular rearrangements were intensely explored by inorganic chemists for the next 20 years [29]. The same year, in work which was little noticed at the time, Halpern began to lay the foundation for understanding the mechanism of homogeneously catalyzed hydrogenations [30]. An event of great technological significance which initiated large industrial research programs was also announced in 1955: the catalytic polymerization of olefins by ill-defined organometallic mixtures, the Ziegler-Natta processes [31]. This work lead to a Nobel Prize in 1963.

In 1956 Longuet-Higgins and L. Orgel predicted that transition metals could stabilize cyclobutadiene [32], and the first stable cyclobutadiene complexes were synthesized by Criegee and Hubel in 1958 [33]. It is rare that theory precedes experiment.

In 1958 the structure of $[CpMo(CO)_3]_2$ showed a covalent metal-metal bond which is unsupported by bridging ligands, and forecast the development of compounds containing metal-metal bonds [34].

In 1959 an early example of an oxidative-addition reaction was described by B. L. Shaw and J. Chatt [35], although the scope and potential of this reaction class (see Chapter 4) was not evident until Vaska's discovery in 1962 of the chemically versatile iridium compound that bears his name [36].

In 1961 the structure of the coenzyme Vitamin $B_{12}$ was announced by Dorothy Crowfoot-Hodgkin [37], work for which she received a Nobel prize in 1964. The remarkable feature of this substance is the existence of a covalent cobalt-carbon $\sigma$-bond, indicating that Nature has been practicing organotransition-metal chemistry for eons.

In 1964 two seemingly unrelated events occurred: E. O. Fischer prepared the first transition-metal carbene complex [38], and Banks discovered catalytic olefin metathesis [39]. We now believe carbene complexes are the active catalysts for olefin metathesis (see Chapter 10).

In 1965 Allen and Senoff [40] reported the first metal $N_2$ complex, $[Ru(NH_3)_5N_2]^{2+}$. Soon many chemists were making $N_2$ complexes, and apparently Nature has done so for many years via the metalloenzyme nitrogenase. The same year G. Wilkinson [41a] discovered the "Wilkinson hydrogenation catalyst," which is one of the most practical organotransition-metal catalysts (see Chapter 6). The same discovery was made nearly simultaneously by chemists at I.C.I [41b].

By this time the discoveries were occurring so rapidly that an analysis is futile. Some commercial discoveries are worth noting: Monsanto's rhodium-based, acetic acid process (Chapter 8); the rhodium-catalyzed oxo reaction (Chapter 8); Dupont's adiponitrile synthesis from HCN and 1,3-butadiene (Chapter 8); Knowles' asymmetric catalytic synthesis of L-Dopa (Chapter 6); Wilke's cyclotrimerization of butadiene (Chapter 12). The future promise of this field is even brighter than these past developments.

It is further instructive to note here the critical role which analytical methodologies have played in the development of this field. Infrared spectroscopy greatly aided characterization of metal-carbonyl complexes. Proton NMR spectroscopy was essential to detect transition-metal hydrides and discover the fluxional behavior of certain organometallic compounds. The emergence of computer-assisted X-ray crystallography as a routine analytical tool lead to the characterization of numerous, hitherto unknown structural classes. Mass spectrometry greatly aided the elucidation of metal clusters. Gas chromatography has been essential to the study of catalytic reactions. It is no accident that the development or commercial availability of these techniques is closely associated with developments in this and other new areas of chemistry.

## Notes.

1. R. H. Abeles and D. Dolphin, Accts. Chem. Res., 9, 114 (1976).

2a. Nitrogen Fixation by Free Living Microorganisms, W. D. P. Steward, ed. (Cambridge Univ. Press, 1975).

2b. A Treatise on Dinitrogen Fixation, Sections I and II: Inorganic and Physical Chemistry and Biochemistry, R. W. F. Hardy, F. Bottomly, and R. C. Burns, eds. (John Wiley, 1979).

3. E. Muetterties, Science, 196, 839 (1977).

4a. F. A. Cotton, Chem. Soc. Rev., 4, 27 (1975).

4b. M. H. Chisholm, F. A. Cotton, M. W. Extine, and L. A. Rankel, J. Amer. Chem. Soc., 100, 807 (1978).

5. J. Halpern, in Organic Synthesis via Metal Carbonyls, I. Wender and P. Pino, eds. (John Wiley, 1977), II, 705-730.

6. M. F. Lappert and P. W. Lednor, Adv. Organomet. Chem., 14, 345 (1976).

7. P. L. Tims and T. W. Turney, Adv. Organomet. Chem., 15, 53 (1977).

8. T. G. Appleton, H. O. Clark, and L. E. Manzer, Coord. Chem. Rev., 10, 335 (1973).

9. F. A. Cotton, Chemical Applications of Group Theory (Interscience, 2d ed., 1971).

10a. J. S. Thayer, Adv. Organomet. Chem., 13, 1 (1975).

10b. J. Organomet. Chem., 100, no. 1, 1-287 (1975). This issue presents perspectives in organometallic chemistry. A similar issue will be published in 1980.

11. W. C. Zeise, Ann. Phys., 9, 932 (1827).

12. M. P. Schutzenberger, Annalen, 15, 100 (1868).

13. L. Mond, J. Chem. Soc., 57, 749 (1890).

14a. L. Mond and C. Langler, J. Chem. Soc., 59, 1090 (1891).

14b. M. Berthelot, C. R. Acad. Sci., 112, 1343 (1891).

15. F. Hein, Ber., 52, 195 (1919).

16a. H. H. Zeiss and M. Tsutsui, J. Amer. Chem. Soc., 79, 3962 (1957).

16b. E. O. Fischer and W. Hafner, Z. Naturforsch., 10b, 665 (1955).

17. F. Fischer and H. Tropsch, DRP 411416 (1922); DRP 484337 (1925).

18. H. Reihlen, A. Gruhl, G. von Hessling, and O. Pfrengle, Leibig. Ann. Chem., 482, 161 (1930).

19. O. Roelen, DRP 849548 (1938); Angew. Chem., 60, 62 (1948).

20. S. Winstein and H. J. Lucas, J. Amer. Chem. Soc., 60, 836 (1938).

21. M. Calvin, Trans. Faraday Soc., 34, 1181 (1938).

22. M. Iguchi, J. Chem. Soc. Japan, 60, 1287 (1939).

23. W. Reppe, Neue Entwicklungen auf dem Gebiete der Chemie des Acetylens und des Kohlenoxyds (Springer-Verlag, 1949).

24a. L. E. Orgel, An Introduction to Transition-Metal Chemistry (Methuen, 1960), pp. 135-143 and references therein.

24b. L. Pauling, The Nature of the Chemical Bond (Cornell Univ. Press, 3d ed., (1960), pp. 331-336 and references therein.

24c. J. W. Richardson, in Organometallic Chemistry, H. Zeiss, ed. (Reinhold, 1960), pp. 12-20 and references therein.

25a. T. J. Kealy and P. J. Pauson, Nature (London), 168, 1039 (1951); G. Wilkinson, J. Organomet. Chem., 100, 273 (1975).
25b. S. A. Miller, J. A. Tebboth, and J. F. Tremaine, J. Chem. Soc., 632 (1952).
26. G. Wilkinson, M. Rosenblum, M. C. Whiting, and R. B. Woodward, J. Amer. Chem. Soc., 74, 2125 (1952).
27. E. O. Fischer and W. Pfab, Z. Naturforsch., 7B, 377 (1952), see also ref. 42.
28a. G. Wilkinson and T. S. Piper, J. Inorg. Nucl. Chem., 2, 23 (1956).
28b. F. A. Cotton, J. Organomet. Chem., 100, 29 (1975).
29a. R. A. Cotton, in Dynamic Nuclear Magnetic Resonance Spectroscopy, L. Jackson and F. A. Cotton, eds. (Academic Press, 1975), pp. 377-440.
29b. R. D. Adams and F. A. Cotton, loc. cit., pp. 489-522.
29c. J. W. Faller, Adv. Organomet. Chem., 16, 211 (1977).
30. J. Halpern, Quart. Rev., 10, 463 (1956).
31a. K. Ziegler, Adv. Organomet. Chem., 6, 1 (1968), and references therein.
31b. G. Natta, Scientific American, 205, 33 (1961), and references therein.
32. H. C. Longuet-Higgins and L. E. Orgel, J. Chem. Soc., 1969 (1956).
33a. W. Hubel and E. H. Braye, J. Inorg. Nucl. Chem., 10, 250 (1958).
33b. R. Criegee and G. Schroder, Liebigs Ann. Chem., 623, 1 (1959).
34. F. C. Wilson and D. P. Shoemaker, J. Chem. Phys., 27, 809 (1958).
35. J. Chatt and B. L. Shaw, J. Chem. Soc., 4020 (1959).
36. L. Vaska and J. W. Kiluzio, J. Amer. Chem. Soc., 84, 679 (1962).
37. P. G. Lenhart and D. C. Hodgkin, Nature, 192, 937 (1961).
38. E. O. Fischer and A. Maasbol, Angew. Chem., Int. Ed. Engl., 3, 580 (1964).
39. R. L. Banks and G. C. Bailey, Ind. Eng. Chem. Prod. Res. Develop., 3, 170 (1964).
40. A. D. Allen and C. V. Senoff, Chem. Comm., 621 (1965).
41a. J. F. Young, J. A. Osborn, F. H. Jardine, and G. Wilkinson, Chem. Comm., 131 (1965).
41b. R. S. Coffey, Imperial Chemical Industries, Brit. Pat. 1,121,642 (filed 1965).
42. G. Wilkinson, J. Amer. Chem. Soc., 74, 6148 (1952)

# 2

# Structure and Bonding: Electronic Bookkeeping

Much of the mystique which inhibits organic chemists from working on organotransition-metal chemistry concerns structure and bonding. At first glance the vast array of organometallic structures seems bewildering and empirical compared with the rather simple theoretical foundation which provides a rational basis for understanding even the most elaborate organic structures. However, we shall see in this chapter that there _is_ a rudimentary system for organizing and cataloging organometallic structures. The underlying rules and principles have "first-order truth" in bonding theory. Classes of isoelectronic and isostructural substances can be simply defined. Moreover the _chemical reactions_ of these molecules can also be related to bonding and structural types. Simplicity is the guiding principle through this discussion. The level of presentation is the minimum that one needs in order to read the primary literature and to plan research concerned with _reactions_ of organometallic substances. Doing so, of course, also requires an elementary understanding of the _stereochemistry_ of organotransition-metal compounds. The simplistic level of bonding and structure presented herein will not satisfy those persons interested in inorganic chemistry _per se_. For a deeper analysis of bonding, more authoritative accounts should be consulted [1].

The concepts of _oxidation state_, _coordination number_, _coordinative saturation and unsaturation_, _d-electron configuration_ ($d^n$), and _coordination geometry_ form the basis of our electron-bookkeeping system and structural organization. These terms are defined, illustrated, and discussed below.

## 2.1. Electronic Configurations of the Transition Metals.

This book is concerned with transition metals, which are defined by Cotton [1a] as "those elements which have a partially filled d-shell in some or all of their compounds."

Typically, transition metals exhibit multiple oxidation states, multiple coordination numbers, and, in some compounds, magnetic and spectroscopic properties which can be accounted for by variable occupancy of the five chemically accessible d-orbitals. As shown in Table 2.1, the first-row transition metals begin with titanium and end with copper by filling the 3d levels. The second-row transition series begins with zirconium and ends with silver by filling the 4d level. Finally, the third-row transition series from hafnium to gold fills the 5d level, but does not include the rare-earth f-level elements.

*Table 2.1. Relationships between oxidation-state and $d^n$ in the transition metals.*

| Group Number | | IVa | Va | VIa | VIIa | VIIIa | | | Ib | |
|---|---|---|---|---|---|---|---|---|---|---|
| First row | 3d | Ti | V | Cr | Mn | Fe | Co | Ni | Cu | |
| Second row | 4d | Zr | Nb | Mo | Tc | Ru | Rh | Pd | Ag | |
| Third row | 5d | Hf | Ta | W | Re | Os | Ir | Pt | Au | |
| Oxidation state | 0 | 4 | 5 | 6 | 7 | 8 | 9 | 10 | -- | $d^n$ |
| | I | 3 | 4 | 5 | 6 | 7 | 8 | 9 | 10 | |
| | II | 2 | 3 | 4 | 5 | 6 | 7 | 8 | 9 | |
| | III | 1 | 2 | 3 | 4 | 5 | 6 | 7 | 8 | |
| | IV | 0 | 1 | 2 | 3 | 4 | 5 | 6 | 7 | |

Elementary chemistry textbooks usually introduce electronic structure by considering the gaseous atoms. This leads to confusion concerning the occupancy of d-orbitals in compounds of the transition elements and the consequent assignment of $d^{\underline{n}}$ values. In the gaseous atoms the ($\underline{N}$ + 1)s level is more stable (lower) than the $\underline{N}$d level. However, the electronic configurations of transition-metal cations tend to revert to a hydrogen-like order; that is, for cations the $\underline{N}$d level is lower in energy than the ($\underline{N}$ + 1)s level, which in turn is more stable than the ($\underline{N}$ + 1)p level. We thus adopt the commonly used convention that in charged (or even neutral) transition-metal compounds, the metal's valence-shell electrons are treated as if they were all in the $\underline{N}$d shell. This convention is a fair description of reality for metals in higher oxidation states and for those elements to the right of the transition series; in neutral complexes of the early transition metals, it is a less-good assumption. ($\underline{N}$ is used here to denote the principal quantum number and to avoid confusion with $\underline{n}$, the number of d electrons.) Thus starting from the argon core of 18 electrons, we proceed: Ar($\underline{18}$) $1s^2$,

$2s^2$, $2p^6$, $3s^2$, $3p^6$; for K, add $4s^1$; for Ca, add $4s^2$; then for Sc, add $4s^2$, $3d^1$. At Ti the configuration might be expected to be $4s^2$, $3d^2$; but compounds or ions of Ti in its lowest common oxidation state, $Ti^{3+}$, have an electronic structure that consists of the Ar core followed by a singly occupied 3d orbital, $3d^1$. That is, the 4s level has now become higher in energy than the 3d level. This situation exists throughout the first transition series to copper in group Ib. Copper is the last element in this row to have a partially filled d-shell in some of its compounds e.g., $Cu^{2+}$ has a $3d^9$ configuration. Thus, by Cotton's definition, copper is the last transition element in the first long row. As shown in Table 2.1, the first triad (vertical group) beginning the transition series is Group IVa (Ti, Zr, and Hf), and the last triad is Group Ib (Cu, Ag, and Au). Throughout the first transition series (row), the 3d level is being filled; for the second transition series it is the 4d level; and for the third transition series it is the 5d level.

Note that within compounds of the entire transition series, the d levels are those which have the highest energy and are consequently those which electrons can be most easily removed from (by oxidation) or added to (by reduction). This generalization is valid to a first approximation, and is the basis from which oxidation-state formalism derives. Thus we shall see that d-electron levels, referred to as $d^{\underline{n}}$, are primarily associated with the metal, and that these metallic d-electrons can be considered nonbonding except for small perturbations brought about by $\pi$-bonding effects and orbital splitting through interaction with the $\sigma$-bonds of the ligand groups (see Section 2.7). It is also necessary to recall that an entire 4f level has been filled by the lanthanides between the second and third transition series.

From the above discussion it should be apparent that a relationship is expected between the d-electron configuration, $d^{\underline{n}}$, and the oxidation state, since it is the $d^{\underline{n}}$ level which electrons are added to or removed from. This will become clear as the oxidation state of the metal, which is the basis of our electron-accounting system, is more precisely defined.

## 2.2. Oxidation-State Formalism.

Oxidation state is defined as "the charge left on the central metal atom when the ligands are removed in their normal, closed-shell configurations" or "the charge left on the central metal atom when each shared electron pair is assigned to the more electronegative atom." The two definitions almost always yield the same result. It is of special importance for organometallic compounds to note that both hydrogen and carbon are more electronegative than transition metals; so, for example, the M-H group is considered as $M^+H^-$ (hydride formalism) for the purpose of determining the formal oxidation state. For more complicated ligands, it is necessary to define their "normal closed-shell configurations" in order to develop widely applicable systems for assigning oxidation states.

Oxidation state is a useful formalism but it is not a physical property. Thus oxidation state, as it will be used throughout this text, can be neither measured nor simply related to any physical property of the compound being discussed. Ligand-oxidation levels are somewhat arbitrarily assigned in order to create the most coherent isoelectronic and isostructural series of compounds. These assigned ligand-oxidation levels lead in turn to the oxidation state of the central metal. The chemical properties of ligands are not necessarily related to their formal oxidation-state assignments. For example, hydrogen attached to the metal is considered a hydride ($H^-$), even though some such "hydrides" may be strong acids, and other M-H bonds may be nonpolar. Oxidation state as used here is always an integer, never fractional, even though modern theory can sometimes calculate a fractional charge distribution between a metal and a ligand in a specific compound. The use of chemical properties or theoretical calculations would lead to an unmanageable system of little heuristic merit. Finally, it should be noted that oxidation state has a sign: positive, negative, or zero.

Before illustrating the use of oxidation level with organometallic compounds, let us review more familiar examples. Consider the halides: NaCl, $AlCl_3$, and $CuCl_2$. We are quite familiar with oxidation-state assignments of Na(+I), Al(+III), and Cu(+II). [Note the use of parentheses, sign, and Roman numerals to denote oxidation state. The positive sign is often omitted.] These salts dissolve in water and form cations which are usually written with the indicated charges as $Na^+$, $Al^{3+}$, $Cu^{2+}$, or more accurately as the hydrated cations $[Na(H_2O)_{\underline{x}}]^+$, $[Al(H_2O)_6]^{3+}$, $[Cu(H_2O)_6]^{2+}$ (ignoring ionization of coordinated water). The charge written for these ions is the same as the oxidation state. Next, note that the familiar copper tetraammine coordination complex $[Cu(NH_3)_4]^{2+}$ also has an oxidation state of (+II). This conforms to our earlier experience about the congruence of cation charges and oxidation-state assignments, but the more formal definition of oxidation state can be applied here with the same result; that is, each ammonia ligand is removed with its bonding electrons, leaving a $2^+$ charge on copper. Next, consider some simple metal carbonyls: $Fe(CO)_5$, $Co_2(CO)_8$, and $Ni(CO)_4$; each has the oxidation state (0). Don't be bothered by the dimeric nature of the $Co_2(CO)_8$, which arises from a covalent cobalt-cobalt bond. Those electrons in the metal-metal bond are evenly divided between the equivalent cobalt atoms without disturbing our oxidation-state calculation. The "normal closed-shell configuration" of the carbon-monoxide molecule has two paired electrons on carbon, which is more electronegative than the transition metals.

## 2.3. D-ELECTRON CONFIRUGATION AND THE 18-ELECTRON RULE.

Before proceeding further with oxidation level, let us next consider the d-configuration. Because of the buildup in d levels as one crosses each transition series from left to right, the group number is the same as the number of d electrons ($d^{\underline{n}}$) for that element in its zero oxidation state. For example, chromium is a member of Group VI, and Cr(0) has configuration $d^6$. This can be discovered by noting that chromium(0) should have six electrons in addition to the argon core. This very useful generalization holds true for the first triad (Fe, Ru, Os) of Group VIII. Since Group VIII has three successive triads, we must then modify this rule of $d^{\underline{n}}$ being equal to the group number, by adding one to $\underline{n}$ for each of the next two triads in Group VIII. Thus Co(0) is $d^9$ (8 + 1) and Ni(0) is $d^{10}$ (8 + 2). The various combinations of oxidation state and $d^{\underline{n}}$ are given in Table 2.1.

Once $d^{\underline{n}}$ is defined for the zero oxidation level of each element, it is simple to subtract the proper number of electrons to arrive at positive oxidation states or to add electrons if negative oxidation states are involved. Consider, for example, the following two series of carbonyls:

| | | |
|---|---|---|
| $[V(CO)_6]^-$ | $Cr(CO)_6$ | $[Mn(CO)_6]^+$ |
| $V(-I)d^6$ | $Cr(0)d^6$ | $Mn(+I)d^6$ |
| | | |
| $[Fe(CO)_4]^{2-}$ | $[Co(CO)_4]^-$ | $[Ni(CO)_4]$ |
| $Fe(-II)d^{10}$ | $Co(-I)d^{10}$ | $Ni(0)d^{10}$ |

Note that the first series are isoelectronic in the sense that all are $d^6$ and have the same bonding pattern (hexacoordinate, octahedral structures). These complexes differ only by the nature of the metal nucleus and the over-all charge. The second series are also isoelectronic, having $d^{10}$ configurations, and these complexes are isostructural as well, having tetrahedral coordination geometries. Thus we begin to see the value of this organization. The formal oxidation state of any particular metal reveals its $d^{\underline{n}}$ configuration, which can be related to the coordination number (CN) and the expected coordination geometry. However, there is even more to this system. Let us consider next the coordination numbers, which can be defined as the number of σ-bonds formed between a metal and its ligands. There is a maximum coordination number (CN) permitted for each $d^{\underline{n}}$ provided we consider only monometallic, diamagnetic compounds ($\underline{n}$ = even number, all electrons paired). We shall see in the following discussion that this relationship between $d^{\underline{n}}$ and CN arises from the unoccupied s-, p-, and d-orbitals which are conceptually used to fashion molecular orbitals to accommodate the ligands.

Thus we shall find that for diamagnetic, mononuclear transition-metal complexes the following rule applies:

$$n + 2(CN)_{max} = 18 \tag{2.1}$$

where $\underline{n}$ is from $d^{\underline{n}}$ and (CN)max is the maximum coordination number. Rearranging this expression, we find (CN)max = $\frac{18 - n}{2}$, which indicates that there is an inverse relationship between the number of nonbonded, metal-centered d-valence electrons and the number of ligands which can be combined with the metal, provided that the electrons are spin-paired. This is logical, since these nonbonding, metal-localized d-electrons have their principle electron density on the metal and in between the ligands. This numerical relationship between (CN)max and $d^{\underline{n}}$ is often referred to as the "18-electron rule." One can see from Equation 2.1 that the sum of d electrons and ligand coordinate-bond electrons cannot exceed 18 (not without involving higher-energy anti-bonding levels, which usually results in unpaired electrons). For example, consider $Cr(CO)_6$. Chromium(0) has six d electrons ($d^6$). When these are combined with six pairs of electrons from the σ-coordinate bonds of the six CO groups, one gets 18 core electrons in the valence shell corresponding to Kr, the next inert gas: 6 + (6 x 2) = 18. This 18-electron rule is similar to Lewis' familiar octet rule for the representative elements.

The 18-electron rule has very few exceptions among well-characterized, stable organometallic or coordination compounds of the transition metals, provided a) only one metal is involved, (b) $\underline{n}$ is even, and (c) the electrons are spin-paired. Thus we expect to find the following maximum coordination numbers for each even value of $d^{\underline{n}}$:

| $d^{\underline{n}}$ | 0 | 2 | 4 | 6 | 8 | 10 |
|---|---|---|---|---|---|---|
| (CN)max | 9 | 8 | 7 | 6 | 5 | 4 |

The maximum coordination number decreases by one for every two electrons that are added to the nonbonded d shell. Such a process is a two-electron reduction of the metal center, which is equivalent to an algebraic decrease of two in the oxidation number. Thus nonbonding electron pairs replace donor, coordinate electron pairs:

$$\underset{\substack{Fe(O)d^8 \\ CN = 5}}{Fe(CO)_5} \underset{-2e^- \; +CO}{\overset{+2e^- \; -CO}{\rightleftarrows}} \underset{\substack{Fe(-II)d^{10} \\ CN = 4}}{Fe(CO)_4^{=}} \tag{2.2}$$

There are some apparent exceptions to the 18-electron rule. An example described in Chapter 3 is $W(PhC{\equiv}CPh)_3(CO)$, a diamagnetic compound which upon superficial analysis seems to be a 20-electron complex and there are other diamagnetic, 20-electron metallocene compounds which we will encounter in Chapter 3.

Later on in this chapter we will find stable compounds in which the maximum coordination number is not attained. These "coordinatively unsaturated" 16- or 14-electron complexes are quite reactive and are important in catalysis. A typical example of a coordinatively unsaturated compound is $RhCl(PPh_3)_3$, a four-coordinate, 16-electron $d^8$ complex of rhodium(I). In Chapter 6 we will encounter this compound as Wilkinson's hydrogenation catalyst.

## 2.4. Spatial Orientation of the d-Orbitals.

The idealized stereochemistries for various organotransition-metal complexes can be correlated with the spatial orientation of the full nonbonding and the empty antibonding d-orbitals. The d-orbital energy-level schemes for various coordination geometries derive from the splitting of the d levels by the spatial arrangement of the ligands about the metal, "the ligand-field." This is the same underlying principle which explains the inverse relationships between CN and $d^n$: that the filled, metal-localized d-orbitals are more stable when directed away from the ligands. This will become clear when we analyze specific molecular-orbital schemes for typical complexes. However, we shall see that the d-electron configuration rarely dictates the preference of one coordination geometry over another.

First we need to consider the spatial orientations of the s-, p-, and d-atomic orbitals. The usual graphical representations of these orbitals, shown in Figure 2.1, give two-dimensional cross sections of the orbital boundary surface along the relevant axes. Each boundary surface is generated from the square of the angular part of a hydrogen wave function. To generate three-dimensional representations, each curve must be rotated around its symmetry axis. The + and - signs arise from the angular

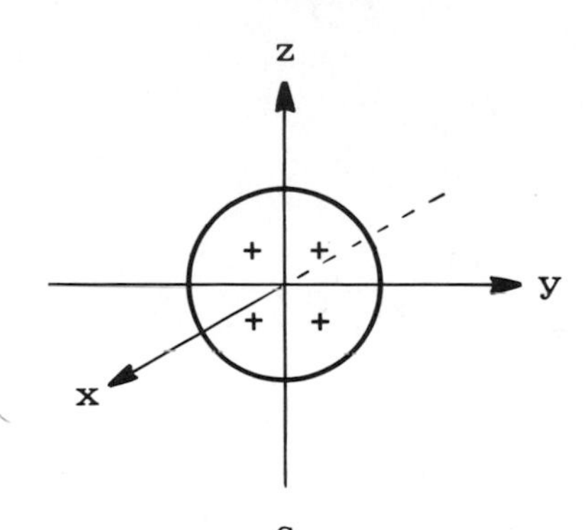


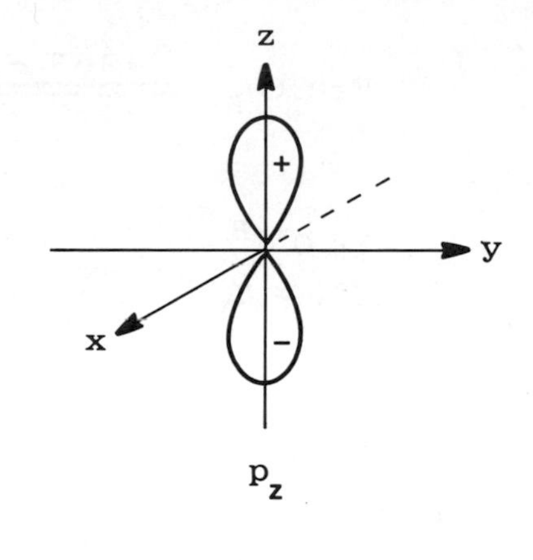


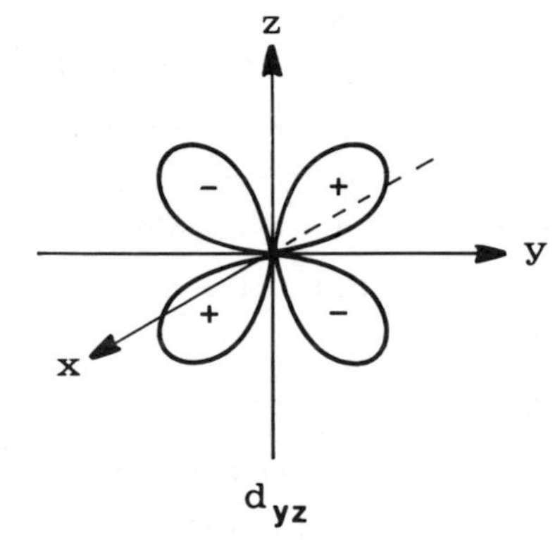


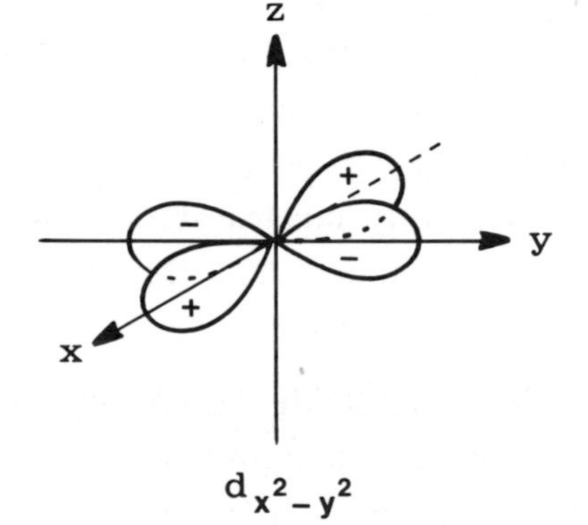


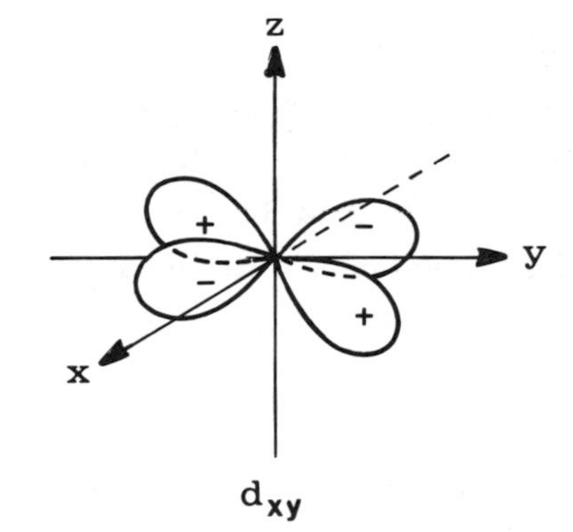


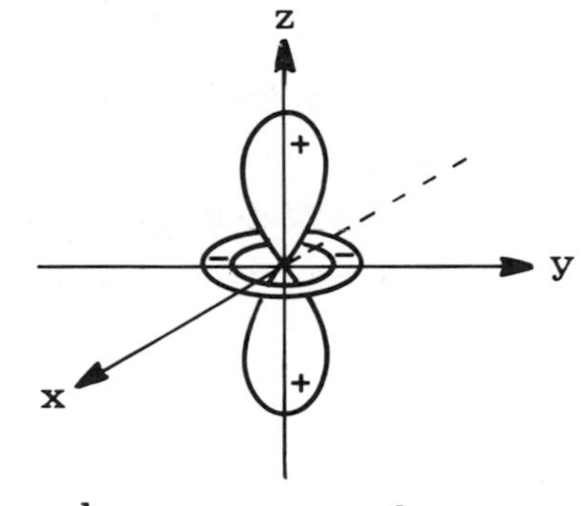


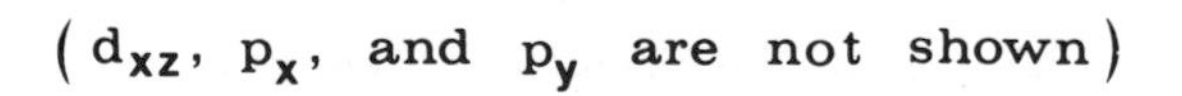


*Figure 2.1. Graphic representation of orbitals.*

part of the wave function itself (rather than its square). We shall find that overlap in regions of like sign between two atomic orbitals (AOs) forms a bonding molecular orbital (MO), whereas overlap of regions having opposite signs affords antibonding MOs (which are usually designated by an asterisk). Each atomic orbital is described by a shorthand notation, such as $4p_x$. The integer 4 describes the major quantum number $N$, the letter p the type of orbital (from the quantum number $l$), and the subscript $x$ the cartesian axis about which the orbital boundary surface is drawn (derived from the quantum number $m$). From introductory chemistry, you may recall the relationships between the quantum numbers $N$, $l$, and $m$: $N = 1,2,\ldots$; $l = N - 1, N - 2, \ldots, 0$; $m = l, l-1, l-2, \ldots, -l$. The values of $l$ denote the type of orbital: $l = 0$,s; $l = 1$,p; $l = 2$,d; etc. From these relationships it follows that; s-orbitals are not degenerate; p-orbitals are three-fold degenerate; and d-orbitals are five-fold degenerate. For a discussion of the underlying mathematics, consult an inorganic text [e.g., 1a,b,d].

The axial designation of an orbital gives the sign of the wave function in various regions of cartesian space. For example, the $3d_{xy}$ orbital has a positive sign for those regions in which both $x$ and $y$ have the same sign, and has a negative sign in those regions where $x$ and $y$ have different signs. That is the symmetry of this wave function denoted by the product of $x$ and $y$. By inspection you can correlate the positive and negative regions in all the orbital representations shown in Figure 2.1. Note that the $d_{xz}$, $p_x$, and $p_y$ orbitals are not shown. It is also useful here to call attention to the symmetry notations $g$ and $u$, which are often shown as subscripts for molecular orbitals but apply as well to individual atomic orbitals and algebraic combinations of ligand atomic orbitals. These symbols refer to gerade (g) and ungerade (u), and denote a center of symmetry or an inversion center with respect to the sign of the wave function about the nucleus of an atom or of the central metal in a complex. From the axial designations in Figure 2.1, it follows that all s- and d-orbitals are $g$; whereas, p-orbitals are $u$. Note that all s-, p-, and d-orbitals have the same angular representation regardless of the major quantum number, $N$. This is because the radial part of the wave function has been omitted. Another point should be explained here. The representation of the $d_{z^2}$ orbital is different from $d_{x^2-y^2}$, $d_{xy}$, $d_{xz}$, and $d_{yz}$; however, mathematically "$d_{z^2}$" is really not different. The problem arises from the fact that for each principal quantum number > 2, there are only five mathematically independent d-orbitals. (For $N \leq 2$, there are, of course, no d-orbitals.) The commonly used $d_{z^2}$ representation results from a linear combination of $d_{z^2-x^2}$ and $d_{z^2-y^2}$. Thus

$d_{z^2}$ is actually $d_{2z^2 - x^2 - y^2}$ (note the smaller negative region about the nucleus). There are other ways of representing five equivalent d-orbitals, but these are difficult to draw and are not used [2].

## 2.5. Molecular Orbital Description of Octahedral Complexes.

With these graphical orbital representations in hand, we can next consider a typical molecular-orbital scheme describing the bonding in an octahedral complex (Figure 2.2). This exercise provides a rationale for correlating $d^n$, CN, and CG, as well as a basis for understanding oxidation-state formalism. This process is a post-facto analysis, since we begin by assuming the number of ligands and their spatial orientation about the metal. This latter property defines the symmetry of the complex. Using a mathematical formalism called group theory, this symmetry can be used to calculate the splitting of d-, s-, and p-metal-orbitals and the algebraic combinations of ligand-donor atomic orbitals which are suitable to form bonding, nonbonding, and antibonding σ-molecular orbitals, as well as π-bonding combinations. Group theory, which is very useful in the analysis of bonding, stereoisomerism, and spectroscopic properties, is outside the scope of the present text. Interested readers are advised to consult Cotton's text [3]. However, in the organometallic literature we will encounter "Mulliken symbols" denoting atomic orbitals which have the appropriate symmetry to form molecular orbitals (*vide infra*). The usual terms are *a*, *b*, *e*, and *t*, which may have subscripts *g* or *u* (as well as a number and primes that we can ignore). Orbitals with representations *a* or *b* are nondegenerate; those with *e* are always doubly degenerate, and those with *t* are triply degenerate. The *u* and *g* subscripts are *ungerade* and *gerade*, as described above. Note that in the octahedral case, the ligands are assumed to lie along the cartesian axes, with the metal at the origin.

Consider, for example, the lefthand side of Figure 2.2. The various metal valence-orbitals (3d, 4s, and 4p) for the octahedral "point group," $O_h$, are labeled with the appropriate "Mulliken symbols" indicating their symmetries. The ordering of these orbitals according to their energies does not derive from group theory. Rather, this level ordering (which is not shown to scale) is that discussed above for coordination complexes: the 3d level is more stable than (below) the 4s level, which is below the 4p level. Further, it is assumed that the d levels which have proper symmetry for σ-bonding with the ligands ($d_{x^2 - y^2}$, $d_{z^2}$) are *destabilized* by interaction with the ligands, since population of these d-orbitals with electrons would interfere with the ligand bonding electrons. It is also assumed that the energy of those d levels which point away from the ligands ($d_{xy}$, $d_{xz}$, $d_{yz}$) is unaffected by the "ligand field." This energy-level concept derives from a bonding theory which is based on point-charges, the "crystal-field theory," and its extension to covalent bonding, the "ligand-field theory." Thus the ordering of metal atomic orbitals shown on the left of Figure 2.2 depicts a "ligand-field perturbation" of the metal-valence orbitals in this symmetry. Later on we will consider the ligand-field splitting diagrams for other symmetries.

On the righthand side of Figure 2.2 are shown the combinations of ligand atomic orbitals (for example, the filled $sp^3$, hybrid orbital-electron pairs in six ammonia ligands) which can interact to form molecular orbitals with the metal-valence orbitals of the same symmetry. These ligand-orbital combinations are arranged, for convenience, according to the expected degree of overlap with the appropriate metal orbitals (vide infra).

The relative energies of molecular orbitals for an octahedral complex are given in the central portion of Figure 2.2. These are not shown to scale. Usual tenets of molecular-orbital construction are employed here. For a combination of $m$ atomic orbitals, one obtains $m$ molecular orbitals (MOs). At least one MO is more stable (bonding) and one MO is less stable (antibonding) than the atomic orbitals which are used to form the MOs. If the atomic orbitals are of the same energy, half the MOs are bonding and half are antibonding. Where no overlap occurs — for example, in the $t_{2g}$ levels in octahedral complexes ($d_{xy}$, $d_{xz}$, and $d_{yz}$) — no net energy change is caused by forming the σ-bonding MO network, so that these particular $t_{2g}$ MOs are metal-localized and nonbonding. We shall consider the consequences of π-bonding later on. A greater degree of molecular-orbital splitting arises from strong metal-ligand-orbital overlap and similar energy levels for the metal and ligand orbitals. That is, more overlap and better energy matching give rise to more stable (lower-energy) bonding and less stable (higher-energy) antibonding molecular orbitals.

As an illustration of this general principle, Figure 2.3 displays the metal and ligand-combination angular wave functions which are combined to form the single $a_{1g}$, the two $e_g$, and the three $t_{1u}$ molecular orbitals of the complex. These represent the six σ-bonding MOs, which are shown in the central part of Figure 2.2. The combination ligand orbitals are represented by Σ. Note that the algebraic combinations of the orbitals are arranged to give matching patterns of + and - signs indicating bonding overlap.

Now let us consider the consequences of the molecular-orbital scheme shown in Figure 2.2 in the context of a typical octahedral complex which satisfies the "18-electron rule," $[Co(NH_3)_6]^{3+}$. For this cobalt(III) $d^6$ complex, the 12 donor electrons

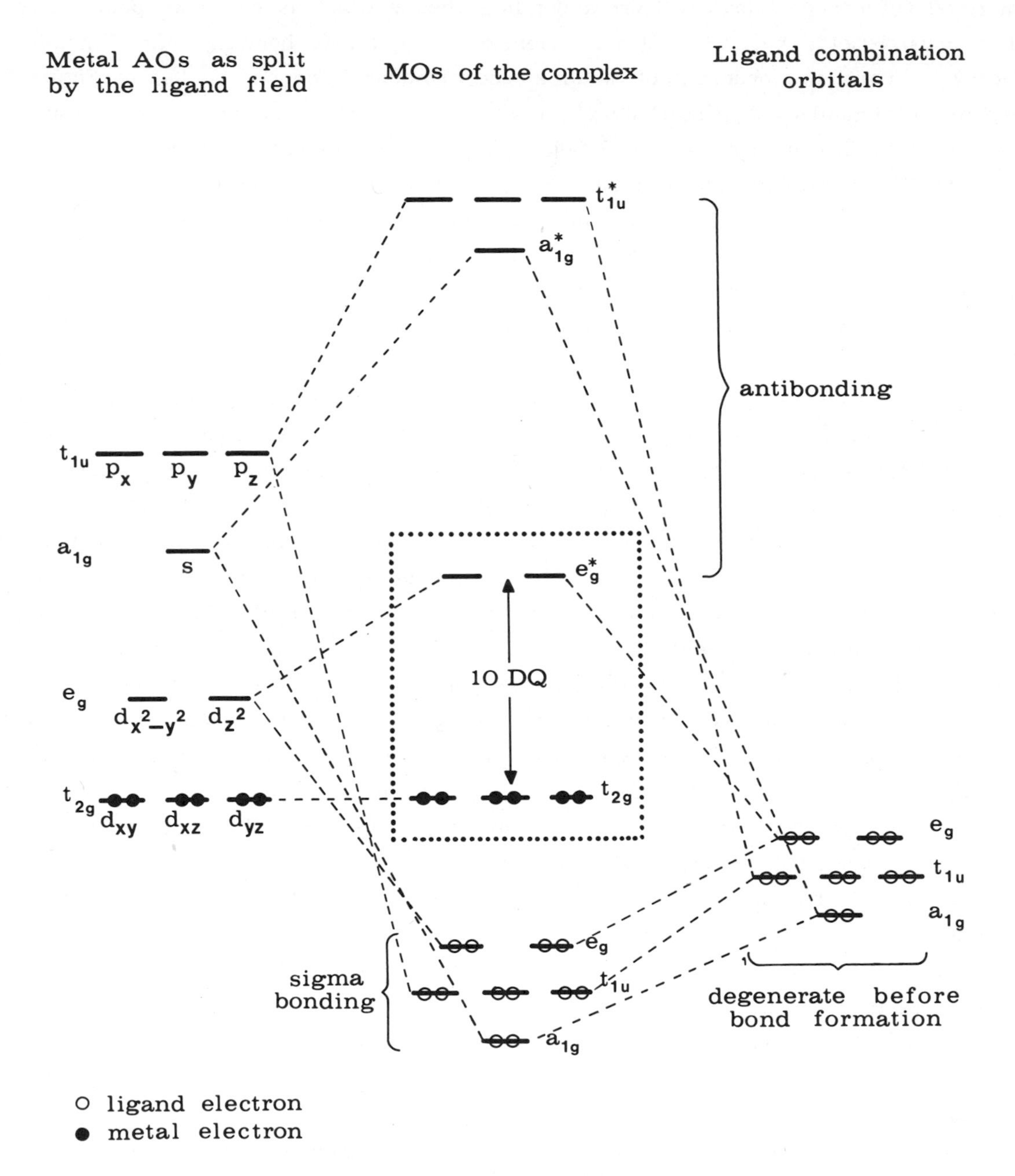


*Figure 2.2. Molecular orbital scheme for an octahedral $d^6$ complex.*

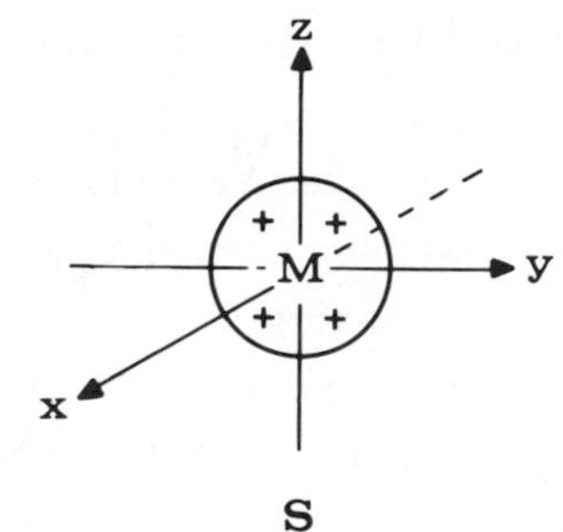

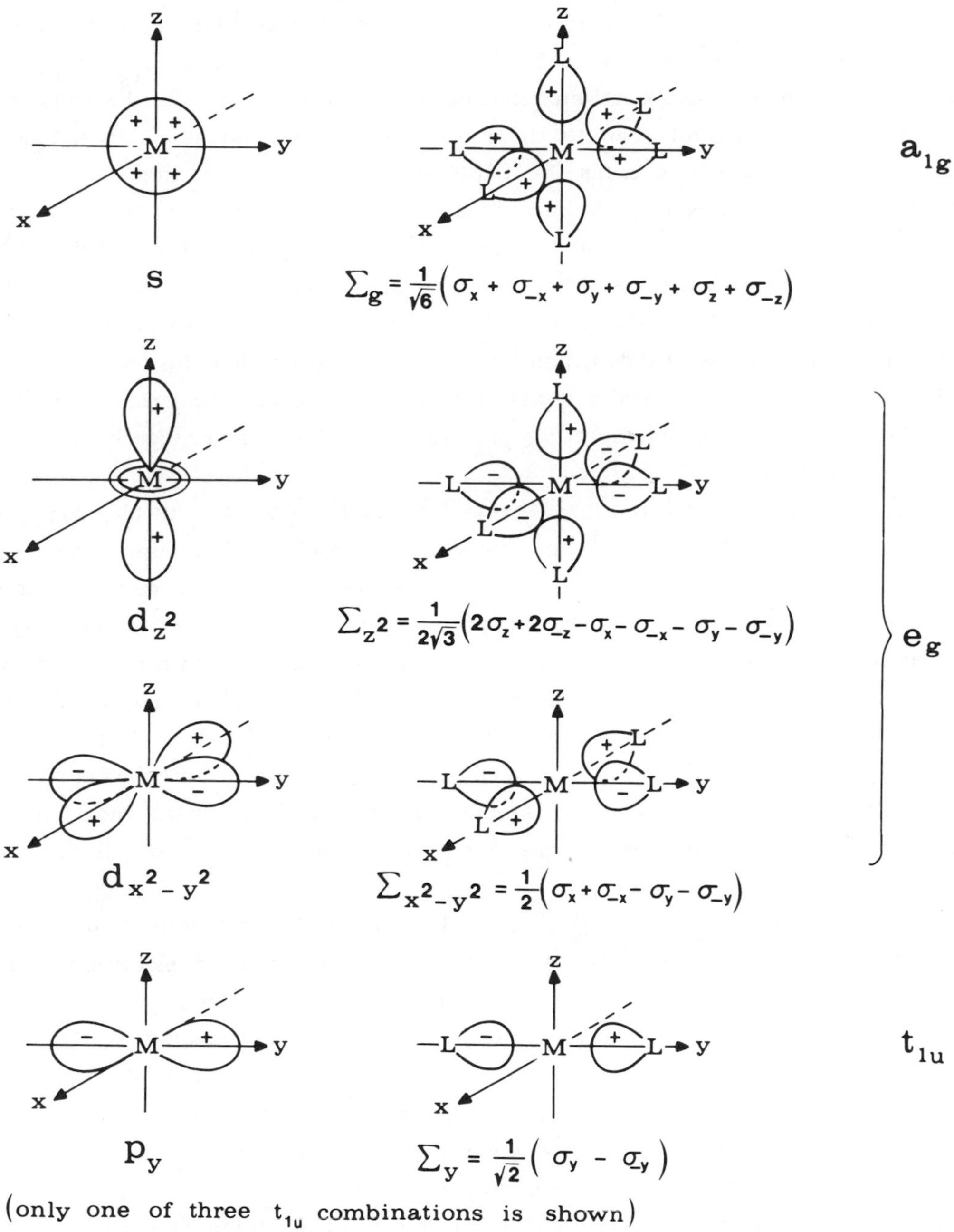

s

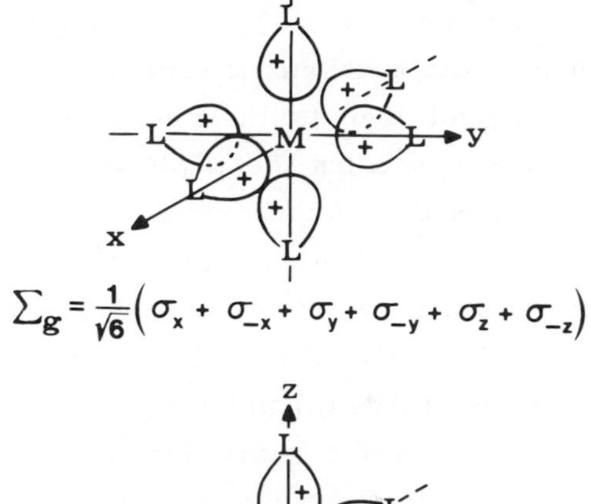


$$\Sigma_g = \frac{1}{\sqrt{6}}(\sigma_x + \sigma_{-x} + \sigma_y + \sigma_{-y} + \sigma_z + \sigma_{-z})$$

$a_{1g}$

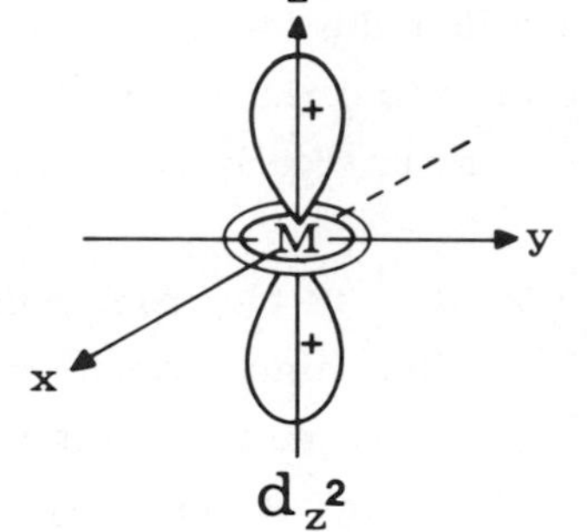


$d_{z^2}$

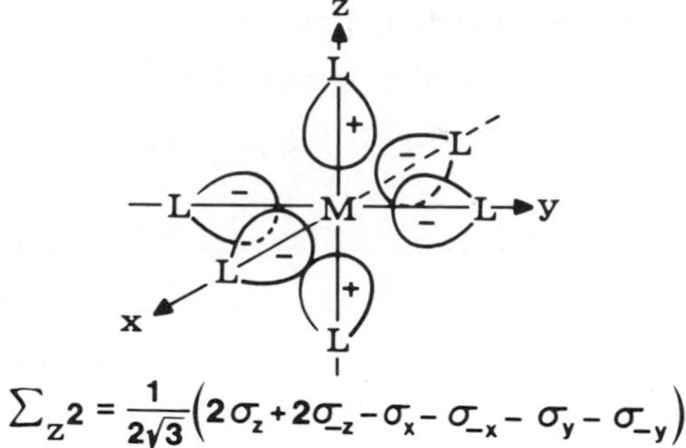


$$\Sigma_{z^2} = \frac{1}{2\sqrt{3}}(2\sigma_z + 2\sigma_{-z} - \sigma_x - \sigma_{-x} - \sigma_y - \sigma_{-y})$$

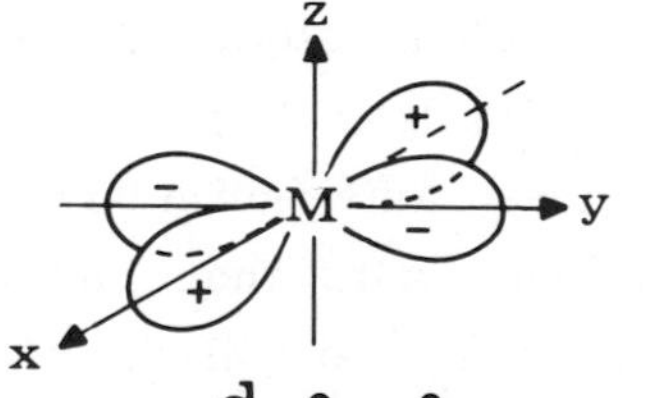


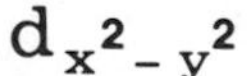

$d_{x^2-y^2}$

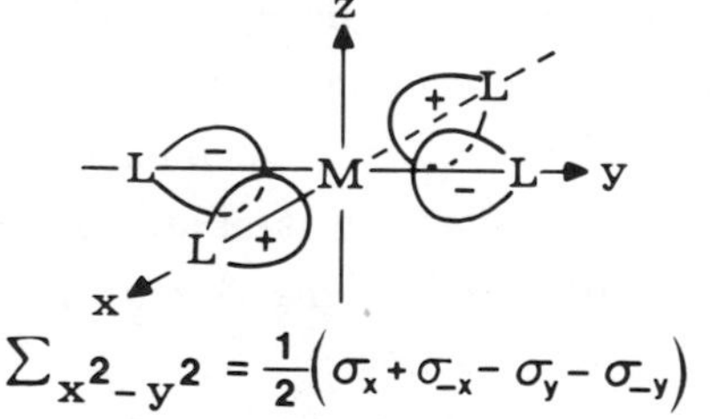


$$\Sigma_{x^2-y^2} = \frac{1}{2}(\sigma_x + \sigma_{-x} - \sigma_y - \sigma_{-y})$$

$e_g$

$p_y$

$$\Sigma_y = \frac{1}{\sqrt{2}}(\sigma_y - \sigma_{-y})$$

$t_{1u}$

(only one of three $t_{1u}$ combinations is shown)

*Figure 2.3. Metal and combination ligand orbitals of the correct symmetry for σ-bonding in octahedral complexes. (Only one of the three $t_{1u}$ combinations is shown.)*

from the six ammonia molecules occupy the six bonding molecular orbitals. Since the energies of the ammonia $sp^3$ lone pairs are much closer to these $a_{1g}$, $t_{1u}$, and $e_g$ molecular orbitals than to the metal atomic orbitals, these electrons can be considered to "belong" more to the ligand than to the metal, in spite of obvious covalent bonding. The remaining six d-electrons enter the triply degenerate, nonbonding $t_{2g}$ levels, and are little perturbed by formation of the complex. Other levels are empty. All electrons are paired and we have satisfied the 18-electron rule. Notice that our oxidation-state formalism and d configuration are based on the relative energies of the metal orbitals, the ligand orbitals, and the molecular orbitals. The $t_{2g}$ levels are called the highest occupied molecular orbitals (HOMOs), and the $e_g^*$ levels are called the lowest unoccupied molecular orbitals (LUMOs). These levels are enclosed within a box in Figure 2.2 and will later be given as d-orbital splitting patterns for other common coordination geometries as shown in Figure 2.7.

We are now in a position to offer a rationale for the 18-electron rule in molecular-orbital terminology. In an 18-electron complex, the nine lowest-energy molecular orbitals arising from the nine metal-valence-shell s-, p-, and d-orbitals are exactly filled. These filled molecular orbitals are usually bonding, since they arise from interaction of metal orbitals with ligand orbitals of the appropriate symmetry. However, in cases where there are no appropriate ligand orbitals, some of the nine filled molecular orbitals are nonbonding (for example, the $t_{2g}$ levels in Figure 2.2); but the filled MOs are never antibonding. Complexes in which these nine molecular orbitals are exactly filled are thus especially stable. A complex with more than 18 valence electrons must generally put some of them into antibonding levels and will therefore be less stable.

Next let us consider $[Ni(H_2O)_6]^{2+}$, a $d^8$, nickel(II), octahedral complex which would violate our 18-electron rule except that it has two unpaired electrons. (Recall, however, the caveat about paramagnetism.) The bonding MO levels and the $t_{2g}$ levels in Figure 2.2 are filled as before, but there are two electrons remaining to be added to doubly degenerate, antibonding, $e_g^*$ molecular orbitals. Because of the coulombic repulsion of two electrons in the same orbital and the stabilizing characteristic ("exchange energy") of singly occupied degenerate orbitals, one electron goes in each $e_g^*$ level.

The separation between the $t_{2g}$ and $e_g^*$ levels for the octahedral complex shown in Figure 2.2 is sometimes designated as 10DQ. This energy gap is especially important for classic coordination complexes of transition-metal cations. If the energy term 10DQ is smaller than the energy required to pair electrons in the same orbital, the d electrons may be distributed equally among the $\underline{t}_{2g}$ and $\underline{e}_g{}^*$ levels with minimum spin-pairing. Consider, for example, two Mn(II), $d^5$ complexes: $[Mn(CN)_6]^{4-}$, having one unpaired electron; and $[Mn(H_2O)_6]^{2+}$, having five unpaired electrons. The former is referred to as a "low-spin" and the latter as a "high-spin" complex. Several factors control the magnitude of 10DQ. "Strong-field ligands" (those which form strong covalent bonds) and π-acid ligands (*vide infra*) increase 10DQ. Third-row, 5d metals have larger 10DQ than 4d metals, which in turn have larger separations than 3d metals. The coordination geometry can also effect 10DQ; i.e., 10DQ is greater for octahedral than for tetrahedral coordination geometries. We shall find that the majority of organotransition-metal complexes are "spin-paired"; that is, 10DQ is large. If such complexes have odd numbers of electrons, there is a marked tendency to pair these by forming covalent metal-metal bonds. This is especially pronounced for second and third row transition-metal derivatives which form stronger metal-metal bonds. Paramagnetic organotransition-metal complexes are known; however, these are seldom involved as catalysts or as synthetic reagents [4].

For classic coordination compounds, the ligand-field splitting of these d levels is very important in understanding optical absorption spectra, magnetic properties, and occasionally esr (electron-spin resonance) spectra. These properties can often be explained by the ligand-field theory. However, we do not need to consider these issues in this book. For example, the weak optical bands arising from d-d transitions in organotransition-metal compounds are usually obscured by intense "charge-transfer" absorptions. In fact, as 10DQ becomes large, the d-d electronic transitions may be displaced into the ultraviolet region. Thus $Cr(CO)_6$ is colorless whereas $[Co(NH_3)_6]^{3+}$ is colored. Organometallic chemists are, therefore, little concerned with optical spectra or magnetic properties.

## 2.6. π-BONDING.

A characteristic feature of the transition metals is their tendency to form complexes with ligands such as CO, RNC, $R_3E$ (E = P, As, Sb), $R_2Z$ (Z = S, Se), NO, olefins, acetylenes, arenes, and other unsaturated organic molecules. Most of these ligands share the property of stabilizing lower metal-oxidation states, even negative ones. These ligands have in common the presence of low-lying *vacant orbitals* of the correct symmetry to form π-bonds by accepting electrons from filled metal d-orbitals, a phenomenon known as *backbonding*. Such ligands are referred to as π-acids. Some of these ligands, such as phosphines, rarely do form π-bonds (see Chapter 3).

Through $\pi$-bonding the degree of electron transfer is highly variable, and in some cases the transfer is nearly complete. This situation can confuse oxidation-state assignments, since our simple oxidation-state, electron-counting scheme ignores $\pi$-bonding. Such $\pi$-bonding stabilizes the otherwise "nonbonding" d-electron pairs, making these less available for chemical reactions. Thus, strong $\pi$-acid ligands lower the reactivity of the metal toward oxidative-addition reactions (Chapter 4). It was once thought that $\pi$-acid ligands play a special role in stabilizing transition-metal-carbon $\sigma$-bonds. This hypothesis is now largely discounted (see Chapter 3).

Cotton divides $\pi$-acids into two broad categories: longitudinal (end-bound) and transverse (side-bound). Carbon monoxide is the prototypical longitudinal $\pi$-acid (see Figure 2.4). Note that for this class the $\sigma$-bonding axis includes the nodal plane of

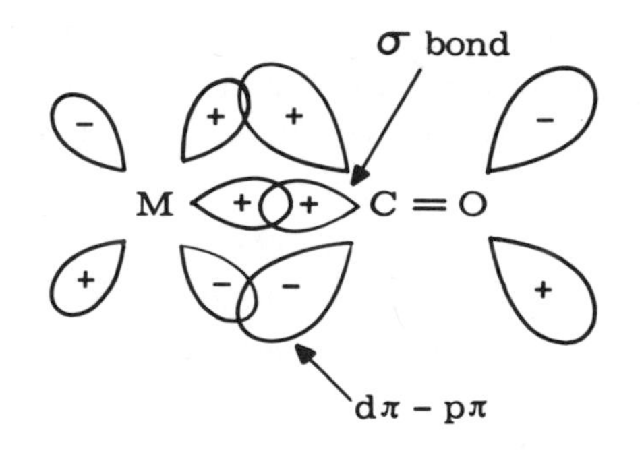


*Figure 2.4.*

the $\pi$-backbond. The algebraic sign of the d-orbital lobes match those of the anti-bonding $p\pi^*$-orbital in the region of orbital overlap, demonstrating that the symmetry is correct for bonding.

Chatt and Dewar [5] proposed the sort of transverse $\pi$-bonding manifested by olefin complexes. In this case the $\sigma$-donor bond is formed by interaction of the filled, bonding $p\pi$ molecular orbital of the olefin with an empty $\sigma$-acceptor orbital on the metal. The $\pi$-backbond involves interaction of a filled metal d-orbital with the vacant anti-bonding $p\pi^*$-orbital on the olefin (see Figure 2.5). Depending on the degree of $\pi$-bonding interaction, this transverse case can vary from a moderately perturbed olefin to a situation indistinguishable from a metallacyclopropane--in which the metal forms two

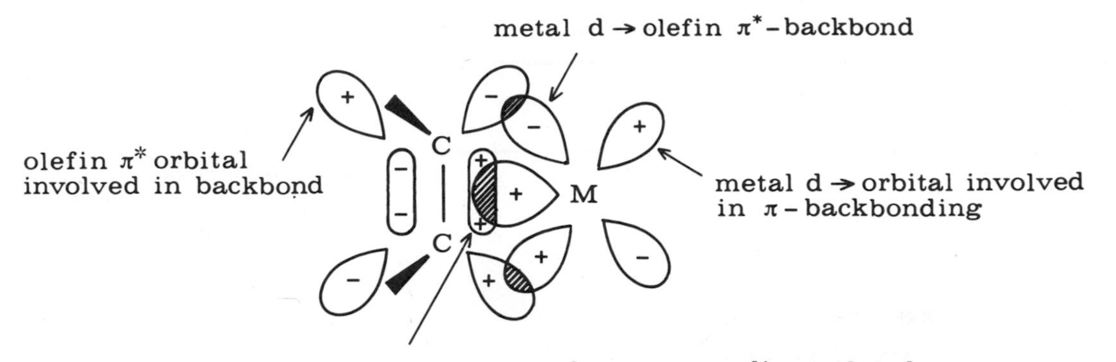


*Figure 2.5.*

σ-bonds to the terminal carbon atoms. These two extremes are simply limiting cases of a continual change rather than distinct bonding forms.

Similar pictures can be constructed illustrating π-backbonding between filled metal d-orbitals and vacant π-symmetry orbitals of many unsaturated organic groups as well as sulfur or phosphorus ligands. The possible π-bonding combinations for various metal stereochemistries are simply derived from group theory.

Figure 2.6 shows the orbital overlap for a typical π-bonding interaction between the π-molecular orbitals of four p-π acceptor ligands and a $d_{xz}$ metal orbital. This situation is found, for example, with $Cr(CO)_6$, a chromium(0), $d^6$-complex. The $t_{2g}$ levels in this octahedral complex have the correct symmetry to form π-bonding molecular orbitals. From the metal d-orbitals, only three net π-bonds are formed (distributed through interactions with the six equivalent ligand-acceptor orbitals). The effect of this π-backbonding is to further stabilize (lower) the metal $t_{2g}$ electrons as shown in the top of Figure 2.6.

If, on the other hand, the ligand π-levels are filled, they are necessarily low in energy and their interaction with the metal $t_{2g}$ orbitals will raise the energy of (destablize) the $t_{2g}$ electrons. For example, this situation occurs with the p-electron pairs on halide ligands.

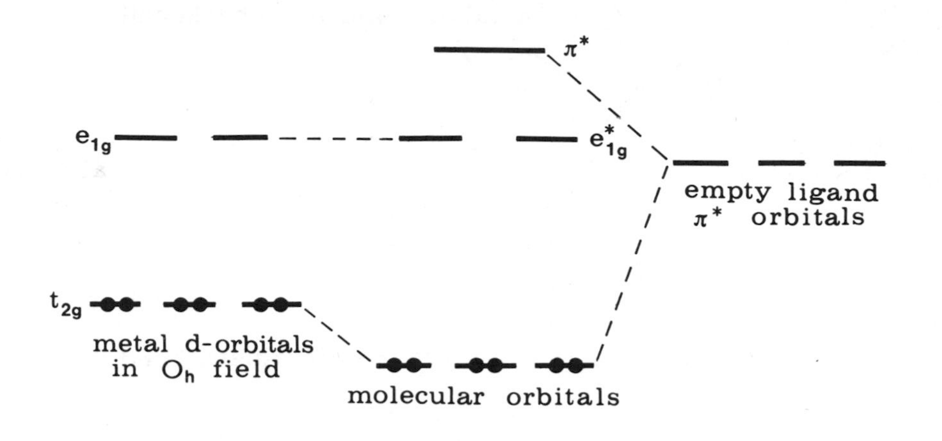


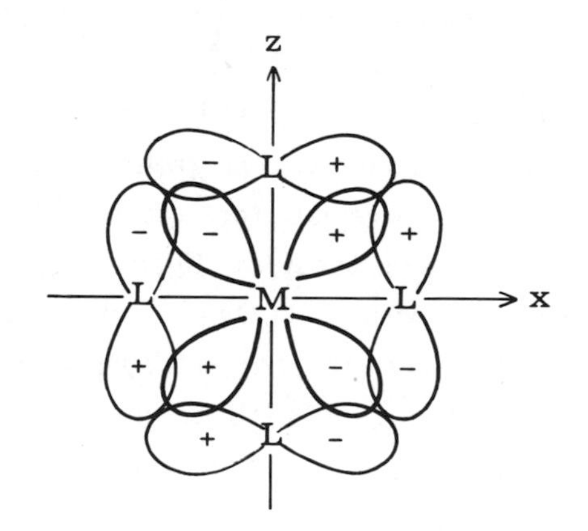


M to L π-bonding involving the $d_{xz}$ orbital of the $t_{2g}$ set and a combination of empty ligand orbitals.

*Figure 2.6. Orbital overlap for π-bonding.*

The former (backbonding) case is far more important with low-oxidation-state organotransition-metal compounds. Such $\pi$-backbonding stabilizes the metal electrons by distributing these over ligand orbitals, thus reducing any excess negative charge on the central metal. Backbonding is diminished by positive charge at the central metal. Consequently metal anions are more prone to backbonding than metal cations. Such $\pi$-bonding requires a close approach of the ligand to the metal, since the overlap of $\pi$-orbitals occurs only at short distances. We shall see that shortening of the bond between a metal and a ligand is one experimental criterion for $\pi$-backbonding. The special affinity which low-oxidation-state transition metals show for the $\pi$-acid ligands is readily explained by $\pi$-backbonding. The distinction between organic derivatives of the transition elements and the nontransition elements can be partly explained by the phenomenon of $\pi$-backbonding, which requires partially filled d-orbitals on the central metal. It should be mentioned that $d\pi$-$d\pi$ backbonding from a filled metal d-orbital to any empty ligand d-orbital is also recognized, for example, with $PF_3$ complexes (see Chapter 3).

Let us return to the MO diagram in Figure 2.2 for octahedral transition-metal complexes and summarize the important points. To a first approximation, the $d^6$ configuration for such complexes consists of a set of the nonbonding d-electron pairs which may be further stabilized by $\pi$-backbonding. This is the HOMO set for the complex, a fact that is consonant with the notion that the most easily removed electrons in a transition-metal complex are metal-centered. This is the principle upon which our oxidation-state formalism is based. Further electrons must be added to degenerate, antibonding metal levels ($e_g^*$). For the octahedral case, this explains the 18-electron rule as it applies to diamagnetic complexes. The $\sigma$-bonding MO framework is occupied by electrons which, on energetic grounds, may be assigned to the ligands in spite of obvious covalency. This is the basis of the idea of a coordinate bond.

An approximate theory such as this is not especially useful for calculating physical properties, such as bond energies, but it does present a qualitative picture of chemical features. As one complex is converted into another, the orbitals that are involved can be correlated, and this can be used as a basis for qualitative discussions of reaction paths.

## 2.7. Ligand-Field d-Orbital Splitting for Various Coordination Geometries.

An essential feature of the above analysis is the splitting of the metal d-orbitals under the influence of a ligand field having a prescribed stereochemistry. This splitting can be derived from group theory, augmented by intuitively reasonable energy arguments. Similar ligand-field-splitting diagrams are given in Figure 2.7 for six common coordination geometries. Each coordination geometry is depicted with the usual assignment of cartesian axes and symmetry-group notation. The ligand-field-splitting pattern is shown beside each structure, but is not drawn to scale. An asterisk is used to label those d-orbitals that are antibonding and would be empty in spin-paired complexes. The appropriate orbitals for σ- and π-bonding are listed beside each splitting diagram. In each case a first-order approximation is made that the σ-bonding electrons are localized on the ligands.

The formation of a σ-bonding MO from a metal d-orbital must produce, in addition to the low-energy bonding MO, a high-energy antibonding MO which is largely d-orbital in character. For spin-paired complexes it turns out that the occupied d-orbitals are usually nonbonding. For these complexes all or part of the d-electrons are in nonbonding orbitals which have arisen from d-orbitals for which there is not a ligand σ-bonding MO of corresponding symmetry.

Next let us consider in turn each of the common coordination geometries displayed in Figure 2.7. The trigonal-planar geometry is uncommon, being found mostly for coordinatively unsaturated $d^{10}$ complexes of platinum(0) and palladium(0). For such cases, $d_{z^2}$, $d_{x^2-y^2}$, and $d_{xy}$ orbitals are filled, and therefore these levels must be only weakly antibonding. This is the geometry expected on the basis of ligand-ligand repulsion when three bulky ligands are bound to a metal having a spherical, filled d shell. All three coordination sites are equivalent, and there is no possibility of isomerism.

The tetrahedral geometry is commonly found for coordinatively saturated, diamagnetic, $d^{10}$ complexes of nickel(0), palladium(0), and platinum(0). For tetrahedral complexes $d^{n}$ where $n < 10$, the separation between the antibonding and bonding d levels is small; so high-spin coordination complexes are usually found. However, such situations are less often encountered with organotransition-metal compounds. This diagram indicates that any tetrahedral $d^8$ complex should exist as a ground-state triplet.

Note that there are more orbitals available than are required to form four σ-bonds. Various mathematical mixes are possible, but this point cannot be predicted *a priori*. There are also more π-bonding orbitals available than can be utilized. Furthermore, in the tetrahedral case and other examples in Figure 2.7, the same atomic orbitals

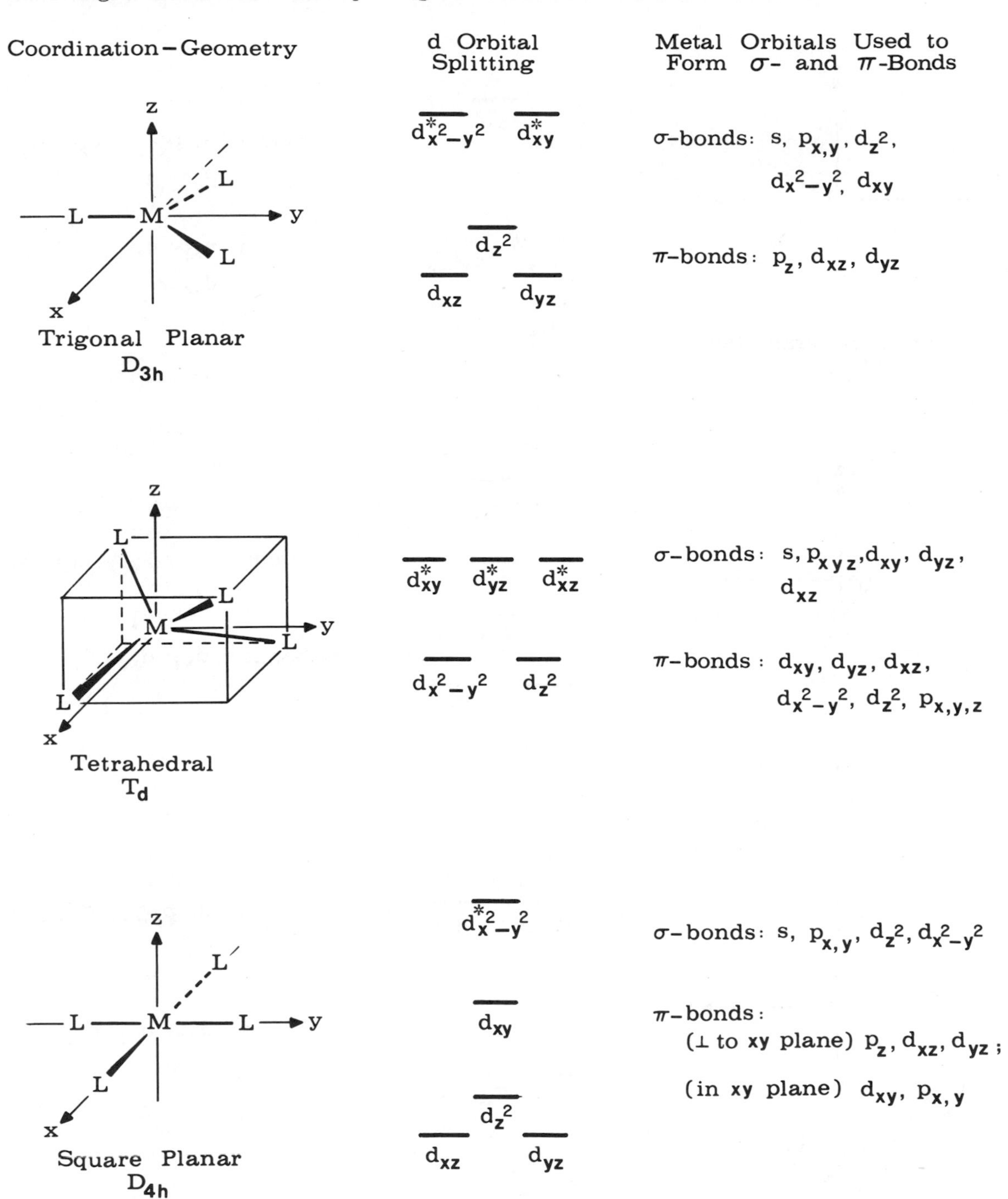


*Figure 2.7. Ligand-field splitting of d-orbitals in selected coordination geometries.*

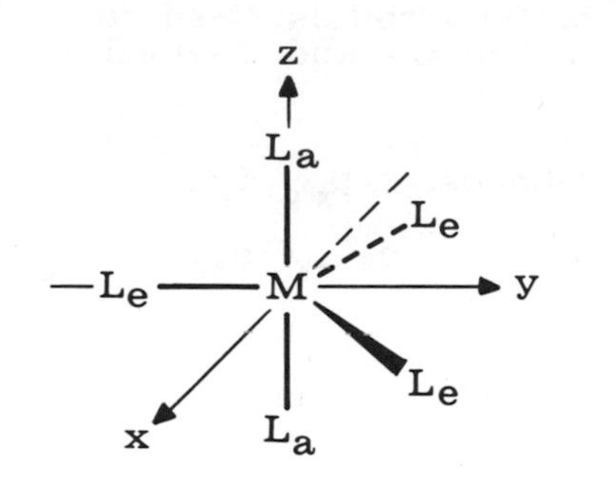


Trigonal Bipyramidal
$D_{3h}$

$d^*_{z^2}$

$d_{xy}$ $d_{x^2-y^2}$

$d_{xz}$ $d_{yz}$

$\sigma$-bonds: s, $p_{x,y,z}$, $d_{z^2}$, $d_{xy}$, $d_{x^2-y^2}$

$\pi$-bonds: $d_{xy}$, $d_{x^2-y^2}$, $d_{xz}$, $d_{yz}$

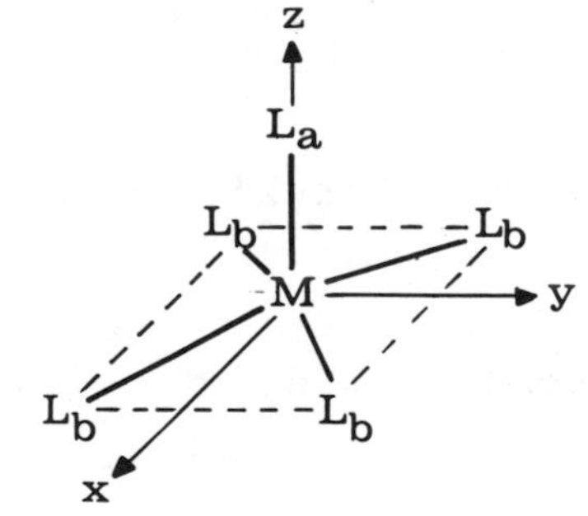


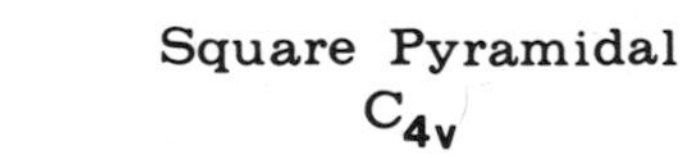
Square Pyramidal
$C_{4v}$

$d^*_{x^2-y^2}$

$d_{z^2}$

$d_{xy}$

$d_{xz}$ $d_{yz}$

$\sigma$-bonds: s, $p_{x,y,z}$, $d_{x^2-y^2}$, $d_{z^2}$

$\pi$-bonds: $d_{xy}$, $d_{xz}$, $d_{yz}$

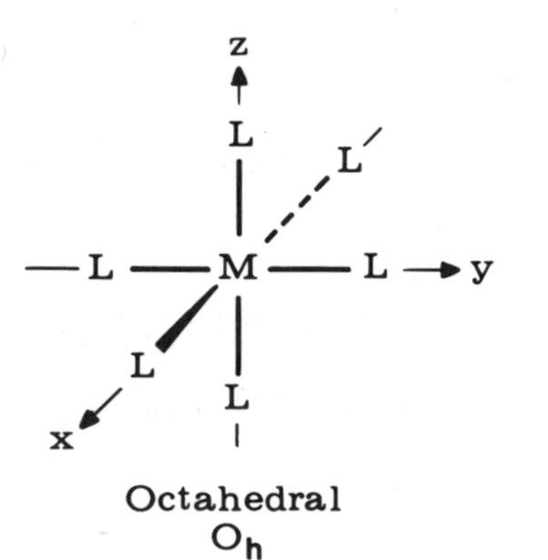


Octahedral
$O_h$

$d^*_{x^2-y^2}$ $d^*_{z^2}$

$d_{xy}$ $d_{xz}$ $d_{yz}$

$\sigma$-bonds: s, $p_{xyz}$, $d_{x^2-y^2}$, $d_{z^2}$

$\pi$-bonds: $d_{xy}$, $d_{xz}$, $d_{yz}$

*Figure 2.7. Continued.*

have the correct symmetry to form both σ- and π-bonds. Since σ-bonds are usually stronger than π-bonds, one assumes that the orbitals which could by symmetry form both kinds of bonds are preferentially used to form σ-bonds.

The tetrahedral geometry minimizes steric interactions between four bulky ligands, as, for example, compared with the square-planar geometry. The situation with isomers for a tetrahedral complex is familiar to all organic chemists and deserves no further comment.

The square-planar geometry is very commonly encountered among coordinatively unsaturated, diamagnetic, $d^8$ complexes of Ni(II), Pd(II), Pt(II), and Co(I), Rh(I), Ir(I). The $d_{x^2-y^2}$ level is strongly antibonding and is usually vacant, but this level is singly occupied for square-planar $d^9$ copper(II) complexes. Note the number of orbitals which may be used for π-bonding either in the $x$, $y$-plane or perpendicular to it. Two relative coordination sites, cis and trans, may be occupied by any pair of equivalent ligands in the square-planar geometry. Thus configurationally stable cis and trans isomers are commonly observed for diamagnetic $d^8$ complexes. Distortion of the square-planar into the tetrahedral geometry for $d^8$ complexes is accompanied by conversion from a $D_{4h}$ diamagnetic, singlet state to a $T_d$ paramagnetic, triplet state. This is occasionally observed for "weak ligand-field" complexes--especially in the case of first-row 3d metals.

On purely steric grounds, the tetrahedral geometry is favored over the square-planar geometry, but the latter occurs with $d^8$ and $d^9$ complexes because of the more favorable electronic situation. This is one of the few instances in which the coordination geometry is dictated by the number of d electrons. Another example is the rare T-shaped three-coordinate $d^8$ geometry, which is not included in Figure 2.7, but which we will encounter as a postulated intermediate in reductive elimination (in Chapter 4).

It should be noted here that square-planar $d^8$ complexes such as platinum(II) or rhodium(I) derivatives undergo facile bimolecular ($S_N2$) ligand-exchange reactions--often without cis-trans isomerism [6].

The trigonal-bipyramidal geometry is commonly found among coordinatively saturated $d^8$ complexes, for example, $Fe(CO)_5$ [7]. This configuration exhibits two distinct sites: two axial and three equatorial positions. The axial bonds make 90° angles with the plane containing the metal and the three equatorial groups, whereas the angle between pairs of equivalent equatorial groups is 120°. Interconversions between axial and equatorial positions occur very readily without dissociation of a ligand, by a process which has been referred to as "pseudorotation" (see Figure 2.8). For example,

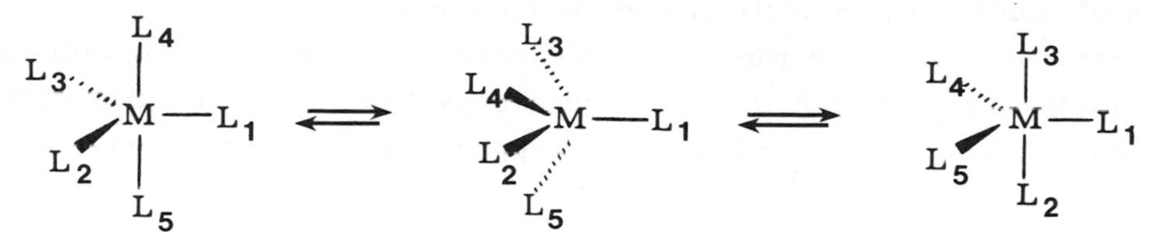


*Figure 2.8. Pseudorotation.*

the $^{13}C$ NMR spectrum of $Fe(CO)_5$ shows a single sharp signal even at low temperatures, implying rapid interconversion of the axial and equatorial CO groups. For a symmetrically substituted complex, the more stable of the two possible $d^8$ trigonal-bipyramidal forms may be estimated by noting that pairs of bulky ligands preferentially occupy axial positions, but a triad of bulky ligands will usually assume three equatorial positions. In $d^8$ complexes, $\pi$-backbonding is strongest for equatorial sites, but strong $\sigma$-donors should prefer axial sites. Five-membered chelate rings tend to span one equatorial and on axial position. However, the energy differences may be quite small; so these guidelines are not reliable. The situation with $d^{10}$ or $d^0$ complexes is different, but lies out of the scope of our discussion [7].

It is important to note that trigonal-bipyramidal $d^6$-complexes should have a triplet ground state, since the two highest-energy electrons would occupy the degenerate $d_{xy}$ and $d_{x^2-y^2}$ levels. This prediction is valid only if the five ligands have the same ligand-field strength. This rule can have important consequences for migratory insertion reactions in which the coordination number is reduced from six to five (see Chapter 5). Trigonal-bipyramidal $d^8$ complexes usually undergo ligand-exchange reactions by dissociative ($S_N1$) paths.

The square-pyramidal coordination geometry is known for a number of transition-metal compounds [7]. The metal is usually located on an axis above the plane defined by four basal ligands, $L_b$, on the same side as the apical ligand, $L_a$ (Figure 2.7). The ordering of the $d_{xy}$, $d_{xz}$, and $d_{yz}$ energy levels shown in Figure 2.7 is variable; it depends on the $L_b$-M-$L_b$ angle and on $\pi$-bonding with apical and basal ligands. This complicated situation, which is outside the scope of our discussion, has been analyzed by Hoffmann [7]. The potential-energy surface connecting the square-pyramidal and trigonal-bipyramidal geometries has low activation-energy barriers. Thus these two geometries are easily interconverted--especially for $d^8$

configurations. However, it should be noted that the square-pyramidal geometry can easily accommodate six d-electrons in a spin-paired configuration. Thus coordinatively unsaturated, diamagnetic, five-coordinate, $d^6$ complexes often exhibit this geometry. Such complexes are often configurationally stable. This has an important consequence in that dissociative, $S_N1$, ligand-exchange reactions of certain octahedral $d^6$ complexes occur without cis-trans isomerism. Such reactions may involve square-pyramidal intermediates, although it is difficult to rule out the possibility that a polar-solvent molecule temporarily occupies the sixth coordination site.

The octahedral configuration is the most common stereochemistry exhibited by transition-metal compounds. It is especially characteristic of kinetically inert complexes having a $d^6$ configuration [Co(III), Rh(III), and Ir(III)], or a $d^3$ configuration [Cr(III)]. In these cases, the triply degenerate $t_{2g}$ level ($d_{xy}$, $d_{xz}$, $d_{yz}$) is filled or half-filled. This level may be further stabilized by $\pi$-backbonding. Octahedral complexes with $d^8$ configurations are paramagnetic. All of the ligand positions in an octahedral complex are equivalent. Pairs of ligands can have a cis or a trans relationship. A triad of ligands can occupy a trigonal face of the octahedron (facial or fac configuration) or meridional positions (mer configuration). It is useful to note that the octahedral configuration is equivalent to a trigonal-antiprism (see Figure 2.9). The

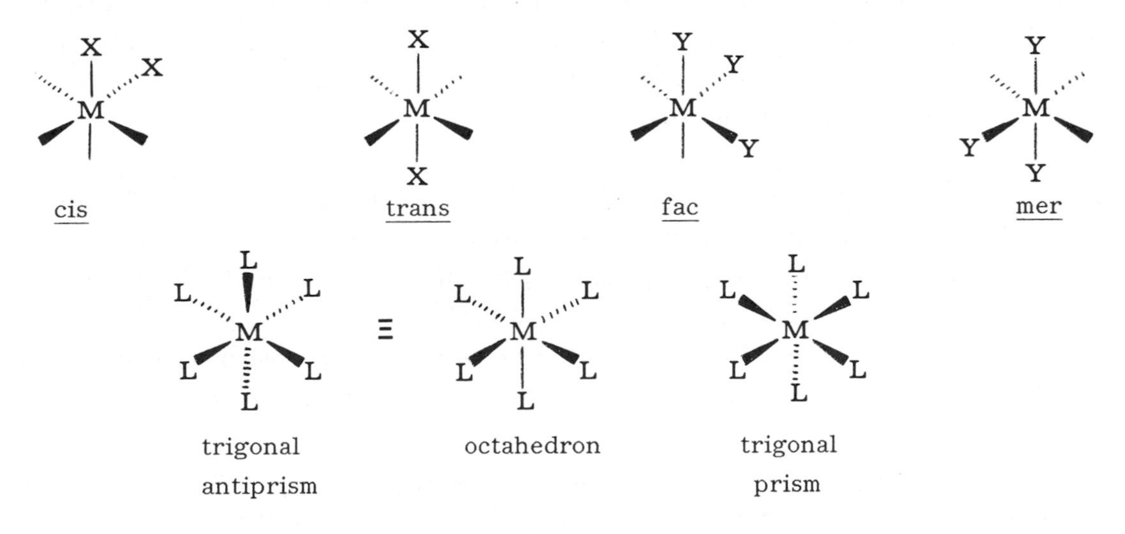


*Figure 2.9.*

trigonal-prismatic configuration [8] is also known for some transition-metal coordination compounds, but is rarely encountered among organotransition-metal compounds. Note the presence of two trigonal and three rectangular faces for this configuration.

Coordination numbers greater than six are occasionally encountered, especially among elements to the left side of the transition series and for $d^{n}$ ($n < 6$). The geometries of these higher coordination numbers are variable, and the interconnecting potential-energy surfaces are quite deformable, resulting in nonrigid or "fluxional" behavior. For these cases we will not discuss d-orbital splitting diagrams, but simply note some of the possible coordination geometries. For 7-coordination, three stereochemistries are encountered: a pentagonal-bipyramid, a capped octahedron (in which a seventh ligand has been added to a triangular face), and a capped trigonal-prism (in which a seventh ligand has been added to a rectangular face). Several polyhedra are found among 8-coordinate compounds, and these are separated by small energy barriers: The cube, the dodecahedron, and the square-antiprism. These higher coordination geometries have little significance in our discussion of organometallic reactions. Further details are given by Huheey [9].

## 2.8. Transition-Metal Complexes of Unsaturated Organic Ligands.

In the preceding discussion, we presented an elementary method for describing the molecular orbitals involved in forming coordination complexes of the transition metals and for understanding the occupancies of the metal-centered d-orbitals which represent the HOMOs and LUMOs. This system rationalizes the "18-electron rule" connecting the coordination numbers, $d^{n}$, and coordination geometries. Next let us generalize these ideas to encompass bonding between extended unsaturated organic molecules and transition metals. The method is actually the same and gives results very similar to those we have already developed.

Consider as an example *bis*-benzene chromium, one of the first "sandwich compounds." Figure 2.10 shows diagrams of selected $\pi$ molecular orbitals of benzene arranged according to their symmetry with respect to a bond from the center of the ring to a metal atom such as chromium.

We now assume $D_{6h}$ symmetry. The two sets of benzene $\pi$-molecular orbitals may be added and subtracted to give twice as many combination molecular orbitals of $g$ and $u$ symmetry, respectively. These have the same symmetries as the metal atomic orbitals. The combinations with symmetry designations $a_{1g}$, $a_{2u}$, $e_{1g}$ and $e_{1u}$ are illustrated in Figure 2.10. Only one of each doubly degenerate $e_{1g}$ and $e_{1u}$ set is shown.

As we did before for an octahedral complex, a molecular-orbital energy-level diagram can be constructed using intuitive concepts to assign qualitatively the relative energies. This is shown in Figure 2.11. Filling the resulting molecular orbitals of *bis*-benzene chromium(0) with the 12 $\pi$-electrons from the two benzene molecules results

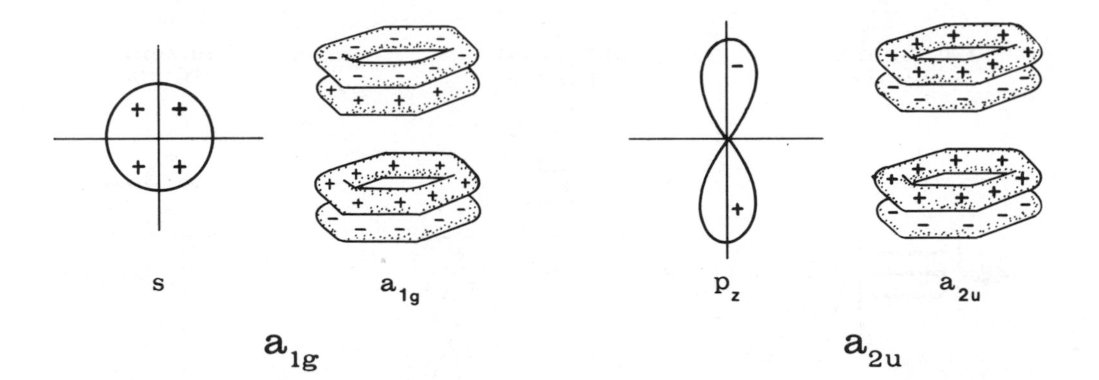


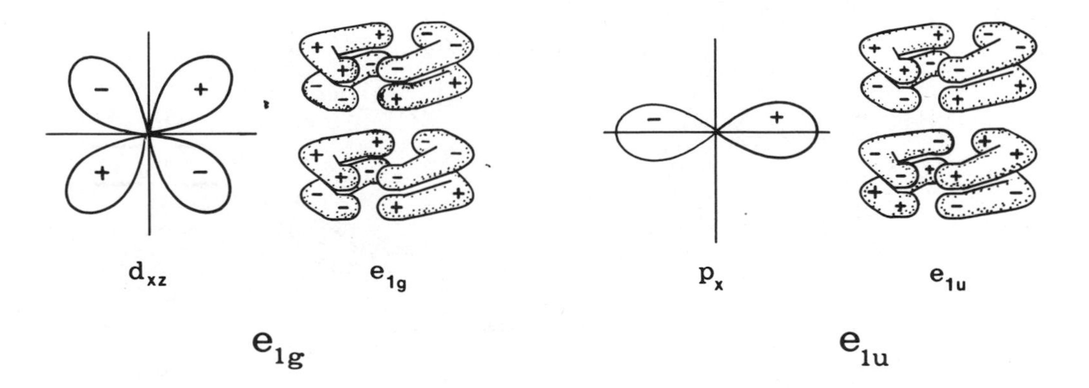


*Figure 2.10. Combinations of metal AOs and benzene MOs to form bonding MOs for bis-benzene chromium.*

in six filled bonding molecular orbitals (one each $a_{1g}$ and $a_{2u}$, and two each $e_{1g}$ and $e_{1u}$). These electrons "belong" primarily to the two benzene ligands, which are each six-electron donors. The six d-electrons from chromium(0) are then used to fill the next three orbitals which are mostly localized at the metal. Two of these, $e_{2g}$ ($d_{x^2-y^2}$

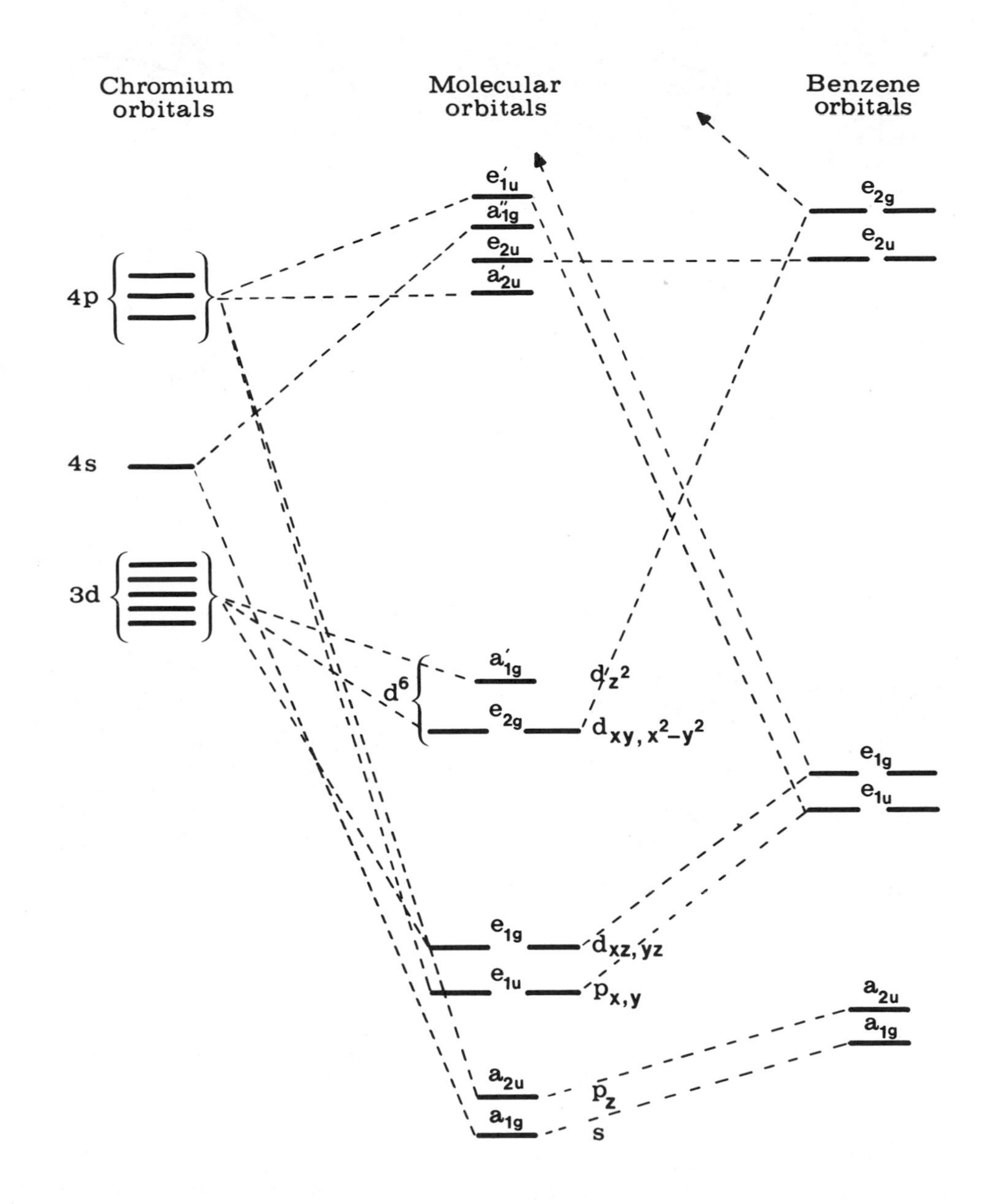


*Figure 2.11. Molecular orbital-energy diagram for bis-benzene chromium.*

and $d_{xy}$) overlap with the empty, antibonding benzene orbitals. These levels are therefore somewhat stabilized by backbonding from the metal to the ligand. The $a_{1g}'$ level is a weakly antibonding $d_{z^2}$ orbital (the complement of the σ-bonding $a_{1g}$ level). The primes in Figure 2.11 are used to distinguish the higher-energy MO of a pair of MOs having the same symmetry. The empty antibonding $e_{1g}'$ and $e_{2g}'$ levels are not shown. Empty benzene $b_{2g}$ and $b_{1u}$ orbitals are also omitted.

Figures 2.10 and 2.11 give a graphic description of the bonding in a coordinatively saturated organometallic π-complex. In spite of the apparent complexity, there is little new information in this bonding scheme beyond that gained from the similar bonding schemes for $[Co(NH_3)_6]^{3+}$ or $Cr(CO)_6$. All three substances may be considered as six-coordinate $d^6$ complexes which satisfy the 18-electron rule. The d-electrons in $(C_6H_6)_2Cr$ are substantially nonbonding, somewhat stabilized by backbonding, as in $Cr(CO)_6$ but not in $[Co(NH_3)_6]^{3+}$. The $a_{1g}'$ level holds the most easily removed (HOMO) electrons in the complex. Rather than three-degenerate d-orbitals, which we found for the purely octahedral complexes in *bis*-benzene chromium, we find two-degenerate and one-unique d-orbitals which are filled. In Chapter 3 we will see that ferrocene, $(C_5H_5)_2Fe$, gives a similar picture.

For other common unsaturated organic ligands that form π-donor complexes with transition metals, Figure 2.12 lists the molecular orbitals capable of forming bonds of σ, π, and δ-symmetries. These ligand orbitals can be used to construct molecular-orbital schemes and energy-level correlation diagrams for complexes between these ligands and transition metals having several $d^{n}$ configurations and geometries. These schemes are analogous to that shown in Figure 2.10. However, we will not give additional examples of these bonding diagrams. Rather, we will introduce and illustrate a very simple electron-counting procedure which can be used for all such unsaturated organic ligands as well as simple ligands. The practicing synthetic chemist needs no further bonding information to deal with the reactions and physical properties of organotransition-metal compounds. The bonding pictures and orbital-correlation diagrams simply make us secure in the knowledge that there are rational analyses for the chemical bonds in organotransition-metal compounds. These MO schemes become important only if we wish to correlate orbitals for molecular rearrangements, such as intramolecular insertions, or to note the nature of empty LUMO or filled HOMO for photochemical and electrochemistry experiments. The latter themes are outside the scope of our presentation.

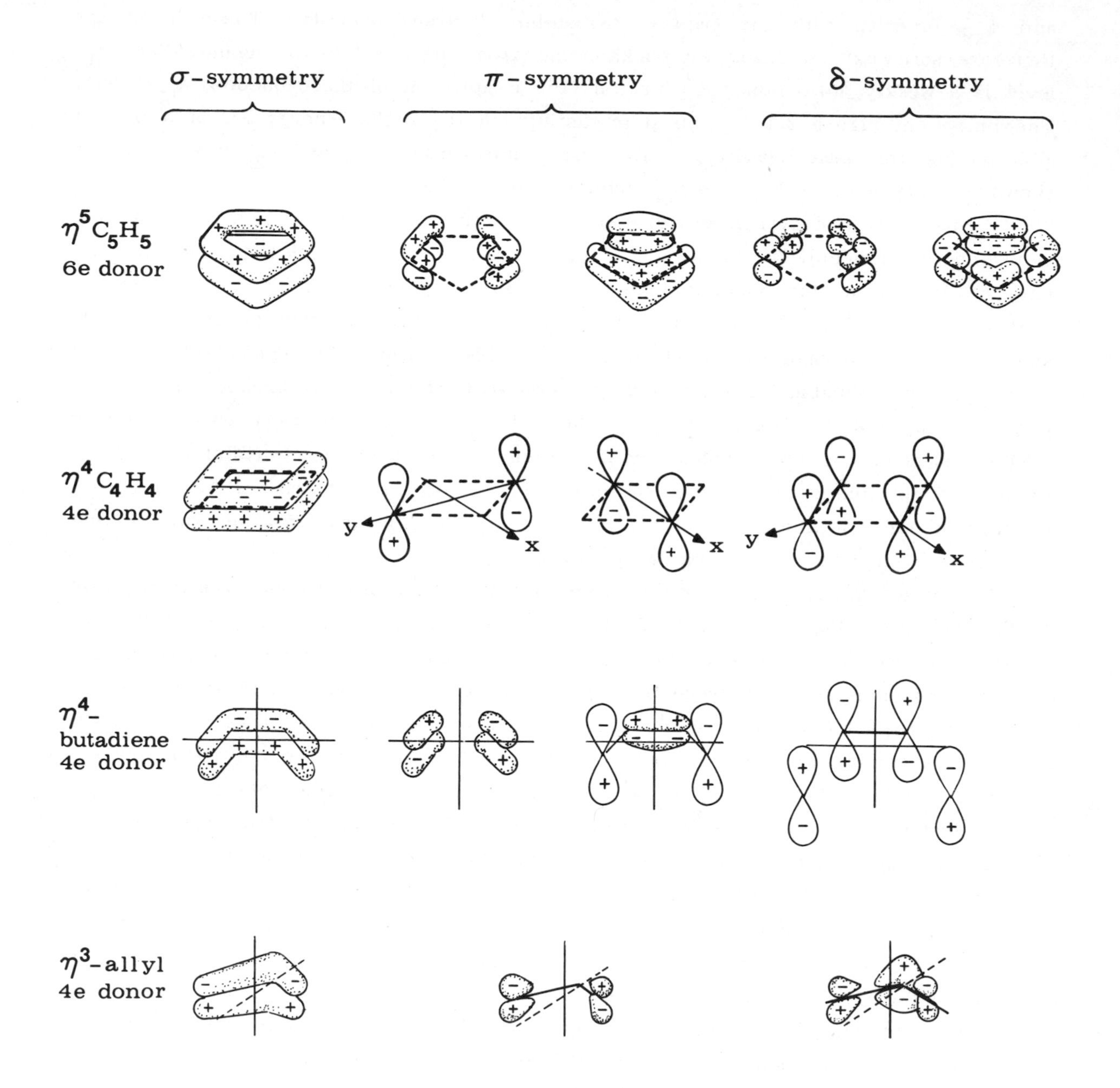


*Figure 2.12. Ligand orbital representations for overlap with metal AOs to form complex MOs having the indicated bonding symmetries.*

## 2.9. A SIMPLIFIED ELECTRON-COUNTING SCHEME.

It is possible to assign a self-consistent set of coordination numbers and charges to most of the ligands we may encounter. These are listed in Table 2.2. For any given complex, we can compute the oxidation state and thus the $d^{\underline{n}}$ value for the central metal by using the values for coordination numbers and ligand charges given in Table 2.2. The number of donor electrons is taken as twice the sum of ligand coordination numbers. The combination of the over all coordination number and $d^{\underline{n}}$ can be used to predict the idealized coordination geometry, whether the complex is coordinatively saturated and, if $\underline{n}$ is even, whether the electrons are expected to be paired.

*Table 2.2. Charges and corresponding coordination numbers for typical ligands.*

| Ligand | Charge[a] | Coordination Number[a] |
|---|---|---|
| X(Cl, Br, I) | -1 | 1(2) |
| H | -1 | 1(2,3) |
| $CH_3$ | -1 | 1(2) |
| Ar | -1 | 1 |
| RCO | -1 | 1(2) |
| $Cl_3Sn$ | -1 | 1 |
| $R_3Z$ (Z = N, P, As, Sb) | 0 | 1 |
| $R_2E$ (E = S, Se, Te) | 0 | 1 |
| CO | 0 | 1(2,3) |
| RNC: | 0 | 1(2) |
| RN: | 0(-2) | 1(2) |
| $R_2C$: | 0(-2) | 1(2) |
| $N_2$ | 0 | 1(2) |
| $R_2C = CR_2$ | 0(-2) | 1(2) |
| $RC \equiv CR$ | 0(-2) | 1(2) |
| CN | -1 | 1 |
| $\eta^4$-cyclobutadiene | 0 | 2 |
| $CH_2 = CHCH_2$- | -1 | 1 |
| $\eta^3$-allyl | -1 | 2 |
| $\eta^6$-benzene ($C_6H_6$) | 0 | 3(2,1) |
| $\eta^5$-cyclopentadienyl ($C_5H_5$) | -1 | 3 |
| $\eta^7$-cycloheptadienyl ($C_7H_7$) | +1 | 3 |
| NO[b] | +1(-1) | 1(2) |
| $ArN_2$[b] | +1(-1) | 1(2) |
| O | -2 | 2 |
| $O_2$ | -2(-1) | 2(1) |

[a] Less common or alternative formulation in parentheses.

[b] Non-innocent ligand (two or more discrete bonding modes).

Some ambiguities arise in the ligand charges, especially in cases of "non-innocent" (amphoteric) ligands, such as NO. We will consider these cases as they are encountered in Chapter 3. In other instances it is a "matter of taste" whether to use one or another set of values; these are denoted by values given in parentheses in Table 2.2. However, most of the compounds we will encounter can be classified within the framework provided in Table 2.2, using the oxidation-state definition and $d^{\underline{n}}$ assignments given earlier. The expected coordination geometries are easily derived from the ligand-field d-orbital splitting patterns presented in Figure 2.7.

The coordination numbers given in parentheses in Table 2.2 are less common (or alternative) values, and encompass both situations in which a ligand is bound to a single metal and examples in which a ligand can bridge. In certain cases coordination numbers greater than one do not imply more than one pair of ligand donor electrons; for example, bridging hydride has a coordination number of two, or even three, but donates a total of only two electrons. These cases will be clarified as we encounter specific examples in Chapter 3.

The simple halide, hydride, alkyl, aryl, and acyl ligands in Table 2.2 require little comment, since all are considered as mono-anions and two-electron donors. For these cases the usual coordination number is one, but higher coordination numbers (given in parentheses) denote bridging ligands. These bridges can form by donation of a second electron pair (bridging halides) or by formation of a three-center bond (bridging hydrides). The $Cl_3Sn^-$ ligand forms compounds which are isostructural with alkyls. Neutral two-electron donors, such as amines, phosphines, organic sulfides, carbon monoxide, and isonitriles, have a coordination number of one; that is, these are "unidentate," and the latter two can bridge. Nitrenes (RN), carbenes, and dinitrogen are analogous to CO. However, nitrene can alternatively be considered a dianion analogous to an oxide group. In Chapter 3 we shall find bridging $N_2$ is structurally different from bridging CO. Olefins can alternatively be considered as doubly negative ligands (metallacyclopropane formalism) with a CN of 2 or as a neutral ligand with a CN of 1. In these cases the olefin is considered as a four- and a two-electron donor, respectively. This situation is discussed further in Chapters 3 and 12. The same dichotomy is found with monometallic acetylene complexes in which the acetylene can be considered a four-electron donor (metallacyclopropene) or a two-electron $\pi$-donor (see Figure 2.13). In dimetallic complexes, acetylenes sometimes display another bonding mode in which each $\pi$-bond of the acetylene is donated to a different metal. An example is $Co_2(CO)_6(HC{\equiv}CH)$.

Cyclobutadiene, which is unstable as the free ligand, forms very stable diamagnetic complexes analogous to those of butadiene. Thus both are considered

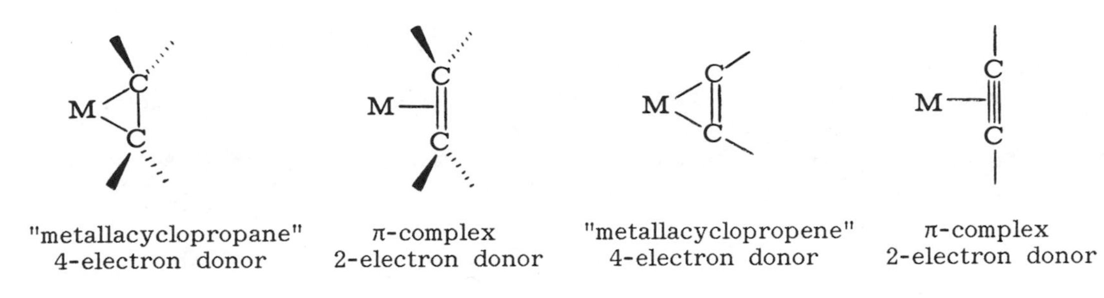


*Figure 2.13.*

neutral bidentate ligands (CN = 2). There are two forms of allyl, the σ- ($\eta^1$) or π-bonded ($\eta^3$) form. The hapto or ηn nomenclature indicates the number of ligand atoms which are attached to the metal. This is discussed in greater detail in Chapter 3. The σ-allyl (a two-electron donor) is analogous to simple alkyls, whereas the π-allyl has a CN = 2 (a four-electron donor). Both forms are treated as mono-anions. The aromatic ring systems $C_6H_6$, $C_5H_5^-$, and $C_7H_7^+$ have the indicated charges, but each may occupy three coordination sites and are considered to be six-electron donors. The non-innocent and amphoteric ligands NO and $ArN_2$ may be considered as positive or negative ligands. These cases are discussed in Chapter 3. Dioxygen, $O_2$, may be considered as a bidentate ($\eta^2$) peroxide (2-) or a monodentate ($\eta^1$) superoxide (1-) ligand.

The isoelectronic six-coordinate complexes in Figure 2.14 further illustrate the use of the information in Table 2.2 in constructing series of analogous, isoelectronic compounds. All of these are $d^6$ and octahedral or quasi-octahedral. A similar set of five-coordinate saturated, $d^8$ complexes are given in Figure 2.15.

Many other structures that illustrate these simple electron-counting rules are given in Chapter 3 and throughout the remainder of the text. These rules provide a very useful guideline for organizing complexes into isoelectronic and isostructural groups. Depending on the relative sizes of ligands (e.g., phosphines versus hydrides) and shapes imposed by rigid ligands (such as the flat tridentate benzene ring), one finds distortions from the idealized geometries of complexes with simple monodentate ligands. Nevertheless, the above organization provides a good approximation of the expected coordination geometries.

As a further exercise in electron-counting rules and the inverse relationship between $d^n$ and coordination number (CN), Table 2.3 illustrates a series of complexes arranged according to $d^n$ ($n$ is even), CN, coordination geometries (CG), and oxidation states, and shows examples of both coordinatively saturated and unsaturated complexes.

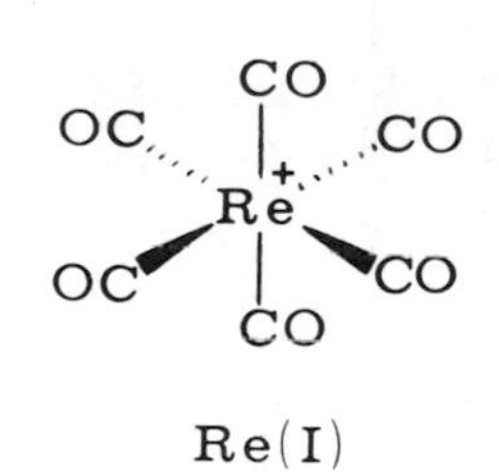


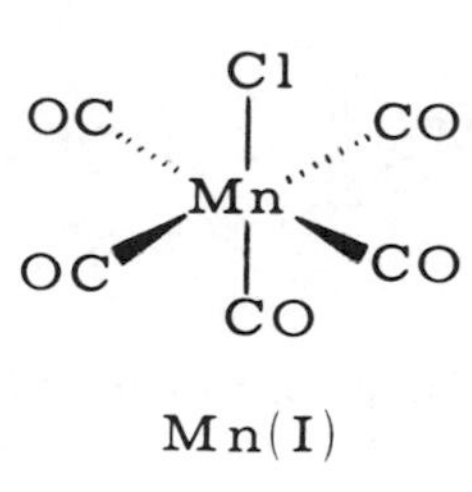


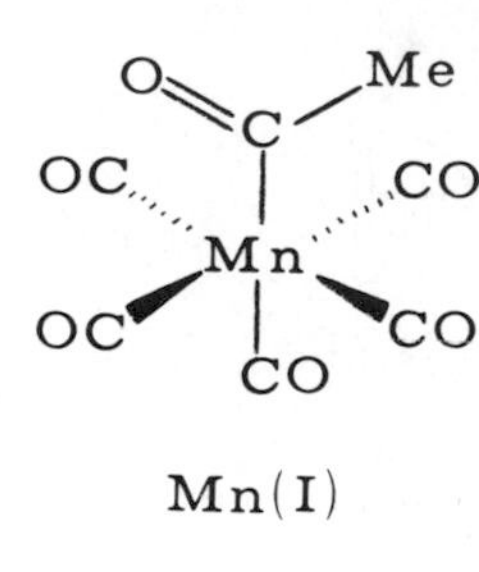


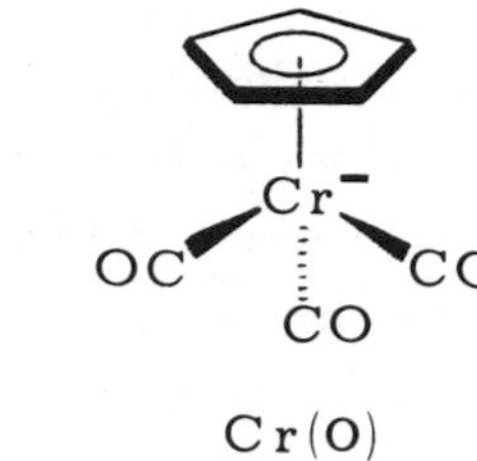


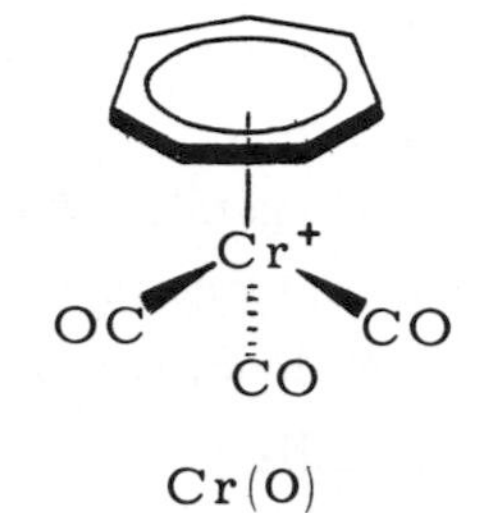


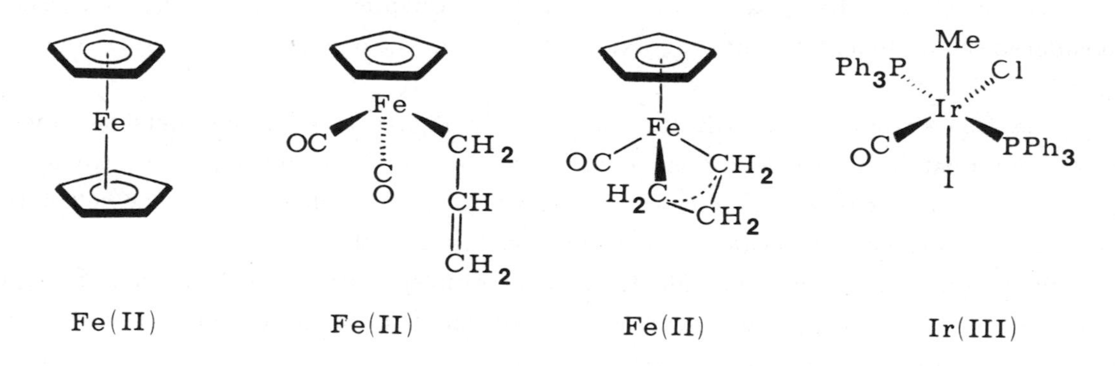


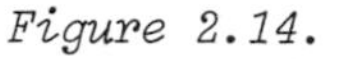

*Figure 2.14.*

In the ensuing chapters we shall encounter many additional examples of these structural electronic relationships.

## 2.10. Paramagnetic Organotransition-Metal Compounds.

Finally, mention should be made of paramagnetic organotransition-metal complexes. It should be obvious that complexes having an odd number of electrons are paramagnetic. Examples are given in Figure 2.16. Note that dmg is an abbreviation for the dianionic bis-dimethylglyoximato ligand (see Chapter 3).

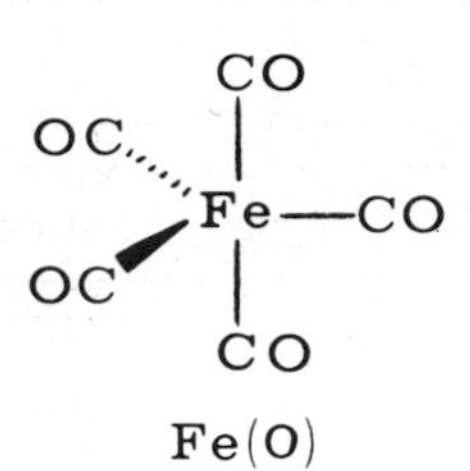


Fe(0)

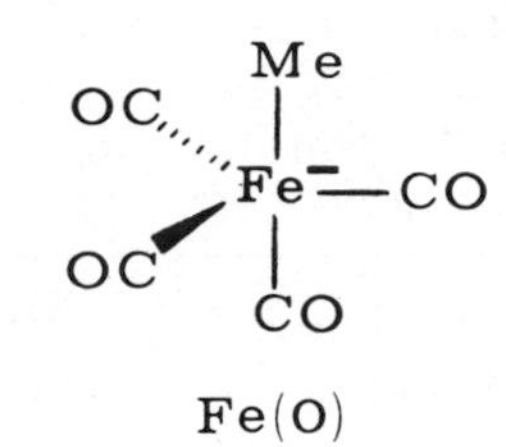


Fe(0)

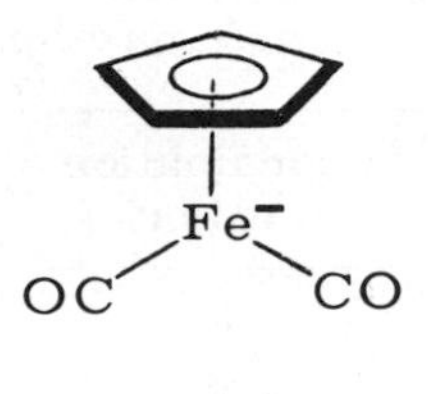


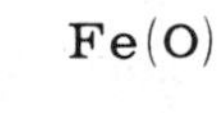


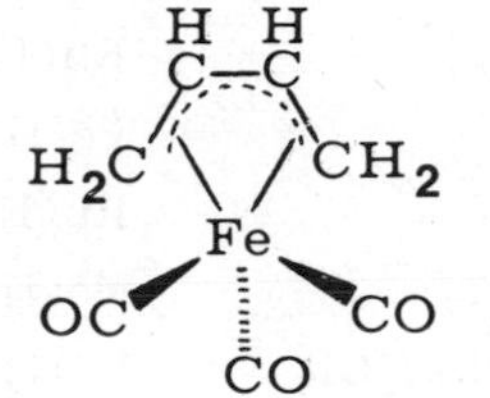


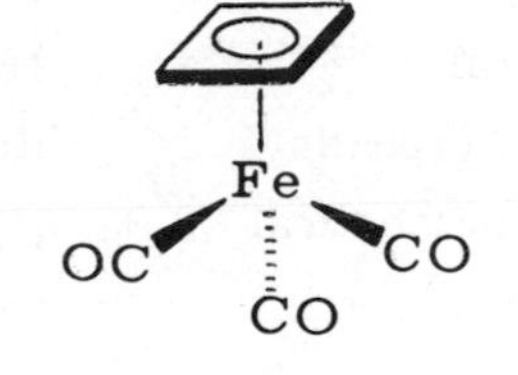


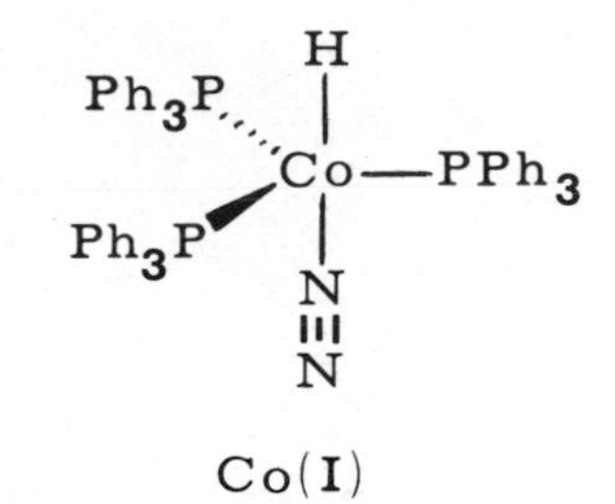


Fe(0)

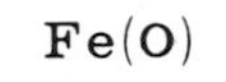


Co(I)

*Figure 2.15.*

$V(CO)_6$

V(0) $d^5$

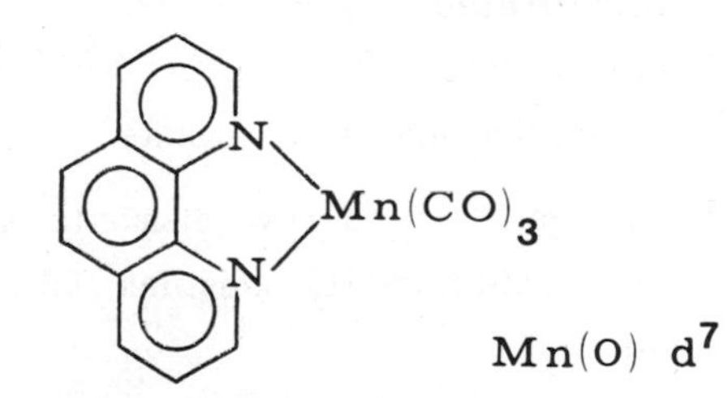


Mn(0) $d^7$

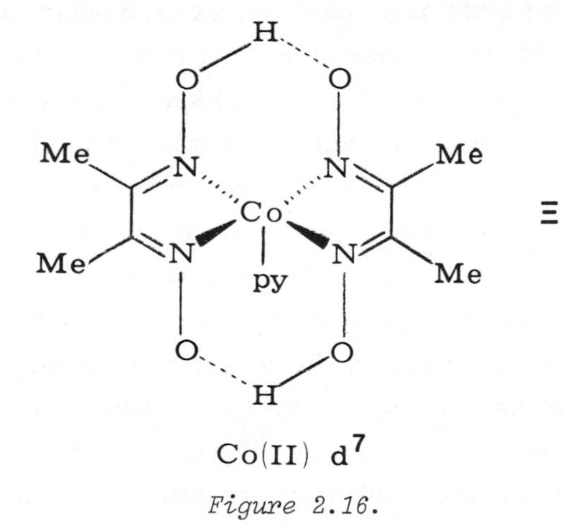


Co(II) $d^7$

*Figure 2.16.*

*Table 2.3. Relationship between $d^n$, coordination number, coordination geometry, and oxidation state of the metal.*

| $d^n$ | Coordination number | Coordination geometry | Example | Oxidation state |
|---|---|---|---|---|
| 10 | 4 | tetrahedral | $Ni(PF_3)_4$ | Ni(0) |
| 10 | 3[a] | trigonal-planar | $Pt(PPh_3)_3$ | Pt(0) |
| 10 | 2[b] | linear | $Au(PPh_3)Cl$ | Au(I) |
| 8 | 5 | trigonal-bipyramidal | $[Co(CNAr)_5]^+$ | Co(I) |
| 8 | 4[a] | square-planar | $[Ir(CO)_2Cl_2]^-$ | Ir(I) |
| 8 | 3[b] | T-shaped[c] | $[Rh(PPh_3)_3]^+$ | Rh(I) |
| 6 | 6 | octahedral | $[Fe(CN)_6]^{4-}$ | Fe(II) |
| 6 | 5[a] | square-pyramidal[c] | $Ru(PPh_3)_3Cl_2$ | Ru(II) |
| 4 | 7 | capped-octahedral[c] | $[Mo(CO)_4Cl_3]^-$ | Mo(II) |
| 4 | 6[a] | octahedral | $W(CO)_2(PPh_3)_2Cl_2$ | W(II) |
| 2 | 8 | square-antiprism[c] | $ReH_5(PPh_3)_3$ | Re(V) |
| 0 | 9 | $D_{3h}$ | $[ReH_9]^{2-}$ | Re(VII) |

[a]Singly unsaturated.

[b]Doubly unsaturated.

[c]These geometries are approximate.

Second- and third-row elements show a strong tendency to pair electrons. For example, the rhodium(II) complex $[Rh(dmg)_2py]_2$ is a diamagnetic dimer joined by a Rh-Rh bond.

Other paramagnetic complexes may have an even number of electrons but exceed the number required to fill the nonbonding d levels. For example, nickelocene, $(\eta^5\text{-}C_5H_5)_2Ni$, is a six-coordinate, $d^8$ complex which, as expected from our electron-counting scheme, has two unpaired electrons (a triplet ground state). In other cases, the paramagnetism is not expected from our elementary bonding theory. Distortions of singlet ($S = 0$) $d^8$ complexes from square-planar geometries can produce triplet ground-state configurations. These cases are not easily predicted a priori, but they are usually restricted to first-row transition metals. For example, $Fe(CO)_4$, an unstable complex which has been detected in a low-temperature matrix, has a triplet ground state and apparently has $C_{2v}$ symmetry [10]. The square-planar $D_{4h}$ symmetry would have been expected for an unsaturated $d^8$ complex with a singlet ground state. The series $(Ph_3P)_3MCl$ shows the following structural, spin-state (S) correlations:

M = Ir, S = 0, square-planar; M = Rh, S = 0, distorted square-planar; M = Co, S = 1, tetrahedral. These situations are often delicately balanced and may be a function of the ligand. For example, the nickel complexes $L_2NiCl_2$ are tetrahedral, S = 1, for L = $Ph_3P$ and square-planar, S = 0, for L = $Et_3P$.

## 2.11. Distinctive Properties of Various Transition Metals.

There are some notable differences between the chemical properties of elements in the various regions of the transition series. We have already noted what appears to be a general trend toward the stronger metal-metal, metal-hydrogen, and metal-carbon bonds formed by 5d and 4d compared with the 3d metals. The 5d complexes usually exhibit slower reactions than either 4d or 3d complexes.

The relative sizes of the transition elements show important trends. The 4d elements are larger than 3d analogues, but about the same size as the 5d elements because of the intervening lanthanide contraction. The size also decreases from left to right for each transition row.

The metals to the lefthand side of the transition series (Groups IV, V and VI) are more electropositive and tend to form strong bonds with nonpolarizable ("hard") donor atoms, such as oxygen [11]. These elements can have fewer d electrons and thus exhibit high coordination numbers. Metal-metal bonding, including multiple bonding, is prevalent in lower oxidation states [12]. $\pi$-Acid ligands such as CO and olefins do not bind as strongly to Group IV and V elements, which means that migratory insertion reactions (Chapter 5) are very facile and lie towards the alkyl or acyl side of the equilibrium.

Group VIII elements are less electropositive and are more easily reduced to the zero oxidation state. The elements themselves are often used as heterogeneous catalysts under reducing conditions. Polarizable ("soft") donor atoms such as phosphorus form strong bonds with Group VIII elements. Many of the complexes derived from these elements have $d^8$ or $d^{10}$ configurations and have a marked tendency to undergo oxidative-addition and reductive-elimination reactions (Chapter 4). These elements exhibit lower coordination numbers and more easily form coordinatively unsaturated complexes. This is especially true for the 4d and 5d Group VIII elements. Group VIII metals tend to form strong bonds with $\pi$-acid ligands such as olefins and CO. This behavior affects the position of the equilibrium in migratory-insertion reactions. Thus we shall find that $\beta$-hydride eliminations and decarbonylations are more common with elements toward the righthand side of the transition series. In the ensuing chapters we will encounter many special characteristics of individual elements, such as the tendency for chromium(0) to form arene complexes and of iron(0) to form diene complexes, and the reactivity of the palladium hydride group.

## NOTES.


1a. F. A. Cotton and G. Wilkinson, Advanced Inorganic Chemistry (Interscience, 4th ed., 1980).

1b. K. F. Purcell and J. C. Kotz, Inorganic Chemistry (W. B. Saunders, 1977).

1c. L. E. Orgel, An Introduction to Transition-Metal Chemistry (Methuen, 1960).

1d. M. C. Day and J. Selbin, Theoretical Inorganic Chemistry (Reinhold, 2d ed., 1969).

1e. G. Henrici-Olive and S. Olive, Coordination and Catalysis (Verlag Chemie, 1977).

2. R. E. Powell, J. Chem. Educ., 45, 45 (1968).

3. F. A. Cotton, Chemical Applications of Group Theory (Interscience, 2d ed., 1971).

4. M. F. Lappert and P. W. Lednor, Adv. Organometal. Chem., 14, 345 (1976).

5. M. J. S. Dewar and G. P. Ford, J. Amer. Chem. Soc., 101, 783 (1979), and references therein.

6a. C. H. Langford and H. B. Gray, Ligand Substitution Reactions (W. A. Benjamin, 1965).

6b. F. Basolo and R. G. Pearson, Mechanisms of Inorganic Reactions (Wiley, 1967).

7a. A. R. Rossi and R. Hoffmann, Inorg. Chem., 14, 365 (1975), and references therein.

7b. B. F. Hoskins and F. D. Whillans, Coord. Chem. Rev., 9, 365 (1972).

8a. E. I. Stiefel, Z. Dori, and H. B. Gray, J. Amer. Chem. Soc., 89, 3353 (1967) and references therein.

8b. R. Eisenberg, Prog. Inorg. Chem., 12, 295, S. J. Lippard, ed., (Interscience, Wiley, 1970).

9. J. E. Huheey, Inorganic Chemistry (Harper and Row, 2d ed., 1978), pp. 401-406.

10. T. J. Barton, R. Grinter, A. J. Thompson, B. Davies, and M. Poliakoff, JCS Chem. Comm., 841 (1977).

11. D. L. Kepert, The Early Transition Metals, (Academic Press, 1972).

12. M. H. Chisholm and F. A. Cotton, Accts. Chem. Res., 11, 225 (1978).

# 3

# Survey of Organotransition-Metal Complexes According to Ligand

Organotransition-metal compounds will be surveyed in this chapter in terms of the various ligand types. This account is intended to be illustrative, but it is by no means comprehensive. Simple ligands will be discussed first, followed by more complex ligands. Certain characteristic physical, chemical, and structural properties of individual ligands are described. In some important cases preparative methods are discussed. Since a majority of organotransition-metal compounds contain more than a single type of ligand, examples of such mixed ligand complexes are used throughout. Compounds having only one type of ligand are called "homoleptic."

The examples cited in this chapter will illustrate further the electron-accounting scheme introduced in Chapter 2. We shall encounter some complexes in which alternative oxidation states seem equally appropriate, and the choice is a matter of "taste." Certain "non-innocent," or amphoteric, ligands having more than one structure and oxidation state will be cited. We shall also see that very reactive, metastable molecules which have electronically unstable ground states may be stabilized by coordination to a transition metal. This chapter will demonstrate the remarkable structural diversity which is characteristic of organotransition-metal chemistry. At the same time, we hope to dispell the confusion which such a bewildering array of compounds presents to chemists first encountering this subject.

## 3.1. Classic Lewis-Base Donors.

There are many ligands that have lone pairs on nitrogen, oxygen, or sulfur, and that bind transition metals into "classical" coordination compounds. We will not present a complete account of these, but note that such groups can and do form bonds with low-oxidation-state transition metals. For instance, labile coordination sites are often occupied by donor solvents such as THF. A more complete discussion of classical donors is given by Cotton and Wilkinson [1].

These classical Lewis-base ligands can be divided into unidentate and multidentate types, according to whether one or more donor atoms are present. Multidentate ligands can form chelate rings, of which the five-membered rings are the most stable. As a general rule, polarizable donor atoms form stronger bonds with third-row transition metals, with metals to the right of the transition series, and with metals in lower oxidation states.

a. Halide Donors. Among the simplest classical donors are the halides I, Br, Cl, and F. Few organotransition-metal fluoride complexes are known. Halides readily form bridges which are easily broken by other ligands. Among the simplest examples is $PdCl_2$, which is, in fact, a square-planar polymer. Iodide is unique among all ligands,

since it is a weak proton base, a strong nucleophile, and a good ligand for transition metals in low oxidation states. In Chapter 8 we shall see that these features are essential to the success of the "Monsanto acetic-acid process." Unfortunately, halide ions are quite corrosive, which creates engineering design problems.

b. Oxygen and Sulfur Donors. Classical unidentate oxygen donors are $H_2O$, MeOH, THF, MeCOMe, $Ar_3PO$, and DMSO. These less-polarizable, weakly basic, "hard" donors are not strongly attached to low-valent transition metals. DMSO may be either S or O bonded and is an example of an ambidentate ligand. Other ambidentate ligands are $CN^-$ and $NCS^-$. Polydentate "crown ethers" are well known for complexing alkali metals [2], but these have been little-studied in the context of organotransition-metal complexes. Thioethers, including crown ethers, are also little-studied but are more promising for low-valent transition metals, since sulfur is a more polarizable, "soft" donor that also has $\pi$-acceptor qualities. Mercaptide anions have a pronounced tendency to bridge metals and are both easily oxidized to disulfides and alkylated, properties which limit their utility as supporting ligands.

Examples of bidentate oxygen-donor ligands are carboxylates, acetylacetonates 1, and tropolonates 2 [3,4].

The last two, monoanionic ligands, undergo quasiaromatic, electrophilic substitution reactions [5].

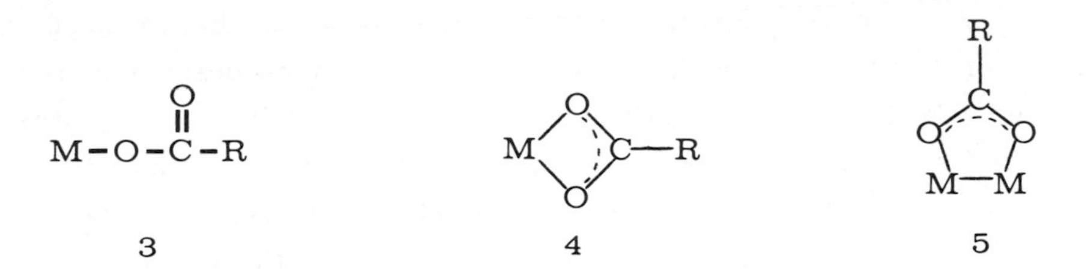


Carboxylates may be unidentate 3, chelating bidentate 4 or bridging bidentate 5. The latter form is especially common in the case of multiple metal-metal bonds [6].

c. Nitrogen Donors. Many polydentate amine complexes are known (e.g., ethylenediamine, en, or tetramethylethylenediamine, tmed), but these have seldom been used with organotransition-metal compounds. Pyridine and its chelating analogues, dipyridyl and phenanthroline, are good ligands for transition metals over a range of oxidation states.

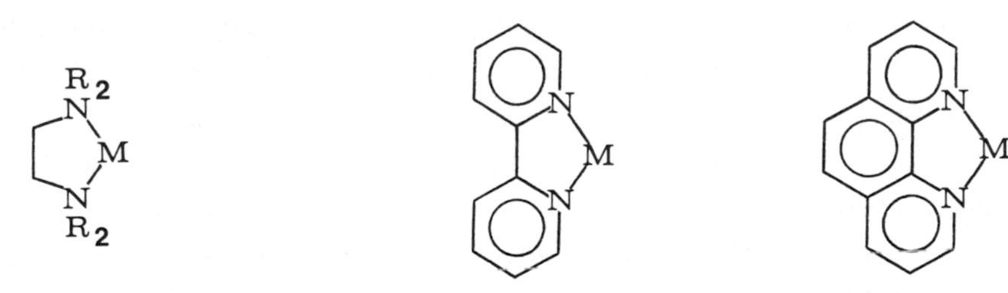

## 3.2. Phosphines and Other Group-Vb Donors.

Transition metals exhibit a pronounced tendency to form complexes with trivalent compounds of phosphorus and arsenic, and to a lesser extent, antimony and bismuth. Among organotransition-metal compounds, tertiary phosphines are the most commonly encountered ancillary ligands. The phosphine ligands in homogeneous catalysts are often modified in order to change the activity or selectivity of the catalyst. It is, therefore, necessary that we consider in some detail the occurrence and properties of these ubiquitous ligands. The number of known transition-metal complexes containing tertiary phosphine ligands is truly immense, and includes monodentate, bidentate, tridentate, and higher chelating phosphines. This subject has been extensively reviewed [7-12].

Organotransition-metal reagents or catalysts containing tertiary arsine, antimony, or bismuth ligands are far less common, and will therefore not be emphasized here. However, some of the trends among the heavy group-Vb donors deserve comment. As we shall see, the significance of $\pi$-bonding in phosphine ligands is controversial. Therefore, it seems likely that the heavier group-Vb donors are predominantly, perhaps exclusively, $\sigma$-donors. Metal-donor steric-repulsion effects increase in the order P < As < Sb < Bi. Steric effects of the substituents decrease in the order P > As > Sb, with Bi being little studied. The bidentate ligands $Ph_2ECH_2CH_2EPH_2$ show a diminishing tendency to chelate in the order P > As > Sb, clearly the result of steric strain. The tendency for these donor atoms to coordinate with metals in higher oxidation states decreases in the order P > As > Sb, which may be the result of increasing polarizability or "softness" of the donor atoms. Bismuth is a very weak donor, and Bi-C bonds are so weak that bismuthine ligands tend to fragment. Many complexes of mixed-donor chelating ligands have been prepared [13], but such exotic, expensive compounds as yet have had little role in homogeneous catalysts.

Tertiary phosphines were once thought to stabilize transition-metal alkyl derivatives through d-$\pi$ backbonding involving filled metal d-orbitals interacting with vacant d-orbitals on phosphorus. The existence of such "backbonding" seemed to be supported by comparisons between the $\nu$CO IR bands of phosphine-carbonyl and of analogous arsine-carbonyl compounds. The possibility that $\pi$-backbonding is important in transition-metal complexes has since been examined with several physical chemical techniques: IR and NMR spectroscopy, and X-ray crystallography for a great variety of phosphine-metal complexes in various oxidation states. Although this issue cannot be said to be fully resolved, for simple tertiary phosphines there seems to be no instance where $d\pi$-$d\pi$ backbonding is required to explain the observed physical properties. However, $\pi$-backbonding is likely to be significant among complexes of $PF_3$ and, to a lesser degree, complexes of phosphites, $(RO)_3P$. Phenomena such as trans effects (trans bond weakening) which were once explained by $\pi$-backbonding can be rationalized in terms of the strong $\sigma$-bonds formed between tertiary phosphines and transition metals. The kinetic stabilization of metal alkyl bonds by phosphine ligands may simply be caused by preserving coordinative saturation at the metal and by steric shielding of the metal by the bulky phosphine groups, thus inhibiting kinetic paths for metal-alkyl bond breaking.

Tertiary phosphine ligands show a pronounced tendency to weaken trans metal-halogen and similar $\sigma$-bonds. This important feature of the phosphine ligand is manifested by an elongated trans metal-halogen bond and an increased kinetic reactivity toward halide replacement reactions. Various physical measurements such as IR (M-Cl), and NMR metal-ligand magnetic coupling have been used to estimate the degree of this trans influence. Remember from Chapter 2 that we are using the terms trans influence and trans effect interchangeably in this text although they are often distinguished [14]. Such trans effects are known for various metals in diverse oxidation states, including higher positive oxidation levels where $\pi$-backbonding is unlikely. Such effects have been extensively studied among platinum phosphine compounds, as the following series illustrates:

$trans$-$[PtHCl(PEt Ph_2)_2]$: Ph$_2$(Et)P–Pt(H)(Cl)–P(Et)Ph$_2$, Pt–Cl 2.42 Å

$trans$-$[PtCl_2(PEt_3)_2]$: Et$_3$P–Pt(Cl)(Cl)–PEt$_3$, Pt–Cl 2.29 Å

$cis$-$[PtCl_2(PMe_3)_2]$: Me$_3$P–Pt(Cl)(Cl)–PMe$_3$ (Cl trans to PMe$_3$), Pt–Cl 2.37 Å

Note that hydride has a more pronounced trans influence on the Pt-Cl bond length than the tertiary phosphine ligand.

A large number of trivalent phosphorus and other group-V ligands have been studied, including monodentate and chelating groups. These ligands encompass a very wide variation in steric and electronic properties. Selected examples of mono- and bidentate ligands are shown in Figure 3.1. The following complexes show the coordination geometries enforced by the multidentate ligands:

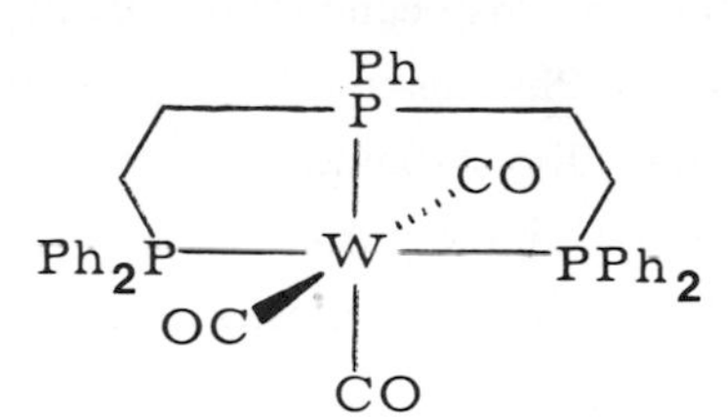


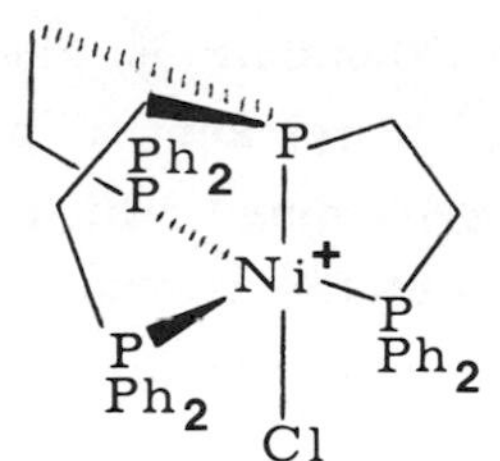


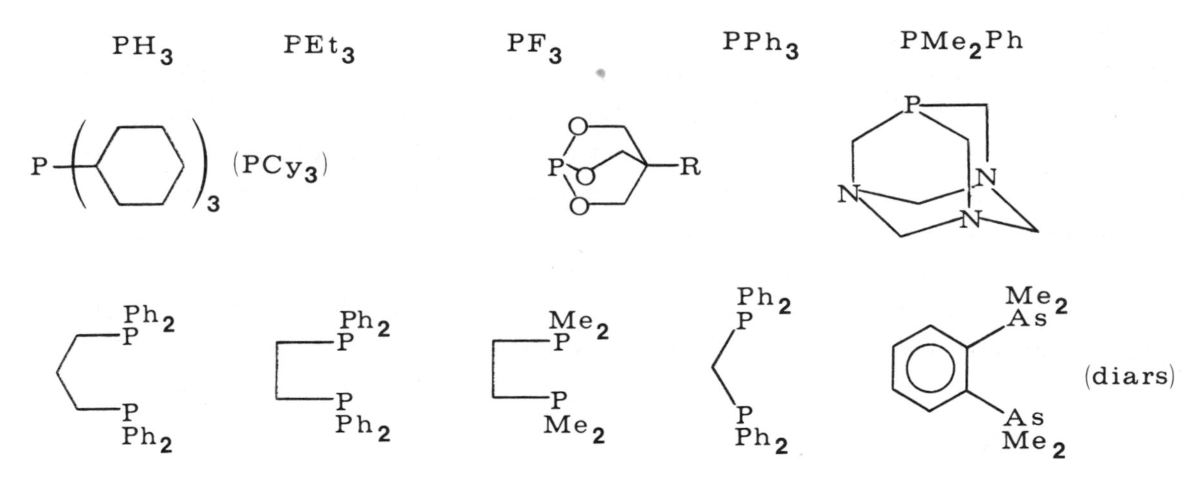


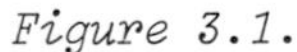

*Figure 3.1.*

Of special note are Verkade's bicyclic bridgehead phosphorus ligands which are sterically constrained to small O-P-O angles [15]. There are also many types of mixed-donor-atom multidentate ligands [16]. Multidentate ligands have been little used in homogeneous catalysts or organotransition-metal reagents because of difficulty and expense in their synthesis, as well as the restrictions which are thus placed on a metal's coordination number and geometry, two factors which must be variable in many catalytic cycles. Even bidentate ligands can markedly alter the course of a catalytic reaction, as is illustrated in Chapter 6, for catalytic hydrogenation.

The relative tendency of phosphine and related ligands to bind to a metal are suggested by the following trend in equilibrium constants:

$$W(CO)_5(\text{amine}) + L \overset{K}{\rightleftharpoons} W(CO)_5L + \text{amine} \tag{3.1}$$

K increases in the order $Ph_3Bi$ < $(PhO)_3P$ ~ $Ph_3Sb$ < $Ph_3As$ < $Ph_3P$ < $(BuO)_3P$ < $CH_3C(CH_2O)_3P$ ~ $Cy_3P$ < $Bu_3P$. In this case [17], steric effects are probably relatively unimportant, so that this trend may reflect the relative intrinsic affinities. However, this may vary from one complex to another.

Altering the substituents on tertiary phosphines can cause substantial changes in the chemical and physical properties of their transition-metal complexes. Some of these effects are electronic, but steric effects are very important. The bulky tertiary phosphine ligands distort the surrounding ligands and shield the metal, thus markedly affecting the chemistry of phosphine complexes. A large number of physical properties are influenced by the bulk of phosphine ligands. These have been reviewed by Tolman [7]. Such steric interactions also have a very important influence on chemical properties, such as the oxidative-addition reactions discussed in Chapter 4. A more subtle sort of steric interaction exerted by chiral phosphines controls the efficacy of asymmetric homogeneous hydrogenation catalysts, which are described in Chapter 6. Tolman has proposed that the cone angle of a phosphine ligand be used as a measure of its steric bulk (see Figure 3.2 and Table 3.1).

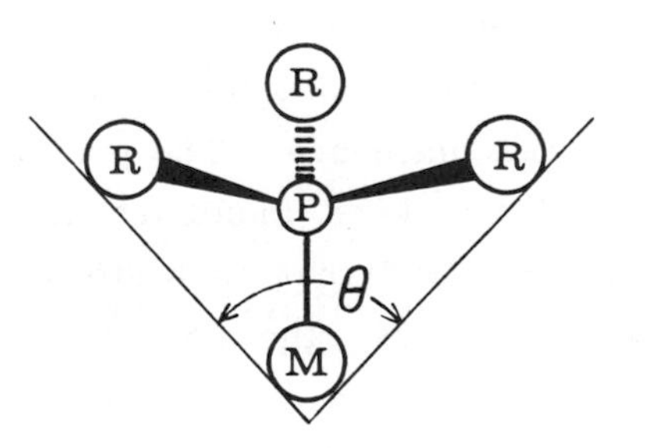


*Figure 3.2. Schematic representation of a phosphine ligand cone angle θ.*

Steric and electronic effects are often strongly interrelated, so that it is difficult to separate these factors. For example, the percent of s-character in the phosphorus donor electron pair should decrease as the cone angle increases. More electronegative substituents on phosphorus produce a shorter M-P bond by inducing more phosphorus s-character into the M-P bond. For example, M-P bonds in phosphites are ~0.06 Å

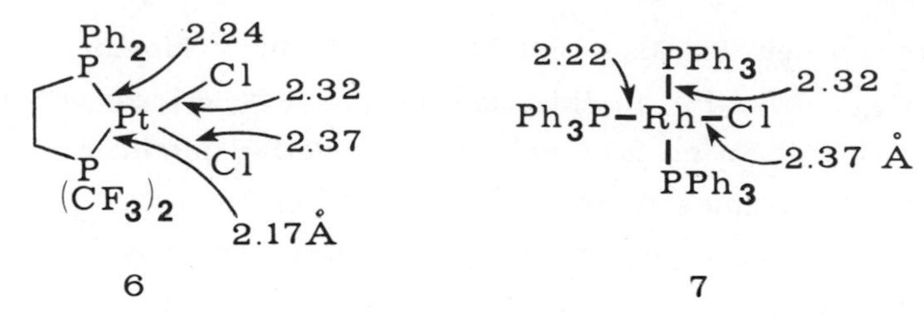


*Table 3.1. Phosphine ligand cone angles.*

| Phosphorus ligand | Cone angle θ° | Phosphorus ligand | Cone angle θ° |
|---|---|---|---|
| $PH_3$ | 87 | $PH_2Ph$ | 101 |
| $P(OCH_2)_3CR$ | 101 | $PF_3$ | 104 |
| $P(OMe)_3$ | 107 | $PMe_3$ | 118 |
| $PMe_2Ph$ | 122 | $Ph_2PCH_2CH_2PPh_2$ | 123 |
| $PEt_3$ | 132 | $PPh_3$ | 145 |
| $PPh_2(t\text{-}Bu)$ | 157 | $PCy_3$ | 170 |
| $PPh(t\text{-}Bu)_2$ | 170 | $P(t\text{-}Bu)_3$ | 182 |
| $P(C_6F_5)_3$ | 184 | $P(mesityl)_3$ | 212 |

shorter than those in phosphines. This electronegativity effect is illustrated in the unsymmetric bidentate phosphine platinum complex 6 [18].

There are many phosphine complexes whose structures show deviations from idealized geometries as the result of steric interactions. For example, Wilkinson's famous hydrogenation catalyst, $RhCl(PPh_3)_3$, 7 (a diamagnetic compound having a formal $d^8$, Rh(I) configuration), has a nonplanar arrangement of donor atoms [19]. Note also that the Rh-P bond trans to chlorine is shorter than the mutually trans Rh-P bonds--surely a manifestation of the trans influence. The presence of several bulky phosphine ligands in the same coordination sphere facilitates ligand dissociation, producing a vacant coordination site. We shall find that such steps can be critical in the functioning of a homogeneous catalyst. Thus, the equilibria among palladium(0) phosphine complexes are dominated by steric effects, the extent of ligand dissociation increasing in the order

$$PMe_3 \sim PMe_2Ph \sim PMePh_2 < PEt_3 \sim PPh_3 < P(i\text{-}Pr)_3 < P(Cy)_3 < PPh(t\text{-}Bu)_2:$$

$$PdL_4 \rightleftarrows PdL_3 + L \rightleftarrows PdL_2 + L$$

Electronic substituent effects on phosphine complexes are most easily studied by varying the para substituents on triarylphosphines, although the magnitude of change is modest. The most dramatic electronic effect is that manifested by $PF_3$ as a ligand. For this ligand, $d\pi$-$d\pi$ backbonding is significant, since the electronegative fluorine atoms are considered to lower the energy of the vacant phosphorus 3d-orbitals. This phenomenon is illustrated by the range and properties of $PF_3$ complexes [20,21]. For example, $\pi$-backbonding undoubtedly stabilizes the negative charge in anionic complexes such as $[M(PF_3)_4]^-$ (M = Co, Rh, Ir). Whereas cationic complexes of the ordinary tertiary phosphines are well known, anionic complexes are rare. The $\pi$-backbonding capability of this ligand is further illustrated by the fact that $PF_3$ forms analogues of most binary carbonyls, including cases for which the carbonyl derivatives have not been detected: $M(PF_3)_4$ (M = Ni, Pd, Pt). The palladium and platinum tetracarbonyls are presently unknown and are probably unstable. In fact, $PF_3$ is one of the few monodentate ligands which can displace all of the CO groups from a binary carbonyl.

There are many other types of phosphine ligands. Some of them will undoubtedly find a role in homogeneous catalysis. Here we need only mention a few of these varieties. Secondary phosphides, $R_2\ddot{P}:^{(-)}$, can serve as ligands [22]. These have a pronounced tendency to bridge two metals:

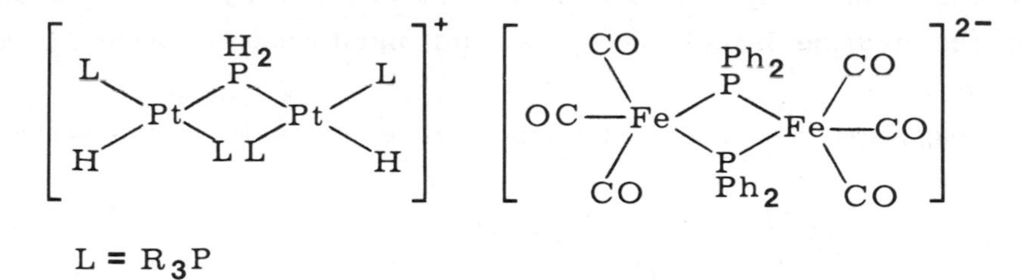


A large number of metal complexes derived from phosphorus ylids have been characterized by Schmidbaur [23]. These include examples of metal-phosphorus bonded chelating ligands, as well as metal-carbon bonded derivatives. To date, this rich structural chemistry has had few catalytic or synthetic applications:

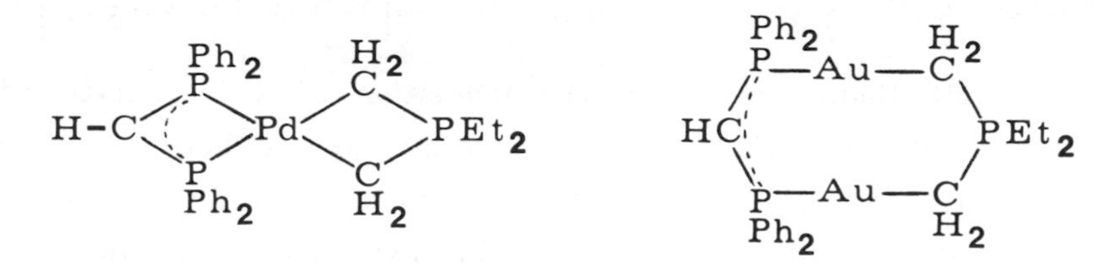


Among the most interesting phosphorus ligands are those which have a formal positive charge in the uncoordinated state. Because of this positive charge, such compounds are expected to exhibit an enhanced $\pi$-acidity. This is consistent with the relatively high C-O stretching frequency observed in the iron(0) derivative, 8, which has been structurally characterized [24]. Note the alternative methods of preparing this substance, especially the use of the Lewis acid $PF_5$ to remove fluoride:

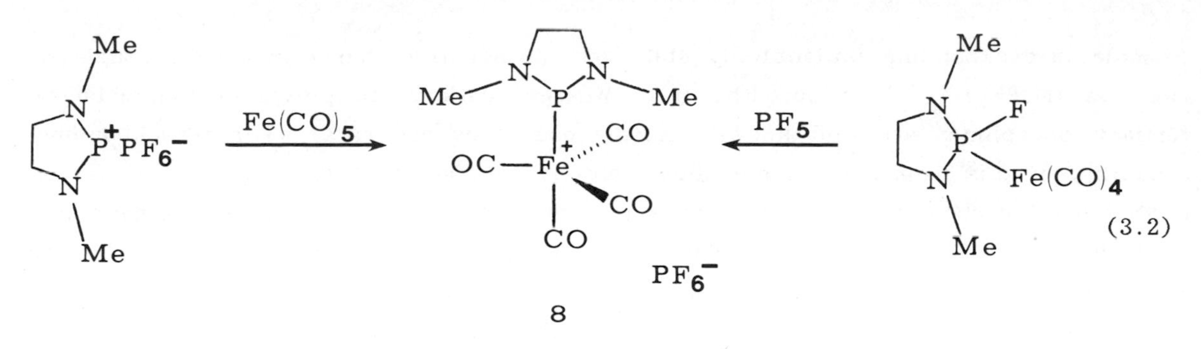


## 3.3. Hydrides.

The term "hydride" is used to describe a hydrogen bound to one or more transition metals. Many catalytic and stoichiometric reactions of transition-metal complexes involve the hydride ligand. Thus, the structural, analytical, and chemical characteristics of the hydride ligand have paramount significance in organotransition-metal chemistry [25-31].

Hydride complexes are known for the entire transition-metal series. Typical examples are:

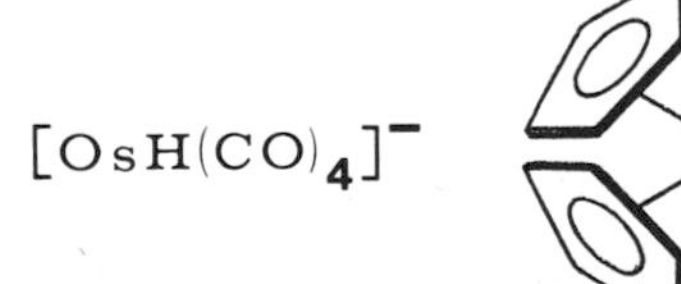

$[OsH(CO)_4]^-$ $(C_5H_5)_2Zr(H)Cl$ $H(Co)(PPh_3)_3$ $WH_6(PMePh_2)_3$

$FeH(NC)(PF_3)_3$ $[HNi[P(OEt)_3]_4]^+$

Towards the upper righthand section of the transition series, hydride complexes are less common and less stable. There is a marked increase in both the kinetic and the thermodynamic stability of metal hydride complexes upon descending a triad. This bond strength trend is in contrast to the situation with hydrides of the post-transition elements. Thus, hydride complexes of nickel and copper are rare and correspondingly reactive. This point is further illustrated by considering the series trans-$(Et_3P)_2M(H)Cl$ (M = Pt, Pd, Ni). The platinum compound can be distilled under high vacuum without decomposition; whereas the palladium compound slowly decomposes in the solid state, and the nickel compound has not been isolated but can be detected spectroscopically. Transition-metal-hydride bond strengths (typically 60 kcal mole$^{-1}$) are, in general, greater than corresponding metal-carbon (alkyl) bond strengths.

a. Structural Characteristics. The hydride ligand may be bound to a single metal (terminal hydrides), or two or more metals (bridging hydrides). Historically, the structures of transition-metal hydride complexes were controversial until clarified by precise X-ray and neutron diffraction studies. Ibers first located the hydride ligand using X-ray diffraction. His review of transition-metal hydride structures is highly recommended [32]. The application of X-ray diffraction to metal hydride complexes is difficult since the X-ray scattering power of an atom is proportional to the square of its atomic number. Thus, the location of a hydride in the presence of a heavy metal by X-ray diffraction is difficult, but in favorable cases the position of a hydride may be measured to ±0.05 Å by X-ray diffraction [33]. Matters are worse with the heavy third-row transition metals and in cases where the hydride is bridging adjacent metals. These problems are further complicated by large thermal motions of the hydride group.

The greater cross section of hydrogen toward neutron scattering makes neutron diffraction a powerful method for accurately mapping transition-metal hydride structures. However, the inherent advantages of neutron diffraction are offset by limited neutron facilities, the requirement for much larger crystals, and often the need for a prior X-ray analysis. Nevertheless, a substantial number of transition-metal hydride structures have been analyzed by neutron diffraction, primarily by Bau [34,35]. These results have greatly increased our understanding of terminal and bridged transition-metal hydride structures.

Although metal-hydrogen bond lengths were once thought to be "long" or "short," modern diffraction studies show that metal-hydrogen lengths are a "normal" sum of covalent radii. Thus, M-H distances fall in the range 1.5 to 1.7 Å. Terminal hydride ligands usually occupy a distinct coordination position. However, bulky ligands may distort the coordination geometry from the ideal as these ligands are displaced toward the small hydride group, thus relieving nonbonded interactions between the large ligands. This is especially found with the voluminous tertiary phosphines.

This point is illustrated in Figure 3.3 by the structures of three monohydride complexes: $HMn(CO)_5$, $HRh(CO)(PPh_3)_3$, and $HRh(PPh_3)_4$. The manganese(I) hydride, a $d^6$ complex, exhibits the expected octahedral ($C_{4v}$) geometry with a 1.601(16) Å Mn-H distance [36]. The cis CO groups are bent toward H, so that the C-Mn-C angles are 167 and 164°, rather than the 180° expected for pure octahedral coordination. The Mn-C bond trans to H is slightly shorter (0.03 Å) than the cis C-Mn-C bonds, probably reflecting the diminished $\pi$-bonding usually seen with trans CO groups which compete for the same d-orbitals.

The structure of $RhH(CO)(PPh_3)_3$ is one in which the hydride ligand occupies the axial position of a trigonal bipyramid [37]. There is only a slight displacement of the phosphines towards the hydride. This may be contrasted with the formally isoelectronic $RhH(PPh_3)_4$ in which the position of the hydride was not located but is thought to be random or to lie on a threefold axis of the tetrahedron of phosphine ligands [38]. In this case the steric bulk of the phosphines totally dominates the structure.

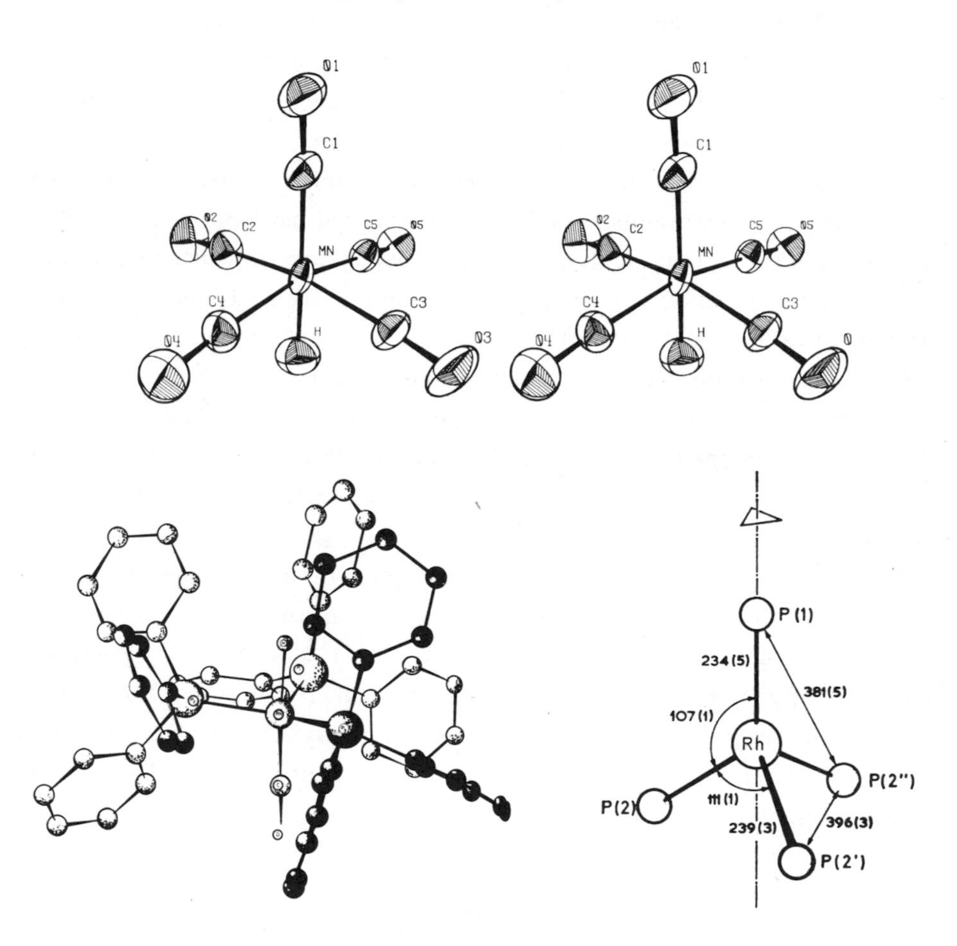


*Figure 3.3. Representative terminal hydride structures.*

There are many compounds having several terminal hydrides bound to a single transition metal; these are $H_6WL_3$, $H_4WL_4$, $H_7ReL_2$, $H_5ReL_3$, $H_3ReL_4$, $H_8OsL$, $H_6OsL_2$, $H_4OsL_3$, $H_2OsL_4$, $H_5IrL_2$, and $H_3IrL_3$. These polyhydride complexes are usually colorless, diamagnetic, and soluble in organic solvents, but not especially reactive or, so far, particularly useful as organometallic reagents. Some polyhydride complexes catalyze $H_2/D_2$ exchange reactions [39]. The seven- and nine-coordinate polyhydrides are typically fluxional, whereas the six-coordinate compounds are rigid, as might be expected [40,41]. Note that although some of these polyhydrides represent high formal oxidation states, they exhibit very little charge separation. All are coordinatively saturated.

The M-H-M bridge bond has been structurally characterized by neutron diffraction. All such doubly bridging hydrides have been found to be bent (i.e., nonlinear) and are best considered as three-center, electron-deficient bonds reminiscent of the B-H-B bridge. However, metal-metal bonding is substantial in bridging transition-metal hydrides, making the concept of bond order ambiguous. Bridging M-H-M bonds are longer than terminal M-H bonds and are of equal length, in contrast to classic bridging hydrogen bonds such as $FHF^-$.

Two or more hydrides may bridge a pair of metals. Hydrides simultaneously bound to three metals (triply bridging, $(\eta^3\text{-H})M_3$, bonds) are known, but these cases are less common. Hydrides bound inside a metal cluster are also known. Doubly and triply bridging and interstitial hydrides may be considered structural models for different types of chemisorbed hydrogen and insoluble binary transition-metal hydrides. For structural detail and examples, consult Bau's review [34,35]. Although the chemical reactions of bridging hydrides have been little studied, these seem to be less reactive than terminal hydrides.

b. Physical Properties. Proton magnetic-resonance spectroscopy is very useful in mapping the presence and the stereochemistry of transition-metal hydride complexes [40]. The transition metal hydrides typically show very high-field chemical shifts in the range $\delta$ = -7 to -24 (or $\tau$ = 17 to 34), well above those of most organic compounds. In rare cases M-H chemical shifts have been reported at much higher (50 ppm above TMS) and much lower values (e.g., $HCo(dmg)_2P(nBu)_3$ at 6$\tau$). The typical high-field M-H chemical shifts have been rationalized as having their origin in magnetic effects arising from both ground-state nonbonded electrons and contributions from excited states. However, these chemical-shift values usually have no obvious correlation with structure or reactivity. An exception involves some *trans* effect relationships which have been noted for square-planar platinum hydrides such as *trans*-$Pt(H)(X)(PEt_3)_2$.

These distinctive chemical shifts are very useful in detecting the presence of a M-H group and have greatly contributed to the development of hydride chemistry. However, the absence of a high-field NMR signal should never be considered as evidence against the presence of an M-H group. These signals can be broad and in some cases have not been observed. The NMR signal of a bridging hydride group is even more difficult to detect.

Magnetic coupling between hydrides and metals, such as $^{103}Rh$, $^{195}Pt$, $^{183}W$, and $^{187}Os$ (J = 38 to 1370 Hz!), provides strong evidence for direct M-H bonds. Coupling between hydride groups and $^{31}P$ in phosphine ligands is also a useful tool. Usually trans $J_{PH}$ (80-150 Hz) values are larger than cis $J_{PH}$ (10-40 Hz); however, exceptions have been found.

Transition metal hydrides generally show an infrared M-H stretching band (νM-H) in the region 1900-2250 $cm^{-1}$. The intensities of these bands vary from weak to moderate. To avoid confusion with νCO and other IR bands, the νM-H bonds may be identified by comparison with the corresponding deuteride, since in most cases $\frac{\nu M\text{-}H}{\nu M\text{-}D} = \sqrt{2}$ in accord with simple theory. This relationship is occasionally not followed in cases of vibrational coupling with another group having a similar frequency and an appropriate symmetry, as, for example, a trans CO group. Bridging M-H-M IR bands may be as low as 1000 $cm^{-1}$ and are usually very difficult to detect. The terminal νM-H band is occasionally solvent-dependent, but is particularly sensitive to structural-electronic trans effects. Substitution of a strongly bonded ligand trans to an M-H bond causes a decrease in the νM-H frequency which reflects the weaker M-H bond. For example:

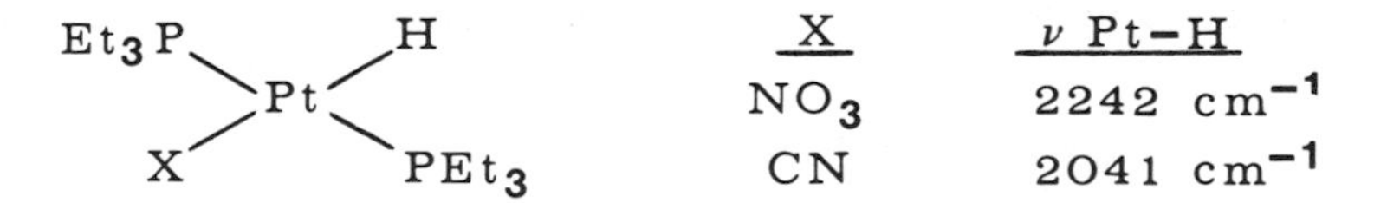


The νM-H frequencies usually increase upon descending a triad. This effect clearly indicates that the M-H bond strength increases with the heavier transition metals.

The deformation mode (νM-H) occurs from 950 to 700 $cm^{-1}$ but is more difficult to assign, although such a band does provide useful supplementary information.

Other spectroscopic techniques have been used to study M-H groups, but such applications are less common. In principle, Raman spectroscopy is useful for measuring νM-H bands. Since Raman intensities depend on changes in polarizability, whereas infrared intensities depend on dipole-moment changes, the two methods are complementary. Thus, low-intensity IR bands are often more intense in the Raman spectrum. However, the need for larger samples, for colorless compounds, and the photolability of CO ligands have discouraged the use of Raman spectroscopy, except for measuring M-M frequencies. Although paramagnetic M-H complexes are rare, electron spin resonance (ESR) has occasionally been used to detect indirectly the presence of a hydride ligand through hyperfine coupling. Details and examples of the application of spectroscopic techniques to metal hydride chemistry are found in Jesson's review [40].

c. Synthesis of Hydrides. There are many methods for synthesizing the transition-metal-hydrogen bond, but few are general. The reader is advised to consult the original literature for each specific case and to hope that explicit, detailed directions have been given. Schunn's review is recommended [42].

There are many examples of hydrogen reacting with metal compounds to afford metal hydrides. High pressures are often required, and a reducing metal, Cu, Ag or Zn, or a ligand, or a proton base are often coreactants. Each example is a special case. Note that metal-metal bonds are usually susceptible to hydrogenolysis. Selected examples are given in Equations 3.3 to 3.6 [43-46].

$$CoS + H_2 + CO \xrightarrow[\substack{200^\circ \\ 250\ atm}]{Cu\ powder} HCo(CO)_4 \tag{3.3}$$

$$OsO_4 + 3{:}1\ CO/H_2 \xrightarrow[160^\circ]{180\ atm} H_2Os(CO)_4 + H_2Os_2(CO)_8 + H_2Os_3(CO)_{12} \tag{3.4}$$

$$(CO)(PPh_3)_2Rh{-}Rh(PPh_3)_2(CO) \xrightarrow[\substack{80\ atm \\ H_2}]{PPh_3} 2\ HRh(CO)(PPh_3)_3 \tag{3.5}$$

$$RuCl_2(PPh_3)_4 + H_2(1\ atm) \xrightarrow[\substack{toluene \\ reflux}]{Et_3N} HRuCl(PPh_3)_3{\cdot}C_6H_5Me \tag{3.6}$$

Oxidative-addition of hydrogen to coordinatively unsaturated complexes, a reaction discussed in Chapter 4, is often a mild method for preparing a cis-dihydride [47]:

$$IrCl(PPh_3)_3 + H_2 \xrightarrow[1\ atm]{25^\circ} H_2IrCl(PPh_3)_3 \tag{3.7}$$

Coordinatively saturated complexes require more vigorous conditions--perhaps to expel initially a ligand [48]:

$$Os(CO)_4(PPh_3) + H_2 \xrightarrow[100^\circ]{80\ atm} H_2O_5(CO)_3(PPh_3) \quad (3.8)$$

If olefins are present in the complex, these are hydrogenated [49]:

$$\overset{+}{Rh}(COD)(PPh_3)_2 + H_2 \xrightarrow[S = acetone]{1\ atm} H_2\overset{+}{Rh}(S)_2(PPh_3)_2 + C_8H_{16} \quad (3.9)$$

We shall find this method used to generate homogeneous hydrogenation catalysts in situ (Chapter 6).

Oxidative-addition reactions of protonic acids are often useful methods for preparing monohydride complexes (see Chapter 4). This method is limited to cases in which the starting metal complex is sufficiently basic and the resulting hydride does not react further with the acid [50]:

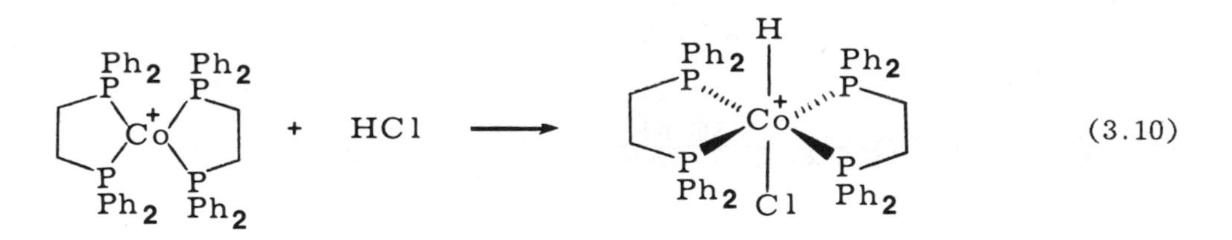


Protonation of anionic metal complexes is a common method for preparing hydrides [51]:

$$\overset{+}{Na}\overset{-}{Re}(CO)_5 + HCl \longrightarrow HRe(CO)_5 \quad (3.11)$$

One of the most useful methods for preparing transition-metal hydride compounds involves replacement of an anionic ligand, such as halide, by a hydride reagent, such as $LiAlH_4$ or $NaBH_4$ [52]:

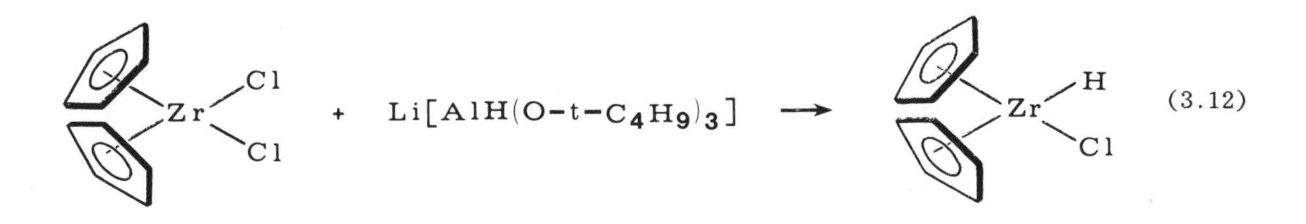

Other compounds, such as hydrazine, formic acid, or alcohol in the presence of a base, can also be used to deliver a hydride ligand [53]:

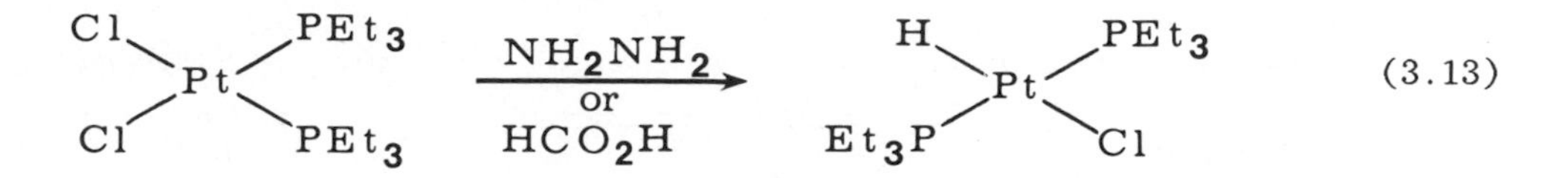

(3.13)

The latter process seems to involve the formation of an alcoholate complex which delivers a hydrogen to the metal, at the same time oxidizing the alcohol to an aldehyde or ketone:

$$IrCl_3(PEt_3)_3 \xrightarrow[CH_3CH_2OH]{KOH} HIrCl_2(PEt_3)_3 + CH_3\overset{O}{\overset{\|}{C}}H \quad (3.14)$$

$$H_3C-\underset{H}{\underset{|}{C}}H-O-Ir(III) \rightleftharpoons H_3C-\overset{O}{\overset{\|}{C}}H + H-Ir(III) \quad (3.15)$$

This reaction is the reverse of metal-hydride reduction of an aldehyde, which we shall encounter in conjunction with hydroformylation (Chapter 8).

Another method for generating hydrides is an insertion reaction (Chapter 5), involving the base hydrolysis of a metal carbonyl:

$$M-CO + OH^- \longrightarrow [M-\overset{O}{\overset{\|}{C}}-OH]^- \longrightarrow M-H^- + CO_2 \quad (3.16)$$

Many other reducing agents have been used to prepare metal hydrides [42].

d. Reactions of the Hydride Ligand. The term "hydride" derives from an oxidation-state formalism. However, this nomenclature should not be confused with chemical reactivity. The chemical properties of transition-metal hydride groups vary enormously. Some M-H groups are, in fact, "hydridic," whereas others are weak or even strong acids, and yet other M-H groups are quite inert to polar reagents. Various reactions of metal hydrides are found in later chapters, such as 6 (Hydrogenation) and 7 (Stoichiometric Reactions of Hydride). However, it is instructive here to mention some typical reactions of the M-H group.

Certain metal hydrides are acidic, but the degree varies enormously with the metal and the ancillary ligands. Examples of acid dissociation constants for selected M-H compounds are gathered in Table 3.2. Note that the strongest M-H acid, $HCo(CO)_4$,

*Table 3.2. Acid dissociation constants of metal hydride compounds* [54].

| Complex | $K_a(H_2O)$ |
|---|---|
| $HCo(CO)_4$ | <2 |
| $HCo(CO)_3P(OPh)_3$ | $1.1 \times 10^{-5}$ |
| $HCo(CO)_3PPh_3$ | $1.1 \times 10^{-7}$ |
| $HCo(CO)_2(PPh_3)_2$ | "very weak acid" |
| $HCo(PF_3)_4$ | strong acid |
| $HMn(CO)_5$ | $8 \times 10^{-8}$ |
| $HRe(CO)_5$ | "very weakly acidic" |
| $H_2Fe(CO)_4$ | $3.6 \times 10^{-5} (K_1)$ |
| | $1 \times 10^{-14} (K_2)$ |

has about the same strength as a mineral acid; but replacement of one CO, an electron-withdrawing ligand, by the electron-releasing $PPh_3$ causes a decrease of more than $10^7$ in acid strength! The effect of the phosphite, $P(OPh)_3$, is intermediate. On the other hand, $PF_3$ is more like CO. In Chapter 4 we see that this acid-base property can be interpreted in terms of oxidative-addition, reductive-elimination.

Reactions of bases with transition-metal hydrides are sometimes used to generate transition-metal anions, as, for example, with the hydridorhenium cluster [55]:

$$H_3Re_3(CO)_{12} + KOH \longrightarrow \overset{+}{K}[H_2Re_3(CO)_{12}]^- \quad (3.17)$$

Strong electrophilic reagents usually "oxidize" the M-H group, sometimes by replacement of $H^+$. Diverse mechanisms may be involved. An example is the reaction with halogens affording hydrogen halides and the corresponding metal halide:

$$HMn(CO)_3[P(OPh)_3]_2 + I_2 \longrightarrow MnI(CO)_3[P(OPh)_3]_2 + HI \quad (3.18)$$

In some cases further oxidation of the metal takes place:

$$H_2IrCl(PEt_3)_3 + Cl_2 \longrightarrow IrCl_4(PEt_3)_2 \quad (3.19)$$

Nitrosating reagents often react with the M-H group in an apparent electrophilic substitution:

$$HRuCl(CO)(PPh_3)_3 + Me{-}\underset{}{\overset{NO}{\overset{|}{N}}}SO_2Ar \longrightarrow RuCl(CO)(NC)(PPh_3)_2 \quad (3.20)$$

Reactions with diazonium salts are formally analogous to nitrosation, except the resulting arylazo derivative is sometimes protonated:

$$HRhCl_2(PPh_3)_2 + FC_6H_4N{\equiv}\overset{+}{N}BF_4^- \longrightarrow RhCl_2(N{=}NC_6H_4F)(PPh_3)_2 \quad (3.21)$$

$$\underline{trans}{-}HPtCl(PEt_3)_2 + XC_6H_4N{\equiv}\overset{+}{N}BF_4^- \longrightarrow [PtCl(NH{=}N(C_6H_4X))(PEt_3)_2]^+BF_4^- \quad (3.22)$$

The reactions of M-H groups with mercury(II) and tin(IV) chlorides can also be considered as electrophilic substitutions, although $SnCl_4$ can lead to further oxidation, and elemental mercury is sometimes formed in reactions with mercuric reagents:

$$HIrCl_2(CO)(PPh_3)_2 + HgCl_2 \longrightarrow IrCl_2(HgCl)(CO)(PPh_3)_2 \quad (3.23)$$

$$H_2Os(CO)_4 + SnCl_4 \longrightarrow \underline{cis}{-}HOs(SnCl_3)(CO)_4 \quad (3.24)$$

$$\underline{trans}{-}HPtCl(PPh_3)_2 + SnCl_4 \longrightarrow PtCl_2(SnCl_3)_2(PPh_3)_2 \quad (3.25)$$

The scope and mechanisms of these processes remain unexplored. Note that for such reactions the reactive metal hydrides can be coordinatively saturated.

Some M-H groups manifest their hydridic character through reaction with proton acids, yielding hydrogen and a complex in which the conjugate base replaces the hydride ligand:

$$HFe(\pi C_5H_5)(CO)_2 + HCl \longrightarrow FeCl(\pi C_5H_5)(CO)_2 + H_2 \quad (3.26)$$

$$H_2Zr(\pi C_5H_5)_2 + 2HOAc \longrightarrow Zr(OAc)_2(\pi C_5H_5)_2 + 2\ H_2 \qquad (3.27)$$

In other cases the hydride is inert, or if several M-H groups are present in the same molecule, their sequential reactivity may diminish. Hydrides trans to a strong ligand, such as phosphine, are often more reactive:

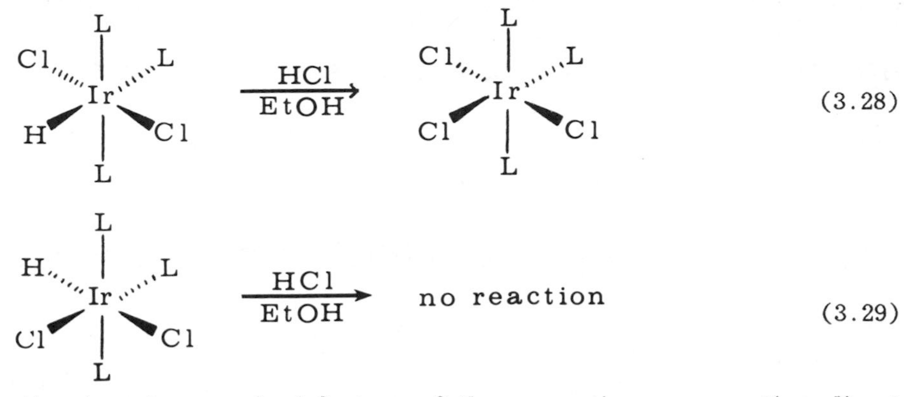


Coordinative unsaturation is not a required feature of these reactions, suggesting direct attack by a proton on the M-H bond. The tendency to react seems to be controlled by the strength of the acid as well as by the reactivity of the hydride.

Transition-metal hydrides are usually good hydrogen-atom donors in apparent radical-chain processes analogous to those of the well-studied tin hydrides. However, in the transition-metal cases the reaction mechanism has been less-studied. Thus, most transition-metal hydrides react with $CCl_4$, affording chloroform (which can often be quantitatively measured by GLC) and the corresponding metal chloride complex. This reaction is sometimes used to confirm the presence of a metal hydride group or even to prepare a stable derivative from a reactive metal hydride. An example of the latter is Norton's characterization of the dinuclear osmium hydride [56]:

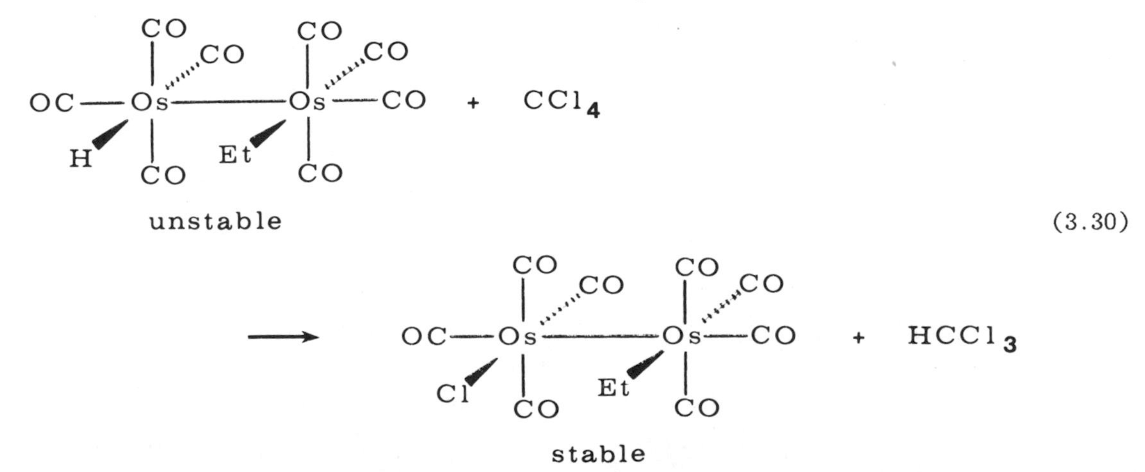

As might be expected from the well-studied radical chemistry of tin hydrides, the following reactivity sequence is observed: $CCl_4 > CHCl_3 > CH_2Cl_2 > CH_3Cl$. Many metal hydride compounds are stable in chloroform as a solvent, but even more are stable in methylene chloride.

Reactions of M-H compounds with alkyl disulfides and with oxygen are perhaps related to these hydrogen-atom transfer processes. Such chain reactions should be subject to initiation and inhibition, but the mechanisms of these reactions are still largely unexplored. Thus, the reactivity of metal hydride complexes toward air varies with the nature of the complex. Many are sensitive, especially in solution. Note the subsequent bridge-forming displacement of CO by coordinated mercaptide [57]:

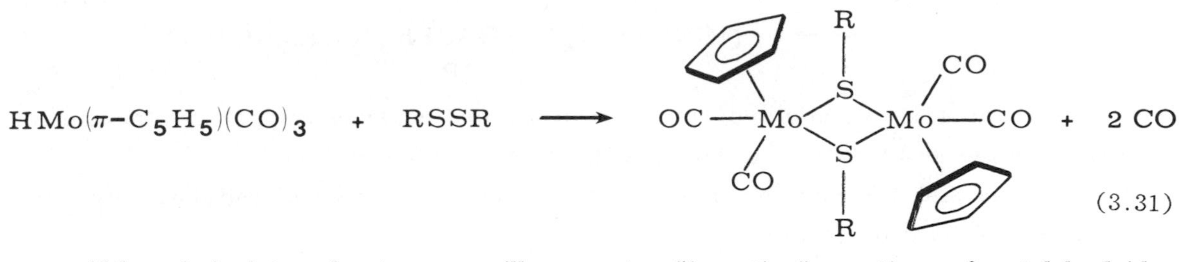


Although in later chapters we will encounter "insertion" reactions of metal hydrides with olefins and acetylenes, it is useful to mention certain of these here, especially in the context of chain-radical processes. Electronegatively substituted olefins readily react with several coordinatively saturated metal hydrides in apparent contrast to ethylene [58]. For example, tetrafluoroethylene reacts with the tungsten hydride [57]:

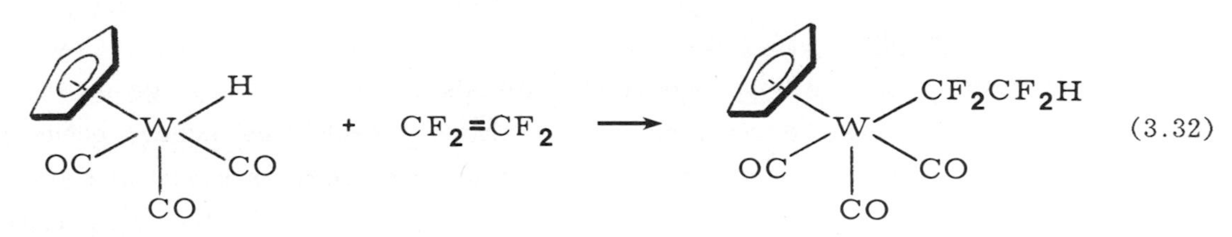


The mechanism of this insertion is presently unknown.

Conjugated dienes also react more readily with metal hydrides than do simple olefins. The products are usually σ- or π-allyl compounds [59,60]:

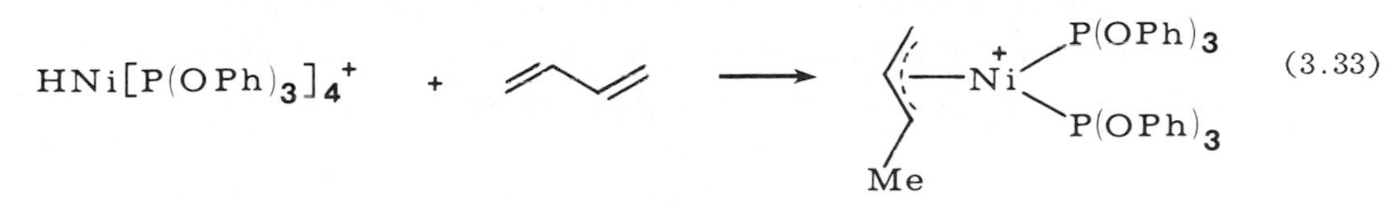


The effects of coordinative saturation and steric factors are unknown.

It is also curious that the regioselectivity of metal hydride additions to acrylonitrile seems to vary from one complex to another [61]:

$$HRhCl_2L_3 + CH_2{=}CHCN \longrightarrow CH_3\underset{\substack{|\\CN}}{C}H{-}RhCl_2L_3 \qquad (3.34)$$

$$L = MeAsPh_2$$

$$HIr(CO)(PPh_3)_3 + CH_2{=}CHCN \longrightarrow \underset{\mathbf{9}}{NCCH_2CH_2Ir(CO)(PPh_3)_2(CH_2{=}CHCN)} \qquad (3.35)$$

The reason for this variation is not understood, but evidence supporting the stereochemistry of the β-substituted iridium complex 9 was not given [62].

A curious and little-studied reaction is that between zinc and cadmium alkyls and transition-metal hydrides [63]:

$$2\ HMn(CO)_4PPh_3 + R_2Zn \longrightarrow 2\ RH + Zn[Mn(CO)_4PPh_3]_2 \qquad (3.36)$$

This sort of process may be related to the binuclear alkyl, hydride eliminations discussed in Chapter 4.

## 3.4. Carbon σ-Bonded Ligands.

a. Transition-Metal Alkyl Compounds. An alkyl, vinyl, or aryl group σ-bonded to a transition metal can be considered as a simple two-electron donor, occupying a single coordination site. As we shall see, there are also some rare examples of bridging transition-metal alkyls. Alkyl derivatives are known for all the transition elements, usually in conjunction with other ligands. Methods for synthesizing metal alkyls are given in Chapter 11. Metal alkyl derivatives are important reagents or intermediates in most stoichiometric or catalytic reactions involving transition-metal compounds. The transition-metal alkyl derivatives vary enormously in their reactivity. We shall find their reactivities depend on mechanistic, rather than thermodynamic, factors. Some typical transition-metal alkyl or aryl compounds are shown:

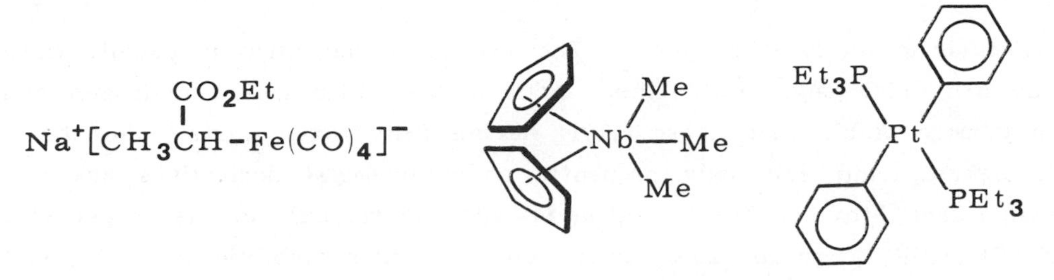


i. <u>Occurrence and Stability</u>. Metal peralkyl, "homoleptic," compounds of the transition metals are usually very reactive. These compounds are often sensitive to oxygen, light, and acids. Such peralkyl complexes were once thought to be thermodynamically unstable. Transition-metal alkyl compounds having other ligands such as CO, π-cyclopentadienyl, or phosphine groups are more common and more stable. Thus, it was proposed by Chatt that π-acid ligands stabilize the transition-metal alkyl bond. It is now clear that the stability of transition-metal alkyl compounds is kinetic, rather than thermodynamic. Thus, these stabilities are determined by the mechanisms available for decomposition. The principal decomposition path of transition-metal alkyl compounds involves either an α- or a β-hydride migration (Equations 3.37 and 3.38), the latter being more facile.

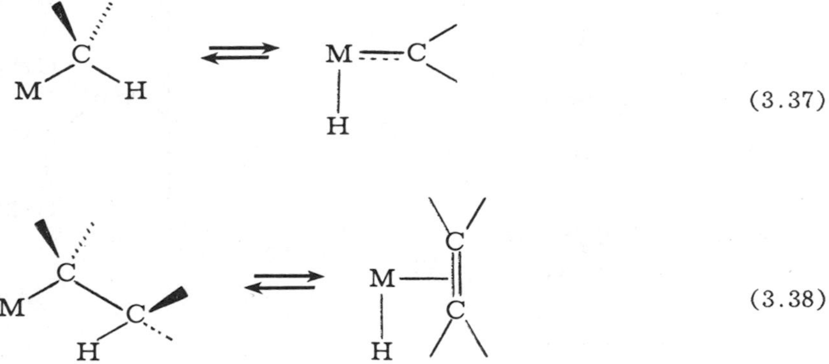


These retrograde insertion reactions are discussed in more detail in Chapter 5. Here we need only note that such hydride migrations usually require a vacant coordination site adjacent to the metal alkyl bond. Thus, we now recognize that most kinetically stable transition-metal alkyl complexes are coordinatively saturated. The stabilizing role of the π-acid ligands thus derives from the fact that these ligands are often substitution-inert.

Once a peralkyl metal undergoes a retrograde insertion, the resulting alkyl hydride complex decomposes very rapidly, forming an alkane by an intra- or inter-molecular reductive-elimination. The β-hydride migration is usually more facile than the α-hydride migration. Thus, metal alkyls which lack β-hydrogen atoms are kinetically more stable than those alkyls having β-hydrogen atoms. For this reason, methyl, benzyl, and especially neopentyl transition-metal derivatives are kinetically more stable than ethyl or butyl compounds. The steric bulk of the neopentyl and the related $(CH_3)_3SiCH_2$ groups may further enhance their stabilities by retarding inter-molecular decomposition pathways.

Metal peralkyls, $MR_n$, where R is incapable of β-hydrogen elimination, are now known for all the early transition metals, for various values of n [64-66]. The second- and third-row peralkyl derivatives tend to exist as dimers having strong (multiple) metal-metal bonds. Some first-row peralkyl derivatives are also dimeric, but certain of these exhibit three-center bonding, which until recently was thought to be an exclusive property of the main-group elements, such as aluminum [67]. Compounds with such alkyl bridges do not seem to possess any significant metal-metal bonding. Examples of the contrasting behavior of 3d versus 4d and 5d metal alkyls are shown in 10 and 11 [68].

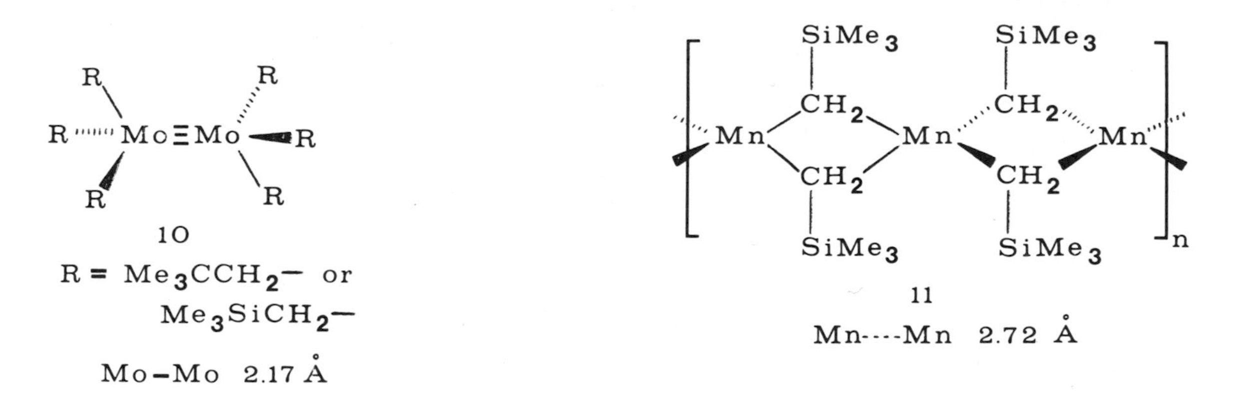


The molybdenum(III) compound, 10, is diamagnetic and possesses a triple Mo-Mo bond, whereas the related manganese(II) alkyls, 11, have unparied electrons, but exhibit temperature-dependent spin coupling. This contrasting behavior illustrates the greater tendency of the heavier transition elements to engage in metal-metal bonding. Some of the most remarkable stable peralkyl transition-metal compounds are those derived from the bridgehead norbornyl group [69]:

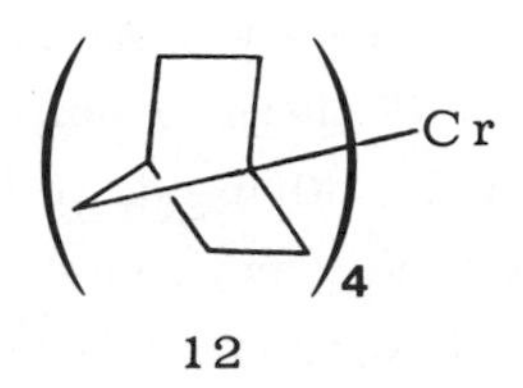


Such complexes cannot decompose through the hydride migration reaction because of the instability of bridgehead olefins (Bredt's rule). For example, the chromium(IV) tetraalkyl, 12, a $d^2$ compound having two unpaired electrons, has a half-life for thermal decomposition of 7.5 hr at 250°C and is stable both in air and in 0.1N $H_2SO_4$!

Metal complexes of fluorinated alkyls such as $CF_3CH_2$- and $CF_3CF_2$- are also quite unreactive--apparently because of the stable β-C-F bonds.

Complexes of cobalt(III), rhodium(III), iridium(III), and chromium(III) are relatively inert to substitution reactions. Thus, it is not surprising to find that $[Rh(NH_3)_5(C_2H_5)](ClO_4)_2$ and $Li_3(CrMe_6)$ [70] are stable enough to have been isolated and characterized.

There are numerous metal alkyl complexes having macrocyclic ligands. Such compounds are analogues of the coenzyme Vitamin $B_{12}$, probably the oldest organometallic compound, (see Figure 3.19). Such metal-carbon bonds are probably not thermodynamically robust, but are kinetically stabilized by the macrocyclic ligand which occupies all the coordination sites adjacent to the metal-carbon bond. These metal-carbon bonds are quite susceptible to both thermal and photochemical homolytic cleavage.

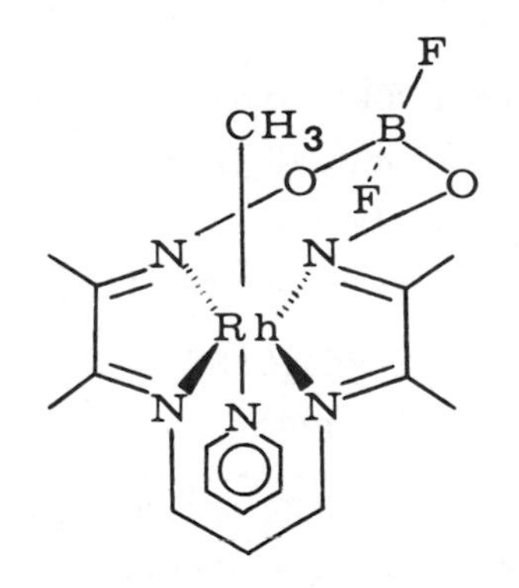

ii. Physical Properties. Transition-metal alkyls exhibit few distinctive physical properties. Both $^{1}H$ and $^{13}C$ chemical shifts at carbons attached to transition metals occur at higher magnetic fields than simple organic derivatives [71]. Direct metal-$^{13}C$ coupling constants may be very large (~500 Hz!). Similar high-field chemical-shift values are found for hydrogens attached to the $\alpha$-carbon in metal alkyls.

iii. Reactions. The most widespread and useful reactions of transition-metal alkyl derivatives occur with groups which are also attached to the same or another metal, insertion (Chapter 5) and reductive-elimination (Chapter 4) reactions. Transition-metal alkyls are only weakly nucleophilic and, for example, do not react readily with aldehydes or ketones, thus differing from main-group metal alkyls. Protons will cleave most transition-metal alkyl bonds, presumably by direct attack on the metal-carbon bond:

$$\geq C{-}M + H^+ \longrightarrow \geq C{-}H + M^+ \tag{3.39}$$

We shall also see that metal alkyls are susceptible to nucleophilic attack; that is, the metal can be a "leaving group":

$$\bar{\ddot{N}} + \geq C{-}\overset{+}{M} \longrightarrow N{-}C\leq + M \tag{3.40}$$

This latter reaction is facilitated by a formal positive charge on the metal in the alkyl compound. Such reactions are often involved when transition-metal alkyls are cleaved under oxidizing conditions [72].

Thus, anionic transition-metal alkyl derivatives are relatively stable under basic, nonoxidizing conditions, whereas cationic alkyls are quite reactive.

b. Aryls and Vinyls. A substantial number of $\sigma$-bonded transition-metal aryl compounds are known. Representative examples are:

| $Rh(Ph)(PPh_3)_3$ | $Cr(Ph)_3(THF)_3$ | $Li(Et_2O)_2Ta(tolyl)_6$ |
|---|---|---|
| [73] | [74] | [75] |

Even though the more electronegative $sp^2$ aryl-metal bonds might be expected to be stronger than corresponding metal-alkyl bonds, the metal aryls are very reactive. Thus binary, "homoleptic" transition-metal aryls are not especially stable, and readily decompose to biphenyl derivatives and other substances. However, the factors which control these degradations are not well-understood [76]. Decomposition paths involving ortho hydrogens have been suggested. We shall see later in this chapter that this decomposition mode has been used to make a benzyne complex. Transition-metal aryl derivatives having ortho substituents are kinetically more stable; for example, mesityl and 1-naphthyl derivatives [77].

Chromium polyaryls have been extensively investigated because they were found to be intermediates in the discovery of $\eta^6$-arene complexes [78]. Polyaryl metal compounds often contain coordinated ether.

Transition-metal-aryl bonds may be formed by simple metathetical reactions of aryl Grignard or lithium reagents with appropriate halides [79]:

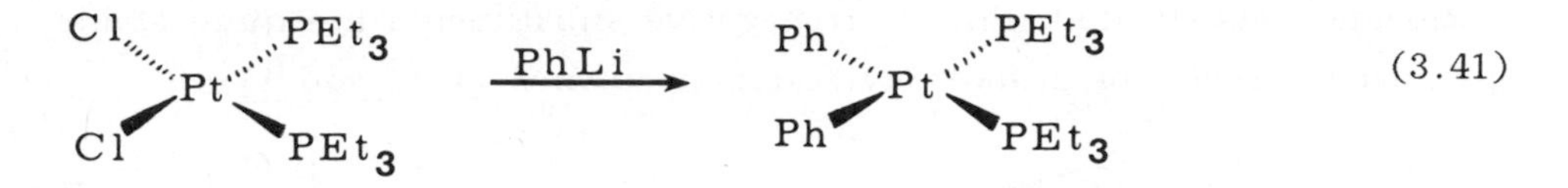


(3.41)

An especially useful method for preparing aryl-metal derivatives involves "transmetallation" between a transition-metal halide and an aryl mercury compound [80], which is conveniently generated by mercuration of a reactive aromatic substrate:

$$ArH + Hg(OAc)_2 \longrightarrow ArHg(OAc) \xrightarrow[L]{Pd(OAc)_2} [ArPdL_2(OAc)] \qquad (3.42)$$

The scope and mechanism of such transmetallations are not known.

As discussed in Chapter 4, oxidative-addition of aryl halides or even aryl C-H bonds to transition metals are useful methods for generating transition metal-aryl σ-bonds.

Transition-metal vinyl groups that are σ-bonded are not a well-studied family of ligands per se. Homoleptic vinyls are quite rare, an example being vinyl silver [81]. The transition metal vinyl group may be formed by decarbonylating acrylyl derivatives:

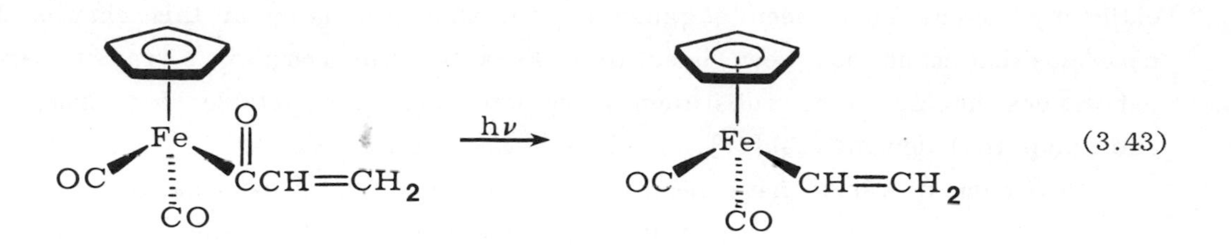


Other methods include oxidative-addition of vinyl halides to $d^8$ and $d^{10}$ derivatives (Chapter 4) and insertion reactions involving coordinated acetylene (Chapter 5). Vinyl compounds substituted with electronegative substituents are quite stable. King [82] has prepared a series of cyano-vinyl derivatives such as 13 and 14.

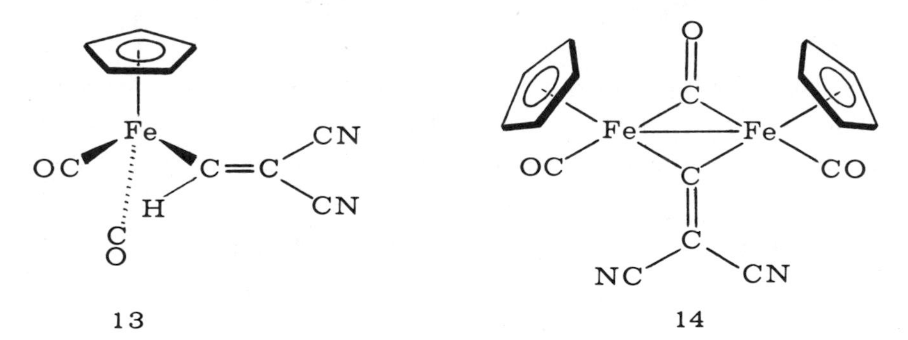


Condensation of vaporized transition-metal atoms with $C_6F_5Br$, and subsequent addition of triphenylphosphine, affords a stable perfluorophenyl derivative, 15 [83]:

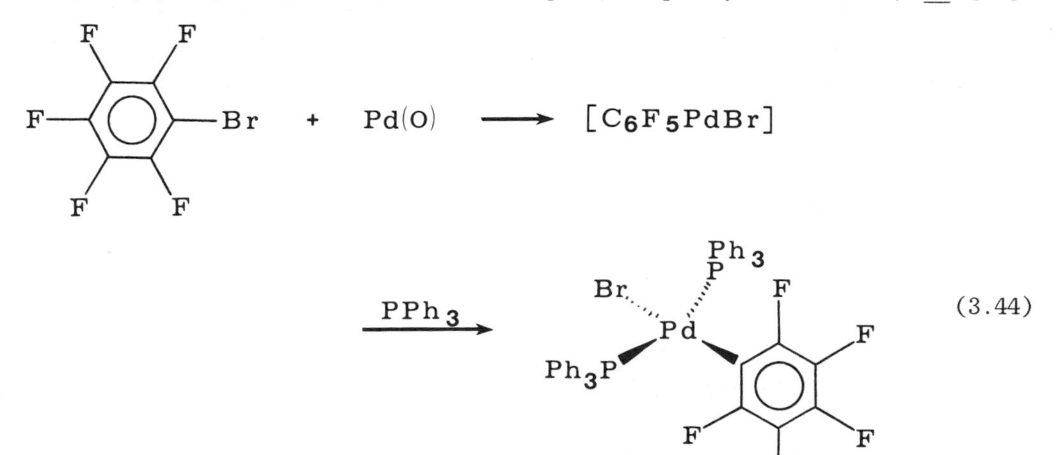

This is a rather general method for preparing many unusual transition metal complexes. However, special equipment is required so that this technique is still not widely used.

c. Acetylides. The acetylide sp-carbon forms σ-alkynyl transition-metal compounds. These may be formed by replacing a halide with an acetylenic anion (Equation 3.45) [84] or by apparent oxidative-addition of the acetylenic C-H bond, followed by reductive-elimination (Equation 3.46) [85]:

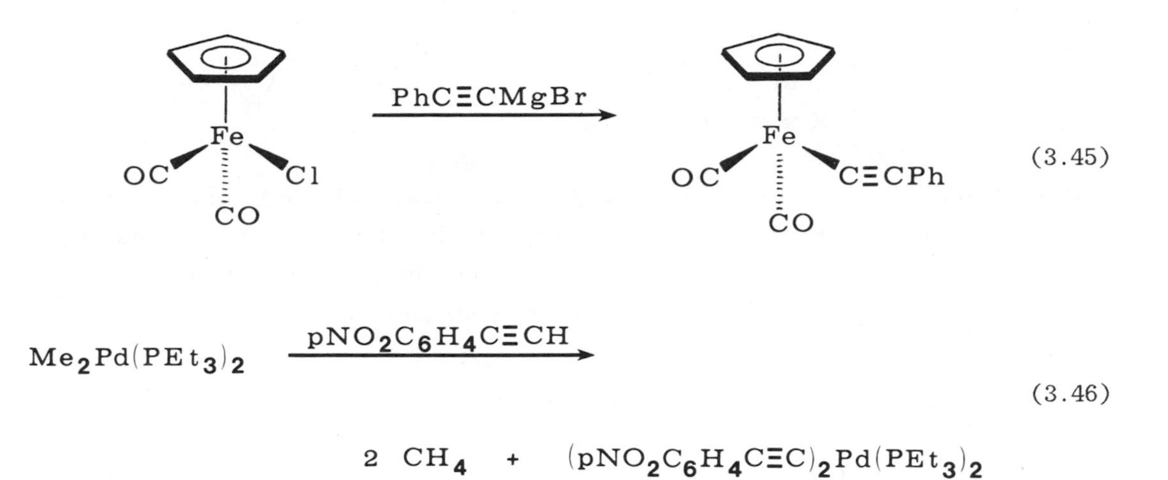


The metal-acetylide bond appears to be less reactive and is probably stronger than metal-alkyl bonds. Polyacetylides such as $K_3[Fe(C≡CH)_6]$ are formally analogous to cyanides, but the former are more readily hydrolyzed. The σ-bonded metal acetylides are capable of π-bonding with neighboring metals thus forming polymeric, insoluble, frequently explosive compounds. This behavior is typical of acetylide complexes of Cu, Ag, and other metals to the right of the transition series [86].

d. Cyanide Complexes. The cyanide ion has a special affinity for and forms complexes with virtually all the transition metals [87,88]. Both homoleptic and mixed complexes are known. Although isoelectronic with carbon monoxide, the cyanide ligand differs from CO in two major ways. The negative charge on cyanide and the greater ligand field strength of nitrogen compared with oxygen renders cyanide a weaker π-acid and a stronger σ-donor ligand than carbon monoxide. Cyanide is thus better able to stabilize metals in higher oxidation states. Cyanide ligands exhibit a greater tendency to bridge two metals, but these bridges are linear: M-C≡N-M. The cyanide ligand has thus far had a limited role in organotransition-metal reactions--the exception being the commercially important hydrocyanation of butadiene that leads to adiponitrile, as discussed in Chapter 8.

e. Acyls and Related Ligands. There are many acyl and carbamoyl transition-metal derivatives. These substances are important in many insertion processes (Chapter 5). Both mono and dihapto acyls are known (Chapter 5).

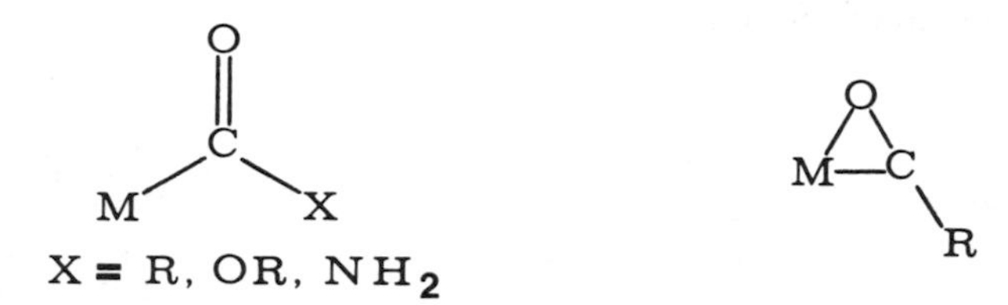


The νCO band in monohapto metal acyl derivatives, ~1650 $cm^{-1}$, is lower than that found in organic ketones, ~1720 $cm^{-1}$. Surprisingly, the reactivity of metal acyl derivatives towards nucleophiles has seldom been directly studied, but as expected, positively charged acyl derivatives appear to be much more reactive than neutral or anionic compounds.

Three reports of stable molecules containing the $MCO_2H$ group have appeared. Pettit's isolation and characterization of an iron(II) carboxylic acid is illustrative [89]:

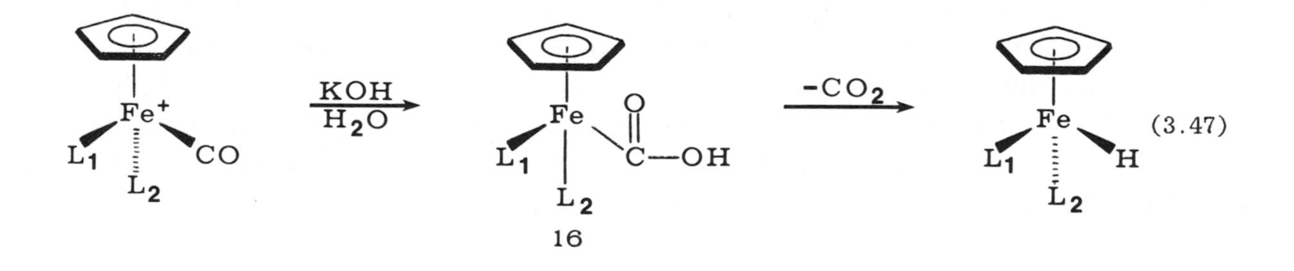


The bis-carbonyl complex (16, $L_1$=$L_2$=CO) is much less stable toward decomposition than the phosphine carbonyl derivative (16, $L_1$=$PPh_3$, $L_2$=CO).

Acyl complexes manifest basic character at the acyl oxygen. Protonation of the acyl oxygen gives rise to a hydroxy-carbene complex with no net change in the formal oxidation state (Equation 3.48) [90]; whereas protonation at the metal would afford the tautomeric acyl hydride and a formal two-electron oxidation of the metal (an oxidative-addition).

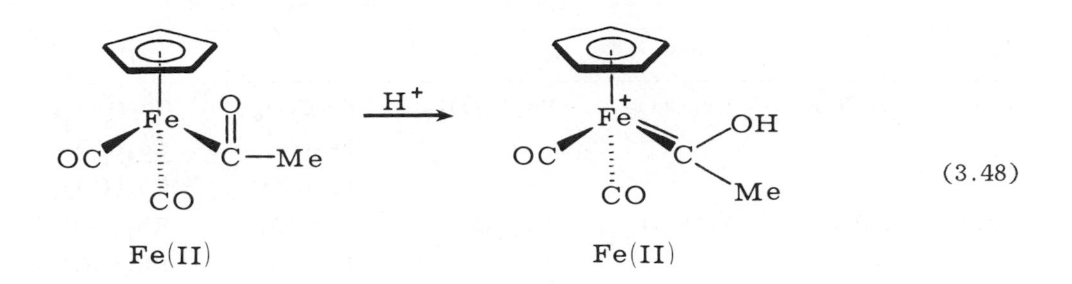


Hydride-acyl complexes are uncommon. These may be intermediates in the formation of the corresponding aldehyde by reductive-elimination:

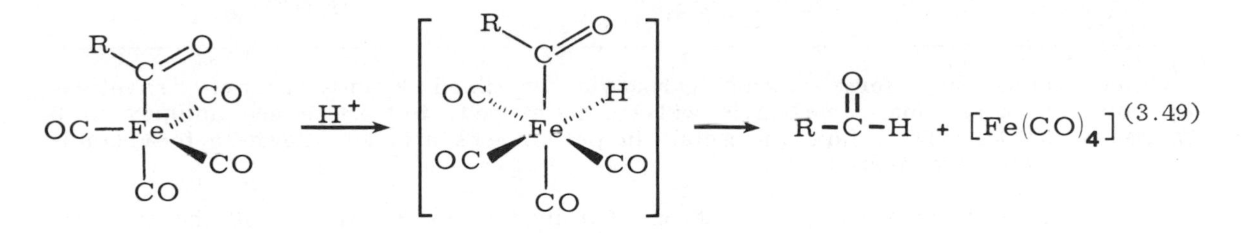


The stability of particular hydroxy-carbene complexes may be accounted for by the relative stabilities of the two oxidation states in the <u>presumed</u> tautomeric equilibrium between the hydroxy-carbene and the hydrido-acyl forms:

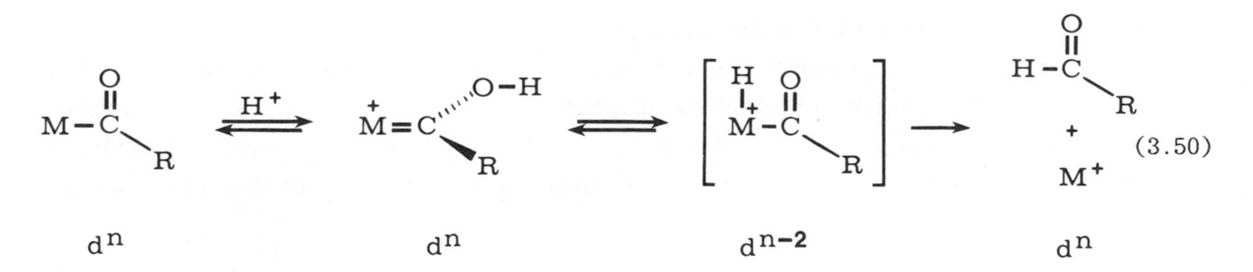


Transition-metal acyl derivatives may be prepared by oxidative-addition of acyl halides, carboxylic anhydrides, and in rare cases aldehydes (Chapter 4) or by insertion reactions (Chapter 5).

## 3.5. End-Bound π-Acid Carbon Ligands.

a. <u>Carbon Monoxide</u>. Metal carbonyls are among the oldest recognized and most widely occurring organotransition-metal compounds. Binary metal carbonyls are known for transition metals in groups V through the nickel triad of group VIII (Table 3.3).

*Table 3.3. Neutral binary metal carbonyls.*[a]

| IV | V | VI | VII | VIII | | |
|---|---|---|---|---|---|---|
| (Ti) | $V(CO)_6$ | $Cr(CO)_6$ | $Mn_2(CO)_{10}$ | $Fe(CO)_5$ | $Co_2(CO)_8$ | $Ni(CO)_4$ |
| | | | | $Fe_2(CO)_9$ | $Co_4(CO)_{12}$ | |
| | | | | $Fe_3(CO)_{12}$ | $Co_6(CO)_{16}$ | |
| (Zr) | (Nb) | $Mo(CO)_6$ | $Tc_2(CO)_{10}$ | $Ru(CO)_5$ | $Rh_2(CO)_8$ | (Pd) |
| | | | | $Ru_3(CO)_{12}$ | $Rh_4(CO)_{12}$ | |
| | | | | | $Rh_6(CO)_{16}$ | |
| (Hf) | (Ta) | $W(CO)_6$ | $Re_2(CO)_{10}$ | $Os(CO)_5$ | $Ir_2(CO)_8$ | (Pt) |
| | | | | $Os_3(CO)_{12}$ | $Ir_4(CO)_{12}$ | |
| | | | | | $Ir_6(CO)_{16}$ | |

[a]Other metals may form anionic carbonyls or mixed ligand-carbonyl derivatives. Undoubtedly more binary carbonyls will be discovered, but these are unlikely to be thermally stable. Thus far, the metals in parentheses are not known to form stable, neutral, binary carbonyls.

Examples of compounds with at least one CO ligand are known for all the transition metals. The homoleptic metal carbonyls are often useful starting materials for organometallic synthesis. They vary widely in physical properties, availability, and cost. Iron carbonyl, a colorless liquid, is the least expensive and is only moderately toxic. Nickel carbonyl is a highly toxic, volatile liquid. The solid brown carbonyl cluster, $Ru_3(CO)_{12}$, is very expensive.

Many polynuclear carbonyls are known. These substances, which are often colored, are held together by metal-metal bonds. Many, though not all, polynuclear carbonyls possess bridging carbonyl groups. With very few exceptions, the bridging carbonyl group requires the presence of a metal-metal bond; complexes with shorter

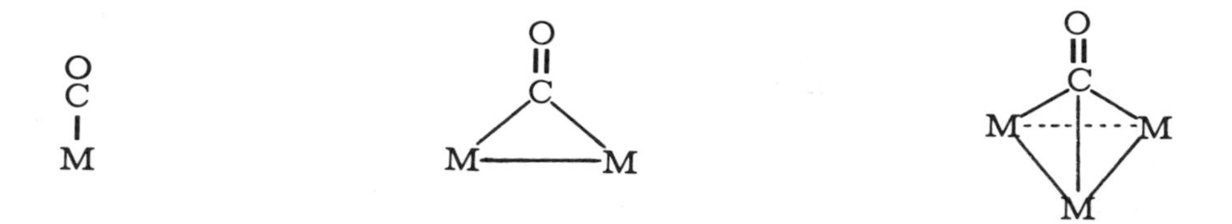


metal-metal bonds have a greater tendency to form carbonyl bridges, and there is often a delicate balance between the bridging and nonbridging carbonyls. For example, $Mn_2(CO)_{10}$ has no CO bridges; $Co_2(CO)_8$ has CO bridges in the solid, but exhibits both bridged and unbridged forms in solution; whereas $Fe_2(CO)_9$ has three bridging CO groups in all phases (Figure 3.4). Heavier congeners, having longer metal-metal bonds, have less tendency to form bridging carbonyls. Triply bridging CO groups are also known, as for example, in $Rh_6(CO)_{16}$. Although most metal carbonyls obey the 18-electron rule, as we noted in Chapter 2, this is not true for some of the higher clusters, such as $Rh_6(CO)_{16}$. Also, note that $V(CO)_6$ has an odd electron.

The structures of polynuclear metal carbonyls are difficult to predict *a priori* [91], and structures of higher clusters are not amenable to simple analysis from the electron-counting rules or the simplified coordination geometries presented in Chapter 2. However, it is true that a CO ligand can conceptually (and often in practice) be replaced by (1) an electron pair, (2) a hydride anion, and (3) two hydrogen atoms. These generalizations are easily derived from our electron-counting rules, and are useful guidelines for predicting the stoichiometry but not the structures of metal carbonyl derivatives, especially clusters. For example, $Fe_3(CO)_{12}$, $[Fe_3(CO)_{11}]^{2-}$, $[Fe_3(CO)_{11}H]^-$, and $Fe_3(CO)_{11}H_2$ have related stoichiometries, but their structures are different.

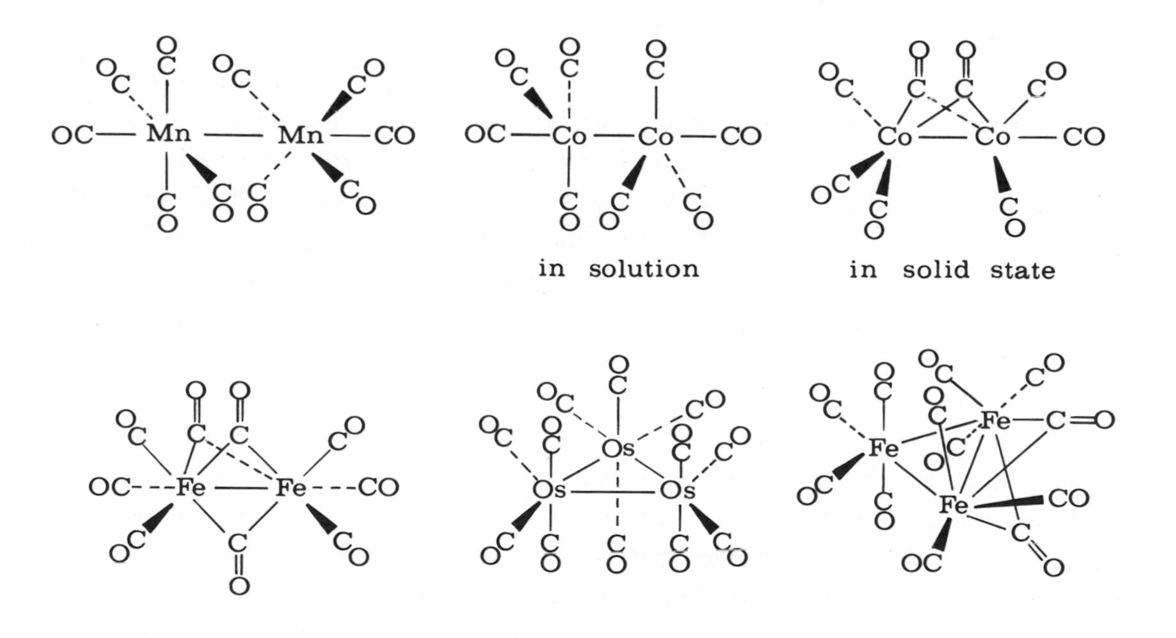


*Figure 3.4.*

In rare cases a carbonyl has been found to bridge two metals which are not joined by a metal-metal bond. These "metalloketones" are thus far structural curiosities, since their reactions have not yet been studied. A case in point is the palladium complex, 17 [92]. Note that the Pd-C-Pd angle is 119°, whereas the corresponding angle in typical bridging carbonyls is <88°. A "normal" Pd-Pd bond is ~2.7 Å, whereas that in 17 is 3.27 Å. The νCO band in 17 is at 1720 $cm^{-1}$.

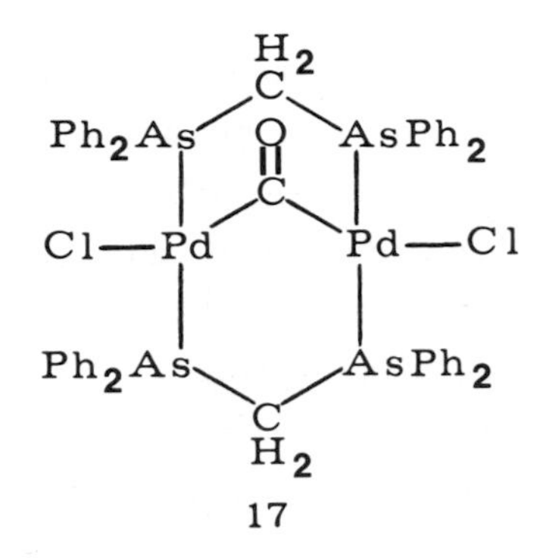

17

As a powerful π-acid, the CO ligand can remove excess charge from a metal atom, thus stabilizing anionic complexes. Indeed, metal carbonyl compounds having from one to four negative charges are known; some examples are $[V(CO)_6]^-$, $[Mn(CO)_5]^-$, $[Co(CO)_4]^-$, $[Fe(CO)_4]^{2-}$, $V(CO)_5{}^{3-}$, $Mn(CO)_4{}^{3-}$, $Co(CO)_3{}^{3-}$, and $[Cr(CO)_4]^{4-}$.

Heavier congeners of these metal carbonyl anions are also known. Those carbonyl anions having negative charges > 2 are very strongly bonded to the counter ion by ion pairing. For example, the insoluble tetranions are best formulated as $[M(CONa)_4]_x$. These materials are so intractable that they have been characterized only by forming soluble organotin derivatives [93]. Many metal carbonyl anions are quite useful reagents for stoichiometric organometallic synthesis [94]. This is discussed in Chapter 9.

As might be expected on the basis of the π-acidity of the CO group, metal carbonyl cations are far less common than carbonyl anions. One example is $[Re(CO)_6]^+$.

Metal-carbonyl bridges are not always symmetric. For example, additional donation of CO π-electrons can occur to a second metal. In some cases the CO may be considered as a four-electron donor [95]. An increasing number of structures have revealed grossly unsymmetrical bridging carbonyls in which a neighboring metal perturbs a terminal carbonyl ligand [96]. These increasingly common structures may be related to the dynamic exchange of carbonyl groups in certain metal carbonyl clusters. An example of an unsymmetrically bridged CO ligand is:

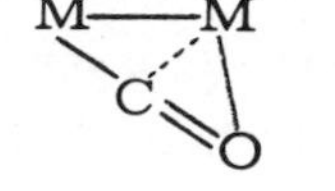


The single most useful spectral method for characterizing the carbonyl ligand is infrared spectroscopy. The intense νCO bands occur in a region of the IR spectrum which is largely free of other strong absorption bands. Possible interfering modes are $\nu N_2$, νNO, and νM-H. The number and frequency of the CO stretching bands are useful in assigning structure and in following the course of a reaction.

Terminal carbonyls exhibit CO stretching frequencies in the region 2140 to 1800 $cm^{-1}$, whereas doubly bridging CO groups are lower (νCO 1850 to 1700 $cm^{-1}$), and triply bridging νCO bands are even lower, down to ~1625 $cm^{-1}$. To a first approximation the νCO frequency reflects the CO bond order which is reduced by π-backbonding. Thus anionic MCO complexes have lower, and cationic MCO complexes higher, νCO values than neutral MCO complexes. Since trans ligands compete with each other for the same two d-orbitals in backbonding, a π-acid ligand often causes an increase in the trans CO stretching frequency. It is instructive to compare the CO stretching frequencies in Figure 3.5.

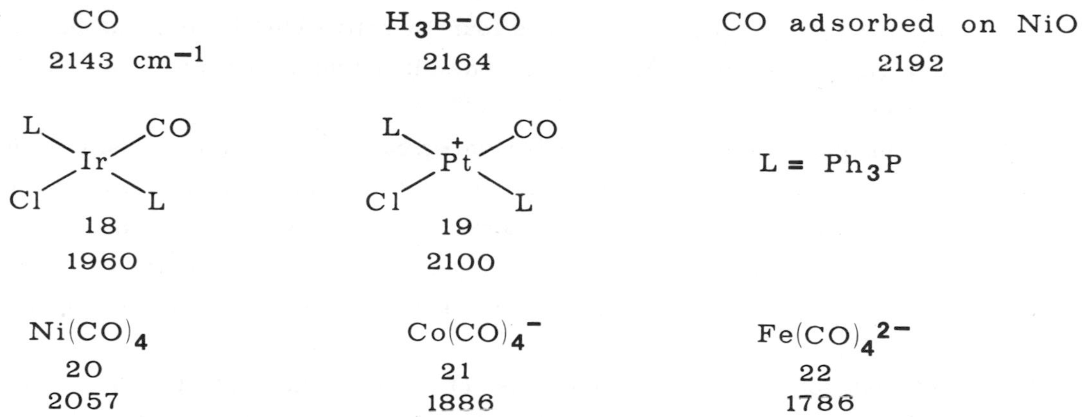

CO
2143 $cm^{-1}$

$H_3B$-CO
2164

CO adsorbed on NiO
2192

L, CO, Ir, Cl, L
18
1960

L, CO, $Pt^{+}$, Cl, L
19
2100

L = $Ph_3P$

$Ni(CO)_4$
20
2057

$Co(CO)_4^-$
21
1886

$Fe(CO)_4^{2-}$
22
1786

*Figure 3.5.*

Because of its small dipole moment, free CO exhibits a very weak νCO band. Cases such as $H_3BCO$ and CO adsorbed on NiO have no backbonding, with the result that the CO frequencies are higher than that of the free ligand. The neutral iridium carbonyl, 18, shows a lower CO frequency than the isoelectric cationic platinum analogue, 19. The tetrahedral $d^{10}$ tetracarbonyls of nickel, cobalt, and iron (20, 21, and 22) show a continual decrease in their principal νCO bands. A formal positive charge is presumed to contract the d-orbitals reducing π-backbonding, whereas a negative charge enhances π-backbonding. This acts to distribute excess negative charge into the π* orbitals of the CO ligands. Such responses in CO stretching frequencies to formal charge and oxidation state are useful guides to structure and reactivity. It is important to note that the relationship between CO frequency and M-C bond order is reciprocal. Those factors which strengthen the metal-carbon bond weaken the carbon-oxygen bond. The weaker νM-C bands of metal carbonyls are difficult to measure and seldom useful.

If a compound contains more than a single CO group, more than one νCO band may be observed. The number of infrared active νCO bands and their relative intensities depend on the local symmetry of the coordination sphere [97,98]. The number of active νCO bands resulting from vibrational coupling can be derived from group theory, a topic outside the scope of this text [99]. For most situations it is easier to look up the result than to carry out the calculation. The active vibrations are those which have a dipole moment change. The number of IR active bands cannot exceed but may be less than the number of CO groups in the complex. For example, an octahedral complex such as *cis*-$L_2M(CO)_4$ ($C_{2v}$ point group) has four IR active νCO bands, but its isomer, *trans*-$L_2M(CO)_4$ ($D_{4h}$ point group), has only one IR active νCO band. Table 3.4 lists some common cases.

It is also important to note that these point-group assignments are only valid for substances in dilute solutions or in the gaseous state. Solids may show additional νCO bands because of solid-state intermolecular vibrational coupling or by having a different symmetry in the solid. The frequencies and bandwidths of CO stretching bands are affected by the polarity of the solvent. Nonpolar noninteracting solvents tend to give narrower, higher frequency νCO bands.

The synthesis of binary carbonyls is of limited interest to the practicing organometallic chemist. Many can be purchased and the preparation of the others often requires special high-pressure equipment, including a compressor for CO. The tank pressure of CO cylinders seldom exceeds 1,000 psi, since the metal tank lining may be

*Table 3.4. The number of infrared active νCO bands expected for common coordination geometries.*

| Point Group | Number of νCO |
|---|---|
| 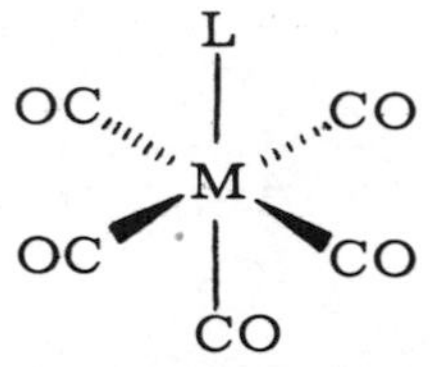 $C_{4v}$ | 3 |
| 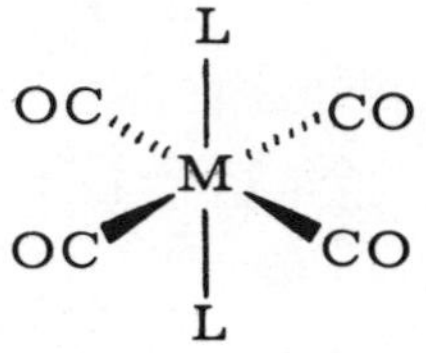 $D_{4h}$ | 1 |
| 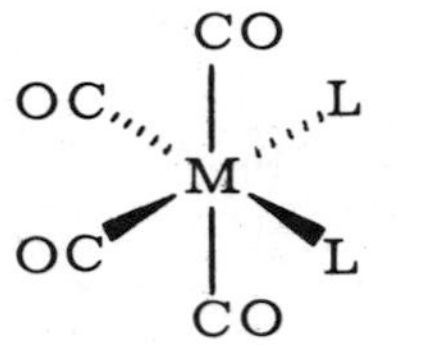 $C_{2v}$ | 4 |
| 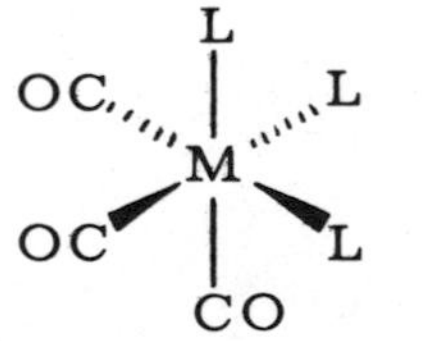 $C_{3v}$ | 2 |
| 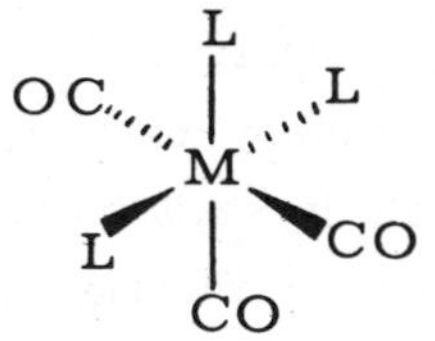 $C_{2v}$ | 3 |
| 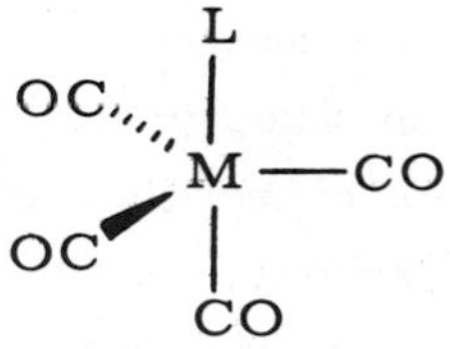 $C_{3v}$ | 3 |

*Table 3.4. Continued.*

| | Point Group | Number of νCO |
|---|---|---|
| CO; OC, OC–M–L; CO | $C_{2v}$ | 4 |
| L; OC, OC–M–CO; L | $D_{3h}$ | 1 |
| L; L–M–CO, CO; CO | $C_s$ | 3 |
| CO; OC–M–CO; CO | $T_d$ | 1 |

attacked. Gaseous CO is often contaminated with traces of $Fe(CO)_5$. Only iron and nickel react directly with CO under moderate conditions to form the binary carbonyls $Fe(CO)_5$ and $Ni(CO)_4$. Nickel was once purified on an industrial scale by this method (Mond process). Iron pentacarbonyl is used to prepare magnet cores by thermal decomposition. Most of the other binary carbonyls are prepared from metal salts under CO pressure in the presence of a reducing agent such as $H_2$ or an active metal.

We shall encounter the reactions of the coordinated CO group in Chapters 5, 8, and 9. However, it should be mentioned here that the MCO group is especially labile toward photodissociation. With few exceptions, irradiation of a metal carbonyl results in loss of a CO group. For example, $Fe_2(CO)_9$, a useful, more reactive starting material than $Fe(CO)_5$, may be prepared by irradiation of the latter, which is readily available but thermally unreactive [100]:

$$2\ Fe(CO)_5 \xrightarrow[CH_3CO_2H]{h\nu} Fe_2(CO)_9 + CO \qquad (3.51)$$

b. Thiocarbonyl and Selenocarbonyls. The first thiocarbonyl complex was discovered by Baird and Wilkinson in 1966 [101,102]. Since that time, several hundred thiocarbonyl compounds have been prepared, involving most of the transition metals; however, selenocarbonyl complexes are still rather rare, and only one tellurocarbonyl complex is now known [103]. Thiocarbonyl complexes have been reviewed through 1973 [104].

The free CS monomer is metastable, although it can be generated by decomposition of $CS_2$ in an electric discharge and isolated in an argon matrix. The free ligand is seldom useful for the preparation of transition-metal CS compounds. Such thiocarbonyl compounds are usually prepared from $CS_2$, $Cl_2CS$, or EtOC(S)Cl. Most of these reactions involve the modification of a coordinated CS precursor. For example, $RhCl(PPh_3)_3$, "Wilkinson's catalyst," forms a $\eta^2$-$CS_2$ complex which is reduced by excess $PPh_3$. The over-all reaction may be carried out in one step:

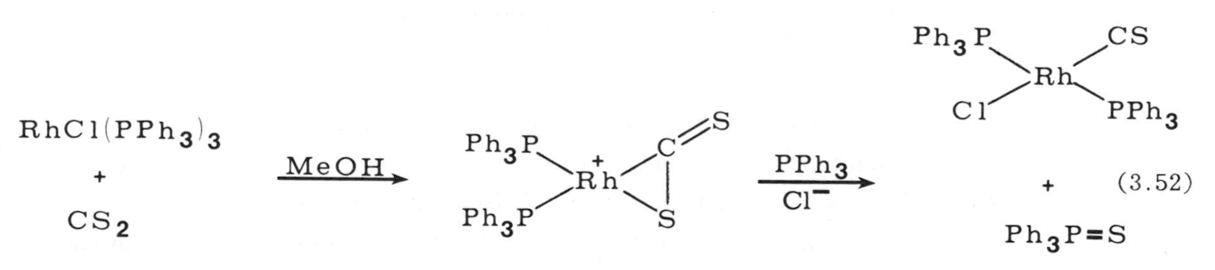


(3.52)

Selenocarbonyl compounds have been prepared in a similar manner starting with $CSe_2$ [105]:

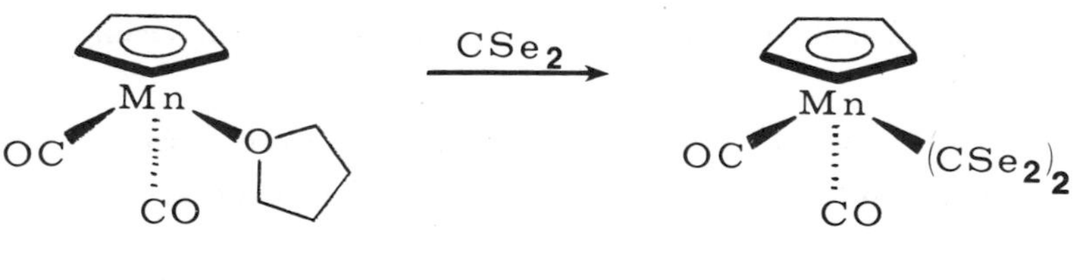


(3.53)

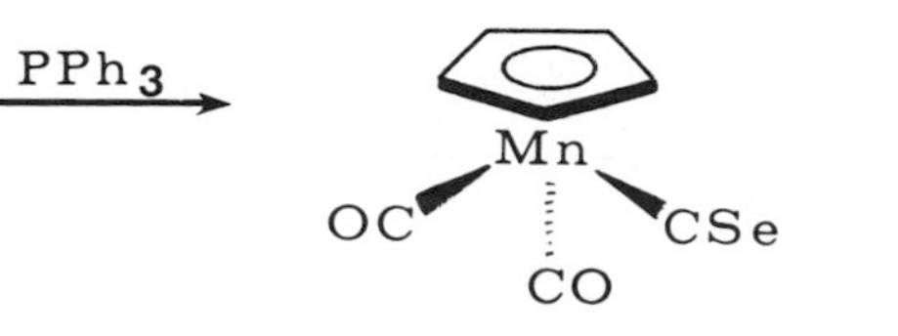

A rational thiocarbonyl synthesis starting from ethylchloroformate involves the reversal of an intermolecular insertion reaction [106]:

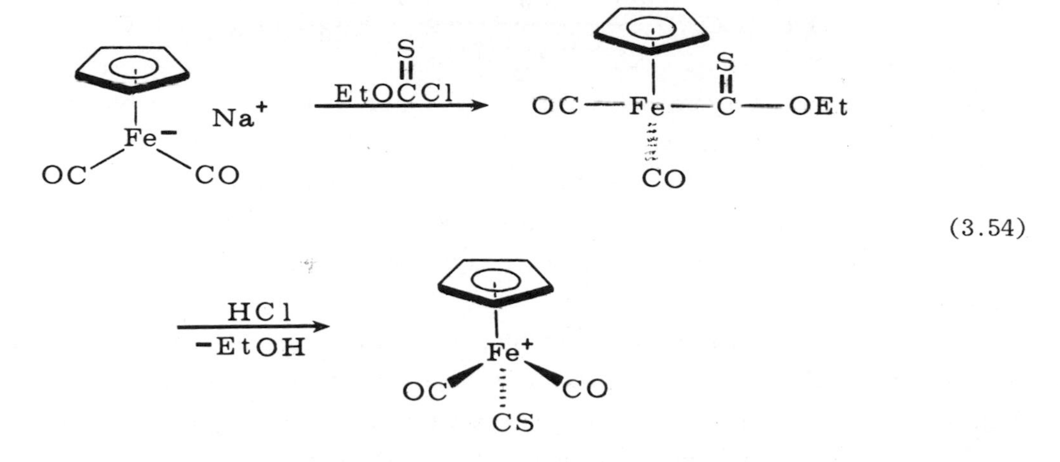


Thiophosgene can be used to prepare CS complexes under reducing conditions [107]:

$$\mathrm{Fe(II)(TPP)} + \mathrm{Cl_2CS} \xrightarrow[\mathrm{py}]{2\,e^-} \mathrm{pyFe(TPP)CS} \qquad (3.55)$$

TPP = tetraphenylporphryrinato

Transition-metal thiocarbonyl complexes exhibit νCS frequencies over the range 1160 to 1410 $cm^{-1}$, depending on the metal, the formal charge, and the other ligands. These values are to be compared with νCS for CS isolated in a $CS_2$ matrix, 1274 $cm^{-1}$. A detailed analysis of the force constants in carbonyl/thiocarbonyl complexes indicates that the CS ligand has a variable capacity to engage in dπ-pπ backbonding. In some cases this effect is greater than and in other cases less than that found for CO [108]. The few reported selenocarbonyl complexes exhibit νCSe in the range 1070 to 1140 $cm^{-1}$ [105,109].

The thiocarbonyl ligand has been found in three forms: terminal, doubly bridging, and "end-to-end" bridging. The latter is not yet structurally characterized [110].

M−CS

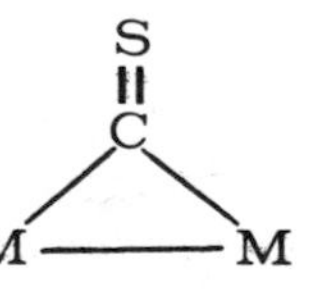


M−C≡S−M

The M-C-S group is linear. The metal-carbon bond lengths in thiocarbonyl complexes are somewhat shorter than those in corresponding carbonyl complexes, as is illustrated by 23 [111]. Bridging CS ligands would seem to be relatively more stable than bridging CO ligands, as is illustrated by the structure of $[CpFe(CO)(CS)_2]$, 24, [112].

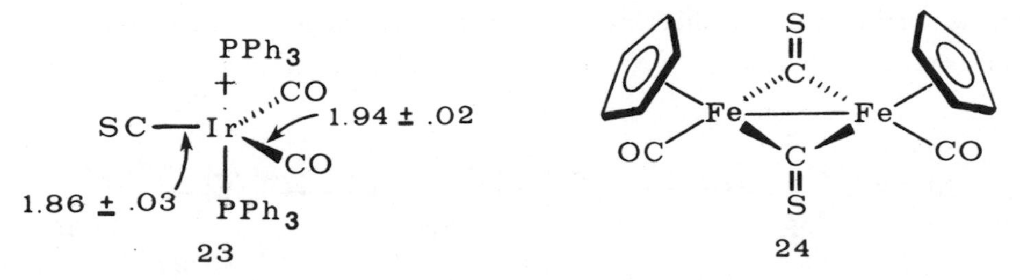


The sulfur atom in MCS complexes is quite basic. Thus, the CS ligand is capable of binding two metals in an end-to-end manner [113]. The νCS band is obscured but thought to be in the region below 1100 $cm^{-1}$.

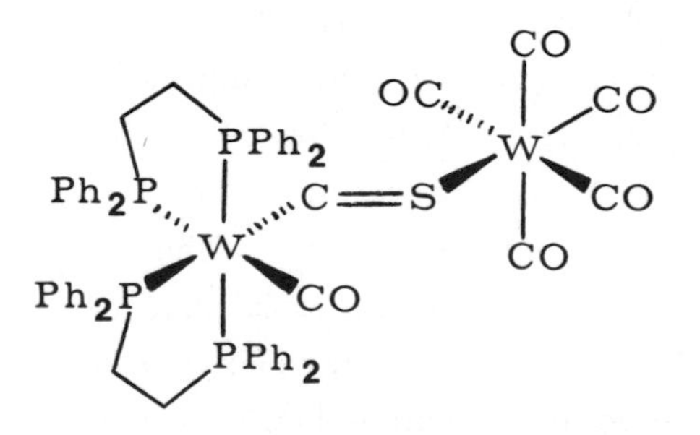


In Chapter 5 we shall see that the thiocarbonyl group is much more susceptible to both inter- and intramolecular insertion reactions than is the carbonyl group [114,115].

c. Carbene Complexes. Compounds in which a divalent carbon is bound to a transition metal, "carbene complexes," are known for the majority of transition elements. The first such compound was isolated and characterized by E. O. Fischer in 1964. Since that time this field developed very rapidly and has been extensively reviewed [116-119].

There are two major classes of carbene complexes: (1) those having one or two heteroatoms bonded to the carbene carbon; and (2) those having only carbon or hydrogen bound to the carbene carbon. The "heterocarbene" type was discovered first and is generally more stable. The reactions of the two types are somewhat different. Because of their different characteristics, we shall consider the two classes of carbene complexes separately. In Chapter 5 we will encounter carbene complexes which are useful reagents for the stereospecific cyclopropanation of olefins.

i. Heterocarbene Complexes. The lone-pair electrons on the heteroatom interact with the metal-carbene group like an amide or an ester. This interaction, depicted in resonance form 26, stabilizes coordinated heterocarbenes. This bonding interaction is manifest by a shortened heteroatom to carbene-carbon distance and in a substantial rotational barrier about that bond. The metal to carbene-carbon distance

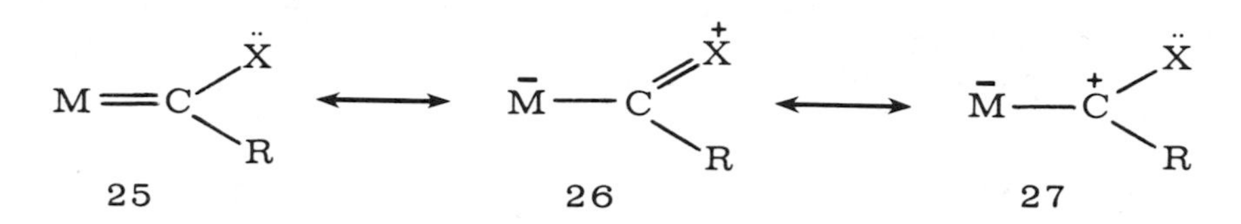


is also short, and that bond has a high rotational barrier, suggesting that resonance form 25 is significant. Finally, certain of the physical and chemical properties of carbene complexes are consistent with a substantial electropositive character at the carbene carbon as depicted by resonance form 27.

The structure of the methoxyphenylcarbene chromium(0) carbonyl, 28, is typical of the heteroatom-carbene complexes. Note the octahedral coordination about chromium expected for the formally $d^6$ configuration. For oxidation-state assignments, the carbene ligand may be considered zerovalent. The carbene carbon-to-chromium distance, 2.04 Å, is substantially shortened from that expected for a chromium-carbon single bond ~2.21 Å. The carbene carbon-to-oxygen bond length, 1.33 Å, is intermediate between that expected for a single bond, ~ 1.43 Å, and a double bond, ~1.23 Å. Low-temperature solution $^1$H NMR shows the presence of another rotational isomer, the "cis" complex, 29. The structural parameters of other heterocarbene complexes are similar [117].

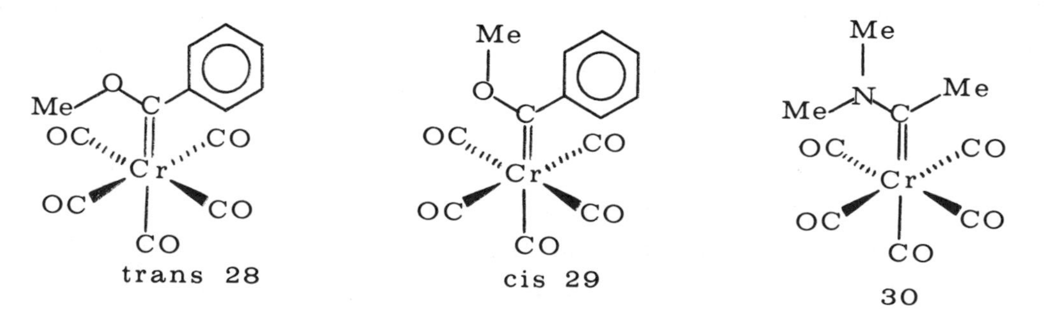

The rotational barrier about the carbon-nitrogen bond in the dimethylamino, methyl carbene complex, 30, is higher than that about the C-O bond in the methoxy compound, 29, just as amide C-NR bonds exhibit higher rotational barriers than ester C-OR bonds. Thus, the $^{1}H$ NMR of 30 shows three methyl signals at 25°C, which are broadened at 100°C. The rotational barrier is estimated to be ~25 kcal $mole^{-1}$.

The $^{13}C$ NMR signals of these heterocarbene carbons occur at very low magnetic fields (highly deshielded) [120]. For example, the chemical shift of the carbene carbon in the methoxy carbene complex 28 occurs at 351 ppm below TMS. This chemical shift is well within the region of a carbocation ("carbonium or carbenium ion"). Not only is this low field signal diagnostic of the presence of a carbene ligand, but the implied positive charge manifests itself in several ways. First, the νCO for the carbonyl trans to the carbene ligand occurs at a substantially lower frequency than those of the cis CO groups, indicating the metal is somewhat negative as implied by resonance forms 26 and 27. Second, the dipole moments of heterocarbene complexes are large (~5 Debye). Third, the hydrogen atoms attached to a carbon α to the carbenoid carbon are estimated to be about as acidic as p-cyanophenol! For example, the methyl group in the methoxy methyl carbene can be deprotonated by n-butyllithium or readily exchanged by basic $CH_3OD$. The resulting anion, which can be alkylated, is not reprotonated by diethylmalonate [121].

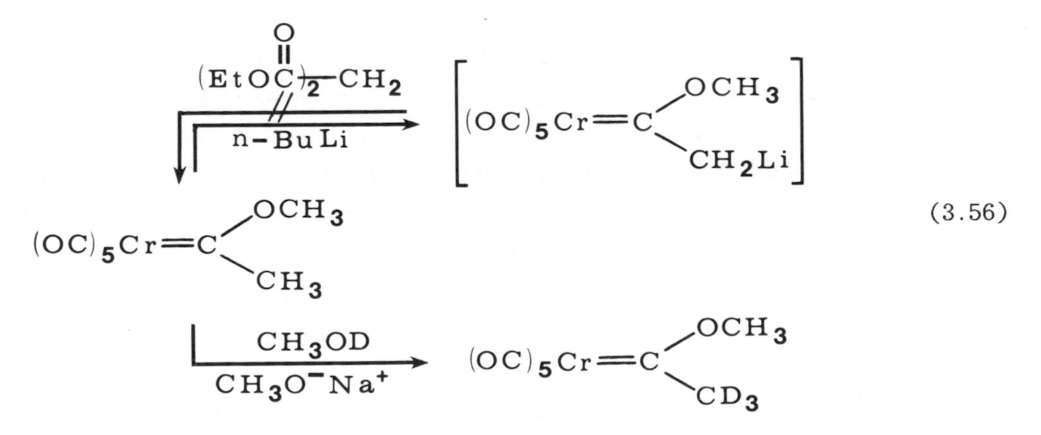


(3.56)

Heterocarbene complexes readily undergo exchange of the heteroatom group in a manner analogous to transesterifications. Such reactions occur via addition-elimination mechanisms involving attack of the Lewis base at the electropositive carbenoid center. In certain cases an intermediate adduct can be detected at low temperature. For example, alkoxy carbene complexes are readily transformed into amino carbene derivatives. This reaction has been used by Fischer to protect the amino group of an amino acid ester. Thus far this sort of procedure has not been widely used.

Carbene complexes are poor sources of free carbenes. This is not surprising since ligands which are unstable in the free state show little tendency to dissociate from a metal. Thermal decomposition of the methoxymethyl chromium carbene does afford E and Z isomers of 1,2-dimethoxy-2-butene:

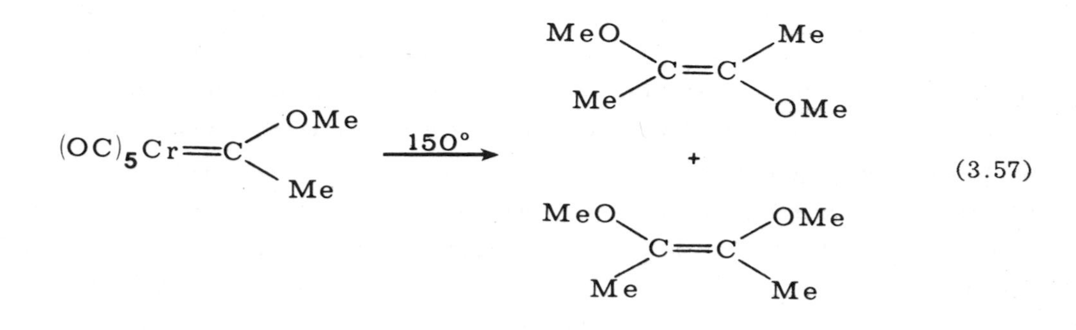

(3.57)

This product probably results from a bimolecular reaction and is typical of both classes of carbene complexes.

Carbene complexes are attacked by various ligands [116]. The resulting reactions are complex, and the mechanisms are presently uncertain. For example, carbon monoxide seems to displace the corresponding ketene, which may be trapped by reaction with an enamide (Equation 3.58). Heterocarbene complexes have been found to act as cyclopropanating agents only with $\alpha,\beta$-unsaturated esters and vinyl ethers. In the presence of vinyl ethers and CO, the isomeric cyclopropanes are formed (Equation 3.59). Since the ratio of isomers depends on the nature of the metal, it appears that the product determining step does not involve a free carbene. The difference between these reactions is puzzling.

Pyridine displaces the carbene and promotes a hydrogen shift affording an enol ether (Equation 3.60):

Electrophilic reagents readily cleave the coordinated carbene. Three quite different examples are illustrated in Equation 3.61 to 3.63. Carboxylic acids afford the corresponding esters (Equation 3.61). Oxygen and the other chalcogens form the ester derivatives (Equation 3.62). The dichlorocarbenoic source affords the olefin (Equation 3.63).

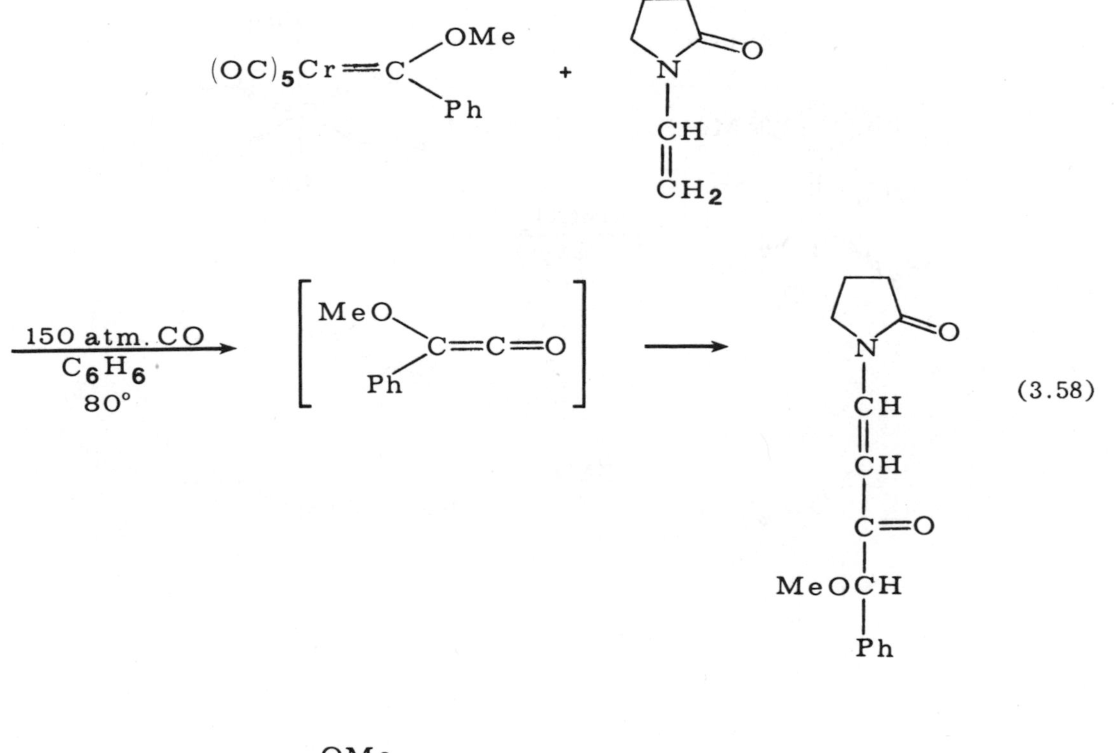


(3.58)

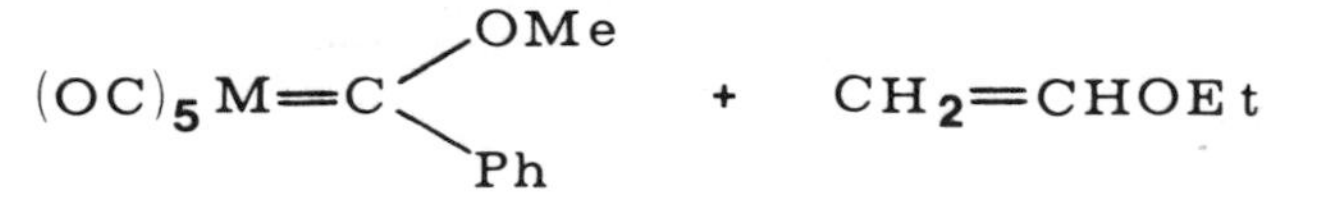


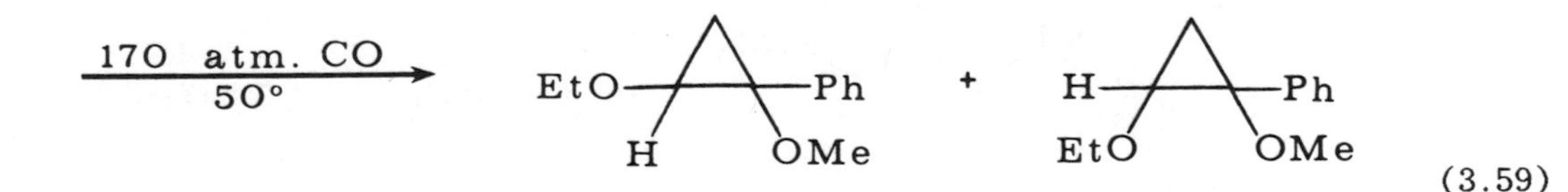


(3.59)

M = Cr, Mo, W

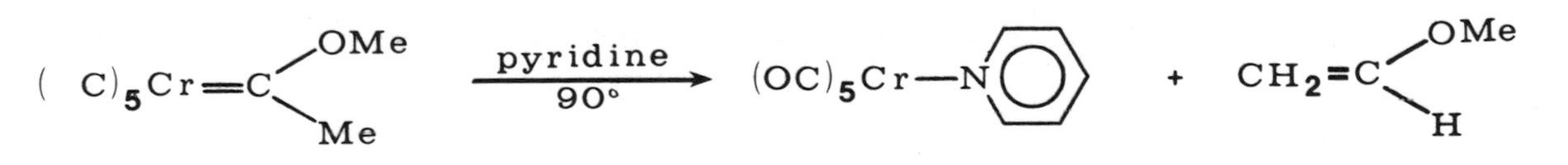


(3.60)

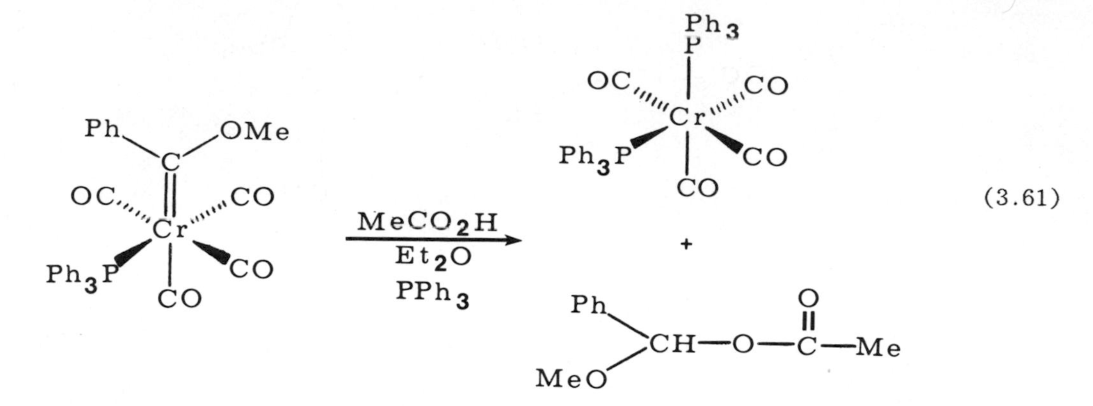


(3.61)

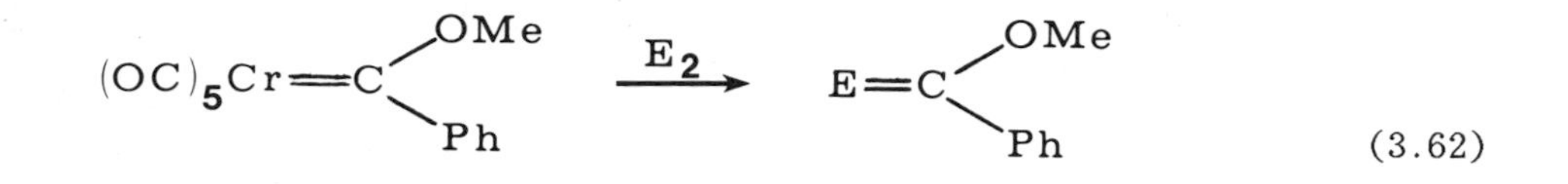


(3.62)

E = O, S, Se

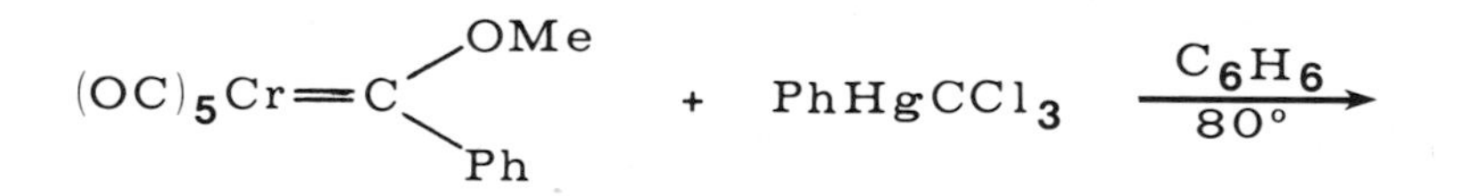


(3.63)

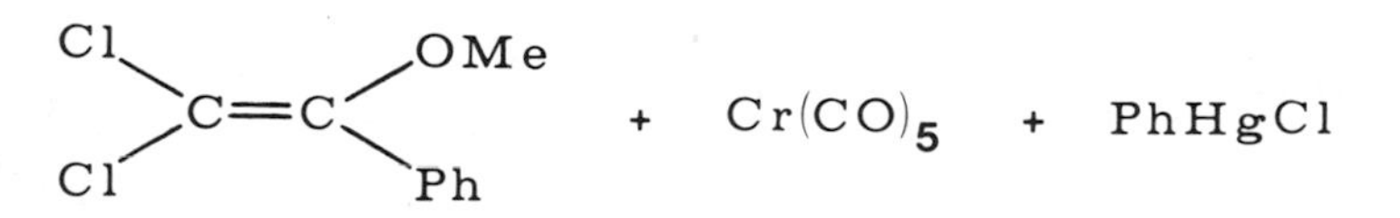


In all cases the mechanisms are unknown.

Most heteroatom carbene complexes are derived from alkylation of the oxygen in anionic acyl complexes. The alkoxy carbene compounds can serve as starting materials for other carbene derivatives through the ester-like heteroatom exchange reactions described above. An example is the formation of the ethoxy phenyl tungsten carbene complex:

$$W(CO)_6 \xrightarrow{PhLi} [(OC)_5W\overset{O}{\overset{\|}{C}}Ph]^- Li^+ \xrightarrow{Et_3O^+} (OC)_5W{=}C(OEt)(Ph) \quad (3.64)$$

Another method for preparing heterocarbene complexes involves direct attack of alcohols and other nucleophiles at certain electropositive coordinated isonitrile ligands:

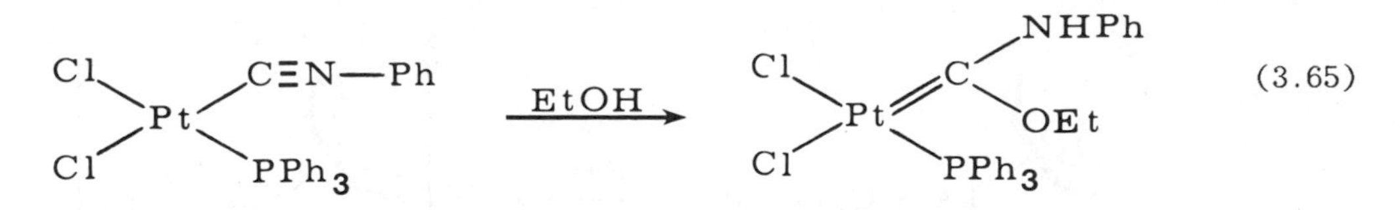

(3.65)

A remarkable, quite general synthesis of diamino substituted carbenes has been developed by Lappert [122]. Treatment of various metal complexes with highly electron rich olefins, such as the biimidazolidizidene, affords carbene complexes derived from cleaving the olefin bond, the reverse of the olefin formation from thermal decomposition of coordinated carbenes. Mono, bis, neutral, and cationic carbene complexes involving a majority of the transition metals have been prepared in this way:

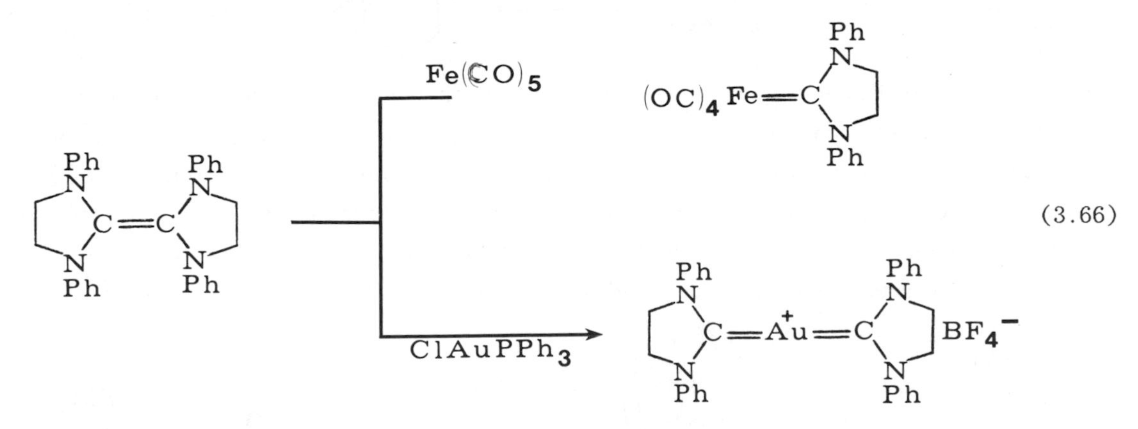

(3.66)

Roper has made good synthetic use of a dichlorocarbene [103], vide infra Equation 3.79.

ii. Alkylidene Complexes. A number of carbene complexes which are not substituted with heteroatoms are known. According to IUPAC nomenclature these complexes are called alkylidenes, as derivatives of alkyls. Many physical and chemical properties of these alkylidene complexes are similar to those of the heterocarbene compounds. However, we shall see that certain of these "simple" carbene compounds exhibit some quite different reactions in which the polarity of the carbene group has been reversed--the carbenoid carbon in the alkylidene compounds may be nucleophilic like the ylid carbon of a Wittig reagent. However, in other cases these carbenes are electron-pair acceptors. This contrasting behavior seems to depend on whether the electronic configuration of the complex is stabilized by the loss or the gain of an electron pair.

The first report of a simple carbene complex was that of the cationic primary carbene complex of iron, which was not isolated but was characterized by the cyclopropanation of cyclohexene to give norcarane [123,124]:

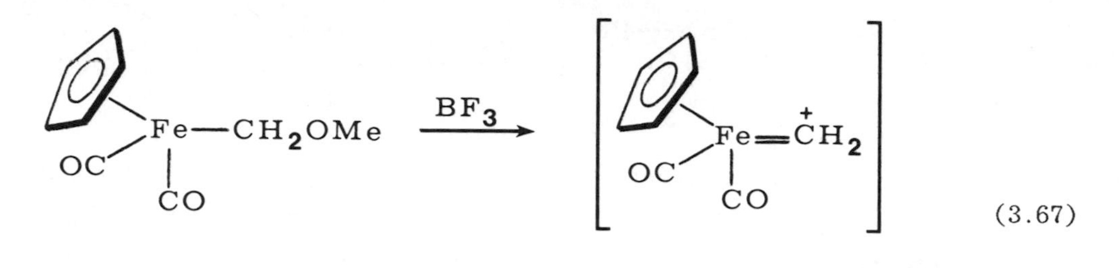


(3.67)

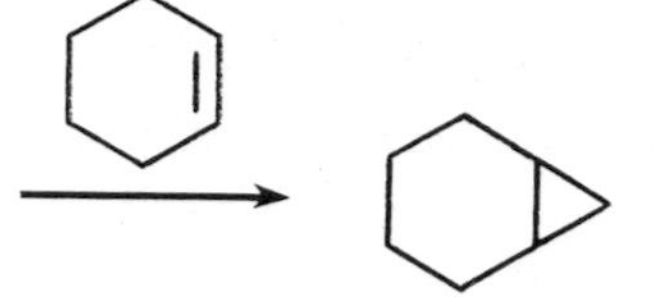

This was followed by the isolation of several nonheteroatom stabilized carbene complexes. These compounds either lacked a C-H group adjacent to the carbenoid carbon, or were so constituted that deprotonation at the α-carbon would form an anti-aromatic system. (The loss of an α-proton is an important pathway for the degradation of such carbene complexes.) Note that the chromium complex is stabilized in a dipolar form by the aromaticity of the diphenylcyclopropenium cation [125,126,127].

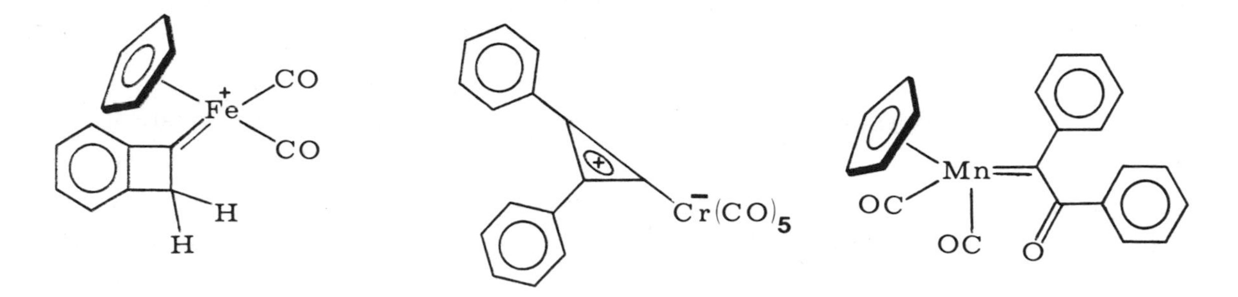


The preparation and detailed structural and chemical investigations of Casey's tungsten [128] and Schrock's tantalum [129] alkylidene complexes have provided a clear picture of this class of carbene complex.

Diphenylcarbene (pentacarbonyl) tungsten(0), 32, is of great interest both because of model studies of the olefin metathesis reaction, in which it is a catalyst for cyclic olefins (see Chapter 10), and because a comparison has been made between its properties and those of the related heteroatom stabilized, alkoxy-phenyl carbene complex, 31. The preparation of the diphenyl carbene derivative from the alkoxy-phenyl carbene compound is another variation of the ester-like addition-elimination reactions we encountered earlier.

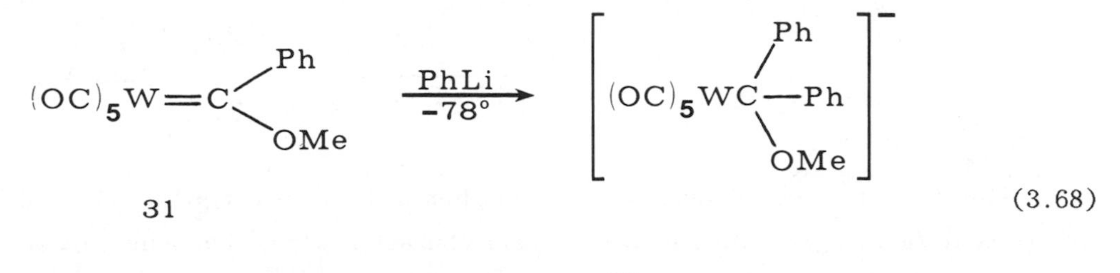


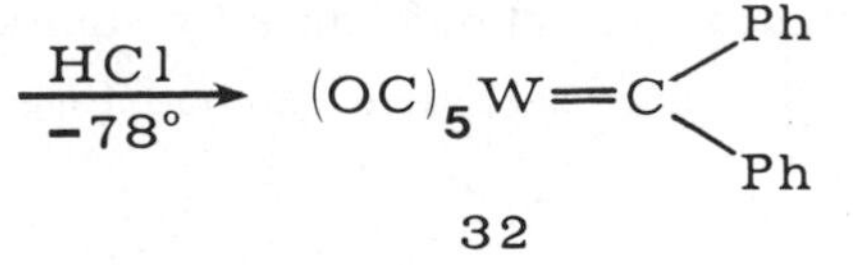


The diphenylcarbene complex, 32, exhibits a low-field $^{13}C$ signal, although not as low as that of the methoxyphenyl complex, 31. Indeed, low-field $^{13}C$ chemical-shift values had been found for all alkylidene carbene carbons, even though some of these have "negative" rather than the "positive" polarities exhibited by most heterocarbene carbons. It is clear that the properties giving rise to the characteristic low-field carbene carbon chemical shifts are not simple. The dipole moment of the diphenyl derivative, 3.48 D, is smaller than that of the methoxyphenyl carbene, 4.39. An analysis of the νCO bands in each indicates that the diphenyl group bears less positive charge than the methoxyphenyl carbene ligand. An X-ray diffraction study of the diphenylcarbene complex reveals a W-C (carbene) distance in the range expected for a double bond.

Many reactions of the diphenyl and methoxyphenyl complexes are similar, although the former is more reactive. For example, various oxidizing agents (Ce(IV), $O_2$, $Me_2SO$) react to form benzophenone, and Wittig reagents give rise to methylene transfer (Equation 3.69).

The major differences between the reactions of hetero substituted carbenes and the alkylidene complexes are revealed in the chemistry of the secondary and primary carbene complexes of tantalum isolated and characterized by Schrock [129]. The methylene complex, the first such isolated complex, is illustrative.

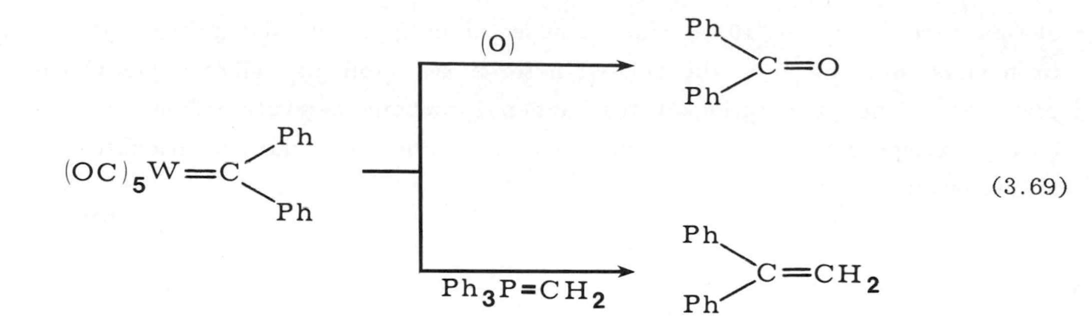


(3.69)

The method used to prepare the tantalum methylene complex, 33, and the reactions of its methylene ligand illustrate the basic character of the carbene carbon. The novel, mild base, $Ph_3P{=}CH_2$, deprotonates the cationic dimethyl tantalum complex, affording the neutral methylene compound, 33. The methylene carbon forms an adduct with the Lewis acid, $Me_3Al$:

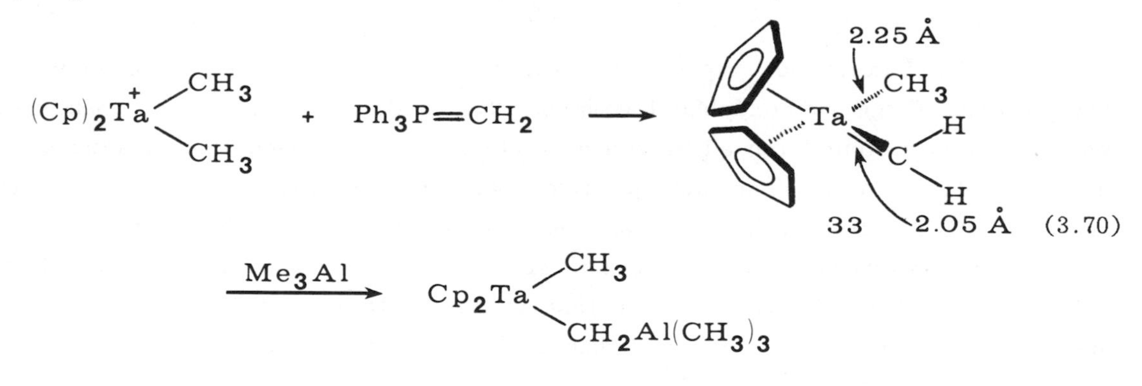


(3.70)

X-ray diffraction shows the C(H)(H) and Ta(C)(C) planes to be perpendicular, with a Ta-$CH_2$ bond length midway between a single and a triple Ta-C bond [130]. Rotation about the Ta=$CH_2$ bond is slow on the $^1H$ NMR time-scale.

The nucleophilicity of this methylene ligand is manifest by its reaction with methyl iodide (Equation 3.71). In this reaction Ta goes from a $d^2$ to a $d^0$ configuration. The presumed ethyl complex loses methane through a β-hydride elimination as revealed by isotopic labeling. Note that the deuterated ethylene (or metallacyclopropane) complex exists as two configurationally stable isomers on the $^1H$ NMR time scale. The bonding scheme which gives rise to these structural features has been discussed [131]. Thermolysis of the methylene complex in the presence of a ligand affords a 1:1 mixture

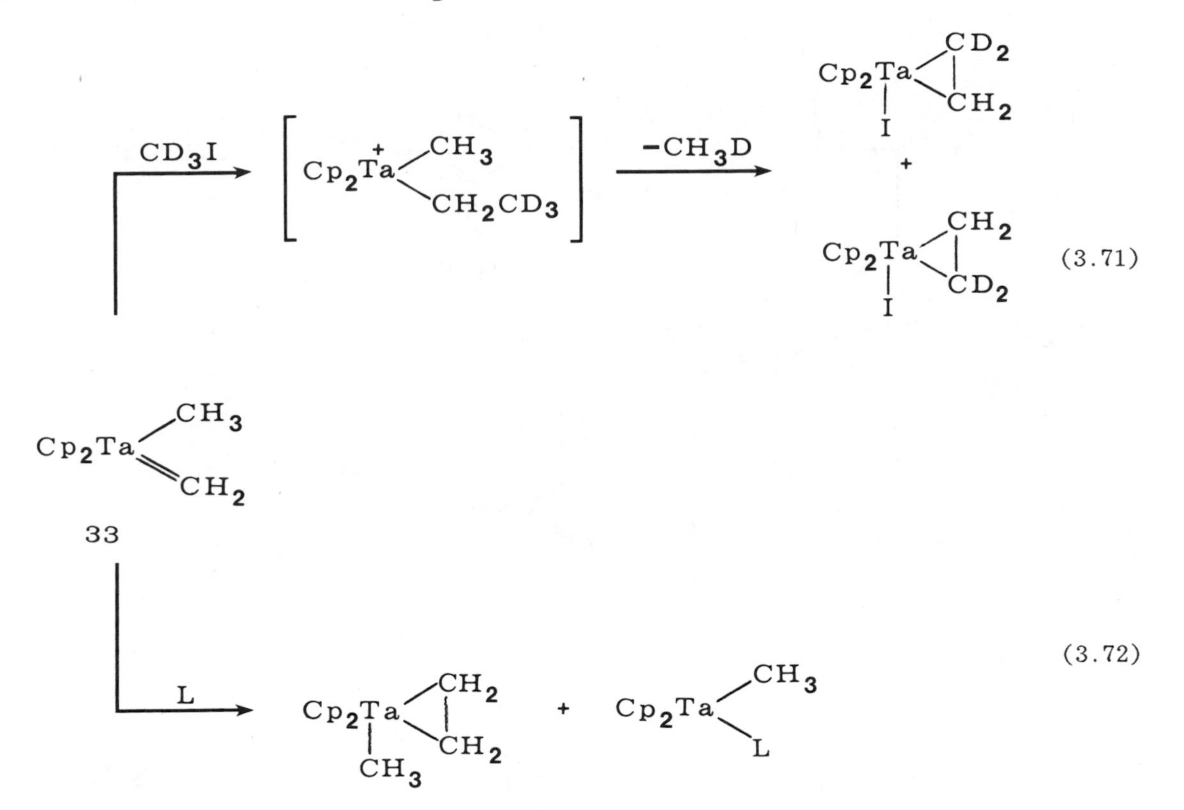


of the ethylene complex and the methyl derivative (Equation 3.72). The formation of coordinated ethylene must occur in an intermolecular step, since the rate law is second-order in Ta and independent of the added ligand L (ethylene or $PMe_3$). This reaction is similar to the attack of the methylene ligand by a Wittig reagent, which is reasonable since these tantalum alkylidene complexes themselves behave as Wittig reagents!

The nucleophilic, "Wittig-like" character of alkylidene complexes is dramatically illustrated by transformations of Schrock's neopentylidene derivative, 34 [132]. Reaction with aldehydes, ketones, esters, amides, and carbon dioxide results in alkylidene replacement of the carbonyl oxygen (Equation 3.73 and 3.74). Several of these substrates rarely react with phosphorus ylides in a "Wittig" sense.

The titanium-aluminum methylene complex has been found to be a practical reagent for the conversion of esters to vinyl ethers [133,134] (Equation 3.75).

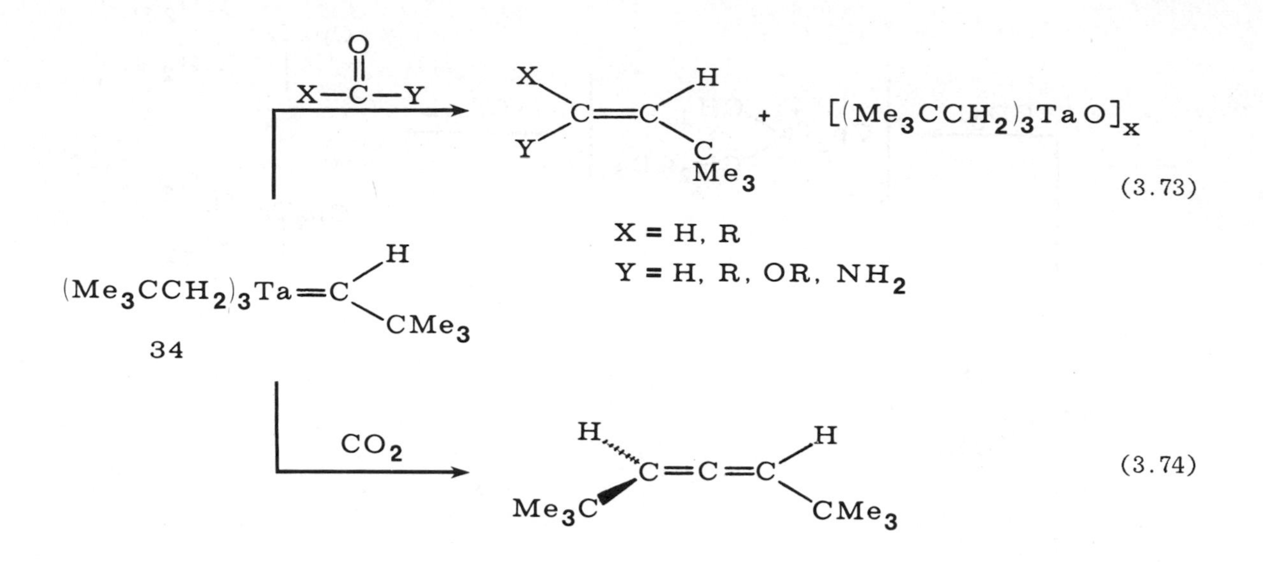


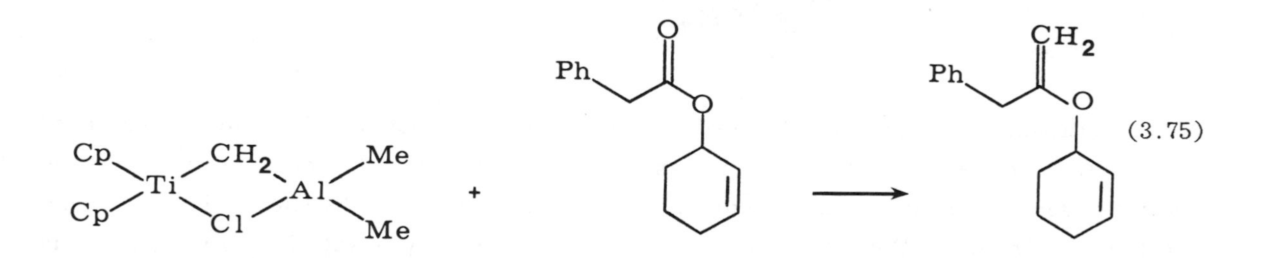


Some heavier group-IVa analogues of carbene ligands are known. Thus far these are structural curiosities. An example is the germylene chromium(0) complex, 35, which is probably kinetically stabilized by the bulky mesityl substituents [135]:

$$(OC)_5Cr{-}GeCl_2(THF) \longrightarrow (OC)_5Cr{-}Ge(C_6H_2\text{-}2,4,6\text{-}Me_3)_2 \qquad (3.76)$$

35

Such complexes have a tendency to form Lewis base adducts at the "carbenoid" atoms. For example, the germylene complex, 35, forms a 1:1 pyridine adduct.

d. Carbyne or Alkylidyne Complexes [136]. Compounds having a monovalent carbon attached to a transition metal have been recently isolated and structurally characterized. Such compounds exhibit very short metal-carbon bond lengths consistent with the assignment of a M≡C triple bond. There seems to be no rational way to categorize such "carbyne or alkylidyne" ligands for the assignment of oxidation states. These compounds are less reactive when the carbyne carbon is bound to a heteroatom with which there may be pronounced multiple bonding. The $^{13}C$ chemical shifts of the carbyne carbon in these complexes occur over a wide range of low field positions (265-340 ppm), overlapping the range of carbene chemical shifts [120]. Representative structures are shown below. Note that the M≡C-R angles approach, but need not attain linearity.

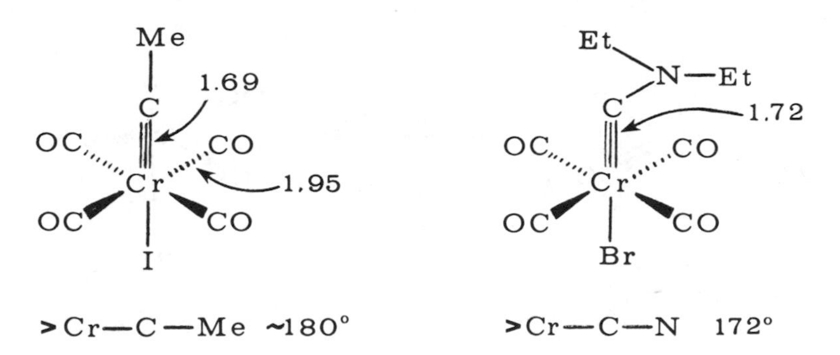


At least three methods have been used to prepare carbyne complexes. Fischer's method involves the abstraction of a Lewis base from a heterocarbene complex [136]:

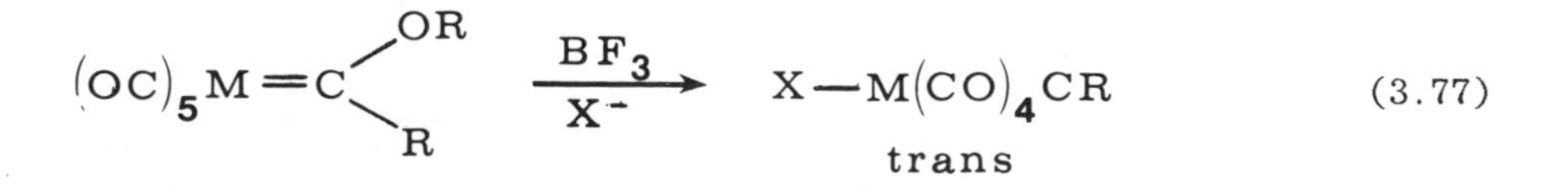


(3.77)

Here M = Cr, Mo, W; X = Cl, Br, I; and R = Me, Et, Ph. During this reaction, a halide replaces the trans CO group.

Schrock has prepared tantalum carbyne compounds, 36, by deprotonating cationic carbene complexes. The latter were generated in situ by reaction with $PMe_3$:

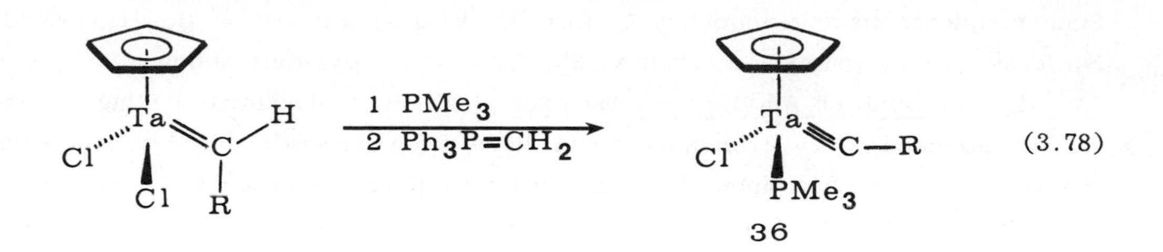


One of these complexes, 36 (R = Ph), has been structurally characterized, revealing a tantalum-carbyne (Ta-C) triple bond length of 1.849(8) Å, some 0.40 Å shorter than a normal Ta-C single bond and 0.2 Å shorter than the carbene Ta-C double bond [137].

Starting from a reactive dichlorocarbene complex, 37, Roper prepared and structurally characterized an osmium carbyne complex, 38 (L = $PPh_3$, Ar = o-tolyl) [138]:

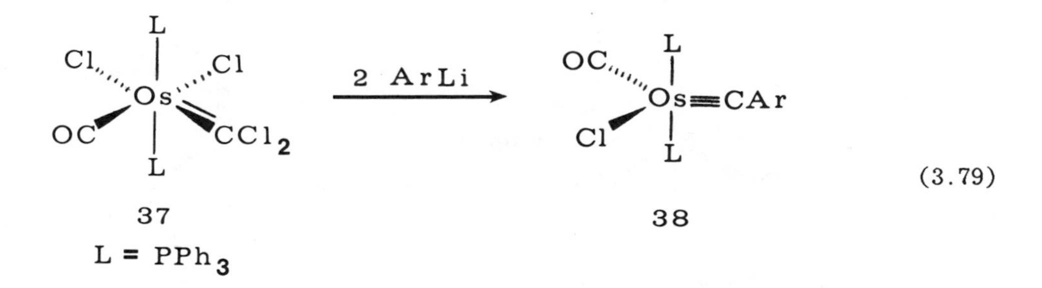


This complex has a very short osmium-carbon bond (1.62 Å), which is reflected by the remarkably high ν Os≡C infrared band (1374 $cm^{-1}$). Reaction with the elemental chalcogens affords the series of $\eta^2$-thioseleno- and telluroacyl complexes, 39 (X = S, Se, Te):

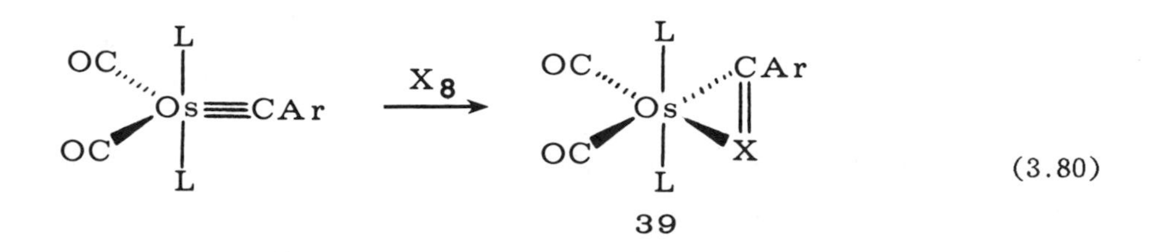

Thermal degradation of some carbyne complexes results in dimerization to alkynes--a reaction analogous to that of the carbene complexes. Thus far, carbyne complexes are little-explored structural curiosities and have not yet found use in organic synthesis.

e. Isonitrile Complexes. The isonitrile ligand, which has a special affinity for transition metals, is formally analogous to carbon monoxide [139]. However, isonitriles are stronger σ-donors and weaker π-acceptors than CO. Thus isonitriles tend to stabilize higher oxidation states. These tendencies are illustrated by the stable cationic complexes $\overset{+}{Mn}(CNAr)_6$ and $\overset{+}{Co}(CNAr)_5$. On the other hand, homoleptic isonitrile analogues of $Mn_2(CO)_{10}$ and $Na_2Fe(CO)_4$ are unknown.

Isonitriles are seldom used as ancilliary ligands in homogeneous catalysts. One reason for this is the proclivity for isonitriles to undergo insertion reactions (Chapter 5).

## 3.6. SIDEBOUND π-ACID CARBON LIGANDS.

a. Olefin Complexes. The olefinic group forms a large number of transition-metal derivatives. Olefin complexes are implicated in many reactions that are catalyzed or promoted by transition-metal compounds. Examples which we shall encounter in later chapters include hydrogenation, dimerization, polymerization, cyclization, hydroformylation, hydrocyanation, isomerization, and certain types of oxidation, such as the "Wacker-Schmidt" process. Typical examples of well-characterized olefin complexes are shown in Figure 3.6. In most cases each coordinated olefin can be considered a two-electron donor, although π-bonding is usually important.

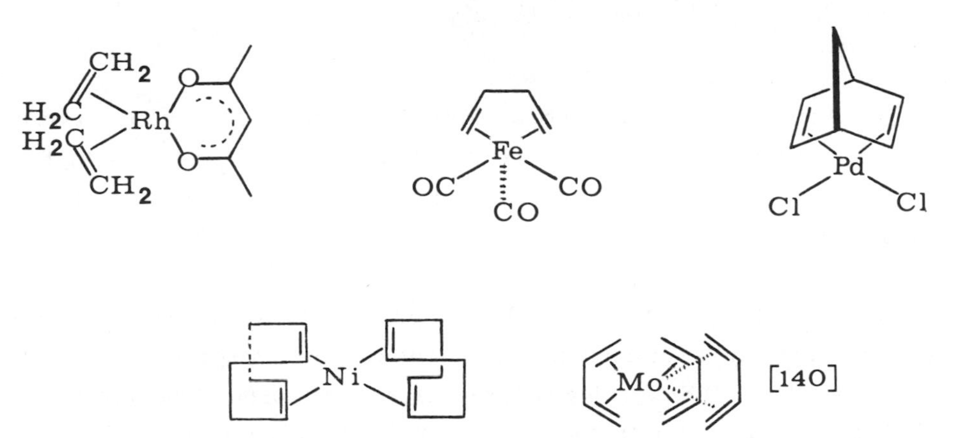


*Figure 3.6.*

Olefin complexation is quite sensitive to steric effects. Thus ethylene forms much more stable complexes than propylene, and complexes of cis olefins are more stable than those of trans olefins. Terminal olefins form more stable complexes than internal olefins. Certain dienes, such as 1,3-butadiene, norbornadiene, and 1,5-cyclooctadiene, form especially stable complexes. Homoleptic olefins complexes (those having only olefin ligands) are exceptional.

It is important here to introduce Cotton's widely used system of notation for olefin and related π-bonding [141]. The number of atoms attached to the metal are specified by a prefix dihapto, trihapto, tetrahapto, etc. This term is derived from the Greek word, haptein, "to fasten." The notation $\eta^n$ is used to denote the number of atoms bound to the metal for a particular ligand. Examples are shown in Figure 3.7.

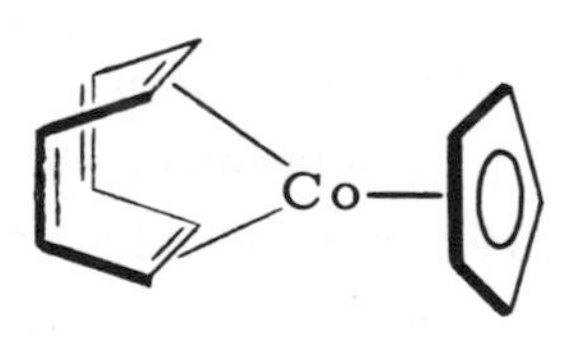


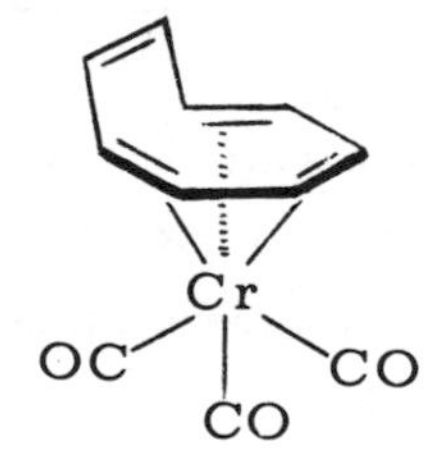


*Figure 3.7.* *$Co(\eta^5-C_5H_5)(\eta^4-C_8H_8)$, (1,2,5,6-tetrahaptocyclooctatetraene)-(pentahaptocyclopentadienyl)cobalt, on the left; and $Cr(CO)_3(\eta^6-C_8H_8)$, (1,3,5-trihaptocyclooctatetraene) tricarbonylchromium, on the right.*

i. Structural Characteristics [141]. Olefins are good π-acid ligands. Thus, we find coordinated olefins to be less common among metals in higher oxidation states or with formal charges greater than +1. In certain cases, the olefinic ligand is strongly distorted by complexation to a degree that the complex may be considered a metallacyclopropane derivative.

The π-bonded olefin and the metallacyclopropane bonding extremes (e.g., 41 a,b and 42 a,b) are not fundamentally different, but can be interpreted in the Dewar-Chatt model as arising from the varying degree of π- and σ-bonding [142]. There is a smooth gradation of one form to the other. Stronger π-bonding results in lengthening the olefinic C-C bond and distorting the coordinated olefin from a planar geometry.

Strongly electron-withdrawing groups on the olefin, such as F or CN, lower the $\pi^*$ level, increasing the strength of the metal-to-ligand $\pi$-bond. The metal is usually located at the midpoint of the olefinic group. The olefinic C-C axis is usually oriented nearly perpendicular to a square-planar complex with three other ligands, 40, but almost in the plane of the square-planar complex with two other ligands, 41a,b. The olefinic axis is approximately in the trigonal plane of a trigonal-bipyramidal metal complex having four other ligands, 42a,b [143]. This latter case is indistinguishable from an octahedral complex in which the coordinated ligand has become a bidentate group (a metallacyclopropane), 42b.

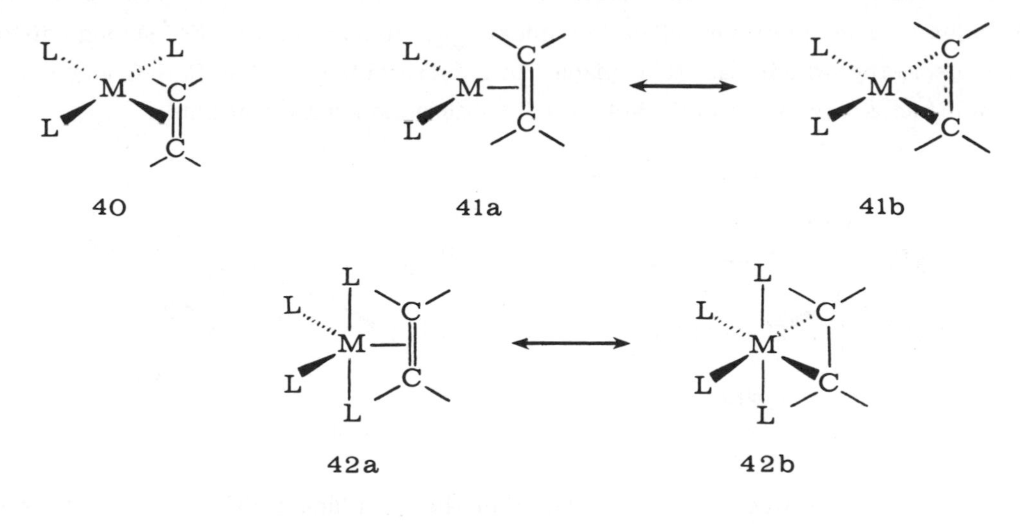


Compare, for example, the structure of Zeise's salt, 43, with that of a nickel ethylene complex, 44 [144-146]. The ethylenic bond in Zeise's salt is 1.37 Å, nearly

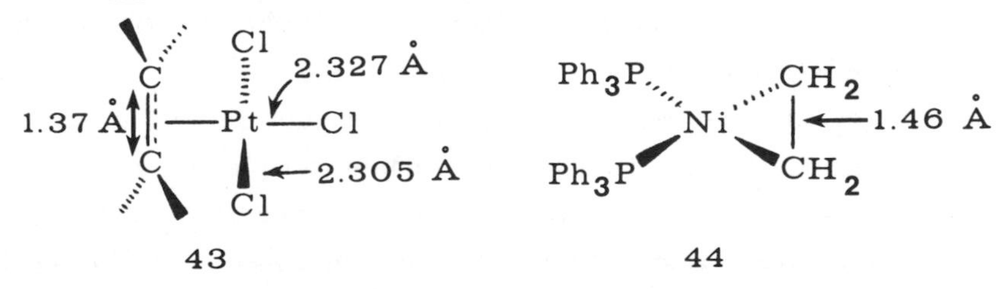


the same as an unperturbed olefin. The carbon-carbon axis is perpendicular to the PtCl coordination plane. This geometry is that expected for a two-electron donor, minimizing nonbonded repulsive interactions. Note the slightly longer Pt-Cl bond trans to the olefin in Zeise's salt, indicating a trans-bond weakening effect of the olefin ligand. In contrast, the ethylenic bond in the nickel complex, 44, is 1.46 Å, intermediate between a double (1.36 Å) and a single (1.54 Å) bond. The carbon-carbon axis makes a dihedral angle of only 5° with the $NiP_2$ plane. This complex might be considered a square-planar compound of nickel(II), although a choice between these formal oxidation states is a matter of "taste."

Electron-withdrawing groups on the olefin tend to favor the metallacyclopropane type of structure. An excellent example of this effect is found in the tetracyano-ethylene complex, 45, in which iridium exhibits the octahedral coordination geometry expected for iridium(III). The coordinated olefinic C-C bond is 1.5 Å, nearly that of a single C-C bond (1.54 Å). The angle NC-C-CN of 110° has moved toward the tetra-hedral value. The platinum allene complex, 46, demonstrates the strong distortion of ligand geometry which can take place upon coordination. The C-C-C angle and the two olefinic bonds are profoundly affected by interaction with platinum.

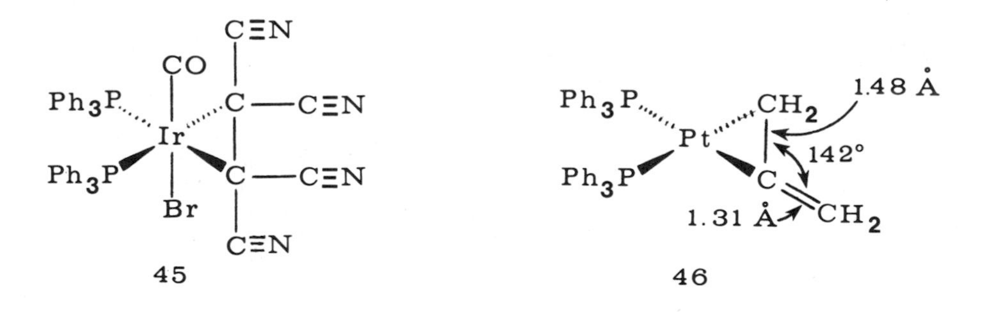


The dihedral angle between the $PtP_2$ plane and the $C_3$ plane is 9°. This complex seems to be best formulated as square-planar platinum(II).

ii. Physical and Chemical Properties. The physical properties of coordinated olefins are not especially distinctive or useful to the synthetic chemist. The coordinated olefin stretching frequency around 1500 $cm^{-1}$ is weak and may contain a significant amount of $CH_2$ deformation. This band is not related in any simple way to the coordinated olefinic C-C bond length. In certain complexes, the coordinated olefin rotates about the metal-olefin bond axis at a rate convenient to study on the NMR time-scale. There have been many studies of this phenomenon which can be related to favorable olefin-metal bonding in two orthogonal orientations. These situations may be important in considering the possible orientation of coordinated olefins and other ligands along various reaction coordinates. An early, well-studied case of olefin rotations is that of $(\eta^5\text{-}C_5H_5)Rh(C_2H_4)_2$, 47. According to $^1H$ NMR analysis, the barrier to ethylene rotation is 15 kcal $mol^{-1}$ [147]. The inner and outer olefinic protons $H_i$ and $H_o$ are sterochemically distinct.

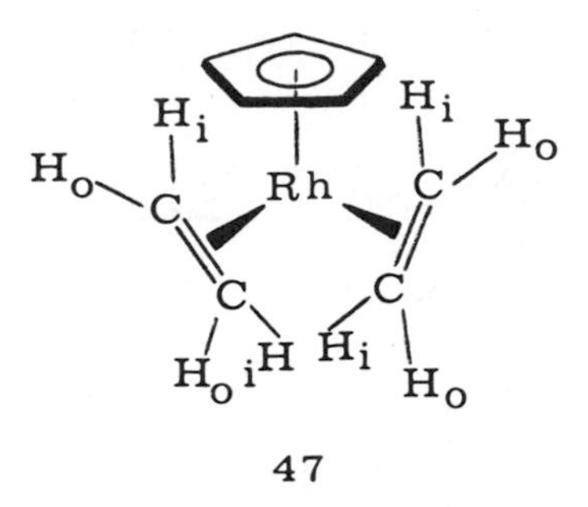


The π-bonded olefin complexes of the later transition elements are electron-deficient and therefore inert towards electrophilic reagents. In fact, these olefin complexes may be attacked by nucleophiles on the face opposite the metal giving trans addition.

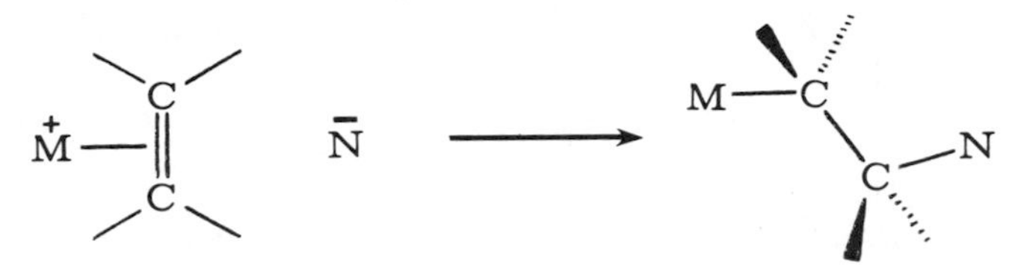


This reactivity towards nucleophilic attack is enhanced by a formal positive charge on the metal. Thus the polarity of π-bonded olefins is the opposite of the free olefins. We will return to this type of reaction in Chapter 5.

By contrast, the olefin complexes of the early transition elements tend to react like metallacyclopropanes, and are attacked by electrophiles such as protons at the electron-rich metal-carbon σ-bonds. The reactivity patterns of olefin complexes which are between these extremes have not been carefully studied.

Coordination of an olefin by a transition metal usually renders the olefin unreactive towards cycloadditions and catalytic hydrogenation if the metal is coordinatively saturated. Thus, the 1,3-butadiene iron(0) tricarbonyl complex, 48, fails to react with maleic anhydride or with hydrogen in the presence of a platinum catalyst [148] (Equation 3.81).

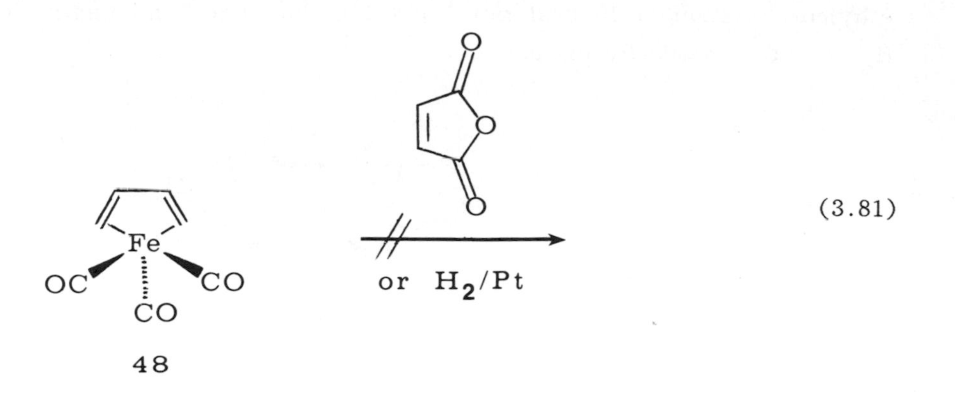


The cyclopentadiene ligand in 49 is bound to the rhenium(I) by only one metal-olefin link. Thus, the uncoordinated olefin may be selectively hydrogenated, and the metal serves as a protective or masking reagent [149]:

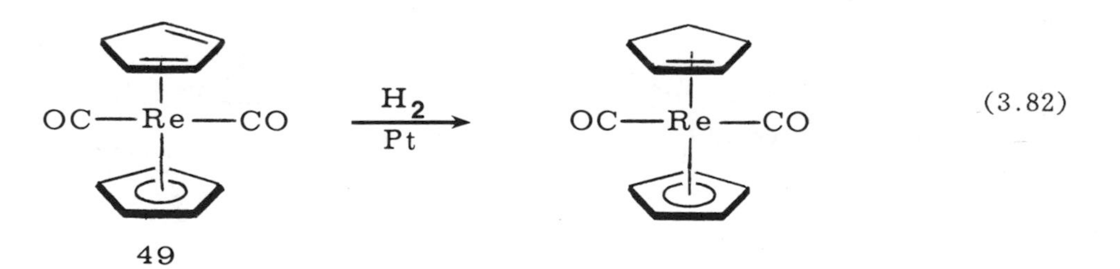


b. Acetylene Complexes. Metal complexes of alkynes are important as intermediates in the cyclooligomerization of acetylenes--forming cyclooctatetraene and benzene derivatives (Chapter 10). Certain acetylenes are also polymerized by transition metals--the most significant case being acetylene itself.

Structurally, the monometallic acetylene complexes are less well-studied than olefin complexes [150]. Upon coordination with a metal, the acetylenic group is distorted toward the geometry of a *cis* olefin. The degree of distortion is highly variable, so that a range of structures is possible. In many cases the structural situation with acetylenes is similar to that we encountered with olefins. In one limiting case the acetylenic group can be considered as a two-electron donor, and in the other as a four-electron donor. The latter "metallacyclopropene" case is favored by electron-withdrawing groups on the acetylene and a tendency for the metal to be more stable in the higher oxidation state. In reality there is no clearcut distinction between the two forms, which are simply different degrees of the same sort of bonding. Consider, for example, the two platinum acetylene complexes, 50 and 51. In the dichloride complex, 50, which is best formulated as Pt(II), the acetylene group is normal to the coordination plane. The angle about the acetylenic axis is 165° and the coordinated acetylenic bond length (1.24 Å) is about midway between that of an uncoordinated acetylene and that of an olefin [151]. The other complex, 51, has a more distorted acetylenic axis (146°) which is nearly in the coordination plane (the dihedral angle is 14°). The acetylenic bond in 51 is somewhat longer (1.32 Å) [152]. The degree to which steric interactions affect the geometries of 50 and 51 is unclear. A striking difference between these two compounds is the position of the IR band assigned to the coordinated acetylenic stretching frequency. In 50, the νC≡C (2020 $cm^{-1}$) is about 150 $cm^{-1}$ below the Raman band for the uncoordinated acetylene. The "metallacyclopropene" complex, 51, exhibits a much lower νC≡C (1750 $cm^{-1}$), about 430 $cm^{-1}$ below that of the free ligand [153]!

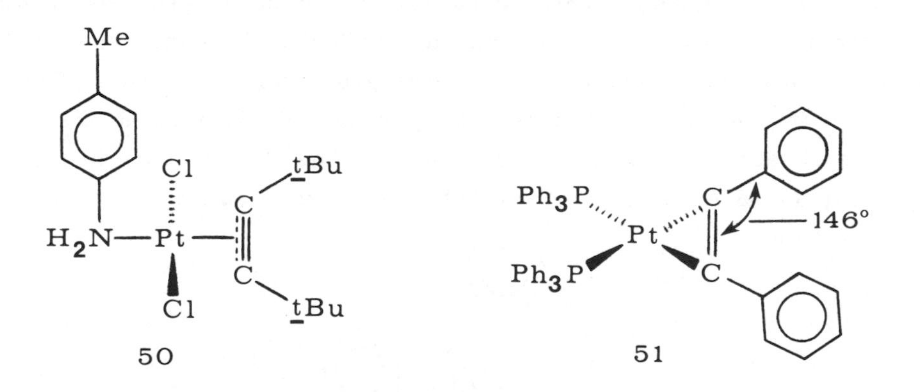


An interesting consequence of the distorted geometry of the coordinated acetylene group is the fact that unstable, small-ring acetylenes are stabilized by complexation with a metal. For example, stable platinum complexes of both cyclohexyne [154] and cycloheptyne [155] have been characterized--even though cyclooctyne is the smallest cyclic free acetylene which can be isolated. The geometry and νC≡C (1721 $cm^{-1}$) of the cycloheptyne complex, 52, are similar to the four-electron donor discussed earlier [155].

The diamagnetic tungsten complex, 53, seems to be a 20-electron system which violates the 18-electron rule [156,157]. The extra electron pair is apparently located on an acetylene-group-combination π-molecular orbital which does not have the correct symmetry to interact with the metal. Thus, this compound does not really violate the 18-electron rule [158]. This case warns us of the limitations to be expected from our oversimplified analysis.

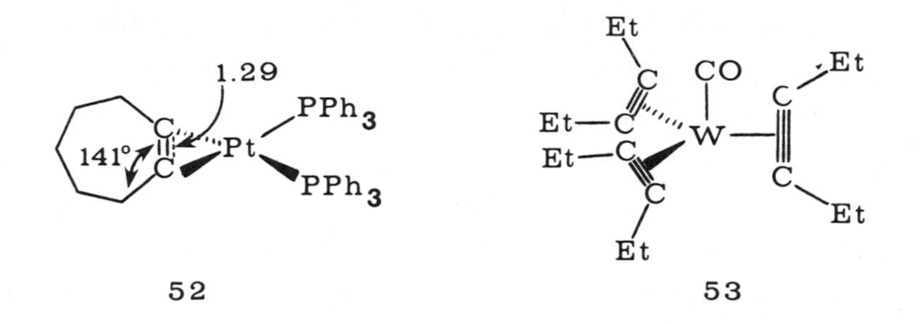


There are a large number of "cluster" complexes in which an acetylenic group is bound to two or more metals. Several of these are structurally characterized and reveal a sharply diminished C-C bond order. In these cases, as in other cluster complexes, simple bonding representations are difficult. The dicobalt complex, 54, is illustrative. It has been used as a protective group for acetylenes. The two cobalt atoms and the two acetylenic carbons are approximately located at the corners of a tetrahedron, each atom bound to the other three. The C-C distance in the coordinated acetylenic group (1.45 Å) is intermediate between a double and a single bond.

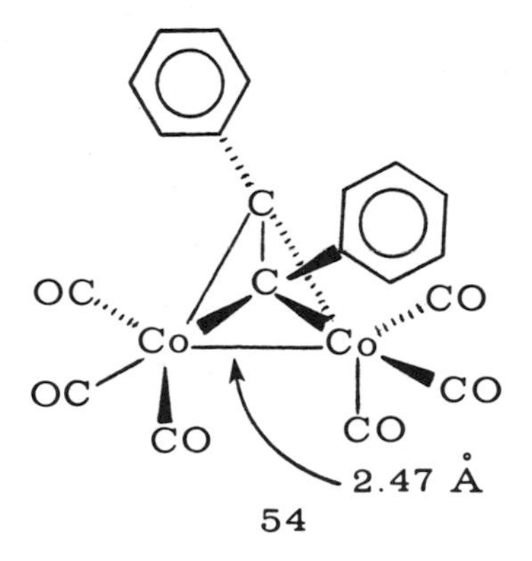

54

Perhaps the most remarkable "acetylene" complex is the tantalum benzyne compound, 55, recently reported by Schrock [159] (Equation 3.83). The tantalum complex, 55, has been characterized by X-ray crystallography. The planar benzyne ring shows $D_{3h}$ C-C bond length alternation; however, the carbyne bond length is

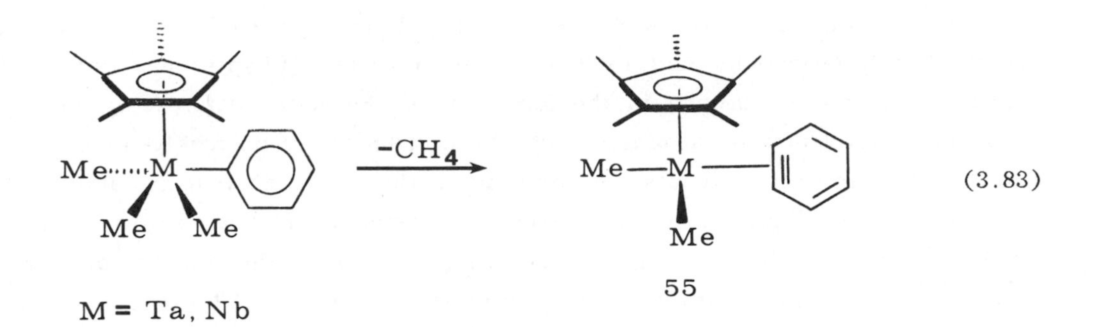


(3.83)

indistinguishable from the other two short C-C bonds. The formulation of these tantalum and niobium compounds as benzyne or o-phenylene complexes is a matter of taste.

c. $\eta^6$-Arene Complexes. Benzenoid aromatics form π-complexes with virtually all the transition metals. A large number of such complexes has been isolated and characterized by a plethora of physical techniques. This subject has been extensively reviewed [160,161].

The first $\eta^6$-arene complexes were prepared by Hein in 1919, but the nature of these compounds was not fully recognized until 1954. The prototypical "sandwich" complex is bis(benzene)chromium, $Cr(C_6H_6)_2$, prepared by E. O. Fischer in 1955 [162].

The arene group usually acts as a six-electron donor (hexahapto ligand), but four- and two-electron forms (tetra and dihapto) are recognized. In the tetrahapto case the aromatic ring is often strongly distorted from the planar, $D_{6h}$ symmetry of benzene. Both diamagnetic and paramagnetic arene complexes are known, and can usually be rationalized from the 18-electron counting scheme. Neutral and cationic π-benzenoid complexes are known, but apparently none having a charge > $2^+$. Anionic arene complexes are rare.

The arene ligand is depleted of its usual electron density upon coordination with a metal. For example, $(C_6H_5NH_2)Cr(CO)_3$ is a far weaker base than free aniline, whereas $(C_6H_5CO_2H)Cr(CO)_3$ is a stronger acid than benzoic acid (see Chapter 14). The electron deficiency of coordinated arenes is manifested by many physical properties and is very important in synthetic applications of arene complexes. For example, nucleophilic substitutions of complexed chlorobenzene are greatly facilitated, but electrophilic substitutions of coordinated arenes are difficult. Nucleophiles readily add to coordinated arenes, especially cationic complexes. Resonance effects between substituents on an arene are diminished by π-bonding with a metal. Similarly the aromatic "ring current" anisotropic magnetic shielding is substantially reduced in a $\eta^6$-arene complex.

Arene complexes are sometimes employed as "precatalysts" in homogeneously catalyzed polymerizations and hydrogenation processes (Chapter 6), but in most cases the arene ligand is removed in the formation of the actual catalyst (vide infra). The pronounced tendency of aromatic rings to coordinate even weakly to many low-valent metals makes aromatic solvents non-innocent in their participation in catalytic processes.

Typical hexahapto-arene complexes are depicted in Figure 3.8. The family of compounds related to bis(benzene) chromium has been the subject of many physical chemical and structural studies, but synthetic applications of these compounds have not been especially important. The parent compound, bis(benzene)chromium, has rigorous $D_{6h}$ symmetry. Electron-releasing substituents invariably stabilize the hexahapto-arene ligand, although steric interactions can counteract this effect. Cationic $d^6$ derivatives of Group-VII and Group-VIII metals with hexamethylbenzene, 56, are illustrated in Figure 3.8. The Group-VI arene tricarbonyls, 57, typified by benzenechromium tricarbonyl, have become important synthetic intermediates. In Chapter 14 synthetically useful reactions of arenechromium tricarbonyl derivatives are discussed in depth. The hexahapto-arene ligand is to be found in combination with many other ligands. Examples are 58, 59, and 60, in Figure 3.8.

The coordinated arene in the Group-VI arene tricarbonyl family is more substitution labile than the CO groups. Free arenes will exchange with the coordinated arene group in arenetricarbonyl complexes. For example, hexamethylbenzene will replace 1,4-diisopropylbenzene in the tungsten carbonyl complex, 61 (Equation 3.84). The hexamethylbenzene complex is apparently more stable because of steric and electronic effects. Such reactions may have complex mechanisms since fractional reaction orders have been reported [160,163]. Phosphine and phosphite ligands will also displace the arene group by an apparent $S_N2$ mechanism (Equation 3.85). On the other hand, a single CO group may be replaced under UV irradiation (Equation 3.86). This is another example of the pronounced tendency for CO ligands to undergo photodissociation.

Tetrahapto-arene complexes are structurally very interesting because the coplanarity and the carbon-carbon bond lengths in the aromatic ring are strongly distorted by complexation of four-arene carbon atoms [164-166]. It is as if the metal, requiring only four additional electrons, has coordinated with the diene portion of a hypothetical "Kekule" structure. The best structurally characterized examples of tetrahapto-arene complexes are the similar, neutral, diamagnetic ruthenium [164,165] and rhodium complexes [166], 62 and 63, in Figure 3.9.

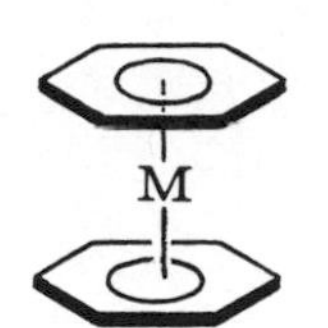


M = Cr, Mo, W

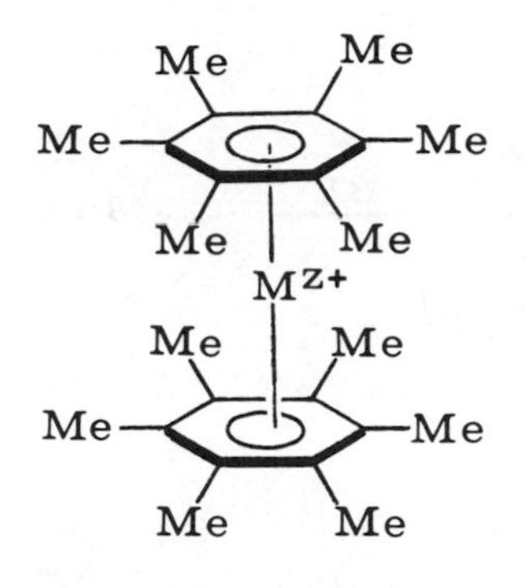


M = Tc, Re; z = 1

M = Fe, Ru, Os; z = 2

56

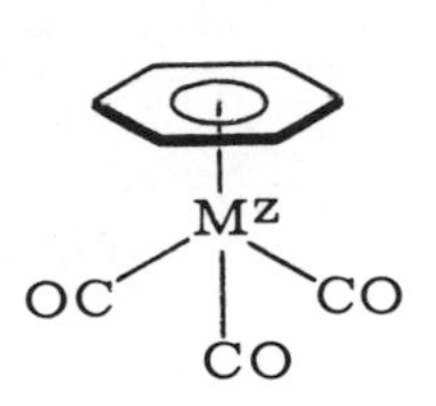


M = Mn; z = +1

M = Cr, Mo, W; z = 0

57

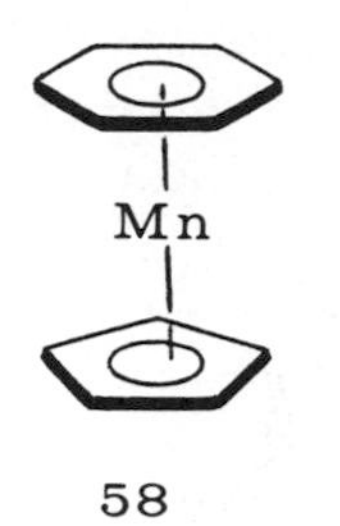


58

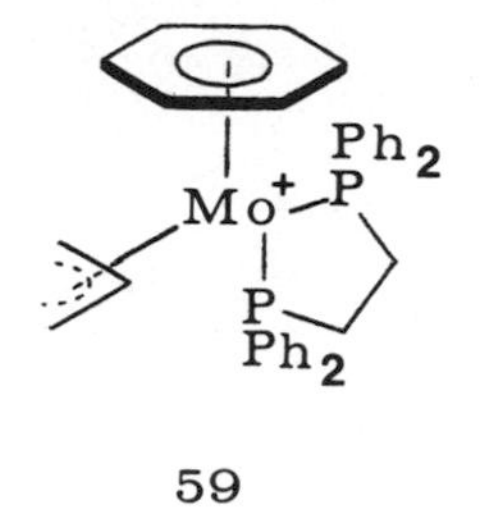


59

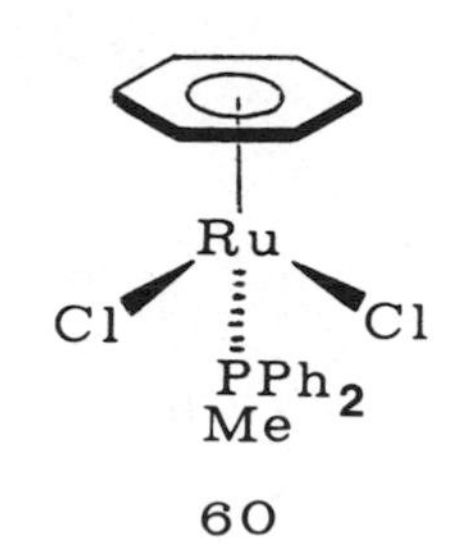


60

*Figure 3.8.*

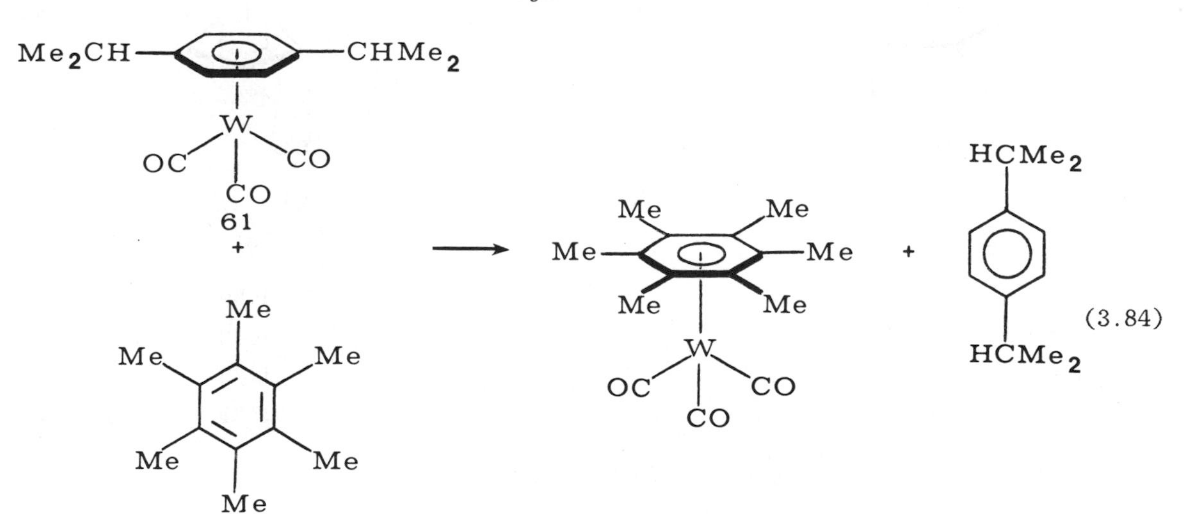

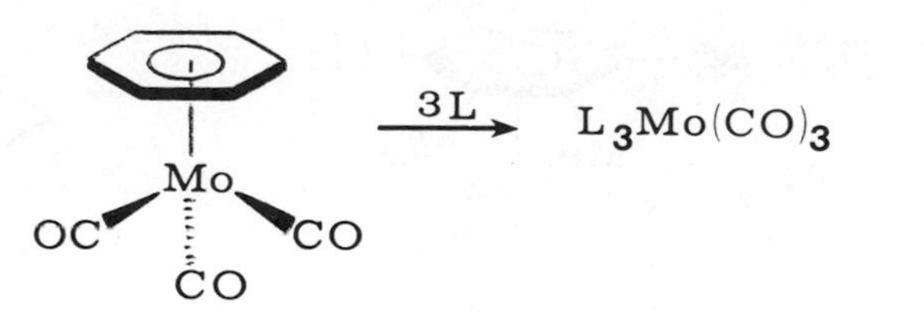


(3.85)

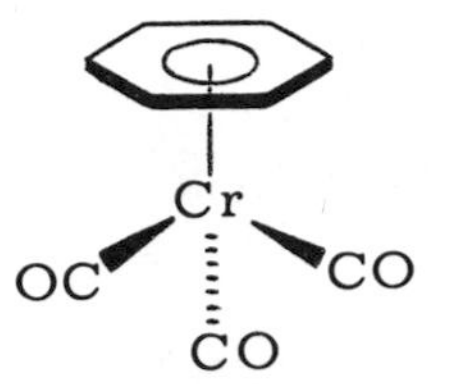


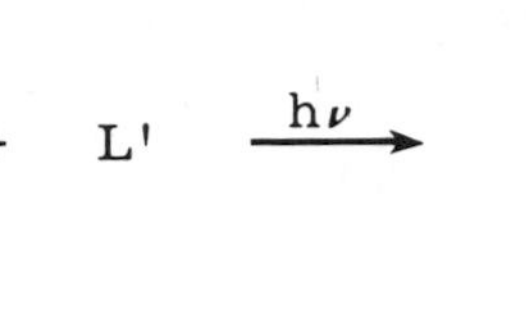


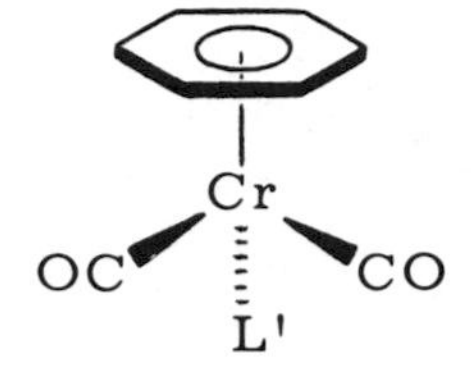


(3.86)

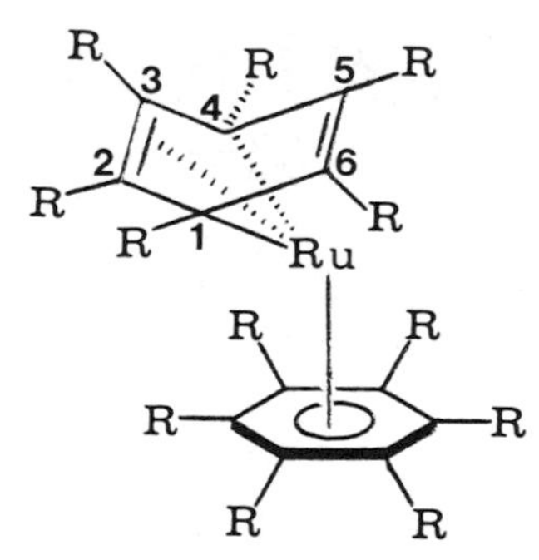


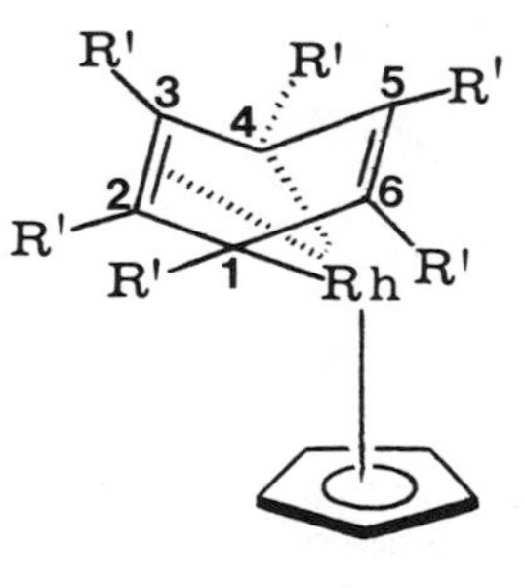

62 63

$R = CH_3$ $R^1 = CF_3$

| | 62 | 63 |
|---|---|---|
| $C_1 - C_2$ | 1.42 | 1.50 |
| $C_2 - C_3$ | 1.41 | 1.42 |
| $C_4 - C_5$ | 1.50 | 1.49 |
| $C_5 - C_6$ | 1.33 | 1.31 |

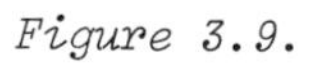

*Figure 3.9.*

In each structure the $C_5$-$C_6$ distance is that of a double bond, whereas the $C_4$-$C_5$ distance is close to that of a single bond. The dihedral angles between the planes intersecting at the $C_{1,4}$ axis are 43° and 48° in the ruthenium and rhodium complexes, 62 and 63, respectively. The complexes differ somewhat in the nature of the bonding at $C_1$-$C_2$. In both cases the metal is clearly not bonded to the distant $C_5$, $C_6$ atoms. The distortion of the arene is clearly related to satisfying the "18-electron rule." Note that the other arene ring in the ruthenium complex, 62, is a normal, planar six-electron donor.

A very interesting dihapto-arene coordination is found in the styrene iron tricarbonyl compounds such as 64 [167]. The vinyl group and two ring carbons are each bound to iron in the dihapto fashion, resulting in a distortion of the bonds in the aromatic ring. Note the pronounced tendency for the $Fe(CO)_3$ unit to form stable diene complexes, resulting in a fixation of a diene unit using one double bond from the hypothetical Kekule" form.

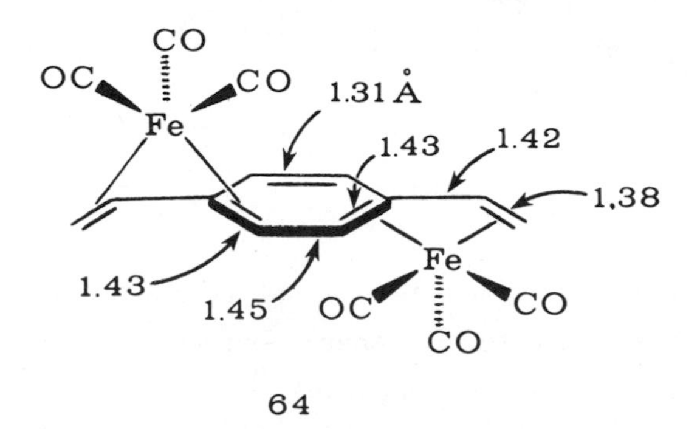


64

Another unusual benzene complex is the sandwich palladium(I) dimer, 65. The quality of the X-ray data does not permit a clear picture of the arene distortion [168].

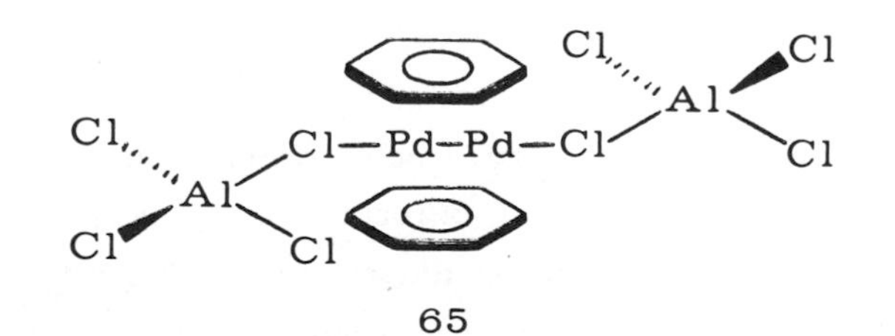


65

The later transition elements have a tendency to form dihapto-arene complexes [169]. Many of these are quite substitution-labile, and are considered to be models for the sort of complexes which must be involved in metal-catalyzed arene reactions such as the Pd(II) acetate-catalyzed dimerization of benzene to biphenyl. The platinum complex of $C_6(CF_3)_6$, 66, has been structurally characterized [170]. The loss of aromaticity is revealed by the alternating double- and single-bond distances. As one might expect, this compound is fluxional on the $^{19}F$ NMR time-scale. That is, a single $^{19}F$ signal is observed down to -90°C, indicating rapid equilibration of the platinum as it apparently moves from one pair of ring-carbon atoms to another.

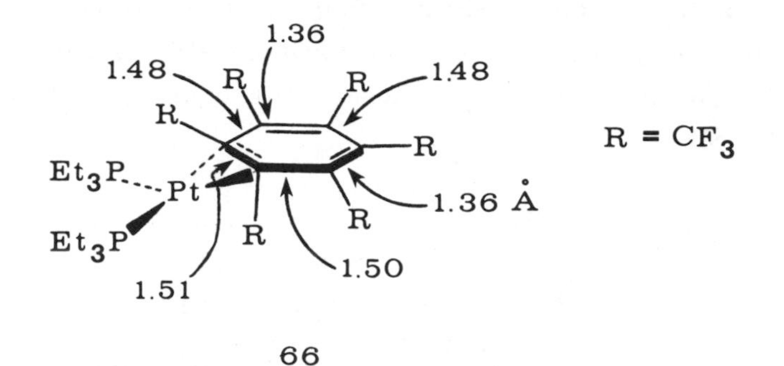


66

Copper(I) and silver(I) form labile arene complexes of various stoichiometries, some of which have been structurally characterized. These all seem to be dihapto. Copper ions in clays also form arene complexes.

Arene complexes are usually prepared by displacement of more weakly held ligands or by creating a highly unsaturated metal by reducing a higher oxidation state in the presence of an arene. The atom vapor technique--although still specialized--is a useful method for preparing arene complexes. These methods are described in more detail in Silverthorne's review [161]. Practical methods for preparing the synthetically useful arenechromium tricarbonyl complexes are outlined in Chapter 14.

Borazines, cyclic trimers which are isoelectronic with arenes, form $\pi$-complexes with the transition metals [171]. The chromium complex, 67, an analogue of the hexahapto-arenechromium(0) tricarbonyls is illustrative. The nitrogens tend to be trans to the CO groups. The borazine ring becomes nonplanar upon coordination.

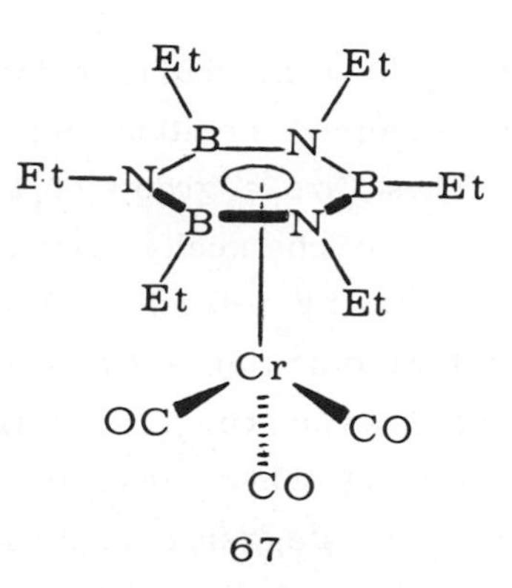


67

d. η⁵-Cyclopentadienyl and Related Compounds. The $\eta^5$-cyclopentadienyl, or "Cp" ligand, is one of the most common we shall encounter. The Cp group is but one member of a class of planar cyclic aromatic ligands, such as benzene. In 1951 the discovery and characterization of $(\eta^5\text{-}C_5H_5)_2Fe$, commonly called ferrocene, catalyzed the development of modern organotransition-metal chemistry [172].

There are three general categories of $\eta^5\text{-}C_5H_5$ complexes: (1) $Cp_2M$, symmetric molecules with parallel cyclopentadiene rings, 68, such "sandwich" compounds being referred to as "metallocenes;" (2) $Cp_2ML_x$, "bent metallocenes," 69, in which the two cyclopentadienylide rings are not parallel and having from one to three additional ligands, L, such as H, R, CO, etc.; and (3) $CpML_y$, "half-sandwich" compounds, 70, in which L represents one to four ligands, including in some cases another aromatic ligand. The cyclopentadienylide group can also be bound to a metal by one carbon, as a unidentate alkyl ligand, but such cases are less common. Note that the symbol Cp is widely used in the literature to represent the $\eta^5\text{-}C_5H_5$ group.

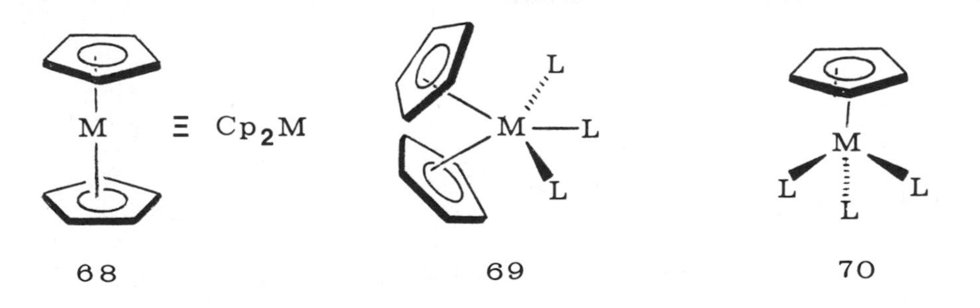


68 69 70

i. Bonding Schemes. The electronic configurations of the symmetric metallocenes, $(\eta^5\text{-}C_5H_5)_2M$, can be accounted for by a molecular-orbital scheme in which there are six strongly bonding, filled molecular orbitals, which have predominantly ligand character. The highest occupied orbitals are of greatest importance to the chemistry of the metallocenes. These are a group of three bonding and two antibonding d-orbitals (Figure 3.10). Other chemically insignificant orbitals are not shown here, but may be found in the text by Purcell and Kotz [173]. The neutral [M = Fe(II), Ru(II), and Os(II)] and cationic [M = Co(III), Rh(III)] diamagnetic metallocenes have the three nonbonding d levels ($e_{2g}$, $a_{1g}$) filled, just as we saw earlier for octahedral $d^6$ complexes. This bonding scheme does not depend on the relative orientation of the two Cp rings (eclipsed or staggered conformation). The rotational barrier between the Cp rings is quite small. Thus, 1,1'-disubstituted ferrocenes exist as a single stereoisomer.

The scheme shown in Figure 3.10 is also useful in discussing those metallocenes with $6 > d^n > 6$. For example, as shown in Figure 3.11, metallocenes having $d^n > 6$, such as cobaltocene and nickelocene, are paramagnetic, since the antibonding d levels, $e^*_{1g}$, are partially occupied. Those metallocenes with $d^n < 6$, such as vanadocene and chromocene, are also paramagnetic; but in such cases the spin state is not easily predicted, since the nonbonding and antibonding levels are sufficiently close in energy that electron-electron repulsion may not be more important than spin-pairing.

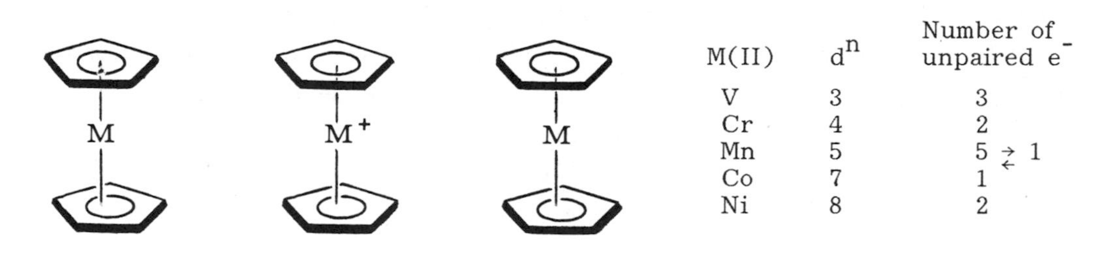


M = Fe, Ru, Os(II) $d^6$

M = Co, Rh(III) $d^6$

diamagnetic $e_{2g}$, $a_{1g}$ levels are filled

*Figure 3.11. Electron spin-pairing in various metallocenes.*

The delicate energy balance between the nonbonding $e_{2g}$, $a_{1g}$ levels and the antibonding $e^*_{1g}$ levels depends on the formal charge, the nature of the metal, and the

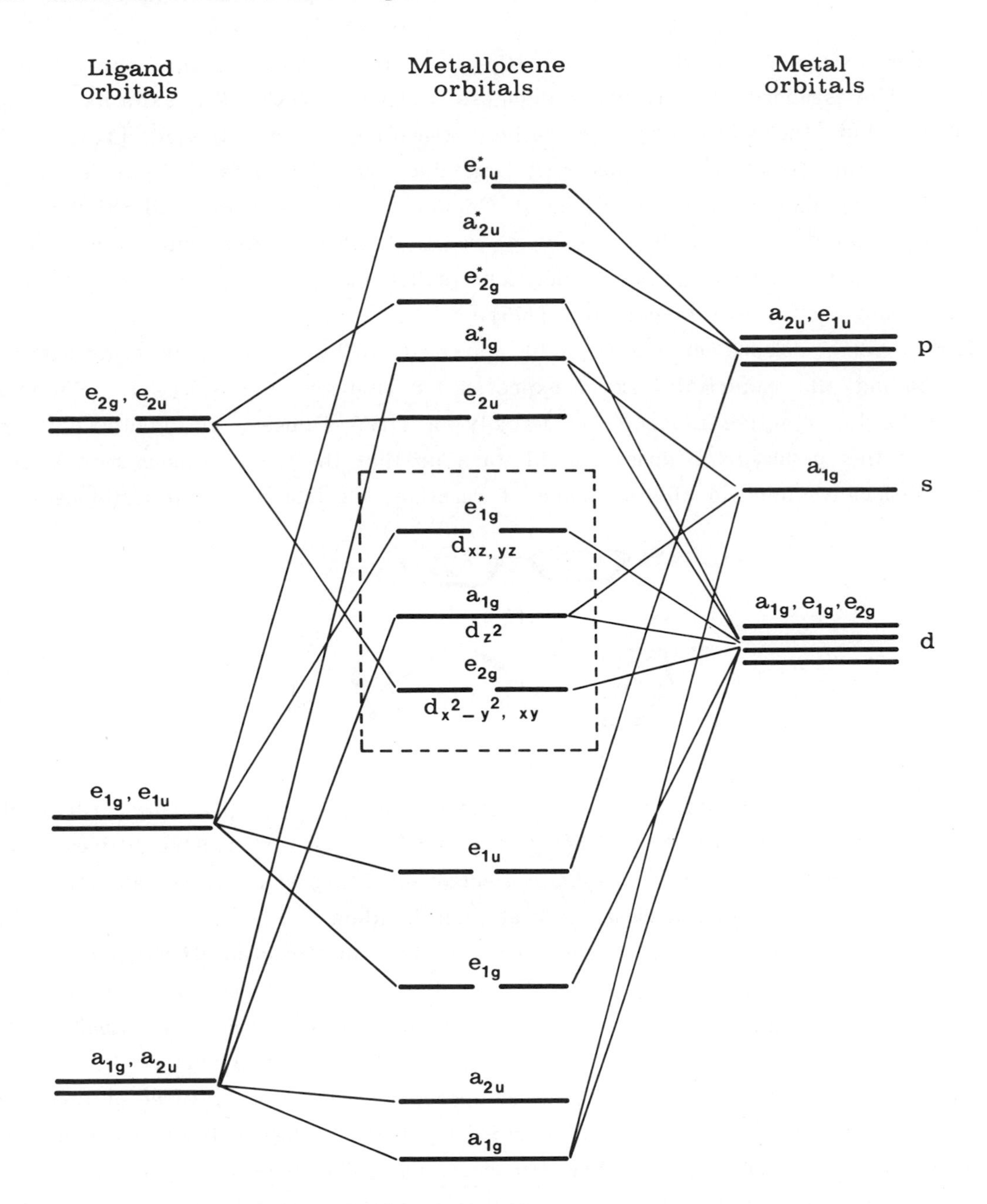


*Figure 3.10. Simplified orbital scheme for metallocenes.*

ligand. For example, the $d^5$ complex, $[Cp_2Fe]^+$, is low-spin, having one unpaired electron. The similar isoelectronic manganese complex $(MeCp)_2Mn$ exhibits a spin equilibrium, but the decamethyl derivative, $(Me_5Cp)_2Mn$ is low-spin [174]. The rhenium analogue $[Cp_2Re]^+$ is apparently unstable, and acquires a hydride ligand, forming the very stable "bent metallocene," $Cp_2ReH$. Other second- and third-row $d^4$ metallocenes, $Cp_2M$ (M = Mo,W), are very reactive intermediates which are usually generated in situ. These electron-deficient substances undergo oxidative-addition reactions analogous to those of carbenes [175].

The bonding scheme in Figure 3.10 appeared to be violated by a diamagnetic complex having the empirical formula expected for titanocene, $(C_5H_5)_2Ti$. However, this exceedingly reactive molecule is actually a Ti(IV) dimer, 71 [176,177]. The formation of this μ-dihydrido dimer should warn us that the CH groups in metallocenes are not chemically inert in the presence of reactive, electron-deficient metal centers.

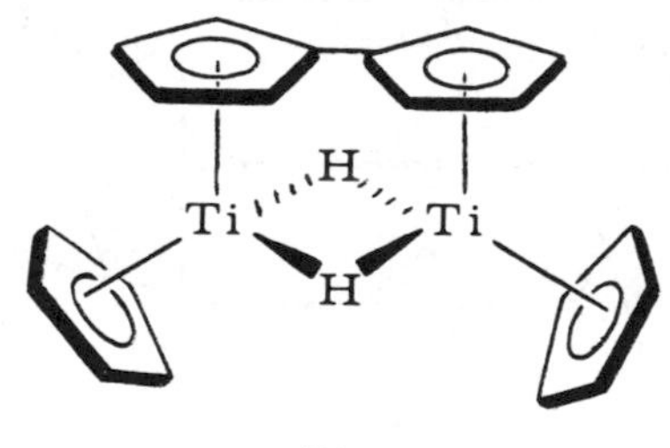


71

When additional ligands bind to the metal, the two $\eta^5$-$C_5H_5$ rings tilt back, resulting in "bent metallocenes." Although the stoichiometries of most such diamagnetic complexes conform to our earlier simplistic electron-counting scheme, certain structural and chemical features require a closer look at their bonding.

These bent metallocenes can accommodate one, two, or three additional ligands (72, 73, and 74) by using bonding molecular orbitals formed from three metal-based d-orbitals and the ligand σ-orbitals. This bonding picture and some of its ramifications for the structures and reactions of bent metallocenes have been discussed in detail by Hoffmann [178]. In the diamagnetic bent metallocenes, one, two, or three of the three highest occupied molecular orbitals are occupied by σ-donor, ligand-based electrons and the remainder by nonbonding metal-based d electrons. The d-electron pairs are always higher in energy, but in favorable cases may be stabilized by backbonding. As Figure 3.12 shows, diamagnetic metallocenes with one unidentate ligand have a $d^4$ configuration; those with two ligands may have either a $d^2$ or a $d^0$ configuration (the latter is coordinately unsaturated); and those with three ligands are usually $d^0$.

| | | |
|---|---|---|
| $[Cp_2FeH]^+$ | $Cp_2MoH_2$ | $Cp_2TaH_3$ |
| Fe(IV) $d^4$ | Mo(IV) $d^2$ | Ta(V) $d^0$ |
| $Cp_2Mo(CO)$ | $Cp_2Zr(Cl)H$* | $Cp_2Nb(C_2H_4)$ (Et) |
| Mo(II) $d^4$ | Zr(IV) $d^0$ | Nb(III) $d^2$ |
| $Cp_2ReH$ | $[Cp_2ReH_2]^+$ | $[Cp_2WH_3]^+$ |
| Re(III) $d^4$ | Re(V) $d^2$ | W(VI) $d^0$ |

*Coordinatively unsaturated, useful reagent.

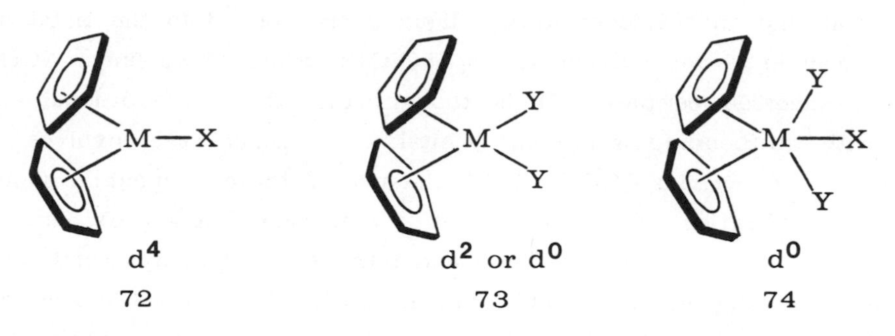


*Figure 3.12. Examples of diamagnetic bend metallocenes.*

This bonding scheme has important consequences for cases in which there is one unidentate ligand and a $d^2$ configuration. If the unidentate ligand is symmetric with respect to the Cp rings, the complex is likely to be high-spin, since the two d electrons must be placed in two closely spaced d-orbitals. An example of such a high-spin $d^2$ complex is $Cp_2VCl$. Insertion reactions (discussed in Chapter 5) involve a decrease of one in coordination number but maintain the same number of d electrons. Thus, insertions involving $d^2$ metallocenes having two ligands are apparently unfavorable, since these would afford a high-energy (probably high-spin) product. This would seem to apply to olefin-hydride, olefin-alkyl, carbonyl-alkyl, and carbene-alkyl insertions, which are discussed in Chapter 5. Thus the olefin-hydride insertion in Equation 3.87 does not seem to occur spontaneously. This restriction may be overcome by simultaneously bringing in an additional ligand.

It is also worth noting that the hydrides in $CpWH_3$ and $CpTaH_3$ show an NMR pattern compatible with that expected for an $A_2B$ set [179]. Thus the two outer and the inner hydrides are stereochemically distinct.

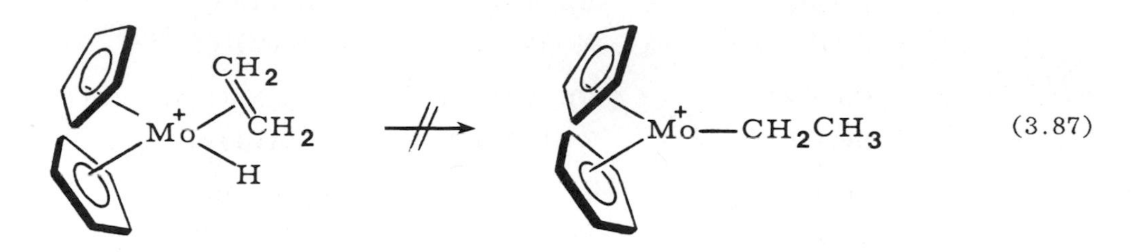


(3.87)

Certain bent metallocenes appear to violate the 18-electron rule by remaining diamagnetic with a formal 20-electron structure [178]. The molecular structures of some of these reveal that the cyclopentadienyl ligands are bonded to the metal in a grossly unsymmetric manner. One example is $Cp_2Mo(NO)R$, which has a linear $NO(NO^+)$ and is formally a 20-electron complex. It is thought that the two additional electrons are associated with antibonding metal-ring orbitals. A similar case involves $Cp_2W(CO)_2$, another 20-electron complex [180]. The existence of these compounds should warn us that the electron-counting scheme in Chapter 2 is an imperfect approximation.

ii. Other Metallocenes. Tris- and tetracyclopentadienyl metal compounds are known. The triscyclopentadienyls are common for the lanthanides and actinides, which are outside of the scope of this text. There are few examples within the transition elements; these are usually $d^0$ situations, and have both $\eta^1$-$C_5H_5$ and $\eta^5$-$C_5H_5$ ligands as does, for example, $Ti(\eta^5\text{-}C_5H_5)_2(\eta^1\text{-}C_5H_5)_2$ [181,182]. Such substances are fluxional, with all four rings equivalent on the NMR time-scale [183].

There are many examples of "half-sandwich" metallocenes. Typical compounds are shown in Figure 3.13. Most of these substances are coordinatively saturated and diamagnetic. In many cases the diamagnetism results from the formation of metal-metal bonds.

A number of "mixed-sandwich $\eta^5$-$C_5H_5$ derivatives have been prepared [184]. The coordinatively saturated series in Figure 3.14 illustrates some of the diverse combinations which are possible. Recall that the symmetric ring systems form an aromatic donor series: $(Ph_3C_3)^+$, a $2e^-$ donor; $(C_4H_4)^0$, a $4e^-$ donor; $(C_5H_5)^-$, $(C_6H_6)^0$, $(C_7H_7)^+$, all $6e^-$donors; and $(C_8H_8)^{-2}$, a $10e^-$ donor. The cyclooctatetraene dianion acts as a planar aromatic 10-electron donor ligand in $(C_8H_8)_2U$ [185], but such planar complexes are rare for d-block metals, except for d complexes of metals in Groups IV and V.

Inasmuch as the $C_3$, $C_4$, $C_7$ and $C_8$ aromatic-ligand systems have limited importance in catalytic and synthetic applications, we have not included a separate section on those compounds.

Another curious variety of Cp complexes are the tri- and tetranuclear sandwich compounds [186]. The diamagnetic nickel complex, 75, appears to violate our simple electron-counting rules; however, the electronic structure of these triple-deck complexes has been explained by Hoffmann [187].

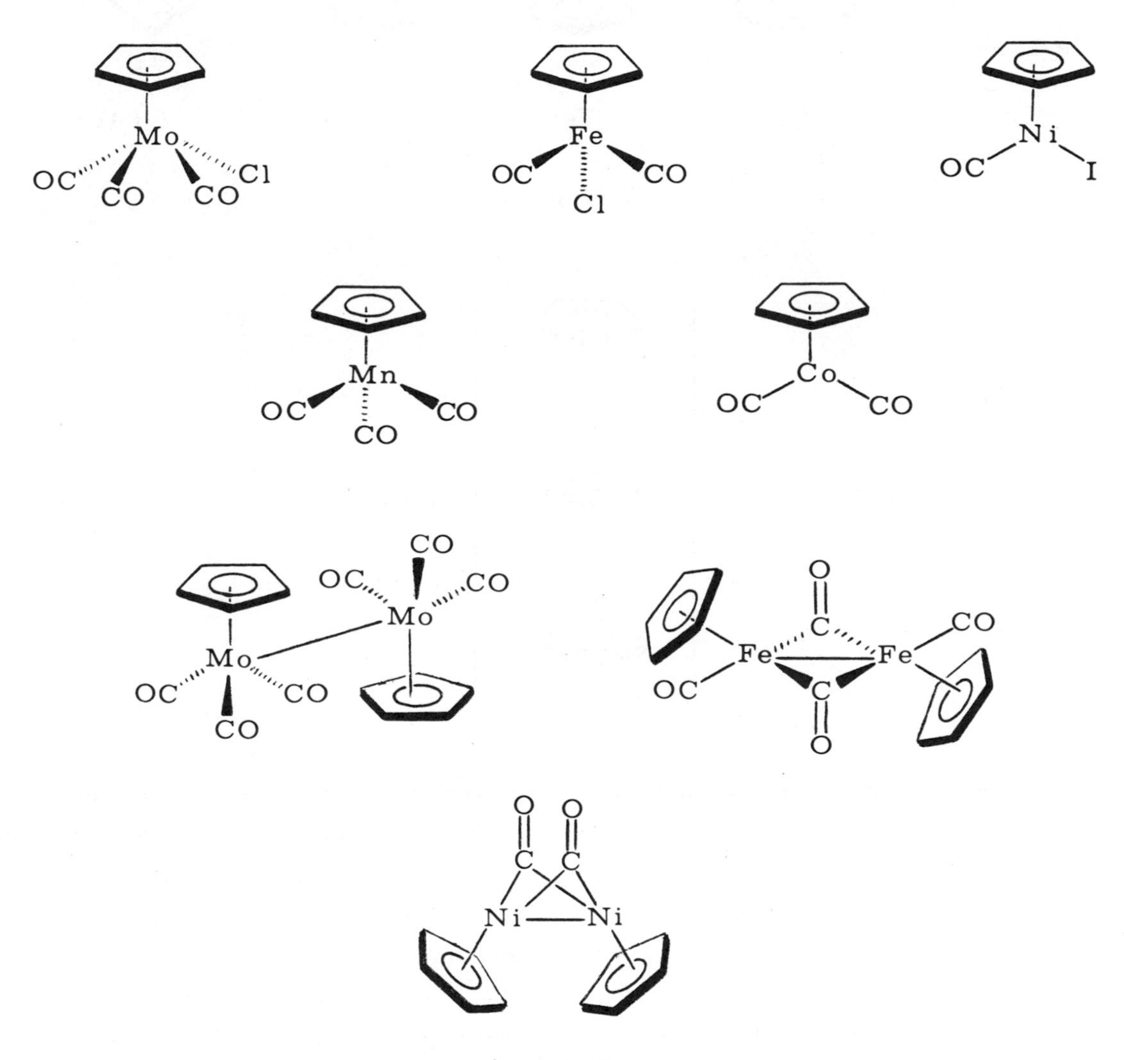


*Figure 3.13.*

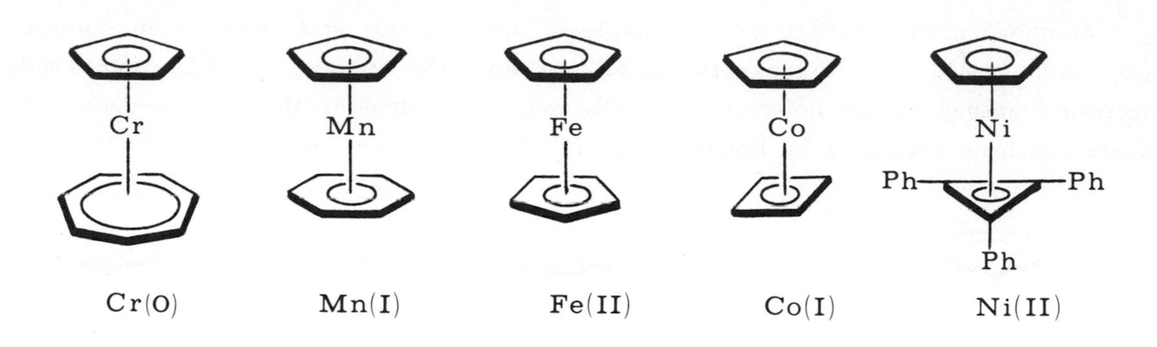


*Figure 3.14.*

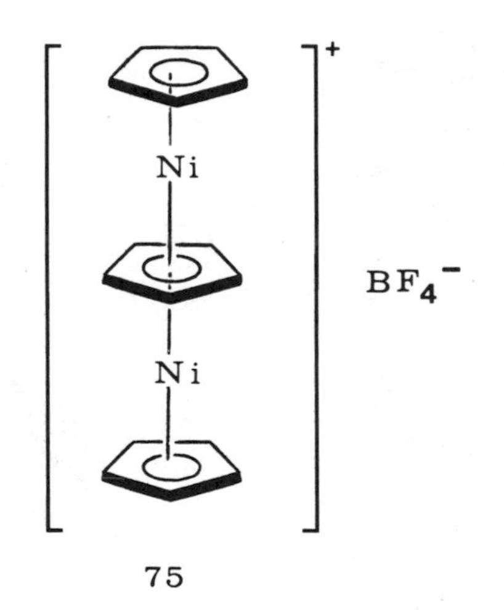


Another class of complexes which has a rich structural chemistry but so far limited practical applications is that derived from various cyclic carboranes [188,189]. For example, the borabenzene anion in the iron complex, 76, is equivalent to the $\eta^5$-$C_5H_5$ anion. A treasure of quasi-arene metal chemistry derives from the polyhedracarboranes. The cobalt complex, 77, is exemplary of this class. In this case $(1,2\text{-}C_2B_9H_{11})^{-2}$ is equivalent to $(C_5H_5)^-$. This vast subject is outside the scope of this text. Interested readers should consult the review by Callahan and Hawthorne [190]. Many other heterometallocene combinations are possible, for example, the chromium and nickel complexes, 78 [191] and 79 [192].

iii. Synthesis of Metallocenes. Metallocene complexes are prepared by taking advantage of the acidity of cyclopentadiene--which may be further enhanced

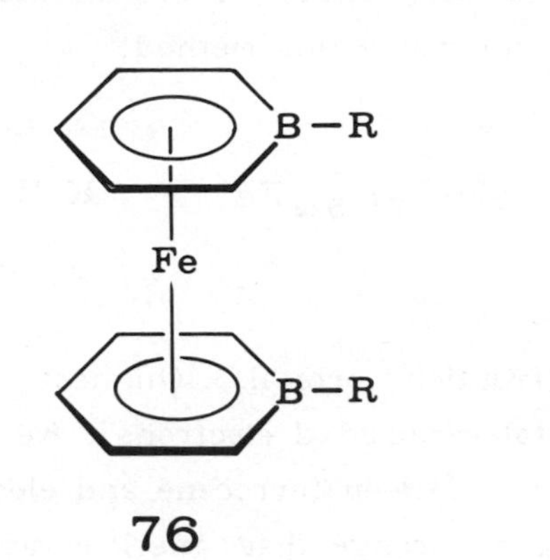


Co

• CH

o BH

77

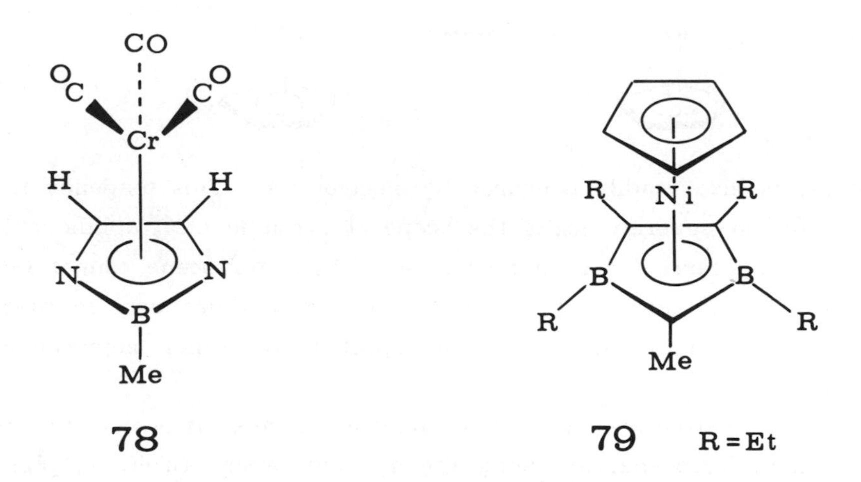


*Figure 3.15. Novel structures among the heterometallocenes.*

upon coordination. Often these compounds are prepared by generating an ionic cyclopentadienylide salt *in situ* from cyclopentadiene and a poorly coordinating proton base in the presence of a suitable metal salt [193]. Recall that cyclopentadiene is conveniently generated by thermal cracking of its Diels Alder dimer--an inexpensive hydrocarbon. The preparation of ferrocene [194] illustrates this method:

$$C_5H_6 + FeCl_2(H_2O) + KOH \longrightarrow (\eta^5\text{-}C_5H_5)_2Fe + KCl \qquad (3.88)$$

iv. *Reactions of Metallocenes*. The frontier orbitals (highest occupied molecular orbitals) of the metallocenes contain metal-localized d electrons. We shall see that these orbitals are very significant in reactions between ferrocene and electrophilic (acidic or oxidizing) reagents. Thus, ferrocene, an orange diamagnetic compound, is easily transformed to the blue, paramagnetic, ferricinium ion by such mild one-electron oxidizing agents as ferric salts, silver(I), halogens, nitric acid, and even sulfuric acid:

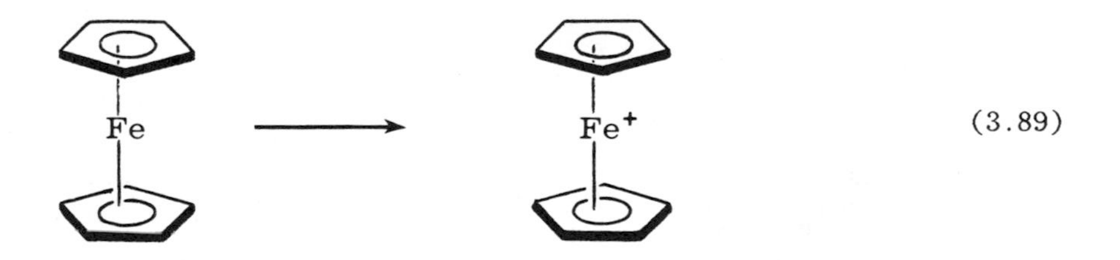


(3.89)

The ferricinium ion is also readily prepared by electrolysis. This tendency to undergo a one-electron oxidation severely limits the scope of aromatic electrophilic substitution reactions which can be carried out on ferrocene. Thus, ferrocene cannot be directly halogenated, nitrated, or sulfonated, although such derivatives can be prepared by indirect methods. These restrictions generally apply to the other metallocenes and to the arene complexes.

The proton is an ineffective one-electron oxidant. Thus, it is not surprising that strong acids protonate ferrocene, affording the hydrido cation, $(\eta^5\text{-}C_5H_5)_2\overset{+}{Fe}\text{-}H$, which is isoelectronic and isostructural with the very stable bent metallocene $(\eta^5\text{-}C_5H_5)_2ReH$. Since the conjugate acid of ferrocene has tilted Cp rings, the short bridging groups in *80* [195] increase its basicity, beyond that of ferrocene:

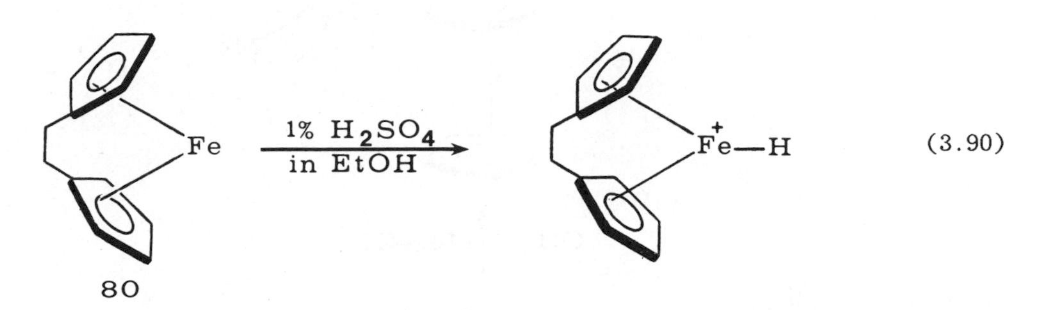


(3.90)

Ferrocene has a rich aromatic chemistry as long as one-electron oxidizing agents are not employed [196]. For example, ferrocene very readily undergoes Friedel-Crafts acylation and alkylation, Vilsmeir formylation, dimethylaminomethylation, and mercuration (Figure 3.16). Ferrocene is $10^6$ more reactive than benzene toward Friedel-Crafts acetylation! The mercury derivatives are useful in the preparation of haloferrocenes.

In view of the basic character of iron in ferrocene, it is plausible that some electrophilic substitution reactions may involve prior attack at iron:

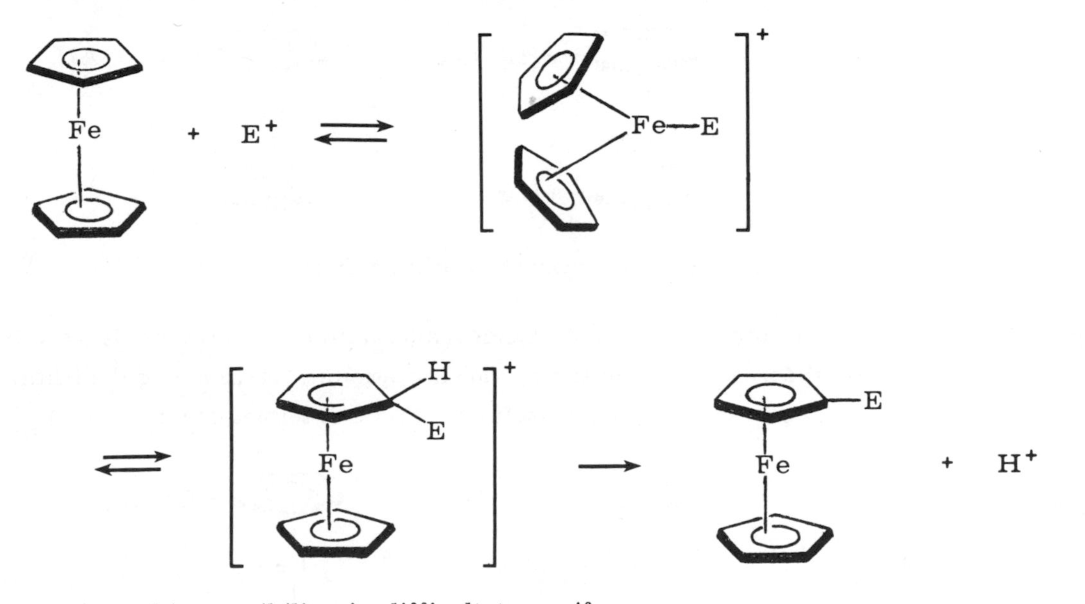


(3.91)

However, this possibility is difficult to verify.

Ferrocene is quite resistant toward catalytic hydrogenation. For example, pressures of 280 atm and temperatures above 300°C over nickel are required for even partial reduction! However, solutions of alkali metals in amines (solvated electrons)

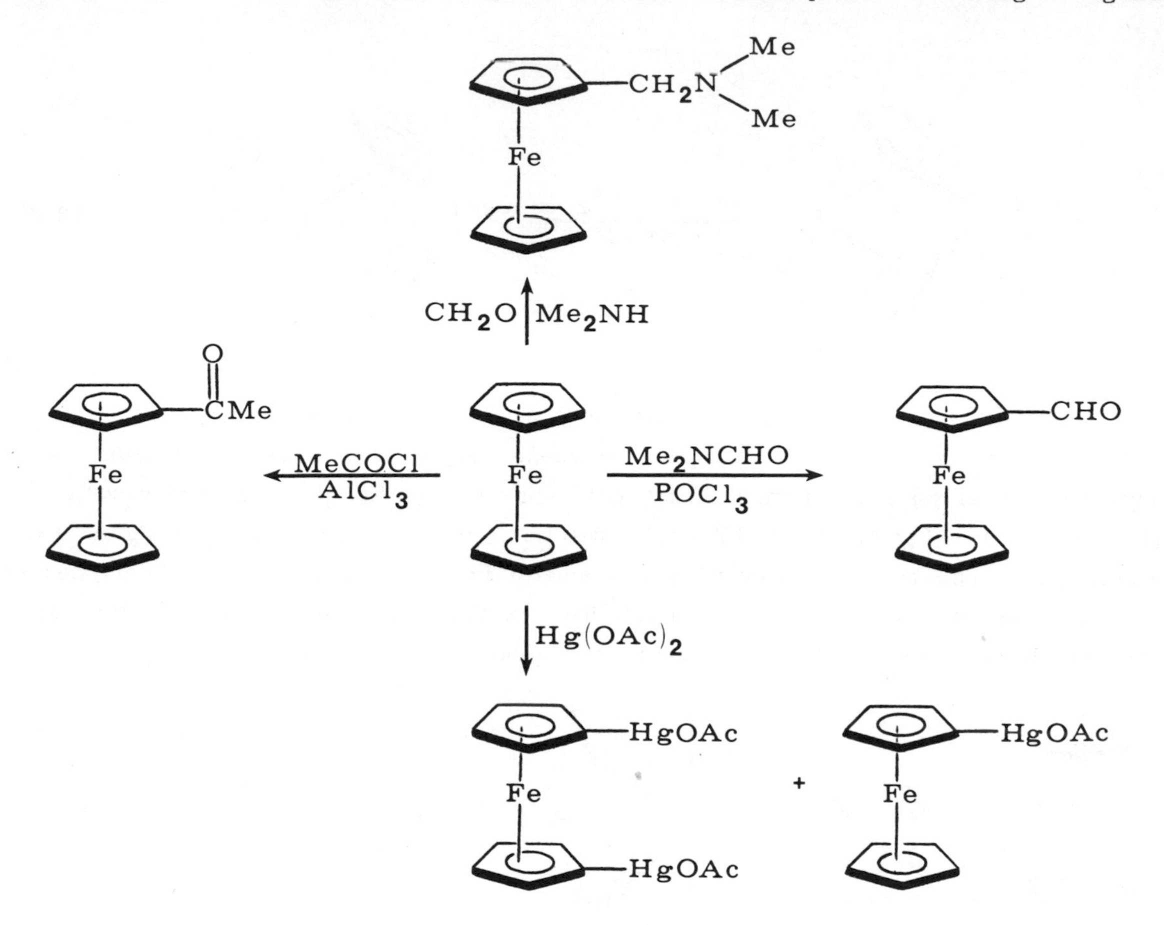


*Figure 3.16. Typical electrophilic substitution reactions of ferrocene.*

readily reduce ferrocene to the cyclopentadienylide anion [197]. Ferrocene is readily lithiated. Depending on the reaction conditions, varying amounts of mono- and di-lithio derivatives are formed. These are especially useful synthetic intermediates:

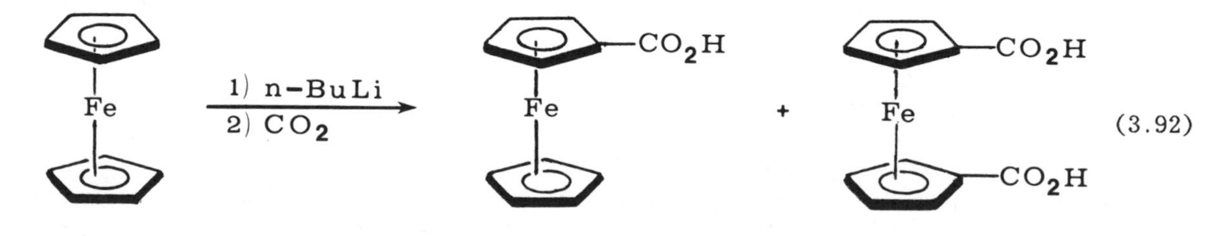


(3.92)

One of the most remarkable aspects of ferrocene chemistry is the facile displacement of bromide from ferrocenyl bromide by a wide variety of nucleophiles in the presence of copper(II). It seems very likely that this activation results from interaction of copper(II) with a basic electron pair on iron, but there is no direct evidence concerning this mechanism. This sort of reaction is useful in preparing derivatives which cannot be derived from electrophilic substitution:

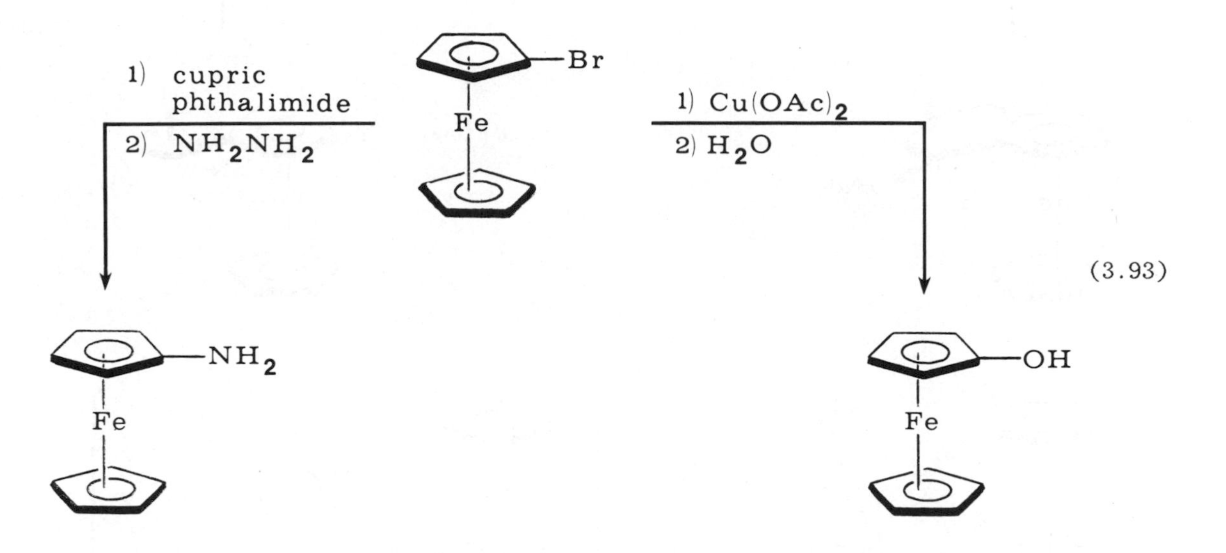


(3.93)

A basic d-electron pair on iron may also be involved in stabilizing carbenium ions α to the ferrocene nucleus. For example, the exo isomer of α-acetoxy-1,2-tetramethylene-ferrocene is solvolyzed more than 2,000 times faster than the endo isomer, but in either case only the exo alcohol is obtained (Equation 3.94).

An unusually facile reaction between ferrocene and aryl diazonium salts provides an excellent route to arylferrocenes. Again, the frontier d-orbitals seem involved. Little has proposed a radical mechanism involving the ferrocinium ion, which has been subsequently shown to be susceptible to radical coupling [198,199] (Equation 3.95).

The tendency for the metal-based frontier orbitals in the metallocenes to engage in redox processes is very pronounced with the $d^7$ metallocenes, cobaltocene, and rhodocene [200]. For example, cobaltocene spontaneously ignites in air. In the reaction with $CCl_4$, the Cp ring in cobaltocene traps the $CCl_3$ radical, forming an exo, 18-electron cobalt(I) complex [201] (Equation 3.96).

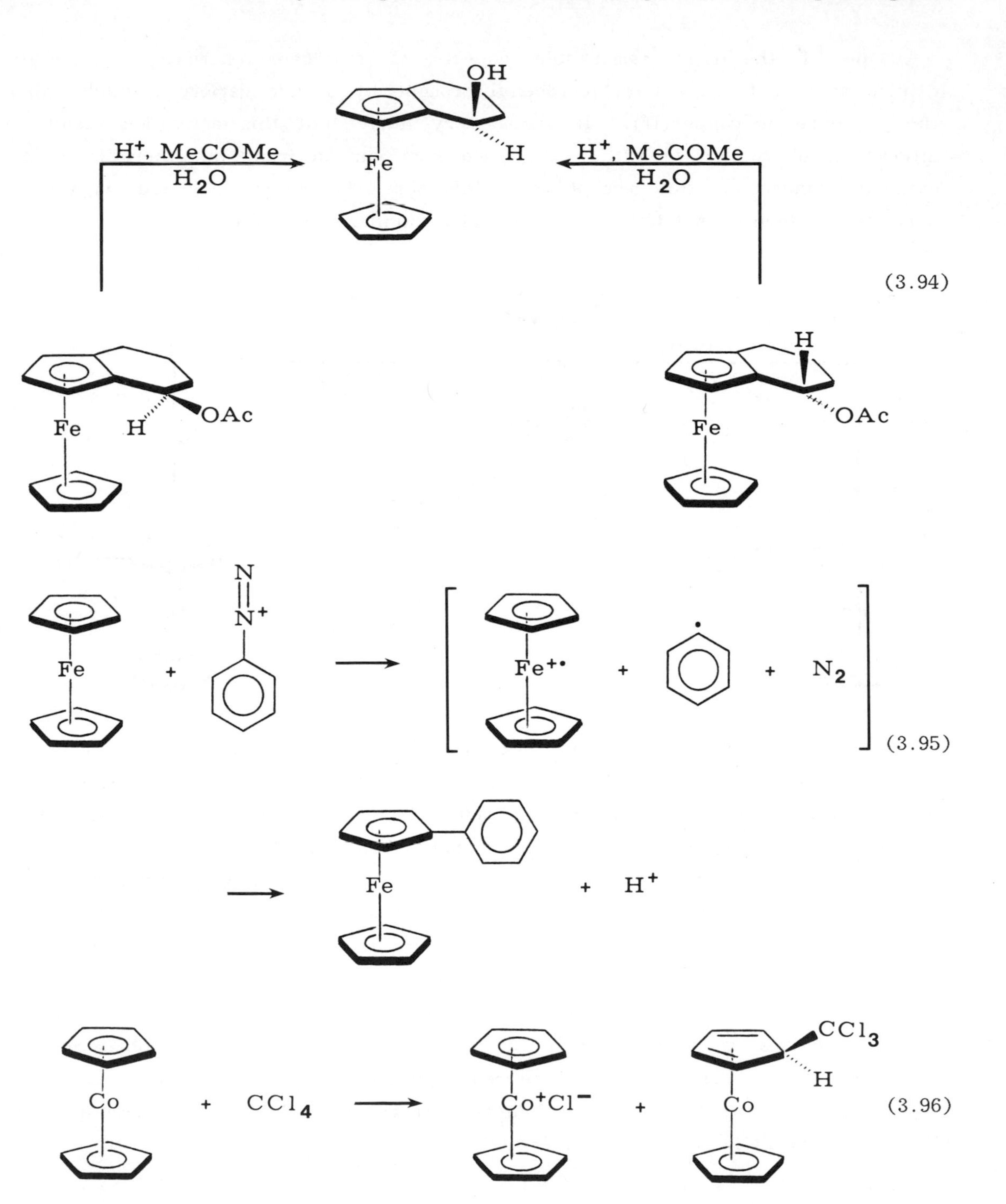
OH
H
Fe
H+, MeCOMe
H2O
H+, MeCOMe
H2O
(3.94)
H
Fe
H
OAc
Fe
OAc
N
N+
Fe
+
Fe+•
+
+
N2
(3.95)
Fe
+
H+
CCl3
H
Co
+
CCl4
Co+Cl−
+
Co
(3.96)

Certain other metallocenes and "half-sandwich" compounds undergo some of the reactions manifested by ferrocene such as Friedel-Crafts acetylation. However, with the exception of the congeners ruthenocene and osmocene, most such Cp derivatives are far less reactive than ferrocene. However, under more vigorous conditions these do react. For example, cyclopentadienylmanganese tricarbonyl can be acetylated and sulfonated [202]:

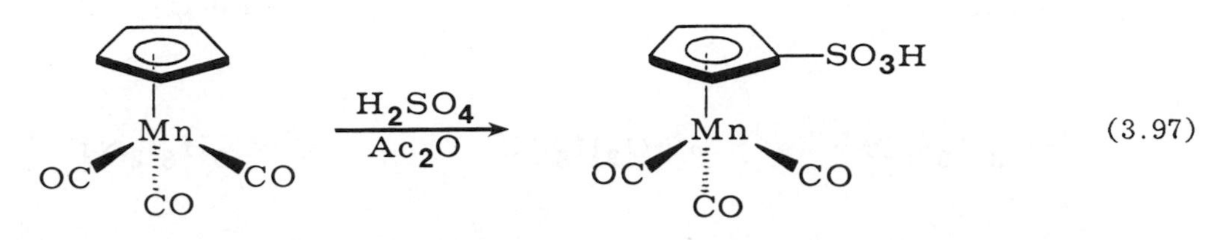


(3.97)

Cyclobutadieneiron tricarbonyl also undergoes a series of characteristic electrophilic substitution reactions, such as acetylation, formylation, and chloromethylation [203]:

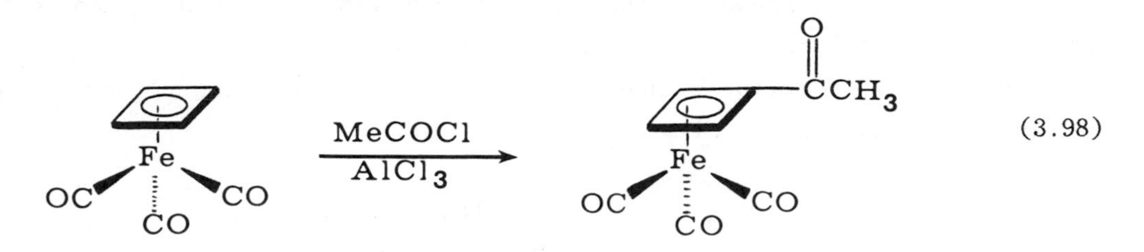


(3.98)

Certain metallocenes are useful catalysts or reagents for reactions we will encounter in later chapters. It is probable that the metal-based orbitals are involved in some of these reactions. For example, "soluble" Ziegler-Natta-type olefin polymerization catalysts are derived from titanocene derivatives, such as $Cp_2TiR^+$ [204,205]. Another example is the "hydrozirconation" of olefins by the sparingly soluble, coordinatively unsaturated $Cp_2Zr(Cl)(H)$. A significant feature of these formally $d^0$, bent metallocenes is that they are not capable of forming dπ-pπ backbonds with olefins or CO. This destabilization of the olefin or CO ligands probably accounts for the facile "migratory insertion" reactions manifested by such compounds. We shall return to this point in Chapter 5.

e. π-Allyl and Related Ligands. The $\eta^3$-allyl, or "π-allyl" group, is an ubiquitous ligand. A great variety of both homoleptic and mixed ligand $\eta^3$-allyl complexes are known (Figure 3.17). The versatile reaction paths and pronounced reactivity of the π-allyl group account for the large number of synthetic and catalytically

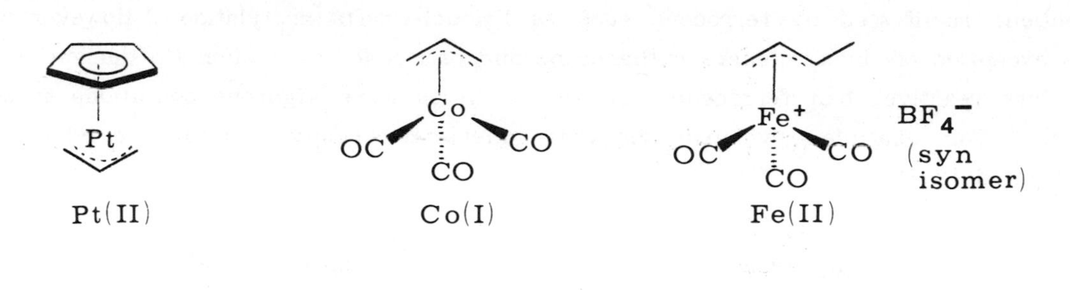


*Figure 3.17. Typical* $\eta^3$*-allyl complexes.*

important reactions manifested by $\eta^3$-allyl complexes. These reactions form the basis of Chapter 15. The synthesis, structure, and occurrence of $\eta^3$-allyl complexes have been reviewed [206], but this review is now out of date.

The typical structural characteristics of a $\eta^3$-allyl ligand are illustrated for bis-($\eta^3$-2-methallyl)nickel, 81 [207].

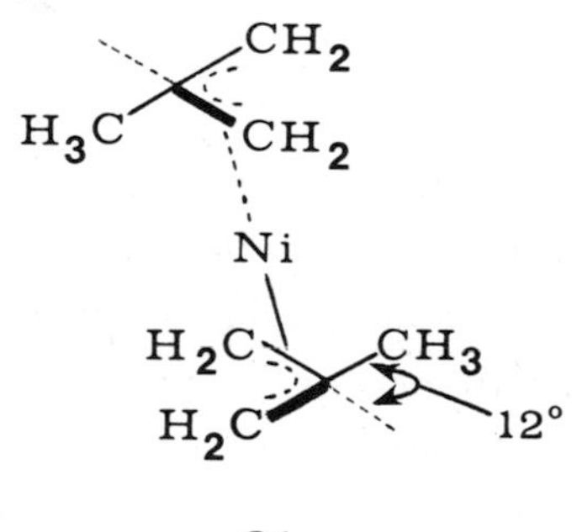


81

Provided that the trans ligands are the same, the three allylic carbon atoms are nearly equidistant from the metal. If the trans groups are different, the terminal allylic carbon-to-metal bonds are not quite equidistant, and the entire $\eta^3$-allyl group can be distorted by such a "trans effect." The allylic carbon atoms and the five substituent atoms are nearly coplanar with the metal on an axis below this plane (a kind of "sandwich" structure). There are thus two types of substituents on the outer carbon atoms: "anti" and "syn" groups, such as $H^3$ and $H^2$ in 82. For simple $\eta^3$-allyl groups which are static on the $^1$H NMR time-scale, the anti and syn hydrogens each exhibit a doublet by coupling with the central hydrogen, which exhibits a multiplet. Typical NMR parameters are: $H^1$ δ 6.5-4, $H^2$ δ 5-2.5, $H^3$ δ 3-1, $J_{12}$ ~ 7 Hz; $J_{13}$ ~ 11 Hz; $J_{23}$ ~ 0 Hz.

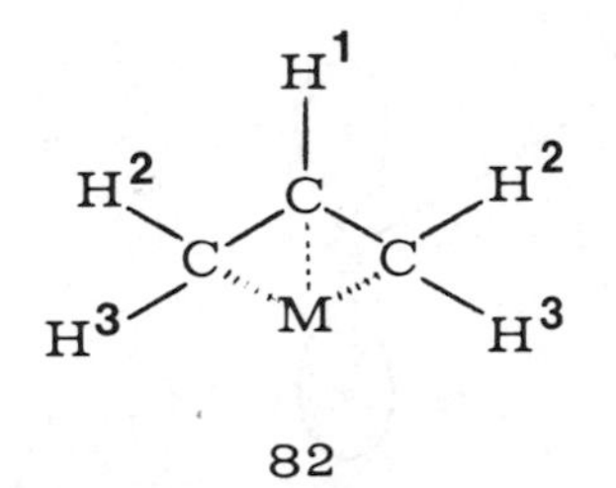

82

The bonding in η$^3$-allyl groups involves two ligand-to-metal donor [206,208] bonds (one σ and the other π) and a metal-to-ligand dπ-pπ backbond. The three ligand molecular orbitals (shown in Figure 3.18) derive from three linear combinations of the allyl anion $p_x$ atomic orbitals. In the terminology of our electron-counting scheme, the π-bonded allyl anion is a four-electron donor which can have some π-acid character.

Chemically the π-allyl group often behaves as an electron-rich ligand, but some of these reactions involve prior dissociation of one terminus to form an intermediate σ-allyl group. This kind of process is often facile and reversible, leading to fluxional rearrangement on the $^1$H NMR time-scale:

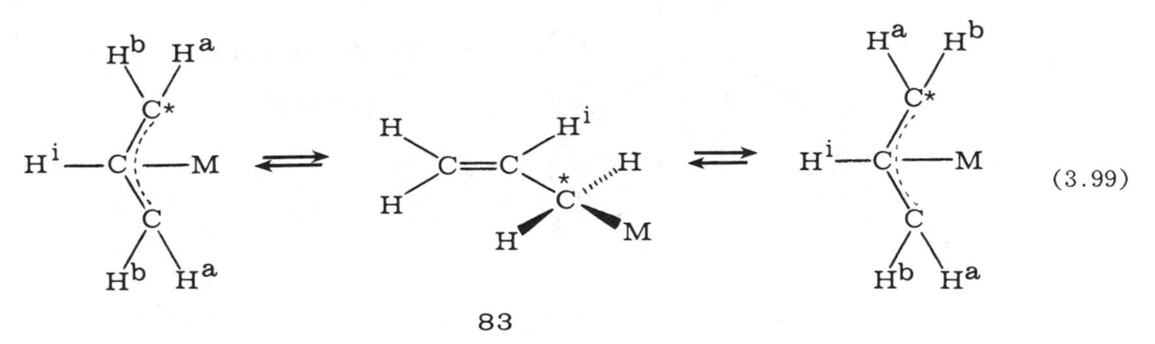

(3.99)

83

Many η$^3$-allyl complexes exhibit such fluxional behavior--especially those homoleptic allyl compounds. In the σ-allyl form there is "free rotation" about the single carbon-carbon bond, so that this facile η$^3$ to η$^1$ rearrangement provides a mechanism for exchanging the syn and anti groups. Note also that since the η$^1$-allyl group cannot act as a π-acid, it is more electron-rich than the η$^3$-allyl group. The transformation η$^3$ to η$^1$ (π to σ) allyl also creates a vacant coordination site on the metal as in 83. The presence

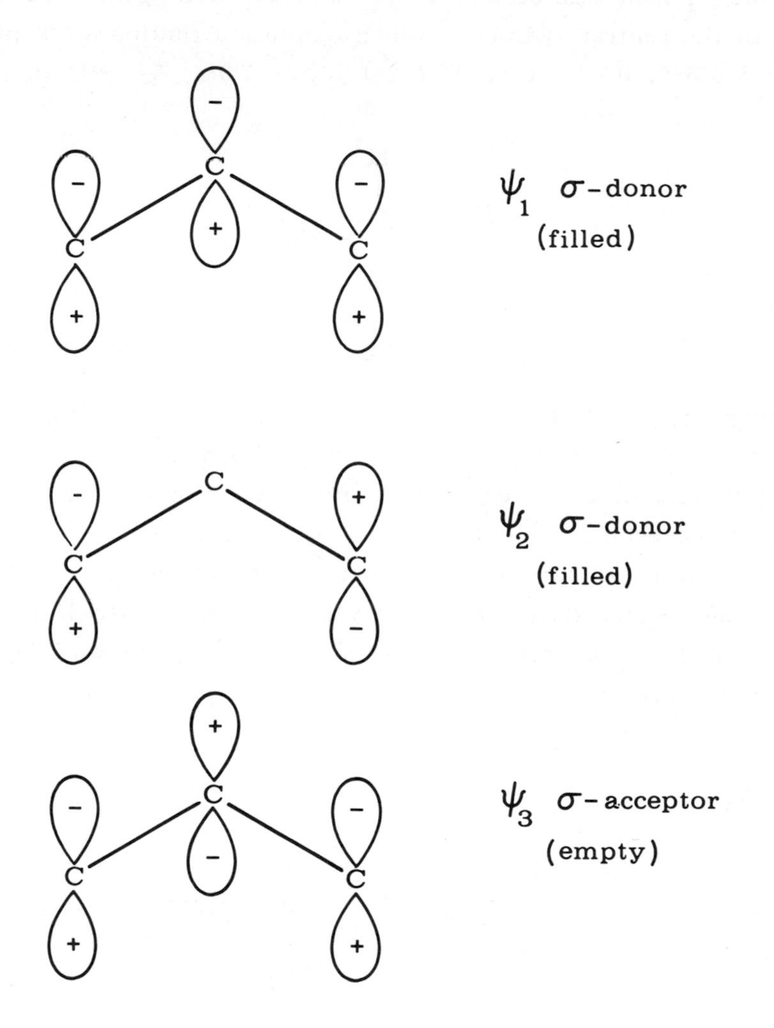


*Figure 3.18. π-Allyl anion orbitals used for $\eta^3$-allyl complexes.*

of this rearrangement of π to σ undoubtedly accounts for the wide range of reactivities exhibited by metal allyl complexes.

Bridging $\eta^3$-allyl complexes such as 84 are also known. It is uncertain whether such substances play a role as intermediates in reactions involving mononuclear transition-metal $\eta^3$-allyl compounds. Note that in Equation 3.100 the $\eta^5$-$C_5H_5$ group is also bridging [209]:

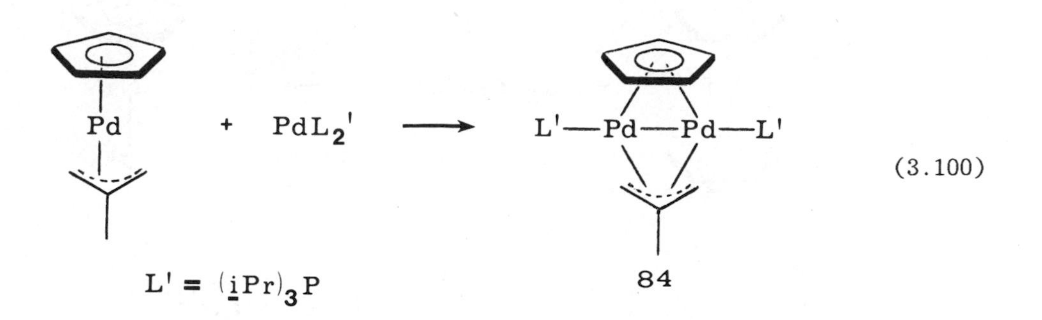


There are many methods for preparing π-allyl complexes. Most involve the preparation of an intermediate σ-allyl derivative which rearranges to the π-allyl form upon the expulsion of another ligand to generate the required vacant coordination site. As shown in Equations 3.101 to 3.104, the σ-allyl derivative may be prepared by adding a metal hydride 1,4 to a conjugated diene, by alkylating a basic metal center with an allyl halide, by displacing a ligand with an allyl Grignard reagent, and by protonating a metal 1,3-diene complex. The latter method is reversible and stereospecific.

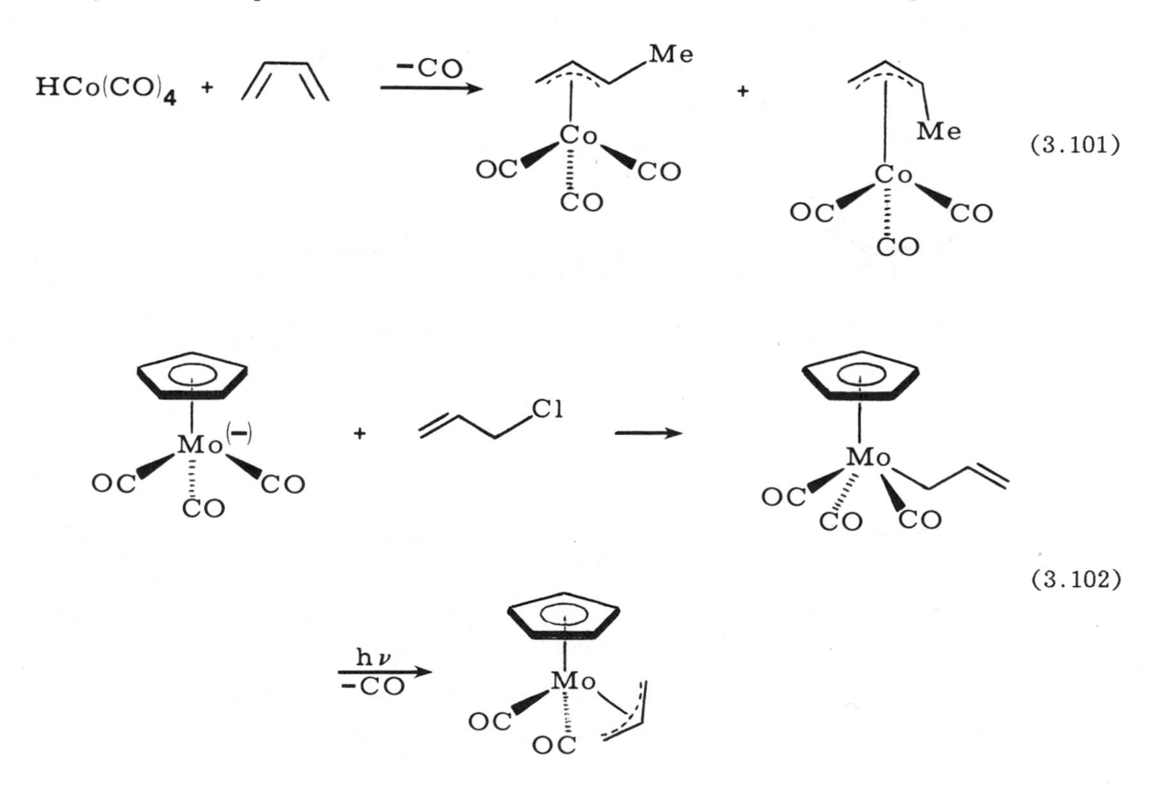

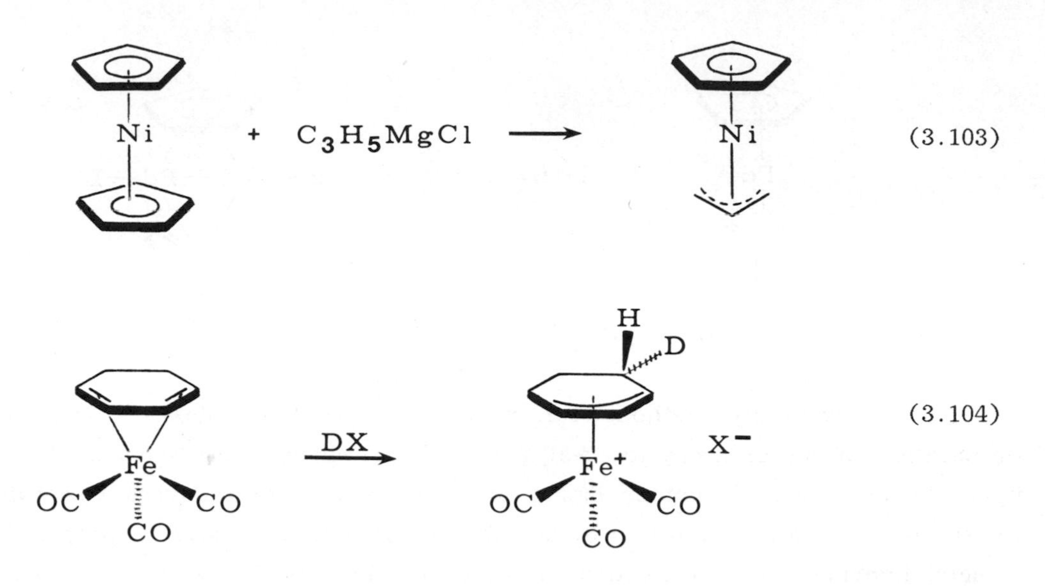


There are other complexes having an odd number of carbon atoms greater than three π-bonded to a metal. These are related to the π-allyl derivatives and are often formed by adding protons to, or removing hydride ions from, transition-metal olefin complexes or by adding nucleophiles to coordinated arenes [210]. Such complexes are often useful synthetic intermediates which we will encounter in the ensuing chapters. Examples are illustrated in Equations 3.105 and 3.106:

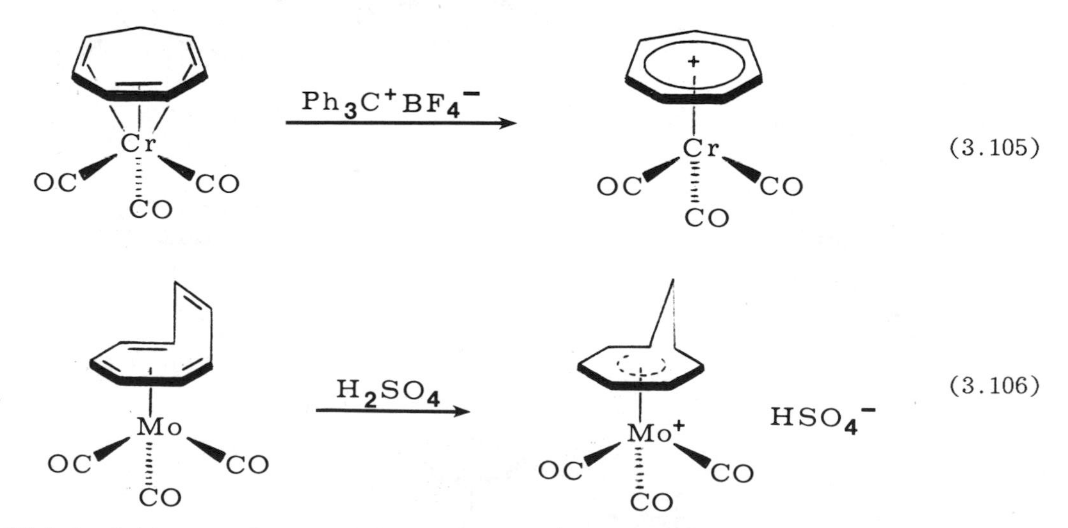


$\eta^3$-Allyl hydride complexes such as <u>85</u> have often been proposed as intermediates in the isomerism of olefins through a type of 1,3-hydride shift:

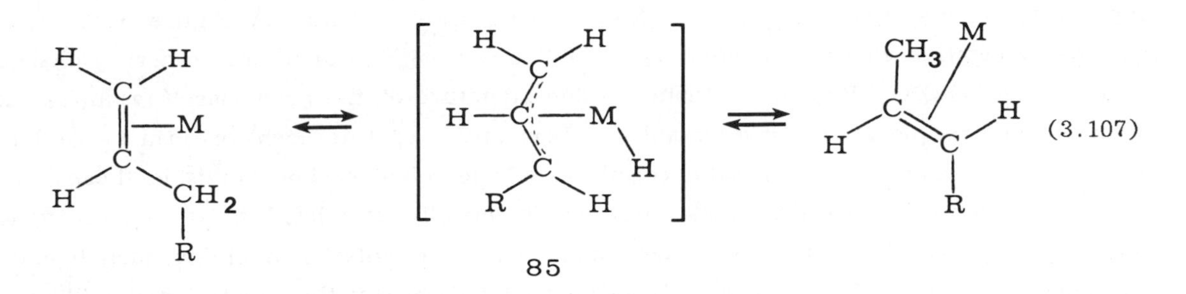


(3.107)

The viability of this mechanism has received support from the preparation and structural characterization of a stable iridium(III) $\eta^3$-allyl hydride, <u>86</u>, from allyl benzene [211]:

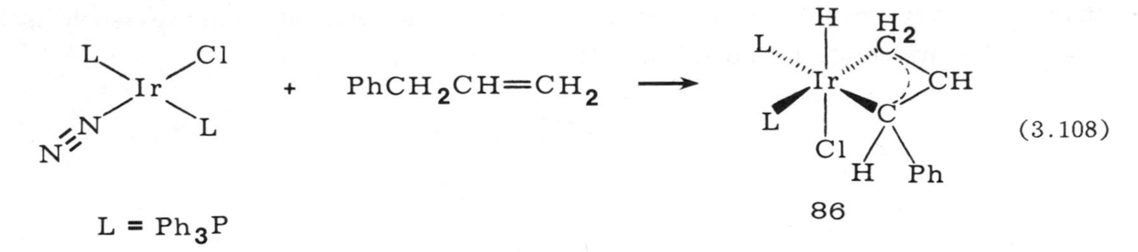


(3.108)

The $\pi$-benzyl ligand is present in well-characterized complexes and is considered to be an intermediate in several organometallic reactions. This ligand can be in equilibrium with the $\eta^1$-benzyl form. Stille [212] has shown that an asymmetric benzyl carbon can retain its configuration during this process.

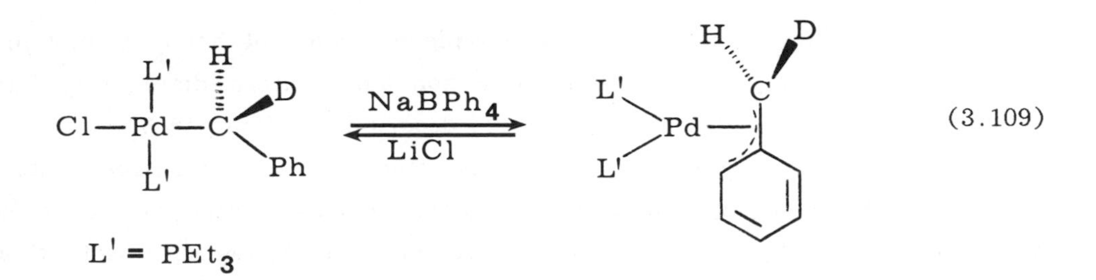


(3.109)

## 3.7. <u>Unsaturated Nitrogen Ligands</u>.

a. <u>Macrocyclic Imines</u>. A very large number of macrocyclic, unsaturated ligands are known. Most are tetradentate and occupy all the coordination positions in a plane containing the metal. Usually this geometry restricts two additional ligands to a <u>trans</u> relationship, preventing mechanisms which require reaction between adjacent sites (for example, insertion reactions). Note, among those ligands in Figure 3.19, that some have negative charges of two [e.g., TPP, Pc, $(dmg)_2$, salen] and others a charge of one, [$(DO)(DOH)_{pn}BF_2$ and corrins]. The structure of the coenzyme Vitamin $B_{12}$ is given as an example of a corrin complex. Derivatives of this coenzyme can be said to be the oldest organotransition-metal complexes. The cobalt-carbon bonds in these complexes are thermodynamically weak, but are kinetically stabilized by the macrocyclic corrin ligand. The best-studied organotransition-metal complexes involving such ligands are those of cobalt. These studies have been stimulated by the unusual and still controversial Vitamin-$B_{12}$-catalyzed rearrangements [213]. Such unsaturated macrocyclic ligands are thought to be capable of $\pi$-bonding with the metal.

As is illustrated in Figure 3.19, quadridentate Schiff-base complexes are usually planar. However, nonplanar arrangements are also known, in complexes such as 87, where the quadridentate ligand is salen [214]:

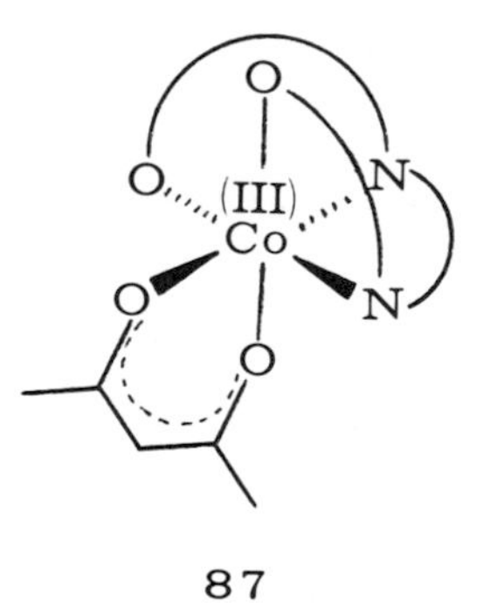


87

So far such nonplanar macrocylic complexes have not been studied in the context of homogeneous catalysis, although such nonplanar intermediates may be involved in certain Schiff-base hydrogenation catalysts described in Chapter 6.

b. Dinitrogen Complexes. For many years it was recognized that, although $N_2$ and CO are isoelectronic and transition-metal carbonyl compounds were bountiful, no metal complexes of $N_2$ were known. Interest in such compounds was further heightened by the knowledge that a family of metalloenzymes (nitrogenases) catalyze the reduction of $N_2$ to $NH_3$ under ambient conditions. Recently the structures of the supposed active site in this molybdenum-iron enzyme has been characterized by EXAFS, and a tentative synthetic analogue has been prepared [215].

In 1965 the first dinitrogen transition metal complex, $[Ru(NH_3)_5N_2]^{2+}$ was serendipitously prepared by Allen and Senoff [216]. Following that discovery a great

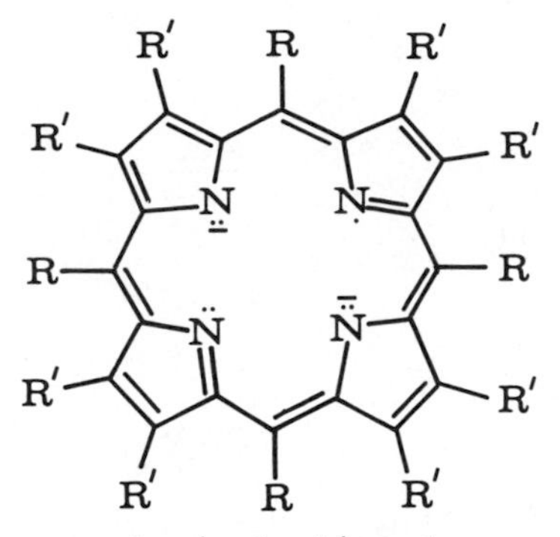

*porphyrinato ligands*
*R = R' = H porphine*
*R = Ph, R' = H meso-tetraphenylporphyrin (TPP)*
*R = H, R' = Et octaethylporphyrin (OEP)*

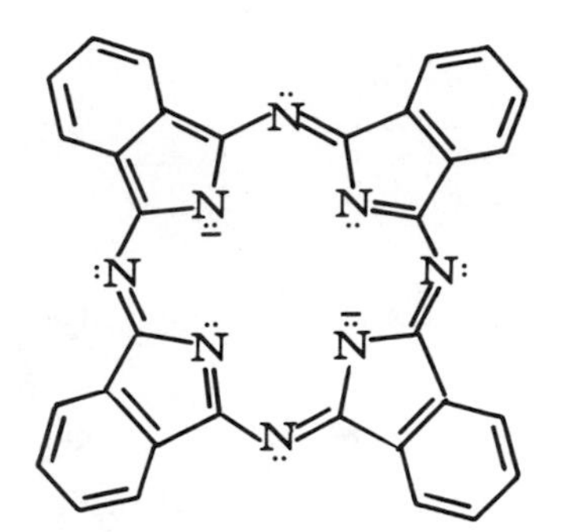

*phthalocyanine (Pc)*

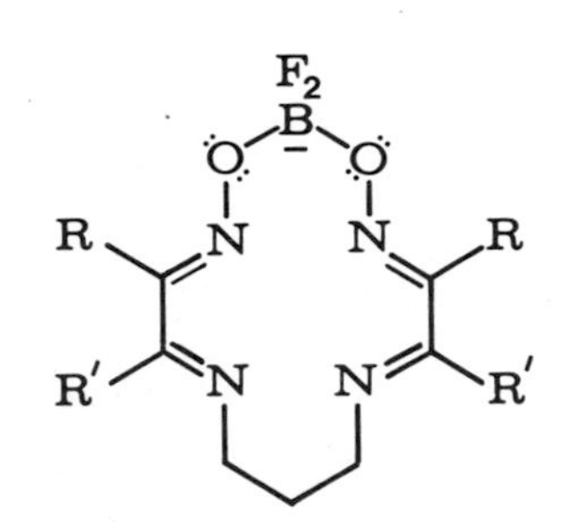

*R = Et, R' = Me:* $C_2(DO)(DOH)pnBF_2$
*R = R' = Me:* $(DO)(DOH)pnBF_2$

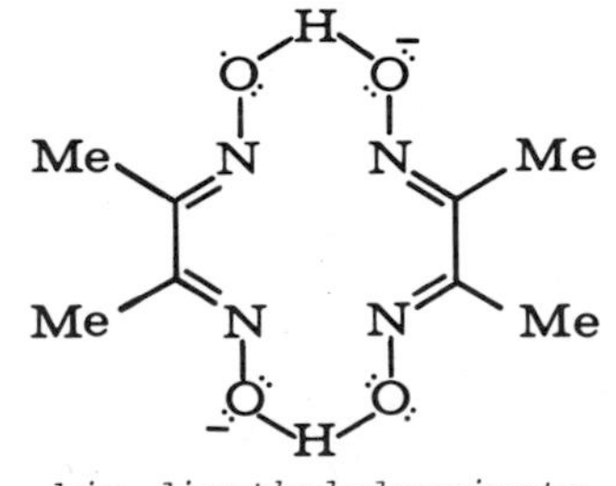

*bis-dimethylglyoximato* $(dmg)_2$

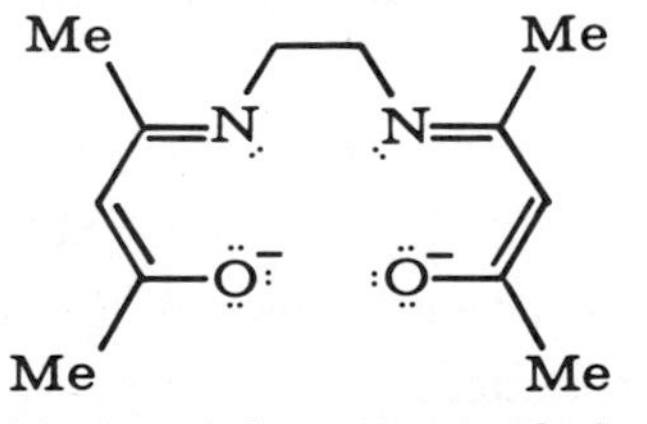

*bis(acetylacetone)ethylenediamine (bae), (acen)*

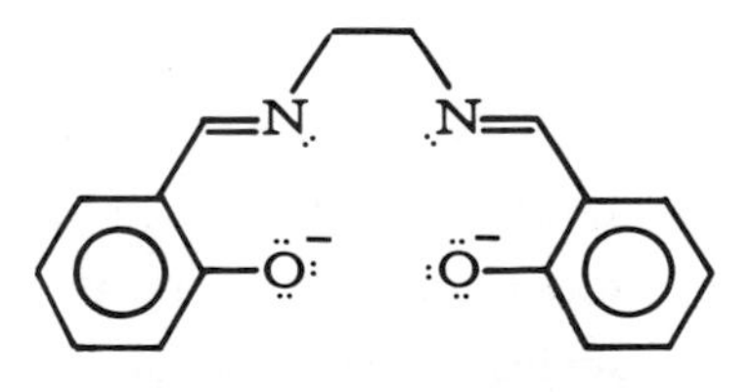

*bis(salicylaldehyde)-ethylenediamine (salen)*

*Figure 3.19. Some representative macrocylcic tetradentate ligands.*

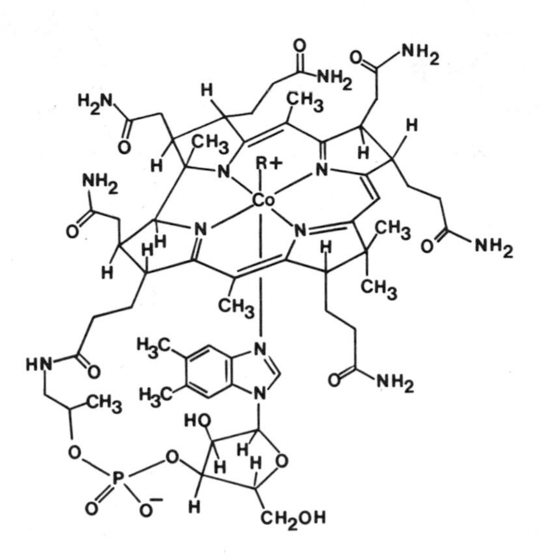

Coenzyme Vitamin B—12 ( a corrin complex )

*Figure 3.19. Continued.*

number and variety of transition-metal dinitrogen complexes have been characterized [217-219]. Examples are now known for most, but not all, of the transition elements. Thus far no useful catalytic processes for reducing $N_2$ have resulted from studies of these synthetic transition-metal dinitrogen complexes, nor have these substances yet achieved a significant role in homogeneous catalysis. We will therefore present only a brief structural discussion of dinitrogen complexes. The reduction of dinitrogen complexes is discussed in several reviews [219-221].

Dinitrogen is both a weaker σ-donor and a weaker π-acid than CO. Nevertheless, dπ-pπ backbonding is very important in stabilizing dinitrogen complexes. Electron-rich metals in lower oxidation states, especially 4d and 5d elements toward the left of the transition series, form the most stable dinitrogen complexes. In a few cases $N_2$ has been shown to displace $H_2O$ or $NH_3$, a reaction necessary for the function of nitrogenase and another indication of the importance of π-bond stabilization of dinitrogen:

$$[Ru(NH_3)_5(H_2O)]^{2+} + N_2 \rightleftarrows [Ru(NH_3)_5N_2]^{2+} \quad (3.110)$$

On the other hand, CO, which is a competitive inhibitor of nitrogenase, effectively displaces $N_2$ from the majority of dinitrogen complexes.

The vast majority of mononuclear dinitrogen complexes are essentially collinear like terminal-metal carbonyls. Structural studies have shown that the N-N bond in such terminal dinitrogen complexes is within about 0.03 Å of that in the free ligand, although the M-N distances are somewhat shorter than the value estimated for a single bond. This modest bond shortening is further evidence supporting dπ-pπ bonding. The $N_2$ stretching frequencies in dinitrogen complexes exhibit shifts toward lower energies similar to those of metal carbonyls. These features are illustrated by the representative selection of monohapto dinitrogen complexes shown in Figure 3.20. Note that bis-dinitrogen complexes are known. The effect on $\nu N_2$ of increasing the oxidation state of the metal reflects diminished backbonding.

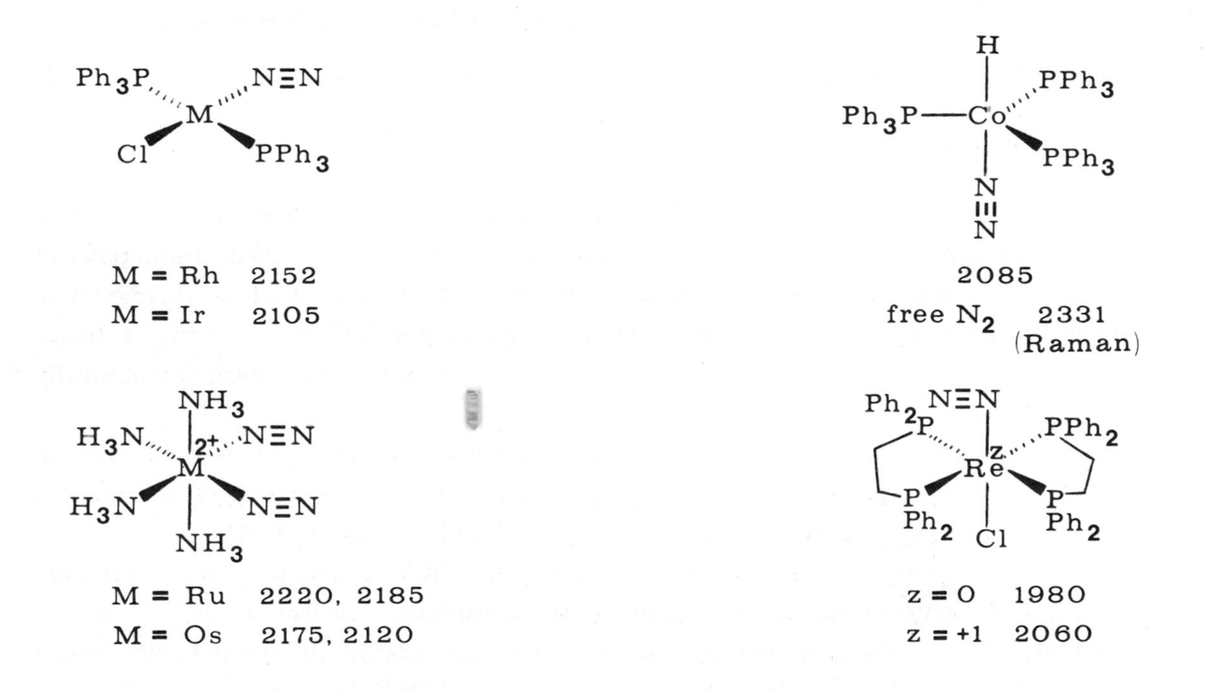


*Figure 3.20. $\nu N_2$ values in $cm^{-1}$ for representative $N_2$ complexes.*

Sidebonded, $\eta^2$, dinitrogen complexes have been claimed [222]. However, a recent structural study at -161°C indicates that these are actually end-on, $\eta^1$ complexes [223]. This evidence is further supported by $^{31}P$ and $^{15}N$ NMR and the high (2100) $\nu N_2$ band.

Bridging dinitrogen complexes are structurally different from most bridging carbonyls in that the former are commonly collinear: M-N≡N-M. In those cases involving two 18-electron metal centers, such as 88, the N-N bond length and the Raman active $\nu N_2$ are little different from those features of the free ligand or of the terminal dinitrogen complexes. However, if one of the metals is an electron acceptor, such as 89, the M-N≡N-M' unit remains collinear, but the N-N distance markedly increases and the $\nu N_2$ frequency falls [224].

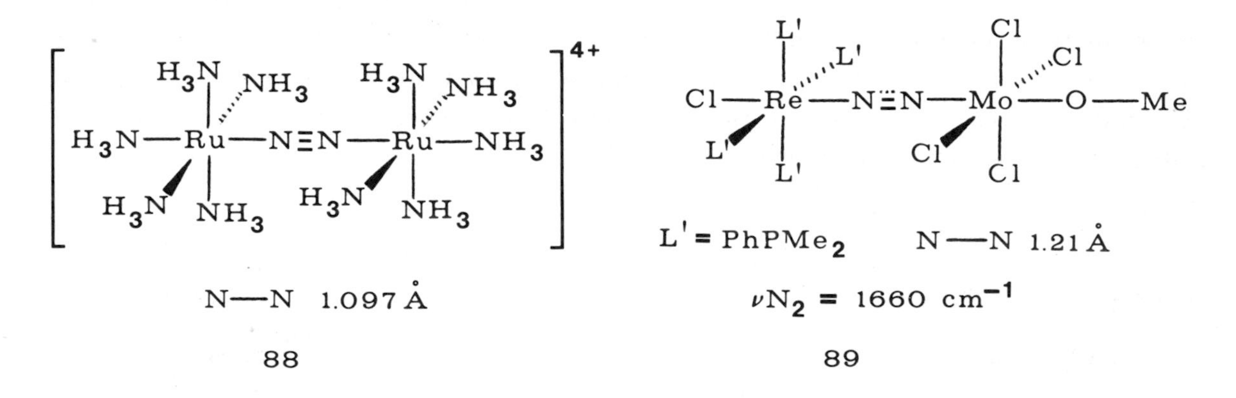


Both of these effects indicate a lowering of the dinitrogen bond order and suggest that such dinitrogen groups should be more susceptible to reduction. Indeed, reductions of coordinated dinitrogen are generally assumed to proceed by way of either bridging or protonated dinitrogen. A trinuclear bridged bis-dinitrogen complex, having a linear ReNNMoNNRe arrangement and long (1.28 Å) N-N bonds, has been structurally characterized [225].

The formation of such Lewis-acid adducts suggests that the terminal nitrogen in linear $\eta^1$-$N_2$ complexes is basic. This basicity is illustrated by alkylation of coordinated dinitrogen, affording an aliphatic diazonium complex [226] (Equation 3.111).

c. Nitrous Oxide. Complexes of $N_2O$ are very little studied. This kinetically inert gas is a very weak ligand which forms complexes similar to $N_2$. Taube's ruthenium(II) $N_2O$ complex is representative of this class [227]. Infrared $N_2O$ stretching frequencies indicate that $N_2O$ is a weak $\pi$-acid: $[(NH_3)_5Ru(N_2O)]^{2+}$, $\nu N_2O$ 2110, 1170 $cm^{-1}$ (free ligand $\nu N_2O$ 2224, 1285) [227].

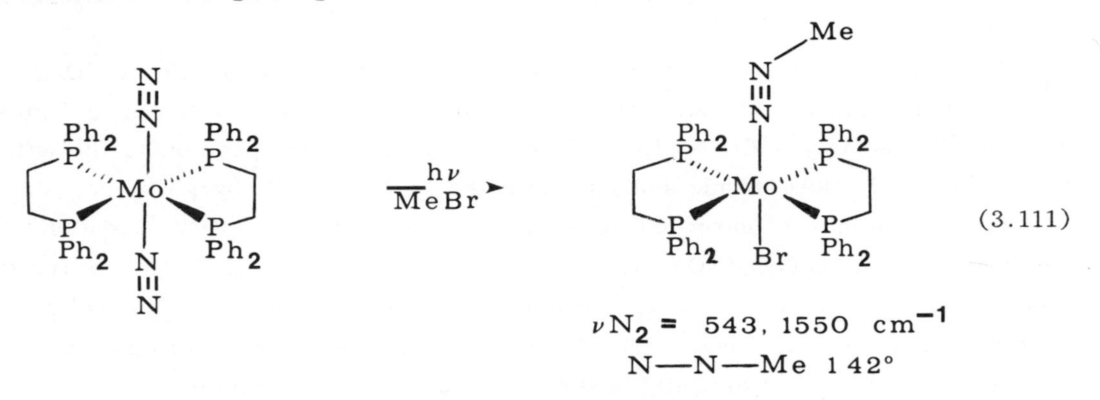


d. Nitroso Arenes. Other very little-studied complexes are those derived from nitroso arenes. An example is the ruthenium complex, 90, whose structure has not been firmly established [228]. Nitroso complexes with $\eta^2$-bonding such as 91 have been prepared. A related $\eta^2$ complex has been structurally characterized [229].

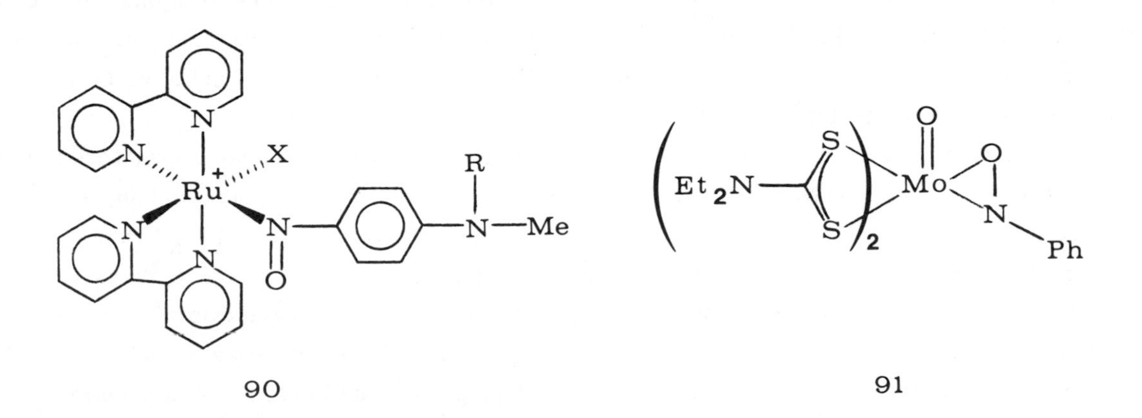


90

91

e. Nitric Oxide Complexes. The nitrosonium ion, $NO^+$, is isoelectronic with CO. It is therefore not surprising that transition metals form NO complexes which are isoelectronic and isostructural with CO derivatives. However, the coordination chemistry of NO shows several additional features, especially the ability to form two distinct bonding modes having different geometries, linear and bent [230]. Thus NO is the protypical example of a "non-innocent" or "amphoteric" ligand ($SO_2$ and $RN_2$ are other cases, vide infra).

Most NO complexes have a terminal, linear structure (M-N-O 180°). For purposes of oxidation-state formalism, in this bonding mode the NO ligand may be considered to be positive, $NO^+$. Complexes containing only NO ligands are very rare (e.g.,

$Co(NO)_3$) [231]. However, complexes having one or two terminal, linear NO groups are rather common and are usually isoelectronic with CO analogues as shown in Figure 3.21. Note that replacing one CO by NO and shifting the metal one position to the left affords an isostructural, isoelectronic analogue with the same over-all charge.

It is common for linear NO complexes to have formal negative oxidation states at the metal, e.g., $Mn(CO)(NO)_3$ (Mn(-III)). However, it is important to recognize in such cases that the actual charge distribution is not related to this formalism, as, for example, it is not in $MnO_4^-$ (Mn(VII)). Nevertheless, this formalism is useful for making first-order predictions about structure and stoichiometry. For example, $Mn(CO)(NO)_3$ is formally $d^{10}$ and has the expected tetrahedral geometry.

The characteristic structural features of linear NO complexes are as follows. The metal-nitrogen bond length is shorter than the sum of covalent radii, which is consistent with the strong $\pi$-backbonding. The linear NO group seems to be more stable in a position which maximizes this intense $\pi$-bonding, e.g., the trigonal plane in $Mn(CO)_4(NO)$ shown in Figure 3.21. If not in an axially symmetric environment, "linear" NO groups may be "slightly bent" (MNO ~170°) [234]. Carbonyl compounds show this same effect, but to a lesser degree.

In certain cases the nitrosyl ligand imparts distinctive chemical characteristics to the complex. The dominant ligand-replacement mechanism for some carbonyl-nitrosyl compounds is often associative ("$S_N2$") rather than the dissociative ("$S_N1$") mechanism exhibited by the isoelectronic carbonyl analogues [235]. Other ligands such as CO in a complex are usually replaced more readily than NO. Ruthenium exhibits a very pronounced tendency to form NO complexes. Chemical attack on coordinated NO occurs by both inter- and intramolecular mechanisms, and leads to complicated redox changes at N [236]. Thus far these complexes have been of little importance in organometallic chemistry, since this reactivity of the NO ligand complicates the use of NO complexes as catalysts. The continuing problem of catalytically removing nitrogen oxides from automobile exhausts will doubtless stimulate further research on reactions of NO complexes.

The most distinctive physical property of the NO ligand is the infrared νNO band. Terminal NO complexes generally exhibit strong νNO bands over a wide frequency range: neutral complexes 1610-1820 $cm^{-1}$, with anionic complexes having νNO as low as 1455 $cm^{-1}$ and cationic complexes as high as 1945 $cm^{-1}$. The above stretching frequencies may be compared with uncoordinated NO 1860 $cm^{-1}$, NO(+) 2220 $cm^{-1}$, and their isoelectronic analogues $O_2^+$ 1876 $cm^{-1}$ and CO 2143 $cm^{-1}$. Comparisons between MNO frequencies and those of the uncoordinated diatomics illustrate the strong metal-to-NO

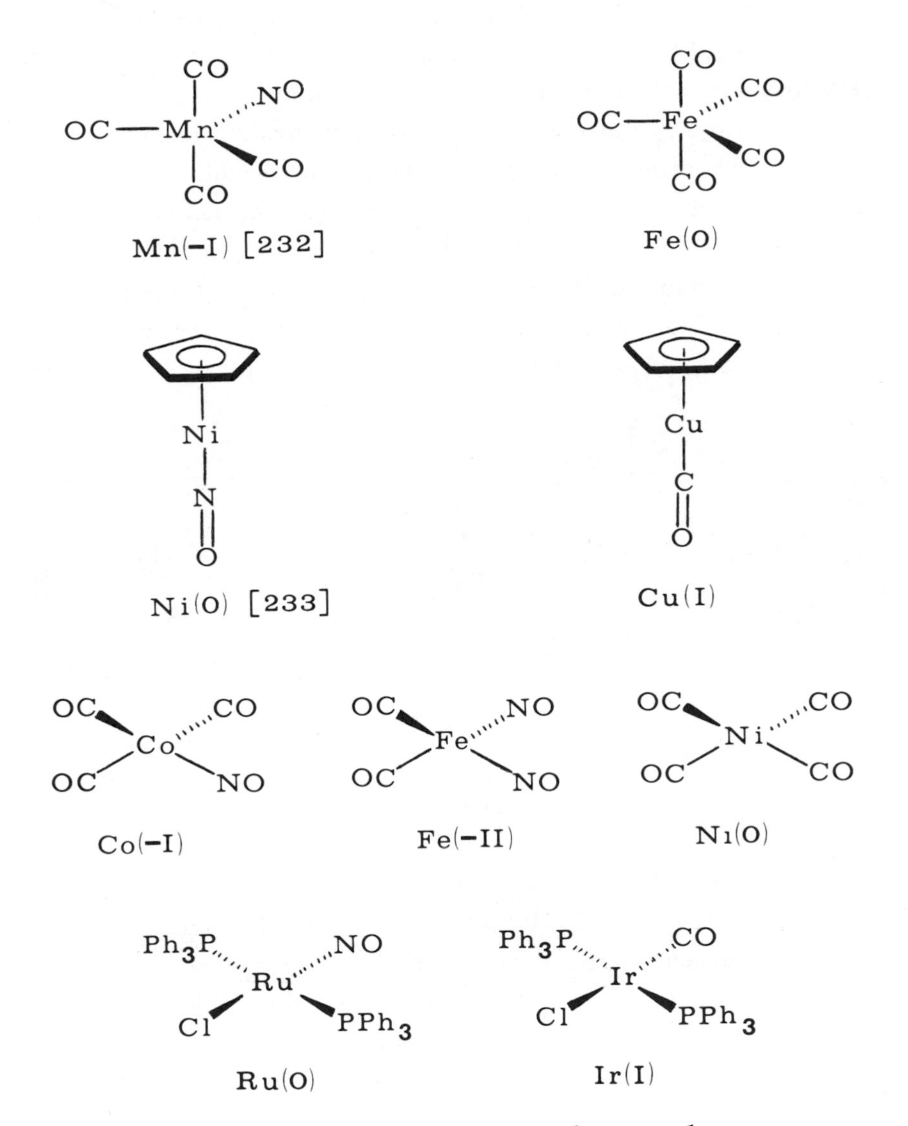


*Figure 3.21. Isoelectronic NO and CO complexes.*

backbonding which populates the ligand $\pi^*$ orbital, thus decreasing the stretching frequency.

Terminal nitrosyls can assume another geometry, the "bent NO," in which the angle MNO is in the range 120-140°. This structural and electronic configuration is a discrete energy minimum. Thus it is possible in certain cases for the linear and bent

NO configurations to coexist as separate "hybridization isomers" having the same over-all stoichiometry [237]. Hybridization isomers of this sort can rarely be observed for the same substance, since in most cases one form is more stable than the other.

Bent nitrosyl groups, which were first structurally characterized by Hodgson and Ibers [238], are encountered as apical ligands in certain square-pyramidal five-coordinate and occasionally in octahedral six-coordinate complexes. Compared with the linear NO group, the M-N bond is slightly longer in the "bent" form. The νNO frequencies of bent NO groups are usually lower (1610-1720 $cm^{-1}$) than those of the linear NO ligands, but the ranges overlap so that the νNO frequency is not a reliable criterion for assigning structures. Below are typical complexes containing "bent" NO ligands:

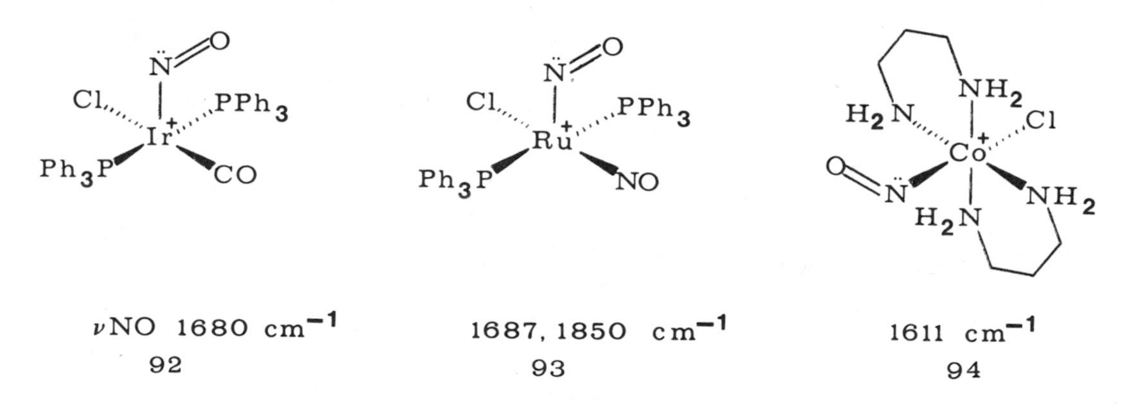


For oxidation-state purposes it is convenient to consider the "bent" NO ligand as $NO^-$. Thus the complexes 92 [238], 93 [239], and 94 [240] are formally $d^6$. This implies that the iridium(III) and ruthenium(III) derivatives, 92, and 93, are coordinatively unsaturated, perhaps reflecting the strong *trans* labilizing effect which appears to be a characteristic feature of the bent NO group. Consistent with this postulated *trans* effect is the weakened Co-Cl bond *trans* to bent NO in the cobalt(III) complex, 94. This bond is ~0.3 Å longer than the sum of covalent radii. The ruthenium *bis*-nitrosyl complex, 93, is exceptional, since it contains both types of terminal NO groups. Its structure allows one to compare both within a common complex. An $^{15}N$ labeling study showed that the two NO groups rapidly interconvert in solution [241].

The bonding situation in bent NO structures has been analyzed by molecular-orbital "Walsh diagrams," which describe the bending process [242,243]. For our purposes the following simplified explanation suffices. The empty $\pi^*$-level of coordinated linear $NO^+$ is about equivalent energetically to the filled nonbonding metal d levels, which accounts for the strong π-bonding. Upon bending the NO group, a d-metal electron pair can be considered to be transferred to the $NO^+$ group occupying a more stable $sp^2$ level localized on nitrogen. Thus the conversion of the linear $NO^+$ to a bent $NO^-$ structure is equivalent to an intramolecular transfer of an electron pair from the metal to the ligand:

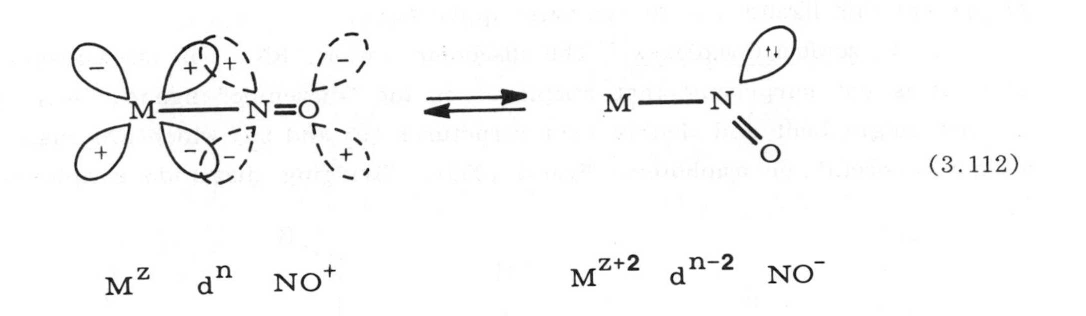


As a result of this electron transfer, the coordination geometry of the metal may change, for example, from a trigonal-bipyramidal $d^6$ configuration to a square-pyramidal $d^8$ configuration. All such amphoteric ligands have a ligand-acceptor orbital which is energetically close to the filled d level and which can become more stable by distorting the ligand. What is unusual is for such situations to give rise to discrete energy minima.

Students interested in NO chemistry may wish to consult references on structures [244], catalytic reactions [245], synthetic methods [246], and reactions [247] of coordinated NO.

Stable compounds having bridging NO groups are also known, but these have been little studied and have so far had no impact on organotransition-metal reactions. As might be expected by analogy with CO compounds, both double- and triple-bridged NO groups are known [248,249]. Bridging NO groups exhibit lower νNO [250,251].

Thionitrosyl complexes have recently been prepared using trithiazyl trichloride [252]:

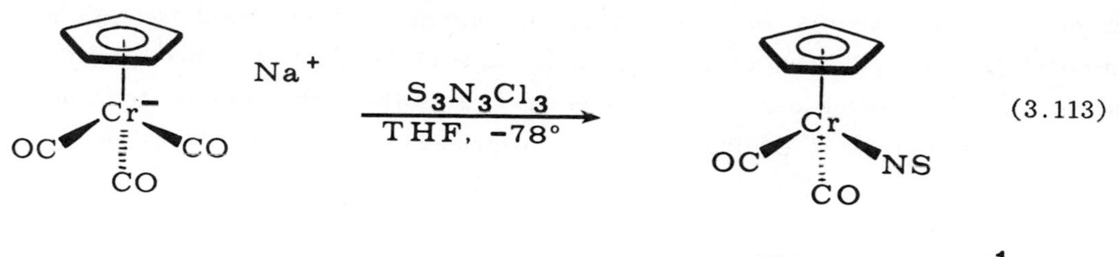

At present this ligand has no synthetic applications.

f. Diazonium Complexes. The diazonium cation, $RN_2^+$, is isoelectronic with NO. Thus it is not surprising that complexes of the "diazenido" ligand, $RN_2$, exhibit the distinct singly bent and doubly bent structures (95 and 96) which are characteristic of a "non-innocent" or amphoteric ligand [253]. Bridging diazenido complexes (97) are

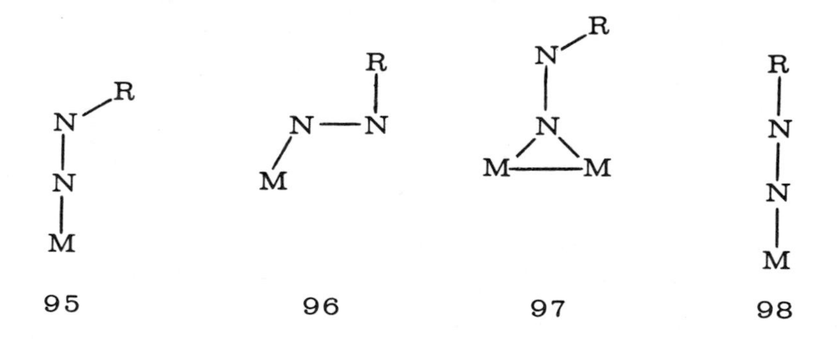


also known; however, completely collinear structures (98) are rare. Thus far only one example is known. Many of these structures have been elucidated by Ibers and his students [254]. Typically these complexes may be prepared from the interaction of aryl diazonium salts with appropriate (usually unsaturated) transition-metal complexes or by alkylation of selected dinitrogen complexes. A novel method for preparing these complexes is the reaction between a primary amine and an electrophilic nitrosyl complex [255] (Equation 3.114).

The structurally characterized six- and five-coordinate, "singly bent" diazenido complexes, 99 [256] and 100 [257], are characteristic of this structural class. For the assignment of formal oxidation states these may be regarded as complexes of $RN_2^+$. The nearly equivalent Ru-Cl distances in 99 indicate that the "singly bent" diazenido ligand does not exert a strong *trans* bond weakening effect.

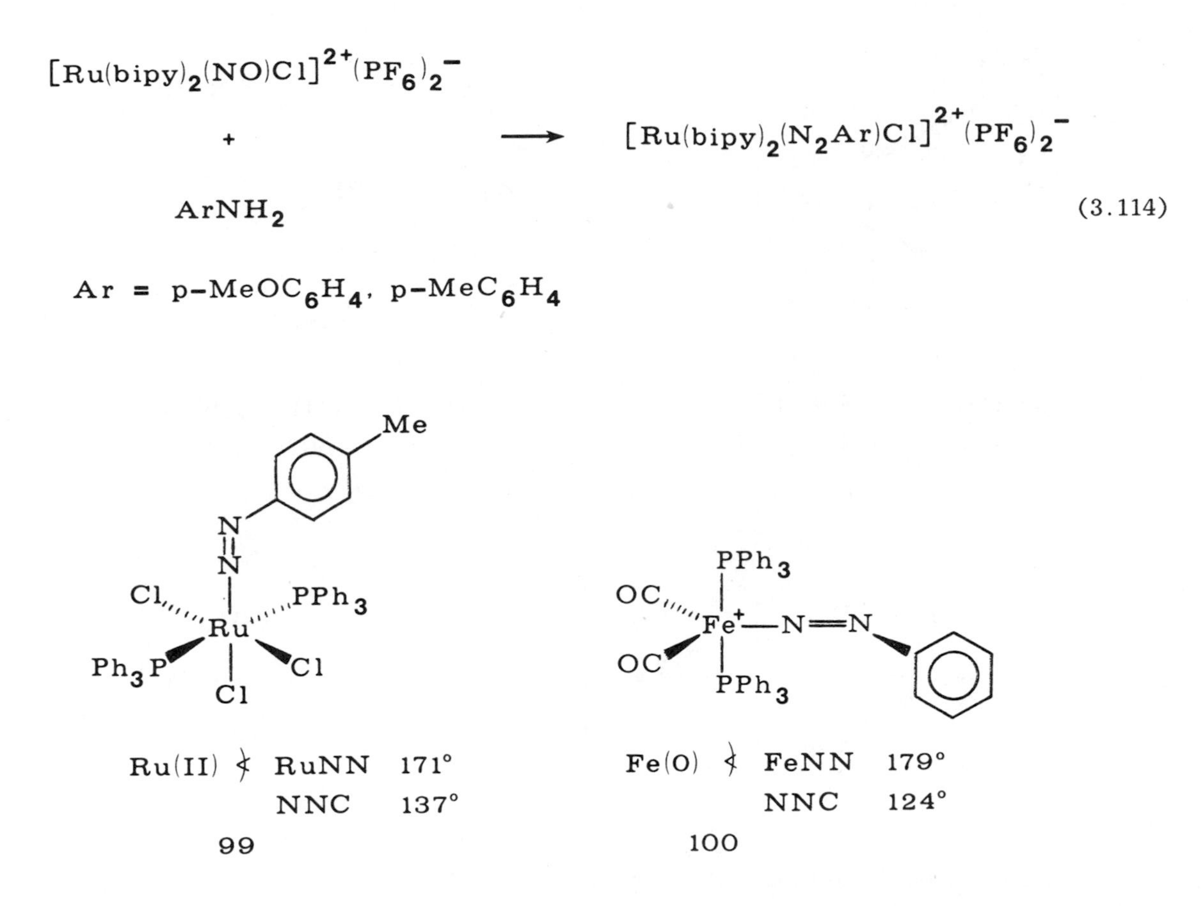


The "doubly bent" diazenido group is analogous to "NO(-)" and therefore might be considered as "$RN_2$(-)." The structurally characterized iridium and platinum complexes in Figure 3.22 are examples of this type. Each of these compounds exhibits a long metal-chlorine bond trans to the doubly bent $ArN_2$ ligand, indicating that this form has a trans bond weakening influence. Those doubly bent diazenido complexes which have been structurally characterized have trans azo structures [258,259].

The use of νN=N infrared frequencies to tentatively assign structure is complicated by several factors, but can be clarified by isotopic substitution and mathematical decoupling procedures [260].

Diazenido complexes can be protonated to form hydrazido (RN=NH-) compounds such as 101, which may be hydrogenated to hydrazino ($RNHNH_2$-) derivatives such as 102 (Equation 3.115).

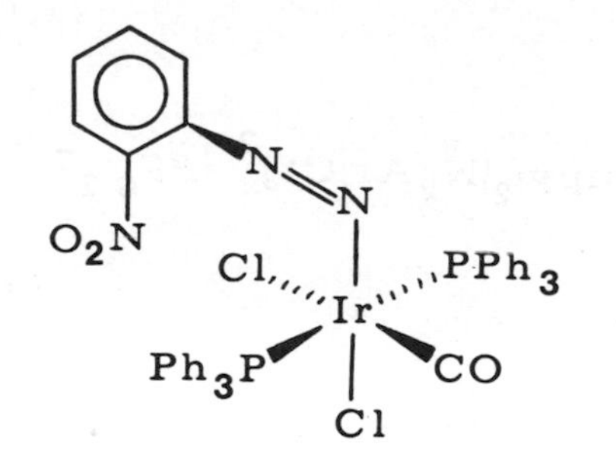


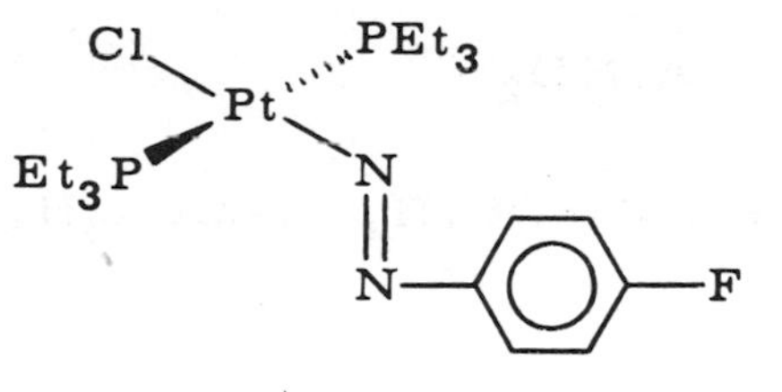


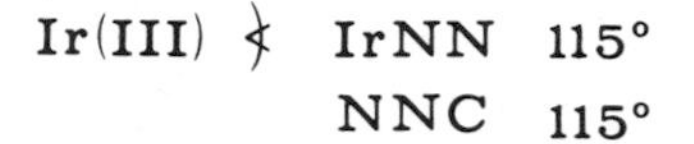


Pt(II) ∢ PtNN 118°
NNC 117°

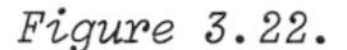

*Figure 3.22.*

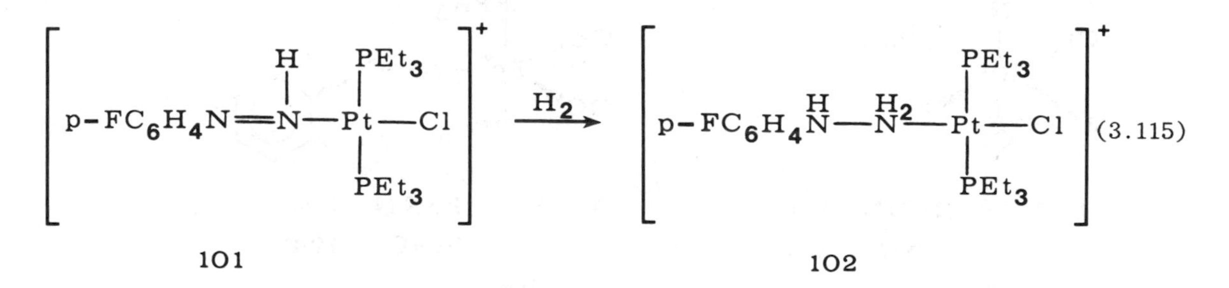


g. <u>Diazoalkane Complexes</u>. Many metal salts including transition-metal complexes catalyze reactions between the very reactive diazoalkanes, $R_2C{=}N{=}N$, and olefins forming cyclopropanes [261,262]. In a few instances stable transition-metal complexes of diazoalkanes have been isolated and structurally characterized. However, this chemistry is still in its infancy. Four structural types have been demonstrated thus far, <u>103</u>, <u>104</u>, <u>105</u>, and <u>106</u>:

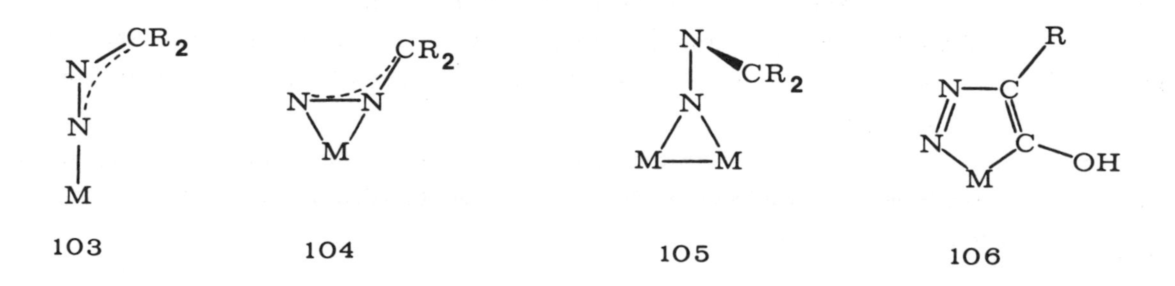

Metal catalyzed cyclopropanation reactions are often proposed to go via another very reactive complex such as 107, which may lose $N_2$, forming an intermediate carbene complex:

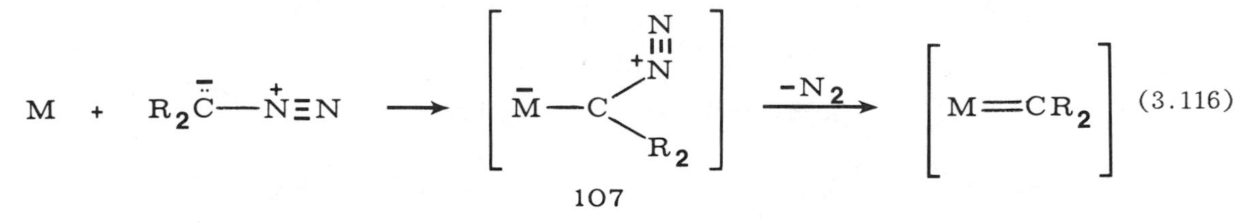


Structures of the η¹-singly bent and the η²-square-planar complexes 108 and 109 have been determined by Ibers [263].

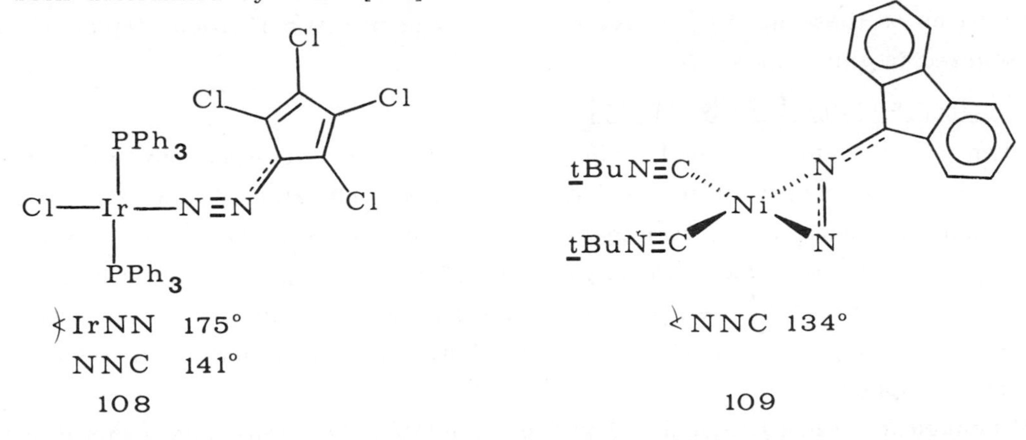


Both doubly and triply bridging compounds of type 105 have been described [264].

h. Nitrile Complexes. A wide variety of metal-nitrile complexes are known [265]. This ligand is quite labile and serves as a useful precursor in the synthesis of other, more stable complexes. The vast majority of nitrile complexes are collinear, although a few are sidebound (dihapto). The νCN band may be shifted to either higher or lower frequencies upon end-on coordination (Figure 3.23).

i. Imines and Nitrides. "Imido" and "nitrido" complexes having double and triple metal-nitrogen bonds are known, as illustrated by 110 [266] and 111 [267]:

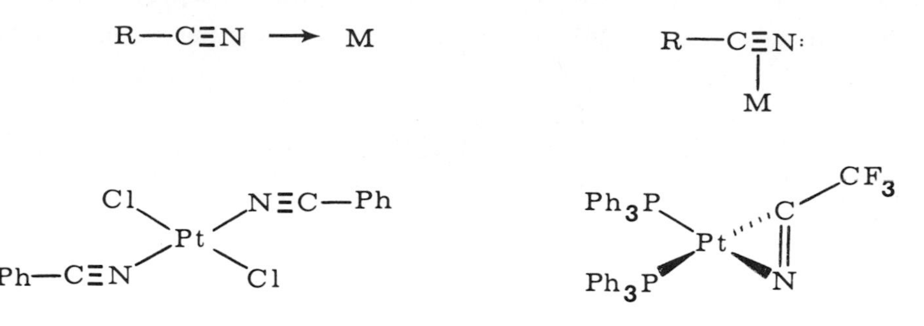


*Figure 3.23.*

These nitrene and nitride ligands are formally analogous to carbenes and carbynes. The reactions of these ligands involve oxidations and nitrogen fixation, topics which are not discussed in this text.

## 3.8. Dioxygen and Its Relatives.

When bound to a metal, the $O_2$ molecule is referred to as dioxygen. Transition-metal dioxygen complexes have been intensively studied [268-272]. Interest in dioxygen complexes stems from their possible role as intermediates in catalytic oxygenation. In fact, most biological oxygenations involve oxygen-binding iron or copper enzymes. There has also been substantial interest in preparing synthetic analogues of the oxygen-carrying metalloproteins: hemoglobin, myoglobin, hemerythrin (iron), and hemocyanin (copper).

Mononuclear transition-metal dioxygen complexes fall into two general classes, bidentate "peroxide" and unidentate "superoxide" compounds. The O-O separation and the $\nu O_2$ IR bands in the $\eta^2$ and $\eta^1$-$O_2$ complexes are sufficiently different from those of the free ligand (1.21 Å and 1556 $cm^{-1}$ (R), respectively) that electron transfer is assumed to have occurred from the metal to the dioxygen ligand. Thus for the assignment of formal oxidation states, we will consider $\eta^2$ and $\eta^1$-$O_2$ complexes as $O_2^{2-}$ and $O_2^-$, four- and two-electron donors, respectively.

a. Dihapto Dioxygen Complexes. The structural, physical, and chemical properties of the $\eta^2$-$O_2$ compounds are compatible with assignment of dioxygen as a four-electron peroxide donor. These bidentate peroxide complexes have isosceles triangular structures with O-O bond lengths in the range 1.46 ± 0.04 Å; comparable to 1.49 Å in hydrogen peroxide. Contrary to earlier reports, there is no significant difference in the O-O distance among complexes which are either reversibly or irreversibly formed by "oxidative-addition" of oxygen to a low-valent transition metal [273]. These complexes are characterized by an IR band near 860 $cm^{-1}$, although this is not purely $\nu O_2$ in character. For comparison, $\nu O_2$ in peroxides is found near 778 $cm^{-1}$. In certain cases acid hydrolysis of such $\eta^2$-$O_2$ complexes affords $H_2O_2$.

In a reaction which seems to be characteristic of the bidentate dioxygen ligand, treatment of $\eta^2$-$O_2$ complexes with $SO_2$ affords chelated sulfate compounds. This is illustrated for the dioxygen adduct, 112, of Vaska's iridium(I) compound [274]:

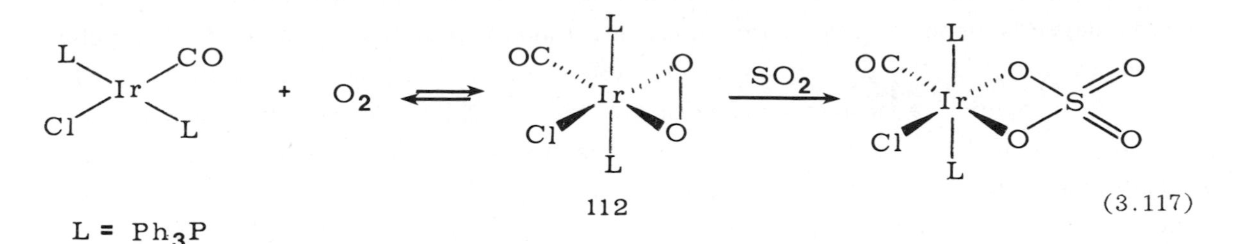


(3.117)

$L = Ph_3P$

Complexes of the peroxide, $\eta^2$-dioxygen class have been prepared by reaction of oxygen with many transition-metal compounds: Fe, Ru, Rh, Ir, Pd, and Pt. Similar complexes may be prepared from hydrogen peroxide and a nonoxidizable metal, for example, the structurally characterized titanium(IV) porphyrinato peroxide complex, 113, in Figure 3.24.

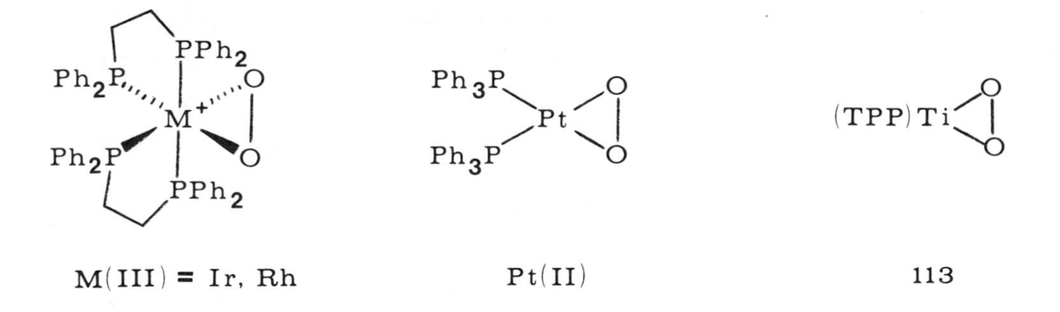


b. Monohapto Dioxygen Complexes. The $\eta^1$-dioxygen complexes have the $O_2$ ligand bound to the metal in an angular manner. The structural characteristics and physical properties of these derivatives are consistent with their formulation as $\eta^1$-superoxide complexes; however, the strengths of and degree of covalency in the metal-$\eta^1$-dioxygen bond are highly variable. Typically such $\eta^1$-dioxygen complexes are formed from metals having a single available coordination site and a favorable one-electron oxidation potential. The appropriate circumstances for forming $\eta^1$-dioxygen are typically encountered with macrocyclic Schiff-base or porphyrin complexes of Cr(II), Fe(II), Ru(II), and Co(II) having a single axial ligand. This is exactly the situation in deoxyhemoglobin (Hb) and myoglobin (Mb), each of which has a five-coordinated, high-spin (S = 2) iron(II) bound to a porphyrinato group, and a single axial imidazole provided by the "proximial histidine" group from the polypeptide globin. Since $\eta^1$-dioxygen complexes are highly reactive and are usually degraded by acids or by mild reducing agents such as another oxidizable metal center, the isolation of such compounds depends upon strategies for minimizing their degradation reactions. The globin protein in Hb and Mb protect dioxygen from further reaction with another iron(II) center by providing a protective enclosure (Figure 3.25). Simple synthetic analogues of oxymyoglobin have been prepared by constructing a similar protective enclosure around one side of the porphyrin ring. Several such metastable iron porphyrin $\eta^1$-dioxygen complexes have been detected in solution. Crystalline iron $\eta^1$-$O_2$ derivatives of the "picket fence" porphyrins have been isolated and characterized by X-ray crystallography [275] (Figure 3.26).

Structural studies of the "picket fence" dioxygen complexes and oxymyoglobin reveal the angular Fe-O$^{\nearrow O}$ structure, but these X-ray structures are not sufficiently accurate to measure the precise O-O separation and the Fe-O-O angle. The most accurate crystallographic data on $\eta^1$-$O_2$ complexes are derived from cobalt macrocyclic $\eta^1$-dioxygen complexes [276]. These structures show a Co-O-O angle of 120° and O-O distances centered around 1.28 ± 0.02 Å, which is close to the value found for free superoxide ion (1.30 Å). All of these $\eta^1$-dioxygen compounds including $HbO_2$ and $MbO_2$ exhibit a $\nu O_2$ stretching frequency in the range 1105 to 1160 $cm^{-1}$ which is similar to that of free superoxide (1145 $cm^{-1}$). It is interesting that this $\nu O_2$ band in $\eta^1$-$O_2$ complexes is insensitive to the nature of the metal, to the trans axial ligand, and to the oxygen binding equilibrium constant.

The magnetic properties of certain $\eta^1$-dioxygen complexes are consistent with the presence of an additional $\pi$-bond between the metal and dioxygen. For example, oxyhemoglobin and the picket-fence iron-dioxygen complexes have a diamagnetic ground state; whereas the chromium-dioxygen complex has two unpaired electrons and the

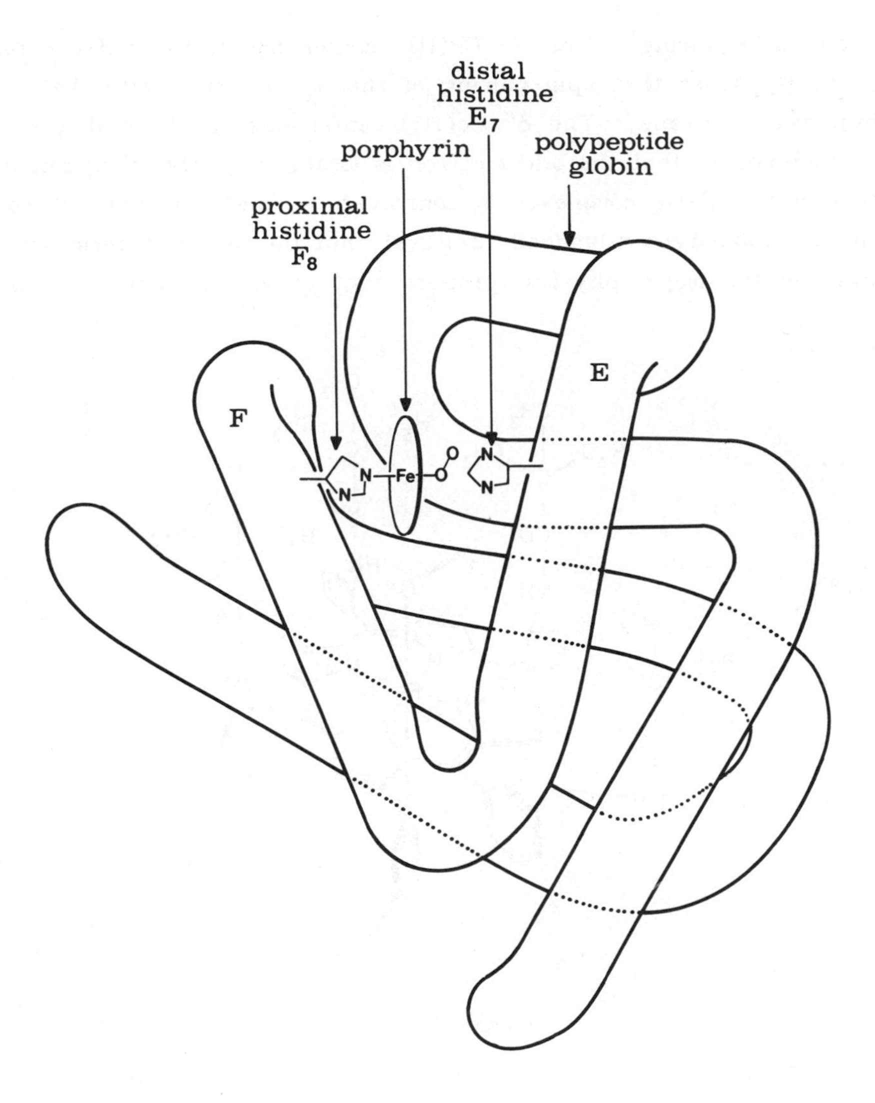


*Figure 3.25. Schematic view of oxymyoglobin, $MbO_2$, showing positioning of the heme in the globin and the proximal and distal histidines.*

cobalt-dioxygen derivatives have a single unpaired electron which ESR studies show to be located primarily on the dioxygen ligand. These results are consistent with an additional bonding component between a $d_{xz}$ orbital on the metal and a singly occupied $\pi$-level in the $\eta^1$-dioxygen ligand [277]. For example, in low-spin Fe(III), the $d_{xz}$ orbital contains a single unpaired electron which couples with the electron on superoxide, giving complete spin-pairing. The $d^3$ Cr(III) center has three half-occupied d levels ($d_{xz}$, $d_{yz}$, and $d_{xy}$), so that spin-pairing of the $d_{xz}$ electron with that on $O_2^-$ would leave two unpaired electrons. The $d^6$ Co(III) center has all three d levels (the $t_{2g}$ d set) doubly occupied, so that the odd electron is localized on the dioxygen ligand. The bonding situation for $\eta^1$-$O_2$ complexes is controversial, and a number of more sophisticated descriptions have been advanced [278,279], but the simple scheme shown in Figure 3.27 accounts for the major physical properties of these compounds. Consistent with

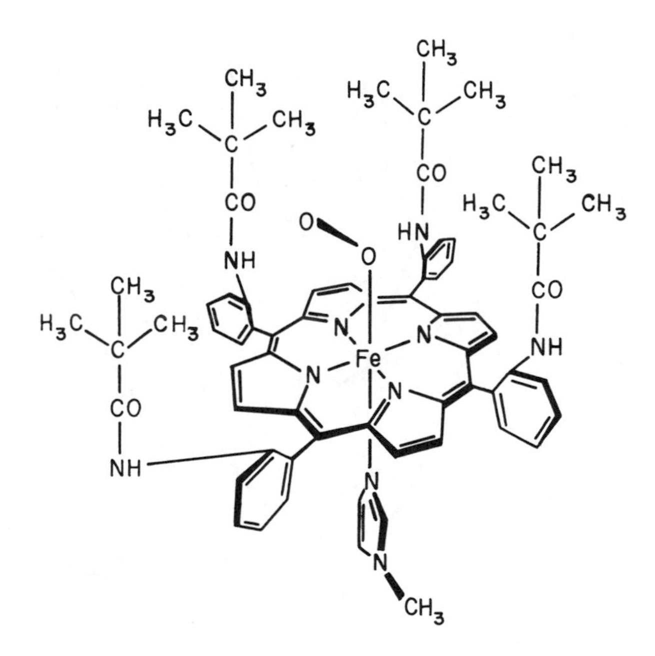

*Figure 3.26.*

this notion of an additional bonding component in the iron-$\eta^1$-dioxygen complexes is a somewhat shortened Fe-O bond length and a higher νFe-O frequency (563 $cm^{-1}$) than would be expected for a simple σ-bond. It is also true that among these porphyrin complexes the oxygen-binding affinity of Fe(II) is greater than that of Co(II). Such affinities can also be correlated with the tendency of the divalent metal to be oxidized, thus Cr(II) > Fe(II) > Co(II). Among $\eta^1$-$O_2$ complexes derived from cobalt(II) macrocyclic complexes, Basolo has shown that a relationship between oxygen binding and the nature of the <u>trans</u> axial base can be similarly related to their redox potentials [280]. It should also be noted that $\eta^1$-$O_2$ porphyrin complexes with closed d-shell ions are known, such as Zn(II), which exhibit $\nu O_2$ bands in the superoxide range. For Zn(TPP)($O_2$), the only reasonable oxidation-state formalism is that involving superoxide ion [281].

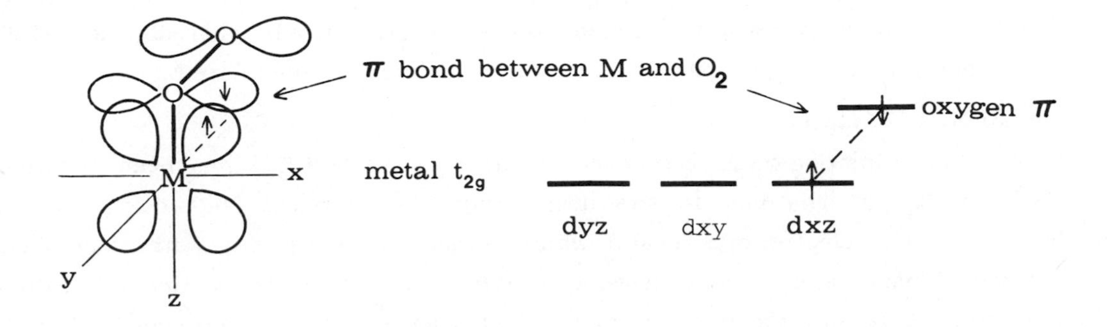


Fe(III) dyz dxy filled - no unparied e

Cr(III) dyz dxy half filled - two unpaired electrons

Co(III) dyz dxy dxz filled - one unparied electron on $O_2^-$

*Figure 3.27. Electron pairing in $\eta^1$-$O_2$ complexes.*

c. Disulfur and Diselenium. Elemental sulfur and selenium form complexes which are analogous to the $\eta^2$-$O_2$, "peroxide," compounds; however, $\eta^1$-$S_2$ complexes are presently unknown [282,283]:

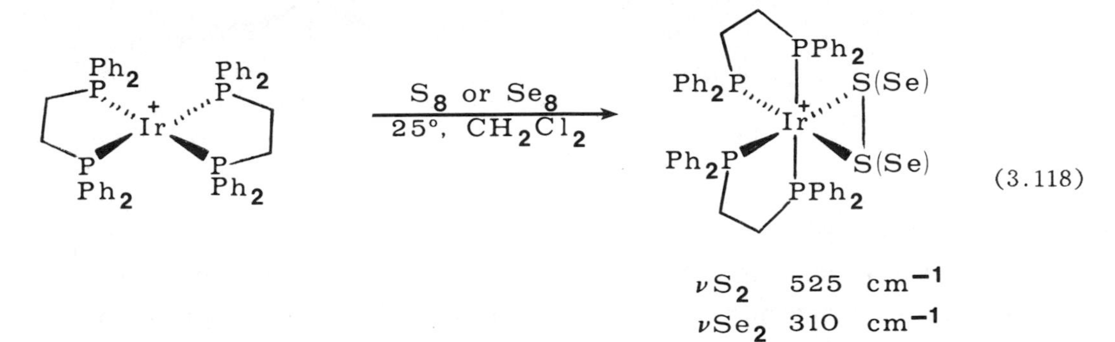

The S-S bond lengths in these $\eta^2$-$S_2$ complexes appear to be highly variable; whereas that in $[Ir(S_2)(dppe)_2]^+$ (2.066 Å) [284] is slightly longer than that of a disulfide (~2.03 Å), that in $[NbCl(Cp)_2(S_2)]$ (1.73 Å) represents the shortest S-S bond yet described [285].

Coordinated disulfur in $Os(S_2)(CO)_2(PPh_3)_2$ has been methylated, giving $[Os(\eta^2\text{-}S_2Me)(CO)_2(PPh_3)_2]SO_3CF_3$, which has been structurally characterized [286]. However, other reactions of coordinated disulfur are thus far unexplored.

## 3.9. Carbon Dioxide.

Interest in transition-metal complexes of carbon dioxide [287] has been stimulated by a search for alternatives to the diminishing petrochemical feedstocks. Carbon dioxide is a weak, electrophilic ligand which forms complexes only with very basic, coordinatively unsaturated transition-metal centers. An example is the 1:1 complex formed between $CO_2$ and the very basic $Na^+[Co(I)salen]^-$ [288]. Such coordinated $CO_2$ can react further with the metal-carbon bonds by an apparent "insertion process." We can expect the chemistry of $CO_2$ complexes to develop rapidly in the next few years.

Several $CO_2$ transition-metal adducts have been reported, and thus far a few have been crystallographically characterized. One such is an $\eta^2$-$CO_2$ Ni(II) complex, 114 [289]. Another contains two entangled $CO_2$ groups, one having an iridium(III) carbon bond, 115 [287].

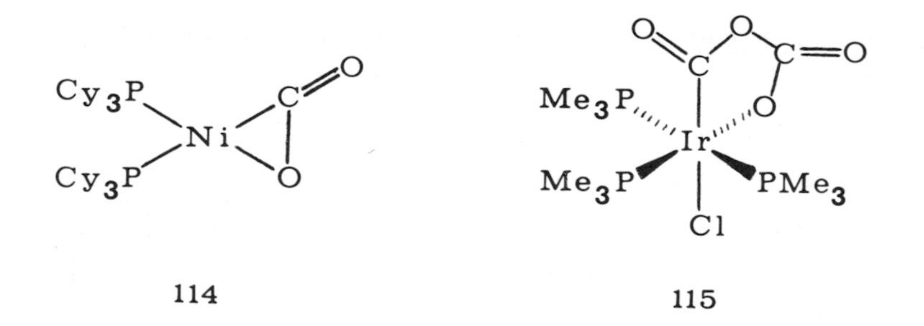


## 3.10. Sulfur Dioxide.

Low valent transition metals form many complexes which contain one or two $SO_2$ ligands. These compounds are of interest as possible intermediates in the removal of $SO_2$ from combustion exhausts, catalytic insertion of $SO_2$ into unsaturated organic substrates, and removal of $SO_2$ from organic sulfonyl compounds ("desulfonation") [290]. The structure, bonding, and reactions of $SO_2$ complexes have been reviewed

[291]. Three distinct bonding modes, $\eta^1$-planar (116), $\eta^1$-pyramidal (117), and $\eta^2$ (118), each having a metal-S bond, have been characterized for terminal $SO_2$ complexes [292,293]. Bridging $SO_2$ is also known in polynuclear compounds [294].

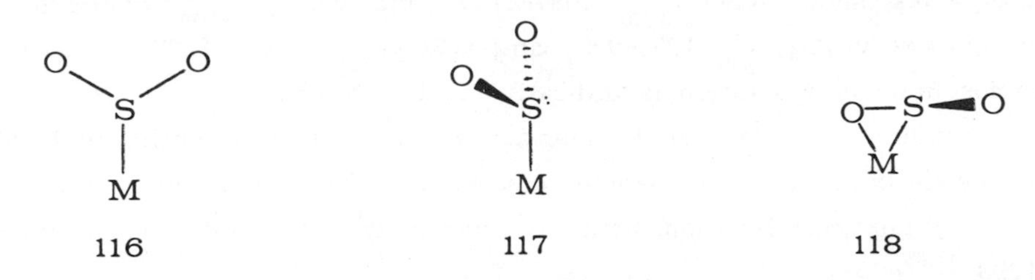


The bonding in S-bonded, $\eta^1$-$SO_2$ complexes is similar to that of the nitrosyls in that two structurally distinct, valence-bond tautomers are known, the $\eta^1$-planar and $\eta^1$-pyramidal forms. Sulfur dioxide has both filled sulfur $\sigma$-donor and empty, delocalized $\pi$-acceptor orbitals. Thus $SO_2$ is a strong $\pi$-acid, and in the extreme a metal-electron pair may be substantially transferred to a sulfur-based orbital, yielding the $\eta^1$-pyramidal form. The structurally characterized pyramidal iridium complex, 119 [295], and the planar ruthenium complexes, 120 [296], are illustrative.

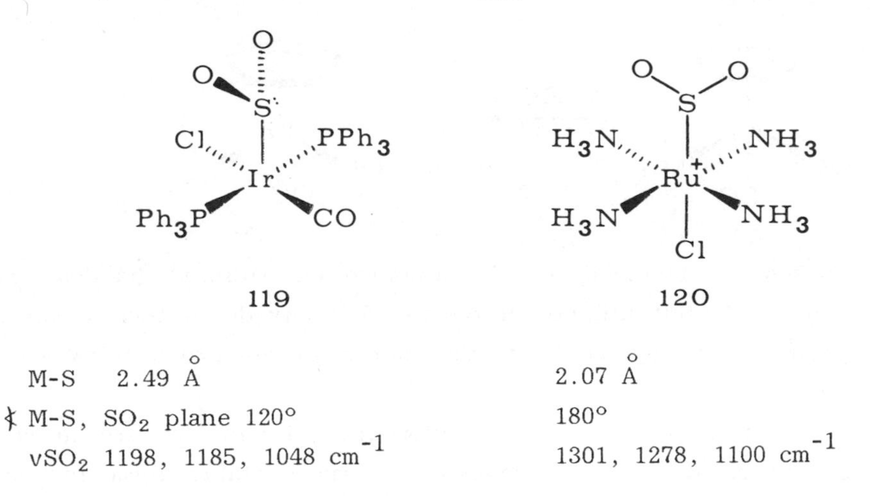


Note that the metal-sulfur bond in the pyramidal form tends to be longer (by ~0.3 Å) than in the planar form, which is reminiscent of the situation with the bent and linear nitrosyls. The angle between the $SO_2$ plane and the metal-sulfur axis is close to 120° in the $\eta^1$-pyramidal complexes and 180° (coplanar) in the $\eta^1$-planar complexes.

The infrared S-O stretching modes are characteristic properties of $SO_2$ complexes, and are reliable diagnostics of the two bonding modes [297]. The free ligand exhibits symmetric and antisymmetric stretching modes at 1151 and 1362 $cm^{-1}$; whereas most $\eta^1$-planar complexes show $\nu(SO_2)_{sym}$ 1087-1130, and $\nu(SO_2)_{asym}$ 1250-1303; and $\eta^1$-pyramidal complexes $\nu(SO_2)_{sym}$ 1065-990, and $\nu(SO_2)_{asym}$ 1150-1237. The lowering of $SO_2$ frequencies upon complexation is indicative of backbonding.

Sulfur dioxide is an unusual ligand, in that it has both nucleophilic and electrophilic Lewis-base and Lewis-acid character. The acid character of $SO_2$ is manifest by its facile complexation with coordinatively saturated basic transition-metal complexes [298]:

$$Ru(CO)_3(PPh_3)_2 + SO_2 \longrightarrow Ru(CO)_2(SO_2)(PPh_3)_2 \qquad (3.119)$$

It appears that $SO_2$ enters the coordination sphere as an electrophile, binding to a nonbonding d-electron pair.

Under very mild conditions, $SO_2$ directly inserts into several transition-metal alkyl complexes affording S-bonding alkyl-sulfinates:

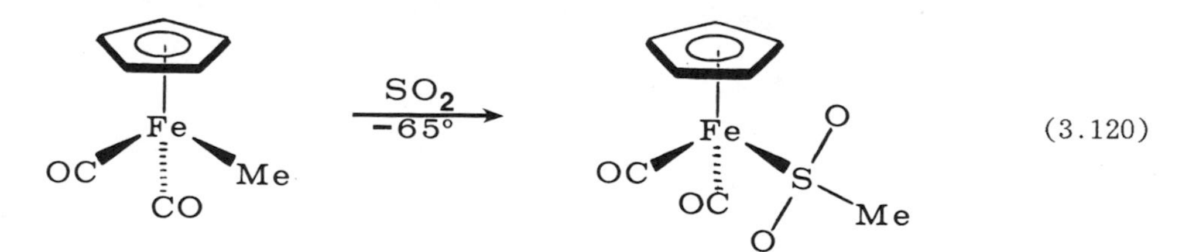


(3.120)

This reaction, which is involved in desulfonation of sulfonyl halides, has been extensively studied [299], but will not be discussed in any detail here. This insertion is highly sterospecific, occuring with retention of configuration at a chiral metal center but <u>inversion</u> at chiral carbon.

Both S- and O-bonded metal sulfinate tautomers, <u>121</u> and <u>122</u>, are known and are probably at equilibrium with the S-bonded the more stable form for low-valent transition-metal complexes:

$$\underset{121}{M\!-\!S(O)(O)\!-\!R} \rightleftharpoons \underset{122}{M\!-\!O\!-\!S(=O)\!-\!R} \qquad (3.121)$$

## NOTES.


1. F. A. Cotton and G. Wilkinson, Advanced Inorganic Chemistry (Interscience, 3d ed., 1972), pp. 620-650.
2. J. M. Lehn, Accts. Chem. Res., 11, 49 (1978), and references therein.
3. J. P. Fackler, Progr. Inorg. Chem., 7, 361 (1966).
4. E. L. Muetterties and C. M. Wright, Quart. Rev., 21, 109 (1967).
5. J. P. Collman, Angew. Chem., 77, 154 (1965).
6. M. H. Chisholm and F. A. Cotton, Accts. Chem. Res., 11, 356 (1977).
7. C. A. Tolman, Chem. Rev., 77, 313 (1977).
8. C. A. McAuliffe, ed., Transition-Metal Complexes of Phosphorus, Arsenic, and Antimony Ligands, (Wiley, 1973), Parts 1 and 2 and references therein.
9. W. Levason and C. A. McAuliffe, Phosphine, Arsine, and Stabine Complexes of Transition Elements, (Elsevier, 1978), in press.
10. W. Levason and C. A. McAuliffe, Accts. Chem. Res., 11, 363 (1978).
11. W. Levason and C. A. McAuliffe, Adv. Inorg. Chem. Radiochem., 14, 173 (1972).
12. G. Booth, Adv. Inorg. Chem. Radiochem., 6, 1 (1964).
13. C. A. McAuliffe, ed., Transition-Metal Complexes of Phosphorus, Aresenic, and Antimony Ligands, (Wiley, 1973), parts 3, 4, and 5.
14. T. G. Appleton, H. C. Clark, and L. E. Manzer, Coord. Chem. Rev., 10, 335 (1973).
15. D. A. Allison, J. Clardy, and J. G. Verkade, Inorg. Chem., 11, 2804 (1972).
16. B. Chiswell, in C. A. McAuliffe, ed., Transition-Metal Complexes of Phosphorus, Arsenic, and Antimony Ligands, (Wiley, 1973), pp. 271-310.
17. R. Colton and I. B. Tomkins, Aust. J. Chem., 19, 1143 (1967).
18. I. Macleod, L. Manojlovic-Muir, D. Millington, K. W. Muir, D. W. A. Sharp, and R. Walter, J. Organometal. Chem., 97, 67 (1975).

19a. P. B. Hitchcock, M. McPartlin, and R. Mason, Chem. Comm., 1367 (1969).

19b. M. J. Bennett and P. B. Donaldson, Inorg. Chem., 16, 655 (1977).

20. T. Kruk, Angew. Chem., Int. Ed. Engl., 6, 53 (1967).
21. J. F. Nixon, Adv. Inorg. Chem. Radiochem., 13, 363 (1970).
22. E. A. V. Ebsworth, H. M. Ferrier, B. J. L. Henner, D. W. H. Ranking, J. F. S. Reed, H. E. Robertson, and J. D. Whiteloch, Angew. Chem., Int. Ed. Engl., 16, 482 (1977).
23. H. Schmidbaur and J. R. Mandl, Angew. Chem., Int. Ed. Engl., 16, 646 (1977), and references therein.

24. R. G. Montemayor, D. T. Sauer, Sr., S. Fleming, D. W. Bennett, M. G. Thomas, and R. W. Parry, J. Amer. Chem. Soc., 100, 2231 (1978).
25. E. L. Muetterties, ed., Transition-Metal Hydrides (Marcel Dekker, 1971).
26. D. M. Roundhill, Adv. Organometallic Chem., 12, 273 (1975).
27. R. Bau, ed., Transition-Metal Hydrides, Adv. in Chem. Series, no. 167, (Am. Chem. Soc., 1978).
28. J. P. McCue, Coord. Chem. Rev., 10, 265 (1973).
29. G. L. Geoffrey and J. R. Lehman, Adv. Inorg. Chem. and Radiochem., 29, 189 (1977).
30. A. P. Ginsberg, Transition-Metal Chem., 1, 111 (1965).
31. M. L. H. Green and D. J. Jones, Adv. Inorg. Chem. and Radiochem., 7, 115 (1965).
32. B. A. Frenz and J. A. Ibers in [25].
33. J. A. Ibers, in [27].
34. R. Bau, "Neutron Diffraction Studies on Transition-Metal Hydride Complexes," in [27].
35. R. Bau, R. G. Teller, S. W. Kirtley, and T. F. Koetzle, Accts. Chem. Res., 12, 176 (1979).
36. S. J. LaPlaca, W. C. Hamilton, J. A. Ibers, and A. Davison, Inorg. Chem., 8, 1928 (1969).
37. S. J. LaPlaca and J. A. Ibers, Acta. Crystallogr., 18, 511 (1965).
38. R. W. Baker and P. Pauling, Chem. Comm., 1495 (1969).
39. E. K. Barefield, G. W. Parshall, and F. N. Tebbe, J. Amer. Chem. Soc., 92, 5234 (1970).
40. J. P. Jesson, in [25], p. 75.
41. P. Meakin, L. J. Guggenberger, E. L. Muetterties, and J. P. Jesson, J. Amer. Chem. Soc., 95, 1467 (1973).
42. R. A. Schunn, in [25], pp. 203-269.
43. W. Hieber, H. Schulten, and B. Marin, Z. Anorg. Allg. Chem., 240, 261 (1939).
44. J. R. Moss and W. A. G. Graham, J. Organometal. Chem., 23, C47 (1970).
45. M. Tagupsky, C. K. Brown, G. Yagupsky, and G. Wilkinson, J. Chem. Soc. A, 937 (1970).
46. R. A. Schunn, Inorg. Syntheses, to be published.
47. M. A. Bennett and D. L. Milner, J. Amer. Chem. Soc., 91, 6983 (1969).
48. F. L'Eplattenier and F. Calderazzo, Inorg. Chem., 7, 1290 (1968).

49. R. R. Schrock and J. A. Osborn, J. Amer. Chem. Soc., 98, 2134 (1976).
50. A. Sacco, M. Rossi, and C. F. Nobile, Chem. Comm., 589 (1966).
51. W. Beck, W. Hieber, and G. Braun, Z. Anorg. Allg. Chem., 308, 23 (1961).
52. P. C. Wailes and H. Weigold, J. Organometal. Chem., 24, 405 (1970).
53. J. Chatt and B. L. Shaw, J. Chem. Soc., 5075 (1962).
54. R. W. Schunn, in [25], p. 239, and references therein.
55. A. Davison and D. L. Reger, J. Organometal Chem., 23, 491 (1970).
56. J. R. Norton, Accts. Chem. Res., 12, 139 (1979).
57. P. M. Treichel, J. H. Morris, and F. G. A. Stone, J. Chem. Soc., 720 (1963).
58. G. W. Parshall and J. J. Mrowca, Advan. Organometal. Chem., 7, 157 (1968).
59. C. A. Tolman, J. Amer. Chem. Soc., 92, 6777 (1970).
60. C. A. Tolman, J. Amer. Chem. Soc., 92, 6785 (1970).
61. K. C. Dewhirst, Inorg. Chem., 5, 319 (1966).
62. W. H. Baddley and M. S. Fraser, J. Amer. Chem. Soc, 91, 3661 (1969).
63. N. A. D. Carey and J. G. Noltes, Chem. Comm., 1471 (1968).
64. G. Wilkinson, Science, 185, 109 (1974).
65. R. R. Schrock and G. W. Parshall, Chem. Rev., 76, 243 (1976).
66. P. J. Davidson, M. F. Lappert, and R. Pearce, Accts. Chem. Res., 7, 209 (1974); Chem. Rev., 76, 214 (1976).
67. H. Holton, M. F. Lappert, G. R. Scollary, D. G. H. Ballard, R. Pearce, J. L. Atwood, and W. E. Hunter, JCS Chem. Comm., 425 (1976).
68. R. A. Andersen, E. Carmore-Guzumbra, J. F. Gibson, and G. Wilkinson, JCS, Dalton, 2204 (1976).
69. B. K. Bower and H. G. Tennent, J. Amer. Chem. Soc., 94, 2512 (1972).
70. J. Drausse and G. Marx, J. Organometallic Chem., 65, 215 (1974).
71. B. E. Mann, Adv. Organomet. Chem., 12, 135 (1974).
72. W. G. Daub, in S. J. Lippard, ed., Progress in Inorganic Chemistry, 22, 375 (1976).
73. W. Keim, J. Organometal. Chem., 14, 179 (1968).
74. H. H. Zeiss and R. P. A. Sneeden, Angew. Chem., Int. Ed. Engl., 6, 435 (1967).
75. V. B. Sarry and P. Velling, Z. Anorg. Allg. Chem., 426, 107 (1976).
76. R. R. Schrock and G. W. Parshall, Chem. Rev., 76, 243 (1976).
77. M. L. H. Green, Organometallic Compounds, Vol. II: The Transition Elements (Methuen, 1968), p. 225.

78. H. H. Zeiss and R. P. A. Sneeden, Angew. Chem., Int. Ed. Engl., 6, 435 (1967).
79. R. J. Cross, Organomet. Chem. Rev., 2, 97 (1962).
80. R. F. Heck, J. Amer. Chem. Soc., 90, 5518 (1968).
81. C. D. M. Beverwijk, G. J. M. Van Der Kerk, A. J. Leusink, and J. G. Noltes, Organometallic Chem. Rev., A, 5, 215 (1970).
82. R. B. King, in Organotransition-Metal Chemistry, Y. Ishii and M. Tsutisui, eds., (Plenum, 1975), p. 36.
83. K. J. Klabunde and J. Y. F. Low, J. Amer. Chem. Soc., 96, 7674 (1974).
84. M. L. H. Green, Organometallic Compounds, (Methuen, 1968), vol. II, p. 200.
85. G. Calvin and G. E. Coates, J. Chem. Soc., 2008 (1960).
86. B. L. Shaw and N. I. Tucker, "Organo-Transition Metal Compounds and Related Aspects of Homogeneous Catalysis," Pergamon Texts in Inorganic Chemistry, vol. 23, p. 810 (1973).
87. R. Rigo and A. Turco, Coord. Chem. Rev., 13, 133 (1974).
88. J. Halpern and R. Cozens, Coord. Chem. Rev., 16, 141 (1975).
89. N. Grice, S. C. Kao, and R. Pettit, J. Amer. Chem. Soc., 101, 1692 (1979), and references therein.
90. M. L. H. Green and C. R. Hurley, J. Organometal. Chem., 10, 188 (1967).
91. J. W. Lauher, J. Amer. Chem. Soc., 100, 5305 (1978).
92. R. Colton, M. J. McCormick, and C. D. Pannan, JCS Chem. Comm., 823 (1977).
93. J. E. Ellis, C. P. Parnell, and G. P. Hagen, J. Amer. Chem. Soc., 100, 3605 (1978).
94. J. P. Collman, Accts. Chem. Res., 8, 342 (1975).
95. R. Colton, C. J. Commons, and B. E. Hoskins, JCS Chem. Comm., 363 (1975).
96. F. A. Cotton, B. A. Frenz, and L. Kruczynski, J. Amer. Chem. Soc., 95, 951 (1973).
97. D. M. Adams, Metal Ligand and Related Vibrations, (St. Martin's Press, 1978).
98. K. F. Purcell and J. C. Kotz, Inorganic Chemistry (Saunders, 1977), pp. 899-905.
99. F. A. Cotton, Chemical Applications of Group Theory (Interscience, 2d ed., 1971).
100. W. L. Jolly, The Synthesis and Characterization of Inorganic Compounds (Prentice-Hall, 1970), p. 472.
101. M. C. Baird and G. Wilkinson, Chem. Comm., 267 (1966).

102. M. C. Baird, G. Hartwell, Jr., and G. Wilkinson, J. Chem. Soc. (A), 2037 (1967).

103. G. R. Clark, K. Marsden, W. R. Roper, and L. J. Wright, J. Amer. Chem. Soc., 102, 1206 (1980).

104. I. S. Butler and A. E. Fenster, J. Organometal. Chem., 66, 161 (1974).

105. I. S. Butler, D. Cozale, S. R. Stobart, JCS Chem. Comm., 103 (1975).

106. L. Busetto, U. Belluco, and R. J. Angelici, J. Organomet. Chem., 18, 213 (1969).

107. D. Mansuy, J. P. Battioni, and J. C. Chottard, J. Amer. Chem. Soc., 100, 4312 (1978).

108. M. A. Andrews, Inorg. Chem., 16, 496 (1977).

109. G. R. Clark, K. R. Grundy, R. O. Harris, S. M. James and W. R. Roper, J. Organometal. Chem., 90, C37 (1975).

110. E. Klumpp, G. Bor and L. Marko, J. Organometal. Chem., 11, 207 (1968).

111. J. S. Field and P. J. Wheatley, JCS Dalton, 2269 (1972).

112. J. W. Dunkler, J. S. Finer, J. Clardy, and R. J. Angelica, J. Organometal. Chem., 114, C49 (1976).

113. B. D. Dombek and R. J. Angelica, J. Amer. Chem. Soc., 96, 7568 (1974).

114. T. J. Collins and W. R. Roper, JCS Chem. Comm., 901 (1977).

115. G. R. Clark, T. J. Collins, D. Hall, S. M. James, and W. R. Roper, J. Organometal. Chem., 141, C5 (1977).

116. E. O. Fischer, Adv. in Organometal. Chem., 14, 1 (1976).

117. F. A. Cotton and C. M. Lukehart, Prog. Inorg. Chem., 16, 487 (1972).

118. D. J. Cardin, B. Cetinkaya, M. J. Doyle, and M. F. Lappert, Chem. Rev., 72, 545 (1972).

119. D. J. Cardin, B. Cetinkaya, M. J. Doyle, and M. F. Lappert, Chem. Soc. Rev., 2, 99 (1973); F. J. Brown, Prog. Inorg. Chem., 27, 1 (1980).

120. M. H. Chisholm and S. Godleski, Prog. Inorg. Chem., 20, 299 (1976).

121. C. P. Casey and R. L. Anderson, J. Amer. Chem. Soc., 96, 1230 (1974).

122. D. J. Cardin, B. Cetinkaya, M. F. Lappert, Jl Manojlovic-Muir, and K. W. Muir, Chem. Comm., 400 (1971).

123. P. W. Jolly and R. Pettit, J. Amer. Chem. Soc., 88, 5044 (1966).

124. R. E. Riley, C. E. Capshew, R. Pettit, W. R. E. David, Inorg. Chem., 14, 408 (1977).

125. A. Sanders, L. Cohen, W. P. Giering, D. Kenedy, and C. V. Magatti, J. Amer. Chem. Soc., 95, 5430 (1973).
126. K. Ofele, Angew. Chem. Int. Ed. Engl., 7, 950 (1968); J. Organometal. Chem., 22, C9 (1970).
127. W. A. Herman, Angew. Chem. Int. Ed. Engl., 13, 599 (1974).
128. P. Casey, T. J. Burkhart, C. A. Bunnell, and J. C. Calabrese, J. Amer. Chem. Soc., 99, 2127 (1977).
129. R. R. Schrock, Accts. Chem. Res., 12, 98 (1979), and references therein.
130. J. L. Petersen and L. F. Dahl, J. Amer. Chem. Soc., 97, 6416 (1975), and references therein.
131. J. W. Lauher and R. Hoffmann, J. Amer. Chem. Soc., 98, 1729 (1976), and references therein.
132. R. R. Schrock, J. Amer. Chem. Soc., 96, 6796 (1974); 98 5399 (1976).
133. F. N. Tebe, G. W. Parshall, and G. S. Reddy, J. Amer. Chem. Soc., 100, 3611 (1978).
134. S. H. Pine, R. Zahler, D. A. Evans, and R. H. Grubbs, J. Amer. Chem. Soc., 102, 3270 (1980).
135. J. Jutzi and W. Steiner, Angew. Chem., Int. Ed. Engl., 16, 639 (1977).
136. E. O. Fischer and U. Schubert, J. Organometal. Chem., 100, 59 (1975).
137. S. J. McLain, C. D. Wood, L. W. Messerle, R. R. Schrock, F. J. Hollander, W. J. Yongs, and M. R. Churchill, J. Amer. Chem. Soc., 100, 5962 (1978).
138. W. R. Roper, private communication, see reference 103.
139. L. Malatesta, Prog. Inorg. Chem., 1, (1959).
140. P. S. Skell, E. M. VanDam, and M. P. Silvon, J. Amer. Chem. Soc., 96, 627 (1974).
141. F. A. Cotton, J. Organometal. Chem., 100, 29 (1975).
142. M. J. S. Dewar and G. P. Ford, J. Amer. Chem. Soc., 101, 783 (1979), and references therein.
143. S. D. Ittel and J. A. Ibers, Adv. Organometal. Chem., 14, 33 (1976).
144. J. A. J. Jarvis, B. T. Kilbourn, and P. G. Owston, Acta Cryst., B27, 366 (1971).
145. M. Black, R. H. B. Mais, and P. G. Owston, Acta Cryst., B25, 1753 (1969).
146. R. A. Love, T. K. Koetzle, G. J. B. Williams, L. C. Andrews, and R. Ban, Inorg. Chem., 14, 2653 (1975).

147. R. Cramer, J. B. Kline, and J. D. Roberts, J. Amer. Chem. Soc., 91, 2519 (1969).
148. H. Reihlen, A. Gruhl, G. von Hessling, and O. Prengle, Ann., 482, 161 (1930).
149. M. L. H. Green and G. Wilkinson, J. Chem. Soc., 4314 (1958).
150. S. D. Ittel and J. A. Ibers, Adv. Organometal. Chem., 14, 55 (1976).
151. G. R. Davies, W. Hewerston, R. H. B. Mais, P. G. Owston, and C. G. Patel, J. Chem. Soc., A, 1873 (1970).
152. F. A. Cotton and G. Wilkinson, Advanced Inorganic Chemistry (Interscience, 3d ed., 1972), p. 751.
153. S. Otsuka and A. Nakamura, Adv. Organometal. Chem., 14, 246 (1976).
154. G. Robertson and P. O. Whimp, J. Organometal. Chem., 32, C39 (1971).
155. M. A. Bennett, G. B. Robertson, P. O. Whimp, and T. Yoshida, J. Amer. Chem. Soc., 93, 3717 (1971).
156. D. Tate, G. M. Augl, W. M. Ritchey, B. L. Ross, and J. G. Grasselli, J. Amer. Chem. Soc., 86, 3261 (1964).
157. R. M. Laine, R. E. Moriarity, and R. Bau, J. Amer. Chem. Soc., 94, 1402 (1972).
158. R. B. King, Inorg. Chem., 7, 1044 (1968).
159. S. J. McLain, R. R. Schrock, P. R. Scharp, M. R. Churchill, and W. J. Youngs, J. Amer. Chem. Soc., 101, 263 (1979).
160. H. Zeiss, P. J. Wheatley, and J. F. S. Winkler, Benzenoid-Metal Complexes, (Ronald Press, 1966).
161. W. E. Silverthorn, Adv. Organometal. Chem., 13, 47 (1975).
162. E. O. Fischer and W. Hafner, Z. Naturforsch, B10, 655 (1955).
163. W. Strohmeier, G. Popp, and J. F. Guttenberger, Chem. Ber., 102, 3608 (1969).
164. G. Huttner and S. Lange, Acta Crystallogr., Sect. B, 28, 2049 (1972).
165. G. Huttner, S. Lange, and E. O. Fischer, Angew. Chem., Int. Ed. Engl., 10, 556 (1971).
166. M. R. Churchill and R. Mason, Proc. Roy. Sr. Ser. A, 292, 61 (1966).
167. R. E. Davis and R. Pettit, J. Amer. Chem. Soc., 92, 716 (1970).
168. G. Allegra, G. T. Casagrande, A. Immizi, L. Porri, and G. Vitulli, J. Amer. Chem. Soc., 92, 289 (1970).
169. J. M. Davidson and C. Triggs, J. Chem. Soc., A, 1324 (1968).
170. J. Browning, M. Green, B. R. Penforld, J. L. Spencer, and F. G. A. Stone, JCS Chem. Chem. Comm., 31 (1973).

171. J. T. Layorvski, Coord. Chem. Rev., 22, 185 (1977).

172. M. Rosenblum, The Iron-Group Metallocenes: Ferrocene, Ruthenocene, and Osmoscene, (Wiley, 1965).

173. K. F. Purcell and J. C. Kotz, Inorganic Chemistry, (Saunders, 1977), pp. 885-886.

174. J. L. Robbins, N. K. Edelstein, S. R. Copper, and J. C. Smart, J. Amer. Chem. Soc., 101, 3853 (1979), and references therein.

175. J. L. Thomas, J. Amer. Chem. Soc., 95, 1838 (1973), and references therein.

176. A. Davidson and S. S. Wreford, J. Amer. Chem. Soc., 96, 3018 (1974).

177. G. P. Pez, J. Amer. Chem. Soc., 98, 8072 (1976), and references therein.

178. J. W. Lauher and R. Hoffmann, J. Amer. Chem. Sco., 98, 1729 (1976).

179. M. L. H. Green, J. A. McCleverty, L. Pratt, and G. Wilkinson, J. Chem. Soc., 4854 (1964).

180. J. L. Atwood, K. E. Stone, H. G. Alt, D. C. Hrncir, and M. D. Rausch, J. Organometal. Chem., 96, C4 (1975).

181. G. W. Halstead, E. C. Baker, and K. V. Raymond, J. Amer. Chem. Soc., 97, 3049 (1975), and references therein.

182. T. J. Marks, A. M. Seyam, and J. R. Kolb, J. Amer. Chem. Soc., 95, 5529 (1973), and references therein.

183. J. L. Calderon, F. A. Cotton, B. G. DeBoer, and J. Takats, J. Amer. Chem. Soc., 93, 3542 (1971).

184. M. D. Rausch, Pure and Appl. Chem., 30, 523 (1972).

185. A. Streitweiser, Jr., U. Muller-Westerhoff, G. Sonnichsen, F. Mares, D. G. Morrell, K. O. Hodgson, and C. A. Harmon, J. Amer. Chem. Soc., 95, 8644 (1973).

186. H. Werner, Angew. Chem., Int. Ed. Engl., 16, 1 (1977), and references therein.

187. J. W. Lauher, E. Elian, R. H. Summerville, and R. Hoffmann, J. Amer. Chem. Soc., 98, 3219 (1976).

188. A. J. Ashe (III), E. Meyers, P. Shu, T. Von Lehman, and J. Bastide, J. Amer. Chem. Soc., 97, 6866 (1975).

189. G. E. Herberich and H. J. Becker, Angew. Chem., Int. Ed. Engl., 14, 184 (1975), and references therein.

190. K. P. Callahan and M. F. Hawthorne, in F. G. A. Stone and R. West, eds., Advances in Organometal. Chem., (Academic Press, 1976), vol. 14, pp. 145-186.

191. G. Schmid and J. Schulze, Angew. Chem., Int. Ed. Engl., 16, 249 (1977).

192. W. Siebert and M. Bochmann, Angew. Chem., Int. Ed. Engl., 16, 468 (1977).
193. R. L. Pruett, "Cyclopentadienyl and Arene Metal Carbonyls" in W. L. Jolly, ed., Preparative Inorganic Reactions, Vol. 2, (Interscience, 1965).
194. W. L. Jolly, Inorg. Synth., 11, 120 (1968).
195. H. L. Lentzner and W. E. Watts, Chem. Comm., 26 (1970).
196. W. F. Little, in A. Scott, ed., Survey of Progress in Chemistry (Academic Press, 1963), vol. 1, p. 133.
197. D. S. Trifan and L. Nicholas, J. Amer. Chem. Soc., 79, 2746 (1957).
198. W. F. Little, K. W. Lynam, and R. Williams, J. Amer. Chem. Soc., 86, 3055 (1964).
199. A. L. J. Bechwich and R. J. Laydon, J. Amer. Chem. Soc., 86, 953 (1964).
200. E. O. Fischer and R. Jira, Z. Naturforsch, 86, 327 (1953).
201. M. L. H. Green, L. Pratt, and G. Wilkinson, J. Chem. Soc., 3753 (1959).
202. M. Cais and J. Kozikowski, J. Amer. Chem. Soc., 82, 5667 (1960).
203. J. D. Fitzpatrick, L. Watts, G. F. Emerson, and R. Pettit, J. Amer. Chem. Soc., 87 3254 (1965).
204. G. Henrici-Olive and S. Olive, Angew. Chem., Int. Ed. Engl. 10, 105 (1971).
205. F. S. Shilova and A. E. Shilov, J. Polm. Sci., Part C, 16, 2336 (1967).
206. M. L. H. Green and P. L. I. Hagy, Adv. Organometal. Chem., 2, 325 (1964).
207. R. Uttech and H. Dietrich, Z. Christ., 122, 60 (1960).
208. N. Rossch and R. Hoffman, Inorg. Chem., 13, 2656 (1974).
209. A. Werner and A. Kuhn, Angew. Chem., Int. Ed. Engl., 16, 412 (1977).
210. S. Winstein, H. D. Kaesz, C. G. Kreiter, and E. C. Friedrich, J. Amer. Chem. Soc., 87, 3267 (1965).
211. T. H. Tulip and J. A. Ibers, J. Amer. Chem. Soc., 101, 4201 (1979), and references therein.
212. Y. Becker and J. K. Stille, J. Amer. Chem. Soc., 100, 845 (1978).
213. R. H. Abeles and D. Dolphin, Accts. Chem. Res., 9, 114 (1976).
214. M. Calligaris, G. Manzini, G. Nardin, and L. Randaccio, J. Chem. Soc., D, 543 (1972), and references therein.
215. T. E. Wolff, J. M. Berg, K. O. Hodgson, R. B. Frankel, and R. H. Holm, J. Amer. Chem. Soc., 101, 4140 (1979).
216. A. D. Allen and F. Bottomly, Accts. Chem. Res., 1, 360 (1968), and references therein.
217. R. Murray and D. C. Smith, Coord. Chem. Rev., 3, 429 (1968).

218. J. E. Fergusson and J. I. Love, Rev. Pure Appl. Chem., 20, 33 (1970).
219. J. Chatt and G. J. Leigh, Chem. Soc. Review, 1, 121 (1972).
220. M. E. Volpin and V. B. Shur, Organometallic Reactions, 1, 55 (1970).
221. E. E. Van Tamelen, Accts. Chem. Res., 3, 361 (1970).
222. C. Busetto, A. D'Alfonso, F. Maspero, G. Perezo, and A. Zazzetta, J. Chem. Soc., D, 1828 (1977).
223. D. L. Thorn, T. H. Tulip, and J. A. Ibers, J. Amer. Chem. Soc., in press (1980).
224. M. Mercer, R. H. Crabtree, and R. L. Richards, JCS Chem. Comm., 808 (1973).
225. P. D. Chadwick, J. Chatt, R. H. Crabtree, and R. L. Richards, JCS Chem. Comm., 351 (1975).
226. A. A. Diamantis, J. Chatt, G. A. Heath, and G. J. Leigh, JCS Chem. Comm., 27 (1975).
227. H. Taube and J. N. Armor, J. Amer. Chem. Soc., 93, 6476 (1971).
228. W. L. Bowden, W. F. Little, and T. J. Meyer, J. Amer. Chem. Soc., 98, 444 (1976).
229. L. S. Liebeskind, K. B. Sharpless, R. D. Wilson, and J. A. Ibers, J. Amer. Chem. Soc., 100, 7061 (1978).
230. J. McCleverty, Chem. Rev., 79, 53 (1979).
231. I. H. Sabherwal and A. B. Burg, Chem. Comm., 1001 (1970).
232. B. A. Frenz, J. H. Enemak, and J. A. Ibers, Inorg. Chem., 8, 1288 (1969).
233. A. T. McPhail and G. A. Sim, J. Chem. Soc. (A), 1858 (1968).
234. L. F. Dahl, E. Rodulfo de Gil, and R. D. Feltham, J. Amer. Chem. Soc., 91, 1653 (1969).
235. D. E. Morris and F. Basolo, J. Amer. Chem. Soc., 91, 2531 (1969).
236. R. Eisenberg and C. D. Meyer, Accts. Chem. Res., 8, 26 (1975), and references therein.
237. C. P. Brock, J. P. Collman, G. Dolcetti, P. H. Farnham, J. A. Ibers, J. E. Lester, and C. A. Reed, Inorg. Chem., 12, 1304 (1973).
238. D. J. Hodgson and J. A. Ibers, Inorg. Chem., 7, 2345 (1968).
239. C. G. Pierpoint and R. Eisenberg, Inorg. Chem., 11, 1088 (1972).
240. D. A. Snyder and D. L. Weaver, Chem. Comm., 1425 (1969).
241. J. P. Collman, P. Farnham, and G. Dolcetti, J. Amer. Chem. Soc., 93, 1788 (1971).
242. R. Eisenberg and C. D. Meyer, Accts. Chem. Res., 8, 26 (1975).

243. R. Hoffmann, D. M. P. Mangos, M. M. L. Chen, and A. R. Rossi, Inorg. Chem., 13, 2666 (1974).

244a. B. A. Frenz and J. A. Ibers, Int. Rev. Sci. Phys. Chem., One, 11 33 (1972).

244b. J. H. Enemark and R. D. Feltham, Coord. Chem. Rev., 13, 339 (1974).

245. G. Dolcetti, N. W. Hoffman, and J. P. Collman, Inorg. Chimica Acta, 6, 531 (1972).

246. K. G. Caulton, Coord. Chem. Rev., 14, 317 (1975).

247. J. Masek, Inorg. Chim. Acta. Rev., 3, 99 (1969).

248. L. Y. Y. Chan and F. W. B. Einstein, Acta. Cryst., B, 26, 1899 (1970).

249. R. C. Elder, F. A. Cotton, and R. A. Schunn, J. Amer. Chem. Soc., 89, 3645 (1967).

250. W. A. Hermann, M. L. Ziegler, and K. Weidenheimer, Angew. Chem., 88, 379 (1976).

251. M. W. Bishop, J. Chatt, J. R. Dilworth, G. Kaufman, S. Kim, and J. Zubieta, JCS Chem. Comm., 70 (1977), and references therein.

252. B. W. S. Kolthammer and P. Legzdins, J. Amer. Chem. Soc., 100, 2247 (1978).

253. P. Sutton, Chem. Soc. Rev., 4, 443 (1975).

254. K. D. Schramm and J. A. Ibers, Inorg. Chem., 16, 2387 (1977).

255. W. L. Bowden, W. F. Little, and T. J. Meyer, J. Amer. Chem. Soc., 95, 5084 (1973).

256. B. L. Haymore and J. A. Ibers, Inorg. Chem., 14, 3060 (1975).

257. Ibid., p. 1369.

258. R. E. Cobbledick, R. W. B. Einstein, N. Farrell, A. B. Gilchrist, and D. Sutton, J. Chem. Soc. (D), 373 (1977).

259. S. Krogsrud and J. A. Ibers, Inorg. Chem., 14, 2298 (1975).

260. B. L. Haymore, J. A. Ibers, and D. W. Meek, Inorg. Chem., 14, 541 (1975), and references therein.

261. P. W. Jolly and G. Wilke, The Organic Chemistry of Nickel, (Academic Press, 1974), vol. 1, pp. 257-340.

262. B. Eistert, G. Heck, M. Regitz, and H. Schwall, Methoden der Organischen Chem., 10/4, (Houben-Weyl, 1968), p. 473.

263. K. D. Schramm and J. A. Ibers, J. Amer. Chem. Soc., 100, 2932 (1978); A. Nakamura, T. Yoshida, M. Cowie, S. Otsuka, and J. A. Ibers, J. Amer. Chem. Soc., 99, 2108 (1977).

264. M. M. Bagga, P. E. Baike, O. S. Mills, and P. L. Pauson, JCS Chem. Comm., 1106 (1967); P. E. Baike and O. S. Mills, JCS Chem. Comm., 1228 (1967).
265. B. N. Stochoff and H. C. Lewis, Jr., Coord. Chem. Rev., 23, 1 (1977).
266. J. Chatt, R. J. Dosser, F. King, and G. S. Leigh, JCS (D) Proc., 2435 (1976).
267a. D. Bright and J. A. Ibers, Inorg. Chem., 8, 709 (1969).
267b. V. V. Tkachev, O. N. Krasochka, and L. A. Atovmyan, Zh. Strukt. Khim., 17k 940 (1976).
268. L. Vaska, Accts. Chem. Res., 9, 175 (1976), and references therein.
269. J. Valentine, Chem. Rev., 73, 235 (1973).
270. J. A. Connor and E. A. V. Ebsworth, Adv. Inorg. Chem. Radiochem., 6, 279 (1964).
271. F. Basolo, B. M. Hoffman, and J. A. Ibers, Accts. Chem. Res., 8, 384 (1975).
272. C. A. Reed in H. Sigel, ed., Metal Ions in Biological Systems, (Marcel Dekker, 1978), vol. 7, p. 277.
273. M. J. Nolte, E. Singleton, and M. Laing, J. Amer. Chem. Soc., 97, 6396 (1975), and references therein.
274. R. W. Horn, E. Weissberger, and J. P. Collman, Inorg. Chem., 9, 2367 (1970).
275. J. P. Collman, Accts. Chem. Res., 10, 265 (1977).
276a. R. S. Gall, J. F. Rogers, W. P. Schaefer, and G. G. Christoph, J. Amer. Chem. Soc., 98, 5135 (1976).
276b. A. Avdeef and W. P. Schaefer, J. Amer. Chem. Soc., 98, 5153 (1976).
277. C. A. Reed and S. K. Cheung, Proc. Natl. Acad. Sci. USA, 74, 1780 (1977).
278. B. H. Huyuh, D. Z. Case, and M. Karplus, J. Amer. Chem. Soc., 99, 6103 (1977).
279. B. D. Olafson and W. A. Goddard, III, Proc. Natl. Acad. Sci., USA, 74, 1315 (1977).
280. F. Basolo, H. M. Hoffman, and J. A. Ibers, Accts. Chem. Res., 384 (1975); F. Basolo, Chem. Rev., 79, 134 (1979), and references therein.
281. M. Nappa and J. Valentine, J. Amer. Chem. Soc., 100, 5075 (1978).
282. A. P. Ginsberg and W. L. Lindsell, Chem. Comm., 232 (1971).
283. P. M. Treichel and G. P. Weber, J. Amer. Chem. Soc., 90, 1753 (1968).
284. W. D. Bonds, Jr., and J. A. Ibers, J. Amer. Chem. Soc., 94, 3413 (1972).
285. L. Dahl, cited in A. P. Ginsberg and W. L. Linsell, Chem. Comm., 232 (1971).
286. G. R. Clark, D. R. Russell, W. R. Roper, and A. Walker, J. Organometal. Chem., 136, C1 (1977).

287. T. Herskovitz, J. Amer. Chem. Soc., 99, 2391 (1977), and references therein.

288. G. Fachinetti, C. Floriani, and P. F. Zanazzi, J. Amer. Chem. Soc., 100, 7405 (1978), and references therein.

289. M. Aresta, C. F. Nobile, V. G. Albano, E. Formi, and M. Manassero, JCS Chem. Comm., 636 (1975).

290. J. Blum, Tetrahedron Lett., 3041 (1966); on removal of $SO_2$.

291. D. M. P. Mingos, Transition Metal Chem., 3, 1 (1978); on $SO_2$ complexes of Pt metals.

292. S. Otsuka, Y. Tatsuno, M. Miki, T. Aoki, M. Matsumoto, H. Yoshioda, and K. Nakatsu, JCS Chem. Comm., 445 (1973).

293. M. Angoletta, P. L. Bellon, M. Manasero, and M. Sansoni, J. Organometal. Chem., 81, C40 (1974).

294. J. Haase and M. Winnewisser, Z. Naturforsch, A, 23, 61 (1968).

295. S. J. LaPlaca and J. A. Ibers, Inorg. Chem., 5, 405 (1965).

296. L. H. Vogt, Jr., J. L. Katz, and S. E. Wiberley, Inorg. Chem., 4, 1157 (1965).

297. G. J. Kubas, Inorg. Chem., 18, 182 (1979).

298. J. Valentine, D. Valentine, Jr., and J. P. Collman, Inorg. Chem., 10, 219 (1971).

299. A. Wojcicki, Adv. Organometal. Chem., 12, 31 (1974).

# 4

# Oxidative-Addition and Reductive-Elimination

"Oxidative-addition" is the term used to describe an ubiquitous class of reactions in which a group, A-B, adds to, and thus oxidizes, a metal complex, M. Both one- and two-electron oxidative-additions are recognized (Equations 4.1 and 4.2, respectively).

$$2M + A{-}B \rightleftharpoons M{-}A + M{-}B \quad \text{or} \quad A{-}M{-}M{-}B \qquad (4.1)$$

$$M + A{-}B \rightleftharpoons \overset{\displaystyle A}{\overset{|}{M}}{-}B \quad \text{or} \quad \overset{\displaystyle A}{\overset{|}{M^{+}}}\ B^{-} \qquad (4.2)$$

Note that for all such reactions, both the oxidation state and coordination number of the metal increases. The reverse reactions are known as "reductive-eliminations." These reactions are very important, since nearly all catalytic and many useful stoichiometric processes involve oxidative-addition and/or reductive-elimination steps. Recall from Chapter 2 that the most easily removed electrons in a transition-metal complex are the d-electrons of the central metal. Thus, oxidative-addition results in a change from a $d^n$ to a $d^{n-2}$ or $d^{n-1}$ formal configuration. <u>It is this ability of the metal to exist in several oxidation states which distinguishes the chemical transformations of organotransition metal from non-transition metal compounds, since the latter usually have closed shell configurations</u>. It is important to emphasize that oxidative-addition is a generic term describing stoichiometry and does not imply any particular mechanism. The same is true for reductive-elimination. These reactions involve changes in the <u>formal</u> oxidation states.

Oxidative-addition reactions constitute the principal method of forming metal-carbon and metal-hydrogen bonds. Conversely, reductive-elimination can lead to the formation of carbon-hydrogen, carbon-carbon, and other C-X bonds. Thus, understanding the scope and mechanisms of this fundamental reaction class is essential to a rational discussion of organotransition-metal chemistry. This chapter will emphasize oxidative-additions which form metal-carbon and metal-hydrogen bonds, because for synthetic organic applications these are the most important examples.

We shall first consider oxidative-addition and subsequently reductive-elimination reactions. The former have been extensively reviewed [1], the latter less so [2].

## 4.1. Oxidative-Addition Reactions with an Over-all Two-Electron Change.

Historically this was the earliest recognized class of oxidative-addition reactions. Prototypical two-electron oxidative-addition reactions shown in Figure 4.1 involve

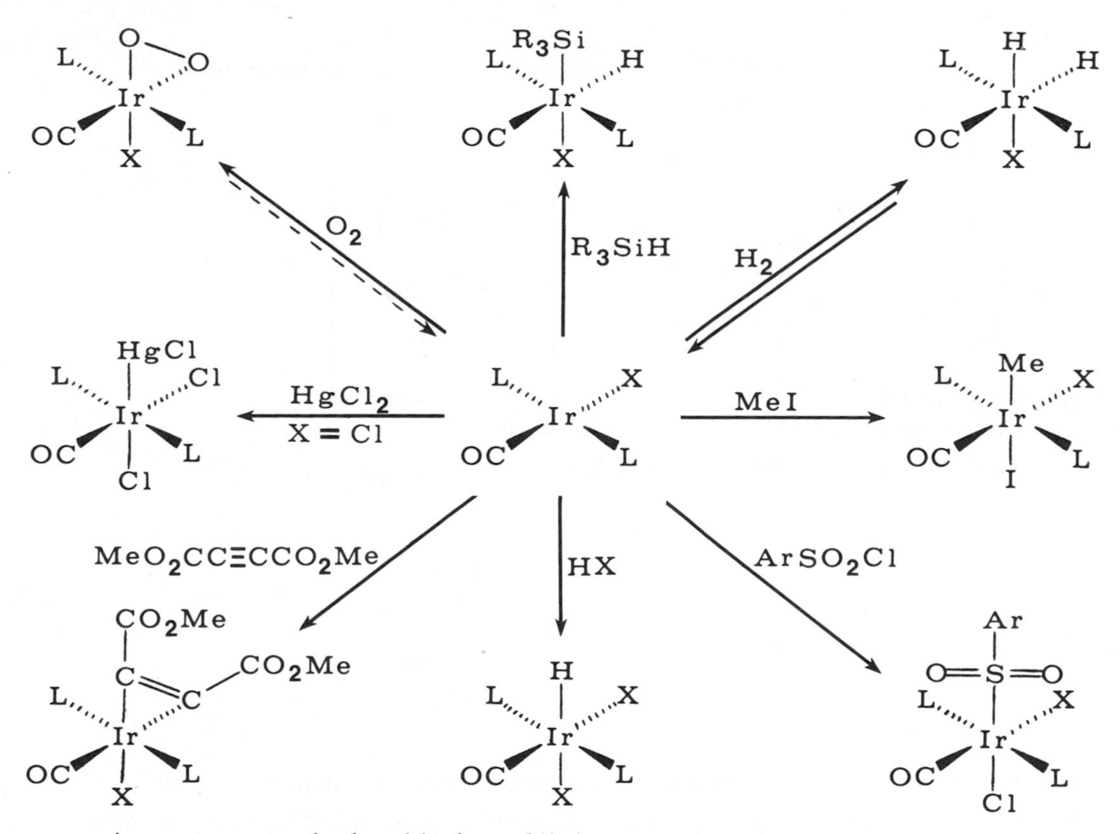


*Figure 4.1. Typical oxidative-addition reactions involving Vask's iridium(I) complex(es).*

"Vaska's complex," a lemon-colored, coordinatively unsaturated iridium(I) compound. Examples of such two-electron, oxidative-additions are known for virtually all even $d^n$ configurations (n = 2, 4, 6, 8, 10), but these reactions are far better studied for the $d^8$ [1a] and $d^{10}$ [1h] compounds found toward the right of the transition series (especially Group VIII).

Both coordinatively saturated and unsaturated complexes undergo oxidative-addition reactions, the latter being more reactive. The general cases for $d^8$ compounds are illustrated in Equations 4.3 and 4.4.

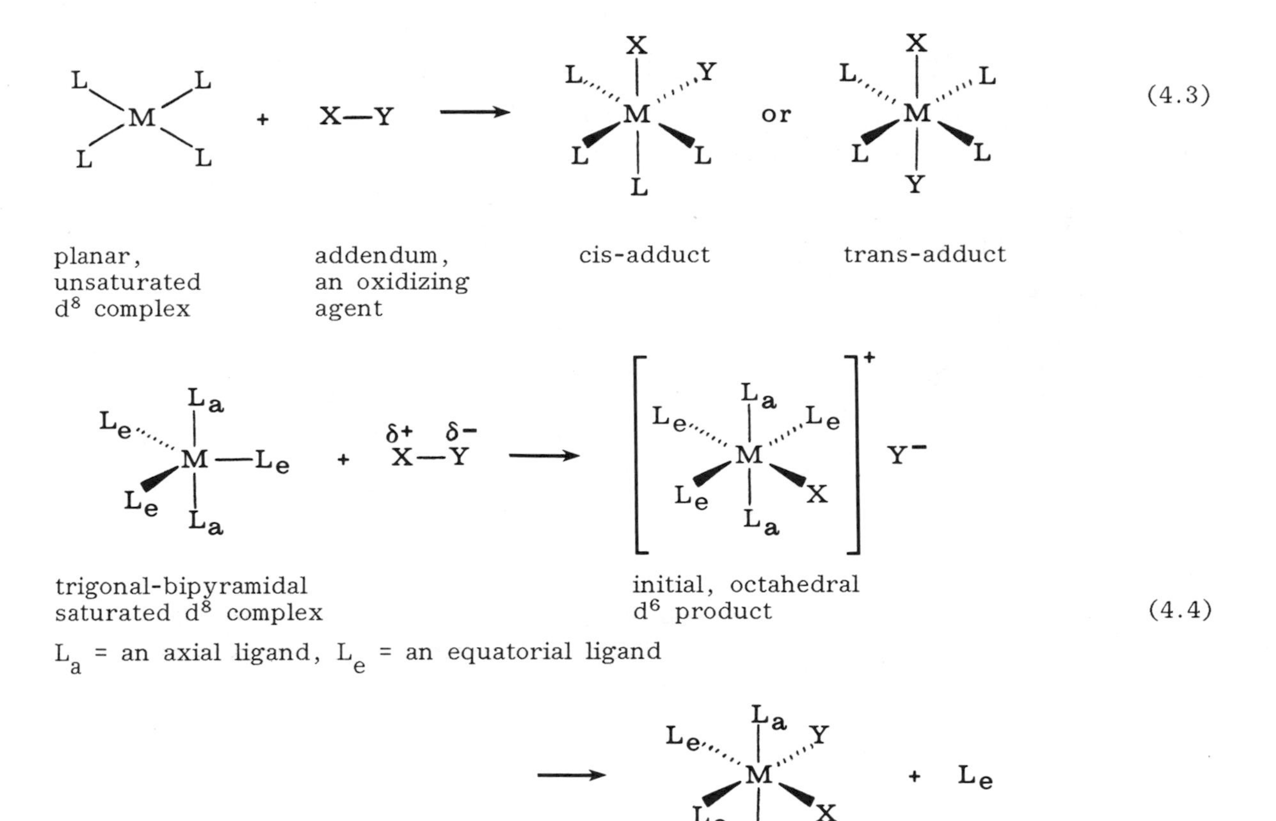


The coordinatively unsaturated $d^8$ complexes may be neutral, anionic, or cationic (Equation 4.3). The "addenda," X-Y, vary from polar electrophiles to nonpolar reagents. As we shall see, ionic, radical, and concerted mechanisms are known.

Additions to unsaturated planar metal complexes are stereospecific. Instances of kinetically controlled *cis* or *trans* additions to the metal are known, depending on the nature of X-Y and the reaction conditions.

Coordinatively saturated $d^8$ compounds are typically less reactive and usually add only the more strongly oxidizing addenda X-Y (Equation 4.4). In certain cases, this reaction has been shown to proceed in a stepwise manner, with the initial attack by $X^+$ forming a coordinatively saturated complex which may subsequently incorporate $Y^-$ by replacement of the ligand $L_e$ (Equation 4.4). Sometimes the final product is that of over-all *cis*-addition, but this depends on the mechanism of the ligand displacement step and is probably not general. The reactivity of the saturated complex seems to depend on its formal charge; anionic compounds, being electron-rich, are more reactive. In certain cases a saturated $d^8$ complex may lose a ligand prior to oxidative-addition. A careful kinetic analysis is required to distinguish this situation.

The addenda X-Y in Equations 4.3 and 4.4 can be divided roughly into three classes: *A*, polar, electrophilic reagents; *B*, those which remain joined by a bond in the adduct; and *C*, nonpolar addenda. Class *A* and *C* addenda fragment upon adding to the metal. Typical examples are as follows.

Class *A* addenda: $X_2$ (halogens), H-Y (Brönsted acids), RSCl, $RSO_2Cl$, R-X (alkylating agents), RCOX (acylating agents), $R_FI$, RCN, $SnCl_4$, and $HgX_2$. The mechanism of the perfluoroalkyl iodide oxidative-additions is presently unknown. These reactions may not be polar. The nonelectrophilic nitriles, RCN, probably react by prior coordination, *vide infra*.

Class *B* addenda: $O_2$, $S_2$, $Se_2$, *o*-quinones, RC≡CR, RN=NR, RCH=CHR, S=C=S, $CH_2$=O, cyclopropanes, $RCON_3$, $RN_3$. 1,3-Dipoles such as $RN_3$ and $RCON_3$ may oxidatively add to metals, forming transitory four-membered metallacyclic adducts which subsequently fragment, giving the final product. Such paths are not well-established.

Class *C* addenda: $H_2$, $R_3SiH$, $R_3Ge$-H, $R_3Sn$-H, RSH, RCHO, Ar-H, R-H. The addition of C-H bonds is usually confined to intramolecular oxidative-additions of "activated" C-H groups, *vide infra*.

Several of the Class *A* addenda add to olefins. There are some analogies between oxidative-addition and the well-known olefin addition reactions. These analogies even apply to reaction mechanisms. For example, oxidative-additions can follow either a polar, two-step mechanism or a chain radical path.

The addenda in Class *B* retain at least one bond in the adducts. Thus, the two new bonds to the metal are necessarily *cis*. This class of reactions is normally limited to coordinatively unsaturated complexes or to complexes which become unsaturated by loss of a ligand before the oxidative-addition step. Whether or not the formation of a complex with an acetylene, an azo compound, or an olefin may be considered an "oxidative-addition" depends on formal oxidation-state assignments which, as we have seen, are often "a matter of taste." We will encounter the addition of the formaldehyde carbonyl group in Chapter 8.

The nonpolar addenda in Class C usually react only with coordinatively unsaturated complexes, and the over-all addition to the metal is almost always stereospecifically cis. An exception to this generalization may be encountered in the hydrogenation of a coordinatively saturated complex in the presence of a proton base, "heterolytic $H_2$ cleavage," which does not involve formal oxidation of the metal [3]. We will encounter reactions of this type in Chapter 6.

a. Reactions with Protons. Perhaps the simplest oxidative-addition involves the reaction of low-valent metal complexes with Bronsted acids, affording transition-metal hydrides. In Equation 4.5, Z = the oxidation number, $M^Z$ is the base, and $H^+$ is the

$$\underset{d^n}{M^Z} + H^+ \rightleftharpoons \underset{d^{n-2}}{\left[M^{(z+2)}\!-\!H\right]^+} \underset{-B^-}{\overset{+B^-}{\rightleftharpoons}} \underset{d^{n-2}}{B\!-\!M^{(z+2)}\!-\!H} \qquad (4.5)$$

acid. Such reactions clearly demonstrate the interrelationship between formal oxidation-reduction and acid-base character which is a common theme throughout chemistry.

Coordinatively unsaturated metals usually capture the conjugate base in a subsequent step, but in certain cases prior coordination of the conjugate base, such as a halide, may preceed reaction of the metal with a proton. For example, kinetic studies on the oxidative-addition reactions of $[Ir(COD)L_2]^+$ and $[IrCl(COD)L]$ with HCl, have shown that nucleophilic attack of $Cl^-$ preceeds the protonation of the complexes [7]. In nonpolar media the addition may be concerted, as shown in Equation 4.6 [4].

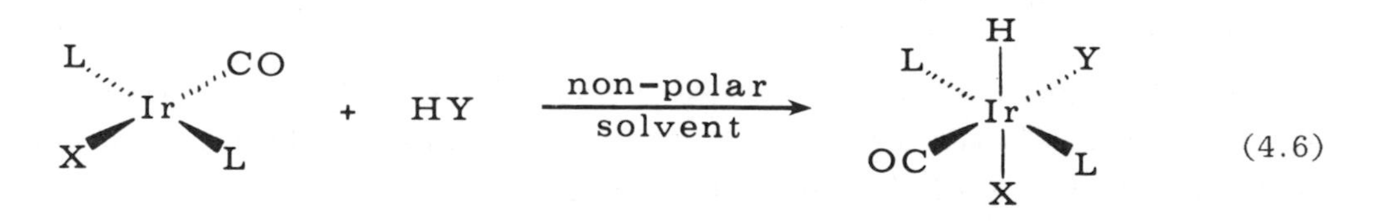

The basicity of saturated transition-metal complexes toward protons is usually more pronounced with anionic complexes (Equation 4.7) [5]:

$$[Re(PF_3)_5]^- + H^+ \longrightarrow HRe(PF_3)_5 \qquad (4.7)$$

but even neutral saturated complexes react (Equation 4.8) [6].

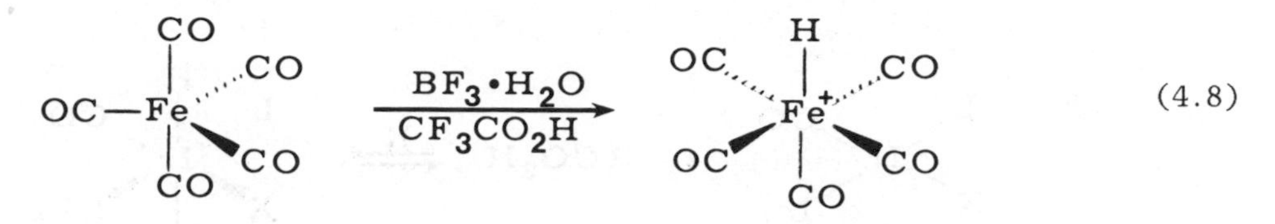


The basicity and thus the reactivity of transition-metal complexes can be gauged from the $K_a$ values of the corresponding hydrides acting as acids (the microscopic reverse of the oxidative-addition step). Unfortunately, there are few accurate $K_a$ values for transition-metal hydrides, because of secondary reactions, air sensitivity, and problems in finding suitable solvents for comparisons. In some cases these acid-base reactions are slow, suggesting the possibility of structural rearrangement during protonation and deprotonation [8]. Those $K_a$ values which have been reported illustrate a strong dependence on both the nature of the metal and the other ligands. Consider, for example, Figure 4.2.

| Compound | $K_a$ |
|---|---|
| $HMn(CO)_5$ | $10^{-6}$ |
| $HRe(CO)_5$ | "very small" |
| $H_2Fe(CO)_4$ | $4 \times 10^{-5}$ |
| $HFe(CO)_4^-$ | $\sim 4 \times 10^{-14}$ |
| $HCo(CO)_4$ | $\sim 1$ |
| $HCo(CO)_3PPh_3$ | $1 \times 10^{-7}$ |

*Figure 4.2.*

The less acidic the hydride, the more basic the conjugate base; thus the third row anion, $[Re(CO)_5]^-$, is more basic than $[Mn(CO)_5]^-$, the first-row, isoelectronic complex. This behavior seems typical, although few quantitative comparisons exist. The greater M-H bond strength of third-row transition-metal hydrides explains this trend.

Next note the substantial estimated difference between the first and second dissociation constant in the dihydride, $H_2Fe(CO)_4$: nine orders of magnitude. Note also the strongly acidic character of $HCo(CO)_4$, which is perhaps the strongest acid among the transition-metal hydrides. Substitution of the electron-releasing $Ph_3P$ ligand for the electron-withdrawing CO $\pi$-acid ligand causes a change in acidity of the hydride by seven orders of magnitude! This is an extreme example of the change in metal basicity (reactivity) which can be brought about by altering the ancillary ligands. The ligand effect on metal basicity is apparent from Shaw's [9] study of the reversible oxidative-addition of carboxylic acids to unsaturated iridium complexes (Equation 4.9).

$$trans\text{-}IrX(CO)L_2 + RCO_2H \overset{K}{\rightleftharpoons} IrH(X)(CO)L_2(OC(O)R) \quad (4.9)$$

Effect on K with
changes in X and L:

X = I > Br > Cl (10 fold)

L = $Me_3P$ > $PhPMe_2$ > $Ph_2PMe$ ~ $Ph_3As$ > $PPh_3$ (3.5-fold, 10 fold, 10 fold)

The extent of reaction is conveniently followed by visible spectroscopy. Plots of $K_a$ of the carboxylic acids versus K for the oxidative-addition are linear. The indicated changes in K as a function of the halide or phosphine ligands illustrates their modest effect on the basicity of iridium.

It should be noted that many transition-metal hydrides formed by oxidative-addition may react further with protons forming hydrogen. In certain cases this hydride protonolysis is facilitated by water. Equation 4.10 illustrates this point [10a,11].

$$Ni[Ph_2P(CH_2)_2PPh_2]_2 \xrightarrow{HX} \begin{cases} \xrightarrow{dry} HNi^+[Ph_2P(CH_2)_2PPh_2]_2X^- \\ \xrightarrow{H_2O} H_2 + Ni^{2+}[Ph_2P(CH_2)_2PPh_2]_2 \end{cases} \quad (4.10)$$

Very weak (nonpolar) protonic acids may undergo oxidative-additions to highly reactive complexes [12a]. In such cases the conjugate base is also a good ligand. Five examples illustrate the broad scope of these reactions (Equations 4.11 to 4.15). The

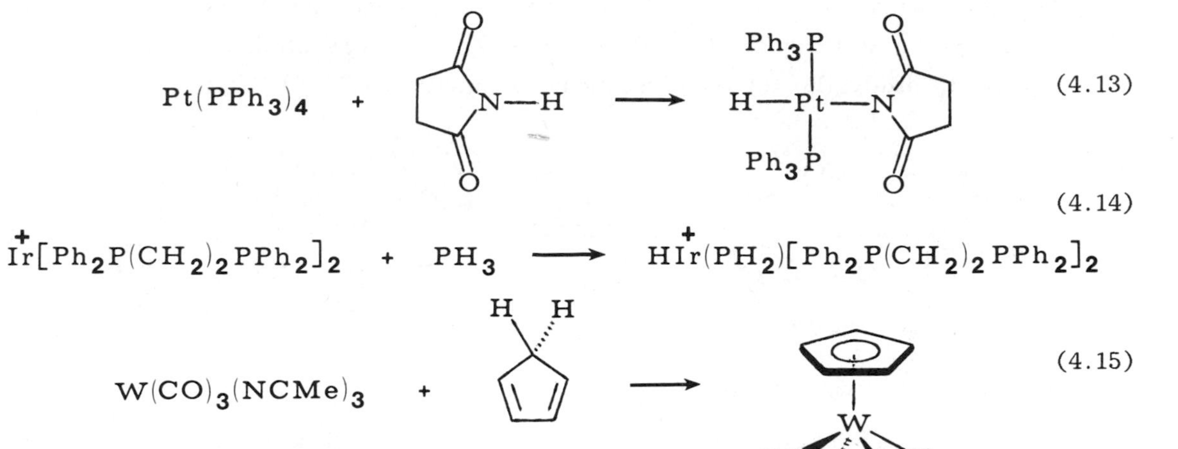


mechanisms of these nonpolar additions are not well-understood, but each case probably involves a coordinatively unsaturated complex. It is likely that some of these reactions have radical mechanisms. Note the less-common oxidative-addition to a $d^6$ complex in Equation 4.15.

The basicity of low-valent transition-metal complexes is manifest in many other ways. For example, one can form Lewis adducts involving transition metals as the base [10b]. Our rules for formal oxidation-state assignments do not catalog reactions such as that in Equation 4.16 as oxidative-addition reactions, even though they are

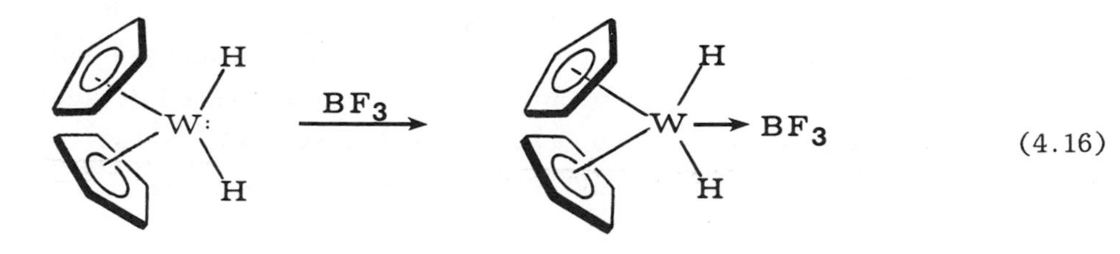

quite equivalent to proton addition. Recall from Chapter 3 that some electrophilic substitutions of ferrocene are thought to involve an intermediate Lewis-base adduct at iron.

Thus $d^8$ compounds are known to form complexes with Lewis acids such as $BX_3$, $AlX_3$, $HgCl_2$, $ZnBr_2$, $TlCl_3$. These Lewis adducts are often difficult to isolate in pure form. If a vacant coordination site is available on the transition metal, a mercuric halide bond cleaves [11], forming an "oxidative adduct" (see, for example, Figure 4.1). The Lewis basicity of transition-metal complexes has been reviewed [12a].

Similarly, metal-metal bonds are basic and may sometimes be protonated without rupture--affording a bridging hydride. An early case was studied by Wilkinson who used $^1H$ nmr to study the following reactions in Equation 4.17 [12b].

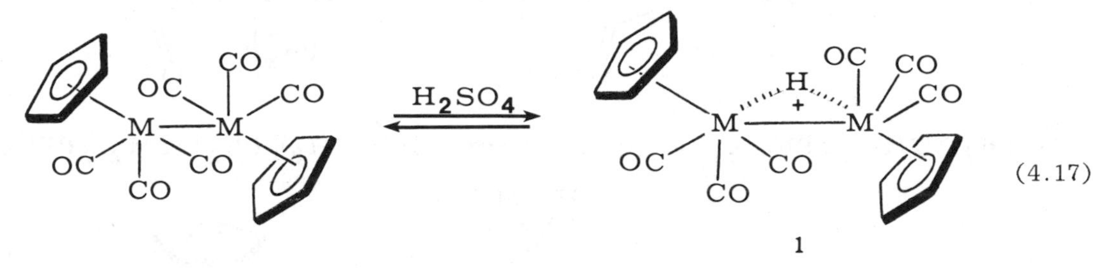


(4.17)

M = Mo, Mo; W, W; W, Mo

2. Reactions Forming Metal-Carbon Bonds. Basic, low-valent transition-metal centers would be expected to be nucleophilic. Thus, it is not surprising to find that alkyl halides and other alkylating reagents oxidatively add to both unsaturated and saturated complexes. These alkylations constitute one of the most important methods of forming metal-carbon bonds. For example, oxidative-addition is considered to be the rate-determining step in the famous Monsanto acetic-acid process (Chapter 8) as well as in stoichiometric reactions involving organocopper reagents (Chapter 11) and $\eta^3$-allyl complexes (Chapter 15). Thus, we will discuss the scope and mechanisms of these "metal alkylations" in considerable detail.

The mechanisms of such metal-alkylations have been extensively studied, but certain cases are still controversial. There is a delicate balance between a polar, two-step, "$S_N2$" path, a radical chain path, and a radical cage path. Perhaps even "three center" mechanisms may eventually be verified. Each individual case must be carefully examined in order to discover the dominant mechanism. Read the original literature and interpretive reviews with caution; much erroneous information has been perpetrated.

Factors such as the formal charge on the complex, the nature of the metal and its accompanying ligands, the nature of the alkylating agent, and experimental conditions such as solvent, temperature, and impurities determine the dominant mechanism of these reactions.

Several criteria have been used to discover the dominant mechanism in these metal alkylations: (1) stereochemistry, both at the metal and at the alkylating carbon; (2) interception of polar or radical intermediates; (3) rearrangement of radical intermediates; (4) the form of the rate law and the nature of the activation parameters; and (5) the effect of solvent polarity and of radical scavengers on the rate. Each of these factors will be illustrated by specific examples.

a. Planar Iridium(I) Complexes. Consider the oxidative addition of methyl halides to derivatives of Vaska's iridium complex (Equation 4.18).

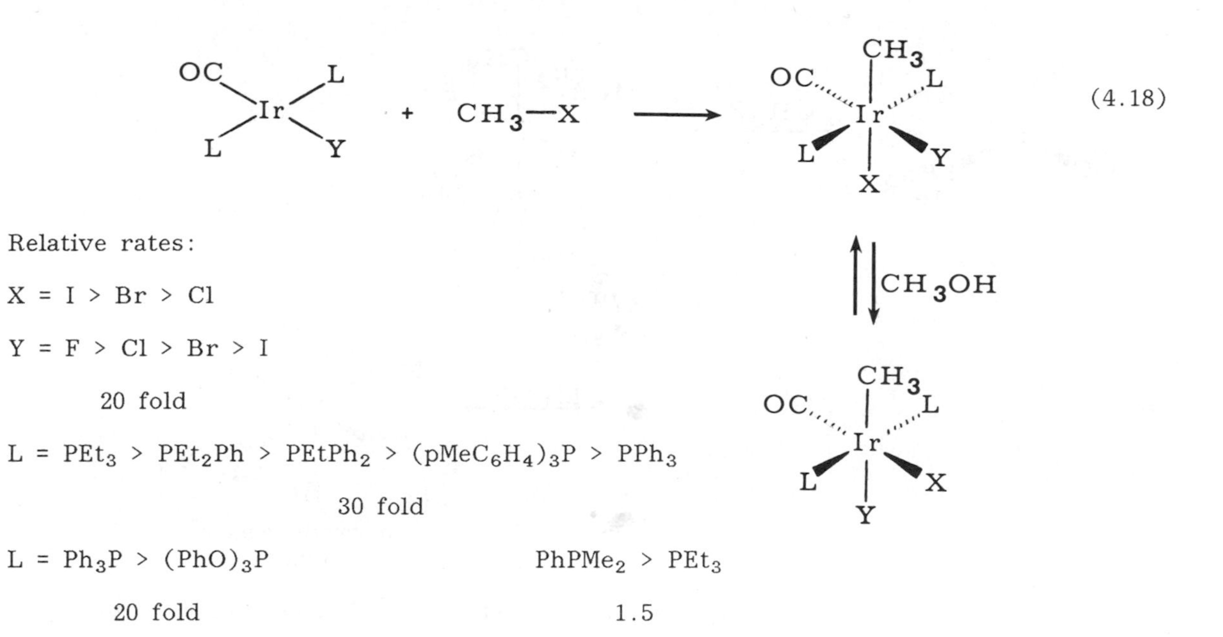


The rate law for this reaction is clearly second-order, first-order each in $CH_3$-X and in iridium(I) [13a]. The reaction is faster in polar solvents, DMF > $CH_3CN$ > THF > $C_6H_5Cl$ > $C_6H_6$. The reaction shows a large negative entropy of activation ($\Delta S^{\ddagger}$ = -43 to -51 eu). These facts are consistent with an $S_N2$ displacement at carbon by iridium(I). One would expect that the two neutral molecules would produce ionic transition states which would be stabilized by polar solvents; however, it has not been possible to intercept the implied cationic, five-coordinate, iridium(III) intermediate with added nucleophiles (Equation 4.19) [13a,b].

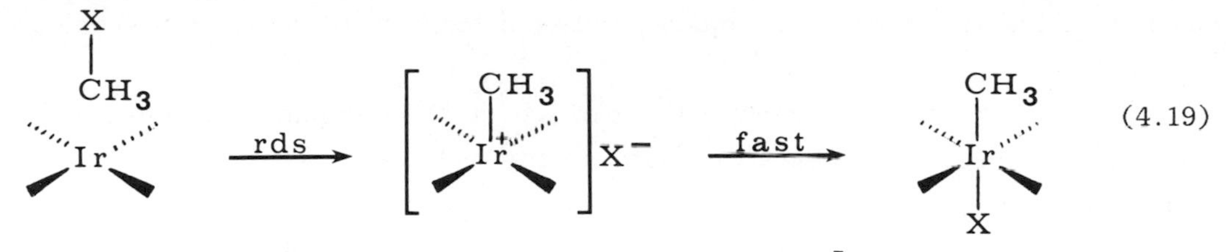


(4.19)

(no detectable exchange with external $X^-$)

In nonpolar solvents, methyl halides add to the iridium(I) complex in the trans manner shown in Equation 4.20 [14]. Methanol causes the trans adduct to rearrange to the thermodynamically more stable cis adducts.

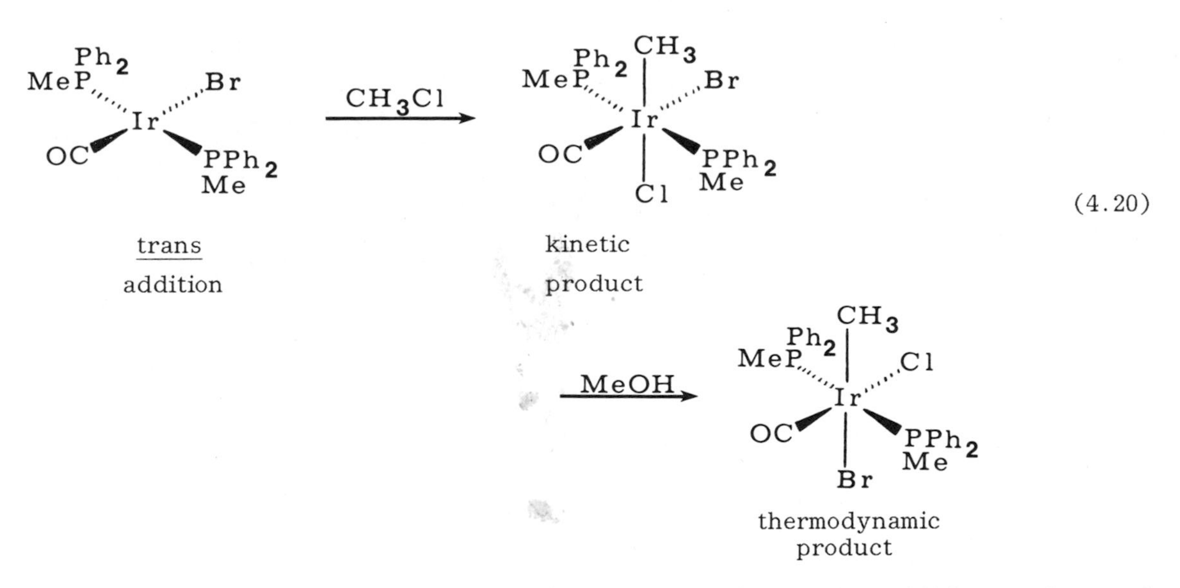


(4.20)

These experiments demonstrate that the sterochemistry of oxidative-addition at the metal is under kinetic control. These stereochemical assignments at the metal were made by a combination of $^1H$ nmr, showing the trans disposition of phosphines, and by the far infrared νIr-Cl band, which is characteristically sensitive to the nature of the trans-substituent.

The rate of methyl-iodide addition is unchanged by galvinoxyl **2** or duroquinone **3**, which might be expected to affect the rate of a radical chain reaction by intercepting the alkyl radical.

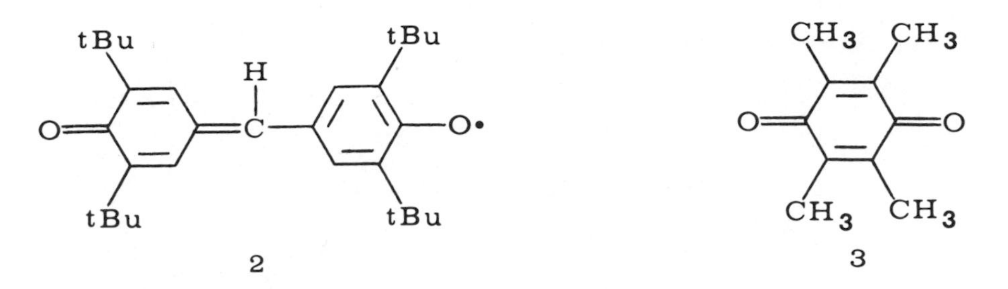


As illustrated in Equation 4.18, more basic phosphines enhance the rate of oxidative-addition to iridium(I), but there is a conflicting effect of steric bulk, since, for example, $PPhMe_2$ > $PEt_3$ [15]. Such a steric effect probably accounts for the relative rate Y = F > I (Equation 4.18). On purely electronic grounds, the polarizable, less-electronegative iodine atom should release a larger amount of electron density to the iridium atom. This is consonant with the protonation of these complexes described in Equation 4.9. Ethyl iodide is much less reactive than methyl iodide, the former not reacting appreciably at ambient temperature. Secondary halides do not appear to react at all if impurities (possible radical initiators) are rigorously excluded. These results are also consistent with steric effects expected for an $S_N2$ process.

With the possible exceptions of the observed trans-oxidative addition and the failure to intercept the expected ionic intermediate [13a,b], the results given above for methyl iodide are consistent with an $S_N2$ polar mechanism (Equation 4.19). A final test corroborating that path would be to discover the stereochemistry at carbon, since an $S_N2$ path should afford inversion of configuration. The initial attempts to examine this point gave conflicting results, but these early experiments [16,17] are now recognized [18,19] to be in error. As it turns out for these iridium(I) complexes, secondary and even primary alkyl halides are insufficiently reactive by the $S_N2$ path which seems to be followed by methyl iodide. Consequently, in the case of these poorer $S_N2$ substrates, a competitive radical chain mechanism takes over (Equations 4.21 to 4.24). This radical chemistry is similar to the one-electron oxidative-addition reactions we shall encounter later in this chapter. Note that addition of a radical to the metal raises its formal oxidation state by one. Many coordinatively unsaturated transition-metal cations are very efficient scavengers of alkyl radicals.

The reactions shown in Equations 4.25 to 4.27 show all the characteristics expected for a chain-radical process: inhibition by galvinoxyl and duroquinone, initiation by AIBN and benzoyl peroxide, and racemization of chiral substrates [19]. The latter stereochemistry at carbon was discovered by an elegant, direct $^1H$ nmr technique which had been developed earlier by G. Whitesides [20].

$$Q\cdot \;+\; Ir(I) \longrightarrow Ir(II){-}Q \tag{4.21}$$

$$Ir(II){-}Q \;+\; R{-}X \longrightarrow X{-}Ir(III){-}Q \;+\; R\cdot \tag{4.22}$$

Initiation steps

$$R\cdot \;+\; Ir(I) \longrightarrow R{-}Ir(II) \tag{4.23}$$

$$R{-}Ir(II) \;+\; RX \longrightarrow R{-}Ir{-}X \;+\; R\cdot \tag{4.24}$$

Propagation steps

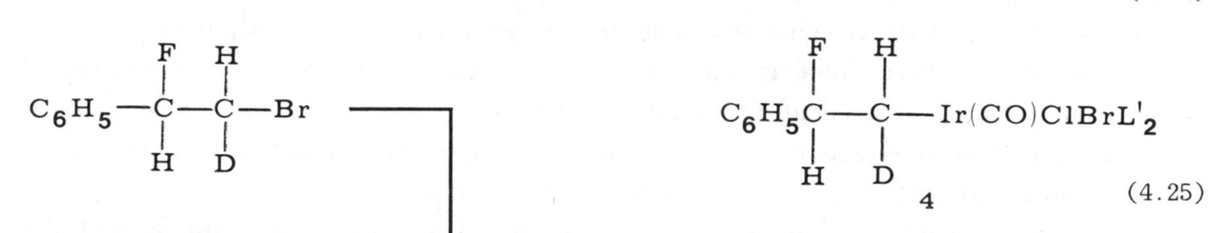


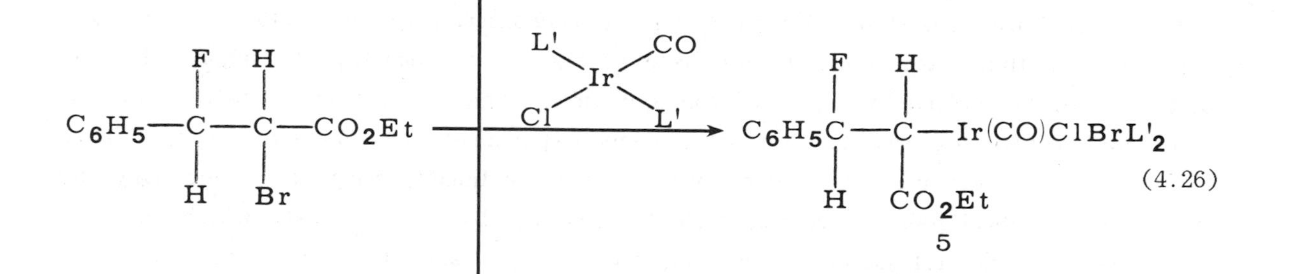

The stereochemistries in 4 and 5 were determined by $^1H$ nmr on racemic diastereomers, by measuring the coupling constant between vicinal hydrogens in the dominant conformation ($J_{HH}$ erythro > 9 Hz; $J_{HH}$ threo < 6 Hz) shown in Figure 4.3.

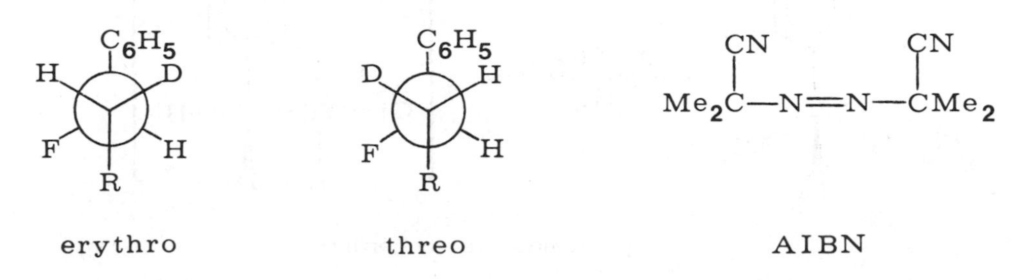


*Figure 4.3.*

The stereochemistry of 6 was studied using resolved enantiomers.

b. $d^{10}$ Complexes. The above alkylations of iridium(I) complexes illustrate the delicate balance between the $S_N2$ and the radical chain path. The extreme steric bulk of the iridium(I) complexes accounts for the relative rates of the two paths with different substrates; with the Vaska complexes only the methyl substrates appear to react by an $S_N2$ path. The situation is more complicated with the $d^{10}$ complexes $M(PR_3)_3$ (M=Pt,Pd; R=Ph,Et). The occasionally conflicting results from several laboratories have been summarized in Lappert's review [1d].

The reaction between $Pd(PPh_3)_4$ and chiral deutero benzyl halides was found to proceed with predominant inversion at carbon (Equation 4.28) [21,22]. The stereochemistry was found by CO insertion (a reaction which precedent indicates should take place with retention; see Chapter 5) and subsequent oxidative degradation to methyl phenylacetate. These results are consistent with an initial $S_N2$ process and would seem to contradict a radical reaction. Note that both primary (Equation 4.28) and secondary benzyl halides (Equation 4.29) appear to react by a polar mechanism.

The observation of a racemic product in oxidative-additions should not necessarily be taken as evidence for a radical path, since racemization could occur after the initial metal-carbon bond is formed. A primary racemization pathway is available in the form of the "self-displacement" reaction (vide infra) which involves the displacement of one metal by another through backside attack at the carbon of a metal-carbon bond [23-25]. Thus far this type of reaction has been little studied. Other racemization pathways are sometimes difficult to identify.

Closely related $d^{10}$ systems have been found to react by radical paths, although the precise mechanisms are controversial. The methods employed to study these mechanisms are instructive, but it is also important to recognize the pitfalls in each technique.

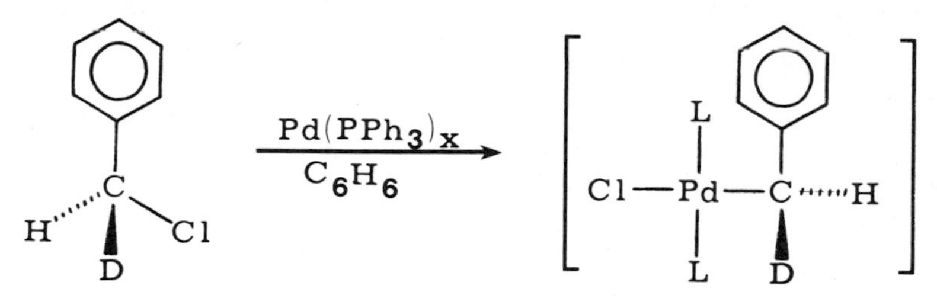


presumed stereochemistry (4.28)

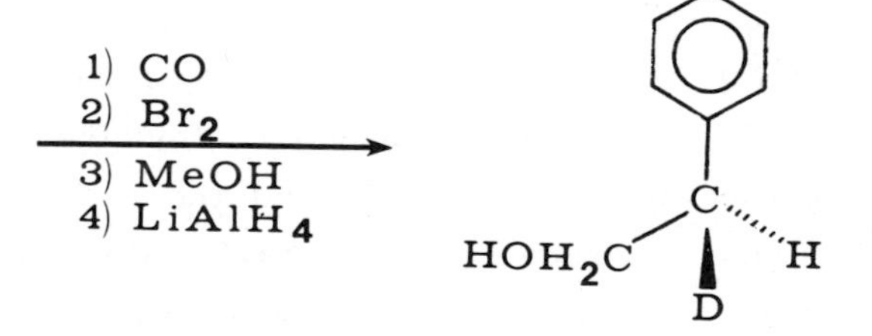


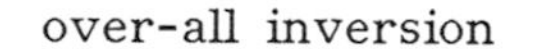

over-all inversion

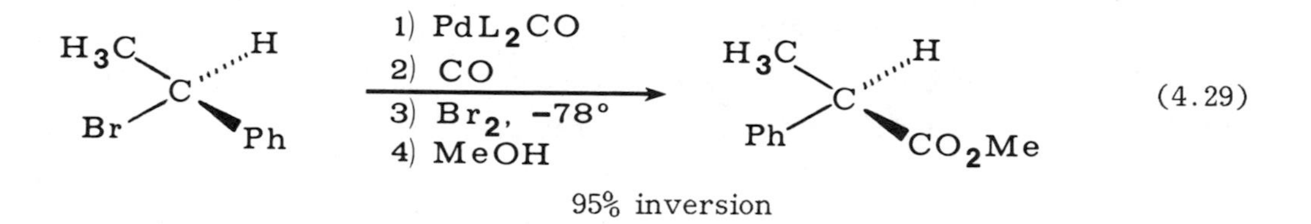


(4.29)

95% inversion

With substrates which are less favorable for $S_N2$ reactions, the palladium(0) complexes afford products which could be the result of one-electron reactions [26]. Examples of products derived from the presumed intermediate radicals are shown in Equation 4.30 and 4.31.

Palladium complexes containing the more basic phosphine, $PEt_3$, are alkylated with over-all inversion at carbon, but the optical yields are lower, and the origin of partial racemization is unclear [26].

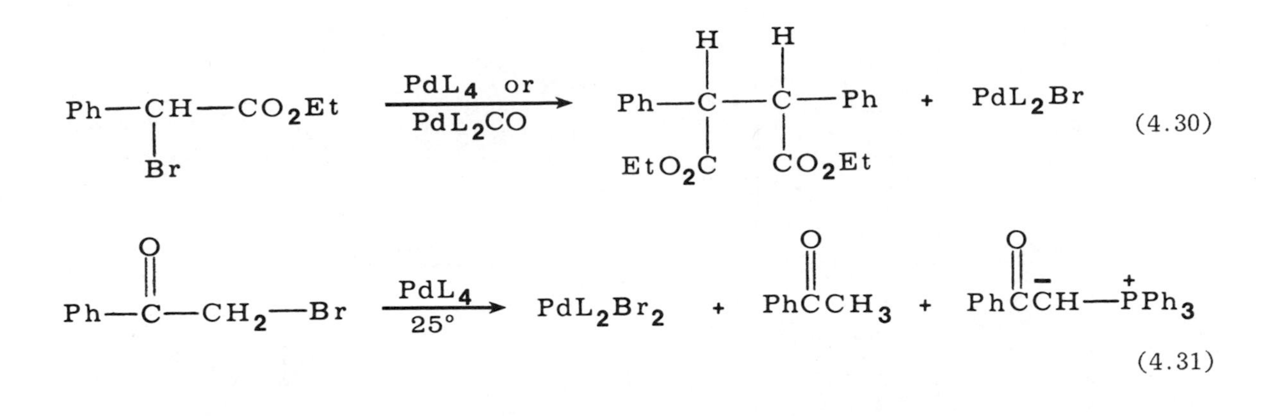


Compared with the above palladium(0) compounds, the corresponding platinum(0) complexes, $PtL_3$ L = $Ph_3P$ and $PEt_3$, seem to have a greater tendency to react by radical paths, but the situation is complicated--both radical-chain and radical-cage paths have been proposed [1d,27,28]. For example, the reaction with n-butyl bromide and $Pt(PEt_3)_3$ is inhibited by galvinoxyl, but it has been pointed out that galvinoxyl reacts with the platinum(0) complex [1d]. Methyl iodide reacts with $Pt(PPh_3)_3$, giving the simple oxidative adduct under conditions where a simple second-order rate law is observed. Introduction of radical spin-trapping reagents such as t-butyl-NO show evidence for the intermediacy of radicals, but it is difficult to quantify this sort of experiment. There is also a danger that the organometallic products may react with the spin-trapping reagent (as does occur with the related, more labile palladium alkyls), but in the platinum system, control experiments appear to rule out this possibility [29]. It has been suggested that this is a bimolecular reaction affording radicals in a solvent cage [1d], involving the two-coordinate platinum(0) complexes (Equations 4.32 and 4.33). Note that t-Bu(NO)R is a stable free radical, easily detected by ESR.

The radical-cage process shown in Equation 4.32 is to be distinguished from a radical-chain process, Equation 4.33, in which the alkyl radical reacts with another platinum(0) forming an alkyl platinum(I) complex, **7**, which reacts in turn with the alkyl halide. These radical-chain reactions should be easier to perturb by radical scavengers (and thus to detect) than the radical-cage reactions.

Another technique which has been used to search for radical intermediates in these reactions is CIDNP (chemically induced dynamic nuclear polarization) [27]. However, since this method is difficult to quantify and exceedingly sensitive, CIDNP results should be interpreted with caution.

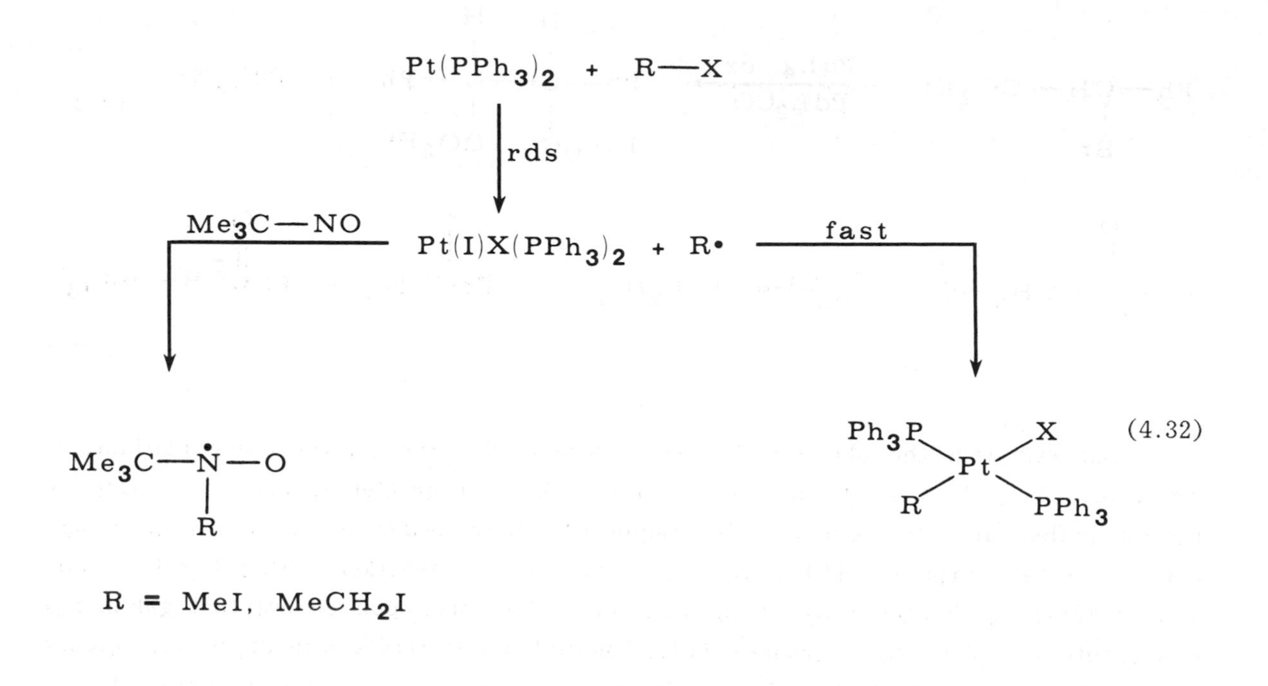


$$\mathrm{Pt(PPh_3)_n \xrightarrow{R\cdot} \underset{7}{[RPt(PPh_3)_n]} \xrightarrow{RX} Pt(PPh_3)_2(R)X + R\cdot} \quad (4.33)$$

In summary, the situation involving radical oxidative-addition mechanisms is still largely unresolved.

c. Macrocyclic Rhodium(I) Complexes. Planar macrocyclic $d^8$ complexes often undergo facile oxidative-addition with alkyl halides. The neutral rhodium(I) macrocycle, **8** in Figure 4.4 has been studied with a wide series of alkylating agents [24]. The observed reactivity profile is consistent with and indicative of an $S_N2$ mechanism. Note that alkyl sulfonates are thought to react by $S_N2$ rather than radical mechanisms.

With the exception of secondary iodides which appear to follow a radical path, the oxidative-addition reactions in Figure 4.4 conform to second-order rate laws: rate = $k_2$[RX][Rh(I)]. It is instructive to compare the second-order rate constant for the

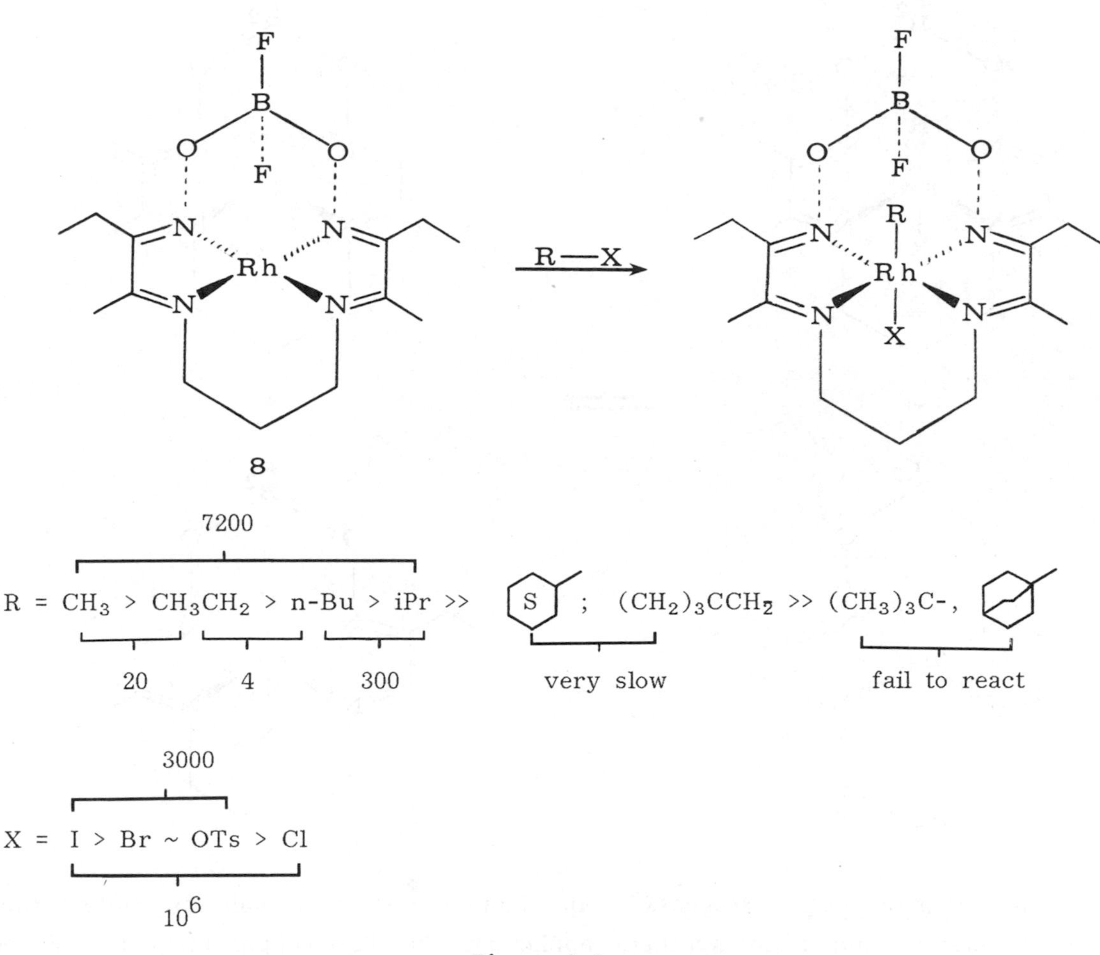


*Figure 4.4.*

macrocycle, 8, and the most reactive iridium(I) Vaska complex (Equation 4.18). (For MeI and the rhodium macrocycle, $k_2 > 10^2\ M^{-1}\ sec^{-1}$, compared with $k_2 = 5 \times 10^{-2}$ for $IrCl(CO)(PMe_2Ph)_2$ at 25°C.) The macrocycle is thus ~$10^4$ faster.

An interesting type of oxidative-addition reaction is the "self-displacement" or metal alkyl exchange reaction. An example is given in Equation 4.34. In such reactions one metal is the leaving group and the other is the nucleophile. The rates of these self-displacement reactions are strongly dependent on steric effects, for example, Me >> Et. The rates of such reactions have been reported in only a few instances [30].

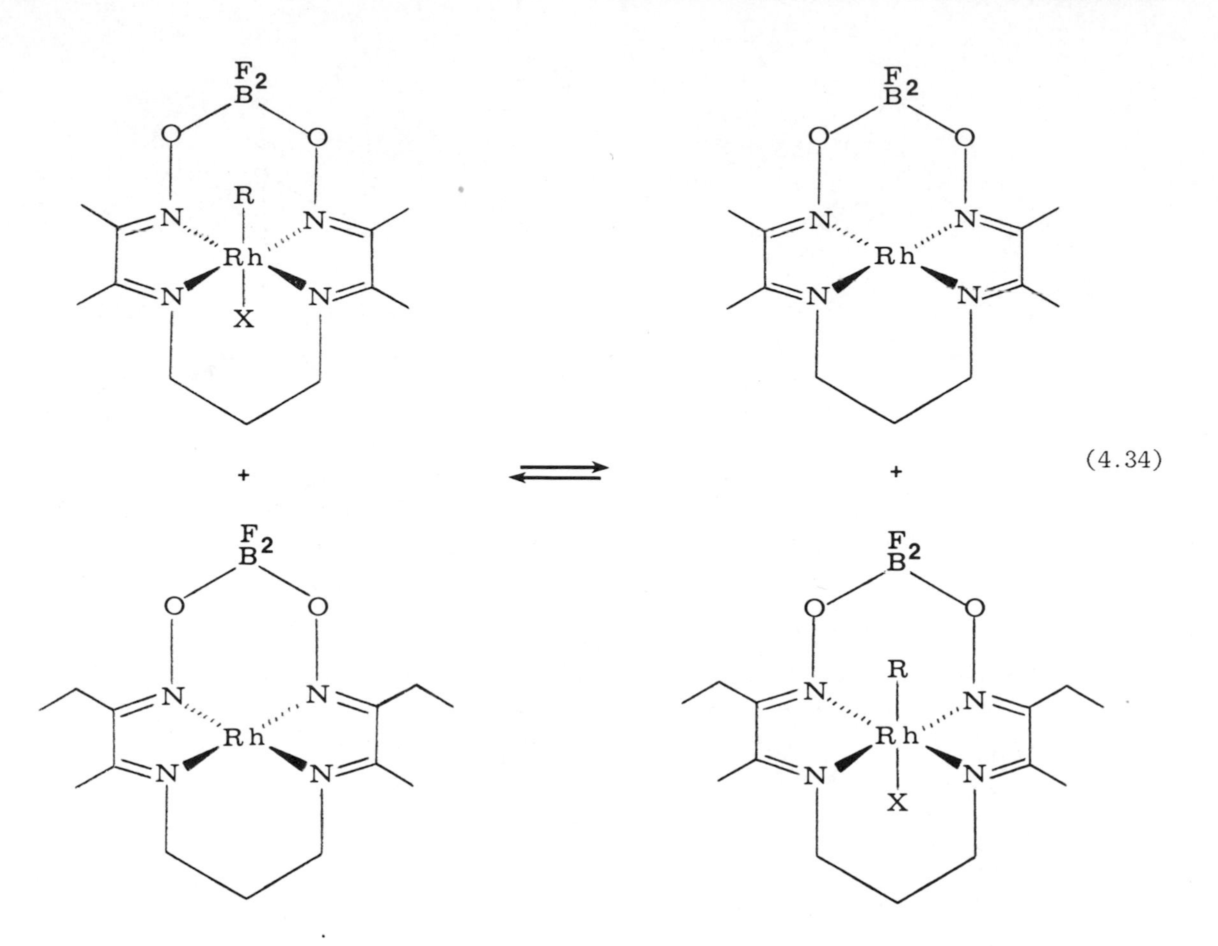


iv. Organo-Copper Reagents. In Chapter 11 we shall encounter the enormously versatile and useful synthetic applications of alkyl copper reagents. Even though the precise nature of these anionic, "ate," complexes remains uncertain, there have been careful studies of the rates and stereochemistries of these reactions. These carbon-carbon bond-forming, alkyl-coupling reactions appear to involve oxidative-addition to a copper(I) ate complex followed by a rapid reductive-elimination from a putative dialkyl copper(III) intermediate. The following results come from G. Whitesides, who was instrumental in developing this chemistry, and subsequently by C. Johnson, who further examined the kinetics and stereochemistry of these reactions [31]. The oxidative-addition step seems to involve inversion at the substrate carbon of the alkylating agent and retention of configuration at both carbon atoms in the supposed reductive-elimination step. In Equation 4.35 the configuration of R in the presumed

$$(R_2Cu(I)Li)_n + R'—X \longrightarrow LiX + [R_2Cu(III)R']_n \longrightarrow R—R' \quad (4.35)$$

intermediate is retained and that at R′ is inverted. The stereochemical course of these reactions is evident from the examples given in Equations 4.36 to 4.40. The

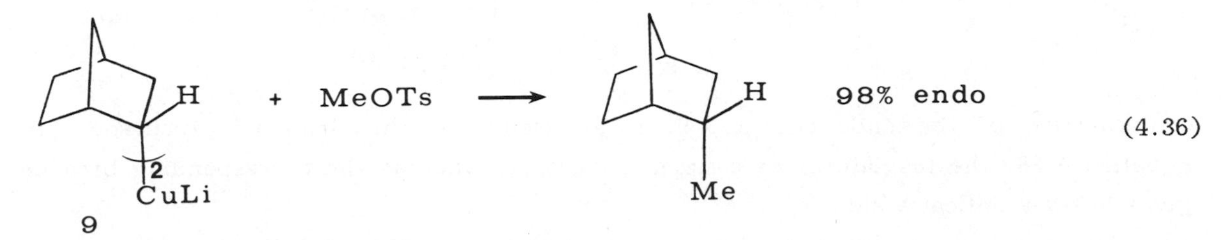


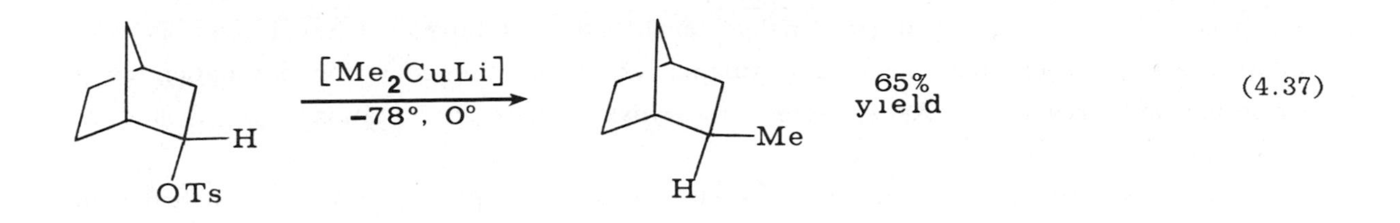


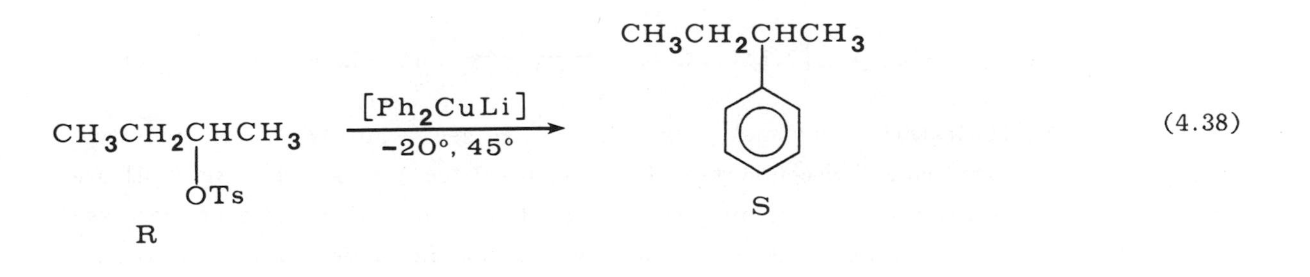


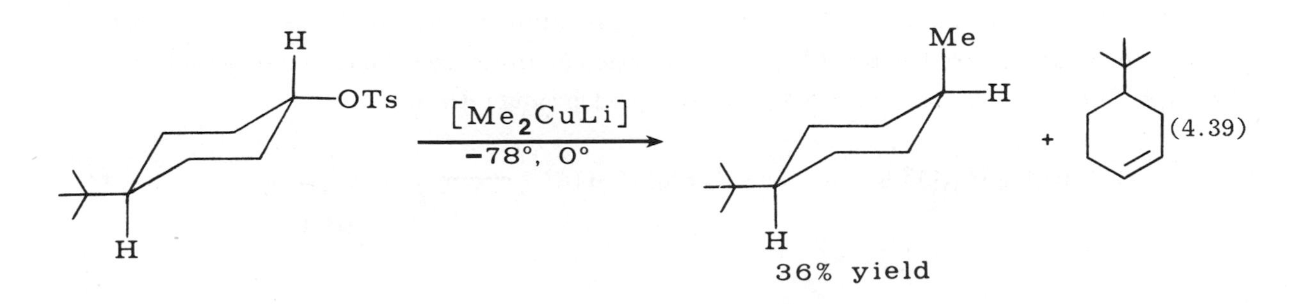

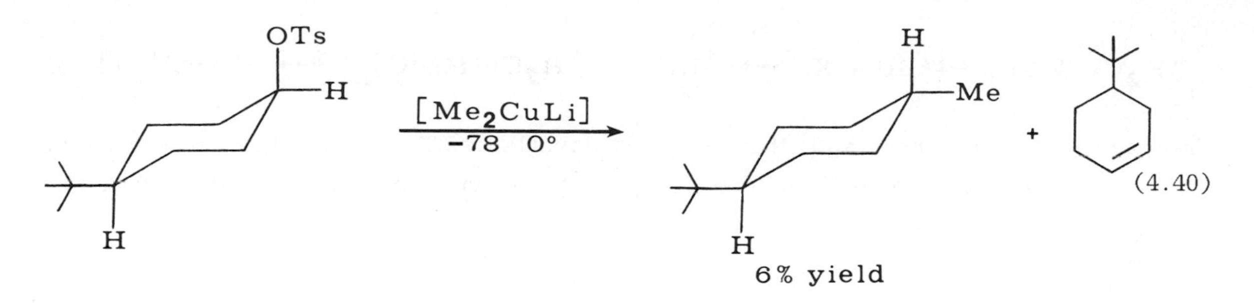


configuration of the endo cuprate, 9, is assumed from the Grignard precursor. In Equation 4.38, the tosylate gives complete inversion, whereas the corresponding bromide gives a lower optical yield.

Note that the stereochemistries shown in Equations 4.36 and 4.37 are under kinetic control and are consistent with the premise of inversion in the oxidative-addition and retention in the reductive-elimination step. The chiral substrate in Equation 4.38 shows the same effect. The conformationally "anchored" 4-t-butylcyclohexyl systems also follow these stereochemical rules and further show the tendency for elimination as a competing side reaction. As expected, the axial tosylate in Equation 4.40 gives more elimination.

The rate-limiting step for the over-all reaction in Equation 4.41 is first-order in both octyl tosylate and the copper reagent. The second-order rate behavior

$$R{-}X + [LiMe_2Cu]_n \xrightarrow[-42^\circ]{Et_2O} R{-}Me \qquad (4.41)$$

(rate = $k_2[RX]^1[LiMe_2Cu]^1$, $k_2(Et_2O) = 2.8 \times 10^{-3}$ M $sec^{-1}$ at -42°C) and substrate reactivity profile [primary > secondary > neopentyl(tosylates); OTs > I ~ Br > Cl] are consistent with an $S_N2$ rate-determining step. The fact that tosylate is even more reactive than iodide is notable. Such a result would not be expected from a radical process.

Note that the high yields in these cross-coupling reactions rule out symmetric three-coordinate intermediates (4.42). A symmetric three-coordinate intermediate would be expected to give a maximum 67% cross-coupled product (4.43).

$$n\text{-}C_5H_{11}OTs + (n\text{-}C_4H_9)_2CuLi \longrightarrow \underset{98\%}{n\text{-}C_9H_{20}} \qquad (4.42)$$

$$R{-}Cu\begin{matrix}\diagup R' \\ \diagdown R'\end{matrix} \longrightarrow 2\ R{-}R' + R{-}R \qquad (4.43)$$

A plausible mechanism for such reactions can be inferred from Kochi's results with isoelectronic organogold reagents, which we will return to in the reductive-elimination section of this chapter.

e. Alkylation of Coordinatively Saturated Complexes. Even though unsaturated $d^8$ or $d^{10}$ complexes are usually more reactive, certain saturated $d^{10}$ and $d^8$ complexes can be alkylated. For example, anionic, coordinatively saturated transition-metal complexes can be especially potent nucleophiles. Alkylation of the metal center in such compounds may be thought of as an oxidative-addition. Perhaps the most thoroughly studied complex of this type is $Na_2Fe(CO)_4$, which is so reactive that it has been referred to as a "supernucleophile" [32,33]. In protic solvents or in the presence of traces of water, this strong Bronsted base forms the markedly less reactive hydride, $NaHFe(CO)_4$. In aprotic solvents there is strong ion-pairing between sodium and carbonyl oxygen atoms. Thus, solvents which solvate sodium cations, such as hexamethylphosphorustriamide (HMPA), dimethylformamide (DMF), or N-methylpyrrolidinone (NMP), greatly enhance the reactivity of $Na_2Fe(CO)_4$. Rate increases of $\sim 10^4$ have been observed.

The observed substrate reactivities, n-$C_{10}H_{21}$-X, X = I > Br > OTs > Cl, and inversion of configuration at the alkylating carbon are consistent with an $S_N2$ reaction. The application of these reactions to organic syntheses is outlined in Chapter 9.

The tremendous range of nucleophilicities manifested by transition-metal centers in alkylation-oxidative-additions is sometimes given by the "Pearson nucleophilicity parameter," [34] which expressed the log of the ratio of rate constants for a given nucleophile and for methyl alcohol reacting with methyl iodide. Consider, for example, the n values in Table 4.1. The iron salt, $Na_2Fe(CO)_4$, in NMP appears to be a "supernucleophile" when compared on this logarithimic scale with a mercaptide salt. Note also that the isostructural monoanion, $Co(CO)_4^-$, is thirteen orders of magnitude less reactive! This result is reasonable when one considers the similarly great difference in $pK_a$ values for the corresponding "hydride" conjugate acids. However, such quantitative comparisons of nucleophilicities have little meaning because of the powerful influence of solvent effects. The high basicity of $Na_2Fe(CO)_4$ limits the range of substrates which may be employed with this reagent because of competing $E_2$ eliminations. For this

*Table 4.1. Pearson nucleophilicity parameter n [34] for transition-metal nucleophiles.*

| Compound | $n = (\log \frac{k_n}{k_{MeOH}})$ | Solvent |
|---|---|---|
| $Na_2Fe(CO)_4$ | 16.7 | NMP |
| Vitamin $B_{12(s)}$ | 14.4 | MeOH |
| $[\eta^5\text{-}C_5H_5Fe(CO)_2]Na$ | 15 | glyme |
| $C_6H_5S^-$ | 9 | MeOH |
| $I^-$ | 7.4 | MeOH |
| $CH_3O^-$ | 6.3 | MeOH |
| $Co(CO)_4^-$ | ~3.5 | glyme |

reason tertiary substrates are useless, and secondary sulfonates are preferable to secondary halides.

A useful technique for examining the relative importance of polar $S_N2$ versus radical paths is to employ a substrate whose radical derivative rearranges at a rate nearly competitive with diffusion-controlled bimolecular reactions at typical concentrations. The cyclopropylcarbinyl radical is perhaps the best substrate for this kind of analysis, since the radical rearranges (Equation 4.44) at a very rapid rate ($k_1 = 1.3 \times 10^8\ sec^{-1}$ at 25°). San Fillippo used this substrate to examine the reaction (4.45)

$$\text{c-}C_3H_5\text{-}CH_2\cdot \underset{}{\overset{k_1}{\rightleftharpoons}} H_2C{=}CH\text{-}CH_2CH_2\cdot \qquad (4.44)$$

between the saturated iron(0) anion, $[\eta^5C_5H_5Fe(CO)_2]^-$, and cyclopropylcarbinyl iodide and bromide [35]. The product(s) were detected by $^1H$ nmr. The iodide gives the product mixture expected from a radical path, but the bromide does not afford a detectable homoallylic product. Thus, there is an apparent change in mechanism in going from the iodide (radical path) to the bromide ($S_N2$ path). It is well-known that halogen-atom abstraction follows the rate profile R-I > R-Br > R-Cl >> R-OTs. During

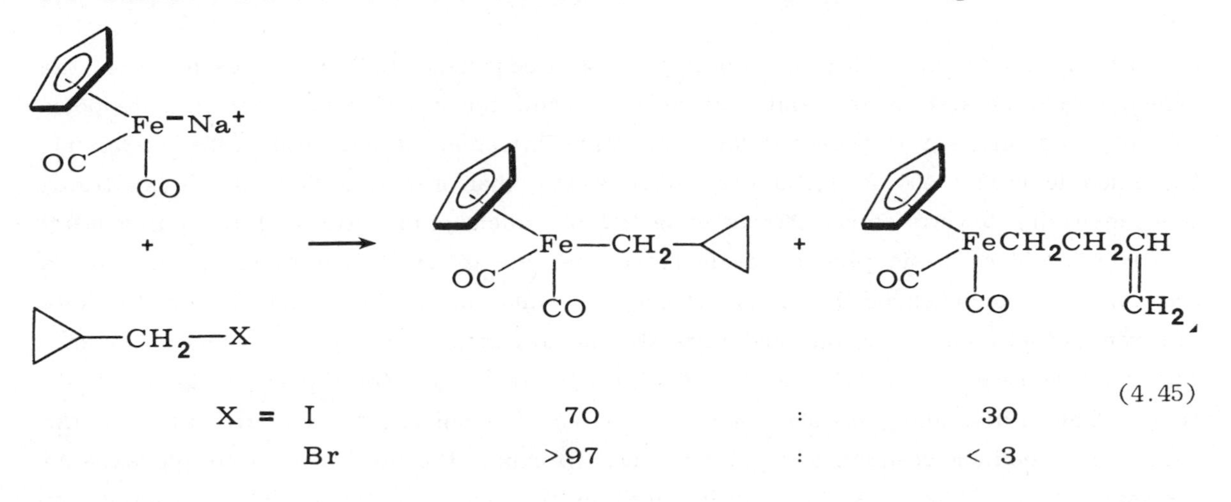


reactions between this iron(0) complex and various alkyl iodides, ESR spectra showed that alkyl radicals were present. The substrates shown in Figure 4.5 were used. The

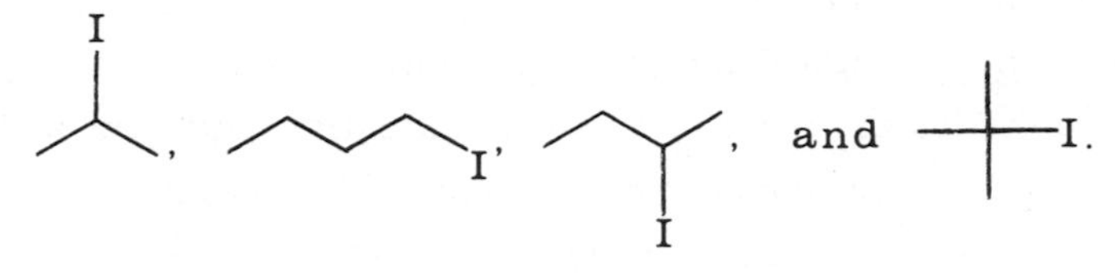

*Figure 4.5.*

corresponding bromides and chlorides did not give such ESR signals. It is important to note that such an ESR experiment is not a reliable measure of how much such radical paths may be involved in these reactions. The technique is extremely sensitive (and thus may detect a minor path), and is difficult to quantify. The failure to detect an ESR signal is meaningless, as are most "negative experiments."

A kinetic analysis of the reaction (Equation 4.46) between alkyl iodides and the coordinatively saturated $d^8$ complexes, $\eta^5$-$C_5H_5M(CO)PR_3$ (M = Co, Rh, and Ir) reveals

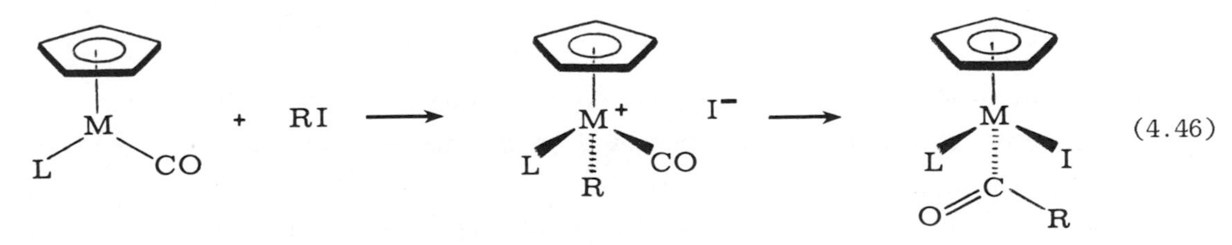

a pattern consistent with a two-step $S_N2$ mechanism [36]. In each case the rate-determining step is over-all second-order, and the reaction is faster in more polar solvents and shows the expected large negative entropies of activation ($\Delta S^{\ddagger}$ = -35 eu). Ethyl iodide reacts 400 to 1,200 times more slowly than methyl iodide, which illustrates the sensitivity toward steric effects expected for nucleophilic attack of the bulky metal at carbon. More basic phosphines increase the rate of oxidative-addition, but this is somewhat counterbalanced by steric effects of bulky phosphines such as $Cy_3P$. Rate differences between the metals are moderate and are influenced by the alkylating agent. The relative rates for $CH_3I$ are Co 1.0, Rh 1.4, Ir 8; but for $C_2H_5I$ are Co 2, Rh 1, Ir 6. The initial ionic oxidative-adduct, 10, can be isolated for iridium; whereas the cobalt and rhodium complexes rapidly insert, affording the neutral acyl complexes. As we shall see in Chapter 5, such differences in the tendency for insertion reactions to occur are typical and can be accounted for by the stronger metal-alkyl bonds formed by the third-row, 5d elements. Benzyl halides show similar behavior with the rhodium derivative. Unfortunately, for this system the stereochemistry at the alkylating carbon has apparently not been examined.

f. Acylation Through Oxidative-Addition. Active acylating agents such as acid chlorides and anhydrides readily undergo oxidative-addition with many saturated and unsaturated low-valent transition-metal complexes. Again, most studies have involved $d^8$ or $d^{10}$ complexes. Initial products are acyl derivatives. As might be expected, such acylations are much more facile than alkylations. Thus, all those complexes which can be alkylated should be susceptible to acylation, although fewer acylations have been examined. Such reactions may be used to form a great variety of acyl complexes; Equation 4.47 is illustrative. Carboxylic esters seem to be rather inert

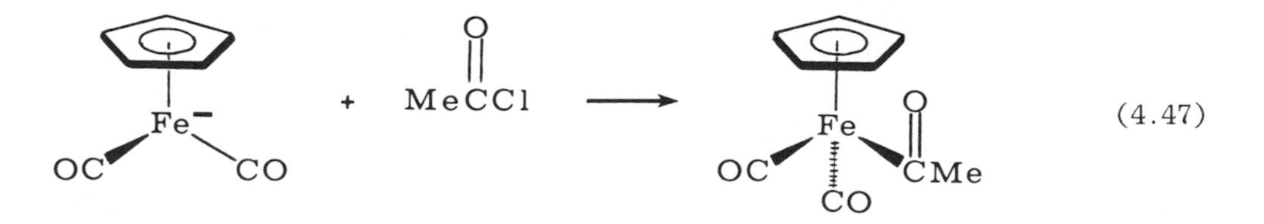


(4.47)

toward oxidative-addition. Alkyl esters do not appear to react; however, activated phenyl and allyl esters have been reported to acylate neutral Ni(0) [37] and anionic iron(0) [38] compounds (Equation 4.48).

In Chapter 8 we shall see that the oxidative-addition of cyclic anhydrides to $Na_2Fe(CO)_4$ is a useful synthesis of half-aldehydes, $HO_2C(CH_2)_nCHO$.

$$Fe(CO)_4^{2-} + R\overset{O}{\overset{\|}{C}}{-}OPh \longrightarrow R{-}\overset{O}{\overset{\|}{C}}{-}\bar{F}e(CO)_4 \qquad (4.48)$$

g. Metal-Carbon-Bond-Forming Oxidative-Addition Reactions Which May Proceed Through Prior Coordination. A number of unsaturated organic halides undergo oxidative-addition under circumstances which indicate that the reaction path involves prior coordination of the organic substrate. Thus palladium(0) adds aryl halides (Equation 4.49). It is interesting that in contrast to most aromatic nucleophilic

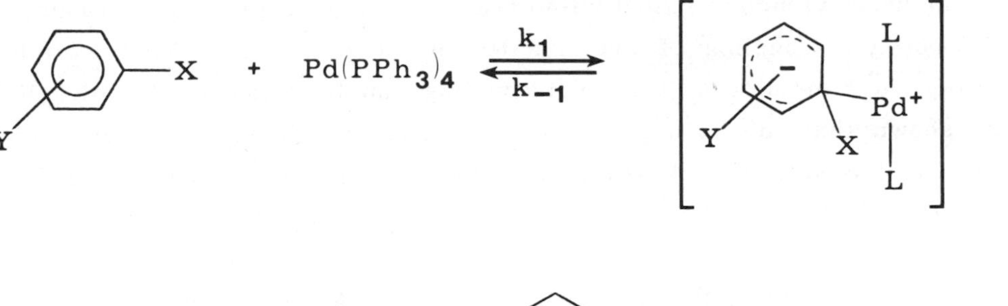


(4.49)

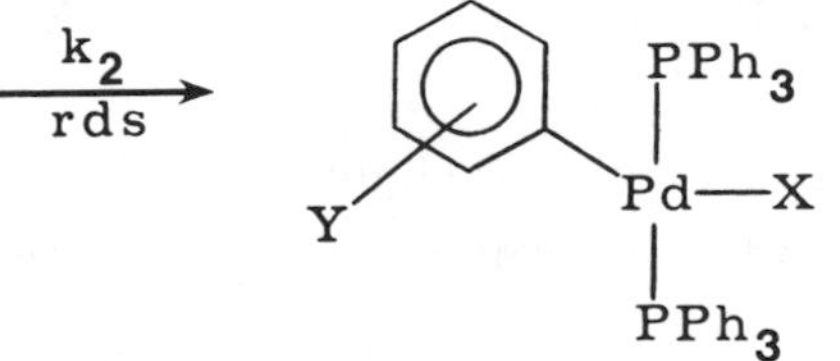


substitutions [39], the order of reactivity for Ar-X is I > Br > Cl (F doesn't react!). This reactivity profile indicates either that the carbon-halogen bond breaking occurs in the rate-limiting step or that electron-transfer is rate-limiting (see below). Electron-withdrawing groups on the aromatic ring increase reactivity [40]. The mechanisms of these reactions are uncertain, but a prior electron-transfer path is as plausible as nucleophilic displacement.

The oxidative-addition of aryl halides to the isoelectronic Ni(0) phosphine complexes affords trans-arylnickel(II) halides and paramagnetic nickel(I) halides (Equation 4.50). This reaction, which is probably related to the palladium(0) case, has been extensively studied by Kochi [41]. An over-all second-order rate-determining step (rate = $k_2$[Ni(0)][ArX]) is proposed to involve electron-transfer from nickel to the aryl halide, perhaps by way of a $\pi$-complex. In Equation 4.50 the postulated

$$NiL_3 + ArX \xrightarrow{rds} \left[\overset{+\bullet}{Ni}L_3\overset{-\bullet}{ArX}\right]_{\mathbf{11}} \begin{cases} \xrightarrow{(a)} ArNiXL_2 + L \\ \xrightarrow{(b)} XNiL_3 + Ar\bullet \end{cases} \quad (4.50)$$

$L = PEt_3$

paramagnetic ion pair, 11, is thought to be a common intermediate which is partitioned into the nickel(II) oxidation-adduct (path a) and the nickel(I) halide complex (path b). The rate-determining step is sensitive to solvent and substituent electronic effects, but not to steric effects [42].

The above aryl halide oxidative-addition to Ni(0) complexes is undoubtedly related to an Ullman-type coupling of aryl halides in the presence of Ni(0) complexes. With zinc powder as a reductant, this aryl coupling can be catalyzed by nickel(II). Kochi [42] has shown that this is a radical-chain process involving intermediate nickel(I) and nickel(III) aryl complexes. A key step is the oxidative-addition of an aryl halide to an intermediate nickel(I) complex (Equation 4.51). Aryl exchange between nickel(III) and

$$Ni(I)X + ArX \longrightarrow ArNi(III)X_2 \quad (4.51)$$

nickel(II) affords a nickel(III) diaryl (Equation 4.52). Reductive-elimination (Equation 4.53) affords the biaryl and regenerates nickel(I). Free aryl radicals are not involved,

$$ArNi(III)X_2 + Ar'Ni(II)X \longrightarrow NiX_2 + ArNi(III)(Ar')X \quad (4.52)$$

$$Ar{-}Ni(III)(Ar'){-}X \longrightarrow Ar{-}Ar' + Ni(I)X \quad (4.53)$$

since 1,4-cyclohexadiene, which should trap aryl radicals, does not affect the biaryl coupling. However, oxidizing agents such as quinones and nitroaromatics inhibit the chain process, apparently by oxidizing the nickel(I) species. This study demonstrates that odd-electron complexes can also undergo two-electron oxidative-additions.

Similar reaction schemes may be involved in transition-metal-catalyzed cross-coupling of Grignard reagents and alkyl halides, which are described in Chapter 11.

Aryl halides also oxidatively add to iridium(I) compounds, but relatively high temperatures (140°) are required [43].

Similar oxidative-additions of vinyl halides to Pt(0), Pd(0), and Ni(0) complexes are related to these aryl halide reactions. Such reactions occur with remarkable stereospecificity--in all cases the metal replaces the halide with retention of E or Z configuration [44]. Since perhaloethylenes have been shown to react with Pt(0) complexes by way of isolable $\eta^2$-olefin compounds which subsequently rearrange to form the σ-vinyl oxidative adducts [45], such a path has been suggested for other vinyl halide oxidative-additions [44f,46].

It is less likely that these vinyl halide oxidative-additions proceed through a free vinyl radical, since such an intermediate vinyl radical should rapidly undergo configurational inversion [47].

Ordinarily carboxylic esters are rather unreactive toward oxidative-addition--by attack either at the acyl carbon (vide infra) or at the alkoxy carbon. However, allylic acetates readily add to palladium(0), affording $\eta^3$-allyl complexes with over-all inversion at carbon. It is probable that prior coordination of the olefin accounts for this facile reaction. The stereochemistry is determined from two consecutive reactions (Equation 4.54), both of which are assumed to go with inversion. This double-inversion sequence has been used by Trost to control the stereochemistry at C-20 in steroids [48].

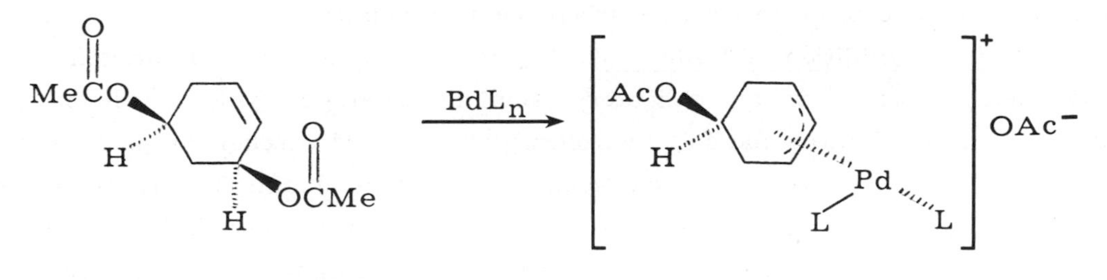


(4.54)

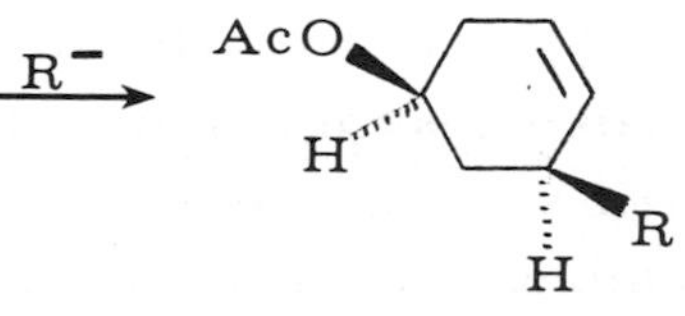


$R^- = C_6H_5SO_2\bar{C}HCO_2Me$

Cleavage of unactivated carbon-carbon bonds by oxidative-addition is rarely observed. However, special cases are recognized. An example is the addition of 1,1,1-tricyanoethane to the platinum(0) complex (Equation 4.55) [49]. The curious cis-stereochemistry at platinum was assigned from IR spectroscopy.

$$\mathrm{MeC(CN)_3} + \mathrm{Pt(PPh_3)_4} \xrightarrow[\text{benzene}]{\text{boiling}} \mathrm{(Ph_3P)_2Pt(CN)(C(CN)_2Me)} \tag{4.55}$$

Strained carbocycles such as cyclopropanes will oxidatively add to reactive transition-metal complexes, affording metallacycles (Equation 4.56) [50]. These

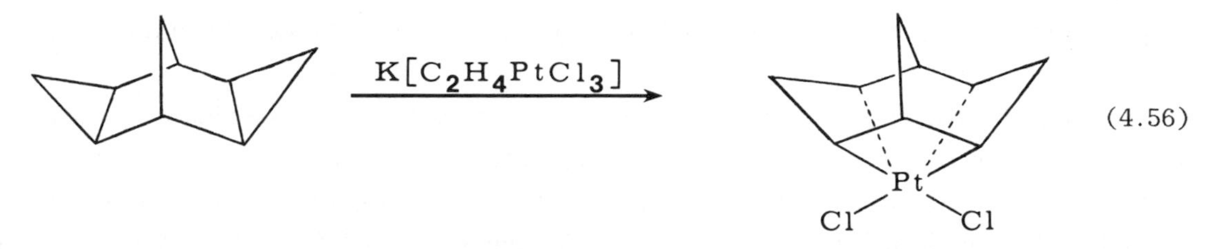


reactions are fairly common. Further examples are discussed in Chapter 10 as an important method of forming metallacyclobutanes. The detailed mechanisms of these oxidative-additions are uncertain. For example, does the metal center form a complex with the cyclopropane prior to carbon-carbon bond breaking?

3. Oxidative-Addition of Hydrogen. The addition of hydrogen to a transition-metal complex is an obligatory step in catalytic cycles such as olefin, acetylene, or arene hydrogenation, hydroformylations, and Fischer-Tropsch reactions. Thus we will encounter this reaction again in Chapters 6 and 8. There are many examples of complexes which add hydrogen in a well-defined, reversible manner. Most have a $d^8$ configuration but a few $d^{10}$ and $d^6$ cases are recognized. Such reactions can be divided into four categories: (a) addition of $H_2$ to coordinatively unsaturated complexes (Equation 4.57); (b) addition of $H_2$ to saturated complexes accompanied by the loss of a ligand (Equation 4.58); (c) homolytic hydrogen splitting by two metal centers, each undergoing a one-electron change (Equation 4.59); and (d) heterolytic splitting of hydrogen by a metal complex in the presence of a base [3] (Equation 4.60). The over-all heterolytic $H_2$ splitting reaction is not an oxidative-addition, but may involve such at an intermediate stage [3].

$$M + H_2 \rightleftharpoons \overset{H}{\overset{|}{M}}{-}H \tag{4.57}$$

$$M{-}CO + H_2 \rightleftharpoons \overset{H}{\overset{|}{M}}{-}H + CO \tag{4.58}$$

$$2M + H_2 \rightleftharpoons 2\ H{-}M \tag{4.59}$$

$$\overset{X}{\overset{|}{M}} + H_2 + \text{base} \longrightarrow M{-}H + H^+{-}\text{base}\ X^- \tag{4.60}$$

The simplest and most thoroughly studied example of category (a) is addition of hydrogen to the family of Vaska's complexes (Figure 4.6). These reactions occur in

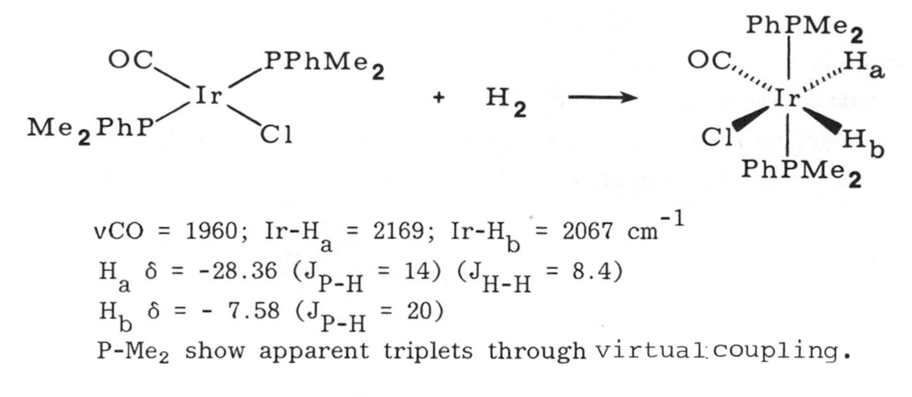


*Figure 4.6.*

solution and in the solid state. In both cases, the $H_2$ addition has been shown to be cis by such NMR and IR data as are shown in Figure 4.6 [52,53b]. Such a concerted hydrogen addition would be expected to afford a cis-adduct. A reaction affording trans oxidative-addition of $H_2$ has been reported, but these results could alternatively be explained by isomerization of an initially formed cis-adduct [53a].

It is reasonable to assume that oxidative-addition of $H_2$ to unsaturated metal complexes is related to the activation of $H_2$ by "chemisorption" to heterogeneous transition-metal hydrogenation catalysts. However, in the chemisorption of $H_2$ on platinum, one H atom becomes attached to each metal atom at the surface, a reaction similar to the homogeneous case in Equation 4.59.

Hydrogenations of $d^{10}$ complexes are rare. The very reactive two-coordinate platinum complexes react with $H_2$ at ambient conditions forming the stable trans dihydride (Equation 4.61). It is uncertain whether the observed trans adducts are the

$$\text{L—Pt—L} + H_2 \xrightarrow[25°]{C_6H_6} \text{trans-}L\text{—Pt(H)}_2\text{—L} \tag{4.61}$$

$$L = PCy_3, P(iPr)_3$$

kinetic products or result from rapid rearrangement of undetected cis-adducts. It is interesting to note that the palladium analogues are inert toward $H_2$ addition at ambient conditions. Three coordinate complexes such as $Pt(PEt_3)_3$ react with $H_2$ only after dissociation of one $PEt_3$. Sufficiently hindered two-coordinate complexes, such as $Pt[P(Ph)t\text{-}Bu)_2]_2$, fail to add $H_2$ [54]. This observation further illustrates the sensivitity of oxidative-additions to steric factors.

Several coordinatively saturated complexes add $H_2$ with the loss of a neutral ligand such as CO, $R_3P$, or $N_2$ (Equations 4.62 to 4.65). It is probable that in these cases

$$Ir^+(CO)_3L_2 \overset{H_2}{\rightleftharpoons} CO + H_2Ir(CO)_2L_2^+ \tag{4.62}$$

$$Ir^+(CO)(PPh_2Me)_4 \xrightarrow{H_2} PPh_2Me + H_2Ir^+(CO)(PPh_2Me)_3 \tag{4.63}$$

$$H_2Ru(N_2)(PPh_3)_3 \xrightarrow{H_2} N_2 + H_4Ru(PPh_3)_3 \tag{4.64}$$

$$Os(CO)_5 \xrightarrow[100-130°C]{80\ atm.\ H_2} H_2Os(CO)_4 + CO \tag{4.65}$$

the ligand is lost prior to $H_2$ addition. Otherwise one need invoke a high-energy, 20-electron intermediate. However, in most instances, this question has not been subjected to a careful kinetic analysis.

Oxidative-addition of hydrogen occasionally requires the presence of a proton base with the resulting over-all heterolytic splitting of the hydrogen molecule into a metal hydride and the protonated base. Some of these reactions undoubtedly proceed through an intermediate dihydride adduct followed by the base-catalyzed removal of a proton. In these cases the over-all reaction (Equation 4.66) is thus an oxidative-addition, reductive-elimination. In other situations a concerted heterolytic splitting of hydrogen may take place (Equation 4.67).

$$M{-}Y + H_2 \rightleftharpoons H{-}\overset{\displaystyle H}{\overset{|}{M}}{-}Y \xrightarrow{B:} H{-}M + H^+{-}B \quad Y^- \tag{4.66}$$

$$M + H_2 + :Base \longrightarrow \overset{-}{M}{-}H + H^+{-}Base \tag{4.67}$$

Equation 4.68 is exemplary of the over-all reaction, without regard to the detailed pathway [59].

$$RuX_2(PPh_3)_3 + H_2 \xrightarrow[C_6H_6]{Et_3N} HRuX(PPh_3)_3 + H{-}\overset{+}{N}Et_3X^- \tag{4.68}$$

$X = Br, Cl$

In other examples of hydrogen activation, a base is not required (Equation 4.69) [60]. This is especially true where Y is a hydrocarbon (Equations 4.70 and 4.71) [61,62]. Related cases in which Pt-Si, Pt-Ge, and Pt-N bonds are hydrogenolyzed are also known. These are probably oxidative-addition, reductive-elimination reaction sequences.

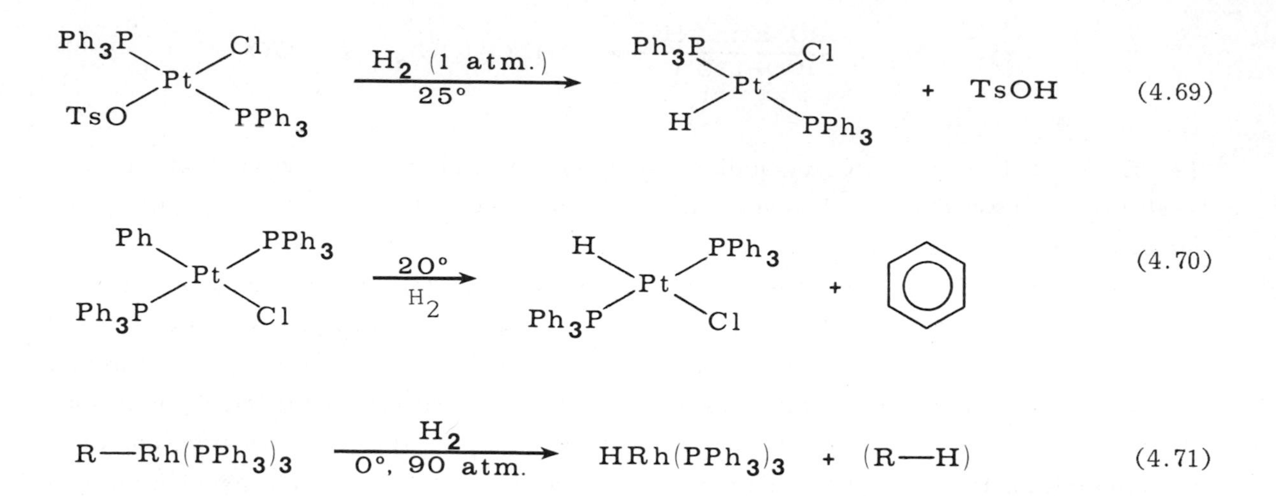


4. Oxidative-Addition of Si-H. Oxidative-addition of the silicon-hydrogen bond is one of the simplest methods of forming transition-metal silyl bonds [62,63]. This reaction is an obligatory step in the homogeneously catalyzed hydrosilation of olefins, acetylenes, and ketones described in Chapter 6, and is known for a number of different transition metals [62,63]. Representative examples of this reaction are shown in Equations 4.72 to 4.74 [65,66,67]. These Si-H additions are usually *cis*, and in some

$$(Ph_3P)_3Co(N_2)H + HSiF_3 \longrightarrow (Ph_3P)_3Co(H)_2(SiF_3) + N_2 \tag{4.72}$$

$$(Ph_3P)_3RhCl + HSiPh_3 \longrightarrow (Ph_3P)_2RhH(SiPh_3)Cl + PPh_3 \tag{4.73}$$

$$(Ph_3P)_2Ir(N_2)Cl + HSi(OEt)_3 \longrightarrow (Ph_3P)_2IrH[Si(OEt)_3]Cl + N_2 \tag{4.74}$$

cases proceed with retention of configuration at chiral-silicon, factors which indicate a concerted reaction.

The reactions in Equation 4.75 illustrate the parallel between $H_2$ and $R_3Si$-H oxidative-addition. Note the formation of μ-dihydride dimers and that the Ge-H bond

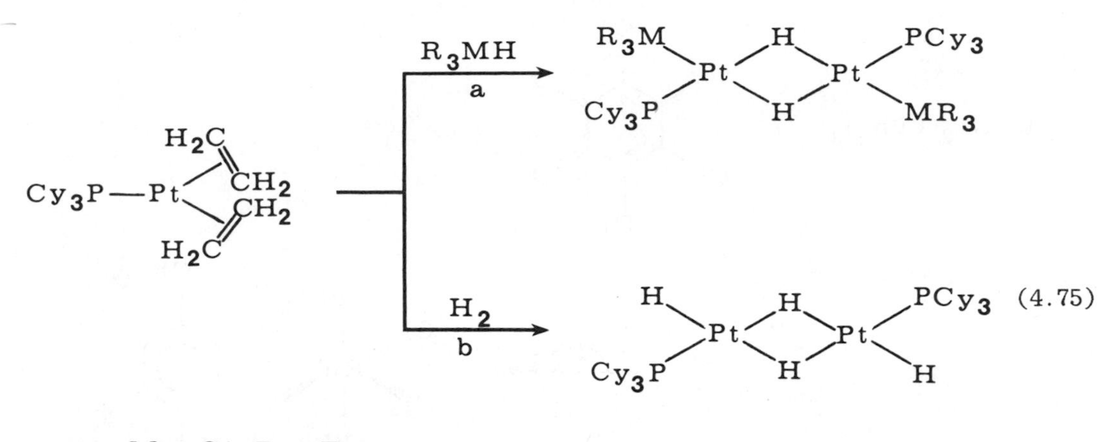


(4.75)

a M = Si; R = Et
M = Ge; R = Me

also adds. In Chapter 6 we will see that these oxidative adducts serve as catalysts for olefin hydrosilation [68].

5. Quinone Oxidative-Additions. Recall that the oxidative-addition of dioxygen to give $\eta^2$-peroxide complexes was described in Chapter 3. A similar oxidative-addition is the reaction of an o-quinone with an unsaturated transition-metal complex affording what may be considered to be a dihydroquinone complex [69]. This is illustrated by a saturated $d^8$ complex (Equation 4.76), and an unsaturated $d^6$ complex (Equation 4.77). It is uncertain whether in the first case a ligand is lost prior to oxidative-addition. Note that the second case involves a less common oxidative-addition to a $d^6$ compound.

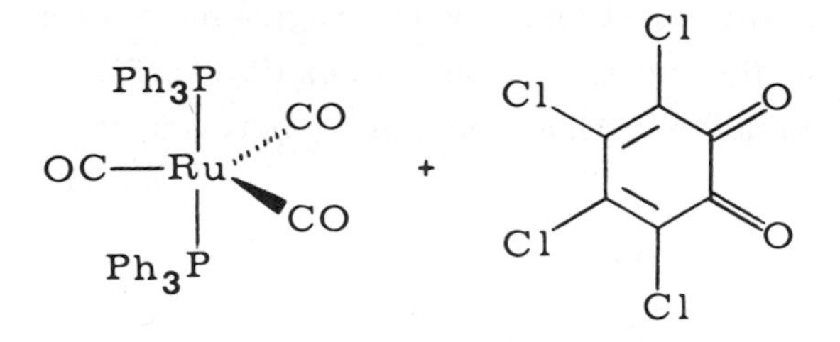


(4.76)

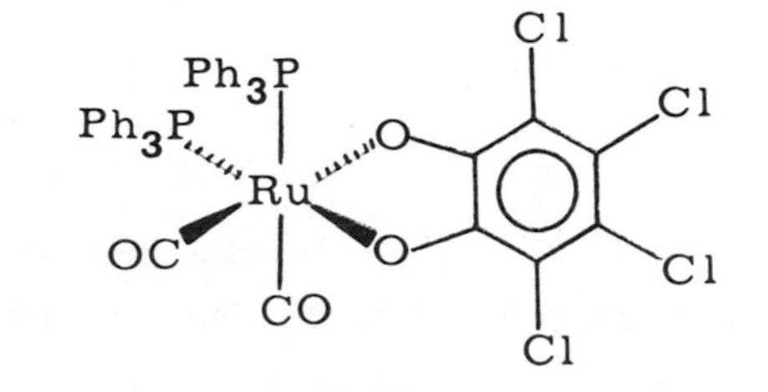

$RuCl_2(PPh_3)_3$ +

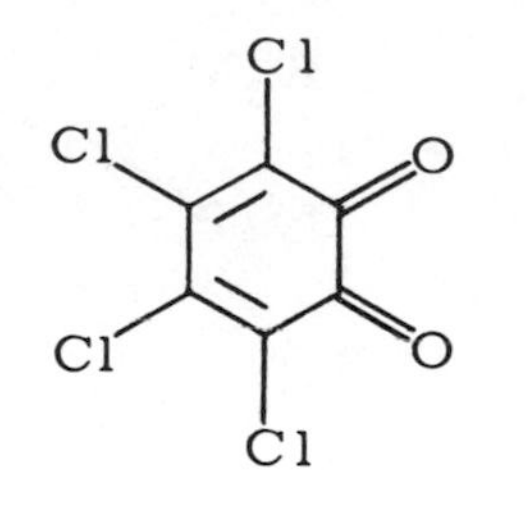


$\xrightarrow{-PPh_3}$

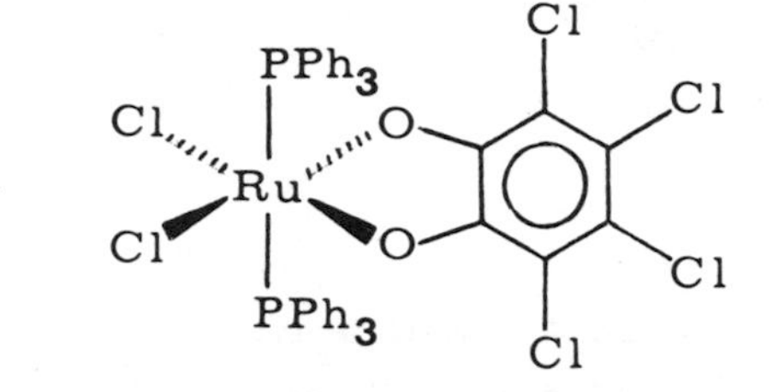


(4.77)

6. Relative Reactivities in Oxidative-Additions. Several factors control the reactivities of metal complexes toward most types of oxidative-additions. The best studied cases involve $d^8$ and $d^{10}$ complexes. As we have seen from examples earlier in this chapter, electron-releasing ligands such as phosphines increase, and $\pi$-acceptors such as CO or olefins decrease, the tendency to undergo oxidative-addition. However, steric inhibition of oxidative-addition is also quite important-especially with the bulky phosphines. Steric and electronic effects may be counterbalanced. Coordinatively unsaturated complexes are invariably more reactive than analogous saturated complexes. This factor is especially evident for $d^{10}$ complexes of the type $ML_4$ (M = Ni, Pd, Pt; and L = $R_3P$ or R'NC). In such cases $ML_2 > ML_3 > ML_4$.

The nature of the metal also influences the reactivity, which might have been expected to reflect the relative tendencies of the metals to become oxidized (Figure 4.7). However, it is usually not possible to make comparisons even between three

| Fe(O) | Co(I) | Ni(II) |
| --- | --- | --- |
| Ru(O) | Rh(I) | Pd(II) |
| Os(O) | Ir(I) | Pt(II) |

Tendency of $d^8$ to be oxidized to $d^6$ ⟶

Tendency of $d^8$ to become 5-coordinate - - - - - >

*Figure 4.7. Characteristics of $d^8$ complexes within Group VIII.*

members of a given triad, nor between metals in adjacent triads. The problem in making such comparisons is that there are few complete isostructural series of complexes. Earlier we saw one such comparison in Graham's study of the coordinatively saturated $\eta^5$-$C_5H_5M(CO)_2$ (M = Co, Rh, Ir) [36]. In most cases, such comparisons are obscured by the marked tendency for $d^8$ complexes to become coordinatively saturated upon ascending a triad and moving towards the left within Group VIII (Figure 4.7). Thus, most Fe(O) complexes are five-coordinate and most platinum(II) complexes are four-coordinate. Therefore, it is difficult to maintain the same degree of saturation within a series. There are, nevertheless, cases in which comparisons can be made between isostructural complexes of second- and third-row elements. The latter are invariably more reactive. Thus, for trans-$MCl(CO)(R_3P)_2$, Ir > Rh (but in this series the first row cobalt(I) derivative is paramagnetic and tetrahedral; so a comparison cannot be made). One complete isoelectronic series has been investigated by Vaska, who reported that the first-row element cobalt is the most reactive (Equation 4.78) [70]. The stereochemistry for the oxidative-adducts is uncertain.

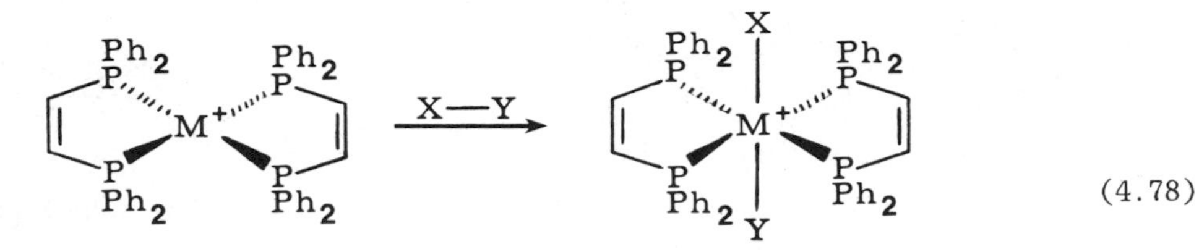


(4.78)

Co > Ir > Rh

X—Y = $O_2$, HCl, $H_2$

In the context of relative reactivities toward oxidative-additions, it is instructive to compare two groups of related complexes in Figure 4.8. The group on the left forms stable hydrogen adducts at ambient temperature and 1 atm. pressure; whereas the group on the right does not. The reactivity relationships described above are implicit in these comparisons. The tendency to form an adduct changes predictably as $\pi$-acids are substituted for electron-releasing ligands, and as first-, second-, and third-row elements are compared. In such reactions kinetic and thermodynamic tendencies are probably parallel, although this point has been little studied, except for Vaska's iridium complex.

7. Oxidative-Addition of C-H Bonds. Activation of saturated hydrocarbons by homogeneous catalysts could lead to the selective functionalization of petrochemical

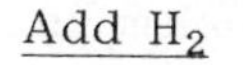


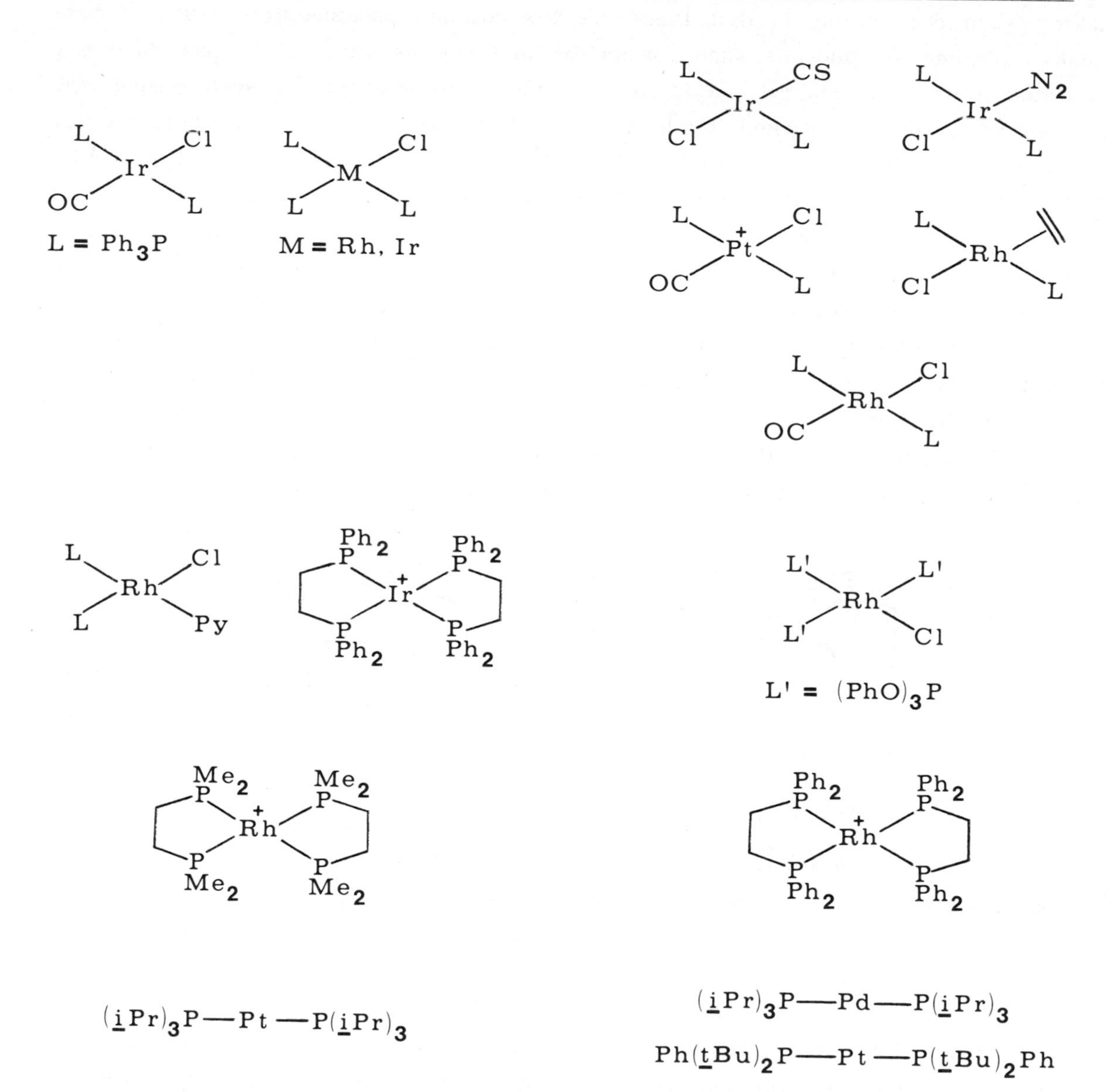


*Figure 4.8. Relative tendencies of complexes to add hydrogen.*

feedstocks. Thus, oxidative-addition of unactivated C-H bonds, especially those in aliphatic hydrocarbons, is a problem of long-standing interest. On thermodynamic grounds such reactions are plausible, since the energetics of C-H oxidative-addition are sometimes not much different from those ubiquitous $H_2$ additions we have previously discussed. However, kinetic paths for C-H additions seem to be far less facile. Intramolecular C-H oxidative-additions, which are called "cyclometallations" or "orthometallations" are in fact common, and have been extensively studied [71,72,73]. Reactions of this type are especially well known for aryl C-H groups. In intramolecular oxidative-additions, the metal and the C-H group are held in close proximity; so the entropy change is more favorable. On the other hand, intermolecular C-H oxidative-additions are more difficult to achieve, especially those involving unactivated aliphatic groups. Such intermolecular hydrocarbon additions have so far not led to synthetically useful procedures. We shall treat the intra- and intermolecular C-H additions separately.

a. Cyclometallations. The first example was discovered by Kleiman in 1963. Nickelocene reacts with azobenzene affording a metallacycle (Equation 4.79).

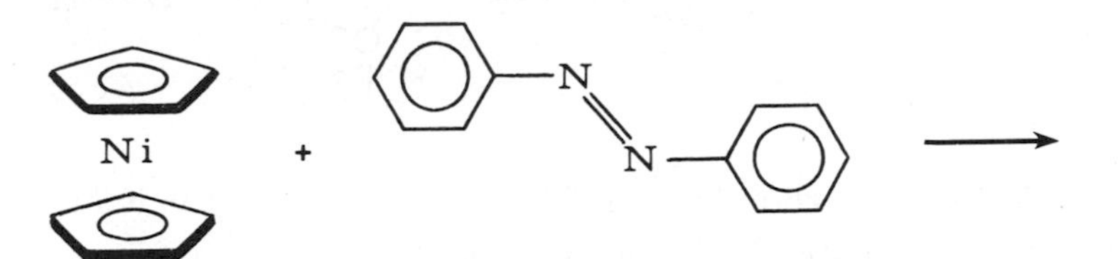


(4.79)

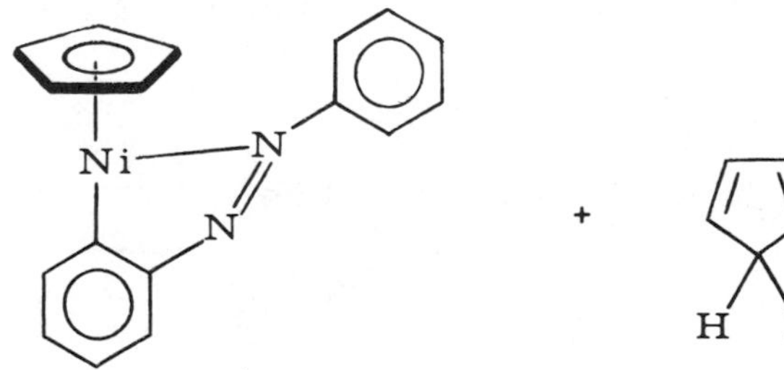


Apparently prior coordination of the azo group leads to oxidative-addition of the ortho C-H group and subsequent reductive-elimination of cyclopentadiene.

Following this discovery, a very large number of such cyclic metallations ("orthometallations") have been reported [71,72]. These cyclometallations follow a similar pattern. The hydrocarbon group is a part of a coordinated ligand and as such is held close to a low-valent metal. The dimensions of the ligand should be such that a ring can be formed. Several sizes are effective, but five-membered rings are normally preferred. For N donor groups five-membered rings are usually formed; whereas both four- and six-membered rings are also formed with the larger P donor groups. In many cases the presence of an adjacent leaving group (an alkyl, hydride, or halide) provides a driving force for the reaction (Equation 4.80) by promoting a subsequent reductive-elimination (path a). In other cases a base may drive the reaction by removing the metal hydride as a proton (path b).

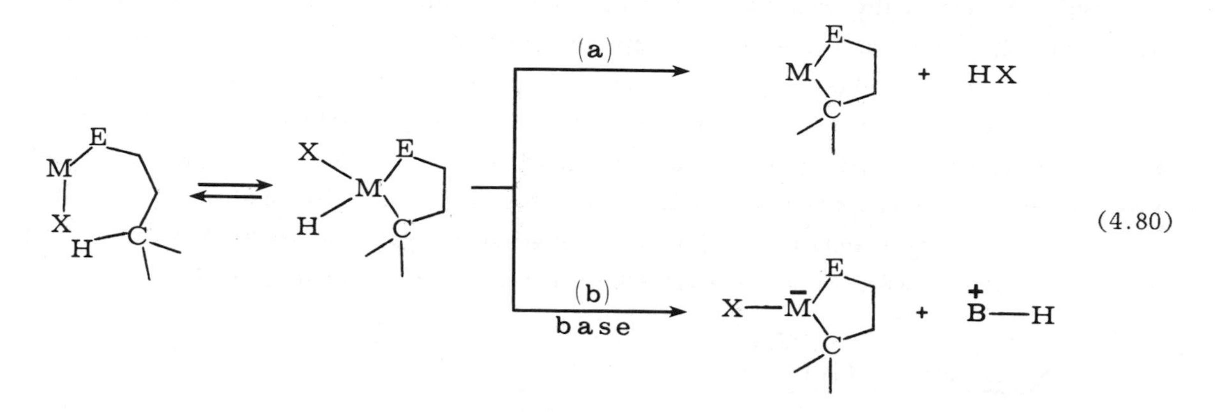


Virtually any donor group, E, may be employed. In spite of the fact that aryl C-H bonds are stronger than aliphatic C-H bonds, aromatic groups are more reactive than aliphatic groups. Reactivity patterns indicate that more than a single mechanism can be involved. Consider Equations 4.81 to 4.86.

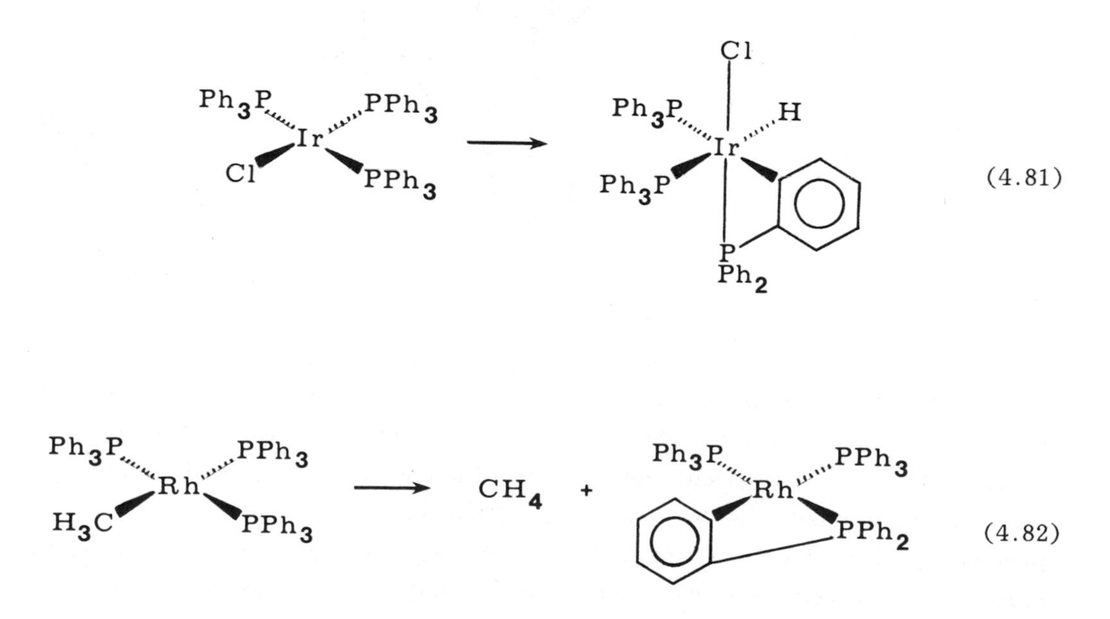

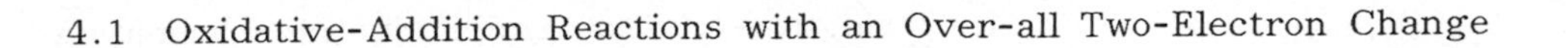

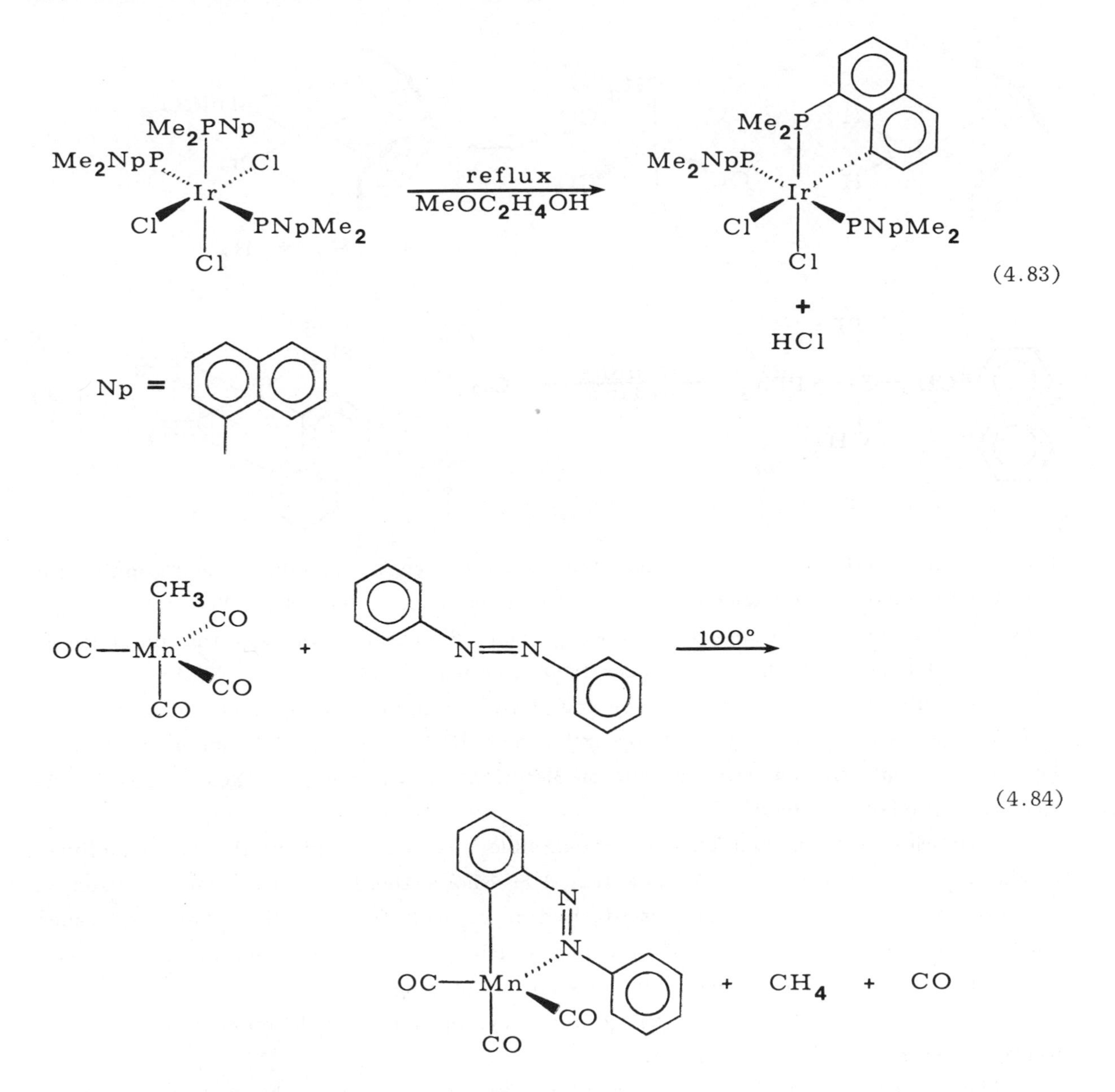


Diverse electronic effects on the reacting ligand and the metal have been observed for orthometallations indicating that more than one mechanism may be involved. For example, in the para-substituted phosphine compounds $[(XC_6H_4)_3P]_3IrCl$, the orthometallation rate decreases in the order: $X = CH_3 > OCH_3 > H >> F$ [72]. These results are consistent with an electrophilic attack by the metal on the aromatic ring. This idea is

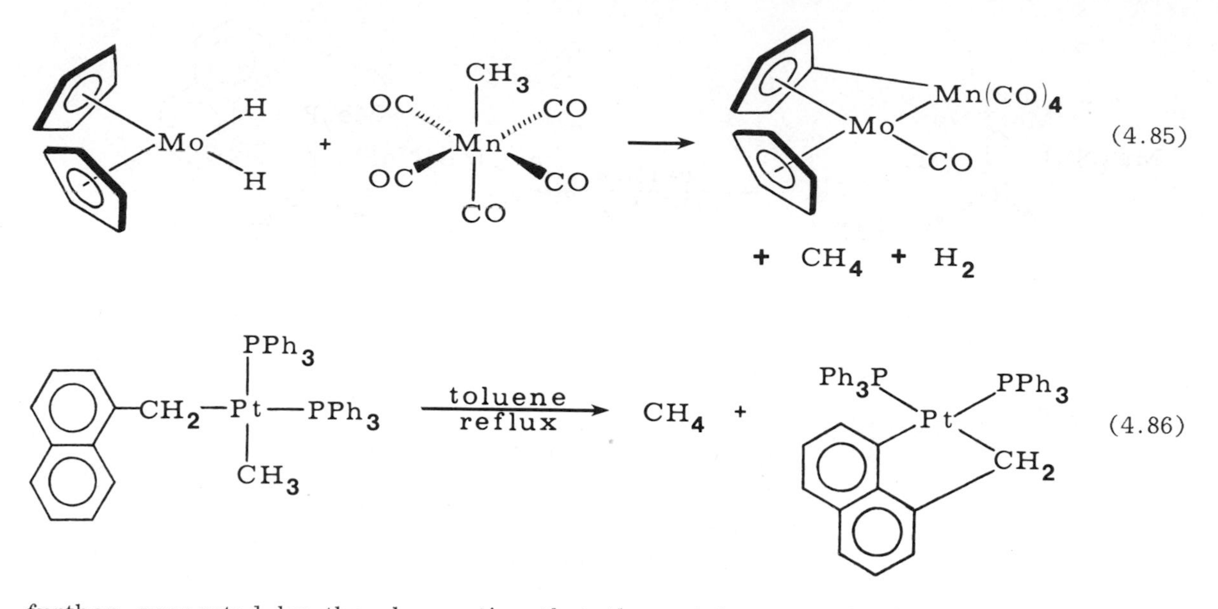


further supported by the observation that the metal center should be electrophilic for reaction to occur. For example, the zero-valent complexes, $M[P(tBu)_2Ph]_2$ (M = Pd, Pt) show no evidence for o-metallation upon prolonged heating, whereas the platinum(II) complex, $PtCl_2[P(tBu)_2Ph]_2$, undergoes facile metallation [74].

The substitution patterns can be found by examining the products formed in the reaction (Equation 4.87) with m-fluoroazobenzene [71]. Palladium(II) chloride shows a substitution pattern opposite to that of $MeMn(CO)_5$. In none of these cases is the mechanism well-understood.

Sometimes orthometallations are reversible, especially when $H_2$ is reductively eliminated (Equation 4.88). As Equation 4.89 shows these reactions can be used to catalyze the exchange of ortho aromatic hydrogens with $D_2$. On the other hand, such exchange can be an irritating side reaction which must be considered when using isotopic labels for studying homogeneous catalytic hydrogenation.

Although less reactive than phosphites, aryl phosphines will undergo similar ortho hydrogen exchange (Equation 4.90).

The greater reactivities of phosphites compared with phosphines are apparently governed by the more favorable metallacycle ring size (5 versus 4), and the more reactive, electron-rich phenoxy groups.

In some cases catalytic exchange is inhibited by excess ligand, indicating the need for a vacant coordination site. For example, in the presence of excess $(PhO)_3P$,

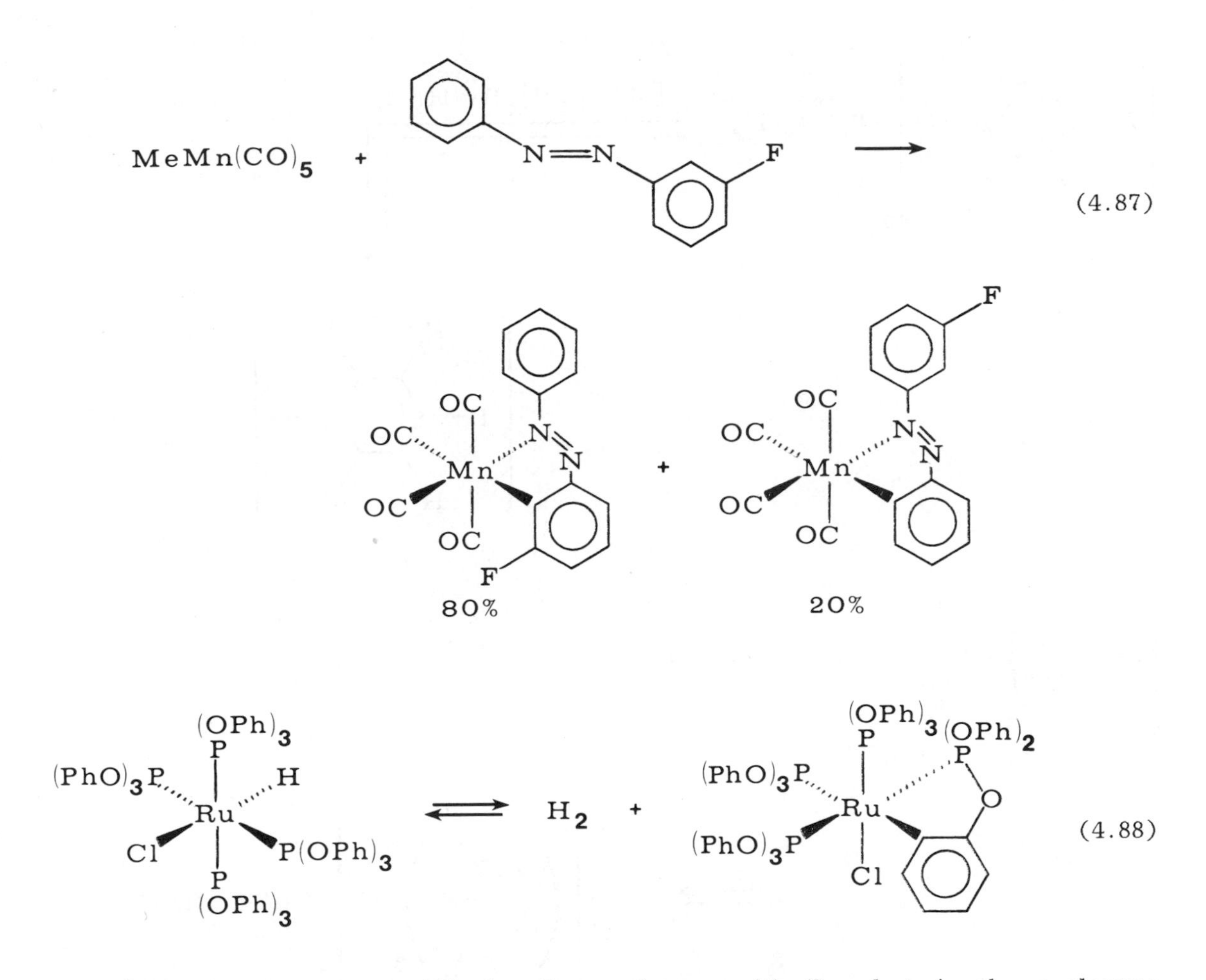


$[(PhO)_3P]_4RhH$ undergoes statistical ortho exchange with $D_2$, but in the analogous cobalt complex such exchange is inhibited by excess $(PhO)_3P$. This is reasonable, because cobalt(I) complexes have a much greater tendency to remain coordinatively saturated than the analogous rhodium(I) compounds.

Cyclometallations can stabilize intermediates which are thought to be involved in analogous intermolecular reactions. Decarbonylation of aldehydes by $RhCl(PR_3)_3$ is an occasionally useful synthetic reaction [75], which has been used in the stereospecific introduction of angular methyl groups. Such reactions go via oxidative-addition of the aldehydic C-H group, followed by retrograde insertion of the acyl to the carbonyl hydride, and subsequent reductive-elimination (Equation 4.91). The over-all process gives retention of configuration at a chiral carbon center [76].

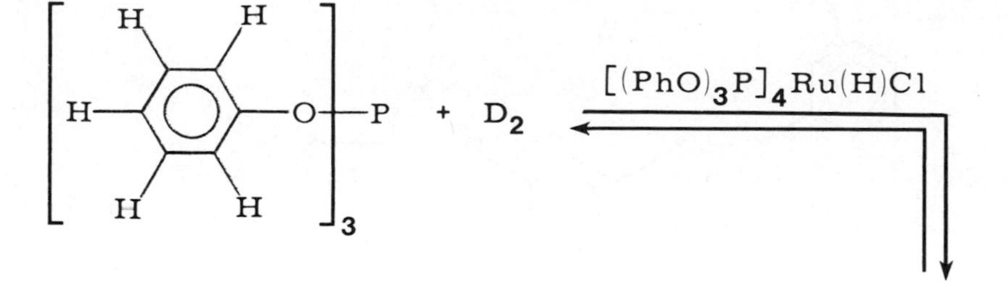

[(PhO)$_3$P]$_4$Ru(H)Cl
D$_2$


(4.89)

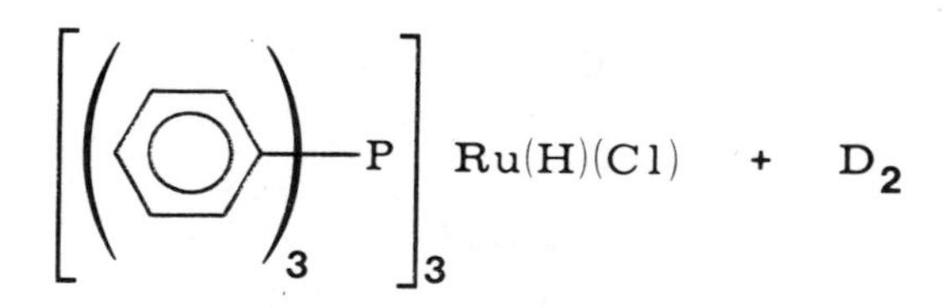

Ru(H)(Cl) + D$_2$


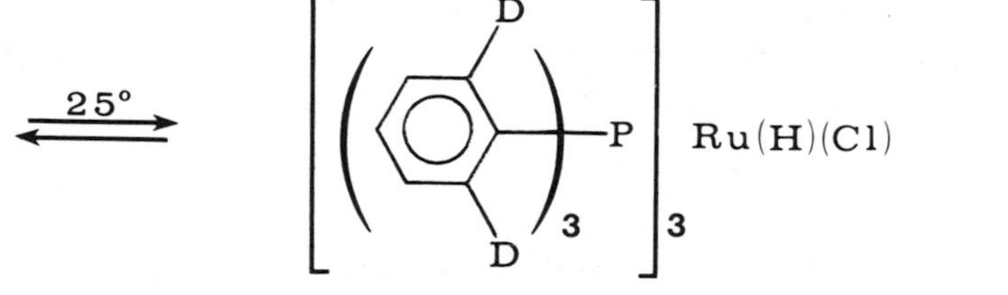

25°
Ru(H)(Cl)


(4.90)

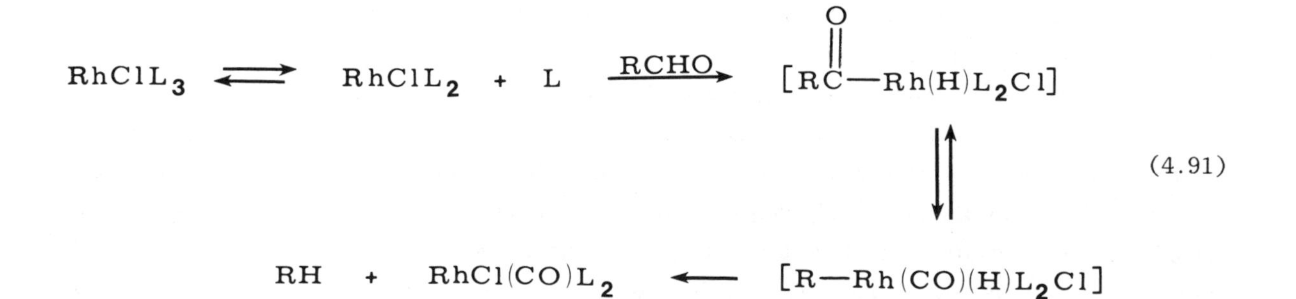

RhClL$_3$
RhClL$_2$ + L
RCHO
[RC(O)—Rh(H)L$_2$Cl]
RH + RhCl(CO)L$_2$
[R—Rh(CO)(H)L$_2$Cl]


(4.91)

Using an aldehyde which contains a ligand, 8-quinolinecarboxaldehyde, the acyl hydride intermediate is stabilized by the five-membered metallacycle and can be isolated (Equation 4.92) [77].

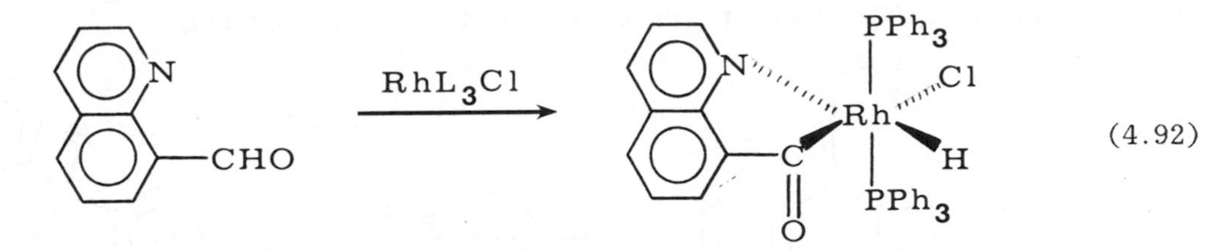


(4.92)

In Chapter 3 we saw that the reversible formation of a $\eta^3$-allyl by intramolecular oxidative addition of an allylic C-H group to the central metal is a common mechanism for olefin rearrangement by a 1-3 hydrogen shift (Equation 4.93). This is a special

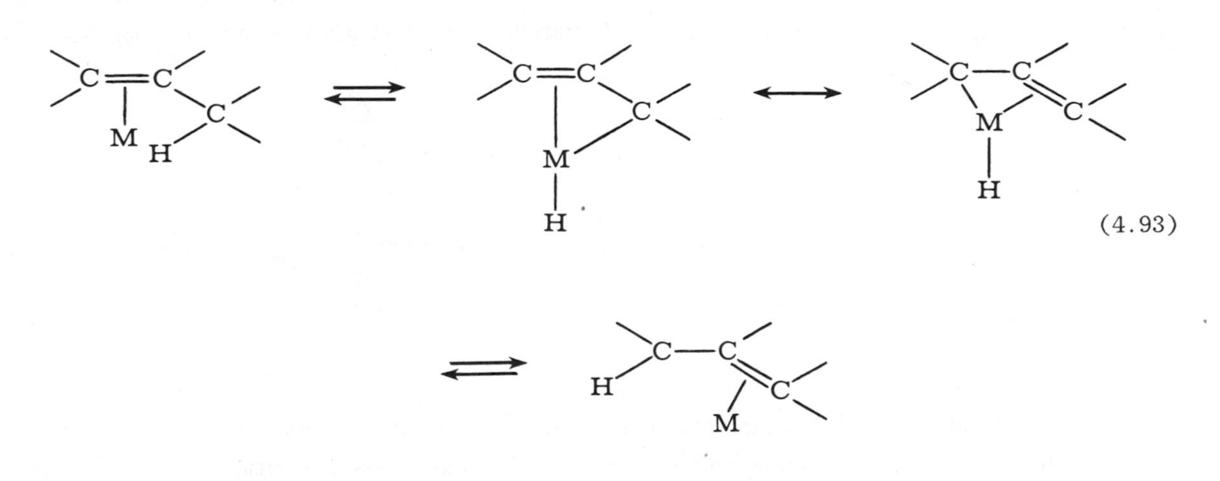


case of the "orthometallation" reaction class. Such reactions are also used (Equation 4.94) to prepare $\pi$-allyl derivatives from olefins in the presence of a base or of a good leaving group on the metal [78].

$$RuH_4(PPh_3)_3 \xrightarrow{MeCH{=}CH_2} (\eta^3\text{-allyl})_2Ru(PPh_3)_2 + 3\,H_2 \qquad (4.94)$$

Arylcyclometallations have been used to prepare heterocycles--especially by carbonylating the intermediate metallacycle [71].

An unusual case is the preparation of oxobenzo[c]thiophene (Equation 4.95) [79]. Note the preferential attack of iron on the more electron-rich ring.

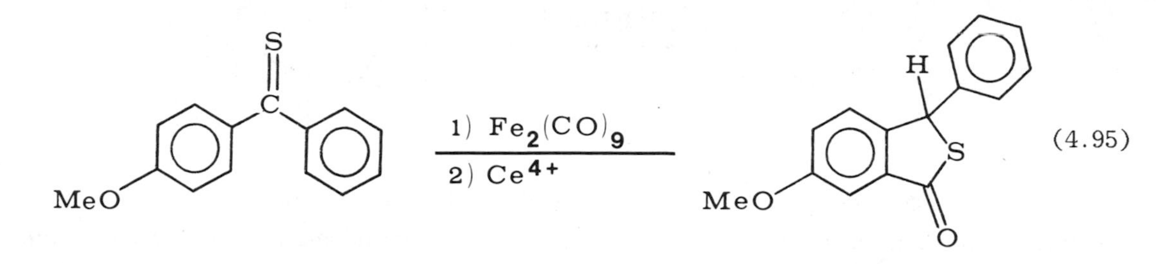


(4.95)

Cyclometallations involving ligands attached to an adjacent metal in a cluster are well-known. Ruthenium and osmium clusters have an unusual propensity to oxidatively add C-H groups. Reactions like that in Equation 4.96 [80] give an added dimension to

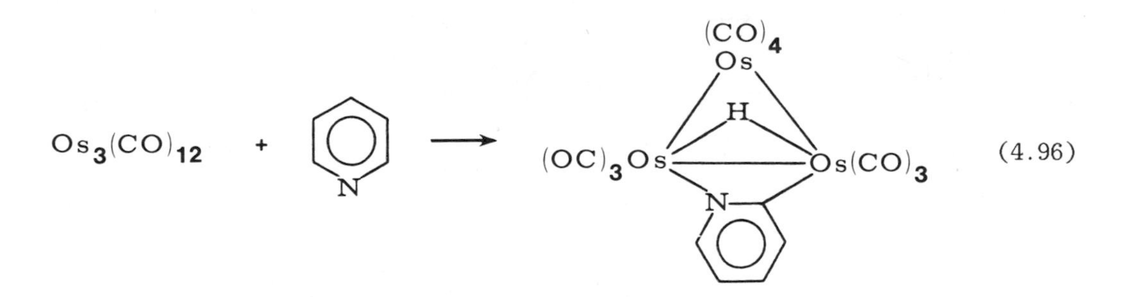


(4.96)

the chemistry of transition-metal clusters, and offer a clue to some of the differences between heterogeneous catalysts and mononuclear homogeneous transition-metal catalysts. Some transition-metal clusters may resemble heterogeneous catalysts [81].

Aliphatic groups also undergo cyclometallation by oxidative-addition of C-H groups. This type of alkane activation has been reviewed [82]. Because of intramolecular cyclometallation and intermolecular C-H additions, highly reactive unsaturated complexes have a tendency to react with one of their own ligands, a ligand on a neighboring complex, or even a solvent molecule. This places a limit on the reactivity of unsaturated, low-valent metal complexes which are stable in solution or which can be isolated. Consider, for example, Chatt's attempted preparation of the very reactive four-coordinate ruthenium(0) tetraphosphine complex. An isolated crystalline product from this reaction (Equation 4.97) is the cyclometallated ruthenium(II) dimer. Note that the

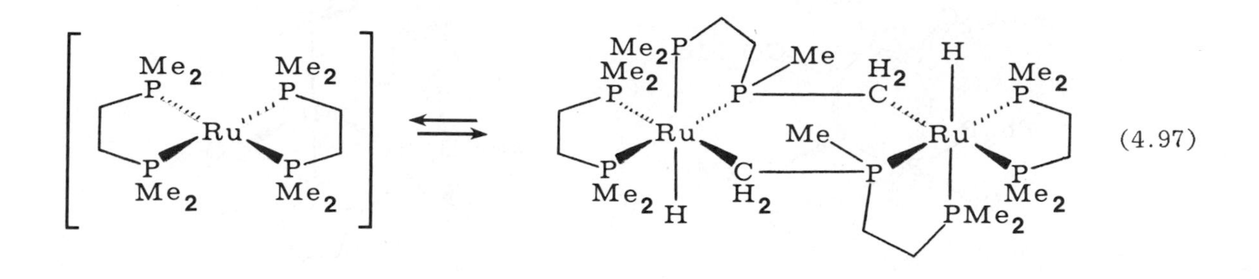


monomolecular metallacycle would have required forming a strained three-membered ring. Several mechanisms are plausible for this reaction, including a mononuclear, intramolecular cyclometallation [83a]. Whitesides has studied the cyclometallation of a neopentylplatinum(II) complex [83b] which involves an unactivated γ C-H bond.

Steric crowding between the metal and bulky ligands appears to facilitate cyclometallation reactions. Consider reactions 4.98 and 4.99, between similar substrates. The more hindered neopentyl phosphine undergoes cyclometallation (4.98) whereas the related n-propyl phosphine does not (4.99).

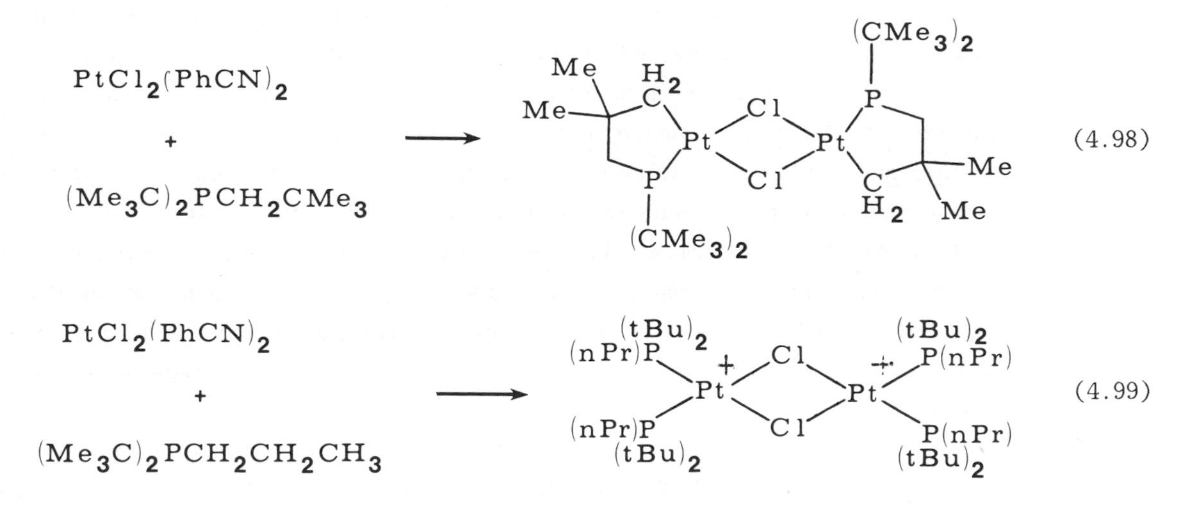


The manner by which steric effects can enforce a weak binding interaction between a metal and a hydrogen atom on a coordinated ligand is evident from some structural studies. Consider the palladium and molybdenum complexes, 12 and 13, in which the Pd-H [84] and Mo-H [85] distances are 2.3 and 2.15 Å, each shorter than the sum of the van der Waals radii (3.1 Å).

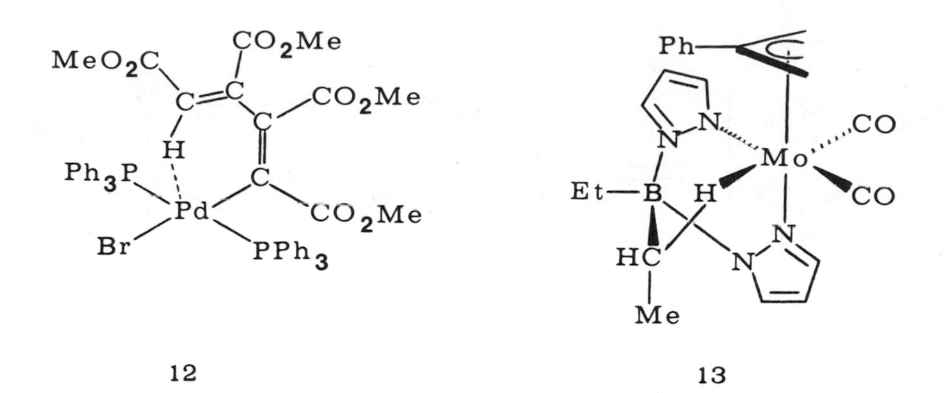


12 13

An especially vivid example of an intramolecular oxidative-addition to an unactivated, aliphatic C-H group is the dehydrogenation shown in Figure 4.9 [86]. The product, 14, has been structurally characterized. The dehydrogenated ligand was freed from the metal by treatment with KCN. The stereospecificity of this reaction (only the trans olefin was detected) can be ascribed to the steric requirements of accommodating the tridentate ligand to the coordination sphere. The reaction sequence is thought to proceed via a trans-chelating diphosphine complex. Recall the strong tendency for such planar $d^8$ complexes to bind two phosphine ligands in a trans-disposition. Each step in the suggested mechanism is plausible since the reverse reactions are well-known for catalytic hydrogenation. Both rhodium(I) and iridium(I) complexes undergo this reaction. Similar reactions have been described by Shaw [87].

b. Intermolecular C-H Additions. Intermolecular C-H oxidative-additions are less facile and less common than the intramolecular "cyclometallations" described above. In fact, most of the intermolecular C-H additions which afford stable adducts involve substrates (such as an aromatic ring) which may form a $\pi$-complex with the metal, prior to C-H oxidative-addition. Such reactions may go through a $\pi$ to $\sigma$ rearrangement and are therefore quite similar to the intramolecular cyclometallations discussed earlier.

An especially instructive example (Figure 4.10) involves very reactive complexes of a highly basic ligand: $M(Me_2PCH_2CH_2PMe_2)_2$ (M = Fe, Ru, and Os). The four-coordinate Fe(0) and Ru(0) complexes, 16, are thought to be generated in situ by heating solutions of the corresponding 2-naphthyl complexes, 15 [88]. In the presence of aromatic molecules, Ar-H, an equilibrium is set up as shown in Figure 4.10. The structures of the octahedral ruthenium and osmium complexes have been established by X-ray diffraction. The equilibrium constants ($K_{eq}$) for path (a) in Figure 4.10

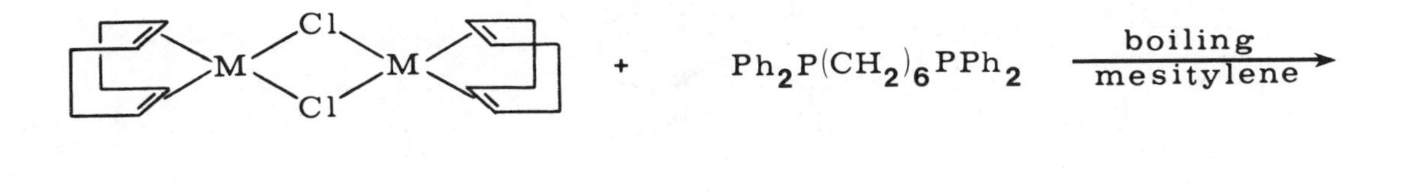


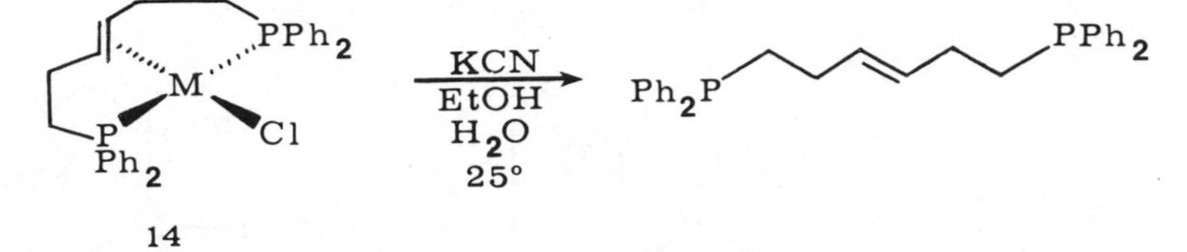


Postulated Mechanism:

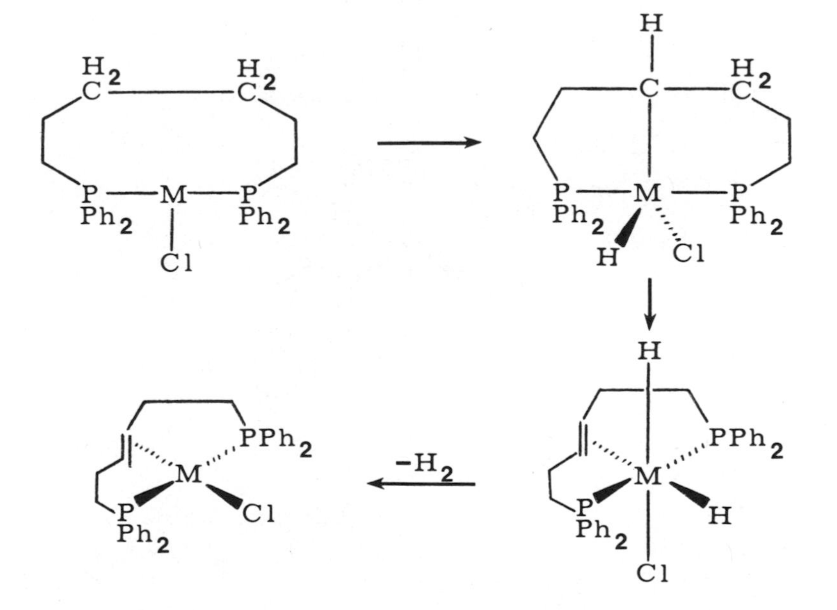


*Figure 4.9. Intramolecular dehydrogenation of a hydrocarbon group.*

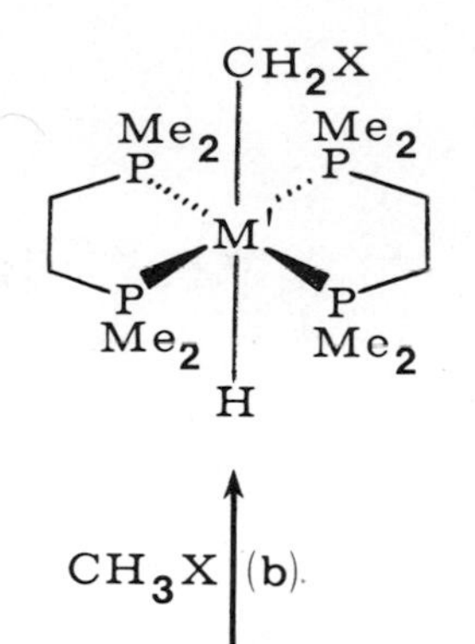


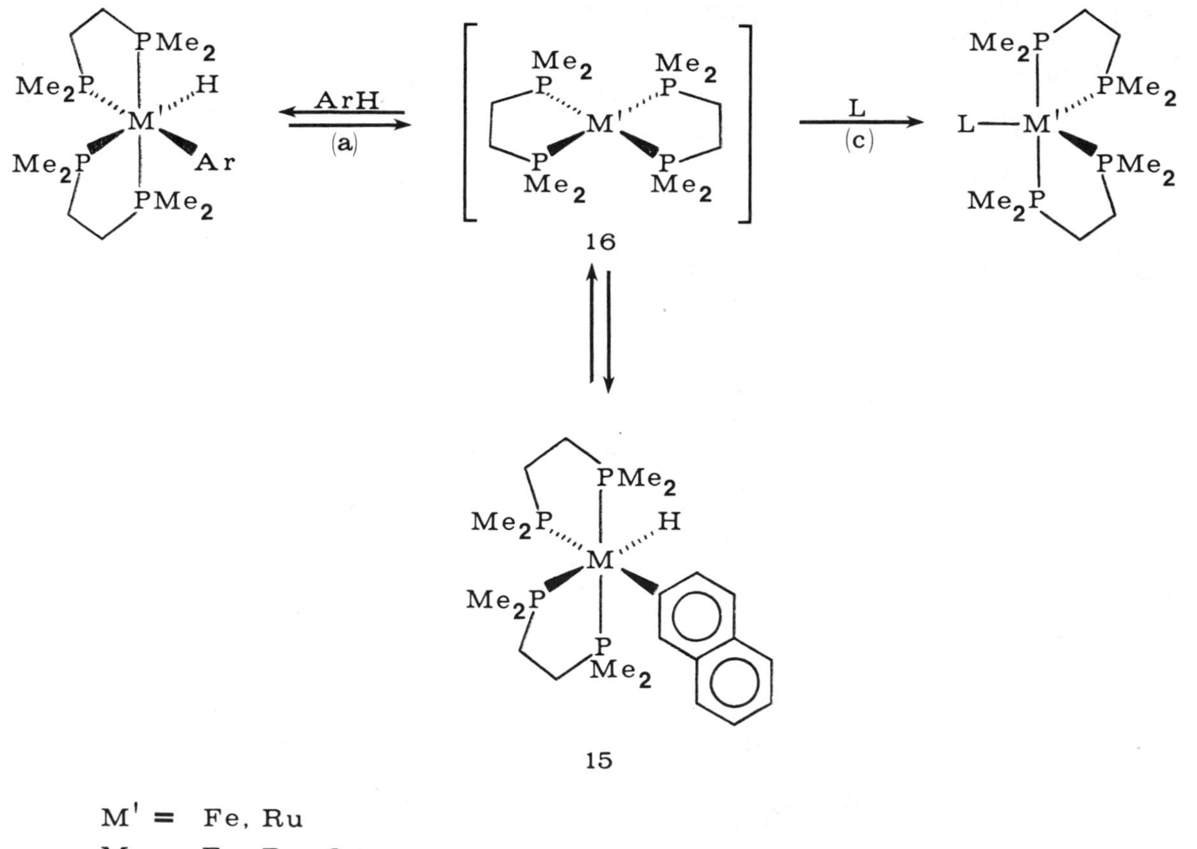


M' = Fe, Ru

M = Fe, Ru, Os

X = CN, COMe, $CO_2Et$, C≡CR

*Figure 4.10. Intermolecular C-H oxidative-addition.*

increase as the ArH molecule is substituted with electron-withdrawing substituents. Substitution of toluene occurs at the meta and para but not at the ortho or the benzylic positions. The naphthyl osmium derivative, 15, does not react because it is apparently too stable to eliminate naphthalene. "Activated" $sp^3$ C-H groups also react (path b). Perhaps it is significant that these hydrocarbon "activating" groups contain potential ligands. π-Arene complexes may be involved. The presumed four-coordinate intermediates, 16, can be intercepted by ligands such as CO, olefins, and phosphines (path c). Although these reactions are complicated by the formation of cis and trans isomers, the nature of these complexes has been thoroughly documented by $^1H$ and $^{31}P$ NMR studies. The very reactive metal centers and the pronounced tendency for the intermediate $d^8$ complexes to add a fifth ligand are important factors governing these C-H additions.

Low-valent, coordinatively unsaturated cyclopentadienyl complexes of Groups IV-VI are extremely reactive. When generated in situ, such complexes tend to form dimers by intermolecular oxidative-addition of a cyclopentadienyl C-H group bound to a second molecule. This is another indication of the limitation which C-H oxidative-additions place on the preparation of highly reactive, coordinatively unsaturated transition metal complexes.

Consider, for example, a reaction which is thought to generate $(\eta^5\text{-}C_5H_5)_2Mo$. This presumed intermediate forms an initial dimer by apparent oxidative-addition of the cyclopentandienyl group on a neighboring molecule (Figure 4.11). Heating the first dimer, 17, at 50° affords a more stable dimer, 18, containing coupled cyclopentadienyl groups. The latter product, 18, was further characterized by forming a stable cationic derivative, 19, whose structure was determined by X-ray crystallography [89].

Similar reactions appear to occur when low-valent Group IV [90] and V [91] metallocenes are generated in situ by reduction of a dihalide or reductive-elimination of R-H or $H_2$.

For example, the coordinatively saturated zirconium(II) derivative, 20, prepared by phosphine-induced reductive-elimination, can serve as a precursor of the coordinatively unsaturated zirconocene, 21, which may have a triplet ground state. The latter spontaneously dimerizes by intermolecular C-H oxidative-addition (Figure 4.12) [92]. Note the marked similarity between the reactions in Figures 4.11 and 4.12. The presumed zirconium(II) intermediate, 21 in Figure 4.12, is very reactive toward oxidative-additions, as demonstrated by the rapid reaction with n-butyl chloride, but especially by addition of the aliphatic C-H group in toluene.

$$[Mo(\eta^5\text{-}C_5H_5)_2(H)Li]_4 \xrightarrow{N_2O} [Mo(\eta^5\text{-}C_5H_5)_2] \longrightarrow$$

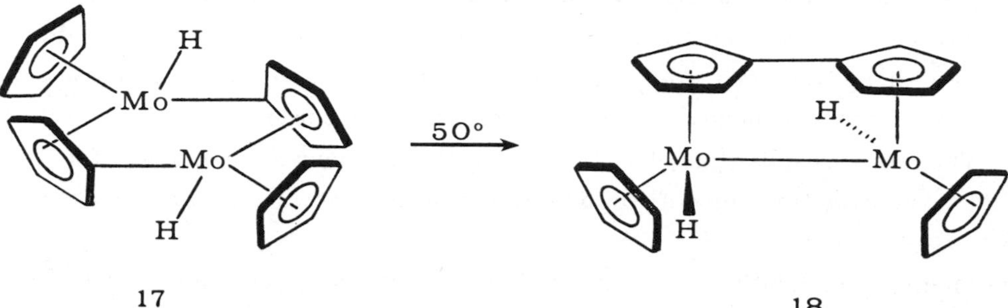


17

cis + trans

18

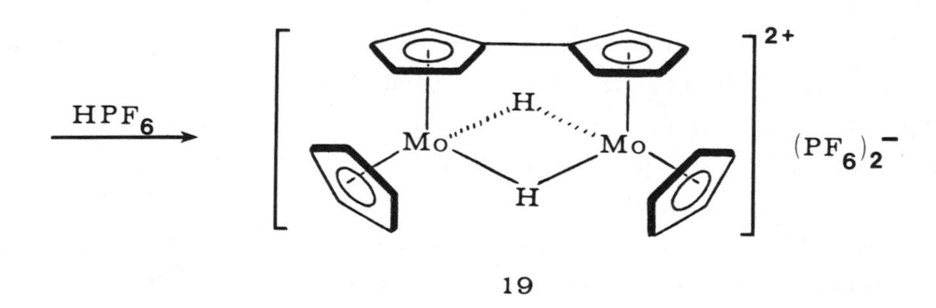


19

*Figure 4.11. Intermolecular oxidative-addition of reactive metallocenes.*

Simple unactivated alkanes will interact with transition-metal salts under certain experimental conditions [82]. Thus far the only evidence of such interaction is the trivial isotopic exchange between $D^+$ in the medium and H atoms of the alkanes. So far it has not proved possible to otherwise intercept or convert the presumed alkyl metal intermediate into a useful product. The conditions in Equation 4.100 are typical of these mysterious reactions. Platinum(II) is catalytically active, whereas platinum(IV) and platinum metal are apparently not (under these conditions). Less polarizable anions such as $Cl^-$ are more effective than polarizable anions such as $I^-$. There is some

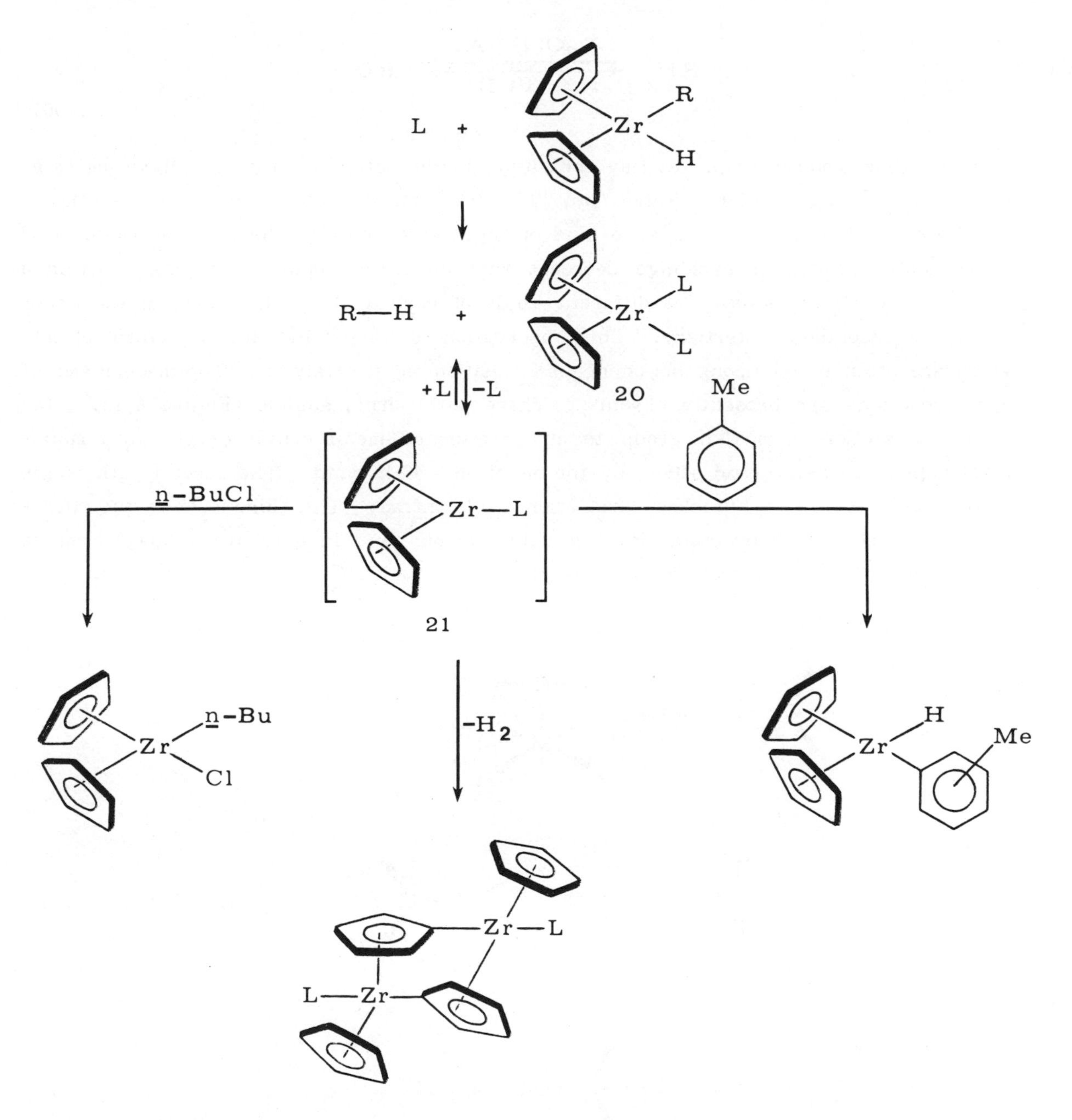


*Figure 4.12. Reactions of a zirconium(II) metallocene.*

$$RH \underset{120°}{\overset{D_2O,\ HOAc}{\underset{HX,\ Pt(II)}{\rightleftarrows}}} RD \qquad (4.100)$$

evidence that dimeric complexes may contribute to the catalytic activity. Maximum rates are observed at particular acidities and Pt(II):$Cl^-$ ratios. Multiple exchange is common for these reactions, indicating some kind of rapid scrambling in the putative metal alkyl intermediate. Rates of exchange decrease with increased alkane branching. Within a given alkane, the reactivity at different kinds of carbon atoms decreases in the order primary > secondary > tertiary. This observation is consistent with the steric effects which are often noted among homogeneous transition-metal catalysts. The mechanisms of these reactions are presently obscure. Three paths are plausible (Figure 4.13): (a) oxidative-addition of a C-H group to an unsaturated metal center or (b) to a metal-metal bond; (c) heterolytic attack by the metal on a C-H bond. The second path is the microscopic reverse of binuclear elimination (vide infra). The third path is the microscopic reverse of the protonolysis of a metal-carbon bond in a saturated alkyl complex and is similar to the heterolytic cleavage of $H_2$.

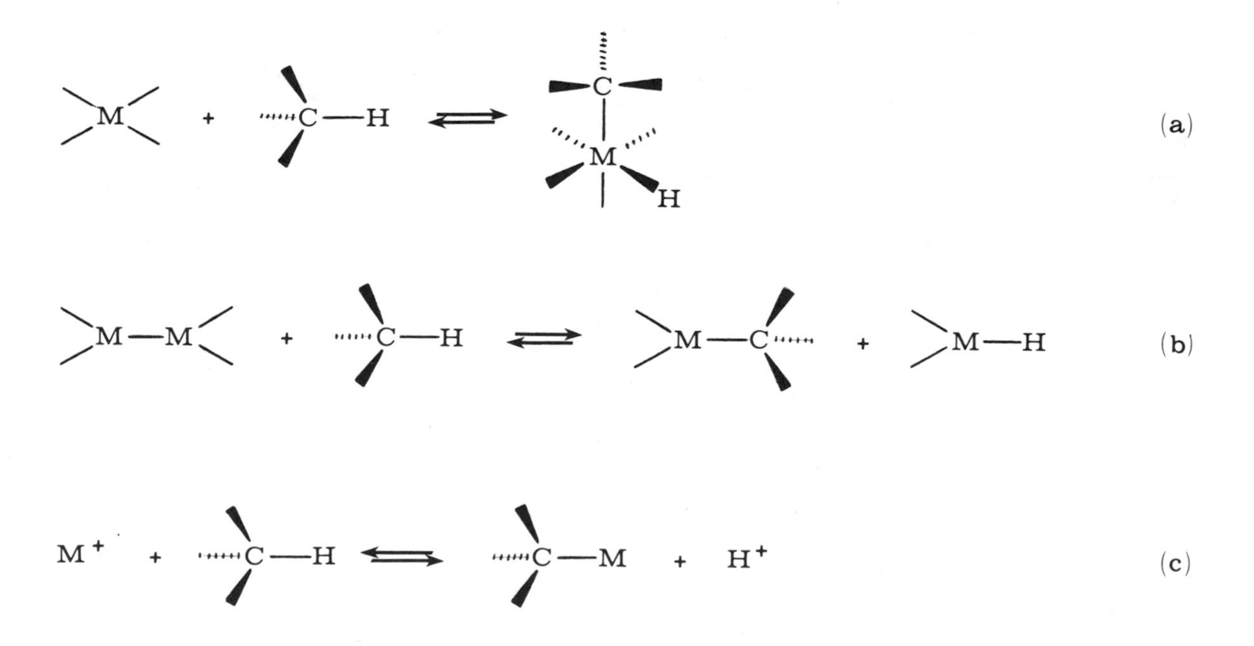


*Figure 4.13. Possible mechanisms for C-H activation.*

## 4.2. OXIDATIVE-ADDITIONS WITH AN OVER-ALL ONE-ELECTRON CHANGE.

There are many examples of oxidative additions in which the metal undergoes a one-electron change. These reactions can be further divided into cases involving monometallic complexes and dimetallic complexes.

1. Mononuclear Systems. Transition-metal complexes having a tendency to undergo one-electron oxidation react with two-electron oxidizing agents according to the stoichiometry shown in Equation 4.101. These reactions usually occur in two steps, the

$$2\ L_nM + A{-}B \longrightarrow L_nM{-}A + L_nM{-}B \qquad (4.101)$$

$$A{-}B = R{-}X,\ Br_2,\ HO{-}OH,\ HONH_2,\ I{-}CN,\ \text{etc.}$$

first being rate-determining. The most thoroughly studied cases involve alkyl halides reacting with complexes of Co(II) [93] and Cr(II) [94].

In these cases the rate-determining step is the extraction of a halogen atom forming an alkyl radical (Equation 4.102) which rapidly adds to another equivalent of the reduced metal (Equation 4.103). The rate law is thus first-order in alkyl halide

$$\underset{d^n}{L_yM} + R{-}X \xrightarrow[k_2,\ \text{rds}]{(a)} \underset{d^{n-1}}{L_yM{-}X} + R\bullet \qquad (4.102)$$

$$L_yM + R\bullet \xrightarrow[\text{fast}]{(b)} \underset{d^{n-1}}{L_yM{-}R} \qquad (4.103)$$

$$L_yM = [Co(CN)_5]^{3-},\ Co(dmg)_2PPh_3,\ [Cr(en)_3]^{2+},\ [Cr(H_2O)_n]^{2+},\ \text{etc.}$$

and in the reduced metal complex; $-d[RX]/dt = 2k_2[RX][LyM]$. The rate depends on the nature of the alkyl halide, following the well-known rate profile for the formation of alkyl radicals, RI > RBr > RCl; R = 3° > 2° > 1° > $CH_3$. Carbon tetrachloride is very reactive and, for this reason, is not a suitable solvent for many low-valent transition-metal complexes.

The alkyl radicals may be scavenged by added reagents, such as hydrogen-atom donors, or may undergo other radical reactions, such as dimerization, disproportionation, or initiation of vinyl-monomer polymerization. The competitive capture of these alkyl radicals by metals (Equation 4.103) is often very fast. For example, the reaction with Cr(II) has a second-order rate constant of $4 \times 10^7\ M^{-1}\ sec^{-1}$ [95]. The alkylated metal oxidative-adduct, LyM-R, may be unstable and undergo subsequent reactions such as β-hydride elimination.

These one-electron oxidative-addition reactions are seldom implicated in homogeneous catalysis, with the exception of hydrogen activation. However, that reaction may involve dimetallic compounds, or Co(I) and Co(III) formed by disproportionation of Co(II) (Equation 4.104).

$$2\ Co(II) \rightleftharpoons Co(I) + Co(III) \tag{4.104}$$

Neighboring heteroatoms enhance the rate of these one-electron oxidative-additions. This is especially true of vicinal dihalides, which are about $10^3$ times more reactive than the monohalides. Olefins are formed (Equation 4.105). The vicinal dihalide must be able to assume a <u>trans</u>-orientation to achieve the maximum rate enhancement [96].

$$ICH_2CH_2I + 2[Co(CN)_5]^{3-} \longrightarrow CH_2{=}CH_2 + 2[ICo(CN)_5]^{3-} \tag{4.105}$$

Metal alkyl self-displacement reactions, which we encountered earlier in this chapter as a formal two-electron exchange between the two metals, are also known as a one-electron exchange. The reaction in Equation 4.106 has been carefully explored by Johnson [97] ($dmg_2$ is <u>bis</u>-dimethylglyoximato and $chg_2$ is <u>bis</u>-cyclohexylglyoximato). It

$$Co(II)(dmg)_2Py + RCo(III)\,chg_2Py \rightleftharpoons RCo(III)(dmg)_2Py + Co(II)(chg)_2Py \tag{4.106}$$

was shown that the configuration at chiral carbon undergoes inversion each time the alkyl is transferred from one cobalt to the other cobalt [97]. Even though these reactions involve formal one-electron exchanges (an alkyl radical may be considered to be transferred), the rate profile for this concerted reaction is like that of an $S_N2$ process: R = Me >> Et > iPr.

2. <u>Dinuclear Systems</u>. Metal-metal bonds are often cleaved by mild oxidizing agents (Equation 4.107). These oxidative-additions which rupture metal-metal bonds are quite analogous to the mononuclear reactions discussed above, since in both cases

each metal undergoes a one-electron oxidation. In other situations metal-metal bonds may be formed by reaction of bimetallic complexes with mild oxidizing agents (Equation 4.108). The first type of dinuclear oxidative-additions (Equation 4.107) is the

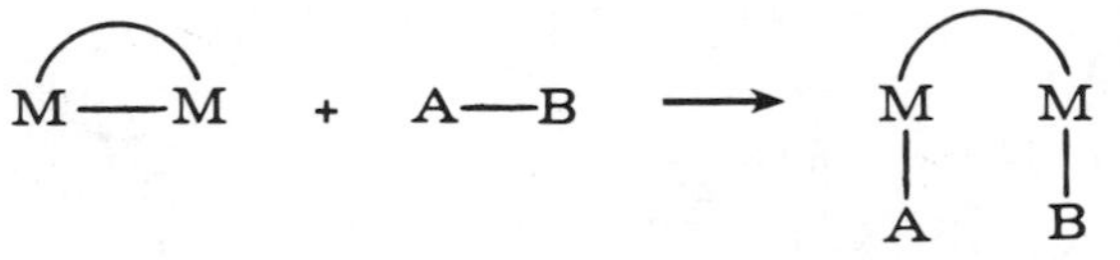

$$\widehat{M{-}M} + A{-}B \longrightarrow \widehat{M\quad M} \text{ with } M{-}A,\ M{-}B \qquad (4.107)$$

$$\widehat{M\quad M} + A{-}B \longrightarrow \widehat{M{-}M} \text{ with } M{-}A,\ M{-}B \qquad (4.108)$$

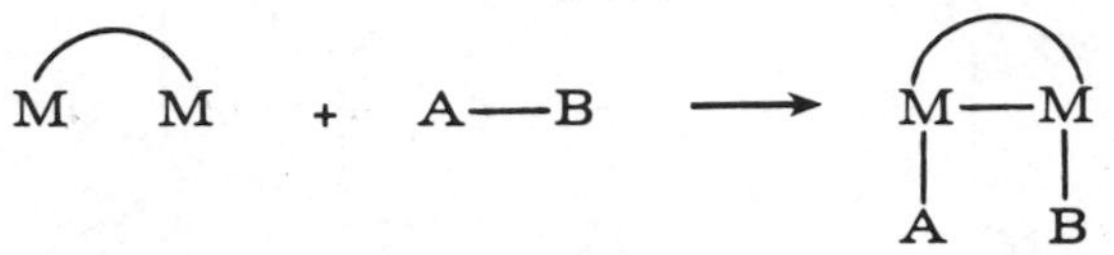

microscopic reverse of dinuclear reductive-elimination (*vide infra*); however, these reactions are very little studied.

Perhaps the best studied dinuclear oxidative-addition is the hydrogenolysis of $Co_2(CO)_8$. This reaction (Equation 4.109), which is an important step in the oxo pro-

$$Co_2(CO)_8 \overset{K}{\rightleftharpoons} [Co_2(CO)_7] + CO \xrightarrow[\substack{H_2 \\ \text{rds}}]{k} 2\ HCo(CO)_4 \qquad (4.109)$$

cess (Chapter 8), is reported to be first order in both $Co_2(CO)_8$ and $H_2$ and inverse order in CO (at low CO pressures) [99]: rate = $kK[Co_2(CO)_8]^1[H_2]^1[CO]^{-1}$. This suggests that the rate-determining step is reaction of $H_2$ with coordinatively unsaturated $Co_2(CO)_7$ which is formed in a pre-equilibrium step. In complexes having a metal-metal bond, a site of unsaturation is the equivalent of an additional metal-metal bond. Thus, in a formal electron-counting sense, the postulated intermediate $Co_2(CO)_7$ in Equation 4.109, could be considered the equivalent of a double Co-Co bond. There are a few studies of the addition of reagents to metal-metal multiple bonds [100], but so far the scope and mechanisms of these reactions are not well-understood.

An interesting type of dinuclear oxidative-addition in which a metal-metal bond is formed has been reported by Gray [101]. The dimeric rhodium(I) complex, 22, adds methyl iodide in such a way that a metal-metal bond is established while the methyl and iodo groups are bound to the opposite rhodiums. Thus, each rhodium undergoes a one-electron change (to rhodium(II)) and simultaneously forms a diamagnetic metal-metal bonded adduct, 23. The geometry of the bis-isonitrile ligands enforces the cofacial

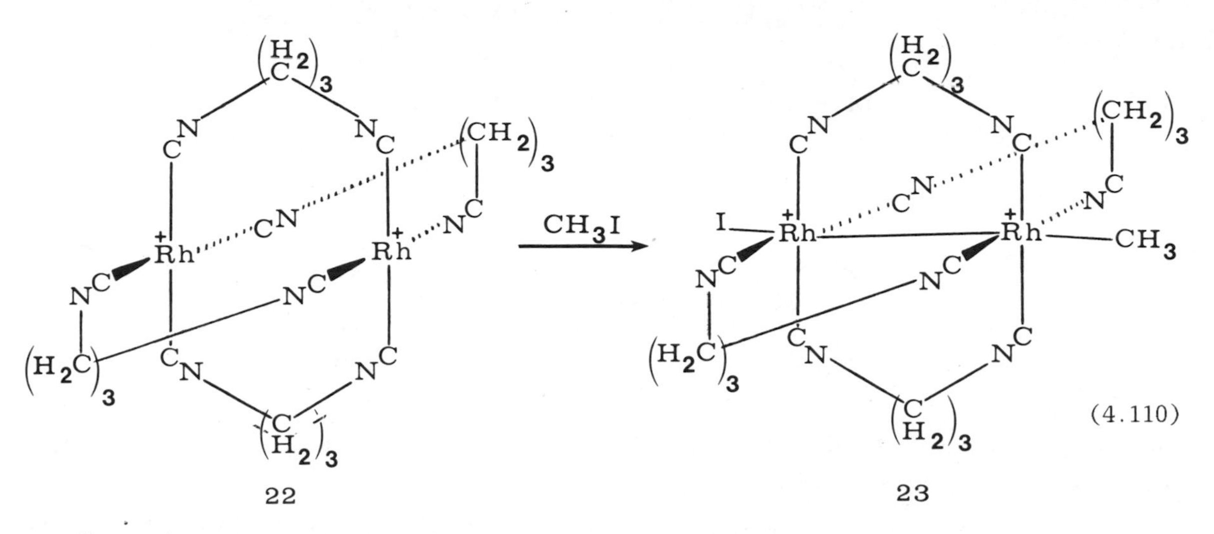


configuration. A similar dinuclear oxidative-addition forming a gold-gold bond (Equation 4.111) has been reported by Schmidbaur [102].

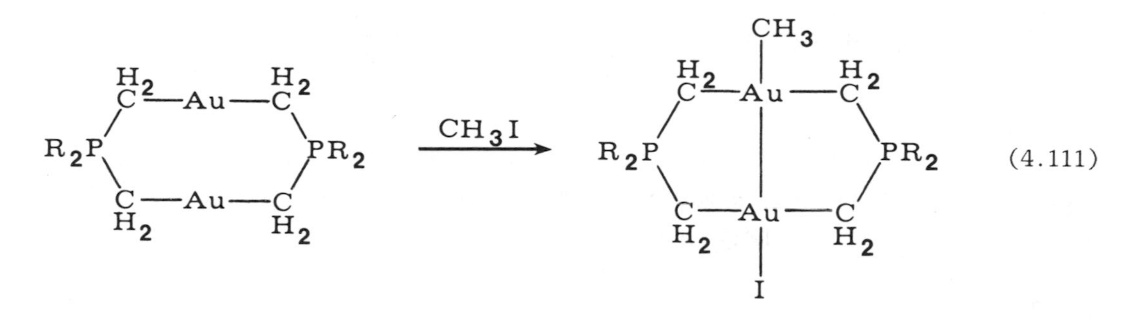


## 4.3. REDUCTIVE-ELIMINATIONS.

Reductive-elimination reactions may be either mononuclear (Equation 4.112) or dinuclear (Equation 4.113). Since reductive-elimination is an obligatory step in

$$L—\overset{\displaystyle Y}{\overset{|}{M}}—X \longrightarrow LM + X—Y \qquad (4.112)$$

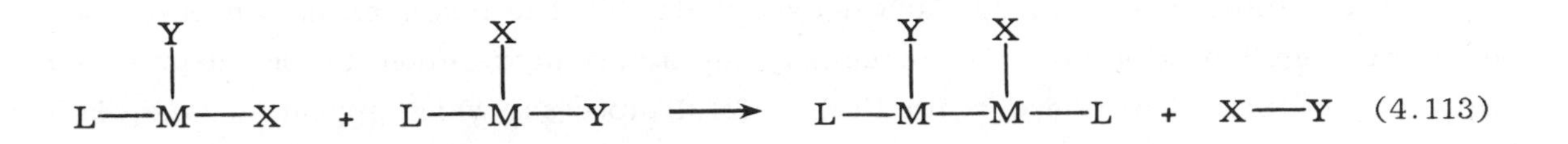


virtually all reactions which are catalyzed by soluble transition-metal complexes, it is very important in our consideration of homogeneous catalysis. Reductive-eliminations are also significant stoichiometric methods of forming carbon-hydrogen and carbon-carbon bonds. Examples of reductive-eliminations are listed in several reviews [2].

For those reversible, concerted, mononuclear oxidative-additions, such as hydrogenation, microscopic reversibility dictates that reductive-elimination occurs by the same concerted path. However, reductive-eliminations as such have been much less studied than oxidative-additions. Those mechanistic studies which have been carried out suggest that reductive-eliminations are just as diverse and complicated as oxidative-additions. Several paths are possible and the dominant mechanism depends on the particular complex, the nature and geometric disposition of the groups undergoing reductive-elimination, the ancillary ligands, the coordination number, the oxidation state, and the formal charge on the complex. Reductive-eliminations may also be oxidatively induced.

Judging from the relatively few well-documented mechanistic studies, we would say the safest way to deal with reductive-eliminations is to avoid generalization and to treat each case separately.

It is difficult to distinguish an intra- from an intermolecular reductive-elimination unless isotopic labels are used to deduce the presence or absence of cross-coupled products. We shall see that complexes which appear to be thermodynamically unstable with respect to reductive-elimination, but are in fact kinetically inert, may exhibit a dinuclear (intermolecular) reductive-elimination path. Typical examples are certain hydrido-alkyl complexes, vide infra.

1. Mononuclear Reductive-Eliminations. Five factors seem to favor intramolecular reductive-eliminations: (1) a high formal charge on the metal; (2) the presence of bulky ligands; (3) cis orientation of the two eliminating groups; (4) an electronically stable product complex; and (5) a coordinatively unsaturated intermediate. Each of these factors should be kept in mind as we examine specific cases. In particular, factor (4) often dictates whether a reductive-elimination is intramolecular or intermolecular. In this section carbon-carbon and carbon-hydrogen eliminations will be discussed.

a. Reactions Forming Carbon-Carbon Bonds. The formation of carbon-carbon bonds by reductive-elimination is potentially an important reaction for metal-promoted organic synthesis. There are synthetically useful processes which appear to take place in this way. Examples are the alkyl cuprate reactions with alkylating agents discussed earlier in this chapter and later in Chapter 11, and metal-catalyzed cross-coupling between an aryl halide and a Grignard or lithium reagent. However, the detailed paths for these processes are uncertain, especially as to whether the carbon-carbon coupling is a concerted intramolecular reaction involving a monometallic complex. Some insight into the possibilities of intramolecular carbon-carbon bond forming reductive-eliminations can be obtained by surveying stoichiometric reactions whose mechanisms have been studied. Gold(III) alkyl compounds will thermally eliminate alkanes, forming a carbon-carbon bond. Kochi [103] has extensively investigated the mechanisms of these thermal reductive-eliminations from planar, $d^8$, trialkyl gold complexes (Equation 4.114). In

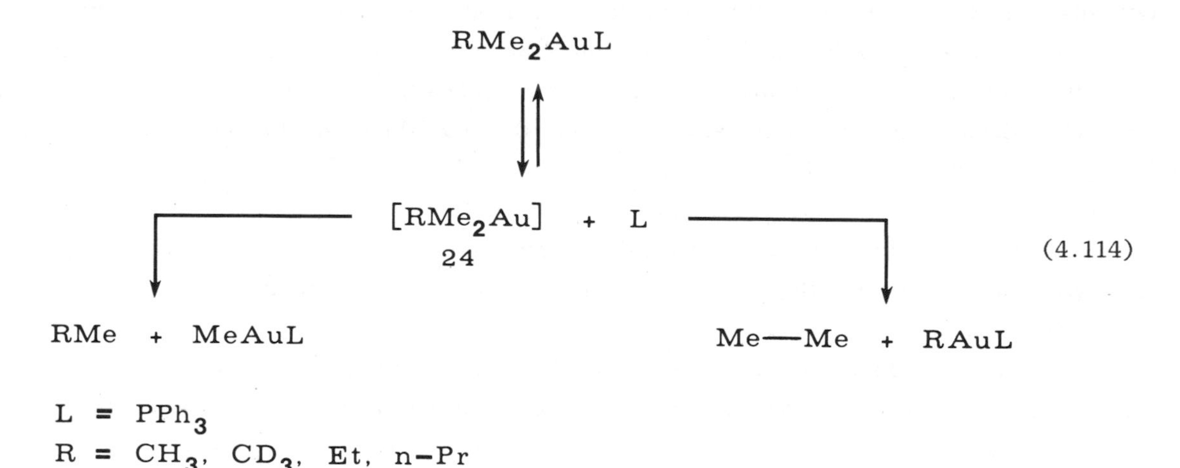


(4.114)

polar solvents such as DMSO or DMF, the dominant reductive-elimination path involves prior dissociation of a phosphine to form a 14-electron intermediate (24) which is presumably T-shaped and may exist as equilibrating cis-trans isomers. Added phosphine inhibits the reaction. Isotopic labeling shows that in polar solvents the reductive-eliminations are intramolecular (no cross-coupling). Both possible carbon-carbon coupled products are formed (Equation 4.114), presumably by cis-elimination from the isomeric three-coordinate intermediates. With alkyl groups other than methyl, these couplings are complicated by side reactions. For example, i-propyl rapidly rearrange to n-propyl groups, probably via a reversible β-hydride-elimination, migratory-insertion sequence (Chapter 5). In nonpolar solvents these reductive-eliminations are apparently complicated by a competing dinuclear path.

Reductive-elimination of alkanes from the related bis-alkyl gold(III) complexes 25 and 26 has also been studied. These complexes were generated by protonolysis of the trialkyl gold compounds (Equation 4.115, a and b). With strong acids the presumed dialkyl intermediate spontaneously eliminates alkane.

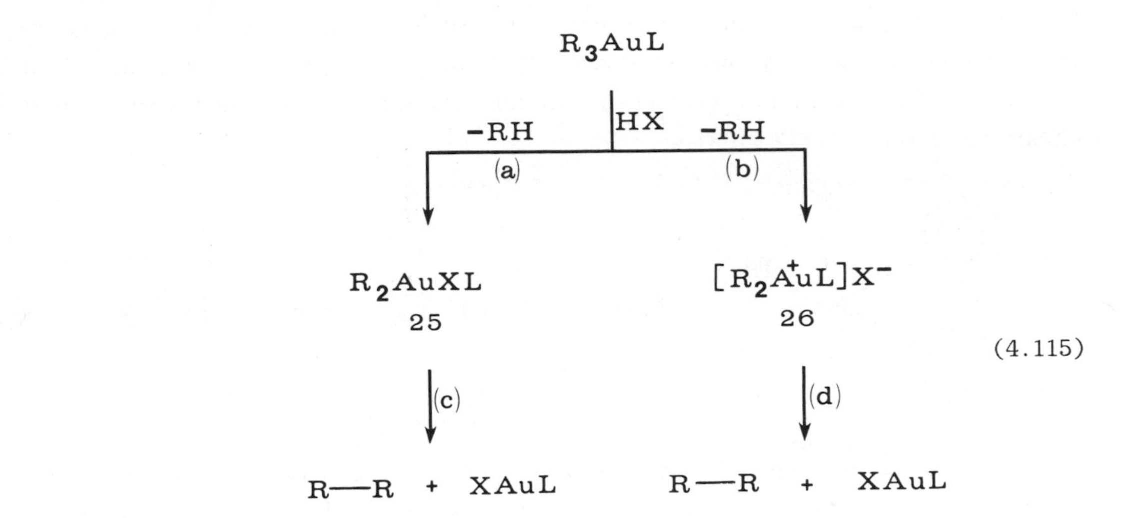


(4.115)

25 $X^- = OAc^-, F^-, Cl^-$
26 $X^- = ClO_4^-, CF_3SO_3^-$

Analogous cationic dialkyl gold(III) complexes, 27, also eliminate alkanes intramolecularly (Equation 4.116) [104]. In these cases radical intermediates have

$$\underset{27}{[R_2AuL_2]^+X^-} \longrightarrow R{-}R + X{-}Au{-}L + L \qquad (4.116)$$

been ruled out. Again free phosphines inhibit the elimination, as do coordinated bidentate phosphines.

Comparison of the reductive-eliminations for all these gold(III) alkyls reveals that: (1) cationic complexes are more prone to elimination; (2) the primary intramolecular path involves three-coordinate, 14-electron intermediates; (3) bulky ligands increase, and more basic ligands decrease, the rate of elimination; and (4) rates of elimination depend on the nature of R: Me < Et < n-butyl.

Electronically and structurally related nickel(II) and platinum (II) dialkyl or diaryl complexes are known. Certain aspects of their stabilities toward thermal reductive-elimination can be ascribed to the same factors which seem to control alkane eliminations in the related gold(III) alkyls. Square planar dialkyl, diaryl, and alkyl aryl nickel(II) complexes are fairly stable toward thermal reductive-elimination provided that a ligand is not lost; apparently three-coordinate complexes are labile. For example, nickel alkyls containing a stabilizing chelating ligand such as 2,2'-bipyridine (bipy) are relatively stable unless they are oxidized (vide infra). Pyrolysis of solid $Et_2Ni(bipy)$ produces products (ethane, ethylene, and butane) which have been attributed to ethyl radicals (Equation 4.117) [105].

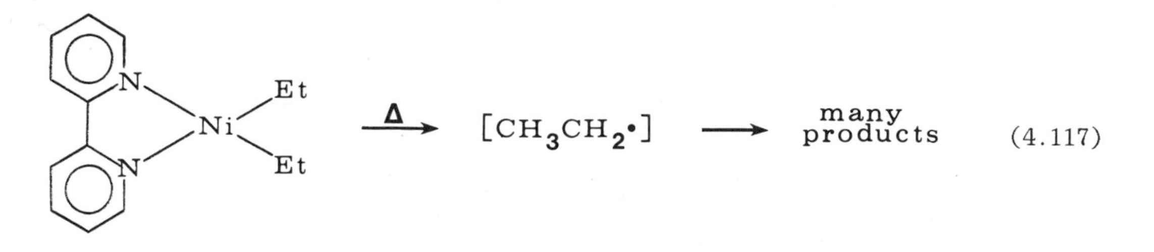


Nickel aryls of the type trans-$Ar_2NiL_2$ ($L = R_3P$) are also moderately stable, particularly where the aryl group has a substituent ortho to the metal-carbon bond [105b]. Such compounds do not decompose in boiling benzene nor in ethanol, and many melt without decomposition, e.g., (o-$CH_3C_6H_4)_2Ni(PEt_3)_2$. The trans-stereochemistry probably contributes to their kinetic stability.

Reductive-elimination of carbon-carbon bonds from these nickel(II) complexes is greatly increased by oxidation. In fact "oxidatively induced reductive-eliminations" have been widely observed, and are thought to be an important facet of several catalytic and stoichiometric carbon-carbon bond-coupling schemes, for example, even in the alkylation of the alkylcuprate compounds discussed earlier in this chapter and in Chapter 11. For instance, oxidative workups are reported to increase the yield of alkyl-aryl coupling products derived from "$R_2CuLi$" and Ar-X [106].

A spectrum of one-electron oxidants ($Co(TFA_3)_4$, $Ce(TFA)_4$, $Tl(TFA)_3$, $CuBr_2$, $Na_2IrCl_6$, $Br_2$) react with trans-$Ar_2Ni(PEt_3)_2$ affording good yields of the corresponding biaryls (Equation 4.118). The coordination number of the reactive intermediate is uncertain. Oxidatively induced metal-carbon bond-breaking reactions are quite common. For example alkyl-aryl insertions (Chapter 5) and nucleophilic metal-alkyl displacements are greatly enhanced by prior oxidation of the metal. Upon oxidation of the metal, the

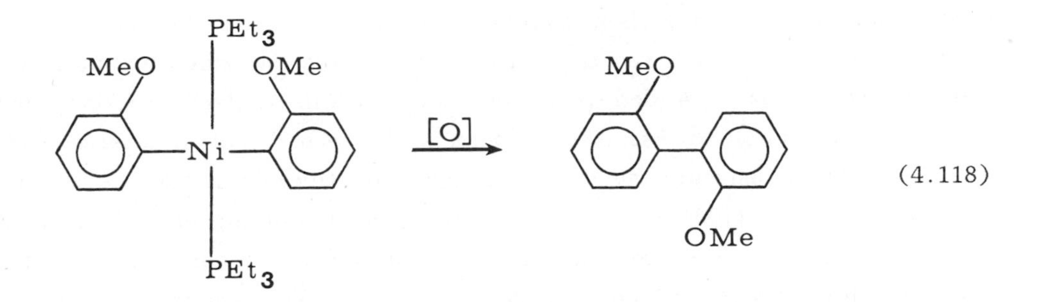


(4.118)

metal-carbon bond apparently becomes much weaker and thus more reactive. In cases of oxidatively induced reductive-elimination, the oxidation may either be electron abstraction (an "outer-sphere" mechanism) or oxidative-addition (an "inner-sphere" mechanism).

Electrolysis of (o-tolyl)$_2$Ni(PEt$_3$)$_2$ in $CH_2Cl_2$ also gives a high yield of coupled product [108]. However, such electrochemical coupling reactions are difficult to interpret, since these processes can be complicated by ligating solvents or the presence of excess ligand. Very bulky aryls such as mesityl do not couple--presumably because these cannot assume the required cis-coordination geometry.

Thermal and oxidatively induced reductive-eliminations have been compared for aryl, alkyl nickel(II) complexes (Equation 4.119). The oxidative path b is much more

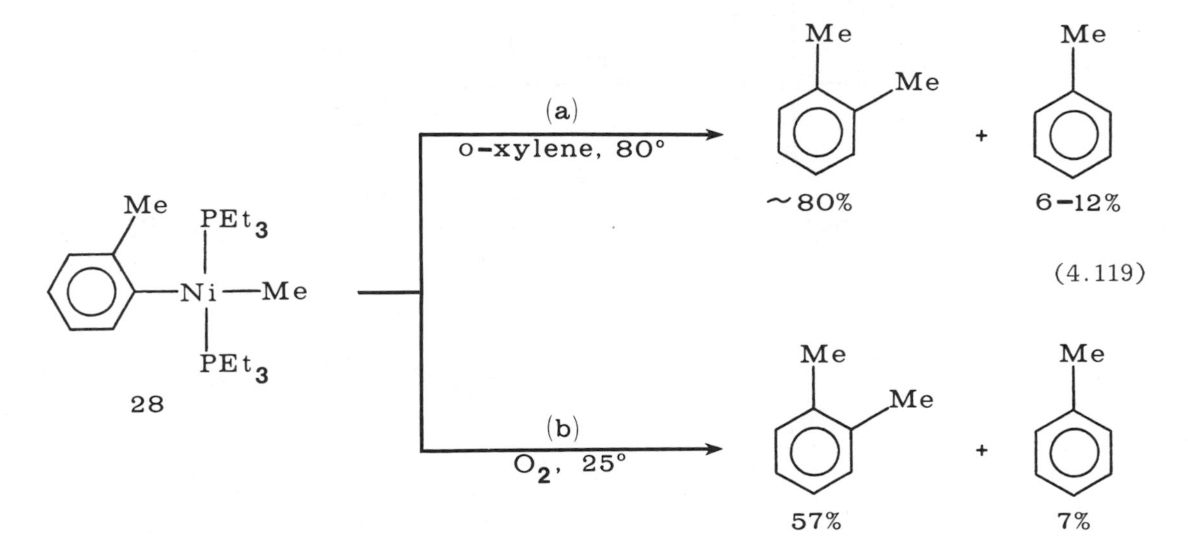


(4.119)

facile, but the carbon-carbon coupling yields depend on the nature of the oxidizing agent. In the thermal reaction (path a) isotopic labeling demonstrated that cross-coupling is minimal. Added phosphine inhibits this thermal reaction, again indicating that prior phosphine dissociation to form a three-coordination intermediate is required.

Even bromobenzene can stimulate the reductive-elimination of 28, perhaps by an oxidative addition [109]. The oxidatively induced coupling of nickel(II) alkyls is probably the basis of cross-coupling reactions between Grignard or lithium reagents and aryl or vinyl halides which are catalyzed by nickel complexes (Figure 4.14) [110].

$$L_2Ni(Ar)(Me) + Ar{-}X \longrightarrow L_2Ni(Ar)(X) + Ar{-}Me$$

$$L_2Ni(Ar)(X) + MeM \longrightarrow L_2Ni(Ar)(Me) + MX$$

$$ArX + MeM \xrightarrow{L_2Ni(Ar)(Me)} Ar{-}Me + MX$$

*Figure 4.14. Reactions involved in nickel-catalyzed coupling of aryl halides and Grignard reagents.*

If only a single aryl is present upon oxidation, the planar nickel(II) compounds tend to reductively eliminate a phosphorus-carbon bond (Equation 4.120).

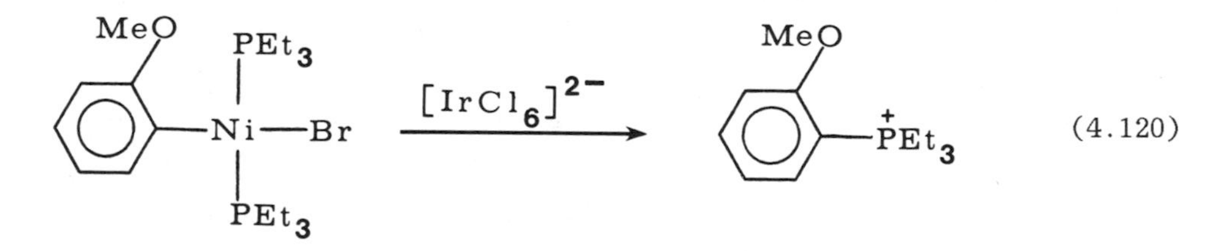


(4.120)

The concerted reductive-elimination of ethane from dimethyl palladium(II) complexes (Equation 4.121) has been carefully examined by Stille [111]. These studies give a

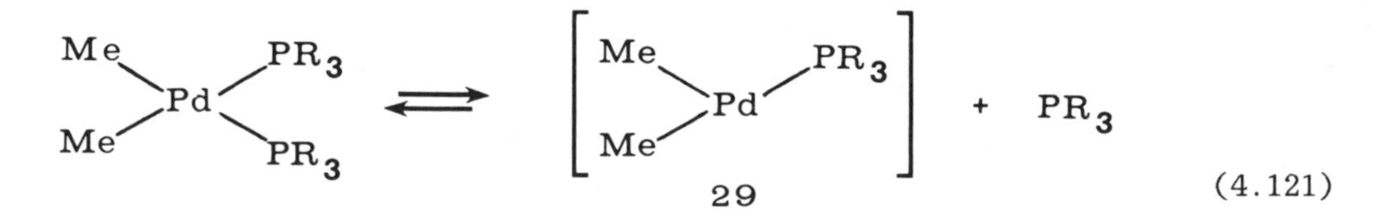


(4.121)

$$\longrightarrow Pd(PR_3)_2 + Me{-}Me$$

clear picture of the reaction mechanism. A cis-orientation of the methyl groups is required for the reaction to occur; trans-dimethyl complexes were shown to rearrange to cis-isomers before elimination takes place. Dimethyl derivatives were chosen to avoid complications arising from β-hydride elimination products which are formed from higher alkyls. Isotopic studies using $CD_3$ groups do not give crossover products, demonstrating that this reductive-elimination is a mononuclear reaction. The palladium(0) product has been characterized by trapping it as a diphenylacetylene adduct, $(PhC{\equiv}CPh)Pd(PPh_2CH_3)_2$.

Addition of excess phosphine slows the reductive-elimination of ethane, indicating that a phosphine is dissociated prior to the rate-determining step, as shown in Equation 4.121; so it is not surprising that the rate of reductive-elimination is 50 to 100 times slower in the case of a chelating phosphine complex, such as cis-$(Me)_2Pd(diphos)$, which is resistant to phosphine dissociation. Polar solvents such as DMSO increase the elimination rate, apparently by promoting phosphine dissociation. It is not clear whether these polar solvents occupy the coordination site vacated by phosphine or if the reactive intermediate, 29, is a T- or Y-shaped three-coordinate complex.

Stille found that transphos dimethylpalladium, 30, does not undergo reductive-elimination of ethane, even at 100° in DMSO; however, addition of $CD_3I$ to this complex rapidly produces $CD_3CH_3$ at 25° (Equation 4.122). The transphos ligand can only

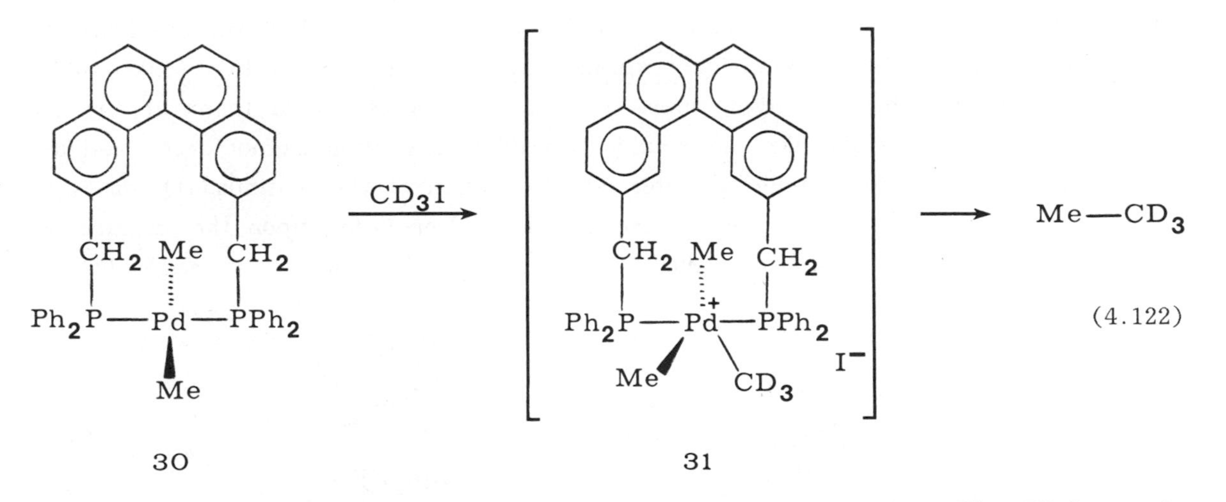


occupy trans-positions; so the methyl groups cannot become cis. The $CD_3I$ reaction appears to take place via a cationic five-coordinate palladium(IV) intermediate, 31. Thus Equation 4.122 is another example of an oxidatively induced reductive-elimination.

The cavity formed by the transphos ligand in 31 is too small to accommodate an iodide ion.

The cis-platinum(II) diaryls behave differently from the similar trans-nickel(II) compounds. Thermal decomposition of the platinum(II) diaryls gives both carbon-carbon and carbon-phosphorus coupled products, unless added phosphine is present, which facilitates biaryl formation [112]. Aryls are not isomerized and the elimination appears to be intramolecular (Equation 4.123). The diphos ligand is reported to retard this

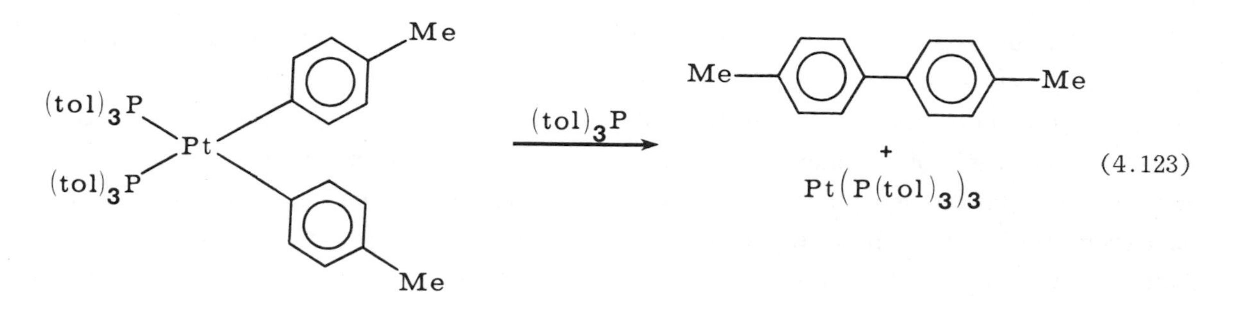


(4.123)

reaction, unless a ten-fold excess of free diphos is present. The basis of this curious effect is not understood.

Dialkyl platinum(II) compounds tend to thermally decompose through an intricate intramolecular mechanism involving ligand-dissociation, β-hydride elimination, and alkyl-hydride reductive-eliminations. Usually 1:1 mixtures of alkene and alkane are formed. These reactions have been extensively studied by G. Whitesides [113]. Like their palladium analogues, the dimethylplatinum complexes are more stable than the higher dialkyls, since with the dimethyl compounds β-hydride migration cannot occur and α-hydride migrations are less facile (Chapter 5). Oxidation of the platinum(II) dimethyl compound (32) does not induce reductive-elimination. Depending upon the nature of the phosphine ligands, a stable platinum(IV) complex (33) is sometimes formed [114]:

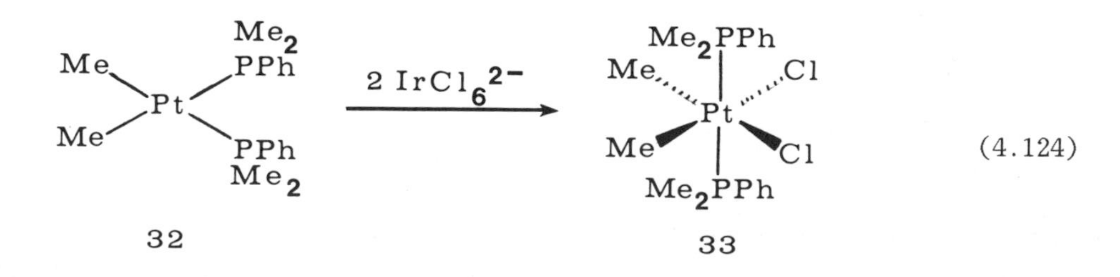


(4.124)

The contrasting behavior of platinum(II) and nickel(II) complexes can be rationalized by noting that platinum(IV) complexes are stable, whereas nickel(IV) compounds are not. Furthermore, Pt-C bonds are expected to be stronger than Ni-C bonds.

However, reductive-eliminations that form carbon-carbon bonds do occur in thermal decompositions of some $d^6$ methyl-platinum(IV) complexes (Equations 4.125 to 4.127)

$$\underline{fac}\text{-}L_2PtIMe_3 \xrightarrow{\Delta} C_2H_6 + \underline{trans}\text{-}L_2PtMeI \quad (4.125)$$

$$L_2PtCl(COMe)(Me)_2 \longrightarrow Me_2CO + \underline{trans}\text{-}L_2PtMeCl \quad (4.126)$$

$$L_2PtIMe_2(CF_3) \longrightarrow C_2H_6 + \underline{trans}\text{-}L_2Pt(CF_3)I \quad (4.127)$$

[115,116]. The ease of elimination, $CH_3CO > CH_3 > CF_3$, appears to be inversely related to the metal-carbon bond strength. Higher alkyls afford a mixture of alkenes and alkanes, presumably via the β-hydride elimination.

A kinetic study of the isotopically labeled trimethyl compounds, 34, 35, 36, and 37 (in 1,4-dioxane at 60-90°C), revealed several interesting features [117]. Methyl

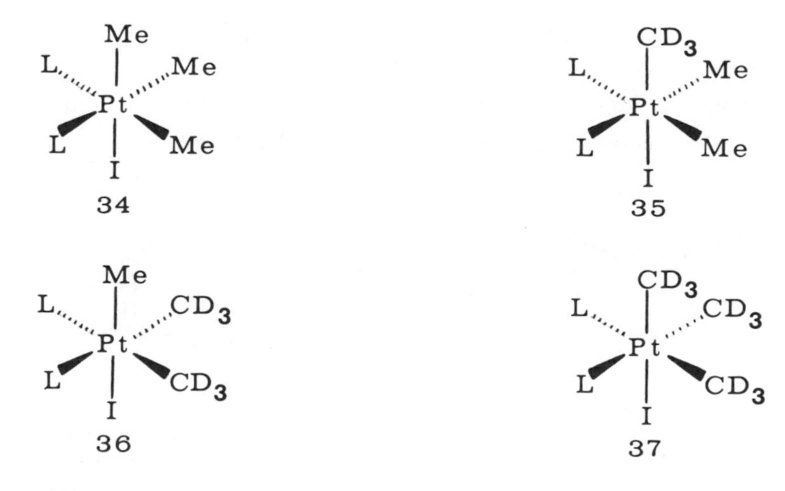


scrambling does not occur prior to reductive-elimination. Methyl groups trans to phosphine are more easily eliminated, and $CH_3$ groups are more easily eliminated than $CD_3$ groups. The eliminations are first-order, but are retarded by added phosphine.

This is consistent with a concerted, rate-determining, intramolecular elimination step $k_2$, from an unsaturated, five-coordinate intermediate (38) formed by loss of a phosphine (Equation 4.128). The more stable trans-platinum(II) products are formed

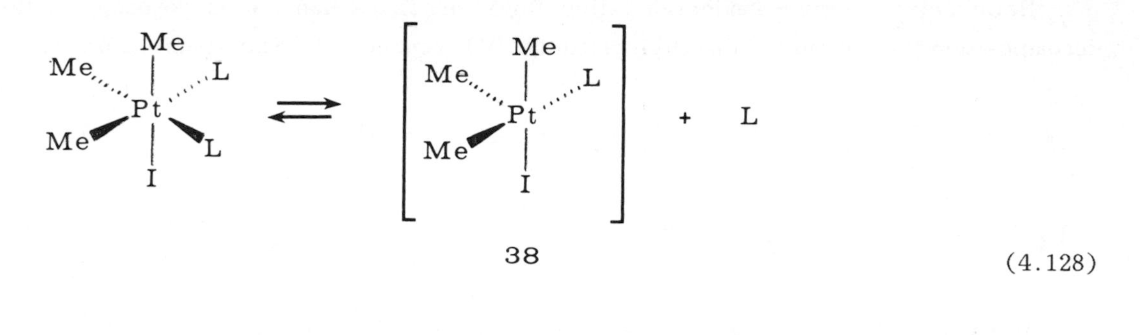


(4.128)

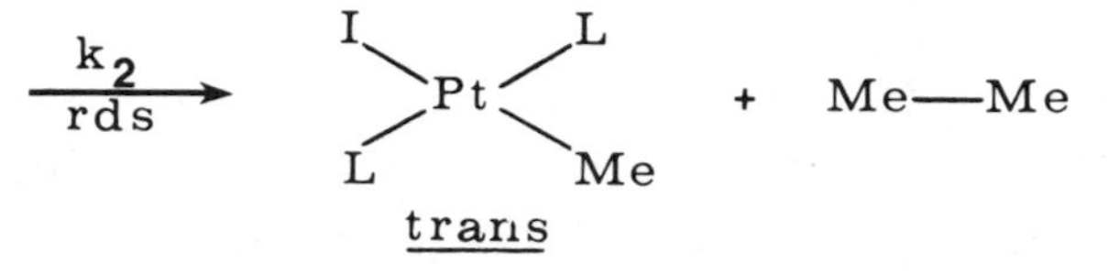


rather than the cis products expected from a concerted elimination. The trans compounds are thought to be formed upon recoordination of the phosphine. These reactions are facilitated by bulky phosphines; however, complexes of chelating phosphines are reported to react by an anomalous mechanism.

The more facile reductive-elimination of the acyl complex 39 illustrates the greater tendency of an acyl to reductively eliminate (Equation 4.129).

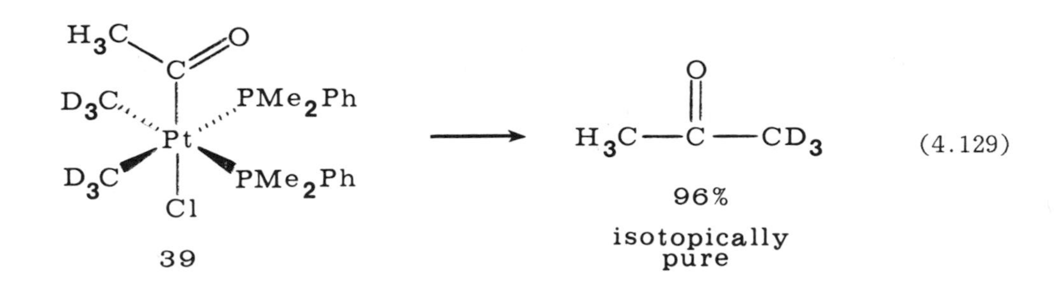


(4.129)

Cationic platinum(IV) methyl compounds seem to be more prone not only toward reductive-elimination, but also to rearrangement [118].

Reductive-elimination reactions that form carbon-carbon bonds have been examined for other $d^6$ complexes--notably those of cobalt(III) and iron(II). Since the metals in these complexes have accessible higher oxidation states, it is possible to examine their oxidatively induced reductive-eliminations, which are again much more facile than the thermal eliminations. Thermal decomposition of the diethyl compound 40 (path a) in solution affords mostly ethylene and ethane, ostensibly via the β-H elimination path [119]; whereas oxidizing agents markedly facilitate the reductive formation of n-butane (path b). The thermal decomposition is inhibited by added bipyridine. Mixtures

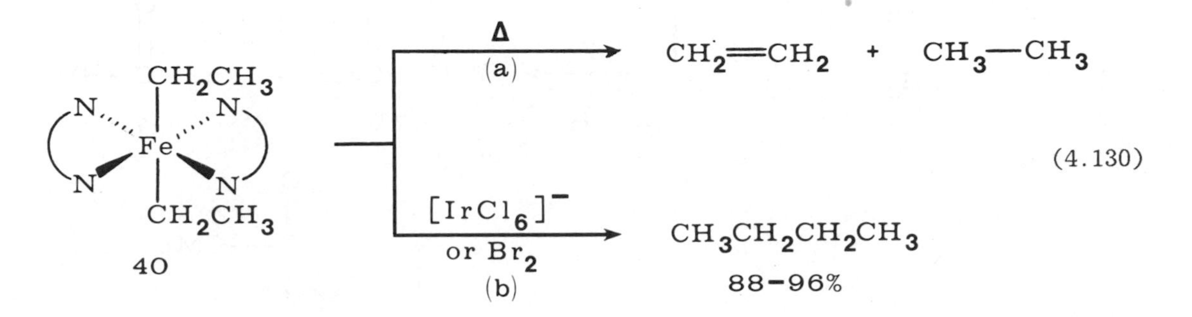


(4.130)

of the diethyl and dimethyl analogues of 40 form ethane and butane but not propane, thus ruling out dinuclear elimination as well as the formation of organic radicals. It is very likely that the oxidatively induced coupling involves a change in coordination geometry so that the alkyl groups can become mutually *cis*.

The anionic manganese(I) diacyl 41 is another $d^6$ complex for which thermal and oxidative reductive-eliminations have been compared as shown in paths (a) and (b) (Equation 4.131) [120]. The thermal path (a) seems to involve prior acyl to alkyl migration: whereas the oxidative path (b) primarily affords a diketone, as the direct reductive-elimination product.

Other examples of reductive-eliminations which form carbon-carbon bonds are given in Chapter 10, which considers the fragmentation of metallacycles.

b. Reactions Forming Carbon-Hydrogen Bonds. Reductive-elimination of a carbon-hydrogen bond is energetically a very favorable reaction. It is clear that *cis*-hydridoalkyl metal complexes are usually kinetically and thermodynamically less stable than their dihydrido and dialkyl analogues. Usually C-H reductive-elimination is so facile that it occurs spontaneously. Consequently intramolecular C-H reductive-eliminations have been rarely observed, although such reactions are often *presumed* steps in a sequence, for example, in protonolysis of metal alkyl compounds, in certain orthometallations, and in catalytic hydrogenation and hydroformylation of olefins.

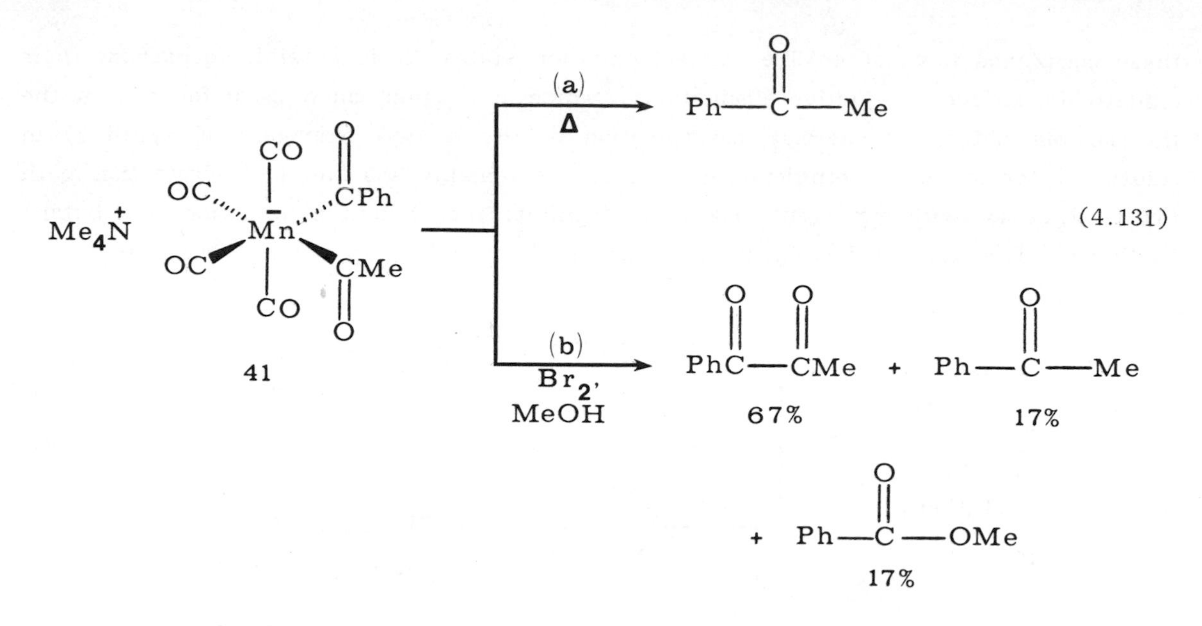


Halpern [121] has studied the direct, intramolecular reductive-elimination of alkanes from cis-hydridoalkylbis(phosphine)platinum(II) complexes such as 42 at -25°C using $^{31}P$ NMR. Isotopic labeling shows that the reaction is intramolecular. The rate

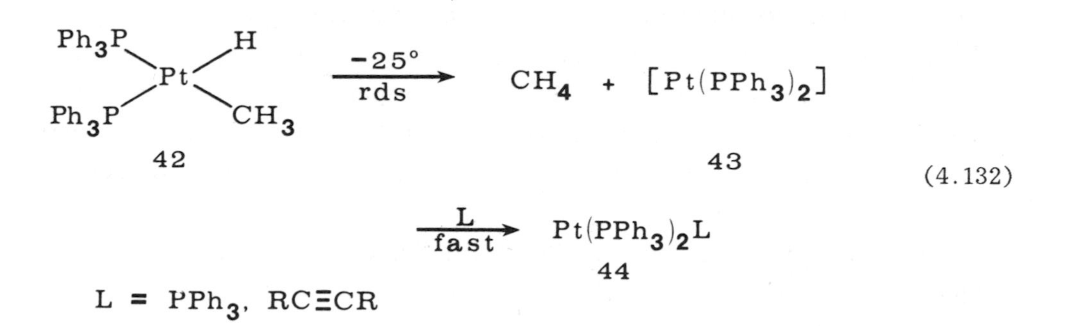


of C-H reductive-elimination is first-order and independent of added phosphine. The starting complex is stable only at low temperatures (below -50°C) and must, therefore, be generated in situ and characterized by low-temperature NMR. The presumed product, a metastable, two-coordinate platinum(0) complex (43), was trapped as various adducts (44). The decomposition of an alkyl- or aryl-hydride, $PtH(R)(PPh_3)_2$, shows the sequence of decreasing reactivity to be: R = $C_6H_5$ > $C_2H_5$ > $CH_3$ > $CH_2CH{=}CH_2$.

The preceding example is highly instructive: here C-H elimination is both thermodynamically and kinetically favorable, which suggests that the reverse process

(C-H oxidative-addition) is precluded on thermodynamic rather than kinetic grounds. Earlier we saw that C-H oxidative-addition does sometimes take place--either by an intramolecular reaction which has a more favorable entropy, or by an intermolecular reaction with alkyl groups having electronegative substituents. The latter are known to stabilize metal-carbon bonds and thus favor the C-H addition.

Equation 4.132 also demonstrates the power of low-temperature NMR spectroscopy as a technique for studying these very facile reactions. In Chapter 6 we will see that Halpern has used this method to observe directly the often-postulated C-H reductive-elimination step in catalytic hydrogenation of an olefin. This technique will undoubtedly be applied to other facile reductive-eliminations.

Presumably, hydride and alkyl ligands must become *cis* before intramolecular C-H elimination can take place. Thus, the planar nickel(II) complex, *trans* $NiH(CH_3)(PCy_3)_2$, is kinetically inert, probably because the very bulky $PCy_3$ ligands must remain *trans*, which enforces the *trans* relationship of the alkyl and hydride groups. *Trans* hydridoplatinum(II) alkyls such as *45* are also kinetically inert. However, addition of another ligand induces rapid reductive-elimination, apparently by promoting *trans* to *cis* isomerization [122]. The resulting C-H reductive-elimination is so facile that insertion reactions (with added CO) or β-H eliminations do not compete.

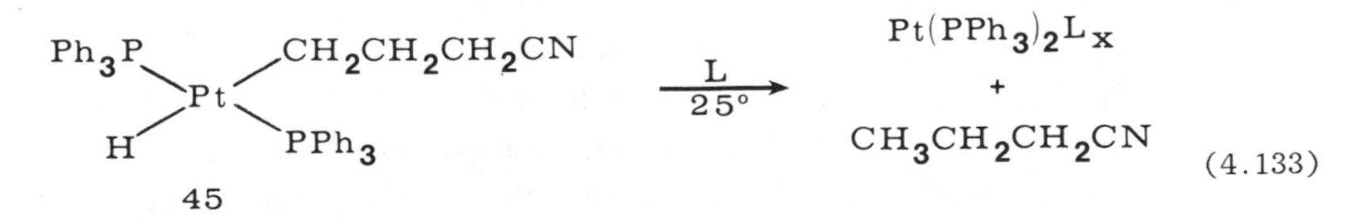


$L = PPhMe_2, PEt_3, CO$, etc.

2. *Dinuclear Reductive Elimination Reactions*. An *intermolecular* reductive-elimination mechanism has been established for a number of reactions. This "dinuclear reductive-elimination" has so far been observed *only for and may be limited to* eliminations involving a hydride. These reactions form a metal-metal bond and a C-H or an H-H bond. Examples are given in Equations 4.134 to 4.136.

$$HMn(CO)_5 + MeAu(PPh_3) \longrightarrow Ph_3P{-}Au{-}Mn(CO)_5 + CH_4 \quad (4.134)$$

$$Os(CO)_4(H)Me + HRe(CO)_5 \longrightarrow HOs(CO)_4Re(CO)_5 + CH_4 \quad (4.135)$$

$$Os(CO)_4(Me)_2 + Os(CO)_4(H)_2 \longrightarrow HOs(CO)_4Os(Me)(CO)_4 + CH_4 \quad (4.136)$$

Although isolated instances had long been recognized, the characteristic features and widespread occurrence of dinuclear eliminations were not appreciated before Norton's mechanism studies [2a]. Using isotopic labels and rate measurements, Norton established some relationships which seem to account for dinuclear hydride eliminations. "Norton's rules" should be used with caution, since they have not been widely tested, nor is the full scope of dinuclear reductive-elimination reactions yet known.

For a dinuclear elimination to occur, two conditions must be met: (1) one metal complex must have a hydride ligand, but may be either coordinatively saturated or coordinatively unsaturated; (2) the other metal may contain a hydride, an alkyl, or an acyl, but an adjacent coordination site must be vacant. The second condition may be achieved by dissociation of a ligand, or by alkyl-to-acyl migratory insertion (see Chapter 5).

In the last subsection we noted that intramolecular reductive-eliminations that form C-H bonds are usually favorable on both thermodynamic and kinetic grounds. Consequently very few stable cis-hydrido-alkyl transition-metal compounds have been prepared. Norton has pointed out that the stability of those hydrido-alkyls which are known can be understood by considering the metal complex which would be formed by reductive-elimination. A concerted intramolecular elimination of a diamagnetic alkyl-hydride should lead initially to a spin-paired reduced metal complex. In those cases in which cis-hydrido, alkyl complexes are stable, the ground state electronic configuration of the reduced product is expected to be a triplet. A concerted reaction path from a singlet starting material should lead to a singlet initial product, but if the ground state of the product is a triplet, the activation energy for the concerted path is necessarily increased. Thus, the concerted reductive-elimination should be unfavorable. The same reasoning holds for the concerted reductive-elimination of any two groups. Complexes 46 to 48 may not be susceptible to simple intramolecular elimination because of the instability of the product complex. A similar argument can be used to explain certain unfavorable migratory insertions (see Chapter 5).

The above concepts are nicely illustrated by Norton's study of the elimination reactions for the series of osmium(II) compounds 49, 50, and 51 whose decomposition temperatures are given below them. Matrix isolation studies indicate that $Os(CO)_4$

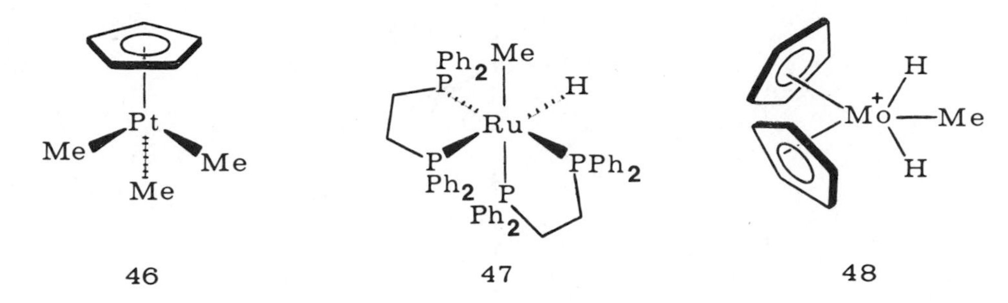


expected from concerted-elimination may be a ground-state triplet since its first-row analog, $Fe(CO)_4$, is paramagnetic [123].

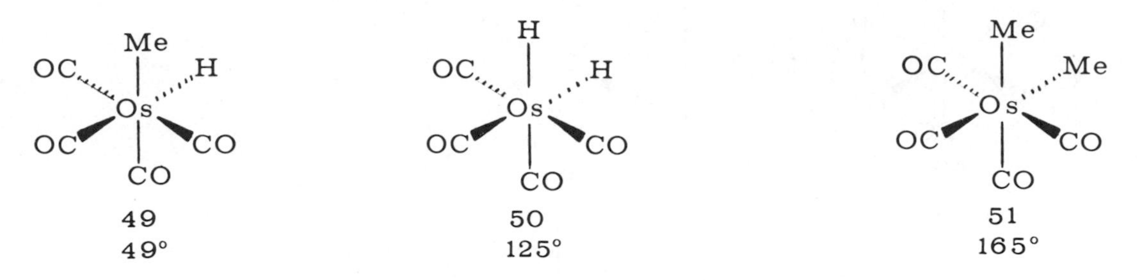


The hydrido-alkyl 49 was shown to react by the cross-coupling mechanism shown in Figure 4.15. The first-order rate-determining step (a) is thought to involve the formation of an intermediate, unsaturated acetyl complex 52. Rapid reaction (step b) of this acyl intermediate 52 with another mole of the hydrido-methyl complex forms methane and the diosmium hydrido-methyl compound 53. The intermediate 52 may be intercepted by nucleophilic ligands, L, affording methane and $Os(CO)_4(PPh_3)$ (step c). The latter reaction occurs at half the rate because of the change in stoichiometry--two moles of 49 are consumed by steps (a) and (b), and one mole by steps (a) and (c). Isotopic labeling demonstrated cross-coupling in step (b) and rules out a concerted elimination. The analogous ethyl complex $Os(CO)_4(H)Et$, was shown to decompose by the same mechanism but at a faster rate. Step (c) depends on the nature and concentration of L; excess L will divert all of the reaction to an intramolecular elimination. Since step (a) is rate-determining, the over-all rate of the reaction is independent of L. The detailed nature of step (c) is not understood.

The structurally characterized dinuclear hydridomethyl compound 53 appears to be stable toward intramolecular methane elimination (which would afford an osmium-osmium double bond), but this diosmium complex will react with the monomeric hydrido-alkyl 49 to form the triosmium complex, 54, by eliminating hydrogen (Equation 4.137).

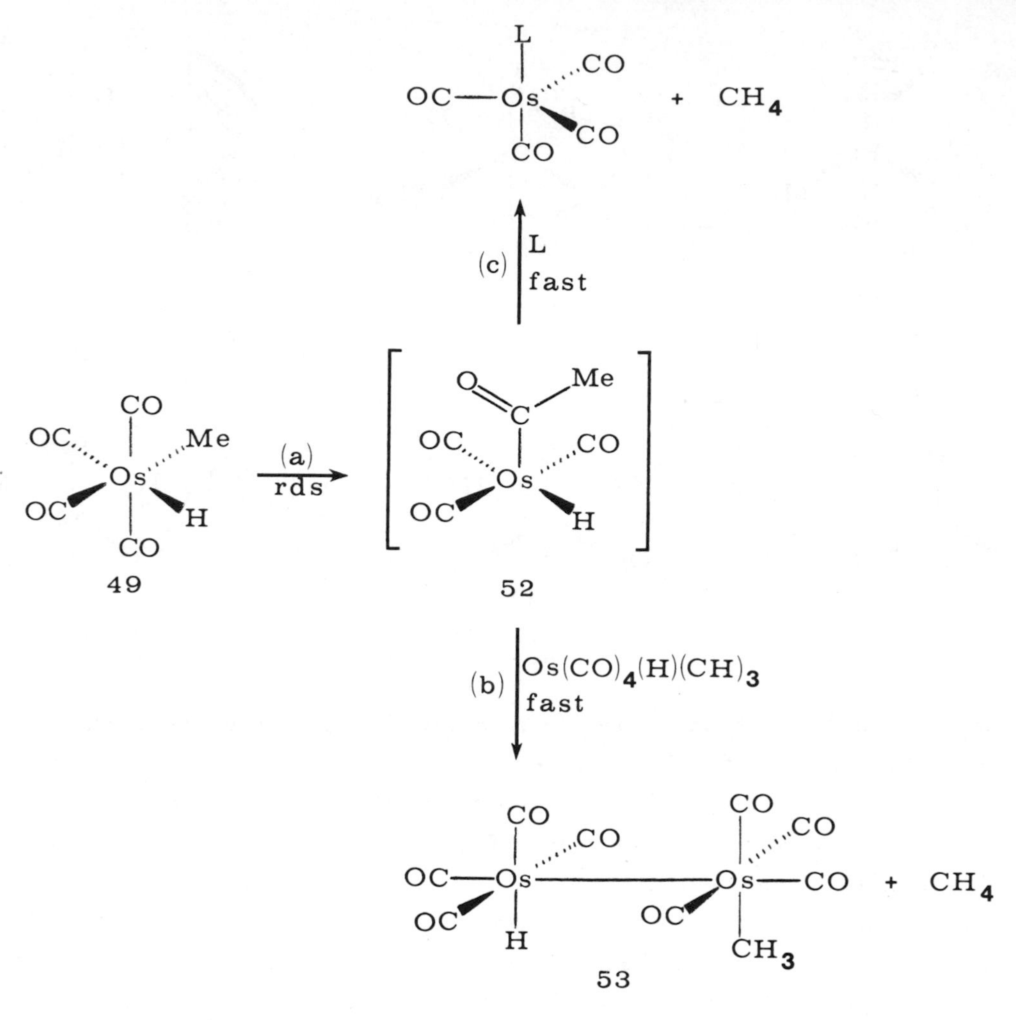


*Figure 4.15. Mechanism of the binuclear eliminations of a hydrido-methyl osmium(II) complex.*

$$\underset{53}{OsH(CO)_4Os(Me)(CO)_4} + \underset{49}{Os(CO)_4(H)(Me)} \longrightarrow \underset{54}{Os_3(CO)_{12}(Me)_2} + H_2 \qquad (4.137)$$

The monomeric dihydride 50 also uses a binuclear elimination path (Figure 4.16). In this case the requisite vacant coordination site is generated by CO dissociation, which is the rate-determining step (a). Rapid reaction (step b) with another molecule

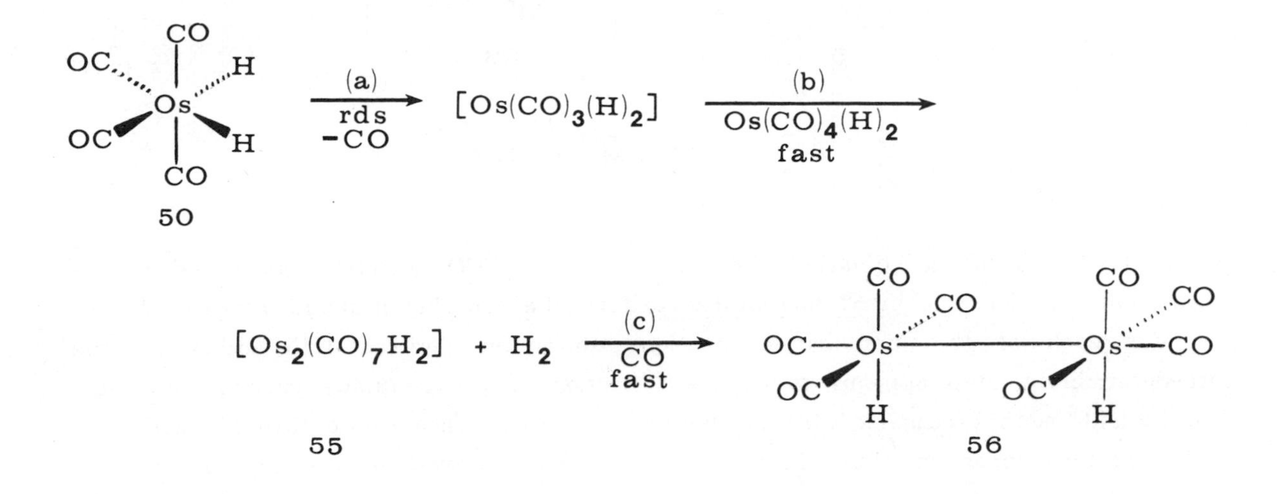


of the dihydride 50 was proposed to form an unsaturated intermediate 55 which rapidly adds CO (step c) yielding the diosmium dihydride 56. The over-all rate can be suppressed under CO pressure. Isotopic labeling showed that the reaction involves cross-coupling.

The dimethyl complex 51 is thermally very stable. Its decomposition appears to involve homolytic rupture of the osmium-methyl bond, forming methyl radicals which extract hydrogen atoms from the solvent:

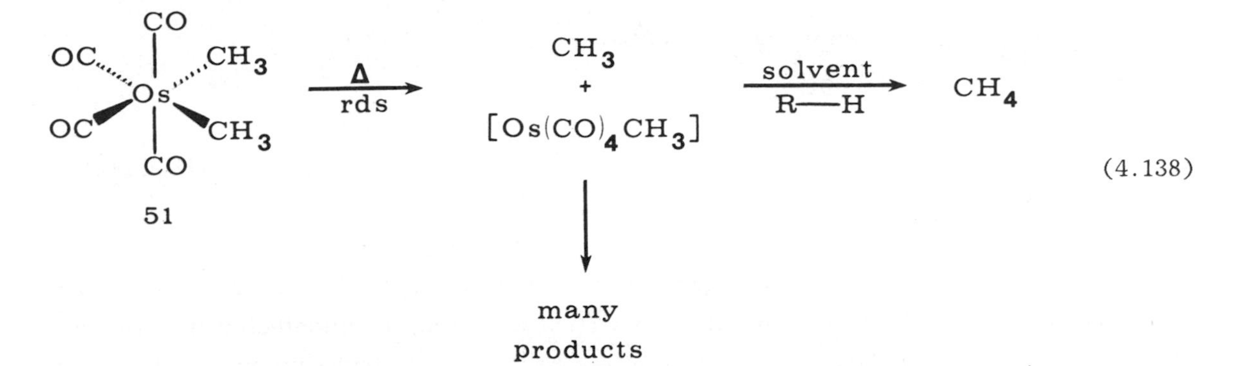


(4.138)

These results and other mixed-metal dinuclear eliminations (*vide infra*) can be accommodated by postulating the formation of a bridging hydride intermediate (58) which rapidly eliminates X-H, forming a metal-metal bond, 59 (Equation 4.139). (M*

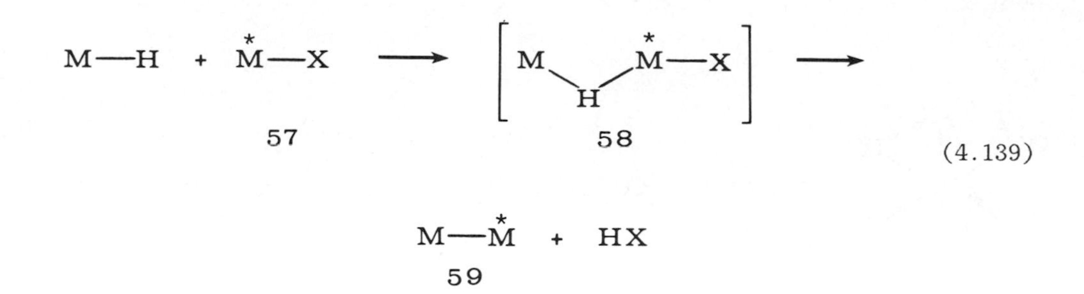


represents a vacant coordination site, X = R-, X- (RCO).) Norton proposed that a vacant coordination site in 57 is required to form the bridging hydride intermediate 58.

For each of the dinuclear eliminations shown in Figures 4.15 and 4.16, the rate-determining step appears to be the formation of the requisite coordination site. The hydrido-methyl complex (49) is thought to form a vacant site through methyl to acyl migratory insertion; but since hydrido-to-formyl migrations are unfavorable (see Chapter 5) the dihydrido complex 50 uses a CO dissociation step to create the vacant site. The dimethyl complex 51 lacks a hydride ligand and, therefore, cannot engage in a binuclear elimination with itself, even though this complex should be capable of generating a vacant coordination site. Thus, its homolytic reaction is understandable. Moreover, this mechanism predicts that the dihydride 50 and the dimethyl compound 51 should be able to consumate a mixed, dinuclear elimination. Norton found just such a reaction (4.140) between the dihydride 50 and the dimethyl compound 51. Both methane and acetaldehyde are formed along with the novel hydrido-methyl diosmium complex 60.

$$\underset{51}{Os(CO)_4(CH_3)_2} + \underset{50}{Os(CO)_4(H)_2} \xrightarrow{\Delta} \underset{60}{HOs(CO)_4Os(CO)_4CH_3} + CH_4 + CH_3CHO \qquad (4.140)$$

In Equations 4.139 and 4.140 it was proposed that the required vacant coordination site is created by methyl-to-CO migratory insertion affording an undetected intermediate acyl. Subsequent dinuclear elimination with other metal hydride unexpectedly forms alkane as the major product; whereas aldehyde is the expected product. These reactions of osmium complexes are thus unusual, since most other acyl-hydride dinuclear eliminations appear to give aldehyde as the major product. An example is the reaction between a manganese benzoyl complex and the corresponding manganese hydride which gives 99% benzaldehyde in 10 minutes at room temperature [124]:

$$C_6H_5C(=O)Mn(CO)_5 + H{-}Mn(CO)_5 \longrightarrow C_6H_5C(=O)H + Mn_2(CO)_{10} \quad (4.141)$$

Dinuclear eliminations may be important steps in reactions which are catalyzed by soluble transition-metal compounds, for example, hydroformylation (see Chapter 8). These dinuclear eliminations are also attractive methods for preparing heterometallic metal-metal bonded compounds; although the scope of such reactions, especially steric effects, is not well-understood.

NOTES.

1a. J. P. Collman and W. R. Roper, Adv. Organomet. Chem., 7, 53 (1968).
1b. J. P. Collman, Accts. Chem. Res., 1, 136 (1968).
1c. J. Halpern, Accts. Chem. Res., 3, 386 (1970).
1d. M. F. Lappert and P. W. Lednor, Adv. in Organometal. Chem., 14, 345 (1976).
1e. R. D. Kemmit and J. Burgess, Inorg. React. Mech., 2, 350 (1972).
1f. A. J. Deeming, MTP Int. Rev. Sci., Inorg. Chem., Ser., 19, 117 (1972).
1g. L. Vaska, Accts. Chem. Res., 1, 335 (1968).
1h. R. Ugo, Coord. Chem. Rev., 3, 319 (1968).
2a. J. Norton, Accts. Chem. Res., 12, 139 (1979).
2b. P. S. Braterman and R. J. Cross, Chem. Soc. Rev., 2, 271 (1973).
2c. J. L. Davidson, Inorg. React. Mech., 2, 398 (1977).
2d. R. D. W. Kemmitt and J. Burgess, Inorg. React. Mech., 2, 350 (1973).
2e. R. D. W. Kemmitt and M. A. R. Smith, Inorg. React. Mech., 3, 451 (1974).
2f. R. D. W. Kemmitt and M. A. R. Smith, Inorg. React. Mech., 4, 3119 (1976).
2g. M. C. Baird, J. Organometal. Chem., 64, 289 (1974).
3. P. J. Brothers, Prog. Inorg. Chem., 28, 1 (1981).
4. D. M. Blake and M. Kubota, Inorg. Chem., 9, 989 (1970).
5. A. Davis, W. McFarlane, L. Pratt, and G. Wilkinson, J. Chem. Soc., 3653 (1962).
6. Th. Kruck and A. Engelmann, Angew. Chem. Int. Ed. Engl., 5, 836 (1966).

7. T. V. Ashworth, J. E. Singleton, D. J. A. de Waal, W. J. Louw, E. Singleton, and E. van der Stork, J. Chem. Soc. (D), 340 (1978).

8. H. W. Walker, C. T. Kresge, P. C. Ford, and R. G. Pearson, J. Amer. Chem. Soc., 101, 7428 (1979), and references therein.

9. A. J. Deeming and B. L. Shaw, J. Chem. Soc. (A), 1802 (1969)

10a. R. A. Schunn, Inorg. Chem., 9, 394 (1970).

10b. D. F. Schriver, Accts. Chem. Res., 3, 231 (1970).

11a. F. Cariati, R. Ugo, and F. Bonati, Inorg. Chem., 5, 1128 (1966).

11b. D. J. Cook, J. L. Dawes, and R. D. W. Kemmitt, J. Chem. Soc. (A), 1547 (1967).

12. J. C. Kotz and D. G. Pedrotty, Organometal. Chem., Rev., Sect. A, 4, 479 (1969).

13. A. Davidson, W. McFarlane, L. Pratt, and G. Wilkinson, J. Chem. Soc., 3653 (1962).

12. P. B. Chock and J. Halpern, J. Amer. Chem. Soc., 88, 3511 (1966).

13. R. G. Pearson and W. R. Muir, J. Amer. Chem. Soc., 92, 5519 (1970).

14. J. P. Collman and C. T. Sears, Jr., Inorg. Chem., 7, 27 (1968).

15. M. Kubota, G. W. Kiefer, R. M. Ishikawa, and K. E. Bencala, Inorg. Chem. Acta, 7, 195 (1973). This reference lists many recent kinetic studies of oxidative-addition.

16. R. G. Pearson and W. R. Muir, J. Amer. Chem. Soc., 92, 5519 (1970).

17. J. A. Labinger, R. J. Braus, D. Dolphin, and J. A. Osborn, J. Chem. Soc. (D), 612 (1970).

18. F. R. Jensen and B. Knickel, J. Amer. Chem. Soc., 93, 6339 (1971).

19. J. S. Bradley, D. E. Connor, D. E. Dolphin, J. A. Labinger, and J. A. Osborn, J. Amer. Chem. Soc., 94, 4043 (1972).

20a. G. M. Whitesides and D. J. Boschetto, J. Amer. Chem. Soc., 93, 1529 (1971).

20b. P. L. Boch, D. J. Boschetto, J. R. Rasmussen, J. P. Demers, and G. M. Whitesides, J. Amer. Chem. Soc., 96, 2814 (1974).

21a. P. K. Wong, K. S. Y. Lau and J. K. Stille, J. Amer. Chem. Soc., 96, 5956 (1974).

21b. K. S. Y. Lau, P. K. Wong, and J. K. Stille, J. Amer. Chem. Soc., 98, 5832 (1976).

21c. J. K. Stille and K. S. Y. Lau, loc. cit., p. 5841.

22. J. K. Stille and K. S. Y. Lau, Accts. Chem. Res., 10, 434 (1977).

23. K. S. Y. Lau, P. K. Wong, and J. K. Stille, J. Amer. Chem. Soc., 98, 5832 (1976).
24. J. P. Collman and M. R. McLaury, J. Amer. Chem. Soc., 96, 3019 (1974).
25. R. W. Johnson and R. G. Pearson, J. Chem. Soc. (D), 986 (1970).
26. Y. Becker and J. K. Stille, J. Amer. Chem. Soc., 100, 838 (1979).
27. A. V. Kramer and J. A. Osborn, J. Amer. Chem. Soc., 96, 7832 (1974).
28. A. V. Kramer, J. A. Labinger, J. S. Bradley and J. A. Osborn, J. Amer. Chem. Soc., 96, 7145 (1974).
29. P. K. Wong, K. S. Y. Lau and J. K. Stille, J. Amer. Chem. Soc., 96, 5956 (1974).
30. D. Dodd and M. D. Johnson, Chem. Comm., 1371 (1971).
31. C. R. Johnson and G. A. Dutra, J. Amer. Chem. Soc., 95, 7783 (1973), and references therein.
32. J. P. Collman, R. G. Finke, J. N. Cawse, and J. I, Brauman, J. Amer. Chem. Soc., 99, 2515 (1977).
33. J. P. Collman, Accts. Chem. Res., 8, 342 (1975).
34. R. G. Pearson, H. Sobel, and J. Songstad, J. Amer. Chem. Soc., 90, 319 (1968).
35. P. J. Krusic, P. J. Fagan, and J. San Fillippo, Jr., J. Amer. Chem. Soc., 99, 250 (1977).
36. A. J. Hart Davis and W. A. Graham, Inorg. Chem., 9, 2658 (1970).
37. J. Ishizu, T. Yamamoto, and A. Yamamoto, Chem. Lett., 1091 (1976).
38. Y. Watanabe, M. Yamashita, M. Igami, T. Mitsudo, and Y. Takegami, Bull. Chem. Soc. Japan, 49, 2820 (1976).
39. J. F. Bunnett, Quart. Rev. Chem. Soc., 12, 1 (1958).
40. P. Fitton and E. A. Rick, J. Organometal. Chem., 28, 287 (1971).
41. T. T. Tsou and J. K. Kochi, J. Amer. Chem. Soc., 101, 6319 (1979).
42. T. T. Tsou and J. K. Kochi, J. Amer. Chem. Soc., 101, 7547 (1979).
43. J. Blum and M. Weitzberg, J. Organometal. Chem., 122, 261 (1976).
44a. P. Fitton and J. E. McKeon, Chem. Comm., 4 (1968).
44b. L. Cassar and A. Giarruso, Gazz. Chem. Ital., 103 793 (1973).
44c. B. E. Mann, B. L. Shaw, and N. I. Tucker, J. Chem. Soc. (A), 2667 (1971).
44d. J. Lewis, B. F. G. Johnson, K. A. Taylor, and J. D. Jones, J. Organometal. Chem., 32, C62 (1974).
44e. B. F. G. Johnson, J. Lewis, J. D. Jones, and K. A. Taylor, J. Chem. Soc., Dalton Trans., 34 (1974).

44f. J. Rajarain, R. G. Pearson, and J. A. Ibers, J. Amer. Chem. Soc., 96, 2103 (1974).

45a. W. J. Bland and R. D. W. Kemmitt, J. Chem. Soc. (A), 127 (1968).

45b. M. Green, R. B. L. Osborn, A. J. Rest, and F. G. A. Stone, J. Chem. Soc. (A), 2525 (1968).

45c. J. Ashley-Smith, M. Green, and D. C. Wood, J. Chem. Soc. (A), 1847 (1970).

46. J. Growning, M. Green, and F. G. A. Stone, J. Chem. Soc. (A), 453 (1971).

47a. O. Sunamura, K. Tokumaru, and H. Yui, Tet. Lett., 5141 (1966).

47b. J. A. Kampmeier and R. M. Fantazier, J. Amer. Chem. Soc., 88, 1959 (1965).

47c. G. M. Whitesides, C. P. Casey, and G. K. Krieger, J. Amer. Chem. Soc., 93, 1379 (1971).

48. B. M. Trost and T. R. Verhoeven, J. Amer. Chem. Soc., 98, 630 (1976).

49. J. L. Burmeister and L. M. Edwards, J. Chem. Soc. (A), 1663 (1971).

50. H. C. Volger, H. Hogeveen, and M. M. P. Gaasbeek, J. Amer. Chem. Soc., 91, 2137 (1969).

51. L. Vaska and M. F. Werneke, Trans. N. Y. Acad. Sci., 33, 70 (1971).

52. F. Moranini and S. Bresadola, Inorg. Chem., 15, 650 (1976).

53a. J. F. Harrod, G. Hamer, and W. Yorke, J. Amer. Chem. Soc., 101, 3987 (1979).

53b. J. P. Jesson, in Transition Metal Hydrides, edited by E. L. Muetterties (Marcel Dekker, 1971), p. 84.

54. T. Yoshida and S. Otsuka, J. Amer. Chem. Soc., 99, 2134 (1977).

55. M. J. Church, M. J. Mays, R. N. F. Simpson, and F. P. Stefanini, J. Chem. Soc. (A), 2909 (1970).

56. A. J. Deeming and B. L. Shaw, J. Chem. Soc. (A), 3356 (1970).

57. W. H. Knoth, J. Amer. Chem. Soc., 98, 7172 (1968).

58. F. L'Eplattenier and F. Calderazzo, Inorg. Chem., 7, 1290 (1968).

59. P. S. Hallman, B. R. McGarvey, and G. Wilkinson, J. Chem. Soc. A, 3143 (1968).

60. R. A. Schunn, in Transition Metal Hydrides, ed. by E. L. Muetterties (Marcel Dekker, 1971), p. 213, and references therein.

61. J. Chatt and B. L. Shaw, J. Chem. Soc., 5075 (1962).

62. W. Keim, J. Organometal. Chem., 14, 179 (1968).

63. C. S. Cundy, B. M. Kingston, and M. F. Lappert, Adv. Organomet. Chem., 11, 253 (1973).

64. S. Murani and N. Sonada, Angew. Chem., Int. Ed. Engl., 18, 837 (1979).

65. N. J. Archer, R. N. Haszeldine, and R. V. Parish, Chem. Comm., 524 (1971).

66. F. de Charentenary, J. A. Osborn, and G. Wilkinson, J. Chem. Soc. (A), 787 (1968).

67. R. N. Haszeldine, R. V. Parish, and R. J. Taylor, J. Chem. Soc. (D), 2311 (1974).

68. M. Green, J. A. K. Howard, J. Pround, J. L. Spencer, F. G. A. Stone, and C. A.Tsipis, J. C. S. Chem. Comm., 671 (1976).

69. A. L. Balch and Y. S. Sohn, J. Organometal. Chem., 30, C31 (1971) and references therein.

70. L. Vaska, L. S. Chen, and W. V. Miller, J. Amer. Chem. Soc., 93, 6671 (1971).

71. M. I. Bruce, "Cyclometallation Reactions," Angew. Chem., Int. Ed., Engl., 16, 73 (1977).

72. G. Parshall, Accts. Chem. Res., 3, 139 (1970); 8, 113 (1975).

73. J. Dehand and M. Pfeffer, Coord. Chem. Rev., 18, 327 (1976).

74. T. Yoshida and S. Otsuka, J. Amer. Chem. Soc., 99, 2134 (1977).

75a. J. Tsuji and K. Ohno, Synthesis, 1, 157 (1967).

75b. B. M. Trost and M. Preckel, J. Amer. Chem. Soc., 96, 7862 (1973).

75c. P. D. Hobbs and P. D. Magnus, Chem. Comm., 856 (1974).

75d. D. J. Dawson and R. E. Ireland, Tet. Lett., 1899 (1968).

76. H. M. Walborsky and L. E. Allen, J. Amer. Chem. Soc., 93, 5465 (1971).

77. J. W. Suggs, J. Amer. Chem. Soc., 100, 640 (1978).

78. D. J. Cole-Hamilton and G. Wilkinson, J. C. S. Chem. Comm., 59 (1977).

79. H. Alper and A. S. K. Chan, J. Amer. Chem. Soc., 95, 4905 (1973).

80a. C. C. Yui and A. J. Deeming, J. Chem. Soc. (D), 2091 (1975), and references therein.

80b. R. B. Calvert and J. R. Shapley, J. Amer. Chem. Soc., 100, 7726 (1978).

81. E. L. Muetterties, T. N. Rhodin, E. Band, C. F. Brucker, and W. R. Pretzer, Chem. Rev., 79, 91 (1979), and references therein.

82. D. E. Webster, Adv. Organomet. Chem., 15, 147 (1977).

83a. F. A. Cotton, D. L. Hunter, and B. A. Frenz, Inorg. Chim. Acta, 15, 155 (1975), and references therein.

83b. P. Foley and G. M. Whitesides, J. Amer. Chem. Soc., 101, 2732 (1979).

84. D. M. Rai, P. M. Bailey, K. Moseley, and P. M. Maitlis, Chem. Comm., 1273 (1972).

85. F. A. Cotton, T. LaCour, and A. G. Stanislowski, J. Amer. Chem. Soc., 96, 754 (1974).

86. P. W. Clark, J. Organomet. Chem., 137, 235 (1977).

87. R. Mason, G. Scollary, B. Moyle, K. I. Hardcastle, B. L. Shaw, and C. J. Moulton, J. Organomet. Chem., 113, C49 (1976).

88. C. A. Tolman, S. D. Ittel, A. D. English, and J. P. Jesson, J. Amer. Chem. Soc., 101, 1742 (1979), and references therein.

89. N. J. Copper, M. L. H. Green, C. Couldwell, and K. Prout, J. C. S. Chem. Comm., 145 (1977).

90. G. P. Pez, J. Amer. Chem. Soc., 98, 8072 (1976) and references therein.

91. L. J. Guggenberger and F. N. Tebbe, J. Amer. Chem. Soc., 93, 5924 (1971); F. N. Tebbe and G. W. Parshall, J. Amer. Chem. Soc., 93 3793 (1971).

92. K. I. Grell and J. Schwartz, J. C. S. Chem. Comm., 244 (1979).

93a. J. Halpern and J. P. Maher, J. Amer. Chem. Soc., 86, 2311 (1964); 87, 5361 (1965).

93b. P. B. Chock and J. Halpern, J. Amer. Chem. Soc., 90, 6959 (1968).

93c. P. W. Schneider, P. F. Phelan, and J. Halpern, J. Amer. Chem. Soc., 91, 77 (1969).

93d. J. Halpern and P. Phelan, J. Amer. Chem. Soc., 94, 1181 (1972).

94a. C. E. Castro and W. C. Kray, Jr., J. Amer. Chem. Soc., 86, 2768 (1963).

94b. J. K. Kochi and J. W. Powers, J. Amer. Chem. Soc., 92, 137 (1970), and references therein.

94c. J. R. Hanson and E. Premuzic, Angew. Chem., Int. Ed. Engl., 7, 247 (1968).

94d. R. S. Nohr and J. H. Espenson, J. Amer. Chem. Soc., 97, 3392 (1975).

95. J. Kochi and J. Powers, J. Amer. Chem. Soc., 92, 137 (1970).

96. D. M. Singleton and J. K. Kochi, J. Amer. Chem. Soc., 89, 6547 (1967).

97. J. Z. Chrzastowski, C. J. Cooksey, M. D. Johnson, B. L. Lockman, and P. N. Steggles, J. Amer. Chem. Soc., 97, 932 (1975).

98. R. Poilblanc, Nouv. J. Chim., 2, 145 (1978).

99. N. H. Alemdaroglu, J. M. L. Penninger, and E. Oltay, Monatsh. Chem., 107, 1043 (1976).

100. M. H. Chisholm, F. A. Cotton, M. W. Extine, and L. A. Rankel, J. Amer. Chem. Soc., 100, 807 (1978), and references therein.

101. N. S. Lewis, K. P. Mann, J. G. Gordon, II, and H. B. Gray, J. Amer. Chem. Soc., 98, 7461 (1976).

102. H. Schmidbaur, J. R. Mandl, A. Frank et Huttner, Chem. Ber., 109, 466 (1976), and references therein.

103. S. Komiya, T. A. Albright, R. Hoffman, and J. K. Kochi, J. Amer. Chem. Soc., 98, 7255 (1976), and references therein.

104a. P. L. Kuch and R. S. Tobias, J. Organomet. Chem., 122, 429 (1976).

104b. S. Komiya and J. K. Kochi, J. Amer. Chem. Soc., 98, 7590 (1976).

105a. T. Yamamoto, A. Yamamoto, and S. Ikeda, J. Amer. Chem. Soc., 93, 3350 (1971).

105b. For a review of nickel-carbon sigma bonds, see D. R. Fahey, Organometal. Chem. Rev., Sect. A, 7, 245 (1972).

105c. M. Uchino, K. Asagi, A. Yamamoto and S. Ikeda, J. Organometal. Chem., 84, 93 (1975).

106. G. M. Whitesides, W. F. Fischer, Jr., G. San Fillippo, Jr., R. W. Bashe, and H. O. House, J. Amer. Chem. Soc., 91, 4871 (1971).

107. T. T. Tsou and J. K. Kochi, J. Amer. Chem. Soc., 100, 1634 (1978).

108. M. Almemark and B. Akermark, J. C. S. Chem. Comm., 66 (1978).

109. D. G. Morrell and J. K. Kochi, J. Amer. Chem. Soc., 97, 7262 (1975).

110. K. Tamao, K. Sumitani and M. Kumada, J. Amer. Chem. Soc., 94, 4374 (1972).

111. A. Gillie and J. K. Stille, J. Amer. Chem. Soc., 102, in press (1980).

112. P. S. Braterman, R. J. Cross, and G. B. Young, J. Chem. Soc., Dalton Trans., 19, 1892 (1977), and references therein.

113. J. X. McDermott, J. F. White, and G. M. Whitesides, J. Amer. Chem. Soc., 98, 6521 (1976).

114. J. Y. Chen and J. K. Kochi, J. Amer. Chem. Soc., 99, 1450 (1977).

115. B. L. Shaw and J. D. Ruddick, J. Chem. Soc. (A), 2969 (1969), and references therein.

116. H. C. Clark and J. D. Ruddick, Inorg. Chem., 9, 2556 (1970).

117. M. P. Brown, R. J. Puddephatt, and C. E. E. Upton, J. Chem. Soc. (D), 2457 (1974), and references therein.

118. T. G. Appleton, H. C. Clark, L. E. Manzer, J. Organomet. Chem., 65, 275 (1974) and references therein.

119. A. Yamamoto, K. Morifuji, S. Ikeda, T. Saito, Y. Uchida, and A. Misono, J. Amer. Chem. Soc., 90, 1878 (1968).

120. C. P. Casey and C. A. Bunnell, J. Amer. Chem. Soc., 98, 436 (1976).

121. L. Abis, A. Sen, and J. Halpern, J. Amer. Chem. Soc., 100, 2915 (1978).

122. R. Ros, R. A. Michalen, R. Bataillard, and R. Roulet, J. Organomet. Chem., 161, 75 (1978), and references therein.

123. T. J. Barton, R. Grinter, A. J. Thompson, B. Davies, and M. Poliakoff, J. C. S. Chem. Comm., 841 (1977).

124. W. Tam., W. K. Wong, and J. A. Gladysz, J. Amer. Chem. Soc., 101, 1589 (1979).

# 5

# Insertion Reactions

There is a substantial body of organometallic reactions which are called "insertions" [1-5]. This term is misleading, in that both intermolecular additions and a wide range of intramolecular rearrangements are combined under this classification. Such reactions are especially important as key steps in homogeneous catalysis as well as in stoichiometric organometallic syntheses. For example, all homogeneously catalyzed "oxo" reactions, olefin-oligomerizations, and olefin-hydrogenations, as well as heterogeneously catalyzed olefin-oligomerizations, olefin-hydrogenations, and Fischer-Tropsch processes, involve insertion reactions. Before we proceed to specific applications of insertion reactions, it is instructive for us to discuss the scope and mechanism of this important reaction class.

Insertion reactions involve the combination of a saturated ligand (X) with an unsaturated ligand (Y), forming a new ligand (Y-X). Two general types of insertion processes are recognized: intramolecular "migratory insertions" (5.1); and

$$\underset{\text{M—Y}}{\overset{\text{X}}{|}} \rightleftharpoons \text{M—Y—X} \underset{}{\overset{\text{L}}{\rightleftharpoons}} \underset{\text{M—Y—X}}{\overset{\text{L}}{|}} \qquad (5.1)$$

intermolecular "nucleophilic additions" (5.2). The intramolecular "migratory insertions"

$$\ddot{\text{X}} + \text{M—Y} \longrightarrow \text{M—Y—X} \qquad (5.2)$$

take place by the combination of X and Y while both are coordinated to the metal. In a separate step, an external ligand L usually binds to the metal, filling the coordination site left vacant by the insertion. No net change in the formal oxidation state of the metal occurs during a migratory-insertion reaction. This aspect helps to account for the broad scope of the migratory-insertion reaction, and for the fact that insertions are well known for nontransition metal compounds.

The intermolecular "nucleophilic-addition" reactions (Equation 5.2) are similar to the intramolecular migratory-insertion reactions (Equation 5.1). However, these nucleophilic additions take place without prior coordination of the nucleophile/ligand X. There may or may not be a change in the formal oxidation state of the central metal during the intermolecular addition reactions (vide infra).

Since the net reactions for both inter- and intramolecular insertions are very similar, it is sometimes difficult to distinguish one from the other. However, for many insertion reactions this distinction is clear. We shall discuss the two processes separately, by first taking up the "migratory-insertion" reaction and subsequently considering "nucleophilic-addition" processes.

## 5.1. Migratory-Insertion Reactions.

These reactions can be subdivided into two classes, one involving longitudinally bonded unsaturated ligands Y, such as carbon monoxide, and the other involving transversely bonded ligands Y such as olefins. We shall consider the longitudinal ligand type first.

1. Migrations to CO: Acyl Formation. The prototypical example of this reaction class is the transformation of an alkyl-carbonyl complex into an acyl complex with the incorporation of an external ligand. Examples of alkyl to acyl insertions are known for most of the transition metals and have been extensively studied.

For example, alkyl manganese pentacarbonyl complexes, 1, rearrange in the presence of ligands (L) such as CO, a tertiary phosphine, an amine, or a halide ion, affording the acyl complex, 2 (Equation 5.3). The mechanism of this particular reaction has been studied in considerable detail by Calderazzo [2]. The mechanisms of several other alkyl carbonyl to acyl migratory insertion reactions have been investigated to a lesser degree. All these mechanistic studies taken together are consistent with a general reaction path. It is, therefore, usually assumed that the mechanisms of various alkyl-to-acyl migratory-insertion reactions are the same or very similar to the well-studied manganese case, but this is an extrapolation of the few definitive studies which

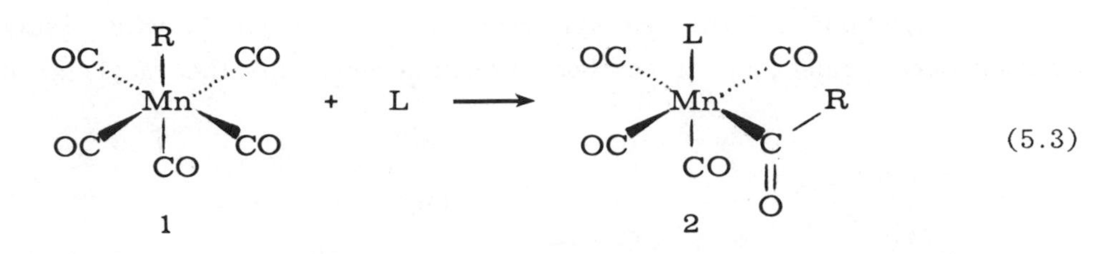


have so far been carried out. In this chapter we review some specific aspects of these mechanistic studies.

a. General features. In so far as generalization based on a few studies is valid, most alkyl to acyl reactions have the following characteristics. The R group migrates to an adjacent carbonyl group, and in a separate, subsequent step, the coordinatively unsaturated intermediate takes up an external ligand in the site vacated by the migrating alkyl group [4]. In most cases which have been examined, this reaction is highly stereospecific [1-7]. Alkyl groups having chirality at the metal-carbon bond migrate with retention of configuration. The rate of these reactions and the position of equilibrium depend on the strengths of the metal-carbon bond and of the metal carbon monoxide bond in the starting alkyl complex [1]. Thus, the migratory aptitudes usually are alkyl > benzyl >> H, perfluoroalkyl, or acyl (the last three groups are so far known to exhibit only the reverse reaction). First-row, 3d-metal alkyls are usually more reactive than second- and third-row 4d and 5d alkyls. Presumably this reactivity profile results from modest differences in metal-carbon bond strengths, the first-row metal-carbon bonds being slightly weaker than those of the 4d and 5d metals. In some cases Lewis acids promote these insertions by binding to the more basic oxygen of the acyl carbonyl group and perhaps stabilizing the unsaturated intermediate acyl complex [8,9]. In certain instances, prior oxidation of the central metal facilitates the forward migratory-insertion reaction [10], probably by lowering the strength of the bond between the metal and the terminal carbonyl group. Metals such as palladium(II) and zirconium (IV), which form rather weak bonds to carbon monoxide, readily undergo the acyl-insertion reaction.

b. The stereochemistry at the metal. The stereochemistry at the metal for the manganese alkyl-acyl insertion was elucidated by Calderazzo [2], who employed $^{13}C$ isotopically labeled CO, infrared spectroscopy, and the concept of microscopic reversibility. Careful infrared spectral analysis of isotopically substituted alkyl and acyl pentacarbonyl complexes 3 and 4 permitted the assignment of isotopically labeled axial or equatorial CO groups as well as the acyl group in both 3 and 4 (Equation 5.4).

Carbonylation of the methyl manganese complex 3 with $^{13}CO$ gives $^{13}CO$ incorporation solely in a carbon monoxide cis to the acyl group in 4. This

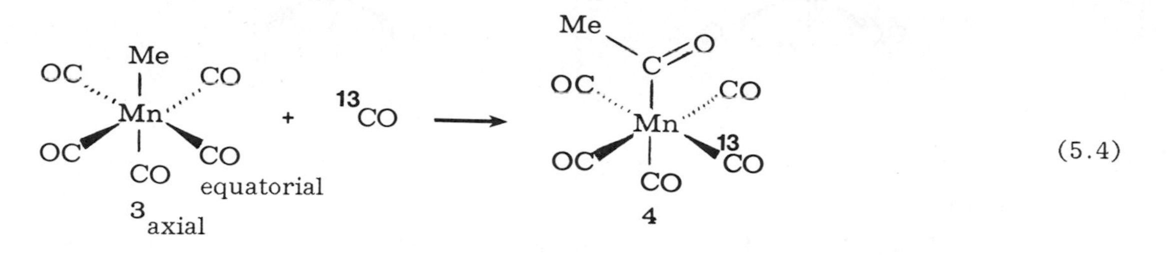


experiment demonstrates two things. First, the entering CO group serves merely as the external ligand L in Equation 5.1 and 5.3. Second, since no trans isomer is observed, the stereochemistry at the metal is under kinetic rather than thermodynamic control. Without such kinetic control we could not study the stereochemistry of this reaction. Unfortunately, the stereochemistries of many other insertion reactions are under thermodynamic control and cannot be so examined.

The reverse of migratory insertion was achieved by thermal decomposition of the isotopically labeled acyl complex 5, yielding exclusively the cis $^{13}CO$ methyl derivative 6 (Equation 5.5). According to the principle of microscopic reversibility this result

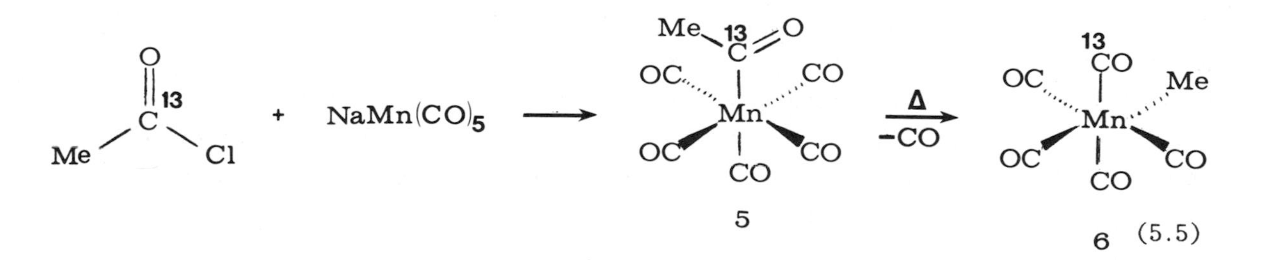


shows that the methyl and CO groups which react to form the acyl must be mutually cis.

A similar thermal decarbonylation of the labeled cis acyl complex 4 affords a 1:2 mixture of the trans and cis methyl derivatives 7 and 8 (Equation 5.6). This result is consistent with a mechanism whose microscopic reverse is methyl migration to a cis carbonyl. Thus, of the two plausible, reversible processes, methyl migratory insertion (Equation 5.7) and carbonyl insertion (Equation 5.8), the former has been established by the above experiments as well as other experiments with the same system.

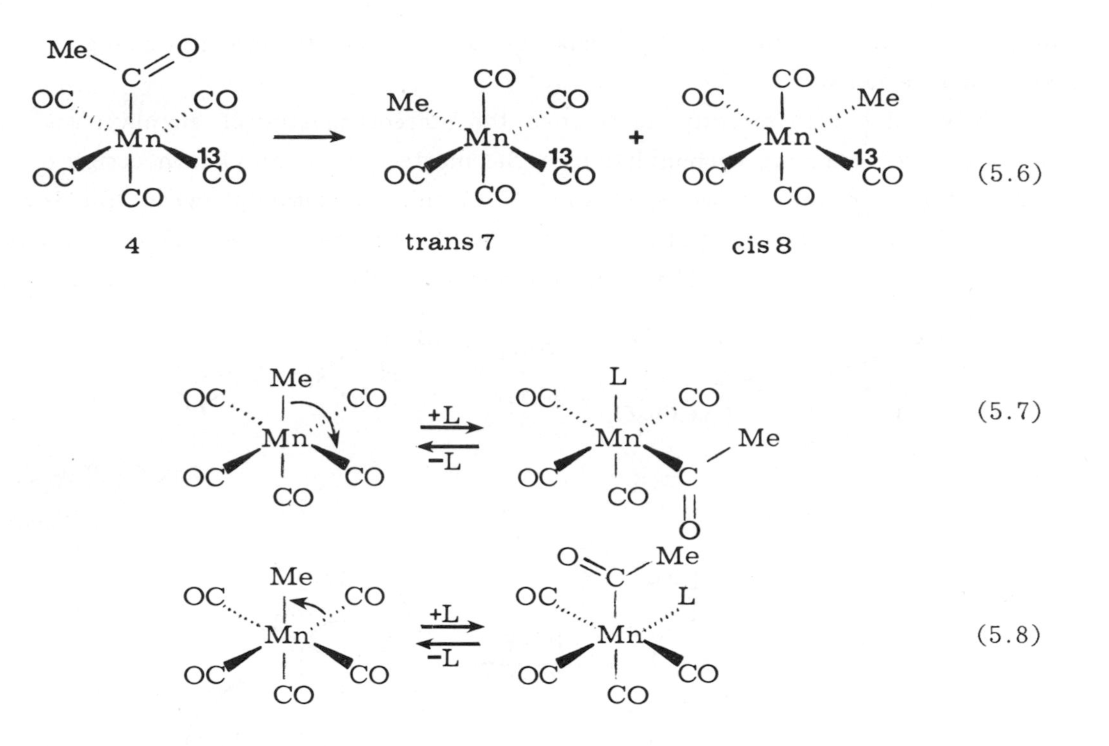


Other alkyl to acyl reactions are presumed to go by the alkyl migration path. Although not enough other cases have been studied in sufficient detail to distinguish these two possibilities, the alkyl migration seems the more likely path.

c. Stereochemistry at the alkyl carbon. Stereochemistry at carbon for alkyl to acyl migratory insertions has been established in a number of situations by the use of various experimental techniques. All of the results reported so far are consistent with retention of configuration at the migrating chiral center. For example, the optically active α-phenethyl manganese acyl 9 was kept under equilibrium conditions of CO pressure and temperature long enough that decarbonylation to the alkyl 10 and migratory insertion back to 9 should have occurred many times (Equation 5.9). The

$$\underset{\mathbf{9}}{C_6H_5{-}CH(CH_3){-}C({=}O){-}Mn(CO)_5} \rightleftharpoons \underset{\mathbf{10}}{C_6H_5{-}CH(CH_3){-}Mn(CO)_5} + CO \qquad (5.9)$$

optical activity of the reaction mixture showed no significant change during this extended period.

Whitesides [11] directly determined the stereochemistry of an alkyl-acyl migratory insertion by examining carbonylation of the diastereomeric alkyl-iron complexes, such as the erythro and threo isomers 11 and 13. Within experimental error, the erythro alkyl 11 affords the erythro-acyl 12 (Equation 5.10) and the threo-alkyl 13 forms the threo-acyl 14 (Equation 5.11). These diastereoisomeric alkyl groups are especially useful in

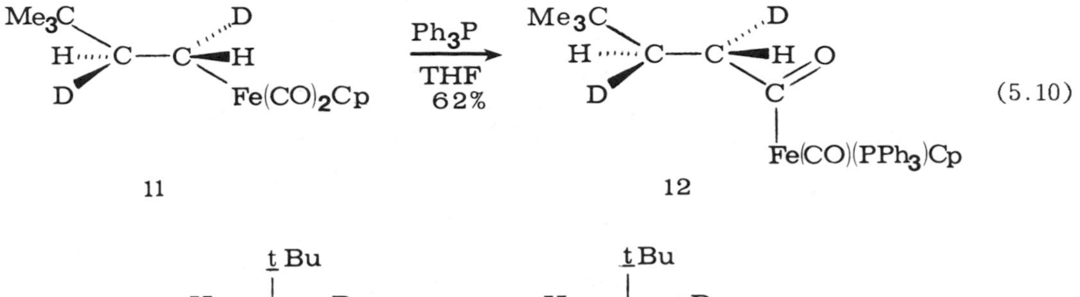


(5.10)

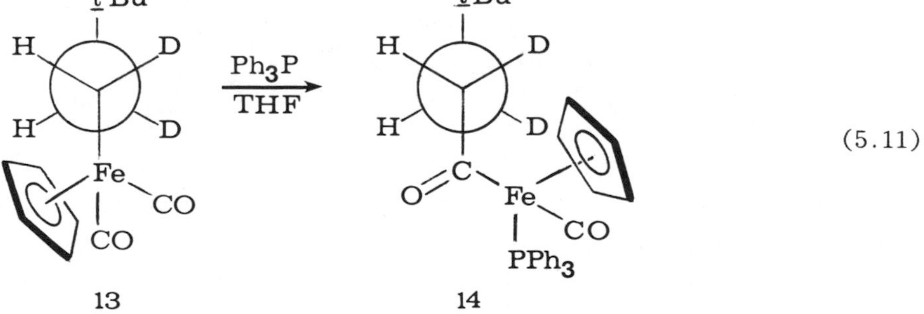


(5.11)

diagnosing reaction mechanism, since both the erythro and threo isomers can be detected simply by measuring their proton NMR spectra. This analysis depends on well-established differences between the $J_{H,H}$ coupling constants for syn and anti hydrogens, and the anchoring of a dominant conformation by the bulky t-butyl and metallo groups. The direct NMR technique used here is thus very powerful. Enantiomers need not be resolved, nor do optical rotations need to be measured. The NMR analyses are quantitative, and both isomers can be distinguished and simultaneously detected. These stereochemistries are more conveniently drawn using "Newman" formulas (Equation 5.11).

Most other studies of the stereochemistry of alkyl-acyl migratory-insertion reactions are indirect, involving a cycle of reactions; so the stereochemistry of one reaction must be assumed in order to assign the stereochemistry of another reaction in the cycle. Working from Whitesides' direct measurements described above, the other cases of alkyl to acyl insertion are usually presumed to proceed with retention of configuration. Consider, for example, Stille's preparation [12] of a palladium-acyl, 16, from the corresponding palladium-alkyl, 15 (Equation 5.12). The palladium-acyl 16

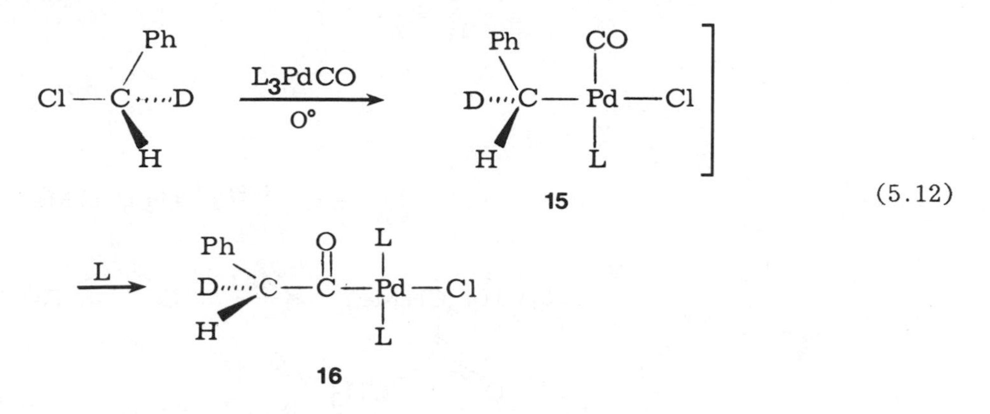


was independently prepared from an acid chloride having a known configuration and enantiomeric purity. The palladium-alkyl was prepared by oxidative addition (see Chapter 4), a reaction presumed in this case to go with inversion of configuration. The isomeric purity of the palladium-acyl was measured by its subsequent degradation in a sequence of reactions which do not involve bonds at the chiral center (Equation 5.13). However, the stereochemical integrity of the reaction cycle is critically

$$\mathbf{16} \xrightarrow[\text{MeOH}]{Cl_2\ -78^\circ} \text{Ph(D)(H)C—C(=O)—OMe} \qquad (5.13)$$

dependent on the reaction conditions. The most reasonable interpretation of over-all inversion at chiral carbon is that given in Equations 5.12 and 5.13.

A similar sequence of reactions involving the iron carbonyl anion 17 has been used to transform (+)-2-octyl tosylate into the corresponding chiral methyl ketone (Equation 5.14). Within the limits of experimental error, the over-all sequence goes with inversion [14]. The migratory-insertion step is assumed to take place with retention of configuration. Thus the first step was claimed to go with inversion. Another stereochemical approach to the migratory-insertion mechanism is to use configurationally stable complexes which are chiral at the metal [15]. For example, photochemical decarbonylation of the (R)-iron acyl, 18, afforded the (S)-iron ethyl derivative, 19, with over-all

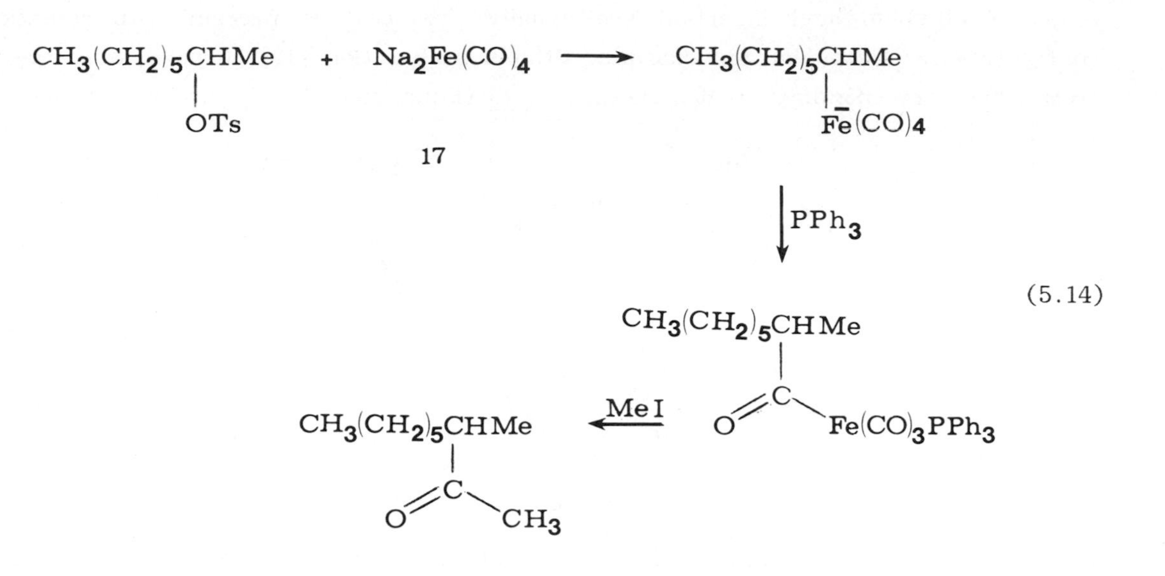


inversion at iron (Equation 5.15). This reaction is consistent with a mechanism in

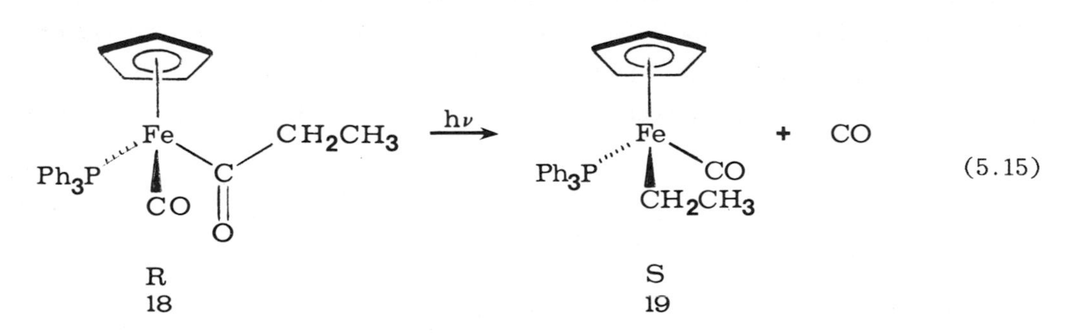


which CO is expelled and ethyl migrates to the coordination site vacated by the CO ligand. These results are fully in accord with alkyl migration to CO for the reverse reaction, since intramolecular CO insertion into the iron-ethyl bond would give retention at the chiral iron center. These results also require that any coordinatively unsaturated intermediate is configurationally stable on the time scale of the reaction.

d. Kinetic studies. Many kinetic studies of alkyl to acyl migratory-insertion reactions have been reported. Most of these can be accommodated by the two-step mechanism shown in Equation 5.16. In the first step there is an equilibrium between the coordinatively saturated alkyl complex, 20, and the coordinatively unsaturated acyl complex, 21. The second step involves the addition of the external ligand L affording

$$R{-}M{-}CO \underset{k_{-1}}{\overset{k_1}{\rightleftharpoons}} \left[ R{-}C(=O){-}M \right] \underset{k_{-2}\ -L}{\overset{k_2\ +L}{\rightleftharpoons}} R{-}C(=O){-}M{-}L \tag{5.16}$$

20 21 22

the product, a saturated acyl complex, 22. Most of the cases have been studied under conditions whereby the equilibrium lies far to the right. That is, the insertion is essentially irreversible under these reaction conditions, and $k_2(L) >> k_{-2}$. Under these conditions a steady-state approximation of the rate of migratory insertion affords the rate law shown in Equation 5.17:

$$\frac{-d[RMCO]}{dt} = k_{obs}[RMCO], \quad \text{where} \quad k_{obs} = \frac{k_1k_2[L]}{k_{-1}+k_2[L]} \tag{5.17}$$

The general mechanism described in Equation 5.16 gives rise to limiting second-order, first-order, or intermediate behavior, depending on the relative magnitudes of $k_1$, $k_{-1}$ and $k_2$.

Consider the situation in which $k_2(L) << k_{-1}$; that is, the second step in Equation 5.16 is rate-limiting. A small steady-state concentration of the assumed unsaturated acyl intermediate, 21, reacts with the external ligand L at a rate much less than that of the reverse step in the prior equilibrium, $k_{-1}$. For this case, the rate law becomes over-all second-order, first-order each in the alkyl complex and the external ligand L, as shown in Equation 5.18:

$$\frac{-d[RMCO]}{dt} = \frac{k_1k_2}{k_{-1}}[RMCO][L] = K_1k_2[RMCO][L] \quad , \tag{5.18}$$

where

$$k_{obs} = k_2K_1[L], \text{ and } K_1 = \frac{k_1}{k_{-1}}$$

For experimental convenience, such cases are usually studied under pseudo-first-order conditions such that the concentration of L, [L], is great enough to be constant. Simple pseudo-first-order rate plots then afford $k_{obs}$. Changing the concentration of L, but still maintaining an excess of L, leads to another value of $k_{obs}$ and verifies the rate dependence on [L]. From these independently measured $k_{obs}$ values, the value of

$K_1k_2$ can be calculated. For such a mechanism, $k_2$ should also be sensitive to the nature of L, more nucleophilic ligands giving rise to increased values of $k_{obs}$. In those cases in which only pure second-order behavior is observed, the two-step mechanism in Equation 5.16 cannot be distinguished from a concerted one-step reaction.

On the other hand, if $k_2[L] >> k_{-1}$, the rate expression in Equation 5.16 reduces to the first-order behavior shown in Equation 5.19. In such cases the rate is over-all first-order, being dependent only on the concentration of the alkyl complex. [RMCO], and independent of either the concentration or of the nature of L. Limiting first-order behavior, $k_2[L] >> k_{-1}$, simplifies the rate law to:

$$\frac{-d[RMCO]}{dt} = k_1[RMCO] \quad , \quad \text{where} \quad k_{obs} = k_1 \quad . \tag{5.19}$$

Occasionally, this limiting first-order rate behavior appears to result from the solvent participating as a ligand.

In migratory-insertion reactions, intermediate-rate behavior is often observed. In other instances an observed second-order rate law can be transformed into an intermediate-rate expression by increasing either the concentration or the nucleophilicity of the external ligand L, thus making the magnitude of $k_2[L]$ similar to that of $k_{-1}$ in Equation 5.17. The use of reciprocal rate plots is a convenient method of expressing such intermediate-rate behavior. For such cases, the external ligand concentration [L] is used in sufficient excess over [RMCO] so that pseudo-first-order rate plots can be obtained, and from these pseudo-first-order rate constants, $k_{obs}$, are calculated:

$$k_{obs} = \frac{k_1k_2[L]}{k_{-1}+k_2[L]} \quad .$$

For L in sufficient excess, [L] is a constant, and therefore $k_{obs}$ is a constant. Plots of $1/k_{obs}$ vs. $1/[L]$ should afford a straight line whose slope and intercepts are $k_{-1}/k_1k_2$ and $1/k_1$ respectively:

$$\frac{1}{k_{obs}} = \frac{k_{-1}}{k_1k_2}\,\frac{1}{[L]} + \frac{1}{k_1} \quad . \tag{5.20}$$

Crude estimates of $k_1$ can be taken from the value of the intercept and used to calculate values for $k_{-1}/k_2$. Since $k_2$ should be affected by the nucleophilicity of L, this can sometimes be confirmed by varying the nature of L.

The well-studied alkyl manganese system [1] exhibits the variations in rate behavior expected from the two-stage path shown in Equation 5.17. For example,

$MeMn(CO)_5$ shows a first-order dependence on [CO] at partial pressures of CO less than 1 atmosphere, but at CO pressures of 15 atmospheres intermediate-rate behavior is observed, giving rise to linear plots of 1/[CO] vs. $1/k_{obs}$. When other more nucleophilic ligands, L, are used, intermediate-rate behavior is again observed, as can be found from linear inverse-rate plots. The $k_1$ values obtained from the intercepts of such plots (Equation 5.20) are independent of the nature of L. However, the values of $k_{-1}/k_2$ taken from the slopes of these linear inverse plots depend on the nature of L. As expected, more nucleophilic ligands L afford smaller values of $k_{-1}/k_2$, since the magnitude of $k_2$ increases with the nucleophilicity of L. The $k_1$ values, however, depend on the nature of the alkyl group.

The role of solvent in these insertion reactions is variable and depends on the particular system [1]. In certain cases a large rate increase for migratory insertions is observed in the presence of a solvent which can serve as a ligand. For example, the reaction between $MeMn(CO)_5$ and cyclohexylamine (Equation 5.21) is accelerated by

$$MeMn(CO)_5 + C_6H_{11}NH_2 \longrightarrow MeC(=O)Mn(CO)_4(C_6H_{11}NH_2) \qquad (5.21)$$

a factor of $10^4$ in DMF as compared with the aromatic hydrocarbon, mesitylene [16]. Such a rate increase induced by a polar solvent is usually ascribed to the solvent's acting as a ligand in such large excess that its concentration is constant. An intermediate having a solvent molecule serving as the incoming ligand may then react further to form the final product, as the stronger external ligand, L, replaces the solvent(S), as in Equation 5.22. In such cases, where the formation of the solvent-bonded

$$R{-}M{-}CO + S \rightleftharpoons [RC(=O)Mn(S)] \xrightarrow[-S]{L} RC(=O){-}M{-}L \qquad (5.22)$$

intermediate is rate-determining, there may result a large negative entropy of activation, "$\Delta S^{\ddagger}$". Such negative $\Delta S^{\ddagger}$ values are occasionally taken as evidence for this sort of pathway [1]. In other instances, large, steady-state concentrations of the solvent-bond intermediate acyl complexes may build up, and can be detected by their characteristic acyl infrared stretching frequency, different from that of the final product [13]. For example, such evidence exists [17] for intermediates of the type $MeCOMn(CO)_4(THF)$ and $PhCH_2COFe(Cp)(CO)(DMSO)$. In other cases, $k_{obs}$ values from reciprocal rate plots are insensitive to the coordinating ability of the solvent. Examples of such systems are $RIr(CO)_2L(Cl)_2$ and RPt(CO)LCl (where L = $Ph_3As$) [1,18]. For such cases, solvents are apparently ineffective as intermediate ligands. It is interesting to note that the capture of an external ligand by a coordinatively unsaturated complex can be very fast and may be solvent-dependent. For example, the photochemically generated chromium pentacarbonyl recaptures CO more rapidly in cyclohexane ($k_{-1} = 3 \times 10^6\ M^{-1}\ sec^{-1}$) than in perfluorocyclohexane ($k_{-1} = 3 \times 10^8$) [19].

$$Cr(CO)_6 \underset{k_{-1}}{\overset{h\nu}{\rightleftharpoons}} Cr(CO)_5 + CO \qquad (5.23)$$

The migratory insertion of the anionic alkyl iron(0) carbonyl compound, 23, has been studied extensively (Equation 5.24). Polar solvents cause a marked decrease in

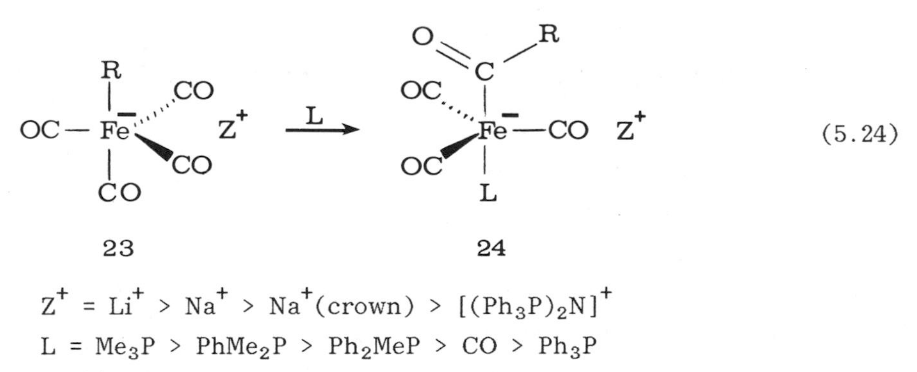


(5.24)

$Z^+ = Li^+ > Na^+ > Na^+(crown) > [(Ph_3P)_2N]^+$
$L = Me_3P > PhMe_2P > Ph_2MeP > CO > Ph_3P$
R = n-alkyl or $PhCH_2$-

the rate of these reactions. For example, changing from THF to NMP (N-methylpyrrolidinone) slows the rate by a factor of 400. This solvent effect is related to ion-pairing phenomena [8]. The tight-ion pair was demonstrated to be the kinetically most reactive species. The rate is thus affected by the nature of the cation. For example, the lithium salt, which forms a strong ion pair, reacts ~$10^3$ faster than do the $[(Ph_3P)_2N]^+$ or crown-ether-complexed sodium salts, which do not form strong ion pairs. Thus in the case of these basic, anionic iron alkyls, the effect of solvent is related to the promotion of the migration by Lewis acids. The acyl oxygen in 24 was shown to be a stronger Lewis base than the terminal carbonyls. IR and $^{13}C$ NMR spectroscopic evidence reveals that the ion pairing in the product is localized at the acyl oxygen.

The rate law observed for the migrations shown in Equation 5.24 was found to be second-order under all feasible experimental conditions. These results were interpreted as the limiting case of the two-step mechanism shown in Equation 5.17, where $k_{-1} >> k_2[L]$. This mechanism cannot be distinguished from a simple, one-step concerted reaction path which would also afford over-all second-order behavior, first-order in $[RFe(CO)_4]^{2-}$ and first-order in L. Either mechanism would also be consistent with the 20-fold increase in rate observed by changing from $Ph_3P$ to $Me_3P$ as the incoming ligand. Such variations seem to derive from both electronic and steric changes in L. The choice of the limiting case of a two-stage mechanism over a synchronous bimolecular mechanism is based on the evidence for the two-step mechanism with other complexes.

Shriver [9] has found that $AlBr_3$ greatly accelerates the methyl to acyl migration, affording a structurally characterized Lewis-acid adduct, 25 (Equation 5.25). A kinetic

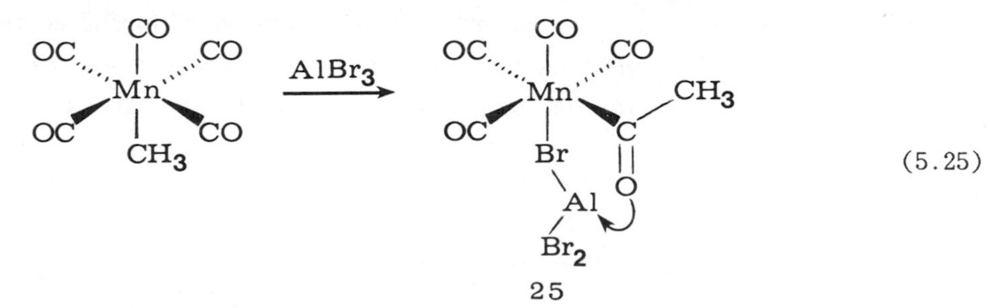


(5.25)

analysis of the above reaction indicates that there is prior complexation with the terminal CO, inducing migratory insertion. This provides a plausible model for Lewis-acid promoted insertions in Fischer-Tropsch processes, which we will encounter in Chapter 8.

e. Migratory aptitudes. For the reaction shown in Equation 5.24, the migratory insertion of n-alkyl was found to be 60 times faster than that of benzyl (at 25°C). This rate difference is consistent with kinetic studies of several other systems which show that electron-withdrawing substituents on the alkyl groups tend to retard the insertion reaction [1,20].

Unfortunately, an extended series of alkyl or aryl groups has not been reported for a single type of metal complex. Nevertheless, the results to date clearly indicate that electronegative groups on R retard the migration reaction, ostensibly by affecting the pre-equilibrium constant $K = k_1/k_{-1}$ (Equation 5.17). This effect probably derives from the increase in the alkyl to metal bond strength brought about by an increase in the electronegativity of R.

Other important results of this electronegativity effect are the failure so far to observe directly any alkyl to acyl migratory insertions involving perfluoroalkyl, perfluoroaryl, hydride, or acyl groups. The latter two cases are especially important.

Direct migration of hydride to carbonyl affording a formyl group is often invoked [21-24] as a mechanism to rationalize the heterogeneously catalyzed "Fischer-Tropsch" reductive oligomerization of CO (Equation 5.26). Formyl metal complexes have been

$$H_2 + CO \xrightarrow{M} [H{-}M{-}CO] \rightleftharpoons [H{-}\overset{\overset{\displaystyle O}{\|}}{C}{-}M] \longrightarrow \tag{5.26}$$

$$CH_3(CH_2)_xCH_3 + CH_3(CH_2)_yCH{=}CH_2 + CH_3(CH_2)_zCH_2OH$$

prepared by other reactions, such as oxidative addition of formic acetic anhydride to a basic metal center [24], or by direct hydride attack on a terminal carbonyl [26,27], as in Equation 5.27. Such formyl complexes are kinetically stabilized, since these go

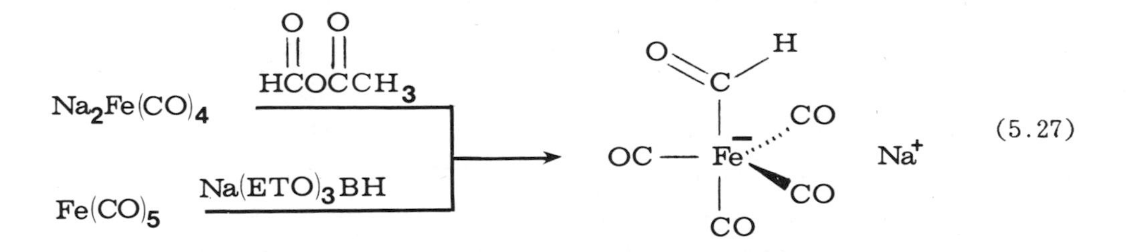


irreversibly to the corresponding hydride upon losing CO. Carbonylation of the hydride, even under forcing conditions, fails to yield any detectable formyl complex [25] (Equation 5.28). Thus the hydride to formyl equilibrium is presumed to lie far to

$$HFe^{-}(CO)_4Na^{+} \underset{\Delta,\ -CO}{\overset{1000\ psi\ CO}{\not\rightleftharpoons}} H\overset{\overset{\displaystyle O}{\|}}{C}Fe^{-}(CO)_4Na^{+} \tag{5.28}$$

the left. Occasionally the migratory insertion of a hydride to a ligand is invoked to explain the kinetics of ligand-replacement reactions [28-30]. However, to date, there is no direct evidence for the migration of a hydride group to CO. We will return to this point in Chapter 8 in the discussion of Fischer-Tropsch processes.

The failure of the acyl group to migrate explains the fact that multiple CO insertions have not been observed [31,32,33]. Casey [31] prepared a kinetically stable α-ketoacyl complex from the corresponding acid chloride (Equation 5.29). Thermal loss

$$CH_3\overset{O}{\overset{\|}{C}}-\overset{O}{\overset{\|}{C}}-Cl \;+\; NaMn(CO)_5 \longrightarrow CH_3\overset{O}{\overset{\|}{C}}-\overset{O}{\overset{\|}{C}}-Mn(CO)_5 \qquad (5.29)$$

of CO results in the formation of the acyl complex, but the reverse carbonylation does not take place (Equation 5.30).

$$CH_3\overset{O}{\overset{\|}{C}}-\overset{O}{\overset{\|}{C}}Mn(CO)_5 \underset{\text{CO, high pressure}}{\overset{\Delta,\,-CO}{\rightleftharpoons}} CH_3\overset{O}{\overset{\|}{C}}Mn(CO)_5 \qquad (5.30)$$

In some instances, products which appear to result from migration of an acyl group, i.e., α-diketo groups, have been isolated [34] (Equation 5.31). It is plausible

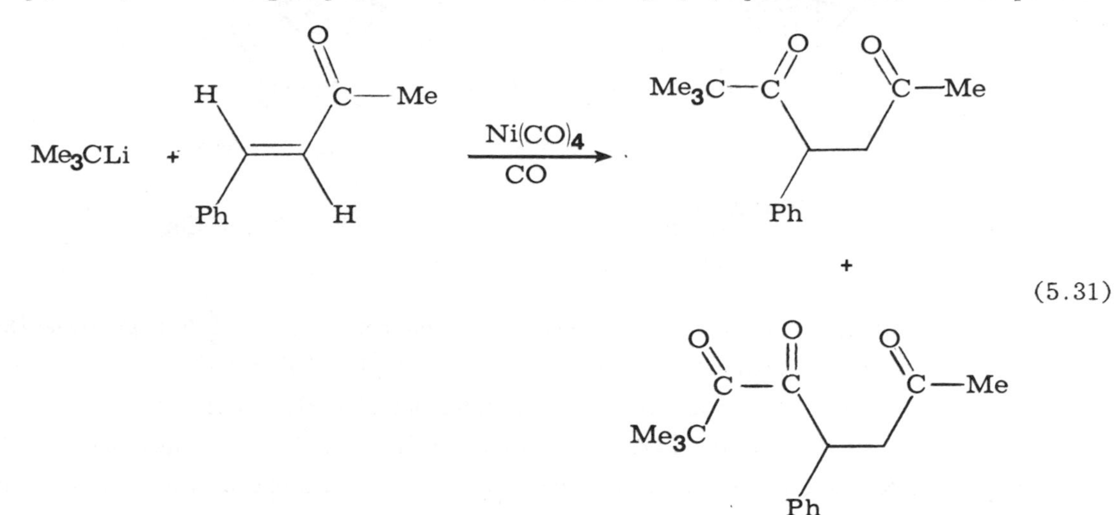


(5.31)

that acyl-carbonyl migration has taken place; however, the detailed mechanisms of such processes are presently unknown.

Retrograde insertions have been used to prepare perfluoroalkyl complexes [35] (Equation 5.32). This sort of procedure is the best general method for preparing perfluoroalkyl metal derivatives. This is a good illustration of the value of looking at reactions in both directions.

$$C_2F_5\overset{\overset{\displaystyle O}{\|}}{C}Cl \xrightarrow{NaMn(CO)_5} C_2F_5\overset{\overset{\displaystyle O}{\|}}{C}Mn(CO)_5 \xrightarrow[-CO]{80^\circ} C_2F_5Mn(CO)_5 \qquad (5.32)$$

f. Relative reactivities of 3d, 4d, and 5d metals. The reactivity of transition-metal alkyls toward carbonyl insertion reactions usually decreases with descent of a triad [1]. That is, alkyls of first-row metals are more reactive than those of second-row metals, which are more reactive than those of third-row metals. Unfortunately, there are few quantitative comparisons, but this generalization seems qualitatively valid in all situations in which the reactivities of isostructural complexes in a given triad have been compared. Consider, for example, the reactivity of the isostructural series Equation 5.33. The ruthenium complex 26b requires higher temperatures to react than the iron derivative 26a, whereas the osmium analog 26c fails to react

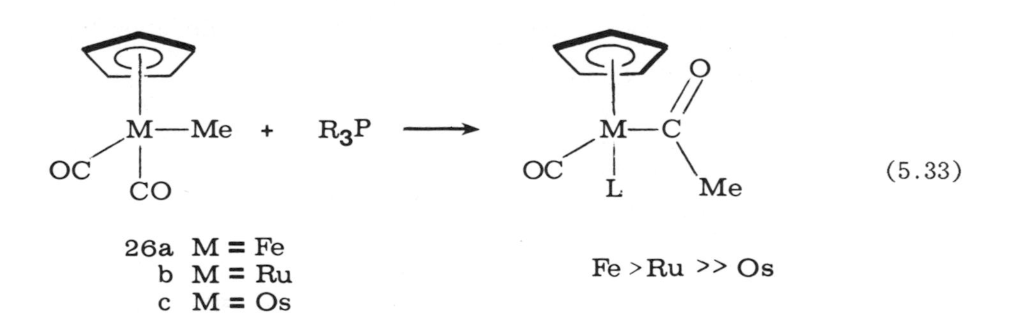


under these conditions. Similar differences have been reported for other systems [36]: $CpM(CO)_3R + Ph_3P$; M = Cr, Mo > W; $RM(CO)_5 + L$; M = Mn > Re. Such reactivity differences are thought to arise from the stronger metal-carbon bond energies for 5d compared with 4d or 3d metals [4]. Note that for nontransition metals, metal-carbon bond energies often follow the reverse trend, the lighter members of the group forming the stronger metal-carbon bonds.

g. Nature of the unsaturated acyl intermediate. The generally accepted two-step mechanism for migratory insertion which is presented in Equation 5.16 invokes a coordinatively unsaturated acyl intermediate, 21. The nature of such intermediates is unclear, but there are several possibilities. For $d^6$ complexes, such an intermediate would be a five-coordinate acyl complex. At least some $d^6$ complexes undergo migratory insertion in a very regioselective manner, such that the migrating alkyl group moves to an adjacent carbonyl, and the incoming ligand takes up the coordination position vacated by the alkyl group. The nature and steric-integrity of such five-coordinate acyl complexes must provide a reason why the observed stereochemistry is kinetically determined. Unsaturated $d^6$ complexes are rare, but it is generally thought that these should have a square-pyramidal structure [4,37]. Among the few five-coordinate $d^6$ complexes which have been structurally characterized, most are square-pyramidal [40], but there are exceptions. From orbital splitting diagrams it has been proposed that trigonal-bipyramidal $d^6$ complexes would be high-spin by having two singly occupied d orbitals—provided that the ligand-field strengths of the five ligands are equivalent [40]. In some cases, a solvent molecule might serve as a weakly held ligand in the otherwise vacant octahedral site [38,39]. In other cases the acyl ligand may have a dihapto structure and thus the complex would be coordinatively saturated (vide infra).

Note that migratory insertion from an octahedral to a square-pyramidal complex would necessarily place this acyl group in a basal position, 27. Hoffmann suggests that

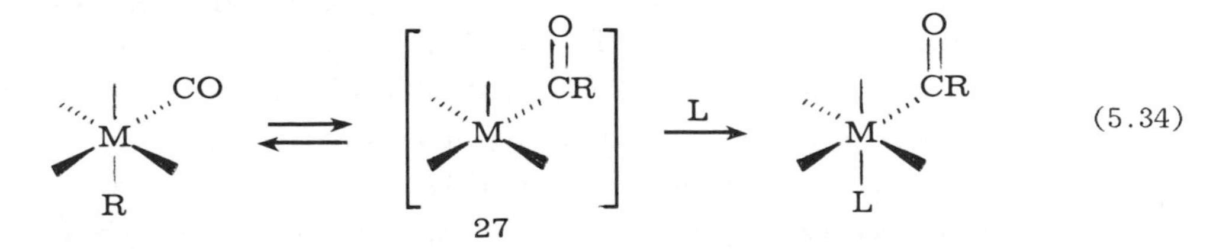


this is also the most stable position for an acyl group [40]. Eisenberg [42] has reported an interesting insertion in which a five-coordinate acyl product was isolated and structurally characterized by X-ray diffraction (Equation 5.35). The acyl group in the isolated product 28 is in the axial position. Furthermore, the metal center in the acyl complex 28 is chiral, a feature which is reflected in the nonequivalence of the $\alpha$-$CH_2$ protons in the NMR. In solution, this nonequivalence is maintained on the NMR time-scale, ruling out rapid interconversion of the two enantiomers of 28. We are left with an unanswered question. Is 28 with the apical acyl group the kinetic product of insertion—a result which implies CO insertion rather than alkyl migration—or, alternatively, did an initial product with a basal acyl group form and subsequently rearrange

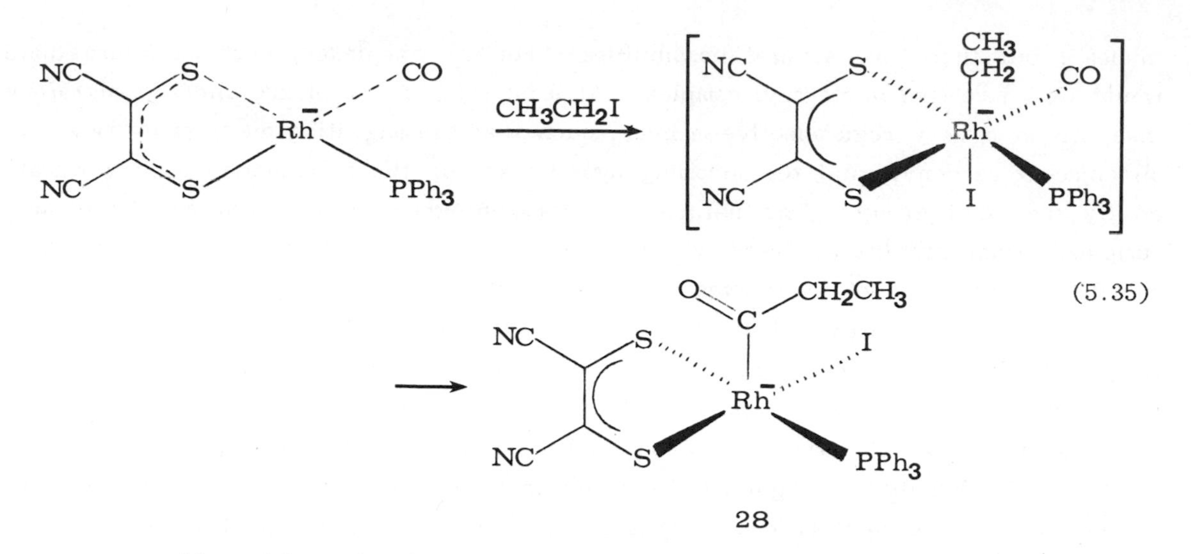


to a more stable axial product? Such questions of kinetically controlled stereochemistry cannot be decided unless both isomers are available.

Five-coordinate $d^6$ acyl complexes do rearrange. For example, Calderazzo has studied the $PPh_3$ induced insertion of the $MeMn(CO)_5$ system, and found that the initially formed cis product 30 (the one expected from the methyl migration) slowly rearranges into a mixture of cis and trans isomers, 30 and 31 [41,43]. Apparently this rearrangement goes through two different five-coordination intermediates, perhaps the basal and apical acyl complexes 29 and 32 (Equation 5.36).

Kubota [44] used the dinitrogen complex 33 to prepare a series of five-coordinate iridium(III) acyl complexes 34. He then studied their decarbonylation to the six-coordinate iridium(III) alkyl, 35 (Equation 5.37). Unfortunately, the unsaturated acyl complexes 30 have not been structurally characterized. However, these are the first direct rate studies of the reverse step $k_{-1}$ in the general two-step mechanism which is usually proposed for migratory insertion (Equation 5.16). Kubota found that electron-withdrawing groups retard this decarbonylation. Other kinetic studies of decarbonylations do not give the same sort of information, since the rate-determining step usually involves the prior loss of a ligand. The remarkable kinetic inertness of 34 can be ascribed to the generally slow reactions which are characteristic of many third-row complexes.

Some other observations of decarbonylations are also consistent with the notion of a coordinatively unsaturated acyl intermediate. An example in which such an intermediate is intercepted involves reaction of the trifluoroacyl complex 36 with "Wilkinson's

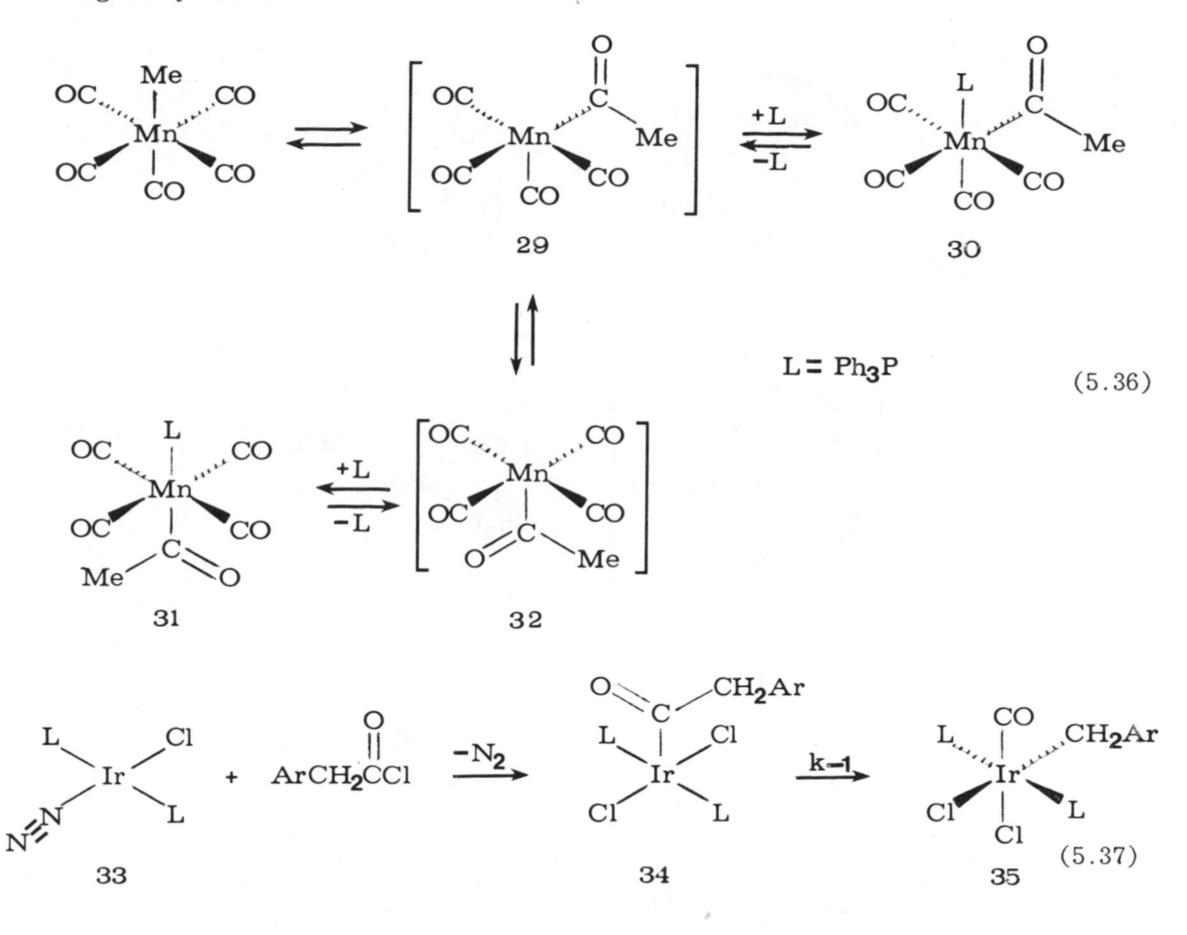


catalyst," 37 [45]. The latter is an excellent reagent for scavenging uncoordinated CO and releasing a phosphine ligand. As shown in Equation 5.38, a portion of the presumed unsaturated trifluoroacyl intermediate 38 combines with the released phosphine, forming the saturated acyl 39, whereas the remainder of 38 undergoes an irreversible reverse migration step, forming the saturated trifluoromethyl complex, 40.

Such "decarbonylation" of acyl complexes can be induced by creating a site of unsaturation. For example, Kubota effected decarbonylation of the platinum(II) acyl 41 by removing a chloride ligand with $Ag^+$ (Equation 5.39) [46]. In the presence of a strong donor solvent, such as acetonitrile, the chloride is simply replaced by acetonitrile; apparently the three-coordinate acyl intermediate 42 reacts rapidly with acetonitrile before the intramolecular reverse migration can take place. However, in a weaker donor solvent such as acetone, both the cationic acyl and alkyl complexes are formed.

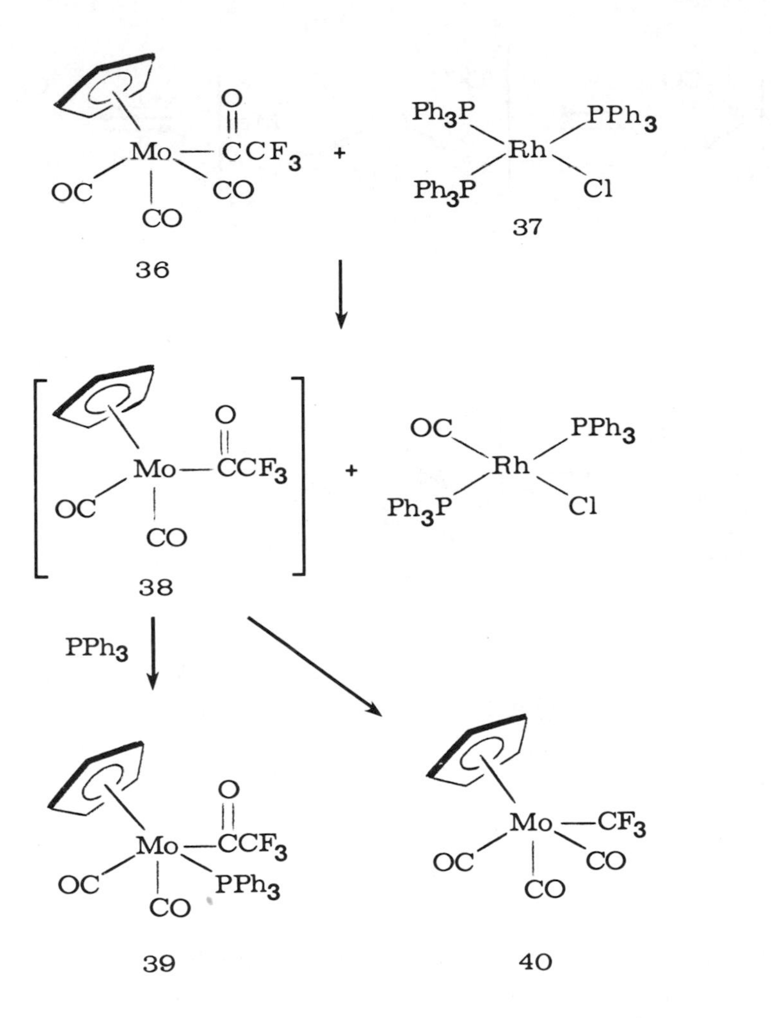


(5.38)

Note here the strong preference for platinum(II) to remain four-coordinate; four-coordinate platinum acyls, even though formally unsaturated, thus do not spontaneously rearrange to the corresponding alkyl.

There is scant information concerning the relative ease of migration of various groups from an acyl ligand to a metal. Thus there is little information concerning the $k_{-1}$ step in Equation 5.16. Some data of this sort has been obtained by examining the decomposition of isotopically labeled unsymmetric diacyl complexes 43 [47]. These studies indicate that the phenyl group migrates in preference to a methyl group (Equation 5.40). However, this particular example is complicated by selective ion pairing with the presumably more basic benzoyl oxygen. When the counter ion $Z^+ = Me_4N^+$,

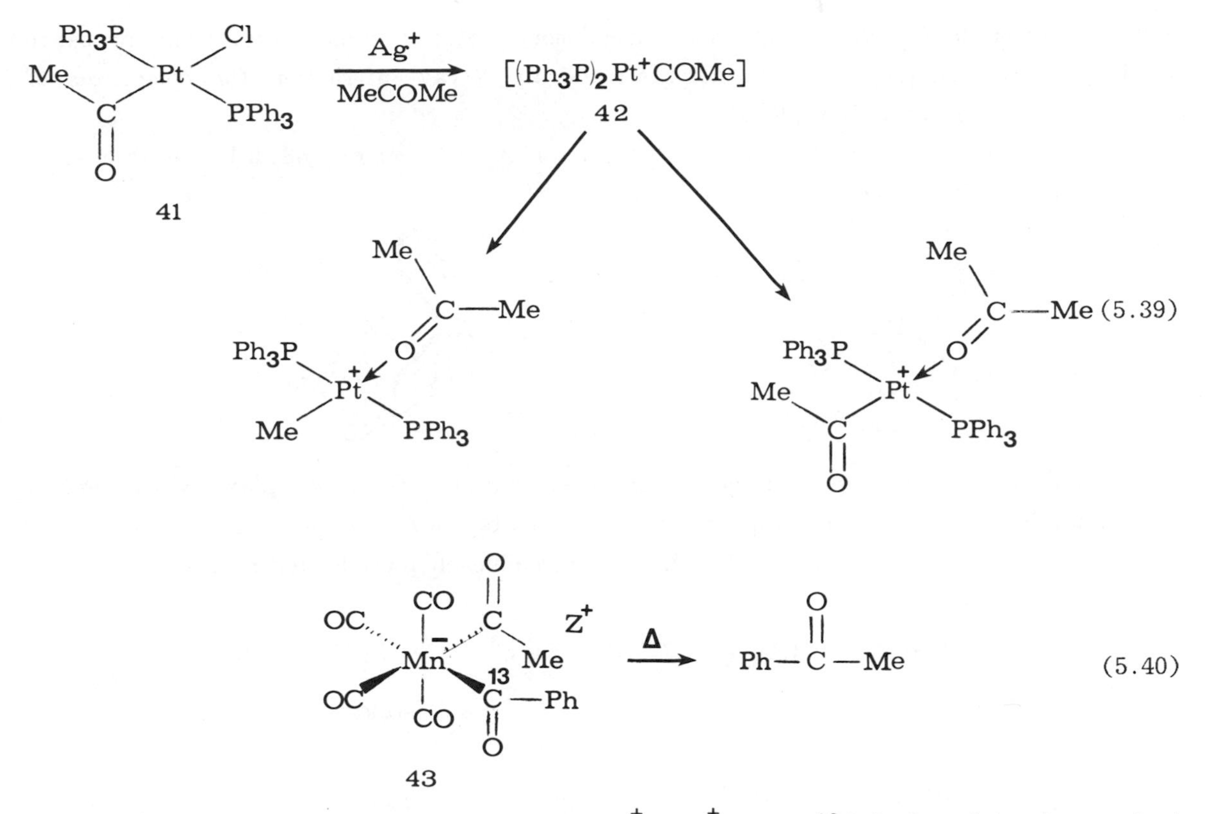


there is no $^{13}C$ in the product; but when $Z^+ = Li^+$, some $^{13}C$ is found in the product. Thus migratory aptitudes in this reaction are altered by the presence of lithium cations. Such acyl stabilization by Lewis acids is quite consistent with the Lewis-acid promotion of migratory insertions in anionic iron alkyls discussed earlier.

In the above case and in many other examples wè shall encounter, several consecutive reactions have taken place. A sequence of steps is usually inferred from our general knowledge of major reaction types. However, such conclusions must be considered tentative unless evidence can be developed for the individual steps. For example, it has been suggested that the above reaction proceeds without prior migration to the metal [47]. There is no direct evidence concerning this proposed mechanism, which is different from a retrograde insertion.

In _each example_ of migratory insertion or the reverse reaction, decarbonylation, one should also consider the possibility that the initially formed intermediate acyl complex has a coordinatively saturated _dihapto_ structure rather than the unsaturated monohapto structure which is usually presumed. _Dihapto_acyl complexes are actually more common than coordinatively unsaturated monohaptoacyl compounds. The presence of dihapto-acyl ligands has been demonstrated by X-ray diffraction for complexes of Groups IV, V, VI, and VIII [48-52].

Dihapto structures have also been characterized for imino- [48,49] and thio-acyls [53]. Typical dihapto acyls are 44 and 45.

Roper [52] has reported an interesting ruthenium(II) complex which exists predominately as the methyl compound in solution, but has a dihapto acyl structure in the solid state (Equation 5.41). The IR spectrum of each form is distinctive, in

$$\underset{CH_2Cl_2}{Ru(CH_3)I(CO)_2(PPh_3)_2} \rightleftharpoons \underset{solid}{Ru(\eta^2\text{-}C(O)CH_3)I(CO)(PPh_3)_2} \quad (5.41)$$

particularly the rather low 1599 $cm^{-1}$ band, which is characteristic of the dihapto acetyl group.

It is probable that dihapto acyl complexes are involved as intermediates in many migratory-insertion reactions. The regiospecific course of these reactions could arise from such configurationally stable dihapto acyl complexes.

The relatively greater stabilities of dihaptothio- and iminoacyl compounds may account for the greater tendency of thiocarbonyls [53-55] and isonitriles [56,57] to undergo migratory insertions (vide infra).

A special situation may arise in the migratory-insertion reactions of certain complexes having rigid multidentate ligands, such as $\eta^5$-$C_5H_5$. In selected cases the coordinatively saturated starting material and the related coordinatively unsaturated product may have different ground-state spin multiplicities (singlet and triplet, respectively). In such situations, since a concerted migratory insertion reduces the coordination number by one, the saturated 18-electron singlet starting complex would form a corresponding unsaturated product which has a triplet ground state. For a concerted mechanism, spin multiplicities should be correlated along the reaction coordinate; so a singlet starting material should form a singlet product; but in those special cases the latter would be an electronically excited state of the product. This should result in an increased activation energy, and such paths would therefore be slow.

This hypothesis has not been experimentally tested and is thus still speculative, but it does seem to explain the fact that several complexes show little tendency to undergo intramolecular migratory insertion in the absence of an external ligand. Similar arguments can be made with regard to reductive-eliminations (Chapter 4) and dissociative ligand replacement reactions.

Consider, for example, the diamagnetic, saturated, $d^2$-tantalum alkyl-carbonyl, 46 [57]. Recall from Chapter 2 that symmetric "bent metallacenes" having one σ-bonding

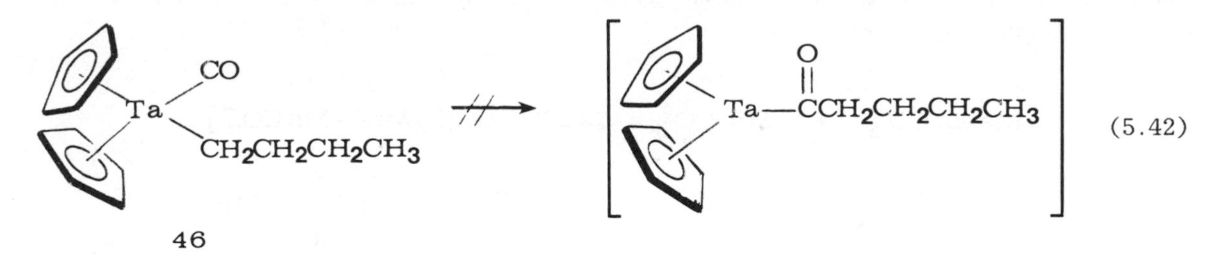


(5.42)

unidentate ligand have an energy-level diagram in which the two highest occupied orbitals are nearly degenerate. In some cases these two orbitals are singly occupied by the two d-electrons; so the ground state for the $d^2$ configuration should be a triplet. Thus a concerted insertion as in Equation 5.42 should be unfavorable. Alternatively, such insertions could take place by a second-order mechanism involving the simultaneous coordination of an external ligand, or by any other mechanism which would provide a singlet product, for example, the formation of a $\eta^2$-acyl complex. On the other hand, reactions such as that in Equation 5.42 may simply be thermodynamically unfavorable, because the terminal carbonyl is stabilized by π-backbonding with the $d^2$ electron pair.

h. The incoming ligand L. Many types of Lewis bases can serve as the incoming ligand, L, in migratory insertions. Examples of such external ligands include CO, $R_3P$, amines, halides, and even basic metal centers. The nature of the products derived from migratory insertions sometimes reveals the incoming ligand L. For example, isolation of the structurally characterized μ-iodo bridged acyl complex 47 from the alkylation-insertion sequence shown in Equation 5.43 suggests that the bridging iodide ligands may have a role in the migratory-insertion stage of the famous Monsanto acetic-acid process [58].

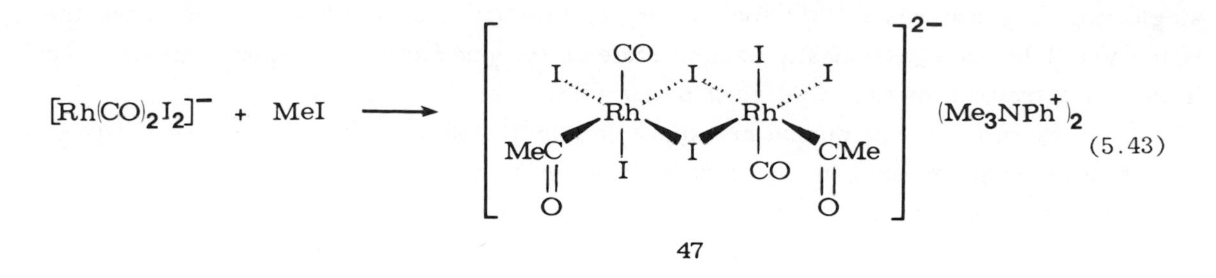


(5.43)

Casey [59] has shown that a metal anion can trap the intermediate unsaturated acyl in migratory insertion by forming a metal-metal bond. The intermediate 48 was stabilized by alkylating the basic acyl oxygen, thus forming a novel binuclear carbene complex, 49 (Equation 5.44).

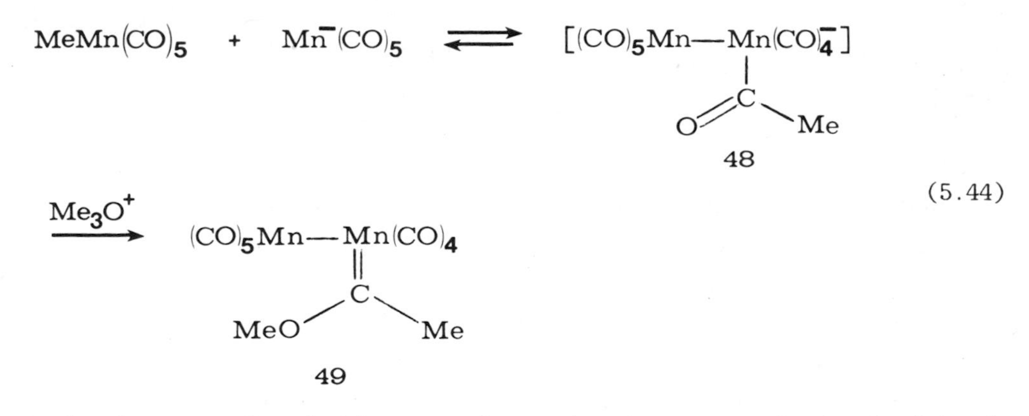


(5.44)

An intramolecular example of this sort of reaction is apparently involved in the alkylation of the anionic binuclear phosphido bridged complex 50 [60]. The first observable product is the acyl dimer 52 (Equation 5.45). This insertion is faster than

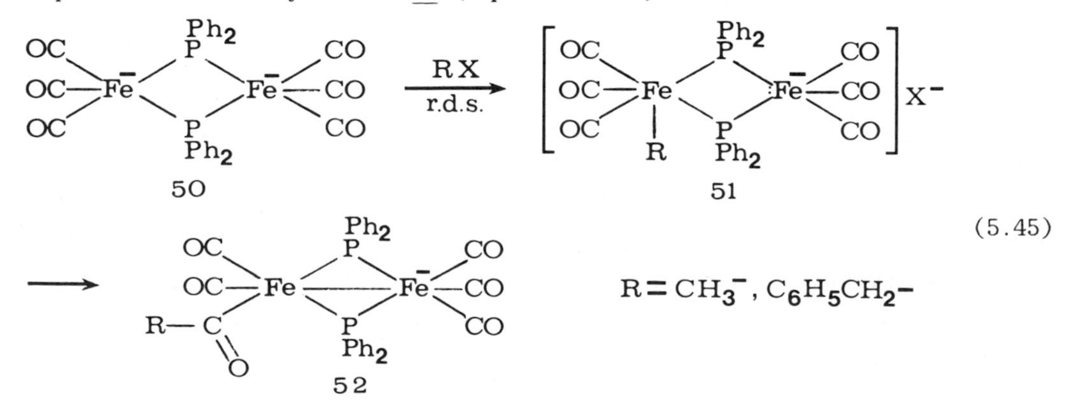


(5.45)

the oxidative-addition. It is probable that the initially formed alkyl dimer 51 subsequently undergoes a rapid migratory insertion. The driving force for the transformation of the putative alkyl 51 into the acyl 52 is the formation of a metal-metal bond. X-ray diffraction studies demonstrate that the dianion 50 does not have a metal-metal bond (none is expected), whereas the acyl 52 does. Thus Equation 5.45 would seem to be an intramolecular analog of Equation 5.44. Note that even the benzyl derivative, usually a sluggish migratory group, forms the acyl 52 very rapidly.

i. Insertions involving external CO. Many noncarbonyl-bearing alkyl complexes react with uncoordinated CO to form an acyl complex. It is commonly presumed that CO must form a bond with the metal before the insertion takes place. This can be difficult to establish, although in most cases there has been no attempt to examine the mechanism. Note that a vacant coordination site, cis to the alkyl group, must be present for migratory insertion to occur.

An interesting example is the insertion of CO into the "16-electron" dialkyl and diaryl zirconium and hafnium bent metallacenes. In solution, the dibenzyl and dimethyl compounds are at equilibrium with the acyl derivatives (5.46). However, the insertion

$$Cp_2MR_2 + CO \rightleftharpoons Cp_2M(COR)R \qquad M = Zr, Hf \quad R = Me \tag{5.46}$$

of CO into the diphenyl compounds is irreversible (5.47). There is no apparent

$$Cp_2MR_2 + CO \longrightarrow Cp_2M(COR)R \qquad M = Zr, Hf \quad R = Ph \tag{5.47}$$

tendency for further insertion to take place (which would form a diacyl). These results appear to reflect the relative thermodynamic migratory tendencies of the alkyl and aryl groups.

The acyl complexes have a dihapto structure of which there are two possible isomers, 54 and 55. A low-temperature spectroscopic study of this reaction [61] shows the prior formation of one dihapto acyl, 54, and its facile rearrangement into another dihapto form, 55, which was shown to have the indicated structure. Thus the reaction appears to go via the unsymmetrical carbonyl 53 (Equation 5.48). However, this

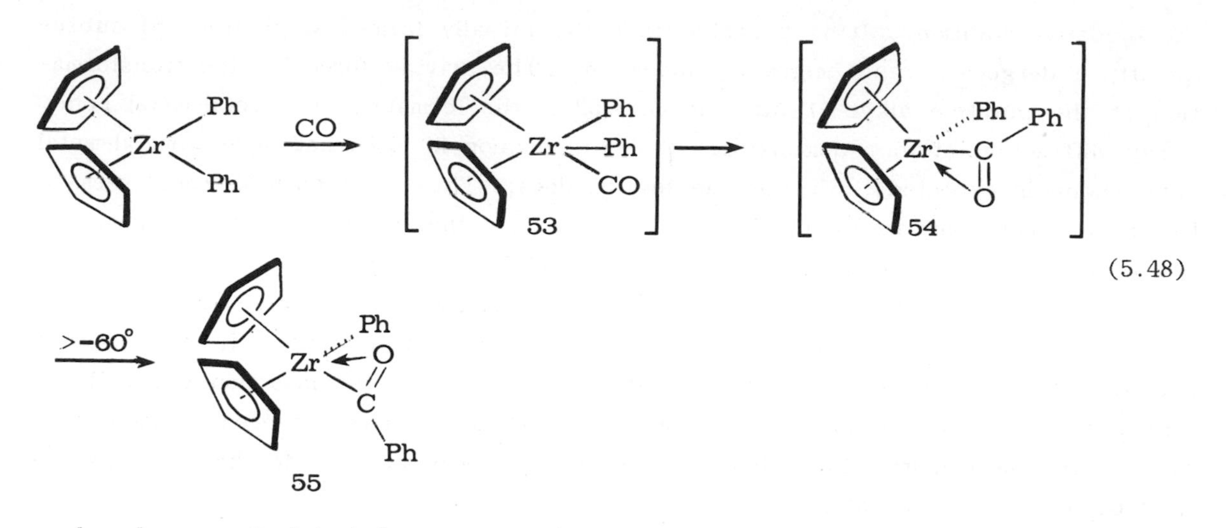


carbonyl was not detected spectroscopically. The rapid insertion exhibited by these compounds may be explained by the relatively weak bond formed between CO and the $d^0$, zirconium(IV), which can not form a $\pi$-backbond.

Another case of assumed migratory insertion (in Equation 5.49) has been described by Sacconi [62]. The five-coordinate nickel(II) methyl complex 56 reacts with CO to

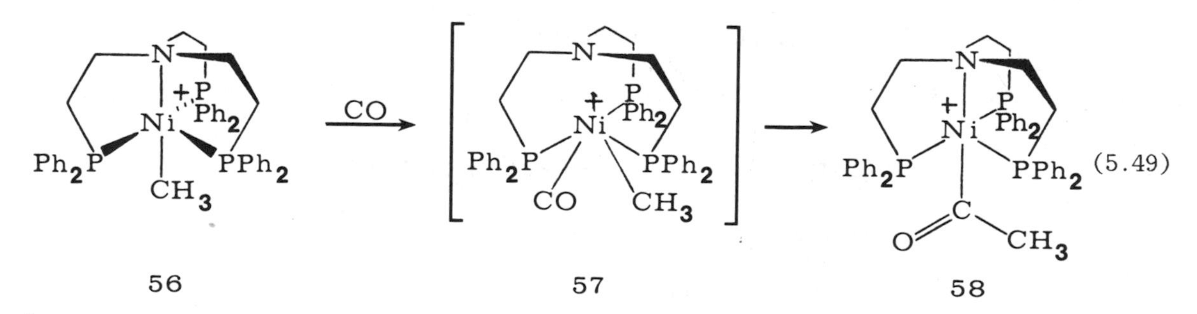


form an acyl 58 which has been structurally characterized as a distorted trigonal bipyramid with an axial acyl group. The presumed intermediate 57 is thought to be a distorted trigonal bipyramid with an axial methyl group, and to have a coordinated CO group with the trans tertiary amine dissociated to avoid a 20-electron system.

Occasionally, such CO intermediates have been detected. An instructive case is that reported by Wilkinson [63], who used low temperatures to favor the CO associative complex formation and infrared spectroscopy as a sensitive probe of coordinated CO in the intermediate 60 (Equation 5.50).

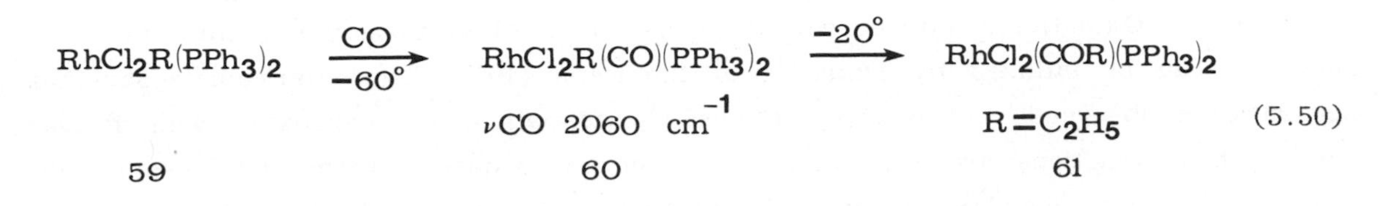


(5.50)

There are two important principles to note here. First is the fact that associative equilibria are favored by lower temperatures. Second is the greater intensity of coordinated CO IR bands compared with free CO. In this example, Wilkinson was also able to examine the tendency of three different groups to migrate in complexes such as 59: R = Et > $CH_2$ = CH- >> $C_2H_4F$ (no reaction). The migratory tendency follows the principle outlined above. Note also that the acyl complex 61 is another example of a coordinatively unsaturated $d^6$ complex, although the sixth coordination site may be occupied by a solvent molecule.

j. The requirement for a cis coordination site. The requirement for a cis coordination site in alkyl to acyl migratory-insertion reactions is nicely illustrated by planar macrocyclic metal complexes. Macrocyclic ligands effectively block the four planar coordination sites and, therefore, do not allow the two axial ligands to become adjacent. This affect also accounts for the kinetic stability of alkyl derivatives of macrocyclic complexes, since the β-hydride elimination is blocked. Thus, both the alkyl and acyl rhodium(III) macrocyclic complexes 62 and 63 are known, stable compounds, but these can neither be interconverted by carbonylation nor by decarbonylation (Equation 5.51). The weakly held acetonitrile ligand in 62 and 63 readily dissociates, creating a vacant coordination site, but it is trans to the alkyl or acyl group [64].

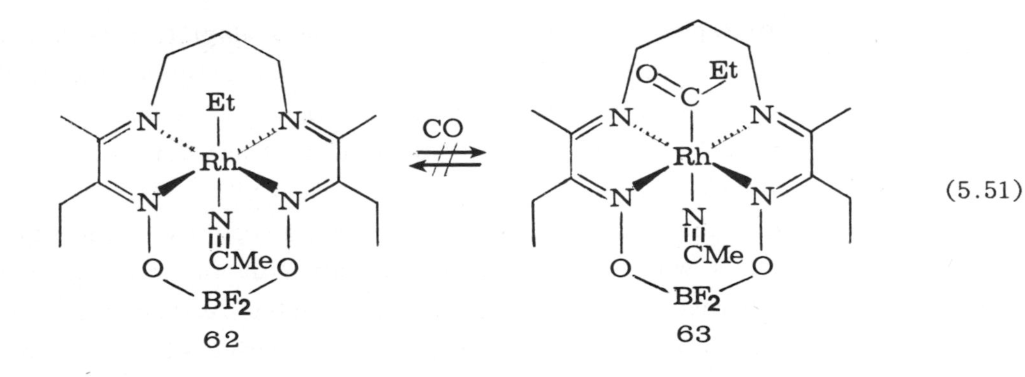


(5.51)

k. Oxidatively induced migratory insertion. Alkyl to acyl migratory-insertion reactions can be induced by oxidation of the metal [10]. The intermediate acyl has seldom been observed, but is inferred from the formation of a carboxylic acid derivative. These reactions are very facile. Apparently oxidation of the metal destabilizes the alkyl-carbonyl complex, probably by reducing the degree of backbonding to the CO group. The presumed acyl is quite reactive, readily solvolyzing like an acid chloride. We shall later see application of this reaction to the synthesis of β-lactams.

The best-studied examples of this sort of reaction are in Rosenblum's work [65] with $\eta^5$-cyclopentadienyl iron(II) alkyls 64. Oxidation with a reagent such as Ce(IV) or $CuCl_2$ in alcohol affords the corresponding ester, 67, presumably by the reaction sequence shown in Equation 5.52. Yields are higher under CO pressure, and the

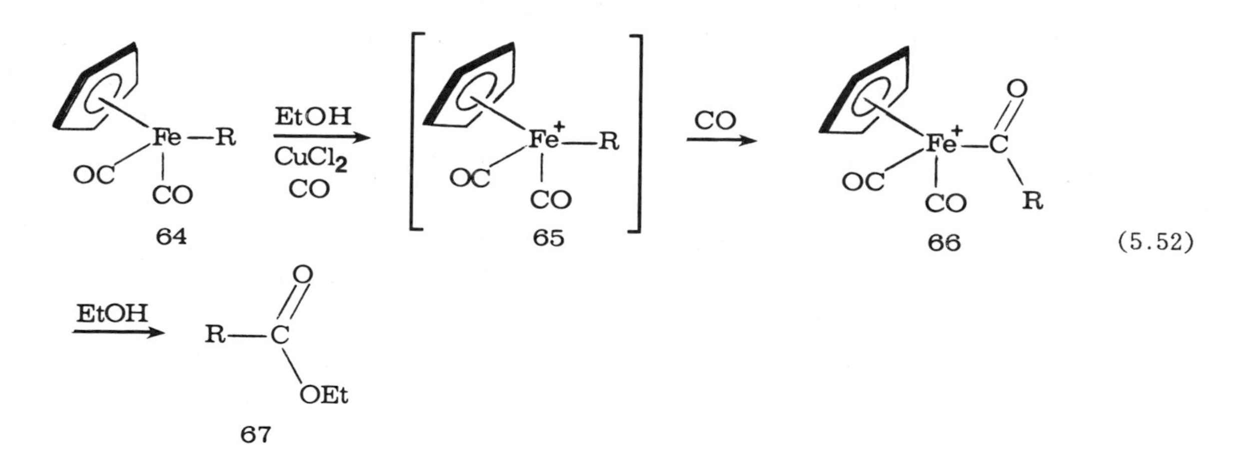


over-all reactions are rapid even at 0°C! The presumed alkyl to acyl transformation, 65 to 66, occurs under remarkably mild conditions, apparently as the result of the initiating oxidation. The intermediate cationic acyl complex 66 seems to be very reactive. Such oxidative solvolysis is occasionally used to transform acyl complexes into carboxylic esters and amides.

These reactions exhibit a high degree of stereoselectivity. As expected, the migratory insertion goes with retention of configuration at the reacting alkyl carbon center. This stereochemistry was deduced from a two-step reaction sequence, such as that shown in Equation 5.53. The initial alkylation (oxidative-addition) is presumed to go with inversion, and the subsequent oxidatively induced insertion with retention, affording the inverted ester product 69. Both stereoisomers were examined. Note that

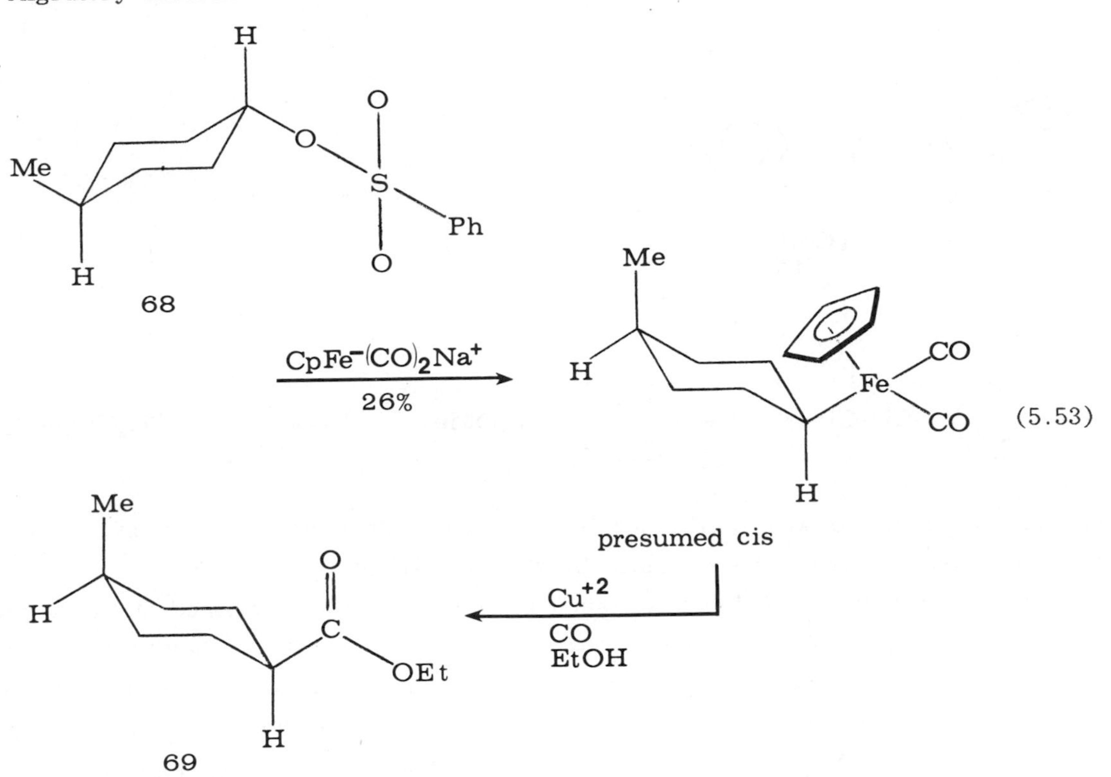


the rather low yield in the first step is probably due to the poor $S_N2$ substrate, an equatorial, secondary arylsulfonate, 68.

These oxidatively induced insertions also reflect the relative tendencies of different alkyl groups to undergo migration. Compare, for example, the oxidative degradation of the two complexes shown in Equations 5.54 and 5.55. The β-p-fluorophenylethyl iron

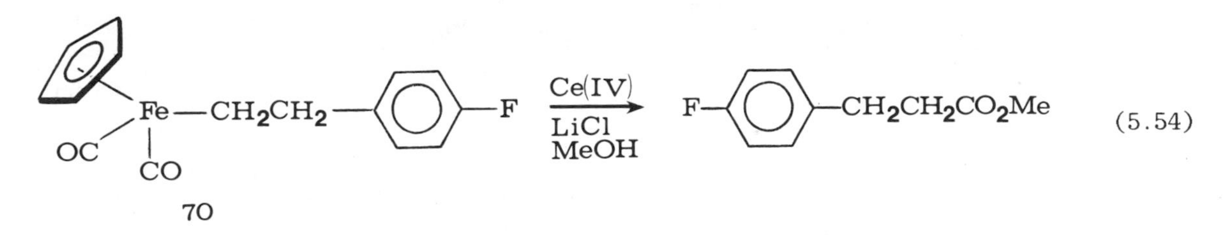


derivative, 70, reacts normally (Equation 5.54); whereas the p-fluorobenzyl compound, 71, forms products derived from both the migratory-insertion path and direct nucleophilic attack on the cationic alkyl intermediate (Equation 5.55) [66]. The oxidatively induced migratory insertion can be rationalized in terms of an increase in formal

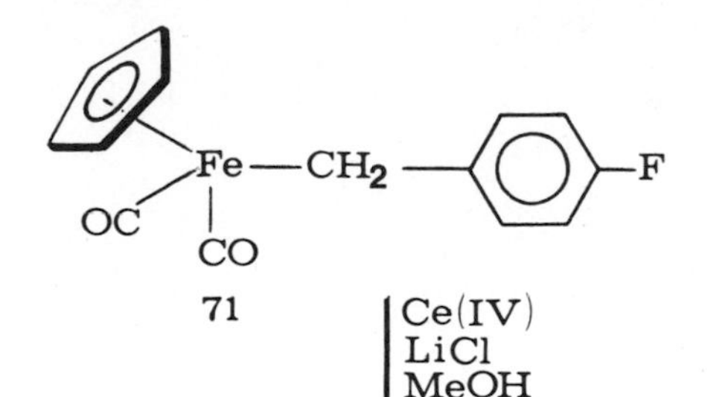


(5.55)

F—C$_6$H$_4$—$CH_2Cl$ + F—C$_6$H$_4$—$CH_2OMe$ + F—C$_6$H$_4$—$CH_2CO_2Me$

charge, which would weaken the metal-carbon monoxide bond. The great reactivity of the acyl intermediate can be explained by the same effect.

2. Migrations to Other Longitudinal Ligands. As was mentioned at the beginning of this chapter, migratory insertion occurs with a great variety of longitudinally bonded ligands in addition to carbon monoxide. Although such reactions have not yet become synthetically important, it is nevertheless instructive to consider some of these cases.

a. Thiocarbonyls. The thiocarbonyl ligand is in many respects similar to carbon monoxide. The most remarkable insertion reaction of the thiocarbonyl group is Roper and Collins's [67] discovery of the hydride migration affording a stable, blue thioformyl complex, 72 (Equation 5.56). Recall that hydride migration to carbon

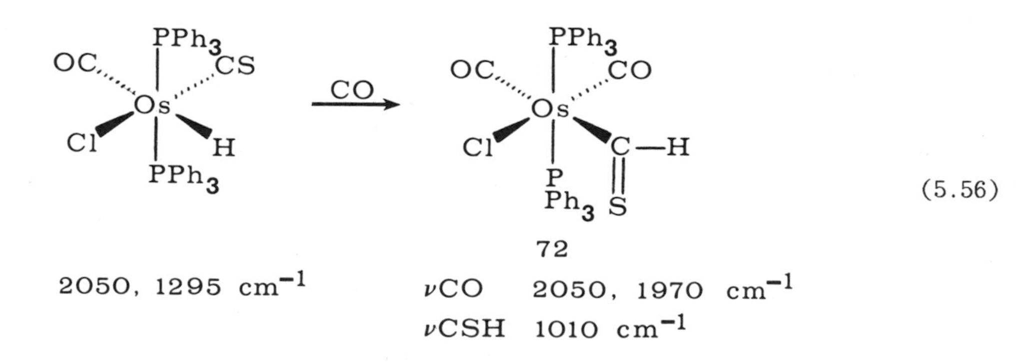


(5.56)

monoxide has not been directly observed although it is occasionally proposed (see Chapter 8). The origin of the differences between the migratory tendencies of thiocarbonyl and carbonyl is not yet understood. Perhaps it is related to the greater tendency for thioacyl compounds to form a dihapto bond [53].

b. Isonitriles. The isonitrile group is similar to carbon monoxide, but undergoes insertions even more readily [68]. In sharp contrast to the carbonyl ligand, for which double insertions have never been directly observed, there is a pronounced tendency for multiple isonitrile insertions to take place. This sort of process probably accounts for the nickel-alkyl-promoted polymerization of isonitriles. A single stoichiometric example is shown in Equation 5.57. The methyl isonitrile derivative, 73, reacts

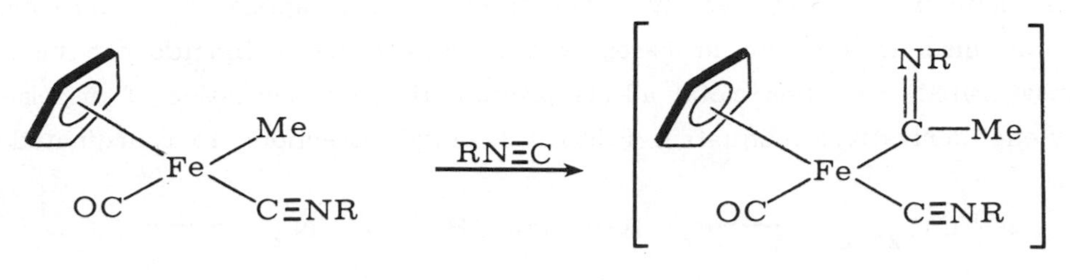


(5.57)

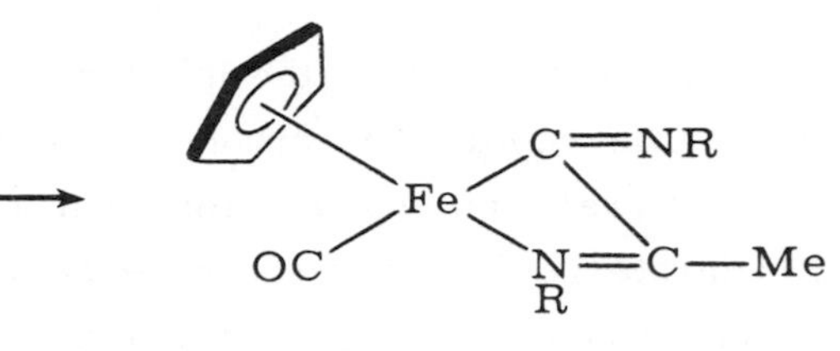


R = t-Bu

with free isonitrile, affording the double insertion production, 75, which is apparently stabilized by the formation of a chelate ring [69]. The initial insertion product, 74, is the probable intermediate, although it was not detected.

c. Carbenes. Insertion reactions involving carbene complexes have been recognized only recently. These reactions are very significant, for they explain several features of carbene chemistry. Migratory insertions of various groups to the carbene ligand are seemingly quite facile, much more so than the corresponding reactions involving carbon monoxide. Consider, for example, the general case in Equation 5.58. In the carbene complex 76, a wide variety of groups X insert, affording

$$\underset{\mathbf{76}}{\overset{}{\underset{\displaystyle X}{\overset{|}{M}}}{=}CH_2} \rightleftharpoons \underset{\mathbf{77}}{M{-}CH_2{-}X} \qquad (5.58)$$

derivatives such as 77. It should be emphasized that the migration of X to carbene is only inferred by analogy with the carbonyl case. In fact insertion by intermolecular addition is very facile with carbene complexes, and is often difficult to distinguish from intramolecular "migratory insertion." Most mechanistic information about reactions of the type shown in Equation 5.58 derives from product analysis in a multistep reaction. Nevertheless, a reactivity pattern is apparent. Groups such as alkyl, hydride, and chloride seem to insert very readily. The alkyl reactions appear to be irreversible, so that multiple insertions occur in cases where fresh carbene ligands can be supplied. Thus, unsaturated transition-metal alkyls promote the polymerization of diazomethane to polymethylene derivatives (Equation 5.59). If alkyl insertions to coordinated carbene

$$R{-}M + CH_2N_2 \longrightarrow R{-}M{=}CH_2 + N_2 \longrightarrow$$

$$RCH_2{-}M \xrightarrow{CH_2N_2} RCH_2M{=}CH_2 \longrightarrow \longrightarrow R(CH_2)_nM \qquad (5.59)$$

were reversible (for example, the reaction to the left for X = R in Equation 5.58), this would provide a scheme for the rupture of carbon-carbon bonds in unsaturated metal alkyl complexes. At present there is no evidence for such a reaction.

Chloride appears to migrate readily to carbene, even though chloride shows little or no tendency to migrate to carbon monoxide. A plausible example of chloride migration to carbene was reported by Mango [70], who found that diazomethane reacts with Vaska's iridium complex 78, affording the chloromethyl derivative 80 (Equation 5.60).

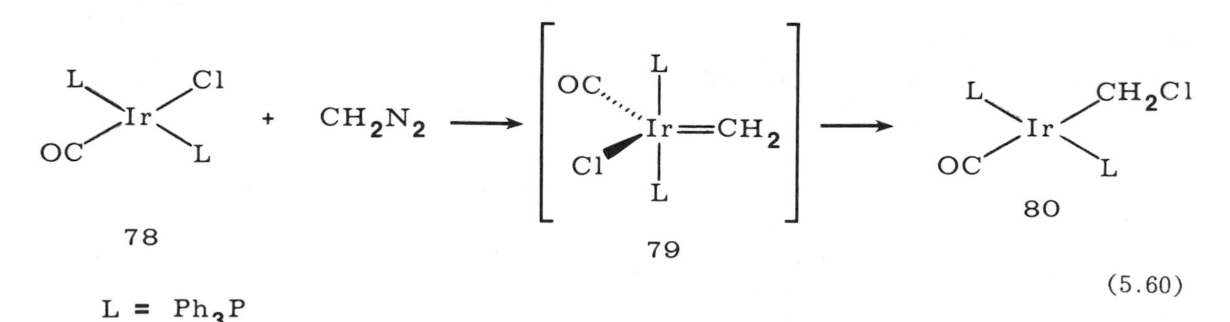


Further multiple insertion is probably prevented by the insolubility of 80, although some polymethylene is formed. Presumably this reaction goes via the generation of an intermediate carbene complex 79. The properties of 80 indicate that it is a chloromethyl derivative, that is, the insertion reaction lies far to the right.

Migration of hydride to coordinated carbene is facile and apparently reversible. An excellent example of this sort of reaction was reported by Shaw [71]. The novel five-coordinate iridium(III) complex 81 loses $H_2$ upon heating under vacuum, affording the structurally characterized carbene complex 83 (Equation 5.61). The reaction is

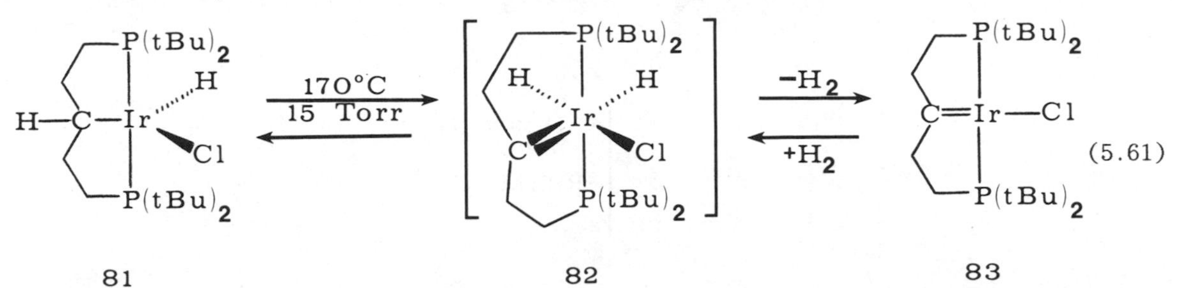


readily reversed by hydrogenation at ambient conditions. It is probable that an undetected intermediate hydrido carbene complex, 82, is formed. On the basis of NMR chemical shifts, the carbene complex 83 appears to have considerable ylid character. Note that the five-coordinate iridium(III) complex 81 is coordinatively unsaturated, a property which should facilitate the α-hydride migration forming the putative intermediate 82. Oxidative-addition of hydrogen to the unsaturated iridium(I) complex 83 should afford 82 by analogy with hydrogenation of the well-studied Vaska complex, 78, which we discussed in Chapter 4.

There is other, indirect evidence for the existence of a reversible hydride-to-carbene insertion reaction [72-75]. Green has proposed such a path, involving an equilibrium between 84 and 85, to explain the formation of the kinetic product 86 and subsequently the thermodynamic product 87. Note that the reversible formation of the kinetic product 86 probably involves an intermolecular nucleophilic attack by the phosphine on the cationic carbene complex 84. This seems to be faster than attack of the phosphine at the metal center in the cationic methyl complex 85, since 86 is formed initially. Note that the postulated methyl intermediate, 85, should be a triplet, unless another ligand is present.

3. Migrations to Olefins and Acetylenes. The intramolecular migration of hydride, alkyl, aryl, and acyl ligands to coordinated olefins and acetylenes is presumed to occur in many synthetic and catalytic reactions. However, these reactions have rarely been directly observed.

a. Olefins. A substantial body of indirect evidence concerning olefin migratory insertions has accumulated. The characteristic features of olefin insertions have been largely deduced from these secondary observations.

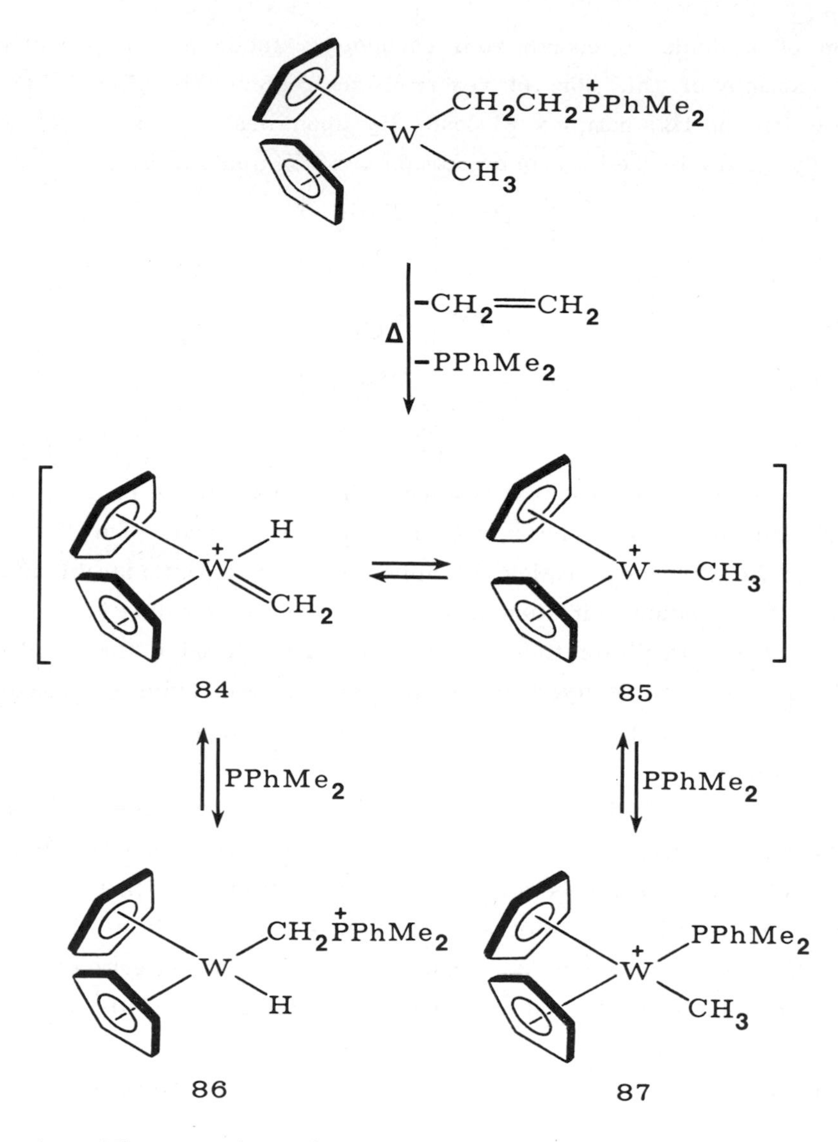


i. Migratory insertions involving hydrides and olefins. The intramolecular insertion of hydride to a coordinated olefin is an obligatory step in olefin hydrogenation (Chapter 6), hydroformylation, and related reactions (Chapter 8). The reverse process, "β-hydride elimination," is the most important path for metal-alkyl decomposition and olefin isomerization.

Intramolecular migration of a hydride to an olefin is quite facile. Consequently, complexes which contain both olefin and hydride groups are rare, and those examples which are known seem to be stabilized by a kinetic barrier. For example, the isolated platinum(II) complex 88 is undoubtedly stabilized by the trans orientation of the hydride and olefin ligands [76].

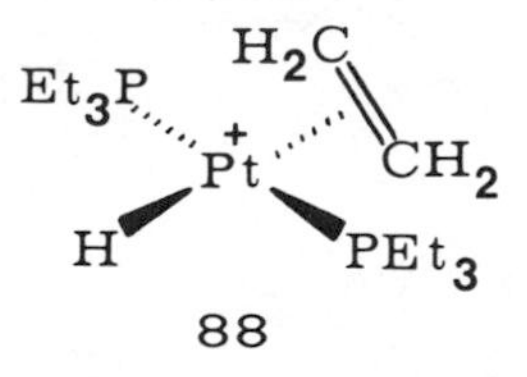


In the absence of external ligands, the molybdenum(IV) complex, 89, is stable in the hydride alkene form [77] (Equation 5.62). The alkyl 90, which should form by

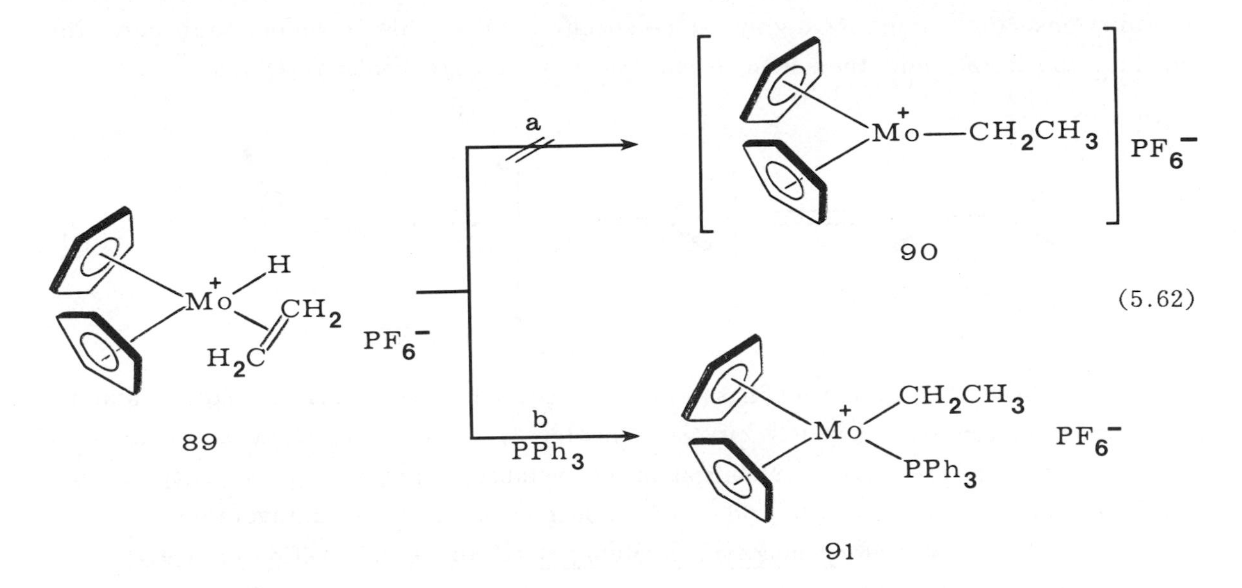


simple migratory insertion (path a), might have a triplet ground state, because the two d-electrons would occupy nearly degenerate orbitals (see the discussion of the orbital scheme for these tilted sandwich compounds in Chapter 2). This suggests that a unimolecular insertion could be energetically unfavorable. However, in the presence of added phosphine, insertion readily takes place, affording the saturated, diamagnetic adduct 91 (path b).

The reaction shown in Equation 5.63 is a rare example of a directly observed hydride-olefin insertion [78]. Using NMR, a hydride-olefin complex 92 is shown to be in equilibrium with the alkyl complex 93. A solvent molecule is thought to fill the coordination site generated by this insertion.

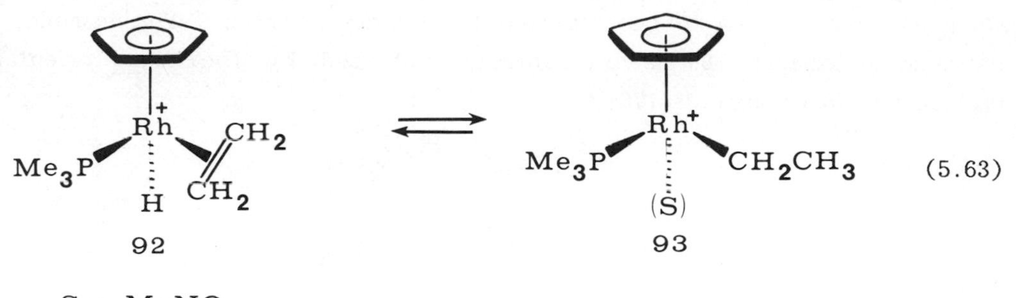


Stereochemical and isotopic labeling studies of metal-alkyl degradations, olefin rearrangements, and olefin hydrogenations clearly indicate that the hydride-olefin migratory-insertion step is highly stereospecific. For this reaction to occur, the hydride, the metal, and the olefin $\pi$-bond must all become coplanar (Equation 5.64).

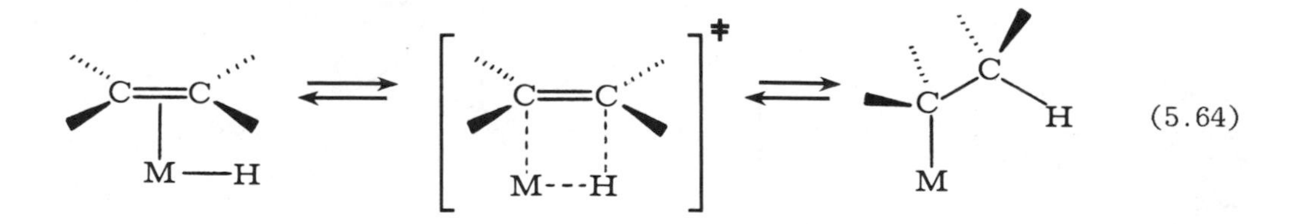


This stereochemical requirement has important consequences. For example, catalytic hydrogenations can be rigorously cis (see Chapter 6). This stereochemical requirement also accounts for the kinetic stabilization of metallacyclopentanes (see Chapter 10), since for that ring size the $\beta$-hydride elimination step is sterically unfavorable.

ii. Migratory insertion involving metal alkyls and olefins or acetylenes. The intramolecular migration of an alkyl ligand to an adjacent coordinated olefin is assumed to occur in several important catalytic processes, such as the dimerization, oligomerization, and Ziegler-Natta polymerization of olefins. There seem to be no unambiguous examples in which a well-characterized metal-alkyl-olefin complex has been directly observed to undergo this reaction. However, Bergman [79] has recently carried out an isotopic labeling study which demonstrates the feasibility of such a migratory-insertion mechanism. Bergman's study was apparently stimulated by Green's provocative suggestion [80] of an alternative mechanism for Ziegler-Natta polymerization, involving a carbene complex derived from $\alpha$-elimination. Reaction between the deuterium-labeled cobalt(III) dimethyl complex, 94, and undeuterated ethylene affords the cobalt(I) complex 95, $CD_3H$, and propene-$d_3$. These are the products expected from the classical insertion mechanism which involves β-H elimination and reductive-elimination (Equation 5.65). Green's α-elimination path is not compatible with these

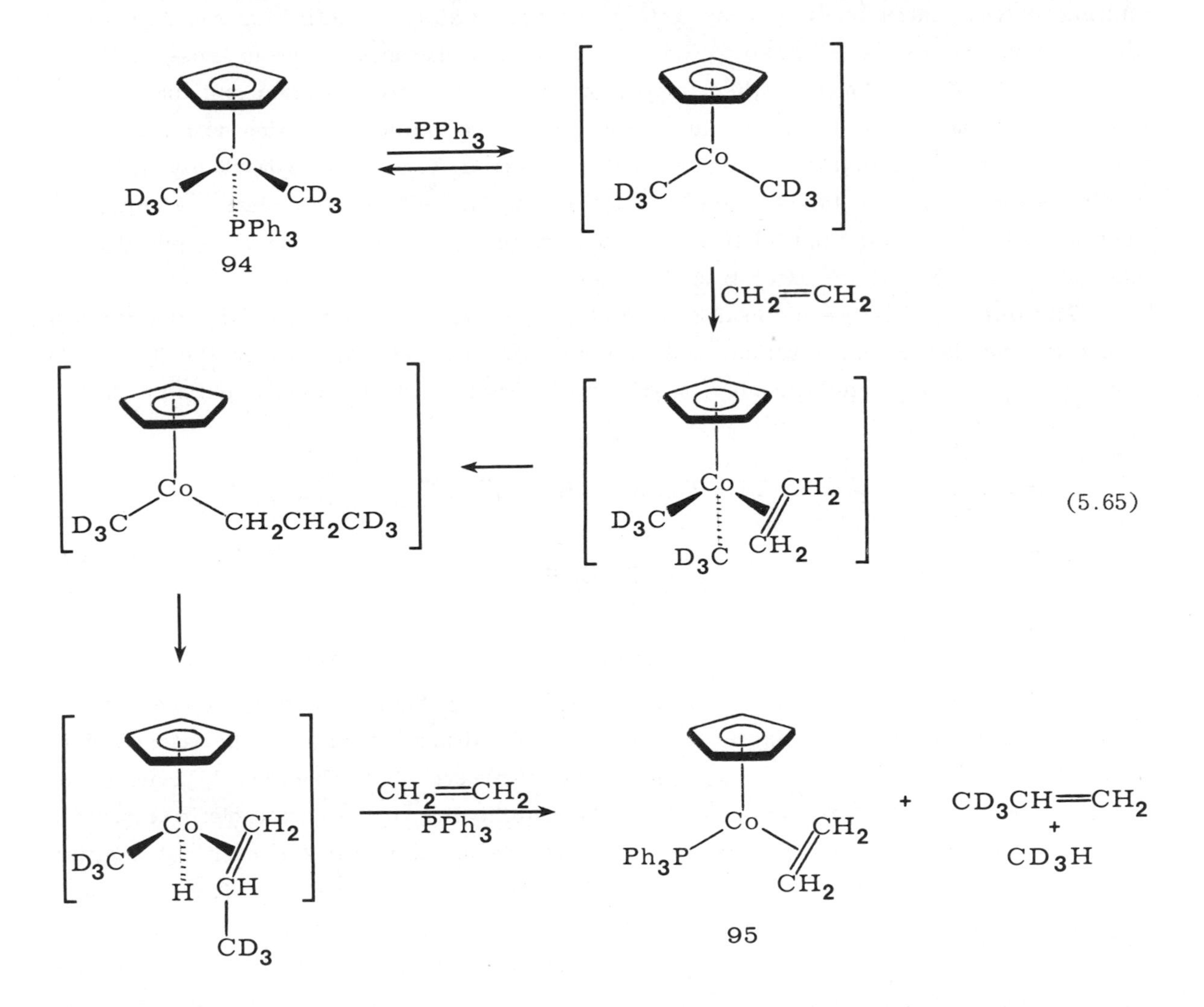


results. Although Bergman's study does distinguish a classical insertion from a carbene path in this specific reaction, it does not provide evidence about the Ziegler-Natta polymerization mechanism. Thus Green's speculation is still debatable.

It is amazing that a reaction as important and intensively studied as Ziegler-Natta olefin polymerization is still an enigma [81]. Almost nothing is known with certainty about either the actual catalyst or the mechanism, although there is considerable speculation. Most of the catalysts are formed in situ from an early transition-metal halide and an aluminum alkyl. A classic example is $TiCl_4$ and $Et_3Al$. In most cases the resulting polymerization catalysts are heterogeneous. Catalytic olefin oligomerization and dimerization are also well-known. The later transition elements, such as those in Group VIII, usually afford dimers rather than high polymers. Polymerization and dimerization are usually limited to ethylene and propylene. More hindered olefins are much less reactive, suggesting that a coordinate bond is involved. Lewis acids markedly influence olefin polymerization catalysts, but the origin of this effect is unclear. Conjugated dienes and other activated olefins are much more readily polymerized, especially by catalysts derived from the later transition metals.

The difference between the degree of olefin polymerization exhibited by the various transition metals has been rationalized in terms of a competition between the β-hydride elimination rate, $k_h$, and the alkyl migratory-insertion rate, $k_p$ (see Equation 5.66).

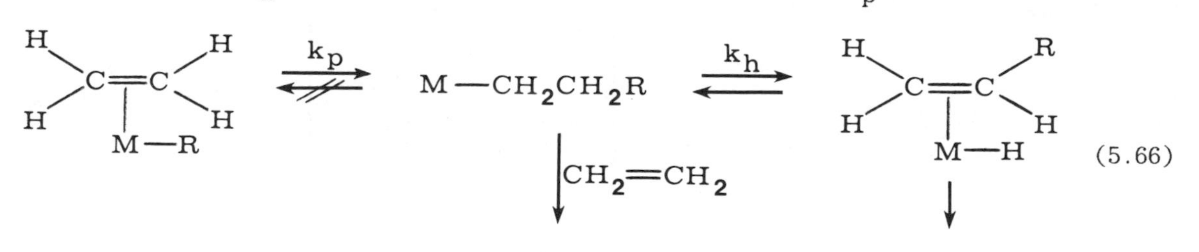


(5.66)

Three situations are plausible: (a) $k_p >> k_h$, giving high polymers, typical of the early transition metals; (b) $k_h >> k_p$, giving olefin dimers, typical of the later transition elements; and (c) $k_h \sim k_p$, affording oligomers. In Chapter 7, concerning transition-metal hydrides, we note that the equilibrium for hydride-olefin insertion seems to be displaced toward the alkyl side of the equation for the earlier transition metals and on the olefin-hydride side for the later transition metals. Thus we might expect more olefin polymerization with catalysts to the left on the transition series.

The regiospecificity of alkyl-olefin insertions is poorly understood. Either Markovnikov addition (path a in Equation 5.67) or anti-Markovnikov addition (path b) occurs, depending in an unpredictable way on a combination of steric and electronic factors.

The nickel-catalyzed codimerization of nobornene and ethylene yields exclusively the exo-product 96 (Equation 5.68). The presence of a chiral phosphine (L = (-)-

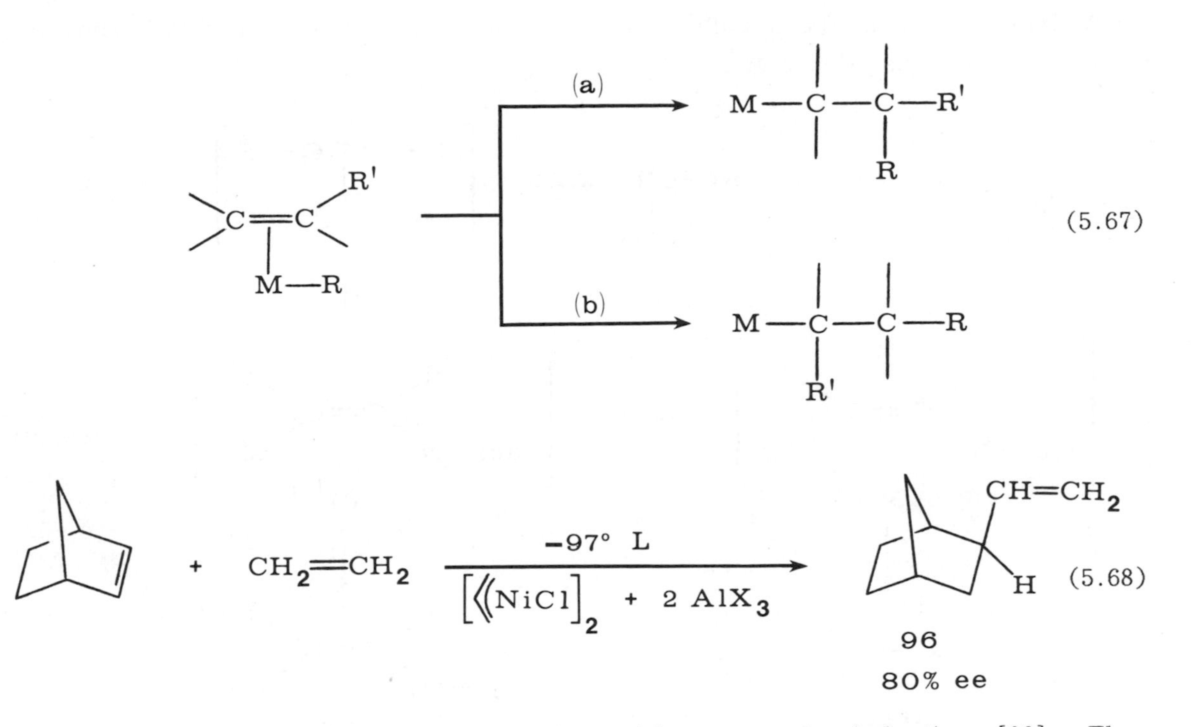


dimenthylisopropylphosphine) affords substantial asymmetric induction [82]. These results imply prior olefin coordination and migratory-insertion at the least-hindered side of norbornene. A less stereo- and regiospecific reaction might have been expected if only one olefin were coordinated to the metal. It should be noted that metallacyclopentanes can also be intermediates in olefin dimerization (see Chapter 10).

Insertions into acetylenic bonds have been much less studied than those involving olefins. The mechanisms of acetylene insertions are correspondingly even less understood. The concerted cis migratory insertion of hydrides and alkyls into acetylenes is the usual result, leading, for example, to stereospecific cis catalytic hydrogenation of acetylenes. However, trans additions have been reported in a number of cases for which the mechanisms are not well-resolved. Bergman [83] has proposed that kinetically controlled trans adducts can derive from an initial cis insertion mechanism. His results serve as a warning that "the observation of a given stereochemical mode of addition, even when the observed complex is found to be the kinetic product of the reaction, does not necessarily mean that the crucial insertion step proceeds with that same stereochemistry." This situation can arise by the intervention of a set of intermediates, such as the initially formed cis adduct 97, which is in equilibrium with an intermediate trans adduct 98.

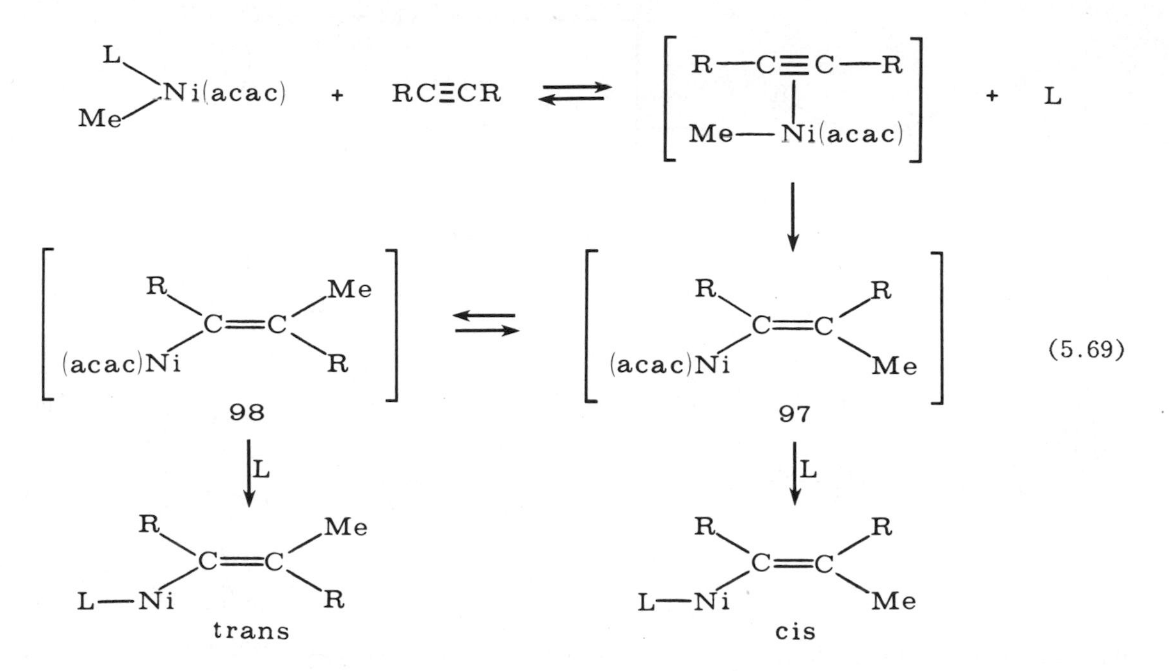


## 5.2. Intermolecular Nucleophilic Additions to Unsaturated Ligands.

There is a class of "insertion" reactions which involve the direct attack of nucleophilic reagents on unsaturated ligands without prior coordination of the nucleophile to the metal. A wide variety of such reactions have been discovered, involving different nucleophilic reagents and many types of unsaturated ligands. Such reactions are important steps in both stoichiometric and catalytic processes, ranging from organic syntheses involving $\eta^3$-allyl (Chapter 15) and olefin (Chapter 12) complexes to catalytic processes such as the Wacker-Schmidt and the water-gas-shift reactions (Chapter 8). These intermolecular reactions are often difficult to distinguish from intramolecular "migratory insertions" involving prior coordination of the nucleophile. Analysis of factors such as the coordinative saturation and substitution inertness or lability of the starting complex, as well as the stereochemistry of the product, are useful in identifying the intermolecular reactions.

We shall see in the examples cited below that three factors facilitate direct nucleophilic reactions at unsaturated ligands: (1) the reactivity of the nucleophile; (2) coordinative saturation of the starting metal complex or a diminished tendency for this complex to take on another ligand; and (3) a formal positive charge on the metal complex. This sort of reaction is known for most unsaturated ligands: carbon monoxide, isonitriles, cyanides, nitrosyls, olefins, chelating dienes, π-arenes, acetylenes, $\eta^3$-allyls, etc. In most cases the over-all reaction does not result in a change in the formal oxidation state, although this does occur in the important $\eta^3$-allyl case which we will encounter in Chapter 15.

1. Attack on Coordinated CO. A variety of nucleophiles may attack the carbon of metal carbonyls without prior coordination to the metal. Examples include Grignard, organolithium, and hydride reagents, aqueous hydroxide, amines, amine N-oxides, and azide ion. The adducts from such intermolecular attack can be stable complexes (e.g., acyls) or they may fragment (e.g., those from amine N-oxides) affording a new complex. *Cis* ligands may be labilized by such adduct formation.

a. Grignard, organolithium, and hydride reagents. Alkyl and aryl lithium reagents, 99, which are very powerful nucleophiles, react directly with coordinatively saturated iron pentacarbonyl, 100 affording anionic acyl lithium reagents, 101. This

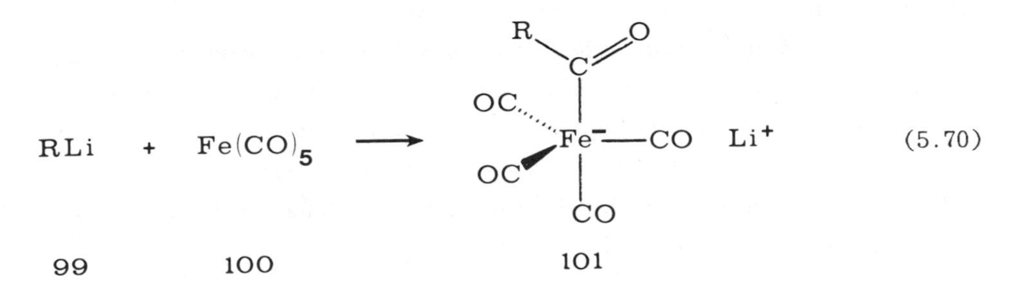


reaction at coordinated CO was discovered by Fischer [84], and has been used as a convenient syntheses of anionic acyl complexes, which can often be isolated as their $[(Ph_3P)_2N]^+$ salts [85]. The reaction occurs at temperatures well below those at which $Fe(CO)_5$ will show exchange with free $^{13}CO$ (a dissociative process, which requires temperatures of about 130°C). Thus the reaction may be considered as a direct nucleophilic attack on coordinated CO; although it might alternatively proceed via a prior electron-transfer path.

Grignard reagents also react with metal carbonyls to form acyl derivatives [86]. Strongly nucleophilic hydride donors react in a similar way. Casey [26] and Gladysz [27] have made use of this as a method for preparing kinetically stable formyl complexes such as 102 (Equation 5.71).

b. Amines. Amines and ammonia attack carbon monoxide in cationic complexes affording carbamoyl compounds, 103. If the amine is primary or secondary,

$$Na^+HB^-(OMe)_3 \quad + \quad Fe(CO)_5 \longrightarrow \underset{\mathbf{102}}{HC(=O)\bar{F}e(CO)_4Na^+} \qquad (5.71)$$

a proton may be lost, affording a stable adduct [87]. On the other hand, with tertiary

$$L_n\overset{+}{M}C{\equiv}O \quad + \quad 2\ HNR_2 \longrightarrow \underset{\mathbf{103}}{L_nM{-}C(=O)NR_2} \quad + \quad R_2\overset{+}{N}H_2 \qquad (5.72)$$

amines and pyridine, the formation of metal-carbamoyl derivatives is reversible. Even though formed in minute amounts, such intermediates are important, since all acyl-type groups appear to strongly labilize cis ligands. Thus pyridine or other nucleophiles, such as hydroxide ion, may cause the stereospecific labilization of ligands which are adjacent to the CO group being attacked [88].

c. Water, hydroxide, and alkoxide. Metal carbonyls may react with water or hydroxide ion [89]. Such a step is involved in "Reppe-type" catalysis and the water-gas-shift reaction (Chapter 8). In some cases, stable metallacarboxylates can be isolated (Chapter 8); in other examples, these intermediates cannot be isolated, but their presence can be deduced from isotopic exchange (Equation 5.73) or from $CO_2$ formation (Equation 5.76).

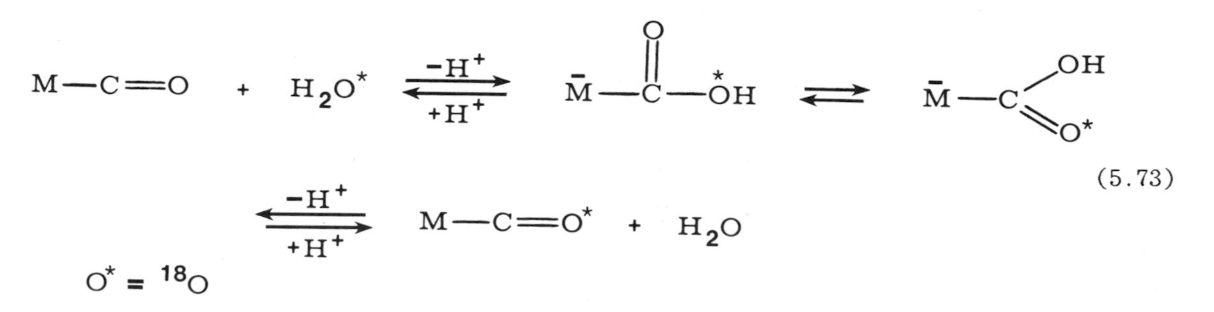

That such direct nucleophilic attack on unsaturated ligands is faciliated by a positive charge on the starting complex is illustrated by the relative rates of isotopic exchange between $H_2{}^{18}O$ and the saturated hexacarbonyls 104 and 105 (Equations 5.74 and 5.75). The cationic $d^6$ rhenium(I) complex 104 has a half-life for exchange of only

$$\overset{+}{Re}(CO)_6 \;(\mathbf{104}) + H_2O^* \rightleftarrows Re(CO)_n(C\overset{*}{O})_{6-n} \tag{5.74}$$

$$W(CO)_6 \;(\mathbf{105}) + H_2O^* \rightleftarrows \text{no reaction} \tag{5.75}$$

30 minutes at 25°, whereas the neutral isoelectronic tungsten complex 105 shows no evidence for exchange after 75 hours [90]. Presumably such exchange involves reversible attack of water on coordinated carbon monoxide, as shown in Equation 5.73.

As we have seen, stronger nucleophiles will react with neutral carbonyls. The attack of saturated metal carbonyls by hydroxide ion is a widespread, important reaction. As shown in Equation 5.76, the intermediate metallacarboxylate 106 loses $CO_2$

$$Fe(CO)_5 + 2\ NaOH \xrightarrow{-H_2O} \left[ \overset{+}{Na}_2 \left[ \begin{matrix} O \\ \vdots \\ O \end{matrix} \!\!>\! C{-}Fe(CO)_4 \right]^{=} \right] \;(\mathbf{106}) \xrightarrow{-CO_2} \overset{+}{Na}_2[Fe(CO)_4]^{=} \xrightarrow{H_2O} \overset{+}{Na}[HFe(CO)_4]^{-} \;(\mathbf{107}) + NaOH \tag{5.76}$$

affording the anionic metal hydride 107. Brown [88] has suggested that this decarboxylation is facilitated by or may require the presence of a vacant coordination site. This $CO_2$ elimination is a key step in the homogeneously catalyzed water-gas-shift reaction. In many cases the anionic hydride product is sufficiently acidic to be deprotonated by excess base. The over-all sequence leads to a formal two-electron reduction of metal, as shown for the general case in Equation 5.77. Thus CO and aqueous base are often employed to reduce a metal complex.

Alkoxide ions form similar complexes, as shown in Equation 5.78, wherein a cationic carbonyl, nitrosyl complex of iron(0), 108, reacts with methoxide ion, forming the stable metallaester 109. Note that in this case carbon monoxide is attacked preferentially to the linear nitrosyl ligand.

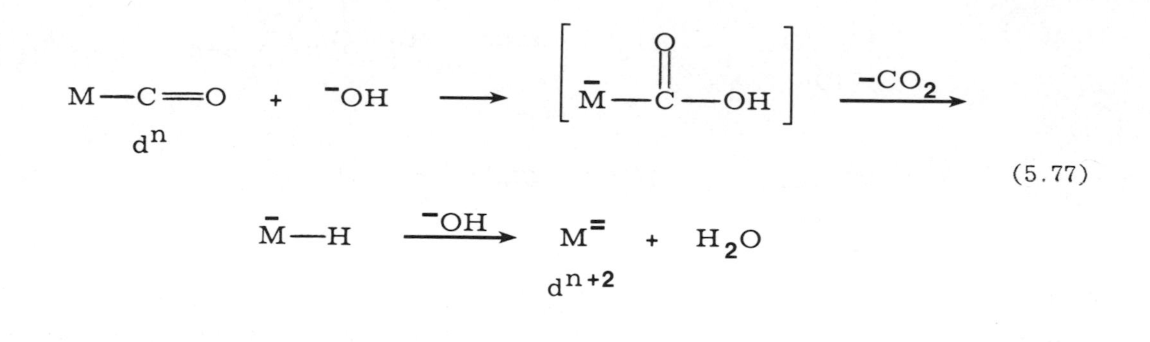

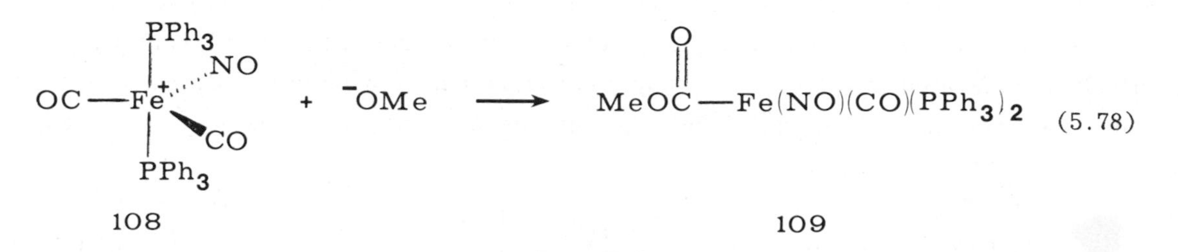

d. Azide ion. Another example of direct attack followed by rapid decomposition of the putative intermediate is the reaction of the strongly nucleophilic azide ion with saturated metal carbonyls [91]. The product isocyanato complex 111 probably derives from a metalloacylazide 110 which loses $N_2$ in a reaction (Equation 5.79) markedly similar to the well-known thermal decomposition of organic acyl azides (Curtius reaction).

$$W(CO)_6 + NaN_3 \xrightarrow{\text{slow}} \left[(CO)_5\bar{W}{-}\overset{\overset{\displaystyle O}{\|}}{C}{-}N_3\right]Na^+ \quad \text{(110)}$$

$$\xrightarrow[\text{rapid}]{-N_2} (CO)_5\bar{W}NCO \quad Na^+ \quad \text{(111)} \qquad (5.79)$$

e. Amine-N-Oxides. An interesting and useful method for removing coordinated CO under mild conditions [92] involves reaction (Equation 5.80) with trimethylamine N-oxide. The mechanism is thought to be an external attack on

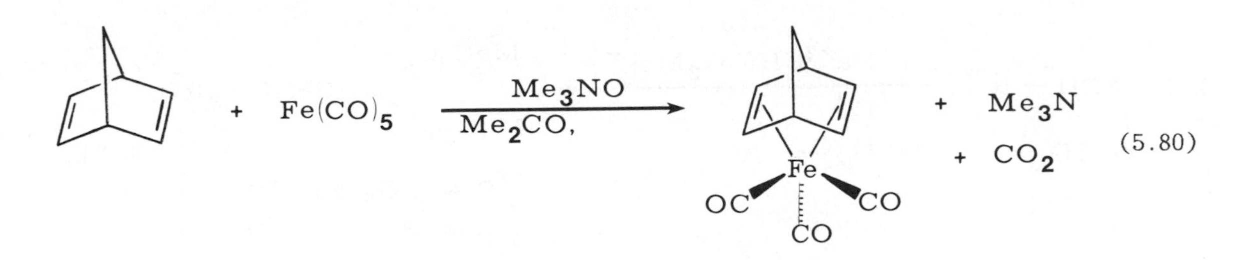


coordinated CO by the nucleophilic N-oxide, followed by fragmentation of the presumed intermediate <u>112</u>, affording $CO_2$, and secondary or tertiary amines [93] (Equation 5.81).

$$M{-}CO + {}^{-}O{-}\overset{+}{N}R_3 \longrightarrow \left[\bar{M}{-}\overset{\overset{\displaystyle O}{\|}}{C}{-}O{-}\overset{+}{N}R_3\right] \quad (5.81)$$

112

$$\longrightarrow M + CO_2 + R_3N$$

Secondary amines may result from a complex reaction involving a second mole of amine oxide [92].

2. <u>Attack on Coordinated Acyls</u>. Many neutral and anionic acyl complexes seem unreactive even in the presence of nucleophilic reagents. However, oxidative decompositions of acyl compounds can result in their facile hydrolysis or solvolysis of the acyl group. Earlier we encountered this as a secondary reaction in oxidatively induced alkyl-migration processes. In most examples the assumed intermediate cationic acyl complex is so susceptible to nucleophilic attack that such groups resemble activated esters or even acid halides. So far, few oxidative decompositions of acyl complexes have been studied <u>directly</u>, and the enhanced reactivity of cationic acyl complexes has seldom been used for organic syntheses. However, this is a promising area of research.

3. <u>Nucleophilic Attack on Coordinated Olefin</u>. Simple monoene complexes seem susceptible to nucleophilic attack, although a limited number of cases have been examined. As shown in Equation 5.82, when an activating group is present, such reactions can be regioselective, the nucleophile going to the β-carbon, as in the Michael addition. In other cases, such as palladium-olefin complexes, regiospecific attack occurs at the more hindered olefinic carbon. We will encounter these in Chapter 12.

$$\underset{\displaystyle Fe(CO)_4}{CH_2{=}CH{-}R} \xrightarrow[\substack{2)\ H^+ \\ 3)\ H_2O_2}]{1)\ Na^+[CH(CO_2Me)_2^-]} (MeO_2C)_2CHCH_2CH_2R \qquad (5.82)$$

$R = CO_2Me$ 90%
$R = H$ 70%

Such reactions are greatly facilitated by a positive charge on the olefin complex or when the nucleophile can be delivered in an intramolecular reaction. As expected in such external attack on a coordinated olefin, the over-all addition of the metal and the nucleophile is trans. These coordinated olefin reactions have been applied to new heterocyclic syntheses and to a β-lactam synthesis, as is illustrated in Chapter 12. With oxygen and nitrogen nucleophiles, this attack on coordinated olefins can be reversible, but with carbon nucleophiles it is irreversible.

Another significant example of this type involves reaction of the norbornadiene palladium complex 113 with sodium diethylmalonate 114 to afford a product 115 having a palladium-carbon σ bond (Equation 5.83). The intermolecular nature of the reaction is

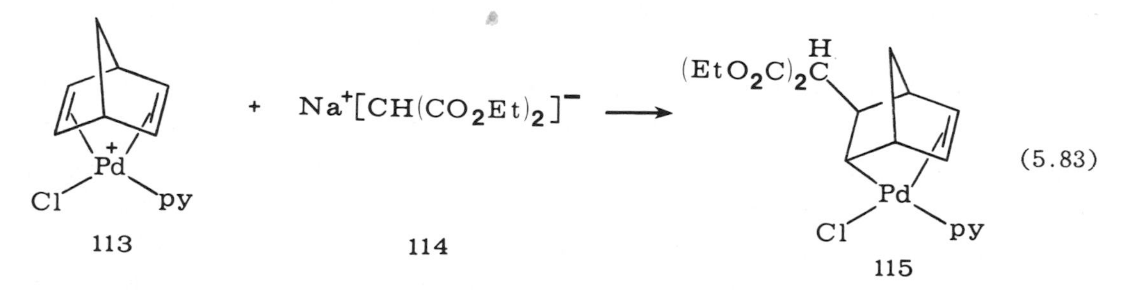


strongly implied by the stereochemistry of the product 115. The exo position of the diethylmalonate group indicates that the addition of carbon and palladium to the coordinated olefin is trans. An intermolecular "migratory insertion" which would have afforded cis addition can therefore be ruled out. Other features of this reaction are instructive. Direct attack occurs in this case even though the four-coordinate palladium complex 113 is coordinatively unsaturated. Both steric hindrance and the general reluctance of $d^8$ palladium(II) to form saturated five-coordinate complexes may explain the intermolecular course of this reaction. The very stable chelating diene in 113 tends to prevent olefin displacement, which can be a serious side reaction in the case of simple monoene palladium complexes. In Chapter 12 we shall find that the intermolecular attack of water on ethylene that is coordinated to palladium(II) is the key step in the Wacker oxidation of ethylene to acetaldehyde.

4. Nucleophilic Attack on Coordinated Acetylene. Cationic platinum(II) acetylene complexes add a variety of nucleophiles. These reactions have been extensively studied by Chisholm and Clark [96], who found that complicated subsequent reactions ensue, especially in the case of terminal acetylene complexes. The reaction of methanol with the coordinated disubstituted acetylenes (generated in situ) affords trans-σ-bonded vinyl ether complexes 116 (Equation 5.84). The trans stereochemistry of the product

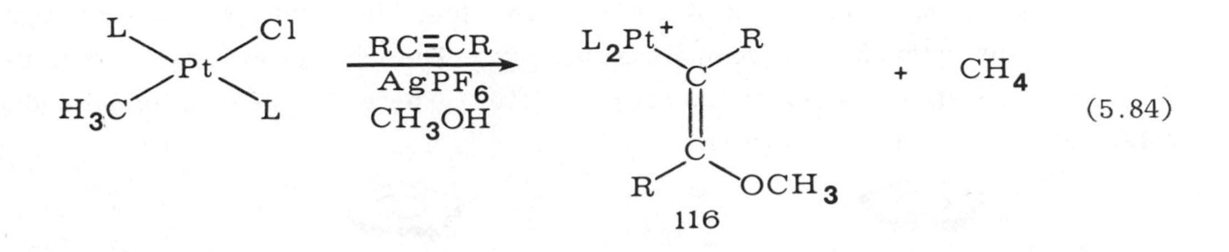


(5.84)

116 indicates that the nucleophilic attack on coordinated acetylene was intermolecular and did not involve prior coordination of methanol.

5. Attack on Coordinated Arene. Cationic arene complexes are especially susceptible to intermolecular nucleophilic additions, such as the example in Equation 5.85. The tropyllium molybdenum complex 117 reacts with methoxide ion affording the $\eta^6$-triene complex 118. Note the endo nature of the adduct 118. In Chapter 14 we shall see that Semmelhack has developed some beautiful synthetic applications of nucleophilic attack by carbanions on neutral arene chromium tricarbonyl complexes.

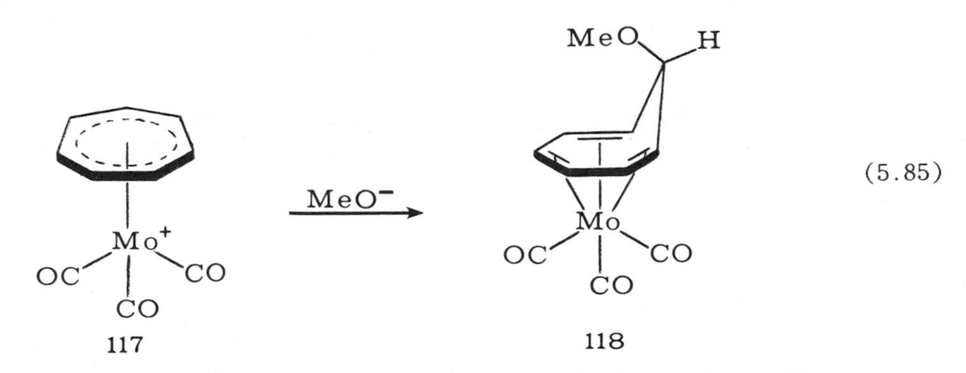


(5.85)

The nucleophilic addition to unsaturated ligands on organometallic cations is very general. This sort of reaction has been reviewed by Davies, Green, and Mingos [97], who have proposed some simple rules to predict the most favorable position for nucleophilic attack.

6. Attack on $\eta^3$-allyl. Nucleophilic backside attack at the terminal carbon of a $\eta^3$-allyl ligand leads to the corresponding olefin complex and a formal two-electron reduction of the metal. The reaction is facilitated by a formal positive charge on the complex. With carbon nucleophiles, carbon-carbon bond formation occurs with inversion of configuration at the allylic carbon. Trost has exploited this reaction in several efficient natural product syntheses. These examples are discussed in detail in Chapter 15.

In the example shown in Equation 5.86, four different unsaturated ligands are present in the cationic molybdenum complex 119. The $\eta^3$-allyl group in 119 is preferentially attacked affording an olefin complex, 120. There has been a formal reduction of molybdenum(II) to molybdenum(O).

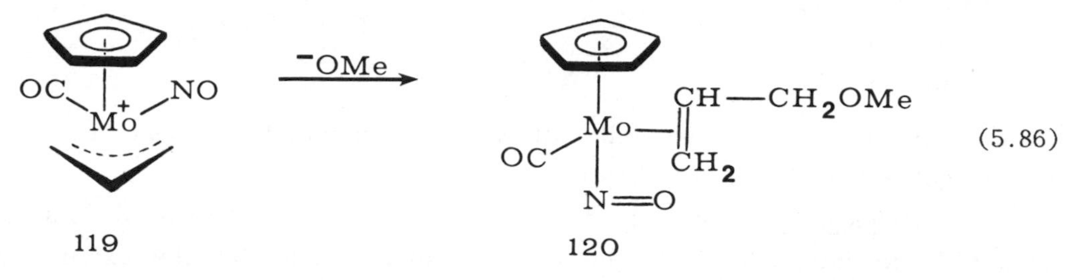


Many direct nucleophilic attacks on coordinated ligands are reversible. Thus, in a process similar to the reverse of the reaction shown in Equation 5.87, allylic substrates can be used to form $\eta^3$-allyl complexes by the abstraction of X by Pd(0). This is one important method for preparing palladium $\eta^3$-allyl complexes such as 121.

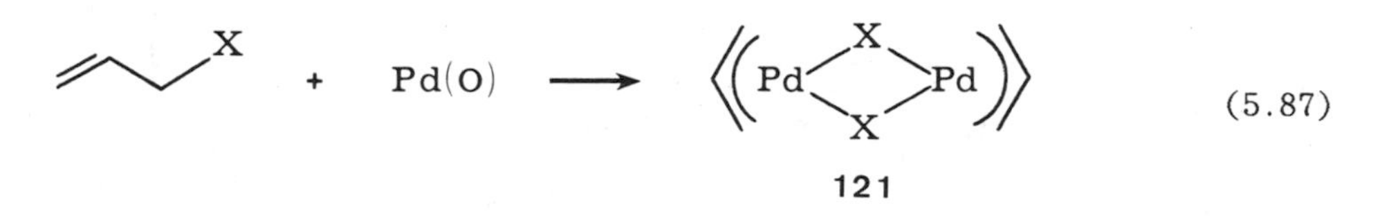


7. Attack on $\eta^5$-$C_5H_5$. Strong nucleophiles may attack $\pi$-cyclopentadiene ligands. Equation 5.88 shows nearly equal attack by methyl lithium on the carbonyl and the $\pi$-cyclopentadienyl ligands affording 122 and 123, respectively. The direction of attack and the degree of selectivity in such processes lies out of the scope of our present understanding. Note that in terms of our oxidation-state formalism, the product 123 has undergone reduction to manganese(-I), a $d^8$ complex. This is because

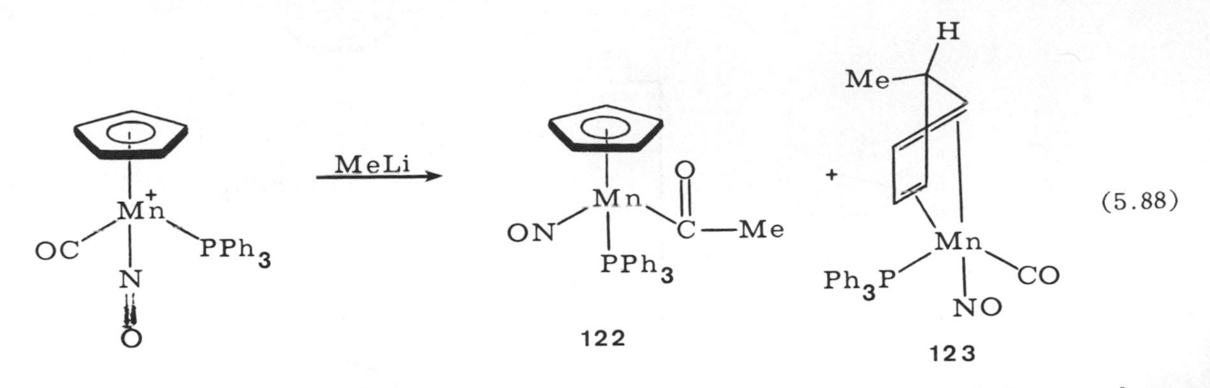


$\eta^5$-cyclopentadienyl groups can be formally considered as an olefin + $\eta^3$-allyl combination.

8. Nucleophilic Attack on Coordinated Carbene and Carbyne. Earlier we noted the propensity for various groups to migrate to coordinated carbene. This tendency carries over to intermolecular attack on coordinated carbene. Such direct nucleophilic attack on coordinatively saturated, positively charged carbene complexes is quite facile. Many examples fit into this category, such as reactions of carbenes with carbanions, amines, thiols, and phosphines. Several of these reactions are reversible as in Equation 5.89.

$$\underset{d^n}{\overset{+}{M}{=}CH_2} + \ddot{\bar{N}} \rightleftharpoons \underset{d^n}{M{-}CH_2{-}N} \qquad (5.89)$$

Gladysz [98] has prepared a cationic rhenium carbene complex, 124, and found it to react with a variety of nucleophiles (Equation 5.90).

In the context of the facile nucleophilic attack on coordinated carbene, it is important to remember that the isonitrile ligand can also be considered as a sort of carbene.

A synthetically useful, irreversible reaction of this class is "cyclopropanation." For example, the well-characterized, coordinatively saturated, cationic carbene complex 125 reacts with propylene at low temperatures to afford the cyclopropanes [99] (Equation 5.91). The tungsten carbene complex 126 reacts similarly [100].

A convenient, stable methylene transfer reagent 127 has been reported to effect the stereospecific cyclopropanation of olefins in high yield [101] (Equation 5.92). This reagent undoubtedly forms a cationic iron carbene complex in situ.

It is sometimes uncertain whether these cyclopropanation reactions involve prior coordination of the olefin or direct external attack of the coordinated carbene by the

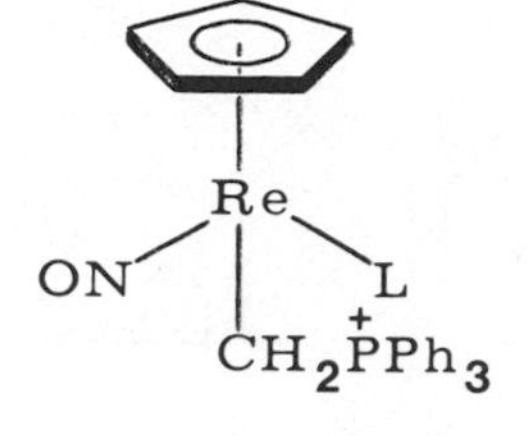


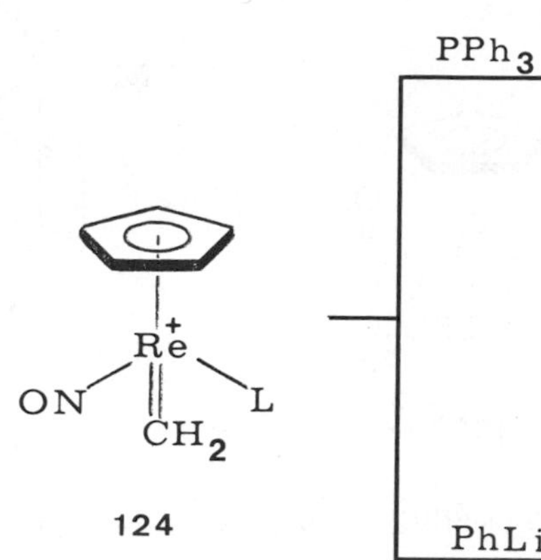


Re
ON
L
CH2Ph

(5.90)

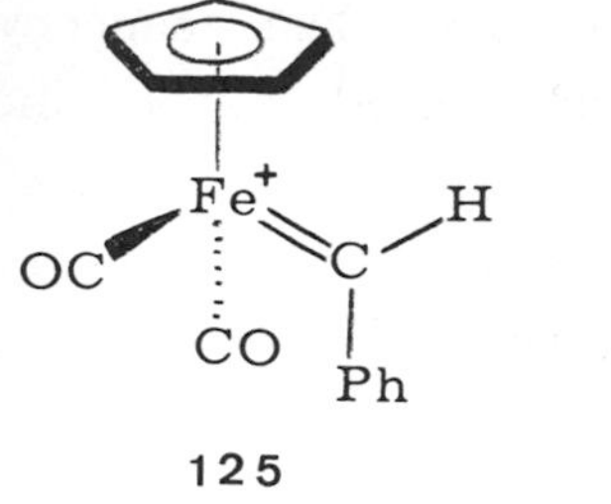


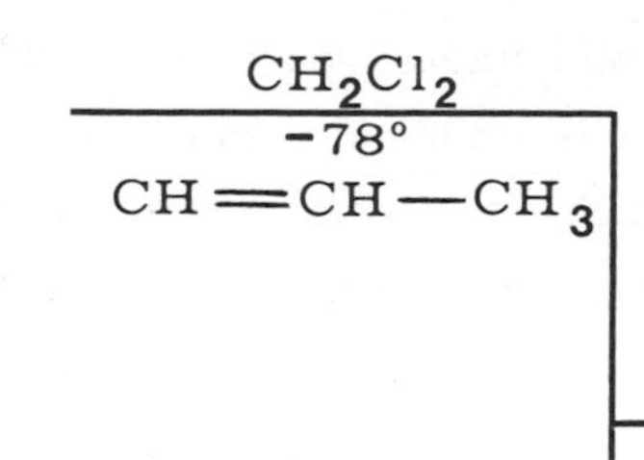


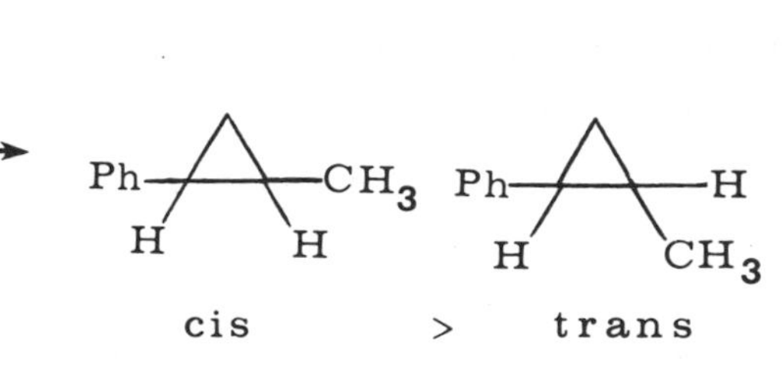


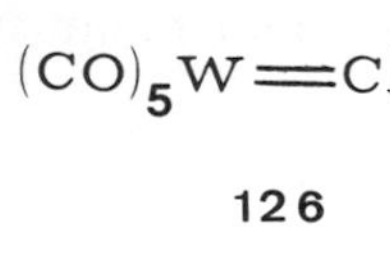


CH2═CH—CH3

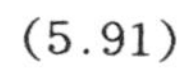

(5.91)

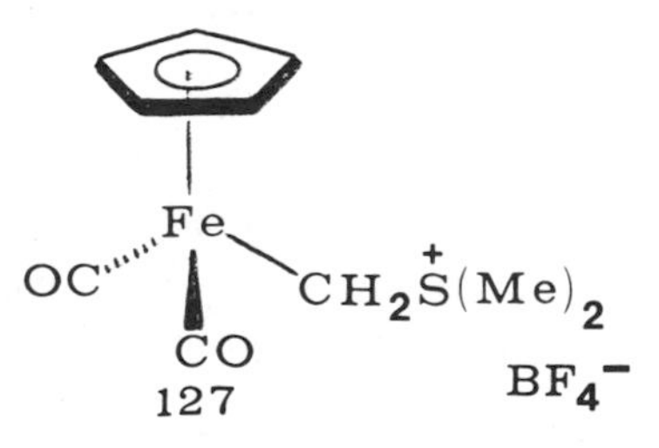


+

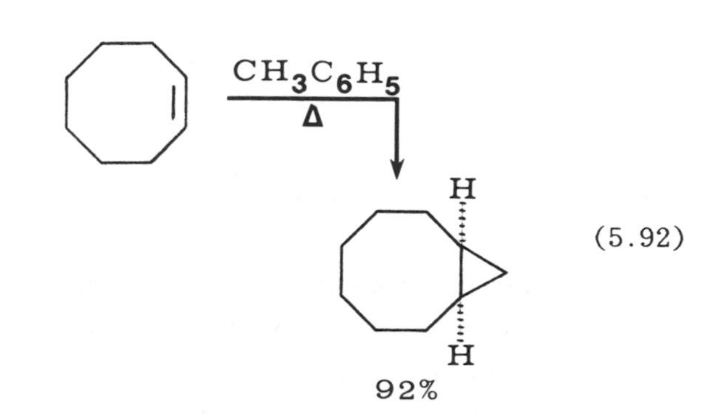


(5.92)

nucleophilic olefin. The latter seems more plausible, because of the coordinatively saturated nature of the metal centers. On the other hand, prior coordination to the metal would afford a metallacyclobutane which could either reductively eliminate cyclopropane (path b in Equation 5.93) or reversibly yield olefin and carbene (path a).

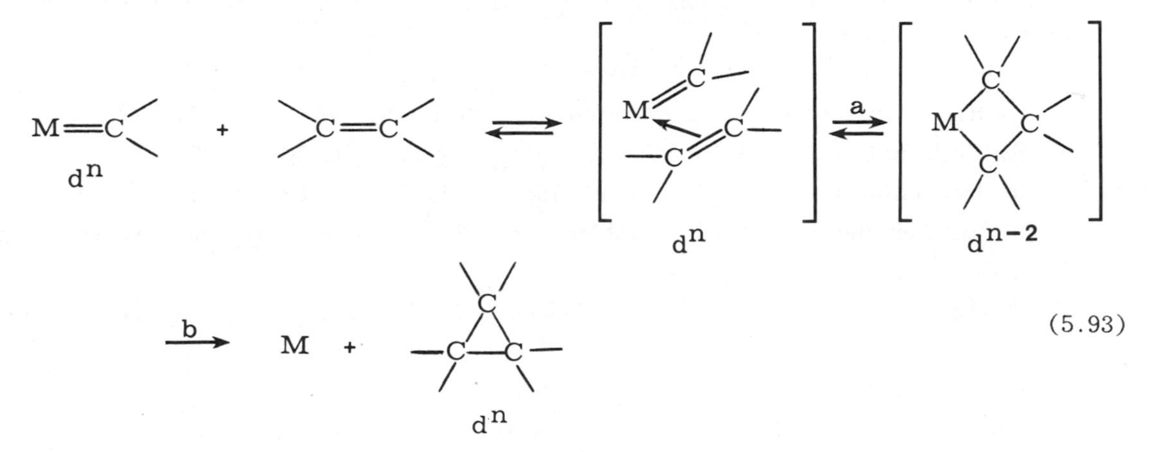


Path (a) is the propagation step in olefin metathesis (see Chapter 10). Cyclopropanes have sometimes been observed during olefin metatheses, but it is not clear whether this is a branch of the main reaction scheme. Extraneous ligands tend to inhibit olefin metathesis, which suggests the need for a vacant coordination site. An important difference between inter- and intramolecular carbene-olefin reactions (Equation 5.91 and 5.93) is that the latter requires a two-electron oxidation of the central metal and the former does not.

In Chapter 8 we will encounter an interesting example of intermolecular nucleophilic attack on coordinated carbene in the discussion of "Fischer-Tropsch" processes.

Although they have been less studied than carbenes, carbyne complexes are also susceptible to direct nucleophilic attack. In fact, this reaction is the reverse of the reaction Fischer used to prepare the first carbyne complex (Chapter 3). An unusual example of this sort of reaction has been discovered by Roper [102], who found that a hydride donor reacts with the cationic *p*-tolyl carbyne complex, 128, by remote attack on the aromatic ring.

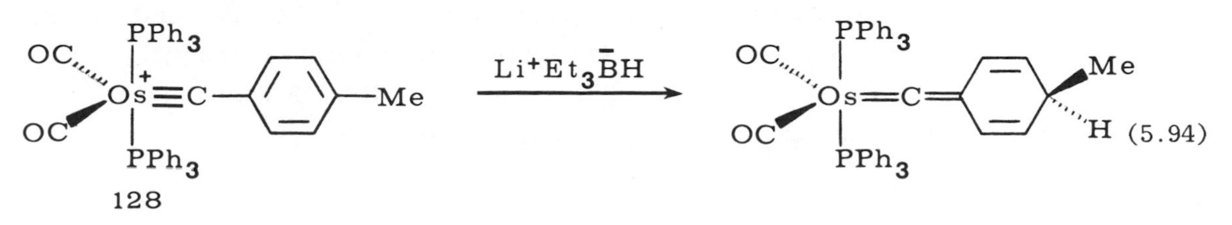

## NOTES.

1a. A. Wojcicki, Adv. Organometal. Chem., 11, 88 (1972).

1b. Ibid., 12, 33 (1974).

2a. F. Calderazzo, Angew. Chem., Int. Ed. Engl., 16, 299 (1977), a review of CO insertion reactions.

2b. M. Kubota and D. M. Blake, J. Amer. Chem. Soc., 93, 1368 (1971).

3a. M. L. H. Green, Organometallic Compounds (Methuen, 1968), II, 261.

3b. M. R. Churchill and S. W. Y. Chang, Inorg. Chem., 14, 1680 (1975).

3c. M. R. Churchill and J. P. Fennessey, Inorg. Chem., 7, 953 (1968).

3d. J. R. Blickensderfer, C. B. Knobler, and H. D. Kaesz, J. Amer. Chem. Soc., 97, 2686 (1975).

3e. C. G. Fachinetti, G. Fochi, and C. Floriani, J. Chem. Soc., Dalton, Trans., 1946 (1977).

4. H. Berke and R. Hoffmann, J. Amer. Chem. Soc., 100, 7224 (1978).

5. G. Henrici-Olive and S. Olive, Transition Met. Chem., 1, 77 (1976).

6. M. M. T. Khan and A. E. Martell, Homogeneous Catalysis by Metal Complexes (Academic Press, 1974), Vol. I.

7a. K. S. Y. Lau, Y. Becker, F. Huang, N. Baenziger, and J. K. Stille, J. Amer. Chem. Soc., 99, 5664 (1977).

7b. H. M. Walborsky and L. E. Allen, Tet. Lett., 823 (1970).

7c. I. C. Douek and G. Wilkinson, J. Chem. Soc., A, 2604 (1969).

7d. R. F. Heck, J. Amer. Chem. Soc., 91, 6707 (1969).

7e. B. L. Shaw, Chem. Comm., 464 (1968).

7f. P. Stoppioni, P. D. Porto, and L. Sacconi, Inorg. Chem., 17, 718 (1978).

8. J. P. Collman, R. Finke, J. N. Cawse, and J. I. Brauman, J. Amer. Chem. Soc., 100, 4766 (1978), and references therein.

9. S. B. Butts, E. M. Holt, S. H. Strauss, N. W. Alcock, R. E. Stimson, and D. F. Shriver, J. Amer. Chem. Soc., 101, 5864 (1979).

10. G. W. Daub, in Progress in Inorganic Chemistry, S. J. Lippard, ed. (Wiley, 1977), 22, 409-423.

11. P. I. Bock, D. J. Boschetto, J. R. Rasmussen, J. P. Demers, and G. M. Whitesides, J. Amer. Chem. Soc., 96, 2814 (1974), and references therein.

12. J. K. Stille and K. S. Y. Lau, Accts. Chem. Res., 10, 434 (1977).

13. F. Calderazzo and F. A. Cotton, Inorg. Chem., 1, 30 (1962).

14. J. P. Collman, Accts. Chem. Res. 8, 342 (1975).

15. A. Davison and N. Martinez, J. Organomet. Chem., 73, C17 (1974).
16. R. J. Mawby, F. Basolo, and R. G. Pearson, J. Amer. Chem. Soc., 86, 3994 (1964).
17. K. Nicholas, S. Raghu, and M. Rosenblum, J. Organomet. Chem., 78, 133 (1974).
18. R. W. Glyde and R. J. Mawby, Inorg. Chem., 10, 854 (1971).
19. E. A. K. von Gusdorf, L. H. G. Leenders, I. Fischler, and R. N. Peruta, Adv. Inorg. and Radiochem., 19, 65 (1976).
20. J. N. Cawse, R. A. Fiato, and R. L. Pruett, J. Organometal. Chem., 172, 405 (1979), and references therein.
21. G. H. Olive and S. Olive, Angew. Chem., Int. Ed. Engl., 15, 316 (1976).
22. I. Wender, Cat. Rev. Sci. Eng., 14, 97 (1976).
23. M. A. Vannice, loc. cit., p. 153.
24. C. Masters, Adv. Organometal. Chem., 17, 61-103 (1979), and references therein.
25. J. P. Collman and S. R. Winter, J. Amer. Chem. Soc., 95, 4089 (1973).
26a. C. P. Casey and S. M. Neuman, J. Amer. Chem. Soc., 98, 5395 (1976).
26b. C. P. Casey, M. A. Andrews, and J. E. Rinz, J. Amer. Chem. Soc., 101, 741 (1979), and references therein.
27a. M. A. Gladysz and W. Tam, J. Amer. Chem. Soc., 100, 2545 (1978).
27b. W. Tam, W. K. Wong, and J. A. Gladysz, J. Amer. Chem. Soc., 101, 1589 (1979).
28. W. J. Miles, Jr., and R. J. Clark, J. Organometal. Chem., 131, 93 (1977).
29. B. H. Byers and T. L. Brown, J. Organometal. Chem., 127, 181 (1977).
30. J. M. Manriquez, R. D. McAlister, R. D. Sanner and J. E. Bercaw, J. Amer. Chem. Soc., 98, 6733 (1976).
31. C. P. Casey, C. A. Bunnell, and J. C. Calabrese, loc. cit., p. 1166.
32. Y. Yamanaoto and H. Yamazaki, Inorg. Chem., 13, 438 (1974).
33. P. M. Treichel and K. P. Wagner, J. Organometal. Chem., 61, 415 (1973).
34. L. S. Hegedus, Ph.D. Thesis, Harvard University (1970).
35. P. M. Treichel and F. G. A. Stone, Adv. in Organomet. Chem., 1, 143 (1964).
36a. K. W. Barnett, D. L. Beach, S. P. Gaydos, T. F. Pollmann, J. Organometal. Chem., 69, 121 (1974).
36b. R. W. Glyde and R. J. Mawby, Inorg. Chim. Acta, 4, 331 (1970); 5, 317 (1971).
36c. D. Egglestone and M. C. Baird, J. Organometal. Chem., 113, C25 (1976).
36d. B. L. Shaw and E. Singleton, J. Chem. Soc., A, 1683 (1967).

36e. K. S. Y. Lau, Y. Becker, F. Huang, N. Baenziger, and J. K. Stille, J. Amer. Chem. Soc., 99, 5664 (1977).
36f. A. J. Hart-Davis and R. J. Mawby, J. Chem. Soc., A, 2403 (1969).
36g. H. F. Klein and H. H. Karsch, Chem. Ber., 109, 2524 (1976).
36h. G. Huttner, O. Orama, and V. Bejenke, Chem. Ber., 109, 2533 (1976).
36i. L. Sacconi, P. D. Dapporto, and P. Stoppioni, J. Organometal. Chem. 116, C33 (1976).
37. P. R. Hoffmann and K. G. Caulton, J. Amer. Chem. Soc., 97, 4221 (1975).
38. J. T. Magne, Inorg. Chem., 9, 1610 (1970).
39. K. S. Y. Lau, Y. Becker, F. Huang, N. Baenziger, and J. K. Stille, J. Amer. Chem. Soc., 99, 5664 (1977).
40. A. Rossi and R. Hoffmann, Inorg. Chem., 14, 365 (1975).
41. F. Calderazzo and F. A. Cotton, Chim. Ind. (Milan), 46, 1165 (1964).
42. C.-H. Cheng, B. D. Spivack, and R. Eisenberg, J. Amer. Chem. Soc., 99, 3003 (1977).
43. K. Noack, M. Ruch, and F. Calderazzo, Inorg. Chem., 7, 345 (1968).
44. M. Kubota, D. M. Blake, and S. A. Smith, Inorg. Chem., 10, 1430 (1971).
45. J. J. Alexander and A. Wojcicki, Inorg. Chem., 12, 74 (1973).
46. M. Kubota, R. K. Rothrock, and J. Geibel, J. Chem. Soc., Dalton, 1267 (1973).
47. C. P. Casey and C. A. Bunnell, J. Amer. Chem. Soc., 98, 436 (1976).
48. G. Fachinetti, C. Floriani, and H. Stoeckli-Evans, J. Chem. Soc. (D), 2297 (1977).
49. G. Fachinetti, G. Fochi, and C. Floriani, J. Chem. Soc. (D), 1946 (1977).
50. U. Franke and E. Weiss, J. Organomet. Chem., 165, 329 (1979).
51. E. C. Guzman, G. Wilkinson, J. L. Atwood, R. D. Rogers, W. E. Hunter, and M. J. Zarvorotko, J.C.S. Chem. Comm., 465 (1978).
52. W. R. Roper, G. E. Taylor, J. M. Waters, and L. F. Wright, J. Organometal. Chem., in press (1980).
53. G. R. Clark, T. J. Collins, K. Marsden, and W. R. Roper, J. Organometal. Chem., 157, C23 (1978).
54. W. R. Roper and L. J. Wright, J. Organomet. Chem., 142, C1 (1977).
55. W. R. Roper, private communication.
56. E. W. Abel and R. J. Rowley, J.C.S. Chem. Comm., 73 (1974).
57. A. H. Klazinga and J. H. Teuber, J. Organomet. Chem., 165, 31 (1979).

58. G. W. Adamson, J. J. Daly, and D. Forster, J. Organomet. Chem., 71, C17 (1974).
59. C. P. Casey and R. L. Anderson, J. Amer. Chem. Soc., 93, 3554 (1971).
60. J. P. Collman, R. K. Rothrock, R. G. Finke, and F. Rose-Munch, J. Amer. Chem. Soc., 99, 7381 (1977).
61. G. Erker and F. Rosenfeldt, Angew. Chem. Int. Ed. Engl., 17, 605 (1978).
62. P. Stoppioni, P. Dapporto, and L. Sacconi, Inorg. Chem., 17, 718 (1978).
63. M. C. Baird, T. T. Magne, J. A. Osborn, and G. Wilkinson, J. Chem. Soc., A, 1347 (1967).
64. M. MacLaury, Ph.D. dissertation, Stanford University (1974).
65. P. K. Wong, M. Madhavarao, D. F. Marten, and M. Rosenblum, J. Amer. Chem. Soc., 99, 2823 (1977).
66. D. H. Ballard, J. Z. Chrzastowski, D. Dodd, and M. D. Johnson, Chem. Comm., 185 (1972).
67. T. J. Collins, and W. R. Roper, J.C.S. Chem. Comm., 1044 (1976).
68a. Y. Yamamoto and H. Yamazaki, Coord. Chem. Rev., 8, 225 (1972).
68b. P. M. Treichel, Adv. Organomet. Chem., 11, 21 (1973).
68c. Y. Yamamoto, K. Aoki, and H. Yamazaki, J. Amer. Chem. Soc., 96, 2647 (1974).
68d. P. M. Treichel and K. P. Wagner, J. Organometal. Chem., 61, 415 (1973).
68e. H. C. Clark, Pure and Appl. Chem., 50, 43 (1978).
69. Y. Yamamoto and Hl Yarmazaki, J. Organomet. Chem., 90, 329 (1975).
70. F. Mango and I. Dvoretzky, J. Amer. Chem. Soc., 88, 1654 (1966).
71. H. D. Empsall, E. M. Hyde, M. Markham, W. S. McDonald, M. C. Norton, B. L. Shaw, and B. Weeks, J.C.S. Chem. Comm., 589 (1977).
72. N. J. Copper and M. L. H. Green, Chem. Comm., 761 (1974).
73. M. L. H. Green, Pure and Appl. Chem., 50, 27 (1978).
74. L. Pu and A. Yamamoto, Chem. Comm., 9 (1974).
75. R. R. Schrock, J. Amer. Chem. Soc., 96, 6796 (1976).
76. A. J. Deeming, B. F. G. Johnson, and J. Lewis, J. Chem. Soc. (D), 1848 (1973).
77. F. W. S. Benfield and M. L. H. Green, J. Chem. Soc. (D), 1324 (1974).
78. H. Werner and R. Feser, Angew. Chem. Int. Ed. Engl., 18, 157 (1979).
79. E. R. Evitt and R. G. Bergman, J. Amer. Chem. Soc., 101, 3973 (1979).
80. K. J. Ivin, J. J. Rooney, C. D. Stewart, M. L. H. Green, and R. Mahtab, J.C.S. Chem. Comm., 604 (1978).
81a. D. G. H. Ballard, J. Poly. Sci., 13, 2191 (1975) and references therein.

81b. J. Boor, Ziegler-Natta Catalysts and Polymerizations (Academic Press, 1979).
82. B. Boganovic, Angew. Chem. Int. Ed. Engl., 12, 954 (1973).
83. J. M. Huggins and R. G. Bergman, J. Amer. Chem. Soc., 101, 4410 (1979).
84. E. O. Fischer and A. Maasbol, Angew. Chem. Int. Ed. Engl., 3, 580 (1964).
85. J. P. Collman, Accts. Chem. Res., 8, 342 (1975), and references therein.
86. D. Drew, M. Y. Darensbourg, and D. J. Darensbourg, J. Organomet. Chem., 85, 73 (1975).
87. R. J. Angelici, Accts. Chem. Res., 5, 335 (1972).
88. T. L. Brown and P. A. Bellus, Inorg. Chem., 17, 3726 (1978), and references therein.
89a. H. C. Kang, C. H. Mauldin, T. Cole, W. Slegeir, K. Cann, and R. Pettit, J. Amer. Chem. Soc., 99, 8323 (1977).
89b. R. M. Laine, R. G. Rinker, and P. C. Ford, J. Amer. Chem. Soc., 99, 252 (1977).
89c. C. Ungermann, V. Landis, S. A. Moya, H. Cohen, H. Walker, R. G. Pearson, R. G. Rinker, and P. C. Ford, J. Amer. Chem. Soc., 101, 5922 (1979).
89d. C. H. Cheng, D. E. Hendrickson, and R. Eisenberg, J. Amer. Chem. Soc., 99, 2791 (1977).
89e. R. B. King, C. C. Frazier, R. M. Hanes, and A. D. King, J. Amer. Chem. Soc., 100, 2925 (1978).
90a. E. L. Muetterties, Inorg. Chem., 4, 1841 (1965).
90b. D. J. Darensbourg and J. A. Froelich, J. Amer. Chem. Soc., 99, 4726 (1977).
91a. R. J. Angelici and L. Busetto, J. Amer. Chem. Soc., 91, 3197 (1969).
91b. W. Beck, H. Werner, H. Engelmann, and H. S. Smedal, Chem. Ber., 101, 2143 (1968).
91c. H. Werner, W. Beck, and H. Engelmann, Inorg. Chim. Acta, 3, 331 (1969).
92. Y. Shuo and E. Hazum, Chem. Comm., 336 (1974).
93. D. J. Blumer, K. W. Barnett, and T. L. Brown, J. Organomet. Chem., 173, 71 (1979), and references therein.
94. J. H. Eekhof, H. Hogeveen, and R. M. Kellog, J.C.S. Chem. Comm., 647 (1976) and references therein.
95. J. K. Stille and D. B. Fox, J. Amer. Chem. Soc., 92, 1274 (1970).
96. M. H. Chisholm and H. C. Clark, Accts. Chem. Res., 6, 202 (1973), and references therein.

97. S. G. Davies, M. L. H. Green, and D. M. P. Mingos, Tetrahedron, 34, 3047 (1978).

98. J. A. Gladysz, private communication.

99. M. Brookhart, private communication.

100. C. Casey, Chemtech, 9, 378 (1979), and references therein.

101. S. Brandt and P. Helquist, J. Amer. Chem. Soc., 101, 6473 (1979).

102. W. R. Roper, private communication, to be published.

# 6

# Homogeneous Catalytic Hydrogenation and Hydrosilation

Homogeneously catalyzed hydrogenation of organic substrates by soluble transition-metal complexes is the most widely studied class of organometallic reactions. Olefins have been the most thoroughly examined substrates, but many other functional groups, such as acetylenes, aldehydes, ketones, nitro groups, and arenes, are also reduced. A very large number of homogeneous hydrogenation catalysts have been discovered; their number and diversity will undoubtedly increase. This subject has been extensively reviewed [1]. Catalytic hydrosilations [2] are similar to hydrogenations and are therefore described at the end of this chapter.

A dominant and very useful feature of homogeneous hydrogenation catalysts is their substrate selectivity, culminating in examples of exceptionally efficient asymmetric hydrogenation [3]. Major disadvantages of such catalysts include their great sensitivity to impurities, particularly traces of oxygen, their proclivity to cause olefin rearrangement, and difficulties in recovering these soluble, often valuable catalysts. Immobilization of molecular hydrogenation catalysts by attachment to insoluble, polymeric supports is a promising method for separating the catalysts from the products.

Table 6.1 lists representative examples of homogeneous hydrogenation catalysts, along with some characteristics of these individual cases. Each "catalyst" listed is actually a "precatalyst," the substance added to the flask to bring about the reaction. This "precatalyst" reacts with hydrogen and/or the substrate to form the actual catalyst, which may be considered the dominant complex present during the catalytic cycle. For example, one or more ligands may dissociate or may be removed by hydrogenation.

*Table 6.1. Representative homogeneous hydrogenation catalysts.*

| "Precatalysts" | $[H_2]$ atm. | Solvents | Substrates | Products | Special features[a] | Ref |
|---|---|---|---|---|---|---|
| $RhL_3Cl$[b] | ~1 | toluene<br>ethanol | simple olefins | hydrocarbons | very selective | [1d] |
| $HRh(CO)L_3$ | <1 | toluene | $RCH=CH_2$ | $RCH_2CH_3$ | internal olefins are unreactive and isomerized | [4] |
| $Ir(CO)ClL_2$ | 1 | toluene | simple olefins | hydrocarbon | 80°, isomerization, H,D exchange | [5] |
| $IrClL_{2-3}$ | 1 | toluene | simple olefins | hydrocarbon | Ir:L 1:2 10 times faster than 1:3, olefin isomerization | [6] |
| $HRuClL_3$ | 1 | toluene | $RCH=CH_2$<br>1,3-diene | hydrocarbon<br>monoene | selective for terminal olefin | [7] |
| $\{M(diene)L_2\}^+$<br>M=Rh>Ir | 1 | THF,<br>acetone | olefin<br>alkyne<br>C=C-COR<br>RCOR | hydrocarbon<br>*cis*-alkene<br>$\overline{CH}$-CHCOR<br>RCH(OH)R' | can be very stereoselective, $H_2O$ required for ketone reduction | [8] |
| $Rh(BH_4)(Cl)_2(DMF)(py)_2$ | 1 | DMF | hindered and unhindered olefins<br>pyridine<br>$ArNO_2$<br>$C=CCO_2Me$ | hydrocarbons<br><br>piperidine<br>$ArNH_2$<br>$CH_2CH_2CO_2Me$ | chiral formamides said to give asymmetric induction, exchange noted between $D_2$ and $BH_4^-$ | [9] |
| $RuCl(H)(CO)L_3$ | ~70 | (none) | aldehydes,<br>ketones | alcohols | 100°C, 94-98% yields | [10] |
| $RhH(DBP)_3$[c] | <1 | $C_6H_6$ | $RCH=CH_2$ | $RCH_2CH_3$ | selective terminal olefins | [11] |
| $PtH(SnCl_3)(PPh_3)_2$ | 4 | $MeOH/C_6H_6$ or $CH_2Cl_2$ | 2-cyclohexenone | cyclohexanone | olefin rearrangement, $D_2$, MeOH exchange, generally poor catalyst, ~90° | [12] |
| $[M(phen)(1,5\text{-hexadiene})]^+$<br>M=Ir>Rh | 1 | acetone | olefin<br>alkyne | hydrocarbon<br>alkene | 1,5-COD Rh complex is inactive | [13] |

*Table 6.1. Continued.*

| "Precatalysts" | $[H_2]$ atm. | Solvents | Substrates | Products | Special features[a] | Ref. |
|---|---|---|---|---|---|---|
| | 1 | MeOH, NaOH | ketone | alcohol | M=Rh | [14] |
| $[Ir^+(COD)L_2]ClO_4^-$ | 1 | acetone | 1,5-cyclo-octadiene | cyclooctene | selective for diene | [15] |
| $[Ir^+(COD)(PCy_3)(py)]PF_6^-$ | 1 | $CH_2Cl_2$ | $Me_2C{=}CMe_2$ | $Me_2CHCHMe_2$ | extremely fast at 0° (~4,000 turnovers per hour), poisoned by polar solvents or excess phosphine | [16] |

[a]All catalysts may be employed at 25° unless otherwise noted.

[b]L = $PPh_3$

[c]DBP =

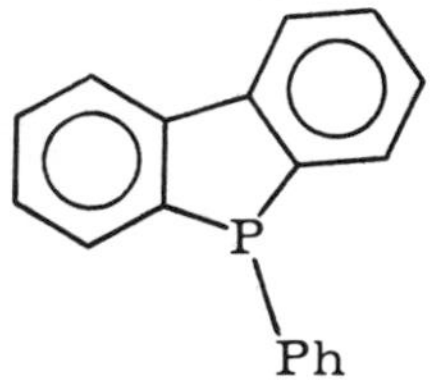

Usually neither the "precatalyst" nor the actual catalyst is recovered after the hydrogenation.

## 6.1. OVERVIEW OF MECHANISM.

The mechanisms of many homogeneously catalyzed hydrogenations have been intensively studied. However, in all but a few cases the complexity of such catalytic cycles has precluded a definitive analysis of the reaction mechanism. The uncatalyzed addition of hydrogen to olefins could theoretically occur by either a concerted process (path a in Equation 6.1) or a stepwise process (path b); however, both paths are

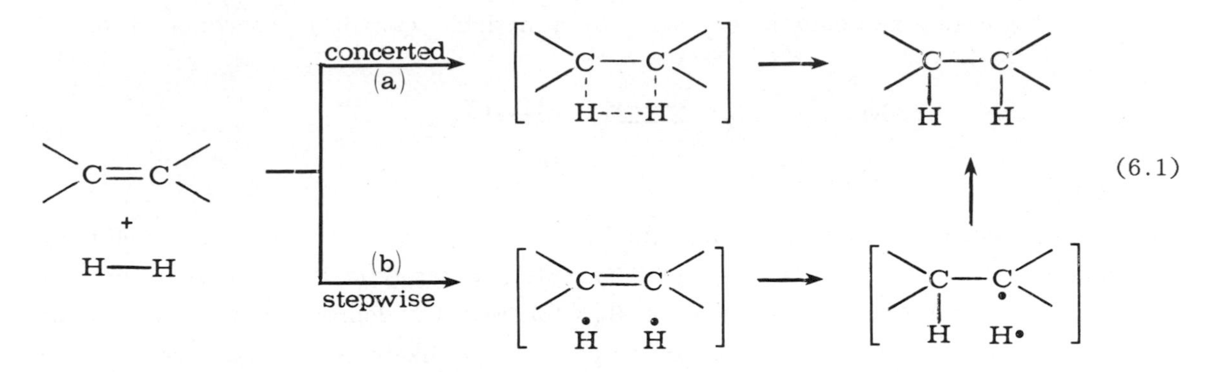


(6.1)

unfavorable. The concerted path is "forbidden" by the Woodward-Hoffman orbital-symmetry rules [18]; whereas the stepwise process would necessarily involve high-energy hydrogen atoms.

The accumulated evidence concerning the mechanism of both homogeneously and heterogeneously catalyzed hydrogenations indicates that intermediate metal hydrides are transferred to the substrate in discrete steps by insertion and reductive-elimination reactions. Thus the formation of metal hydrides from molecular hydrogen is an obligatory step in homogeneously catalyzed hydrogenations. The most common mode of activating molecular hydrogen is by oxidative-addition (Chapter 4). Both monometallic (Equation 6.2) and bimetallic (Equation 6.3) $H_2$ oxidative-additions are known. Both oxidative-additions seem to require the presence of at least one vacant coordination site ("site of unsaturation") on the metal.

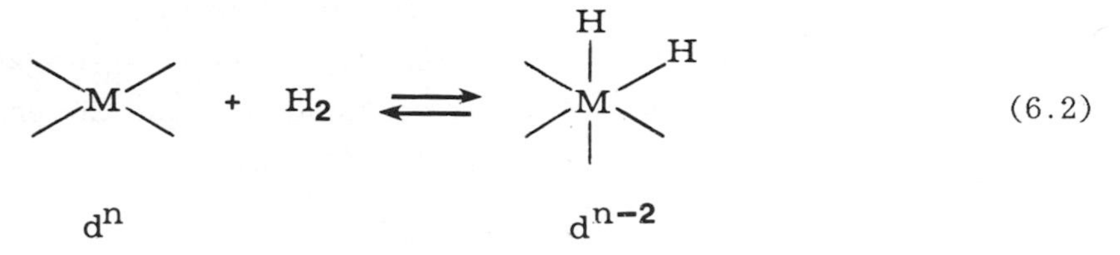


(6.2)

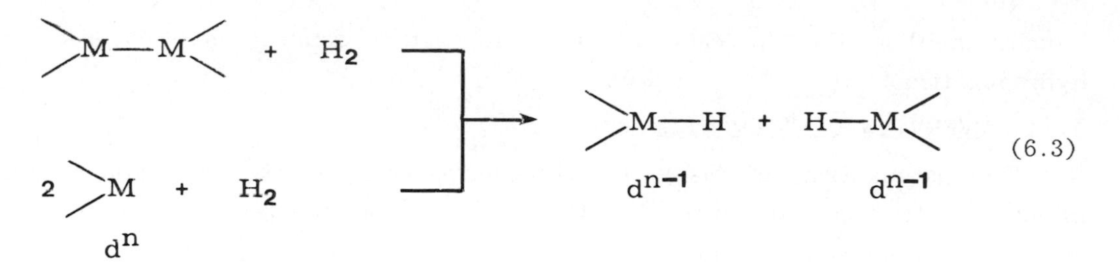


There is another mode of metal-hydride formation which does not require formal oxidation of the metal, the "heterolytic" activation of hydrogen [19]. In this process (6.4) the hydrogen molecule is cleaved into a hydride (which is coordinated to the

$$M + H_2 \rightleftharpoons M{-}H^- + H^+ \qquad (6.4)$$
$$d^n \qquad\qquad d^n$$

metal) and a proton (which is usually stabilized by a base). The reverse reaction is well-known as the protonolysis of a metal hydride. Heterolytic $H_2$ cleavage is difficult to distinguish from oxidative-addition of $H_2$, followed by deprotonation of a hydride with a base (6.5). Heterolytic $H_2$ activation is more likely with metal complexes in higher oxidation states where oxidative-addition is less feasible.

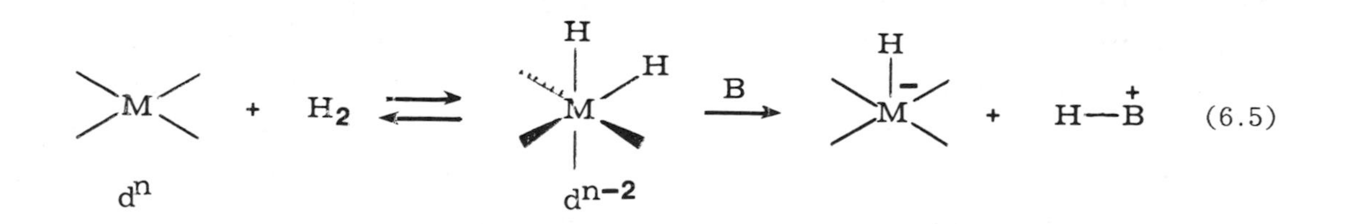


Note that heterolytic $H_2$ cleavage also requires the presence of a vacant coordination site on the metal. In some situations a homogeneous catalyst is formed from a "precatalyst" having a metal in a high oxidation state by a two-step process as shown in Equation 6.6: (a) heterolytic $H_2$ cleavage; and (b) deprotonation of the hydride affording the actual hydrogenation catalysts.

In other cases heterolytic $H_2$ splitting appears to be an integral step in the catalytic cycle. This seems to be best substantiated for the enzyme hydrogenase [20], and for the hydrogenase model system, Pd(II)(salen) [21].

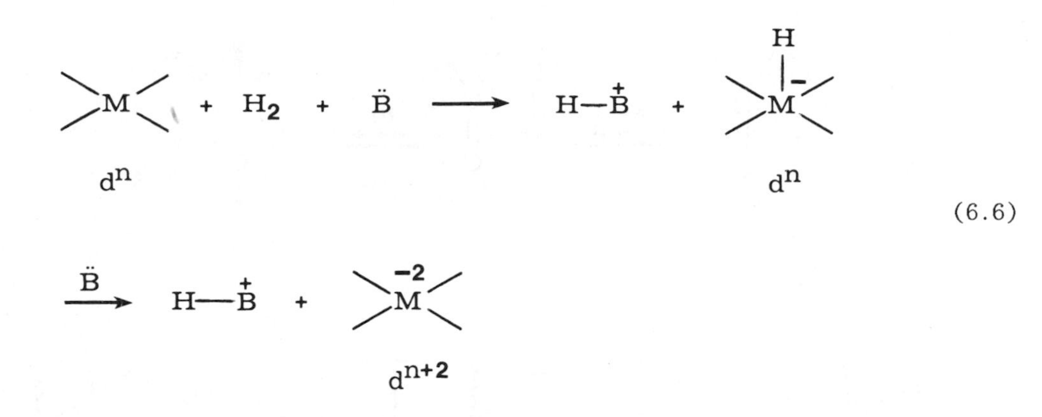


Other homogeneous catalysts derive their hydrogen from sources other than $H_2$. These "hydrogen transfer" catalysts form the requisite metal hydrides from alcohols, from nontransition-metal hydrides such as $NaBH_4$, or from hydrolysis of coordinated CO (see the water-gas shift reaction in Chapter 5). Since these catalysts are otherwise similar to homogeneous hydrogenation catalysts, both types are discussed in this chapter.

The majority of mononuclear, homogeneous olefin hydrogenation catalysts can be roughly divided into the following two classes:

(A) Homogeneous catalysts having a single M-H group present at some characterized stage of the catalytic cycle; such catalysts usually promote the isomerization of olefins and often exhibit selectivity toward the hydrogenation of conjugated olefins, and of terminal versus internal olefins.

(B) Homogeneous catalysts having two adjacent hydrides (cis-$MH_2$) present in one stage of the catalytic cycle; such complexes often catalyze the cis addition of hydrogen and under favorable circumstances olefin isomerization is not a serious side reaction.

1. Monohydride Catalysts. A proposed generalized mechanism for the monohydride catalysts is illustrated in Figure 6.1. The mechanisms of monohydride catalysts are much less studied than those of the dihydride catalysts; so certain gross features of the mechanism of the former have not been well-established. A major uncertainty concerns the hydrogenolysis of the metal alkyl intermediate to form the products. This may involve simple oxidative-addition of hydrogen, followed by alkyl-hydride reductive-elimination, or reaction with a second metal hydride affording a metal-metal bond (binuclear reductive-elimination). Either path is plausible, but so far neither has

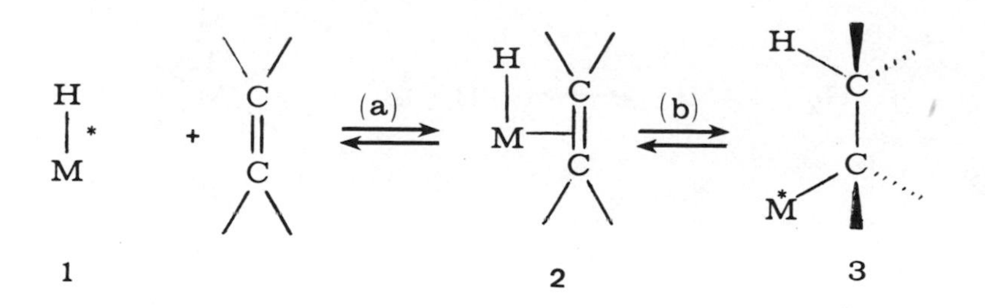


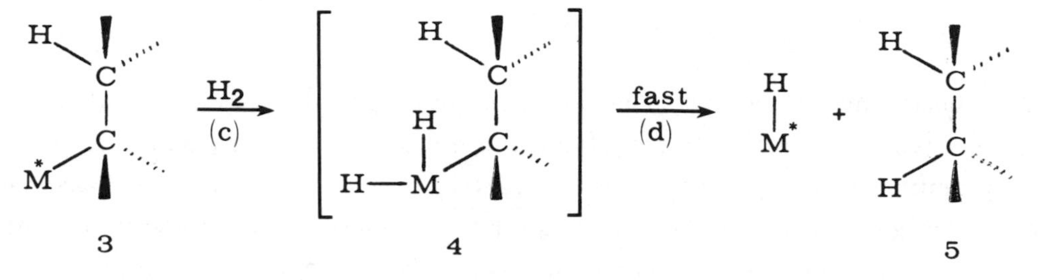


such intermediates have
not been detected

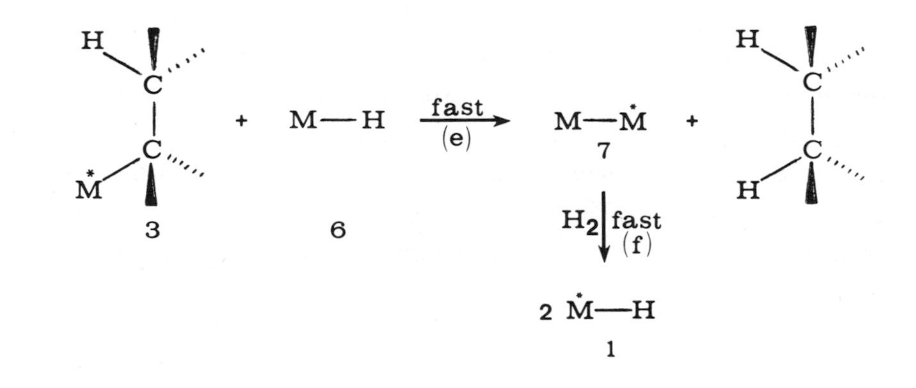


*Figure 6.1. The probable mechanism for "monohydride" hydrogenation catalysts. An asterisk indicates a site of unsaturation which is required for the indicated reaction to take place. Extraneous ligands may inhibit these steps. However, see the discussion of conjugated olefins. Accompanying ligands are not depicted.*

been well-established in any individual case. With simple, nonconjugated olefins, the first step (path a in Figure 6.1) requires complexation of the olefin to a coordinatively unsaturated metal hydride 1. Bulky ligands on the metal often limit useful substrates to the least-hindered olefins, such as terminal monosubstituted olefins. Free ligands should inhibit this step by competing for the coordination site on the hydride 1. Conjugated olefins react with certain coordinatively saturated metal hydrides by an unestablished mechanism; in some cases a radical path seems probable. The reversibility of the two steps (a and b) provides for the isomerization of simple olefins and the isotopic exchange between the initial hydride 1 and hydrogens on the olefin, reactions which are characteristic of the monohydride catalysts.

Product formation from the alkyl intermediate 3 appears to occur by paths c and d or e and f, although so far neither case seems to have been well-established for any monohydride catalyst.

The product-forming path c and d involves simple oxidative-addition of $H_2$ to the metal alkyl intermediate 3, and requires the presence of a vacant coordination site. Subsequent rapid reductive-elimination of the alkyl and one hydride ligand from the postulated intermediate 4 (step d) should afford the product and regenerate the coordinatively unsaturated active catalyst 1. Dihydride alkyl intermediates such as 4 have not been observed.

The product-forming path e avoids the difficult oxidative-addition step and also removes the need for an intramolecular reductive-elimination, which, as we have seen in Chapter 4, can in some cases be an unfavorable path. Thus path e involves a binuclear reductive-elimination which seems to require a coordinatively unsaturated alkyl 3 and a metal hydride 6, which could be either coordinatively saturated or coordinatively unsaturated. Product would be formed in this fast step, along with a metal-metal dimer 7 having at least one site of unsaturation. Hydrogenolysis of the singly unsaturated dimer 7 (step f) regenerates two moles of the active catalyst, the unsaturated monohydride 1. This last step appears to require a vacant coordination site. Path e could be inhibited by steric bulk at the metal sites, especially at the binuclear elimination stage. Isolation of metal centers by high dilution or by true immobilization on insoluble supports should inhibit path e. If either paths c and d or b and e are very fast, the over-all reduction could take place with cis stereochemistry and minimal olefin rearrangement and/or isotopic hydrogen exchange, since the back-reactions in steps a and b would then be less significant.

Let us next consider two examples of monohydride catalysts. Equation 6.7 shows the selective hydrogenation of terminal olefins catalyzed by $HRh(CO)(PPh_3)_3$, 8, under

$$RCH{=}CH_2 + H_2 \xrightarrow[25°,\ <1\ atm.\ H_2]{HRh(CO)(PPh_3)_3\ \ 8} RCH_2CH_3 \qquad (6.7)$$

mild conditions. The pronounced substrate selectivity exhibited by this catalyst is illustrated by its failure to promote the reduction of cyclohexene. Internal olefins are isomerized, but are not competitively reduced. Other groups such as RCHO, RCN, RCl, and $RCO_2R'$ are not reduced. Added $PPh_3$ inhibits the reduction of terminal olefins. From these results we can be certain that a site of unsaturation is required at least in one segment of the catalytic cycle. In Chapter 8 we shall see that this same complex catalyzes the "oxo" reaction. Oxo catalysts can usually also serve as monohydride hydrogenation catalysts. Consequently, olefin hydrogenation is a competing side reaction in the oxo process; and with certain conjugated olefin substrates, such competitive hydrogenation becomes very significant. Oxo catalysts also usually promote the hydrogenation of aldehydes to alcohols, but more strenuous $H_2$ pressure and temperature are required—seemingly because the aldehyde carbonyl group is a relatively ineffective substrate.

The second example of such monohydride catalysts is $HRuCl(PPh_3)_3$, 9, a rare case of a coordinatively unsaturated, five-coordinate, $d^6$ complex. Apparently the three bulky phosphine ligands inhibit coordination of other large ligands. This catalyst shows remarkable substrate selectivity by reducing terminal olefins (Equation 6.8) more rapidly than internal olefins by a factor of $10^3$. The rate of olefin reduction by this

$$RCH{=}CH_2 + H_2 \xrightarrow[9]{HRuCl(PPh_3)_3} RCH_2CH_3 \qquad (6.8)$$

ruthenium(II) catalyst is a linear function of $K_{Ag+}$ (the equilibrium binding constant between Ag+ and the olefin). It has thus been proposed that olefin coordination is the rate-determining step. Rapid exchange occurs between $D_2$ and protons on terminal olefins as well as protons at the _ortho_ positions of the triphenylphosphine ligands. The latter is another case of "orthometallation" or intramolecular oxidative-addition, which, as we have seen in Chapter 4, is a rather common reaction that complicates isotopic hydrogen-labeling experiments.

2. Dihydride Catalysts. The second major class of homogeneous hydrogenation catalysts, dihydride complexes, are much better studied, for both historical and practical reasons. A generalized mechanism for this class is illustrated in Figure 6.2. In

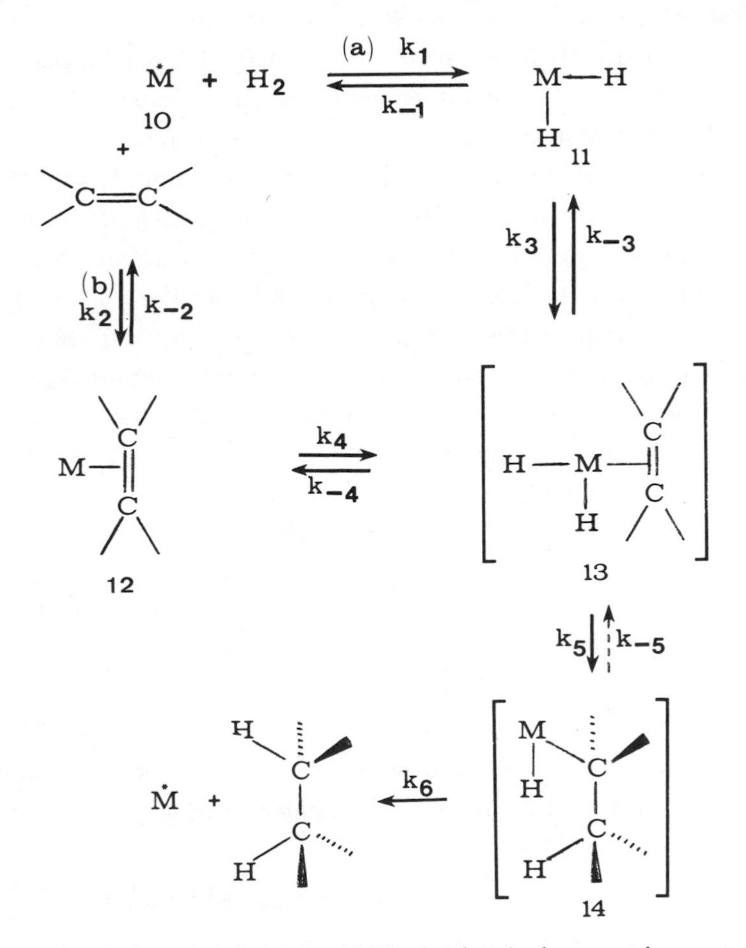


*Figure 6.2. Generalized mechanisms for "dihydride" hydrogenation catalysts. The accompanying ligands are not shown. In compound 10, at least one site of unsaturation must be present, although there are occasionally two sites. Complex 11 must be unsaturated for the subsequent olefin coordination step to occur. This unsaturation can result either from ligand dissociation or from addition of $H_2$ to a doubly unsaturated complex. Complex 14 would be unsaturated unless a ligand has been taken up from the solution, which is the usual case.*

path a, a singly or doubly unsaturated metal complex 10 oxidatively adds $H_2$ in step $k_1$, affording the cis dihydride 11, which can usually be independently characterized.

Coordination of an olefin on the same face of the coordination sphere as the two hydrides affords coordinatively saturated complex, 13. The transformation of 11 to 13 ($k_3$) requires a vacant coordination site, accounting for some of the inhibition of the catalytic cycle brought on by competing extraneous ligands.

In an alternative path (b in Figure 6.2) the olefin is coordinated first, forming an unsaturated olefin complex 12, which then adds $H_2$ ($k_4$), affording the dihydride olefin complex 13. We shall see that the relative importance of path a versus path b can be governed by *trans* effects in the accompanying ligands. For example, chelating phosphines seem to favor the olefin path b, whereas *bis*-monodentate phosphine ligands, which can assume mutually *trans* coordination sites, seem to favor the hydrogen addition path a. This occurs because *cis* phosphines decrease the value of the equilibrium constant, $k_1/k_{1}$. In other cases olefins, which have a very high affinity for the metal, may react by path b, because the equilibrium constant $k_2/k_{2}$ is increased.

The migratory insertion of 13 to 14 ($k_5$) goes by *cis* addition of the M-H bond to the olefin, and requires the olefinic bond axis, the metal, and the hydride to be coplanar. This step $k_5$ is usually assisted or consummated by the incorporation of an external ligand. Alkyl-hydride intermediates such as 14 have rarely been observed, because the following step, $k_6$, is usually very rapid, permitting no buildup of 14. The high stereoselectivity (over-all *cis* addition) and the absence of olefin rearrangements and hydrogen-isotope scrambling which are usually observed for "dihydride" catalysts can be accounted for by the relative rates: $k_6 >> k_{5}$.

Two examples of "dihydride" catalysts will now be discussed. First, we will examine the general features and mechanism of the famous Wilkinson catalysts $RhClL_3$ (L = a tertiary phosphine), and then the similar cationic catalysts of the type $[Rh(H)_2(L)_2(S)_2]^+$.

## 6.2. General Features of Wilkinson's Hydrogenation Catalyst.

In 1964 Wilkinson and Coffey independently and nearly simultaneously discovered that solutions of the rhodium(I) complex, $RhCl(PPh_3)_3$, 16, would catalyze the hydrogenation of olefins. Wilkinson fully explored the scope, selectivity, and characteristics of this complex, which is now commonly referred to as "Wilkinson's catalyst." He also carried out the first extensive study of the mechanism by which this complex catalyzes the hydrogenation of olefins. Since that time many other studies have contributed to our understanding of this remarkably selective and synthetically useful catalyst, which is routinely employed by organic chemists.

There are two convenient methods for preparing this catalyst. The first in Equation 6.9, is limited to triphenylphosphine and some other triaryl phosphines. The

$$\underset{15}{RhCl_3(H_2O)_x} + PPh_3 \xrightarrow[\Delta]{CH_3CH_2OH} \underset{16}{RhCl(PPh_3)_3} + Ph_3P{=}O \quad (6.9)$$

burgundy-red, mildly air-sensitive, crystalline catalyst 16 is formed by reduction of the ethanol-soluble rhodium(III) hydrate 15. Triphenylphosphine acts as the reducing agent, but the mechanism is uncertain. The chloride in 16 may be exchanged for bromide or iodide by metathesis with LiBr or LiI. Such facile ligand displacements are expected for planar $d^8$ complexes.

A more general and versatile synthesis of Wilkinson's catalyst is shown in Equation 6.10. First, $RhCl_3(H_2O)_x$ is reduced and converted into a dimeric rhodium(I)

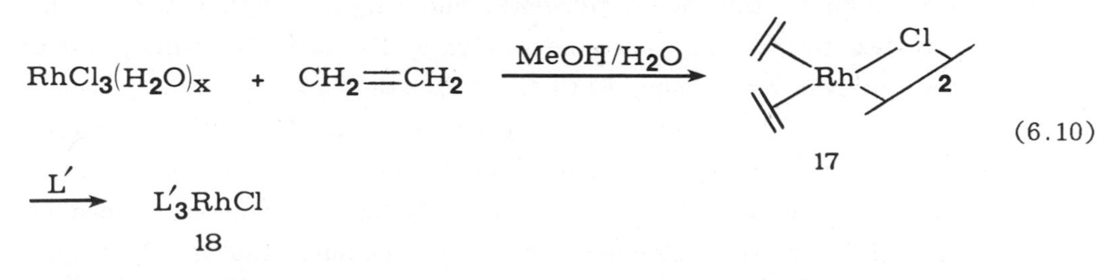


bis-olefin complex 17. Water is required for this reaction, in which one of the olefins is oxidized in a step similar to that in the Wacker process (see Chapter 12). Ethylene affords a sparingly soluble and thus easily isolated complex, 17, which is not stable for prolonged periods. In a second step, the rhodium(I) olefin complex is treated with virtually any tertiary phosphine, L', affording a variety of Wilkinson catalysts, 18. The L/Rh ratio can be varied by means of this method. Hydrogenation catalysts having varying degrees of reactivity can thus be generated *in situ*.

The general characteristics of the Wilkinson catalysts, 18 are as follows. Unconjugated olefins and acetylenes are hydrogenated at ambient temperature and low $H_2$ pressures (<760 Torr). Relative substrate reactivities for unconjugated olefins tend to parallel their tendencies to coordinate to rhodium, and thus reflect the steric crowding afforded by the bulky phosphine ligands. The over-all rate difference among the following olefins is about fifty-fold!

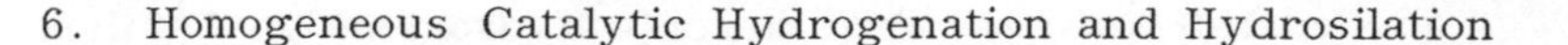

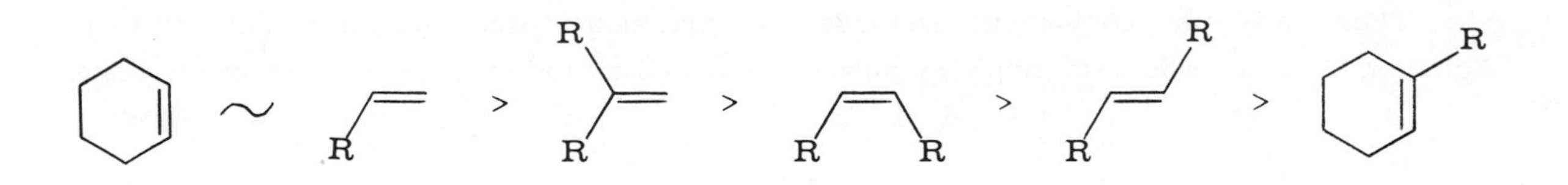

Certain olefins which are especially good ligands appear to inactivate the catalyst and to inhibit the reduction of the substrates listed above. Examples are ethylene and 1,3-butadiene. These substrates are reduced at higher temperatures or by using $RhI(PPh_3)_3$ in place of $RhCl(PPh_3)_3$. Examination of the reaction mechanism (vide infra) provides an explanation for this effect of the more strongly bonding substrates in terms of an alteration in mechanism. The selectivity of the Wilkinson catalysts is illustrated by the number of unsaturated groups which are not reduced. Examples include: arenes, ketones, carboxylic esters, carboxylic acids, amides, nitriles, azo compounds, and nitro compounds. The Wilkinson catalysts are poisoned by competing ligands, such as $Ph_3P$, thiophene, pyridine, and $CH_3CN$. Other compounds poison or inhibit this system by reacting with the catalyst. For example, $CHCl_3$ and $CCl_4$ slowly deactivate the catalyst by forming $RhCl_3L_2$, but some catalytic activity can be restored by passing a stream of $H_2$ through the solution--evolving HCl. This reactivation probably involves heterolytic $H_2$ cleavage. Aldehydes and to a lesser extent primary alcohols slowly but irreversibly react with the Wilkinson catalyst, affording the very stable rhodium(I) carbonyl complex, 19, the rhodium analog of Vaska's iridium compound. This stoichiometric reaction can be used preparatively to decarbonylate

$$RCHO + RhClL_3 \longrightarrow \underset{\mathbf{19}}{RhCl(CO)L_2} + L + RH \qquad (6.11)$$

aldehydes but at great expense (see 6.11). A similar catalytic reaction is known, but is less effective (see Chapter 8).

In benzene, polar cosolvents such as ethanol, acetone, and ethyl acetate cause an approximately two-fold increase in the rate of hydrogenation compared with the reaction in pure benzene. These effects are probably related to the rate-determining migratory-insertion step (vide infra).

A remarkable feature of the Wilkinson catalyst is the lack of isotopic scrambling between $D_2$ and protons on the solvent or of scrambling between $H_2$ and $D_2$. For example, reduction of 1-hexene with a mixture of $H_2$ and $D_2$ at 15 atm. gives 94%

reduction without $D_2/H_2$ scrambling as demonstrated by mass-spectral analysis of the products.

The stereochemistry of hydrogenation clearly affords cis addition of hydrogen provided that very pure catalyst and solvent are employed. This was first illustrated by the addition of deuterium to maleic acid to afford the meso-1,2-dideuterosuccinic acid, 20, as in Equation 6.12. Similarly fumaric acid yields d,ℓ-1,2-dideuterosuccinic

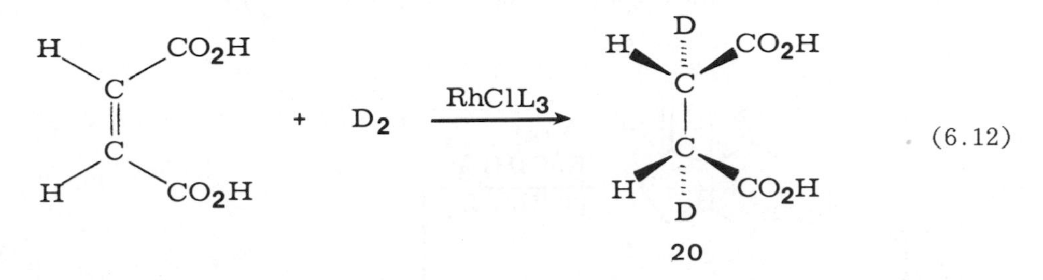


(6.12)

acid. This is a convenient reaction for examining stereochemistry, because the diastereoisomeric 1,2-dideuterosuccinic acids can be distinguished by their infrared spectra. However, these olefinic substrates lack α-hydrogens and, therefore, do not offer the opportunity for a competing olefin rearrangement pathway.

Both regio- and stereoselectivity of such hydrogenations were illustrated by Birch [1d] in the selective reduction of ergosterol acetate, 21, to 5α,6α-$D_2$-ergost-7-en-3-β-ol acetate, 22. This result illustrates a number of features. The least-hindered olefin is

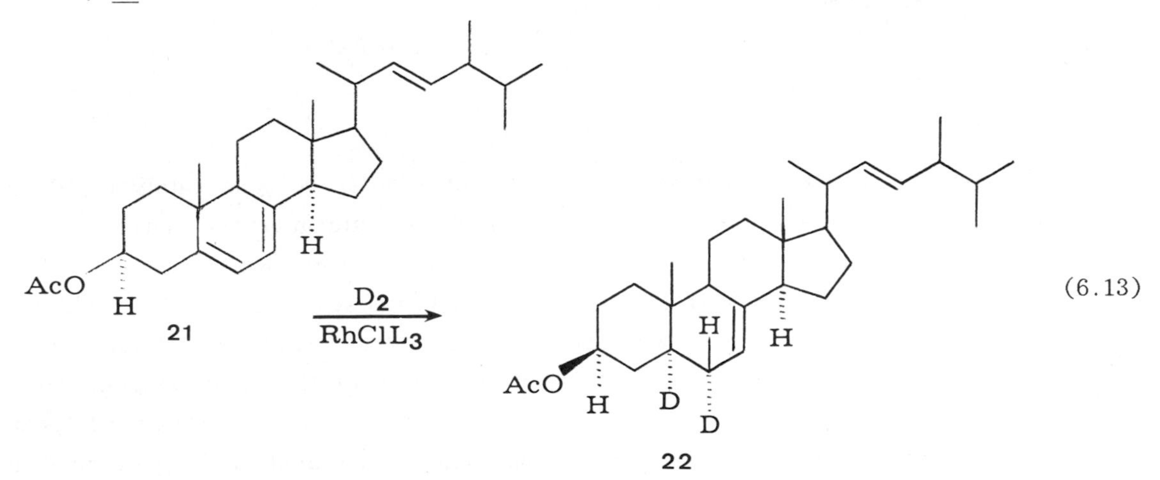


(6.13)

reduced from the less-crowded face of the steroid. Neither the remaining trisubstituted 7-ene nor the trans-disubstituted 21-ene are affected under these conditions. The conjugated diene is expected to be a superior ligand compared with these simple olefins. Furthermore, isotopic hydrogen-deuterium exchange has not taken place.

The selectivity of the Wilkinson catalyst toward less-hindered olefins [1d] is illustrated by the reduction of the two steroid dienes shown in Equation 6.14. Note that the ruthenium catalyst shows a similar selectivity (path a).

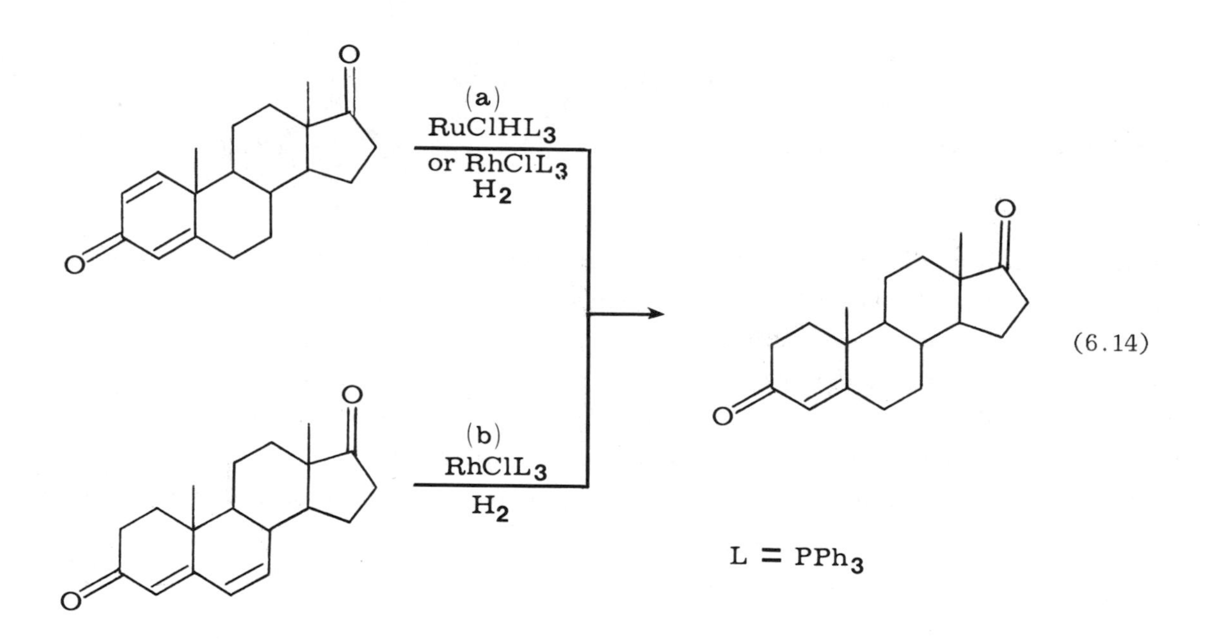


(6.14)

The comparison between Wilkinson's catalyst and a heterogeneous catalyst [1d] is illustrated by the selective reduction of the less-hindered, unconjugated olefin with the former and by reduction of the more-hindered, conjugated olefin with the latter (6.15).

In other examples, Birch [1d] has shown the delicate selectivity of the Wilkinson hydrogenation catalyst in contrast with most heterogeneous catalysts. Reduction of 1,4-dihydrobenzenes, such as 23, proceeds as in 6.16 without the disproportionation to cyclohexane and benzene derivatives, which characterizes most heterogeneous catalysts. A direct comparison between the Wilkinson homogeneous catalyst and heterogeneous platinum or palladium catalysts in the reduction of dihydrobenzene derivatives is illustrated in Equation 6.17. A further example involving a very sensitive spiroether is

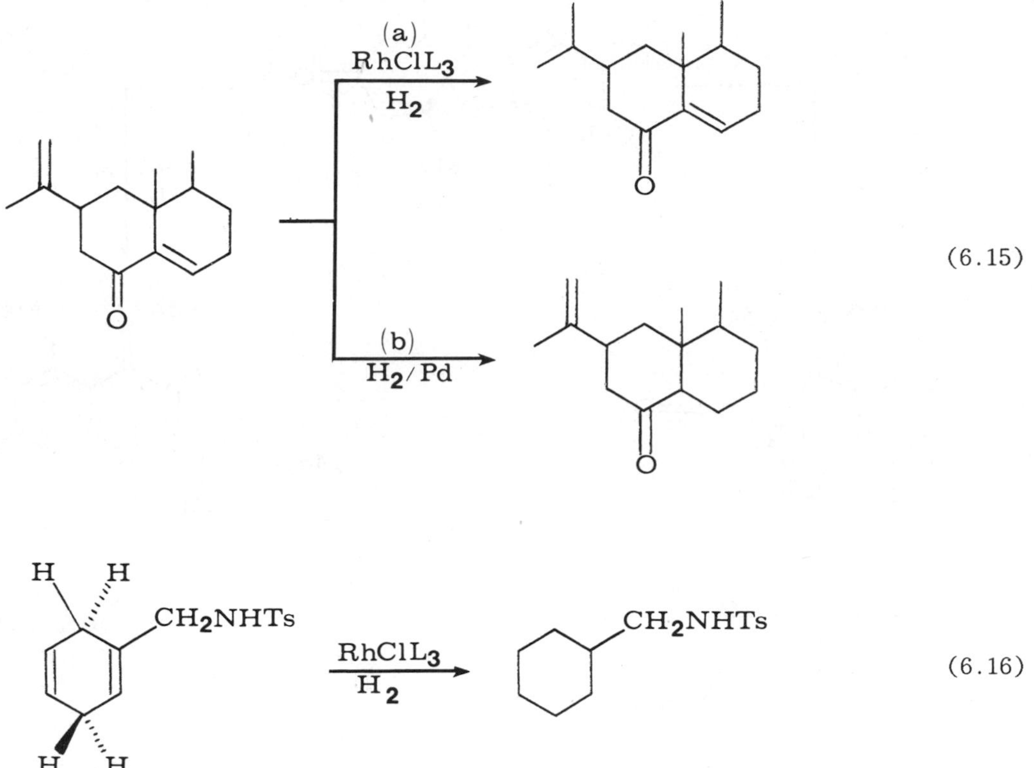


(6.15)

H H
$CH_2NHTs$
$RhClL_3$
$H_2$
$CH_2NHTs$

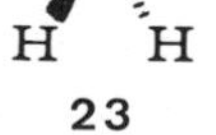


(6.16)

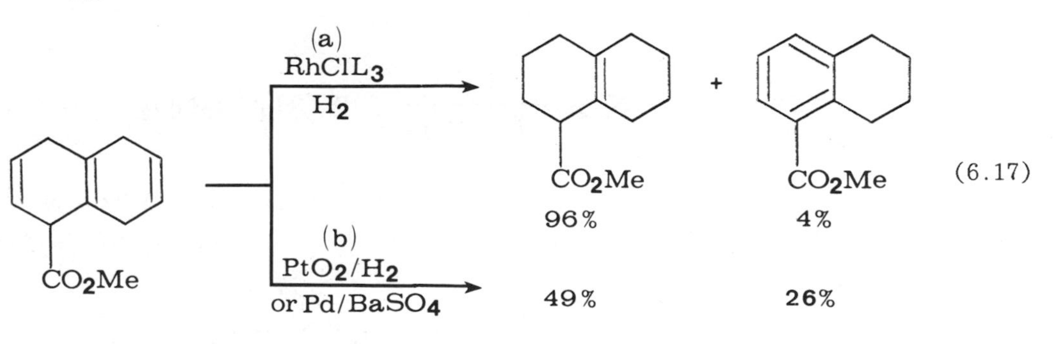


(6.17)

shown in Equation 6.18. The Wilkinson catalyst reduces only one olefin, whereas heterogeneous catalysts degrade the ether and dehydrogenate the product affording the diarylketone [1d]. Reduction of a nitro olefin, 24, proceeds as in 6.19 without affecting the nitro group.

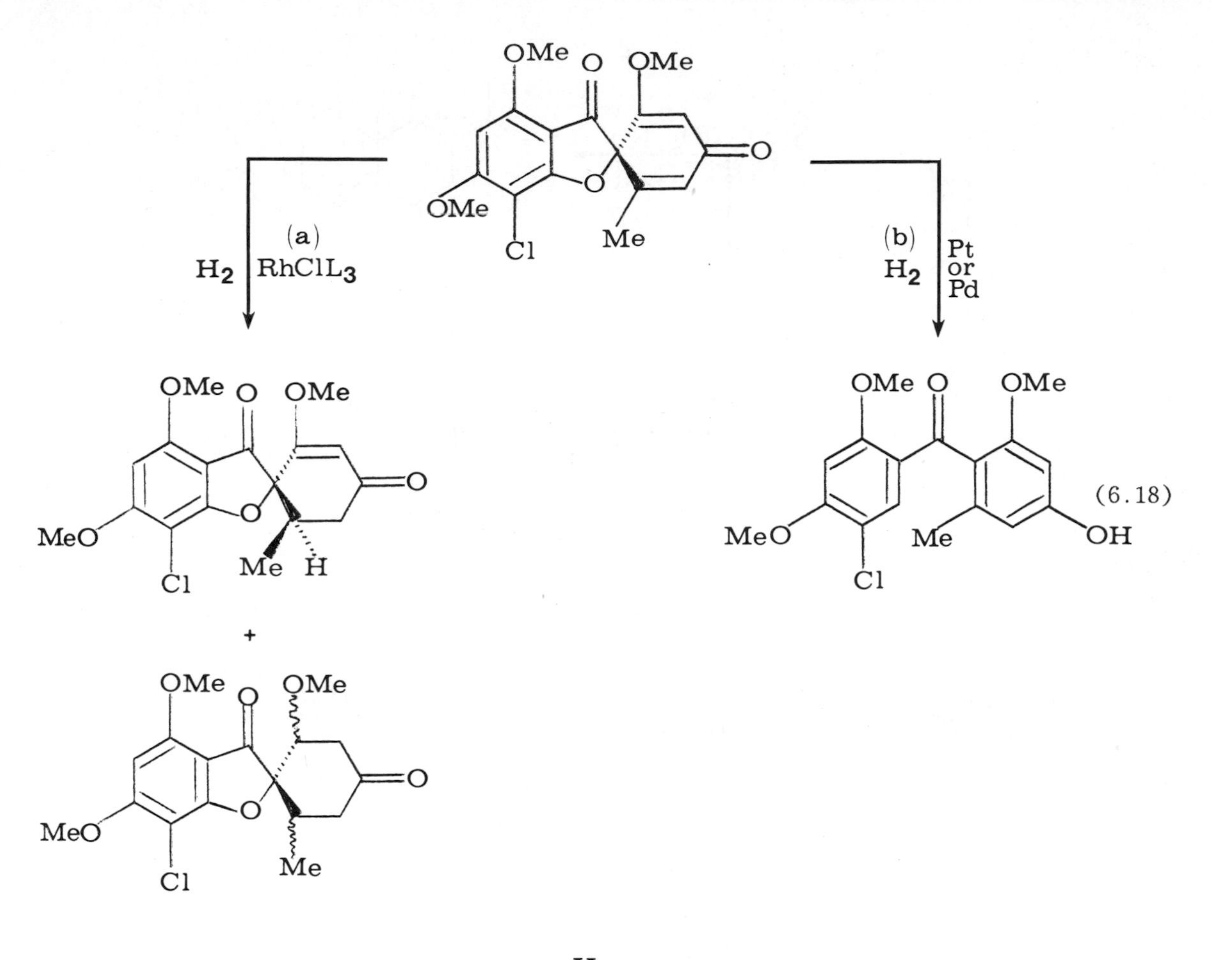


$$\underset{\mathbf{24}}{PhCH{=}CHNO_2} \xrightarrow[RhClL_3]{H_2} PhCH_2CH_2NO_2 \qquad (6.19)$$

Not all substrates exhibit such selectivity in reactions with Wilkinson's catalyst. For example, the α-methylene-lactone, damsin, 25, rearranges to isodamsin, 26,

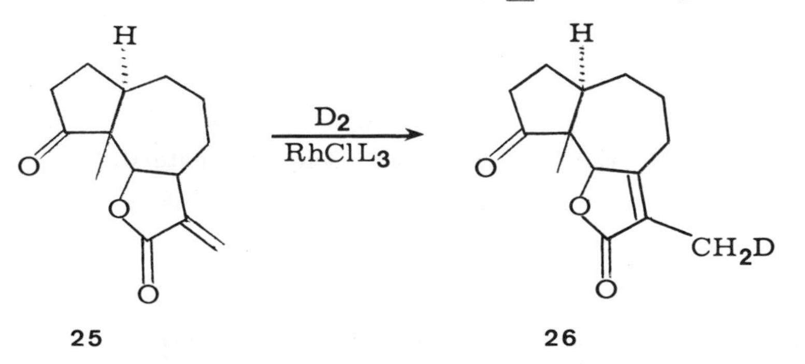

exchanging one deuterium with $D_2$. Apparently the usually rapid sequential addition of the second hydrogen is intercepted in this case.

The Wilkinson catalyst can be internally delivered, giving rise to remarkably stereoselective hydrogenations [22]. For example, the sodium alkoxide 27 reacts to form the cis fused decalin 29, presumably via an oxide-coordinated intermediate 28.

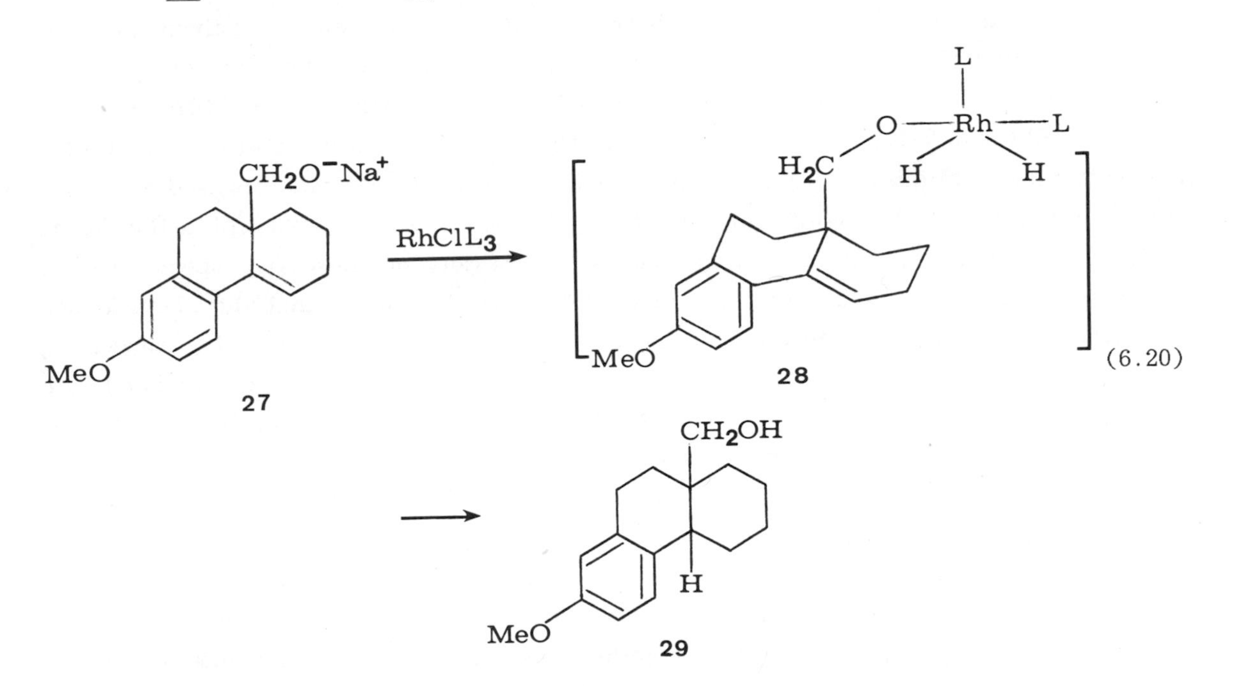


## 6.3. Mechanisms of Hydrogenation by Rhodium(I) Dihydride Catalysts.

Since catalysis is a kinetic phenomenon, kinetic measurements are essential to the elucidation of catalytic mechanisms. However, the number and complexity of reactions associated with a catalytic cycle are often sufficiently great that an over-all kinetic analysis is a nearly useless diagnosis of the mechanism. The number of variables in a multistep process is simply too large to distinguish one path from another. A more meaningful evaluation of the mechanism can be achieved by examining individual stages of the over-all reaction cycle. It is occasionally possible to measure kinetic and equilibrium parameters for each step in the catalytic cycle. When these parameters can be combined to calculate the experimentally measured rate of the catalytic reaction, the over-all mechanism can be considered to be "understood". Halpern [17,23] has successfully carried out just such analyses for the Wilkinson catalyst, $RhClL_3$, and for the cationic catalyst $[Rh(diphos)(S)_2]^+$ (where S is a solvent molecule) [24]. Because of their heuristic value, we will examine the details of both mechanisms.

1. Wilkinson's Catalyst, $RhClL_3$. The over-all catalytic cycle and side reactions for Wilkinson's catalyst are outlined in Figure 6.3. This mechanism results from Wilkinson's earlier studies [25] and Halpern's recent stepwise kinetic analysis [17,23]. Within this catalytic system, five rhodium complexes have been directly observed and characterized [17,26], either by isolating solids or by solution $^{31}P$ NMR studies: $RhClL_3$, $RhClL_2(C{=}C)$, $Rh_2Cl_2L_4$, $Rh(H)_2ClL_3$, and $Rh_2(H)_2Cl_2L_4$. However, Halpern's kinetic studies [17,23,27] reveal that none of these complexes is directly involved in the kinetically significant catalytic cycle (which is contained within the dotted circle in Figure 6.3). In fact, the accumulation of any of these five species should reduce the over-all reaction rate [17]. The "take-home lesson" is that the identification of a dominant or even a detectable species in a catalytic system may lead to misleading interpretations of the reaction mechanism. Only when such studies are associated with kinetic measurements defining the role of these characterized complexes as well as "invisible" complexes can the actual mechanism be defined.

Consider the cycle of complexes which Halpern found in the $RhClL_3$-catalyzed hydrogenation of cyclohexene (Figure 6.3). The mechanism has been elucidated by dividing the over-all reaction into two parts: (a) hydrogenation of $RhClL_3$; and (b) reaction of $RhH_2ClL_3$ with cyclohexene. The combination of parts (a) and (b) corresponds to a "hydride path." Taken together, the kinetic and equilibrium parameters measured for parts (a) and (b) allow one to calculate with good agreement the experimentally established over-all rate of cyclohexene hydrogenations.

Part a. Studies of $RhClL_3$ hydrogenation were carried out by monitoring the UV absorbance at 390 nm. A conventional UV instrument was used in the presence of excess ligand L, which suppresses the rate; but in the absence of added L the reaction is so rapid that it was necessary to use a stopped-flow spectrophotometer. Halpern's kinetic analysis is best accommodated by a path involving dissociation of a phosphine from 30 (Figure 6.3) affording the 14-electron intermediate 31 (reaction 1), followed by very fast hydrogenation of 31, forming the unsaturated dihydride 33, which is in equilibrium with free triphenylphosphine, giving 32. This circuitous route from 30 to 32 is much faster than the direct hydrogenation of 30 (reaction 2), which was independently measured in the presence of excess triphenylphosphine. In fact $RhClL_2$, 31, reacts with $H_2$ at least $10^4$ times faster than $RhClL_3$, 30! Wilkinson originally proposed such a

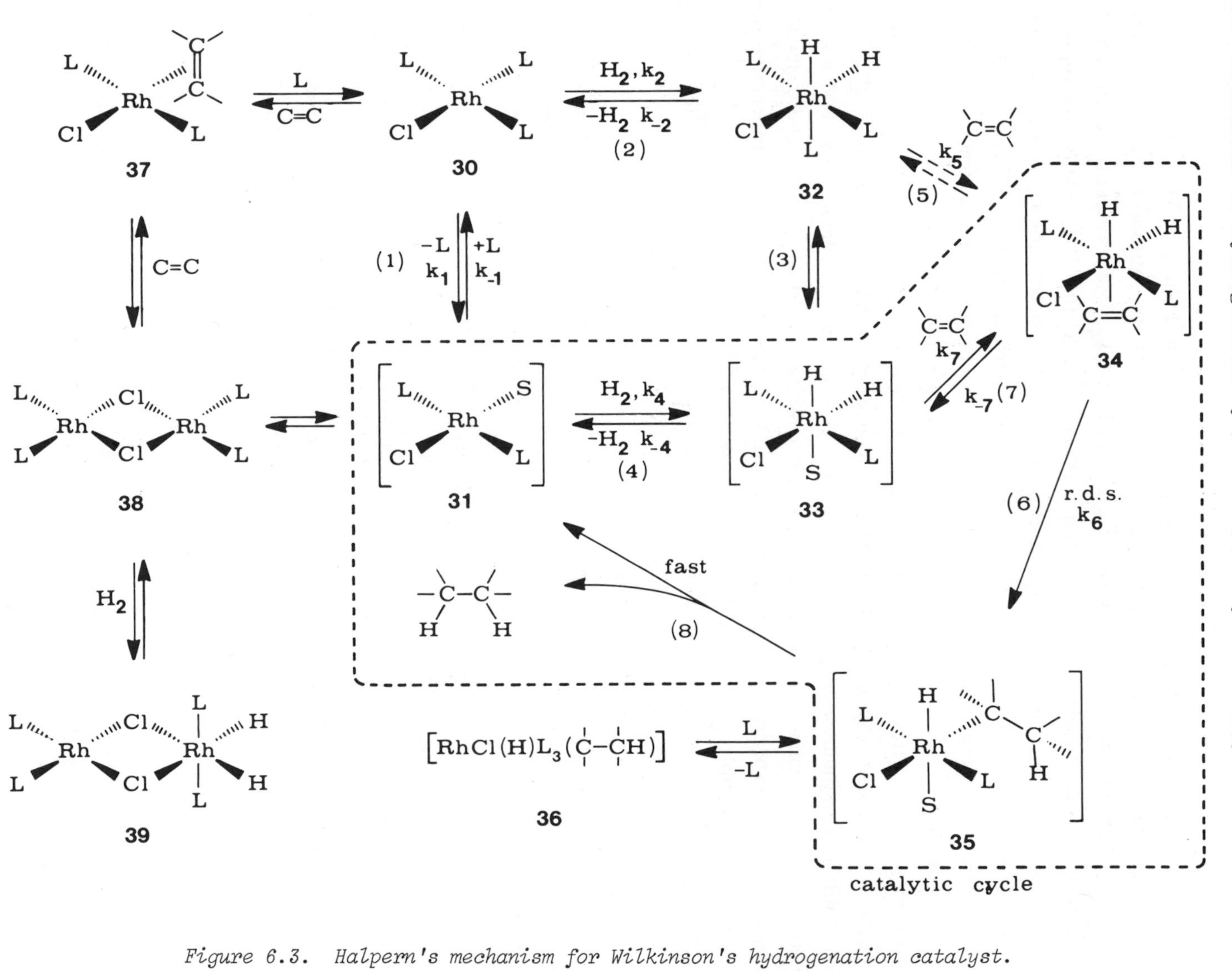


*Figure 6.3. Halpern's mechanism for Wilkinson's hydrogenation catalyst.*

path for hydrogenating $RhClL_3$ (reaction 1,4, and 3 in Figure 6.3). However, his proposal was based on the erroneous presumption that the dissociative equilibrium, $K = k_1/k_{-1}$, lies far toward the 14-electron complex, 31. More recent measurements have shown this equilibrium to lie to the left ($K_1 < 1 \times 10^{-5}$ M). This discrepancy derives from the great air sensitivity of $RhClL_3$, which reacts with $O_2$, affording dissociated $Ph_3P{=}O$ and giving values of low apparent molecular weight. This is a serious problem with many organometallic compounds; molecular weight measurements on air-sensitive complexes are treacherous and often unreliable. Actually the "invisible" 14-electron complex, $RhClL_2$, has a pronounced tendency to dimerize (31 $\rightleftarrows$ 38, Keq ~ $10^6$ $M^{-1}$). The dimer (39 in Figure 6.3) has been characterized by Jesson *et al*. [28], using NMR techniques. The dimer, 38, adds $H_2$ on one rhodium atom forming 39 (Figure 6.3) at a rate similar to the parent precatalyst $RhClL_3$. However, in the presence of $H_2$ these reactions are not significant, since $H_2$ effectively intercepts the three-coordinate complex 31 before it can dimerize.

In the absence of $H_2$, coordinatively unsaturated olefin complexes such as 37 (Figure 6.3) have been characterized; however, these are not kinetically significant in the catalytic hydrogenation of cyclohexene and similar substrates.

Part b. A separate kinetic study of the reaction between the dihydride 32 (Figure 6.3) and a large excess of cyclohexene (pseudo-first-order conditions) afforded the rate law shown in Equation 6.21. A steady-state treatment of the transformation 32 to

$$-d[RhH_2ClL_3]/dt = k_{obs}[RhH_2ClL_3] \qquad (6.21)$$

34 ($K_5$) and to 35 ($k_6$) predicts the dependence of $k_{obs}$ shown in Equation 6.22.

$$k_{obs} = \frac{K_5k_6[C{=}C]}{L + K_5[C{=}C]} \qquad (6.22)$$

Values of $k_{obs}$ were measured over a wide range of olefin and phosphine concentrations. The reciprocal of Equation 6.22 (Equation 6.23) predicts that plots of $1/k_{obs}$ vs. $L/[C{=}C]$ should be linear, having a slope of $1/K_5k_6$ and an intercept of $1/k_6$. Such linear plots

$$\frac{1}{k_{obs}} = \frac{1}{K_5k_6}\frac{L}{[C{=}C]} + \frac{1}{k_6} \qquad (6.23)$$

were obtained, and from these, values of $K_5$ and $k_6$ were calculated. The migratory-insertion step $k_6$ turns out to be the slowest step in stage b, as well as in the over-all catalytic hydrogenation cycle. The subsequent product-determining step (reaction 8 in

Figure 6.3) is apparently so rapid that the reverse of reaction 6 is kinetically insignificant. (Recall that this is why isotopic exchange does not occur between $D_2$ and the olefinic substrate.)

A very interesting aspect of this mechanism is that the rate-determining step (reaction 6) shows no dependence on free $Ph_3P$, although a subsequent rapid formation of a saturated unstable alkyl-hydride complex 36 cannot be ruled out. One final point: the fact that step $k_6$ involving Rh-H bond breaking is rate-determining suggests that the reaction between 32 and cyclohexene should show a kinetic isotope effect. Comparison between reaction 32 and that of the analogous dideuteride, $Rh(D)_2ClL_3$, does show a small isotope D effect: $k_{obs}(H)/k_{obs}(D) = 1.15$.

In all these structures, the sites of unsaturation (indicated by (S) in Figure 6.3) could be occupied by a solvent molecule. The complexes shown in brackets in Figure 6.3 are "invisible" in the sense that these have not been directly observed, but their existence and stoichiometry were deduced from kinetic and equilibrium measurements. The stereochemistry suggested for the postulated alkyl hydride intermediate 35 is based on the trans effect, leaving a vacant coordination site trans to the hydride ligand. Note that Tolman et al. [29], used NMR to show that the phosphine trans to hydride in 32 is substitution-labile and to assign the stereochemistry of 32.

The combined rate laws from parts a and b successfully account for the observed catalytic cyclohexene hydrogenation rate. Since $k_6$ is the slowest step in the catalytic cycle, at constant $H_2$ pressure and high olefin concentrations (~1M) with no added phosphine, the rate reaches a limiting value determined by this slow step. The over-all rate law shows that little advantage can be gained by modifying the catalyst so that it is in the form of $RhClL_2$, 31, for example, by immobilizing and thus site-isolating this 14-electron complex on a rigid support such as silica (which could prevent its dimerization).

In spite of the fact that this mechanism seems to be well-understood, slight modifications can change the dominant mechanism. For example, styrene (which is a better ligand than cyclohexene) shows a rate behavior which requires a second parallel path involving the coordination of two styrene molecules. Wilkinson's earlier studies [25] indicate that other strongly binding substrates such as ethylene or 1,3-butadiene show different kinetic behavior--a path involving the olefin complex 37 (Figure 6.3) may be important. Furthermore, substantial changes in the phosphine (for example, using alkyl-substituted phosphines) strongly affect dissociative equilibria and may thus alter the hydrogenation mechanism. Thus we are dealing with a multistep reaction of such complexity that modification of the catalyst or substrate may alter the rate in an unpredictable way. Nevertheless, the mechanism shown in Figure 6.3 contains no real surprises, but is a stepwise combination of the kinds of reactions encountered in Chapters 4 and 5 and the sort of organometallic compounds discussed in Chapter 3.

2. The Cationic Catalyst, $[Rh(diphos)(S)_2]^+$. There is a family of cationic, dihydride rhodium catalysts of the type $[RhL_2(S)_2]^+$ in which S is a polar solvent ligand, such as THF or MeCN, and $L_2$ represents two tertiary phosphine ligands. These complexes merit special attention, since all the catalysts which afford high enantioselectivity in the asymmetric hydrogenation of amidocinnamic acids are of this type (vide infra). Such catalysts may be generated in situ by hydrogenation of the readily accessible cationic diene complexes, $[RhL_2(diene)]^+$. This is a convenient method for generating sites of unsaturation; however, the dominant products of this reaction depend on the nature of the phosphine ligands. The complex having two monodentate phosphines, 40, absorbs three moles of $H_2$, affording the dihydride 41 [30,24], whereas those having a chelating, bidentate phosphine 42 absorb two moles of $H_2$, yielding the highly unsaturated complex 43 [31,32] (Equations 6.24 and 6.25).

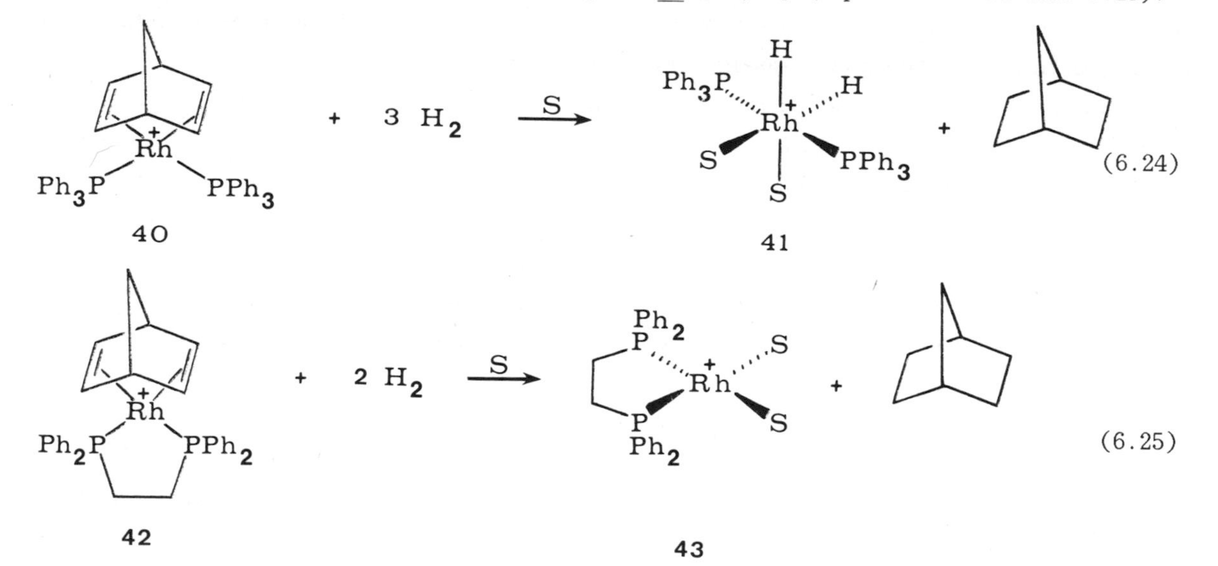


The difference between these two reactions can be explained by the trans interactions between phosphine and hydride groups. Phosphines avoid becoming trans to hydrides whenever possible. Since oxidative-addition of hydrogen is cis, and the relative disposition of chelating phosphines is also cis, the diminished tendency of complexes such as 43 to add hydrogen can be rationalized. Therefore the relative

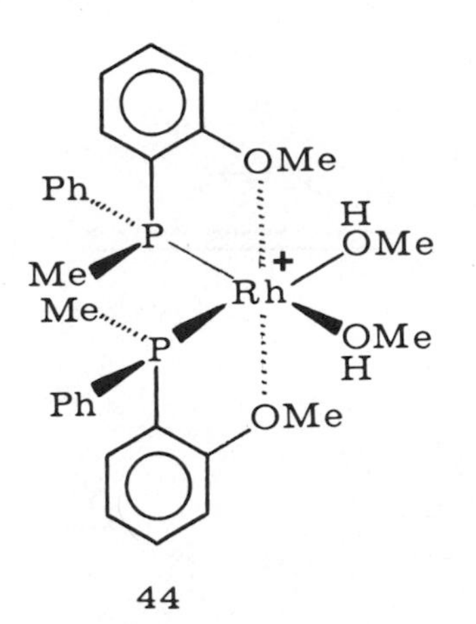


44

orientation of the phosphine influences the equilibrium for oxidation-addition of hydrogen. This thermodynamic effect changes the dominate kinetic path; complexes having two monodentate phosphines use a "hydride" hydrogenation path, whereas those having a chelating phosphine seem to react by an "olefin" path. This generalization is not always followed. For example, the complex (44) derived from (o-methoxyphenyl) methylphenylphosphine has the two monodentate phosphines cis according to $^{31}P$ NMR spectra. The o-methoxy groups are thought to be weakly coordinated, thus stabilizing this unexpected coordination geometry [32].

The mechanism of hydrogenations involving the diphos catalyst 43 has been elucidated in considerable detail by Halpern [24]. This very instructive mechanism (outlined in Figure 6.4) provides an interesting contrast to that described above for Wilkinson's catalyst.

The dominant mechanism for the diphos complex 43 involves an olefin rather than a hydride path. An olefin complex 45 is in equilibrium with the catalyst 43. Halpern has measured equilibrium constants $K_{eq}$ for a series of olefin substrates. At ambient temperatures the rate-determining step, $k_1$, involves oxidative-addition of $H_2$ to the unsaturated rhodium(I) complex, 45, giving 46 which rapidly forms 47 by migratory insertion and subsequently hydrocarbon product by reductive-elimination.

Three complexes in the catalytic cycle (43, 45, and 47) were thoroughly characterized by multinuclear NMR measurements, and two by x-ray diffraction (43 and 45). At low temperatures (<-40°C), $k_2$, a dissociative reaction, becomes slower than $k_1$, an intermolecular reaction, so that under these condition, 47 accumulates and can

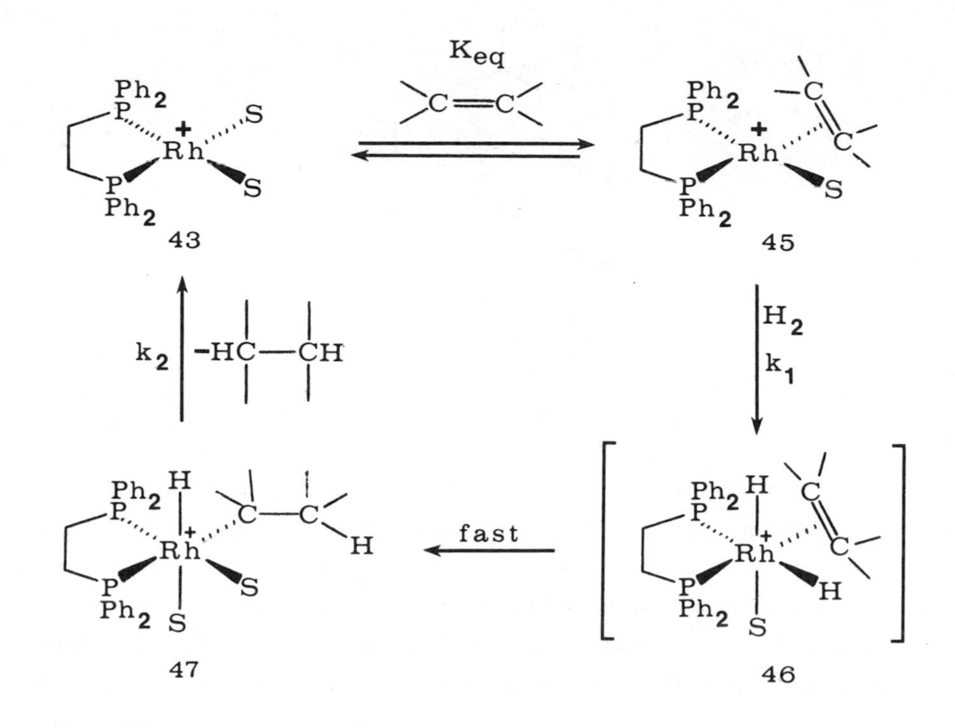


$$\text{Rate } (25^\circ) = \frac{k_1 K_{eq} [\text{Rh}]_{tot} [\text{C=C}][\text{H}_2]}{1 + K_{eq}[\text{C=C}]} \quad ;$$

$$\frac{[\text{Rh}]_{tot}[\text{H}_2]}{[\text{Rate}]} = \frac{1}{k_{obs}} = \frac{1}{k_1 K_{eq}[\text{C=C}]} + \frac{1}{k_1} \quad .$$

Below -40°C, $k_1 > k_2$ and 47 can be observed.

*Figure 6.4. Mechanism of olefin hydrogenation catalyzed by* $[Rh(diphos)_2(S)_2]^+$.

be characterized by NMR ($^1$H, $^{31}$P, $^{13}$C, $^{15}$N for S = MeCN[15]). Although such intermediates have often been proposed, 47 is the first hydride alkyl complex to be directly observed in a catalytic hydrogenation. The labile intermediate hydride olefin complex 46 rapidly forms 47 over a wide temperature range and therefore cannot be detected.

The affinity of 43 for olefins, manifested by $K_{eq}$, strongly depends on the structure of the olefin [30] as shown in Equation 6.26. Even aromatic rings coordinate

$$[Rh(diphos)]^+ + unsat. \underset{}{\overset{K_{eq}}{\rightleftharpoons}} [Rh(diphos)(unsat)]^+ \qquad (6.26)$$

| unsat : | 1-hexene | benzene | styrene | toluene | α-acetamidocinnamic acid |
|---|---|---|---|---|---|
| $K_{eq}$ (25°C) | 2 | 18 | 20 | 97 | $8 \times 10^3$ |

with the highly unsaturated complex 43. Thus in this system benzene, toluene, and especially xylenes are competitive inhibitors of olefin hydrogenation. The $^1H$ NMR of the benzene complex has been determined, indicating $\eta^6$-arene coordination like that revealed in the x-ray structure of the dimer, 48.

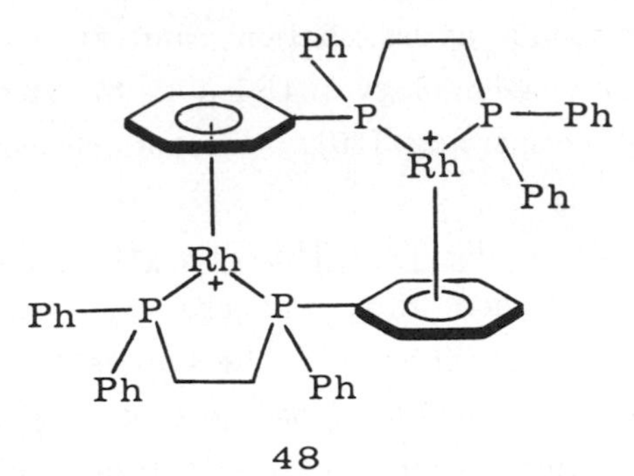


48

The amidocinnamic acid, PhCH=C(NHAc)(COOH), forms a very strong complex with this rhodium(I) center. As we shall see in the section on asymmetric hydrogenation, an x-ray structure of this complex, determined by Halpern [30], accounts for its large association constant ($K_{eq} = 8 \times 10^3\ M^{-1}$) and provides some insight into the induction of chirality.

The rate law derived from a steady-state treatment of this mechanism (Figure 6.4) was confirmed by linear plots $(k_{obs})^{-1}$ vs. $(C{=}C)^{-1}$. Since $K_{eq}$ can be independently measured, $k_1$ values are overdetermined.

The NMR studies of 47 (where S = MeCN) reveal the stereochemistry shown in Figure 6.4 [24]. In passing from 46 to 47, the hydride migrating to the olefin is trans to a phosphine. The trans destabilization of this hydride probably accounts for the speed of this transformation, which, under limiting conditions, is in the range of an enzyme (a turnover number of $\geq 100\ sec^{-1}$).

## 6.4. Asymmetric Homogeneous Catalytic Hydrogenation.

One of the most significant developments in the recent history of organotransition-metal chemistry is the discovery of soluble complexes which catalyze the asymmetric addition of hydrogen to unsaturated prochiral substrates. These asymmetric syntheses are achieved by homogeneous catalysts bearing chiral ligands. Such chiral catalysts form diastereomeric complexes with prochiral substrate ligands such as olefins. In certain circumstances one diastereomeric intermediate is favored; stereoselective metal-to-substrate hydrogen transfer by migratory insertion and reductive-elimination then results in the formation of a chiral product. In favorable cases, such as the amidocinnamic acids, nearly quantitative asymmetric synthesis is realized, thus mimicking the stereoselectivity of an enzyme! Asymmetric homogeneous catalytic hydrogenation has been extensively reviewed [3]. By contrast, heterogeneous asymmetric hydrogenation is much less effective and has been less well studied [33].

The discovery of Wilkinson's hydrogenation catalyst and the nearly simultaneous development of chiral phosphine technology [34] led to the first asymmetric homogeneous hydrogenations of prochiral compounds [35]. Intense research in this area has continued to the present.

1. Asymmetric Catalysts of the Type $[Rh(PR_3)_2{*}(S)_2]^+$. Most of the asymmetric catalysts which have been studied thus far are rhodium complexes derived from monodentate and bidentate chiral phosphines. The highest enantioselectivity reported to date (>95% ee) has been achieved from cationic precatalysts of the type $[Rh(PR_3){*}_2(diene)]^+$, where diene = norbornadiene (nor) or 1,5-cyclooctadiene (cod), and $(PR_3){*}_2$ is a chiral chelating diphosphine. Per cent enantiomeric excess (ee) is defined as [R]-[S]/[R]+[S] x 100 where R is the major isomer. This is to be distinguished from per cent optical purity = $\alpha$ sample/$\alpha$ reference x 100, where $\alpha$ is the specific rotation at a given wavelength.

In certain instances chiral monodentate phosphines will give nearly as high enantioselectivity (>90% ee), but the more rigid chelating ligands appear to be superior, and, as we have seen, the chelated complex appears to use a different mechanism. These high optical yields have so far been restricted to hydrogenation of prochiral $\alpha$-acetylaminocinnamic acid derivatives. Thus the difference in enantioselectivity derived from catalysts with chelating and those with monodentate chiral phosphines probably arises from the different stereochemistries of the olefin complexes 49 and 50, which

are intermediates in the two catalytic cycles. The stabilization of one diastereomeric olefin complex also depends on an additional weakly coordinating group which is usually present on the olefin.

From this research Knowles [3g,35a] (Monsanto) developed a significant process (Equation 6.27) for the manufacture of L-dopa, used in the treatment of Parkinson's disease.

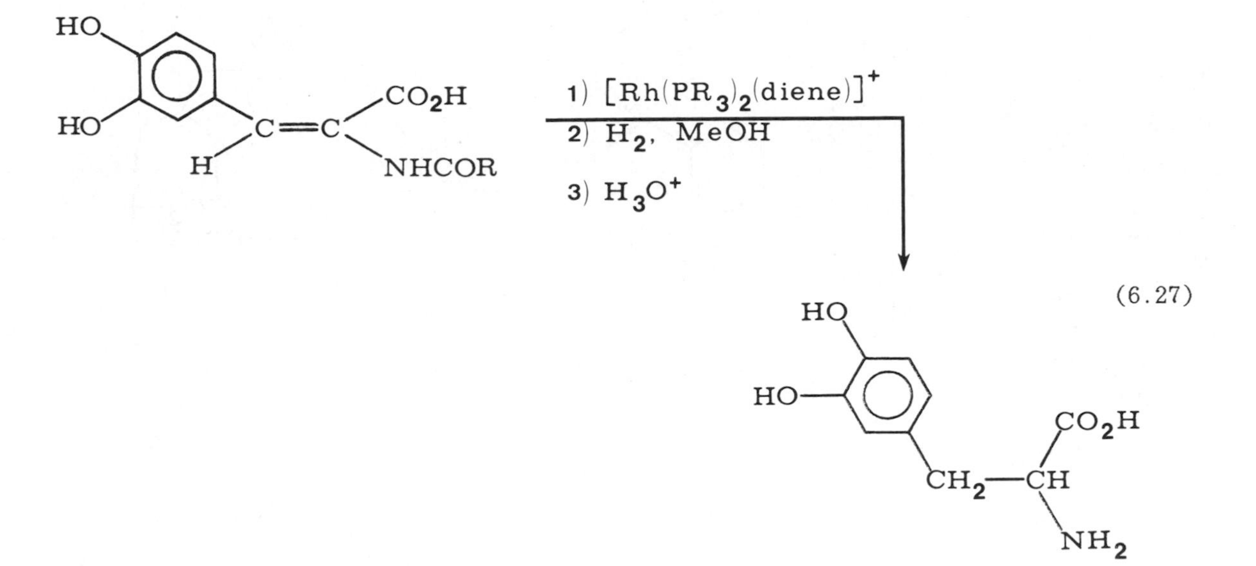


(6.27)

The investigations of asymmetric hydrogenation catalysts have depended on the synthesis of chiral phosphines. In fact, Knowles' development of the L-dopa process stemmed from procedures which Mislow and coworkers [36] had developed for the efficient preparation of tertiary phosphines which are chiral at phosphorus. A very large number of chiral phosphines have now been prepared and used for asymmetric catalytic hydrogenations [3]. Representative examples of these chiral ligands are listed in Figure 6.5, along with their usual abbreviations. In Table 6.2 selected chiral ligands are given along with the per cent enantiomeric excess and the major product configuration (R or S) for the reduction of α-acetylaminocinnamic acid. In all cases the usual cationic rhodium catalysts were used. Note that several of these ligands are chiral at carbon rather than at phosphorus. Notable among ligands which are chiral at carbon are "DIOP", prepared by Kagan [3a] from the readily available diethyl ester of (-)-tartaric acid, and "chiraphos," recently described by Bosnich [37].

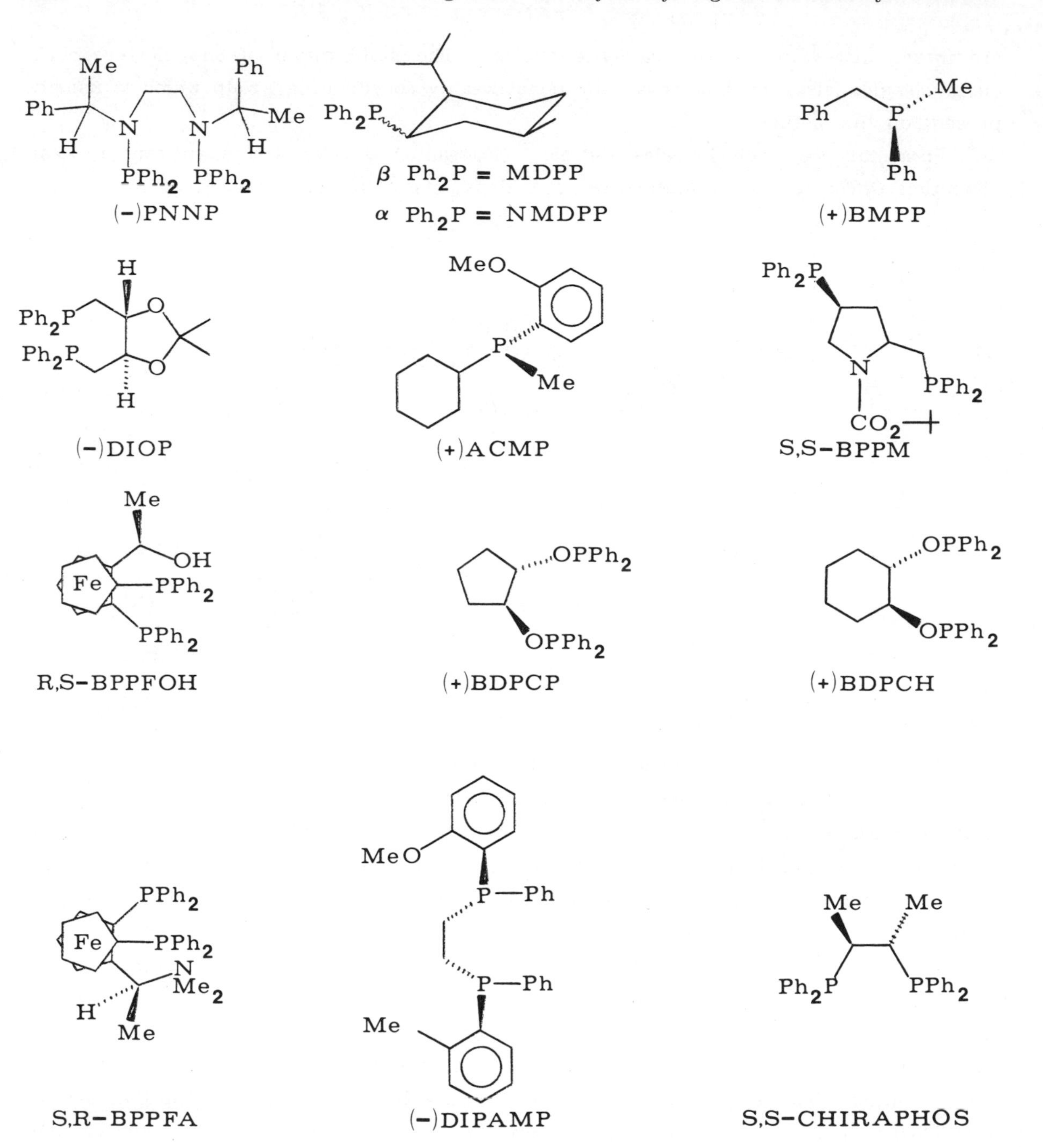


*Figure 6.5. Representative chiral ligands in asymmetric hydrogenation catalysts.*

*Table 6.2. Homogeneous asymmetric hydrogenation of α-acetylaminocinnamic acid by catalysts of the type* $[Rh(PR_3)^*{}_2(S)_2]^+$.

| Ligand ($PR_3^*$) | Percent ee | Product Configuration | Reference |
|---|---|---|---|
| S,S-Chiraphos | 99 | R(D) | [37] |
| (-)DIPAMP | 96 | S | [38] |
| S,R-BPPFA | 93 | S | [39] |
| S,S-BPPM | 91 | R | [40] |
| (+)ACMP | 85 | S | [3g] |
| (-)PNNP | 84 | R | [41] |
| (+)DIOP | 81 | S | [3a] |
| (+)BDPCP | 68 | S | [42] |
| (+)BDPCH | 12 | S | [43] |

Specific preparations and relative efficiencies of the various chiral phosphines are found in several reviews [3]. We need note two interesting factors. (1) Substrates other than acetamidocinnamic acid derivatives (Table 6.3) give at best modest enantioselectivity (usually <60% ee, but 83% in one case). (2) There is no obvious relationship between the % ee of the product, the structures of the chiral phosphines, and the substrate. It is as if one problem, the asymmetric reduction of acetamidocinnamic acids, has been solved several times, but the asymmetric hydrogenation of many other substrates is not solved, nor is there any obvious systematic way of searching for a solution. It is, however, probable that substrates which can act as bidentate ligands should give the best results.

Fortunately, a great deal is known about the mechanism and intermediates involved in hydrogenations of the α-acetamidocinnamic acid substrates with chiral cationic rhodium complexes, which are good models for the chiral catalysts. From these studies, one can obtain a better understanding of the factors which could lead to asymmetric induction in other substrates. We shall see that very strong stereoselective coordination of the substrate to a highly unsaturated catalyst by a minimum of two binding sites is required. For this reason different results may be obtained if a good coordinating anion such as chloride is present.

Consider again Halpern's [24,30] mechanism for the catalytic hydrogenation of α-acetamidocinnamic acid by the cationic rhodium(I) complex, outlined in Figure 6.4.

*Table 6.3. Hydrogenation of various substrates with chiral homogeneous catalysts $[Rh(PR_3)^*_2(S)_2]^+$.*

| Ligand ($PR_3$*) | Substrate | Product | Percent ee | Ref. |
|---|---|---|---|---|
| (-)DIOP | $=C(Ph)CO_2H$ | $MeCH(Ph)CO_2H$ (S) | 63 | [3a] |
| (-)DIOP | $=C(Ph)CO_2Me$ | $MeCH(Ph)CO_2Me$ (R) | 7 | [3a] |
| (+)ACMP | $=C(Ph)CO_2H$ | $MeCH(Ph)CO_2H$ | 12 | [3g] |
| NMDPP | $PhCH{=}C(Me)CO_2H$ | $PhCH_2CH(Me)CO_2H$ (R) | 60 | [44] |
| MDPP | $PhCH{=}C(Me)CO_2H$ | $PhCH_2CH_2(Me)CO_2H$ (S) | 17 | [44] |
| (+)BDPCP | $=C(Ph)Et$ | $CH_3CH(Ph)Et$ (R) | 60 | [45] |
| (+)BDPCH | $=C(Ph)Et$ | $CH_3CH(Ph)Et$ (R) | 33 | [42] |
| (-)DIOP | $=C(Ph)Et$ | $CH_3CH(Ph)Et$ (S) | 24 | [45] |
| (+)DIOP | PhCOMe | PhCH(OH)Me(R) | 51 | [46] |
| R,S-BPPFOH | PhCOMe | PhCH(OH)Me(R) | 43 | [47] |
| R,S-BPPFOH | $MeCOCO_2H$ | $MeCH(OH)CO_2H$(R) | 83 | [47] |
| (-)DIOP | $MeCOCO_2H$ | $MeCH(OH)CO_2H$(R) | 6 | [47] |
| ACMP | $MeCOCO_2Me$ | $MeCH(OH)CO_2Me$ | 71 | [48] |

As shown in Equation 6.26, the equilibrium constant $K_{eq}$ for the substrate binding step is much larger for α-acetamidocinnamic acid than that for simple olefins, arenes, or even styrenes. Measurements of $K_{eq}$ for a series of acrylic acid derivatives suggests that the acetamido and arene groups are particularly important in increasing the affinity of the catalyst for the substrate. The nature of this tight binding is revealed by Halpern's X-ray crystallographic analysis of the racemic rhodium(I)-cinnamate complex having the achiral diphos ligand [24]. Only one of the two enantiomeric forms of 51 is shown. Both the olefin and the amide oxygen are coordinated to rhodium.

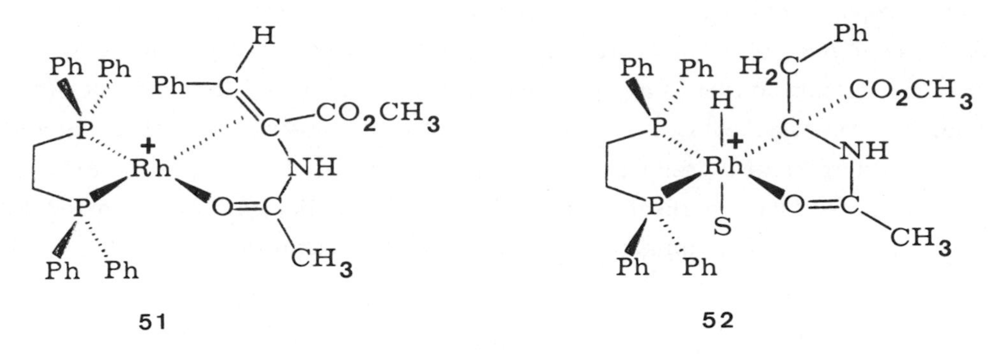


Thus far, it has not been possible to characterize structurally the similar complex of the chiral DIPAMP ligand. However, it can be inferred that the high degree of ee derives from preferential formation of one of the two possible diastereomers which would result from substitution of chiral DIPAMP for the achiral diphos ligand in 51. The energy difference appears to depend on steric interactions between the aryl groups on DIPAMP and on nonbonded interactions between these groups and the aryl and ester functions on the substrate. The strength of these interactions is apparently greater because of the rigid chelates formed by the bidentate phosphine and the bidentate substrate. Halpern has established a relationship between the substrate-catalyst affinity ($K_{eq}$) and the observed ee. For example, the Z ("cis") isomer exhibits a larger $K_{eq}$ and ee than the E ("trans") isomer.

The subsequent hydride-olefin insertion step is highly regio-and stereoselective. The coordination geometry of the product, 52, has been established by low-temperature multinuclear NMR studies [24]. The amide oxygen remains coordinated, and the regiospecificity of hydride addition is apparently controlled by the formation of a more favorable five-membered chelate ring. The rhodium hydride to olefin intramolecular insertion is assumed to take place in the usual *cis* manner. Inasmuch as the olefin binding step appears to dominate the stereochemical course of these asymmetric hydrogenations, it is interesting to measure directly the stereochemical purity of the diastereomers formed between Z-α-benzamidocinnamic acid and chiral rhodium chelates. This has been accomplished by examining the $^{31}P$ NMR spectra of cationic rhodium DIOP complexes generated _in situ_ [49]. For the more tightly binding Z isomer, the $^{31}P$ NMR spectrum is consistent with the presence of only one of the two possible diastereoisomers.

2. _Other Asymmetric Homogeneous Catalysts_. Although there have been relatively few reports of asymmetric hydrogenation catalysts which employ chiral ligands other than phosphines, this approach will undoubtedly become more thoroughly investigated. In the section under supported catalysts, mention will be made of Whiteside's protein-bound asymmetric catalyst. Some successful asymmetric induction has also been achieved by adding chiral amines to catalysts derived from bis-dimethylglyoximato cobalt(II) complexes. The highest optical yield to date, 78% ee, was reported [50] for the reduction of benzil using quinine and benzylamine in mesitylene at -10° (see 6.28).

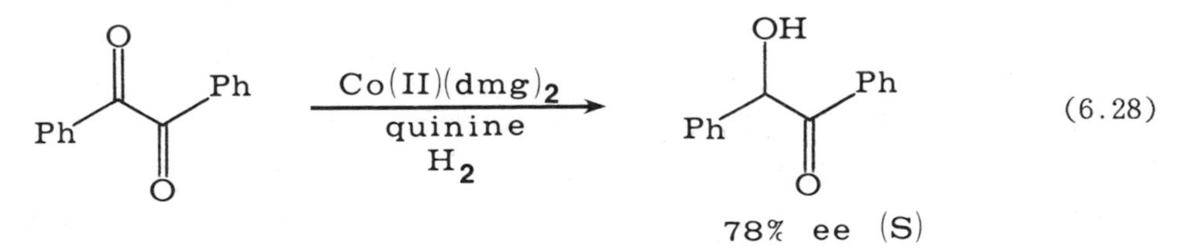


(6.28)

The way in which this asymmetric induction occurs is unclear.

Chiral amides have been used to successfully induce modest asymmetric hydrogenation with a catalyst derived from borohydride reduction of a rhodium pyridine complex [9]. A 57% ee was achieved in the hydrogenation of (E)-methyl-β-methylcinnamate (6.29).

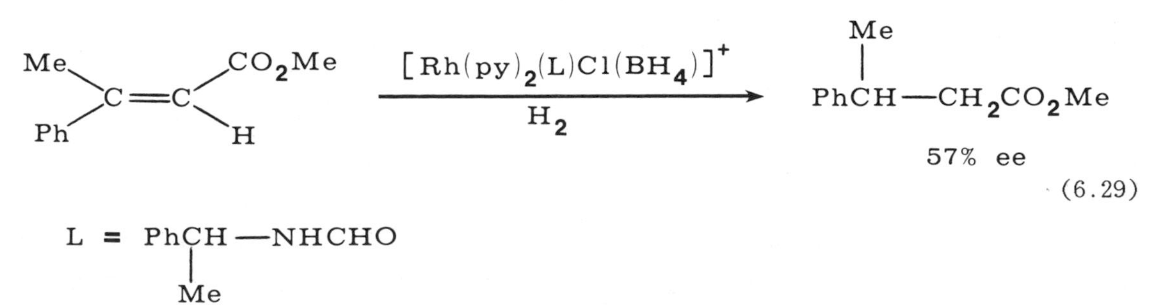


(6.29)

Other homogeneously catalyzed asymmetric reactions have been reported, but these are vastly outnumbered by asymmetric hydrogenation. These asymmetrically catalyzed reactions represent a truly embryonic field. In this context, the formation of chiral cyclopropanes from olefins and diazo compounds in the presence of chiral copper complexes (ee ~8%) [51], rhodium-catalyzed asymmetric hydroformylations (ee 44%) [52], and metal-catalyzed coupling of ethylene with norbornene (ee 80%) should be mentioned [3h]. Such asymmetrically catalyzed reactions have been surveyed up to 1978 [3f]. Examples of asymmetric hydrosilation are discussed later in this chapter.

## 6.5. Other Hydrogenation Catalysts.

1. Catalysts for Olefin and Acetylene Reduction. Many reports have appeared concerning soluble complexes which catalyze olefin and acetylene hydrogenations. Beyond the Wilkinson catalysts, few of these systems have been widely used by synthetic organic chemists. A representative list of olefin and acetylene "precatalysts" is given in Table 6.1, along with the usual experimental conditions and special characteristics. Note that most of these precatalysts contain rhodium, ruthenium, or iridium, and represent rather minor variations with respect to changes in ligand. Most precatalysts contain phosphine ligands, but some have pyridine and phenanthroline ligands. Various substrate selectivities are indicated, notably terminal versus internal olefinic reactivities, relative tendencies to catalyze olefin rearrangement, and reactivity toward other unsaturated groups such as ketones.

One of the most unusual catalysts is the last entry in Table 6.1 [16]. This precatalyst would appear to be very similar to the two preceding entries; however, under high dilution in $CH_2Cl_2$, this complex forms a presumably homogeneous catalyst which is very different from the other examples.

The features which usually distinguish homogeneous from heterogeneous hydrogenation catalysts are reported to be completely absent in these catalysts derived from hydrogenation of 1,5-cyclooctadiene in $CH_2Cl_2$. The resulting uncharacterized catalyst reduces very hindered olefins. For example, most homogeneous catalysts completely fail to reduce tetra-substituted olefins; however, one catalyst in this series, 53 (L = $PCy_3$, L' = Py), reduces $Me_2C=CMe_2$ at a very high rate (4,000 turnovers per hour at 0°C), as in Equation 6.30. Under comparable conditions, this catalyst is about 100 times more reactive than the Wilkinson catalyst toward 1-hexene reduction. In contrast to the usual sensitivity which homogeneous catalyst show toward impurities, this iridium catalyst is reported to be uneffected by either methyl iodide or by oxygen!

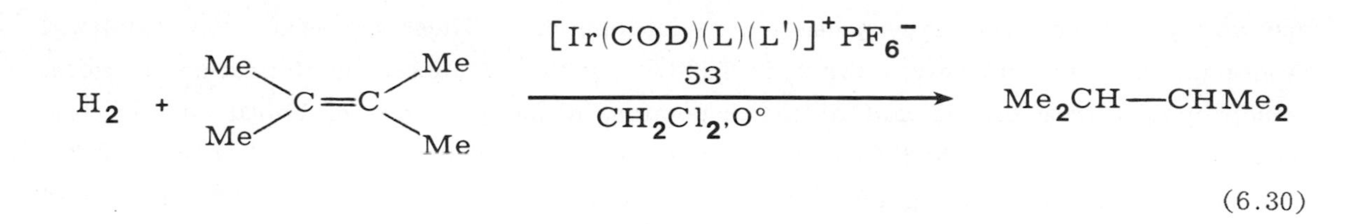


(6.30)

Although most second-row (4d) catalysts are more reactive than their third-row (5d) analogues, in this case the rhodium system is less reactive than the iridium complex by a factor of $10^6$! The iridium catalyst, 53, is extraordinarily sensitive to the nature of the solvent; nearly all catalytic activity is lost when acetone is used in place of methylene chloride. High dilution and the presence of excess olefinic substrate seem necessary to sustain activity; otherwise, a catalytically inactive dimer, $[Ir_2H_2(\mu\text{-}H)_3(PPh_3)_4]PF_6$, is formed [16].

It is difficult to judge the utility of these iridium catalysts until they have been employed for organic synthesis in other laboratories on a wider range of substrates, but the initital reports are very interesting. Preliminary attemps to support these highly reactive iridium catalysts with polymeric ligands have so far been unsuccessful. The exact nature of this iridium catalyst remains a mystery.

2. Catalysts for the Hydrogenation of Conjugated Olefins. There are several homogeneous hydrogenation catalysts which exhibit marked selectivity toward dienes and conjugated olefins [53]. An interesting example is $(\eta^5\text{-}C_5H_5)_2MoH_2$, which reduces cyclopentadiene to cyclopentene at 180°, 160 atm. in the absence of solvent (Equation 6.31). Other dienes are similarly reduced to monoenes which are nearly inert toward

(6.31)

further reduction. Yields range from 40 to 90%. It is surprising that norbornadiene is reduced, whereas 1,5-cyclooctadiene is not.

The olefinic groups in α,β-unsaturated carbonyl compounds such as methyl acrylate are also reduced. Aromatic solvents inhibit these reactions, presumably by competition for coordinated sites. Selective 1,2-metal hydride addition (as opposed to 1,4-addition) has been proposed on the basis of product analysis. For example, dimethyl muconate, 54, affords exclusively α-dihydromuconate, 56, which was shown not to arise from β-dihydromuconate, 55, the 1,4-addition product. It was proposed that only one

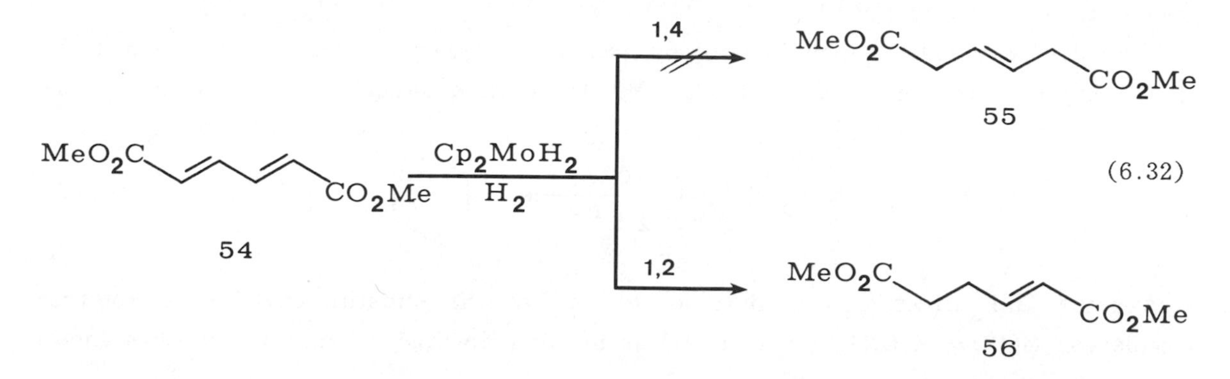


(6.32)

olefinic group in the diene is bound to the metal. At this point, the detailed mechanism of these diene reductions remains obscure. The major commercial interest in such reactions is the conversion of cyclopentadiene to cyclopentene. This is usually accomplished using heterogeneous catalysts.

Other catalysts show selectivity toward the 1,4-hydrogenation of conjugated dienes which can achieve a S-cis configuration, indicating the probable formation of an intermediate chelated diene complex (6.33). A well-studied example involves precatalysts

S−trans ⇌ S−cis (6.33)

which are derived from arene chromium carbonyl complexes. These catalyze the hydrogenation of methyl sorbate, 57, to methyl 3-hexenoate. Deuterium labeling verified the

57 ($CO_2Me$) $\xrightarrow[D_2,\ DMF]{(ArH)Cr(CO)_3}$ (D D) $CO_2Me$ (6.34)

exclusive 1,4-hydrogen addition [54]. This catalyst is remarkable for its ability to catalyze the conjugation of 1,4-dienes rapidly prior to their reduction, but not to catalyze the cis-trans monoene isomerization at an appreciable rate. This 1,4-diene conjugation does not require hydrogen. The severe conditions required for the function of this system (160°/30 atm. $H_2$) suggest that the actual catalyst may be quite different from the precatalyst. This supposition is further supported by the observation that substitution-labile chromium arene complexes are active at much lower temperatures.

Similar catalysts which selectively reduce conjugated dienes are formed by irradiating solutions of $M(CO)_6$ (M = Cr, Mo, W) as in Equation 6.35. However, a more

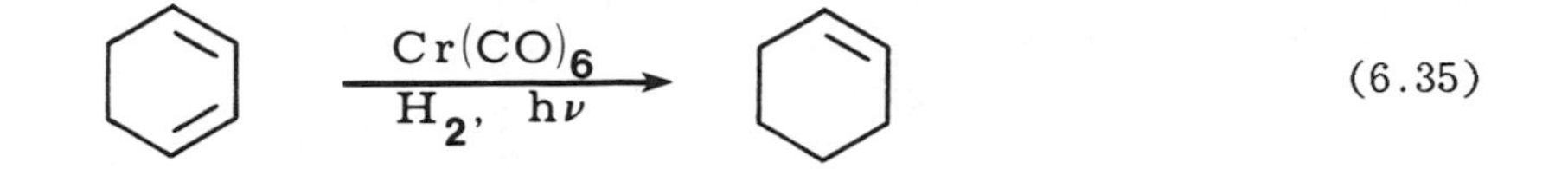


convenient and effective procedure is to employ the substitution-labile acetonitrile complexes, $M(CO)_3(MeCN)_3$ (m = Cr, W) as in Equation 6.36. Wrighton [55] has shown

$D_2$ / $Cr(CO)_3(MeCN)_3$ — D D (6.36)

that such catalysts add $D_2$ to S-cis dienes in a 1,4-manner. $Mo(CO)_6$, $W(CO)_6$ are also effective, but these catalyze monoene isomerization.

Selective reduction of nitro groups, azo groups, and olefins and carbonyl groups which are conjugated to electron-withdrawing systems have been reported to be catalyzed by bis(dimethylglyoximato)(pyridine)cobalt(II) [56] (Equations 6.37 and 6.38). It

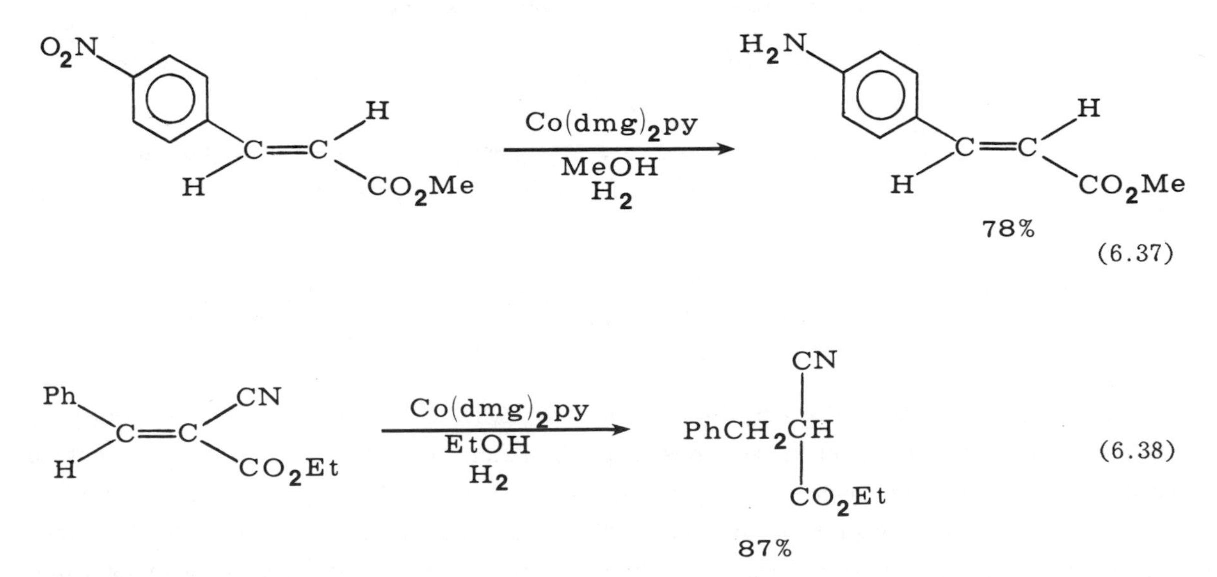

seems doubtful that these substrates form a complex with the metal center, but that a reactive cobalt hydride adds directly to the substrate without prior coordination. Rather large (2-10) ratios of cobalt to substrate are required, indicating low turnover numbers. Note that nitro groups can be selectively reduced.

The mechanism of the reduction of conjugated olefins by this catalyst has been investigated by substrate reactivities, and isolation of intermediates [57]. These studies indicate that the reaction (6.39) proceeds in two steps: (a) the formation of an α-cobalt alkyl; and (b) the reduction of this complex by another cobalt(II) derivative.

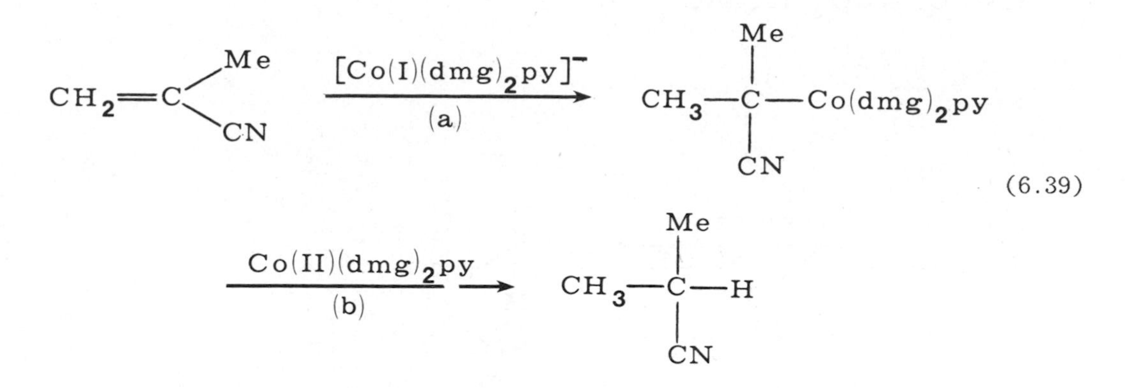


The first step is favored by electron-withdrawing substituents and neutral conditions. It has been found that the regiospecific addition of such cobalt complexes to conjugated olefins depends on the pH. Under basic conditions β-addition is the preferred path (apparently a "Michael addition" by nucleophilic cobalt(I), 59), whereas under acidic conditions α-addition takes place (apparently cobalt(III) hydride (58) addition). The α-alkyl (60) has been isolated and demonstrated to be a reactive intermediate (6.40, 6.41).

$$\underset{58}{[HCo(III)(dmg)_2py]} \underset{+H^+}{\overset{-H^+}{\rightleftharpoons}} \underset{\underset{59}{(basic)}}{[Co(I)(dmg)_2py]^-} \qquad (6.40)$$

The second step is an ill-defined reduction of the α-cobalt alkyl by another cobalt complex and seems to require basic conditions. This step is facilitated by the presence of another substituent on the α-carbon; thus acrylonitrile is not reduced, but methacrylonitrile is reduced in high yield (6.42). The use of optically active bases such as

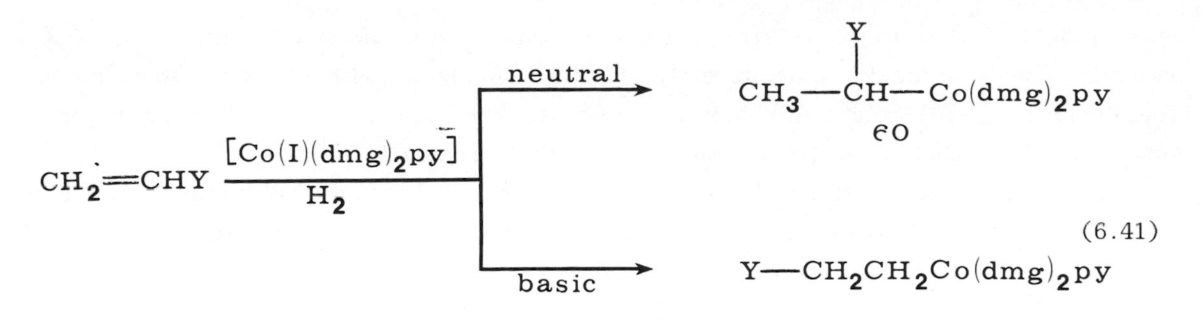


Y = CN, $CO_2Me$, etc.

$Co(dmg)_2py$ / $H_2$, MeOH — CN — CN — H — 87% (6.42)

quinine with these cobalt catalysts gives asymmetric induction; however, the enantioselectivity is low (~50% ee) (Equation 6.43). These results indicate that the chiral base

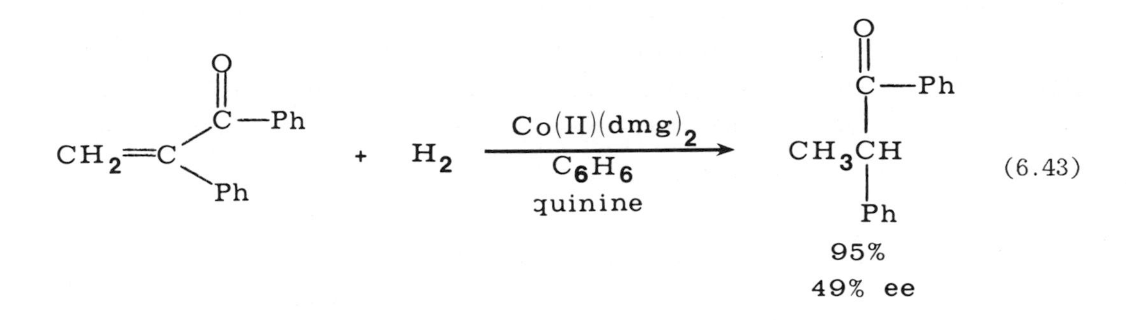


must be in the vicinity of the reaction site--the glyoximinato hydroxyl groups or the substrate carbonyl are logical positions. The second hydrogen comes from the solvent rather than from $H_2$ or $NaBH_4$, which can be used in place of $H_2$ (Equation 6.44). Thus far, reactions of this type are not especially useful as synthetic procedures, nor are their mechanisms well-understood, but such novel reaction paths may eventually be put to good use.

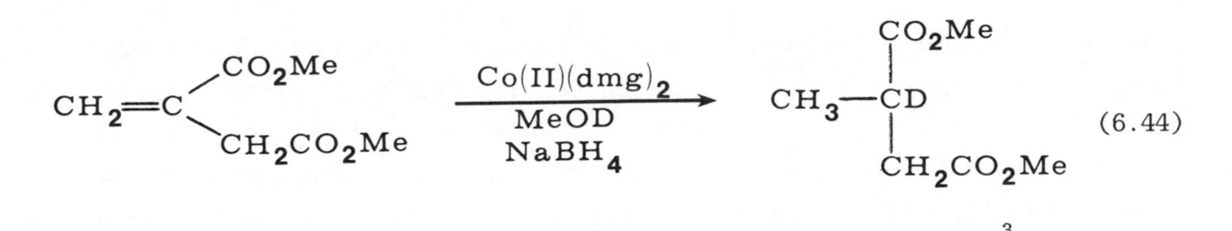


(6.44)

Cobalt(II) ions in the presence of excess cyanide afford $Co(II)(CN)_5^{3-}$, which is a catalyst for the selective hydrogenation of conjugated dienes to monoenes in aqueous or methanolic solutions. This $Co(CN)_5^{3-}$ catalyst system, which shows some similarities to $HCo(CO)_4$ and $Co(dmg)_2$, has been extensively studied [58]. Hydrogen activation of $Co(CN)_5^{3-}$ is similar to that of $Co_2(CO)_8$ in that the rate is second-order in cobalt (Equation 6.45); however the mechanism is complex [59]. A typical diene reduction is

$$2\,[Co(CN)_5]^{3-} + H_2 \rightleftharpoons 2\,[CoH(CN)_5]^{3-} \qquad (6.45)$$

illustrated in Equation 6.46. The intermediates are mixtures of <u>mono</u> and <u>trihapto</u> allyl cobalt(III) cyano complexes <u>61</u> and <u>62</u>. This catalyst is not especially stereoselective;

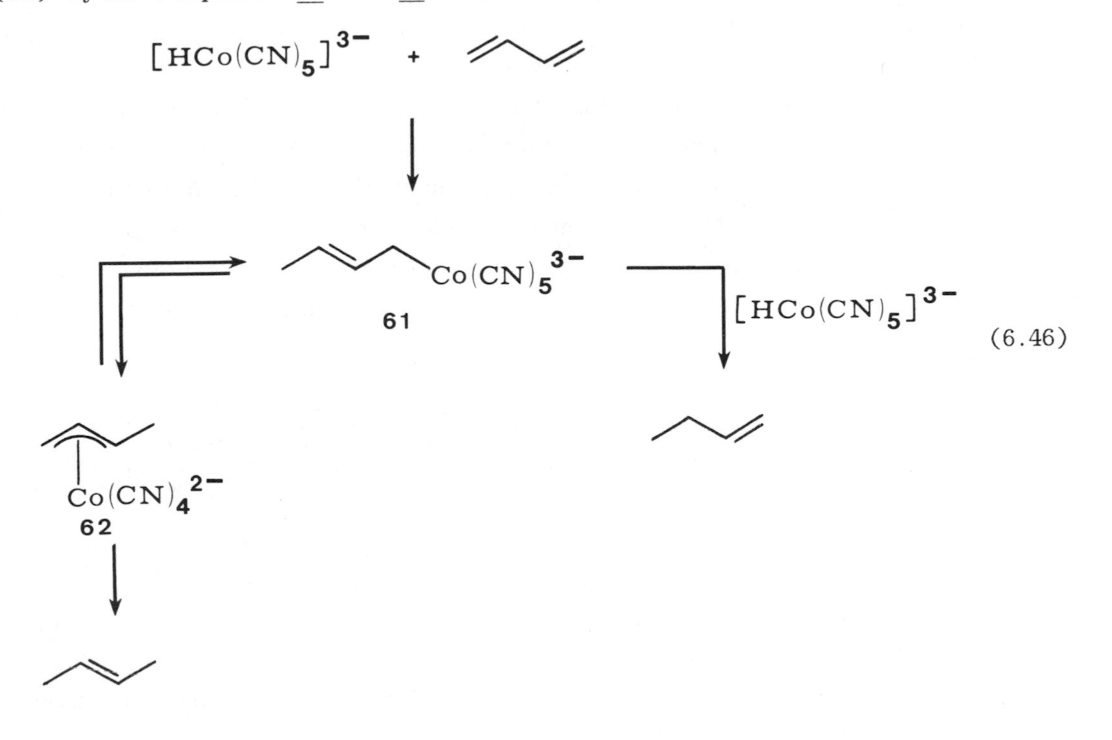


(6.46)

proportions of isomeric butenes depend on various factors (solvent, cyanide to cobalt ratios, and base concentration). Addition of deuterium to conjugated dienes with this catalyst is not at all specific. Certain α,β-unsaturated olefins are reduced, whereas others are not. For example, acrylic acid does not react, but methacrylic acid is reduced. Note this similarity to the $Co(dmg)_2$ system. Other groups such as azo, nitro, acetylenes, and oximes are also hydrogenated in the presence of this catalyst. The oxime reduction (6.47) is synthetically the most useful.

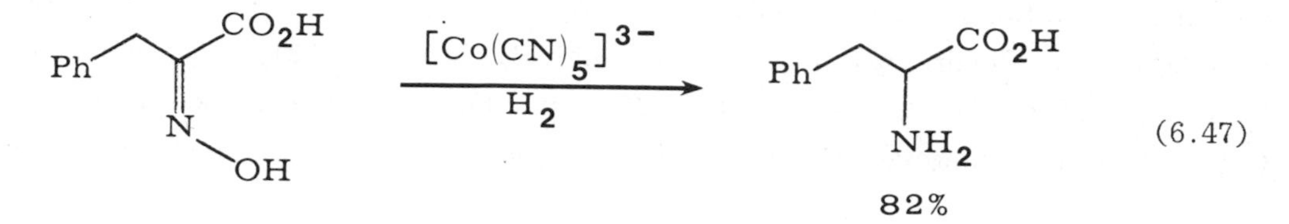


(6.47)

Hydridotetracarbonylcobalt(I), which is formed either by hydrogenation of $Co_2(CO)_8$ or acidification of $NaCo(CO)_4$, is occasionally used for hydrogenations, but this catalyst is only stable under CO, and hydroformylation is therefore a potential side reaction. The high temperatures and especially the high pressures which are required inhibit routine laboratory applications. This catalyst is, however, useful in certain situations, such as the reduction of conjugated dienes. Like the other cobalt dimethylglyoxime complex and cyano hydrides, the first step is addition to the conjugated dienes affording π-allyl complexes. Subsequent reduction probably occurs by hydrogen transfer from another mole of $HCo(CO)_4$ (a binuclear elimination), although direct hydrogenolysis of the cobalt-carbon bond has also been proposed. Unconjugated dienes are thought to be conjugated prior to reduction. A typical application of this catalyst is the hydrogenation of α,β-unsaturated carbonyl compounds, for example, the furyl derivative (6.48).

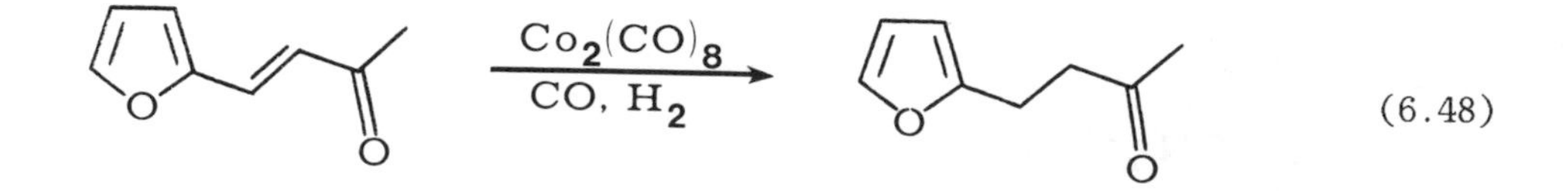


(6.48)

3. Catalysts for the Hydrogenation of other Functional Groups. Organic derivatives such as aldehydes, ketones, esters, amides, nitriles, and anhydrides are not very good ligands for low-valent transition-metal compounds. It is therefore not surprising that thus far there are few synthetically useful homogeneous catalyst for the hydrogenation of these substrates.

Esters, amides, nitriles, and carboxylic acids are especially resistant toward homogeneously catalyzed hydrogenations. Most instances of such reactions require severe conditions or involve catalysts which may be heterogeneous. Examples up to 1972 may be found in James' comprehensive book [1b].

The catalytic hydrogenation of succinic anhydride [60] is a rare example of such a reaction (6.49). The 50 percent yield may be accounted for by the formation of water,

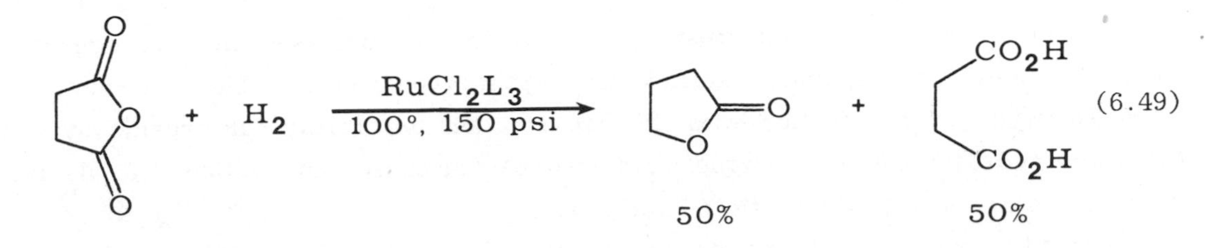


which opens the anhydride to succinic acid, which is inert. Note that the lactone is not further reduced under these conditions. With unsymmetrical cyclic anhydrides, the reduction is regiospecific. The least-hindered carbonyl group is reduced [60b].

Cyclic anhydrides also afford aldehydic acids with $HCo(CO)_4$, but yields are modest (6.50).

$Co_2(CO)_8$, $H_2$, CO → CHO / $CO_2H$    32%    (6.50)

The homogeneous catalytic hydrogenation of ketones is much more limited than the reduction of olefins. One of the best-known homogeneous catalysts for ketone hydrogenation derives from cationic precatalysts of the type $[Rh(diene)L_2]^+$. For the hydrogenation of aldehydes and ketones, these are much better than neutral catalysts. However, water or a proton base is required for reduction of the organic carbonyl compounds, perhaps because the active catalyst is a monohydride generated by the loss of a proton. Such catalysts have been used on conjunction with chiral ligands to afford asymmetric induction (see Figure 6.5 and Table 6.3). Simple ketones afford low optical yields (<50% ee), but 1,2- and 1,3-dicarbonyl compounds, which can act as bidentate ligands, give better results [61] (Equation 6.51).

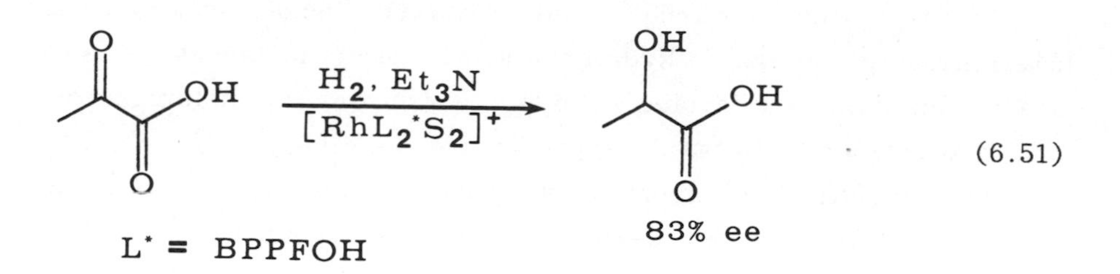


It should be noted that most oxo catalysts will promote the hydrogenation of aldehydes and ketones under sufficiently forcing conditions. These reductions are often employed in conjunction with the oxo reaction but rarely for organic syntheses. Catalytic hydrosilation of aldehydes and ketones (discussed in Section 6.8, 4) is more general, and probably will be more widely used.

Benzyl alcohols are hydrogenolyzed by $HCo(CO)_4$ (Equation 6.52). Hydrogenolyses

$$\mathrm{Ph\overset{O}{\overset{\|}{C}}CH_3} \xrightarrow[\mathrm{CO,\ H_2}]{\mathrm{Co_2(CO)_8}} [\mathrm{PhCH(OH)CH_3}] \longrightarrow \mathrm{PhCH_2CH_3} \qquad (6.52)$$

are rarely catalyzed by homogeneous complexes. This example probably depends on the high acidity of the hydride (Chapter 3) to make the alcoholic OH into a good leaving group. A useful feature of the $Co_2(CO)_8$ catalyst system is that it is not poisoned by sulfur compounds. Consider for example the sequential reduction of α-acetyl thiophene (6.53). Note that the thiophene ring is reduced in a second stage.

Imines are much better ligands and are more susceptible to reduction than aldehydes or ketones. A few complexes have been reported to catalyze the hydrogenation of imine derivatives, for examples: $[HCo(CN)_5]^{3-}$ [62], $RhCl_2(BH_4)(DMF)py_2$

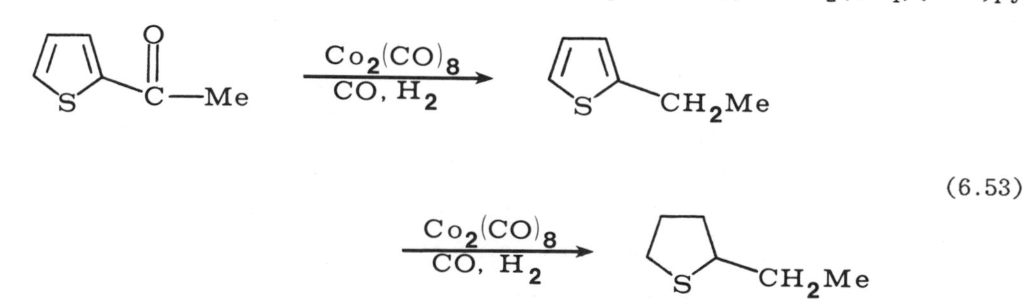

[63], and $[Rh(nbd)L_2]^+ClO_4^-$ ($L_2$ = DIOP) [64]. However, these substrates have been comparatively little studied.

Nitro groups are rather easily reduced by many reagents as well as by heterogeneous catalytic hydrogenation. Homogeneous complexes will often catalyze nitro hydrogenations, but are seldom used synthetically for this purpose. These homogeneously catalyzed nitro hydrogenations are not well-understood mechanistically. Furthermore, such reductions are sometimes complicated by the formation of azo compounds. Both rhodium and cobalt oxo catalysts promote the reduction of aromatic nitro compounds. The versatile McQuillin catalysts, 63, efficiently catalyzes the hydrogenation of aryl nitro compounds. This reaction occurs even with crosslinked polystyrene substrates, 64. This result strongly supports the contention that the active catalyst is

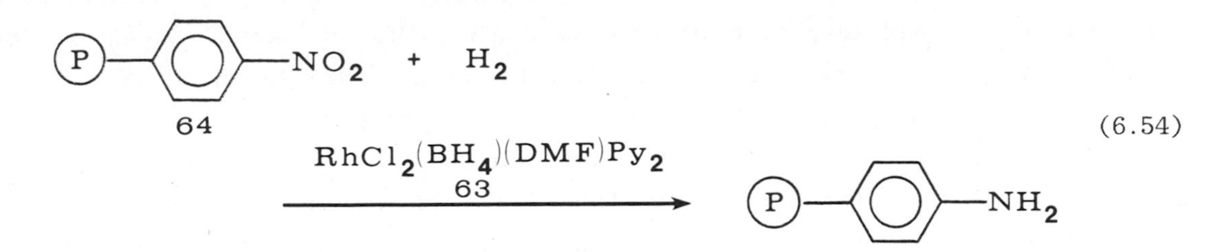


(6.54)

homogeneous, because heterogeneous particles do not appear to affect such crosslinked polymeric substrates [65].

Homogeneous catalytic hydrogenation of simple cyclopropanes does not seem to have been described, although such a stereospecific reduction would be synthetically interesting. Hydrogenation of cyclopropyl alkenes has been reported [66].

A few cases of the homogeneous catalytic hydrogenation of epoxides have been reported, mostly by electrophilic catalysts [67]. Little is known about the mechanism.

4. Catalytic Hydrogenation of Arenes. Many transition-metal complexes form solutions which will catalyze the hydrogenation of aromatic substrates [1a,1b,68-80]. Representative examples are listed in Table 6.4. At present the mechanisms of these reations are ill-defined. In particular, it has been difficult to establish the nature of the active catalyst or even whether these catalysts are soluble monometallic complexes rather than suspended metal particles ("sols"). Previously we saw how difficult it is to establish the catalytically active species with the well-defined homogeneous olefin hydrogenation catalysts. In the instance of arene hydrogenation catalysts, this is even more difficult.

Thus far, "homogeneous" arene hydrogenation catalysts have not proven to be simultaneously both selective and reactive enough to supplant the well-entrenched heterogeneous catalysts. However, the latter suffer from a lack of stereospecificity and from an unfortunate tendency to catalyze H-D exchange reactions. It is in these two areas that ostensibly homogeneous arene hydrogenation catalysts have their greatest promise, but we shall find the most stereoselective arene catalysts are usually not very reactive and, in several instances, are unstable. Further difficulties with the preparation of the requisite "precatalysts" have so far discouraged their use by synthetic chemists.

Among the first "homogeneous" arene hydrogenation catalysts to be reported were ill-defined substances produced in situ by the reaction of reducing agents, such as $LiAlH_4$, alkyl-aluminum derivatives, or organolithium reagents, with organic-soluble transition-metal complexes, such as carboxylic acid salts and acetylacetonates. Such "Ziegler-type" systems have been described by Sloan [69], by Kroll [70], and by Lapporte [71a,b], who studied these catalysts more thoroughly.

Typically these catalysts require high pressure ($\geq$ 1,000 psi) and moderate temperatures (100°), e.g., entry A in Table 6.4. The reactivity profile Ni > Co > Fe is characteristic of such systems. Stereospecific cis-reduction is observed, as Equation 6.55 shows for the reduction of dimethylphthalate. The nickel catalyst exhibits chemical

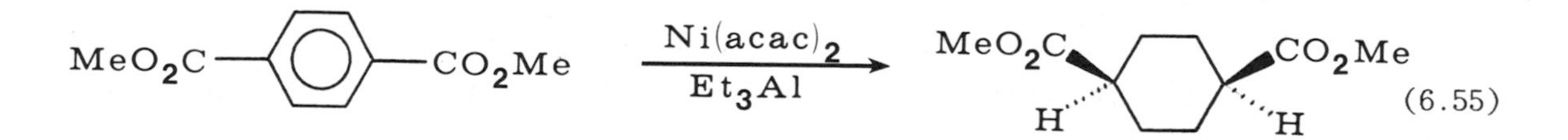

(6.55)

properties similar to Wilke's "naked nickel" homogeneous organometallic catalysts. For example, this species will catalyze the cyclotrimerization of 1,3-butadiene and is deactivated by nitrobenzene. Olefins are hydrogenated and isomerized at 25° under a few psi $H_2$. An iron catalyst of this type exhibits a Mossbauer spectrum different from that expected for elementary iron [71c].

Soluble polystyrene is hydrogenated by these catalysts. Similar catalysts have recently been examined for the hydrogenation of other soluble polymers [82]. In the course of this work, the physical properties of the black catalyst solution were examined. The catalyst solution was passed through a 0.5 μ micropore filter (under $N_2$). An electron micrograph of the filtered material revealed no particulate structure at

*Table 6.4. Homogeneous catalysts for arene hydrogenation.*

| Precatalyst | Conditions | Substrate | Product | Characteristic Features | References |
|---|---|---|---|---|---|
| A. | | | | | |
| $Et_3Al$ + Ni(II) 2-ethyl-hexenoate | 1,000 psi 150-210° | o-xylene | cis-1,2-dimethyl-cyclohexane | Al/Ni ~3:5:1 most active black solutions | [69,70,71, 72] |
| or $Et_3Al$ + $(RCO_2)_2M$ | " | dimethyl terphthalate | cis-1,4-dicarbomethoxy-cyclohexene | $Ph_3P$ poisons reaction Ni $\geq$ Co > Fe > Cr > Cu | |
| B. | | | | | |
| $\eta^3$-$C_3H_5Co[P(OMe)_3]_3$ | 25°, ~14 psi | $C_6D_6$ | $C_6H_6D_6$ all cis | ~20 turnovers in 48 hr. $NO_2$, F, CN groups not tolerated | [73,68] |
| | | $C_{10}H_8$ | $C_{10}H_{18}D_6$ | $C_6H_6$>$C_6H_5Me$>$C_6H_4Me_2$ $C_6H_6$ > naphthalene | |
| C. | | | | | |
| Rh(I) N-phenyl-anthranilic acid complex in DMF | 20°, 14 psi DMF | $C_6H_6$ | $C_6H_{12}$ | anthracene>naphthalene> benzene | [74] |
| | | $C_{10}H_8$ | $C_{10}H_{18}$ | | |
| D. | | | | | |
| $RhCl_2(BH_4)(DMF)py_2$ | DMF | pyridine | piperidine | quinoline is reduced isoquinoline is not | [9,75] |
| E. | | | | | |
| $[(\eta^5$-$C_5Me_5)\ RhCl_2]_2$ | 50°, 70 psi 15 eq. $Et_3N$ i-propanol | $C_6D_6$ | $C_6D_6H_6$ nearly all cis | 225 turnovers in 36 hr. in i-propanol; 400 in benzene, benzene>toluene | [76] |
| F. | | | | | |
| $(\eta^6$-$C_6Me_6)(\eta^4$-$C_6Me_6)Ru$ | 90° 24-42 psi | $C_6H_6$ | $C_6H_{12}$ | quantitative catalyst recovery, rates decrease with alkyl substitution, significant cyclohexene formation, extensive H-D exchange | [77,68,78] |
| | | p-xylene | cis-1,4-dimethyl cyclohexane | | |

*Table 6.4. Continued.*

| Precatalyst | Conditions | Substrate | Product | Characteristic Features | References |
|---|---|---|---|---|---|
| G. | | | | | |
| $(\eta^6\text{-}C_6Me_6)Ru(H)Cl(PPh_3)$ | 50°<br>700 psi | $C_6H_6$ | $C_6H_{12}$ | turnover rate 0.8 $min^{-1}$<br>no cyclohexenes detected | [79a,b] |
| H. | | | | | |
| $[(\eta^6\text{-}C_6Me_6)Ru)_2$<br>$(\mu\text{-}H)_2(\mu Cl)]Cl$<br>(generated in situ<br>from $\{RuCl_2(\eta^6\text{-}C_6Me_6)\}_2$<br>and $Na_2CO_3$) | 50°<br>700 psi | $C_6D_6$ | $C_6D_6H_6$<br>>95% | 9,000 turnovers in 36 hr.<br>activity suppressed by thiophene<br>hydrogenolysis of halobenzenes | [80] |

500-fold magnification; however, light scattering indicated some 3500 Å colloidal particles. Whether these particulate "sols" are the active catalysts is very difficult to determine. These solutions are similar to the ill-characterized "metal hydrides" prepared from transition-metal halides and hydride reducing agents which are described in Chapter 7. In all these cases, the active catalysts may either be reactive, unstable, highly unsaturated metal complexes or small metal particles in the 10-40 Å range. Such "naked metal" compounds should be stabilized by coordination with substrate molecules which are removed by hydrogenolysis (as ligands such as 1,5-cyclooctadiene are). Metal particles are intrinsically unstable, tending to aggregate into larger particles, with a corresponding loss in surface area and diminished catalytic activity. In commercial heterogeneous catalysts, the major function of supports such as $Al_2O_3$ or carbon is to stabilize metal particles against aggregation.

It is very difficult, if not impossible, to distinguish as the active catalyst a homogeneous metal complex from a metastable metal particle. The latter may often be present in trace quantities in catalytic systems involving high hydrogen pressures and/or strong hydridic reducing agents--especially with precatalysts which are known to decompose during the catalytic reaction. Precipitates isolated from such reactions cannot be used as a gauge of the actual catalyst, since these large particles may have such small surface areas that they do not exhibit a measurable degree of catalytic activity. Catalytic activity is inherently amplified, since one catalyst molecule may produce thousands of product molecules. Thus, very small quantities of a very active catalyst may be "invisible" in a solution containing spectroscopically observable organometallic complexes, which are themselves catalytically inactive. A qualitative, imperfect guide to whether a catalyst is homogeneous or heterogeneous is the limited reactivity with bulky substrates and the tendency to exhibit a high degree of stereospecificity, which are characteristics of the former.

Another puzzling feature of the mechanism used by monometallic arene hydrogenation catalysts concerns how the proper orientation of the aromatic ring, the metal, and the hydride is attained for migratory insertion. This problem has been discussed in the context of chelating diene hydrogenations [82]. It would seem that a *dihapto* or at most a *tetrahapto* coordination of the arene may be required, since *hexahapto* coordination might not lead to the correct geometry for intramolecular hydrogen transfer.

Keeping these issues in mind, consider the representative homogeneous arene hydrogenation precatalysts listed in Table 6.4. We have already discussed the ill-defined "Ziegler" type hydrogenation catalysts (entry A). Note the high stereoselectivity reported by Lapporte suggestive of homogeneous catalysis.

The cobalt allyl phosphite systems (entry B) have been extensively studied by Muetterties [68]. These catalysts exhibit very low activities toward arene reduction under mild conditions (1 atm, 25°). The remarkably stereospecific hydrogenation of $C_6H_6$ to all cis-$C_6D_6H_6$ without H-D exchange has been contrasted with the lower specificity and extensive H-D exchange found with heterogeneous catalysts [73]. Xylenes are deuterated in a similar stereospecific cis manner by the homogeneous cobalt precatalyst. Catalytic activity is enhanced by bulkier phosphites $P(OiPr)_3 > P(OEt)_3 > P(OMe)_3$, but this increasing activity is accompanied by decreasing catalyst lifetimes. The low reactivity and limited lifetimes of this system will preclude any synthetic applications and will complicate further mechanism studies.

The rhodium N-phenylanthranilate complexes, entry C, appear to be the most reactive homogeneous arene hydrogenation catalysts yet reported. We shall see that a polymer-bound form of this catalyst has also been found to be quite active for hydrogenation of arenes [74].

The rhodium borohydride complex, entry D, reduces pyridine, but this may be a special case, since aromatic hydrocarbons are apparently not affected.

The air-stable $[Rh(\eta^5\text{-}C_5Me_5)Cl_2]_2$ precatalyst, entry E, has been studied extensively by Maitlis. A variety of substituted benzenes are hydrogenated (50°, 50 atm $H_2$) under apparently homogeneous conditions (no metal was visually apparent). A base such as $Et_3N$ is required, presumably to take up HCl and so promote the formation of a rhodium hydride by $H_2$ oxidative-addition and deprotonation. The reactivity is modest: 225 turnovers of benzene molecules per rhodium atom were observed in 36 hours in isopropanol. $D_6$-Benzene is hydrogenated to $D_6$-cyclohexane. High but not completely stereospecific cis hydrogenation was reported. Alkyl substituents sterically retard the rate. Some functional groups are tolerated (ether, ester, ketone, amide), whereas others are not (phenol and carboxylic acid). These factors support, but do not establish, the purported homogeneous nature of this catalyst. The iridium analog is significantly less active.

The catalytic activity of a ruthenium arene hydrogenation catalyst (entry F) was discovered by Muetterties: $(\eta^6\text{-}C_6Me_6)Ru(\eta^4\text{-}C_6Me_6)$. This complex was originally prepared and structurally characterized by Fisher [77]. This precatalyst is moderately active (five benzene turnovers per hour at 90° and 2-3 atm $H_2$). The soluble complex was quantitatively recovered after a ten-hour reaction, and no free hexamethylbenzene could be detected in the reaction product. Hydrogenation rates decrease with increasing alkyl substitution on the arene. Stereospecificities are modest and approximate those reported for heterogeneous catalysts: o- and p-xylenes afford cis to trans product ratios of 9:1 and 9:2, respectively. This catalyst affords extensive H-D exchange for arene $+D_2$ and perdeuteroarene $+H_2$ reactions. There is evidence that the permethylated arene ligand on the soluble precatalyst is involved in the reaction. The methyl groups in this ligand contained deuterium when the precatalyst was recovered from reaction of $D_2$ with benzene. The recovered benzene substrate was also deuterated. Cyclohexenes are significant reaction products. For example, 55 percent of p-xylene was converted to 1,4-dimethylcyclohexenes during one hydrogenation experiment. The exact nature of the active catalyst and the reaction mechanism have not yet been established. An interesting question is whether the permethylbenzene ligand is exchanging with the substrate during the reaction; otherwise it is difficult to understand how the coordination sites necessary for catalysis are formed.

Bennett has described other, apparently more reactive ruthenium arene hydrogenation catalysts containing the hexamethylbenzene ligand. For example, the hydride $[RuHCl(\eta^6-C_6Me_6)PPh_3]$, which is generated in situ, is a stable, long-lived precatalyst for the hydrogenation of arenes and olefins (entry G in Table 6.4). Under moderate conditions (50°C, 50 atm $H_2$), this air-sensitive complex catalyzes the reduction of neat benzene at a good rate (turnover 0.8 $min^{-1}$). Slow decomposition occurs during the hydrogenation, affording a catalytically inactive black solid. About 45 percent of the precatalyst was isolated after a 36-hour reaction. Hex-l-ene is more readily reduced than benzene, and no cyclohexenes were detected in benzene hydrogenation products.

Another ruthenium hexamethylbenzene complex (entry H) is claimed to be the most active and stable arene hydrogenation complex discovered so far. Under relatively mild conditions (50°C, 50 atm $H_2$), 9,000 moles of benzene are reduced per mole of catalyst during 36 hours, corresponding to a turnover number of four per minute. Hydrogenation of $C_6D_6$ gives more than 95 percent $C_6D_6H_6$. Methyl benzoate, phenol, and anisole are reduced to cyclohexyl derivatives without reaction of the functional groups. Nitrobenzene may be selectively reduced to aniline. Sulfur and thiophene reduce but do not completely suppress the catalytic activity. Sulfur poisoning is well-known for heterogeneous catalysts.

Finally, it should be mentioned that homogeneous catalytic arene hydrogenation can take place by free-radical mechanisms that extract hydrogen atoms. Feder and Halpern [84] have presented evidence that $HCo(CO)_4$ catalyzed hydrogenation of anthracene derivatives takes place by a two-step hydrogen atom transfer (6.56). Although this

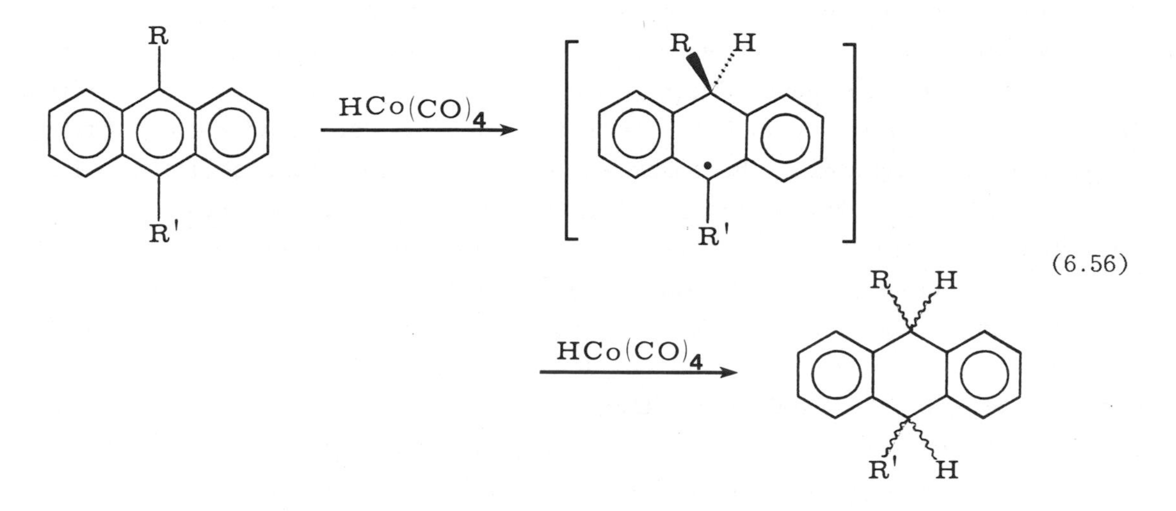


(6.56)

case is not synthetically useful, it does indicate that such free-radical hydrogenations are viable possibilities in arene and olefin hydrogenation [85].

5. A Model for Hydrogenase. Catalysis of the hydrogenation of 1-hexene by Pd(II) and Ni(II) complexes of a number of ligands has been studied by Henrici-Olive and Olive [21]. The most active of these Pd(salen) was studied in considerable detail. This complex shows many similarities to the enzyme hydrogenase [20], which catalyzes the exchange between $H_2$, $H^+$, and a biological electron carrier, E (see Equation 6.57).

$$H_2 + E_{ox} \xrightleftharpoons{\text{hydrogenase}} 2\,H^+ + E_{red} \tag{6.57}$$

Suspensions of the planar complex Pd(salen), 65, in ethanol activate molecular hydrogen thereby catalyzing H-D exchange as well as olefin isomerization and hydrogenation:

$$\underline{n}\text{-}C_4H_9CH{=\!=}CH_2 \xrightarrow[H_2,\ 20^\circ]{Pd(II)(salen)} \underline{n}\text{-}C_6H_{14} \tag{6.58}$$

Catalytic activity is completely inhibited in acidic media. In alkaline media an increase in activity was noted, and a rate maximum was observed at a particular concentration of NaOH. This rate-pH profile is analogous to that of hydrogenase. Under $D_2$ in the presence of an olefin, many deuterated complexes were observed, but none were detected using $H_2$ in EtOD. Homogeneous hydrogenation and exchange reactions were observed with DMF solutions of this catalyst. Rates were measured as a function of hydrogen, olefin, and catalyst concentration.

The postulated mechanism of $H_2$ activation involves dissociation of basic phenolic oxygen of the salen ligand (see Chapter 3) and promotion of a "heterolytic" cleavage of hydrogen [19] of the type discussed in Chapter 4. In this particular case it has been suggested that the phenolate oxygen acts as the basic site, leaving the coordination sphere on accepting a proton, but remaining attached to the complex (6.59). The

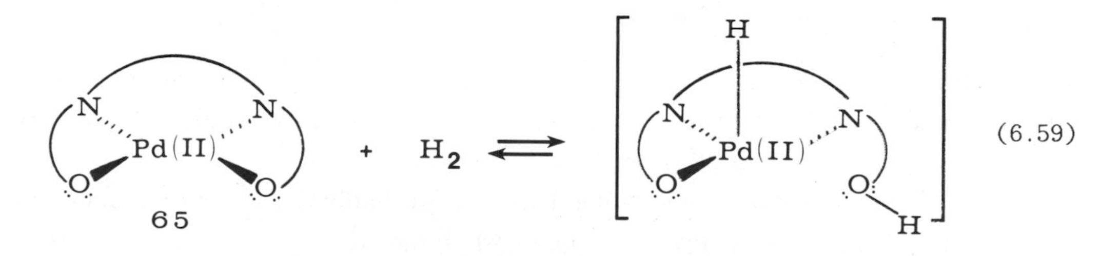


reported kinetic analysis is consistent with the mechanism in Equations 6.60 to 6.62.

$$PdL + H_2 \overset{K_1}{\rightleftharpoons} HPd(LH) \tag{6.60}$$

$$HPd(LH) + \text{>C=C<} \overset{K_2}{\rightleftharpoons} R\text{—}Pd(LH) \tag{6.61}$$

$$R\text{—}Pd(LH) \xrightarrow{k} RH + PdL \tag{6.62}$$

$$L = (salen)$$

Values were estimated for $K_1$, $K_2$, and k ($3.6 \times 10^3$ ℓ $mol^{-1}$, 4.2 ℓ $mol^{-1}$, and $6.2 \times 10^{-3}$ $s^{-1}$ at 20°C).

6. Free Radical Hydrogenation. Free-radical mechanisms have been implicated in some olefin hydrogenations that are catalyzed by transition-metal hydrides. These are similar to the free-radical arene hydrogenation. The mechanism for the hydrogenation of α-methyl styrene by $HMn(CO)_5$ shown in Equations 6.63 to 6.65 has been established by Halpern [85] using CIDNP (chemically induced dynamic nuclear polarization). For

$$CH_2{=}C(Me)Ph \quad + \quad HMn(CO)_5 \xrightarrow{k_a} (Me)_2\dot{C}{-}Ph \quad + \quad \bullet Mn(CO)_5 \qquad (6.63)$$

$$Me_2\dot{C}{-}Ph \quad + \quad HMn(CO)_5 \xrightarrow{fast} Me_2C(H){-}Ph \quad + \quad \bullet Mn(CO)_5 \qquad (6.64)$$

$$2\ \bullet Mn(CO)_5 \xrightarrow{fast} Mn_2(CO)_{10} \xrightarrow[H_2]{k_b} 2\ HMn(CO)_5 \qquad (6.65)$$

the stoichiometric reaction the rate-determining step, $k_a$, is hydrogen-atom transfer from the metal hydride to the olefin. The catalytic reaction requires high temperatures (~200°) to regenerate $HMn(CO)_5$ by hydrogenolysis of $Mn_2(CO)_{10}$, $k_b$.

Free-radical hydrogenation mechanisms are probably more common than is generally appreciated. High temperatures, conjugated substrates, and coordinatively saturated metal hydrides are conditions under which such free-radical mechanisms are more favorable.

7. Catalysis of Hydrogen Transfer Reactions. For some hydrogenation reactions other reducing agents may be used in place of $H_2$; e.g., $NH_2NH_2$, $NaBH_4$, alcohols, and $HCO_2H$. References to such cases are collected in Birch's review [1d]. For example, formic acid may be used with a ruthenium hydride catalyst to reduce terminal olefins [86] (Equation 6.66). Under these conditions $H_2$ is ineffective! Such systems have not been thoroughly developed and merit further study.

$$RCH{=}CH_2 \quad + \quad HCO_2H \xrightarrow[DMF]{Ru(H)ClL_3} RCH_2CH_2 \quad + \quad CO_2 \qquad (6.66)$$

An interesting homogeneous catalyst for hydrogen transfer has been reported by Henbest [1d]. The combination of an iridium chloride salt, a hydrogen donor such as wet isopropanol, and a ligand such as trimethylphosphite in acidic medium will reduce cyclohexanones to predominately axial alcohols. Dimethyl sulfoxide (DMSO) or phosphorus acid may be used in place of trimethylphosphite. In some cases the phosphite or DMSO appear to be the primary reducing agents. Examples are shown in Equations 6.67 and 6.68.

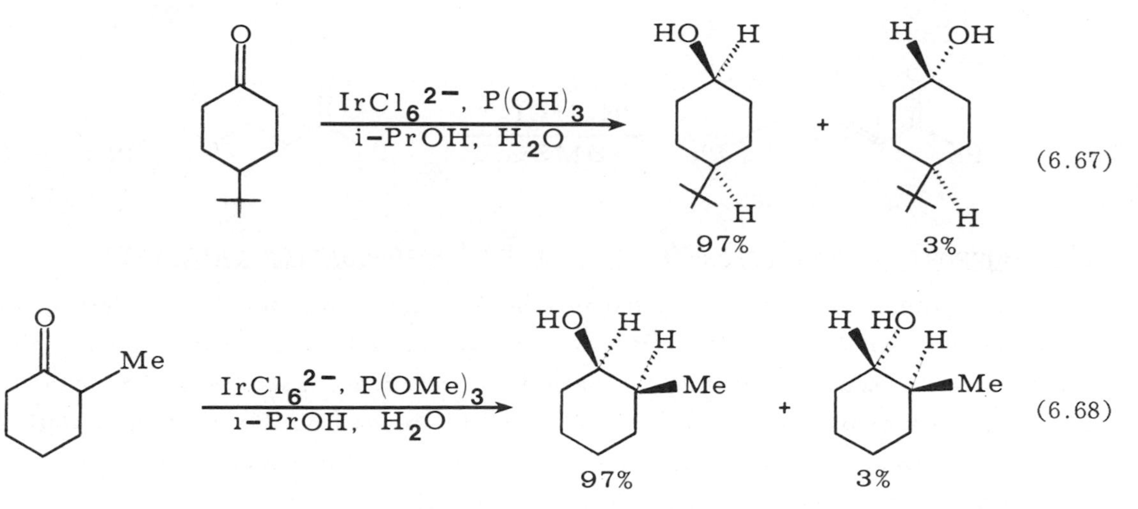


This reagent is sensitive to steric hindrance in the substrate. Thus in steroid substrates 2- and 3-keto groups are reduced under conditions where 11-, 17-, and 20-keto groups are unaffected (Equations 6.69 and 6.70). Aldehydes are reduced, as are certain conjugated olefins (6.71); in this example the keto group is not reduced under these conditions.

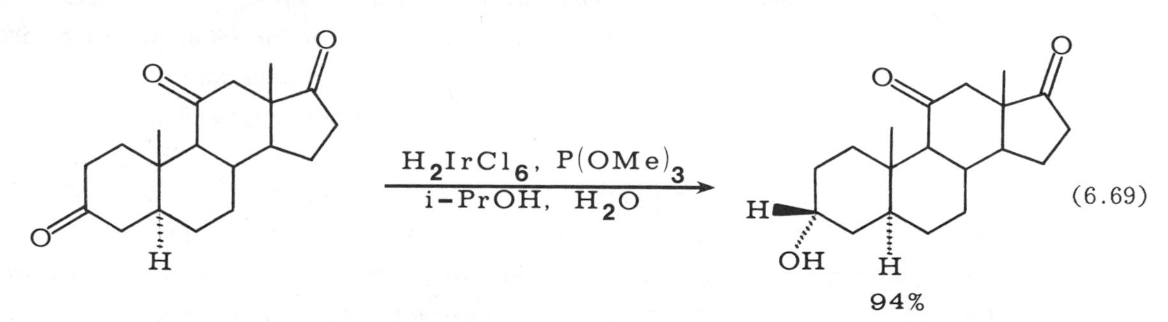

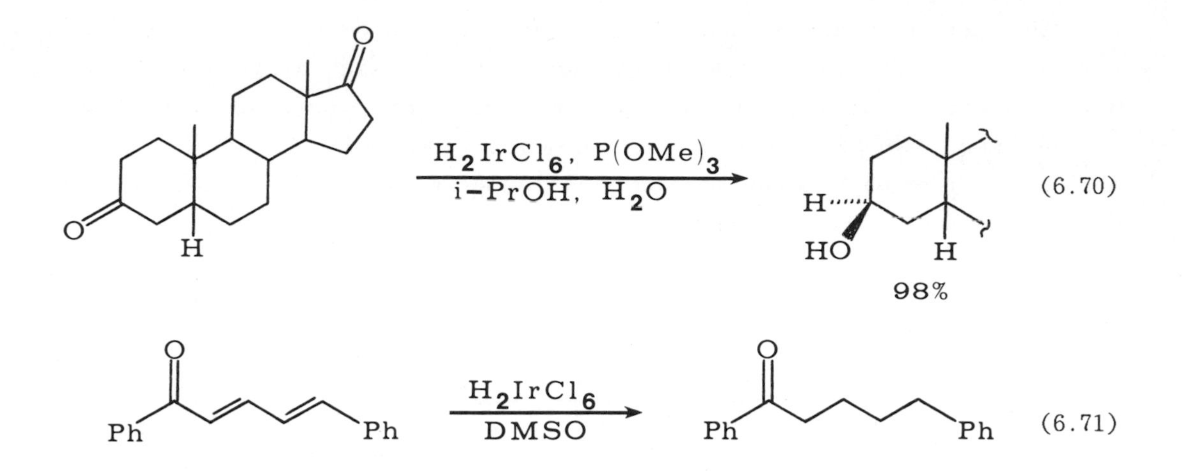


## 6.6. SUPPORTED TRANSITION-METAL COMPLEX HYDROGENATION CATALYSTS.

During the past decade extensive research has been devoted to homogeneous transition-metal catalysts which are attached to insoluble supports. Such "hybrid catalysts" have been employed for a variety of reactions, such as hydroformylation, Wacker ethylene oxidation, olefin hydrogenation, hydrosilation, dimerization, metathesis, and acetylene cyclo-oligomerization. This subject has been reviewed [87].

The most widely studied hybrid catalysts are olefin hydrogenation catalysts. For this reason supported catalysts will be discussed here, in the context of hydrogenation catalysts.

In principle, hybrid catalysts could combine the advantages of soluble catalysts—homogeneous catalyst sites, utilization of all metal atoms, fast reaction rates, and high selectivity (including asymmetric induction)—with certain valuable characteristics of heterogeneous catalysts: facile separation of the products from the catalyst along with recovery and reuse of the expensive catalyst. Membrane systems and water-soluble catalysts derived from sulfonated phosphines are alternative new approaches to catalyst recovery [88].

In other respects hybrid catalysts might even be superior to their soluble analogues. For example, homogeneous catalysts are usually coordinatively unsaturated and thus have a tendency to aggregate, blocking the coordination sites necessary for catalysis. Immobilization of highly reactive, unsaturated metal complexes on a rigid support could theoretically result in a higher concentration of the active catalyst than could be attained in solution. Site-isolated catalysts should also be incapable of undergoing binuclear reductive-elimination steps. If such a step were involved in a homogeneously catalyzed reaction cycle, the heterogenized catalyst might exhibit different properties from its homogeneous analog (see Chapter 8). Another possible advantage is that complex catalysts bound to solid inorganic oxide supports such as silica can in principle be used with virtually any solvent or even for gas-phase reactions. Thus far, this prospect has not been widely examined.

The hypothetical advantages of hybrid catalysts have seldom been realized in practice. The use of anchored catalysts is complicated by their preparation and characterization and by special properties of the support material. Nevertheless, this technique should eventually become very significant.

The majority of hybrid catalysts reported in the literature are attached to the support by covalent rather than ionic or van der Waal's bonds. Herein we will focus on the two most widely studied support materials: cross-linked polystyrene and silica. Each has contrasting advantages and disadvantages which should be considered by anyone planning to work with hybrid catalysts.

1. Analysis of Supported Catalysts. One experimental problem which is common to both cross-linked polystyrene and silica-supported catalysts is analysis of the attached catalytic complex. Such anchored complexes are usually present in low concentrations compared with the large amount of support material. To characterize the structure, homogeneity, and concentration of an anchored complex is a difficult problem. A battery of methods have been employed to analyze hybrid catalysts, but one can never hope to map their structures or measure their purities as thoroughly as one can with crystalline, soluble complexes. The most commonly employed technique is elemental analysis, especially the measurement of heteroatom ratios; however, accuracy is limited by the large support background. Another very useful method is Fourier transform infrared spectroscopy (FT-IR) using the spectral-difference technique, which subtracts out the spectrum of the support to reveal the spectrum of the anchored complex [89]. The latter may be compared with the spectrum of a well-characterized soluble analog. UV-visible spectroscopy is seldom very useful, even with the diffuse-reflectance method [90].

In a few cases X-ray photoelectron spectroscopy (ESCA) has been used to discover the oxidation state of the metal [91,92] and to establish whether bridging ligands are present [93]. It is important to recall that ESCA is sensitive to the first few layers near the surface.

In cases where it is suspected that a supported metal complex has decomposed to the metal, electron microscopy has been useful in detecting metal crystallites [94,95].

Electron microprobe analysis has been used to measure the uniformity of the catalyst distribution within a hybrid catalyst bead or particle [96-98].

Fourier transform $^{31}P$ NMR can be used to learn about the nature of phosphorus ligands within a cross-linked polystyrene resin--especially if the resin is solvent-swollen [96]. Solid-state, "magic angle" NMR techniques should also be useful for the analysis of these polymeric catalysts.

The interatomic distances and the identity of certain ligand atoms in the complex coordination sphere can be studied by the new, but difficultlyaccessible EXAFS technique (extended X-ray absorption fine structure analysis [99].

2. Polymer-Supported Catalysts. Most of the reported hybrid catalysts have been attached to cross-linked and thus insoluble organic polymers (resins). Polystyrene resins are by far the most commonly used catalyst supports. Two main types of polystyrene resins have been used: microporous ("microreticular") and macroporous ("macroreticular"). Each is derived from copolymerization of divinylbenzene and styrene in various ratios.

Gel type, microporous resins have been more commonly employed, typically with 1 or 2 percent divinylbenzene crosslinks. The interior of microreticular resins is chemically accessible only when they are swollen by solvents. Under these circumstances, the gel-type polymer is so internally mobile that ligands which are widely spaced along the polymer chain can coordinate with a single metal [99,100]. This characteristic renders these microreticular resins poor supports for site isolation of complexes which tend to dimerize or otherwise aggregate. The reaction shown in Equation 6.72 (where

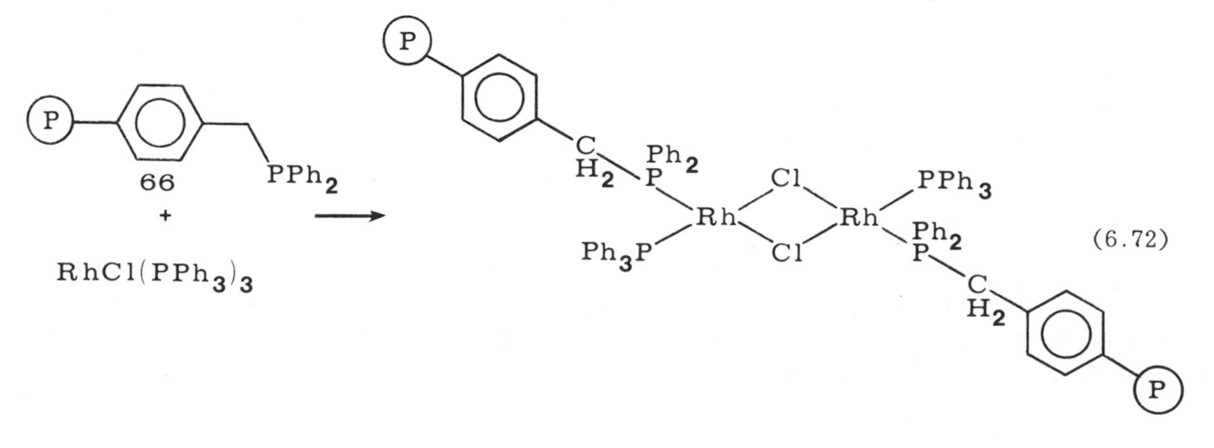

66 is a polymeric ligand) illustrates this phenomenon. In the absence of a solvent which will swell microporous resins, metal sites are inaccessible to substrates, and catalytic activity may be completely lost.

Macroreticular resins are prepared by copolymerizing styrene and divinylbenzene in the presence of a third component which is a good solvent for the monomers but does not swell the cross-linked polymer [101]. The resulting resin has large diameter pores, but nominal shrinking and swelling properties. With macroreticular resins the degree of mobility or flexibility of the polymer backbone is less than with microreticular resins; however, the interior benzene rings in the former are chemically much less accessible. In many instances macroreticular resins can be functionalized only at the surface of the polymer bead or in very large pores, thus limiting the concentration of ligands or metal sites which can be incorporated into the polymer support.

The degree of site isolation within functionalized polymer resins is controversial [89,102]. Polymer-bound functional groups can sometimes be isolated, especially with highly crosslinked, macroreticular resins [99], [103,104]; however, this phenomenon is time-dependent [105], difficult to predict, and difficult to control.

Functionalized resins may be prepared either by introducing ligand groups into benzene rings in commerical resins or by copolymerizing a functionalized monomer with styrene and divinylbenzene. Chloromethylated and phosphine substituted polystyrenes can be purchased. An extensive washing procedure has been recommended to remove impurities from commercial polystyrene resins before introducing ligand groups [87a]. Numerous methods have been developed for introducing ligand groups into polystyrene [87] (Equation 6.73). Care must be taken to employ high-yielding reactions, since byproducts which may remain attached to the polymer can interfere with catalyst activity and characterization. Diphenylphosphino functionalized resins have been the most commonly used polymeric ligands. Note that $MeOCH_2Cl$ is a carcinogen which can be used only with extreme caution. The use of $EtOCH_2Cl$ and $BF_3 \cdot OEt_2$ is preferred for this reason [106].

Metal complexes are usually introduced into these ligand-substituted polymers by ligand-exchange reactions such as those shown in Equations 6.74 and 6.75. In view of the problem of site isolation and internal mobility, the exact nature of the complexes obtained by this method may be obscure. Subtle differences may also arise from using polymeric ligands prepared by different methods.

In most cases the activity of a polymer-bound hydrogenation catalyst is less than that of its soluble analogue. For example, a polymer-supported analogue of Wilkinson's

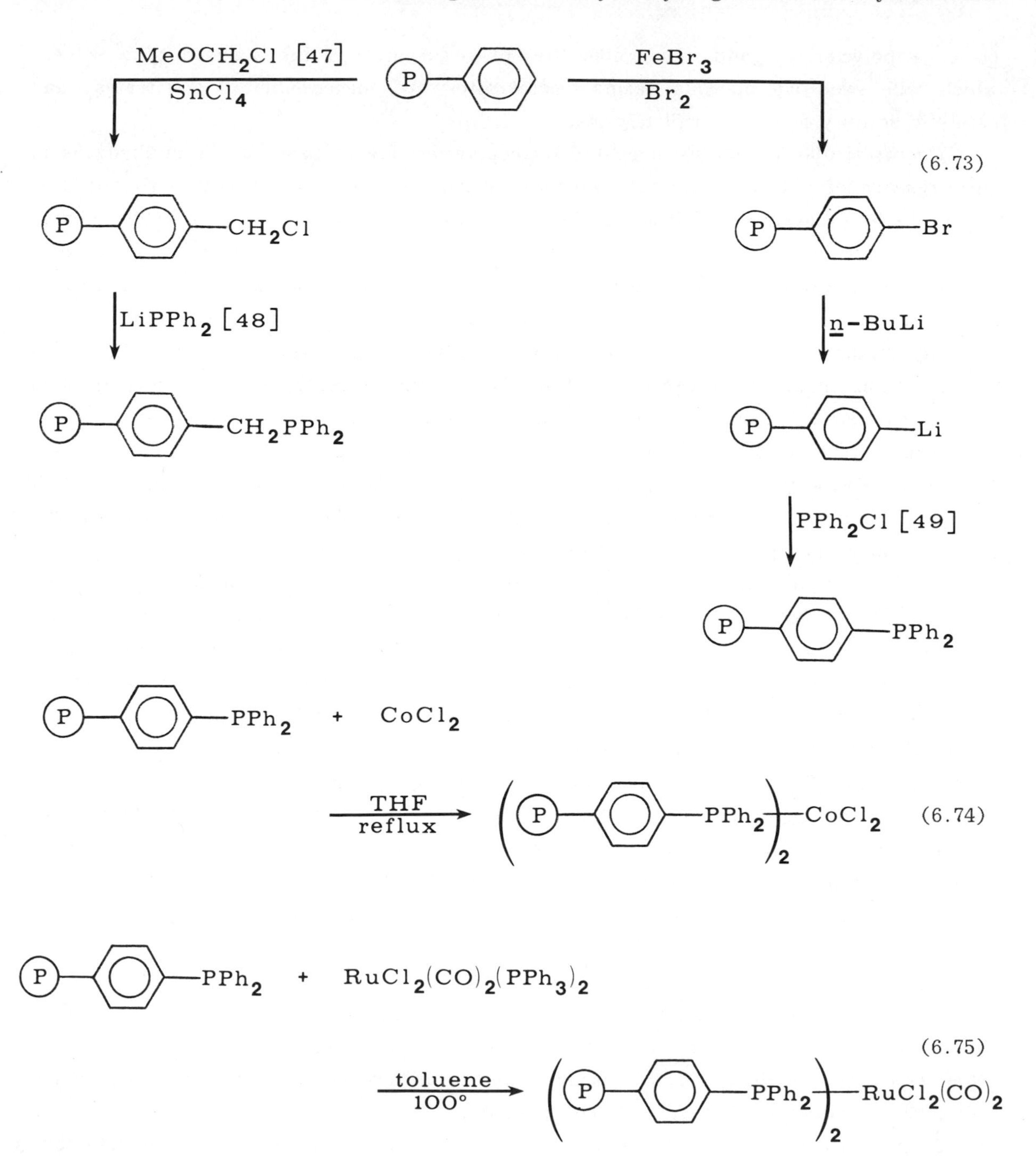

MeOCH2Cl [47]
SnCl4
P
FeBr3
Br2
(6.73)
P
CH2Cl
P
Br
LiPPh2 [48]
n-BuLi
P
CH2PPh2
P
Li
PPh2Cl [49]
P
PPh2
P
PPh2
+
CoCl2
THF
reflux
P
PPh2
2
CoCl2
(6.74)
P
PPh2
+
RuCl2(CO)2(PPh3)2
(6.75)
toluene
100°
P
PPh2
2
RuCl2(CO)2

catalyst is 1/16 as active as the soluble complex [107]. This may be due to the relatively slow diffusion of substrate olefins to the catalyst sites or to inhibition by free ligands along the flexible polymer chain. However, some site isolation and greater catalytic activity can be achieved by using highly crosslinked (18 to 20 percent DVB) polystyrene. For example, Grubbs has prepared a titanocene catalyst anchored to macroreticular 20 percent crosslinked polystyrene. This polymeric catalyst is 60 to 120 times more active for olefin hydrogenation than the corresponding soluble catalyst derived from $Cp_2TiCl_2$ [108]. Because reduced, soluble titanocene compounds tend to form catalytically inactive dimers, the enhanced activity of the supported metallocene was attributed to immobilization of the monomeric complexes on the polymer bead. Similar enhanced activity (100-fold) was obtained with polymeric catalyst, 67, derived from anchored $TiCpCl_3$ complexes [109].

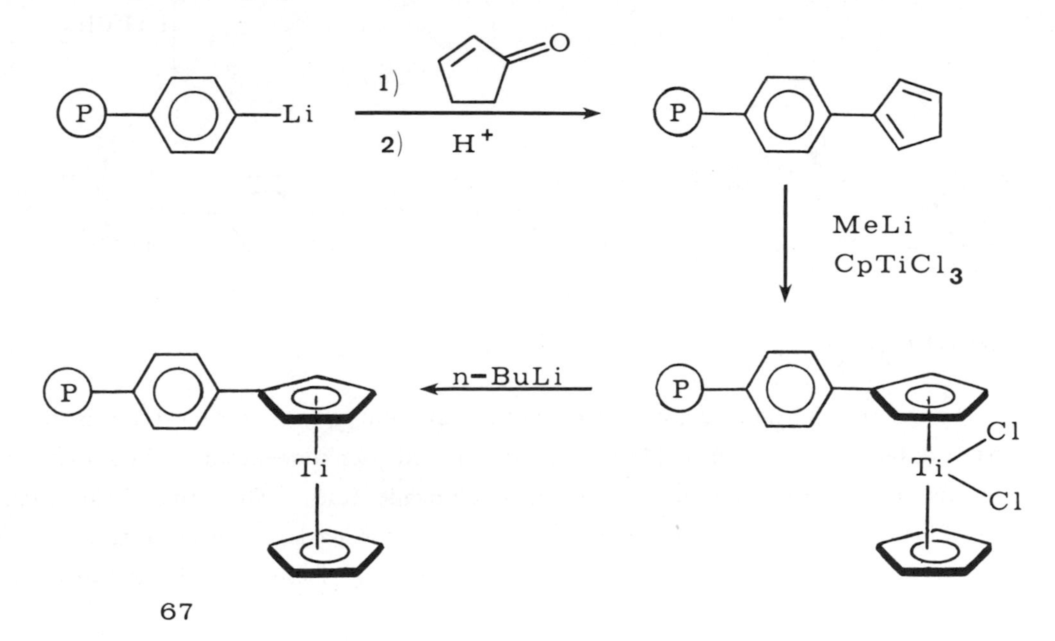


Modest changes have been noted in substrate reactivities for polystyrene anchored analogues of Wilkinson's catalyst [87b].

The availability and ease of modification of polystyrene resin complexes having a wide degree of crosslinking has contributed to their widespread use. However, these resin-bound catalysts are limited by certain physical properties. For example, these polymers have poor mechanical and thermal stability, with a maximum temperature limit of about 160°C [87j].

Since substrates must be able to penetrate the polymer matrix in order to reach catalyst sites, the choice of reaction medium is limited to solvents having good swelling or polymer-penetrating properties. This solvent compatibility problem is illustrated by the following example.

The polymer-anchored analogue of the rhodium DIOP catalyst (Equation 6.76) is

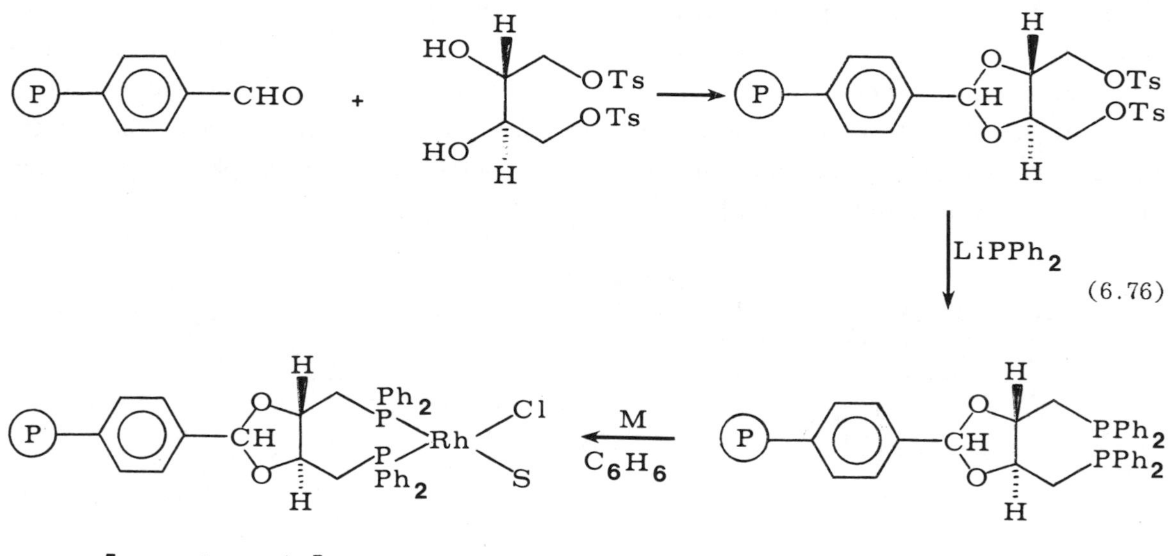


active for the hydrogenation of non-polar olefins in benzene at 25° and 1 atm $H_2$, and for the hydrosilation of ketones [110]. However, in benzene-ethanol this catalyst is *inactive* toward the hydrogenation of α-acetamidocinnamic acid. The ethanol cosolvent is required to dissolve the polar substrate and to afford the cationic form of this catalyst. However, the polymer is virtually collapsed by this polar cosolvent, making the catalytic sites inaccessible.

This problem has been overcome by preparing a polar DIOP-bearing resin which swells in alcoholic solvents [111]. Catalysts prepared from this copolymer *68 in situ*, using 1:3 benzene-ethanol, are effective in the catalytic asymmetric hydrogenation of α-acetamidocinnamic acid and related compounds (in 1:5 benzene/ethanol, 1-2.5 atm, 25°C). Both reaction rates and the degree of asymmetric induction are similar to the soluble analogue. This example illustrates that polymer chemistry is very important in the application of hybrid catalysts. To achieve significant advances in heterogenized catalysts, one must be willing to prepare new functionalized polymers.

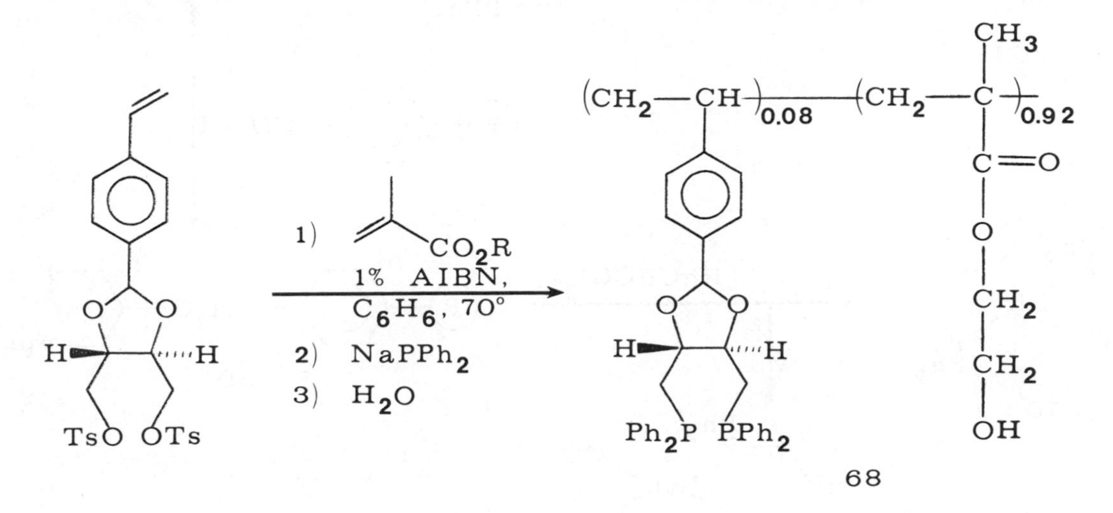


The polymeric DIOP catalysts are unusual in the sense that a chelating diphosphine is involved; most polymeric catalysts contain monodentate phosphines. However, it is important to recognize that bidentate phosphines may not form chelated complexes within a flexible polymer. Consider, for example, the polymeric bidentate rhodium(I) complex [112] derived from the chelating phosphine 69 as shown in Equation 6.77. Infrared studies [113] indicate trans phosphine coordination, 70, rather than the expected cis coordination, 71.

Even though few supported catalysts have so far been derived from chelating phosphines, the latter appear to offer some advantage over monodentate phosphines. Supported catalysts prepared from monodentate phosphines are more susceptible to two problems: (1) reduction of the complex to elemental metal particles; and (2) leaching of the active catalyst from the support by other ligands, such as CO or soluble phosphines.

When highly unsaturated catalysts are produced on a support, for example, by hydrogenation of a coordinated diene, there is a marked tendency to form metal particles, which are themselves very active catalysts of the classic heterogeneous type. For example, the latter may hydrogenate arenes. These particles can sometimes be detected by visual inspection; however, electron microscopy is a more reliable method. Such metal particles have been reported in both polystyrene [95b,114] and silica-supported

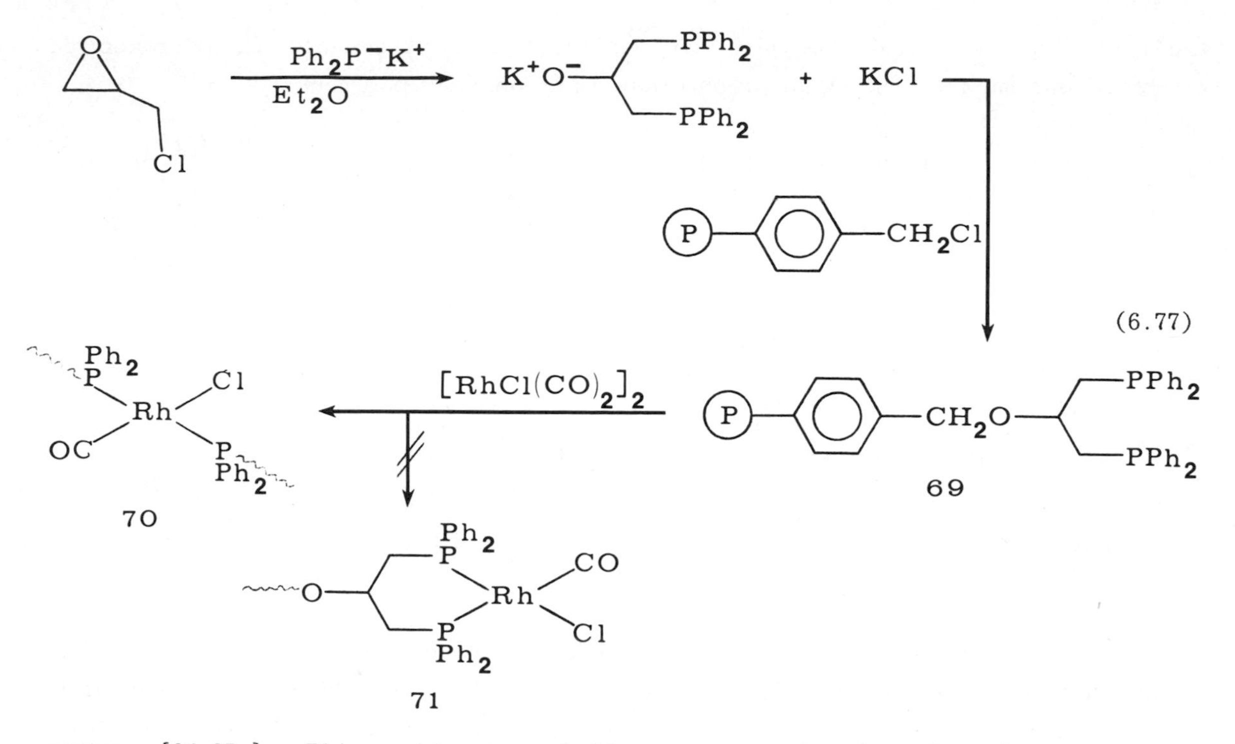


systems [94,95a]. This problem is probably more pervasive than those few cases would suggest.

Other evidence indicating that catalyst bound to supports by monodentate ligands may have poor stability comes from leaching studies. Instances of metal elution have been reported from organic [114-116] and inorganic [117,118] supports functionalized with various monodentate ligands ($-NR_2$, $-PPh_2$, -SH). Complexes attached by multiple links appear to be more stable [87j,116-118], but loss of metal still can occur with some bidentate ligands [114,116-118]. A small amount of metal leaching is very difficult to measure.

A recent development in polymer-bound catalysts should be mentioned here. That is the technique of using a soluble polymer which can be separated from the reaction mixture by membrane filtration [119]. For example, Whitesides [120] has attached the bis-phosphine to a water-soluble protein, 72, and formed a rhodium catalyst which was used as a polymeric catalyst for asymmetric induction. A modest enantioselectivity was obtained.

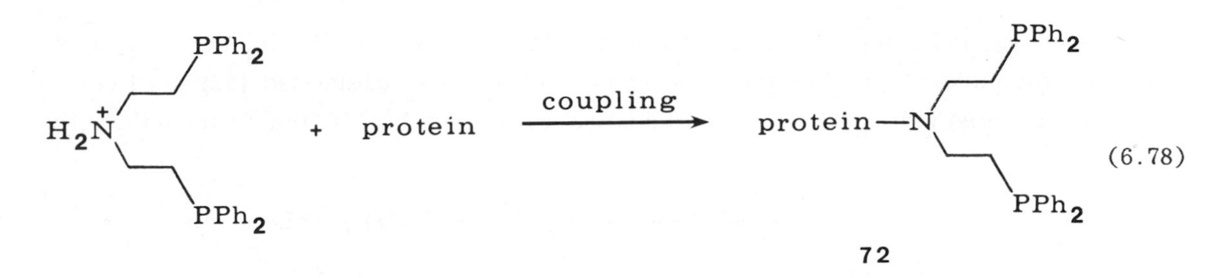


(6.78)

3. Silica-Supported Catalysts. Homogeneous catalysts bound to silica surfaces appear to offer advantages over the catalysts supported on resins. However, these potential advantages are offset by more severe problems involving synthesis and characterization, with the result that silica is still much less commonly used as a catalyst support. Silica is an attractive support because of several factors: spherical and porous silicas having large surface areas are commercially available; silica surfaces are rigid matrices which are not subject to swelling by solvents; and they have very high thermal stability. There is a large literature concerning the physical and chemical nature of silica surfaces [121]. Three species are present on silica surfaces as shown in Figure 6.6: (a) free hydroxy (silanol) groups; (b) pairs of hydrogen bonded silanol

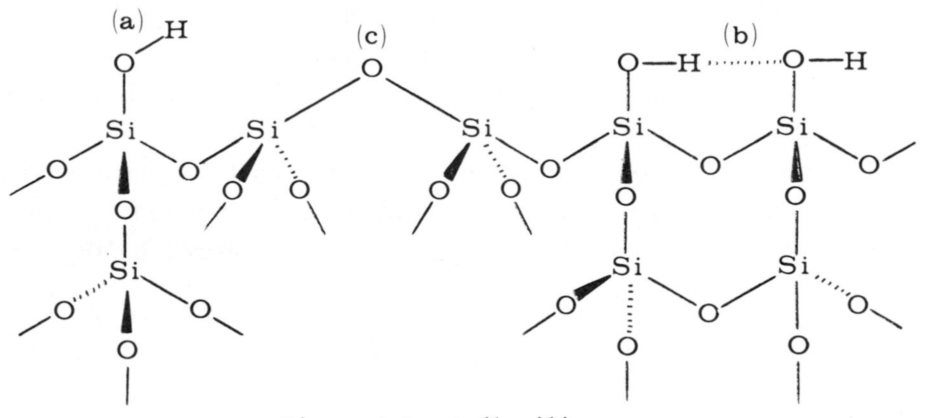


*Figure 6.6. Bulk silica.*

groups; and (c) siloxane bridges. The vicinal hydroxyl pairs (b) are much less reactive than the free silanols (a), which are believed to be the principal sites at which silica surfaces may be functionalized. We shall represent these reactive silanol groups by the symbol Si—OH. The maximum concentration of groups such as (a) is roughly one per 100 $Å^2$ on a surface area of about 300 $m^2$ per gram.

Various methods have been developed for binding ligands (L) to silica surfaces, principally for chromatographic supports. These have been summarized [122]. The two most important methods are shown in Equation 6.79a and b. In the first method the

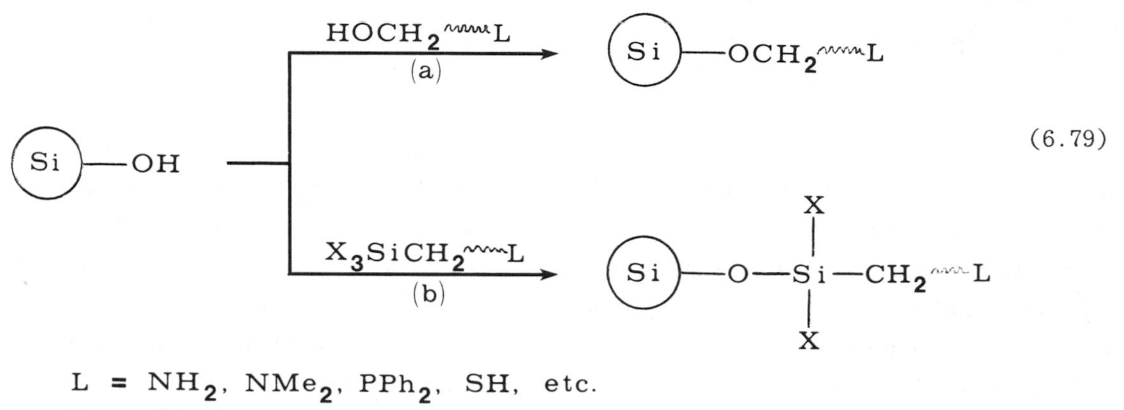


L = $NH_2$, $NMe_2$, $PPh_2$, SH, etc.
X = Cl, OR, $RCO_2$, etc.

silica surface is esterified with an appropriately substituted alcohol. In the second method, a condensation is carried out using a suitably functionalized organosilane. In order for this condensation to occur, at least one substituent on the organosilane should be reactive: -Cl, $-OCH_3$, or $-O_2CCH_3$. The second method is preferred because of the strong Si-O-Si bond and the greater thermal and hydrolytic stability of the Si-C link [123].

Several ligand substituted organosilanes are commercially available (L = $NH_2$, -py, -SH, etc.), but others cannot be purchased and must be prepared. A useful example containing the diphenylphosphino group 73 is readily synthesized by the method shown [124].

$$(EtO)_3SiCH{=}CH_2 + HPPh_2 \xrightarrow{h\nu} \underset{73}{(EtO)_3SiCH_2CH_2PPh_2} \qquad (6.80)$$

Unfortunately, many other potentially interesting ligand-substituted organosilanes are very difficult to prepare. These synthetic problems have impeded research on silica-supported catalysts. Another approach to the functionalization of silica surfaces involves prior coating with a rigid-ladder polymer polyphenylsiloxane. These low-molecular-weight units are further condensed by heating [125] (Equation 6.81). The

$$\text{(Si)—OH} + \left(\begin{matrix}\text{Ph} \\ \text{—O—Si—} \\ \text{—O—Si—} \\ \text{Ph}\end{matrix}\right)_x \xrightarrow[\text{2) 280°C, 10 min.}]{\text{1) THF}} \text{(Si)—C}_6\text{H}_5 \qquad (6.81)$$

aromatic rings may then be functionalized like polystyrene. This multistep process is not completely reproducible [95a].

Metal complexes may be anchored on functionalized silica surfaces as on polystrene substituted ligands. An example is the bridge-splitting reaction (6.82). However, in reactions of this sort, assumptions must be made about the nature and purity of the surface-bonded complex. These assumptions are difficult to verify.

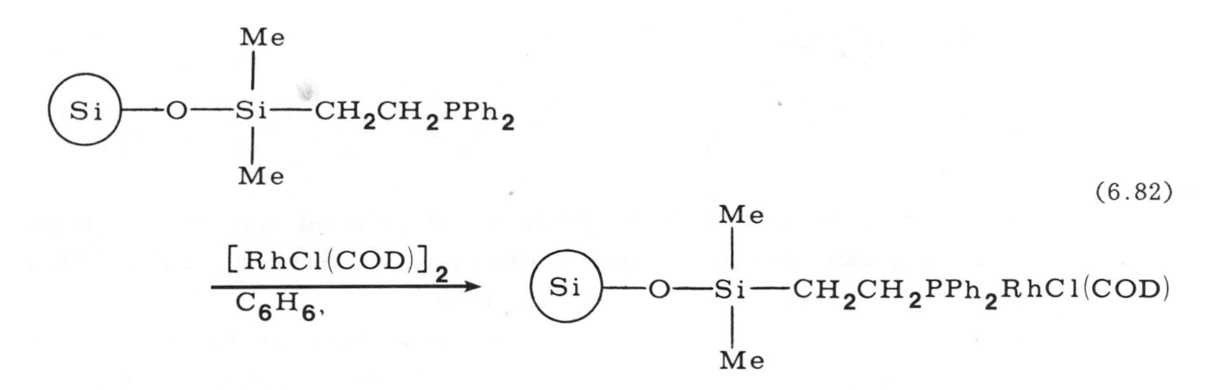


An alternative approach first reported by Allum [94] is now widely employed [87j, 116-118, 124a, 126]. A metal complex of a ligand-substituted silane, 74, is first prepared and subsequently bound to the silica surface. This approach should lead to a fairly well-defined surface species, 75, and it also permits one to control the ratio of anchored ligand to metal (L/M). In fact, L/M ratios greater than 1.0 would be difficult to achieve using the ligand-functionalized silica technique because of the widely spaced groups on the rigid silica surface. An example of this technique [122] is the preparation of the silica-bonded trisphosphine complex 76.

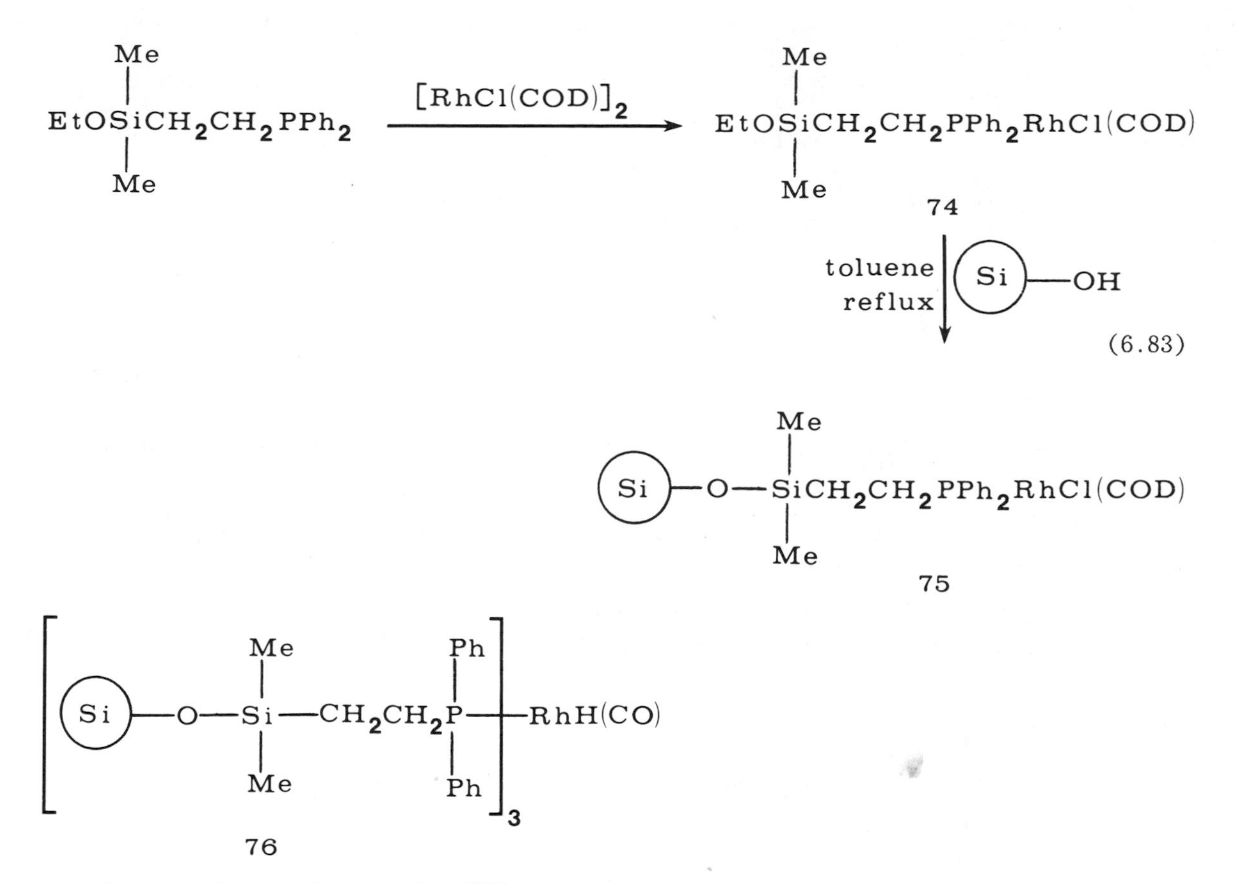


In actual practice, it is difficult to obtain the silane-metal complexes as pure, crystalline solids, and these are used without purification. Furthermore, it is not clear that all the silated ligands become bound to the silica surface.

The catalysts anchored on silica should be truly immobilized and freely accessible to solvent, and therefore should be superior to the polystyrene systems. Thus far, there have been relatively few quantitative comparisons between silica-bound catalysts and their soluble analogues. The relative reactivities of the hydrogenation catalysts derived from the silica-anchored complex 77 and its soluble analogue 78 were found to be comparable [127]. In contrast to similar silica-bonded complexes having a single monodentate phosphine, the complex derived from the chelating phosphine showed less tendency to form metal particles upon exposure to hydrogen in the presence of olefin substrate.

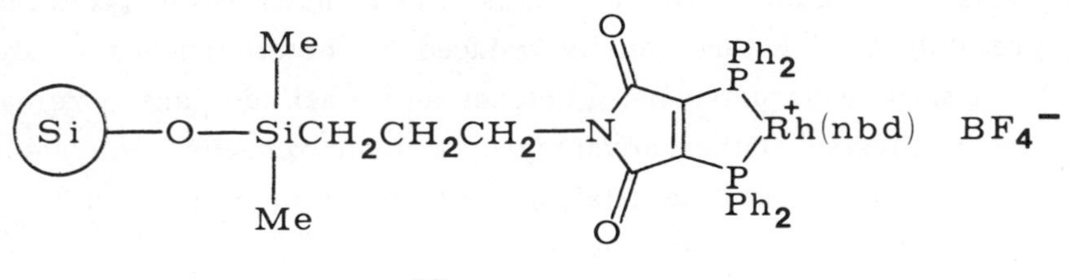


77

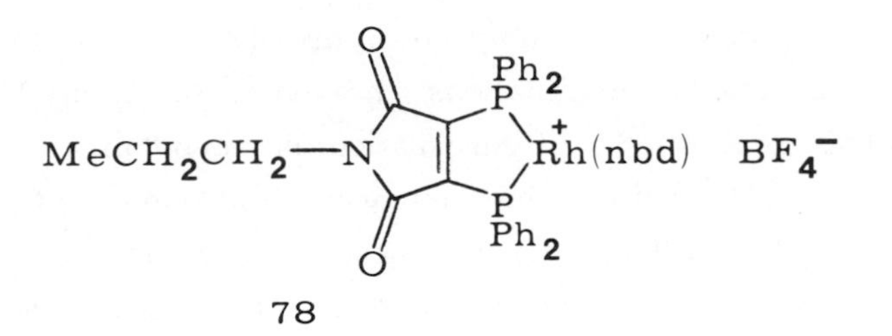


78

## 6.7. RELATIVE ADVANTAGES OF HOMOGENEOUS VERSUS HETEROGENEOUS HYDROGENATION CATALYSTS.

Homogeneous catalysts are typically more sensitive toward steric effects and are thus much more selective in the reduction of less-hindered olefins, such as terminal versus internal, cis versus trans, 1,1-versus 1,2-disubstituted, and disubstituted versus trisubstituted olefins. Certain homogeneous catalysts afford selective deuteration: cis-dideuteration of monoenes and 1,4-dideuterations of S-cis-1,3-dienes. Homogeneous catalysts of the dihydride type can often be employed with neither olefin rearrangement nor disproportionation of 1,4-dihydroaromatics. Hydrogenolysis reactions are rarely observed among homogeneous catalysts, whereas heterogeneous catalysts frequently cleave carbon-heteroatom bonds. Olefins and other groups in polymeric substrates are more easily reduced by homogeneous catalysts, especially in gels and solvent-swollen, cross-linked polymers.

The ultimate selectivity of homogeneous catalysts is the high degree of asymmetric reduction which is so far limited to specific types of prochiral olefins. Nothing approaching this selectivity has so far been discovered among heterogeneous catalysts. In fact heterogeneous catalysts usually have several active sites, differing in reactivity and selectivity. Heterogeneous catalysts are usually much more reactive. For example, heterogeneous catalysts reduce aromatic rings under much milder conditions than do homogeneous precatalysts. Esters can be reduced by heterogeneous catalysts but apparently not by homogeneous catalysts. Ketones and aldehydes are more easily reduced by heterogeneous catalysts. Other advantages of heterogeneous catalysts are the ease of separating the products from the catalysts and the greater range of temperature and hydrogen pressures which may be employed.

## 6.8. HOMOGENEOUSLY CATALYZED HYDROSILATION.

The addition of Si-H bonds to olefins, acetylenes, aldehydes, and ketones, "hydrosilation," is catalyzed by many soluble transition-metal complexes [2,128]. Both homogeneously and heterogeneously catalyzed hydrosilations are known. These reactions are similar to catalytic hydrogenations; however, olefin hydrosilations are more exothermic by about 5 kcal $mol^{-1}$. Hydrosilation is important in the manufacture of some silicon derivatives [129]. In a few instances, additions of Ge-H bonds have also been reported [130], but we will not discuss these rare examples.

Catalytic hydrosilations are very complex, and their mechanisms are not well-understood. Plausible reaction schemes can be written which rationalize the homogeneous hydrosilations. However, in no individual case is a hydrosilation mechanism well-documented. These reactions are troubled by many factors: induction periods, irreproducible kinetics, radical chain paths, olefin rearrangement and dimerization, exchange reactions at silicon, and acute sensitivity of the products to the nature of the hydrosilating reagent. In some situations it is not even clear whether a hydrosilation catalyst is homogeneous. One _must_ consult the original experimental papers before attempting catalytic hydrosilation.

Quite diverse transition-metal complexes act as precatalysts for apparently homogeneous hydrosilation. The same complexes will sometimes also catalyze olefin hydrogenation.

In spite of these complexities, catalytic hydrosilation promises to become an important synthetic method. This expectation is enhanced by examples of asymmetric inductions (_vide infra_).

1. _Mechanism_. A plausible but speculative mechanism for homogeneous olefin hydrosilation is outlined in Figure 6.7. Oxidative-addition of silicon hydrides, which is described in Chapter 4, is presumed to be an obligatory step in catalytic hydrosilations [131]. However, it is unclear whether the olefin substrate is coordinated prior to or subsequent to the Si-H oxidative-addition step, raising the same question of an "olefin path" or a "hydride path" that we found in catalytic hydrogenation. The olefin path

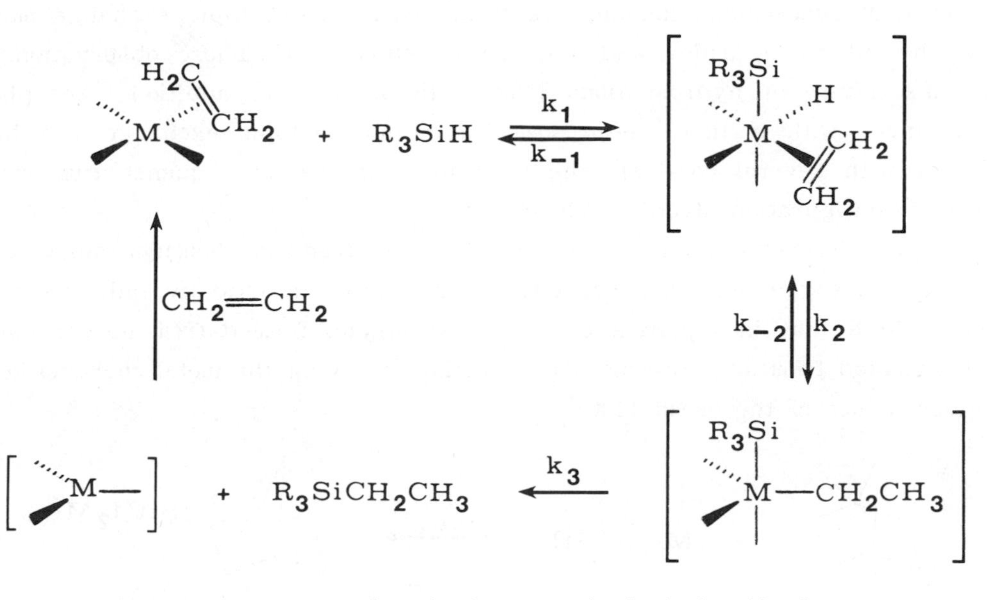


*Figure 6.7. Proposed mechanism for olefin hydrosilation.*

has been suggested in some instances, since olefin complexes are often effective hydrosilation precatalysts.

Consider the mechanism in Figure 6.7. Reversible, cis, Si-H oxidative-addition, $k_1/k_{-1}$, is followed by reversible hydride migratory-insertion, $k_2/k_{-2}$, and then irreversible reductive-elimination of the silane occurs. The oxidative-addition step activates the Si-H bond, and the insertion step establishes the regiospecificity of the addition to the olefin. The reversibility of the first two steps accounts for isotopic exchange involving the Si-H bond, which is occasionally observed, and olefin rearrangement, which is a common side reaction. Little is known about the reductive-elimination step $k_3$. In certain cases both the oxidative-addition and reductive-elimination reactions may be binuclear, for example, in hydrosilations which are catalyzed by $Co_2(CO)_8$. In these cases radical-chain oxidative-addition paths are plausible. The regiospecificity of insertion into unsymmetric olefins and the tendency for olefin rearrangement is quite sensitive to the number and nature of the substituents on silicon. The means by which the silyl group exerts this control over the migratory-insertion step is unknown.

In some cases it has been shown that the $H_2PtCl_6$-catalyzed Si-H addition to olefins is cis [132]. This is the expected result from a reaction sequence like that shown in Figure 6.7.

When deuterium-labeled silanes are used, extensive isotopic exchange may occur between the silicon hydride, and the olefinic substrate. These observations reflect the complex nature of hydrosilations [133]. In other cases addition takes place in a simple manner, with neither olefin isomerization nor isotopic exchange. It has been shown that with several catalysts the over-all hydrosilation sequence can occur with retention of configuration at chiral silicon [134].

That a silyl-metal complex is probably involved in homogeneously catalyzed hydrosilation is suggested by the fact that some silyl-metal oxidative adducts are active catalysts. Note that in Equation 6.84 the exo adduct formed from norbornene is the product expected from an intramolecular insertion involving the metal coordination at the less-hindered face of the olefin [130].

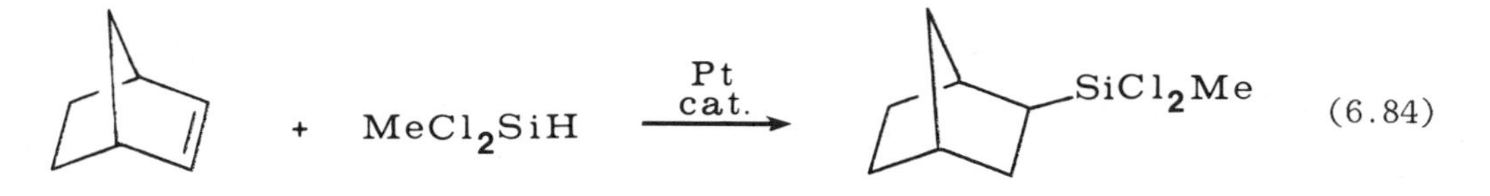


2. General Features of Olefin Hydrosilation. Representative catalysts, silating reagents, and conditions for hydrosilation of olefins are listed in Table 6.5.

Several features of these olefin hydrosilation catalysts are notable. The rate, yield, and product obtained from reaction with 1-hexene are essentially the same when either $H_2PtCl_6$ or $[PtCl_2(C_2H_4)]_2$ is used as a precatalyst. These catalysts are usually more stable when excess olefin is present, and the olefin is typically used as the solvent. Prior treatment of the precatalyst with the silicon hydride tends to deactivate the catalyst. Cyclic olefins are often unreactive, and terminal olefins tend to form n-alkyl silanes (6.85). Depending upon the silane, yields can be diminished by rapid,

$$\mathrm{RCH{=}CH_2} + \mathrm{R_3'SiH} \xrightarrow[\mathrm{C{=}C\ complex}]{\mathrm{Rh(I)\ or\ Pt(II)}} \mathrm{RCH_2CH_2SiR_3'} \qquad (6.85)$$

competing olefin isomerization (see Table 6.5). Branched olefins tend to give several products.

Dicobaltoctacarbonyl exhibits an induction period after which hydrosilation is catalyzed even at 0°C, provided that excess olefin is present. This catalyst system

*Table 6.5. Examples of homogeneously catalyzed olefin hydrosilation.*

| Catalyst | Silane | Substrate | Typical conditions | Major product | Special features | Reference |
|---|---|---|---|---|---|---|
| $H_2PtCl_6$ | $MeCl_2SiH$ | 1-heptene | olefin is solvent 120° | 1-heptyl-$SiMeCl_2$ | 93% yield | [135] |
| $[PtCl_2(C_2H_4)]_2$ | $(MeO)_3SiH$ | 1-hexene | | 1-hexyl-$Si(OMe)_3$ | no isomerization >90% yield | [136] |
| | $Cl_3SiH$ | 1-hexene | olefin is solvent at reflux | 1-hexyl-$SiCl_3$ | extenstive isomerization of excess olefin 50-80% yield | |
| | $Et_3SiH$ | 1-hexene | | 1-hexyl-$SiEt_3$ | extensive isomerization, deactivation of catalyst, poor yields < 20% | |
| $[(C_2H_4)_2RhCl]_2$ | $EtCl_2SiH$ | 1-hexene | olefin is solvent at reflux | 1-hexyl-$SiEtCl_2$ | 50% yield | [136] |
| $Co_2(CO)_8$ | $(MeO)_3SiH$ | 1-octene | olefin is solvent <60°C | 1-octyl-$SiR_3$ | rapid isomerization | [137] |
| | $Et_3SiH$ $PhCl_2SiH$ | | | | | |
| $Pt(C_2H_4)(PPh_3)_2$ | $MeCl_2SiH$ | 1-hexene | excess olefin | 1-hexyl-$SiMeCl_2$ | internal olefins do not react | [138] |
| $[Pt(SiR_3)(\mu\text{-}H)(PCy_3)]_2$ | $Me_2EtSiH$ | styrene | 20° | $PhCH_2CH_2SiMe_2Et$ 92% | mild conditions, cyclic olefins react, Ge-H addition also catalyzed | [130] |
| $RhCl(PPh_3)_3$ | $Et_3SiH$ | 1-hexene | excess olefin, 60° ~2 hr | 1-hexyl-$SiEt_3$ 80% | olefin isomerization | [139,140] |
| | $PhMe_2SiH$ | cis-2-pentene | excess olefin, 25° 2 days | 1-pentyl-$SiPhMe_2$ 84% | trans isomer is unreactive | [140] |
| $\{RhCl(CO)_2\}_2$ | $(EtO)_3SiH$ | $C_2H_4$ | 60 atm., 65° 20 hr. | Et-$Si(OEt)_3$ 95% | ethylene dimerization observed | [140] |

*Table 6.5. Continued.*

| Catalyst | Silane | Substrate | Typical conditions | Major product | Special features | Reference |
|---|---|---|---|---|---|---|
| $RhH(CO)(PPh_3)_3$ | $PhMe_2SiH$ | $CH_2=CHCN$ | 15 min, reflux | $PhMe_2SiCH(Me)CN$ 80% | note regiospecificity | [140] |
| $Pd(PPh_3)_4$ | $Cl_3SiH$ | 1-octene | 100°, 5 hr., excess olefin | n-octyl-$SiCl_3$ 90% | internal olefin unreactive | [141] |
| $NiCl_2(dmpf)$ | $MeCl_2SiH$ | 1-hexene | 120°, 20 hr. | n-hexyl-SiMeClH + n-hexyl-$SiMeCl_2$ (dmpf=1,1-bisdimethyl phosphinolferrocene) | note H,Cl interchange at Si | [142] |
| $Ni(acac)_2$ + $AlEt_3$ | $(EtO)_3SiH$ | 1,3-pentadiene | 3 hr., 20° | $EtCH=CHCH_2Si(OEt)_3$ 96% | best of "Ziegler catalysts" | [143] |

causes extensive olefin rearrangement, even with trialkoxysilanes, which do not induce isomerism in the platinum-catalyzed system.

With ethylene used as a substrate, the relative hydrosilation rates for three diverse catalysts have been estimated [144]; however, these comparisons may depend on the silane and the substrate (Equation 6.86).

$$(XC_6H_4)Me_2SiH \quad + \quad CH_2{=}CH_2 \xrightarrow{\text{cat.}} XC_6H_4Me_2SiCH_2CH_3 \qquad (6.86)$$

| | cat.: | $H_2PtCl_6$ > | $RhCl(PPh_3)_3$ > | $Co_2(CO)_8$ |
|---|---|---|---|---|
| relative rates | | 4000 | 200 | 1 |

Olefin isomerization and a pronounced tendency to form n-alkyl silanes accounts for the rhodium-catalyzed formation of the n-pentyl derivative from cis-2-pentene (Equation 6.87). It is interesting that the more poorly coordinating trans-2-pentene is unreactive under these conditions [140].

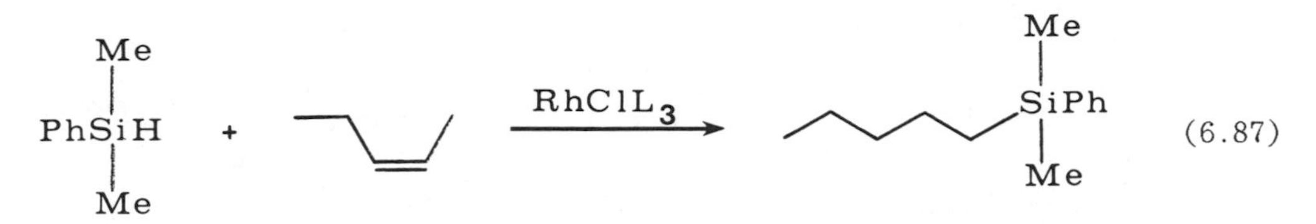


Hydrosilation of acrylonitrile is catalyzed by several Group VIII complexes. The addition of the Si-H bond is highly regiospecific, the silane going to the α-carbon [140] (Equation 6.88). Caution: acrylonitrile is a carcinogen.

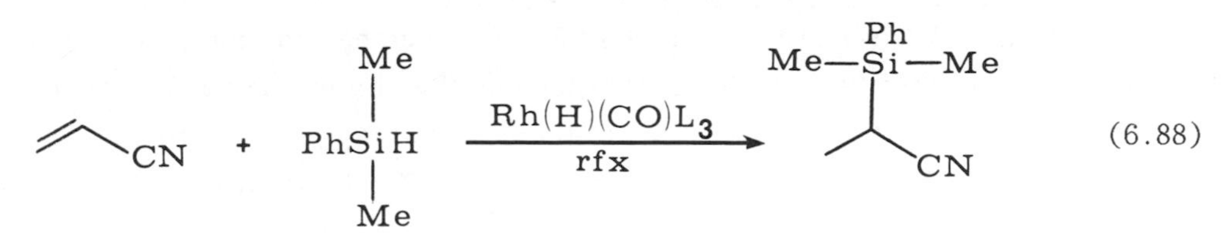


3. Acetylene Hydrosilation. Hydrosilation of acetylenes is even more complex than that of olefins. With terminal acetylenes, the Si-H addition is apparently not regiospecific, so that two monoadducts are obtained [2]. The terminal silane 79 in Equation 6.89 results from cis Si-H addition.

Another difficulty is that both single and double addition can take place as in Equation 6.90, and hydrogen is evolved [145]. An intermediate disilylnickel complex is thought to be involved. Such double silations may become synthetically useful.

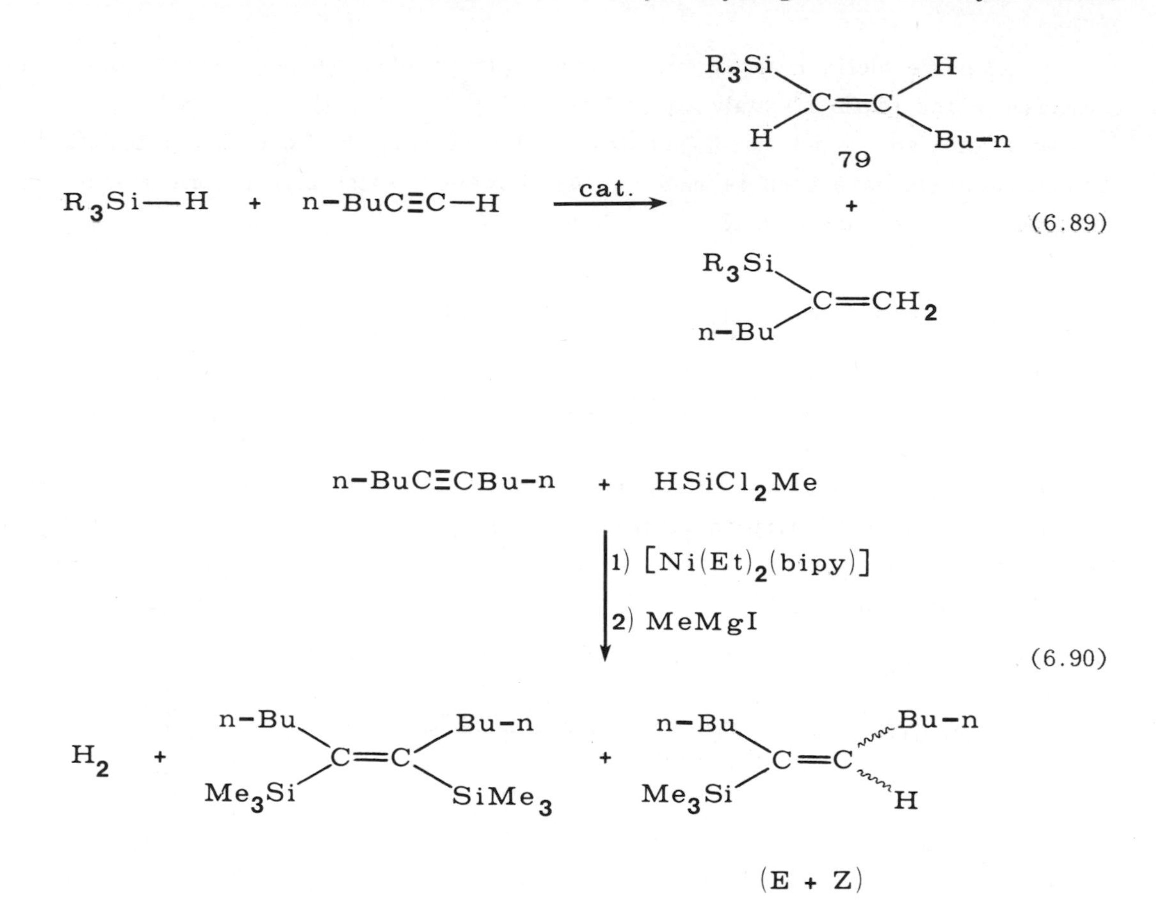


4. Hydrosilation of Aldehydes and Ketones. Because of the avidity which silicon has for oxygen, hydrosilation of aldehydes and ketones is more facile than hydrogenation of other substrates. It is not surprising that hydrosilation of carbonyl compounds is completely regiospecific, the silicon binding to oxygen and the hydride to carbon. Subsequent hydrolysis of the silyl ether product affords an alcohol [146]. Thus hydrosilation of aldehydes and ketones can be considered an alternative to catalytic hydrogenation. Plausible mechanisms have been suggested for homogeneously catalyzed hydrosilation of ketones.

When α,β-unsaturated ketones are used as substrates, either the olefinic or the ketone function may be reduced. It is curious that monohydrosilanes are reported to react exclusively by 1,4-addition (olefin reduction), whereas dihydrosilanes apparently give predominantly 1,2-addition (ketone reduction), as shown in Equation 6.91.

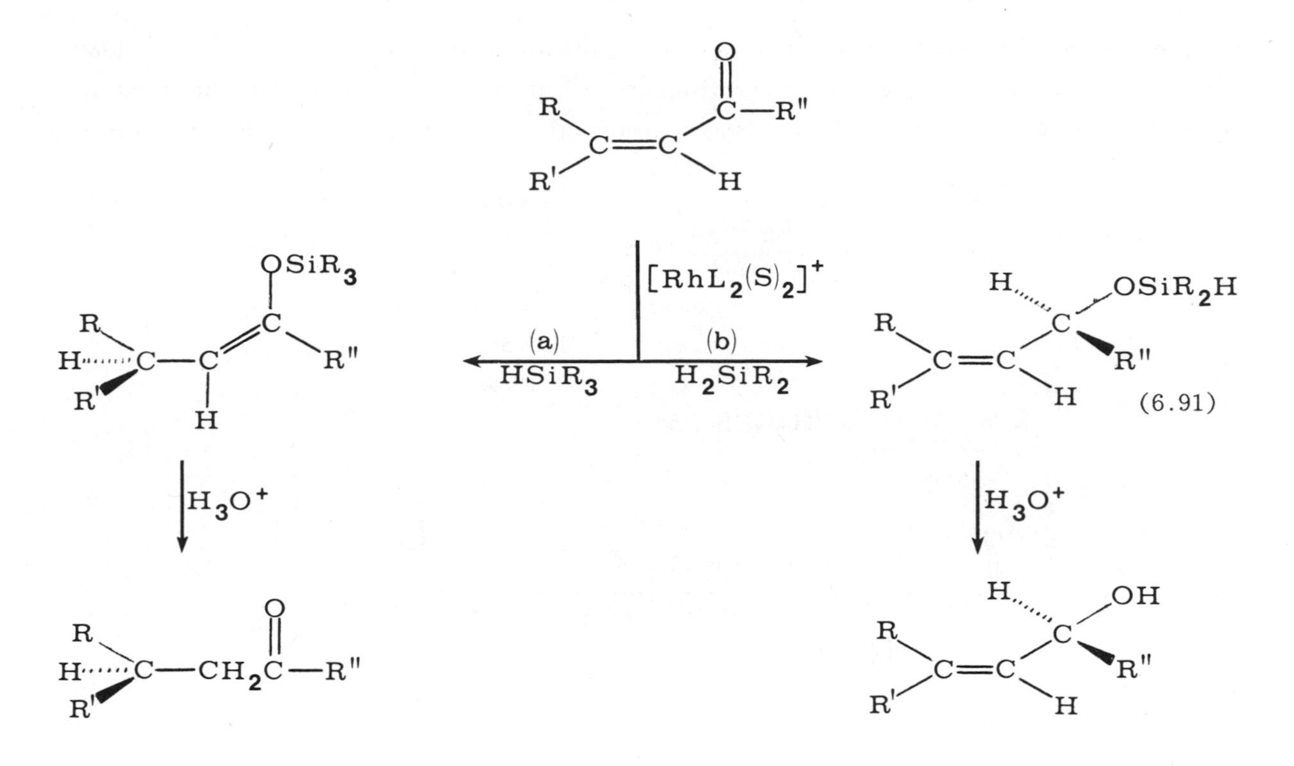


5. Asymmetric Hydrosilation. When chiral phosphine ligands are employed asymmetric induction is observed. Selected examples are listed in Table 6.6. For

*Table 6.5. Asymmetric hydrosilation of ketones using ligands shown in Figure 6.5.*

| Ligand | Silane[a] | Substrate[b] | Percent ee | Product configuration | Reference |
|---|---|---|---|---|---|
| (+)BMPP | $PhMe_2SiH$ | A | 62 | S | [148] |
| (+) BMPP | $EtMe_2SiH$ | A | 56 | R | [149] |
| (+)DIOP | $NpPhSiH_2$ | B | 58 | S | [150] |
| S,R-BPPFA | $Ph_2SiH_2$ | B | 49 | R | [151] |
| (-)BMPP | $PhMe_2SiH$ | B | 43 | R | [152] |
| (+)BMPP | $PhMe_2SiH$ | B | 32 | S | [153] |
| (+)DIOP | $NpPhSiH_2$ | C | 82 | R | [154] |
| (-)DIOP | $NpPhSiH_2$ | D | 42 | R | [155] |

[a] Np = α-naphthyl.

[b] A is PhCO(t-Bu); B is PhCOMe; C is $MeCOCO_2$(n-Pr); D is EtCOMe.

simple ketones and α,β-unsaturated ketones, optical yields are modest, the ee seldom exceeding 50% [146] (Equation 6.92). Higher optical yields are realized in the catalytic hydrosilation of 1,2- and 1,4-dicarbonyl compounds [147] (Equations 6.93 and 6.94).

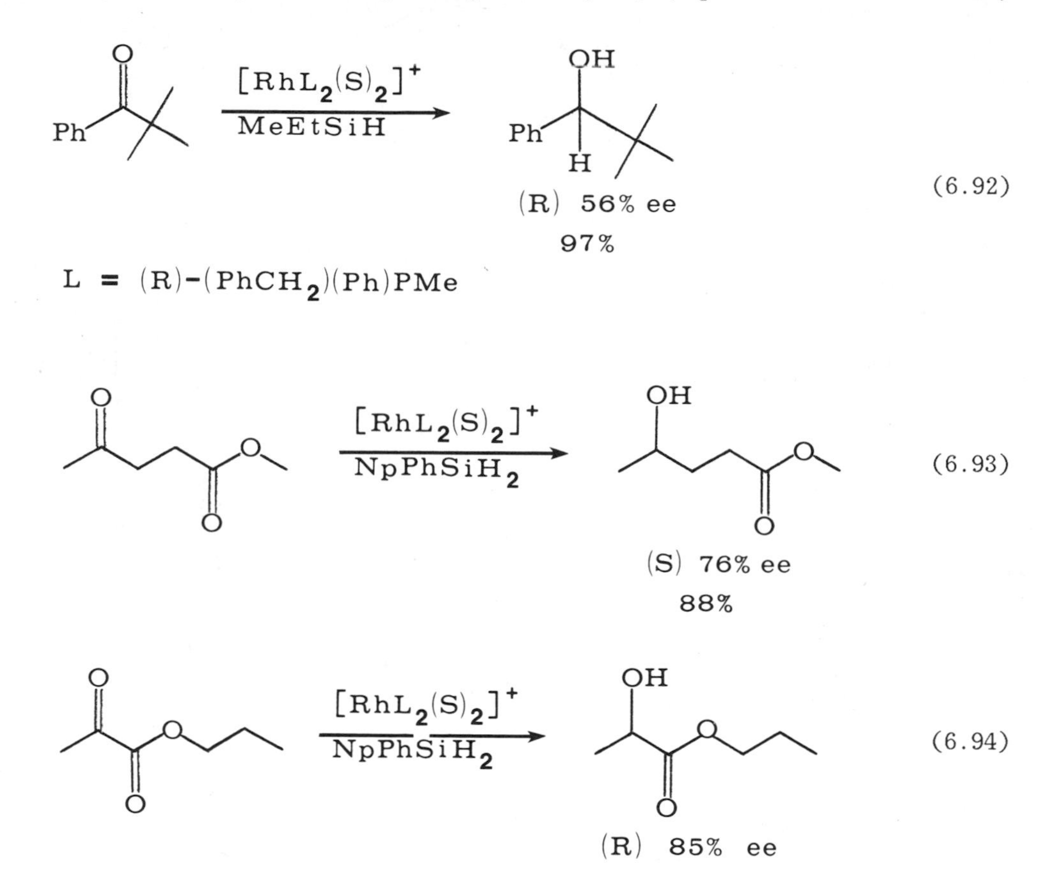


It is probable that the second carbonyl group acts as a ligand, helping to orient the ketone at the chiral catalytic center. This situation is reminiscent of that observed in the asymmetric catalytic hydrogenation of the amidocinnamic acids. Both the degree of enantioselectivity and the dominant configuration of the product depend on the nature of the silane. In practice it is difficult to choose among the large number of experimental variables: the chiral phosphine, the substrate, and the silane. For particular cases, these reactions will doubtless become important, but at present it is unwise to generalize.

## NOTES.


1a. B. R. James, Homogeneous Hydrogenation, (Wiley, 1974).

1b. B. R. James, Adv. Organomet. Chem., 17, 319 (1978).

1c. J. Halpern, T. Okamoto, and A. Zakhariev, J. Mol. Cat., 2, 65 (1977).

1d. A. J. Birch and D. H. Williamson, Organic Reactions, 24, 1 (1976).

1e. F. J. McQuillin, Homogeneous Hydrogenation in Organic Chemistry, (D. Reidel, 1976).

1f. F. R. Hartley and P. N. Vezey, Adv. Organomet. Chem., 15, 189 (1977).

2. J. L. Speier, Adv. Organomet. Chem., 17, 407 (1978).

3. For reviews on asymmetric homogeneous catalytic hydrogenation, see the following.

3a. H. B. Kagan and J. C. Frian, "New Approaches in Asymmetric Synthesis," in Topics in Stereochemistry, N. L. Allinger and E. L. Eliel, eds., 10 (1977).

3b. J. D. Morrison, W. F. Masler, and M. K. Neuberg, in Advances in Catalysis, 25, 81 (1976).

3c. H. B. Kagan, Pure and Appl. Chem., 43, 401 (1975).

3d. J. D. Morrison, W. F. Masler, and S. Hathaway in Catalysis in Organic Synthesis, P. N. Rylander and H. Greenfield, eds., (Academic Press, 1976).

3e. L. Marko and B. Heil, Catalysis Rev., 8, 269 (1974).

3f. D. Valentine and J. W. Scott, Synthesis, 329 (1978).

3g. W. S. Knowles, M. J. Sabacky, and B. D. Vineyard, Adv. Chem. Ser., No. 132, 274 (1974).

3h. B. Bogdanovic, Angew. Chem., Int. Ed. Engl., 12, 954 (1973).

4. Reference 1a, p. 250.

5. Ibid., p. 296; reference 1d, p. 29.

6. Reference 1a, p. 304.

7. Reference 1d, p. 65.

8a. J. R. Shapley, R. R. Schrock, and J. A. Osborn, J. Amer. Chem. Soc., 91, 2816 (1969).

8b. R. R. Schrock and J. A. Osborn, J. Amer. Chem. Soc., 93, 3089 (1971).

9. C. J. Love and F. J. McQuillin, J. Chem. Soc., Perkin I, 2509 (1973), and references therein.

10. R. A. Sanchez-Delgado and D. L. DeOchoa, J. Mol. Cat., 6, 303 (1979).

11. D. G. Holah, I. M. Hoodless, A. N. Hughes, B. C. Hui, and D. Martin, Can. J. Chem., 52, 3758 (1974), and references therein.

12. Reference 1b, p. 359; reference 1d, p. 41.
13. G. Zassinovich, G. Mestroni, and A. Camus, J. Mol. Cat., 2, 63 (1977).
14. G. Mestroni, G. Zassinovich, and A. Camus, J. Organomet. Chem., 140, C3 (1977).
15. R. H. Crabtree, H. Felkin, T. Khan, and G. E. Morris, J. Organomet. Chem., 144, C15 (1978).
16. R. Crabtree, Accts. Chem. Res., 12, 331 (1979), and references therein.
17. J. Halpern, Trans. Am. Crystallogr. Assoc., 14, 59 (1978).
18. R. B. Woodward and R. Hoffmann, Angew. Chem. Int. Ed. Engl., 8, 781 (1969), and references therein.
19. P. J. Brothers, Prog. in Inorg. Chem., 29, (1980), in press, and references therein.
20. R. H. Holm, in Biological Aspects of Inorganic Chemistry, A. W. Addison et al., Eds. (Wiley, 1977), Chap. 3.
21. G. Henrici-Olive and S. Olive, J. Mol. Catal., 1, 121 (1975/76).
22. H. W. Thompson and E. McPherson, J. Amer. Chem. Soc., 96, 6232 (1974).
23. J. Halpern, T. Okamoto, and A. Zakhariev, J. Mol. Cat., 2, 65 (1976).
24. A. S. C. Chan and J. Halpern, J. Amer. Chem. Soc., 102, 838 (1980), and references therein.
25. J. A. Osborn, F. H. Jardine, J. F. Young, and G. Wilkinson, J. Chem. Soc. (A), 1711 (1966).
26. J. Halpern and C. S. Wong, Chem. Comm., 629 (1973).
27. J. Halpern, in Organotransition Metal Chemistry, Y. Ishida and M. Tsutsui, eds. (Plenum, 1975), p. 109.
28. P. Meakin, J. P. Jesson, and C. A. Tolman, J. Amer. Chem. Soc., 94, 3240 (1972).
29. C. A. Tolman, P. Z. Meakin, D. L. Lindner, and J. P. Jesson, J. Amer. Chem. Soc., 96, 2762 (1974).
30. J. Halpern, D. P. Riley, A. S. C. Chan, and J. J. Pluth, J. Amer. Chem. Soc., 99, 8055 (1977).
31. R. R. Schrock and J. A. Osborn, J. Amer. Chem. Soc., 98, 2134 (1976).
32a. J. M. Brown and P. A. Chaloner, J.C.S. Chem. Comm., 321 (1978).
32b. J. M. Brown, P. A. Chaloner, and P. N. Nicholson, J.C.S. Chem. Comm., 646 (1978).
33. Y. Izumi, Angew. Chem., Int. Ed. Engl., 10, 871 (1971).

34. L. Horner, H. Winkler, A. Rapp, A. Mentrup, H. Hoffmann, and P. Beck, Tet. Lett., 161 (1961).

35a. W. S. Knowles and M. J. Sabacky, Chem. Comm., 1445 (1968).

35b. L. Horner, H. Buthe, and H. Siegel, Tet. Lett., 4023 (1968).

36. W. B. Farnham, R. K. Murray, Jr., and K. Mislow, J. Amer. Chem. Soc., 92, 5810 (1970), and references therein.

37. M. D. Fryzuk and B. Bosnich, J. Amer. Chem. Soc., 99, 6262 (1977).

38. W. S. Knowles, M. J. Sabacky, B. D. Vineyard, and C. J. Winkauff, J. Amer. Chem. Soc., 97, 2567 (1975).

39. T. Hayashi, T. Mise, S. Mitachi, and W. Marconi, J. Mol. Cat., 1, 451 (1976).

40. K. Achiwa, J. Amer. Chem. Soc., 98, 8265 (1976).

41. M. Fiorini, G. M. Giongo, F. Marcati, and W. Marconi, J. Mol. Cat., 1, 451 (1975/76).

42. M. Tanaka and I. Ogata, J.C.S. Chem. Comm., 735 (1975).

43. T. Hayashi, M. Tanaka, and I. Ogata, Tet. Lett., 295 (1977).

44. A. M. Aguiar, C. J. Morrow, J. D. Morrison, R. E. Burnett, W. F. Masler, and N. S. Bhacca, J. Org. Chem., 41, 1545 (1976).

45. T. Hayashi, M. Tanaka, and I. Ogata, Tetrahedron Lett., 295 (1977).

46. B. Heil, S. Toros, S. Vastag, and L. Marko, J. Organomet. Chem., 94, C47 (1975).

47. I. Ojima, T. Kogure, M. Kumagai, S. Horiuchi, and T. Sato, J. Organomet. Chem., 122, 83 (1976).

48. J. Solodar, Chemtech, 421 (1975).

49. J. M. Brown and P. A. Chaloner, J. Amer. Chem. Soc., 102, 3040 (1980).

50. Y. Ohgo, Y. Natori, S. Takeuchi, and J. Yoshimura, Chem. Lett., 1327 (1974).

51. Y. Kiso, K. Yamamoto, K. Tamao, and M. Kumada, J. Amer. Chem. Soc., 94, 4373 (1972), and references therein.

52. M. Tanaka, Y. Ikeda, and I. Ogata, Chem. Lett., 1115 (1975).

53a. A. Nakamura and S. Otsuka, Tet. Lett., 4529 (1973), and references therein.

53b. A. Andreetta, F. Conti, and G. F. Ferrari, Aspects Homo. Cat., 1, 204 (1970).

54. E. N. Frankel, E. Selke, and C. A. Glass, J. Amer. Chem. Soc., 90, 2446 (1968).

55. M. Wrighton and M. A. Schroeder, J. Amer. Chem. Soc., 95, 5764 (1973).

56. Y. Ohgo, S. Takeuchi, and J. Yoshimura, Bull. Chem. Soc. Japan, 44, 283 (1971).

57a. S. Takeuchi, Y. Ohgo, and J. Yoshimura, Bull. Chem. Soc. Japan, 47, 463 (1974).
57b. M. N. Ricroch and A. Gaudemer, J. Organomet. Chem., 67, 119 (1974).
58. Reference 1d, p. 47.
59. Reference 1a, pp. 107-114.
60a. J. E. Lyons, J.C.S. Chem. Comm., 412 (1975).
60b. P. Morand and M. Kayser, J.C.S. Chem. Comm., 314 (1976).
61. T. Hayashi, T. Mise, and M. Kumada, Tet. Lett., 4351 (1976), and references therein.
62. M. Murakami and J. Kang, Bull. Soc. Chem. Japan, 36, 763 (1963).
63. I. Jardine and F. J. McQuillin, Chem. Comm., 626 (1970).
64. A. Levi, G. Modena, and G. Scorrano, J.C.S. Chem. Comm., 6 (1975).
65. J. P. Collman, M. Bressan, and K. Kosydar, unpublished results.
66. C. H. Heathcock, and S. R. Poulter, Tet. Lett., 2755 (1969).
67. I. Mochida, S. Shirahama, H. Fujitsu, and K. Takeshita, Chem. Lett., 421 (1977).
68. E. L. Muetterties and J. R. Bleeke, Accts. Chem. Res., 12, 324 (1970).
69. M. F. Sloan, A. S. Matlock, and D. S. Breslow, J. Amer. Chem. Soc., 95, 4014 (1963).
70. W. R. Kroll, J. Catal., 15, 281 (1969).
71a. S. J. Lapporte, Ann. N.Y. Acad. Sci., 158, 510 (1969).
71b. S. J. Lapporte and W. R. Schuett, J. Org. Chem., 28, 1947 (1963).
71c. K. A. Klindinst and M. Boudart, J. Catal., 28, 322 (1973).
72. V. C. Lipovich, F. K. Schmidt, I. V. Kalechits, Chem. Abstr., 68, 59185w (1968).
73. L. S. Stuhl, M. Rakowski Du Bois, F. J. Hirsekorn, J. R. Bleeke, A. E. Stevens, and E. L. Muetterties, J. Amer. Chem. Soc., 100, 2405 (1978).
74. N. L. Holy, J. Org. Chem., 44, 239 (1979), and references therein.
75. P. Abley, I. Jardine, and F. J. McQuillin, J. Chem. Soc. C, 870 (1971).
76a. J. W. Kang, K. Moseley, and P. M. Maitlis, J. Amer. Chem. Soc., 91, 5970 (1969).
76b. M. J. Russell, C. White, and P. M. Maitlis, J.C.S. Chem. Comm., 427 (1977).
77. For the preparation and characterization of this complex, see E. O. Fischer and C. Eischenbroich, Chem. Ber., 103, 162 (1970); G. Huttner and S. Lange, Acta Crystallogr., Sect. B, 28, 2049 (1972).
78. J. W. Johnson and E. L. Muetterties, J. Amer. Chem. Soc., 99, 7395 (1977).

79a. M. A. Bennett and A. K. Smith, J. Chem. Soc. Dalton, 233 (1974).
79b. M. A. Bennett, T.-N. Huang, A. K. Smith, and T. W. Turney, J.C.S. Chem. Comm., 582 (1978).
80. M. A. Bennett, T.-N. Huang, and T. W. Turney, J.C.S. Chem. Comm., 312 (1979).
81. R. L. Augustine, Catalytic Hydrogenation (Marcel Dekker, 1965).
82. J. C. Falk, in Catalysts in Organic Synthesis, P. N. Rylander and H. Greenfield, eds. (Academic Press, 1976), pp. 305-324.
83. R. H. Crabtree, H. Felkin, T. Khan, and G. E. Morris, J. Organomet. Chem., 144, C15 (1978).
84. H. M. Feder and J. Halpern, J. Amer. Chem. Soc., 97, 7187 (1975).
85. R. Sweany and J. Halpern, J. Amer. Chem. Soc., 99, 8335 (1977).
86. I. S. Kolomniko, D. Y. Koreshov, V. P. Kukolev, V. A. Mosin, and M. E. Volpin, Izvest. Akad. Nauk, SSSR, Ser. Khim., 175 (1973).
87a. F. R. Hartley and P. N. Vezey, Adv. Organomet. Chem., 15, 189 (1977).
87b. R. H. Grubbs, Chemtech, 512 (1977).
87c. Z. M. Michalska and D. E. Webster, Chemtech, 117 (1975).
87d. J. C. Bailar, Jr., Catal., Rev., Sci. Eng., 10, 17 (1974).
87e. E. M. Cernia and M. Graziani, J. Appl. Polym. Sci. 18, 2725 (1974).
87f. C. U. Pittman, Jr., and G. O. Evans, Chemtech, 560 (1973).
87g. P. Hodge, Chem. Brit., 14, 237 (1978).
87h. J. Manassen, Platinum Metals Rev., 15, 142 (1971).
87i. A. L. Robinson, Science, 194, 1261 (1976).
87j. L. L. Murrell, in Advanced Materials in Catalysis, J. J. Burton and R. L. Garten, eds. (Academic Press, 1977), pp. 235-265.
88a. M. T. Westaway and G. Walker, U.S. Patent 3,617,553 (1971).
88b. Y. Dror and J. Manassen, J. Mol. Cat., 2, 219 (1977).
88c. F. Joo, Z. Toth, and M. T. Beck, Inorg. Chimica Acta, 25, L61 (1977).
89. J. I. Crowley and H. Rapoport, Accts. Chem. Res., 9, 135 (1976), and references therein.
90. B. A. Sosinsky, W. C. Kalb, R. A. Grey, V. A. Uski, and M. F. Hawthorne, J. Amer. Chem. Soc., 99, 6768 (1977).
91. M. Terasawa, K. Kaneda, T. Imanaka, and S. Teranishi, J. Catal., 51, 406 (1978).
92. N. Takahashi, I. Okura, and T. Keii, J. Mol. Catal., 3, 277 (1977/78).

93. T. H. Kim and H. F. Rase, Ind. Eng. Chem., Prod. Res. Dev., 15, 249 (1976).
94. K. G. Allum, R. D. Hancock, I. V. Howell, T. E. Lester, S. McKenzie, R. C. Pitkethly, and P. J. Robinson, J. Organomet. Chem., 107, 393 (1976).
95a. M. Bartholin, Ch. Graillat, A. Guyot, G. Coudurier, J. Bandiera, and C. Naccache, J. Mol. Catal, 3, 17 (1977/78).
95b. A. Guyot, Ch. Graillat, and M. Bartholin, J. Mol. Catal., 3, 39 (1977/78).
96. R. H. Grubbs and S. C. H. Su, in Organometallic Polymers, C. E. Carraher, Jr., J. E. Sheats, and C. U. Pittmen, Jr., eds. (Academic Press, 1978), pp. 129-134.
97. F. Dawans and D. Morel, J. Mol. Catal., 3, 403 (1977/78).
98. Th. G. Spek and J. J. F. Scholten, J. Mol. Catal., 3, 81 (1977/78).
99. J. Reed, P. Eisenberger, B.-K. Teo, and B. M. Kincaid, J. Amer. Chem. Soc., 100, 2375 (1978).
100. J. P. Collman, L. S. Hegedus, M. P. Cooke, J. R. Norton, G. Dolcetti, and D. N. Marquardt, J. Amer. Chem. Soc., 94, 1789 (1972).
101a. N. W. Frisch, Chem. Eng. Sci., 17, 735 (1962).
101b. R. Kunin, E. F. Meitzner, J. A. Oline, S. A. Fisher, and N. Frisch, Ind. Eng. Chem., Prod. Res. Dev., 1, 140 (1962).
102. D. C. Neckers, Chemtech, 108 (1978).
103. G. Gubitosa, M. Boldt, and H. H. Brintzinger, J. Amer. Chem. Soc., 99, 5174 (1977).
104. G. A. Crosby and M. Kato, J. Amer. Chem. Soc., 99, 278 (1977).
105a. L. T. Scott, J. Rebek, L. Ovsyanko, and C. L. Sims, loc. cit., p. 625.
105b. S. Mazur and P. Jayalekshmy, J. Amer. Chem. Soc., 101, 677 (1979).
106. J. T. Sparrow, Tet. Lett., 4637 (1975).
107. R. H. Grubbs, L. C. Kroll, and E. M. Sweet, J. Macromol. Sci. Chem., 7, 1047 (1973).
108. W. D. Bonds, Jr., C. H. Brubaker, Jr., E. S. Chandrasekaran, C. Gibbson, R. H. Grubbs, and L. C. Kroll, J. Amer. Chem. Soc., 97, 2128 (1975).
109. E. S. Chandrasekaran, R. H. Grubbs, and C. H. Brubaker, Jr., J. Organomet. Chem., 120, 49 (1976).
110. W. Dumont, J.-C. Poulin, T.-P. Dang, and H. B. Kagan, J. Amer. Chem. Soc., 95, 8295 (1973).
111. T. Masuda and J. K. Stille, J. Amer. Chem. Soc., 100, 268 (1978); N. Takaishi, H. Imai, C. A. Bertelo, and J. K. Stille, loc. cit., p. 264.
112. I. Tkatchenko, C.R. Acad. Sci., Ser. C, 282, 229 (1976).

113. A. R. Sanger and L. R. Schallig, J. Mol. Catal., 3, 101 (1977/78).
114. C. U. Pittman, Jr., S. K. Wuu, and S. E. Jacobson, J. Catal., 44, 87 (1976).
115. W. H. Lang, A. T. Jurewicz, W. O. Haag, D. D. Whitehurst, and L. D. Rollman, J. Organomet. Chem., 134, 85 (1977).
116. K. G. Allum, R. D. Hancock, I. V. Howell, R. C. Pitkethly, and P. J. Robinson, J. Catal., 43, 322 (1976).
117. K. G. Allum, R. D. Hancock, I. V. Howell, T. E. Lester, S. McKenzie, R. C. Pitkethly, and P. J. Robinson, J. Mol. Catal., 43, 331 (1976).
118. I. V. Howell, R. D. Hancock, R. C. Pitkethly, and P. J. Robinson, in Cataysis, Heterogeneous and Homogeneous, B. Delmon and G. Jannes, eds. (Elsevier, 1975), pp. 349-359.
119. V. Schurig and E. Bayer, Chemtech, 212 (1976).
120. M. E. Wilson and G. M. Whitesides, J. Amer. Chem. Soc., 100, 306 (1978).
121a. R. K. Iler, The Colloid Chemistry of Silica and Silicates (Cornell University Press, 1955).
121b. M. L. Hair, Infrared Spectroscopy in Surface Chemistry (Marcel Dekker, 1967), pp. 79-139.
121c. A. V. Kiselev and V. I. Lygin, Infrared Spectra of Surface Compounds (Wiley, 1975), pp. 75-236.
121d. V. L. Snoeyink and W. J. Weber, Jr., in Progress in Surface Membrane Science, Vol. 5, J. D. Danielli, M. D. Rosenberg, and D. A. Cadenhead, eds. (Academic Press, 1972), pp. 63-119.
122. K. G. Allum, R. D. Hancock, I. V. Howell, S. McKenzie, R. C. Pitkethly, and P. J. Robinson, J. Organomet. Chem., 87, 203 (1975).
123. C. Eaborn, Organosilicon Compounds (Butterworths Scientific Publications, 1960).
124a. A. K. Smith, J. M. Basset, and P. M. Maitlis, J. Mol. Catal., 2, 223 (1977).
124b. Z. M. Michalska, Ibid., 3, 125 (1977/78).
125. J. Conan, M. Bartholin, and A. Guyot, Ibid., 1, 375 (1975/77).
126. Cyclopentadienyl metal carbonyl complexes: F. R. W. P. Wild, G. Gubitosa, and H. H. Britzinger, J. Organomet. Chem., 148, 73 (1978).
127. K. Neuberg, Ph.D. dissertation, Stanford University (1978).
128. M. M. T. Khan and A. E. Martell, Homogeneous Catalysis by Metal Complexes, Vol. II (Academic Press, 1974), p. 66.
129. W. Noll, Chemistry and Technology of Silicons (Academic Press, 1968).

130. M. Green, J. A. N. Howard, J. Proud, J. L. Spencer, F. G. A. Stone, and C. A. Tsipsi, J.C.S. Chem. Comm., 671 (1976).
131. A. J. Chalk and J. F. Harrod, J. Amer. Chem. Soc., 87, 16 (1965).
132. T. G. Selin and R. West, J. Amer. Chem. Soc., 84, 1863 (1962).
133. J. W. Ryan and J. L. Speier, J. Amer. Chem. Soc., 86, 895 (1964).
134. L. H. Sommer, E. W. Pietruszar, and F. C. Whitmore, J. Amer. Chem. Soc., 91, 7051 (1969).
135. J. C. Saam and J. L. Speier, J. Amer. Chem. Soc., 80, 4104 (1958).
136. A. J. Chalk and J. F. Harrod, J. Amer. Chem. Soc., 87, 16 (1965).
137. J. F. Harrod and A. J. Chalk, J. Amer. Chem. Soc., 87, 1133 (1965).
138. K. Yamamoto, Y. Hayashi, and M. Kumada, J. Organomet. Chem., 28, C37 (1971).
139. R. N. Haszeldine, R. V. Parish, and R. Taylor, J. Chem. Soc. (D), 2311 (1974).
140. A. J. Chalk, J. Organomet. Chem., 21, 207 (1970).
141. M. Hara, K. Ohno, and J. Tsuji, Chem. Comm., 247 (1971).
142. M. Kumada, Y. Kiso, and M. Umeno, Chem. Comm., 611 (1970).
143. M. F. Lappert, T. A. Nile, and S. Takahashi, J. Organomet. Chem., 72, 425 (1974).
144. P. Svoboda, M. Dapka, J. Hetflejs, and V. Chvalovsky, Collct. Czech. Chem. Commun., 37, 1585 (1972).
145. K. Tamao, N. Miyaki, Y. Kiso, and M. Kumada, J. Amer. Chem. Soc., 97, 5603 (1975).
146. I. Ojima, K. Yamamoto, and M. Kumada, Aspects Homo. Catal., 3, 186 (1977).
147. I. Ojima, T. Kogure, and M. Kumagai, J. Org. Chem., 42, 1671 (1977).
148. T. Hayashi, K. Yamamoto, K. Kasuga, J. Omizu, and M. Kumada, J. Organomet. Chem., 113, 127 (1976).
149. I. Ojima and Y. Nagai, Chem. Lett., 223 (1974).
150. W. Dumont, J. Poulin, T. Dang, and H. B. Kagan, J. Amer. Chem. Soc., 95, 8295 (1973).
151. Y. Hayashi, K. Yamamoto, and M. Kumada, Tet. Lett., 4405 (1974).
152. I. Ojima and T. Kogure, Chem. Lett., 541 (1973).
153. K. Yamamoto, Y. Hayashi, and M. Kumada, J. Organomet. Chem., 54, C45 (1973).
154. I. Ojima, T. Kogure, and Y. Nagai, Chem. Lett., 985 (1975).

155a. R. J. Corriu and J. J. Moreau, J. Organomet. Chem., 64, C51 (1974).
155b. Ibid., 85, 19 (1975).

# 7

# Stoichiometric Reactions of Transition-Metal Hydrides

Virtually all the transition metals form "hydride" complexes, although the stabilities of these vary widely among the various metals. The hydride ligand occupies a discrete coordination site on the metal and behaves primarily as a $\sigma$-donor ligand. Transition-metal hydrides are characterized by very high-field $^1H$ NMR absorptions (typically -7 to -24 $\delta$) and by infrared M-H stretching bands in the 1900-2250 $cm^{-1}$ range. The classification of M-H complexes as "hydrides" arises from the formalisms discussed earlier, and does not imply chemical behavior. In fact, transition-metal "hydrides" are frequently quite acidic.

Transition-metal hydrides are central to several important catalytic processes, including hydrogenation and hydroformylation reactions of olefins. Additionally, some transition-metal hydrides effect useful stoichiometric reactions, often with a high degree of regio- and stereoselectivity.

One of these reagents is $KHFe(CO)_4$, generated *in situ* by treatment of $Fe(CO)_5$ with ethanolic potassium hydroxide. These solutions are effective reducing systems for the conversion of conjugated enones to saturated carbonyl compounds:

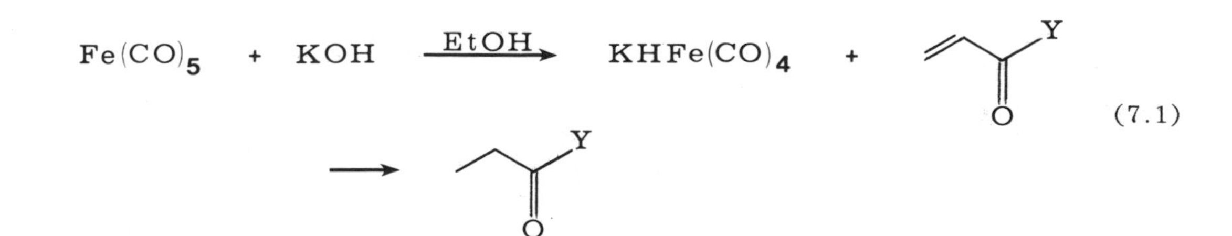


(7.1)

Unsaturated aldehydes, ketones, esters, lactones, nitriles, and diesters undergo clean reduction of the double bond with no reduction of the carbonyl group [1]. Although other polynuclear iron carbonyl-hydride complexes are present, it appears the $HFe(CO)_4^-$ is responsible for the observed reductions. Steric hindrance about the conjugated enone drastically slows the rate of reaction, and the system is not of use with base-sensitive substrates. The basic conditions necessary to generate the reducing agent can be used to an advantage. For example, ketones having α-methyl or α-methylene groups undergo a clean reductive α-alkylation when treated with $KHFe(CO)_4$ and a variety of aldehydes [2,3]. This reaction is thought to proceed by an initial base-catalyzed condensation of the ketone with the aldehyde, followed by reduction of the thus-produced conjugated enone:

$$RCH_2C(=O)R' + R''CHO \xrightarrow{\text{base}} \left[ RC(=CHR'')-C(=O)R' \right] \xrightarrow{KHFe(CO)_4} RCH(CH_2R'')C(=O)R' \qquad (7.2)$$

Active methylene compounds such as β-ketoesters react in a similar fashion [4,5], whereas indole undergoes clean reductive alkylation at the 3 position [6].

A careful study of the mechanism [7] by which $NaHFe(CO)_4$ reduces ethyl acrylate indicates that the key steps are as shown in (7.3). $NaHFe(CO)_4$ adds rapidly and

$$H\overset{(-)}{Fe}(CO)_4 \rightleftharpoons CO + H\overset{(-)}{Fe}(CO)_3 + CH_2{=}CHC(=O)OEt \longrightarrow CH_2(H)-CH(\overset{(-)}{Fe}(CO)_4)C(=O)OEt \xrightarrow{H^+} CH_3CH_2C(=O)OEt \qquad (7.3)$$

irreversibly to ethyl acrylate to produce (in the absence of a proton source) an insoluble σ-alkyliron intermediate $[CH_3CH(Fe(CO)_4)CO_2Et]^-$. This reaction is strongly inhibited by excess CO, indicating a dissociative first step. This alkyliron species undergoes a rapid protonolysis to produce the reduced product.

Several other functional groups are reduced by $HFe(CO)_4^-$. Acid chlorides are reduced to aldehydes in excellent yield [8]. Although mechanistic studies have not been reported, the reaction is claimed to involve initial displacement of halide by $Fe^-$ to form $RCOFe(H)(CO)_4$, followed by reductive elimination of aldehyde. Carbon-nitrogen double bonds are also readily reduced by this reagent. Enamines [9] are reduced to tertiary amines via the iminium salt, by both $KHFe(CO)_4$ and $KHFe_2(CO)_8$:

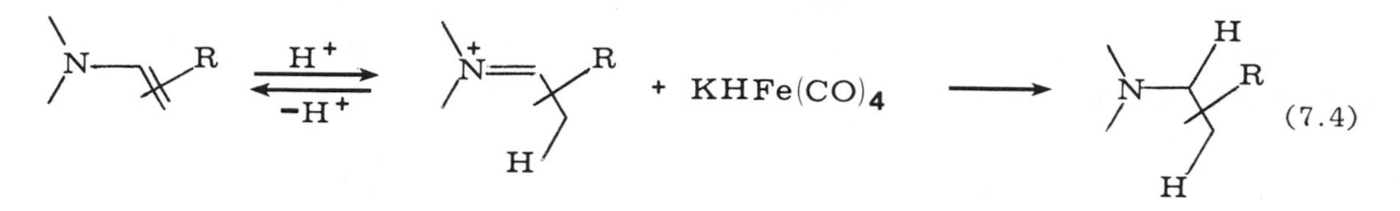


(7.4)

This forms the basis of a useful reductive amination procedure [10,11] in which aldehydes and ketones react with primary amines and $Fe(CO)_5$ in 1 N alcoholic KOH to produce secondary amines:

$$RCHO + R'NH_2 + Fe(CO)_5 \xrightarrow[EtOH]{1\ N\ KOH} [RCH{=}N{-}R'] \longrightarrow RCH_2NHR' \qquad (7.5)$$

Use of glutaraldehyde [12] allows the synthesis of N-alkyl pyrrolidines in excellent yield:

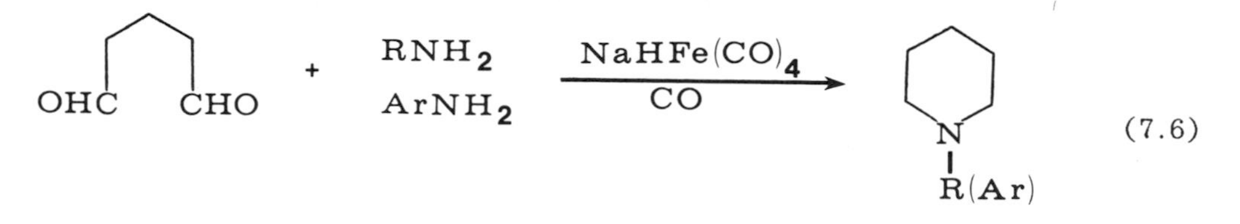


(7.6)

All these reactions involve reductions of imines, formed by the condensation of 1° amines and aldehydes.

Another very useful reducing agent, developed by Collman, is the binuclear cluster $NaHFe_2(CO)_8$. This complex is prepared by the reaction of $Na_2Fe(CO)_4$ with $Fe(CO)_5$, followed by acidification with acetic acid. This procedure results in a red-brown acetic acid solution of the complex (NMR +8.47 ppm upfield from TMS), which is used directly in reductions. This reagent is a very effective reducing agent for conjugated enones of all types. Conjugated esters, ketones, aldehydes, nitriles, amides, and lactones all undergo clean reduction of the carbon-carbon double bond, with no competing carbonyl reduction. Epoxides are not effected by this system, and isolated double bonds are neither reduced nor rearranged. Hindered systems react in only low yield (i.e., testosterone, 10%), and two or more equivalents of $NaHFe_2(CO)_8$ are required for complete reaction. Both added $NH_4Cl$ and added $Fe(CO)_5$ increase the yield of reductions. A study of relative reduction rates of a variety of substrates shows that cis double bonds (e.g., cyclohexenone-type enones) reduce rapidly, that substitution $\alpha$ or $\gamma$ to the carbonyl slows the rate, and that there is no correlation between rates of reduction and the reduction potential of the enone, indicating that electron-transfer processes are probably not involved [7].

On the surface, this chemistry is similar to that of $NaHFe(CO)_4$. However, a massive mechanistic study by Collman, Brauman, and coworkers [7] of the reaction of $NaHFe_2(CO)_8$ with conjugated enones indicates that, mechanistically, these two complexes react quite differently. Recall that $HFe(CO)_4^-$ reacts with ethyl acrylate in a dissociative manner, with a rapid irreversible, regiospecific addition to the double bond, followed by rapid protonolysis of the iron-carbon bond (Equation 7.3). In contrast, $NaHFe_2(CO)_8$ reacts with ethyl acrylate in an associative fashion, undergoing rapid regiospecific reversible addition to the double bond, followed by relatively slow protonolysis of the iron-carbon bond. The dinuclear iron complex does not fragment, but adds intact to the double bond. It is at least 26 times more reactive than the mononuclear $NaHFe(CO)_4$, and the reaction rate is first order in both dimer and substrate. The resulting alkyliron species decomposes to product by two pathways as shown in Equation 7.7, one involving direct protonolysis of the dinuclear species, the other

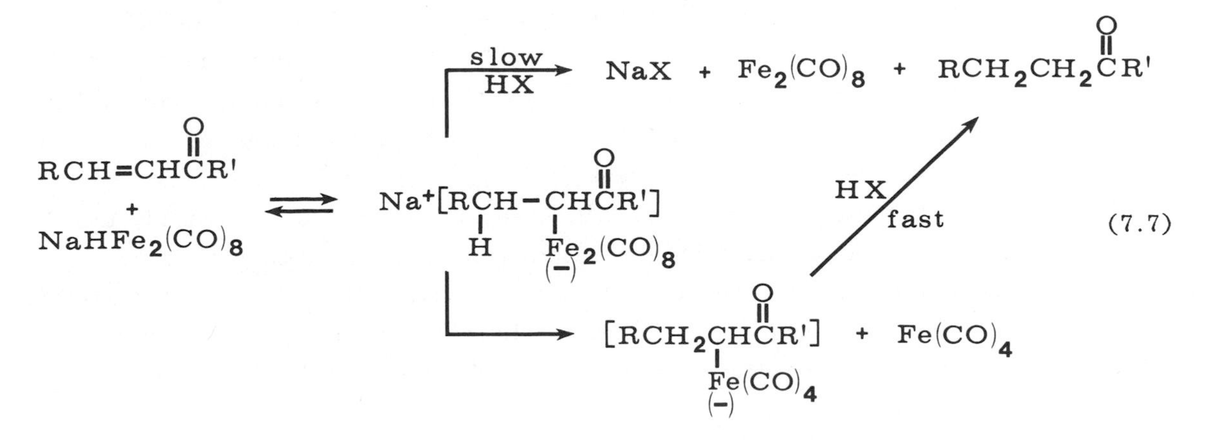

being an acid-independent path involving slow fragmentation (loss of $Fe(CO)_4$), followed by rapid protonolysis. The chemistry of the two species is complementary, since the mononuclear species reacts best under basic conditions, whereas the dinuclear complex requires acid conditions.

The related iron hydridocarbonyl cluster, $Et_4N^+HFe_3(CO)_{11}^-$, reduces aromatic nitro compounds to amines. The yields in this reaction are quite high, and functional groups such as aldehydes, ethers, and chlorides are tolerated [13]. Similarly, the chromium hydridocarbonyl cluster, $NaHCr_2(CO)_{10}$, is a very efficient reducing agent for conjugated ketones, aldehydes, esters, and nitriles, resulting in clean reduction of the carbon-carbon double bond [14]. The mechanism of this reaction has not been examined, but is expected to be similar to that of the related iron systems.

The extremely useful chemistry of alkylcopper reagents (Chapter 11) has served as a model for the development of several related copper-hydride reducing agents. The mixed copper-hydride complex $[n\text{-}C_3H_7C{\equiv}CCuH]^-Li^+$, when used in fourfold to sixfold excess, cleanly reduces conjugated esters and ketones to saturated carbonyl compounds in excellent yield. The stereochemistry (with fused-ring systems) is comparable to that obtained using catalytic hydrogenation. With polyunsaturated compounds, only the conjugated double bonds are reduced, whereas acetylenic esters are inert [15]. The mixed cuprate $[n\text{-}BuCuH]^-Li^+$ selectively transfers the hydride (rather than the alkyl) group, and is also an efficient reagent for the 1,4-reduction of conjugated enones at low temperatures (-40°). This reagent even reduces highly substituted substrates such as 3,5,5-trimethylcyclohex-2-en-l-one. At higher temperatures (25°) it reduces aldehydes and ketones to alcohols, and 1°, 2°, and even 3° halides, mesylates, and tosylates to hydrocarbons in excellent yield. Elimination reactions with 3° halides are not a problem. Esters are inert [16]. An even more effective reducing agent for halides is produced by reacting cuprous iodide with two equivalents of $LiAlH(OMe)_3$ in THF. This complex, whatever its structure, reduces 1°, 2°, 3°, neopentyl, allyl, vinyl, and aryl halides as well as gem dihalocyclopropyl compounds to the corresponding hydrocarbons in excellent yields in all cases. This is indeed a remarkable range of substrates. Additionally, epoxides are reduced to alcohols [17].

Let us notice two other examples of structurally ill-defined, but chemically useful, reducing agents. One is produced by the reaction of cuprous bromide with two equivalents of $LiAlH(OMe)_3$ in THF, hereafter referred to as the "Li complex"; the other is produced by the treatment of cuprous bromide with one equivalent of $NaAlH_2(OCH_2CH_2OCH_3)_2$, hereafter known as the "Na complex" [18]. Both of these reagents are obtained as heterogeneous mixtures in THF, which are not purified or characterized before use. (They react quite differently from LiCuH prepared by an alternative route.) Although both reagents are efficient in 1,4-reductions of conjugated enones, the "Li complex" is superior for 2-cyclohexenone and related systems. Use of three equivalents of this reagent effects complete reduction of cyclohexenone to cyclohexanone in 0.3 hr. The more highly substituted 5,5-dimethyl- and 3,5,5-trimethylcyclohexenones are also reduced, but somewhat more slowly. In contrast, the "Na complex" is much more effective for acyclic systems, leading exclusively to 1,4-reduction, whereas the "Li complex" leads to substantial 1,2-reduction as well. The "Na complex" is also more effective for the reduction of conjugated esters and acetylenic esters. In all cases, base-catalyzed condensation of the conjugated carbonyl compounds is a competing reaction when the "Na complex" is used, and leads to both oligomeric and reduced products. The "Na complex" is also much more reactive toward other functional groups than the "Li complex," which is specific for α,β-unsaturated systems. The "Na complex" reduces saturated ketones and aldehydes, and alkyl bromides, about as rapidly as it reduces a typical conjugated ketone. Ester and nitrile groups are inert to both reagents.

Although the structures of these complexes are not known, several pieces of experimental evidence provide some mechanistic information. Reduction of an α,β-unsaturated ketone with the reagent prepared from CuBr and $LiAlD(OMe)_3$ results in exclusive deuteration at the β-position (Equation 7.8). This implies a normal

$$2\ LiAlD(OMe)_3 + CuBr + RCH{=}CH\overset{\overset{O}{\|}}{C}CH_3 \longrightarrow$$

$$[R\underset{\underset{D}{|}}{C}H{-}\overset{-}{C}H{-}\overset{\overset{O}{|}}{C}CH_3]M^+ \xrightarrow{H_2O} R\underset{\underset{D}{|}}{C}HCH_2\overset{\overset{O}{\|}}{C}CH_3 \qquad (7.8)$$

1,4-addition process. Whether the intermediate before the aqueous quench is a Li or a Cu enolate has not been discovered. The role of a one-electron transfer process versus a straightforward 1,4-addition of hydride (formally a two-electron process) in this system is also unknown. (This point is more fully discussed for the related 1,4-alkylation reactions of alkylcuprates in Chapter 11.) However, the isolation of large amounts of dimethyl-3,4-diphenyladipate from the reaction of methyl cinnamate with the

"Na complex" suggests the intermediacy of radical anions, and hence a mechanism involving a one-electron transfer, at least for the substrate in Equation 7.9. (The

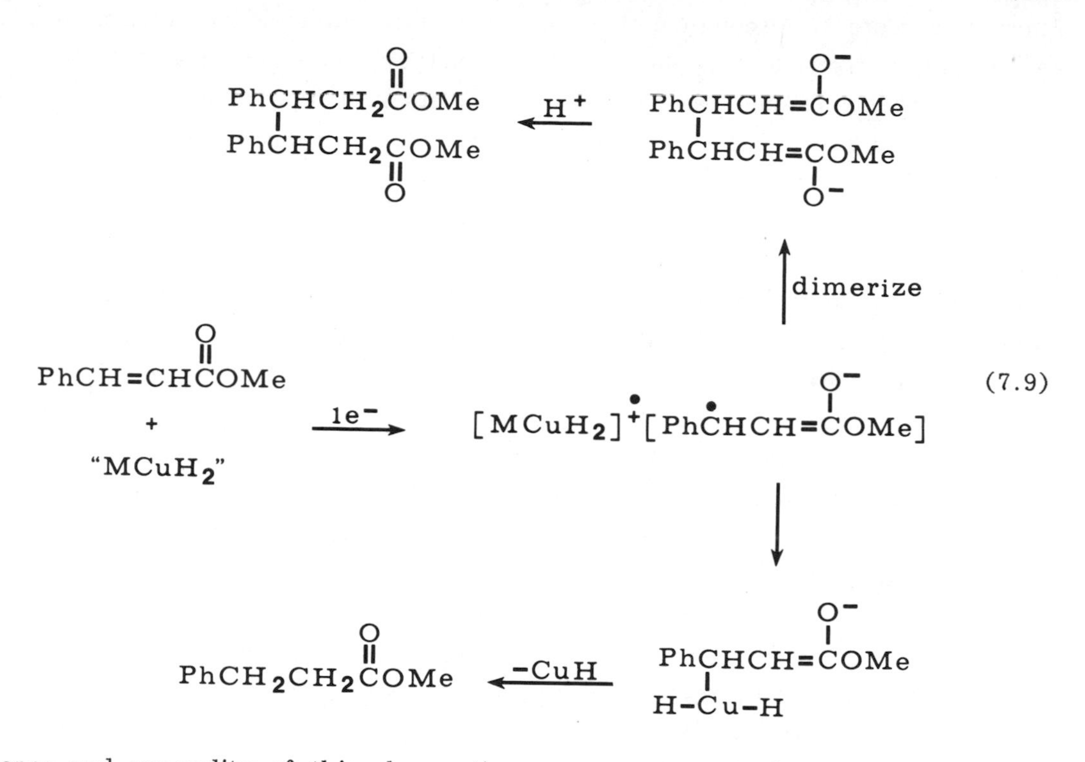

significance and generality of this observation are open to question, however, since in most systems radical-derived products are not observed.) Again, it has not been demonstrated that the radical-ion intermediates responsible for dimerization are also responsible for reduction.

The behavior of these "copper hydride" species depends largely on how they are prepared. The reagent resulting from the reaction of $LiAlH_4$ with four equivalents of cuprous iodide is very effective in the 1,4-reduction of a large number of conjugated ketones, including those with extensive $\alpha$ and/or $\beta$ substitution. However, this reagent fails to react with cyclohexenone, a substrate normally quite reactive toward 1,4-reduction by "CuH" species. In this instance, however, Ashby has shown the active reducing agent is $AlH_2I$, not some copper species [19]. Discrete, characterized copper hydride "ate" complexes with the compositions $Li_nCuH_{n+1}$, where n = 1 to 5, can be made by reacting $LiAlH_4$ with the corresponding lithium alkylcuprate complexes, $Li_nCu(CH_3)_{n+1}$. The chemistry of these complexes depends very much on their specific constitution. For example, $Li_2CuH_3$ reduces conjugated ketones exclusively 1,4 to produce saturated ketones, whereas $Li_4CuH_5$ reduces the same substrates 1,2, giving allylic alcohols. Although all these complexes reduce decyl bromide to n-decane, $Li_4CuH_5$ is the best reagent for this transformation [20].

A complex of unspecified constitution resulting from the reaction of cuprous iodide with two equivalents of n-butylmagnesium bromide reduces alkynes to cis-alkenes very efficiently. The complex is thought to be some sort of "CuH" resulting from β-elimination from butyl copper [21]. However, the same clean cis reduction of alkynes to alkenes is effected by the reagent produced by mixing equal amounts of $MgH_2$ with CuI or t-BuOCu [22]. Alkenes are unreactive towards both reagents. Vinyl cuprates are clearly intermediate in these reductions, since they can be trapped with electrophiles (Equation 7.10). Acetylenic sulfides are cleanly reduced to cis vinyl

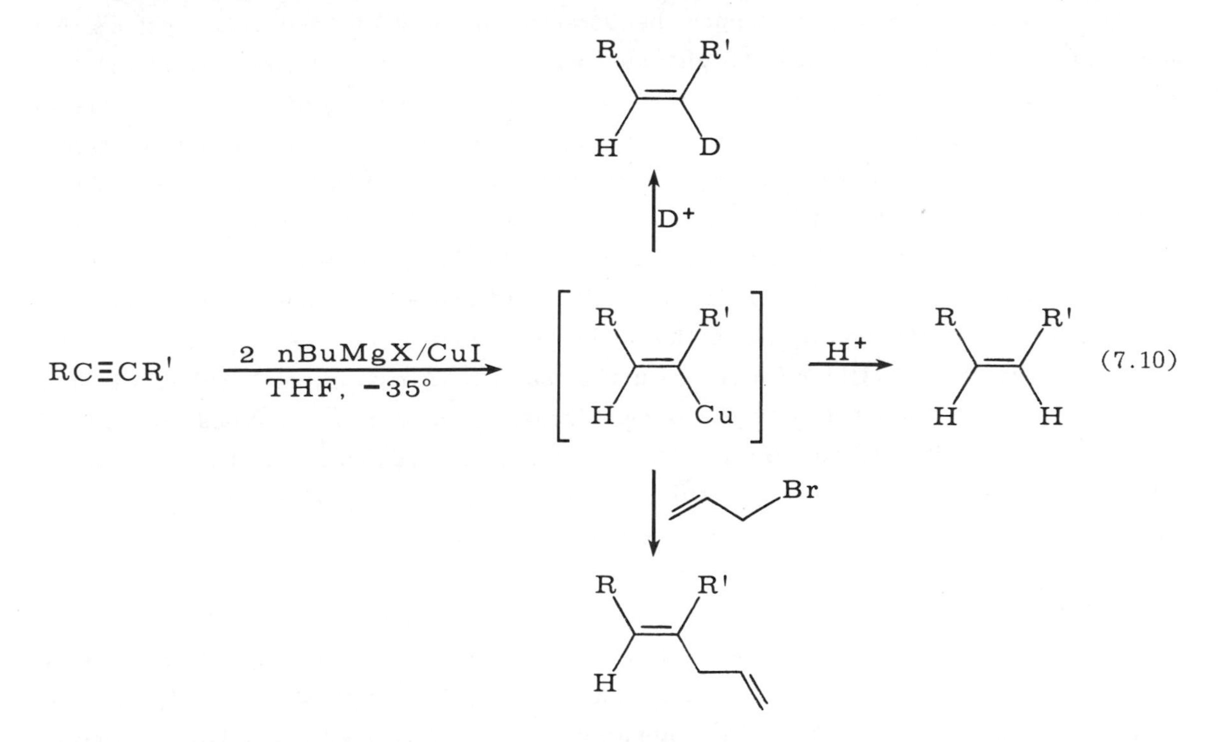


sulfides [23] by the complex formed by the treatment of cuprous bromide with one equivalent of $LiAlH(OMe)_3$ as shown in Equation 7.11. In the absence of cuprous bromide, exclusive trans reduction is observed.

A somewhat different copper-based reducing agent is $(Ph_3P)_2CuBH_4$. This is an easily prepared white, crystalline, air-stable solid, soluble in a range of organic

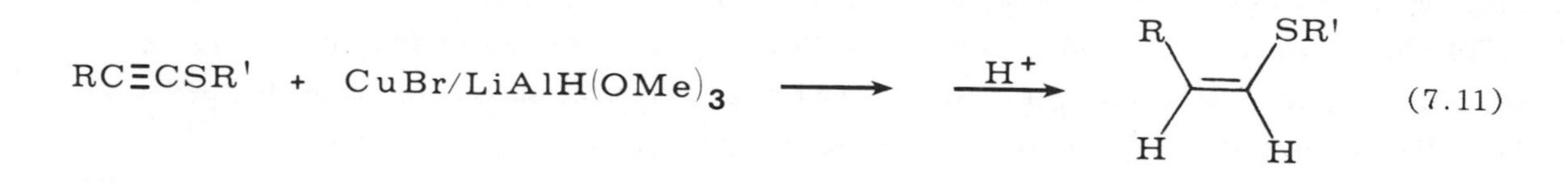


solvents, and insoluble in water and ethanol. It is a very specific reducing agent for the conversion of acid chlorides to aldehydes. The reduction is usually carried out in acetone, and aromatic, heteroaromatic, $\alpha,\beta$-unsaturated, and simple aliphatic acid chlorides are all reduced cleanly and in high yield to the corresponding aldehydes. Esters, ketones, nitriles, epoxides, imines, alkenes, and alkynes are stable to this complex [24,25]. Clearly complexation of the $BH_4^-$ to copper alters its reducing ability, since aldehydes and ketones are normally reduced by $NaBH_4$.

The relative success of copper halide/aluminum hydride reducing systems led Ashby to extend this approach to virtually all the first-row transition-metal halides. The $LiAlH_4/NiCl_2$ systems reduce 1°, 2°, cyclic, and aromatic halides and tosylates to the corresponding hydrocarbons in essentially quantitative yield. Similar reagents from $FeCl_2$, $CoCl_2$, and $TiCl_3$ perform similar reductions but less efficiently. Those produced from $LiAlH_4$ and $VCl_3$, $CrCl_3$, $MnCl_2$, and $FeCl_3$ are not effective for this type of reduction [26]. Similar mixtures of $LiAlH_4/MCl_x$ reduce alkenes--including cyclic, terminal, and internal olefins as well as styrene--to alkanes. The order of reactivity for the first-row metals in this alkene reduction is Co > Ni > Fe(II) > Fe(III) > Ti(III) > Cr(III) > V(III) > Mn(II) > Cu(I) > Zn(II). A mixture of $LiAlH_4/NiCl_2$ in a 10:1 ratio reduces internal alkynes to cis alkenes, and terminal alkynes to terminal alkenes, stereospecifically and quantitatively, with no overreduction of the olefin to the alkane [27]. A mixture of $FeCl_3$ with three equivalents of sodium hydride is less efficient in this reduction, but reduces ketones and aldehydes to alcohols in good yield [28].

Caubere has studied even more complex reducing systems that result from mixtures of sodium hydride, sodium alkoxide, and various transition-metal halide salts in varying proportions. Since the mode of action of this system is not understood at all, the best mixtures for specific reductions have apparently been developed by trial and error. The best mixture of this type for the reduction of ketones is sodium hydride, sodium t-amyloxide, nickel(II) acetate, and the ketone in a 4:1:4:1 ratio in THF at 40-60°. Under these conditions alkyl and aryl ketones as well as cyclohexanones are reduced to the corresponding alcohols. This mixture also reduces alkynes to alkenes. The mixture of sodium hydride, sodium t-butoxide, and nickel acetate, in a ratio of 4:2:1, reduces olefins to alkanes in anisole as solvent. Terminal, internal, and geminally disubstituted olefins, as well as cyclohexene and cyclooctene, are reduced in good yield under these conditions. Trisubstituted olefins react only in low yield [29]. A better reducing system for alkenes and alkynes is produced by the mixture of sodium hydride, sodium t-butoxide, ferric chloride, and the alkene or alkanes in THF in the ratio of 4:2:1:0.5. This mixture reduces alkenes and alkynes to alkanes the presence of ketones. 4-Vinylcyclohexene is reduced cleanly to 4-ethylcyclohexene, indicating a preference for less-substituted double bonds [30]. The same system reduces organic halides to alkanes [31]. The reduction of organic halides by mixtures of NaH, NaOR, and the halides of Ti, V, Cr, Mn, Fe, Co, Ni, Cu, Zn, Cd, Mo, Pd, and W has been studied in detail, but no generalizations have emerged [32].

Although all the above systems are of some use for synthesis, we are monumentally ignorant about what exactly is going on. Not only are few structures known, but in most cases even the elemental constitution of the reactive complex is unknown. Clearly the transition metal is involved in some fashion, since in the absence of metal the chemistry is significantly different. However, the direct involvement of transition-metal hydride species, as opposed to aluminum or boron hydride species complexed to the metal, has not been unequivocally demonstrated. The specific sequence of steps involved in any given reduction, and the modes of interaction of substrate with reducing agent, are even more unclear. This situation is not likely to change soon, because of the potential number of species involved. The Grignard reaction has been studied for almost a century, and it still is not completely understood; so patience is definitely in order.

A refreshing contrast to this state of uncertainty is provided by Bergman in his study of the chemistry of $[(\eta^5\text{-}C_5H_5)V(CO)_3H]^-$. This complex, made by the reduction of $\eta^5\text{-}C_5H_5V(CO)_4$ with sodium, followed by protonation by water, is a very effective reducing agent for virtually all types of organic bromides and iodides (Equation 7.12). Primary, secondary, and tertiary alkyl bromides, benzyl bromide, bromobenzene, and bromostyrene are reduced to the corresponding hydrocarbons in good to excellent yields; acid chlorides are reduced to aldehydes; _gem_-dibromocyclopropanes are reduced to monobromocyclopropanes, which are then inert to further reduction; and _vic_-dihalides are cleanly dehalogenated in a stereospecific _trans_ fashion, like that of other radical reactions. Ketones and esters do not react with this complex. Several features of these reactions suggest that a radical chain mechanism is involved. The relative

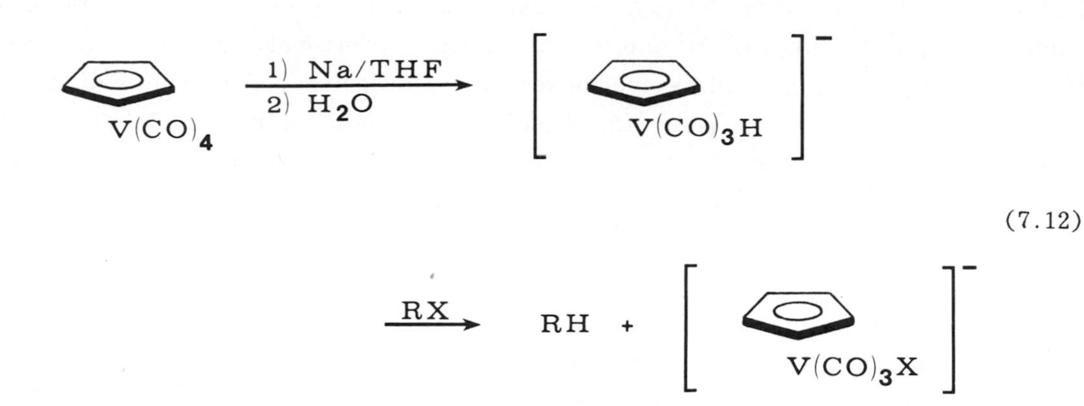


(7.12)

reactivities of the various substrates parallel those observed for reduction by trialkyltin hydrides, a known radical process. Particularly striking is the observation that primary, secondary, and tertiary halides are reduced at about the same rates, and that alkyl tosylates and chlorides are very unreactive. This clearly eliminates any $S_N2$-type processes involving vanadium-complex anions. Chiral secondary halides are reduced to racemic hydrocarbons, and both cis and trans vinyl halides lead to the same mixture of cis and trans olefins. This loss of stereochemistry is typical for processes involving alkyl or vinyl radicals. Reduction of bromoethyl allyl ether leads to substantial amounts of cyclized reduced product (Equation 7.13). This is considered to be firm evidence

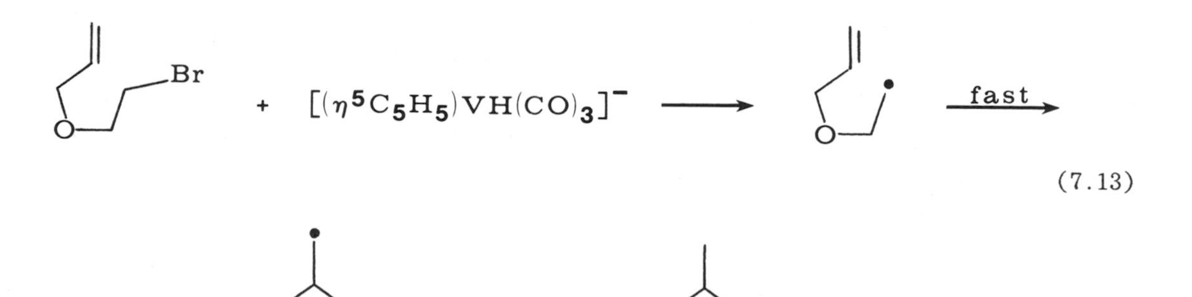


(7.13)

for the intermediacy of radicals, since the rate of cyclization of these unsaturated radicals is at least $10^6$ $sec^{-1}$, fast enough to compete with most intermolecular radical trapping processes. The mechanism which is most consistent with all the data is presented in Figure 7.1 [33]. Note its similarity to that proposed for radical

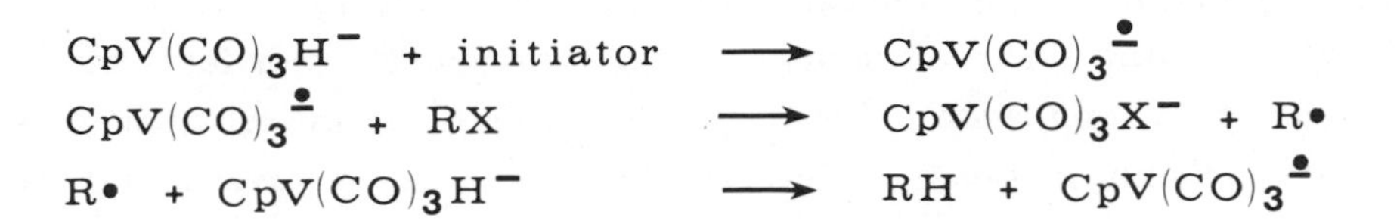


*Figure 7.1. Mechanism of the reduction of halides by $CpV(CO)_3H^-$.*

chain-oxidative additions (Chapter 4) and to the reaction of $\eta^3$-allylnickel halides with organic halides (Chapter 15).

All the reactions discussed above are reductions, and most undoubtedly pass at some point through a σ-alkyl metal complex, which is subsequently cleaved by hydrogen in some form ($H^+$, $H^{\cdot}$, or $H^-$) to lead to the reduced product. However, σ-alkyl transition-metal complexes have a very rich chemistry in their own right (i.e., oxidative-addition/reductive-elimination, insertion, oxidative cleavage) and have tremendous potential for use in organic synthesis, provided they survive long enough to undergo the desired chemistry. Particularly important and useful would be the ability to generate σ-alkyl metal complexes by the reaction of a transition-metal hydride with simple unactivated olefins or alkynes (Equation 7.14). This would allow direct

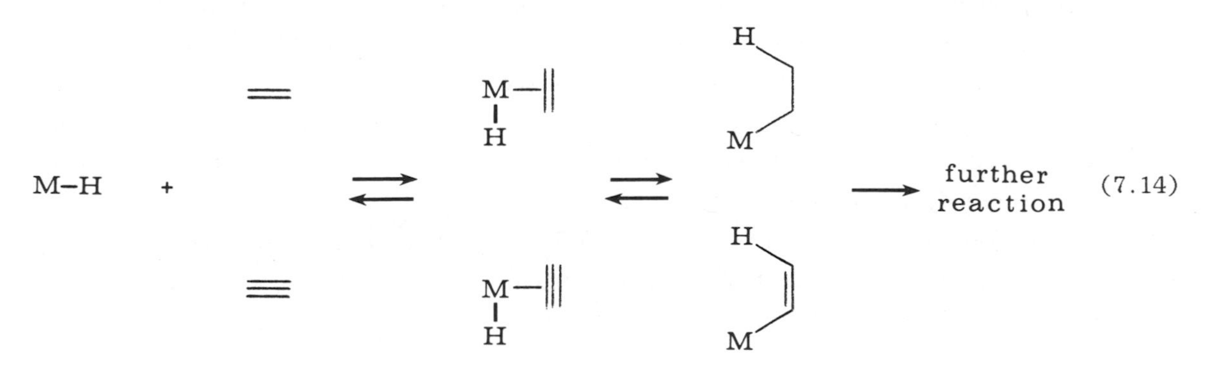


functionalizations of these unsaturated species not normally possible by traditional synthetic organic methods. To achieve this, either the equilibrium in Equation 7.14 must be far to the right, strongly favoring the σ-alkyl metal species over the hydride-olefin complex, or the reagent used to cleave the σ-alkyl metal species must react rapidly and selectively with the σ-alkyl metal complex while ignoring the hydride-olefin complex. In the catalytic hydrogenation of olefins and alkynes, the latter situation obtains; although σ-alkyl metal complexes are not detected, the hydrogenation proceeds. However, reagents other than hydrogen do not meet these criteria. Attempts to functionalize olefins by using typical hydrogenation catalysts such as $Rh(Cl)(PPh_3)_3$ are not very successful. The difficulty is that, with hydride complexes of the relatively electron-rich transition metals ($d^6$-$d^{10}$), the equilibrium in Equation 7.14 lies very far to the left, strongly in favor of the hydride-olefin complex; presumably because olefin ($\pi$-acceptor) complexes of electron-rich metals are more stable than the corresponding alkyl ($\sigma$-donor) complexes. This suggests that electron-poor transition metals ($d^0$-$d^4$) from the lefthand portion of the series should favor the $\sigma$-alkyl complex in Equation 7.14. This is indeed the case for a number of early transition metals in high oxidation states. The most extensively studied of these (by Schwartz) is the $d^0$ Zr(IV) complex $(\eta^5\text{-}C_5H_5)_2Zr(H)Cl$, which reacts with a number of alkenes and alkynes to give stable, isolable alkyl complexes of the type $(\eta^5\text{-}C_5H_5)_2Zr(R)Cl$.

This reaction is remarkable in several respects. The addition occurs under mild conditions, and the resulting alkyl complexes are quite stable. Regardless of the initial position of the double bond in the substrate, the Zr ends up at the sterically least-hindered accessible position of the olefin chain as a whole (Equation 7.15) [34]. This "contrathermodynamic" rearrangement of the olefin to the less stable terminal position occurs by a Zr-H elimination, followed by readdition in a fashion to place Zr at the less-hindered position of the alkyl chain in each instance. This migration proceeds in this fashion because of the steric congestion about Zr caused by the two cyclopentadienyl rings. The order of reactivity of various olefins towards hydrozirconation is: terminal olefins > cis-internal olefins > trans-internal olefins > exocyclic olefins > cyclic olefins and terminal > disubstituted > trisubstituted olefins. Tetrasubstituted olefins and trisubstituted cyclic olefins fail to react.

Alkynes also react with $(\eta^5\text{-}C_5H_5)_2Zr(H)Cl$, adding Zr-H in a cis manner. With unsymmetric alkynes, addition of Zr-H gives mixtures of vinyl Zr complexes with the less-hindered complex predominating. Equilibration of this mixture with excess $Cp_2Zr(H)Cl$ leads to mixtures greatly enriched in less-hindered complex (Equation 7.16).

Finally, 1,3-dienes add ZrH to the less-hindered double bond to give $\gamma$-$\delta$-unsaturated alkylzirconium complexes (Equation 7.17).

With the development of this chemistry, the long-sought reagent for the functionalization of olefins seems to have been found. The usefulness of this process for synthesis depends on the reactivity of the alkyl zirconium complexes formed above and will be discussed in Chapter 11.

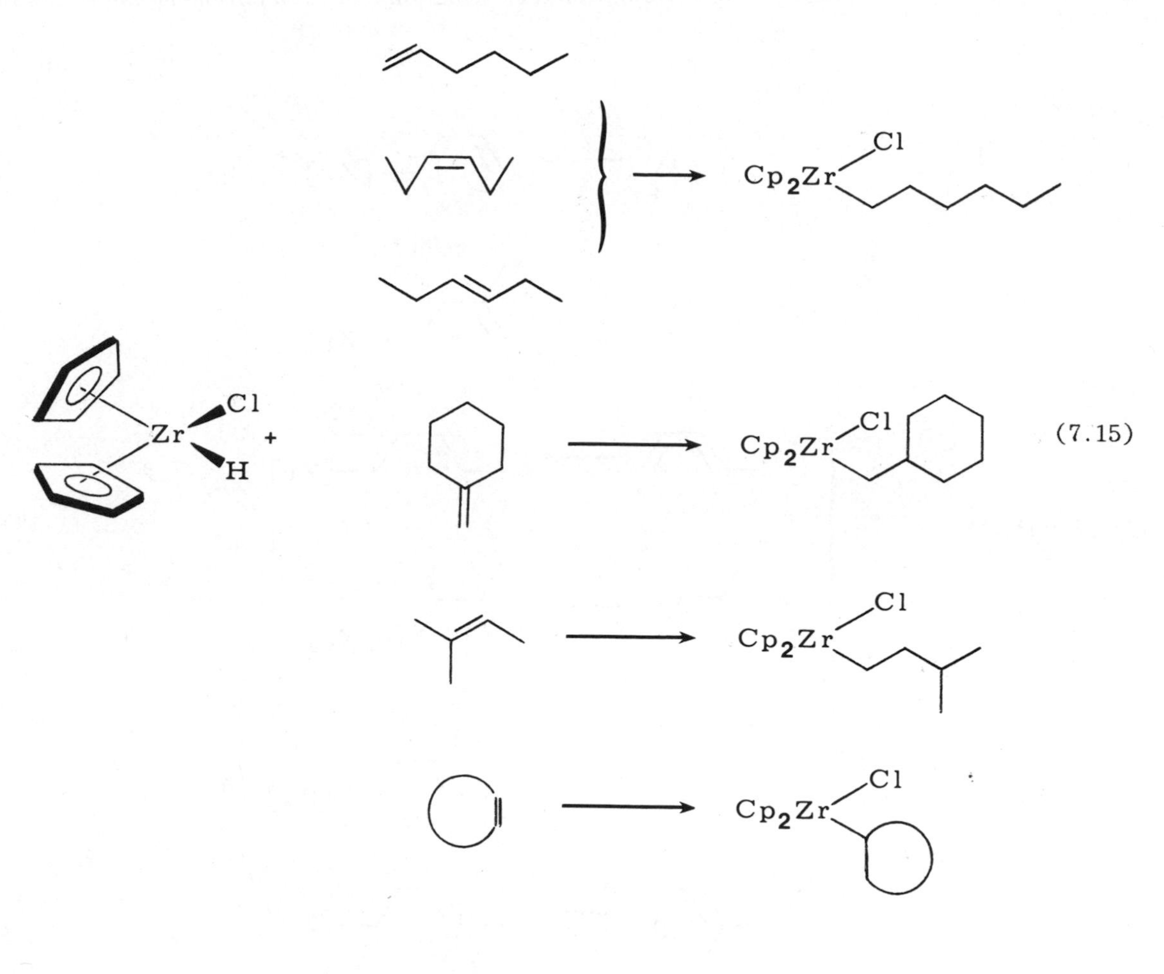


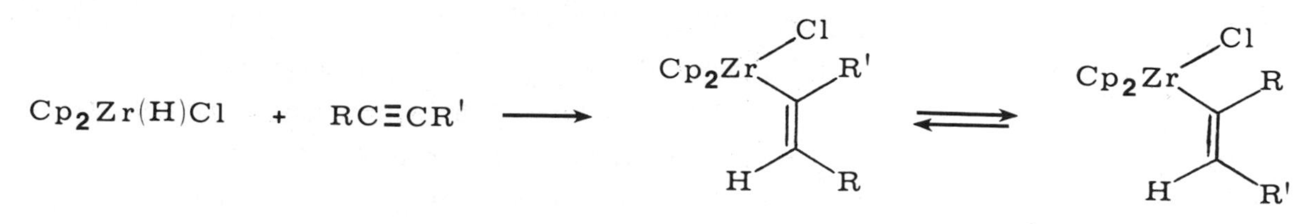


(7.16)

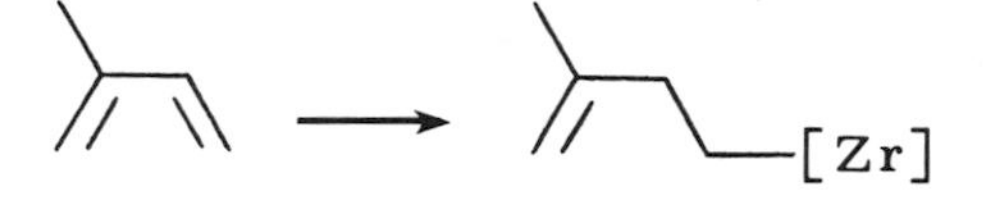


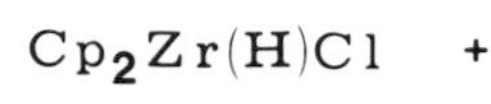


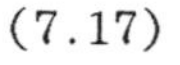


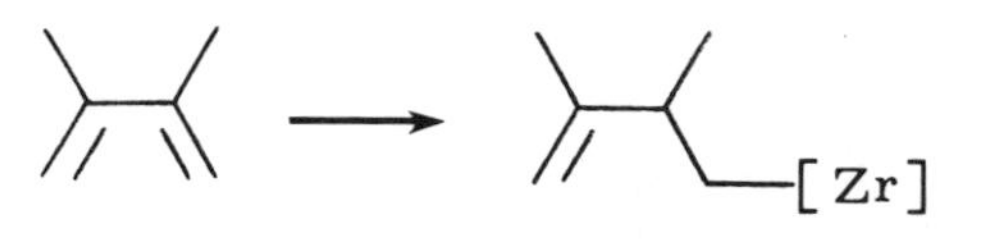


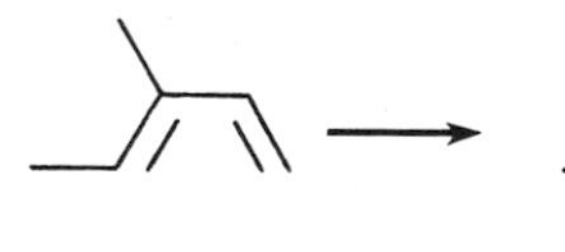

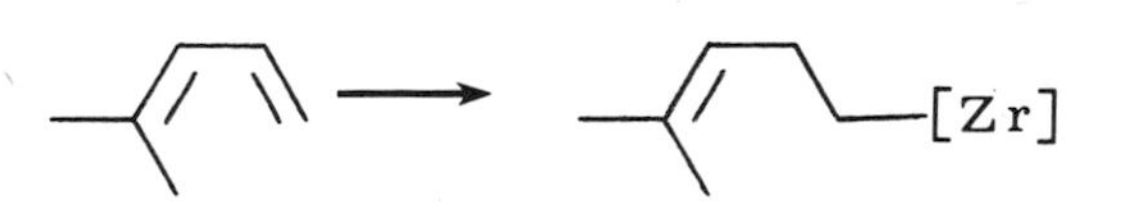


## Notes.


1. R. Noyori, I. Umeda, and T. Ishigami, J. Org. Chem., 37, 1542 (1972).
2. G. Cainelli, M. Panunzio, and A. Umani-Ronchi, Tetrahedron Lett., 2491 (1973).
3. G. Cainelli, M. Panunzio, and A. Umani-Ronchi, JCS Perkin I, 1273 (1975).
4. M. Yamashita, Y. Watanabe, T-a. Mitsudo, and Y. Takegami, Tetrahedron Lett., 1867 (1975).
5. M. Yamashita, Y. Watanabe, T-a. Mitsudo, and Y. Takegami, Bull. Chem. Soc. Japan., 51, 835 (1978).
6. G. P. Boldrini, M. Panunzio, and A. Umani-Ronchi, J.C.S. Chem. Comm., 359 (1974).
7. J. P. Collman, R. G. Finke, P. L. Matlock, R. Wahren, R. G. Komoto, and J. I. Brauman, J. Amer. Chem. Soc., 100, 1119 (1978).

8. T. E. Cole and R. Pettit, Tetrahedron Lett., 781 (1977).
9. T-a. Mitsudo, Y. Watanabe, M. Tanaka, S. Atsuta, K. Yamamoto, and Y. Takegami, Bull. Chem. Soc. Japan, 48, 1506 (1975).
10. Y. Watanabe, M. Yamashita, T-a. Mitsudo, M. Tanaka, and Y. Takegami, Tetrahedron Lett., 1879 (1974); Bull. Chem. Soc. Japan, 49, 1378 (1976).
11. G. P. Boldrini, M. Panunzio, and A. Umani-Ronchi, Synthesis, 733 (1974).
12. Y. Watanabe, S. C. Shim, T-a. Mitsuda, M. Yamashita, and Y. Takegami, Bull. Chem. Soc. Japan, 49, 2302 (1976).
13. G. P. Boldrini, A. Umani-Ronchi, and M. Panunzio, J. Organometal. Chem., 171, 85 (1979).
14. G. P. Boldrini, A. Umani-Ronchi, and M. Panunzio, Synthesis, 596 (1976).
15. R. K. Boeckman, Jr., and R. Michalak, J. Amer. Chem. Soc., 96, 1623 (1974).
16. S. Masamune, G. S. Bates, and P. E. Georghiou, J. Amer. Chem. Soc., 96, 3686 (1974).
17. S. Masamune, P. A. Rossy, and G. S. Bates, J. Amer. Chem. Soc., 95, 6452 (1973).
18. M. F. Semmelhack, R. D. Stauffer, and A. Yamashita, J. Org. Chem., 42, 3180 (1977).
19. E. C. Ashby, J. J. Lin, and R. Kovar, J. Org. Chem., 41, 1939 (1976).
20. E. C. Ashby, J. J. Lin, and A. B. Goel, J. Org. Chem., 43, 183 (1978).
21. J. K. Crandall and F. Collonges, J. Org. Chem., 41, 4089 (1976).
22. E. C. Ashby, J. J. Lin, and A. B. Goel, J. Org. Chem., 43, 757 (1978).
23. P. Vermeer, J. Meijer, C. Eylander, and L. Brandsma, Rec. Trav. Chim., 95, 25 (1976).
24. G. W. J. Fleet, C. J. Fuller, and P. J. C. Harding, Tetrahedron Lett., 1437 (1978).
25. T. N. Sorrell and R. J. Spillane, Tetrahedron Lett., 2473 (1978).
26. E. C. Ashby and J. J. Lin, J. Org. Chem., 43, 1263 (1978).
27. E. C. Ashby and J. J. Lin, J. Org. Chem., 43, 2567 (1978).
28. T. Fujisawa, K. Sugimoto, and H. Ohta, J. Org. Chem., 41, 1667 (1976).
29. J. J. Brunet, L. Mordenti, B. Loubinoux, and P. Caubere, Tetrahedron Lett., 1069 (1977).
30. J. J. Brunet and P. Caubere, Tetrahedron Lett., 3947 (1977).
31. B. Loubinoux, R. Vanderesse, and P. Caubere, Tetrahedron Lett., 3951 (1977).

32. J. J. Brunet, R. Vanderesse, and P. Caubere, J. Organometal. Chem., 157, 125 (1978).

33. R. J. Kinney, W. D. Jones, and R. G. Bergman, J. Amer. Chem. Soc., 100, 635; 7902 (1978).

34. These arguments are advanced in a review by J. Schwartz and J. A. Labinger, Angew. Chem. Int. Ed. Engl., 15, 333 (1976).

# 8

# Catalytic Reactions Involving Carbon Monoxide and Hydrogen Cyanide

The catalytic incorporation of CO into organic molecules has been intensively studied during the past four decades. Major industrial processes have evolved from this research [1,2]. Nevertheless, CO chemistry is currently receiving even greater attention because of the rapidly increasing price of petroleum, which has been the principal chemical feedstock. In the future, coal is destined to become a major source of many organic chemicals that are produced on a large scale. The principal carbon source from coal is "syn-gas," a mixture of CO and $H_2$ derived from combustion of coal in the presence of steam (Equation 8.1) combined with the water-gas shift reaction (Equation 8.2):

$$C + \tfrac{1}{2} O_2 \longrightarrow CO \qquad (8.1)$$

$$CO + H_2O \rightleftharpoons H_2 + CO_2 \qquad (8.2)$$

The "oxo," "Reppe," "Fischer-Tropsch," Monsanto, and related processes are discussed in this chapter. We shall find that certain steps in these catalytic cycles are familiar as oxidative-addition, reductive-elimination, and insertion reactions. Often one can propose a plausible mechanism which is consistent with the experimental facts. Well-characterized organometallic complexes can sometimes be detected and isolated from these reactions. However, major questions about the actual reaction mechanism usually remain unanswered, especially for the heterogeneously catalyzed Fischer-Tropsch process. Significant innovations and additional improvements in these reactions are both possible and probable. Thus, mechanisms of these catalytic CO reactions are subjects of great current interest. Pedagogically, these processes provide practical examples of reactions and principles which we encountered in earlier chapters. Hydrocyanation reactions are also included in this chapter.

## 8.1. The Oxo Reactions.

Otto Roelen discovered the "oxo," or hydroformylation, reaction [3] in 1938 while studying the Fischer-Tropsch reaction, an existing industrial process. The hydroformylation reaction is used to convert olefins to the saturated aldehydes via reaction of the olefin with syn-gas (8.3). The product composition can be considered to arise from

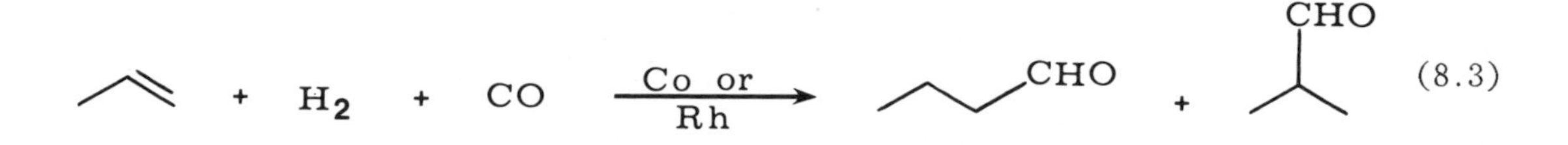


the formal addition of the elements of formaldehyde across the C-C double bond; however, formaldehyde is not an intermediate species in the reaction. The reaction is homogeneously catalyzed by a number of transition-metal carbonyl complexes. Cobalt is the most widely used catalytic element for this reaction, although the use of rhodium is beginning to become important with the introduction of new process chemistries based upon rhodium. Rhodium catalysts increase both the reaction rate and the selectivity of the process. However, since rhodium is very expensive, the problems of catalyst recovery are crucial to the economics of the process.

Commercial use of the hydroformylation reaction is on a large scale. About 6.5 million tons of "oxo" products are produced per year with propylene as the principal feedstock (Equation 8.3). The ultimate commercial products are alcohols, either 1-butanol or 2-ethylhexanol, which are formed by the hydrogenation of the aldehyde (8.4). Although the "oxo" catalysts can also catalyze the hydrogenation of both the

$$\text{CHO} + H_2 \xrightarrow[\text{Cu}]{\text{Ni or}} \text{OH} \qquad (8.4)$$

product aldehyde and the starting olefin, the aldehyde hydrogenation requires higher temperatures and pressures to proceed at an acceptable rate. Hence, these processes are usually conducted separately. In addition, the plasticizer component,

2-ethylhexanol, requires an intermediate aldol condensation of 1-butanal before reduction of the aldehyde function (8.5).

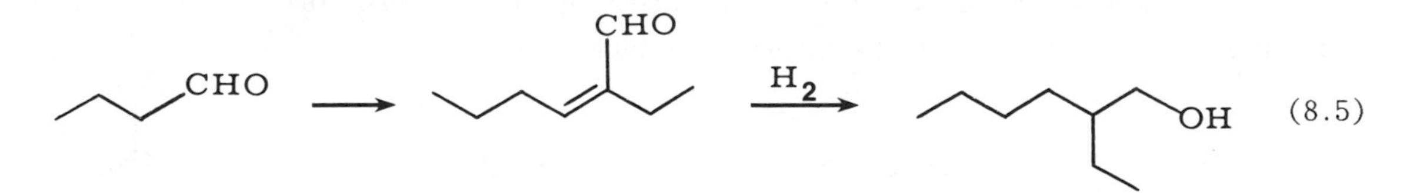


(8.5)

At present, the straight-chain alcohols such as 1-butanol are the preferred commercial products. Hence, the maximization of the selectivity of the hydroformylation reaction to the n-isomer has been a major concern in the commercialization of this chemistry. It should be noted, however, that other aspects of the reaction are equally important in optimization of the process for commercial operation. For example, the rate of the reaction will determine the reactor size for a given production rate; here rhodium enjoys a distinct advantage over cobalt, since the reaction rates achieved with rhodium are much higher. This advantage of rhodium is somewhat offset by its high price; the decision on the specific catalyst used will be made on the basis of economics.

The normal-to-iso ratio can be dramatically increased by the addition of bulky phosphines, such as trioctyl phosphine, to the reaction media. However, with cobalt catalysts these ligands tend to reduce the over-all rate of the reaction, thus reducing the productivity of the plant, and requiring a compromise between selectivity and production rate. Olefin rearrangement and hydrogenation also represent undesirable side reactions; however, only the latter is a concern with propylene. Here again, rhodium has an advantage over cobalt, since olefin hydrogenation is not a problem in processes using rhodium catalysis. Olefin hydrogenation can be suppressed in cobalt-catalyzed processes by increasing the CO partial pressure. In addition, substantial CO partial pressures are required to prevent catalyst decomposition. However, high CO partial pressures will reduce the over-all reaction rate, again requiring a compromise.

Separation of the products from the catalysts, a problem intrinsic to the whole field of homogeneous catalysts, is especially troublesome with the oxo reaction because these catalysts are often volatile and/or expensive. Further oxo reactions of the aldol products can afford heavy "bottoms." In fact, some of the earlier oxo literature is misleading because of imprecise product characterization and questions concerning material balance.

Olefin reactivities in hydroformylation reactions follow a pattern like that found for homogeneously catalyzed hydrogenation. That is, more hindered olefins are less reactive. A typical rate profile for the cobalt-carbonyl-catalyzed oxo process is shown in Figure 8.1. Similar results are found with rhodium catalysts.

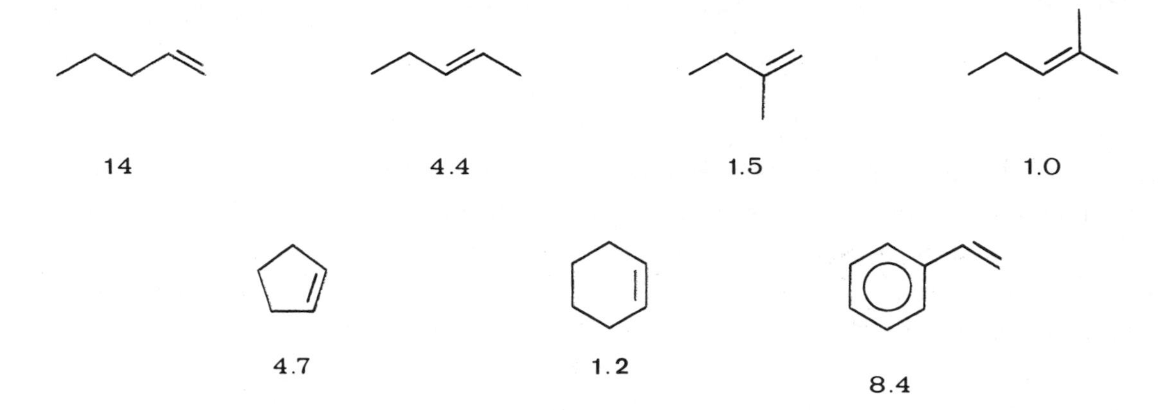


*Figure 8.1. Relative substrate reactivities for cobalt-catalyzed hydroformylation.*

The data in Figure 8.1 suggest two things about the oxo mechanism. First, coordination of the olefin by the transition-metal catalyst must occur as a step in the over-all reaction, followed by a subsequent metal hydride intramolecular migratory insertion. Second, this olefin coordination step must sometimes take place at or before the rate-determining step in the over-all catalytic cycle; olefin structure should not otherwise affect the relative reaction rates. It is usually assumed or implied that the oxo reaction mechanism does not change when different olefins are employed, but this premise is probably not justified. Recall from Chapter 6 that the dominant mechanism of homogeneous hydrogenation can depend on the nature of the olefin.

Examination of the oxo process, especially by IR spectroscopy, has indicated the nature of some metal carbonyl complexes which are present during the reaction cycle. Certain of these plausible intermediates have been isolated and characterized, and shown to undergo stoichiometric reactions which are probable steps in the catalytic oxo process. The rate laws and equilibria for the individual steps have not been carefully deduced and then reassembled to gauge the veracity of proposed catalytic reaction mechanisms. That is, the oxo reaction has not been dissected and subjected to the sort of thorough kinetic analysis that Halpern has carried out for the two homogeneous catalytic hydrogenations which are discussed in Chapter 6. Thus, we should regard certain aspects of the oxo mechanisms with caution.

a. Cobalt Catalysts. First, we will consider the cobalt-catalyzed oxo reaction. This process requires rather severe conditions (100-120°C, and 100-300 atm of CO + $H_2$). Heck and Breslow [4] proposed a mechanism for this reaction which has been widely cited. Their mechanism is outlined in Figure 8.2. The active catalyst in this cycle is thought to be the four-coordinate cobalt carbonyl hydride, $HCo(CO)_3$, 3, formed by dissociative loss of CO from the highly acidic saturated hydride, $HCo(CO)_4$, 2, step b in Figure 8.2. The latter has been detected under oxo conditions (vide infra). In fact, cobalt-catalyzed oxo reactions do not commence until cobalt octacarbonyl, 1, has been converted into this hydride through hydrogenolysis of the cobalt-cobalt bond, step a. In actual practice the hydride, 2, is sometimes formed in situ from cobalt salts or by protonation of $Co(CO)_4^-$ salts. Note that each of these steps (a and b) should exhibit a reciprocal rate dependence on the CO partial pressure $(p_{CO})^{-1}$. Under certain conditions the over-all reaction rate is proportional to $(p_{CO})^{-n}$ $(n = 1...2)$.

The unsaturated cobalt hydride, 3, is proposed to coordinate the olefin, step c in Figure 8.2. This step must be reversible, since under certain conditions isomerized olefinic substrate has acquired deuterium from $D_2$ or $DCo(CO)_4$. The proposed olefin-hydride complex, 4, which has not been observed, is thought to undergo a reversible migratory insertion, affording an undetected unsaturated alkyl complex, 5 (step d). A reversible reaction such as step d seems necessary to account for the n/i product ratio, olefin rearrangement, and H,D scrambling within the substrate (vide infra). The postulated unsaturated alkyl, 5, is thought to acquire a CO group, forming the saturated alkyl, 6 (step e). Alkyls such as 6 have been prepared, but under operating oxo conditions (because of interfering absorptions) the presence of 6 cannot be determined by IR analysis. The following step, f, involves migratory insertion, first to an unobserved, unsaturated acyl, 7, and then a saturated acyl, 8, reaction g. The latter acyl, 8, has been observed under reaction conditions by IR analysis. The product and rate-determining step, h, was proposed by Heck and Breslow to be the oxidative-addition of $H_2$ to the unsaturated acyl, 7, affording postulated dihydrido acyl, 9, which presumably would undergo rapid intramolecular reductive-elimination (step i) affording aldehyde and regenerating the hydride catalyst, 3. All the steps are proposed to be reversible except the product-forming reaction (step i). This is consistent with the fact that aldehydes do not appear to reenter the catalytic cycle. The Heck-Breslow

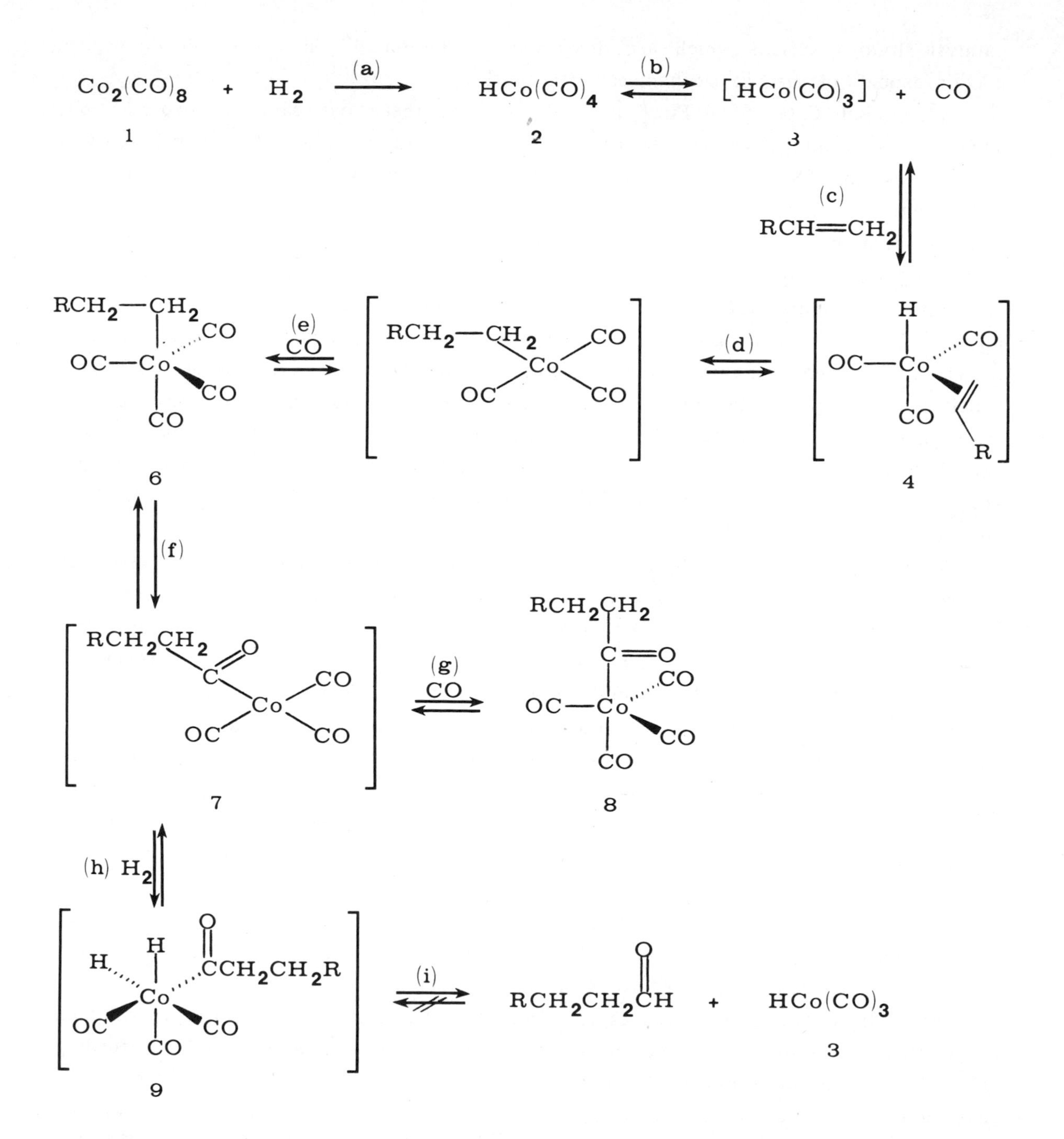


*Figure 8.2. Heck-Breslow mechanisms for oxo reaction.*

mechanism qualitatively accounts for many characteristics of the cobalt-catalyzed oxo reaction. Several of the postulated intermediates have been prepared and are known to give the indicated reactions. However, for this complex, over-all catalytic process, it is not feasible to test this mechanism from kinetic data. There are simply too many variables. The postulated rate-determining step, oxidative-addition of $H_2$ to the cobalt (I) complex, which is substituted with the electronegative acyl group (step h in Figure 8.2) would be expected to be very slow. The presence of phosphines such as $(n\text{-}Bu)_3P$ might facilitate step h unless the saturated acyl 8 were further stabilized by phosphine substitution. Actually, in the presence of such phosphines, the rate of the cobalt-catalyzed oxo reaction is diminished, although the n/i product ratio increases, and the thermal stability of the catalyst is improved.

The Heck-Breslow mechanism has been challenged on the basis of IR studies of the reaction under actual operating conditions (100 atm, 125°C) [5a,b]. The equilibrium between $Co_2(CO)_8$ and $HCo(CO)_4$ was first established, and the terminal olefinic substrate was then injected into the high-pressure, high-temperature IR cell under actual operating conditions. The concentrations of various cobalt complexes, olefin substrate, and the aldehyde product were quantitatively monitored as a function of time during the period required for the catalytic reaction to reach a steady-state condition. These results clearly show that step h in the Heck-Breslow scheme cannot be the rate-determining step. Instead, these IR studies are consistent with a fast binuclear-elimination reaction as the product-determining step (8.6). This nicely accounts for the

$$\underset{\mathbf{10}}{RC(=O)\text{—}Co(CO)_4} + HCo(CO)_4 \xrightarrow{\text{fast}} RC(=O)H + Co_2(CO)_8 \qquad (8.6)$$

observed initial surge in the concentration of $Co_2(CO)_8$ and the decrease in $HCo(CO)_4$ upon injection of olefin into a solution of preformed catalyst. The slowest step in the catalytic cycle is the reaction between $HCo(CO)_4$ and the olefin leading to the acyl complex, presumably in the manner proposed by Heck and Breslow, except that at some stages in their reaction scheme, Alemdaroglu *et al.* [5a] suggest twenty-electron intermediates. Their product-forming binuclear-elimination step (Equation 8.6) has been observed independently, but details regarding the coordination number of the acyl, 10, are not clear. Recall that $HCo(CO)_4$ is a strong acid. Acyl and alkyl complexes are subject to simple protonolysis without oxidative-addition, and it is possible that such a reaction path is involved.

The most interesting aspect of the modified oxo mechanism is that the oxidative-addition and reductive-elimination steps are binuclear, and, in the over-all catalytic cycle, neither is rate-limiting. Whether this sort of mechanism will be found to be significant in the rhodium-catalyzed oxo reaction or in several other homogeneously catalyzed reactions is now an open but vital question! Two immediate implications of this binuclear mechanism are: (a) heterogenized catalysts may alter the rate or nature of oxo reactions, since site isolation of the monomeric oxo catalysts should block binuclear steps; and (b) oxo catalysts involving two different metals are possible and could show very different catalytic behavior.

Olefin isomerization occurs in the presence of cobalt oxo catalysts, but these side-reactions are not well-understood, nor are the factors which lead to higher n/i product ratios. One enigmatic result [6] has been reported concerning the hydroformylation of ω-trideutero terminal olefins under relatively low CO and $H_2$ pressures, conditions which typically result in substantial olefin rearrangement. The n-aldehyde product was converted into the corresponding methyl ester and then analyzed by mass spectrometry (8.7). The average deuterium substitution at each carbon was found to

$$CD_3CH_2CH_2CH_2CH{=}CH_2 \xrightarrow[\text{2) [O]; 3) MeOH, H}^+]{\text{1) } H_2,\ CO,\ Co_2(CO)_8} \overset{7}{C}H_3\overset{6}{C}H_2\overset{5}{C}H_2\overset{4}{C}H_2\overset{3}{C}H_2\overset{2}{C}H_2\overset{1}{C}O_2Me \qquad (8.7)$$

be as follows: for carbon 7 (in Equation 8.7), D content was 2.14 per cent; for carbons 5 and 6, D was 3.17; for carbon 4, D was 1.66; for carbon 3, D was 1.48; and for carbon 2, D was 1.55. This analysis showed no over-all deuterium loss in the product, but that deuterium is almost statistically spread along the carbon chain! This rearrangement apparently occurs without exchange of $DCo(CO)_4$ with the catalyst pool, since such exchange would lead to a loss of D to the reaction pool and to incorporation of D in the aldehydic group. The aldehydic deuterium would have been removed by the subsequent oxidation to a carboxylic acid. These experiments should be independently examined and extended. However, it appears that substantial rearrangement occurs along the entire carbon chain through intermediates which lead to n-aldehydes. It is not clear whether both acyl and alkyl or only alkyl cobalt intermediates take part in this extensive isomerization. Higher CO pressures, which are known to inhibit olefin rearrangement, tend to favor the formation of acyl complexes.

Halpern has suggested an alternative free-radical mechanism (Figure 8.3) [7] to account for the cobalt-catalyzed hydroformylation of styrene and possibly other conjugated olefins. The cobalt hydride is proposed to donate a hydrogen atom to styrene, generating a radical pair (path a in Figure 8.3). Collapse of the radical pair affords the α-phenethyl cobalt complex 11 (path b). Note that the regiospecificity of the over-all cobalt-hydride addition is thus controlled by the stability of the benzyl radical. In a side reaction (path c) the benzyl radical extracts a hydrogen atom from another cobalt hydride, affording the hydrogenation product. This step represents a free-radical hydrogenation (see Chapter 6).

Migratory insertion to CO affords the acyl, 12 (path d). At this point homolysis of the cobalt-acyl bond affords an acyl radical (path e) which can extract a hydrogen atom from another cobalt hydride (path f), affording the aldehyde. Dimerization of cobalt carbonyl radicals yields $Co_2(CO)_8$ which is in equilibrium with the hydride (paths g and h).

b. Rhodium Catalysts. The remarkably active rhodium catalysts are troubled by competing equilibria involving the very stable but catalytically inactive tetrameric and hexameric clusters, 13 and 14 in Equation 8.8. Addition of $H_2$ and CO is

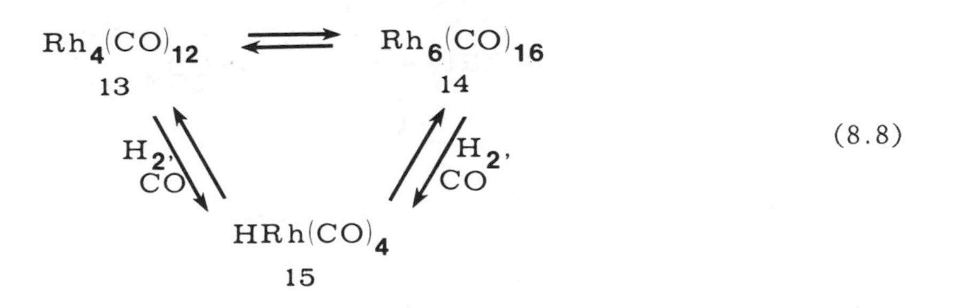


(8.8)

necessary to form the mononuclear hydrido carbonyl, $HRh(CO)_4$ (15), which is thought to be the active catalyst--probably in the form of the coordinatively unsaturated intermediate, $HRh(CO)_3$. This rhodium carbonyl oxo catalyst tends to hydrogenate and isomerize olefins, and affords lower n/i product ratios (1:2) than the analogous cobalt carbonyl. However, addition of phosphine ligands yields a modified rhodium oxo catalyst with several desirable characteristics. Thus, both industrial and laboratory

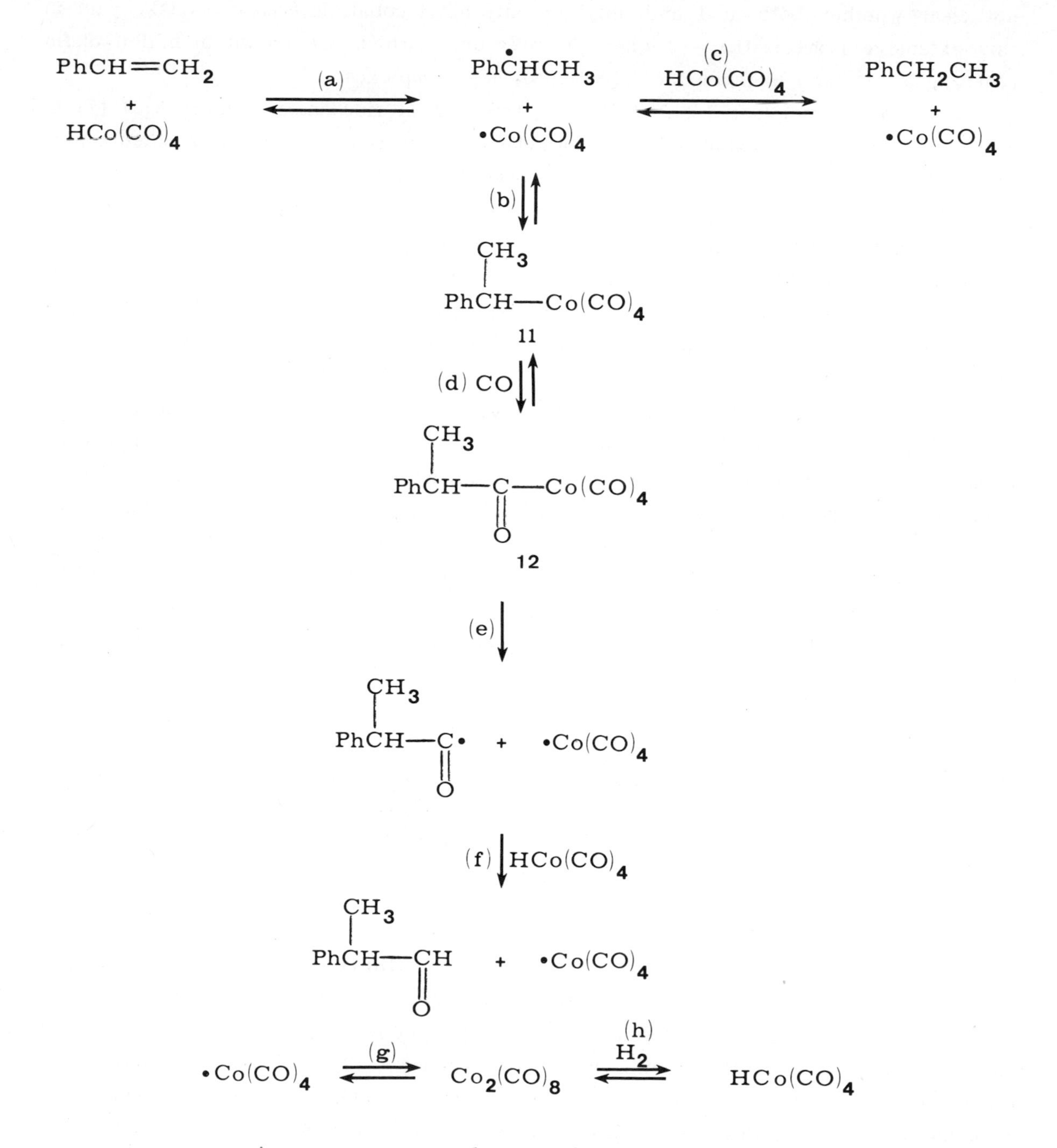


*Figure 8.3. Proposed free-radical oxo mechanism.*

applications of rhodium oxo catalysts typically employ added phosphine or phosphite. These ligands stabilize the monomeric rhodium complexes toward cluster formation, markedly increase n/i product ratios (up to 30:1), suppress olefin hydrogenation and rearrangement, and yet exhibit rates like those of the parent rhodium carbonyl catalyst. Additionally, the phosphine-modified rhodium oxo catalysts are reactive even at ambient temperatures and pressures. Thus, these catalysts can be used in the synthetic laboratory without requiring the specialized high-pressure equipment which is rarely available in academic laboratories.

Although rhodium oxo catalysts can be generated in situ from $Rh_2O_3$ or even rhodium metal, the phosphine catalysts may be introduced in the form of an easily prepared rhodium complex, $HRh(CO)(PPh_3)_3$. Such catalysts have been extensively studied, especially by Wilkinson [8] and by Pruett [9]. In many respects the over-all kinetic responses and side reactions of the cobalt and the rhodium-catalyzed reactions are similar, except that the latter is much faster with phosphine ligands present. For the rhodium-catalyzed reactions, reaction sequences similar to the Heck-Breslow cobalt mechanism have been suggested. Wilkinson has proposed both dissociative and associative paths (Figure 8.4).

Although the mechanisms shown in Figure 8.4 for the rhodium oxo catalysts are speculative and undoubtedly flawed, these can serve as a basis for our discussion of this process. Wilkinson proposed that the five-coordinate bis-phosphine rhodium hydride, 16, is the active catalyst. In the "dissociative" reaction cycle, loss of a phosphine (path a in Figure 8.4) is proposed to afford the unsaturated hydride, 17, which coordinates an olefin forming 18 (path b). Insertion (path c) affords the unsaturated alkyl, 19, which adds a phosphine to form the saturated alkyl, 20, (path d). Wilkinson speculates that in this step the opposite mode of addition may be favored, with this acidic carbonyl hydride complex giving a 2°-alkyl-rhodium bond. Reversible alkyl to carbonyl migratory insertion (path e) affords the unsaturated acyl, 21. Oxidative-addition of hydrogen to the unsaturated acyl, 21 (path f), is proposed to be the rate-determining step. This is plausible in the sense that the over-all reaction rate is first order in $[H_2]$, and such an oxidative-addition would be more facile for rhodium than cobalt and should also be accelerated by phosphine ligands. However, the available evidence does not rule out a binuclear elimination as the product-forming step (8.9).

The final step in Wilkinson's proposed mechanism is a fast, irreversible, intramolecular, reductive-elimination of 22 (path g in Figure 8.4), affording 17a, which

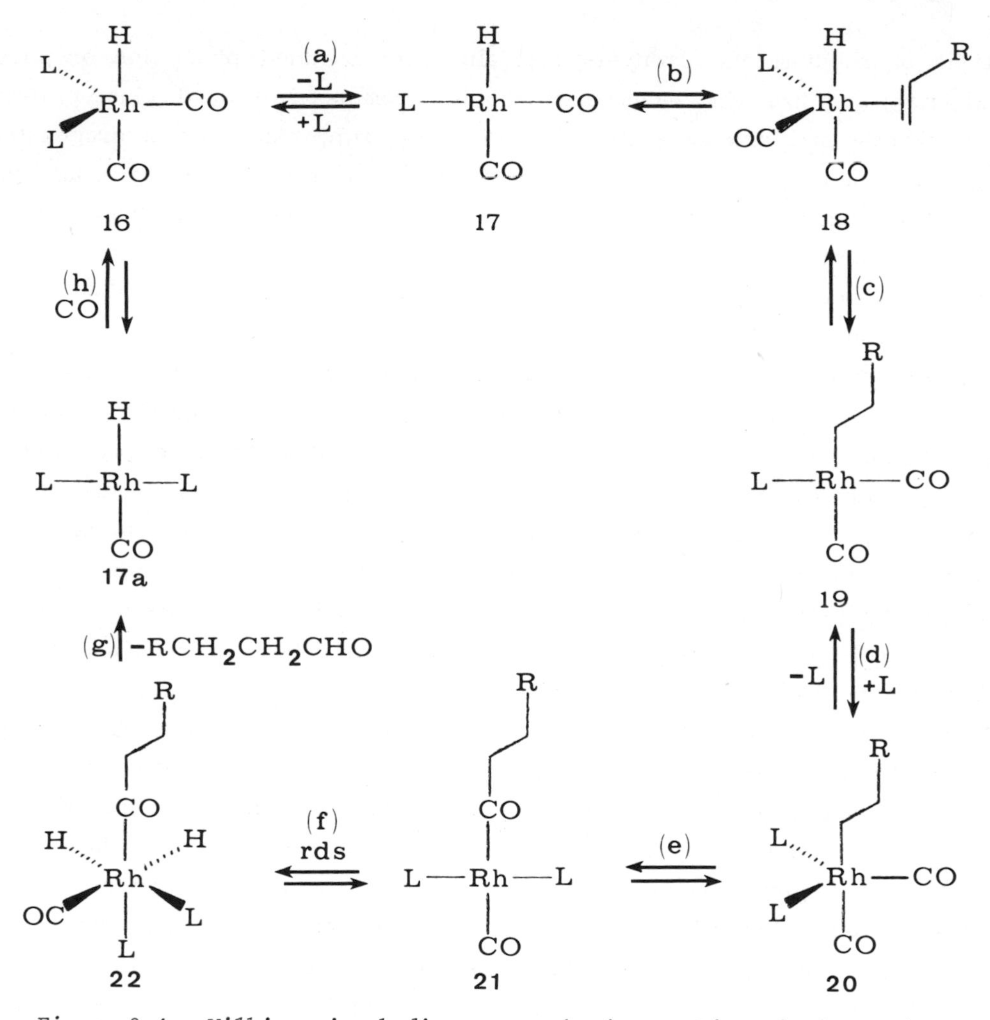


*Figure 8.4. Wilkinson's rhodium oxo mechanisms. Dissociative path.*

acquires CO regenerating 16 (path h), completing the catalytic cycle. Note also the unusual cis orientation of phosphines, which Wilkinson postulates for this cycle.

Wilkinson has also proposed a questionable associative mechanism (Figure 8.4, continued), path i, which involves an unprecedented 20-electron intermediate, 23. The remaining steps (j, k, and l) are nearly the same as those in the dissociative-reaction cycle.

The stereochemistry of the rhodium-catalyzed oxo reaction has been examined [10]. The salient results showing stereospecific cis addition of D and DCO to Z- and E-3-methyl-2-pentene are illustrated in Equations 8.10 and 8.11. The E olefin

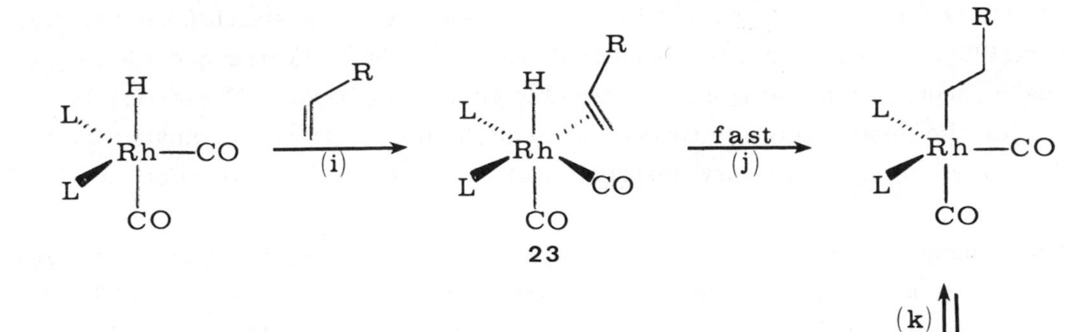


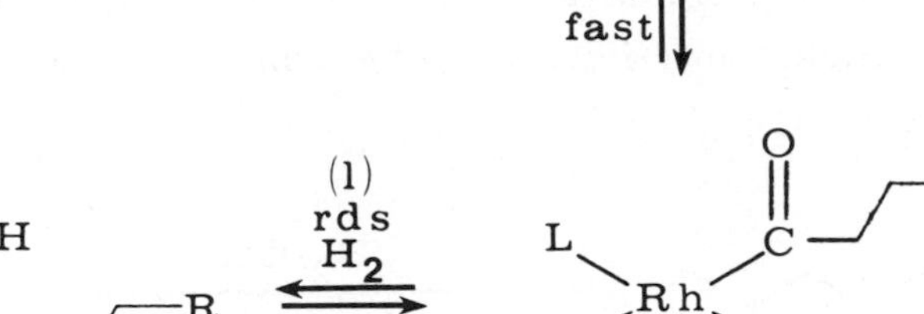


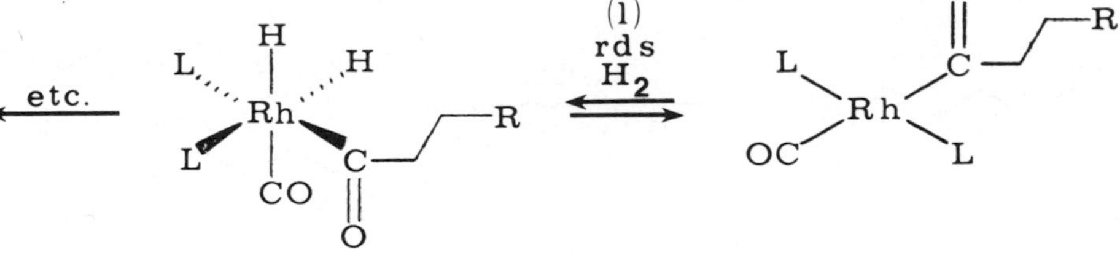


*Figure 8.4. Continued. Associative path.*

$$\mathrm{HRhL_n(CO)_{3-n}} + \mathrm{R\overset{O}{\overset{\|}{C}}{-}RhL_m(CO)_{3-m}} \xrightarrow{?} \mathrm{Rh_2L_{n+m}(CO)_{(6-m-n)}} + \mathrm{R\overset{O}{\overset{\|}{C}}H} \quad (8.9)$$

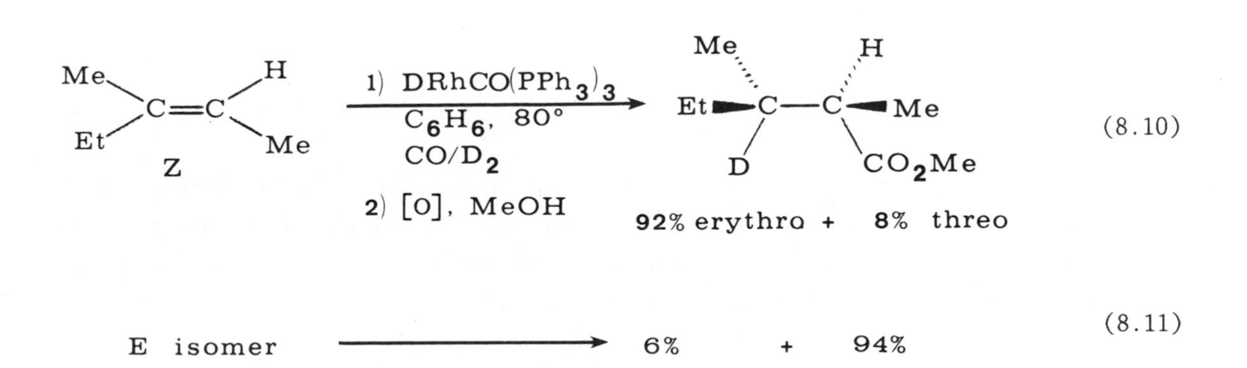

correspondingly affords the threo aldehyde. The small loss of stereospecificity was shown to result from cis-trans rearrangement of the olefin. Rearranged olefin was found to have incorporated one D atom, ostensibly from $DRh(CO)L_3$. These results are in accord with the expected cis insertion of metal hydride into a coordinated olefin, followed by an alkyl-acyl migratory insertion with retention of configuration at alkyl carbon.

Tertiary amines have been occasionally used as modifying ligands in rhodium oxo reactions. For example, the isolated tribenzylamine-substituted rhodium hydride, 24, catalyzes hydroformylation of 1-octene under ambient conditions [11] as in Equation 8.12. High selectivity can be achieved, for example, up to 93% n-aldehyde at low

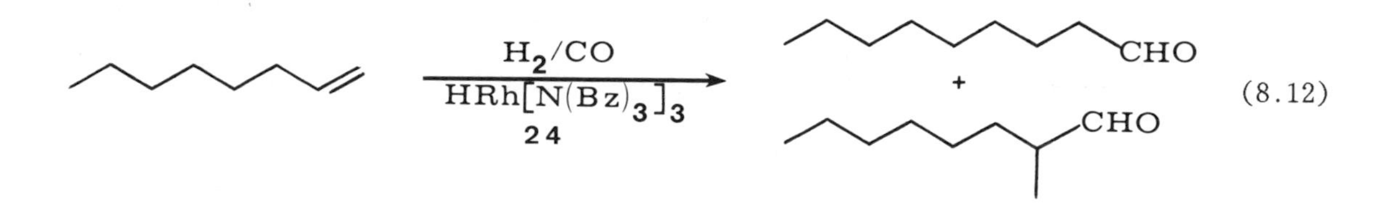


(8.12)

conversion. However, at higher conversion, olefin isomerization is a problem, and over-all yields are modest. Other tertiary amines such as N-methylpyrrolidine and tetramethylethylenediamine have been used as in situ additives to rhodium-catalyzed oxo reactions. Compared with secondary amines or phosphines, tertiary amines are resistant to oxidation. This sort of ligand merits further study.

c. Other Oxo Catalysts. Complexes of several other transition metals have been employed as oxo catalysts. Ancillary ligands are often required to increase reactivity. The reactivities of various metal carbonyls as oxo catalysts relative to cobalt are shown in 8.13. Although these are presently inferior to the rhodium catalysts, it is,

$$\begin{array}{ccccccccccc} Rh & > & Co & > & Ru & > & Mn & > & Fe & > & Cr, Mo, W, Ni \\ 10^3\text{-}10^4 & & 1 & & 10^{-2} & & 10^{-4} & & 10^{-6} & & 0 \end{array} \quad (8.13)$$

nevertheless, important that we note these other systems. The coordinatively saturated ruthenium(O) catalyst, 25, appears to function by dissociation of a CO group and subsequent oxidative-addition of $H_2$, which is thought to be the rate-limiting step in the catalytic cycle [12] shown in Equations 8.14 and 8.15.

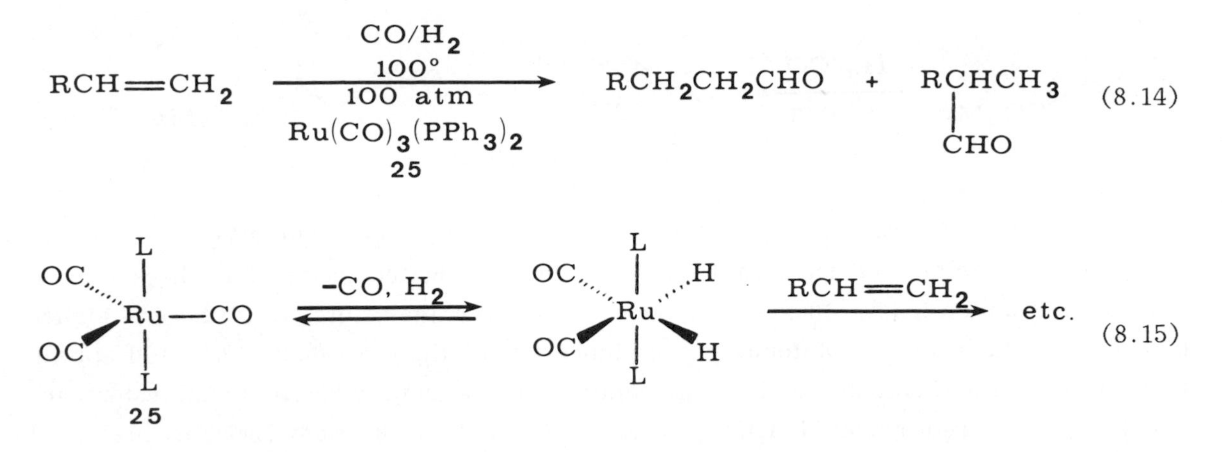


L = $PPh_3$

At 100°C and 100 atm $H_2$ and CO, this catalyst affords a n/i product ratio of approximately 3:1.

Certain platinum complexes are oxo catalysts, but these have low activity and poor stability [13,14]. For example, $[(Ph_3P)_3PtH]^+PF_6^-$ affords low yields of aldehyde and alcohol from 1-hexene at 15° and 100 atm $H_2$ and CO. The presence of the $SnCl_3^-$ ligand (formed in situ from $SnCl_2$ addition to a platinum chloride complex) increases catalyst activity. Thus, $PtH(SnCl_3)(CO)(PPh_3)_2$ is thought to be an active but relatively unstable catalyst which affords 95% hexanol from 1-pentene at 100° and 200 atm CO and $H_2$.

Iridium oxo catalysts are known, but are less reactive and less well-studied than rhodium analogues. A notable example is $HIr(CO)_3P(iPr)_3$, which catalyzes the hydroformylation of ethylene at 200 psi and 50°C. An IR analysis of this sytem has identified successive hydride, alkyl, and acyl stages in the catalytic reaction [15]. These results suggest a mechanism whose early steps are very similar to those proposed by Heck and Breslow for the $Co_2(CO)_8$ catalyzed reaction. Recall from Chapter 5 that the alkyl-to-acyl migratory-insertion step should be much slower for the third-row iridium alkyl.

Oxo reactions of conjugated olefins give complex product mixtures, and so far these reactions have not been useful [16]. For example, 1,3-butadiene does not give a clean product with cobalt catalysts, and rhodium catalysts afford mixtures of mono-and dialdehydes. The formation of $\eta^3$-allyl intermediates is a problem with this substrate. Olefin hydrogenation is a major competing reaction with $\alpha,\beta$-unsaturated esters. Furthermore, mixtures of $\alpha$ and $\beta$ oxo products are formed (8.16). These product ratios

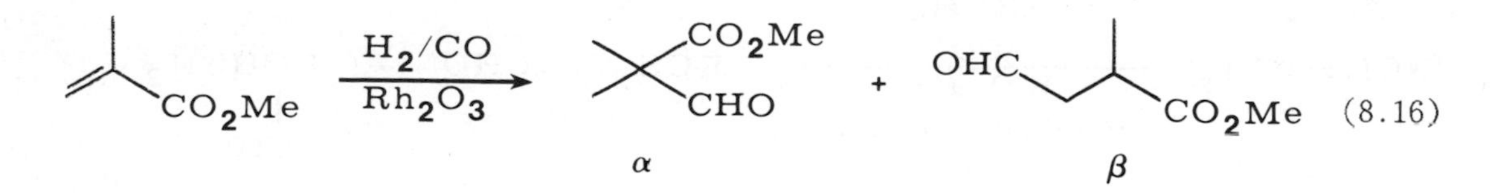


are apparently determined by either kinetically or thermodynamically controlled regioselective addition of rhodium hydride to the conjugated olefin, but these factors are not well-understood. Increased temperatures favor the β product, whereas higher pressure or the presence of tertiary phosphines favors the α product. It would appear that the α product (which results from addition of the metal adjacent to the ester carbonyl) is the predominant kinetic product, whereas the thermodynamically preferred rhodium intermediate is the β isomer. The latter may be formed by a rearrangement of either a rhodium alkyl or a rhodium acyl prior to the irreversible product-forming hydrogenolysis. This topic deserves further study. More extensive rearrangement is observed at elevated temperatures. For example, a γ-formyl product is formed from reaction of crotonate esters at 180°C.

## 8.2. THE FISCHER-TROPSCH REACTION.

The reductive polymerization of carbon monoxide to afford straight-chain hydrocarbons, olefins, and alcohols (8.17) is an old process which has recently attracted great interest [17-21]. These processes offer the possibility of obtaining chemical feedstocks and liquid fuels from coal by way of syn-gas. The Fischer-Tropsch

$$CO + H_2 \xrightarrow[\text{cat.}]{\Delta,\ p} CH_3(CH_2)_nCH_2OH + CH_3(CH_2)_nCH{=}CH_2 + CH_3(CH_2)_mCH_3 \quad (8.17)$$

process may be traced to German patents in 1913 and 1914 describing the formation of hydrocarbon, alkenes, and oxygenated compounds over heterogeneous cobalt and osmium catalysts. Franz Fischer and Hans Tropsch, after whom these processes are named, discovered the alkali-iron-catalyzed reaction in 1923-1925. In 1936 the first commercial plant was opened in Germany, which subsequently depended upon the cobalt-catalyzed Fischer-Tropsch process for a substantial portion of its wartime liquid hydrocarbon fuels.

South Africa has operated a large (~230,000 tons per year) iron-catalyzed Fischer-Tropsch plant (SASOL) since 1957. The United States is considering building such plants as world petroleum prices soar. However, the present economics of preparing either liquid hydrocarbon fuels or chemical feestocks from syn-gas are unattractive. A principal problem is the lack of product selectivity, both in terms of product type (olefin or alcohol) and in the statistical molecular-weight range of the products. Additionally, a large quantity of the carbon content in coal must be discarded as $CO_2$ to form the necessary hydrogen. The use of Fischer-Tropsch reactions to make liquid fuels (gasoline and diesel) is, therefore, a political decision requiring government subsidy. On the other hand, the synthesis of small, valuable molecules retaining at least one oxygen of CO from efficient and selective Fischer-Tropsch reactions would have great economic merit. We shall see that the Union Carbide ethylene glycol process is the premier example of such a reaction.

a. Heterogeneous Fischer-Tropsch Catalysts. Until very recently, all Fischer-Tropsch reactions were heterogeneously catalyzed. Although there has been much speculation, the mechanism of these reactions remains uncertain. Recently, reports have described homogeneous reactions which model certain steps in the Fischer-Tropsch process. Some claims of weakly catalytic, homogeneous Fischer-Tropsch reactions have also been described (*vide infra*). First we shall consider mechanisms proposed for the heterogeneously catalyzed Fischer-Tropsch process.

The hydrogenation of CO to methane, "methanation," is a reaction which is heterogeneously catalyzed by several transition metals, such as nickel, ruthenium, and rhodium. Methanation seems to involve the formation of surface carbides which, in some instances, do not appear to participate in chain growth [22]. It is uncertain at present whether methanation is fundamentally different from Fischer-Tropsch reductive polymerization.

However, Fischer and Tropsch's original dissociative mechanism, involving a metal carbide intermediate, has been given new life by the discovery that a soluble metal carbide can be carbonylated, affording a structurally characterized carbomethoxy cluster [23], as in Figure 8.5. This result also lends support to the argument that the adjacent metal atoms in clusters may provide pathways which are not possible with mononuclear homogeneous catalysts [24]. However, it is difficult to find a cluster catalyst which is sufficiently robust to withstand high temperatures and $H_2/CO$ pressures.

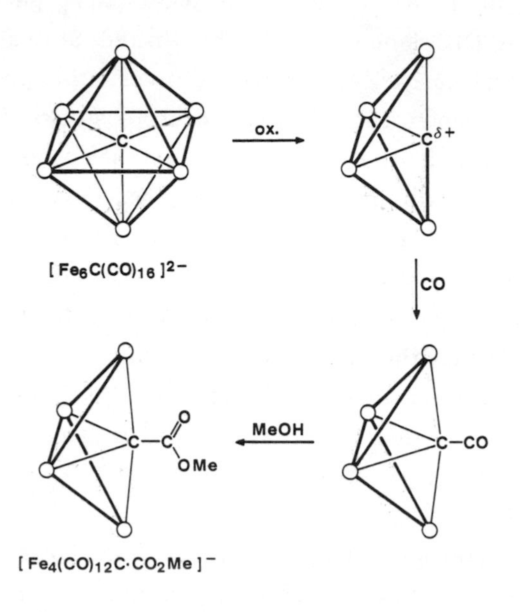


*Figure 8.5.*

A major problem in formulating a reasonable Fischer-Tropsch mechanism is the initiation of the chain--from a surface carbide or by hydride migratory insertion to CO, affording formyl. As mentioned in Chapter 5, this hydride insertion appears to be thermodynamically unfavorable. Further on in this chapter, we will discuss soluble formyl complexes and their possible role in CO reductions. It has been suggested that bonding of the formyl oxygen to an adjacent metal or an acidic center on a heterogeneous catalyst support might facilitate the formation of the formyl (8.18). Further

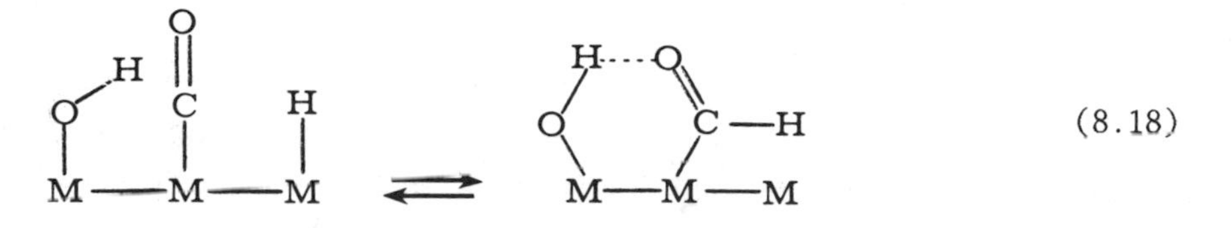


(8.18)

reduction of the formyl is more reasonable, the initial product being a hydroxycarbene complex (8.19). Further reaction might proceed by forming a surface alkyl (8.20), inserting another acyl and subsequent reduction of this acyl (8.21).

Thus, an alkyl chain might grow by the continued insertion of CO groups and reduction of the acyl intermediates. At random times, the alkyl groups could be

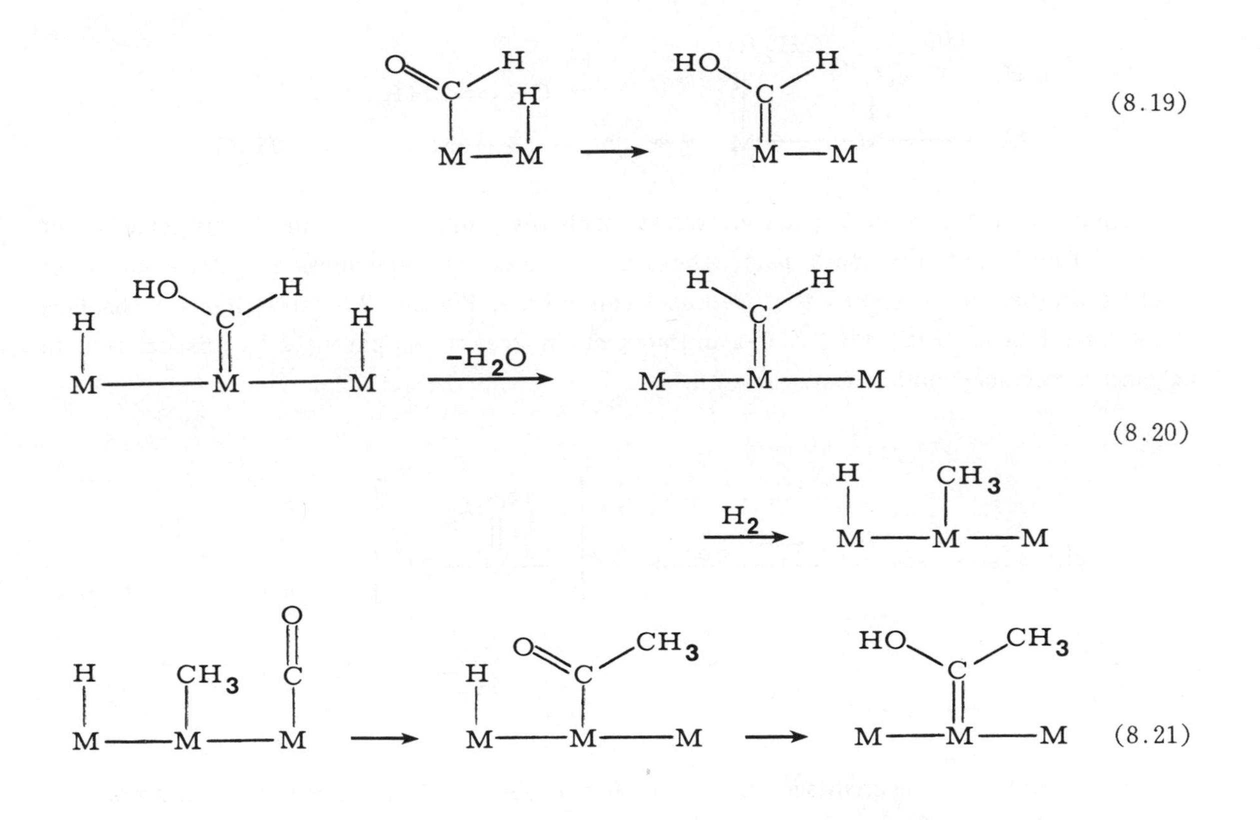

(8.19)

(8.20)

(8.21)

reduced to hydrocarbons (8.22), or the acyl groups to aldehydes and subsequently to

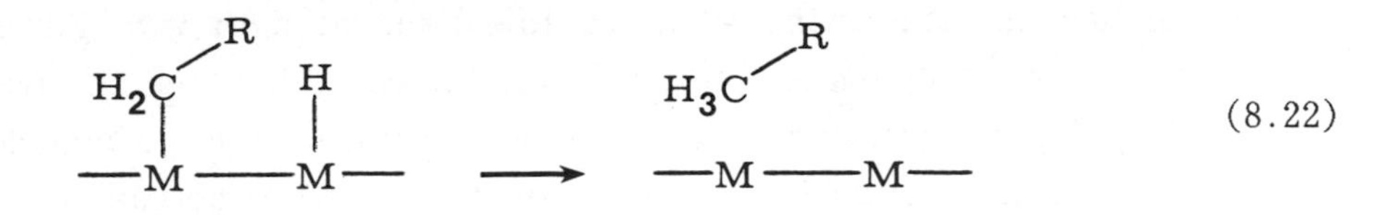

(8.22)

alcohols (8.23). Dehydration of α-hydroxyalkyl groups might afford olefins (8.24).

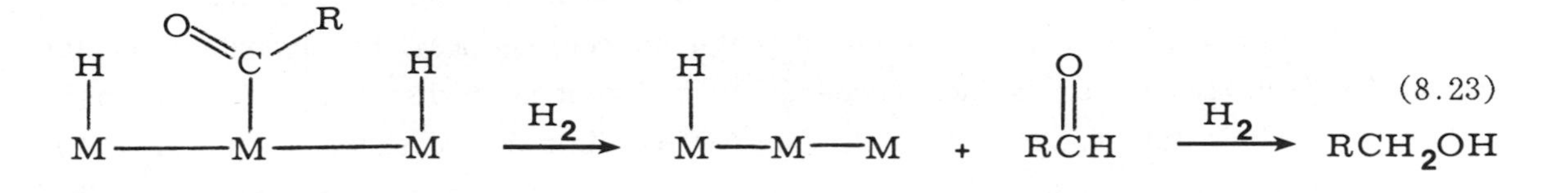

(8.23)

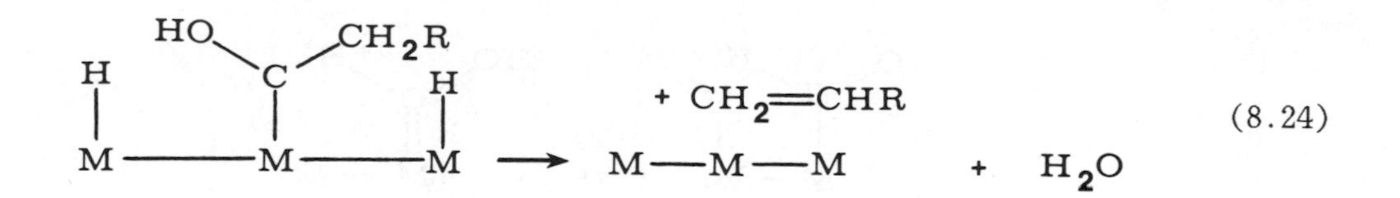


(8.24)

The final saturated-hydrocarbon products probably derive from the hydrogenation of these olefins. For the most part, these mechanisms are speculation. However, from labeling studies it is known that methanol can initiate Fischer-Tropsch chains. Similarly, ketene labeled with $^{14}C$ at C-2 initiates chain growth--apparently by dissociation to CO and a surface-bonded methylene (8.25).

$$^{14}CH_2{=}C{=}O \xrightarrow{\text{F.T. cat.}} \left[ \begin{array}{c} ^{14}CH_2 \\ \| \\ -M-M \end{array} \right] \xrightarrow[H_2]{CO} {}^{14}CH_3(CH_2)_nR \qquad (8.25)$$

Note that the mechanisms proposed for both the homogeneously catalyzed oxo reaction and the heterogeneously catalyzed Fischer-Tropsch reaction have one intermediate in common: a metal acyl derivative. Under reducing conditions, soluble acyl complexes are normally transformed into aldehydes, and very seldom (*vide infra*) give rise to further chain growth, which is the dominant path with Fischer-Tropsch catalysts. It has been suggested [20] that the difference between the two processes might be accounted for by noting that the soluble acyls can undergo binuclear elimination by reaction with another hydride molecule (Equation 8.26); whereas this path is not open to the heterogeneous Fischer-Tropsch catalyst, so that reductive chain growth becomes dominant (8.27). This interpretation would be weakened by the existence of either a truly site-isolated oxo catalyst or by a homogeneous Fischer-Tropsch catalyst.

The hypothesis that site-isolation of catalyst centers may be important in the Fischer-Tropsch mechanism gained credence from Vollhardt's discovery of a polymer-anchored cobalt catalyst, **26** [25]. Under $CO/H_2$ (initial pressure, 75 psi at 25°), octane suspensions of **26** at 190°C afford methanation and linear hydrocarbons (8.28).

$$M{-}C(=O)R \xrightarrow{M'{-}H} M{-}M' + RCHO \qquad (8.26)$$

$$M{-}C(=O)R \xrightarrow{H_2,\ CO} R(CH_2)_nCH_3 \qquad (8.27)$$

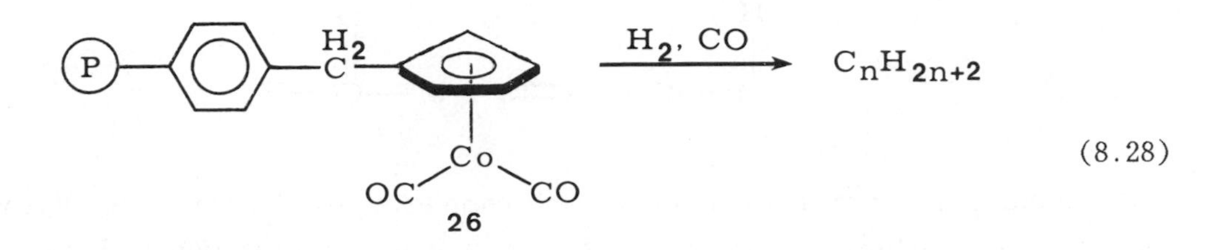


n = 1–20

A steady turnover of CO was maintained for many cycles. Activity was increased by removal of CO from 26 by prior vacuum pyrolysis. Control experiments support the contention that the active catalyst is a metal complex derived from 26. Soluble $CpCo(CO)_2$ and $Co_2(CO)_8$ showed no activity, but poisoned the polymer-bound catalyst, 26. Cobalt was not leached from 26 during the Fischer-Tropsch reaction, since treatment of the active catalyst with CO quantitatively regenerated the precatalyst 26 as judged by IR analysis. The polymer in the active catalyst, 26, is a cross-linked macroreticular resin. An analog of 26 bound to a more flexible, microreticular resin showed no Fischer-Tropsch activity, but upon vacuum pyrolysis this flexible resin complex formed CO-bridged species, in contrast to 26. These results indicate that site isolation, and perhaps other specific features of the polymer environment, are important factors in controlling catalytic activity. In Chapter 6 we encountered many polymer-supported hydrogenation catalysts. The rates and product distributions of these polymeric catalysts may differ from their soluble analogues, and may also depend on the pore size, degree of cross-linking, swelling properties of the resin, and the solvent. However, the above case is the first observation of new catalytic activity upon immobilization of a homogeneous precatalyst. It should be emphasized that the actual catalyst formed from 26 is presently unknown.

b. Homogeneous Model Reactions. Next we will examine some homogeneous model reactions for steps which are thought to occur in the Fischer-Tropsch process. The Fischer-Tropsch synthesis requires the reduction of coordinated carbon monoxide. The formation of the first C-H bond is a key step in such a process. The alternative to a carbide or carbene mechanism is the formation of a formyl complex, either by insertion of a hydride from the same metal to which the CO is attached or, as discussed above for the heterogeneous case, by intermolecular hydride addition from an adjacent metal (8.29).

$$\mathrm{H{-}M{-}CO} \rightleftharpoons \mathrm{M{-}C({=}O){-}H} \qquad (8.29)$$

Actually, a large number of formyl complexes are known. In all cases, these appear to be kinetically stabilized. That is, the forward reaction in Equation 8.29 has never been directly observed, but the reverse reaction is well established. The first formyl complex was prepared by formylating the strongly basic, coordinatively saturated iron carbonyl anion [26] as in 8.30.

$$\mathrm{Fe(CO)_4^{2-}} + \mathrm{H{-}C({=}O){-}O{-}C({=}O){-}Me} \longrightarrow \mathrm{H{-}C({=}O){-}Fe(CO)_4^{-}} \qquad (8.30)$$

A more general route to formyl complexes involves reduction of metal carbonyls by trialkyl or trialkoxyborohydrides $Et_3BH^-$, $(MeO)_3BH^-$, $(iPrO)_3BH^-$ [27] as in 8.31. Recall from Chapter 5 that this reaction is only one example of a general class of

$$\mathrm{Na^+H\bar{B}(OR)_3} + \mathrm{Fe(CO)_5} \longrightarrow \mathrm{HC({=}O){-}\bar{F}e(CO)_4} \qquad (8.31)$$

intermolecular nucleophilic additions to unsaturated ligands. In this way, Casey, Gladysz, and Winter have been able to generate many unstable formyl complexes in situ [28-32]. These formyls were detected by their characteristic low-field aldehydic $^1H$ NMR signal in the region 15-16 δ. In some instances, crystalline formyl complexes were isolated and fully characterized. An example is the air-stable dinuclear formyl, $Et_4N^+Re_2(CHO)(CO)_9^-$, 27 in Equation 8.32. These anionic formyl complexes owe their kinetic stability to the fact that they are coordinatively saturated. The loss of a neutral ligand seems to be the rate-determining step in their apparently irreversible formation of the corresponding hydride.

All these formyl complexes are potent hydride donors toward electrophiles such as $H^+$, ketones, aldehydes, alkyl halides, and even other metal carbonyls. The reactions of the dinuclear rhenium hydride (27 in 8.32) illustrate this point. The reactions in

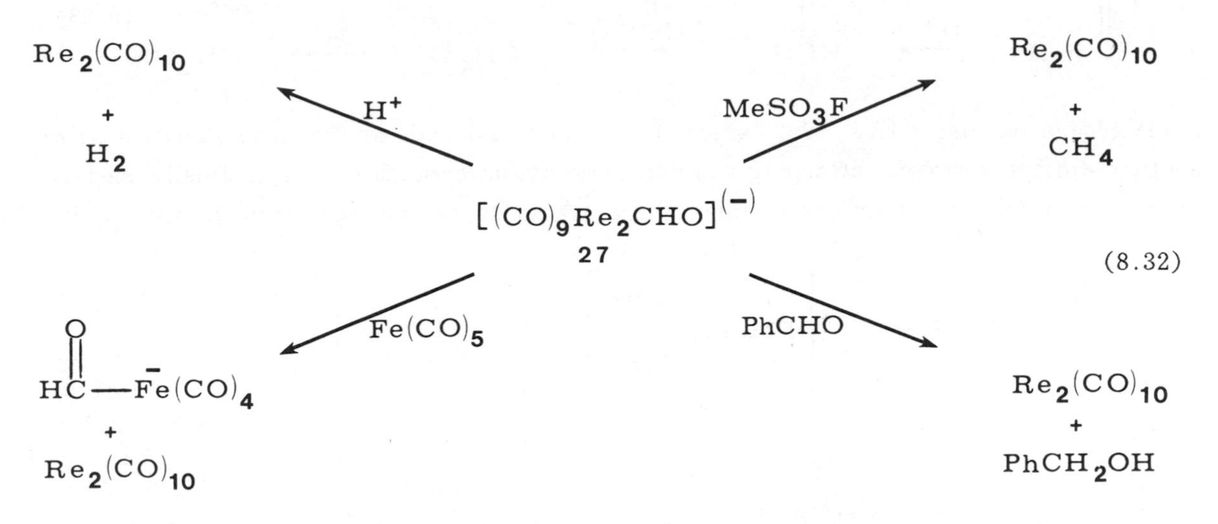


(8.32)

Equation 8.32 have been shown to go by direct hydride donation from the coordinated formyl and not through an intermediate metal hydride derived from the formyl [33]. This hydride donation may be aided by the formation of a stable, neutral metal carbonyl (8.33).

$$\bar{M}-\overset{O}{\overset{\|}{C}}-H \longrightarrow M-C{=}O + H^- \qquad (8.33)$$

An important consequence of this hydride donation is that formyl complexes can undergo self-reduction to form methanol. For example, treatment of the iron formyl, 28, with 10 equivalents of $CF_3CO_2H$ gives a 55 percent yield of methanol and a trace of formaldehyde (8.34). This reaction may take place by protonolysis of the iron-formyl bond to generate formaldehyde, which is subsequently reduced by a second mole of the

$$L'(CO)_3Fe(CHO) + CF_3CO_2H \longrightarrow CH_3OH + (CH_2O) \quad (8.34)$$

**28**

$L' = (PhO)_3P$

formyl complex (8.35). Alternatively, protonation of the formyl oxygen could produce a

$$Fe{-}C(=O){-}H \xrightarrow{H^+} HC(=O)H \xrightarrow{[HC(=O)\bar{Fe}]} CH_3O^- \xrightarrow{H^+} CH_3OH \quad (8.35)$$

hydroxycarbene derivative, 29, which is further reduced by hydride donation from another formyl complex, affording a hydroxymethyl intermediate that is finally cleaved by a proton (8.36). Carbene complexes are known to be susceptible to hydrogenation

$$\bar{Fe}{-}C(=O){-}H \xrightarrow{H^+} \left[Fe{=}C(OH)(H)\right] \quad (8.36)$$

**29**

$$\xrightarrow{[HC(=O)Fe]^-} [Fe{-}CH_2OH]^- \xrightarrow{H^+} CH_3OH$$

[34], further supporting their plausible role in Fischer-Tropsch reactions (8.37).

$$(CO)_5Cr{=}C_4H_6O \text{ (2-oxacyclopentylidene)} \xrightarrow[H_2]{\Delta \text{ or } h\nu} \text{tetrahydrofuran} \quad (8.37)$$

This carbene mechanism is given further credence by the recent work of Gladysz [35]. Addition of electrophiles such as $MeSO_3F$, $Me_3SiCl$, and $CF_3CO_2H$ was shown to induce disproportionation of the rhenium formyl. By independently generating all the intermediates in this reaction (8.38), the reduction of an intermediate carbene complex, 30, was substantiated.

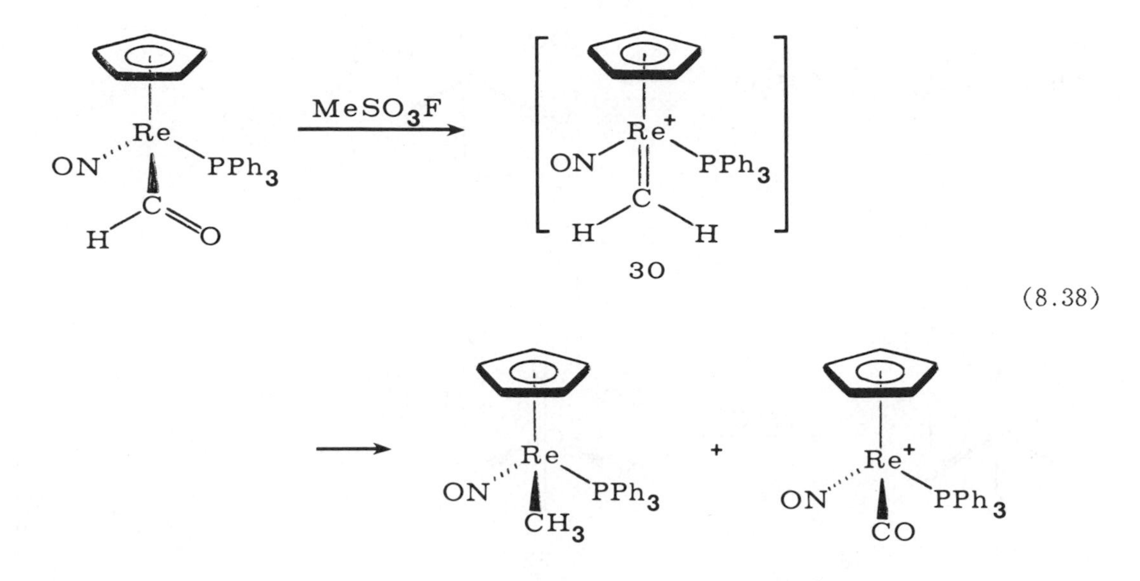


Even though, in the cases that have been examined, the insertion of hydride to formyl (Equation 8.29) appears to be thermodynamically unfavorable, there are several factors which might promote that reaction. A Lewis acid could shift the equilibrium, either by coordinating with and destabilizing the starting carbonyl or by binding the formyl oxygen which would stabilize the product. Such Lewis-acid effects are well-documented in rate studies of alkyl-to-acyl migratory insertions, which we discussed in Chapter 5. The formation of a $\eta^2$-formyl (see Chapter 5) would also stabilize that side of the equilibrium. A $d^0$ complex in which the carbonyl has no possibility of engaging in $\pi$-backbonding would also help to stabilize the formyl side of the equilibrium in Equation 8.29. The latter two properties are found among complexes of the early transition series, such as those in Group IV. We will return to this point further on.

The formation of formaldehyde from CO and $H_2$ is thermodynamically unfavorable. Consequentally, any catalytic process for hydrogenating carbon monoxide cannot pass through formaldehyde or even a complex of formaldehyde unless formaldehyde is greatly stabilized by coordination of the metal or by interaction with a strong Lewis acid. This situation at ambient pressure is illustrated by Roper's preparation [36] of a structurally characterized $\eta^2$-formaldehyde complex (31 in 8.39). Upon heating, the solid $\eta^2$-formaldehyde complex, 31, spontaneously forms the hydridoformyl compound, 32, which is unstable in solution at 40°C, affording CO, $H_2$ and 33 (in 8.40).

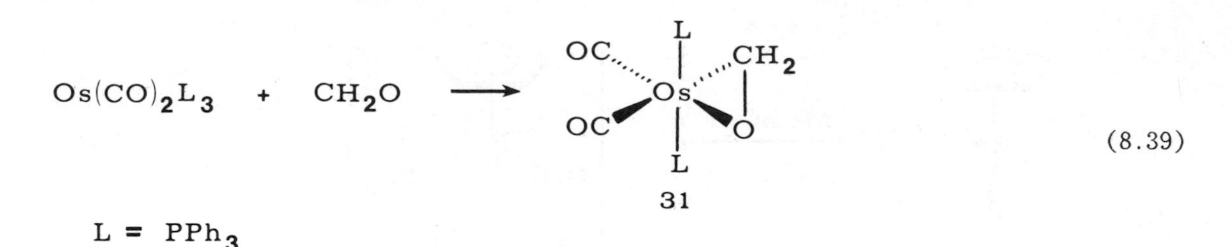


(8.39)

L = $PPh_3$

31 —75°C→ 32 —40°C→ 33 + $H_2$

(8.40)

After formyl formation, the next plausible stage in the reduction of CO is a hydroxymethyl complex. The related α-hydroxyalkyl ligands are also probably involved in subsequent chain-growth stages. The first well-characterized hydroxymethyl compound, 34, was prepared by Casey [37] as in 8.41. An α-hydroxyalkyl metal complex,

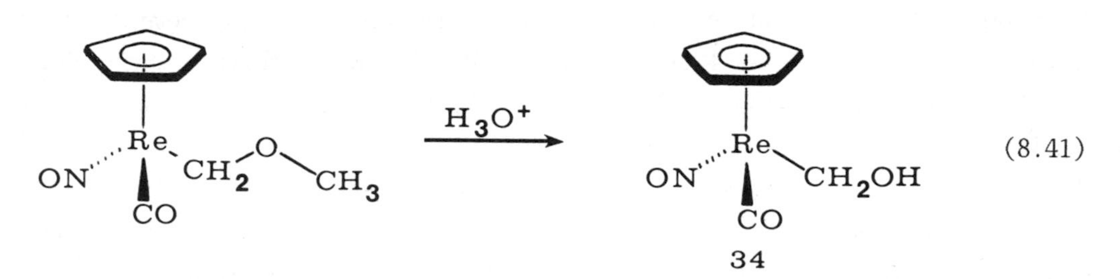


(8.41)

$CpFe(CO)_2(C(CF_3)_2OH)$, had been reported earlier [38]. Note that the hydroxymethyl ligand could alternatively be formed by hydroxide addition to a carbene complex. Such a reaction should be reversible, so that acid-catalyzed removal of the hydroxyl group can generate a cationic carbene (8.42). Thus the hydroxymethyl and the carbene

$$M{-}CH_2{-}OH \underset{OH^-}{\overset{H^+}{\rightleftharpoons}} \overset{+}{M}{=}CH_2 \qquad (8.42)$$

ligands can be interconnected. This is significant for several reasons. A reaction such as that shown in Equation 8.42 might explain the role of $K_2O$ promotors in the heterogeneous Fischer-Tropsch process. Carbene ligands should be easily reduced to alkyl groups by the inter- or intramolecular addition of a hydride. The alkyl can subsequently insert CO in a Fischer-Tropsch chain-growth step. It is also possible that the carbene could be derived from an alternative path which does not involve the thermodynamically questionable formyl intermediate. Further on, we shall see that a hydroxymethyl complex is a plausible intermediate in the formation of ethylene glycol from $H_2$ and CO.

The first direct reduction of a carbonyl to an alkyl (8.43) was reported in 1968

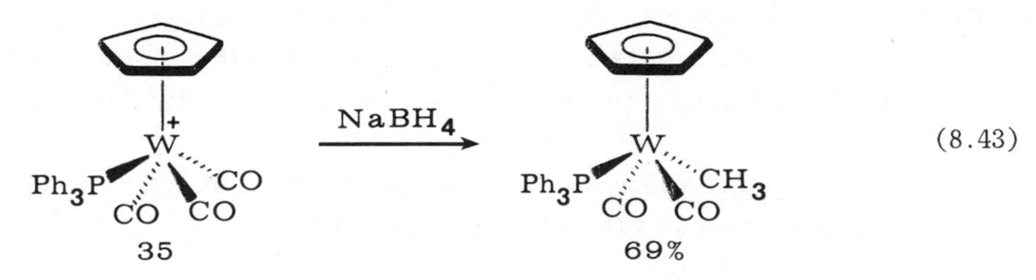


[39]. Presumably a formyl intermediate is involved. Note that the formal positive charge on the tungsten carbonyl complex, 35, would be expected to increase its reactivity toward intermolecular hydride addition.

The reduction of a neutral metal acetyl to an ethyl derivative has also been achieved (8.44). In this case, $BH_3$, a Lewis acid, is an effective reducing agent, but $NaBH_4$ fails to react [40].

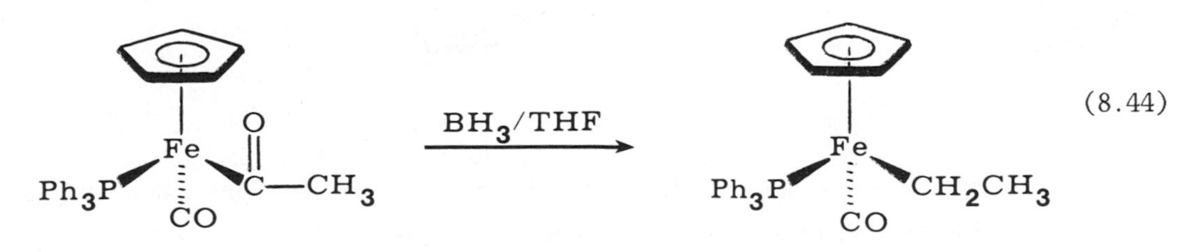

An example of the way in which a Lewis acid could facilitate the reduction of an acyl is found in reaction (8.45) between the cationic carbene complex, 36, and $NaBH_4$,

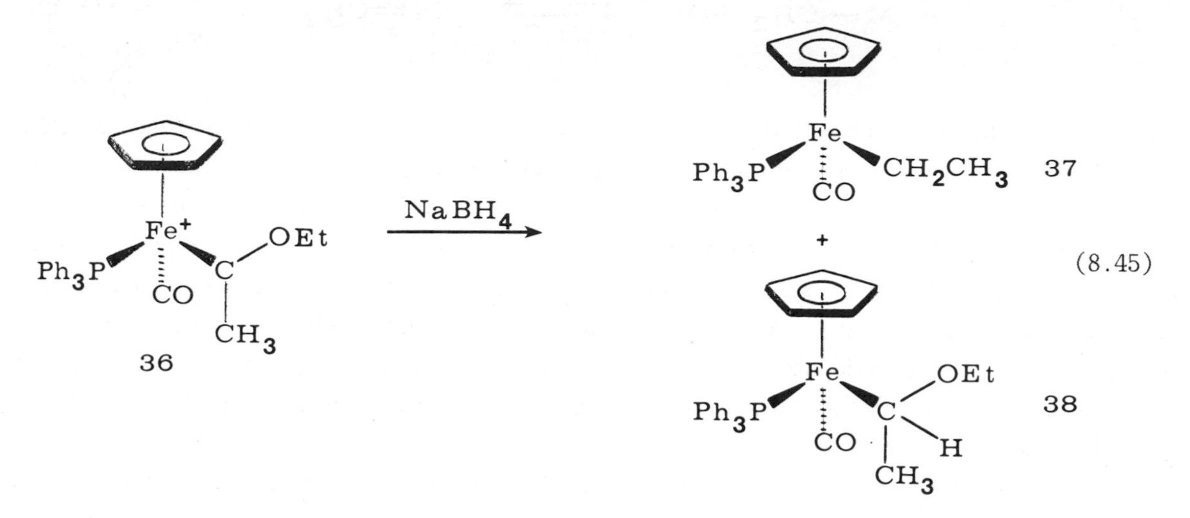


affording the ethyl complex, 37, and the unusual α-ethoxyethyl derivative, 38, which is presumably an intermediate in the formation of the ethyl derivative [41]. It can be argued that the ethoxycarbene complex, 36, is similar to the Lewis-acid adduct of an acyl complex.

The stoichiometric reduction (in 8.46) of the zirconium(II) carbonyl, 39, was the first reported homogeneous CO hydrogenation [42].

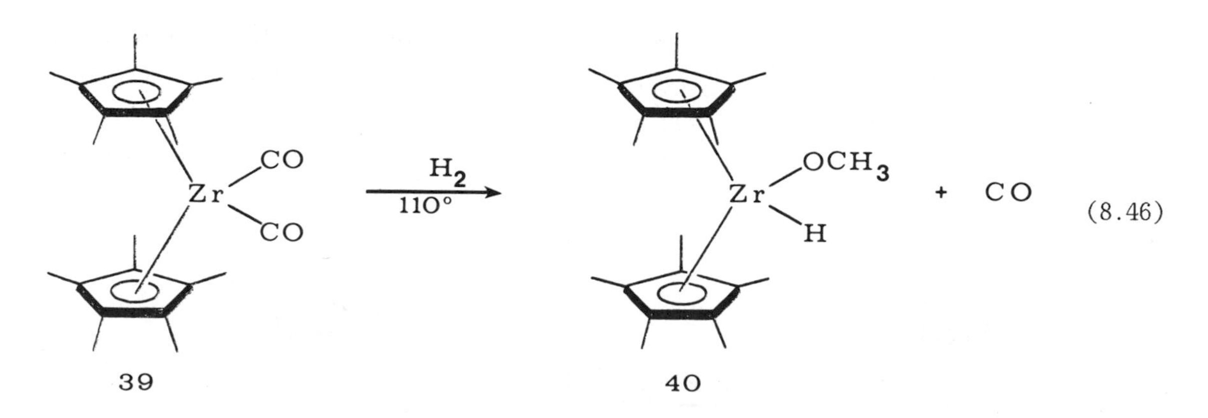


In spite of Bercaw's careful study of this and related reactions [43], it has so far not been possible to distinguish an intramolecular (Equation 8.47) from an intermolecular

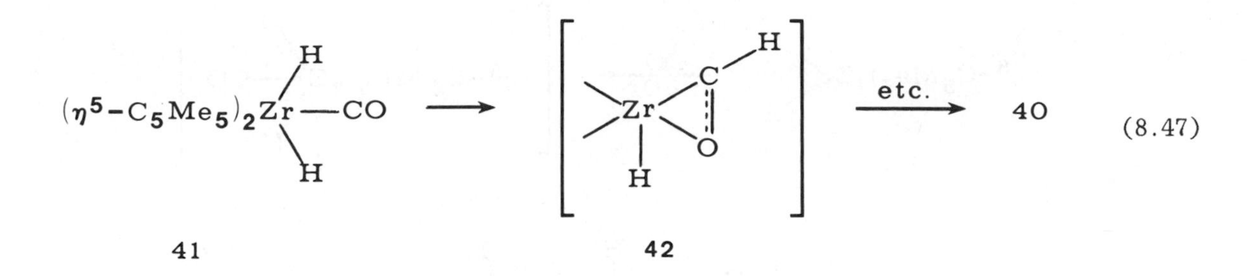


(Equation 8.48) hydride transfer to coordinated CO. The intermolecular path avoids

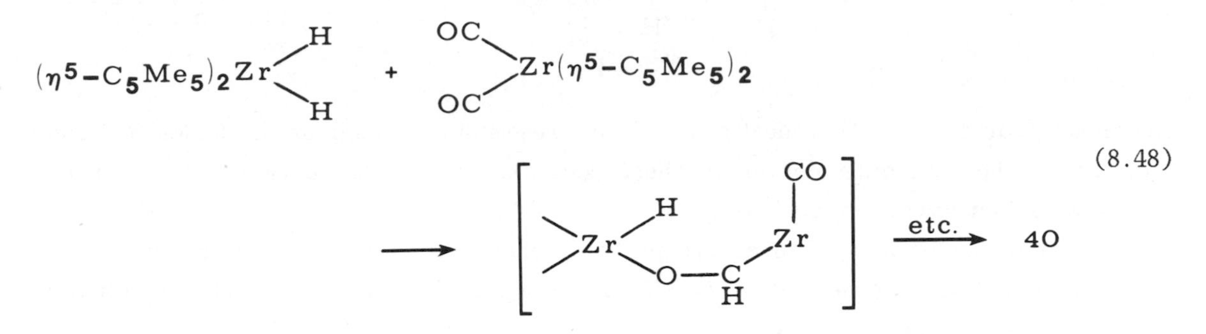


the need to invoke a hydride to formyl migratory insertion, and has been unequivocally demonstrated for a heterometallic system (8.49).

$$(C_5Me_5)_2ZrH_2 + W(CO)_6 \longrightarrow (C_5Me_5)_2Zr(H){-}O{-}C(H){=}W(CO)_5 \quad (8.49)$$

The dihydridocarbonyl, 43, is thought to be an intermediate in the formation (8.50) of the curious dimer 44. This reaction serves as a plausible model for a Fischer-Tropsch carbon-carbon bond-forming step or the formation of ethylene glycol in the Union Carbide process, vide infra. The formation of the dimer, 44, may reflect the carbene-like character of a $\eta^2$-formyl intermediate, 42.

There are several special features of the zirconium system which facilitate CO reductions under such mild conditions. The zirconium carbonyls are very reactive toward alkyl-to-acyl insertion, the acyl being stabilized by a dihapto structure and the CO destabilized by the lack of a $\pi$-backbond in $d^0$ Zr(IV) complexes such as 41. Thus even the intramolecular hydride-to-formyl insertion seems possible for this system.

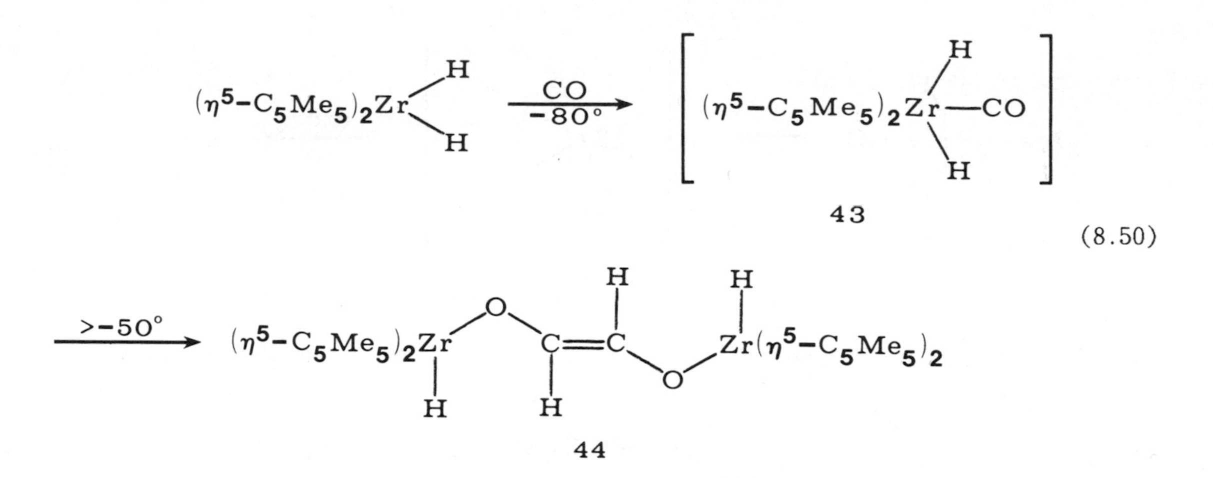


(8.50)

Zirconium hydrides are chemically hydridic, resembling aluminum hydrides in their reactivity. The zirconium center in these complexes is also an excellent Lewis acid--especially toward oxygen-donor groups.

The above-mentioned features are probably important in the zirconicene-dichloride-catalyzed reductive oligomerization of CO by $\underline{i}$-$Bu_2AlH$ [44] in 8.51. This reaction is

$$\begin{matrix} iBu_2AlH \\ + \\ CO \end{matrix} \xrightarrow[2)\ H_3O^+]{1)\ Cp_2ZrCl_2} \underset{1.0}{MeOH} + \underset{0.1}{EtOH} + \underset{0.15}{n\text{-}PrOH} + \underset{0.03}{n\text{-}BuOH} \quad (8.51)$$

related to Bercaw's stoichiometric hydrogenation. Apparently, $\underline{i}$-$Bu_2AlH$, being both a strong hydride-transfer agent and a Lewis acid can promote continued reduction and insertion steps, affording a weakly catalytic synthesis of methanol and higher alcohols. Note also that these $d^0$ systems are not likely to form alkanes or aldehydes by mononuclear reductive-elimination.

c. Homogeneous CO Hydrogenation. Very recently, claims of homogeneously catalyzed CO hydrogenation have appeared. Most of these reports are complicated by severe experimental conditions, slow rates, low conversions, and the difficulty in distinguishing homogeneous from heterogeneous catalysis.

For example, cobalt carbonyl, $Co_2(CO)_8$, has been reported [45] to be a weak catalyst for the hydrogenation of CO to form methanol and other minor products under extreme conditions (300 atm, 200°C) as in 8.52. This reaction was proposed to be homogeneous on the basis of limited kinetic data. Only ten turnovers were observed

$$CO + H_2 \xrightarrow[C_6H_6]{Co_2(CO)_8\ [HCo(CO)_4]} MeOH + [\ MeOC(O)H + EtOH + EtOC(O)H + n\text{-}PrOH\ ] \quad (8.52)$$

while the solution remained homogeneous! Under more severe conditions (306 atm, 240°C), $Mn_2(Co)_{10}$ also affords a small quantity of methanol. These results are significant in the sense that mononuclear metal hydrides may be the actual catalysts. A free-radical path has been proposed for the cobalt-catalyzed reaction [45].

At 225-275°C and 1300 atm, several ruthenium complexes serve as catalyst precursors for the homogeneous hydrogenation of CO to methanol and methyl formate (8.53); however, $Ru(CO)_5$ is the only detectable ruthenium complex in the reaction

$$CO + H_2 \xrightarrow{Ru(CO)_5} MeOH + MeOC(O)H \quad (8.53)$$

mixture [46]. Whereas increased CO pressure suppresses the rate and increases the formate yield, addition of triphenylphosphine increases the methanol selectivity to >95 per cent. The hydrogenation of CO to form methanol is a highly exothermic reaction with a negative entropy change. Therefore this reaction is thermodynamically more favorable at lower temperatures. The present heterogeneous catalysts for this reaction require relatively high temperatures to achieve reasonable rates. At these temperatures high pressure is required to obtain the appropriate equilibrium state. However, if the temperature is not controlled and becomes too high, these heterogeneous catalysts lose their activity. Homogeneous catalysts dissipate heat more easily than heterogeneous catalysts do. Thus the discovery of a homogeneous catalyst for selective hydrogenation of CO to methanol, which would be efficient at moderate temperatures and pressures, would be an important technological development. Such a catalytic system could substantially reduce the cost of methanol produced from coal or methane via syn-gas. The activation energy for the $Ru(CO)_5$ catalyst is 32 ± 3 kcal $mol^{-1}$, which is lower than that reported for $HCo(CO)_4$, 41 kcal $mol^{-1}$. These results are significant since they show that a mononuclear complex is capable of homogeneously catalyzing the hydrogenation of CO. Metal clusters are not required for such reactions.

Earlier claims of methane and Fischer-Tropsch hydrocarbon products derived from the precatalyst $Ru_3(CO)_{12}$ were shown to arise from ruthenium metal formed by decomposition of the cluster [46,47]. Other claims of CO hydrogenation to methane catalyzed by $Os_3(CO)_{12}$, and $Ir_4(CO)_{12}$ at 140°C and 2 atm are probably due to heterogeneous catalysts [48]. Methanation is a characteristic product derived from many heterogeneous transition-metal catalysts [49]. In the absence of other evidence, the formation of hydrocarbons from CO hydrogenation is diagnostic of metal particles arising from the degradation of homogeneous precatalysts.

The implication of cluster complexes and Lewis acids in homogeneously catalyzed Fischer-Tropsch reactions led Muetterties to invent a catalyst involving $Ir_4(CO)_{12}$ in the molten salt $NaCl{\cdot}2AlCl_3$ [50]. Under moderate pressures CO is hydrogenated to low-molecular-weight hydrocarbons (8.54). Ethane is the major initial product.

$$CO + 3\,H_2 \xrightarrow[1\text{ atm.}\;\; Ir_4(CO)_{12}]{180°\;\; NaCl{\cdot}\,AlCl_3} CH_4 + C_2H_6 + CH_3CH_2CH_3 + i\text{-}C_4H_{10} \qquad (8.54)$$

However, the number of turnovers is quite low. It is interesting to note that $Ir_4(CO)_{12}$ is the most chemically robust of all the known transition-metal cluster complexes. Other carbonyl clusters such as $Rh_4(CO)_{12}$, $Rh_6(CO)_{16}$, and $Ru_3(CO)_{12}$ visibly decompose to the metals in this corrosive molten salt. Since water is formed from CO hydrogenation, the $AlCl_3$ is constantly degraded as the reaction proceeds. Note that $AlCl_3$ could stabilize formaldehyde as an intermediate.

The most intriguing and economically attractive homogeneous syn-gas reaction is Union Carbide's rhodium-carbonyl-catalyzed ethylene glycol synthesis [51,52] in 8.55. This reaction requires high pressures to bias the equilibrium toward ethylene glycol rather than methanol. High reaction temperatures seem to be necessary to obtain reasonable rates. The virtual absence of methane would appear to rule out heterogeneous

$$CO + H_2 \xrightarrow[3{,}000\ \text{atm.}]{200^\circ,\ [Rh_{12}(CO)_{30-34}]^{-n}} HOCH_2CH_2OH + [HOCH_2CH(OH)CH_2OH] + [CH_3OH] \quad (8.55)$$

rhodium catalysts. The product selectivity in this process is quite impressive--up to 70% ethylene glycol in the products. This is unlike the other Fischer-Tropsch catalysts, which give statistical product distributions.

Infrared evidence (under high pressure and temperature) indicates the presence of anionic clusters containing 12 rhodium atoms (formed from two rhodium octahedra). The active catalyst is thought to be one of these anionic clusters, but which one is uncertain. Ion pairing affects the product selectivity--perhaps by influencing equilibria between various anionic rhodium carbonyl clusters. Thus cationic "promoters," such as $Cs^+$, and quaternary ammonium ions are beneficial. The published literature on this process is minimal [51]; the primary reports are confined to patents [52]. At present it is not clear whether this reaction, which requires severe temperatures and pressures and an expensive catalyst, and which has special product-separation problems, will become a commercial process. Nevertheless, the probable role of cluster catalysts and the mysterious product selectivity make this a fascinating reaction. A hydroxymethyl rhodium complex, 45, is a plausible intermediate in the Union-Carbide process (8.56).

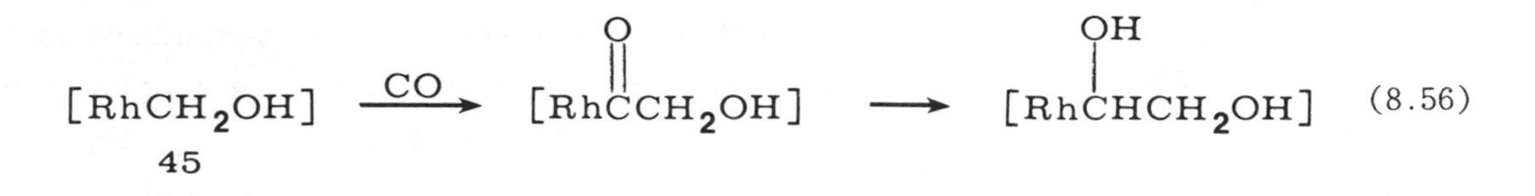


Related α-acyloxymethyl manganese compounds, such as 46, have been reported [53] to insert CO and to form ethylene glycol derivatives (8.57).

Another selective coupling of carbon monoxide is oxidative rather than reductive. This is a complicated palladium(II)-chloride-catalyzed reaction affording diethyloxylate and diethylcarbonate as major and minor products (8.58). A copper cocatalyst is

$$(CO)_5MnCH_2OC(=O)C(CH_3)_3 \ (\mathbf{46}) \xrightarrow{H_2} HC(=O)CH_2OC(=O)C(CH_3)_3 \longrightarrow HOCH_2CH_2OC(=O)C(CH_3)_3 \quad (8.57)$$

$$EtOH + CO + O_2 \xrightarrow[\substack{CuCl_2 \\ (Et_3O)_3CH}]{PdCl_2} EtOC(=O)\text{—}C(=O)OEt + EtOC(=O)OEt \quad (8.58)$$

required, apparently to reoxidize palladium(0) which is formed during the catalytic cycle. Typical conditions are 1,000 psi, 125°C, affording a 93% total yield of diethyloxalate and diethylcarbonate in a 2.4:1 ratio [54].

## 8.3. Reppe Reactions.

A number of homogeneously catalyzed reactions are known which involve three components: CO, an olefin or an acetylene, and a nucleophile such as $H_2O$, ROH, $RNH_2$, RSH, or $RCO_2H$. These processes, which form carboxylic acid derivatives, were first developed by Reppe and are sometimes collectively called "Reppe reactions" [55]. These reactions seem to make use of a step similar to the water-gas shift reaction to create a metal hydride, which adds to an olefin to form an intermediate metal alkyl which subsequently inserts CO. The resulting acyl then reacts with a nucleophile. These are all reaction paths we have previously encountered. Water is required to form the essential metal-hydride intermediate. It is often easier to generate a metal hydride by hydrolysis of coordinated CO than by direct hydrogenation, because the former process does not require the presence of a vacant coordination site, but the latter does. Very little is known about the detailed mechanism of most of these reactions. However, it is evident that free hydrogen gas is usually not an active component in the reaction. Therefore, although these reactions resemble the water-gas shift process in forming a metal hydride, this intermediate is thought to be captured by olefin insertion and diverted away from $H_2$ formation. An example of such a Reppe reaction is the hydroformylation of olefins using CO and $H_2O$ catalyzed by $Fe(CO)_5$, which is normally a very poor oxo catalyst (8.59). Pettit [56] has noted such reactions are thermodynamically more favorable than processes employing $H_2$ gas. He has shown that reactivity of the $Fe(CO)_5$ system depends on the pH, which controls the rate of formation, and the equilibrium concentrations of the iron-hydride intermediates (8.60 and 8.61).

$$RCH{=}CH_2 + CO + H_2O \xrightarrow[OH^-,\ 90^\circ]{Fe(CO)_5} CO_2 + RCH(CHO)CH_3 + RCH_2CH_2CHO \quad (8.59)$$

$$Fe(CO)_5 + OH^- \longrightarrow HFe(CO)_4^- + CO_2 \quad (8.60)$$

$$HFe(CO)_4^- \underset{}{\overset{H_2O}{\rightleftharpoons}} H_2Fe(CO)_4 + OH^- \quad (8.61)$$

The dihydride is thought to be the most reactive complex. This system is capable of reducing the intermediate aldehyde to alcohols (8.62). In Chapter 6 we saw that the

$$RCH_2CH_2CHO + CO + H_2O \xrightarrow[OH^-]{Fe(CO)_5} RCH_2CH_2CH_2OH + CO_2 \quad (8.62)$$

combination of a metal carbonyl and an aqueous base could be used to hydrogenate aryl nitro compounds to amines, and conjugated enones to saturated derivatives.

Most Reppe reactions involve the formation of an acid, an ester, or an amide, presumably from solvolysis of an intermediate metal acyl with $H_2O$, ROH, or $RNH_2$, respectively (8.63). A number of transition-metal salts and carbonyl complexes can serve as precatalysts: Ni, Co, Fe, Rh, Ru, Pd, etc. Phosphine-substituted carbonyls are expecially reactive. The mechanism by which the putative acyl intermediate undergoes solvolysis is uncertain. The Reppe reactions in Equation 8.63 differ from the oxo

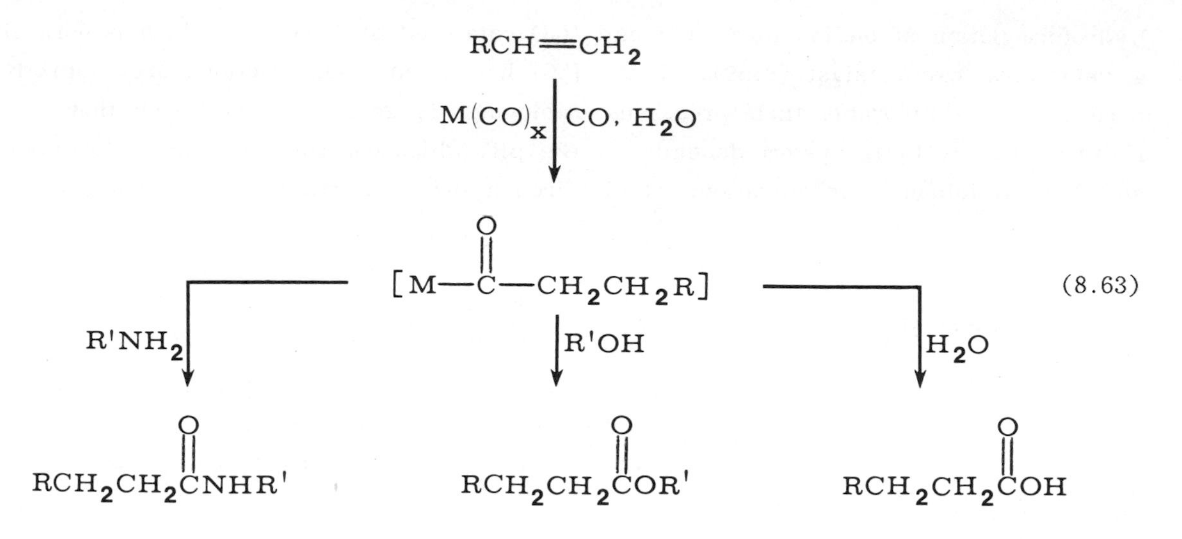


(8.63)

process in that the acyl in the latter is cleaved by a reduction, forming an aldehyde. In both cases isomeric products are formed which depend on the regiospecificity of the metal-hydride insertion step.

Stoichiometric Reppe reactions often require milder conditions than corresponding catalytic reactions. For example, the $Ni(CO)_4$-catalyzed formation of acrylic acid from acetylene requires 120-220°C and 30 atm; whereas the stoichiometric reaction occurs at 35-80°C and 1 atm (8.64).

$$HC{\equiv}CH + CO + H_2O \xrightarrow{Ni(CO)_4} H_2C{=}CHCO_2H \qquad (8.64)$$

## 8.4. Monsanto's Acetic-Acid Process.

The carbonylation of methanol to form acetic acid (8.65) stands out as the technologically most successful example of homogeneous catalysis [57]. This highly

$$MeOH + CO \xrightarrow[180°C,\ 30-40\ atm.]{Rh\ cat.\ /\ I^-} \underset{99\%}{MeCO_2H} \qquad (8.65)$$

selective reaction is extremely rapid. Yields are >90%, and rates are in the enzymatic range. Current annual acetic-acid production by this process is estimated at $10^6$ tons. This and related processes are especially interesting, since one-carbon chemicals derived from coal via syn-gas are expected to become increasingly important chemical feedstocks. Few chemical reactions are entirely new. Thus the rhodium-based Monsanto acetic-acid process bears some resemblance to an earlier BASF cobalt-based process. Both require an iodide "promoter," but Monsanto's rhodium-catalyzed reaction is much faster and takes place at considerably lower temperatures and pressures than the cobalt-catalyzed reaction.

Much is known about the mechanism of the Monsanto process. Under moderate conditions (80°C; 3 atm CO) the reaction has been studied with NMR and IR, using a well-characterized "precatalyst," $Ph_4As^+[Rh(CO)_2I_2]^-$ and methyl iodide. Intermediate reaction stages have thus been characterized. These studies are consistent with the mechanism shown in Figure 8.6.

The rate-determining step a is thought to be the oxidative-addition of methyl iodide to the unsaturated anionic complex 47. Methyl iodide is formed by the reaction of HI with methanol. The HI is derived from hydrolysis of acetyl iodide, the penultimate organic product. The proposed catalytic mechanism involves a cycle of four steps shown in Figure 8.6: (a) oxidative-addition, (b) migratory insertion of alkyl to carbonyl, (c) coordination of CO, and (d) reductive-elimination of acetyl iodide. Step a is thought to be rate-determining on the basis of the rate law observed under steady-state conditions: rate = $k[Rh]^1[I]^1[CO]^0[CH_3OH]^0$. Direct evidence supports the reductive-elimination step (d). A solid, acetyl μ-iodo dimer, 48, has been isolated and structurally characterized by X-ray diffraction. Treatment of dimer 48 with CO affords acetyl iodide (observed by NMR) and the starting rhodium(I) catalyst 47, which does not appear to react with acetyl iodide--that is, step f is nearly irreversible. Even if this step were reversible, hydrolysis or methanolysis would consume acetyl iodide. The putative acetyl intermediate 49 is the logical product of CO reacting with the dimer 48 by splitting the μ-iodo bridge.

Iodide is a necessary component in the over-all reaction scheme. Virtually any source of iodide will serve to activate the catalyst. The mechanism of the Monsanto process depends on the special chemical characteristics of iodide, which is simultaneously a good nucleophile, a good leaving group, an effective ligand for rhodium, and a very weak proton base. Iodide also appears to activate rhodium(I) toward oxidative-addition. Bromide can carry out these functions, but is much less effective. The rate is reduced tenfold. Pentachlorothiophenoxide, $C_6Cl_5S^-$, can also be substituted for iodide, but the rate is greatly diminished [58].

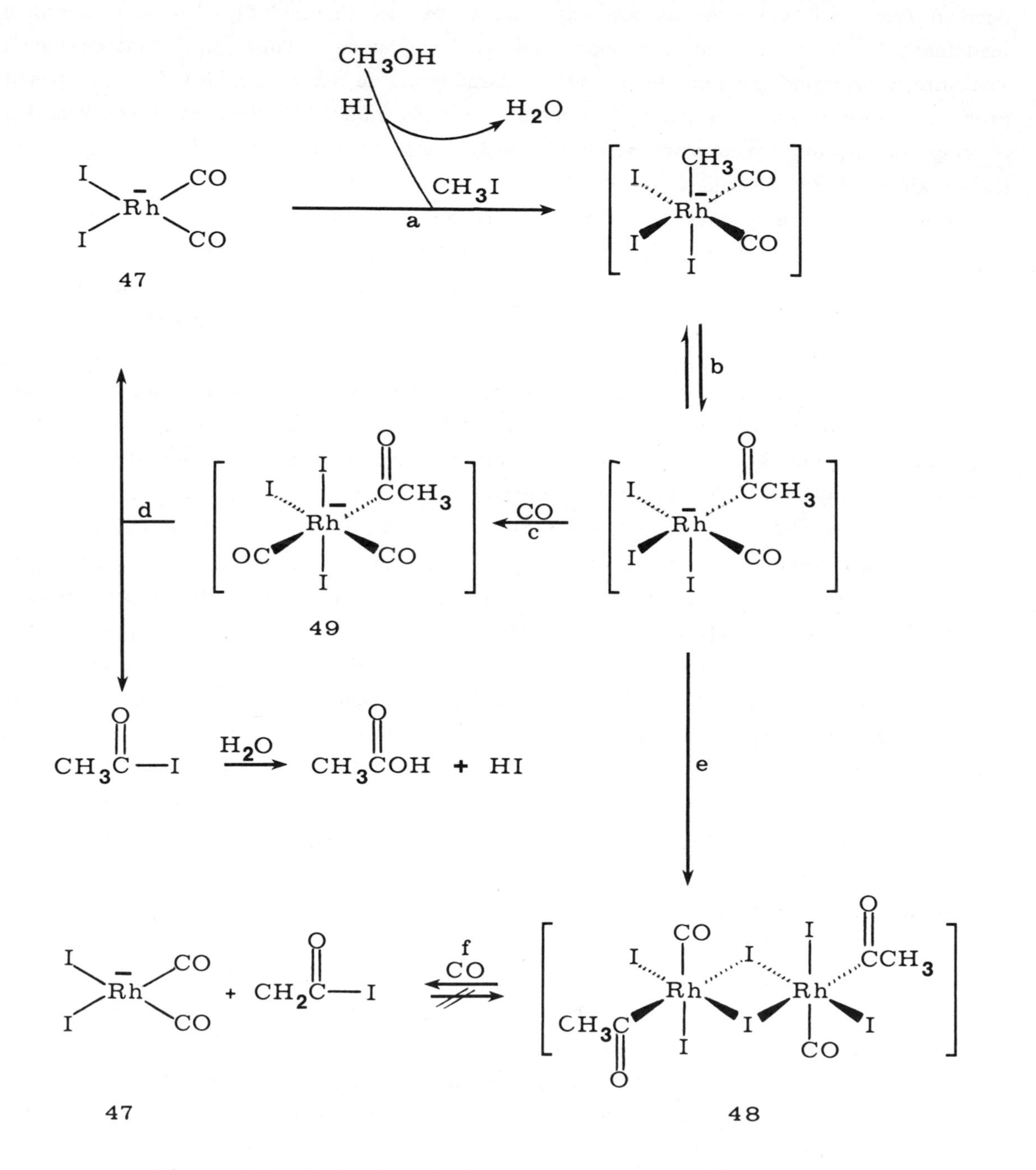


*Figure 8.6. Mechanism of the Monsanto acetic-acid process.*

Cobalt is less reactive than rhodium, seemingly because of a slower oxidative-addition step, although the nature of the cobalt(I) complexes may be different. Iridium catalysts are also active, but less selective and mechanistically more complex than the analogous rhodium system. Forster [59] has shown that the iridium system exhibits two principal catalytic cycles for methanol carbonylation, one involving neutral and the other anionic iridium species. Under many conditions a competitive water-gas shift reaction is observed with iridium catalysts.

Attempts to immobilize the Monsanto acetic-acid catalyst have been mostly unsuccessful. Retrospectively, this is not surprising. Added ligands such as phosphines not only slow the catalytic reaction, but once dissociated, these ligands are irreversibly alkylated by methyl iodide. Thus any ligands which might be used to bind rhodium to a solid support should be subject to alkylations. An imaginative attempt to immobilize a Monsanto-type catalyst used a polymer-bonded perchlorothiophenylate in place of iodide ion, but the rate and the catalyst stability were unsatisfactory [60]. Of course iodide ion cannot be attached to a support. Alkylation of iodide affords a key reaction intermediate, methyl iodide. Finally, it should be said that the successful containment of such valuable (Rh and I) and corrosive (I) components in a continuous homogeneous process is a major engineering achievement.

## 8.5. Other Iodide-Promoted CO Reactions.

Some other catalytic reactions using iodide promoters have been reported. Even though these have not yet led to commercial processes, the chemistry is nevertheless instructive and thus worth comment.

Aqueous formaldehyde can be carbonylated to glycolic acid in the presence of a rhodium salt and an iodide promoter [61] as in 8.66. Acetic acid is a significant

$$(CH_2O)_3 + 3\,H_2O + CO \xrightarrow[150°C,\ 1{,}000\ atm.]{Rh(I),\ I^-} \begin{array}{ll} HOCH_2C(=O)OH & 86\% \\ + & \\ CH_3CO_2H & 14\% \end{array} \quad (8.66)$$

byproduct. A plausible mechanism has been suggested for this reaction (Figure 8.7). The analogy between this mechanism and that shown above for the acetic-acid process is evident.

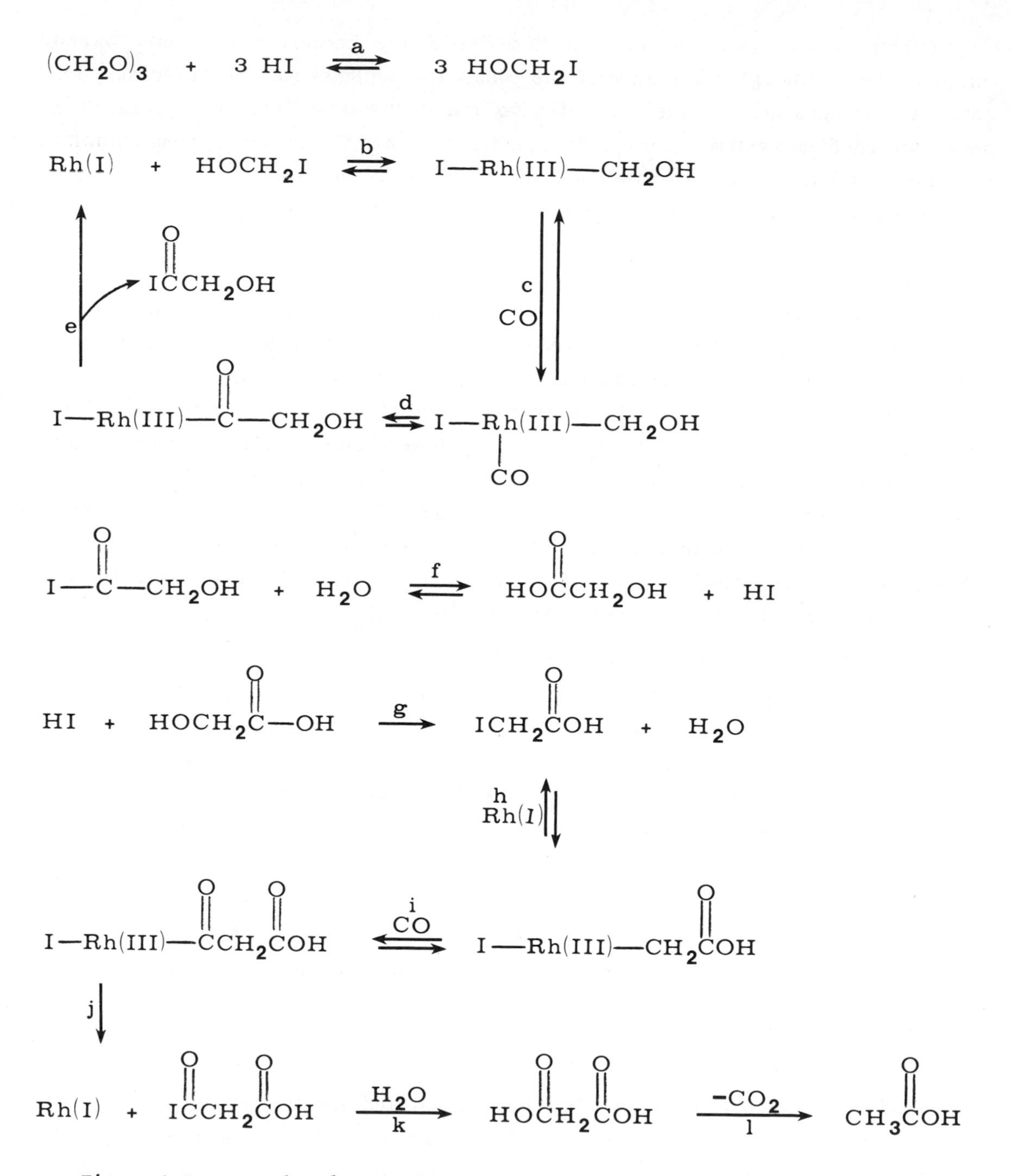


*Figure 8.7. Postulated mechanism for Chevron's glycolic-acid synthesis.*

In Figure 8.7, the reactive organic intermediate is proposed to be $ICH_2OH$, formed by reaction between paraformaldehyde and HI (step a). This undergoes alkylation by rhodium(I) (step b), followed by carbonylation and CO insertion (steps c and d). Reductive-elimination of the acyl iodide (step e) completes the primary catalytic cycle. The product, glycolic acid, is formed by hydrolysis of the acyl iodide (step f). A secondary catalytic cycle is proposed to afford acetic acid and carbon dioxide through formation of malonic acid (steps h-l).

Another modification incorporates ethylene and yields methylene-bis-propionate [61] as in 8.67. In this process, formaldehyde does not appear to react with the metal

$$(CH_2O)_3 + CO + CH_2{=}CH_2 \xrightarrow[\text{125°C, 5 hr}]{CH_3CH_2I,\ [Rh(CO)_2Cl]_2} \underset{84\%}{(CH_3CH_2C(=O){-}O)_2CH_2} + \underset{16\%}{CH_3CH_2CO_2H} \quad (8.67)$$

catalyst, but rather scavenges propionic anhydride, which is the principal reaction product. The yield is based on ethylene. Figure 8.8 depicts the proposed mechanism.

In this scheme, HI is proposed to oxidatively add to rhodium(I) (step a). Insertion of ethylene affords the ethyl rhodium derivative (steps b and c), followed by CO insertion (step d) and the familiar reductive-elimination (step e) affording propionyl iodide, the proposed penultimate product. The formaldehyde simply serves to capture propionyl iodide or propionic anhydride (steps f and g).

A third iodide-promoted process involves the catalytic synthesis of tetrahydrofuran from ethylene oxide and ethylene [61] as in 8.68. This reaction is catalyzed by several complexes: $Pd(PPh_3)_4$, $RuCl_2(PPh_3)_3$, and $Ru(CO)_3(PPh_3)_2$ in the presence of HI. Unfortunately, the yields and rates are apparently insufficient to make a commercially viable process in spite of the valuable product, THF. The proposed mechanism (shown in Figure 8.9) invokes β-iodoethanol formed from action of HI on ethylene oxide (step a). Oxidative-addition (step b), ethylene insertion (step c), and reductive-elimination (step d) yields the penultimate product, 4-iodobutanol. The latter is known to cyclize, forming THF and regenerating HI (step e).

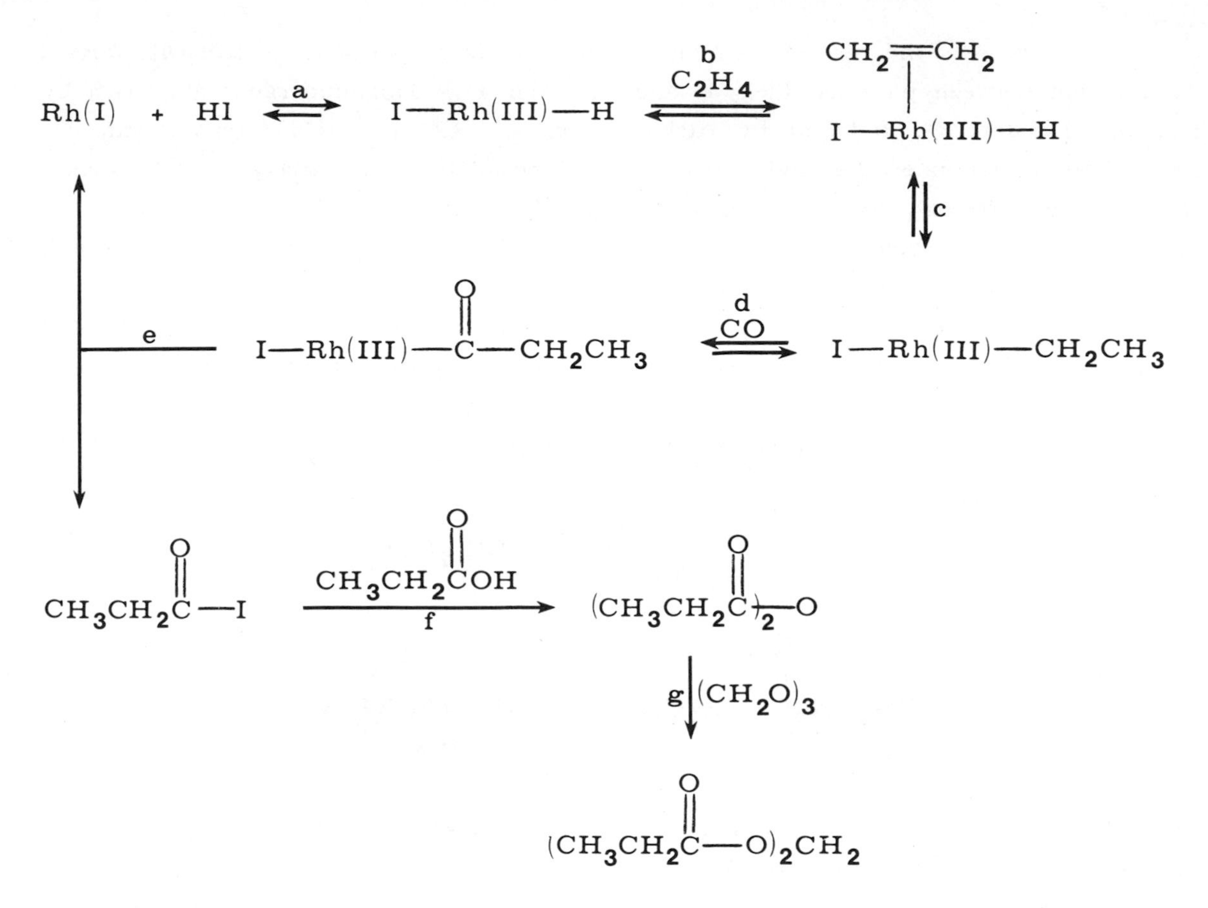


*Figure 8.8. Proposed mechanism for bis-methylene propionate synthesis.*

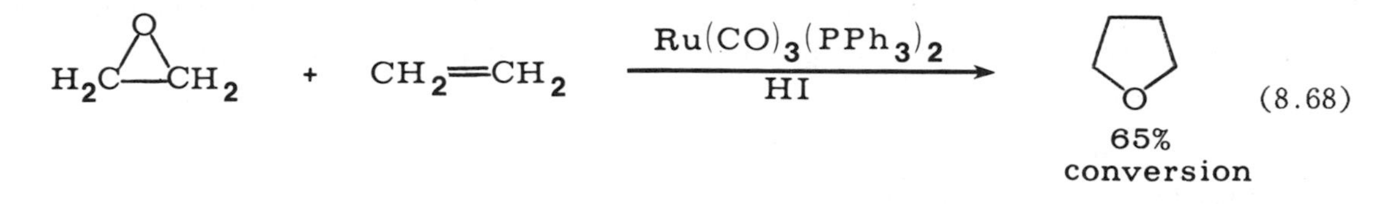


(8.68)

In Figures 8.7 to 8.9, the other ligands are not shown. Actually, these mechanisms are conjecture, proposed to account for the observed reactions. Nevertheless, the individual steps have sound analogies, and the schemes are plausible. Moreover, such hypothetical schemes are often employed to invent new catalytic reactions. Such was probably true for the above three reactions.

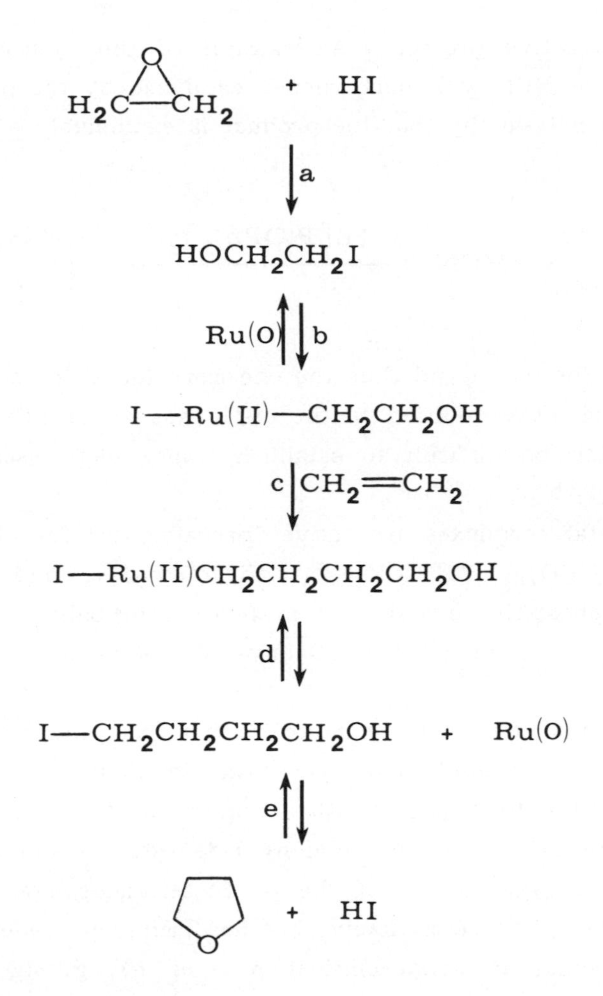


*Figure 8.9. Iodide-promoted tetrahydrofuran synthesis.*

## 8.6. Hydrocyanation of Olefins.

Hydrogen cyanide does not react with simple unconjugated olefins. However, certain transition-metal complexes will homogeneously catalyze such reactions [62]. Apparently DuPont is using this type of reaction to produce adiponitrile, a nylon-6,6 intermediate, from 1,3-butadiene and HCN as in 8.69. In terms of raw materials, this

$$CH_2{=}CH{-}CH{=}CH_2 + HCN \xrightarrow{\text{Ni cat.}} NC(CH_2)_4CN \qquad (8.69)$$

is an economically attractive process. An example of the hydrocyanation of a simple olefin is the reaction of HCN with norbornene, catalyzed by the palladium(0) phosphite complex (8.70). It is noteworthy that the product is exclusively exo, since that is the

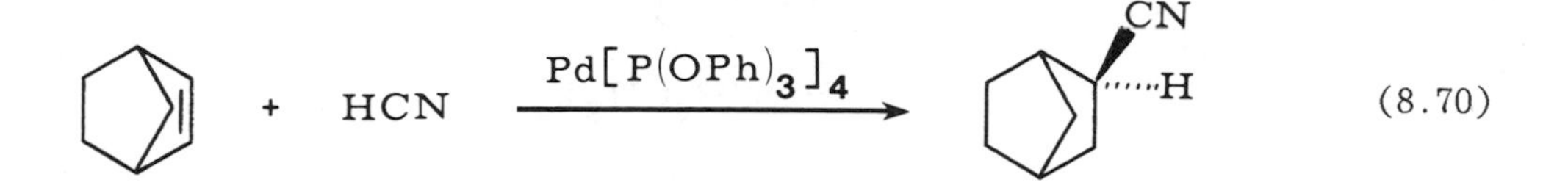


less hindered side of the olefin and thus the one more likely to coordinate to the metal. The stereochemistry of HCN addition to the olefin has apparently not been examined, although it would often be difficult to establish, since olefin isomerization is a major side reaction.

A number of metal complexes are active "precatalysts" for olefin hydrocyanation. Examples include $Co_2(CO)_8$, $Pd[P(OPh)_3]_4$, $Ni[P(OPh_3)]_4$, and $Na_4[Ni(CN)_4]$. For palladium catalysts, phosphite ligands are preferred, probably because these ligands stabilize palladium-hydride intermediates. Catalyst deactivation by formation of $Pd(CN)_2$ is a problem.

The mechanism of olefin hydrocyanation is not well understood--at least in the published literature. A plausible path is shown in Figure 8.10. Step a involves oxidative-addition of HCN to a palladium(0) complex--a well-precedented reaction (see Chapter 4). Olefin insertion into the metal hydride (step b and c) is very likely. A major question concerns the nature of the product-forming step. An intramolecular reductive-elimination (step d) seems likely, but has limited precedence. Another possibility is an intermolecular reductive-elimination (step e), giving an undefined palladium(I) dimer, 50. Other paths can be envisaged, and perhaps different metal catalysts use different reaction paths.

The regiospecificity of HCN addition to unsymmetric olefins depends on the nature of the metal catalyst. The factors which control the direction of addition are not understood. Consider, for example, the three reactions shown in Equation 8.71 to 8.73.

A major problem with the adiponitrile synthesis (Equation 8.69) is the possibility of olefin rearrangement between the addition of the first and second mole of HCN. Rearrangement to the more stable conjugated olefin, $CH_3CH_2CH{=}CHCN$, would seem to block the formation of adiponitrile. The method of controlling the olefin rearrangement and regiospecificty of HCN addition has not been published.

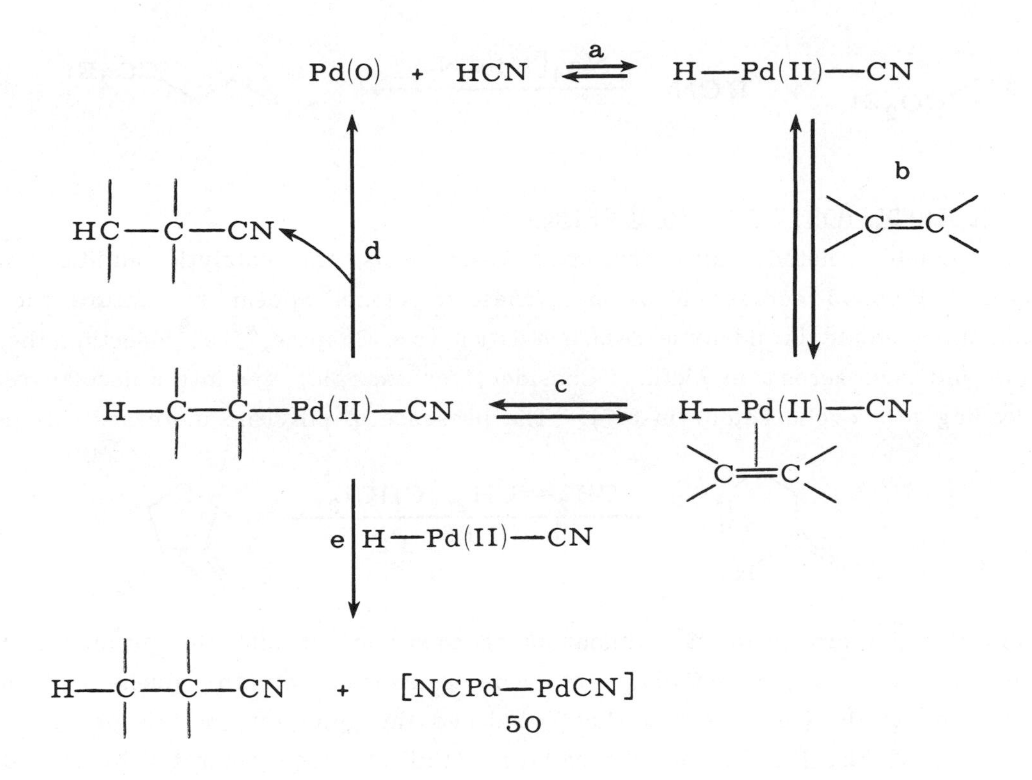


*Figure 8.10. Possible mechanisms for palladium-catalyzed hydrocyanation.*

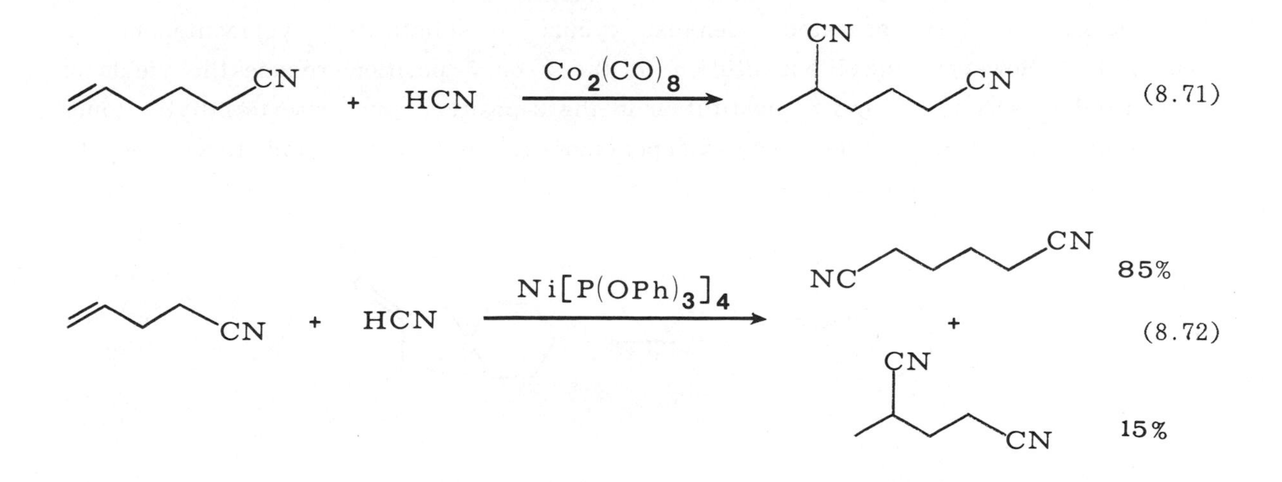

$$CH_2{=}CHCO_2Et \;+\; HCN \xrightarrow{Na_4[Ni(CN)_4]} NC\!-\!CH_2CH_2\!-\!CO_2Et \qquad (8.73)$$

## 8.7. Aldehyde Additions to Olefins.

A vaguely related, little-developed reaction is the catalytic addition of the aldehyde C-H bond across an olefin. These reactions appear to capture the acyl-hydride intermediate in aldehyde decarbonylation (see Chapter 7) by inserting the metal hydride and acyl across an olefin. Consider, for example, the intramolecular reaction [63] leading to a cyclic ketone in 8.74. The presence of ethylene increases the yield to

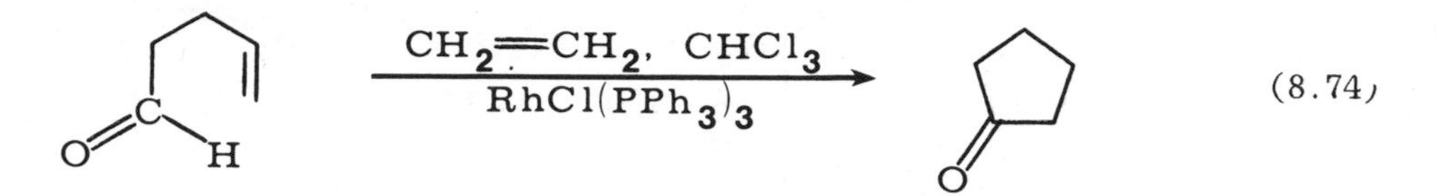


72% and the conversion to 96% without being incorporated into the product. With a different catalyst, $Rh(acac)(C_2H_4)_2$, ethylene is inserted into the aldehydic bond. A more detailed study [64] has somewhat broadened the generality of this process. Of a wide variety of Rh, Ir, Ru, and Pd catalysts studied, those generated by the addition of two equivalents of tri-p-tolylphosphine, tri-p-anisylphosphine, or tris-(p-dimethylaminophenyl)phosphine to $[(cyclooctene)_2RhCl]_2$ were the most reactive. With these catalysts, 4,5-unsaturated aldehydes cyclize to substituted cyclopentanones in good yield. However, alkyl substitution in the 2 or 5 position reducesthe yields of cyclic product somewhat, and disubstitution in the 2 position gives rise to ethyl ketones (by ethene insertion) rather than cyclopentanones. Both spiro and fused bicyclic ketones can be made by this method (8.75 and 8.76).

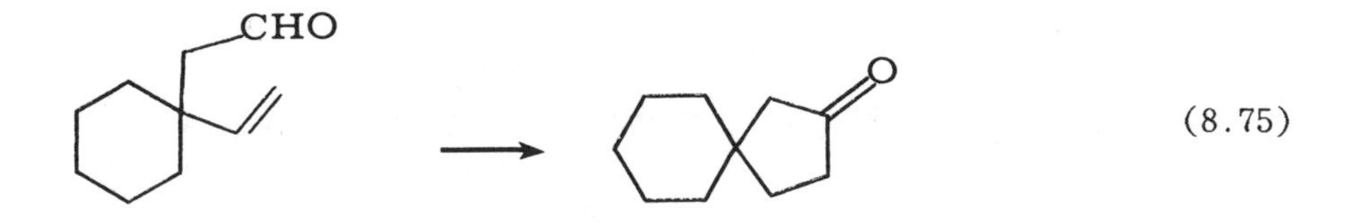

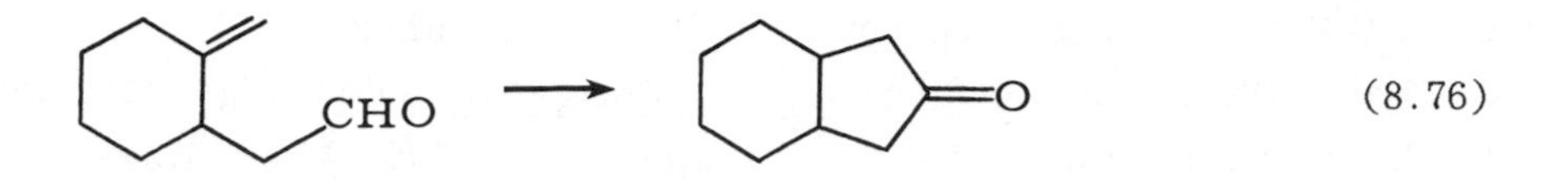


## Notes.

1. G. Parshall, Homogeneous Catalysis, (Wiley-Interscience, 1980).
2. K. Weissermel and H. J. Arpe, Industrial Organic Chemistry (Verlag Chemie, 1978).
3. R. L. Pruett, Adv. Organometal. Chem., 17, 1 (1979).
4. R. F. Heck and D. S. Breslow, J. Amer. Chem. Soc., 83, 4023 (1961).
5a. N. H. Alemdarogly, J. L. M. Penniger, and E. Oltay, Monatsheft für Chemie, 107, 1153 (1976).
5b. M. van Boven, N. H. Alemdaroglu, and J. M. L. Penniger, Ind. and Eng. Chem. (Pdt. Res.), 15, 279 (1975).
6. M. Bianchi, F. Piacenti, P. Frediani, and U. Matteoli, J. Organometal. Chem., 137, 361 (1977).
7. J. Halpern, Proceedings of the First International Symposium on Homogeneous Catalysis (Plenum Press).
8a. G. Yagupsky, C. K. Brown, and G. Wilkinson, J. Chem. Soc. A, 1392 (1970).
8b. C. K. Brown and G. Wilkinson, J. Chem. Soc. A, 2753 (1970).
9. R. L. Pruett and J. A. Smith, J. Org. Chem., 34, 327 (1969).
10. A. Stefani, G. Consiglio, C. Botteghi, and P. Pino, J. Amer. Chem. Soc., 95, 6504 (1973).
11. B. Fall and E. Muller, Monatsheft für Chemie, 103, 1222 (1972).
12. R. A. Sanchez-Delgado, J. S. Bradley, and G. Wilkinson, J. Chem. Soc. D, 399 (1976).
13. I. Schwager and J. F. Knifton, J. Catal., 45, 256 (1976).
14. C. Hsu and M. Orchin, J. Amer. Chem. Soc., 97, 3553 (1975).
15. R. Whyman, J. Organometal. Chem., 94, 303 (1975).
16a. Y. Takegami, Y. Watanabe, and H. Musada, Bull. Chem. Soc. Japan, 40, 1459 (1967).
16b. J. Falbe and N. Huppes, Brennst.-Chem., 48, 46 (1967).
17. M. A. Vannice, Catal. Rev.-Sci. Eng., 14, 153 (1976).

18. V. Ponec, Catal. Rev.-Sci. Eng., 18, 151 (1978).
19. C. Masters, Adv. Organometal. Chem., 17, 61-103 (1979), and references therein.
20. Henri-Olive and S. Olive, Angew. Chem., Int. Ed. Engl., 15, 136 (1976).
21. E. L. Muetterties and J. Stein, Chem. Rev., 79, 479 (1979).
22a. H. Pichler, Adv. Catal., 4, 271 (1952).
22b. K. A. Kini and A. Lahiri, J. Sci. Ind. Res., 34, 97 (1975).
22c. A. Jones and B. D. McNicol, J. Catal., 48, 384 (1977).
23. J. S. Bradley, G. B. Ansell, and E. W. Hill, J. Amer. Chem. Soc., 101, 7417 (1979).
24. E. L. Muetterties, Bull. Soc. Chim. Belg., 84, 959 (1975).
25. P. Perkins and K. P. C. Vollhardt, J. Amer. Chem. Soc., 101, 3985 (1979).
26. J. P. Collman and S. R. Winter, J. Amer. Chem. Soc., 95, 4089 (1973).
27. C. P. Casey and S. M. Neumann, J. Amer. Chem. Soc., 98, 5395 (1976).
28. Ibid.
29. J. A. Gladysz and W. Tam, J. Amer. Chem. Soc., 100, 2545 (1978).
30. J. A. Gladysz and J. C. Selover, Tetrahedron Letters, 319 (1978).
31. J. A. Gladysz, G. M. Williams, W. Tam, and D. L. Johnson, J. Organometal. Chem., 140, C1-C6 (1977).
32. S. R. Winter, G. W. Cornett, and E. A. Thompson, J. Organometal. Chem., 133, 339 (1977).
33. C. P. Casey and S. M. Neumann, J. Amer. Chem. Soc., 100, 2544 (1978).
34. C. P. Casey, and S. M. Neumann, J. Amer. Chem. Soc., 99, 1641 (1977); Inorganic Compounds with Unusual Properties, II, Advances in Chemistry Series, no. 173 (American Chemical Society, 1979), Chapter 12.
35. W-K. Wong, W. Tam, and J. A. Gladysz, J. Amer. Chem. Soc., 101, 5440 (1979).
36. K. L. Brown, G. R. Clark, C. E. L. Headford, K. Marsden, and W. R. Roper, J. Amer. Chem. Soc., 101, 503 (1979).
37. C. P. Casey, M. A. Andrews, and D. R. McAlister, J. Amer. Chem. Soc., 101, 3371 (1979).
38. T. Blackmore, M. I. Bruce, P. J. Davidson, M. Z. Igbas, and F. G. A. Stone, J. Chem. Soc. A, 3153 (1970).
39. P. M. Treichel and R. L. Shubkin, Inorg. Chem., 6, 1328 (1968).
40. J. A. Van Doorn, C. Masters, and H. C. Volger, J. Organometal. Chem., 105, 245 (1976).
41. A. Davison and D. L. Reger, J. Amer. Chem. Soc., 94, 9237 (1972).

42. J. M. Manriquiez, D. R. McAlister, R. D. Sanner, and J. E. Bercaw, J. Amer. Chem. Soc., 98, 6734 (1976); 100, 2716 (1978).
43. P. T. Wolczanski, R. S. Threlkel, and J. Bercaw, J. Amer. Chem. Soc., 101, 218 (1979).
44. L. I. Shoer and J. Schwartz, J. Amer. Chem. Soc., 99, 5831 (1977).
45. J. W. Rathke and H. M. Feder, J. Amer. Chem. Soc., 100, 3623 (1978).
46. J. S. Bradley, J. Amer. Chem. Soc., 101, 7419 (1979).
47. M. J. Doyle, A. P. Kouwenhover, C. A. Schaap, and B. Van Oort, J. Organometal. Chem., 174, C55 (1979).
48. M. G. Thomas, B. F. Beier, and E. L. Muetterties, J. Amer. Chem. Soc., 98, 1296 (1976).
49. M. A. Vannice, J. Catal., 37, 61 (1979).
50. G. C. Demitras and E. L. Muetterties, J. Amer. Chem. Soc., 99, 2796 (1977).
51. R. L. Pruett, Ann. N.Y. Acad. Sci., 295, 239 (1977).
52. C. Masters, Adv. Organometal. Chem., 17, 80-83 (1979), and references therein.
53. B. D. Dombek, J. Amer. Chem. Soc., 101, 6466 (1979).
54. D. M. Fenton and P. J. Steinwand, J. Org. Chem., 39, 701 (1974).
55a. W. Reppe and H. Vetter, Annalen, 582, 133 (1953).
55b. J. Falbe, Carbon Monoxide in Organic Synthesis (Springer Verlag, 1970), pp. 78-122.
56. N. Grice, S. C. Kao, and R. Pettit, J. Amer. Chem. Soc., 101, 1692 (1979), and references therein.
57. D. Forster, Adv. Organometal. Chem., 17, 255-267 (1979).
58. K. M. Webber, B. C. Gates, and W. Drenth, J. Catal., 47, 269 (1977).
59. D. Forster, J. Chem. Soc. D, in press (1980).
60. K. M. Webber, B. C. Gates, and W. Drenth, J. Mol. Cat., 3, 1 (1977-78). See also A. Krzywicki and M. Marczewski, J. Mol. Catal., 6, 431 (1979).
61. S. Lapporte and V. P. Kurkov, in Organotransition-Metal Chemistry, Y. Ishii and M. Tsutsui, eds. (Plenum Press, 1975), p. 199.
62. E. S. Brown, Aspects of Homogeneous Catalysis (Reidel, 1974), II, 57-76.
63. C. F. Lochow and R. G. Miller, J. Amer. Chem. Soc., 98, 1282 (1976).
64. R. C. Larock, K. Oertle, and G. F. Potler, J. Amer. Chem. Soc., 102, 190 (1980).

# 9

# Stoichiometric Reactions of Transition-Metal Carbonyl Complexes

The chemistry of transition-metal carbonyls dates to the discovery of nickel carbonyl by Mond in 1890. Carbon monoxide coordinates to virtually all the transition metals, and a prodigious number of complexes with a bewildering array of structures are known. Because carbon monoxide is among the best of $\pi$-acceptor ligands, it stabilizes low oxidation states. Many quite stable metal carbonyls have the metal in a very low (0 to -2 or -3) oxidation state. Hence many metal carbonyls are quite electron-rich, and react as nucleophiles and as electron-transfer reducing agents. Additionally, the ease of the migratory-insertion reaction, in which alkylmetal carbonyl complexes convert to acylmetal complexes, allows metal carbonyls to be effective carbonylation (acylating) agents for a variety of organic substrates. Important catalytic processes involving metal carbonyls have already been discussed. In this chapter, useful stoichiometric reactions of metal carbonyls are considered [1]. These fall into three major categories: (1) coupling reactions; (2) carbonylation (acylation) reactions; and (3) decarbonylation reactions.

## 9.1. Coupling Reactions.

As early as 1948 the conversion of allyl chloride to biallyl (1,5-hexadiene) by treatment with nickel carbonyl had been reported [2]. Subsequent studies by Corey showed this to be a general reaction, and led to the development of several useful synthetic procedures. The coupling reaction requires coordinating solvents such as DMF, HMPA or N-methyl pyrrolidone, and proceeds readily at 25°-50°, with evolution of carbon monoxide (9.1). The process is thought to entail the chemistry presented in Figure 9.1 [3]. Coordinatively saturated $Ni(CO)_4$ is quite labile; loss of one CO creates a

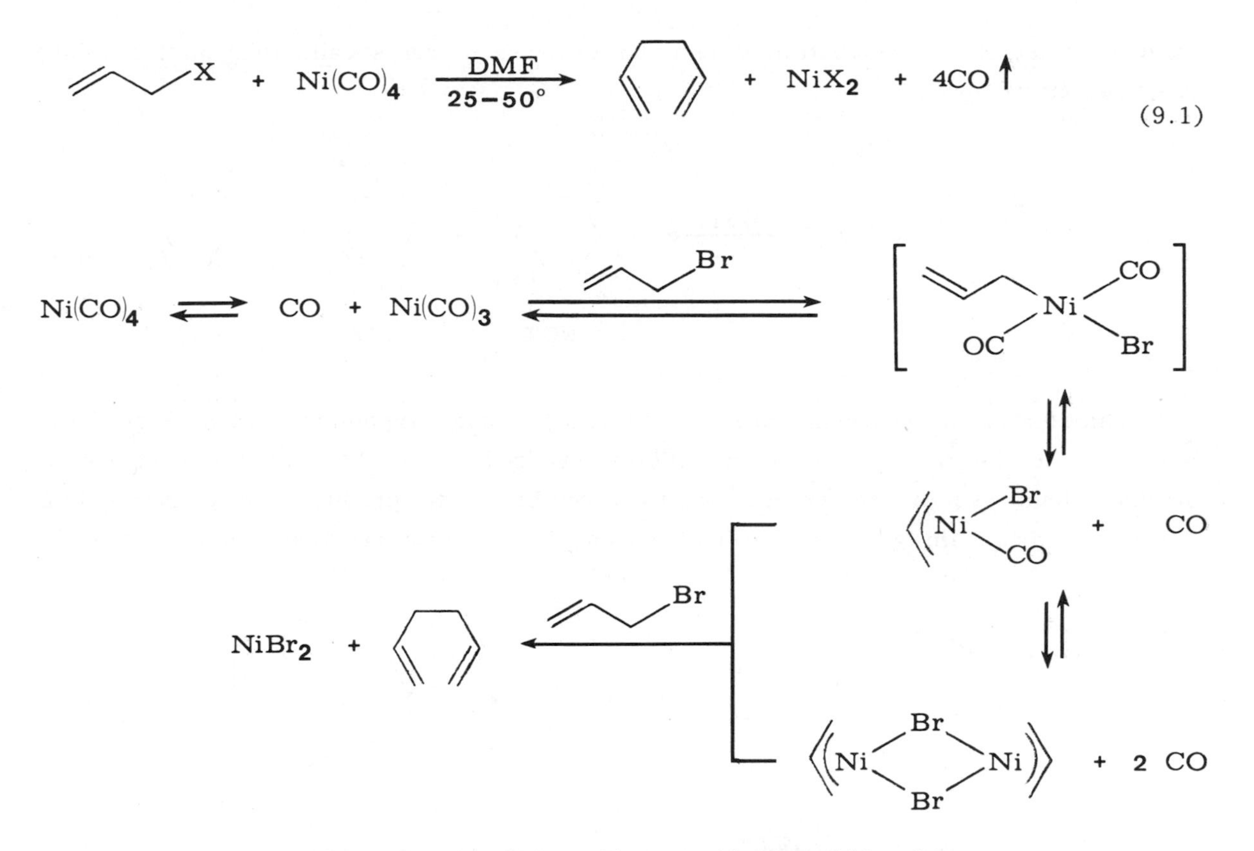


*Figure 9.1. $Ni(CO)_4$ coupling of allylic halides* [3].

vacant coordination site, and allows an interaction with allyl bromide to proceed, producing a $\eta^3$-allylnickel(bromo)carbonyl complex. Dimerization with loss of CO produces the well-known bis-$\eta^3$-allylnickel bromide complex. In polar solvents both of these complexes react irreversibly with allyl bromide to give biallyl and nickel bromide. (In nonpolar solvents, coupling does not occur, and the bis-$\eta^3$-allylnickel bromide complex is isolated in excellent yield.) The intermediacy of bis-$\eta^3$-allylnickel halide complexes is inferred from the deep red color, characteristic for these complexes, which develops, then dissipates, during this coupling reaction. The $\eta^3$-allylnickel(bromo) carbonyl complex can be detected spectroscopically ($\nu_{CO}$ 2060 $cm^{-1}$) and is easily generated by treatment of the dimer with CO in nonpolar solvents. Both complexes react with allyl bromide in DMF to produce biallyl rapidly. Both react with CO in DMF to produce some allyl bromide initially, and ultimately biallyl. Hence the equilibria in Figure 9.1 are well-established. Unsymmetric allylic halides couple predominantly to exclusively at the less-substituted end, in contrast to almost all other allyl coupling reactions (9.2).

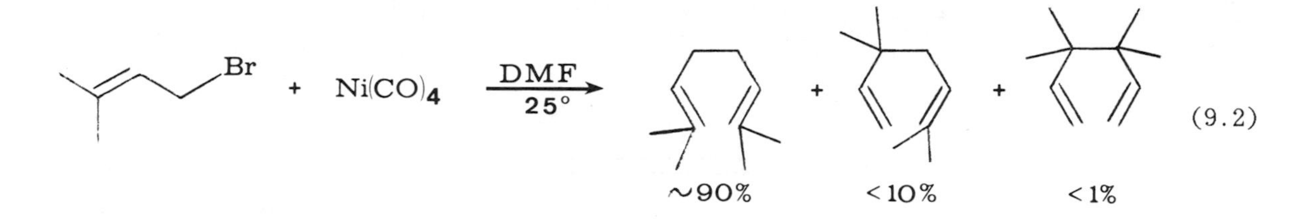


This feature is especially useful for forming cyclic compounds from α,ω-bis-allylic halides. In the initial studies [4], Corey cyclized simple 12-, 14-, and 18-carbon straight-chain systems under high dilution conditions to produce the corresponding cyclic 1,5 dienes in good to excellent yields (9.3). This reaction has subsequently

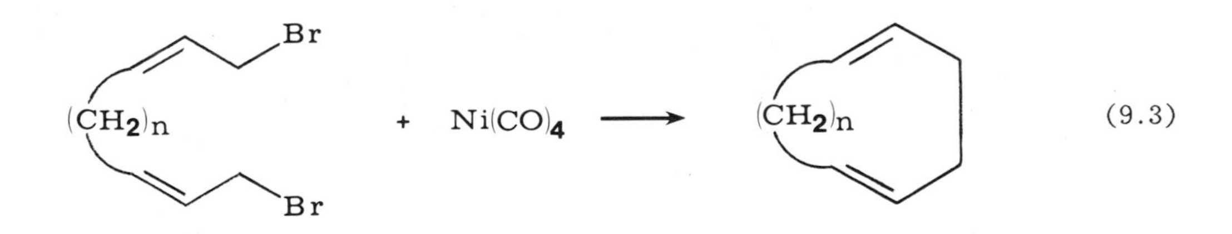


been used to synthesize several macrocyclic compounds, including humulene (1) [5], cembrene (2) [6], and macrocyclic lactones (3) [7]. Note that coupling always occurs at the less-substituted allyl terminus.

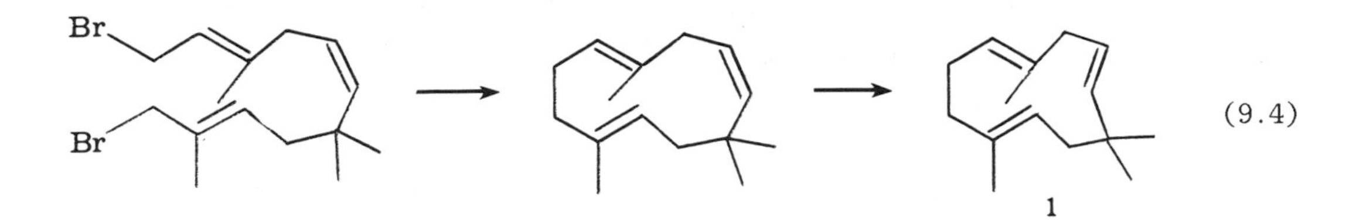


Although by far the most work has been directed toward allylic halide couplings, Tsutsumi has shown that a variety of organic halides with systems similar to the allyl system also couple. For example, α-haloketones, which are essentially oxygen analogs of allyl halides, couple readily with nickel carbonyl [8]. The ultimate product is

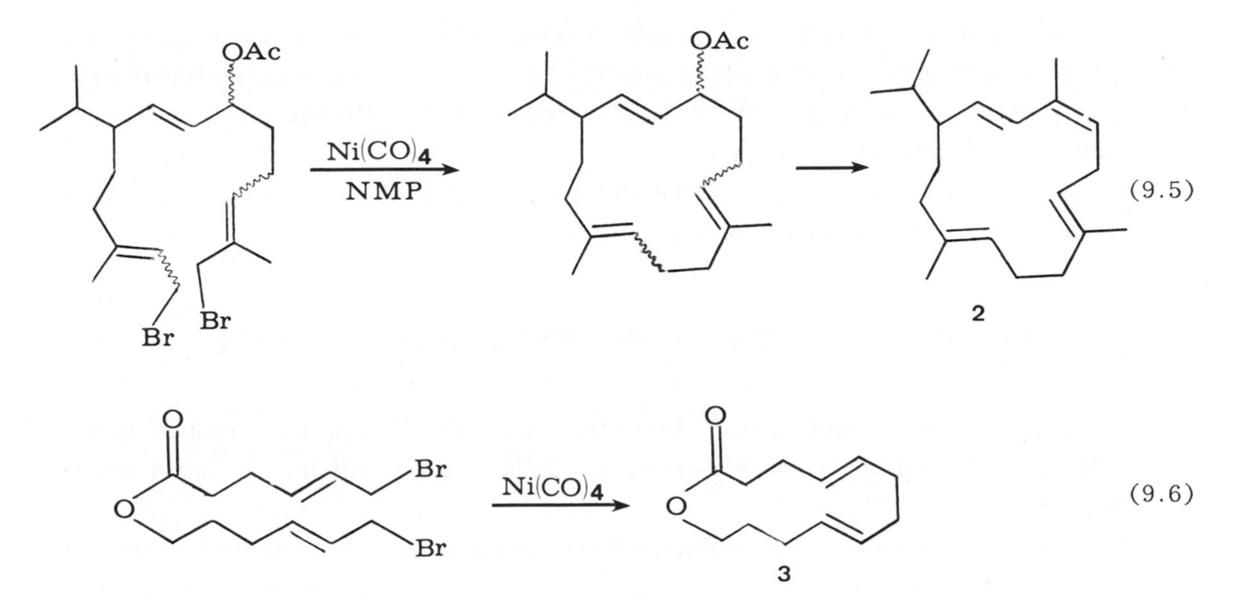


strongly dependent upon the solvent used. In THF, the symmetric 1,4-diketone is the sole product, corresponding to the formation of "biallyl." However, when the reaction is carried out in DMF solvent, the β-ketoepoxide is isolated in excellent yield. (This converts, under acid conditions, to the furan, a common reaction of this class of compounds.) This unexpected coupling product can be explained by invoking the attack of an "oxallyl" nickel species on the carbonyl group of the bromoketone, followed by epoxide formation with ejection of the bromide (Figure 9.2). Note that under

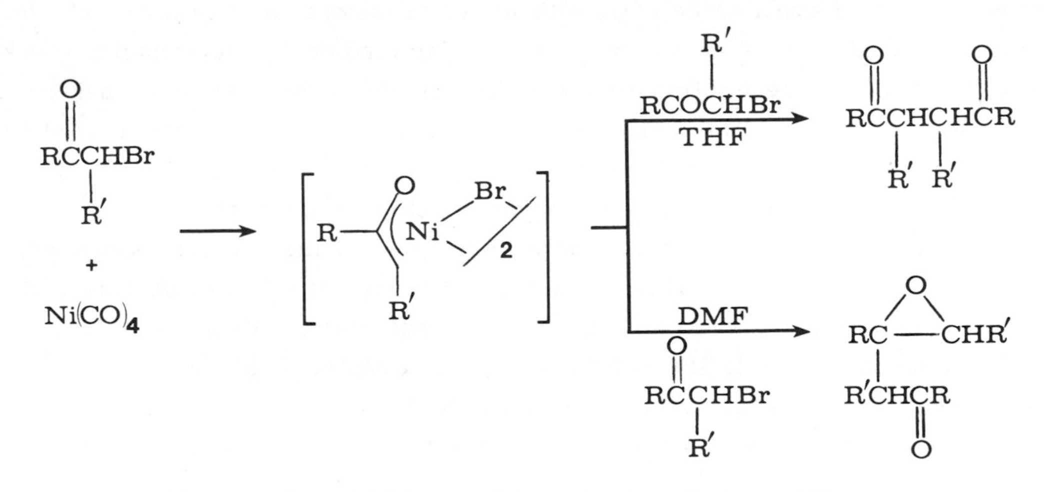


*Figure 9.2. $Ni(CO)_4$ coupling of α-bromoketones [8].*

appropriate conditions, π-allylnickel halides do react with ketones and aldehydes at the carbonyl carbon to produce homoallylic alcohols; so this behavior is not without precedent (Chapter 15). Obviously the solvent exerts a great influence on the course of this particular coupling reaction.

Nickel carbonyl also couples α-halosulfides to give 1,2-dithioethanes, whereas α-haloethers do not lead to identifiable products [9]:

$$RSCH_2Cl + Ni(CO)_4 \longrightarrow RSCH_2CH_2SR + NiCl_2 \qquad (9.7)$$

Nickel carbonyl also couples benzyl bromide. However, the major product here is dibenzylketone, formed by carbonyl insertion. This process will be discussed later in this chapter.

All the above reactions are "reductive" couplings, in the sense that the low-valent Ni(0) is oxidized to Ni(II) during the coupling process. Noyori has developed a related "reductive" coupling of α,α'-dihaloketones with a variety of organic substrates to provide a facile entry into a number of normally difficult to synthesize seven-membered ring compounds [10]. This chemistry is based on the product resulting from the reaction of $Fe_2(CO)_9$ with α,α'-dihaloketones (Equation 9.8), believed to be an iron-stabilized oxallyl cation (5). Recall that $Fe(CO)_5$ itself is coordinatively saturated, and that thermal or photochemical ejection of one CO (producing $Fe(CO)_4$) is required to initiate the reaction. $Fe_2(CO)_9$ is a photodimer of $Fe(CO)_5$, and can be considered a convenient source of reactive $Fe(CO)_4$ without the requirements of heat or light to produce it. The formation of 4 can be viewed as proceeding by nucleophilic attack (or oxidative addition) of the electron-rich $Fe^0(CO)_4$ on the highly reactive α-haloketone to produce the $Fe^{2+}$-stabilized α-bromoenolate. Loss of bromide gives the iron-stabilized oxallyl cation, 5, which is formally a dipolar species. However, complexation (coordination) of the $O^-$ to iron produces a complex which is functionally a simple oxygen-substituted allyl cation. Although a π-type interaction of this oxallyl species with iron is possible, the chemistry of this species can be understood without this additional complexity. This, in conjunction with the fact that stable cationic $\eta^3$-allyliron complexes are unknown, leads to the formulation of this complex as 5.

The steps proposed in its formation, as well as the cationic nature of 5, are supported by the following experimental observations [11]. The first step, a formal

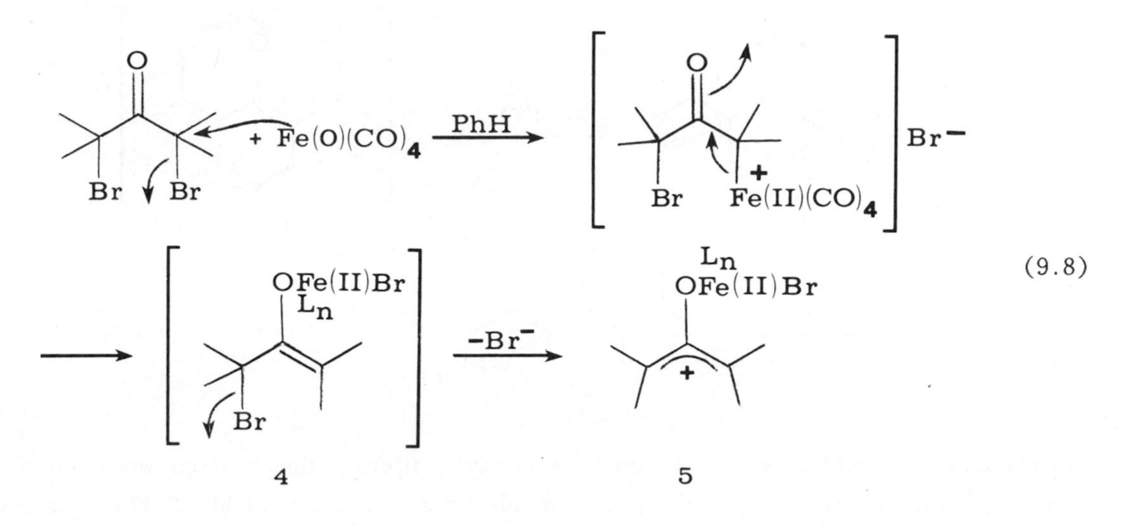


(9.8)

two-electron reduction of substrate, finds precedent in the reduction of endo-α-bromocamphor by $Fe_2(CO)_9$ in aqueous DMF (9.9). This reduction proceeds

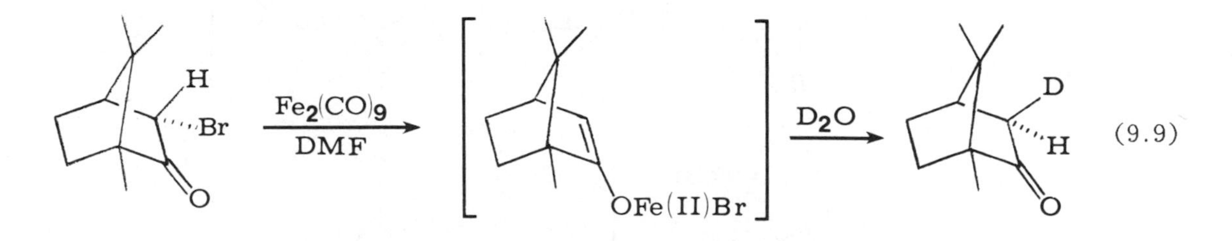


(9.9)

stereoselectively in an exo fashion. Since protonation of the enolate of camphor is known to occur from the exo side of the bicyclic system, the presence of an iron enolate is indicated. The second step, loss of halogen, was confirmed by monitoring the reduction of α,α'-dibromoketones with 0.5 equivalents of $Fe_2(CO)_9$ in aqueous DMF. Under these conditions, the yield of mono bromoketone reaches a maximum after several hours, then decreases, while the yield of the parent (unbrominated) ketone increases steadily. The cationic nature of complex 5 (when formed in nonprotic solvents such as benzene) is inferred from several experimental observations. Reactions of α,α'-dibromoketones with $Fe_2(CO)_9$ in the presence of nucleophiles such as acetate or methanol lead to products arising from nucleophilic attack on a cationic species (9.10).

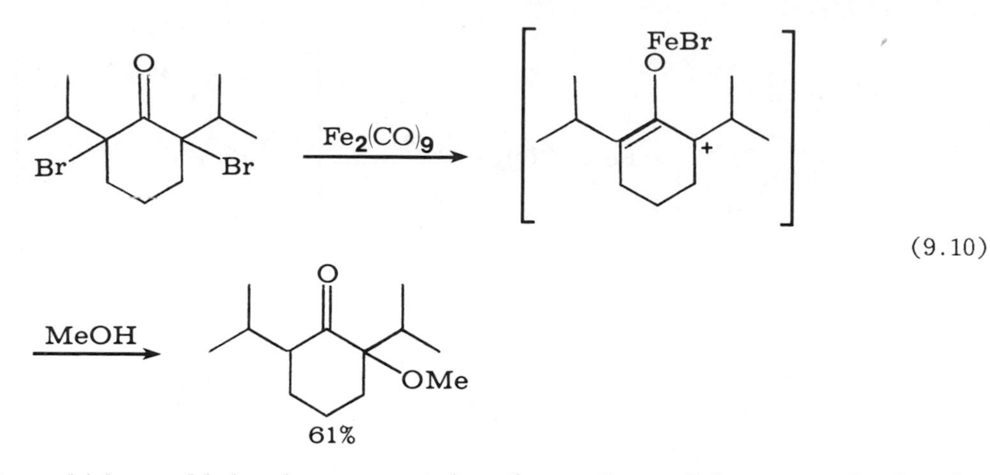


(9.10)

With substrates which would involve neopentyl cations, the well-known carbonium-ion rearrangement of this type of system is observed in high yield (9.11). In systems

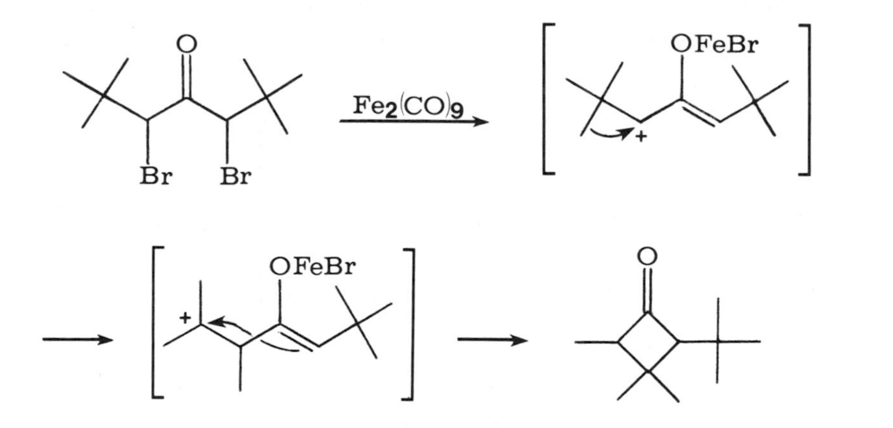


(9.11)

containing additional sites reactive toward cations, internal trapping is observed (9.12). Finally, $\alpha,\alpha'$-dibromoacetone itself does not react to give products arising from the oxallyl intermediate, since it lacks carbocation-stabilizing substituents at the sites which would bear formal positive charges.

Confirmation of the allyl cation nature of complex 5 is also found in its behavior in cycloaddition reactions. Allyl cations are three-carbon, two-electron systems which should undergo a 3 + 4-cyclocoupling reaction (formally a $\pi^2 + \pi^4$ cycloaddition) to produce seven-membered rings. Complex 5 does indeed engage in this kind of reaction, and herein lies its usefulness for synthesis. Thus, reaction of a number of open-chain dienes, including 1,3-butadiene and isoprene, with tertiary or secondary

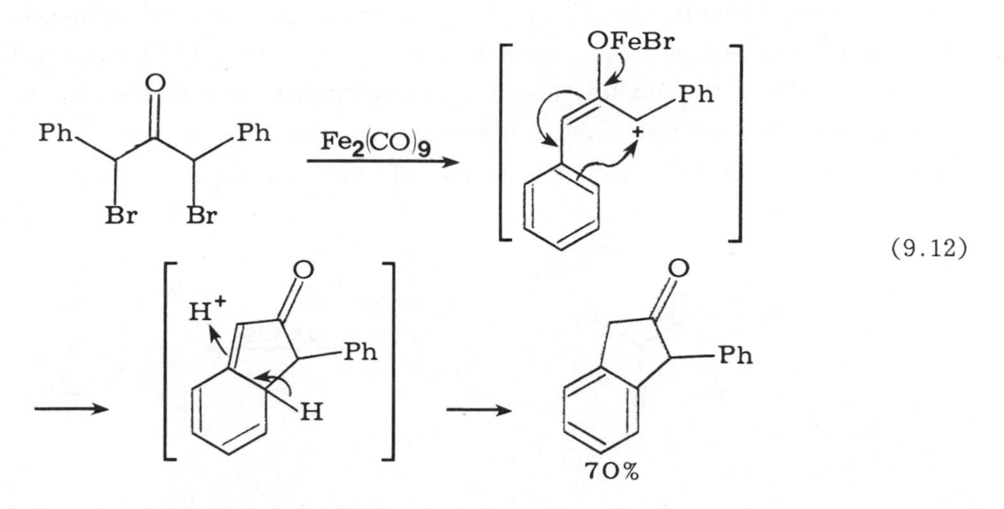


(9.12)

α,α'-dihaloketones (9.13) produces substituted 4-cycloheptenones in fair to excellent yields [12]. This provides a very facile entry into seven-membered ring systems which

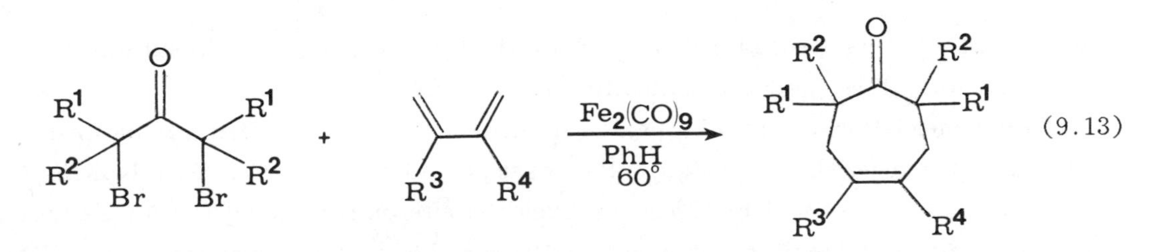


(9.13)

are normally both difficult and cumbersome to synthesize. As might be expected for a cycloaddition reaction, dienes which have a high equilibrium concentration of the S-cis isomer are the most reactive systems. Particularly reactive are cyclic diene systems such as cyclopentadiene, furan, and N-acetylpyrrole. These react in excellent yields to give the corresponding bridged bicyclic systems (9.14).

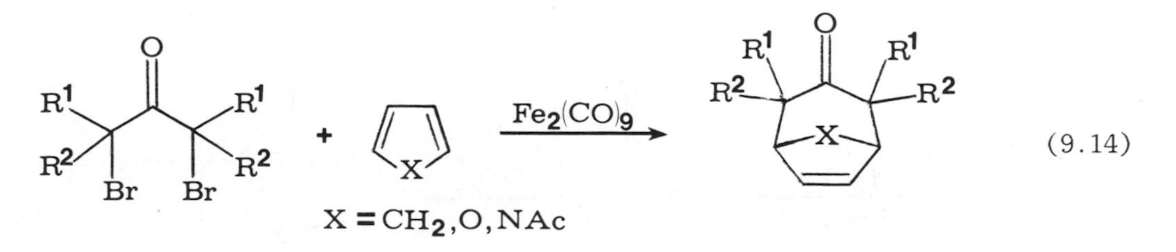


(9.14)

A major difficulty in all of these reactions is that α,α'-dibromoacetone itself, as well as dibromides of other methyl ketones, fail to give cycloadducts in the above reaction, making the unsubstituted cycloheptenones inaccessible by this route. This problem can be circumvented, however, by the use of tri- or tetrabromoketones, followed by reductive dehalogenation of the cyclized product (9.15). With the

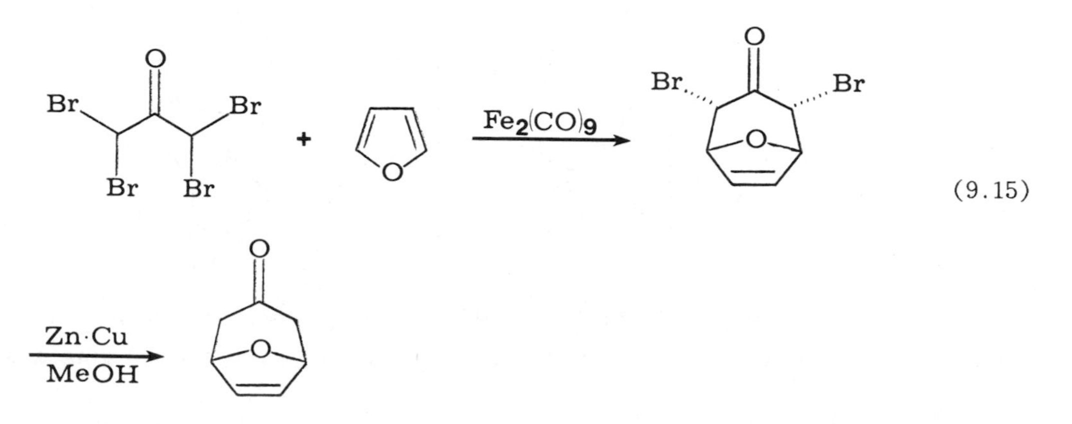


(9.15)

development of this procedure, all substitution patterns (i.e., unsubstituted, mono-, 1,1-di-,1,3-di-, tri-, and tetrasubstituted) are accessible.

This chemistry offers the most direct route to a number of natural products containing seven-membered rings. For example, nezukone (6), β-thujaplicin (7), and α-thujaplicin (8) are available from the cycloadducts of furans [13]. An elegant stereo-controlled approach to C-nucleosides (9.19) also uses this chemistry [14]. Depending

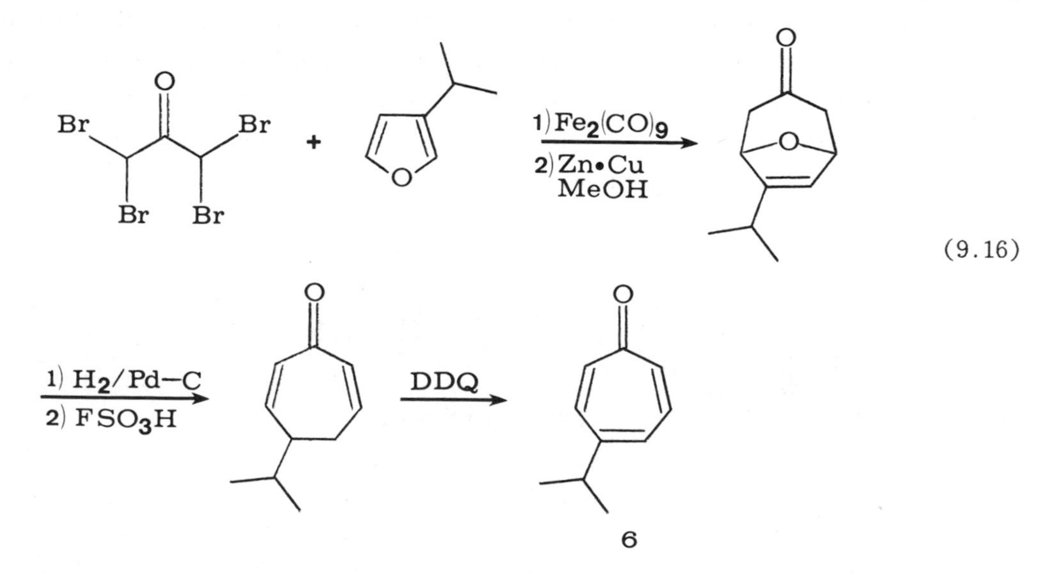


(9.16)

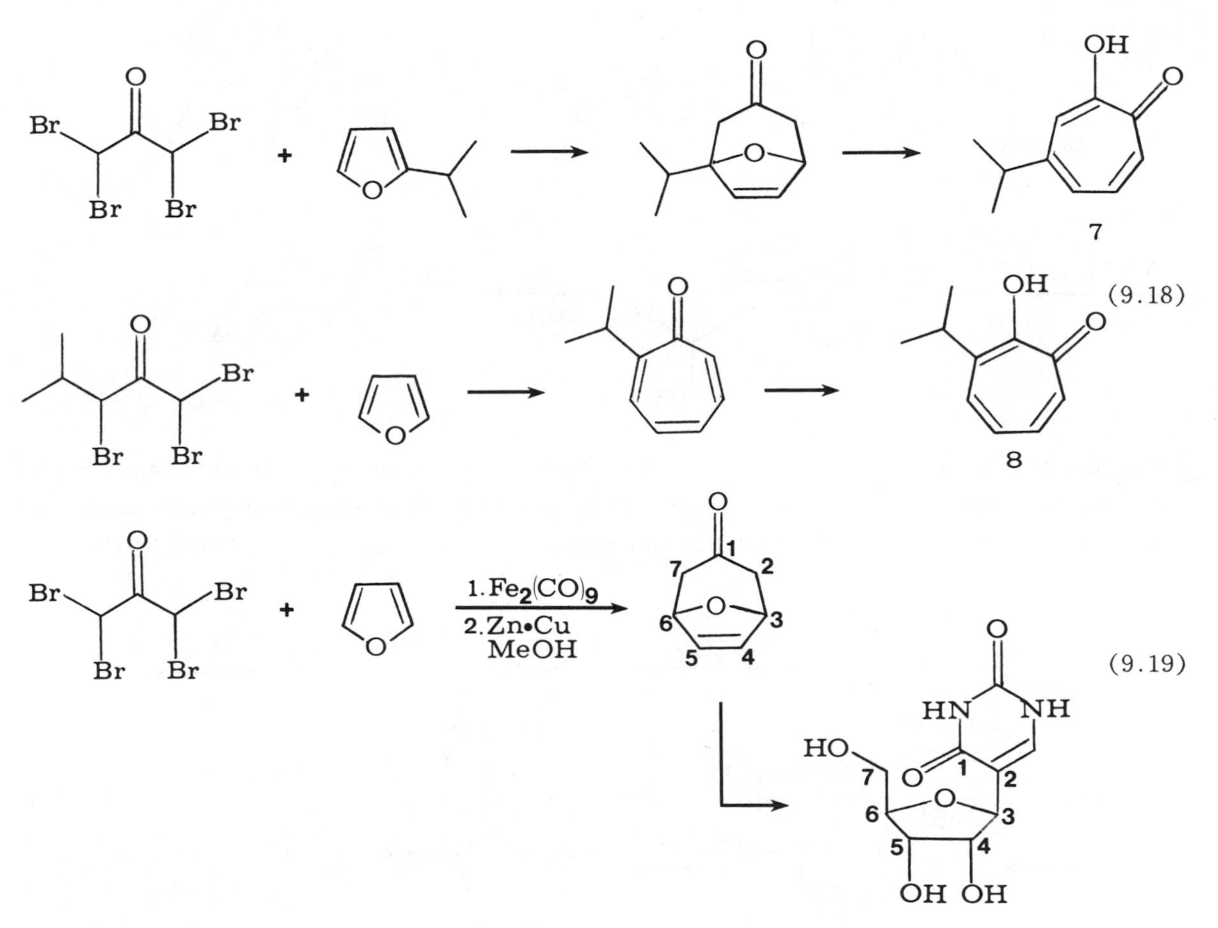


on the substrate used in the final condensation step, a number of C-nucleosides are available in excellent yields by this route. It should be noted that this is one of the few syntheses of nucleosides that does not use carbohydrates as precursors. Finally, this cycloaddition reaction with pyrroles offers a general synthetic approach to the tropane alkaloids, including tropine (9) in 9.20, scopine, and hyoscyamine [15]. These 3 + 4-cycloadditions are thought to proceed by a concerted process, with the regioselectivity controlled by the frontier molecular orbitals of the cycloadducts [16].

Iron-stabilized oxallyl cations are even more useful for synthesis because they react in a formal [3+2] cycloaddition with aromatic olefins [18] and enamines [19] to give 3-arylcyclopentanones and cyclopentenones, respectively. The concerted thermal

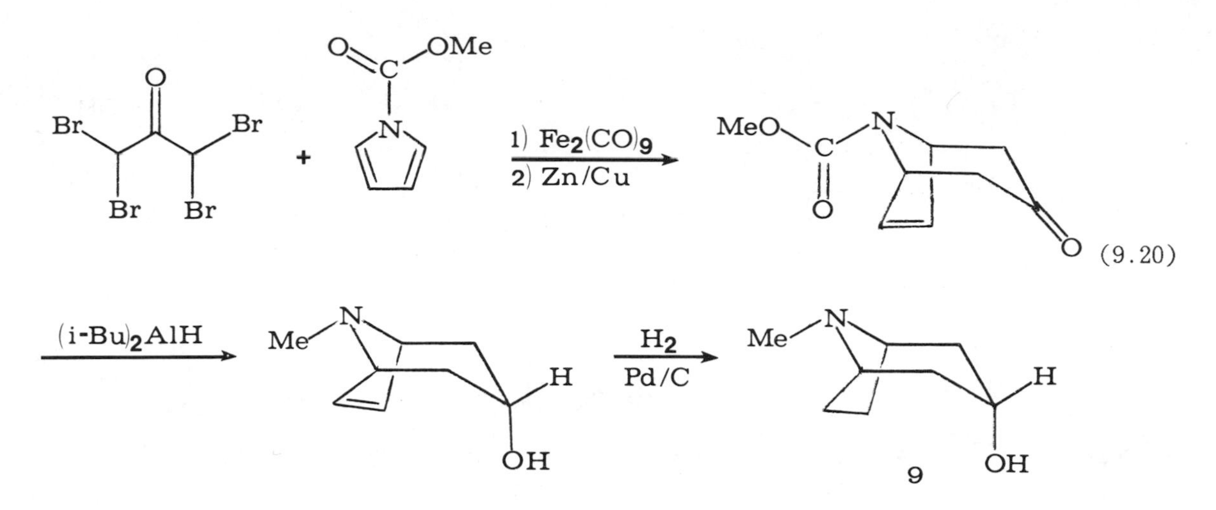


[3+2] cycloaddition of the allyl cation to ethylene is symmetry-forbidden, and ample experimental evidence [16] indicates that this cycloaddition is indeed stepwise, with the orientation controlled by the relative stabilities of the ionic intermediates (9.21).

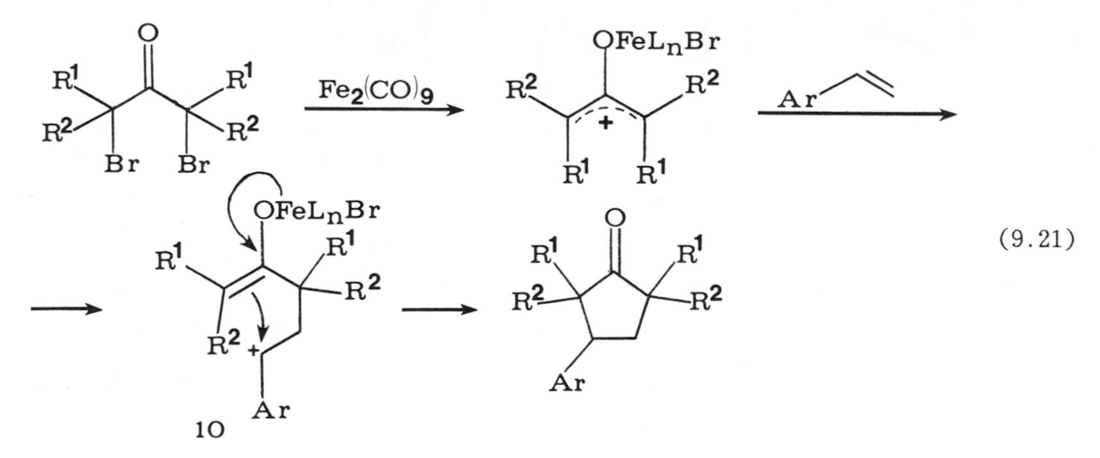


Again, α,α'-dibromoacetone fails to undergo this reaction, and in this case so does α,α,α',α'-tetrabromoacetone. Hence unsubstituted products remain inaccessible by this route. Although this is a stepwise reaction, it is stereospecific, with over-all retention of configuration. Apparently free rotation in 10 is prevented by strong attractive interactions between the cationic part and the enolate [17]. The reaction of enamines is thought to go by a similar pathway (9.22).

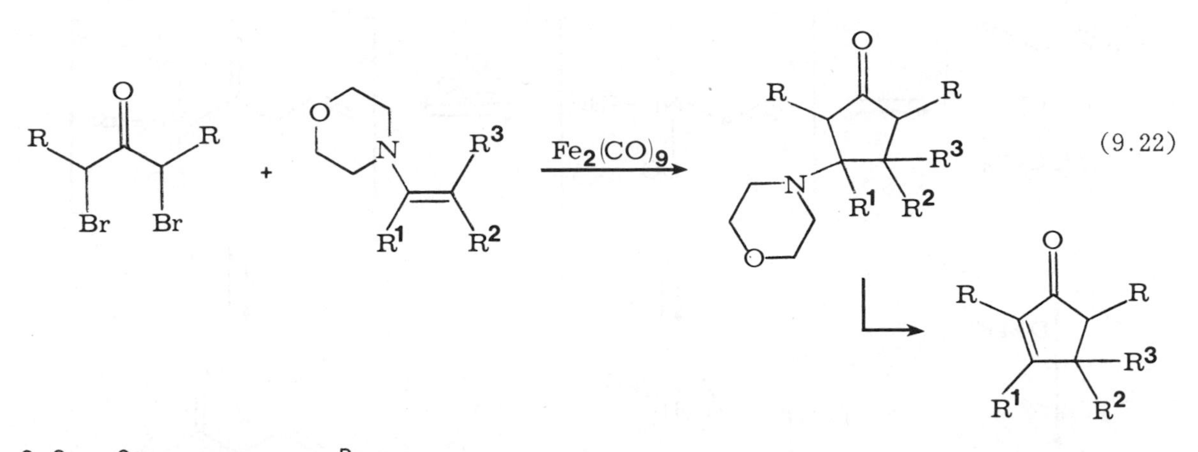

(9.22)

## 9.2. CARBONYLATION REACTIONS.

Although all the reactions in Section 9.1 involve metal carbonyls as reactive species, the carbon monoxide served simply as a ligand, providing the complex with the necessary reactivity and/or stability to allow the reaction to ensue. However, because of the facility of the migratory-insertion reaction (Chapter 5), reactions of metal carbonyls which proceed through alkylmetal carbonyl complexes can be made to produce acyl-metal complexes instead, often by minor changes in reaction conditions. This can result in the incorporation of carbon monoxide in the final product, and is the basis for several useful synthetic approaches to organic carbonyl compounds.

A typical example of this is observed when the nickel carbonyl coupling of allylic halides is carried out in solvents of moderate polarity and in the presence of excess carbon monoxide (Figure 9.3). Here the carbonylation product (either the β-γ-unsaturated acid halide or the ester) is the sole product [20]. Furthermore, treatment of preformed $\eta^3$-allylnickel halide complexes with carbon monoxide under the same conditions produces the same product. This strongly implies the intermediacy of a $\eta^1$-allylnickel carbonyl complex which undergoes a migratory insertion followed by a reductive elimination of acid halide. Figure 9.3 summarizes the reactions of nickel carbonyl with allylic halides, and vividly illustrates the sensitivity of the reactive intermediates to reaction conditions. In benzene, $\eta^3$-allylnickel halide complexes are isolated in high yields. In very polar solvents such as DMF, coupling occurs to form biallyl exclusively. In solvents of intermediate polarity (e.g., $Et_2O$) migratory insertion ensues, to produce the acyl halide, which reacts with added alcohol to form the unsaturated ester. Note that all products proceed through the same set of reactive intermediates.

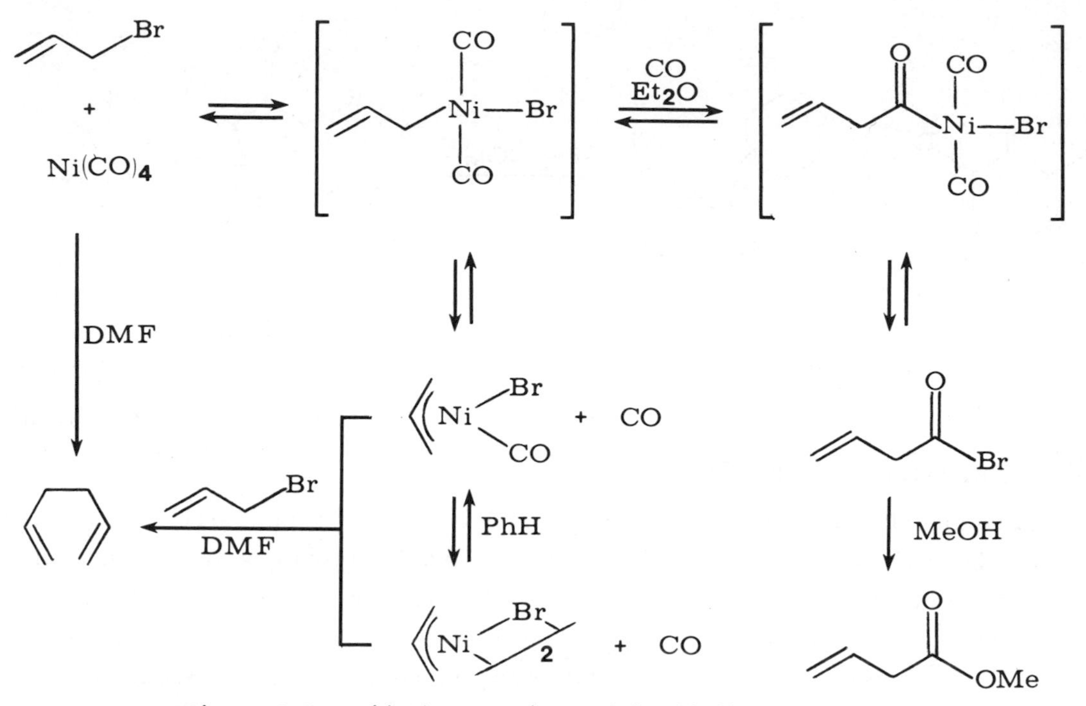


*Figure 9.3. $Ni(CO)_4$ reactions with allylic halides.*

A variety of other unsaturated species, including alkynes, alkenes, conjugated enones, and 1,5-dienes, can be made to insert with $\eta^3$-allylnickel complexes, to produce a number of more complex products (Equations 9.23 to 9.25). Again, the specific sequence of reactions that ensue is quite sensitive to reaction conditions, and careful attention to experimental detail is required to insure production of only one of the many possible products [21].

Nickel carbonyl also carbonylates other organic halides which are sufficiently reactive to form σ-alkylnickel complexes. Aryl iodides in particular react in alcohol solvents to give benzoate esters in high yield.

Although other neutral metal carbonyls effect isolated carbonylations of organic substrates, anionic metal carbonyl complexes have found much greater application for this type of conversion. The major reason for this is that most anionic metal carbonyls are quite strong nucleophiles, and are highly reactive toward a variety of organic substrates subject to nucleophilic attack. Similarly, being highly reduced species, anionic metal carbonyl complexes are quite reactive in oxidative addition reactions (not all of which are nucleophilic displacements) with organic substrates. Probably the most

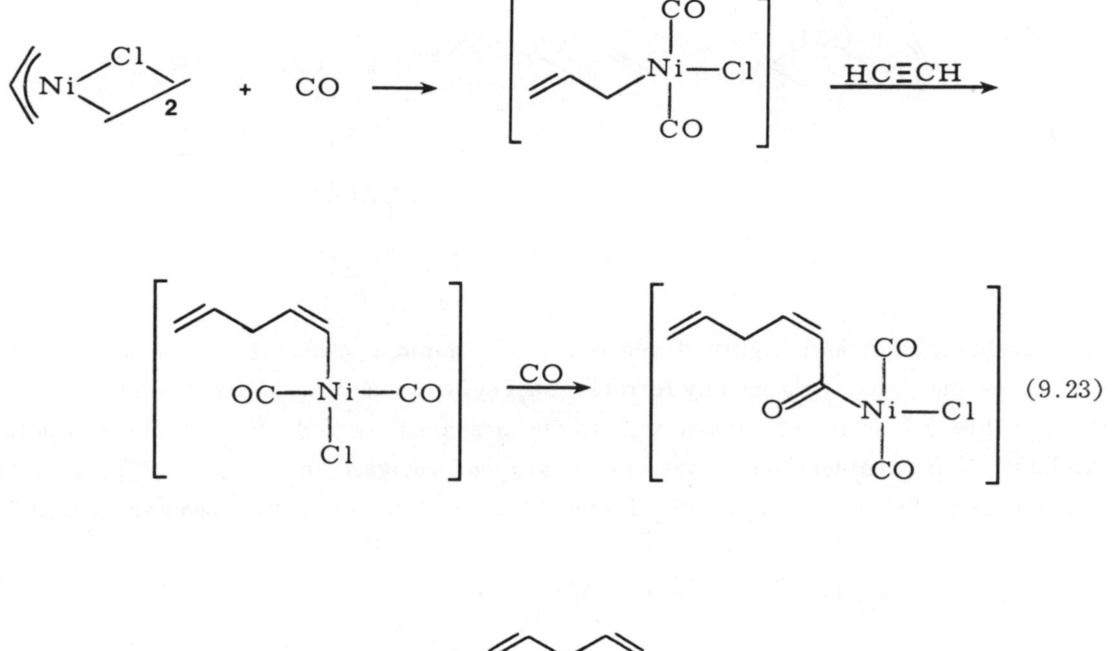
Ni
Cl
2
+ CO
CO
Ni—Cl
CO
HC≡CH
OC—Ni—CO
Cl
CO
CO
O
Ni—Cl
CO

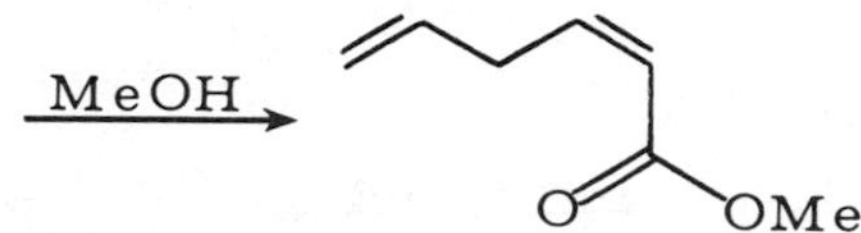
MeOH
O
OMe

(9.23)

Ni
Cl
2
+
+ CO + NaOAc

(9.24)

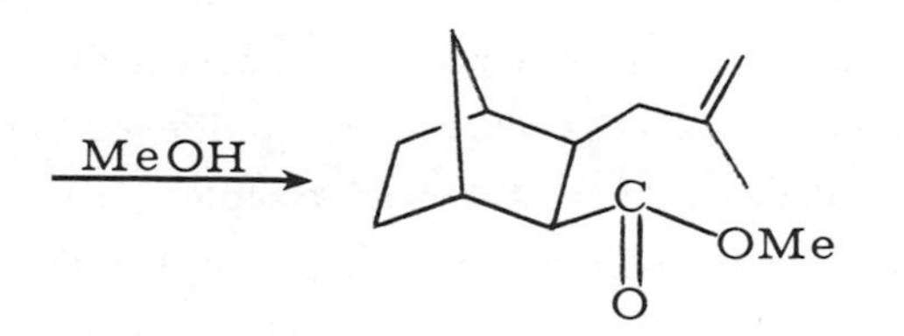
MeOH
C
O
OMe

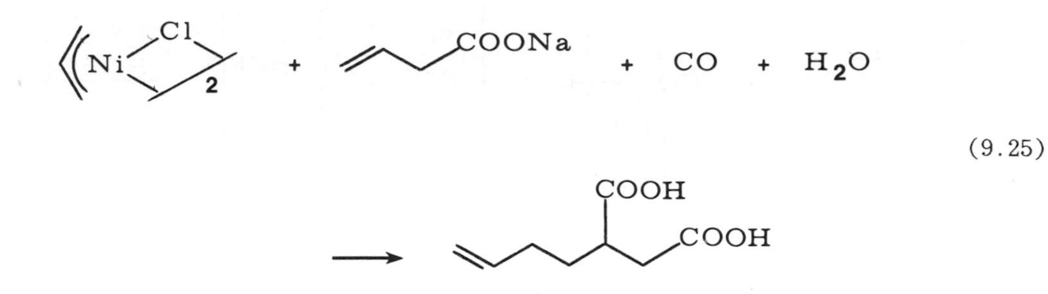


(9.25)

intensively studied and highly developed (for organic synthesis) anionic metal carbonyl species is disodium tetracarbonylferrate, $Na_2Fe(CO)_4$, developed by Cooke and Collman [22]. This $d^{10}$ iron (-2) complex is easily prepared by reduction of the commercially available iron pentacarbonyl by either sodium amalgam in THF (9.26) or sodium/benzophenone ketyl in refluxing dioxane (9.27). Although the amalgam method is

$$Fe(CO)_5 + Na(Hg) \xrightarrow{THF} Na_2Fe(CO)_4 \tag{9.26}$$

$$Fe(CO)_5 + Na + PhCOPh \xrightarrow{dioxane} Na_2Fe(CO)_4 \cdot 1.5\ dioxane \tag{9.27}$$

convenient on a small scale, the product $Na_2Fe(CO)_4$ is contaminated with mercury salts and polymeric iron species. Hence the sodium/benzophenone ketyl route is preferred. (This complex is also available commercially.)

This complex is useful for synthesis specifically because of its high nucleophilicity and the ease of the migratory-insertion reaction (Chapter 5) in this system. Figure 9.4 summarizes this chemistry. Organic halides and tosylates react with $Na_2Fe(CO)_4$ with typical $S_N2$ kinetics (second order), stereochemistry (inversion), and order of reactivity ($CH_3$ > $RCH_2$ > R'RCH; RI > RBr > ROTs > RCl; vinyl, aryl, inert) to produce coordinatively saturated anionic $d^8$ alkyl iron (0) complexes (a). These undergo protolytic cleavage (b) to produce the corresponding hydrocarbon, an over-all reduction of the halide. In the presence of excess carbon monoxide or added triphenylphosphine, migratory insertion takes place to form the acyl complex (c), accessible directly by the reaction of $Na_2Fe(CO)_4$ with acid halides (d). This acyl complex reacts with acetic acid to produce aldehydes (e), providing a high-yield conversion of alkyl and acyl halides to

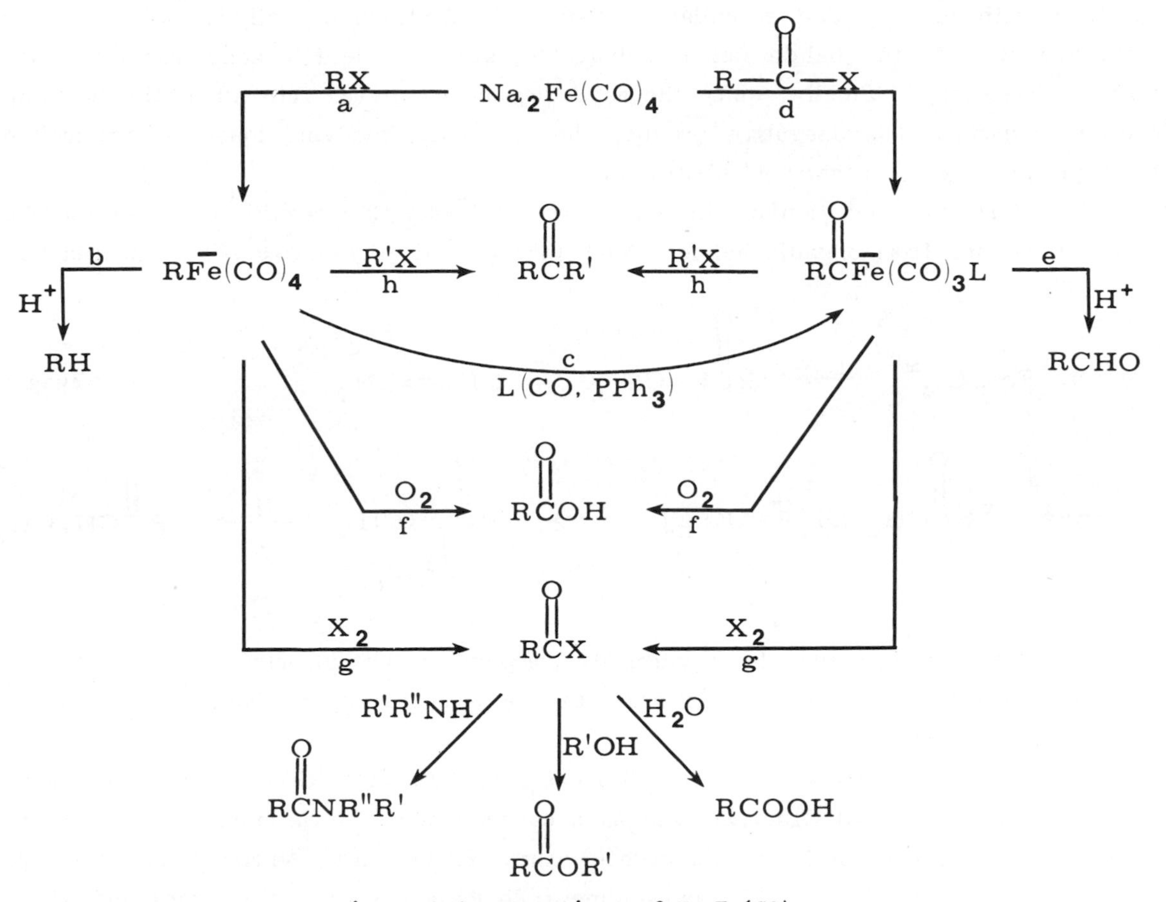


*Figure 9.4. Reactions of $Na_2Fe(CO)_4$.*

aldehydes. Oxidative cleavage of either the acyl or alkyliron (0) complex by oxygen (f) or halogen (g) produces carboxylic acid derivatives. Finally, the acyl iron(0) complex itself is still sufficiently nucleophilic to react with reactive alkyl iodides to give unsymmetrical ketones in excellent yield (h). Interestingly, the alkyl iron(0) complex reacts in a similar fashion with alkyl iodides to produce unsymmetric ketones inserting CO along the way (h). Hence this chemistry provides routes from alkyl and acid halides to alkanes, aldehydes, ketones, and carboxylic acid derivatives. The reagent is quite specific for halides. Ester, ketone, nitrile, and olefin functionality is tolerated elsewhere in the molecule. In mixed halides (i.e., chloro-bromo compounds) the more reactive halide is the exclusive site of reactivity. The major limitation is the high basicity ($pK_b$ about that of $OH^-$) of $Na_2Fe(CO)_4$, which results in competing elimination reactions with tertiary and secondary substrates. Additionally, allylic halides having alkyl groups δ to the halide fail as substrates, since stable 1,3-diene iron complexes form preferentially. Finally, since the migratory insertion (c) fails when the R group contains adjacent electronegative groups, the syntheses involving insertion are limited to simple primary or secondary substrates.

An interesting variation of this chemistry by Cooke [23] results in the production of ethyl ketones from organic halides, $Na_2Fe(CO)_4$, and ethylene (9.28). The reaction

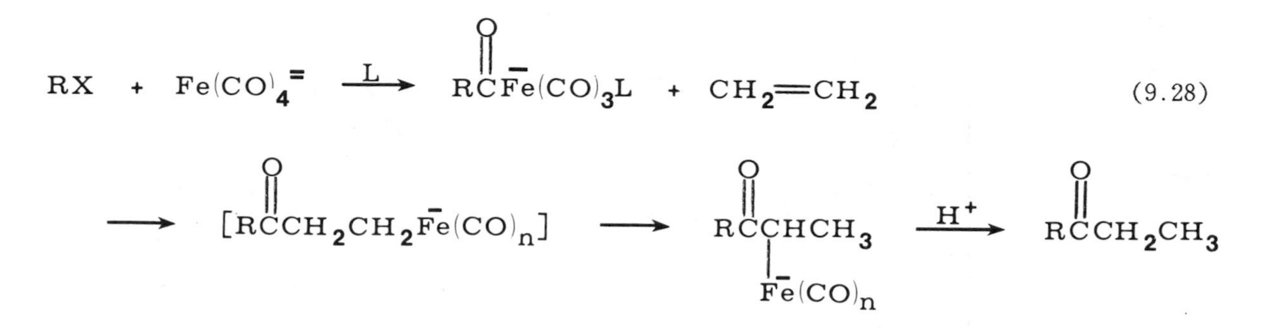

is thought to proceed via the acyliron(0) complex, which inserts ethylene into the acyl-metal bond and rapidly rearranges to the α-metalloketone, which then cannot further insert ethylene.

The chemistry described in Figure 9.4 has been subjected to very close experimental scrutiny by Collman and Brauman, and as a consequence, the mechanistic features are understood in detail. Both the alkyliron(0) complex, 11, and acyliron(0) complex, 12, have been isolated as air-stable, crystalline $[(Ph_3P)_2N]^+$ salts, fully characterized by elemental analysis and IR and NMR spectra, and shown to undergo the individual reactions detailed in Figure 9.4. Careful kinetic studies show that ion-pairing effects dominate the reactions of $Na_2Fe(CO)_4$, and account for the observed $2 \times 10^4$ rate

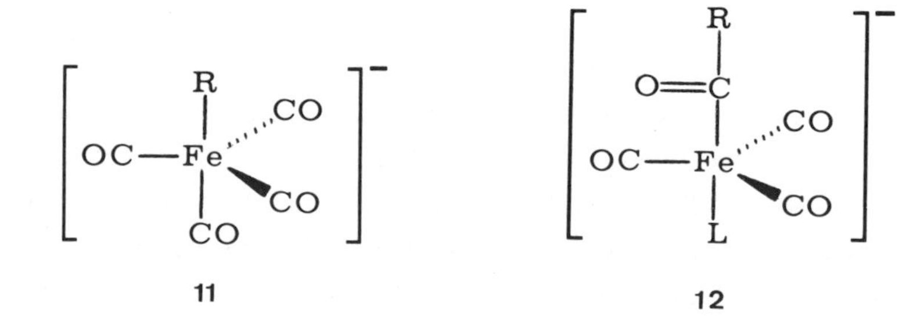

11 12

increase (tight ion pair $[NaFe(CO)_4]^-$ vs. solvent-separated ion pair $[Na^+{:}S{:}Fe(CO)_4]^-$ as the kinetically active species) in going from THF to N-methylpyrrolidone. The rate law, substrate order of reactivity, stereochemistry at carbon, and activation parameters (particularly the large negative entropy of activation) are all consistent with an $S_N2$-type oxidative-addition, with no competing one-electron mechanism [24]. The migratory-insertion step (c) is also subject to ion-pairing effects, and is accelerated by tight ion pairing [$Li^+ > Na^+ >$ Na-crown$^+ > (Ph_3P)_2N^+$]. The insertion reaction is over-all second-order, first-order in both $NaRFe(CO)_4$ and added ligand, with about a twenty-fold difference in rate over the range of ligands $Ph_3P$ (slowest) to $Me_3P$ (fastest) [25].

Heck has studied another anionic transition-metal carbonyl complex which effects a variety of useful carbonylation reactions, $NaCo(CO)_4$ [26]. This $d^{10}$ cobalt(-I)complex is easily prepared by the reduction (Na/Hg) or disproportionation (base) of $Co_2(CO)_8$. The sodium salt is a colorless crystalline solid, which is quite air-sensitive. The anion is moderately nucleophilic. Reaction with organic halides normally reactive in $S_N2$ processes (1°, 2°, allyl, benzyl halides, α-haloesters) produces an alkylcobalt tetracarbonyl complex, which, in the presence of one atmosphere of carbon monoxide, rapidly converts to the corresponding acylcobalt tetracarbonyl. Treatment with alcohol produces the ester (9.29). Since $HCo(CO)_4$ is a product of the alcohol cleavage of

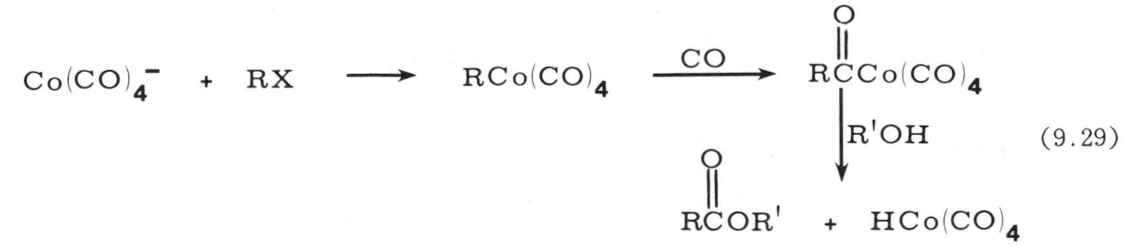

(9.29)

the acyl cobalt species, the reaction can be made catalytic in cobalt by converting this complex back to $Co(CO)_4^-$. This is best accomplished, in practice, by carrying out the reaction at 50° in the presence of a hindered tertiary amine, such as dicyclohexylethylamine. Under these conditions, alkyl iodides, sulfates, sulfonates, and bromides undergo carboalkoxylation in moderate to good yield.

β-Hydroxyalkylcobalt carbonyls are produced by the reaction of $HCo(CO)_4$ or $NaCo(CO)_4$ with epoxides. These complexes similarly insert carbon monoxide, producing acylcobalt complexes which are cleaved by alcohols to produce β-hydroxyesters (9.30). Although the β-hydroxyacyl complexes are isolable if desired, the system can also be

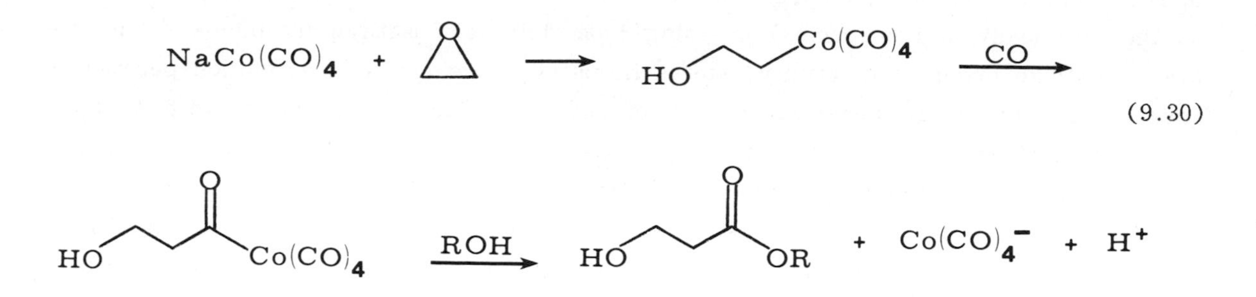


(9.30)

made catalytic. As the substitution about the epoxide increases, yields drop, an observation consistent with a nucleophilic attack of $Co(CO)_4^-$ on the epoxide. The less hindered terminus of the epoxide is always the site of acylation.

In addition to their reaction with alcohols to produce esters, acylcobalt carbonyl complexes, formed by either of the above methods, react with 1,3-dienes to form acylated $\eta^3$-allylcobalt carbonyl compounds. These can be cleaved by base to produce acyldienes (9.31). Again the reaction can be made catalytic by preparing the acylcobalt

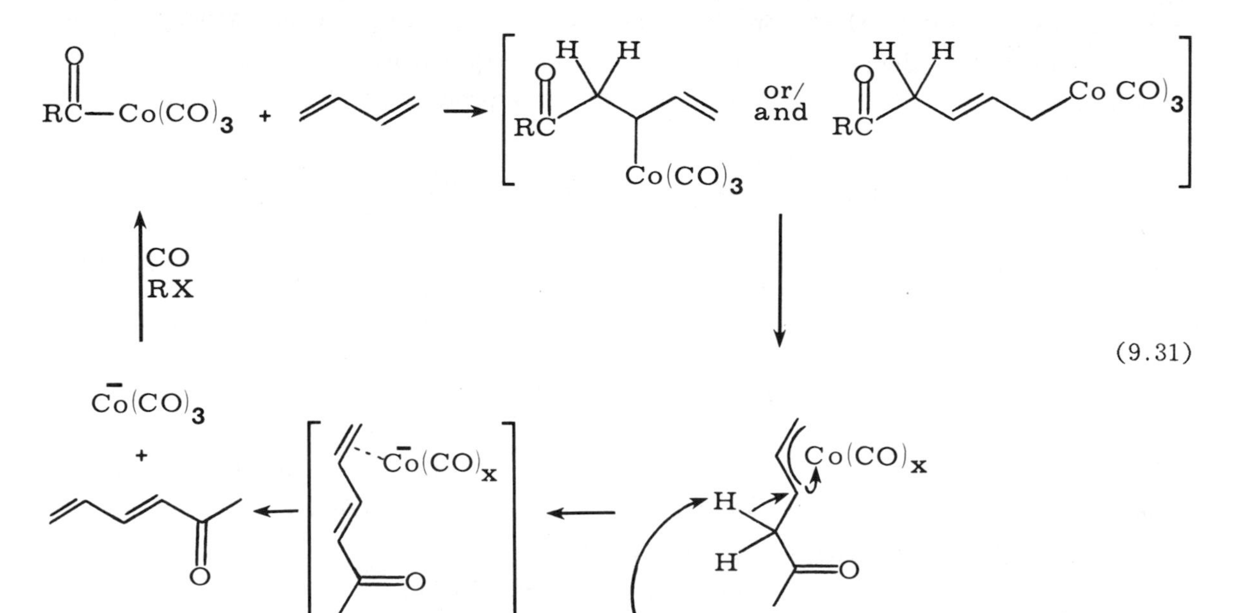


(9.31)

carbonyl complexes from $Co(CO)_4^-$ and alkyl or acyl halides in the presence of the diene, carbon monoxide, and base. The reaction is likely to occur by 1,2- or 1,4-addition of the acylcobalt complex to the diene, followed by collapse of the $\eta^1$-allyl complex formed. With unsymmetric dienes, mixtures of products result. The usefulness of this system for synthesis is difficult to assess, since the products were not purified, and the reported yields were based on the amount of carbon monoxide absorbed. However, this appears to be a potentially useful transformation if one is willing to optimize conditions and work out isolation procedures [26].

Both the iron and the cobalt carbonyl chemistry discussed above rely on the reaction of the metal carbonyl anion (produced by reduction of the neutral metal carbonyl) with an organic halide ($S_N2$) to produce a σ-alkylmetal complex, which inserts CO to form the σ-acylmetal complex. Another approach to acylmetal carbonyl complexes is the reaction of the neutral metal carbonyl with an organolithium compound. The properties of the resulting "acylmetal carbonylate" depend strongly on the properties of the specific transition metal involved. For example, group VI transition-metal carbonyls (Cr, Mo, W) react with a large number of organolithium reagents, undergoing attack at a carbonyl to produce an "acylmetal carbonylate" in which much of the charge resides on the carbonyl oxygen. Methylation with $(Me_3O)^+BF_4^-$ produces a very stable "carbene" complex (9.32). The chemistry of these complexes has been studied

$$Cr(CO)_6 + RLi \longrightarrow [(CO)_5\bar{Cr}\overset{O}{\overset{\|}{C}}R \longleftrightarrow (CO)_5Cr{=}\overset{O^-}{\overset{|}{C}}R] \xrightarrow{Me_3OBF_4} (CO)_5Cr{=}\overset{OMe}{\overset{|}{C}}R \tag{9.32}$$

extensively. Although the stable (heteroatom-substituted) ones are of little use in organic synthesis, unstable complexes of this type are implicated in the potentially useful olefin metathesis reaction (Chapter 10).

In contrast, the acylmetal carbonylate complexes of iron and nickel are highly reactive and perform a variety of transformations of use for organic synthesis. The complexes resulting from the reaction of organolithium reagents with iron carbonyl are the same as those produced by the reaction of $Na_2Fe(CO)_4$ with acyl halides, and are well-characterized structurally. The chemistry of these complexes has been presented above, and requires no further comment. The complexes resulting from the reaction of organolithium reagents with nickel carbonyl are much less well-characterized structurally, because they are much less stable. However, the general formula $[RCONi(CO)_3]^-$ has been assigned to them, and much of their chemistry is consistent with this assumption.

Tsutsumi [27] has shown that treatment of acylnickel carbonylates with ethanolic HCl produces α-diketones from aryl complexes and symmetrical ketones from alkyl complexes (9.33). The complexes produced from alkyllithium reagents are considerably

$$RLi + Ni(CO)_4 \xrightarrow{-70^\circ} [RC(=O)Ni(CO)_3]^- Li^+ \xrightarrow[EtOH]{HCl} ArC(=O)-C(=O)Ar \text{ or } RC(=O)R \qquad (9.33)$$

more reactive than those from aryllithiums, and react with allylic halides at -50° to produce high yields of the β,γ-unsaturated ketones (Hegedus) [28], as in 9.34. With

$$RLi + Ni(CO)_4 \xrightarrow[-50^\circ]{Et_2O} [RC(=O)Ni(CO)_3]^- Li^+ + R'CH=CHCH_2Br \longrightarrow R'CH=CHCH_2C(=O)R \qquad (9.34)$$

substituted allylic halides, acylation occurs at the halogen-bearing carbon in all reported cases, except for the C-12 dibromide, with which products arising from both terminal and internal attack are obtained in the ratio 5:4 as in 9.35. With trans-geranyl bromide, a substrate susceptible to cis-trans isomerization, the configuration of the allylic double bond remains unchanged by the acylation reaction, making the intermediacy of $\eta^3$-allylnickel species unlikely (9.36). Aryl iodides, benzoyl chloride, and aliphatic halides, including 1° iodides, are unreactive toward these acylnickel carbonylates under conditions sufficiently severe (~0°C) to decompose the nickel complex. In contrast, vinyl halides are quite reactive. Thus, 1-bromo-2,2-diphenylethylene reacts

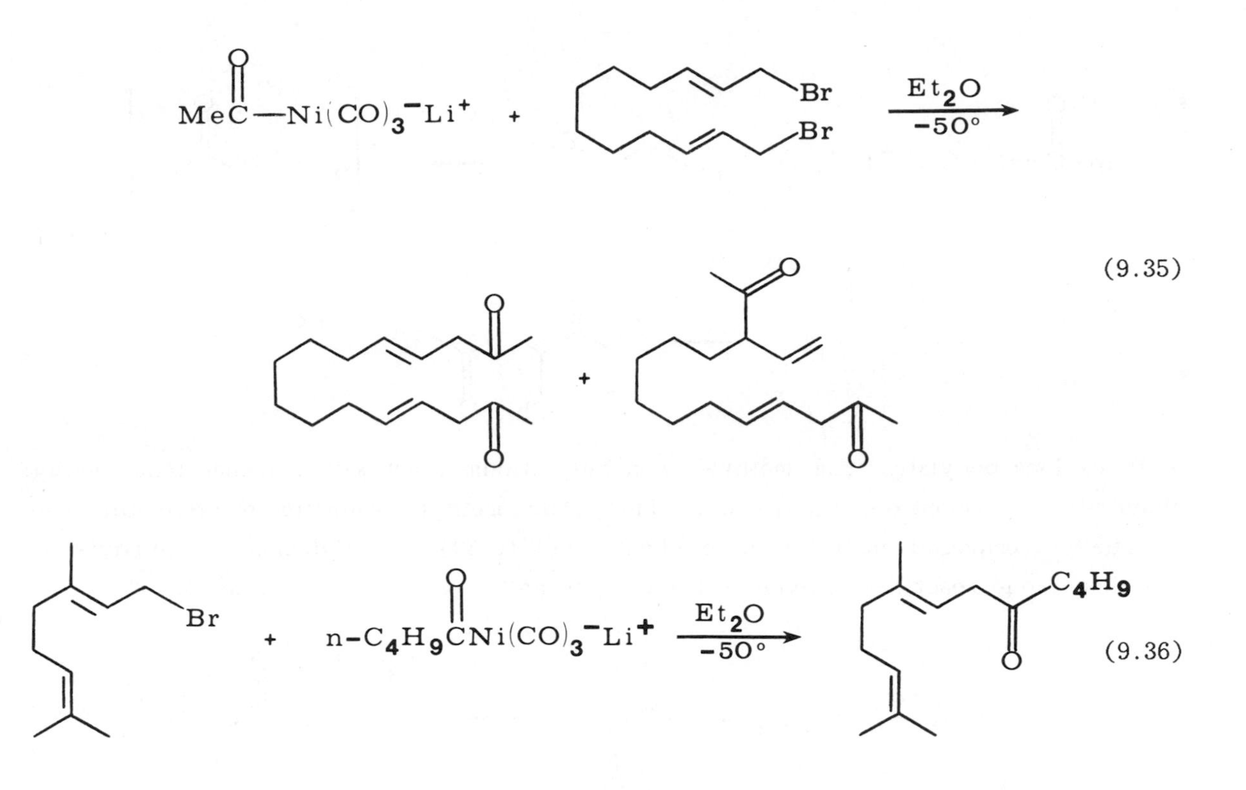


with the methyllithium-nickel carbonyl complex to produce the expected unsaturated ketone (9.37). However, trans-β-bromostyrene reacts to produce 3-phenyl-2,5-

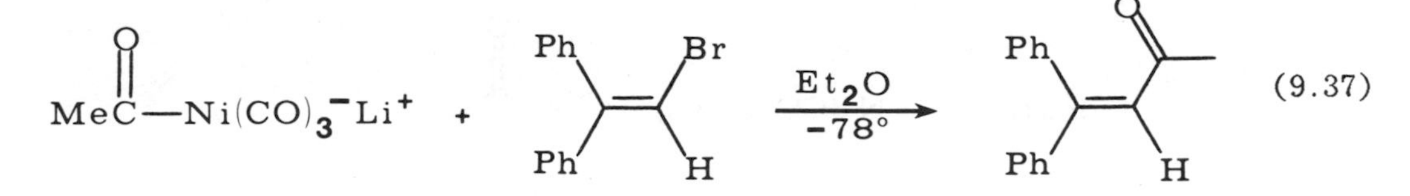


hexanedione as the major product (9.38). This dione arises from a 1,4-addition of the acylnickel carbonylate to the initially formed conjugated ketone. Thus the acylation of vinyl bromides is restricted to those substrates which are too hindered to undergo the conjugate addition.

However, this 1,4-acylation of conjugated enones is of interest in its own right, since this type of direct acylation is not possible using classical synthetic reactions, and since the product 1,4-diketones are important starting materials for a variety of heterocyclic syntheses. Corey [29] showed that the reaction is quite general.

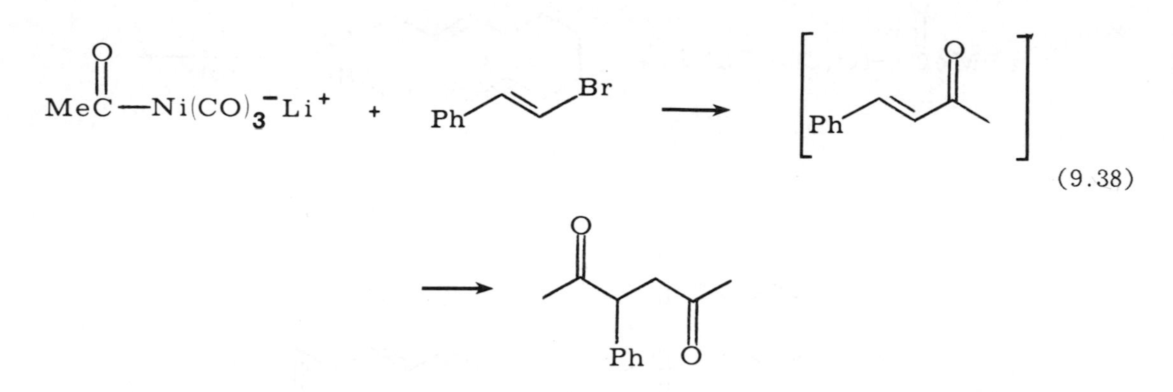


Acylnickel carbonylates from methyl- or n-butyllithium react with benzalacetone, methyl cinnamate, 2-cyclohexenone, mesityl oxide, and methyl crotonate to give the 1,4-dicarbonyl compound in good to excellent yield (9.39). The dienone, 1-phenyl-1,3-hexadien-5-one reacts exclusively 1,4 and gives no 1,6-adduct. The reaction of

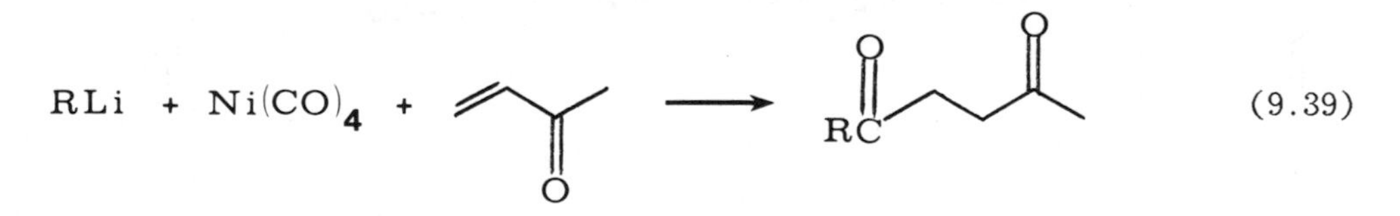


t-butyllithium, nickel carbonyl and benzalacetone produces the expected 1,4-dione as well as substantial amounts of a 1,4,5 trione (9.40). The yield of the trione increases

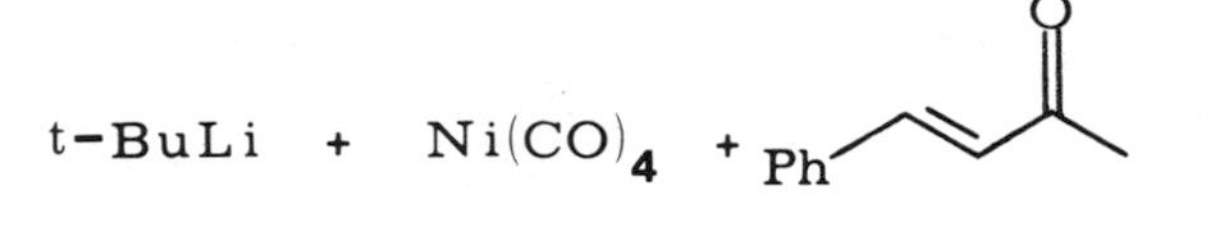


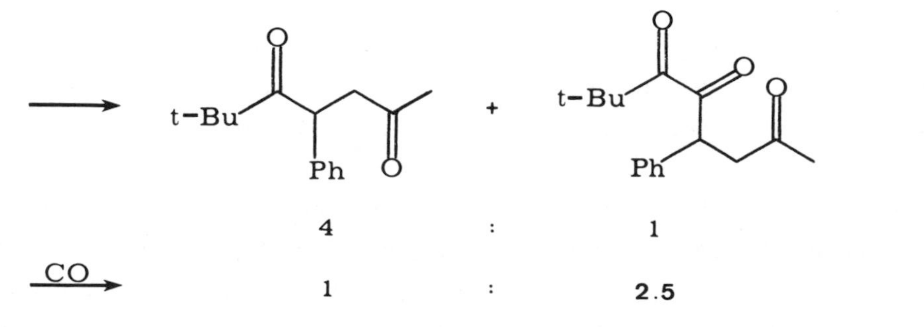

when the reaction is run under an atmosphere of carbon monoxide, implying the highly unusual migration of an acyl group to a nickel-bound carbon monoxide. This is one of the very few examples of this type of reaction and requires further study before definitive mechanistic statements can be made.

Monosubstituted alkynes also react with acylnickel carbonylates to produce 1,4-diketones. Tsutsumi [30] obtained the best yields with aroyl complexes (9.41).

$$ArC(=O)Ni(CO)_3^-Li^+ + RC{\equiv}CH \longrightarrow [RCH{=}CH{-}C(=O){-}Ar] \longrightarrow ArC(=O)CH(R)CH_2C(=O)Ar \qquad (9.41)$$

In the reaction of phenylacetylene with $[CH_3CONi(CO)_3]^-Li^+$, small amounts of benzalacetone ($PhCH{=}CHCOCH_3$) are isolated, implicating conjugated enones as intermediates in this reaction, and linking it to the 1,4-acylation of conjugated enones just discussed. Disubstituted acetylenes do not react in this system.

Nucleophiles other than organolithium reagents react with nickel carbonyl to produce synthetically useful complexes. Vinyl bromides react with complexes formed from nickel carbonyl and sodium methoxide in methanol to produce α,β-unsaturated esters in excellent yield (Corey) [31], as in 9.42. With careful control of reaction conditions, the geometry of the double bond is maintained.

$$MeO^-Na^+ + Ni(CO)_4 + RCH{=}CHBr \xrightarrow{MeOH} RCH{=}CHCO_2Me \qquad (9.42)$$

With sensitive substrates, this may be the method of choice for the carboxylation of vinyl halides. For example, in Corey's synthesis of α-santalol [32], conversion of the vinyl iodide to the conjugated ester was less than 10% when the $Mg/CO_2$ method was used, whereas $CH_3O^-/Ni(CO)_4$ raised the yield to 85% (9.43).

Aryl iodides and allylic bromides are also readily carboxylated by this complex, although these substrates are also carboxylated (although more slowly) by nickel carbonyl in the absence of methoxide [31]. Surprisingly, simple primary alkyl iodides do

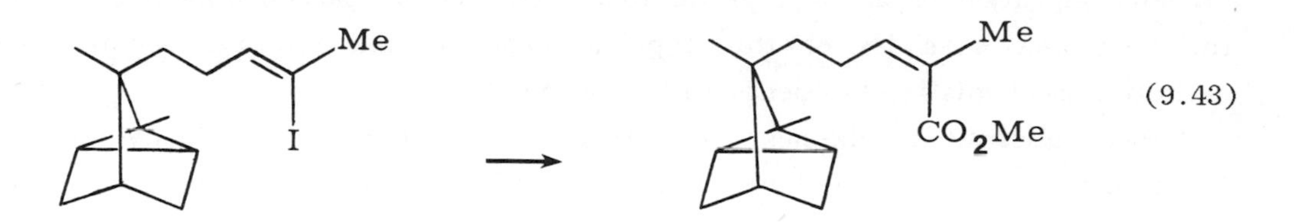


(9.43)

not react with sodium methoxide/nickel carbonyl under conditions sufficiently severe to decompose the nickel complex. A more reactive complex is formed by the reaction of potassium t-butoxide with nickel carbonyl. This reagent converts primary alkyl halides to the corresponding t-butyl esters in quite good yield. With very reactive vinyl bromides such as β-bromostyrene, amino-carbonylation results from treatment with nickel carbonyl and pyrrolidine at 60°. However, this reaction is sluggish and of little use with less reactive substrates. In a reaction related to this chemistry and the chemistry of aroylnickel carbonylates, the reaction of lithium dimethylamide with nickel carbonyl produces a complex which converts phenylacetylene to 2-phenyl-tetramethyl succinamide in moderate yield [33], as in 9.44. However, the same reagent reacts with β-bromostyrene to give the unsaturated amide in excellent (96%) yield [34].

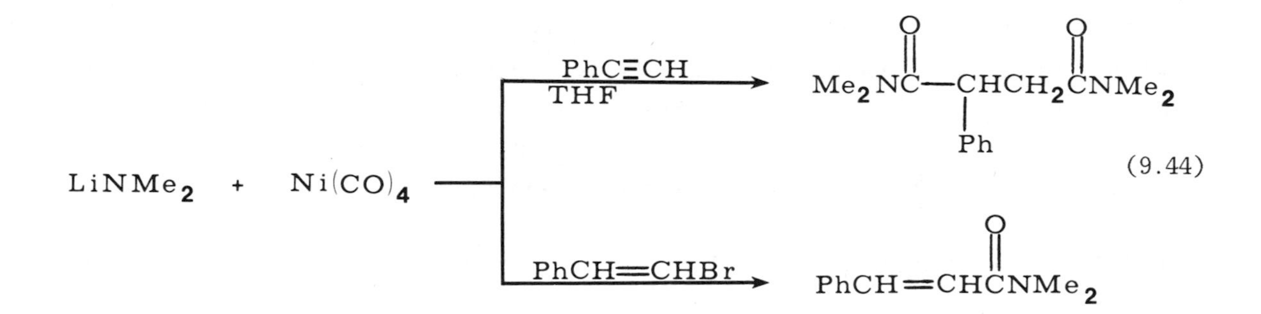


(9.44)

Although the chemistry performed by the above acylnickel carbonylates is quite useful, neither the structures of the complexes nor their mechanisms of reaction have been examined to any extent. The specific complexes involved may well be dimeric or polymeric, since highly colored solutions are formed upon treatment of nickel carbonyl with the bases involved, and since nickel carbonyl is known to form colored clusters [e.g., $Ni_3(CO)_8^{=}$] in the presence of base [35]. Whatever the mechanism of reaction, it must be significantly different from that of the acyliron carbonylates, since the reactivity pattern toward halides (vinyl, aryl > allyl >> 1° alkyl) is opposite to that of the iron complex (1° > 2° > 3° alkyl; vinyl, aryl inert). The 1,4-addition reaction of acylnickel carbonylates with conjugated enones is reminiscent of the chemistry of lithium dialkylcuprates (Chapter 11) for which an electron-transfer mechanism has been proposed but not proven. The reactions with organic halides are adequately explained by a mechanism involving an oxidative-addition step of the type involved in the reaction of $L_4Pd(0)$ complexes with aryl iodides rather than that of the $S_N2$ type followed by iron (9.45). Just why alkyl iodides are so unreactive is not clear.

$$Ni(CO)_4 + B^- \longrightarrow [B\overset{O}{\overset{\|}{C}}Ni(CO)_3]^- \rightleftarrows [B\overset{O}{\overset{\|}{C}}Ni(CO)_2]^- \xrightarrow{RX}$$

$$\left[ B-\overset{O}{\overset{\|}{C}}-Ni(R)(X)(CO)_2 \right]^- \longrightarrow B\overset{O}{\overset{\|}{C}}R + [Ni(CO)_2X]^- \quad (9.45)$$

Alkynes are carbonylated by carbon monoxide and alcohols using a number of different catalysts. Two general mechanisms are recognized for this process (9.46 and 9.47). The first (9.46) involves a "metal-hydride" pathway, in which the key step is addition (insertion) of a metal hydride to an alkyne to produce a vinyl metal complex. Insertion of CO into the metal-carbon bond produces a metal acyl complex. Cleavage by the alcohol produces the unsaturated ester and regenerates the metal hydride. The second mechanism (9.47) involves coordination of CO to the metal, followed by

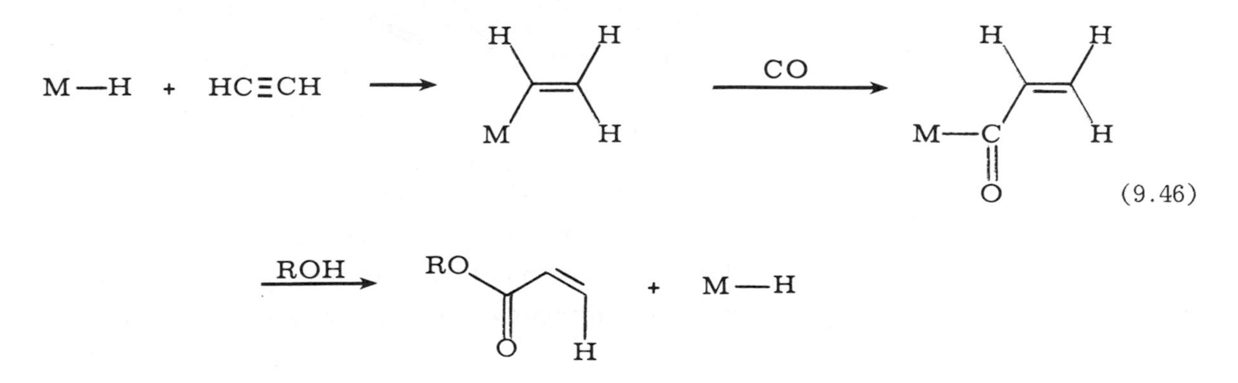

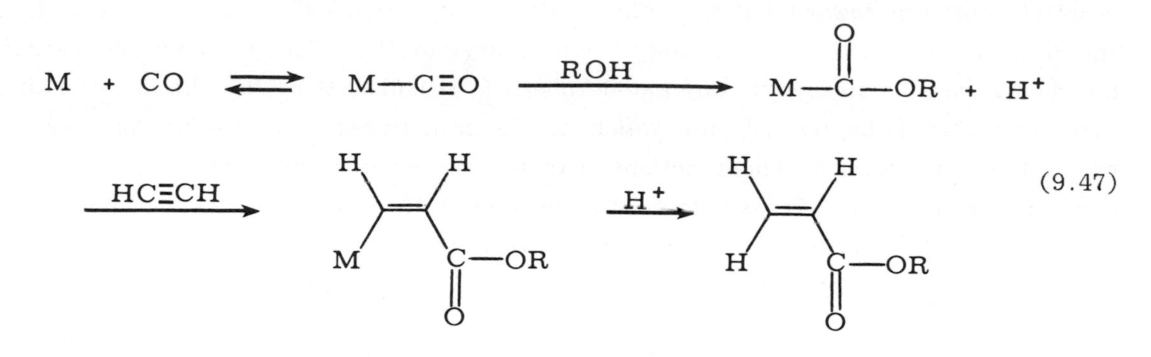

(9.47)

nucleophilic attack of alcohol on the metal-bound CO to produce a metal carboxylate complex. Insertion of the acetylene into the metal-CO bond produces a vinylmetal complex which undergoes protolytic cleavage, freeing the conjugated ester and the original metal catalyst. The pathway followed depends on the catalyst system used.

Recently Norton [36] developed an intramolecular version of this alkyne carbonylation to produce α-methylene lactones from β-hydroxyalkynes (9.48). A careful

$$HC{\equiv}CCH_2CH_2OH \quad + \quad CO \xrightarrow{PdI_2/Bu_3P/CH_3CN}$$ (9.48)

kinetic study showed a rate law for which CO coordination was the rate-limiting step (rate = $k_{obs}$ [Pd][CO]), and a labeling study showed that the proton of the alcohol was found exclusively on the terminal olefinic carbon (9.49). The fact that no

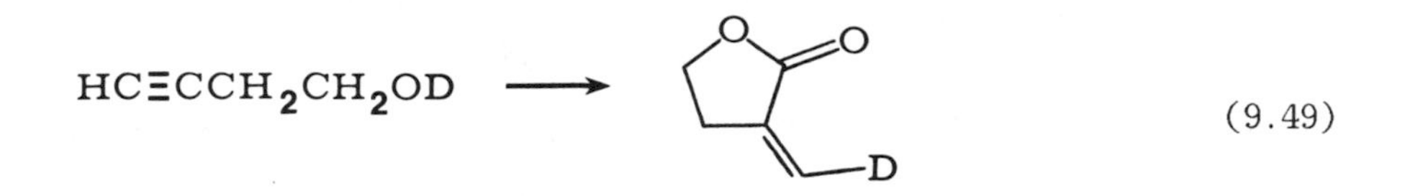

(9.49)

rearrangement of the exocyclic double bond to the more stable endocyclic position was observed, an easy process in the presence of metal hydride species, led to the mechanism shown in 9.50. The requisite intramolecular insertion of the alkyne into the Pd-CO bond was demonstrated by preparing the proposed intermediate independently, and

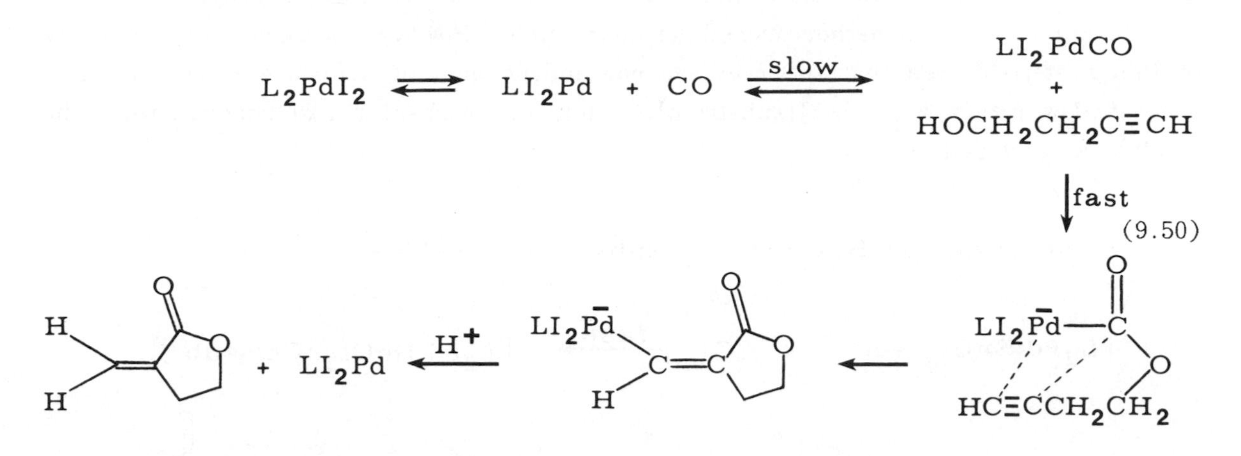


studying its conversion to α-methylene lactone (9.51). These results strongly support the proposed mechanisms.

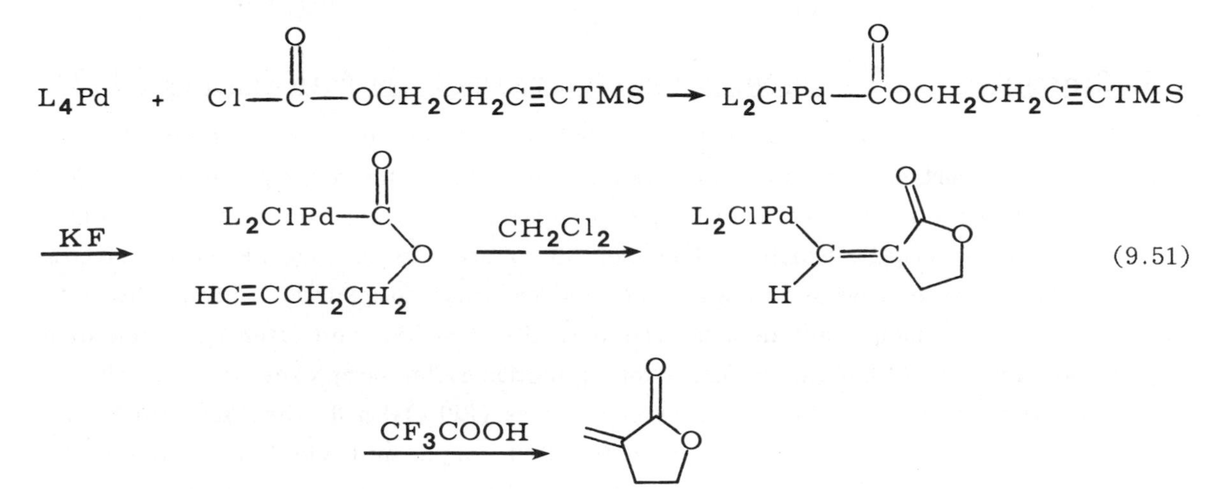


Because CO coordination is the rate-limiting process in the above reaction, Pd is being used inefficiently, since only the small amount that coordinates CO reacts. To increase the catalytic efficiency of this system, one must make the coordination of CO more efficient. To achieve this, $SnCl_2$ was added to the system. $SnCl_2$ reacts with Pd-Cl complexes to produce $PdSnCl_3$ complexes. The $SnCl_3^-$ ligand is a strong π-acceptor, and it labilizes the ligand in the position <u>trans</u> to it. It was reasoned that this <u>trans</u> labilizing effect should facilitate CO uptake by making a vacant coordination site more available. Indeed, the new catalyst system, consisting of $PdCl_2$, $Ph_3P$, and

$SnCl_2$ in acetonitrile, was more than 100 times as reactive as the previous catalyst system and resulted in higher over-all yields as well. However, a careful kinetic study of this system showed that the rate was now independent of CO concentration and had the rate-law rate = $k_{obs}$ [Pd][substrate]. Thus the mechanism had changed when the catalyst was altered (9.52).

$$[L_2Pd(SnCl_3)(CH_3CN)]^+ + \text{substrate} \xrightarrow{\text{slow}}$$

$$[L_2Pd(SnCl_3)(sub)]^+ + CO \xrightarrow{\text{fast}} [L_2Pd(SnCl_3)(CO)(sub)]^+ \quad (9.52)$$

$$\longrightarrow L_2(SnCl_3)PdC(=O)CH_2CH_2C{\equiv}CH \longrightarrow \longrightarrow \text{(α-methylene-γ-butyrolactone)}$$

## 9.3. Stoichiometric Decarbonylation of Aldehydes and Acid Chlorides [37].

Much of this chapter has been devoted to consideration of methods for the introduction of carbonyl groups into organic substrates via a "migratory insertion" reaction, in which a metal alkyl or aryl group migrates to an adjacent coordinated CO to produce a σ-acyl complex which is then cleaved to produce the organic carbonyl compound. The reverse process, in which organic carbonyl compounds, specifically aldehydes and acid chlorides, are decarbonylated is also possible, and often quite useful in organic synthesis. Although a number of transition-metal complexes will function as decarbonylation agents, by far the most efficient is $(PPh_3)_3RhCl$, the Wilkinson hydrogenation catalyst. This complex reacts with alkyl, aryl, and vinyl aldehydes under mild conditions to produce the corresponding hydrocarbon and the very stable $(Ph_3P)_2Rh(Cl)CO$, as in 9.53. The course of the decarbonylation can easily be

$$RCHO + (Ph_3P)_3RhCl \xrightarrow{20-100^\circ} (Ph_3P)_2Rh(Cl)CO + PPh_3 + RH \quad (9.53)$$

monitored by the change in color of the solution, from the deep red of $(Ph_3P)_3RhCl$ to the canary yellow of $(Ph_3P)_2Rh(Cl)CO$. *Trans*-α-alkylcinnamaldehydes are decarbonylated with retention of olefin geometry, producing *cis*-substituted styrenes [38], as in 9.54.

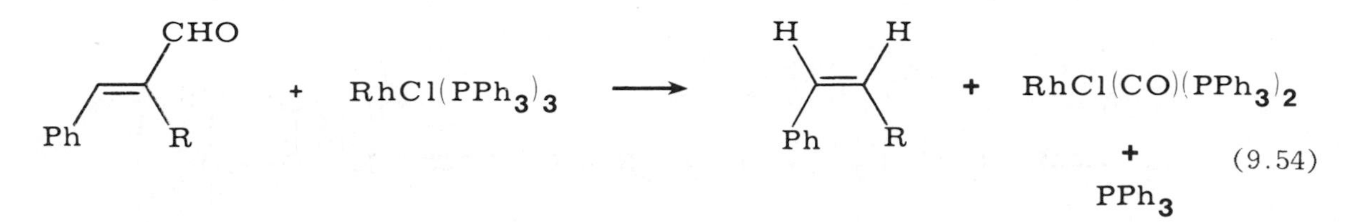


More impressively, chiral aldehydes are converted to chiral hydrocarbons with over-all retention and with a high degree of stereoselectivity [39,40], as in 9.55 and 9.56.

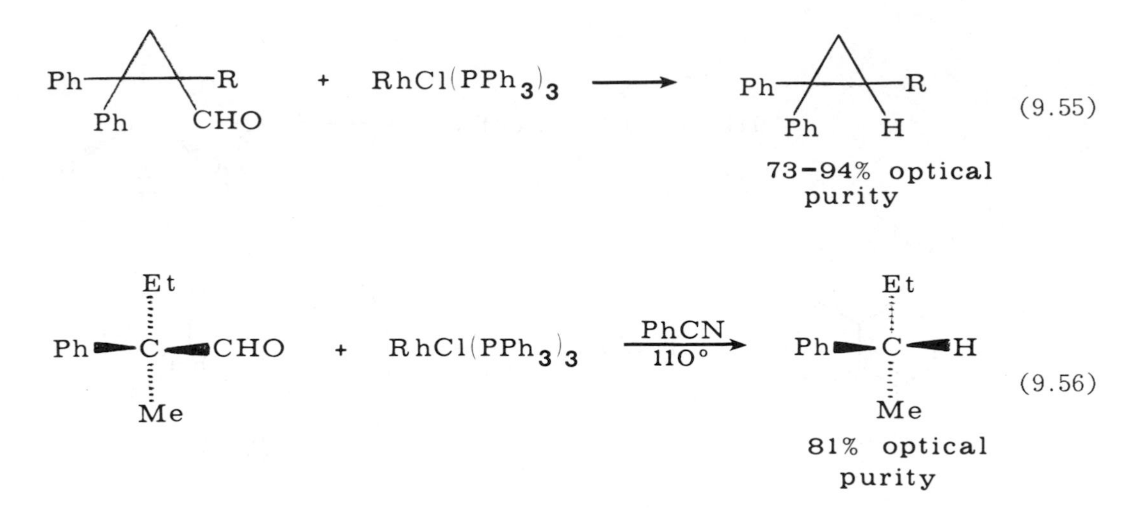


Use of readily accessible C-1 deuterated aldehydes [from $Fe(CO)_4^{=}$ chemistry, for example] makes previously inaccessible deuterated optically active hydrocarbons available. The mechanism proposed for this decarbonylation is that in 9.57. It involves, as a first step, oxidative addition of the aldehyde to the rhodium complex. This is not unreasonable, in light of the high reactivity of $L_3RhCl$ toward oxidative addition. Although it is not possible to isolate this initial oxidative adduct from simple aldehydes, Suggs [41] used 8-quinoline carboxaldehyde, which reacts to give an acylhydride rhodium complex, which is stabilized by chelation and hence more easily isolated and characterized (9.58). The next step is the reverse of the carbonyl insertion reaction, that is, migration of the alkyl group from carbonyl to metal. It is well-known that migratory insertion is a reversible process, and can be driven in either direction by appropriate choice of reaction conditions. In addition, it has been shown in other cases that both forward and reverse reactions proceed with retention of stereochemistry of the migrating alkyl group; hence the retention of stereochemistry in the decarbonylation

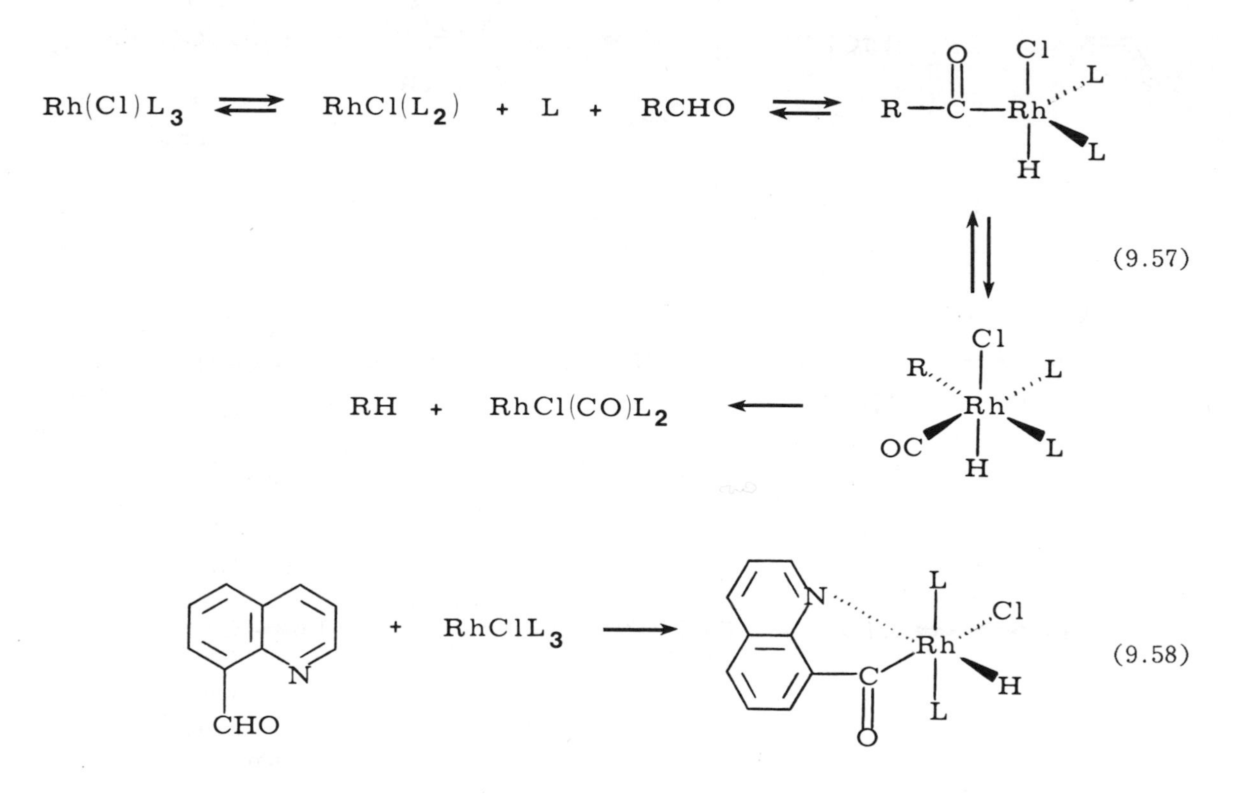


of chiral aldehydes. The last step is the reductive elimination of alkane (RH) and the production of $RhCl(CO)L_2$. This last step is irreversible [alkanes do not oxidatively add to $RhCl(CO)L_2$], and drives the entire process to conclusion. Since $RhCl(CO)L_2$ is much less reactive in oxidative addition reactions than $RhClL_3$ (CO is a $\pi$-acceptor and withdraws electron density from the metal), it cannot react with aldehydes under these mild conditions, and the decarbonylation as described is stoichiometric. When the reaction is carried out at temperatures in excess of 200°, both $RhClL_3$ and $RhCl(CO)L_2$ function as catalytic decarbonylation agents, presumably because $RhCl(CO)L_2$ will oxidatively add aldehydes under these severe conditions.

This decarbonylation reaction is quite useful in organic synthesis, since it removes a carbonyl (aldehyde) group that is often necessary to activate a position for functionalization, but not required in the final product. Its tolerance toward a variety of other functional groups and its usefulness when applied to complex molecules is illustrated by its use in the synthesis of $\alpha$-linked disaccharides [42], desoxypodocarpate [43], occidentalol [44], and grandisol [45], as shown in 9.59 to 9.62.

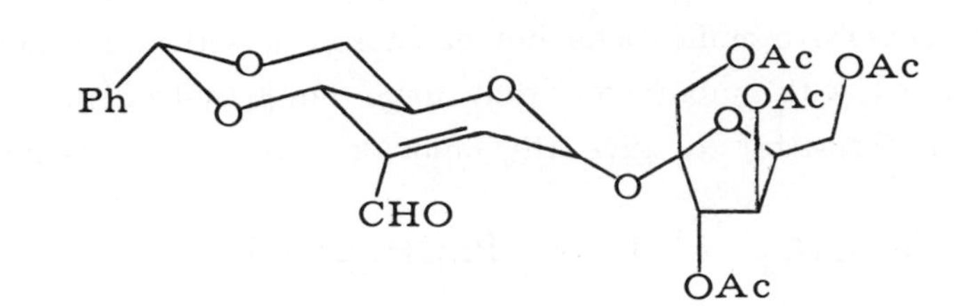
Ph
OAc
OAc
OAc
CHO
OAc

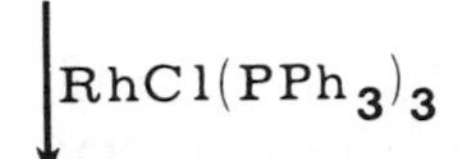
RhCl(PPh3)3

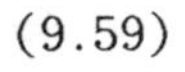
(9.59)

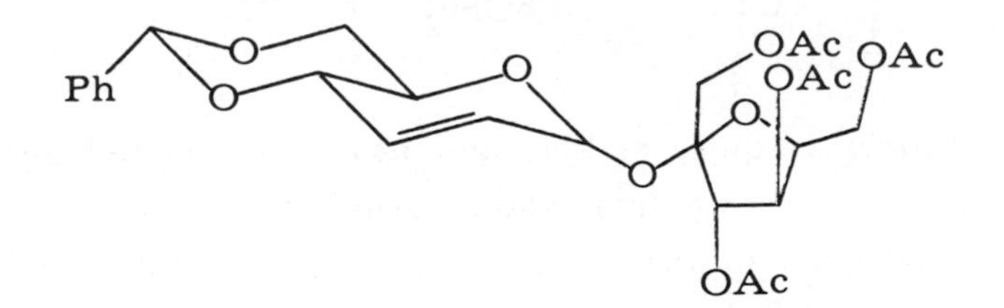
Ph
OAc
OAc
OAc
OAc

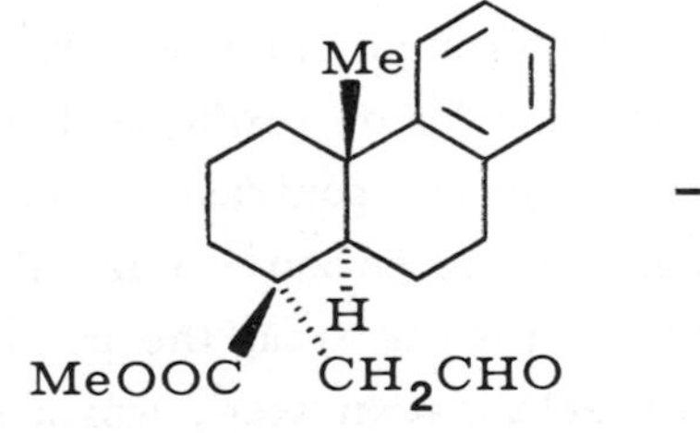
Me
H
MeOOC
CH2CHO

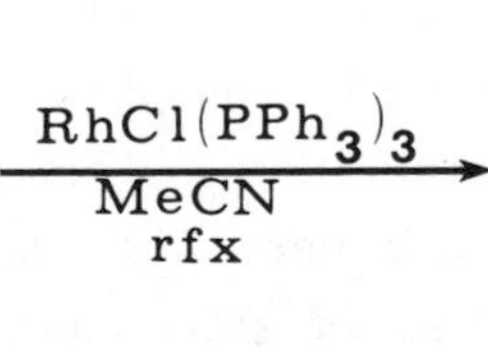
RhCl(PPh3)3
MeCN
rfx

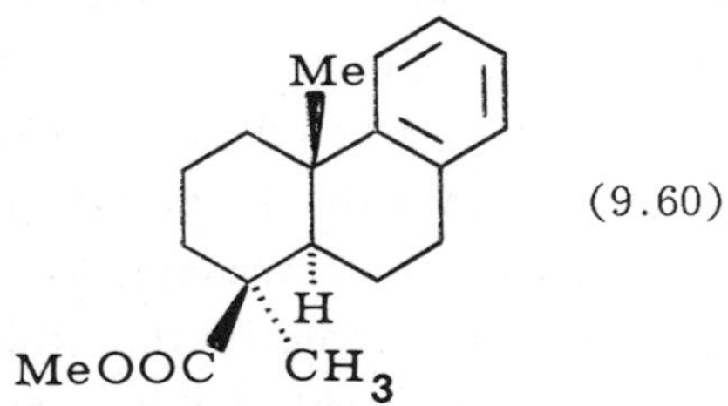
Me
H
MeOOC
CH3
(9.60)

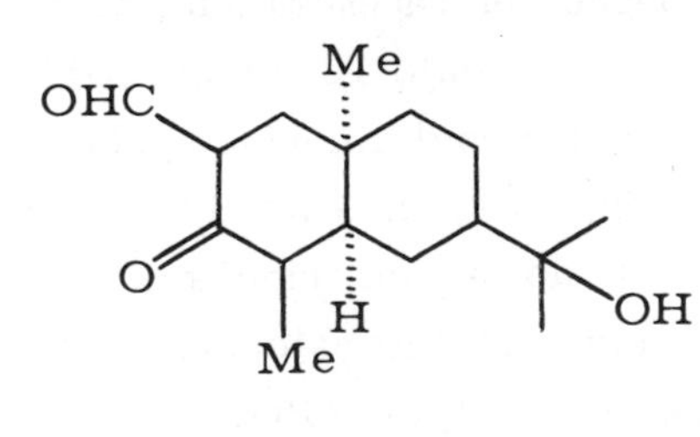
Me
OHC
O
H
Me
OH

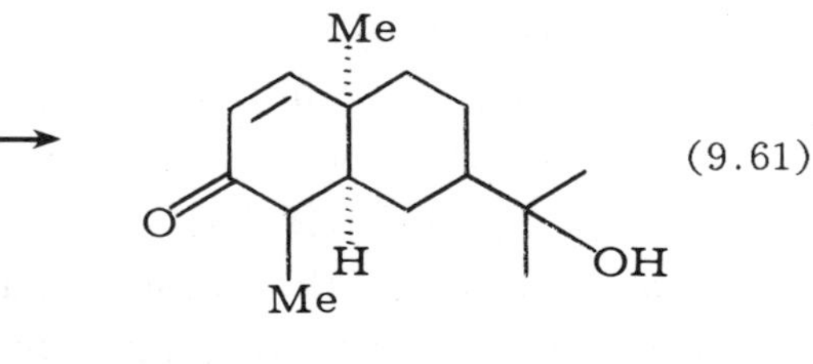
Me
O
H
Me
OH
(9.61)

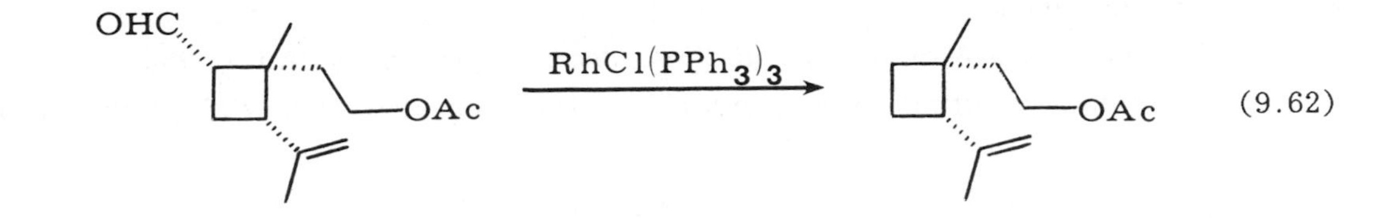
OHC
OAc
RhCl(PPh3)3
OAc
(9.62)

Acid chlorides also undergo decarbonylation upon treatment with $RhCl(PPh_3)_3$, but the reaction suffers several complications not encountered with aldehydes. The reaction is most straightforward with substrates that have no β-hydrogens. In these cases, decarbonylation occurs smoothly to give the chloride (9.63). In contrast to aldehydes,

$$RhClL_3 \rightleftharpoons RhClL_2 + L + PhCH_2COCl \tag{9.63}$$

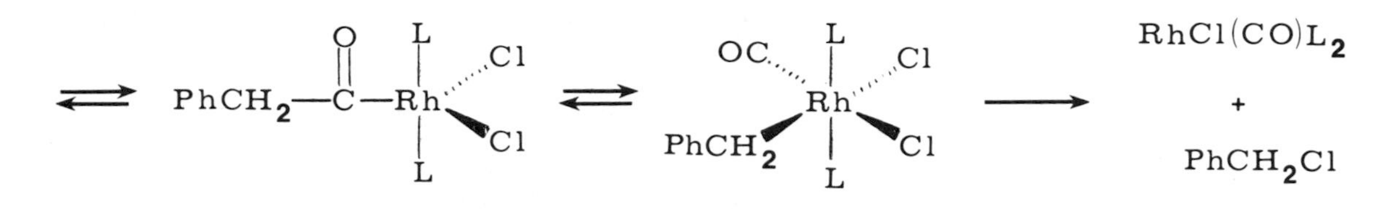


the initial oxidative adduct is quite stable, and has been isolated and fully characterized in several instances. Heating this adduct leads to decarbonylation, via a mechanism thought strictly analogous to that involved in aldehyde decarbonylation. However, the decarbonylation of acid halides differs from that of aldehydes in several respects. Decarbonylation of optically active (S)(-)-α-trifluoromethylphenylacetyl chloride with $RhCl(PPh_3)_3$ produces completely racemic α-trifluoromethylbenzyl chloride [46]. Similarly, the decarbonylation of (S)(+)-α-deuteriophenylacetyl chloride produces benzyl-α-d chloride with only 20 to 27% net retention. This is in direct contrast to aldehydes, which undergo decarbonylation with a very high degree of retention. Careful studies by Stille [47] indicate that this loss of stereochemistry results from the many acyl $\rightleftarrows$ alkyl rearrangements prior to the rate-limiting reductive-elimination step, which must be slow relative to the aldehyde case. Even if only a small degree of racemization occurs with each acyl → alkyl → acyl conversion, significant racemization could result by attrition. Use of $Rh^{36}Cl(PPh_3)_3$ as the decarbonylation agent with benzoyl chloride leads to an equal distribution of labelled $^{36}Cl$ in the product benzyl chloride, unreacted benzoyl chloride, and $RhCl(CO)(PPh_3)_2$. This complete scrambling of labeled chlorine indicates that both chlorines in the oxidative adduct are equivalent, and that the oxidative addition also reverses many times prior to the rate-limiting loss of $PhCH_2Cl$. Hence this is a very dynamic system.

Acid halides with β-hydrogens undergo decarbonylation by $RhCl(PPh_3)_3$ to produce primarily olefins, resulting from β-hydride elimination from the σ-alkyl intermediate, rather than reductive elimination (9.64). The decarbonylation of erythro-2,3-diphenylbutanoyl chloride gives exclusively trans-methylstilbene (9.65), whereas the threo acid chloride gives exclusively cis methylstilbene (9.66) [48]. This is entirely

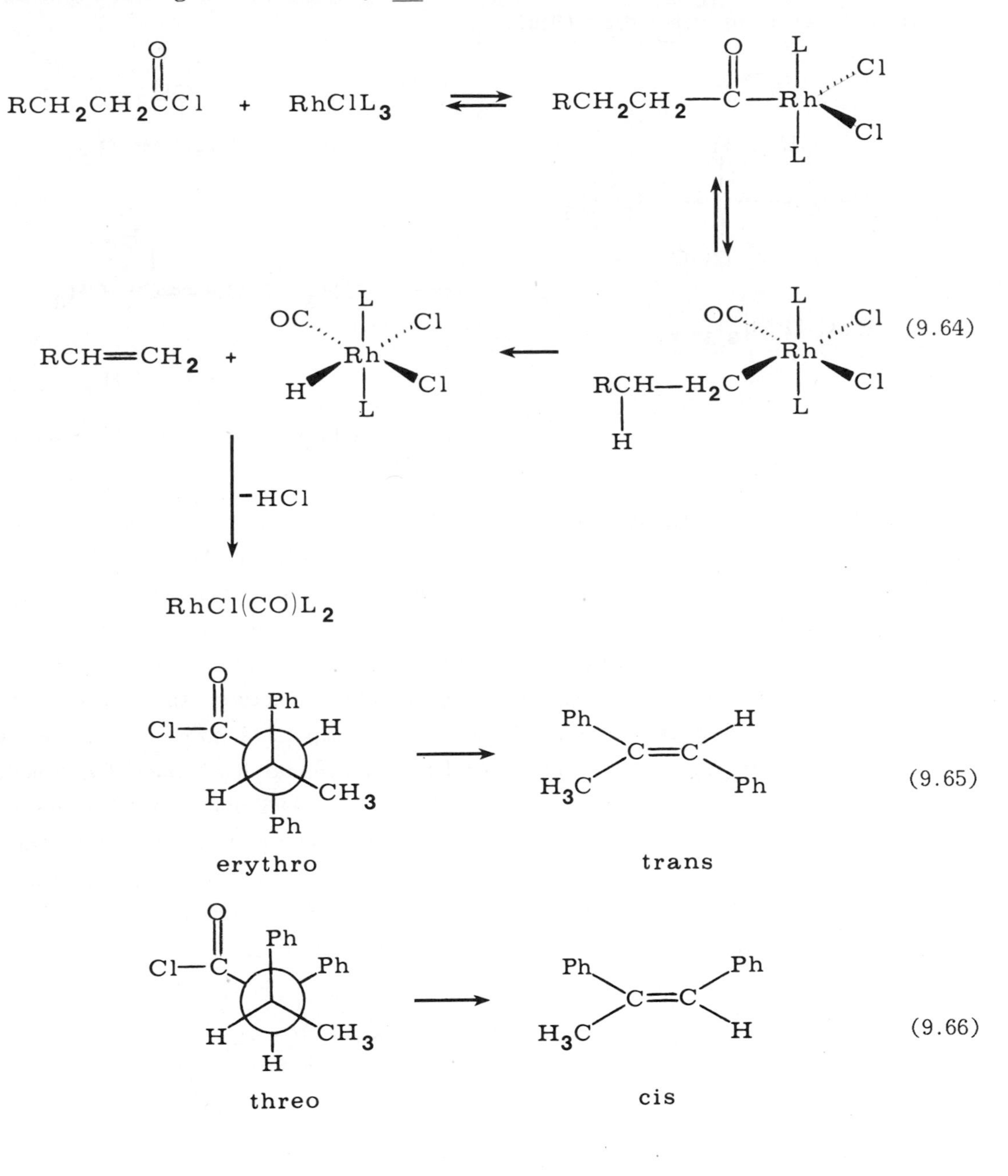

consistent with migratory deinsertion (acyl → alkyl) with retention of configuration, followed by the expected syn β-hydride elimination. Branched acid chlorides that can undergo β-elimination in several directions give mixtures of products with the most substituted olefins predominating (9.67).

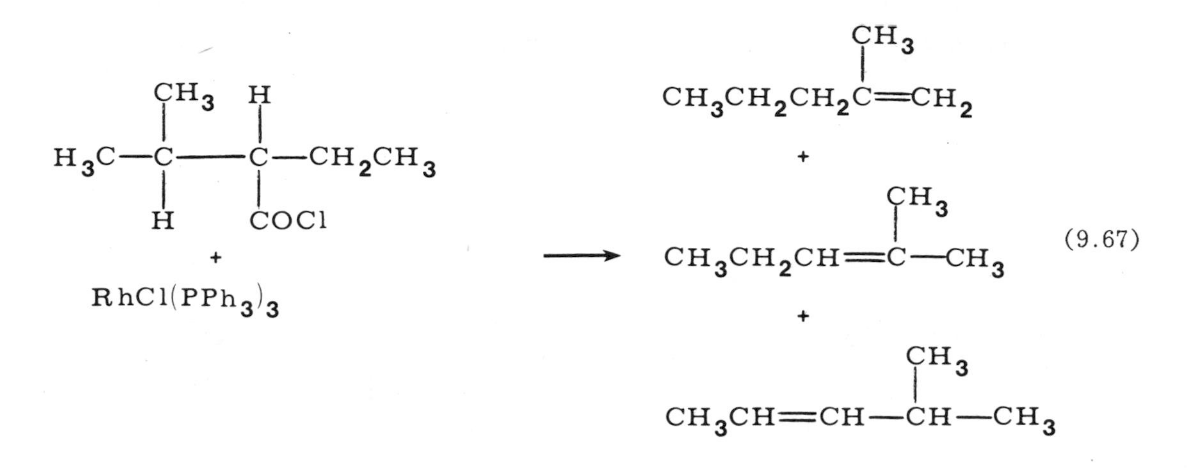


All the above decarbonylations are stoichiometric reactions, carried out in aromatic solvents such as benzene or toluene at temperatures between 20 and 100°C. As with aldehydes, the decarbonylation of acid chlorides can be made catalytic in Rh complex by going to higher temperatures (200-300°C). Under these more severe conditions, other transition-metal complexes, most notably $PdCl_2$ or Pd/C, also catalyze this reaction. Although catalytic decarbonylations are economically attractive, the severe conditions required preclude their use with sensitive organic compounds commonly intermediate in sophisticated organic syntheses. This problem has been at least partially remedied by the development of a procedure to reconvert $Rh(Cl)(CO)(PPh_3)_2$ back to the reactive $RhCl(PPh_3)_3$ by treatment with benzyl chloride and triphenylphosphine [49]. However, this regeneration is done in a separate step, and decarbonylation of large amounts of substrate requires either the running of several small-scale reactions or the purchase of a large amount of $RhCl(PPh_3)_3$, which is ultimately recovered.

## Notes.

1. For general reviews on the use of metal carbonyls in synthesis, see I. Wender and D. Pino, eds., Organic Synthesis via Metal Carbonyls, (Wiley-Interscience, vol. I, 1968; vol. II, 1977); M. Ryang, Organometal. Chem. Rev. A, 5, 67 (1970); M. Ryang and S. Tsutsumi, Synthesis, 55 (1971); H. Alper, J. Organometal. Chem. Library, 1, 305 (1976).
2. I. D. Webb and G. T. Borcherdt, J. Amer. Chem. Soc., 73, 2654 (1951).
3. E. J. Corey, M. F. Semmelhack, and L. S. Hegedus, J. Amer. Chem. Soc., 90, 2416, 2417 (1968).
4. E. J. Corey and E. Wat, J. Amer. Chem. Soc., 89, 2757 (1967).
5. E. J. Corey and E. Hamanaka, J. Amer. Chem. Soc., 86, 1641 (1964); 89, 2758 (1967).
6. W. G. Dauben, G. H. Beasly, M. D. Broadhurst, B. Muller, D. J. Peppard, P. Pesnelle, and C. Sutter, J. Amer. Chem. Soc., 96, 4724 (1974).
7. E. J. Corey and H. A. Kirst, J. Amer. Chem. Soc., 94, 667 (1972).
8. E. Yoshisato and T. Tsutsumi, Chem. Comm., 33 (1968).
9. M. F. Semmelhack, Ph.D. thesis, Harvard University, 1967, p. 74.
10. R. Noyori, Accts. Chem. Res., 12, 61 (1979).
11. R. Noyori, Y. Hayakawa, H. Takaya, S. Muria, R. Kobayashi, and N. Sonada, J. Amer. Chem. Soc., 100, 1759 (1978).
12. H. Takaya, S. Makino, Y. Hayakawa, and R. Noyori, J. Amer. Chem. Soc., 100, 1765 (1978).
13. H. Takaya, Y. Hayakawa, S. Makino, and R. Noyori, J. Amer. Chem. Soc., 100, 1778 (1978).
14. R. Noyori, T. Sato, and Y. Hayakawa, J. Amer. Chem. Soc., 100, 2561 (1978).
15. Y. Hayakawa, Y. Baba, S. Makino, and R. Noyori, J. Amer. Chem. Soc., 100, 1786 (1978).
16. R. Noyori, F. Shimizu, K. Fukuta, H. Takaya, and Y. Hayakawa, J. Amer. Chem. Soc., 99, 5196 (1977).
17. Y. Hayakawa, K. Yokoyama, and R. Noyori, Tetrahedron Lett., 4347 (1976).
18. Y. Hayakawa, K. Yokoyama, and R. Noyori, J. Amer. Chem. Soc., 100, 1791 (1978).
19. Y. Hayakawa, K. Yokoyama, and R. Noyori, J. Amer. Chem. Soc., 100, 1799 (1978).

20. R. F. Heck, Accts. Chem. Res., 2, 10 (1969); F. Guerrieri and G. P. Chiusoli, J. Organometal. Chem., 15, 209 (1968).
21. G. P. Chiusoli, Accts. Chem. Res., 6, 422 (1973).
22. M. P. Cooke, J. Amer. Chem. Soc., 92, 6080 (1970); J. P. Collman, Accts. Chem. Res., 8, 342 (1975).
23. M. P. Cooke and R. M. Parlman, J. Amer. Chem. Soc., 97, 6863 (1975).
24. J. P. Collman, R. G. Finke, J. N. Cawse, and J. I. Brauman, J. Amer. Chem. Soc., 99, 2515 (1977).
25. J. P. Collman, R. G. Finke, J. N. Cawse, and J. I. Brauman, J. Amer. Chem. Soc., 100, 4766 (1978).
26. R. F. Heck, in I. Wender and P. Pino, eds., Organic Synthesis via Metal Carbonyls, (Wiley-Interscience, 1968), Vol. I, 373-404.
27. S. K. Myeong, Y. Sawa, M. Ryang and S. Tsutsumi, Bull. Chem. Soc. Japan, 38, 330 (1965).
28. L. S. Hegedus, Ph.D. thesis, Harvard University, 1970, pp. 85-87.
29. E. J. Corey and L. S. Hegedus, J. Amer. Chem. Soc., 91, 4926 (1969).
30. Y. Sawa, I. Hashimoto, M. Ryang, and S. Tsutsumi, J. Org. Chem., 33, 2159 (1968).
31. E. J. Corey and L. S. Hegedus, J. Amer. Chem. Soc., 91, 1233 (1969).
32. E. J. Corey, H. A. Kirst and J. A. Katzenellenbogen, J. Amer. Chem. Soc., 92, 6314 (1970).
33. S. Fukuoka, M. Ryang and S. Tsutsumi, J. Org. Chem., 33, 2973 (1968).
34. S. Fukuoka, M. Ryang and S. Tsutsumi, J. Org. Chem., 36, 2721 (1971).
35. H. W. Sternberg, R. Markby, and I. Wender, J. Amer. Chem. Soc., 82, 3638 (1960).
36. T. F. Murray and J. R. Norton, J. Amer. Chem. Soc., 101, 4107 (1979).
37. For a review on this subject, see J. Tsuji, in I. Wender and P. Pino, eds., Organic Synthesis via Metal Carbonyls (Wiley-Interscience, 1977), Vol. II, 595-654.
38. J. Tsuji and K. Ohno, Tetrahedron Lett., 2173 (1967).
39. H. M. Walborsky and L. E. Allen, Tetrahedron Lett., 823 (1970).
40. H. M. Walborsky and L. E. Allen, J. Amer. Chem. Soc., 93, 5465 (1971).
41. J. W. Suggs, J. Amer. Chem. Soc., 100, 640 (1978).
42. D. E. Iley and B. Fraser-Reid, J. Amer. Chem. Soc., 97, 2563 (1975).
43. B. M. Trost and M. Preckel, J. Amer. Chem. Soc., 95, 7862 (1973).
44. M. Sergent, M. Mongrain, and P. Deslongchamps, Can. J. Chem., 50, 336 (1972).

45. P. D. Hobbs and P. D. Magnus, Chem. Comm., 856 (1974).

46. J. K. Stille and R. W. Fries, J. Amer. Chem. Soc., 96, 1514 (1974).

47. K. S. Y. Lau, Y. Becker, F. Huang, N. Baenziger, and J. K. Stille, J. Amer. Chem. Soc., 99, 5664 (1977).

48. J. K. Stille, F. Huang, and M. T. Regan, J. Amer. Chem. Soc., 96, 1518 (1974).

49. R. W. Fries and J. K. Stille, Synth. Inorg. Met. Org. Chem., 1, 295 (1971).

# 10

## Formation and Fragmentation of Metallacycles

Metallacycles are organotransition-metal complexes in which the transition metal is part of a carbocyclic ring. Metallacycles containing a single metal are known for ring sizes ranging from three to seven members. Since three-membered metallacycles can be regarded as valence bond or resonance structures of olefin or acetylene complexes, they will not be further considered here. The formation and fragmentation of metallacycles is an important class of organotransition-metal reactions, since these processes are involved in a number of significant catalytic transformations such as olefin metathesis and cyclooligomerization of both alkenes and alkynes. Metallacycles often behave differently from their open-chain analogues (i.e., cis-dialkylmetal complexes $L_nMR_2$), and it is these differences in reactivity that account for much of the interesting and useful chemistry of this class of compounds.

Metallacycles can be prepared in a number of ways (Figure 10.1). These include oxidative-addition of strained ring compounds to low-valent transition metals, migratory insertion of alkenes and alkynes into metal carbene complexes, metathetical reactions of metal dihalides with $\alpha,\omega$-dilithio compounds, cyclometallation reactions of metal alkyl complexes, and cyclodimerization of alkenes and alkynes. Additionally, metallacyclobutanes can be prepared by nucleophilic attack on the central carbon of a $\eta^3$-allylmetal complex. Many of these reactions which form metallacycles are reversible, accounting for a number of important catalytic processes such as olefin metathesis and the cyclooligomerization of alkenes and alkynes. Many of these catalytic reactions formally violate Woodward-Hoffman rules for orbital-symmetry correlation. Initially it was postulated that involvement of the transition metal, with its partially filled d orbitals,

Metallacyclobutanes

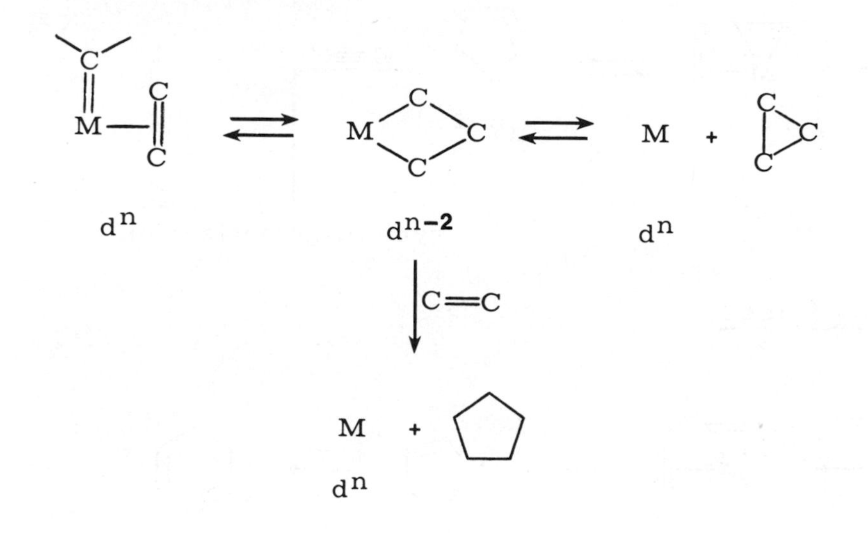


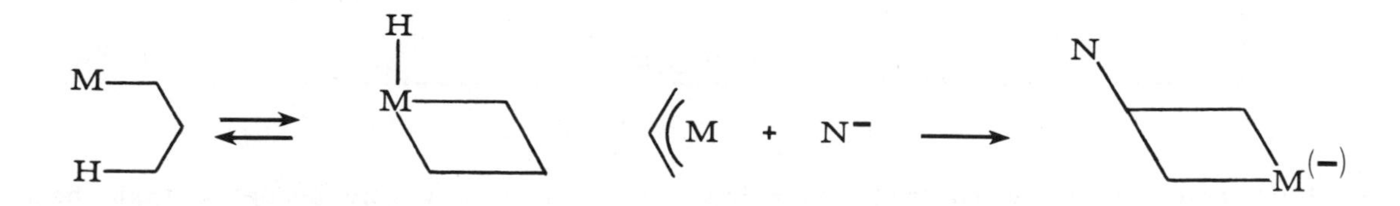


Metallacyclopentanes

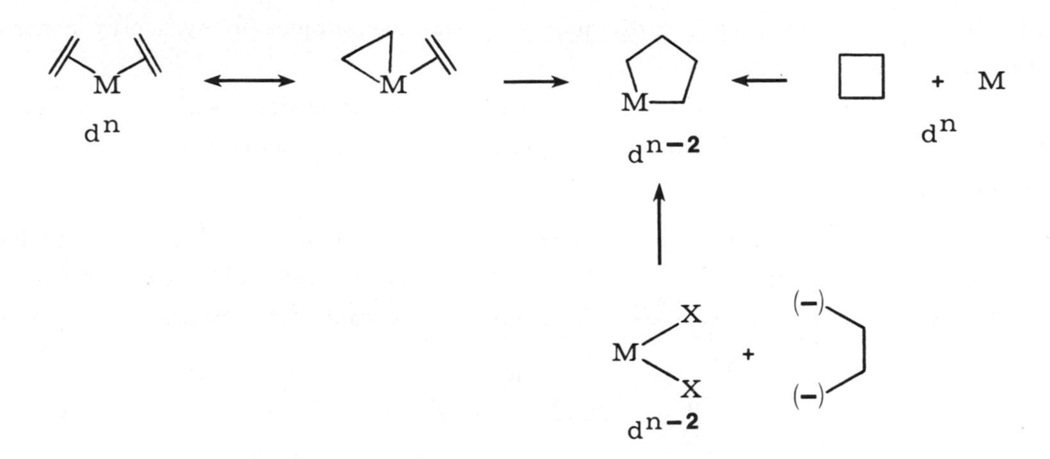


*Figure 10.1. Preparation of metallacycles.*

Metallacyclopentenes

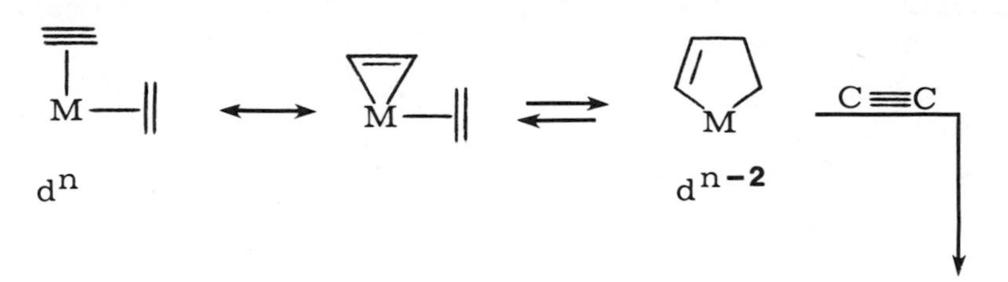


Metallacyclopentadienes

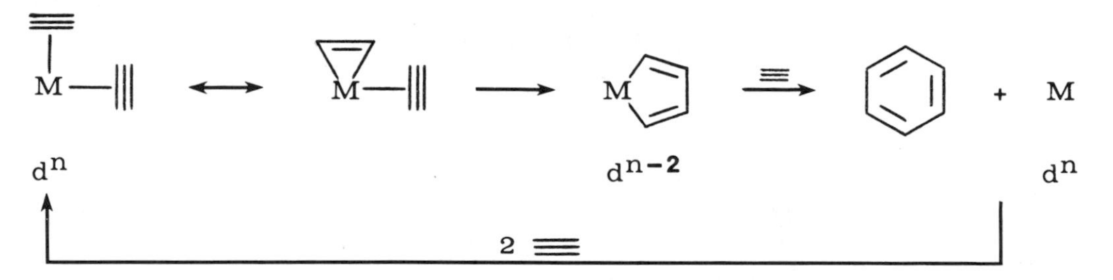


*Figure 10.1. Continued.*

obviated these symmetry restrictions. However, it is now widely accepted that these reactions are not concerted, but rather occur in steps with discrete intermediates, and hence the over-all reactions are not subject to the symmetry control originally postulated (vide infra). However, each individual step may be subject to symmetry control and can be treated accordingly.

In this chapter metallacycles will be considered by ring size, which provides a useful (if somewhat arbitrary) format for the presentation of the material.

## 10.1. METALLACYCLOBUTANES.

Perhaps the most interesting and useful chemistry which is thought to involve metallacyclobutanes as reactive intermediates is the "olefin-metathesis" reaction (10.1) [1,2,3]. This reaction equilibrates alkylidene units of olefins, for example, converting

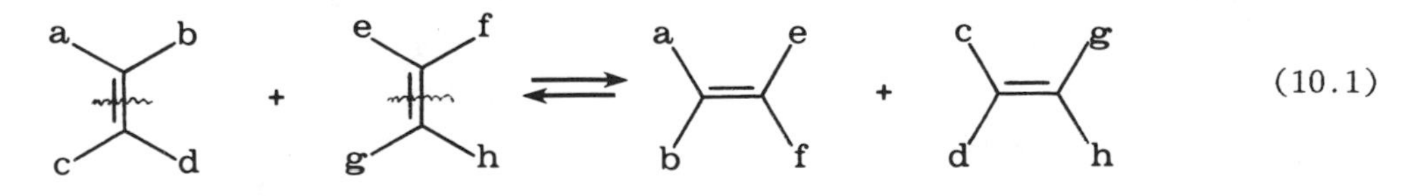


(10.1)

propene to a mixture of ethene and butenes (10.2). With cyclic olefins, metathesis produces polyalkenemers, often of high molecular weight and stereoregularity (all cis),

$$CH_3CH{=}CH_2 + CH_3CH{=}CH_2 \rightleftharpoons CH_2{=}CH_2 + CH_3CH{=}CHCH_3 \quad (10.2)$$

as in Equation 10.3. A variety of olefins react in this manner, although those

(10.3)

substituted with functional groups other than alkyl usually do not. This olefin metathesis reaction is characterized by low activation barriers, fast rates, thermoneutrality, and approximate first-order reaction rates in both metal catalyst and olefin.

The most common olefin metathesis catalysts are produced from the reaction of tungsten or molybdenum halides with reducing or alkylating agents, such as aluminum hydride. Both homogeneous and heterogeneous catalyst systems are known, but the exact nature of the catalytically active species is rarely understood. Most catalyst systems are quite sensitive to oxygen, with small amounts activating the catalyst and larger amounts poisoning it. This feature has made accurate kinetic studies difficult. The mechanism of this olefin metathesis reaction has been hotly debated, with Grubbs, Katz, Casey, and others being major contributors.

Originally, a concerted four-center mechanism (10.4) was proposed for olefin metathesis [3]. Calculations showed that this concerted process should have a low

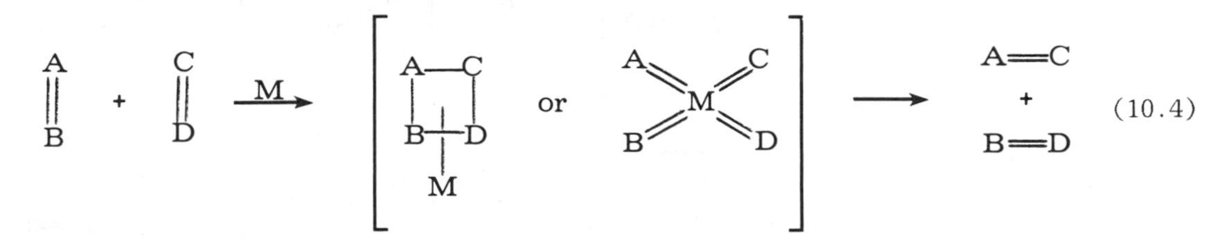

(10.4)

activation barrier, since interaction of the olefins with metal d orbitals removed the symmetry restrictions from this formally symmetry-forbidden reaction. However, this mechanism suffers from several deficiencies, including the absence of precedence for the proposed intermediate complexes, the fact that cyclobutanes are neither produced nor reactive under metathesis conditions, and the requirement of a pairwise exchange of alkylidene units (not observed).

Another mechanism (10.5) involves metallacyclopentanes, formed by oxidative addition of olefins. Although complexes of this type are known, this mechanism also

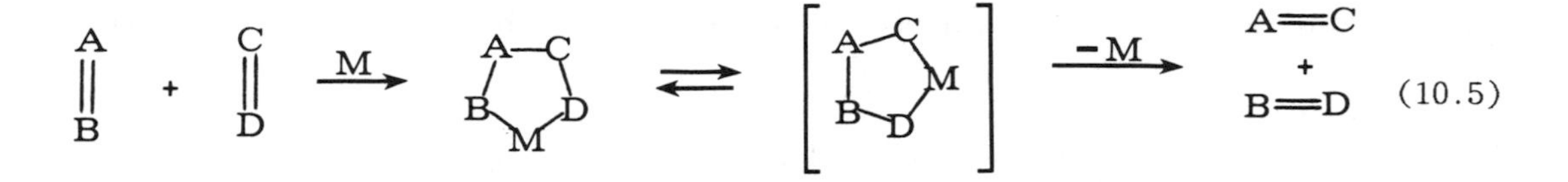


results in pairwise exchange of alkylidene units. Since pairwise exchange is not observed in metathesis, this mechanism can also be dismissed.

The currently accepted mechanism for olefin metathesis (first proposed by Chauvin and Herisson [4]) is a one-carbene chain mechanism, proceeding through metallacyclobutane intermediates (Figure 10.2). This reaction differs from the two above in

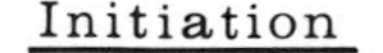


M $\xrightarrow{A{=}B}$ M=B + (M=A) (or M—$CH_2$R ⟶ M=CHR, with H on M)

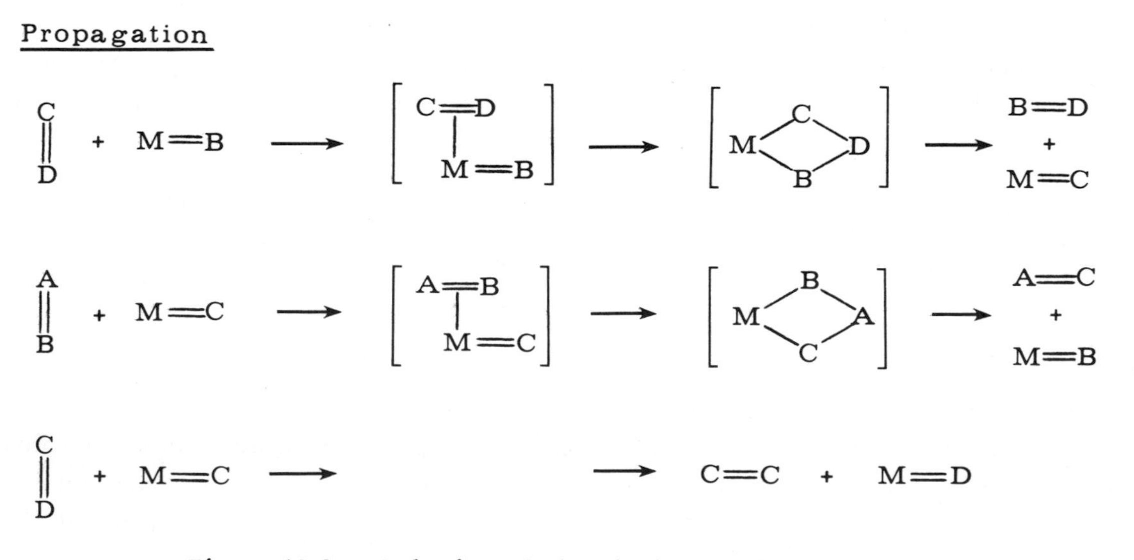


*Figure 10.2. Mechanism of the olefin metathesis reaction.*

predicting that the initial products of metathesis of two olefins would be a statistical mixture of all possible alkylidene units, rather than specific pairwise exchange. This mechanism now rests on a firm basis, since nonpairwise exchange has been demonstrated, and since models for each step of this process have been discovered. (It is interesting that when this mechanism was originally suggested, it found little acceptance, since the known chemistry of metal carbenes at that time was inconsistent with such a species being an intermediate in olefin metathesis.) The olefin-carbene metallacyclobutane mechanism requires a reversible, two-electron, oxidation-reduction of the central metal. The association of reversible redox reactions with transition-metal catalysis is a recurring theme throughout organotransition-metal chemistry.

Nonpairwise exchange has been demonstrated in several ways, primarily by Grubbs [1] and by Katz [2]. One of the simplest experiments to understand involves the metathesis of a mixture of 1,7-octadiene and 1,7-octadiene-1,1,8,8-$d_4$ [1] (Equation 10.6). It has previously been shown that 1,7-octadiene metathesizes to cyclohexene and

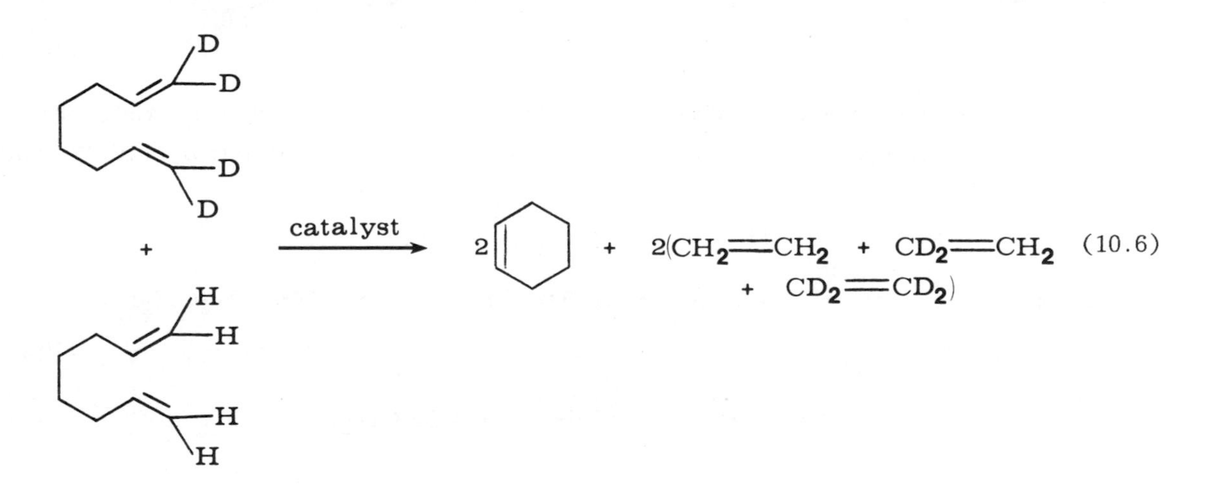


ethylene in nearly quantitative yield, and that the cyclohexene does not undergo further metathesis. Furthermore, neither the product ethylenes nor the octadienes are scrambled in the reaction. Metathesis of a 1:1 mixture of labeled and nonlabeled octadienes by a pairwise exchange would produce a 1:1.6:1 mixture of ethylene $d_0$/ethylene $d_2$/ethylene $d_4$, whereas a nonpairwise exchange would give a 1:2:1 ratio. In fact, a 1:2:1 ratio is observed experimentally, confirming nonpairwise exchange. A number of other experimental tests confirm the nonpairwise exchange in olefin metathesis [1,2].

The metathesis reaction described in Figure 10.2 is a chain reaction, and has both initiation and propagation steps. The initiation step requires the generation of a metal carbene complex. Recall that most metathesis catalysts are generated by treating metal halides with alkylating agents. Thus metal alkyls are likely precursors to metal carbenes by an α-elimination reaction. This type of reaction has been proposed by Muetterties [5] in the reaction of $WCl_6$ with dimethylzinc and observed by Schrock [6] in the formation of stable tantalum alkylidene complexes (Equations 10.7 and 10.8). For catalysts

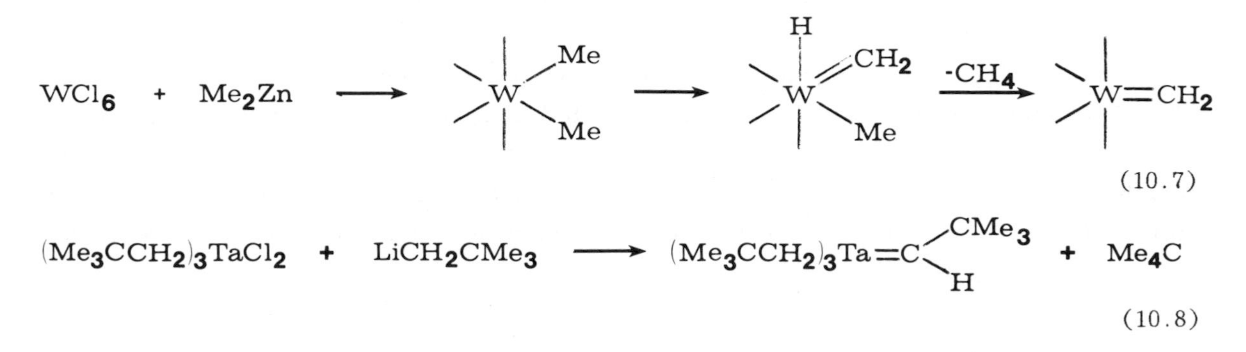


not involving alkylating agents, a variety of approaches to metal carbenes exist. These include metal hydride addition to an olefin, followed by α-elimination (10.9), and

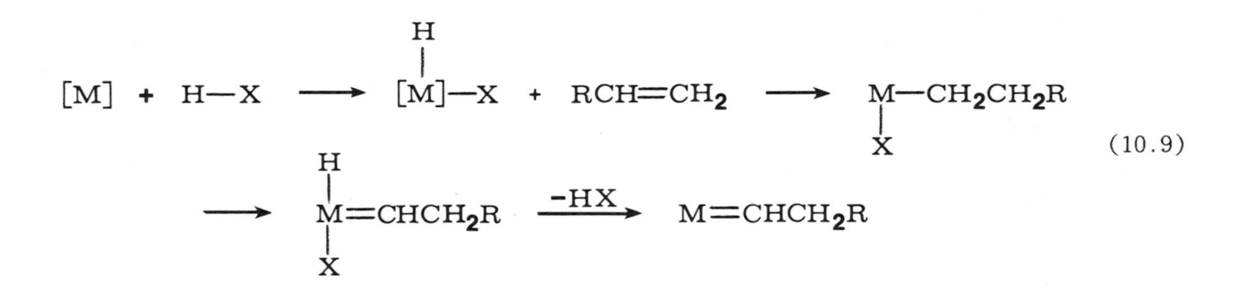


formation of a metallacyclopentane followed by α-elimination and reductive-elimination (10.10). Thus, many reasonable chain-initiation processes are available.

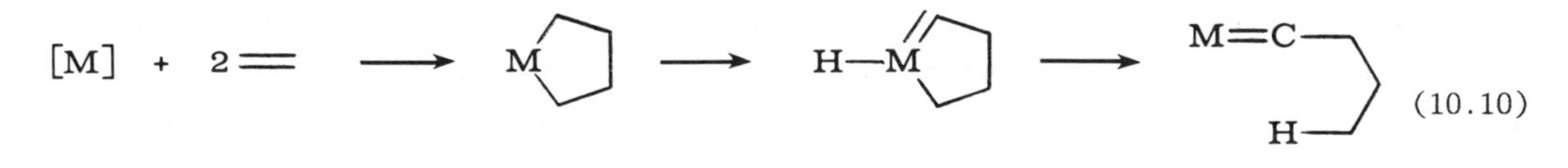

The chain-propagating steps involve exchange of alkylidene groups between the olefin and the metal carbene catalyst. This reaction (10.11) has been observed independently [7]. Since the $(CO)_5W{=}CPh_2$ complex also functions as a metathesis catalyst

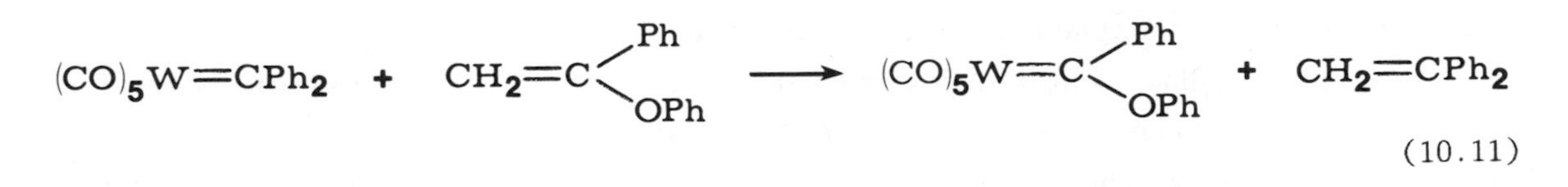


(10.11)

for alkenes [8] and a polymerization catalyst for alkynes [9], these chain-carrying steps rest on a firm precedent in the literature.

Very recently Schrock [10] has prepared a series of active niobium, tantalum, and tungsten metathesis catalysts, which not only confirm the proposed mechanism but also shed light on some of the features of the "black box" catalysts effective in metathesis. For example, consider the reaction (10.12) between the W(VI) d(0) complex

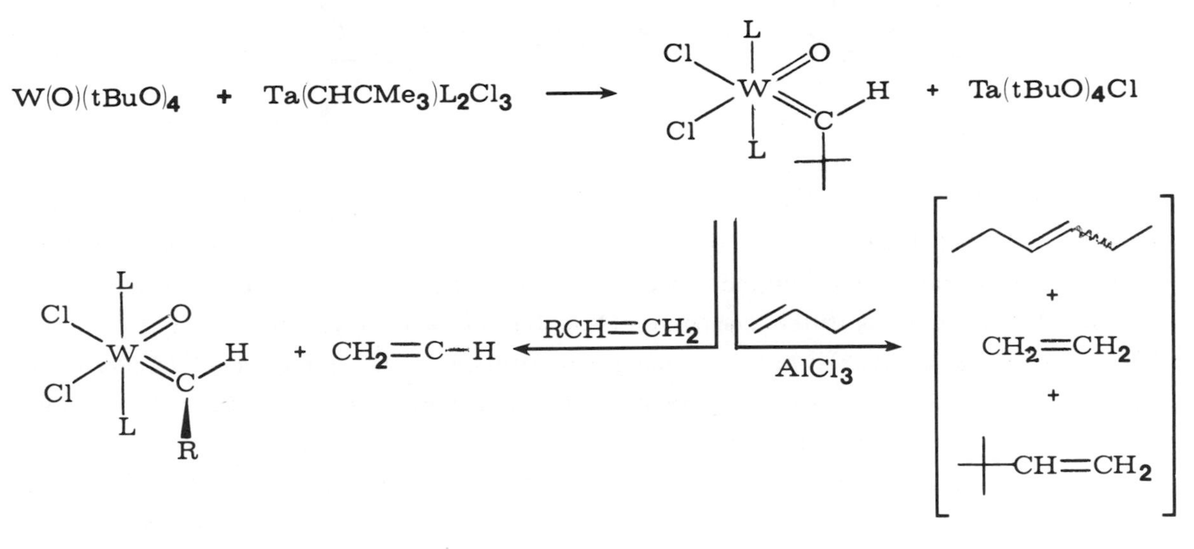


(10.12)

$W(O)(\underline{t}\text{-}BuO)_4$ and the tantalum carbene complex $Ta(CHCMe_3)(PEt_3)_2Cl_3$ to form the tungsten-carbene complex. This complex reacts slowly at 25° (1 day) with ethene, propene, or styrene to exchange alkylidene groups, much as in Equation 10.11. In the presence of traces of $AlCl_3$, this complex catalyzes the metathesis of 1-butene to 3-hexenes at 25°C. The results of these experiments and others suggest that Group VI d(0) carbene complexes are probably the most common type of metathesis catalyst; that at least one oxo ligand is required to keep the Group VI metal in a d(0) state (hence the accelerating effects of small amounts of oxygen or oxygenated solvents on metathesis catalysts prepared from Group VI halides and alkylaluminum reducing agents); and that aluminum halide greatly accelerates the metathesis reaction (hence the activity of catalysts prepared using alkylaluminum reducing agents).

Although metallacyclobutane intermediates are proposed in the metathesis reaction, they have never been observed with metals which are typical metathesis catalysts. Their involvement in the metathesis reaction is inferred in part from stereochemical studies beyond the scope of this discussion [2].

Stable metallacyclobutanes <u>are</u> known for a number of Group VIII metals. Their reactions and structural features may have some bearing on the olefin metathesis reaction. Some of the best-characterized complexes have been prepared by the reaction (10.13) of Pt(II) complexes with cyclopropanes (oxidative addition). These Pt(IV)

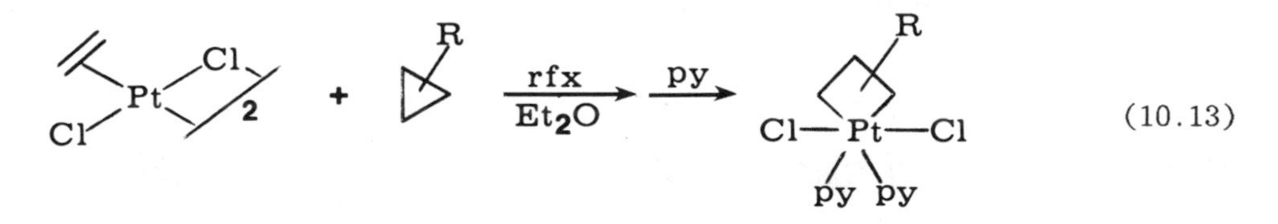


(10.13)

metallacycles are quite stable, and have been characterized by X-ray crystallography, primarily by Ibers [11]. These studies show that the cyclopropane ring has been fully opened, and that little if any residual bonding between $C_1$ and $C_3$ exists. That is, these are true metallacyclobutane complexes, not metal-coordinated cyclopropane complexes. In the unsubstituted compound (R = H) the metallacyclobutane ring is almost flat (dihedral angle 168°), but increasing substitution leads to increased puckering of the ring (to 160°-150°). Cyclobutane itself has a dihedral angle of 150°. The degree of puckering in the substituted metallacyclobutanes and, as a consequence, the amount of interaction between substituents on $C_1$ and $C_3$ of the ring is central to many arguments concerning the stereochemistry of olefin metathesis. In these arguments it is

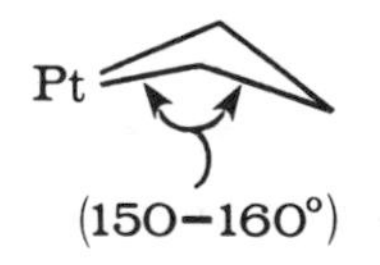


assumed that the structural features of these stable platinacyclobutanes are similar to those of the presumed molybdenacyclobutanes involved in metathesis.

These stable platinacyclobutanes do not engage in olefin metathesis reactions. Thermal degradation at relatively high temperatures produces cyclopropane and propene, from competitive reductive elimination and β-hydride elimination [12] (10.14). The

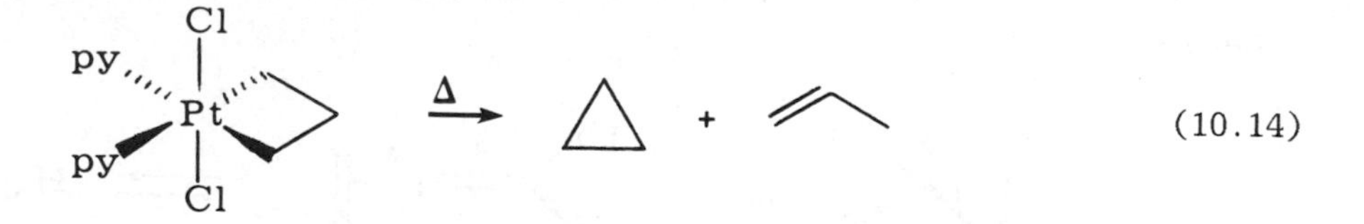


phenyl-substituted complex undergoes a remarkably facile rearrangement (10.15). Crossover experiments show that this rearrangement does not occur by elimination of

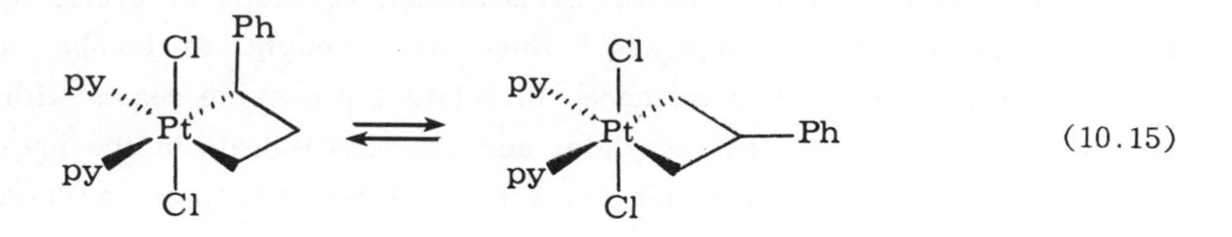


phenylcyclopropane followed by readdition, nor by fragmentation to free styrene and a $Pt{=}CH_2$ complex. The observation that the rearrangement goes with clean retention of stereochemistry eliminates mechanisms involving metal carbene-alkene complexes as intermediates, since some loss of stereochemistry would be expected from these intermediates. Hence Puddephatt [13] proposes a concerted mechanism (10.16) to account

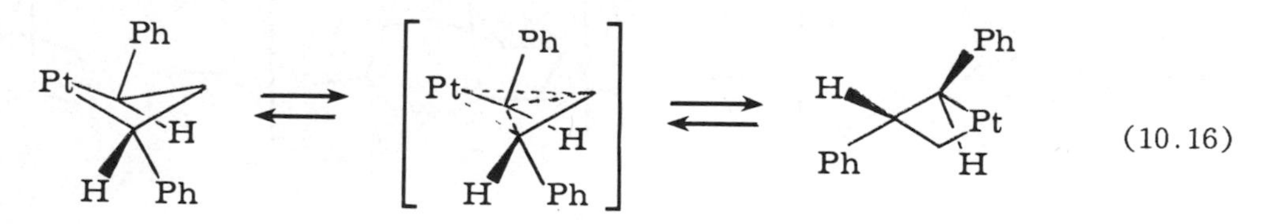


for this rearrangement. Casey [14] offers two closely related views of this rearrangement. One involves formation of an edge-metallated cyclopropane which undergoes an edge-to-edge isomerization (10.17). The other involves a disrotatory ring

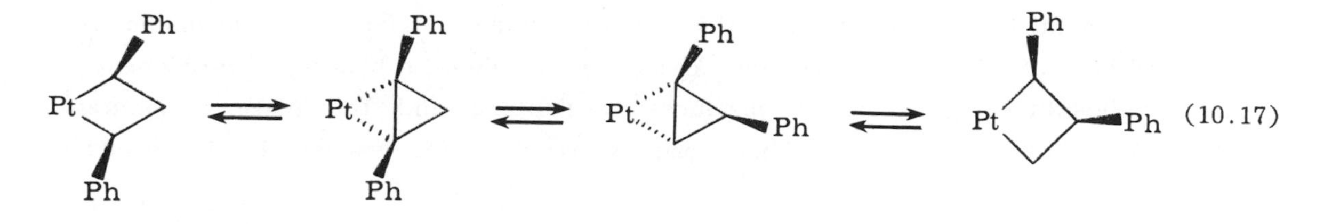

opening to give an incipient metal carbene-olefin complex in which the plane of the alkene ligand and the plane of the carbene ligand become perpendicular. Concerted rotation can then lead back to starting material or isomerized product with retention of stereochemistry (10.18). All three explanations are consistent with the available experimental data.

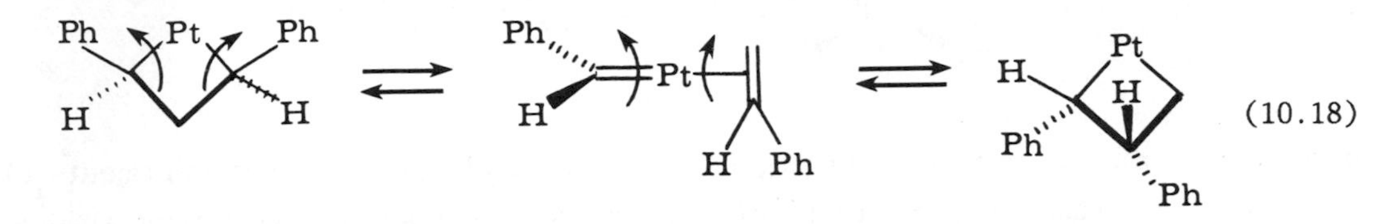


Noyori studied several useful cycloaddition reactions of cyclopropanes and olefins catalyzed by nickel(0) complexes, which are thought to involve metallacyclobutane intermediates [15,16]. For instance, bicyclo[2.1.0]pentane reacts with activated olefins such as acrylonitrile, methyl acrylate, and dimethyl maleate in the presence of nickel(0) catalysts to produce norbornanes ([2.2.1]bicycloheptanes), as in Equation 10.19. The

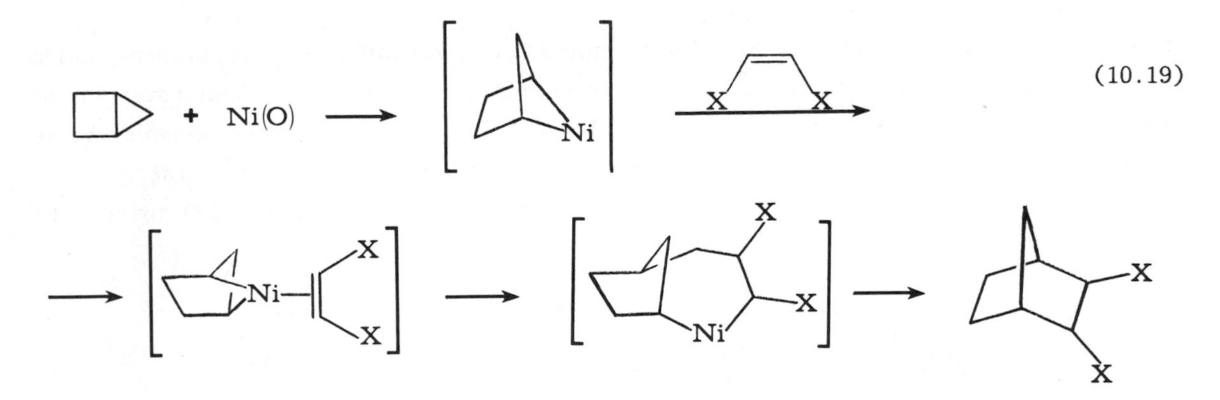


reaction is thought to involve oxidative addition of the strained cyclopropane to nickel(0), giving the metallacyclobutane. Insertion of the olefin into the carbon-nickel bond followed by reductive-elimination completes the process. Although this mechanism is quite reasonable, little experimental proof is available.

Stable metallacyclobutanes of the entire nickel triad are available by cyclometallation of bis-neopentylmetal complexes. Thus, heating bis-neopentyl-bis(triethylphosphine) platinum(II) produces neopentane and the platinacyclobutane complex [17] (Equation 10.20). This complex reacts with excess $I_2$ to produce

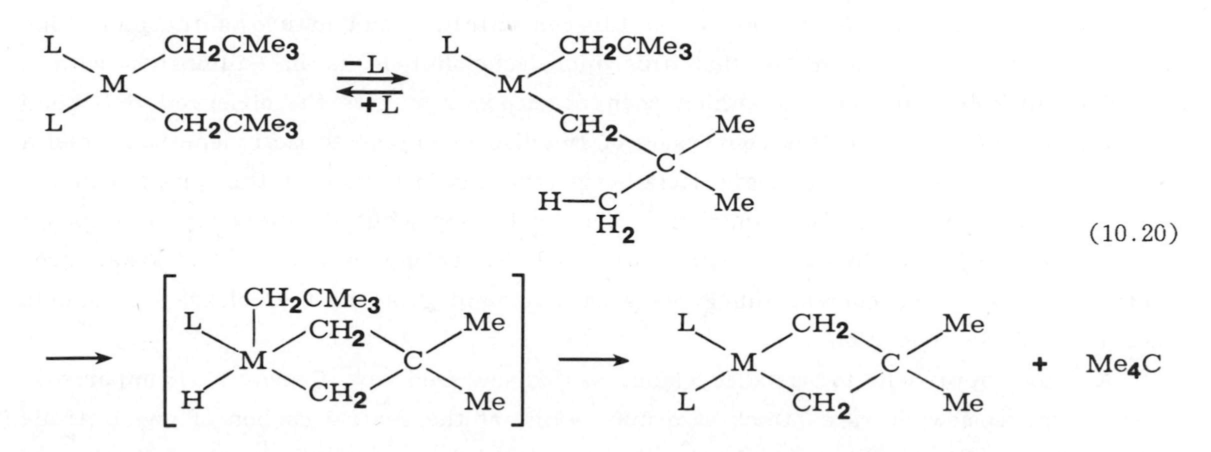


(10.20)

1,1-dimethylcyclopropane, and with DCl to produce predominantly $(DCH_2)_2C(CH_3)_2$, confirming its structure. The corresponding metallacyclobutanes of nickel and palladium form in a similar fashion, and undergo identical reactions with oxidizing agents and DCl [18]. Thermal decomposition of the nickel complex is particularly interesting, in that significant carbon-carbon bond cleavage is observed (Figure 10.3). The relative

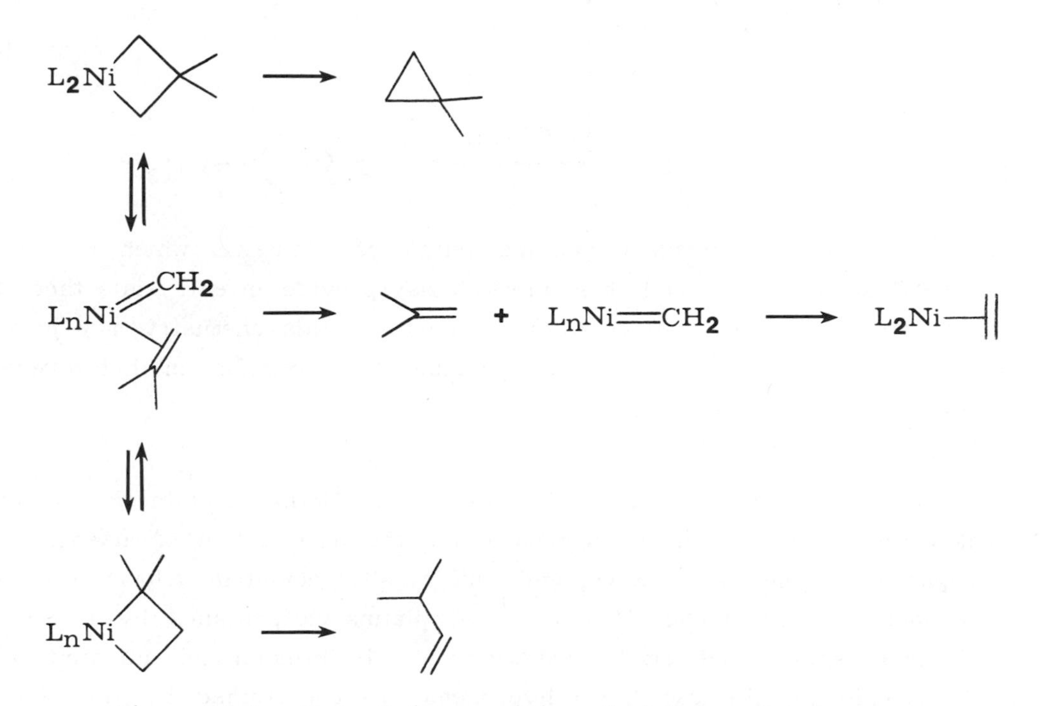


*Figure 10.3. Thermal decomposition of nickelacyclobutanes.*

amount of each product is dependent on the concentration and nature of the phosphine ligand. What is significant is that this nickelacyclobutane is in equilibrium with a methylene-nickel-olefin complex, which then decomposes to give the observed C-C bond cleavage products. Since this conversion of metallacyclobutane to methylene-metal-olefin is proposed as a key step in olefin metathesis, the demonstration of this process in the nickel system is particularly significant. The palladiacyclobutane thermally decomposes to give primarily dimethylcyclopropane, with only minor amounts of C-C cleavage products observed. The corresponding platinum compound gives no C-C cleavage products upon thermal decomposition.

A final approach to metallacyclobutanes discovered by Green [19] apparently involves the somewhat rare attack of a nucleophile at the central carbon of an electron-rich $\pi$-allylmetal complex. Stable tungstenacyclobutanes result from the reaction of methyllithium or $NaBD_4$ with a cationic $\eta^3$-allyltungsten complex (10.21). This is of

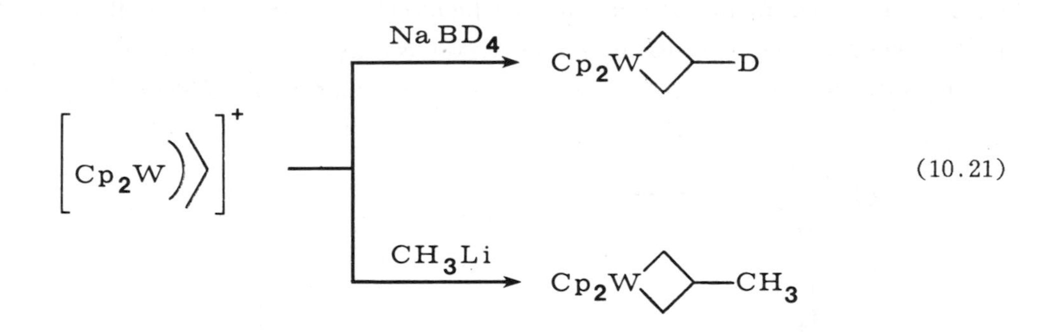


(10.21)

some importance, because metallacyclobutane complexes of metals which are efficient metathesis catalysts are rare, and this approach may provide an entry into these types of complexes for use in mechanistic studies. Further, this chemistry may provide a route for the formation of complexes responsible for initiation in heterogeneously catalyzed olefin metathesis reactions.

## 10.2. METALLACYCLOPENTANES.

While metallacyclobutanes are involved in olefin metathesis reactions, metallacyclopentanes are of central importance for the dimerization of alkenes. They can be prepared in a number of ways, and, like metallacyclobutanes, have some unique chemical properties. Whitesides [20] prepared platinacyclopentanes by the reaction (10.22) of bis-phosphine platinum(II) halides with 1,4-dilithiobutane and studied their thermal decomposition. Because the $\beta$-hydrogens are constrained by the ring from

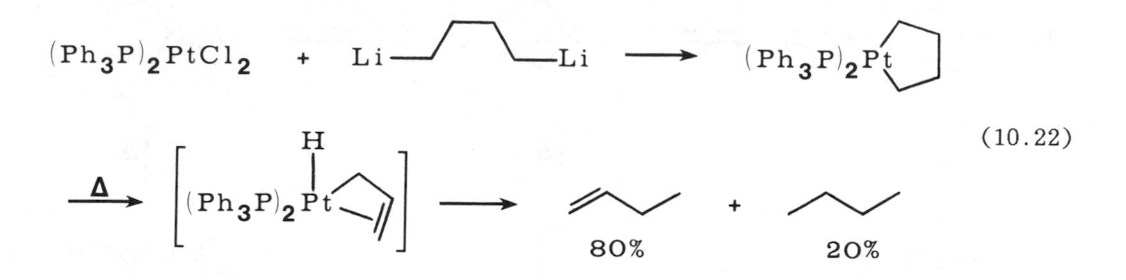


(10.22)

achieving a Pt-C-C-H dihedral angle of 0° (required for β-elimination), they are quite stable compared to acyclic dialkyl species. For example, the platinacyclopentane decomposes $10^4$ more slowly than the corresponding di-n-butyl compound, giving mixtures of 1- and 2-butenes. The six-membered platinacycle shows similar stability, but the seven-membered system does not. The course of this thermal reaction is quite sensitive to the nature of the phosphine, however (a feature particularly pronounced with nickelacyclopentanes; see below). For example, the corresponding bis-tri-n-butylphosphine complex decomposes in methylene chloride to give primarily (60%) cyclobutane, by a reductive-elimination, with only small amounts of β-hydride elimination products (butene) being formed. However, these results are not absolutely clear-cut. The formation of cyclobutane has a strong dependence on solvent, and may, in fact, involve oxidation of the platinacyclopentane to Pt(IV) by methylene chloride, followed by reductive elimination of cyclobutane. Regardless, this is unique to the metallacycle, since the corresponding diethylplatinum complex decomposes in methylene chloride to produce only ethene and ethane, but no butane.

The related metallacyclopentane complex, $(diphos)Pd(CH_2)_4$, has been made and is similarly more stable than its acyclic analog [21]. The metallacycle thermally decomposes to give a mixture of 1- and 2-butenes, reacts with bromine to give 1,4-dibromobutane, and reacts with acid to give mixtures of butane and butenes.

Grubbs [22] has extensively studied the metallacyclopentanes arising from the reaction of 1,4-dilithiobutane with $L_2NiCl_2$ complexes. The mechanism (10.23) of thermal decomposition of these complexes is strongly dependent on the coordination state of the metal. The monophosphine complex is the major species formed when the phosphine is the bulky tricyclohexylphosphine. This species decomposes exclusively to linear butenes, presumably by a β-hydride elimination mechanism. This is reasonable because a vacant coordination site is required for β-elimination, and the three-coordinate

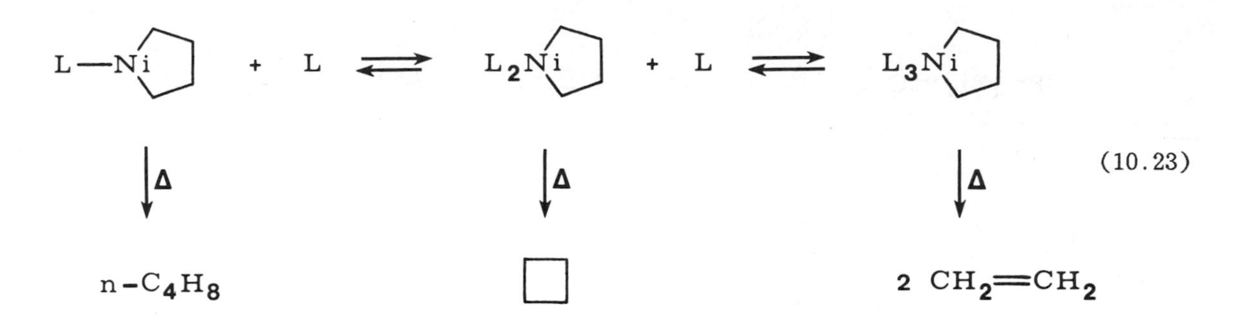


(10.23)

complex is quite unsaturated. In contrast, the bis-phosphine complex decomposes (by a reductive-elimination mechanism) to yield cyclobutane as the main product. Finally, the tris-phosphine complex decomposes to ethylene, the reverse of a cyclodimerization of ethylene [23]. All three complexes produce predominately cyclobutane when oxidatively decomposed.

The formation of ethylene from the tris-phosphine complex implies that this complex is in equilibrium with a bis-olefin complex. Grubbs showed this to be the case [24] by preparing the 2,2,5,5-tetradeuteronickelacyclopentane and observing the scrambling of the deuterated positions (10.24). The scrambled products were characterized by bromi-

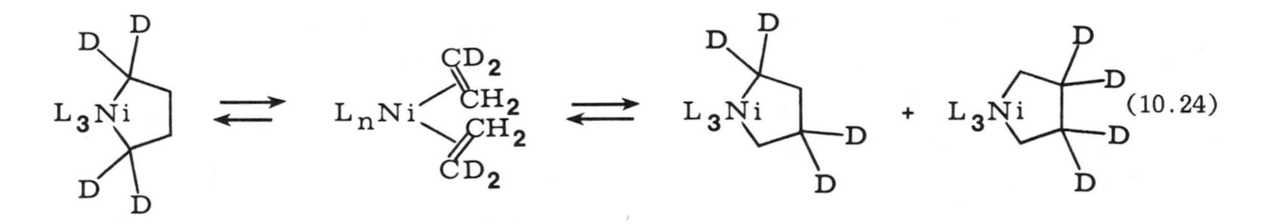


(10.24)

native cleavage to the deuterated 1,4-dibromobutanes. In the bis-olefin complex, the olefins remain tightly coordinated to the metal, since equilibration of a mixture of nickelacyclopentane and tetradeuteronickelacyclopentane produces only bromobutanes $d_4$ and $d_0$ upon cleavage. The bis-phosphine nickelacyclopentane, which decomposes to cyclobutane upon thermolysis, does not equilibrate as in Equation 10.24, and is thus unlikely to be in equilibrium with a bis-olefin complex. There is one awkward feature in this study. Should the tris-phosphine complex directly form the bis-olefin complex $L_3Ni(CH_2CH_2)_2$, a 20-electron system would be produced. Since this is not likely, a prior loss of phosphine is invoked, and is supported by the observation that excess phosphine does indeed slow the rate of deuterium scrambling. However, loss of phosphine from the tris-phosphine complex must produce a bis-phosphine complex different from the square planar $L_2Ni(CH_2)_4$ in Equation 10.23, since this complex does not undergo deuterium scrambling and decomposes to cyclobutane, not ethylene. The details of this issue remain unresolved.

Despite this problem, the significance of the equilibrium in Equation 10.24 is that it suggests that appropriate metallacyclopentanes (or complexes derived from them) should be able to dimerize olefins. This is indeed the case. The tris-phosphinenickel-acyclopentane reacts with 1,7-octadiene to give a high yield of the trans-2-nickelahydrindane. This can be converted to a number of useful organic compounds by standard methods [25], as in Equation 10.25. The same initial complex catalyzes the dimerization

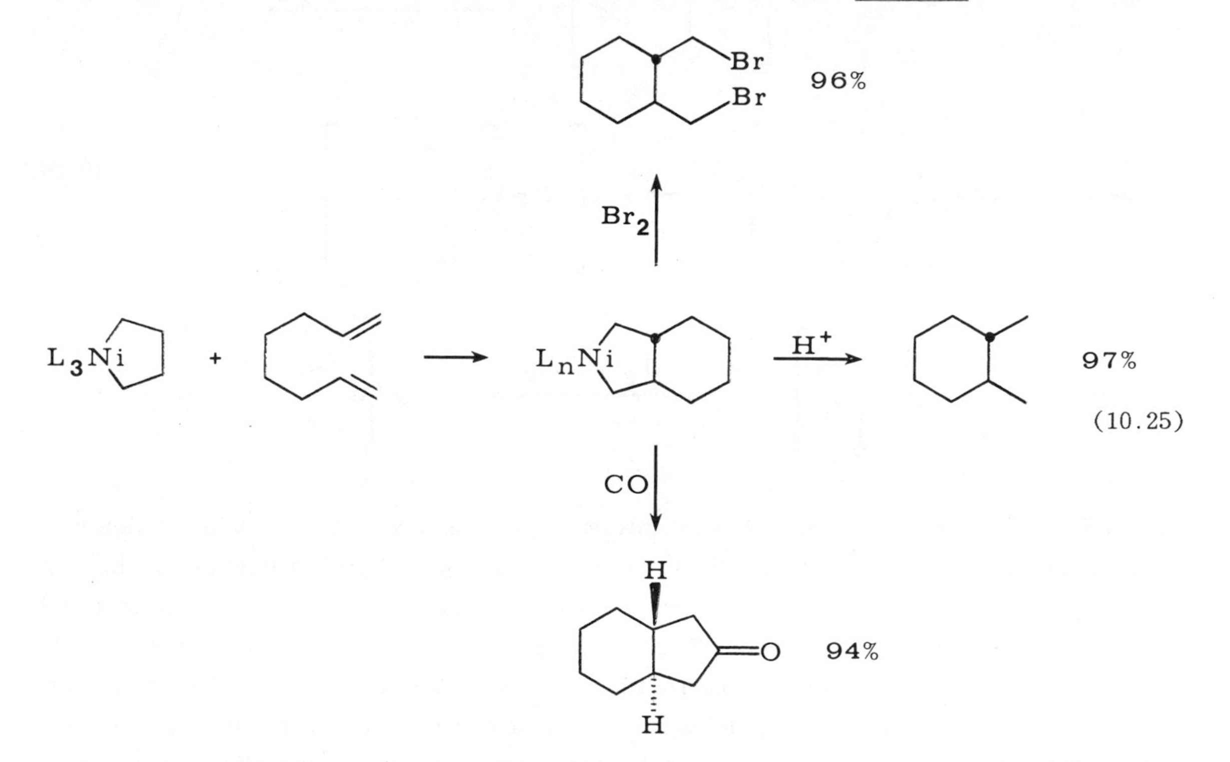


of ethylene to mixtures of cyclobutane and linear butenes. The product distribution is quite dependent on catalyst concentration and temperature, and only modest (23 catalyst turnovers) activity is observed [26].

This type of chemistry is not restricted to the Group VIII metals. Metallacyclopentane complexes of titanium [27,28] and zirconium [29] also react with olefins to form new metallacyclopentanes incorporating the olefins. Tantalum complexes, studied extensively by Schrock, account for some of the most interesting chemistry of this type. $CpTa(CH_2CMe_3)Cl_2$ reacts with olefins to form a tantalacyclopentane and neopentylethylene in excellent yield. The reaction is thought to proceed via the carbene and metallacyclobutane complex which ultimately β-hydride eliminates and reductively eliminates to produce the olefin complex. This reacts rapidly with external olefin to form the metallacyclopentane [30], as in Equation 10.26. With the simple

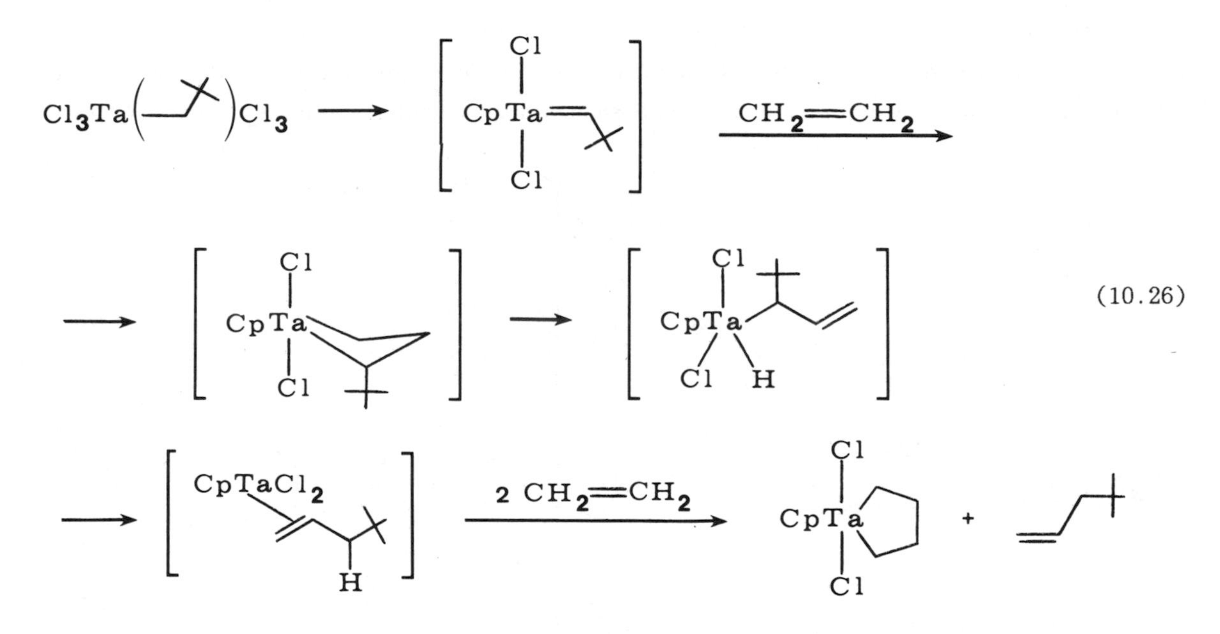


cyclopentadienyl ligand, none of the intermediate complexes are isolable. However, using a pentamethylcyclopentadienyl (Cp') ligand instead, olefin complexes of the type $Cp'TaCl_2$(olefin) can be isolated. These react reversibly with olefins to form metallacyclopentanes [31] as in Equation 10.27. These complexes catalytically dimerize terminal olefins to mixtures of linear dimers when exposed to excess olefin. The mechanism of decomposition of the metallacycle to the product olefins appears at first glance to be a straightforward sequence of β-hydride elimination and reductive-elimination. However, a labeling study showed the decomposition to proceed through a metallacyclobutane (Figure 10.4). This further demonstrates the complexities encountered in metallacyclic chemistry.

A rather different approach to metallacyclopentanes involves the oxidative-addition of metals to strained cyclobutanes [32]. The best-documented case of this type of

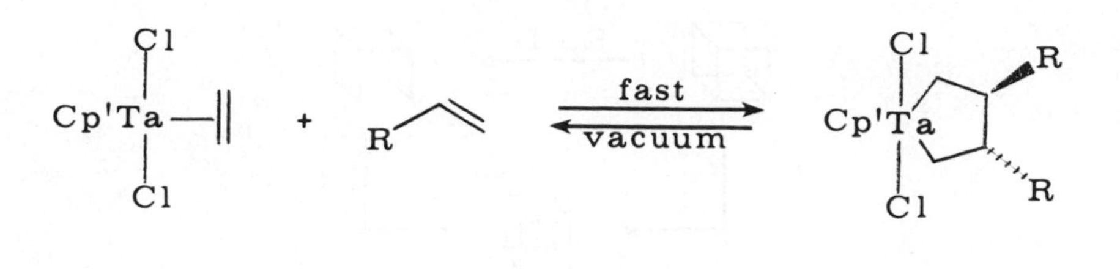


(10.27)

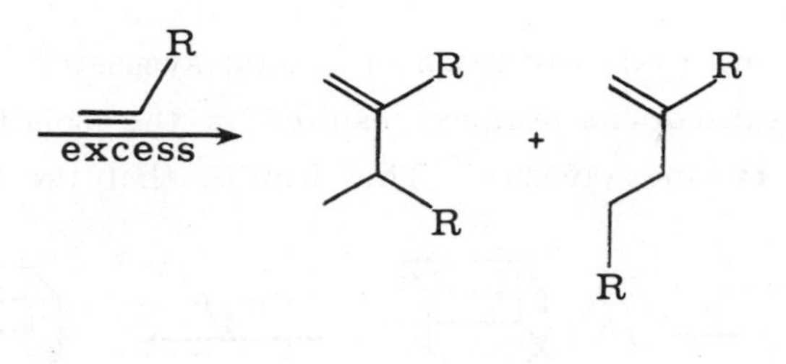


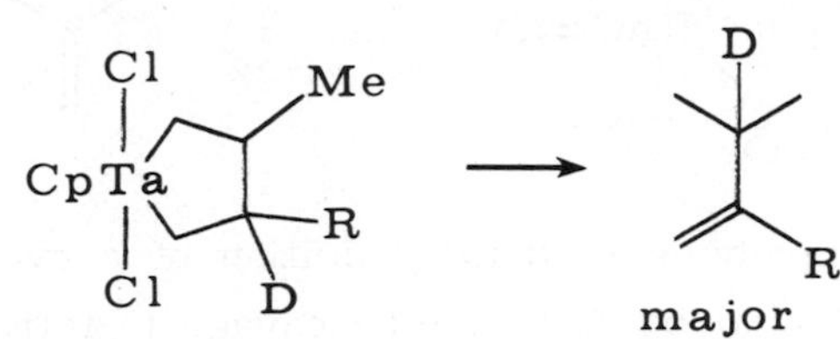


<u>via</u>

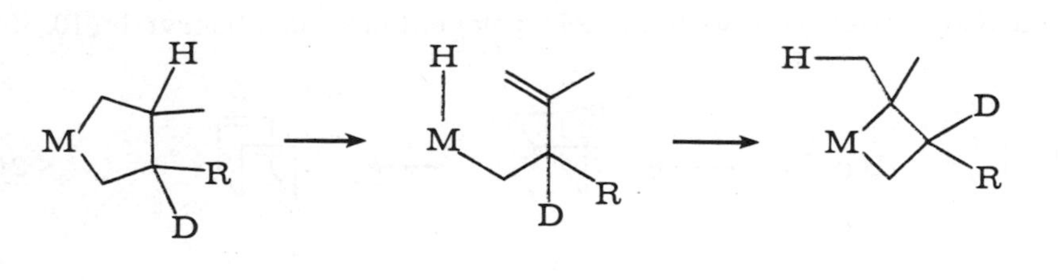


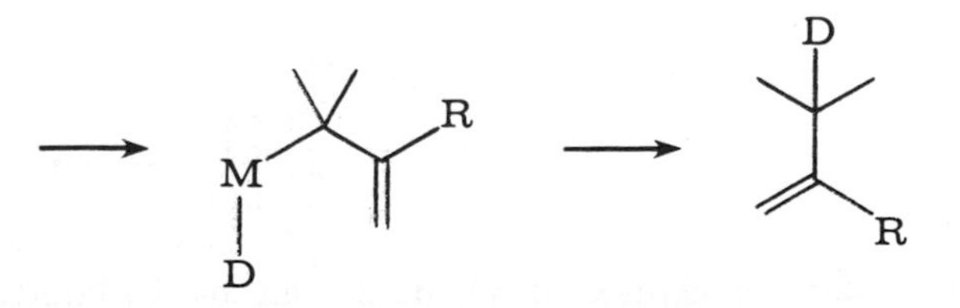


*Figure 10.4. Decomposition of tantalacyclopentanes.*

reaction is the Rh(I) catalyzed rearrangement (as in 10.28) of cubane to <u>syn</u>-tricyclooctadiene studied by Halpern and Eaton. Originally this reaction was claimed

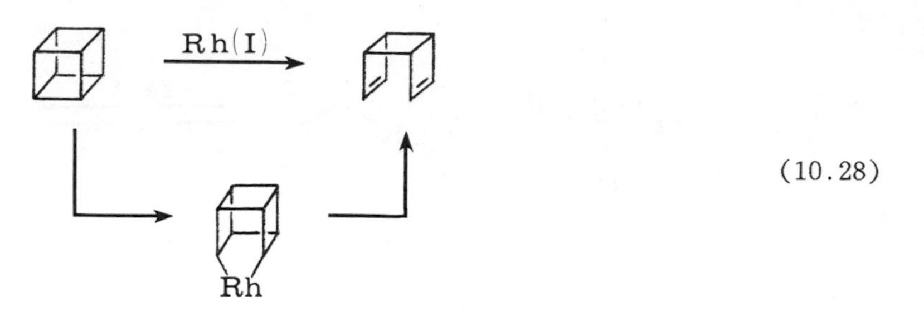


(10.28)

to go by a concerted mechanism, with circumvention of orbital symmetry restrictions by the transition metal. However, subsequent studies resulted in the isolation of a stable Rh(III) oxidative adduct of the cubane (10.29). This implies that the rearrangement

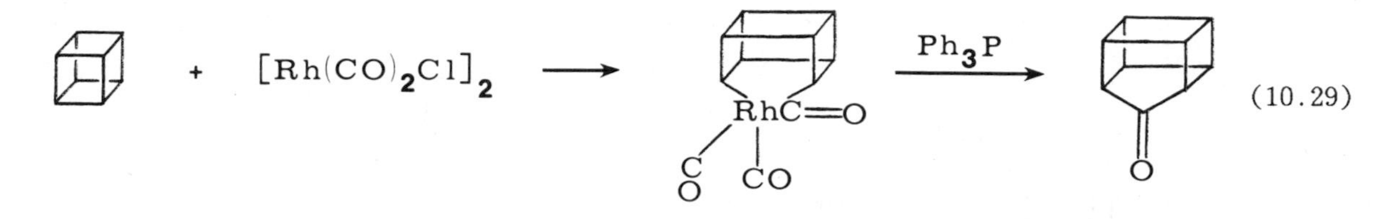


(10.29)

follows a similar nonconcerted pathway involving oxidative addition of a cyclobutane to Rh(I) to form a metallacyclopentane, followed by C-C bond cleavage to form the diene. A number of other Rh(I) catalyzed rearrangements of strained rings are thought to go by this process. In some cases the same competition between β-hydride elimination and C-C bond cleavage observed with nickelacyclopentanes is observed (10.30). In ideal

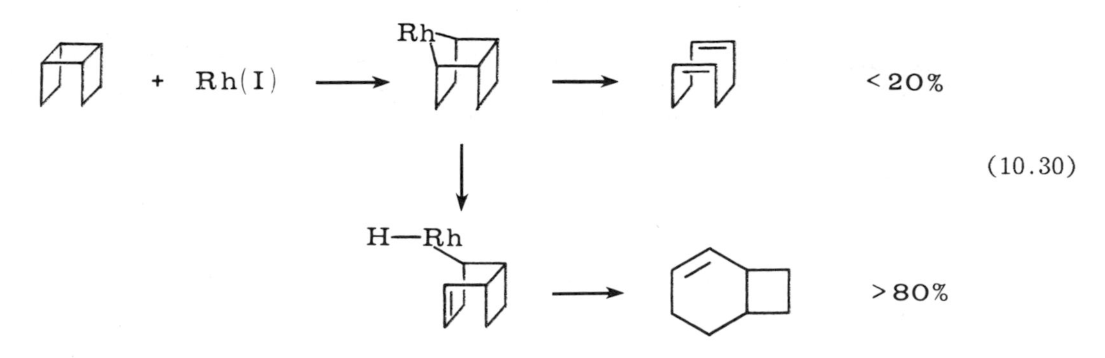


(10.30)

cases, the diolefin complex of rhodium can be isolated, and the ring opening reversed by adding excess olefin [33] (Equation 10.31).

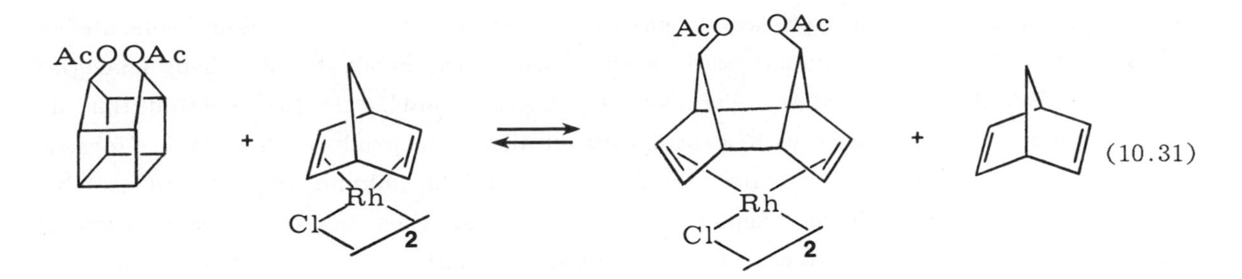


## 10.3. METALLACYCLOPENTADIENES AND METALLACYCLOPENTENES.

In much the same fashion that metallacyclopentanes are formed from two olefins, metallacyclopentadienes are formed by the reaction of a variety of metals with alkynes. Since this reaction is central to the synthetically important metal-catalyzed cyclotrimerization of alkynes to aromatics discussed in Chapter 13, it will be considered in some detail.

Although a number of transition metals catalyze the cyclotrimerization of alkynes to aromatics, $CpCoL_2$ complexes are among the most efficient. The proposed mechanism is described in Figure 10.5 and is thought to involve coordination of two molecules of

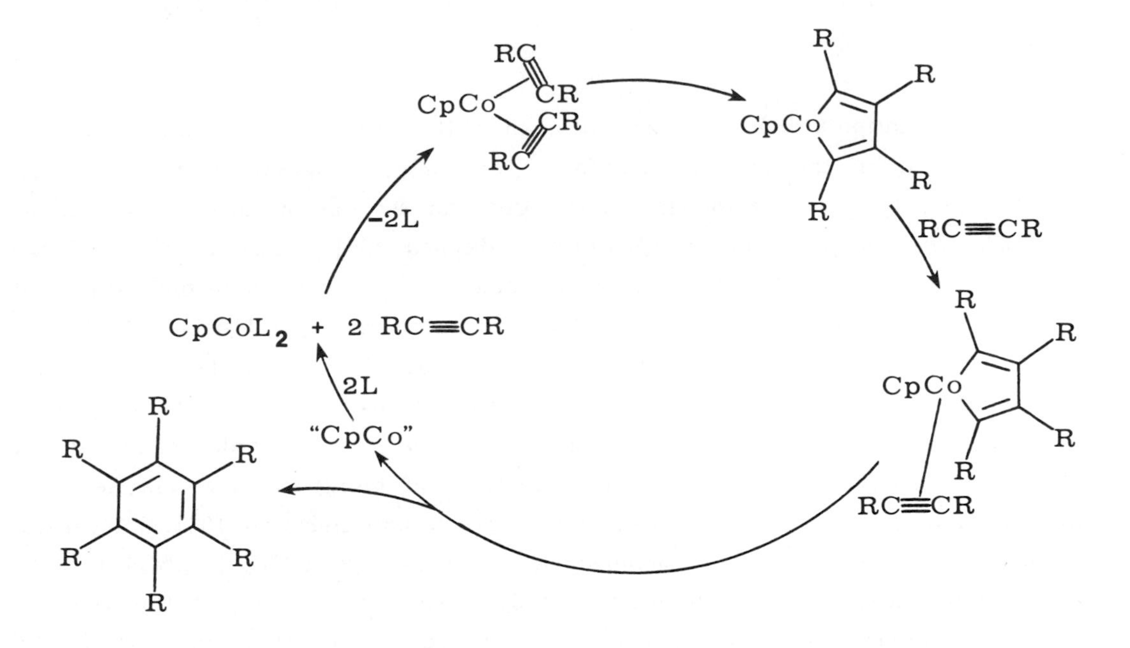


*Figure 10.5. Co catalyzed cyclotrimerization of alkynes.*

alkyne, formation of the metallacyclopentadiene, coordination of another molecule of alkyne, and collapse to products, with regeneration of the catalyst. Excellent analogies exist for all but the last step. For example, Collman and Little [34] showed that an Ir(I) nitrogen complex reacts with a disubstituted alkyne to produce an isolable (formally) Ir(III) alkyne complex. This reacts with additional alkyne to give an isolable Ir(III) metallacyclopentadiene complex, which itself catalyzes the cyclotrimerization of alkynes at temperatures in excess of 110° (Equation 10.32).

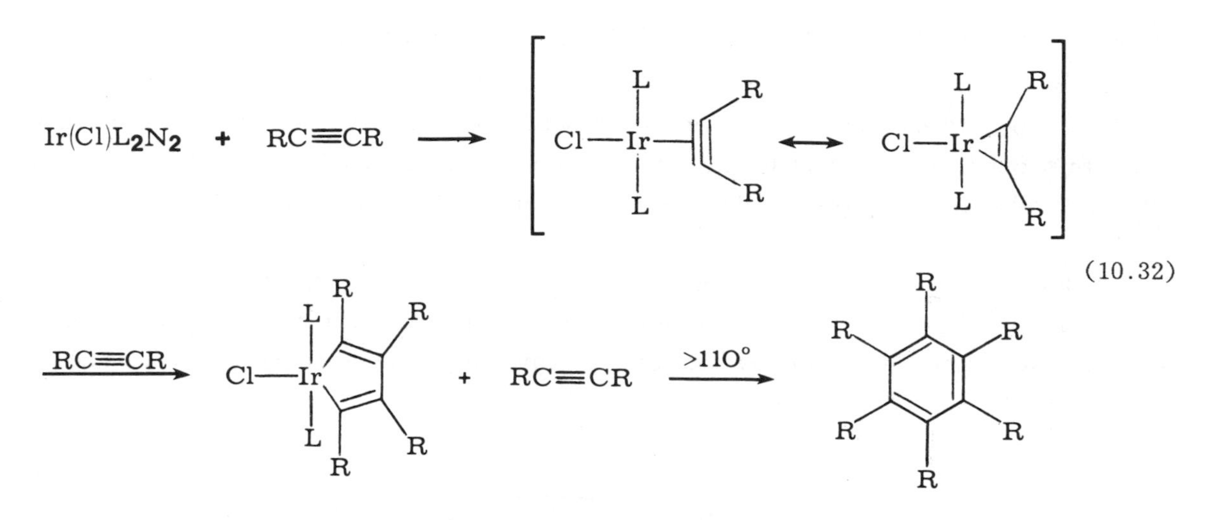


The rhodium complex behaves similarly. In both cases, added ligands such as phosphine or CO inhibit the catalytic reactions. Similarly, Yamazaki showed the cobalt complex $CpCo(PPh_3)_2$ reacts with alkynes to give an isolable metallacyclopentadiene complex which also catalytically cyclotrimerizes alkynes [35]. The reaction of the metallacyclopentadiene with alkyne to produce arenes can proceed by several different mechanisms: (1) insertion to give a metallacycloheptatriene; (2) "Diels-Alder" type cycloaddition within the coordination sphere of the metal; (3) "Diels-Alder" type cycloaddition without prior coordination of alkyne (Figure 10.6). It has not yet proved possible to distinguish among these pathways. However, some evidence for direct cycloaddition or insertion (without prior coordination) of highly reactive alkynes with the cobaltacyclopentadiene has been obtained by Bercaw and Bergman [36]. A careful kinetic study of the reaction of the metallacyclopentadiene $CpCo(PPh_3)\{C_4Me_4\}$ with 2-butyne to produce hexamethylbenzene (catalytically) clearly indicates that this reaction proceeds by rate-determining loss of phosphine, then coordination of alkyne, and finally

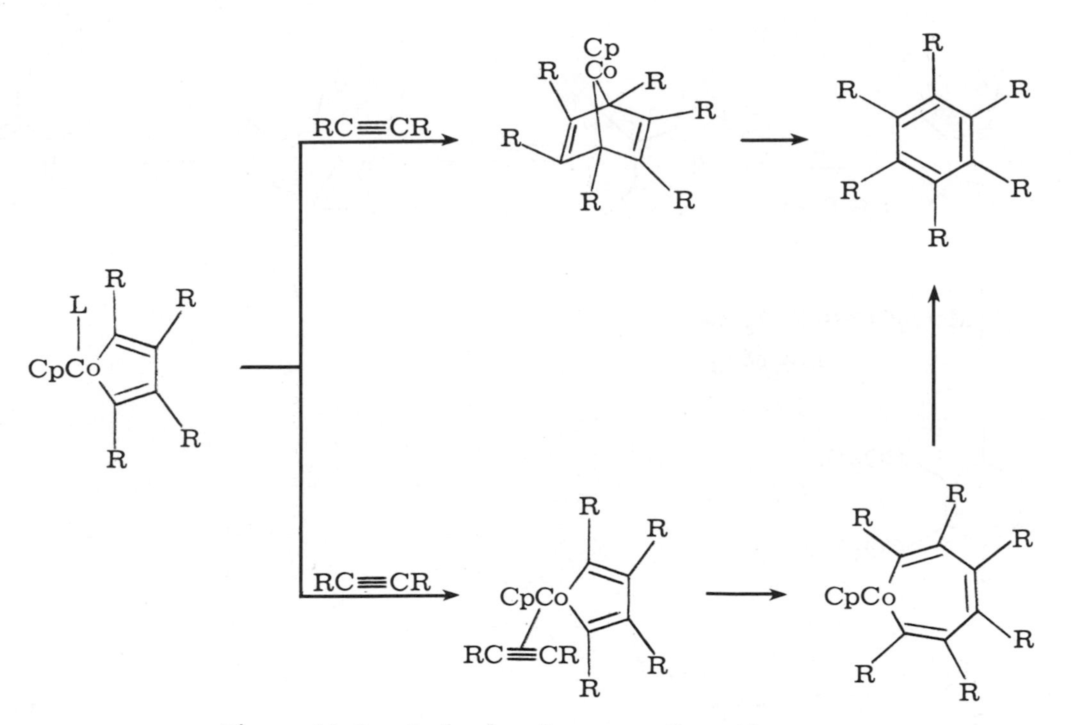


*Figure 10.6. Mechanism for arene formation.*

collapse to form product. Replacement of triphenylphosphine by the much more strongly coordinating trimethylphosphine prevents prior coordination of alkyne, and completely inhibits the cyclotrimerization of 2-butyne (several days at 120°). However, this same trimethylphosphine complex reacts quickly at 20° with dimethylacetylene dicarboxylate to produce 3,4,5,6-tetramethylphthalate dimethyl ester, in a process that does not involve dissociation of trimethylphosphine and coordination of dimethylacetylene dicarboxylate at all (10.33).

Finally, it should be noted that metallacyclopentadiene and cyclobutadiene complexes are valence tautomers [37] (10.34). However, the fact that cyclobutadiene metal complexes do not catalyze cyclotrimerization reactions, and in fact are often isolated as unreactive side products from cyclotrimerization reactions, indicates they are not intermediates in these reactions.

Some of the same complexes that catalyze the cyclotrimerization of alkynes to arenes also catalyze the cocyclotrimerization of alkynes and alkenes to cyclohexadienes (10.35). These reactions may proceed via metallacyclopentadienes or metallacyclopentenes, depending on the relative coordinating abilities of the unsaturated species

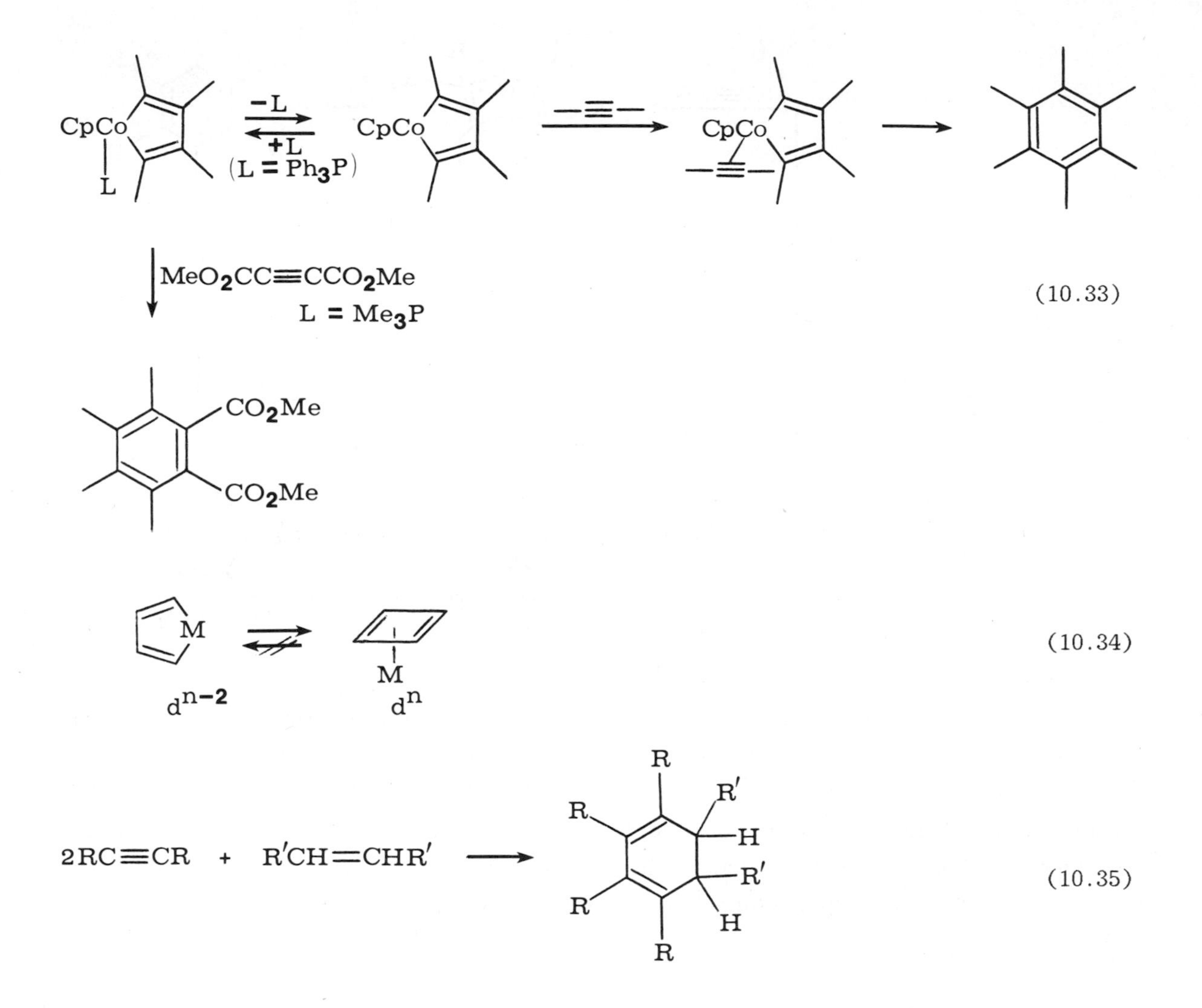


involved. Yamazaki showed that the complex $CpCo(PPh_3)_2$ reacts sequentially with two different alkynes to give a mixed cobaltacyclopentadiene complex, which reacts with added olefin to produce cyclohexadiene complexes or cyclohexadienes as in Equation 10.36, depending on the nature of the substituents [38]. Since each of the intermediate complexes in this reaction has been isolated and characterized, the over-all proposed mechanism is very reasonable. However, the last step is still in question. Kinetic studies show this reaction to be inhibited by added phosphine, indicating prior

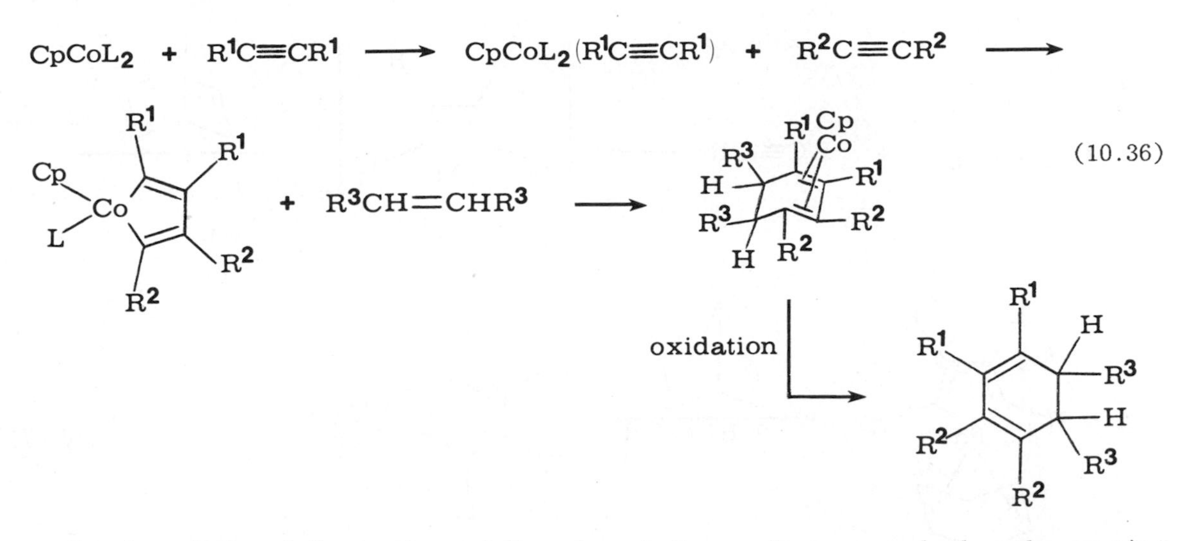


coordination of the olefin to the metallacyclopentadiene. However, whether the ensuing step is an insertion to form a metallacycloheptadiene, or a Diels-Alder addition within the coordination sphere of the metal, is a question that remains unanswered.

The palladiacyclopentadiene complex resulting from the reaction of the stable Pd(0) complex $Pd_2$(dibenzylideneacetone)$_3$ with dimethyl acetylenedicarboxylate followed by norbornadiene has been characterized by X-ray crystallography by Ibers. This complex also catalyzes the cocyclotrimerization of norbornadiene with dimethyl acetylenedicarboxylate [39], as in Equation 10.37.

Cocyclotrimerization can, however, proceed by a different path, one involving metallacyclopentenes (Figure 10.7) [40]. Electron-deficient (electrophilic) olefins such as dimethyl maleate and fumarate, fumaronitrile, and crotononitrile react with $CpCo(PPh_3)(RC{\equiv}CR)$ complexes to form metallacyclopentenes, one of which has been characterized by X-ray crystallography. Kinetic studies by Yamazaki [40] indicate that this reaction proceeds through an intermediate in which both the alkene and alkyne are coordinated to the metal. The thus-formed metallacyclopentene reacts subsequently in two different ways. With methyl phenylpropiolate or diphenylacetylene, cyclohexadiene complexes form. The reaction is inhibited by added phosphine, suggesting prior coordination of the alkyne is important. The formation of the cyclohexadiene complex almost certainly proceeds by insertion of alkyne into the metallacyclopentene, since a Diels-Alder type of cycloaddition is impossible in this instance.

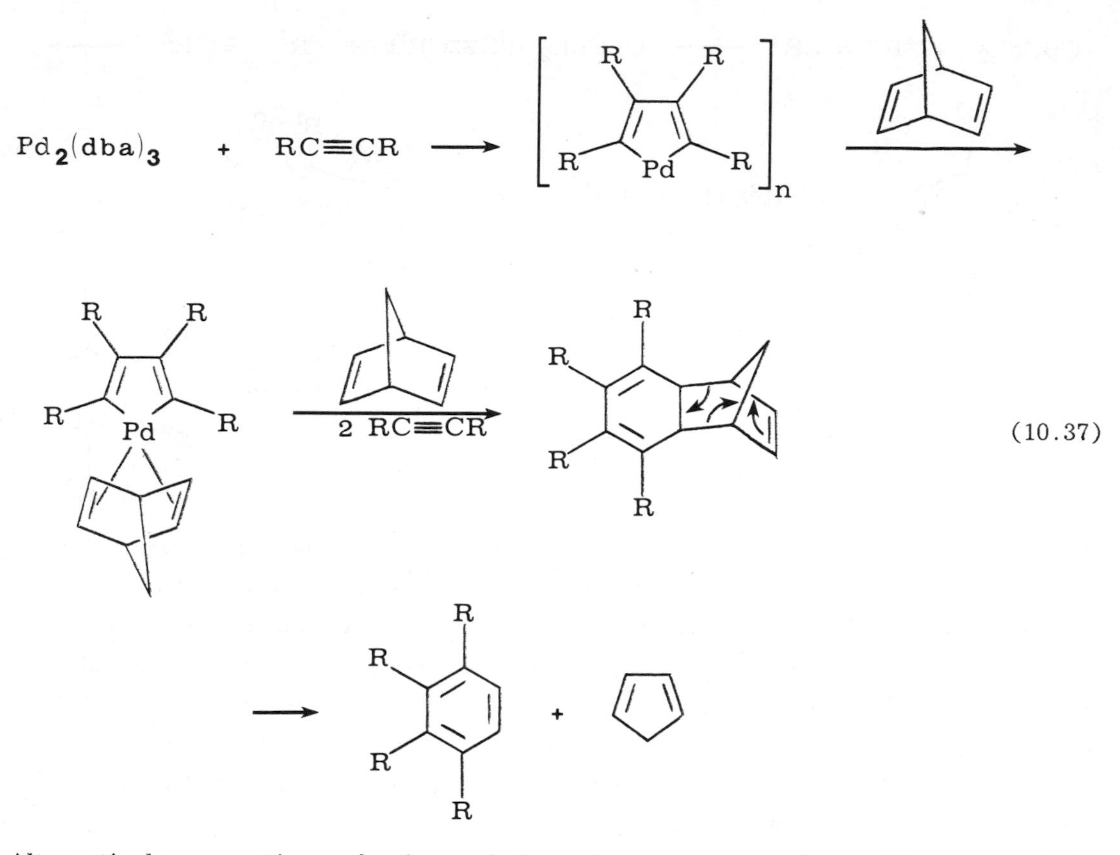


Alternatively, reaction of the cobaltacyclopentene complex with acrylonitrile produces open-chain diene complexes. This is thought to result from insertion to form a metallacycloheptene, followed by β-hydride elimination and reductive elimination. Which process ensues depends on the relative coordinating ability of the alkene in comparison with the alkyne.

## 10.4. Other Metallacycles.

Six- and seven-membered metallacycles for Pt and Ni are known. The bis-phosphine platinacyclohexane is resistant to thermal degradation, decomposing at a rate roughly equal to that of the corresponding platinacyclopentene, which is $2.5 \times 10^4$ slower than the rate of the corresponding open chain compounds. As discussed above, this rate results from the constraints imposed by the small ring, preventing the metal from becoming completely cis to a β-hydrogen, as required for β-hydride elimination. In contrast, the corresponding bis-phosphine platinacycloheptane thermally decomposes

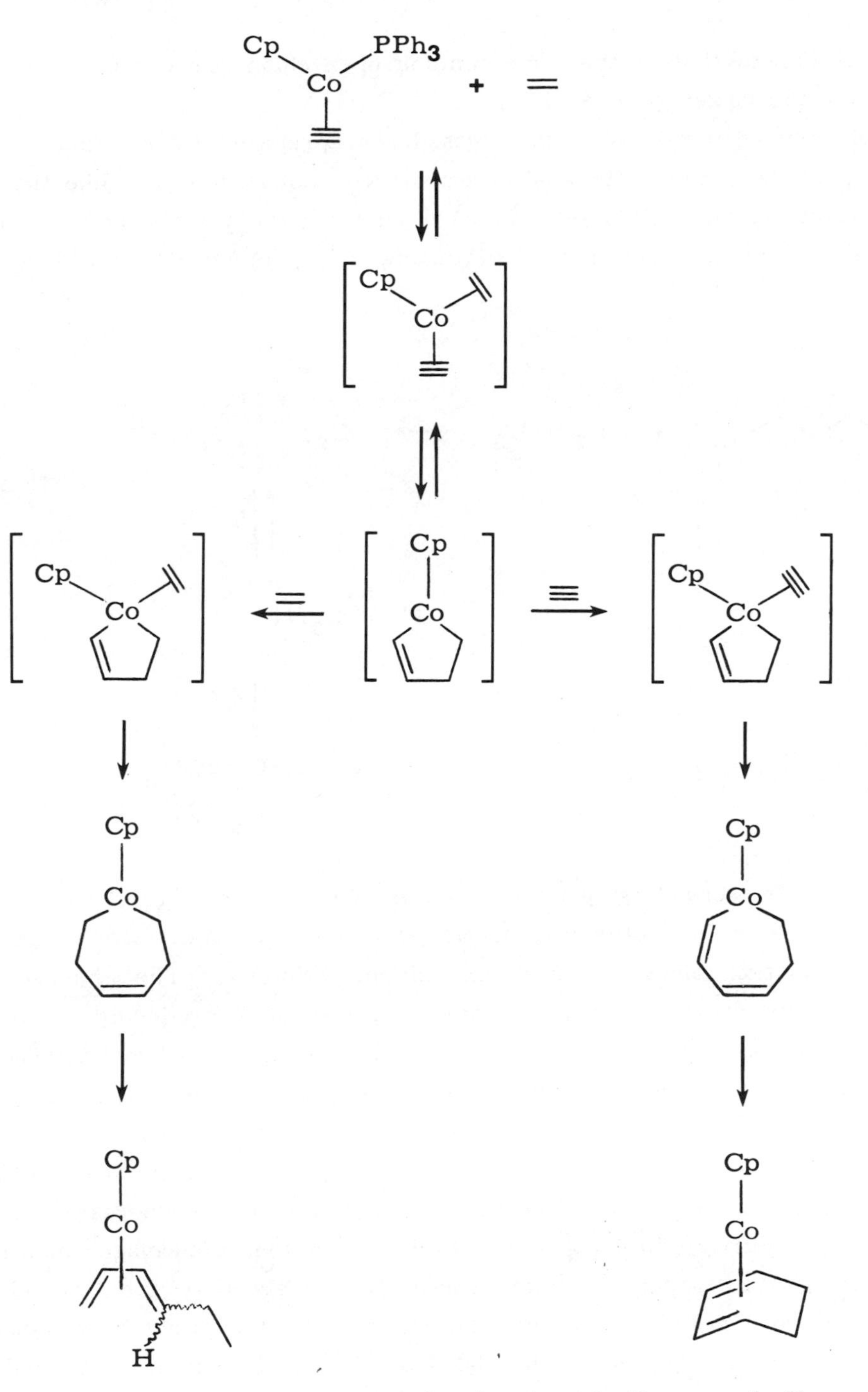


*Figure 10.7. Co catalyzed cocyclotrimerization of alkenes and alkynes.*

at a rate similar to that of the corresponding open-chain compounds, indicating greater flexibility for the larger ring.

Bis-phosphine nickelacyclohexane complexes, prepared by reaction of $L_2NiX_2$ with 1,5-dilithiopentane, undergo thermal decomposition (Equation 10.38) like the corresponding nickelacyclopentanes (Equation 10.23). That is, the bis-phosphine complex undergoes primarily reductive elimination, producing cyclopentane, along with a small amount

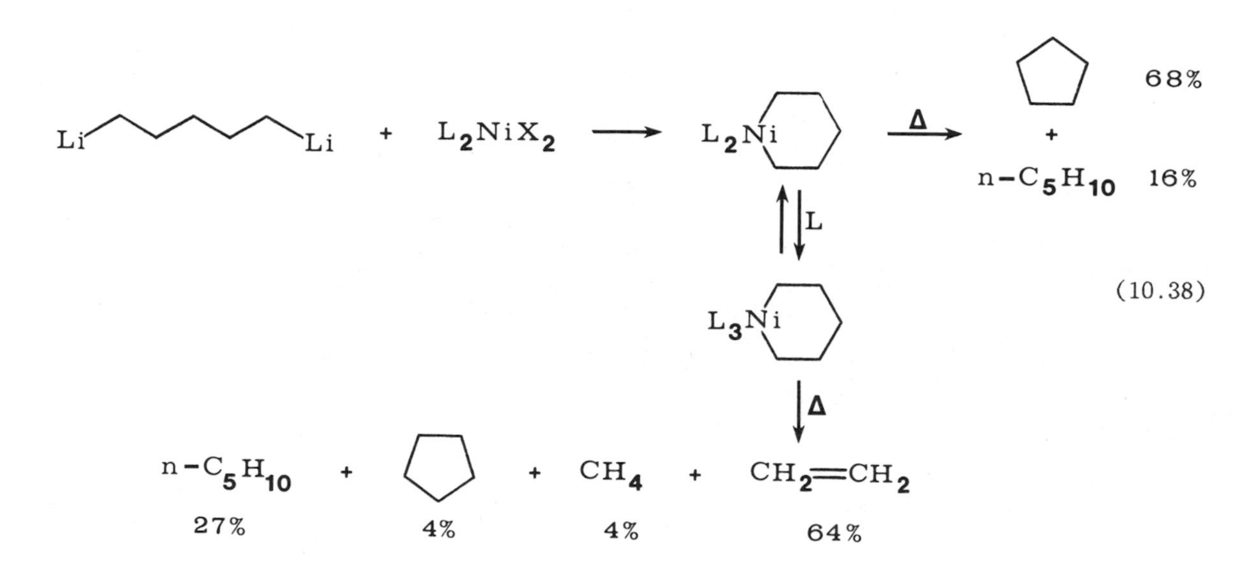


(10.38)

of pentenes from competing β-hydride elimination. In contrast, the tris-phosphine complex produces mainly ethylene, from C-C bond cleavage, with β-elimination and reductive elimination being minor decomposition pathways. The observed C-C bond cleavage products could result either from a β-cleavage, producing an olefin-metallacyclobutane complex, or from an α-cleavage, producing a carbene-metallacyclopentane complex (10.39). Although β-cleavage is the expected pathway, labeling studies by Grubbs [41] indicate that α-cleavage is the major pathway followed.

Decomposition of the 2,2,6,6-tetradeutero complex in the presence of cyclohexene as a trap for the metal carbene complex produces dideuterated norcarane with 70 to 90 percent isotopic purity (10.40). This clearly shows that α-cleavage must have occurred, producing the carbene complex almost exclusively from the labelled α-carbons. Had β-cleavage occurred, the carbene carbon would result from the metallacyclobutane having only one of its two methylenes labeled. Hence the norcarane formed could have only 50% (maximum) isotopic purity.

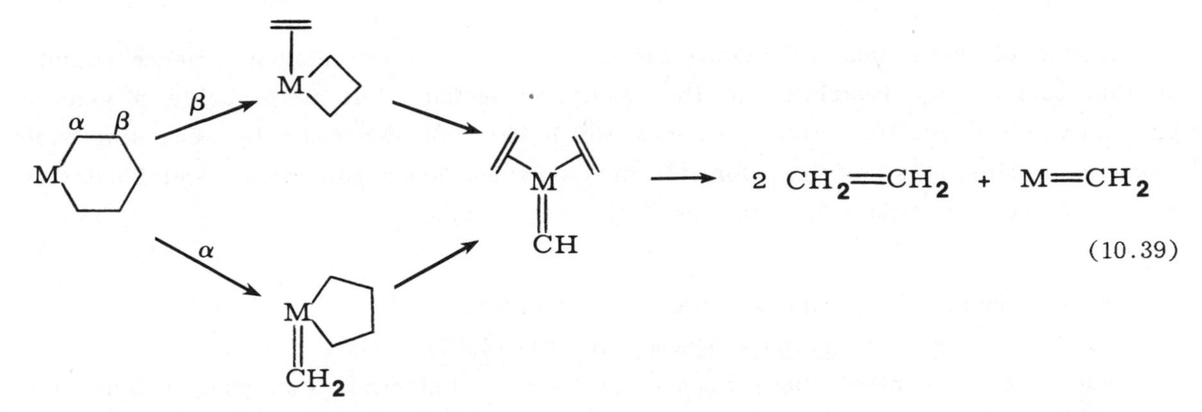


(10.39)

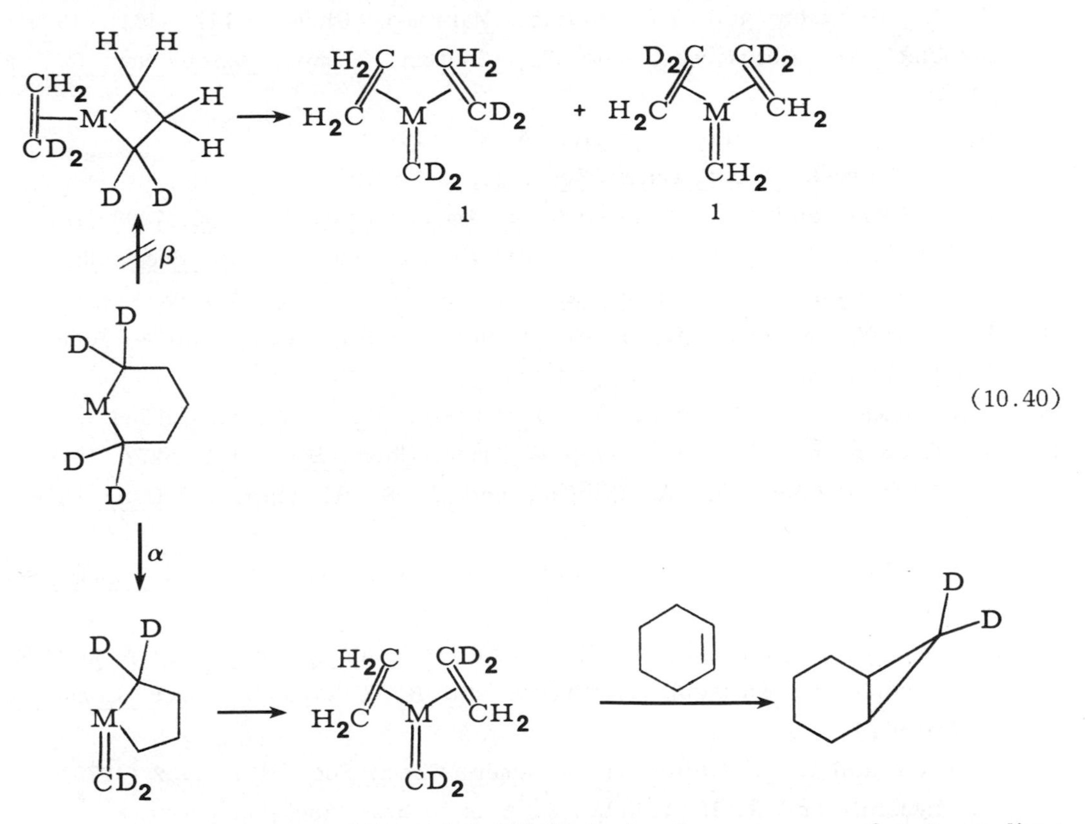


(10.40)

More surprisingly, decomposition of $L_3Ni(CH_2)_5$ in the presence of 1,7-octadiene results in the catalytic (560 percent, based on Ni) production of cyclohexane and ethylene, olefin metathesis products. Apparently the nickel carbene formed by thermal degradation of the metallacyclohexane can act as a metathesis catalyst. Hence activity of this sort is not restricted to the Group VI metals, but is probably a general phenomenon for reactive metal carbenes which are not consumed by competing side reactions. This is gratifying, for it brings our discussion full circle, and illustrates the coherence of metallacycle chemistry [42].

## NOTES.

1. R. H. Grubbs, Prog. Inorg. Chem., 24, 1 (1978).
2. T. J. Katz, Adv. Organomet. Chem., 16, 283 (1977).
3. For a discussion of these mechanisms, see J. Halpern, "Catalysis of Symmetry Restricted Reactions by Transition Metal Compounds" in Organic Synthesis via Metal Carbonyls, Vol, II, I. Wender and P. Pino, eds. (Wiley, 1977), pp. 705-730.
4. J. L. Herisson and Y. Chauvin, Makromol. Chem., 141, 161 (1970); J. P. Soufflet, D. Commereuc, and Y. Chauvin, Compt. Rendus Ser. C., 276, 169 (1973).
5. E. L. Muetterties, Inorg. Chem., 14, 951 (1975).
6. R. R. Schrock, Accts. Chem. Res., 12, 98 (1979).
7. C. P. Casey and T. J. Burkhardt, J. Amer. Chem. Soc., 96, 7808 (1974).
8. J. McGinnis, T. J. Katz, and S. Hurwitz, J. Amer. Chem. Soc., 98, 605 (1976).
9. T. J. Katz and S. J. Lee, J. Amer. Chem. Soc., 102, 422 (1980).
10. R. Schrock, S. Rocklage, J. Wengrovius, G. Rupprecht, and J. Fellman, J. Mol. Catal., in press.
11. J. Rajaram and J. A. Ibers, J. Amer. Chem. Soc., 100, 829 (1978).
12. T. H. Johnson and S. S. Cheng, J. Amer. Chem. Soc., 101, 5277 (1979).
13. R. J. Puddephatt, M. A. Quyser, and C. F. H. Tipper, J.C.S. Chem. Comm., 626 (1976).
14. C. P. Casey, D. M. Scheck, and A. J. Shusterman, J. Amer. Chem. Soc., 101, 4233 (1979).
15. R. Noyori, Y. Kumagai, and H. Takaya, J. Amer. Chem. Soc., 96, 634 (1974).
16. R. Noyori, Y. Kumagai, I. Umeda, and H. Takaya, J. Amer. Chem. Soc., 94, 4018 (1972).
17. P. Foley and G. M. Whitesides, J. Amer. Chem. Soc., 101, 2732 (1979).
18. A. Miyashita and R. H. Grubbs, J. Amer. Chem. Soc., in press.
19. M. Ephritikhine, M. L. H. Green, and R. E. Mackenzie, J.C.S. Chem. Comm., 619 (1976); M. Ephritikhine, B. R. Francis, M. L. H. Green, R. E. Mackenzie, and M. J. Smith, J.C.S. Dalton, 1131 (1977).

20. J. X. McDermott, J. F. White, and G. M. Whitesides, J. Amer. Chem. Soc., 98, 6521 (1976).
21. P. Diversi, G. Ingrosso, and A. Lucherini, J.C.S. Chem. Comm., 735 (1978).
22. R. H. Grubbs, A. Miyashita, M. Liu, and P. Burk, J. Amer. Chem. Soc., 100, 2418 (1978).
23. These results have been rationalized with orbital symmetry arguments, P. S. Braterman, J.C.S. Chem. Comm., 70 (1979).
24. R. H. Grubbs, and A. Miyashita, J. Amer. Chem. Soc., 100, 1300 (1978).
25. R. H. Grubbs, and A. Miyashita, J. Organomet. Chem., 161, 371 (1978).
26. R. H. Grubbs, and A. Miyashita, J. Amer. Chem. Soc., 100, 7416 (1978).
27. J. X. McDermott, M. E. Wilson, and G. M. Whitesides, J. Amer. Chem. Soc., 98, 6529 (1976).
28. R. H. Grubbs, and A. Miyashita, J.C.S. Chem. Comm., 864 (1977).
29. G. Erker and K. Kropp, J. Amer. Chem. Soc., 101, 3659 (1979).
30. S. J. McLain, C. D. Wood, and R. R. Schrock, J. Amer. Chem. Soc., 101, 4558 (1979).
31. S. M. McLain, J. Sancho, and R. R. Schrock, J. Amer. Chem. Soc., 101, 5451 (1979).
32. L. Cassar, P. E. Eaton, and J. Halpern, J. Amer. Chem. Soc., 92, 3515 (1970).
33. P. E. Eaton and D. R. Patterson, J. Amer. Chem. Soc., 100, 2573 (1978).
34. J. P. Collman, Accts. Chem. Res., 1, 136 (1968); J. P. Collman, J. W. Kang, W. F. Little, and M. P. Sullivan, Inorg. Chem., 7, 1298 (1968).
35. Y. Wakatsuki, T. Kiramitsu, and H. Yamazaki, Tetrahedron Lett., 4549 (1974).
36. D. R. McAllister, J. E. Bercaw, and R. G. Bergman, J. Amer. Chem. Soc., 99, 1666 (1977).
37. H. Yamazaki and N. Hagihara, J. Organomet. Chem., 7, P22 (1967).
38. Y. Wakatsuki and H. Yamazaki, J. Organomet. Chem., 139, 169 (1977).
39. H. Suzuki, K. Itoh, Y. Ishii, K. Simon, and J. A. Ibers, J. Amer. Chem. Soc., 98, 8494 (1976).
40. Y. Wakatsuki, K. Aoki, and H. Yamazaki, J. Amer. Chem. Soc., 101, 1123 (1979).
41. R. H. Grubbs and A. Miyashita, J. Amer. Chem. Soc., 100, 7418 (1978).
42. A variety of other metallacycles have been implicated in a number of catalytic oligomerization and cyclooligomerization reactions. For a review, see G. Wilke, Pure & Appl. Chem., 50, 677 (1978).

# 11

# Transition-Metal Alkyl Complexes

Transition-metal alkyl complexes are very common and are of great importance in organic synthesis. Like the hydride ligand, the alkyl group fills a regular coordination site on a transition metal, and behaves primarily as a strong σ-donor. Although the protons on a carbon directly bonded to a transition metal can appear almost anywhere in the $^1H$ NMR spectrum, frequently the signals appear upfield from the same group in a hydrocarbon (i.e., the metal is often shielding relative to carbon). Transition-metal alkyl complexes are frequently not very stable, having at least three relatively low-energy decomposition pathways available to them. By far the most common mode of decomposition is the β-hydride elimination pathway. Alkyl groups lacking β-hydrogens (i.e., $CH_3$, $CH_2SiMe_3$) often form relatively stable alkyl complexes. Homolytic decomposition, forming alkyl radicals, is much less common under normal conditions. Decomposition by α-elimination, forming metal (hydride) carbene complexes, is also infrequently observed.

Transition-metal alkyl complexes rarely react as if they are carbanionic in character. The covalent metal-to-carbon σ bond strongly mediates the reactivity of the alkyl group, restricting its reactions to those mechanistically accessible to the transition metal (i.e., oxidative-addition, migratory insertion). That is, the transition metal is more than just a sophisticated counter ion for the alkyl group; it is, in fact, the major determinant of the chemical behavior of that alkyl group.

There are several general methods for the synthesis of transition-metal alkyl complexes, summarized in Table 11.1. This variety of approaches makes virtually every class of organic compound a potential source of the alkyl group and hence chemically

*Table 11.1. Preparation of transition-metal complexes.*

| Type of complex | Preparation |
|---|---|
| Metal hydride-olefin | M-H + C=C → M-C-C-H |
| Carbanion-metal halide | R(-) + MX → R-M + $X^-$ |
| Metal anion-organic halide | M(-) + RX → M-R + $X^-$ |
| Transmetallation | RM' + MX → RM + M'X (M' = Hg, Sn, Al) |
| Oxidative addition | $L_nM$ ($d^n$) + RX → $L_nM(R)X$ ($d^{n-2}$) |
| Metal olefin-nucleophile | $(CH_2=CH_2)$ M + Nuc → $NucCH_2CH_2M$ |
| Cyclometallation (orthometallation) | L⌒C(H)-R + M → cyclic L–M(H)–C(R) metallacycle |

accessible to the array of reactions in which σ-alkylmetal complexes participate. All of these approaches have found some application in organic synthesis, and the remainder of this chapter will discuss this chemistry following the classification presented in Table 11.1.

## 11.1. Metal Alkyls Produced from Olefins and Metal Hydrides.

As was mentioned in Chapter 7, a particularly appealing approach to alkylmetal complexes is the addition of a metal hydride to a simple, unactivated olefin, since this would allow functionalizations of unactivated olefins not available by standard methods. Early attempts based on Group VIII transition-metal hydrides met with very limited success because with these metals the equilibrium lies strongly to the olefin-metal-hydride complex side rather than the metal-alkyl complex side. However, the development by Schwartz of the chemistry of $Cp_2Zr(H)Cl$, which adds cleanly and in high yield to a wide variety of simple alkenes and alkynes to produce very stable alkyl complexes, provides the basis for considerable useful chemistry [1]. For example, these zirconium alkyls are reactive toward a number of electrophiles. Protonation produces the hydrocarbon, and halogenation produces the terminal halide [2], as in Equation 11.1. This reagent adds regioselectively and stereospecifically cis to alkynes to produce vinylzirconium complexes. Equilibration of the initial reaction mixture results in enrichment in the isomer in which Zr occupies the less-hindered carbon. Treatment with

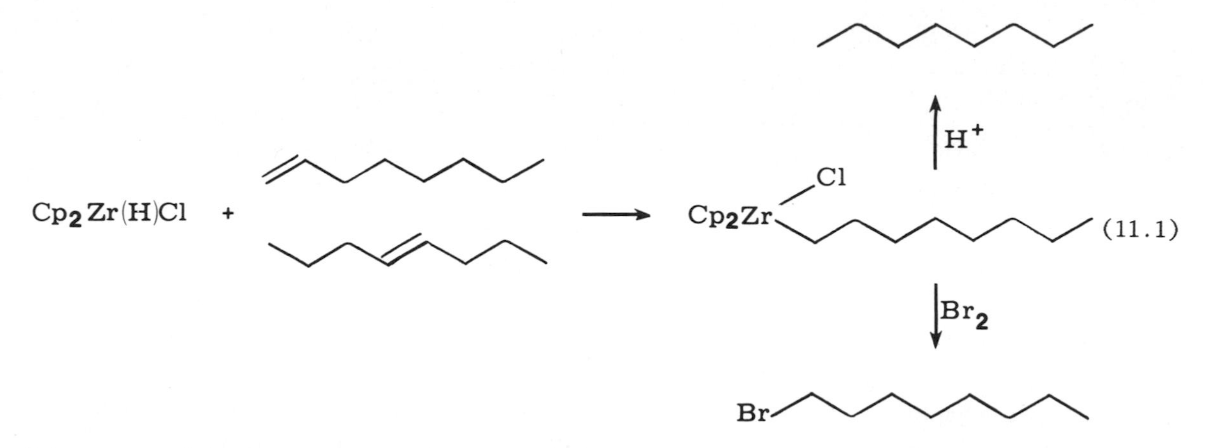


N-bromosuccinimide produces the vinyl halide with retention of olefin geometry, making Equation 11.2 a useful procedure for the stereospecific synthesis of vinyl halides [3].

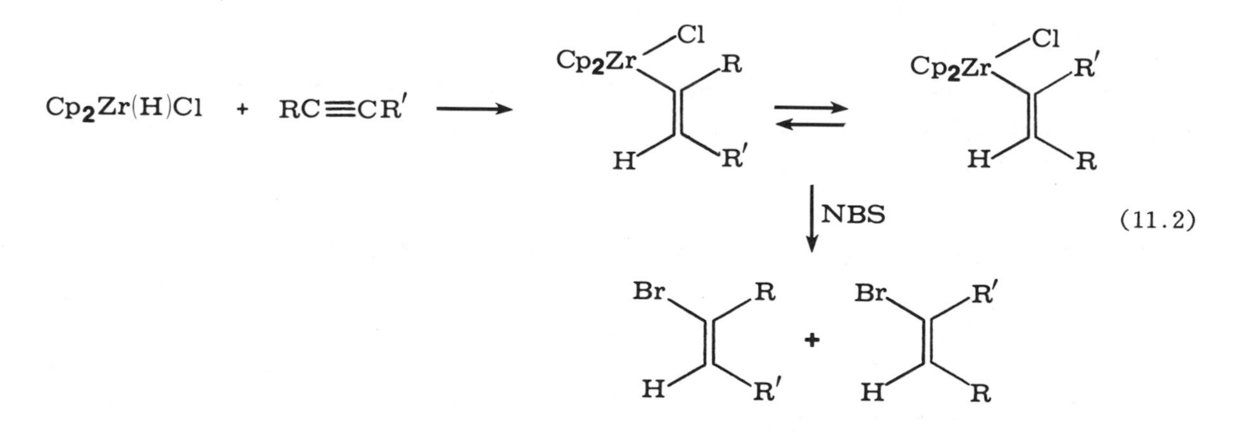


The mechanism of this electrophilic cleavage is of some interest. Often electrophilic cleavages of transition-metal alkyl complexes occur with inversion of configuration at carbon, and are thought to proceed by oxidation of the metal by the electrophile, followed by an $S_N2$ displacement of the metal [4]. However, the zirconium in $Cp_2Zr(R)Cl$ is a $d^0$ metal, and further oxidation is difficult. Hence this mechanism cannot operate in this system. Using the NMR technique developed by Whitesides (Chapter 5), cleavage of the Zr-C bond by bromine was shown to proceed with retention of configuration at carbon, confirming the assumption of a change in mechanism. This cleavage is proposed to proceed by prior coordination of the electrophile to the metal by donation of a pair of electrons to a vacant low-lying metal orbital, facilitating frontside attack by the electrophile on the zirconium-carbon bond [5].

Whenever an alkyl-metal complex is produced, one is almost obliged to treat it with carbon monoxide to see if it will undergo migratory insertion. Indeed, $Cp_2Zr(R)Cl$ complexes insert CO under very mild conditions (25°, 20 psi) producing the corresponding acyl complexes in excellent yield. This is somewhat surprising, since $Cp_2ZrR_2$ complexes are quite unreactive toward insertion. Not only is CO insertion facile, the acyl complexes react with dilute HCl to produce aldehydes in excellent yield. Overall, this process corresponds to a "hydroformylation" of the olefin, a well-known catalytic process. However, in contrast to the catalytic system, this system (in 11.3) produces exclusively terminal aldehyde regardless of the original position of the olefin [6].

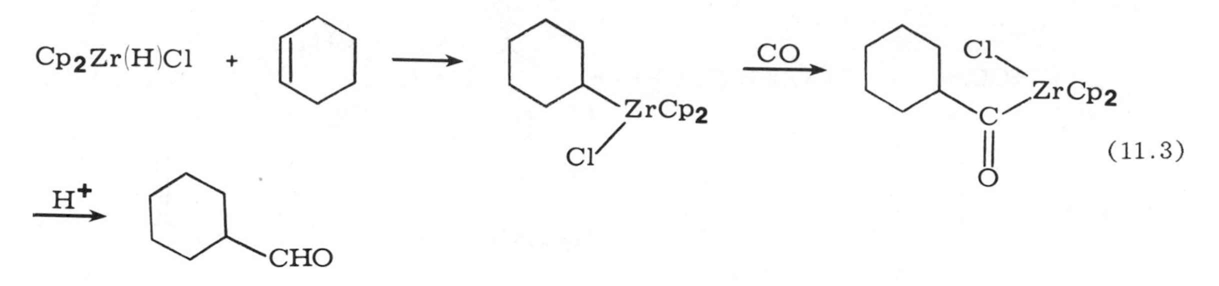


(11.3)

Application of this "hydroacylation" to 1,3-dienes is particularly interesting, in that it produces γ,δ-unsaturated aldehydes in good yield [7], as in 11.4.

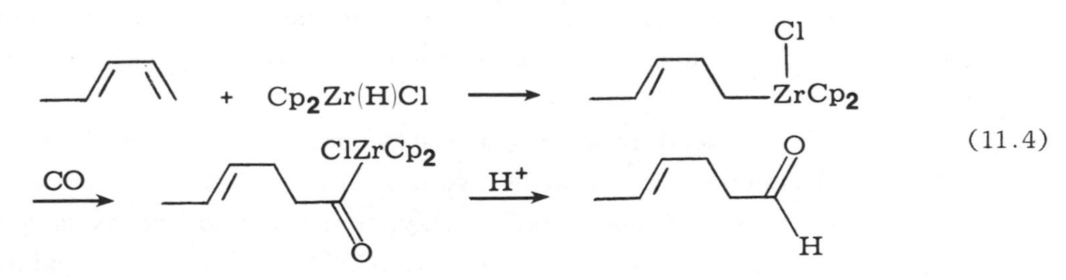


(11.4)

Acylzirconium complexes can also be cleaved directly to carboxylic acid derivatives under oxidizing conditions. Thus, aqueous $H_2O_2$ produces carboxylic acids, $Br_2$ in methanol produces methyl esters, and N-bromosuccinimide produces acid bromides. Oxidation of the alkyl zirconium complexes by a number of oxidizing agents, including $H_2O_2/H_2O$, t-butylhydroperoxide, and m-chloroperbenzoic acid produces the corresponding alcohols in moderate yield [8].

Although these zirconium alkyls are thus of potential synthetic utility, the very stability that allows their ready formation also prevents them from having sufficient reactivity to effect a number of other useful transformations. An elegant solution to this problem is to "transmetallate" the zirconium alkyls; that is, to exchange zirconium for another metal which has a greater reactivity in the desired reaction than does zirconium. In principle, this would allow the activation of olefins by chemistry unique to zirconium, "hydrozirconation," and the transfer of this reactive alkyl group to another metal to utilize that metal's inherent reactivity. In practice, this is precisely what happens. Although $Cp_2Zr(R)Cl$ complexes react slowly with acid halides to produce ketones, this reaction is difficult, and rarely proceeds in good yield. However, treatment with $AlCl_3$ results in a transfer of the alkyl group from Zr to Al, and the resulting aluminum alkyl complex reacts quickly with a variety of acid halides to produce ketones in excellent yield [9] (Equation 11.5). Since the specific aluminum alkyls

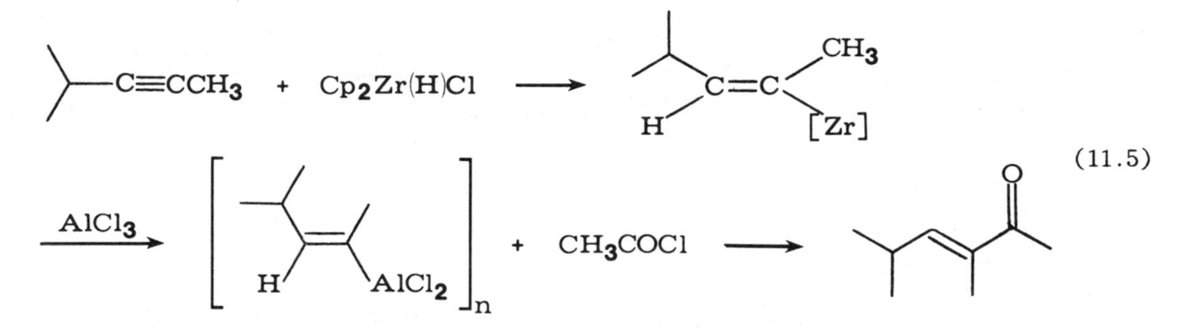


are not directly accessible from the corresponding olefin and an aluminum hydride species, this combined process allows each metal to manifest its own inherent reactivity to produce chemical transformations impossible for a given single type of organometallic complex.

Of all the alkyl metal complexes available, organocopper complexes are far and above the most heavily used in organic synthesis (vide infra), because of their unique ability to alkylate organic halides, and to alkylate conjugated enones in a 1,4 fashion. Unfortunately, transfer of an alkyl group from Zr to Cu does not produce a reagent that is very efficient in the above reactions [10]. However, successful 1,4-alkylations are achieved by transmetallations to Ni instead of Cu [11], as in 11.6. Although internal alkynes react in only low (6 to 10%) yield, terminal alkynes including those containing oxygen functionality (a protected OH) react well. Both cyclic and acyclic conjugated ketones (lacking β-substitution) undergo alkylation. This chemistry has been applied to the synthesis of prostaglandin-related compounds, which have long side-chains bearing both oxygen functionality and unsaturation. As is common with most 1,4-additions, the initial product is an enolate (Ni or Zr) which can be further alkylated by electrophiles. In this case, the second alkylation was incomplete, resulting in a mixture of products [12] (Equation 11.7).

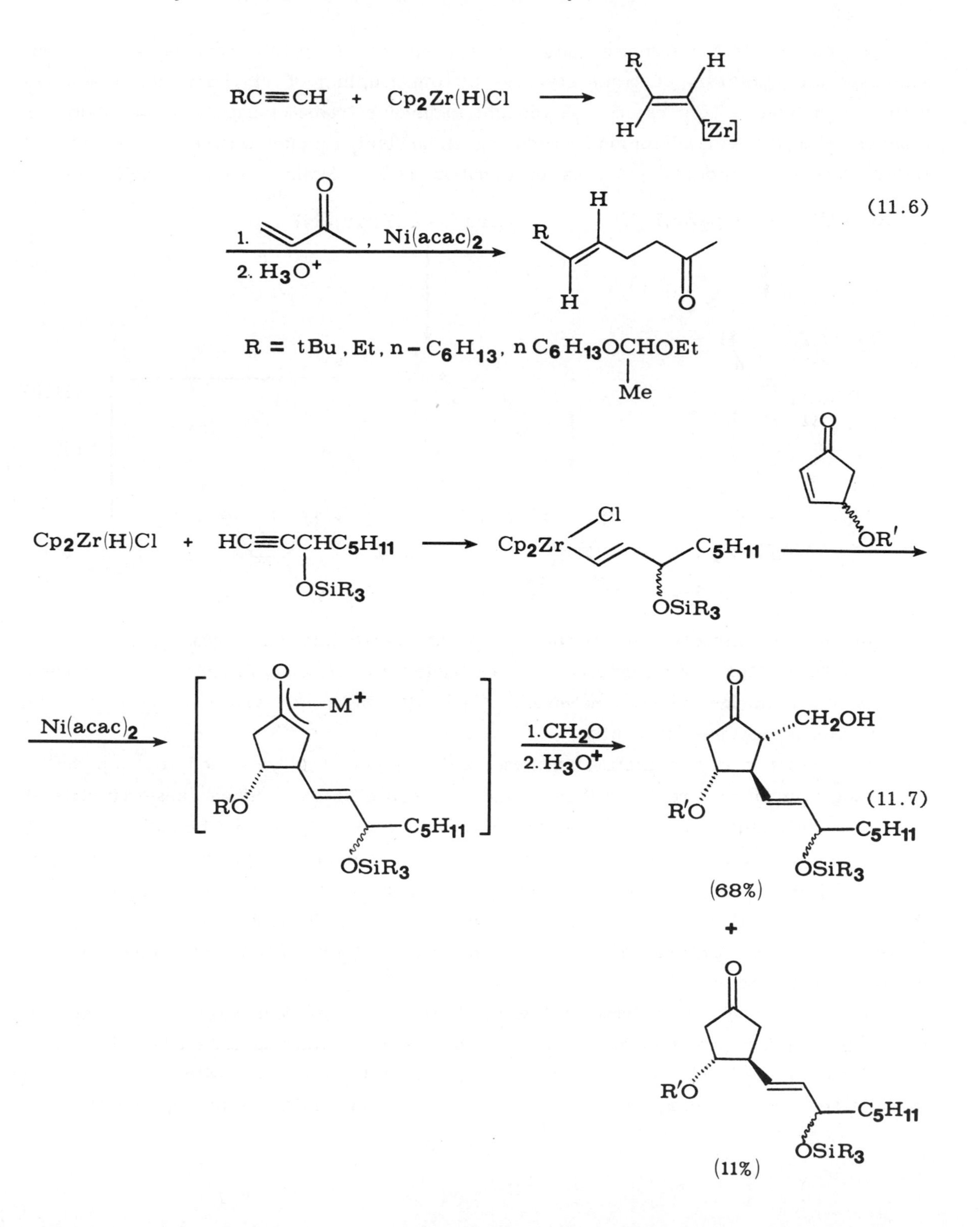
RC≡CH + Cp2Zr(H)Cl
R
H
H
[Zr]
1.
, Ni(acac)2
2. H3O+
(11.6)
R = tBu, Et, n-C6H13, nC6H13OCHOEt
Me
Cp2Zr(H)Cl + HC≡CCHC5H11
OSiR3
Cp2Zr
Cl
C5H11
OR′
Ni(acac)2
M+
R′O
1. CH2O
2. H3O+
CH2OH
(11.7)
(68%)
+
(11%)

A related Zr-Ni transmetallation developed by Negishi provides an efficient stereospecific synthesis of trans styrenes by the coupling of vinylzirconium complexes with aryl halides. This reaction is thought to involve transmetallation of the arylnickel complex with the vinylzirconium, producing an arylvinyl nickel complex, which reductively eliminates products [13], as in Equation 11.8. Again, oxygen functionality is

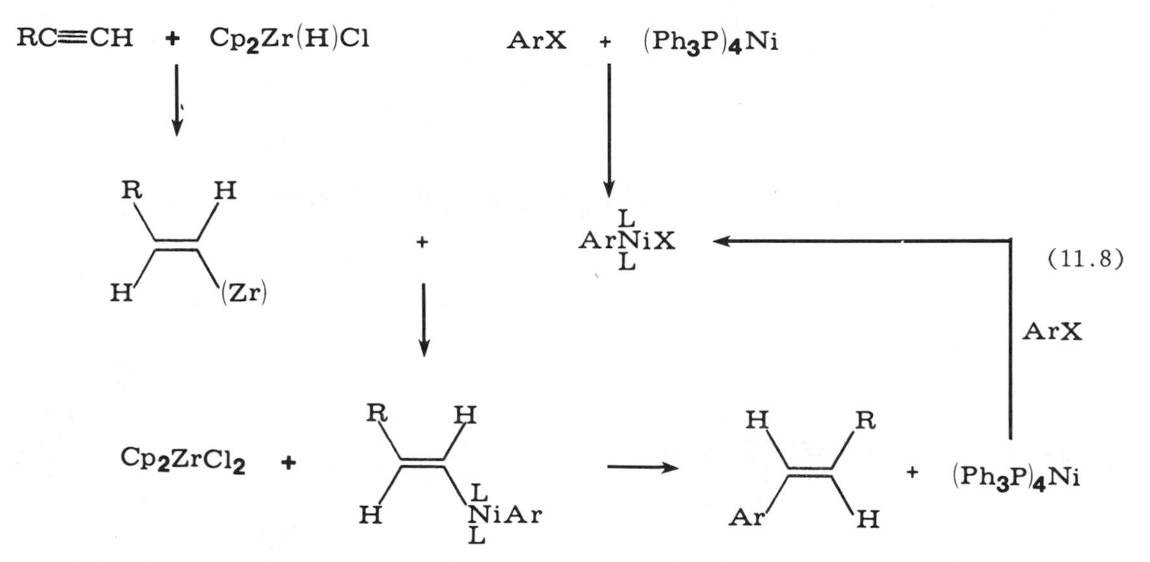


tolerated in the vinylzirconium complex, and the aryl halide can contain Cl, OMe, CN, and COOMe groups. This approach is restricted to terminal alkynes, since internal alkynes react in low yield. However, a slightly different system (vide infra) will handle internal alkynes efficiently.

The reaction of trimethylaluminum with alkynes is catalyzed by $Cp_2ZrCl_2$, producing vinylalanes regiospecifically and stereospecifically. Since these vinylalanes are themselves quite reactive, a number of trisubstituted olefins can be made with very high regio- and stereoselectivity [14], as in 11.9. This route has been used to synthesize the terpenes geraniol, and ethyl geranate, in excellent yield [15] (11.10). The course of this reaction is not clear. Since $Cp_2Zr(CH_3)Cl$ itself does not add to alkynes, the active methylating species must involve both Zr and Al. However, the details remain obscure.

None of the above systems are very effective with internal alkynes, and only low yields of coupling products are obtained using these as starting materials. Since hydrozirconation goes well with internal alkynes, the problem must be with the transmetallation step. To overcome this problem, the use of an additional transmetallation step

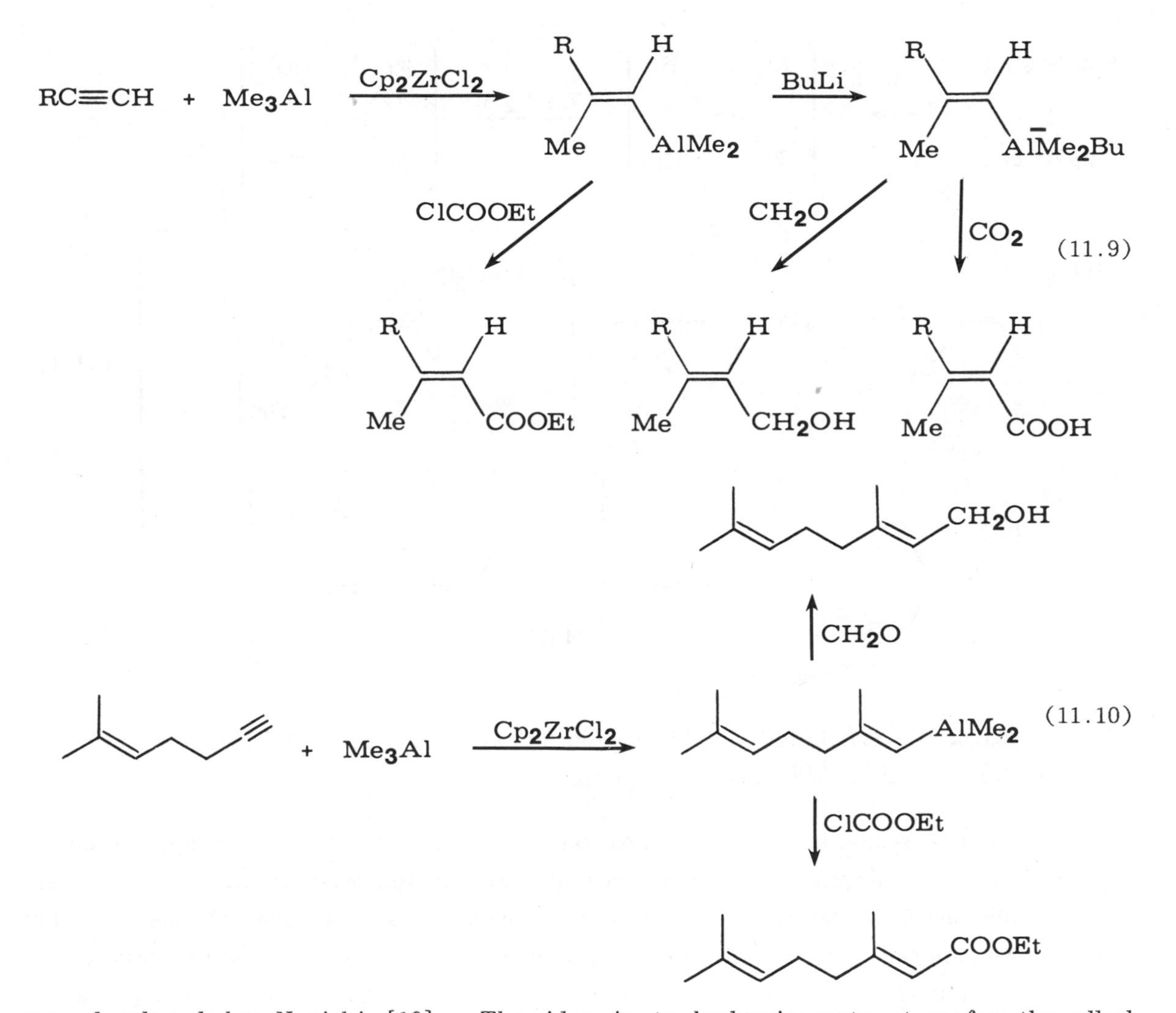


was developed by Negishi [16]. The idea is to hydrozirconate, transfer the alkyl group from Zr to some other metal that undergoes this particular transmetallation easily, then transfer the alkyl group from this intermediary to the final metal (Ni, Pd, or Al) to effect the desired coupling reaction. This leads to a rather complicated system, and the specific choice of intermediate metal is not an obvious one. However, experimentally it is found that $ZnCl_2$ greatly facilitates the coupling of substituted vinyl zirconates (or alanes) with alkenyl, aryl, or alkynyl halides using Pd(0) or Ni(0) catalysts. Since the stereo- and regiospecificity of the Zr and Ni(0) or Pd(0) chemistry is maintained in this process, some very useful synthetic procedures are available by this method (11.11).

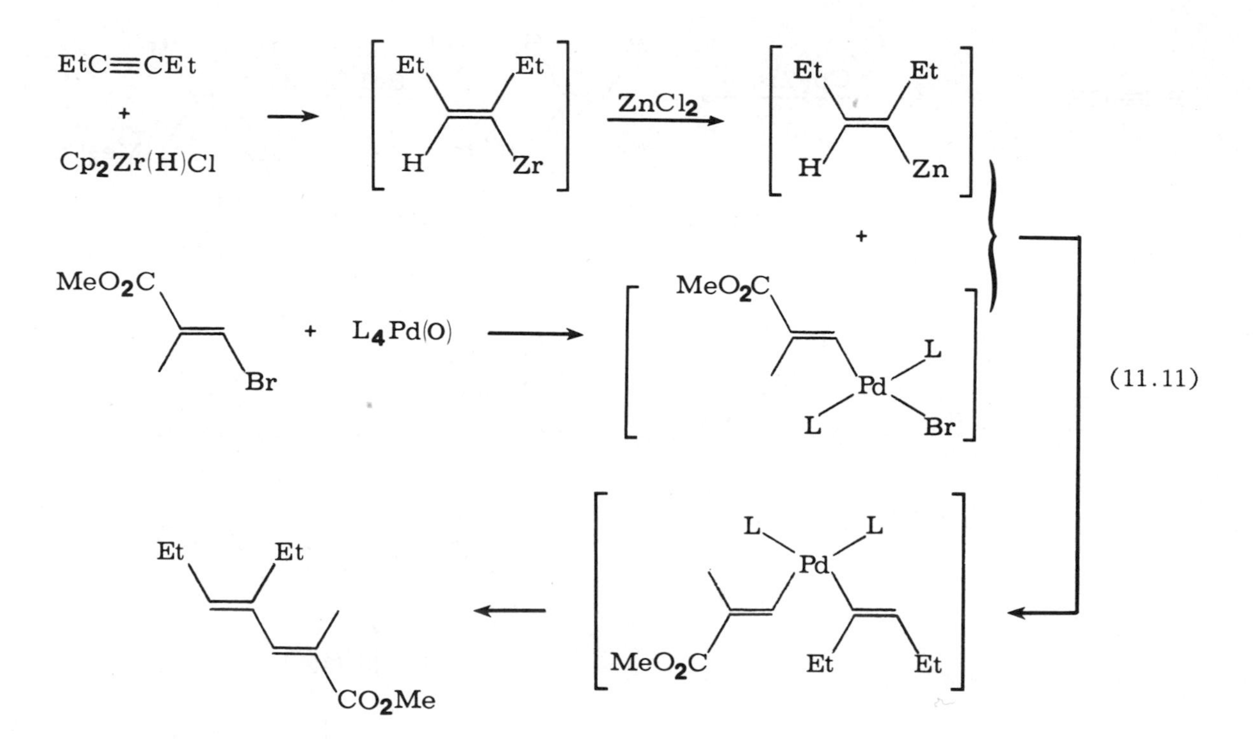


## 11.2. Metal Alkyls Produced from Carbanions and Metal Halides; Organocopper Complexes.

Of all the transition-metal organometallic reagents developed for application to organic synthesis, organocopper complexes are by far the most heavily used and enthusiastically accepted by the synthetic organic chemist, as evidenced by the 80 to 100 papers that appear annually in this area. There are many reasons for this (well-deserved) popularity, not the least of which is the ease with which these reagents are prepared and the remarkably useful chemistry they effect. Additionally, ever since the development of copper-catalyzed Grignard reactions more than 30 years ago, organocopper chemistry has been a standard synthetic technique for the organic chemist. Hence new advances in the field have been rapidly accepted and applied to synthetic problems.

Because of the activity in this field, and the vast number of new complexes developed in recent years, only a skeletal treatment of the subject is possible here. The different types of organocopper complexes useful in synthetic chemistry are summarized in Table 11.2 [17,18]. Much of the early work in the area evolved from the

*Table 11.2. Summary of organocopper reagents.*

| Types of organocopper reagents | General name |
|---|---|
| RM + CuX(catalytic) | Catalytic Organocopper Compound |
| RCu | Stoichiometric Organocopper Compound |
| RCu·Ligand | (Ligand = $R_3P$, $(RO)_3P$, $R_2S$) |
| $R_2CuM$ (M = Li, MgX) | Homocuprate |
| $R_2CuM$·Ligand | |
| RR'CuM | Mixed Homocuprate |
| R(Z)CuM (Z = OR', SR', CN, Cl, Br) | Organo(hetero)cuprate or Heterocuprate |
| $R_nCuLi_{n-1}$ (n>2) | Complex Organocopperlithium Compounds |
| RCu·$BF_3$ | Organocopper-borane Complexes |
| RCu·$BR_3$ | |

observation of Kharasch [19] that Grignard reagents added 1,4 to conjugated enones in the presence of catalytic amounts of cuprous salts (11.12). Although this is a very

$$RMgX + CH_2{=}CH\overset{O}{\overset{\|}{C}}CH_3 + CuX \longrightarrow RCH_2CH_2\overset{O}{\overset{\|}{C}}CH_3 \qquad (11.12)$$

useful process, it is quite sensitive to the particular copper salt used, the presence of trace impurities in the magnesium and/or copper, and even the order of addition of reagents. Hence variable results are often obtained. Although the intermediacy of discrete organocopper complexes was always assumed in these reactions, it wasn't until the middle 1960s that it was conclusively demonstrated that stoichiometric organocopper reagents are indeed the reactive species in these copper-catalyzed Grignard reactions. With this realization, it became possible to better control these reactions, and organocopper chemistry developed with explosive rapidity. Major contributors to this development are Corey, House, Normant, Posner, and Whitesides.

The most intensively studied and extensively used organocopper complex is the symmetrical organocuprate, formed by the reaction of two equivalents of organolithium (or Grignard) reagent with one equivalent of anhydrous cuprous iodide. The resulting complex is of tremendous utility for two types of transformations: the alkylation of organic halides [20] and the 1,4-alkylation of conjugated enones [21]. The organocuprates are clearly the reagents of choice for both processes.

Although a number of other organometallic reagents will alkylate organic halides, they frequently suffer from competing reactions, such as halogen-metal exchange, elimination reactions, and coupling reactions to form symmetrical dimers. In contrast, organocuprates, formed <u>in situ</u> by the reaction of two equivalents of organolithium complex with one equivalent of cuprous iodide, cleanly alkylate a number of organic halides (11.13). The range of reactivity is very broad indeed. The R group of the

$$2\,RLi + CuX \xrightarrow{<0^\circ} [R_2Cu]Li \xrightarrow{R'X} RR' \qquad (11.13)$$

copper complex can be primary, secondary, or tertiary alkyl, vinyl, propenyl, aryl, and heteroaromatic. Although diallylcuprates can be made, they only react cleanly with alkyl halides and epoxides (see below). Dialkynylcuprates do <u>not</u> react well with halides, although simple copper acetylides (RC≡CCu) do (see below). Dialkenylcuprates <u>retain</u> the stereochemistry of the double bond upon reaction, leading to stereospecific coupling. These organocuprates are considerably less nucleophilic than the organolithium reagents whence they arose, and will react with halides in the presence of remote ester, cyano, and keto functionality. Aldehydes, however, cannot be present, since they react faster than halides.

The range of reactive halide substrates is also very broad. The order of reactivity of alkyl ($sp^3$) halides is $1^\circ > 2^\circ \gg 3^\circ$, with iodides more reactive than bromides and chlorides. Alkyl tosylates also undergo substitutions by diorganocuprates. Alkenyl halides are also very reactive toward lithium organocuprates, and react with essentially complete retention of configuration of the double bond. This feature has also proved useful in organic synthesis. Aryl iodides and bromides are alkylated by lithium organocuprates, although halogen-metal exchange is sometimes a problem, and RCu reagents are sometimes more efficient in this reaction. Benzylic, allylic, and propargylic halides also react cleanly. Although propargyl halides produce primarily allenes (via $S_N2'$ type reactions), allylic chlorides and bromides react without allylic transposition. In contrast, allylic acetates react mainly with allylic transposition. Although acid halides are converted to ketones by lithium dialkylcuprates, alkyl hetero (mixed) cuprates are more efficient at this process. Finally, epoxides undergo ring opening, resulting from alkylation at the less-substituted carbon. Consideration of a great deal of experimental data leads to the following order of reactivity of lithium diorganocuprates with organic substrates: acid chlorides > aldehydes > tosylates $\approx$

epoxides > iodides > ketones > esters > nitriles. This is only a generalization; many exceptions are caused by unusual features in either the organocopper reagent or the substrate.

As is the usual case in organometallic chemistry, mechanistic studies lagged far behind synthetic studies involving lithium diorganocuprates, and considerable controversy currently exists regarding both the nature of the copper reagent and the mechanisms by which the above alkylations ensue. Although $R_2CuLi$ is written for convenience, to denote the stoichiometry of the reagents involved, it is clearly not indicative of the structure of the reagent in solution. Studies by Pearson [22], including vapor-pressure depression data, indicate that "$R_2CuLi$" is dimeric in solution. The Cu-Cu distances estimated from X-ray scattering data are most consistent with a cyclic structure in which Cu and Li alternate with the methyl groups (1); $^1$H NMR studies are also

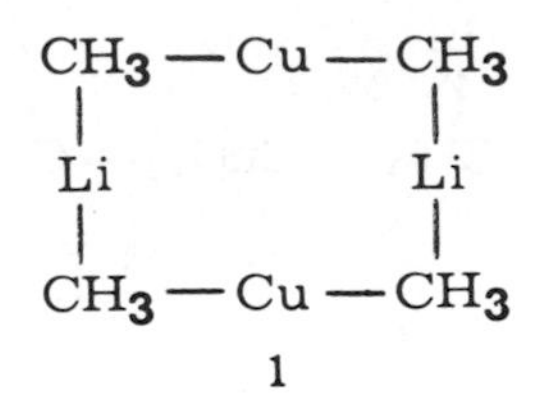

consistent with structure 1. This structure, if correct, goes a long way toward explaining some otherwise difficult-to-rationalize observations of the reactions of $R_2CuLi$ reagents with organic halides. A careful study of the reaction of lithium diorganocuprates with alkyl halides by Johnson strongly implicated an $S_N2$-type reaction mechanism [23]. The rate of this reaction was found to be second order, first order in substrate, and first order in cuprate. The order of reactivity is 1° > 2° >> 3°, and OTs > I > Br > Cl, the same as observed for straightforward $S_N2$ reactions. With optically active secondary tosylates, clean inversion of configuration is observed. A simple $S_N2$ process is possible, but the role of copper is then hard to specify (11.14). Alternatively,

$$R-\bar{Cu}(I)-R \; + \; >C-X \longrightarrow R-C\!\!< \; + \; RCu(I) \tag{11.14}$$

an $S_N2$ type oxidative addition (i.e., nucleophilic attack by Cu) was proposed (11.15). The major objection to this oxidative-addition pathway is the obligatory involvement of Cu(III) species, which are rather unlikely. Additionally, selective cross coupling is hard to explain, since it appears that the reductive elimination should produce symmetrical as well as cross-coupled products with equal facility. Both difficulties are

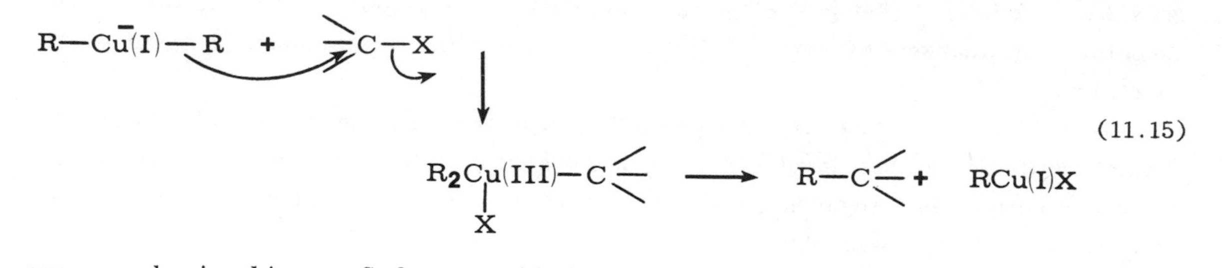


(11.15)

overcome by invoking an $S_N2$-type oxidative addition to the intact dimer (11.16). In this

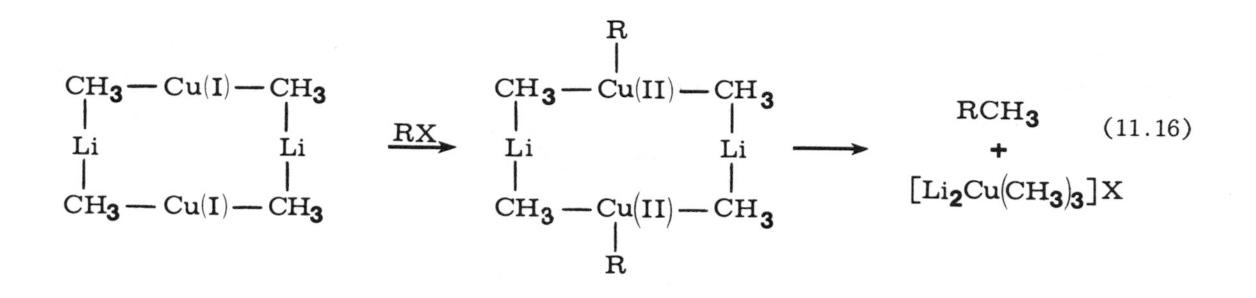


(11.16)

scheme, each Cu of the dimer contributes only one electron for the oxidative-addition process, resulting in the production of two Cu(II)s rather than Cu(III). In the dimeric oxidative adduct, the alkyl groups from the cuprate reagent are distinguishable from that of the substrate, and hence selective cross coupling results.

The mechanism of the reaction of $R_2CuLi$ complexes with vinyl and aryl halides has not been examined in any detail. Clearly, direct $S_N2$ reactions are unlikely. However, an oxidative addition reaction mechanism similar to that involved in the reaction of the isoelectronic Pd(0) complexes with these substrates is likely (Chapter 4). Again, involvement of both coppers of an intact dimer would obviate the involvement of Cu(III) intermediates.

Whatever the mechanism of reaction, lithium diorganocuprates have been extensively utilized for the alkylation of complex molecules [24-28]. A few representative examples are listed in Equations 11.17 to 11.21.

Another extensively used process in the synthesis of complex organic molecules is the conjugate addition of lithium diorganocuprates to conjugated enones (11.22). The range of R groups available for this useful reaction is similar to that for the alkylation reaction. Virtually all types of $sp^3$ hybridized dialkylcuprate reagents react cleanly,

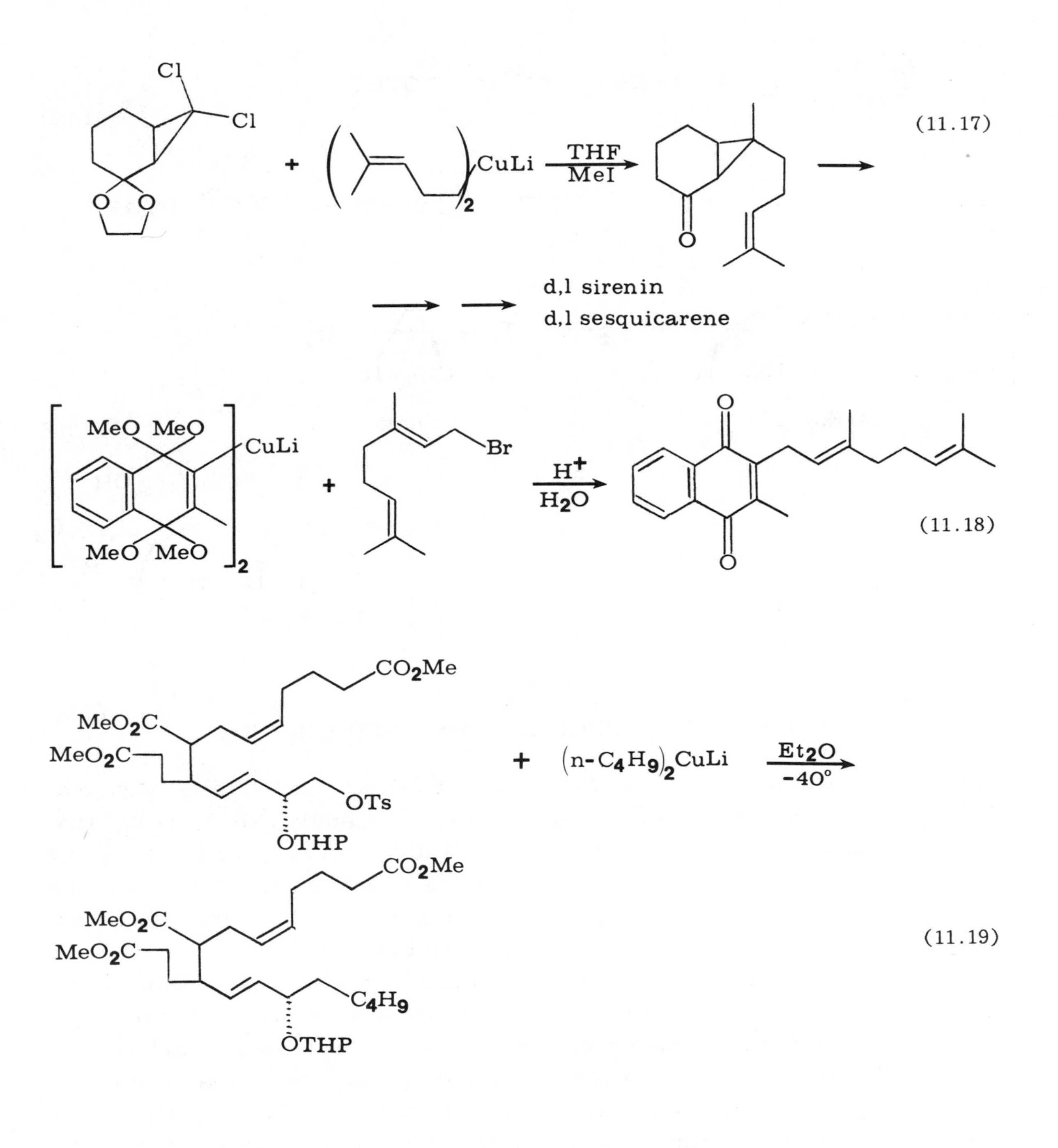
Cl
Cl
O O
+
CuLi
2
THF
MeI
O
(11.17)
d,l sirenin
d,l sesquicarene
MeO MeO
CuLi
MeO MeO
2
+
Br
H+
H2O
O
O
(11.18)
CO2Me
MeO2C
MeO2C
OTs
OTHP
+
(n-C4H9)2CuLi
Et2O
-40°
CO2Me
MeO2C
MeO2C
C4H9
OTHP
(11.19)

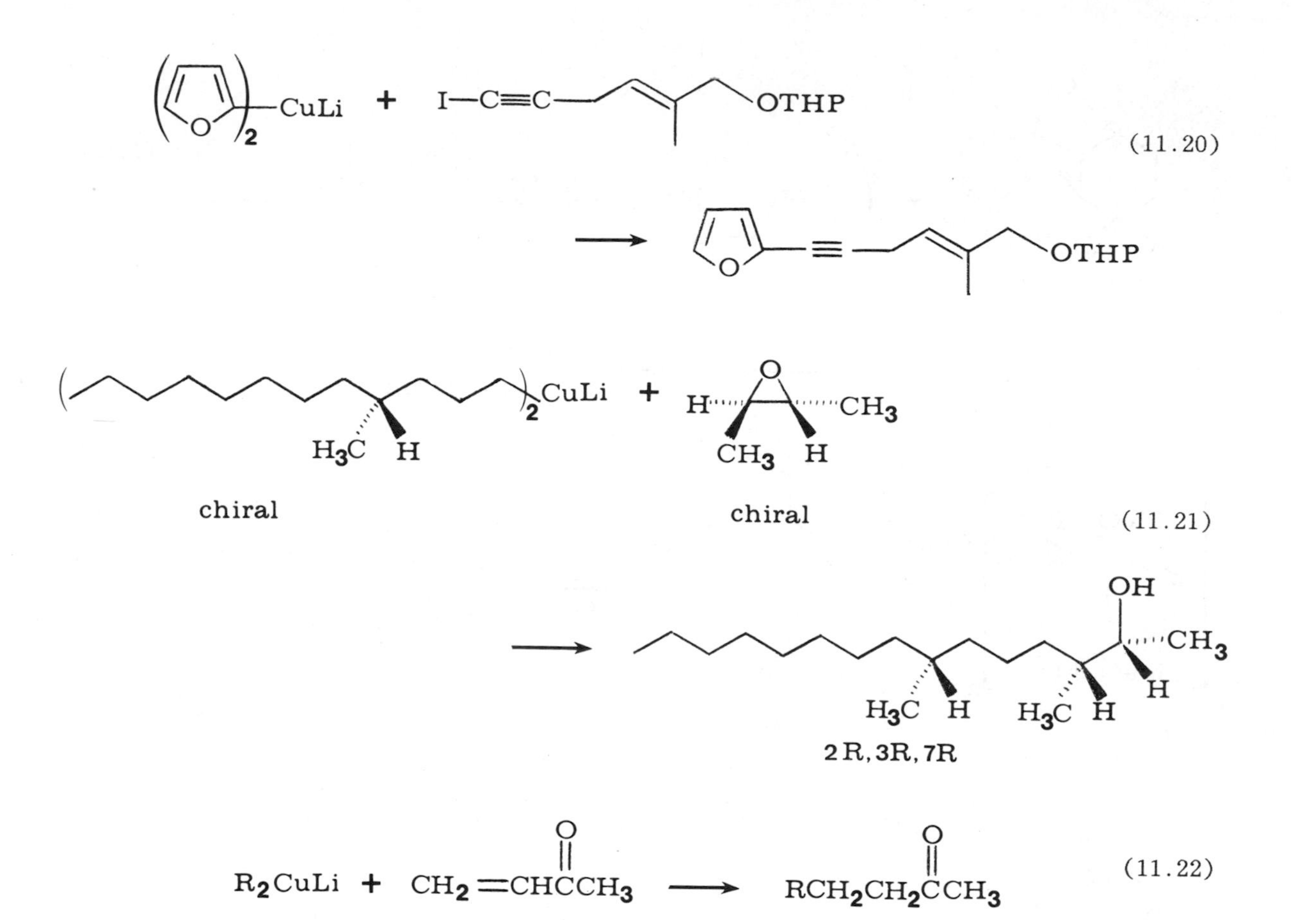


including straight-chain and branched primary, secondary, and tertiary alkyl complexes. Phenyl and substituted arylcuprates react in a similar fashion, as do vinylcuprates. Allylic and benzylcuprates also add 1,4 to conjugated enones, but they are no more efficient than the corresponding Grignard reagents. Acetylenic copper reagents do not transfer an alkyne group to conjugated enones, a feature used to advantage in the chemistry of mixed cuprates discussed below.

The range of α,β-unsaturated carbonyl substrates that undergo 1,4-addition with diorganocuprates is indeed broad. Although the reactivity of any particular system depends on many factors, including the structure of the copper complex and the substrate, sufficient results are available to allow the formulation of some generalities. Conjugated ketones are among the most reactive substrates, reacting very rapidly with diorganocuprates (the reaction is complete in less than 0.1 sec at 25°) to produce the

1,4-adduct in excellent yield. Alkyl substitution at the α,α', or β positions of the parent enone causes only slight alterations in the rate and course of this conjugate addition with acyclic enones, although this same substitution, as well as remote substitution, does have stereochemical consequences with cyclic enone systems. For example, with 4-substituted cyclohexenones, the trans isomer of the adduct always predominates, but as the steric bulk of the 4-substituent increases, the amount of trans product also increases (11.23). With the 1-methyloctalone system, alkylation produces the cis product (11.24).

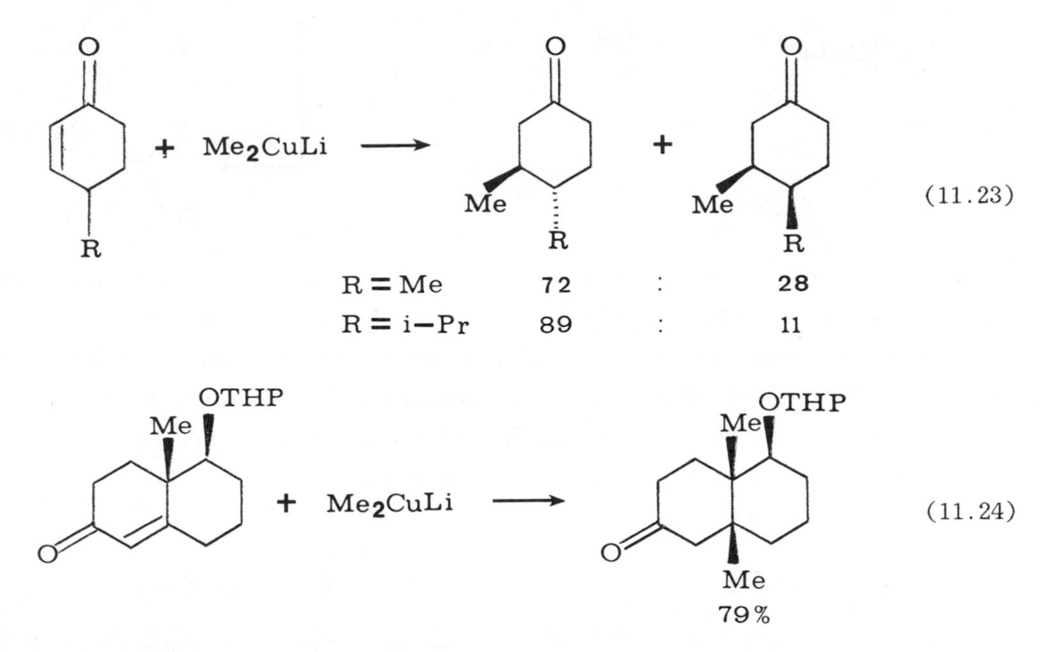


Conjugated esters are somewhat less reactive than conjugated ketones. With these substrates, α, β, and β,β-disubstitution decrease reactivity drastically, making these substrates of less use synthetically. Conjugated carboxylic acids do not react with lithium diorganocuprates, and conjugated aldehydes suffer competitive 1,2-additions. Conjugated anhydrides and amides have not been studied to any extent.

Acetylenic esters are considerably more reactive with diorganocuprates than are the corresponding ethylenic esters, and this reaction has found significant use in organic synthesis. The reaction is thought to proceed through a vinylcopper species which is formed by a stereospecific cis addition of cuprate. Reaction of this vinylcopper complex with a number of electrophiles (Equation 11.25) permits the stereospecific

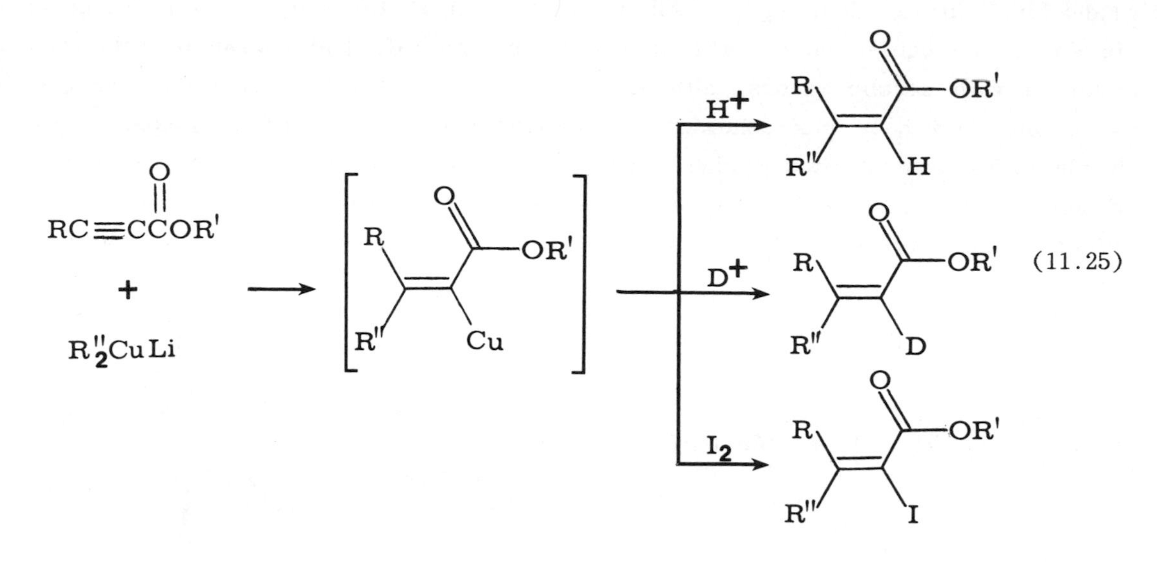


synthesis of trisubstituted olefins. Similar reactions occur with acetylenic ketones, acids, and amides, although they have been studied much less [29].

Again, mechanistic studies of this reaction lag far behind synthetic studies, and significant differences of opinion exist concerning the specifics of this reaction. It is currently (1979) thought that conjugate addition proceeds either by an electron-transfer process or by a direct nucleophilic-addition process [30]; see Figure 11.1. (Note that

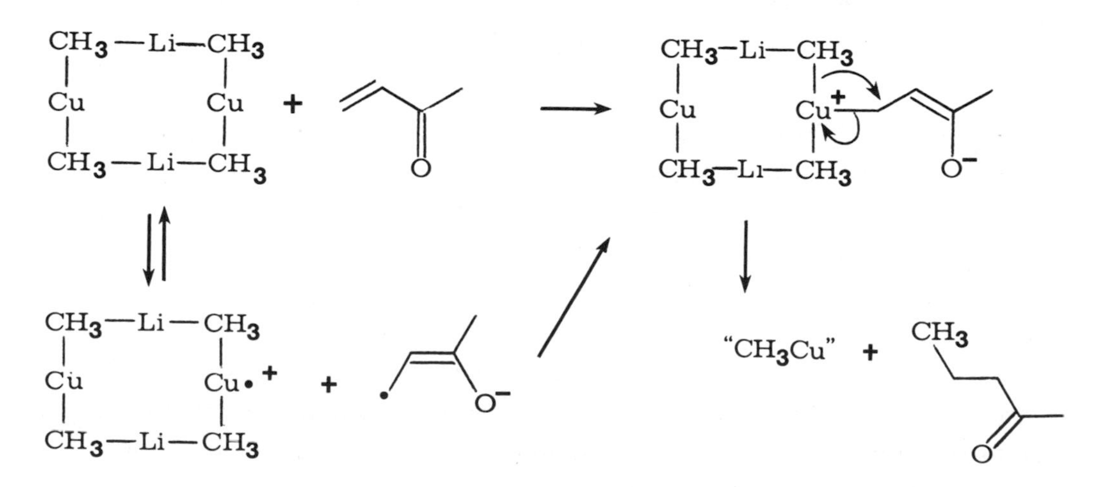


*Figure 11.1. Mechanism for alkylcuprate conjugate additions.*

both ultimately involve a Cu(III) species for which there is no direct evidence. Note also that direct transfer of the alkyl from copper to enone, in a process not involving a copper-enone intermediate, cannot be ruled out.)

The nucleophilic-addition mechanism is supported by the demonstrated ability of organocuprates to act as nucleophiles. Recall that alkyl halides, alkyl tosylates, and epoxides all react with organocuprates with the clean inversion of stereochemistry characteristic of direct nucleophilic displacements (vide supra). In this view, the conjugate addition of organocuprates to enones is similar to the familiar Michael addition of nucleophiles to the same substrates.

On the other hand, a number of experimental results (primarily the work of House) are consistent with the electron-transfer mechanism [31]. First, there is a strong correlation between reactivity of conjugated enones toward conjugate addition and their polaragraphic one-electron reduction potentials. Thus, with enones whose reduction potentials are less than ~-1.5 volts (easily reduced), reduction of the double bond (a net 2 $e^-$ process) without alkylation results. With enones having reduction potentials between -1.5 V and -2.35 V, conjugate addition occurs. With enones even more difficult to reduce (reduction potentials >-2.35 V), no reaction occurs. Opponents of electron transfer remark that reduction potential simply measures the electron affinity of the enone, and should correlate with both a one-electron transfer process and a nucleophilic (two-electron) addition process. Electron-transfer buffs respond that other Michael additions to conjugated enones show no apparent correlation with reduction potential.

Second, if enone radical anions are obligatory intermediates, reactions characteristic of these species should be observed with appropriate substrates. For example, the less stable cis enone 2 very rapidly equilibrates to the trans enone 3 when converted to the corresponding radical anion (11.26).

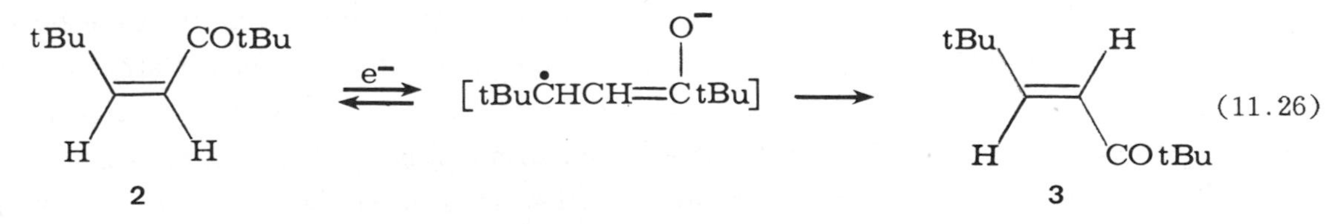

(11.26)

Reaction of excess 2 with reagents known to react by a direct nucleophilic-attack mechanism (i.e., $LiAlH_4$ or $CH_3Li$) leads to no isomerization of recovered starting material, whereas reaction of 2 with $Me_2CuLi$ leads to complete isomerization of recovered enone to the trans isomer. However, because electron exchange between radical anions and their precursors is very rapid, only small amounts of radical anion are necessary for complete isomerization to occur. Thus it is unclear whether the isomerization is due to an unrelated side reaction, or to the intermediacy of radical anions in the conjugate addition of organocuprates to enones.

Finally, the observation that β-cyclopropyl-α,β-unsaturated ketones undergo competitive alkylative ring opening along with conjugate addition has been attributed to the intermediacy of radical anions (11.27). However, this mechanism passes through an

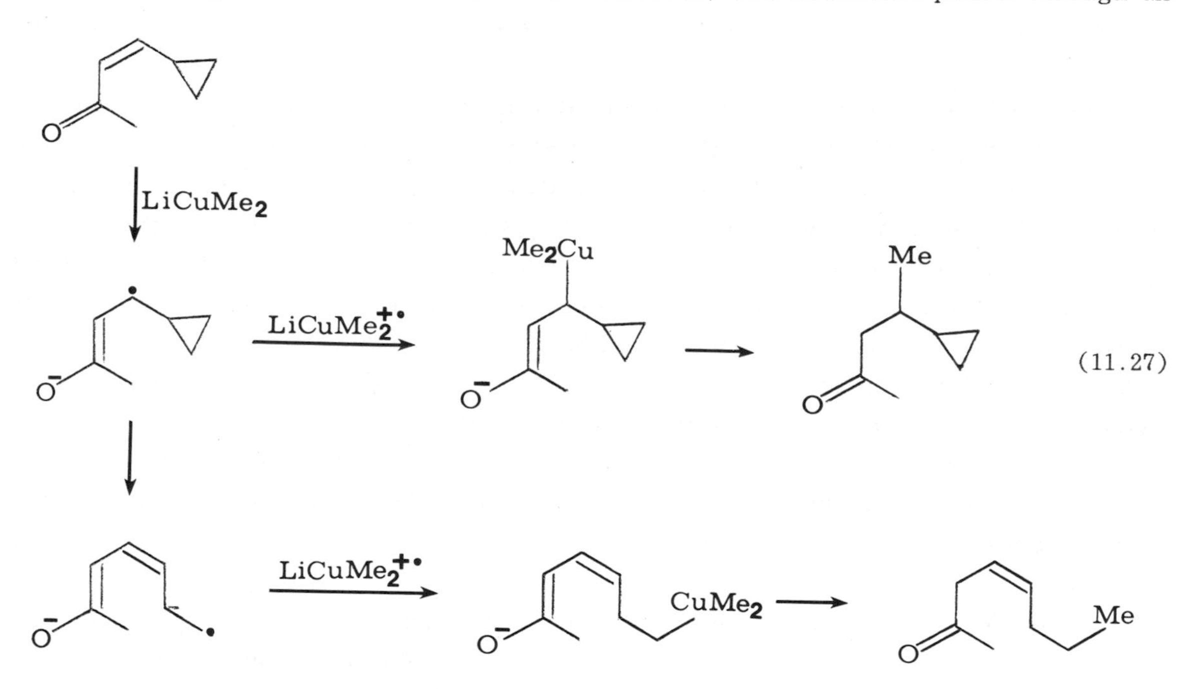


intermediate in which the cyclopropyl carbon which is alkylated is a radical. If this were the case, it should lose any stereochemistry it had. A careful labeling study by Casey [30] has shown that the cyclopropyl group undergoes alkylative cleavage by alkylcuprates stereospecifically (11.28; although the absolute stereochemistry is not known, inversion is assumed). Thus this reaction appears to go by direct nucleophilic cleavage rather than by an electron-transfer mechanism. The dispute continues.

Although the mechanism of conjugate addition is still subject to some question, applications in synthesis continue unabated. Prostanoid syntheses [32,33] have relied particularly heavily on this conjugate-addition reaction (11.29, 11.30). Note the degree of functionality tolerated in both the substrate and the organocopper complex. Many

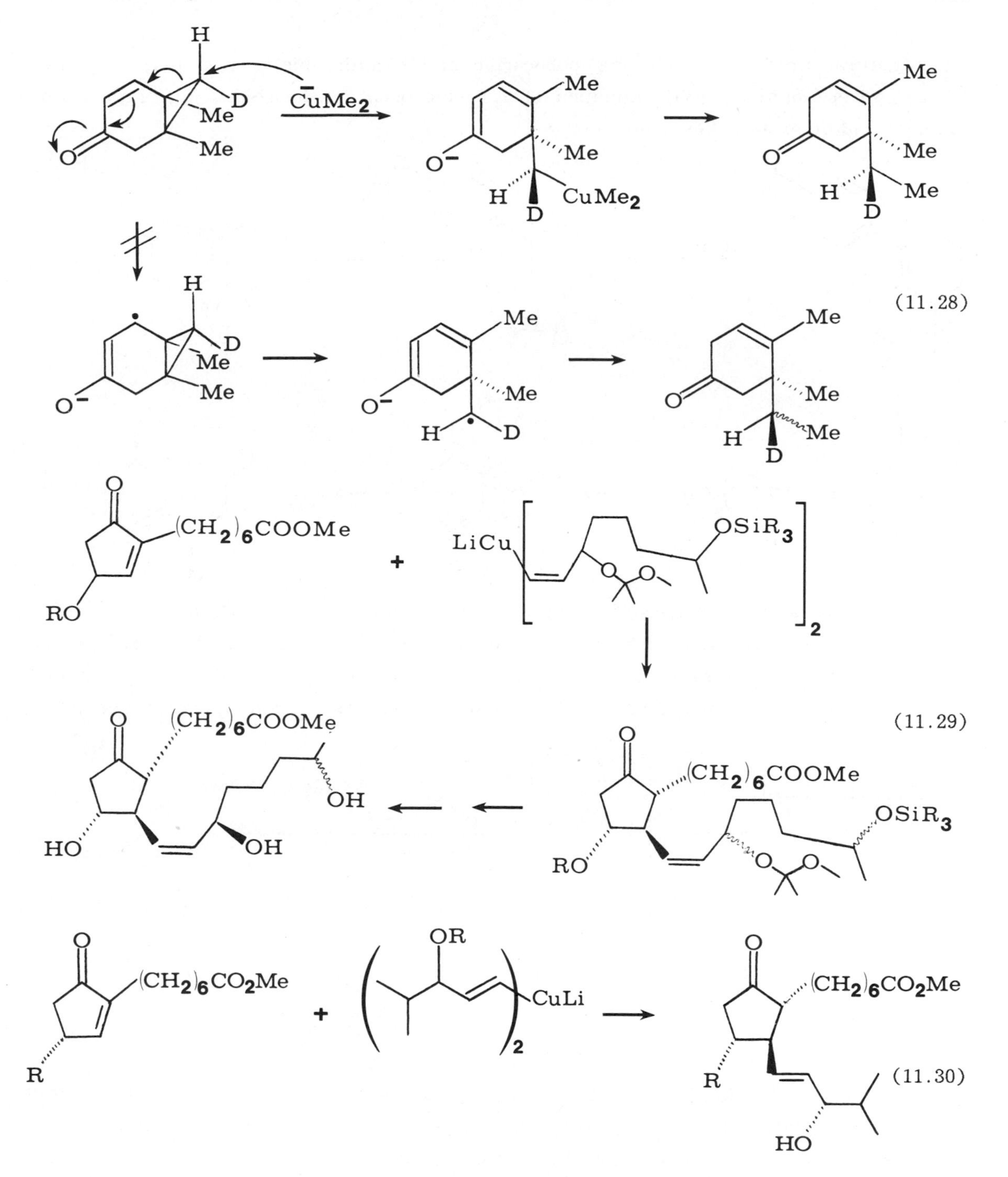
H
CuMe2
D
Me
Me
O
O−
CuMe2
H
D
(11.28)
(CH2)6COOMe
LiCu
OSiR3
RO
2
(11.29)
HO
OH
(CH2)6CO2Me
OR
CuLi
R
(11.30)
HO

other natural products, including podocarpic and dehydroabietic acid [34] (Equation 11.31) and eremophilone [35] (Equation 11.32) have been synthesized using organocopper conjugate addition as a key step.

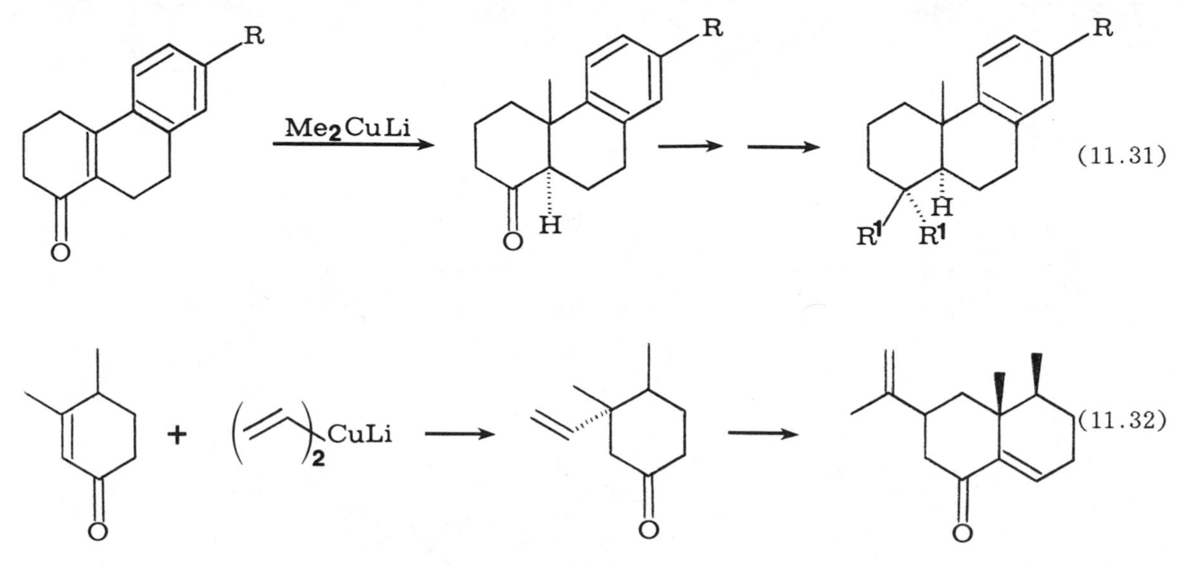


The initial adduct, prior to aqueous quench, is either a copper or a lithium enolate [36], and further reaction with electrophiles is possible. This has been used to an advantage in several syntheses. For example, a deoxyprostaglandin $E_2$ model compound has been made in this fashion [37] (Equation 11.33). A very clever annellation

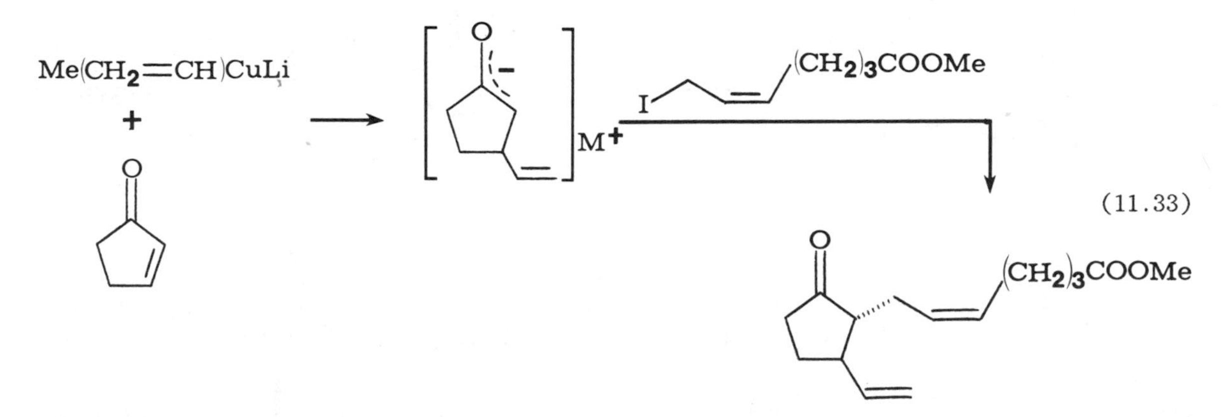


procedure has been developed, utilizing a conjugate addition of the enolate to a vinyl silane [38] (Equation 11.34). Similarly, the bicyclic product, valarane, is produced by an intramolecular version [37] of this reaction (11.35). A particularly impressive

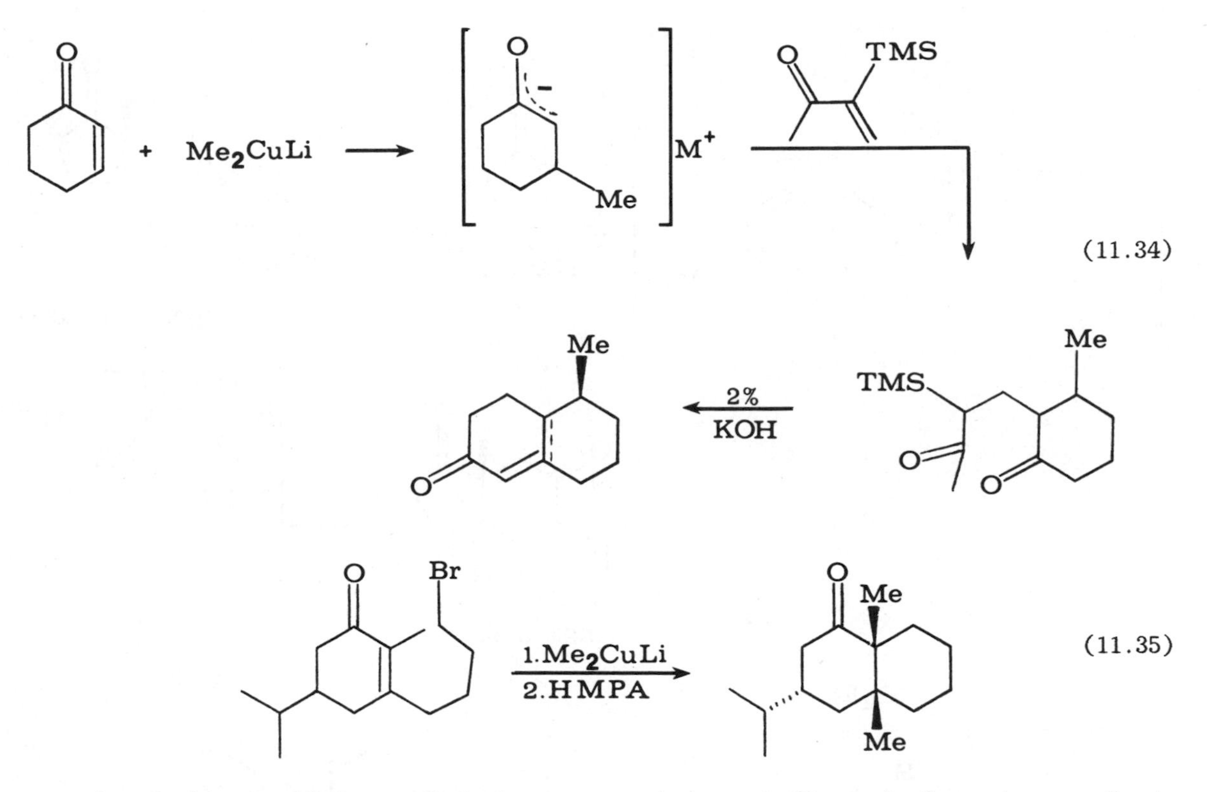


example of this β-addition-α-alkylation process is seen in Posner's four-step synthesis [39] of equilenin (11.36). The one-pot copper-assisted process proceeds in very high yield and is essentially stereospecific. Saponification of the ester, protection of the keto group, and acid-catalyzed ring closure completes the synthesis. Allenic esters also undergo this conjugate addition-alkylation sequence. Lavandulol has been synthesized [40] in this manner (11.37).

Although lithium dialkylcuprates are very useful reagents, they suffer several serious limitations, the major one being that only one of the two alkyl groups is transferred. Because of the instability of the reagent, a threefold to fivefold excess (a sixfold to tenfold excess of R) is often required for complete reaction. When the R group is large and complicated, as it often is in prostaglandin chemistry, for example, this lack of efficiency is intolerable. Cuprous alkoxides, mercaptides, and acetylides are considerably more stable than cuprous alkyls, and, furthermore, are relatively unreactive toward the substrates normally used with dialkylcuprates. By making mixed alkyl heterocuprates, in which one group is a nontransferrable, stabilizing alkoxide,

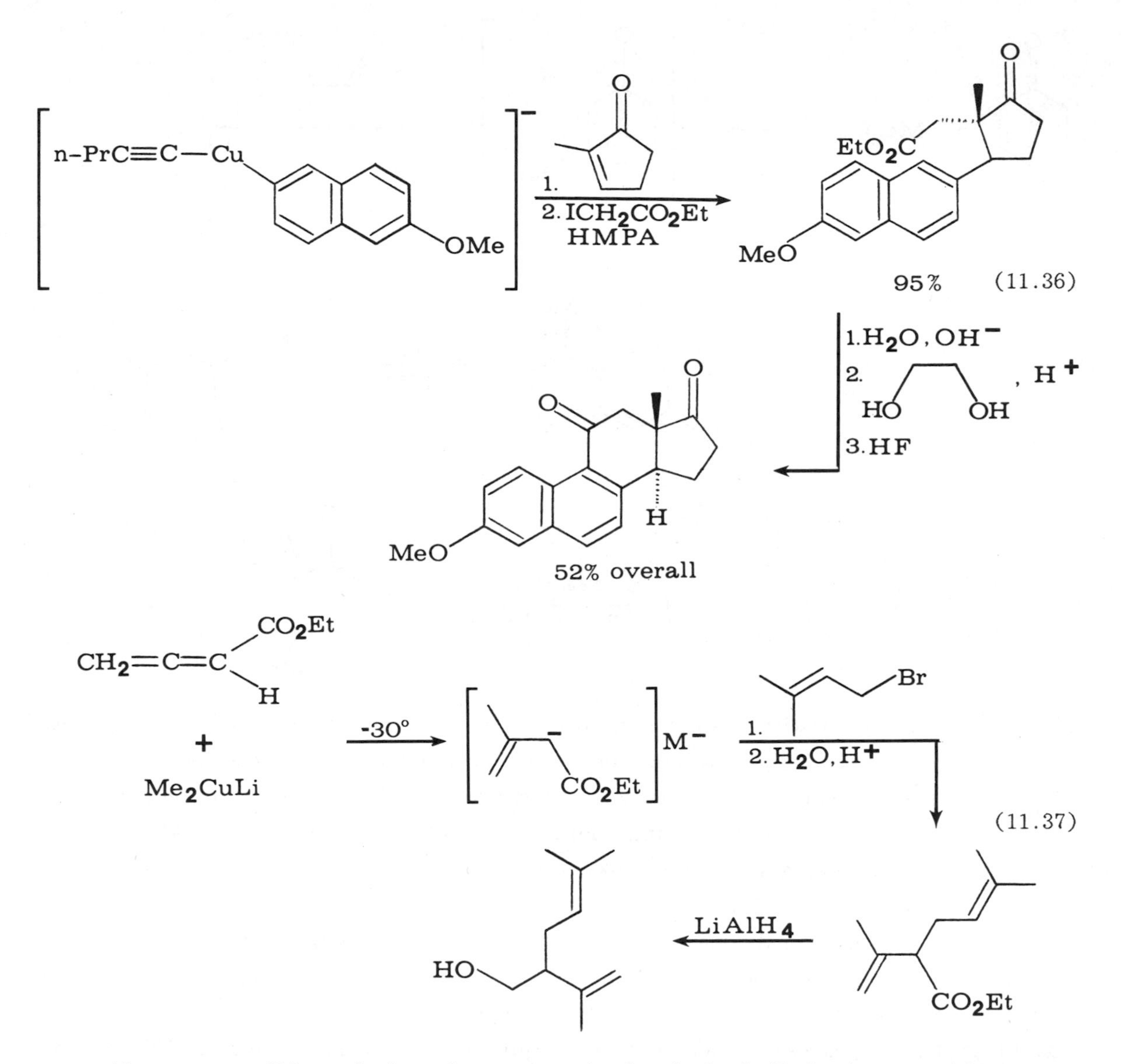


mercaptide, or acetylide and the other group is the desired alkyl, these problems were largely solved by Posner [41]. Among the heterocuprates, lithium phenylthio(alkyl)-cuprates and lithium t-butoxy(alkyl)cuprates are the most useful.

For example, these are the most efficient reagents for the conversion of acid chlorides to ketones. Although large excesses of lithium dialkylcuprates are required for this reaction, and secondary and tertiary organocuprates never work well, a 10 percent excess of these mixed cuprates is sufficient to lead to high yields of ketones, even in the presence of remote halogen, keto, and ester groups. In addition

$$Br(CH_2)_{10}COCl + tBuO(tBu)CuLi \longrightarrow Br(CH_2)_{10}COtBu \quad 78\% \tag{11.38}$$

$$n\text{-}BuCO(CH_2)_4COCl + tBuO(secBu)CuLi \longrightarrow n\text{-}BuCO(CH_2)_4COsecBu \quad 86\% \tag{11.39}$$

$$EtO_2CCH_2CH_2COCl + PhS(tBu)CuLi \longrightarrow EtO_2CCH_2CH_2COtBu \quad 65\% \tag{11.40}$$

although aldehydes are unstable toward lithium dialkylcuprates even at -90°, equal portions of benzaldehyde and benzoyl chloride react with one equivalent of lithium phenylthio(t-butyl)cuprate to produce pivalophenone in 90 percent yield with 73 percent recovery of benzaldehyde. The lithium phenylthio(t-alkyl)cuprate is also the reagent of choice for the alkylation of primary alkyl halides with t-alkyl groups. Secondary and tertiary halides fail to react with this reagent, however.

Perhaps most importantly, these mixed cuprate reagents, especially the acetylide complexes, are extremely efficient in the conjugate alkylation of $\alpha,\beta$-unsaturated ketones. Since it is this reaction that is used most extensively in the synthesis of complex molecules, this was an important advance which was rapidly applied to a variety of total syntheses, including the prostaglandins [42,43] (Equations 11.41 and 11.42).

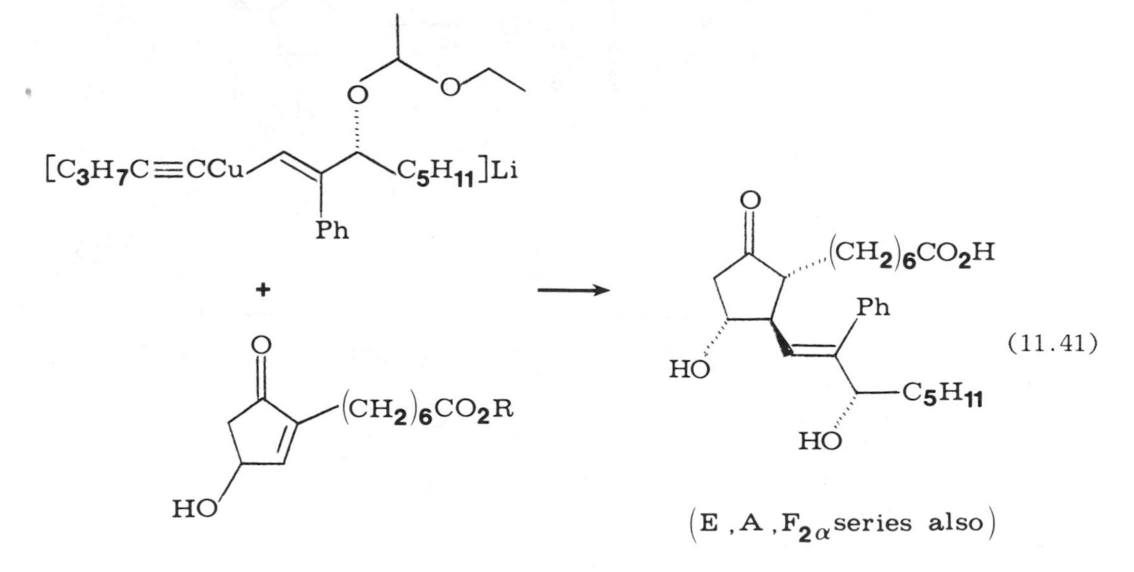

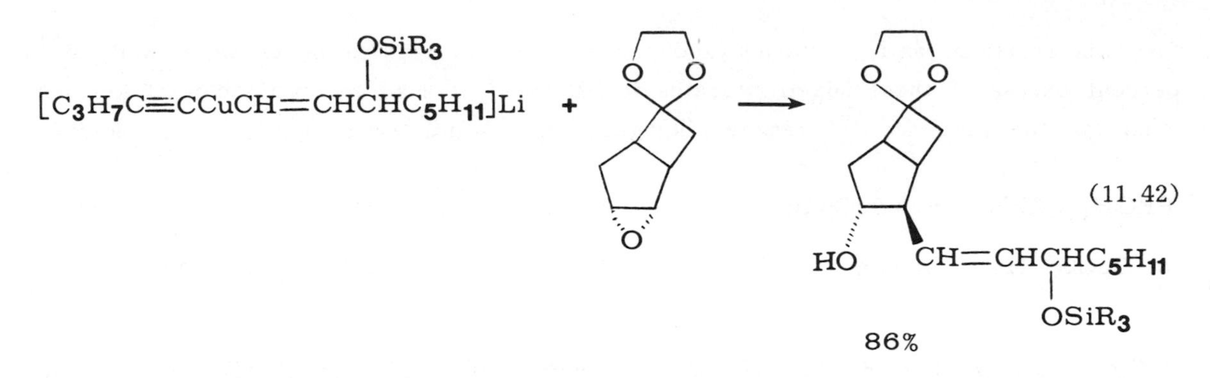


The ability to introduce complex, functionalized groups efficiently via these mixed cuprates has also been applied [44-46] to a number of other interesting systems (11.43 to 11.45). As already discussed, alkynylcuprates do not transfer the alkynyl group to

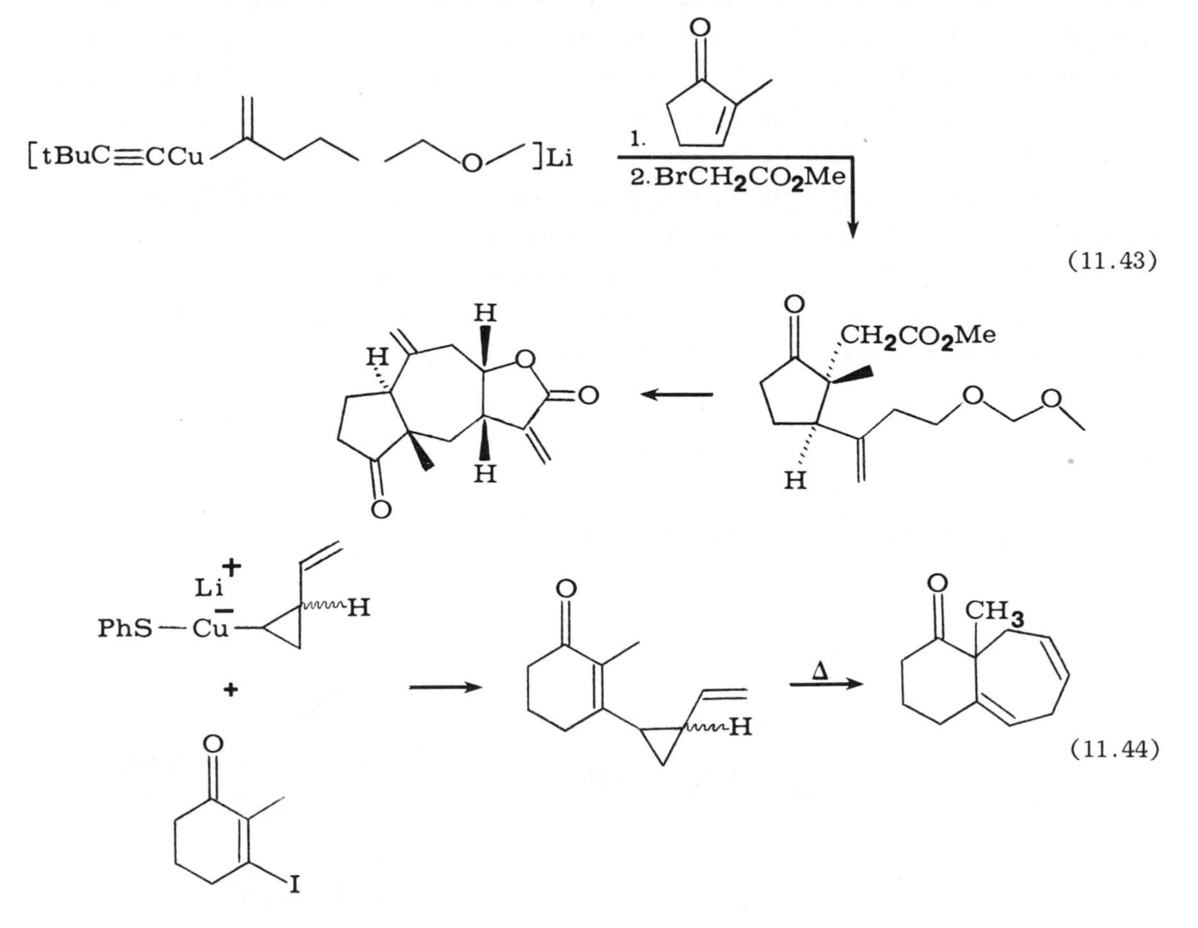

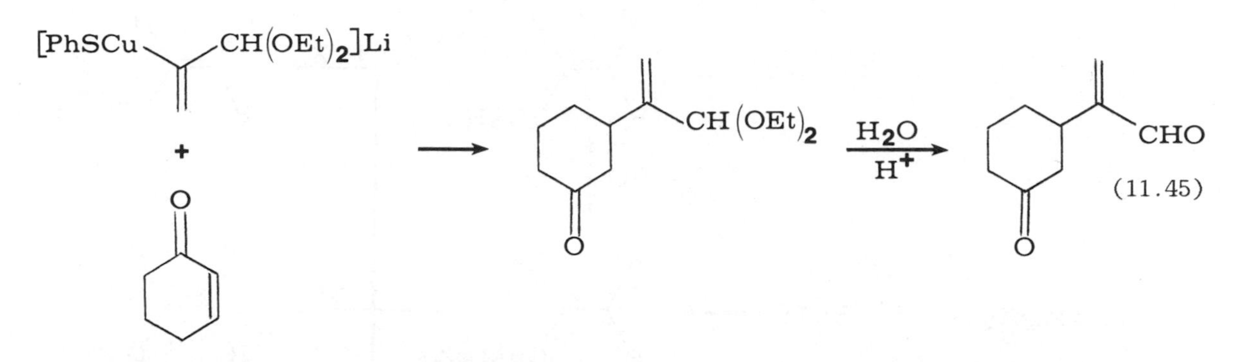

(11.45)

conjugated enones. However, a mixed cuprate containing a transferrable ethynyl-group equivalent permits the facile introduction of an angular ethynyl group into fused ring systems [47] (Equation 11.46).

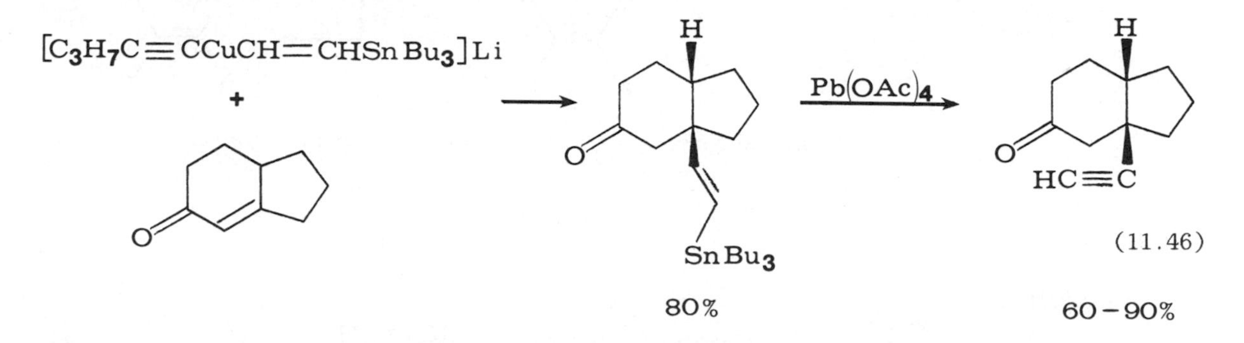

(11.46)

Although dialkyl or alkyl heterocuprates are by far the most extensively used organocopper reagents, the simple organocopper complexes RCu or RCuMX perform much of the same chemistry, and are clearly superior for certain types of reactions, particularly the addition of $RCuMgX_2$ to terminal alkynes. Although $R_2CuLi$ complexes merely abstract the acidic acetylenic proton from terminal alkynes, and pure RCu or RCuLiX do not add at all, the complexes $RCuMgX_2$, generated from Grignard reagents and copper(I) iodide, cleanly add syn to simple terminal alkynes to produce vinylcuprates both regio- and stereospecifically. These vinyl complexes enjoy the reactivity common to organocopper species and react with a variety of organic substrates (Figure 11.2) [17, 48-51]. The range of reactive electrophiles can be extended by the use of slightly different conditions [52,53] (Equation 11.47). These reactions are of particular value for the synthesis of trisubstituted double bonds, and should find extensive application in the syntheses of terpenoids and insect hormones.

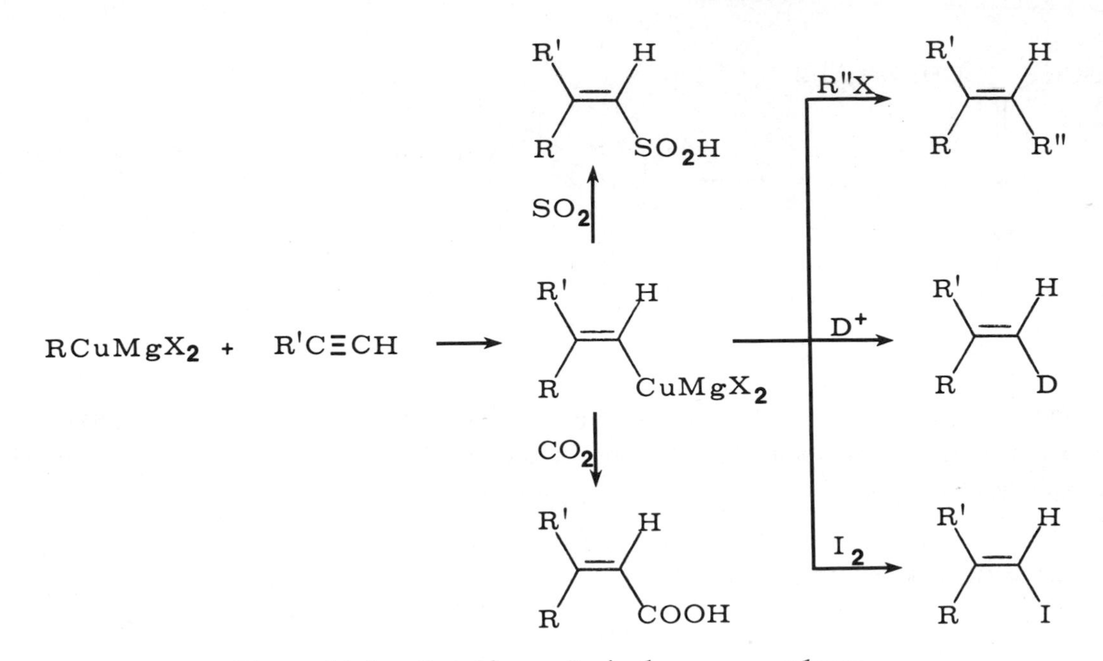


*Figure 11.2. Reactions of vinylcopper complexes.*

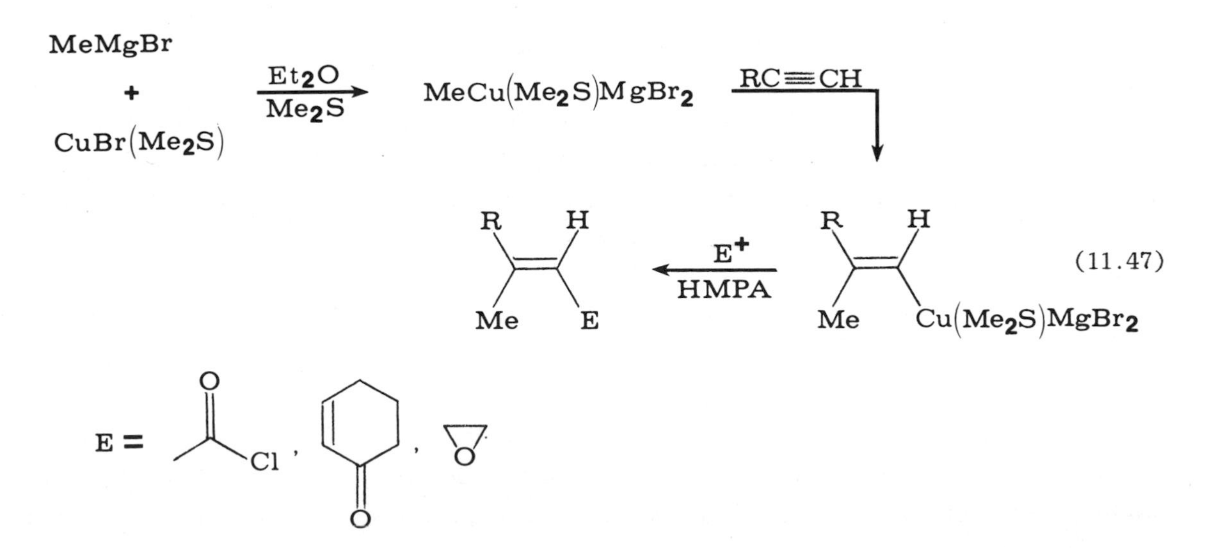


With heteroatom-substituted alkynes, the regiochemistry is somewhat more complex, and appears to be controlled by both the initial polarities of the substrates [17] and the ability of the heteroatom to coordinate to the copper and deliver it to a specific site [54] (Equation 11.48). Alkynes having good leaving groups in the propargyl position

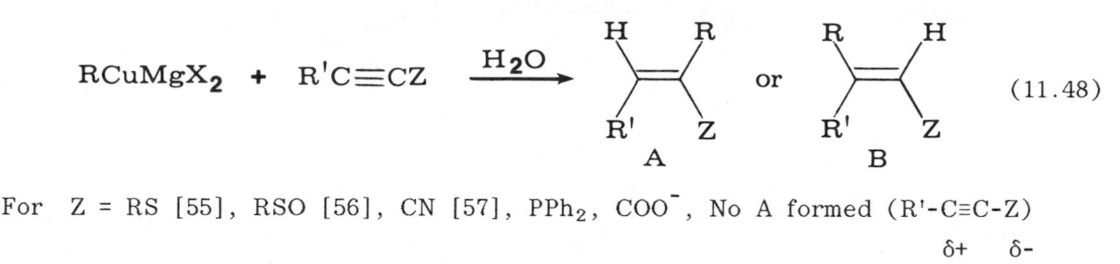


For Z = RS [55], RSO [56], CN [57], $PPh_2$, $COO^-$, No A formed (R'-C≡C-Z) ($\delta+$ $\delta-$)

For Z = OEt [55], $NEt_2$ [55], $NPh_2$ [54], No B formed (R-C≡C-Z) ($\delta-$ $\delta+$)

undergo an $S_N2'$ type displacement to produce the corresponding allene [58]. Chiral propargyl sulfones produce chiral allenes in excellent yield [59] (Equation 11.49).

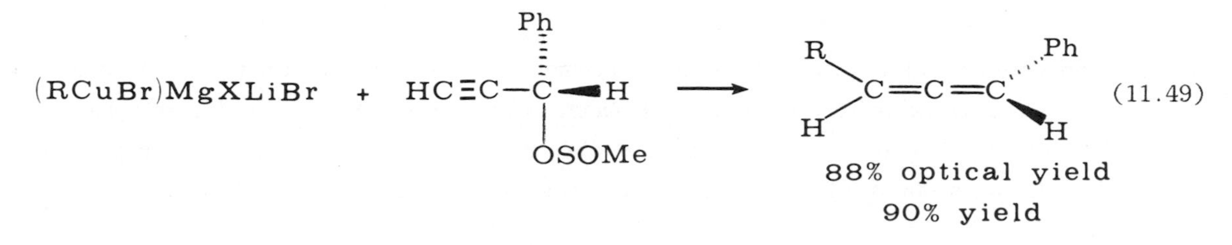


It was the observation that copper salts catalyzed the 1,4-addition of Grignard reagents to conjugated enones that sparked investigations into the behavior of stoichiometric organocopper complexes. In turn, current work on stoichiometric copper reagents has led to a resurgence of interest in copper-catalyzed Grignard reactions, and the utility of this type of reaction has been greatly expanded [17,54]. The conjugate addition reaction, already well-established, has been improved for use with vinyl Grignards by the addition of ligands, such as diisopropyl sulfide, to the reaction mixture. Propargyl alcohols undergo an anti alkylation by Grignards when the reaction is catalyzed with 3 percent CuI (as in 11.50), whereas preformed organocopper complexes add syn.

Substitution reactions of primary alkyl halides and tosylates by Grignards are also catalyzed by copper salts, and $\alpha,\omega$-dihalides undergo monoalkylation exclusively, at $-78^\circ$, as in 11.51. Vinyl iodides (available from the alkylation of alkynes by $RCuMgX_2$) similarly undergo clean alkylation by Grignard reagents in the presence of catalytic

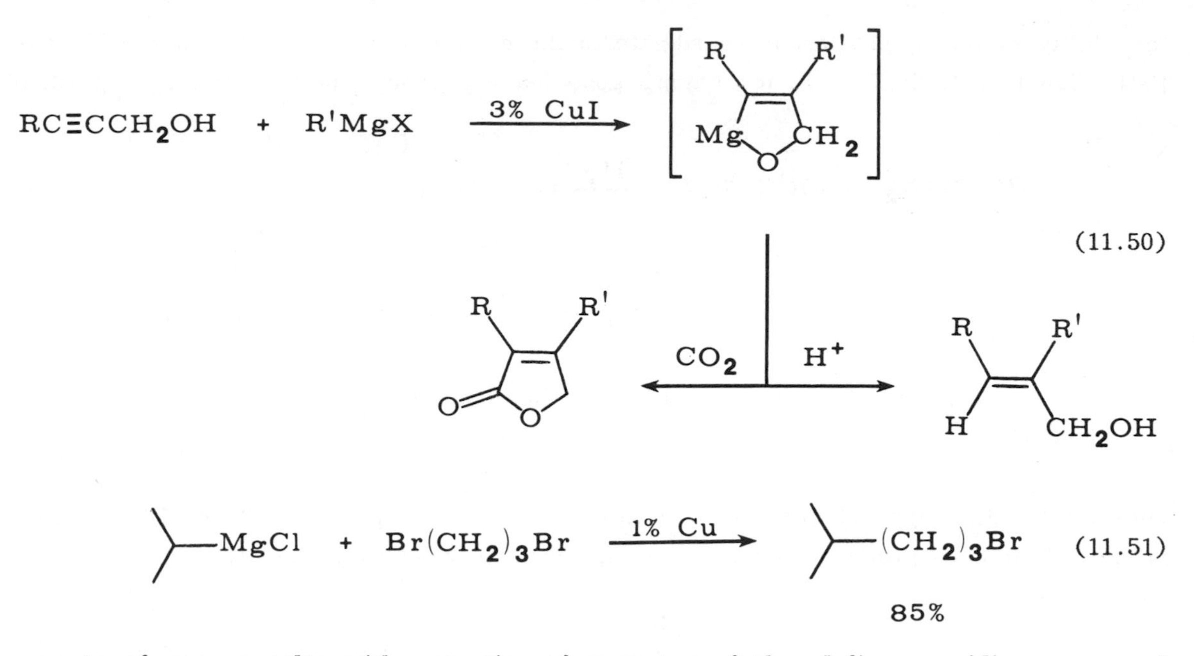


amounts of copper salts with retention of geometry of the olefin, providing a general route to di- and trisubstituted olefins. Primary allylic acetates are regio- and stereoselectively substituted by Grignards in the presence of catalytic amounts of copper(I) salts (11.52), whereas allyl ethers give mixtures of products resulting from both $S_N2$

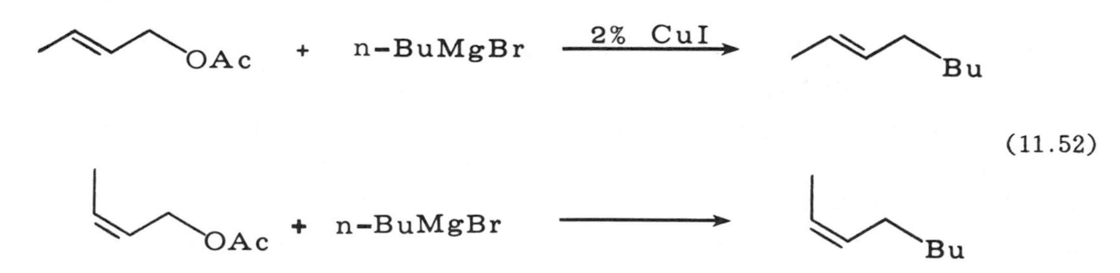


and $S_N2'$ reactions, depending on the substitution pattern of the substrate (11.53). Allyl acetals react cleanly to produce enol ethers which are easily hydrolyzed to the corresponding aldehydes (11.54).

As was mentioned at the outset of this section, organocopper chemistry is a complex and rapidly changing field in which new reactions are developed rapidly, but mechanistic understanding only slowly. Hence what is presented here is a brief summary of this field as it now stands.

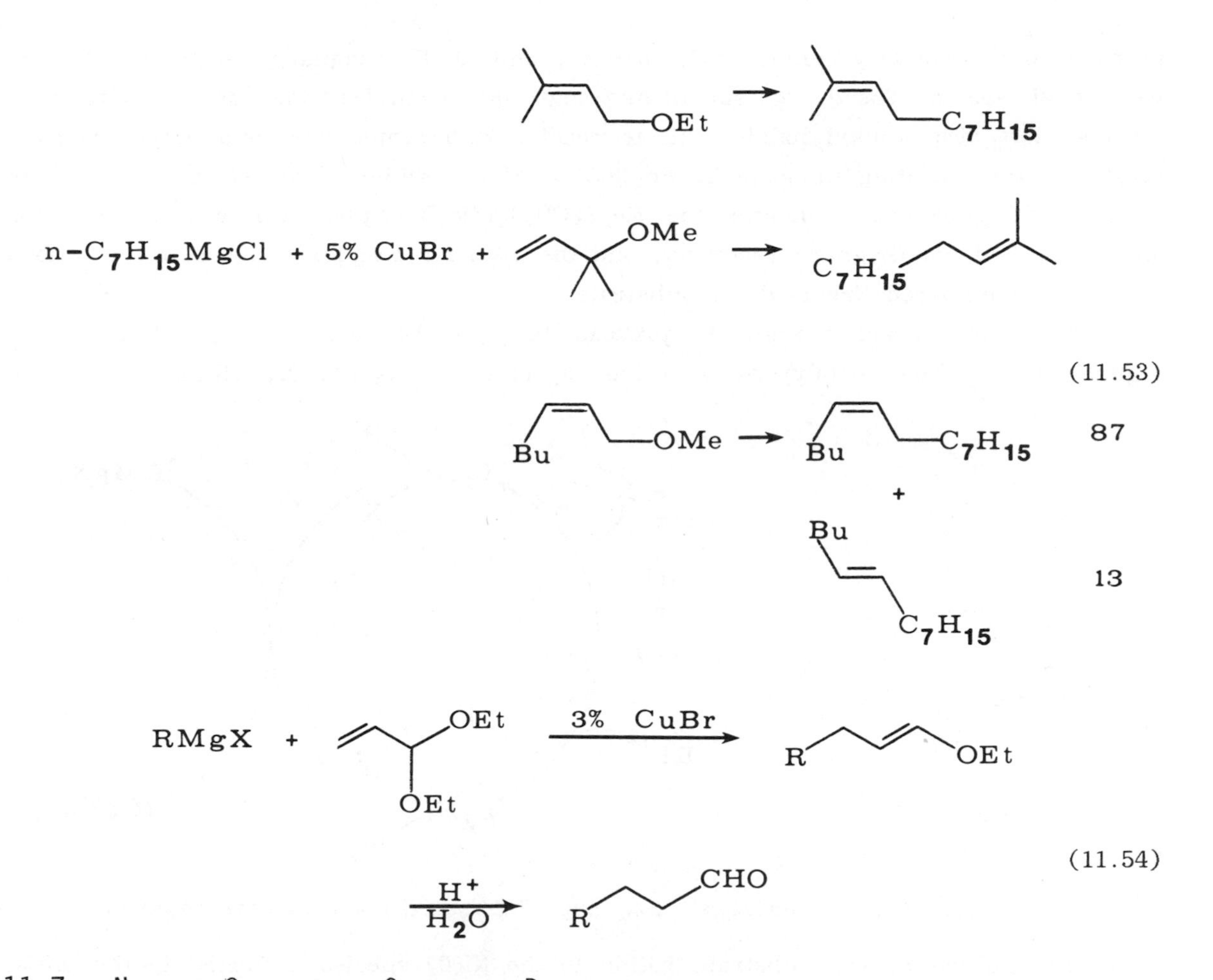


## 11.3. Nickel-Catalyzed Grignard Reactions.

The successful use of copper salts to catalyze Grignard reactions has led to the investigation of other transition-metal salts for this same purpose. Although many transition-metal salts do effect the course of Grignard reactions with various substrates [60], nickel complexes have been the most extensively studied, primarily by Kumada [61]. Nickel-phosphine complexes catalyze the selective cross coupling of Grignard reagents with aryl and olefinic halides (11.55). The reaction works equally well with

$$\text{RMgX} + \begin{matrix}\text{ArX}' \\ \text{or} \\ \text{CH}_2\text{=CH–X}\end{matrix} \xrightarrow{\text{NiL}_2\text{Cl}_2} \begin{matrix}\text{RAr} \\ \text{or} \\ \text{CH}_2\text{=CH–R}\end{matrix} + \text{MgXX}' \qquad (11.55)$$

primary and secondary alkyl, aryl, alkenyl, and allylic Grignard reagents. A wide variety of simple, fused, or substituted aryl and alkenyl halides are reactive substrates, but simple alkyl halides fail to react. With simple aryl and alkyl Grignard reagents, the chelating diphosphine $Ph_2P(CH_2)_3PPh_2$ complex of nickel chloride was the most effective catalyst, whereas the $Me_2P(CH_2)_2PMe_2$ complex was most suitable for alkenyl and allylic Grignard reactions, and the bis $Ph_3P$ complex was best for reactions of sterically hindered Grignards or substrates.

The reactions are thought to proceed through the catalytic cycle described in Figure 11.3. The catalytically active species is $L_2Ni(R)X$, formed by the

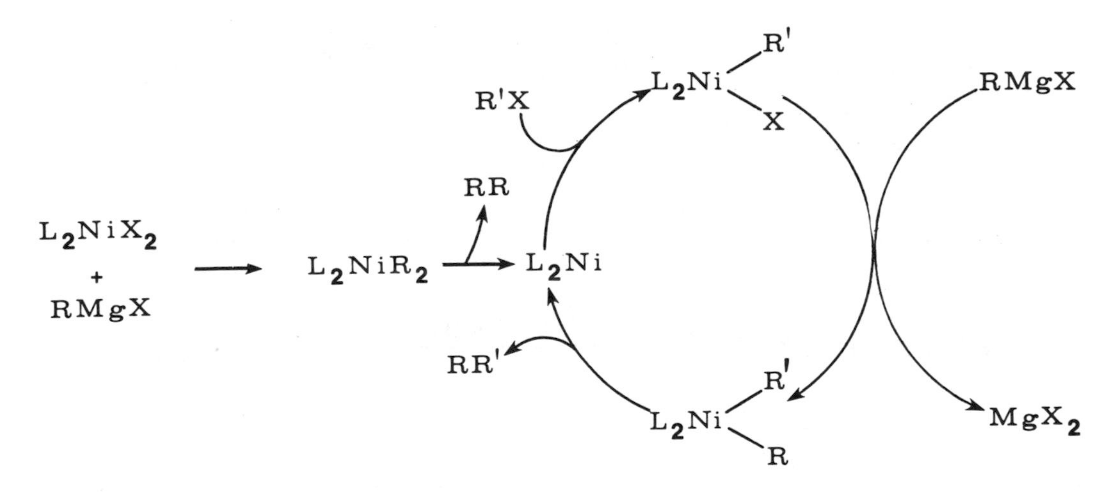


*Figure 11.3. Nickel-catalyzed coupling of Grignard reagents with halides.*

oxidative addition of the substrate halide to the Ni(0) species produced in the initial step. Alkylation of this complex by the Grignard reagent produces an unsymmetrical dialkylnickel complex, which reductively eliminates the cross-coupled product and regenerates $L_2Ni(0)$ to carry the catalytic cycle. If this is the correct mechanism, it would suggest that phosphine-nickel(0) complexes would also be catalysts in this system, and that the reaction is mechanistically related to some Pd(0) coupling reactions discussed below.

Recent studies by Kochi [62] of the nickel-catalyzed coupling of aryl halides to biaryls (a process closely related to that of Figure 11.3) show this to be a radical chain process in which the key steps are oxidative addition of RX to a Ni(I) species, and reductive elimination from a diaryl Ni(III) species. Similar steps may be involved in the nickel-catalyzed Grignard coupling reactions (11.56). Either mechanism must be

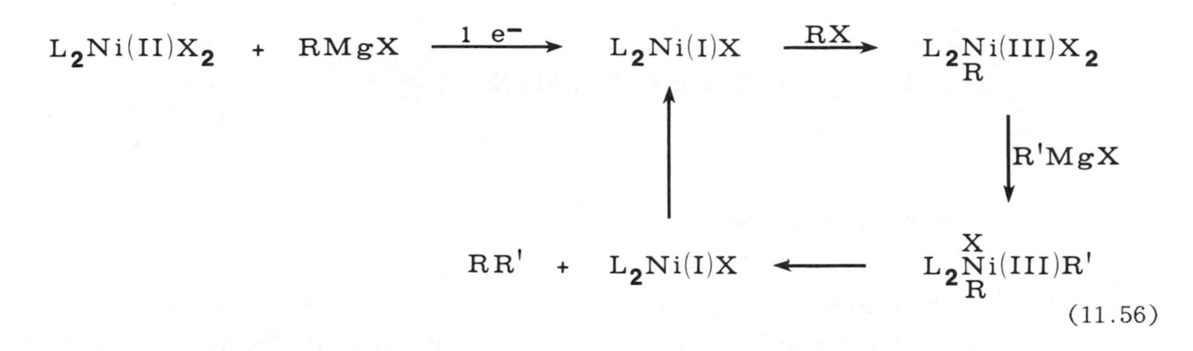


(11.56)

incomplete, since it fails to account for the lack of reactivity of organolithium reagents in this system. Clearly, the magnesium is playing a role in these coupling reactions, but that role remains to be clarified.

Despite the mechanistic ignorance shrouding this reaction, it has been used in a number of interesting and synthetically useful reactions. For example, 2-chloroquinolines [63] and halopyridines [64] are cleanly alkylated by Grignard reagents under these conditions (11.57 and 11.58). Bromoenol ethers react similarly, resulting in a

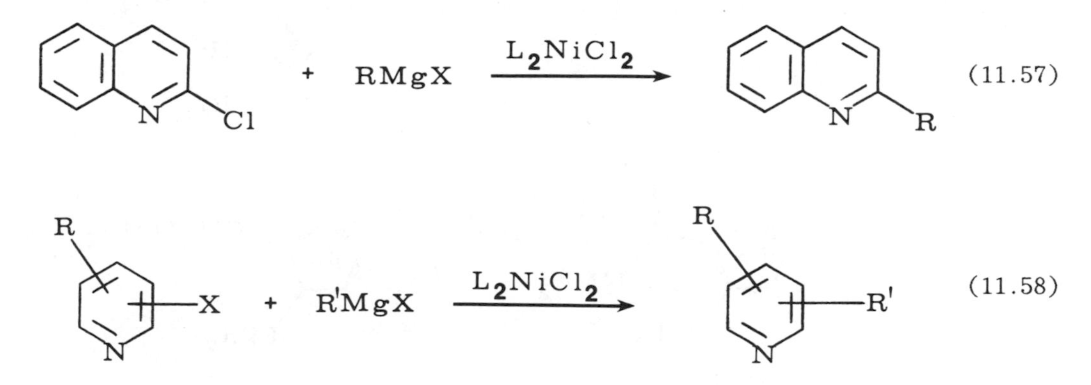


two-carbon homologation of the Grignard reagent [65] (Equation 11.59).

$$\text{RMgBr} + \text{BrCH{=}CHOEt} \xrightarrow{(\text{diphos})\text{NiCl}_2} \text{RCH{=}CHOEt} \qquad (11.59)$$

The use of bis Grignard reagents with aromatic dihalides results in a cyclocoupling to produce metacyclophanes. Although the yields are only modest (10 to 30 percent), this one-step procedure (11.60) is still competitive with other multistep approaches [66]. In this fashion muscopyridine (4) was prepared in 20 percent yield in one step.

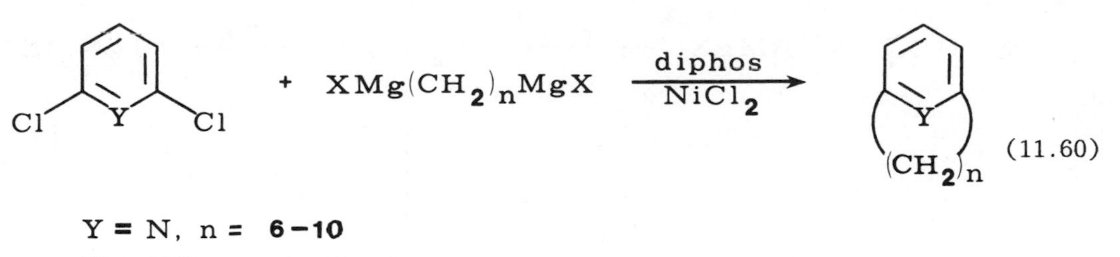


(11.60)

Y = N, n = 6–10

Y = CH, n = 8–10, 12

Since phosphine-nickel complexes are clearly involved in the carbon-carbon bond-forming step in this reaction, the use of optically active (chiral) ligands results in the production of optically active coupling products in appropriate systems. The chiral ligands used in these reactions are the unusual (aminoalkylferrocenyl)phosphines 5-7.

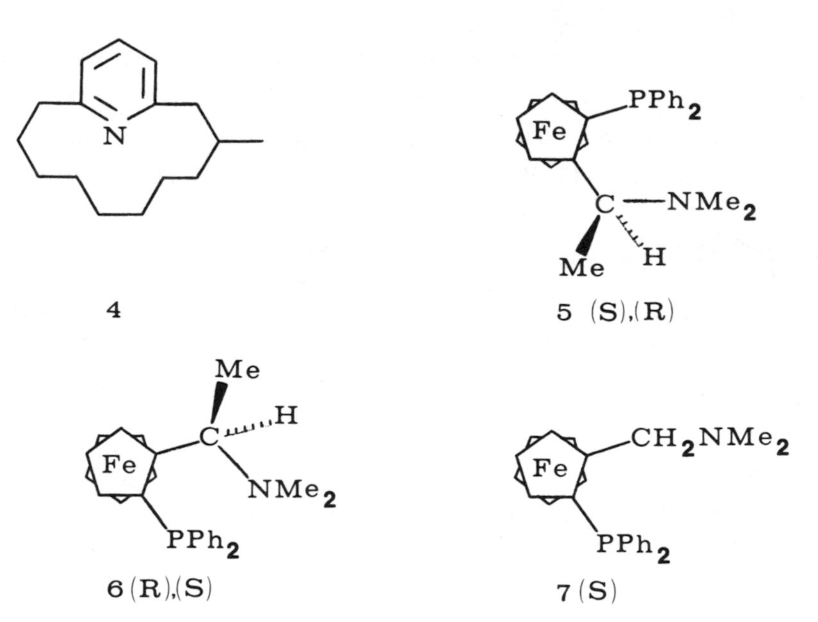


Reaction of 1-phenylethyl Grignard reagent with vinyl bromide using a 1:2 mixture of anhydrous nickel chloride and a chiral ligand results in the formation (11.61) of

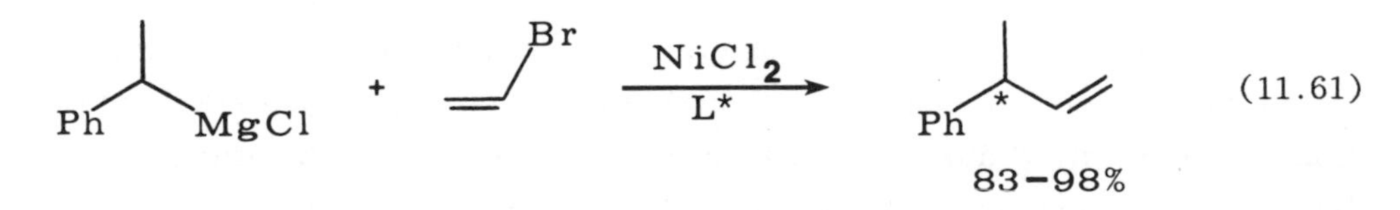


(11.61)

optically active 3-phenyl-1-butene in high chemical yield and optical purity [67]. With 5, which has (S),(R) chirality, the product has the (R) configuration, but with ligands

6 (R), (S) and 7 (S), the product has the (S) configuration. It should be noted that this reaction is actually a kinetic resolution of the racemic Grignard reagent. That is, one enantiomer of the Grignard reagent reacts with the chiral nickel alkyl intermediate faster than the other enantiomer, resulting in the observed optical induction. Since the Grignard reagent undergoes inversion rapidly relative to coupling, high yields of chiral product can be obtained.

## 11.4. Alkylrhodium(I) Complexes.

The reaction of $RhCl(PPh_3)_2CO$ with organolithium reagents or Grignard reagents (developed by Hegedus) produces unstable σ-alkylrhodium(I) complexes which are very efficient in the alkylation of acid chlorides to produce ketones [68] (Equation 11.62).

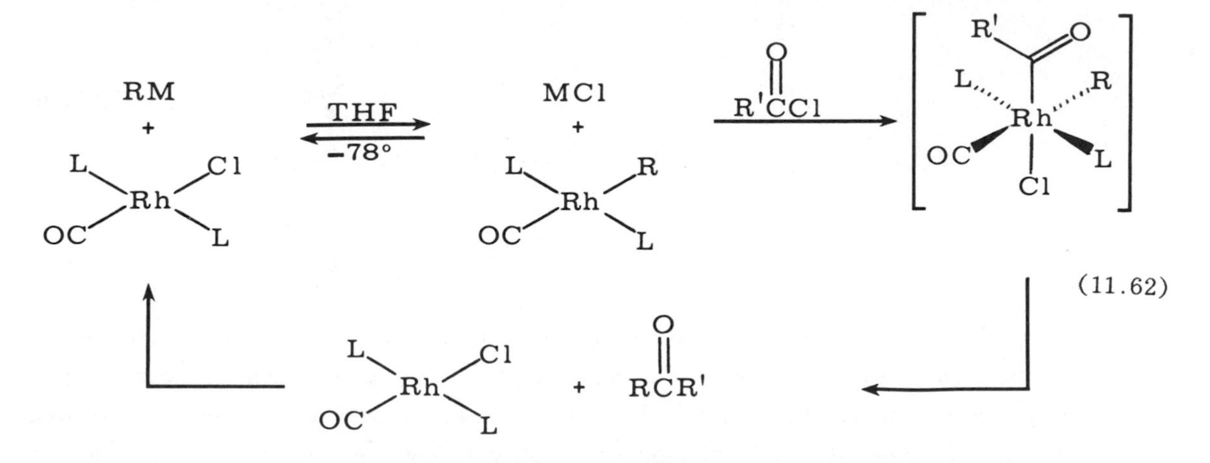


(11.62)

Although the chlororhodium complex is inert to acid halides, conversion to the electron-rich alkylrhodium complex significantly increases its reactivity, and permits facile oxidative addition even at -78°. The oxidative adduct rapidly reductively eliminates ketone and regenerates the starting chlororhodium complex for reuse in this system. Although this mechanism rests on firm literature precedent, experimental verification is sparse. The alkylrhodium(I) intermediate could not be isolated, but the shift in the CO stretching frequency from 1982 $cm^{-1}$ in the chlorocomplex to 1962 $cm^{-1}$ upon treatment with the organolithium reagent is consistent with replacement of the chloro group by the more electron-donating alkyl group. The proposed Rh(III) complex resulting from oxidative addition was never detected, despite much effort. Rather, upon addition of acid chloride, the soluble, red alkylrhodium complex disappears and the insoluble chlororhodium complex forms, along with the product ketone. Thus, alkyl, aryl, vinyl, and α-chloro acid chlorides are converted to the corresponding methyl, n-butyl, phenyl, or allyl ketones in moderate to high yield. The reaction is specific for acid halides, and can be carried out in the presence of aldehydes, esters, amides, nitriles, and other organic halides without complication. Optically active α-methyl acid halides are alkylated without racemization of the chiral center, as evidenced by the high-yield conversion of (S)-(+)-2-methylpentanoyl chloride to (S)-(+)-4-methyl-3-heptanone (the principle alarm pheremone of fungus-growing ants of the genus Atta) with complete retention of stereochemistry (11.63).

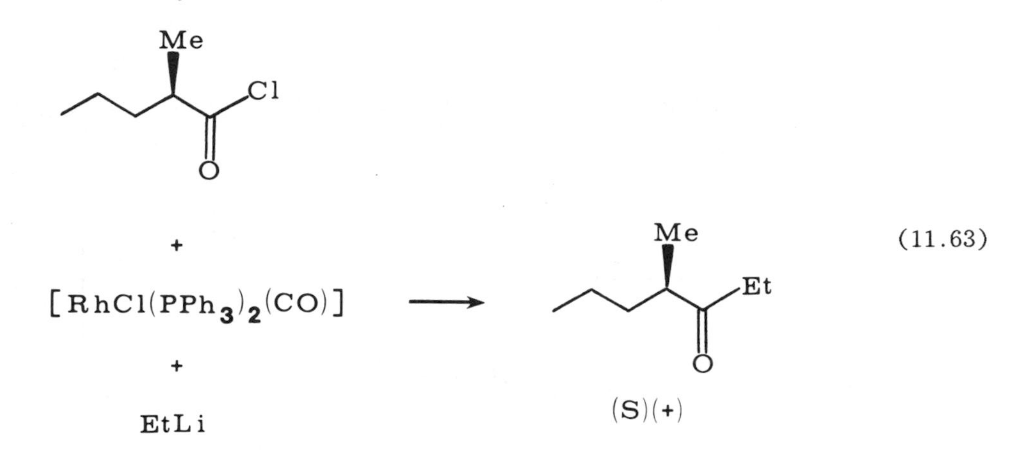


(11.63)

Secondary and tertiary organolithium and Grignard reagents cannot be used in this system, since the resulting alkylrhodium complexes β-hydride-eliminate more rapidly than they react with the acid chloride. Thus, reaction of sec-butyllithium, $RhCl(CO)(PPh_3)_2$, and benzoyl chloride gives only a low yield of ketone in which the major product is the n-butyl ketone rather than the sec-butyl ketone. When this reaction is carried out in the presence of excess hexene, the sole product is the n-hexyl ketone. These results are consistent with Figure 11.4, in which the initially formed sec-alkylrhodium species undergoes a rapid β-hydride elimination, producing 1-butene and a hydridorhodium(I) complex. Since this equilibrium is likely to lie to the rhodium-hydride-olefin side (Chapter 7), the low over-all yield of ketone is explained by loss of alkylrhodium(I) complex by this route. However, readdition occurs to a small extent, with the metal bonding preferentially to the primary position, producing the n-butyl complex and ultimately the n-butyl ketone. If excess hexene is present, it successfully traps the hydridorhodium(I) complex, producing n-hexyl ketone.

An attractive feature of this reaction is that the starting rhodium complex is regenerated in usable form in the last step, and hence there is no net consumption of

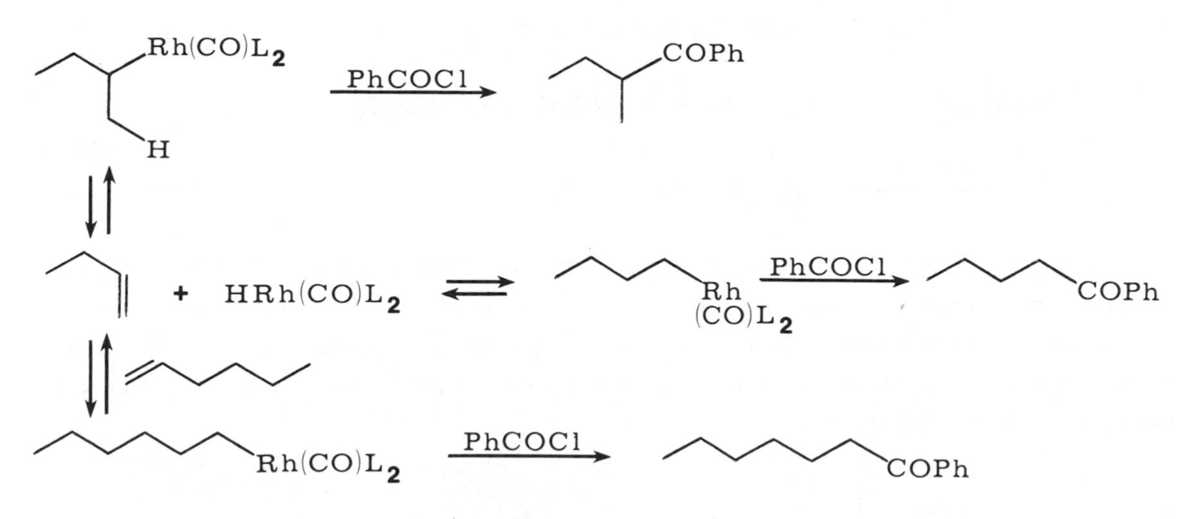


*Figure 11.4. Rearrangements of σ-alkylrhodium complexes.*

the expensive rhodium complex. However, for any particular reaction, a stoichiometric am⌃⌃nt of rhodium complex is required, since the alkylrhodium(I) complex must be preformed prior to addition of the acid halide to prevent direct reaction of organolithium and acid halide to produce tertiary alcohols.

## 11.5. Reactions of Metal Alkyls from Metal Anions and Alkyl Halides.

Most of the synthetically useful alkylmetal complexes arising from the reaction of anionic transition-metal complexes with alkyl halides involve anionic transition-metal carbonyl complexes. This is primarily because carbon monoxide is a strong π-acceptor and stabilizes negative charge. The most useful complexes in this regard are those of $Fe(CO)_4^{=}$ and $Co(CO)_4^{-}$. Both of these have been discussed in detail in Chapter 9, and will not be further considered here.

## 11.6. Reactions of Metal Alkyls from Transmetallation Reactions.

The reaction of organolithium or organomagnesium compounds with some metal halides, particularly palladium chloride, results in reduction of the transition metal rather than formation of organometallic species, thereby severely limiting the synthetic utility of this type of reaction. The desired transition-metal organometallics can be formed from less-reactive main-group organometallics, particularly organomercury and tin compounds. Several very useful synthetic transformations rely on this "transmetallation" from mercury or tin to palladium as a key step.

The earliest use of this chemistry was in the synthesis of symmetrical biaryls by Heck, and involved the palladium(II) coupling of arylmercuric halides [69]. The reaction (11.64) is thought to involve transmetallation from mercury to palladium followed

$$\mathrm{ArHgX} + \mathrm{PdX_2} \longrightarrow \mathrm{ArPdX} + \mathrm{HgX_2} \xrightarrow{\mathrm{ArHgX}} \mathrm{Ar_2Pd} + \mathrm{HgX_2} \longrightarrow \mathrm{Ar_2} + \mathrm{Pd(O)} \quad (11.64)$$

by reductive elimination of the two aryl groups. Larock extended this reaction to vinylmercuric halides, which are readily available from the corresponding alkyne via hydroboration-mercuration. Depending on the conditions chosen, either head-to-head or head-to-tail dimerization can be achieved [70-72]. (The mechanism for head-to-tail dimerization is not known.)

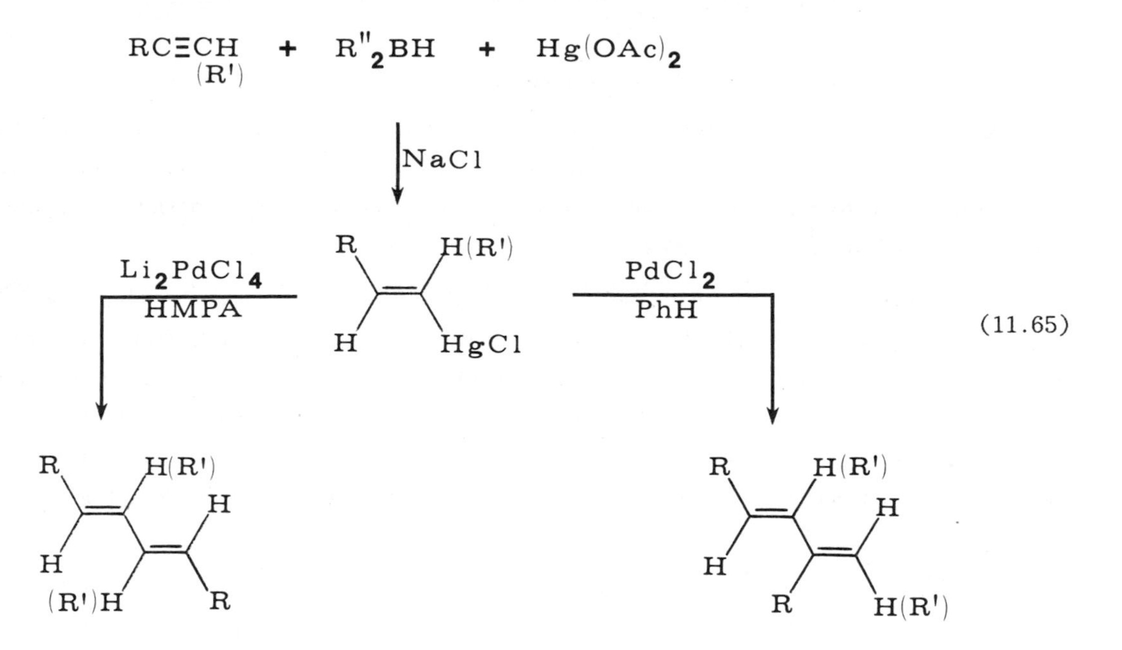


(11.65)

Aryl- and vinylpalladium complexes formed from organomercuric halides undergo a variety of other useful reactions characteristic of organopalladium species, regardless of the method of formation (i.e., insertion and oxidative addition). For example, allylic halides react with aryl- or vinylmercuric halides in the presence of Pd(II) to produce allylbenzenes or 1,4-dienes, depending on starting material. Allylic transposition is always observed, and the reaction (11.66) is thought to involve addition (insertion) of RPdX to the olefin, followed by $PdX_2$ elimination [73]. Alkenes, particularly

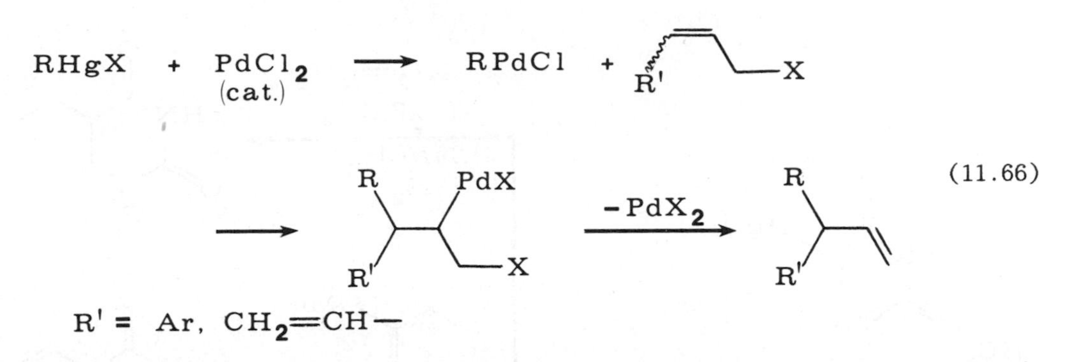


electrophilic ones, undergo a similar insertion-elimination sequence with arylmercuric halides, allowing the palladium-assisted alkylation of olefins by organomercuric halides (11.67).

$$ArHgX + PdX_2 \longrightarrow ArPdX + HgX_2$$

$$\xrightarrow{CH_2{=}CHY} ArCH_2{-}CHY(PdX) \xrightarrow{-PdHX} ArCH{=}CHY + Pd(O) + HX \qquad (11.67)$$

This reaction has found extensive application in nucleoside chemistry, of all places, because the pyrimidine ring is easy to mercurate in the five position [74,75]. Thus, a number of 5-substituted nucleosides, normally difficult to synthesize, were prepared in quite good yield (11.68). With mercurated pyrimidines, even electron-rich olefins such as dihydropyran and other enol ethers react, forming the C-C bond between $C_5$ of the pyrimidine and $C_1$ of the enol ether or acetate [76,77].

In contrast, Trost [78] showed that vinyl sulfides react with phenylmercuric chloride in the presence of a Pd-Cu catalyst system to produce arylation at the 2 (β) position (11.69). Surprisingly, intramolecular versions of this reaction using arylmercuric halides fail completely. However, the corresponding aryltin compounds do react nicely. These systems alkylate at the α position, and result in the elimination of the thioalkyl group (11.70). The system becomes catalytic in Pd(II) by the addition of a reagent to scavenge the thio groups (in this instance, $HgCl_2$) which otherwise would coordinate to the Pd(II) and inactivate it. The change in regiochemistry must be due

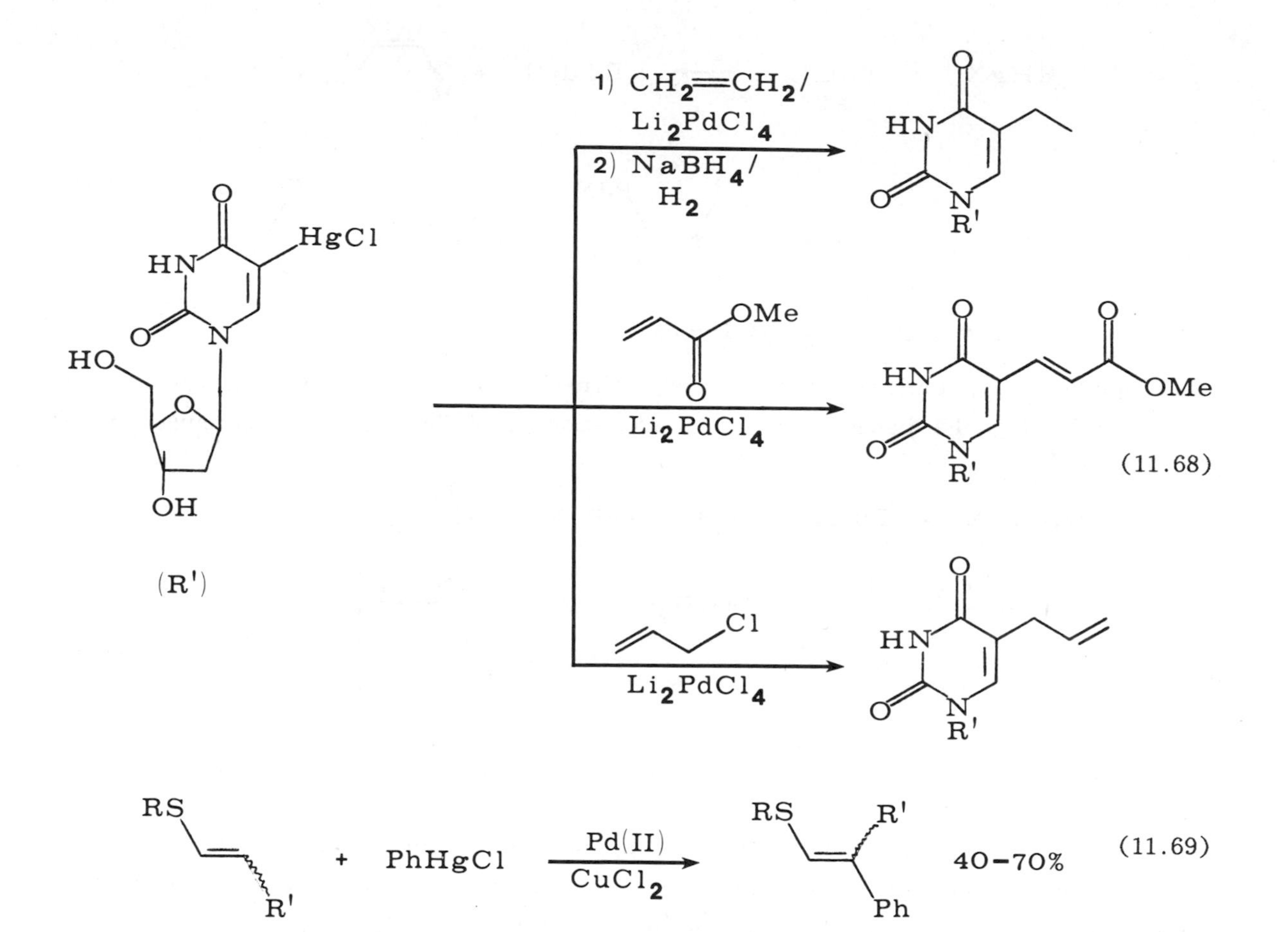


to the intramolecular nature of this reaction and to the constraints of ring size. Currently, a more precise explanation is not available.

Surprisingly, vinylmercuric halides react with olefins in the presence of Pd salts somewhat differently from arylmercuric halides, producing quite high yields of functionalized $\eta^3$-allylpalladium halide complexes rather than the alkylated olefin. Presumably, readdition of PdH to form the $\eta^3$-allyl complex occurs very efficiently in these cases [79]. Reaction 11.71 is very useful in its own right, since it allows the synthesis of of unusually substituted $\eta^3$-allylpalladium halide complexes for use in organic synthesis in other ways (Chapter 15).

Finally, palladium catalyzes the carbonylation of organomercuric halides to produce esters. This involves the well-known insertion of carbon monoxide into σ-alkylpalladium complexes (Chapter 9) as a key step. Vinylmercuric halides are particularly

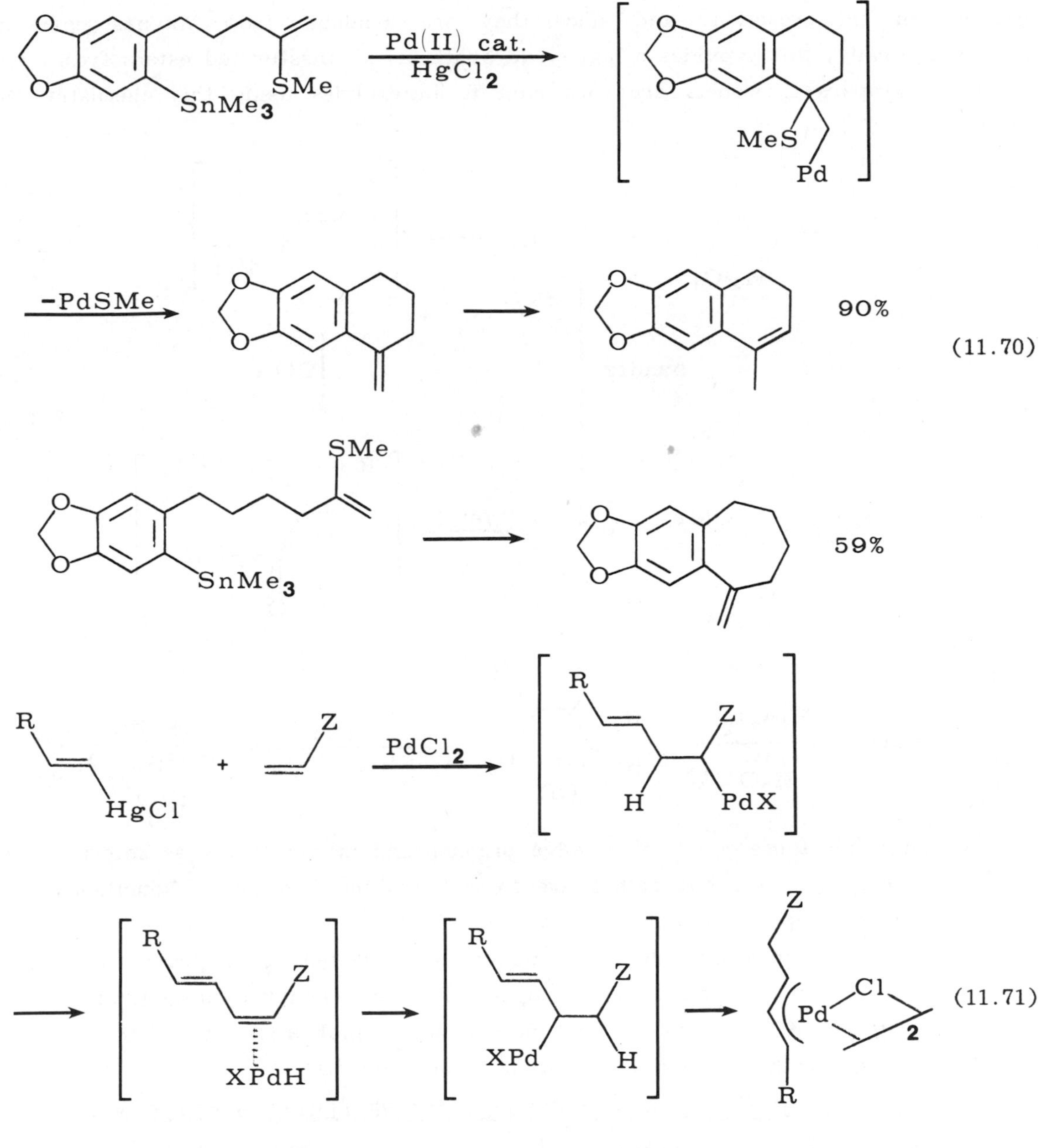


R = Ph, n–Bu, t–Bu

Z = COOEt, CN, COMe, H, alkyl

efficient in this reaction, and since they are available from alkynes regio- and stereospecifically, this provides a high-yield synthesis of unsaturated esters from alkynes (11.72). Propargyl alcohols are converted to butenolides using this chemistry [80] (Equation 11.73).

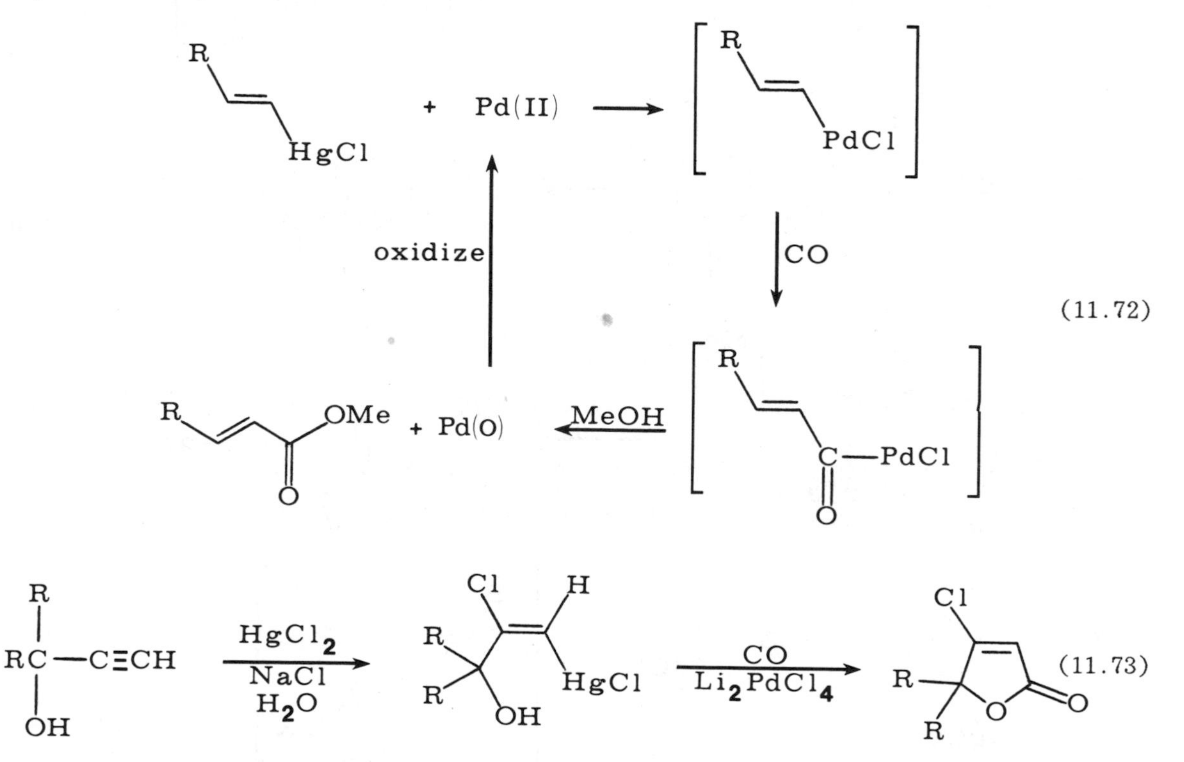


Although all these reactions involve organopalladium complexes as intermediates, they are rarely isolated, but rather are formed and used in situ. Nonetheless, the organopalladium complex must have a degree of stability for the above reactions to succeed. All the chemistry discussed above is restricted to R groups lacking β-hydrogens (i.e., aryl, vinyl, benzyl, neopentyl), since when β-hydrogens are present in the alkyl group, β-elimination of Pd-H occurs more rapidly than any other chemistry, producing alkenes rather than coupling or insertion products.

## 11.7. Reactions of Metal Alkyls from Oxidative Addition Reactions.

Section 11.6 presented chemistry of σ-organopalladium(II) complexes produced by the reaction of organomercuric halides with Pd(II)salts. Unfortunately, many desirable organomercuric halides are not readily accessible, and alternative routes to organopalladium complexes are desirable. An approach (developed primarily by Heck and by

Stille) which has found a great deal of application involves the oxidative addition of organic substrates to Pd(0) compounds. As before, this system is restricted to organic groups lacking β-hydrogens, since β-hydride elimination is a competitive process. The types of reactions utilizing oxidative addition as the key step are summarized in Figure 11.5 and involve insertion of small molecules into the Pd(II)-C bond, followed by a

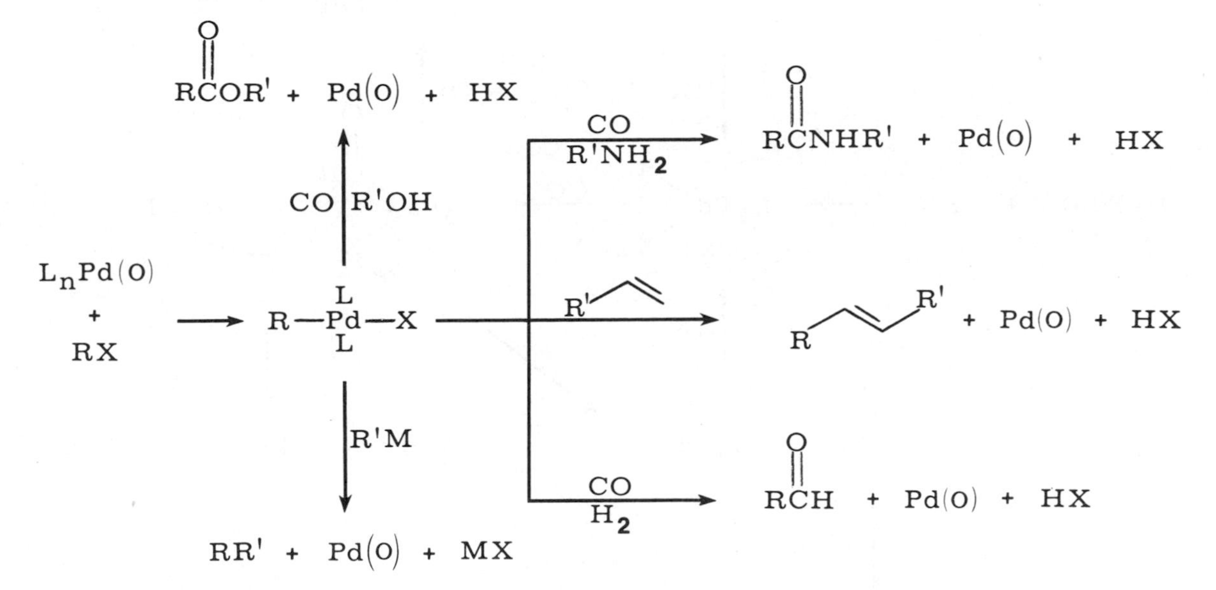


*Figure 11.5. Palladium(0) reactions with organic halides.*

reductive elimination to give organic product and to regenerate a Pd(0) species. Hence all these systems can be made catalytic, at least in principle [81].

One of the first systems to be developed was the Pd(0)-catalyzed carbonylation of organic halides. Thus aryl, benzyl [82], vinyl, and heteroaromatic halides [83] are carbonylated to esters in the presence of methanol, to amides in the presence of primary amines, and to aldehydes in the presence of $H_2$. Either $Pd(PPh_3)_4$ or $PdCl_2(PPh_3)_2$ can be used as the catalyst, the Pd(II) species being reduced to a Pd(0) species by CO rapidly under the reaction conditions. In all cases, a tertiary amine is necessary to absorb the HX generated. The reaction is thought to proceed as in Figure 11.6, but specific details of the processes occurring after CO insertion are not available. Clearly, the first step is an oxidative addition of the substrate to the Pd(0), followed by CO insertion to form a Pd(II) acyl complex. The mode of decomposition of this complex is less clear. It may undergo a reductive elimination of RCOX, which reacts with alcohol (or amine) to produce the acid derivative. Alternatively, it may undergo an

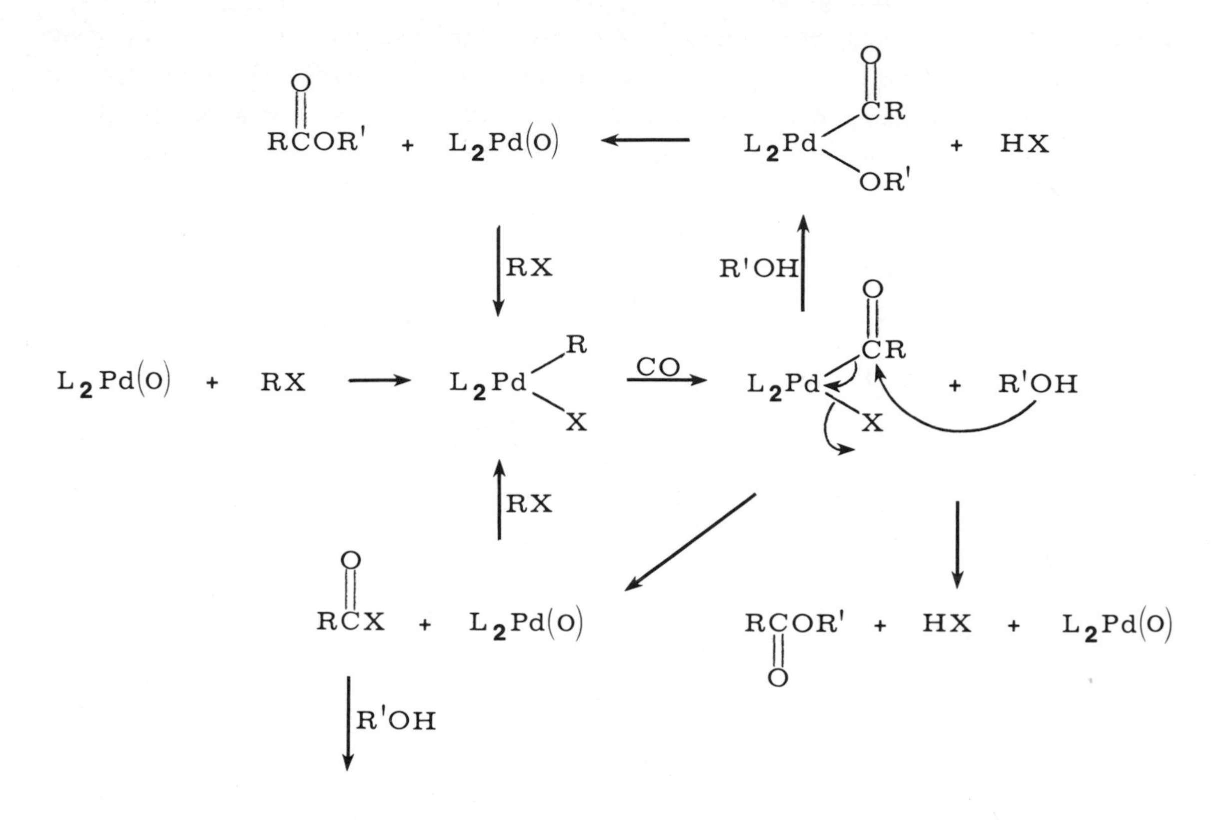


*Figure 11.6. Mechanism of palladium(0)-catalyzed carbonylation reactions.*

alkoxide-halide exchange, followed by reductive elimination of ester. Finally, it may undergo direct attack of the acyl by alcohol, producing ester and Pd(0). The mechanism of the conversion of the acylpalladium complex to aldehydes by $H_2$ is similarly obscure, and may involve either an oxidative-addition reductive-elimination or a binuclear reductive-elimination (11.74; see Chapter 4).

Intramolecular versions of this catalytic carbonylation of alkyl halides lead to a number of useful heterocyclic systems. Stille [84] showed that propargyl alcohols are readily converted to the corresponding Z vinyl iodides, which in turn react with CO (2 atm) in the presence of a Pd(0) catalyst at 25° to produce $\Delta^{\alpha,\beta}$-butenolides in excellent yield (11.75). Since both the oxidative addition and the CO insertion proceed with retention of configuration at carbon, the Z geometry required for cyclization to the

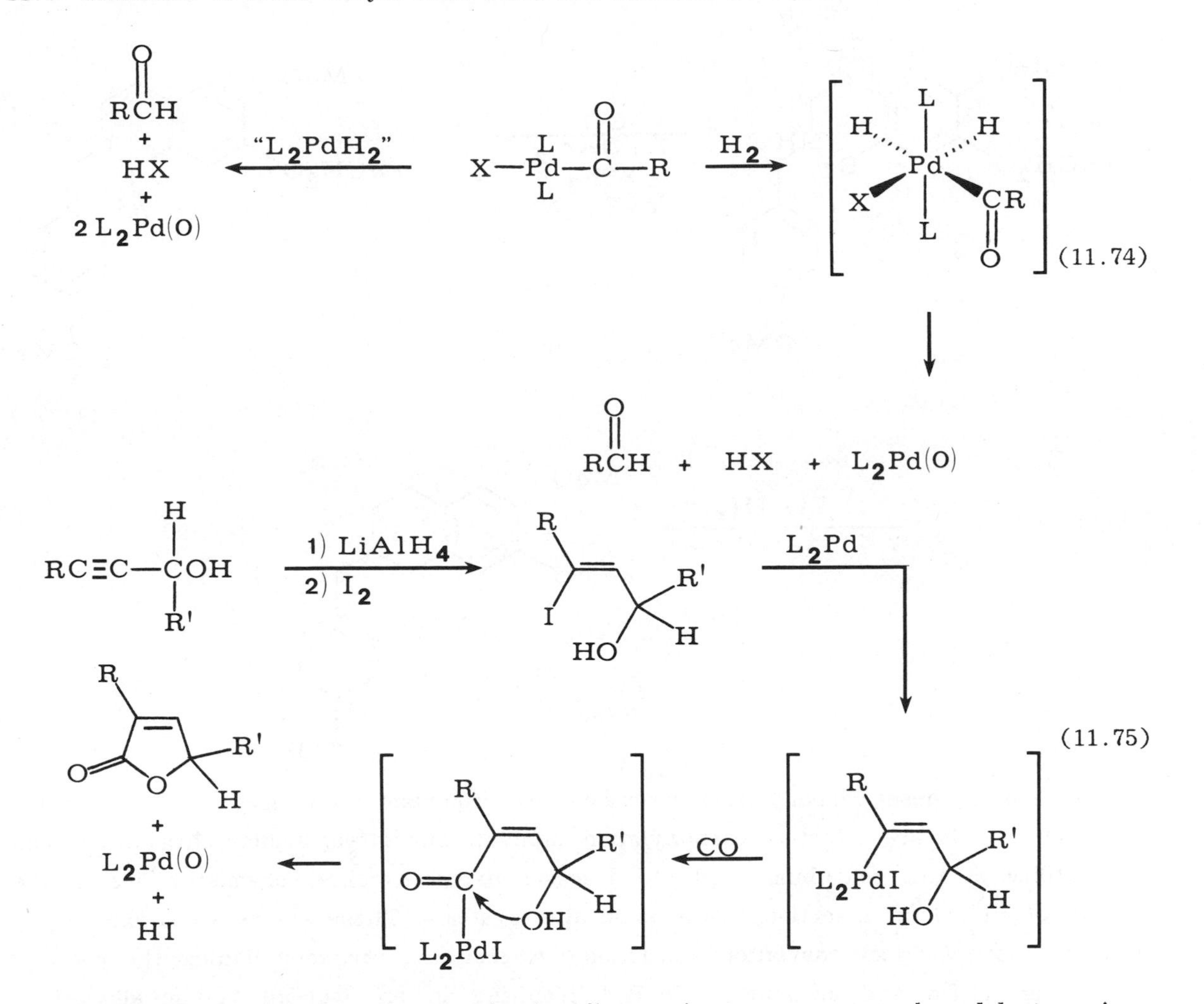


lactone is maintained. Benzolactams having five-, six-, or seven-membered lactam rings are available from the corresponding o-bromo-alkylaminobenzenes by a similar route (11.76), developed by Ban [85]. This has been used as a key step in the total synthesis (11.77) of sendaverine [86].

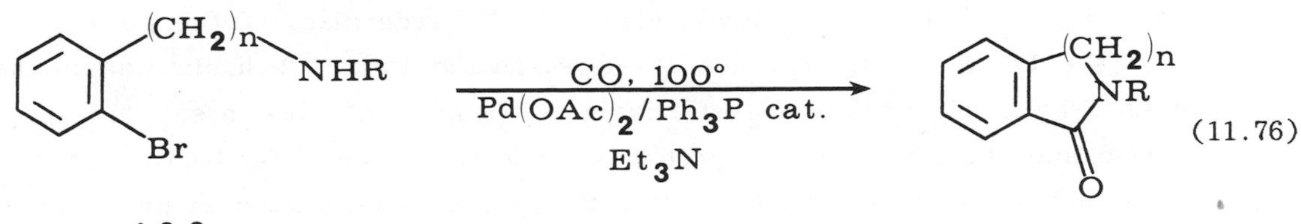

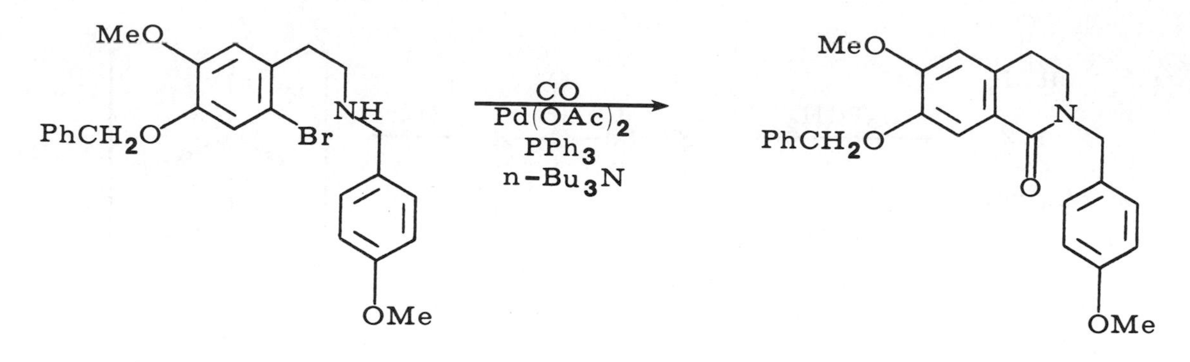


(11.77)

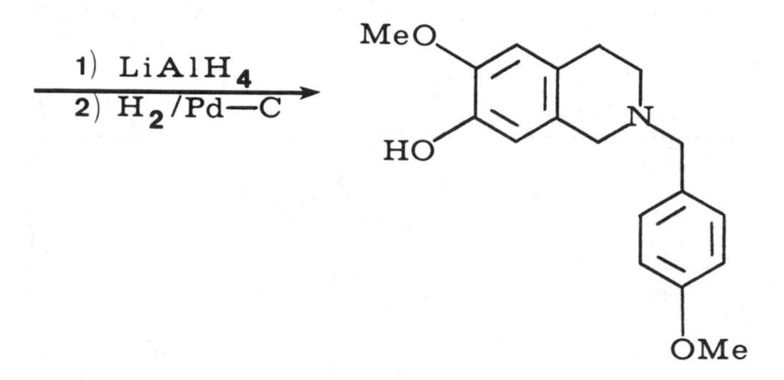


Although these carbonylation reactions are important and useful, several other transition metals are effective carbonylation catalysts and effect similar chemistry. The real utility of this palladium (and, to a lesser extent, nickel) chemistry lies in the facility of the olefin insertion reactions in this system. These are all variations of the well-established "Heck arylation" reaction, wherein an organopalladium(II) complex inserts an olefin and eliminates "Pd-H," resulting in an over-all vinylic alkylation. Although the precise mechanism of this reaction is not known (a familiar refrain by now), there is experimental evidence for a reaction sequence involving coordination of the olefin to the alkylpalladium(II) complex, cis insertion of the olefin, and cis elimination of "H-Pd-X" [81]. The R group adds exclusively or predominantly to the least-substituted carbon of the double bond (Figure 11.7), regardless of the polarization of the olefin (see below). The reaction is stereospecific with 1,2-disubstituted alkenes and proceeds by a cis addition-cis elimination sequence, with the most stable product being formed and the most hydridic proton being lost. If the olefin bears a good leaving group, such as a halide or an acetate, this group is eliminated in preference to a hydride. This reaction is quite general, and has been used to prepare a wide variety

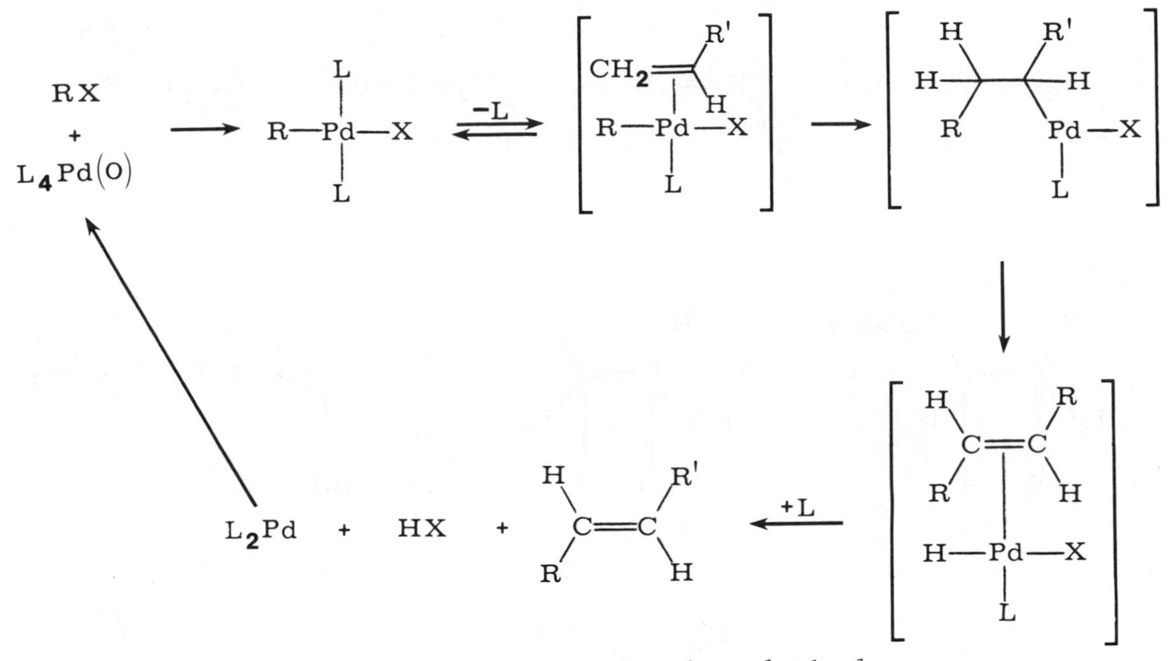


*Figure 11.7. Olefin insertions into Pd-C bonds.*

of compounds. Substituted aromatic [87], heteroaromatic [88], and pyrimidinyl [89] halides react with ethylene [87], disubstituted olefins [90], conjugated enones [91], and N-vinyl amides [92] in the presence of palladium catalysts to produce the corresponding alkylated olefins in fair to excellent yields. With allylic alcohols as the olefinic substrate, reductive elimination occurs toward the alcoholic carbon, producing carbonyl compounds as products [93,94] (Equation 11.78). In these cases, the catalyst

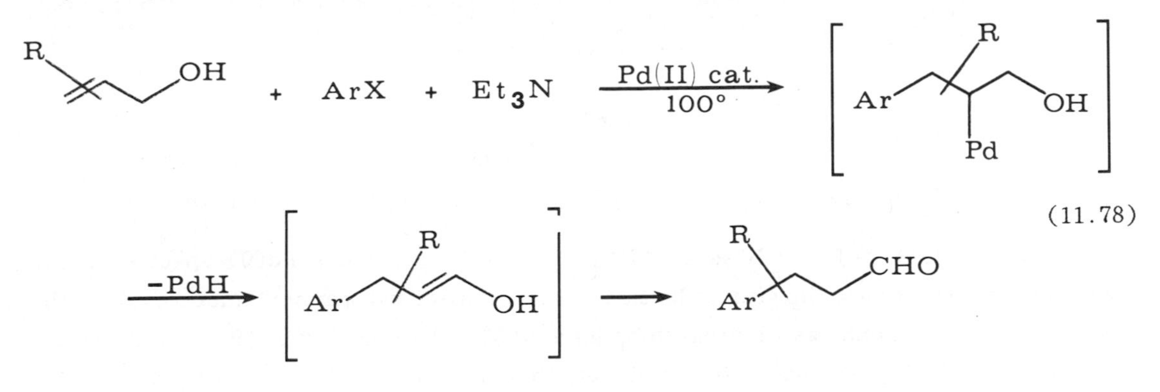

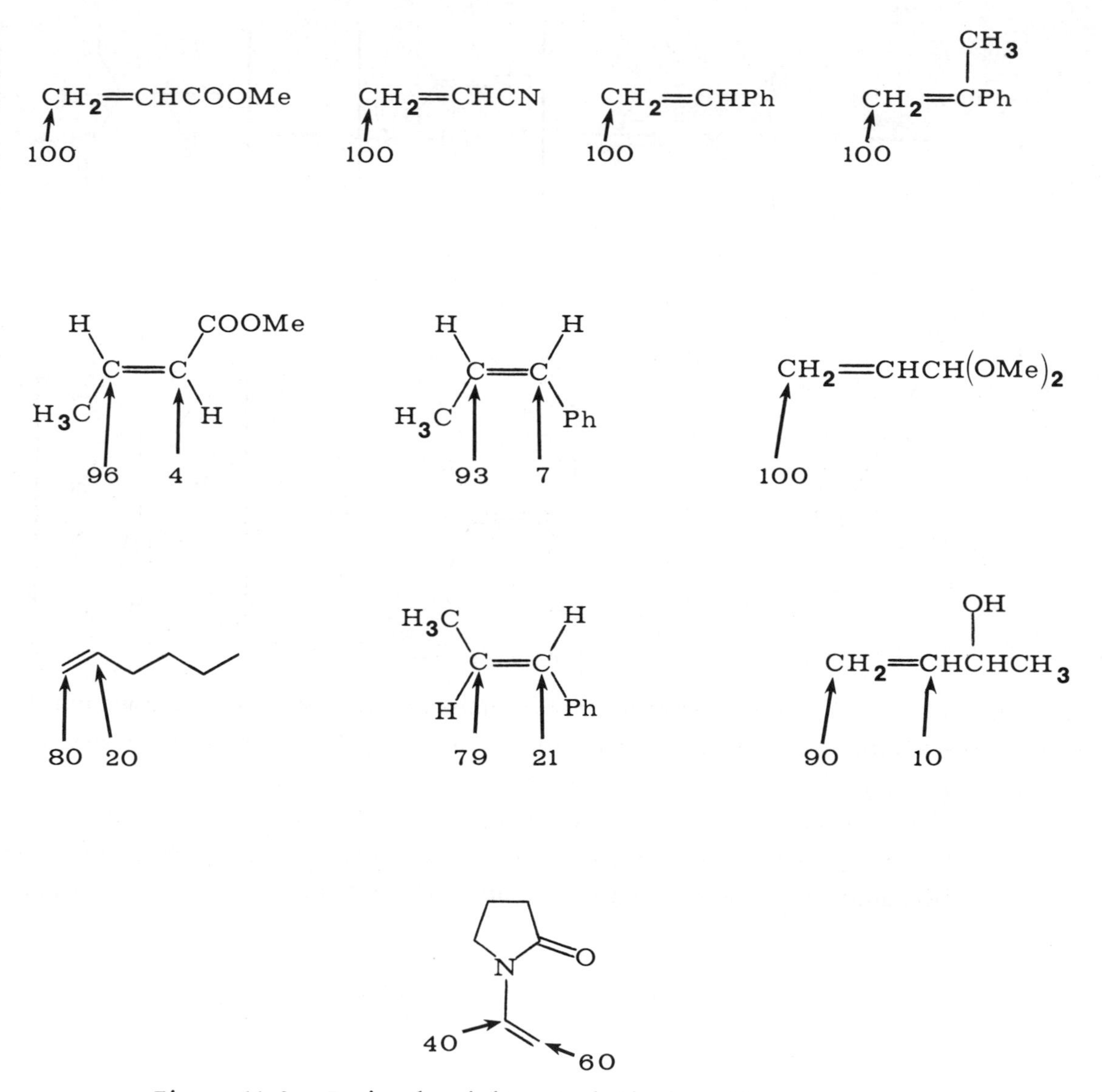


*Figure 11.8. Regioselectivity of olefin insertion reactions.*

added is a Pd(II) salt, such as $Li_2PdCl_4$ or $Pd(OAc)_2$. Since Pd(0) species are clearly required for catalysis, in situ reduction in an unspecified manner must result. Heteroaromatic halides such as 2-bromothiophene [95], 2-bromofuran [96], and 2- and 4-bromopyridine [97,98] also work well in this reaction, allowing the direct, specific introduction of carbonyl-containing side chains into these heterocyclic compounds.

Ban [99] showed that intramolecular oxidative-addition olefin-insertion reactions are also quite facile and lead to heterocyclic systems, such as indoles and isoquinolines (11.79 and 11.80). Similar cyclizations to form indoles or oxindoles using Ni(0)

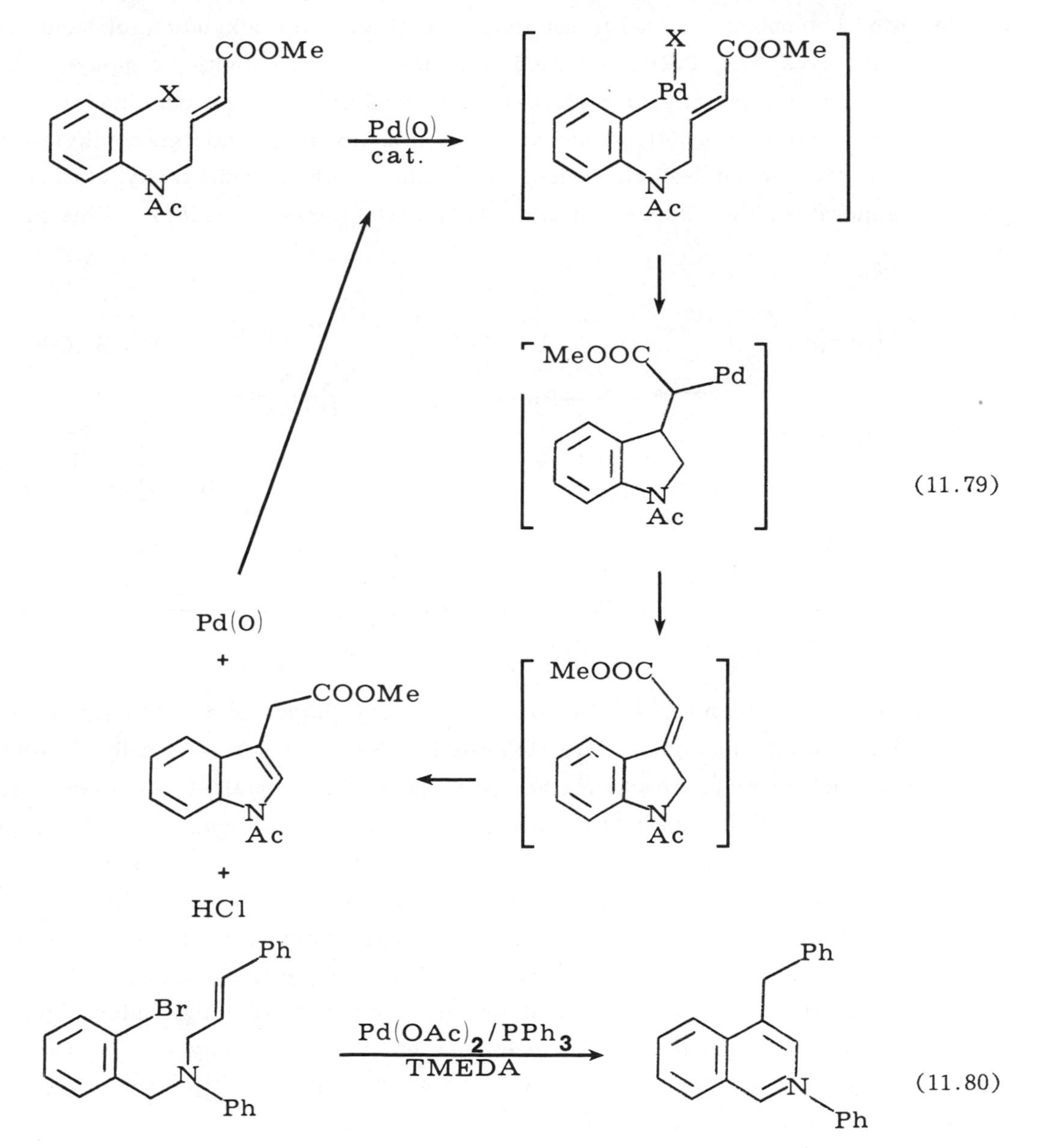

catalysts rather than Pd(0) catalysts are also successful [100,101] and have been used in the synthesis of natural products [100,102].

Finally, oxidative addition combined with transmetallation has been developed by Stille into a number of useful coupling reactions. An alkylation of acid halides to ketones involves the Pd(0)-catalyzed reaction of tetraalkyltin complexes with acid chlorides, and is thought to involve oxidative-addition of the acid halide to Pd(0) to produce an acylpalladium(II) complex. This complex then undergoes alkylation by the alkyltin to give an acyl-alkylpalladium(II) complex, which reductively eliminates ketone and regenerates the Pd(0) catalyst [103,104] (Equation 11.81). This reaction has

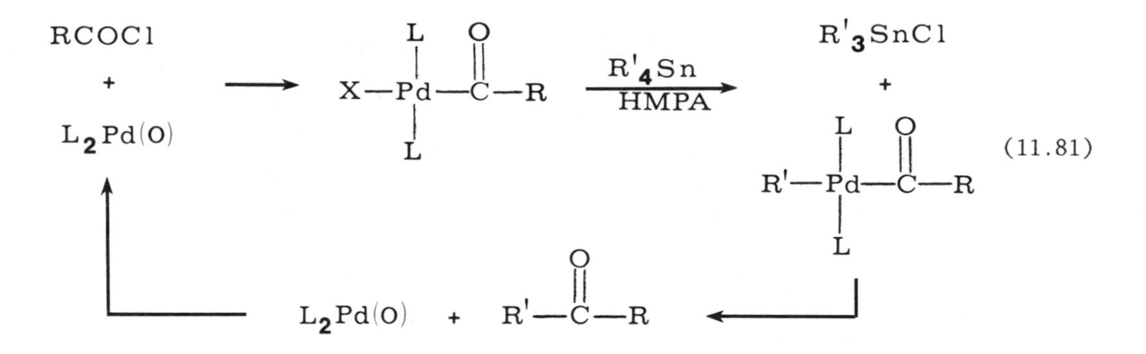


(11.81)

several attractive features. The yields are very high, and virtually all functional groups, including aldehydes, are tolerated. Sensitive or sterically hindered acid chlorides react cleanly; reaction times are short; high catalyst turnover (~20,000) is observed; and virtually no side products are produced. Although two of the four alkyl groups can be transferred, the reaction of $R_3SnCl$ in this system is five times slower than that of $R_4Sn$. $R_2SnCl_2$ and $RSnCl_3$ react with HMPA; so a maximum of two of the tin's alkyl groups can be utilized. Methyl, phenyl, vinyl, benzyl, and allyltin reagents work well, but more complex tin reagents have not yet been studied.

In a related reaction, vinyl halides are alkylated by vinyl, alkyl, phenyl, and alkynyl Grignard reagents in the presence of a Pd(0) catalyst. In this case the Grignard reagent is thought to alkylate the alkylpalladium(II) halide formed by the oxidative addition of the vinyl halide to the Pd(0) complex [105].

## 11.8. Reactions of Metal Alkyls from Nucleophilic Attack on Metal Complexed Olefins.

A characteristic reaction of a number of metal olefin complexes, particularly those of Pd(II), Pt(II), and $[Fe(II)]^+$, is nucleophilic attack on the complexed olefin, producing a σ-alkylmetal complex (11.82). Often these alkylmetal complexes are not very

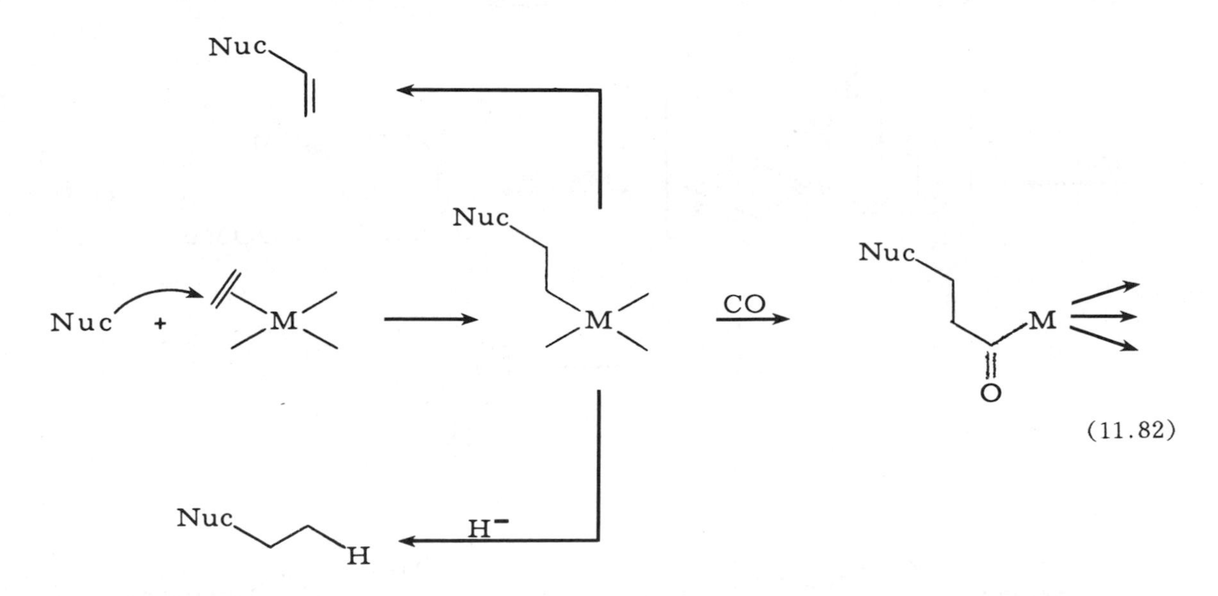


(11.82)

stable and undergo spontaneous β-hydride elimination. However, many of these complexes have sufficient lifetimes to undergo further reactions of the types discussed above (insertion reactions), allowing the difunctionalization of the original olefinic substrate.

The earliest well-characterized reactions of this type were those of the very stable, isolated, and well-characterized alkyl complexes resulting from nucleophilic attack on chelating olefin complexes. The cyclooctadiene complex of palladium chloride reacts with methoxide to form a methoxyalkylpalladium complex, which decomposes on exposure to carbon monoxide in methanol to produce the trans β-methoxyester in 77 percent yield. The process clearly involves the insertion of CO into the Pd-C bond, followed by decomposition of the thus-formed acyl complex, as presented above, and results in an over-all methoxycarbonylation of the olefin [106] (Equation 11.83). Addition of Cu(II) salts to reoxidize palladium makes this system function catalytically. A similar reaction (11.84) occurs with N,N-dimethylallylamine [107]. For this methoxycarbonylation reaction, the intermediate alkyl complex need not be isolable, since CO

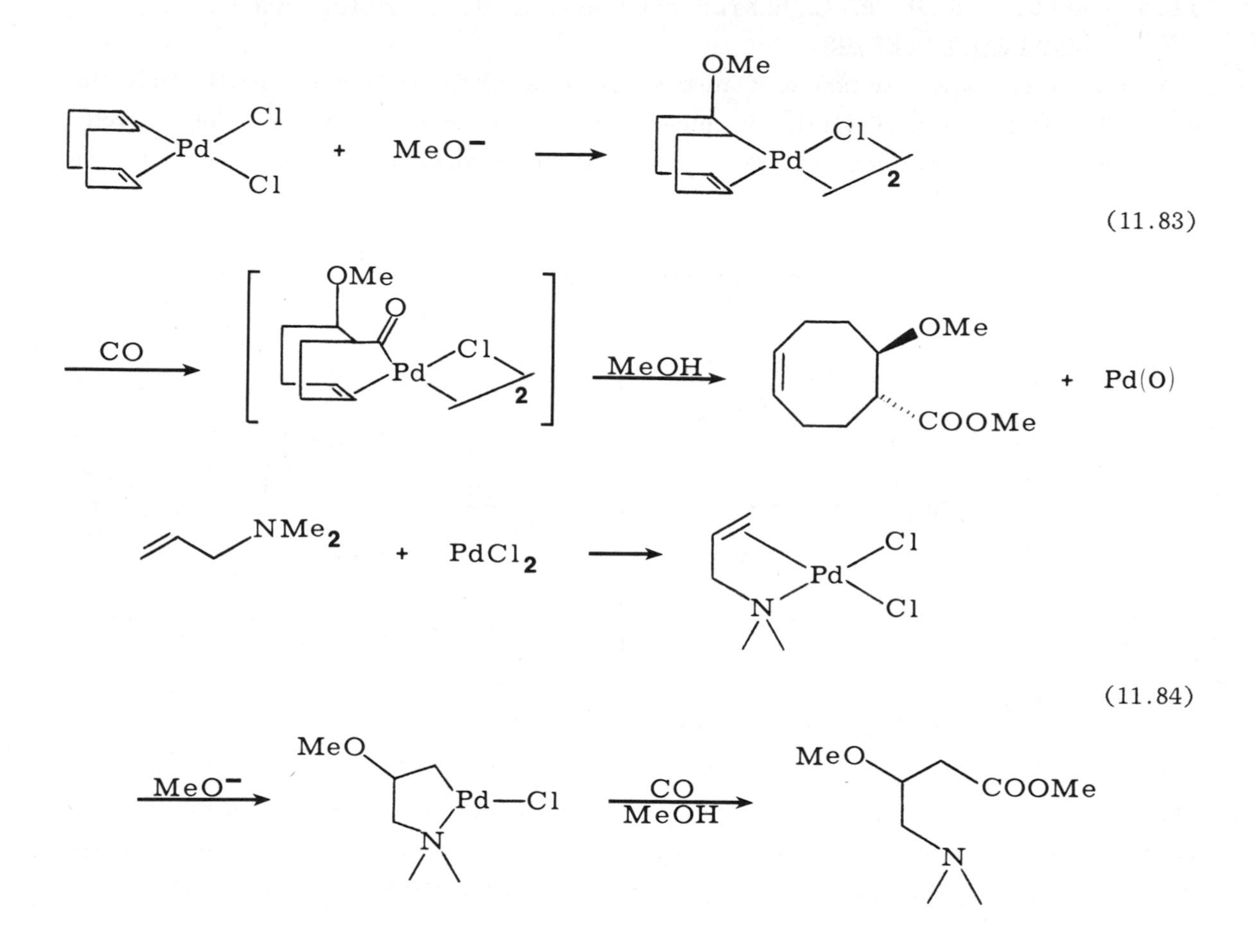


appears to insert under conditions sufficiently mild to preclude β-elimination. Hence, simple olefins can be methoxycarbonylated using $PdCl_2$ to promote the reaction as shown in 11.85 [108]. In contrast, the insertion of conjugated enones requires more severe

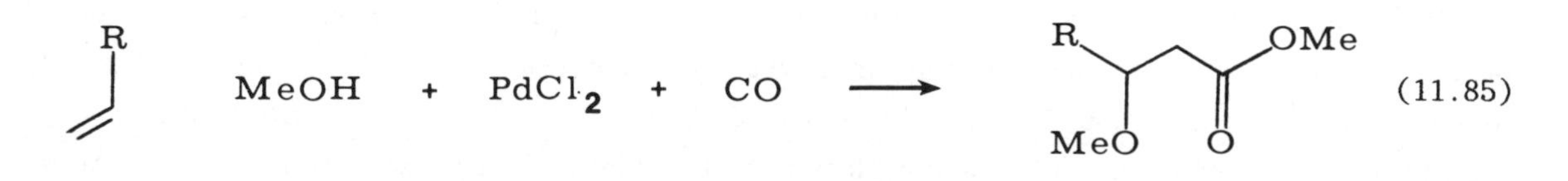


conditions, and only relatively stable alkyl complexes have been successfully added to these substrates. Holton reacted the complex from "methoxypalladation" of

N,N-dimethyl-2-allylamine with conjugated ketones lacking β-substituents to produce the β-substituted enone in excellent yield [109] (Equation 11.86). This is a very valuable

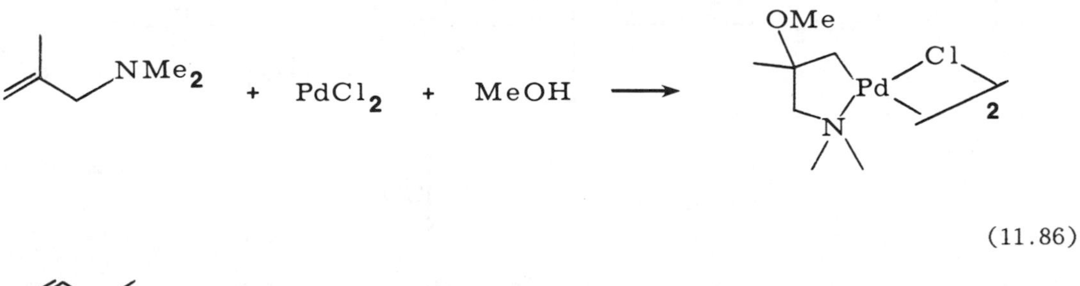


(11.86)

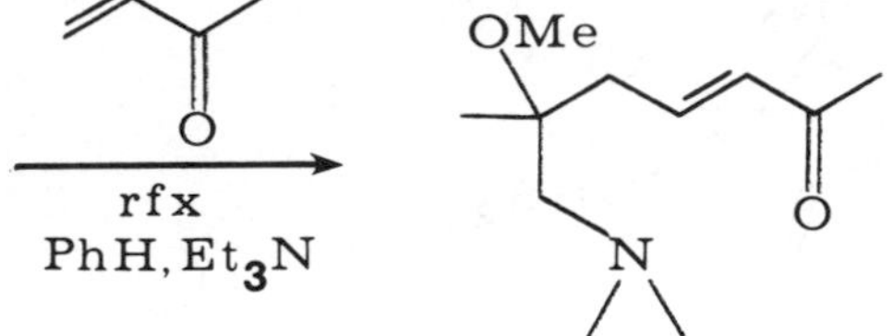


reaction, since it allows the difunctionalization of the olefin substrate in a systematic fashion, and has been used in a very neat synthetic approach to the prostaglandins (Chapter 12).

Hegedus showed that the alkylmetal complexes from the amination of olefin-palladium complexes also undergo insertion reactions under appropriate conditions. Thus, amination of an olefin in the presence of Pd(II) followed by treatment with CO results in the production of a very stable β-aminoacyl complex. β-Aminoacid derivatives are produced by decomposition of this complex by excess diphos or $Br_2/CH_2Cl_2$ in the presence of a nucleophile [110] (Equation 11.87). Similar chemistry utilizing a cationic iron-olefin complex and a primary amine was used by Rosenblum in a synthesis of β-lactams (Chapter 12). Backvall subjected the unstable σ-alkylpalladium(II) complex resulting from amination of the olefin-palladium complex to an oxidative cleavage to produce 1,2-aminoalcohols or 1,2-diamines, depending on the conditions of the reaction (11.87) [111]. This reaction is likely to proceed by oxidation of Pd(II) to a higher oxidation state (Pd(III) or Pd(IV)), making it a better leaving group, followed by displacement of this oxidized metal by $H_2O$ or $R_2NH$. The sterochemistry of introduction of the OH or second $R_2N$ group (inversion) is consistent with this mechanism, and accumulating evidence in other systems is also supportive of this interpretation [112]. With a primary amine used as the nucleophile, oxidation of the unstable

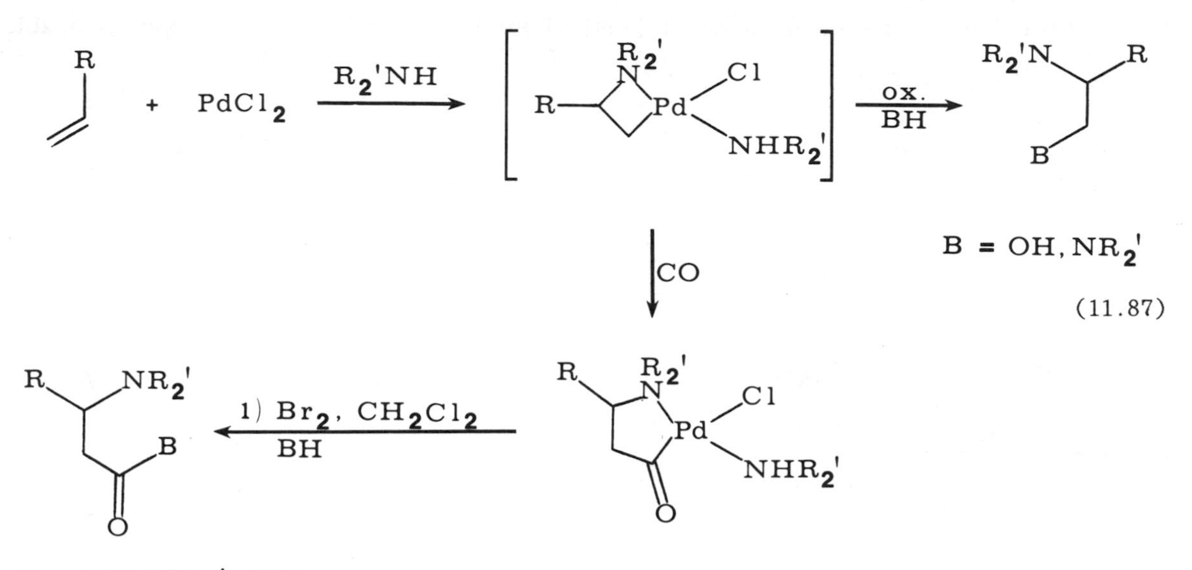


σ-alkylpalladium(II) complex with bromine produces aziridines, by intramolecular displacement of oxidized Pd by the (now) secondary amine [113] (Equation 11.88).

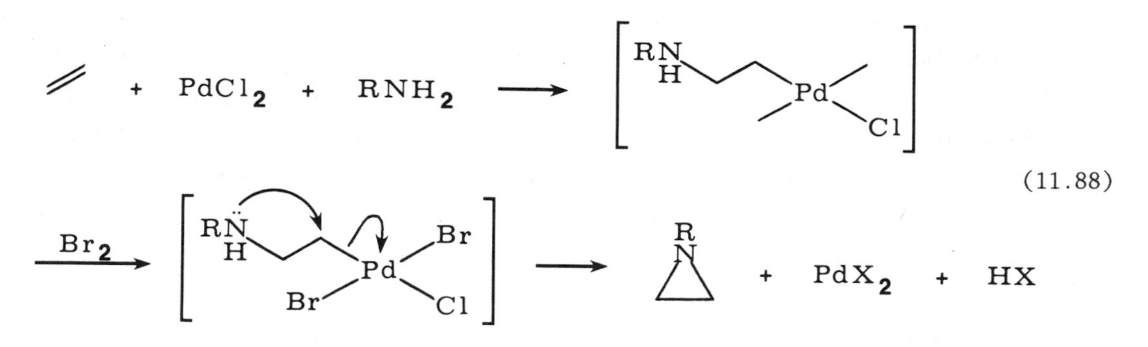


Hegedus synthesized a variety of heterocyclic compounds by palladium-assisted intramolecular amination of olefins (Chapter 12). These too proceed through relatively unstable σ-alkylpalladium complexes, which can be made to undergo insertion reactions under appropriate conditions. Some of these are summarized in Figure 11.9 [114]. With olefinic addends, alkylation always occurs at the less-substituted carbon, regardless of the polarization of the olefin. This closely parallels the chemistry of alkylpalladium complexes produced in other ways, discussed above. Surprisingly, when the

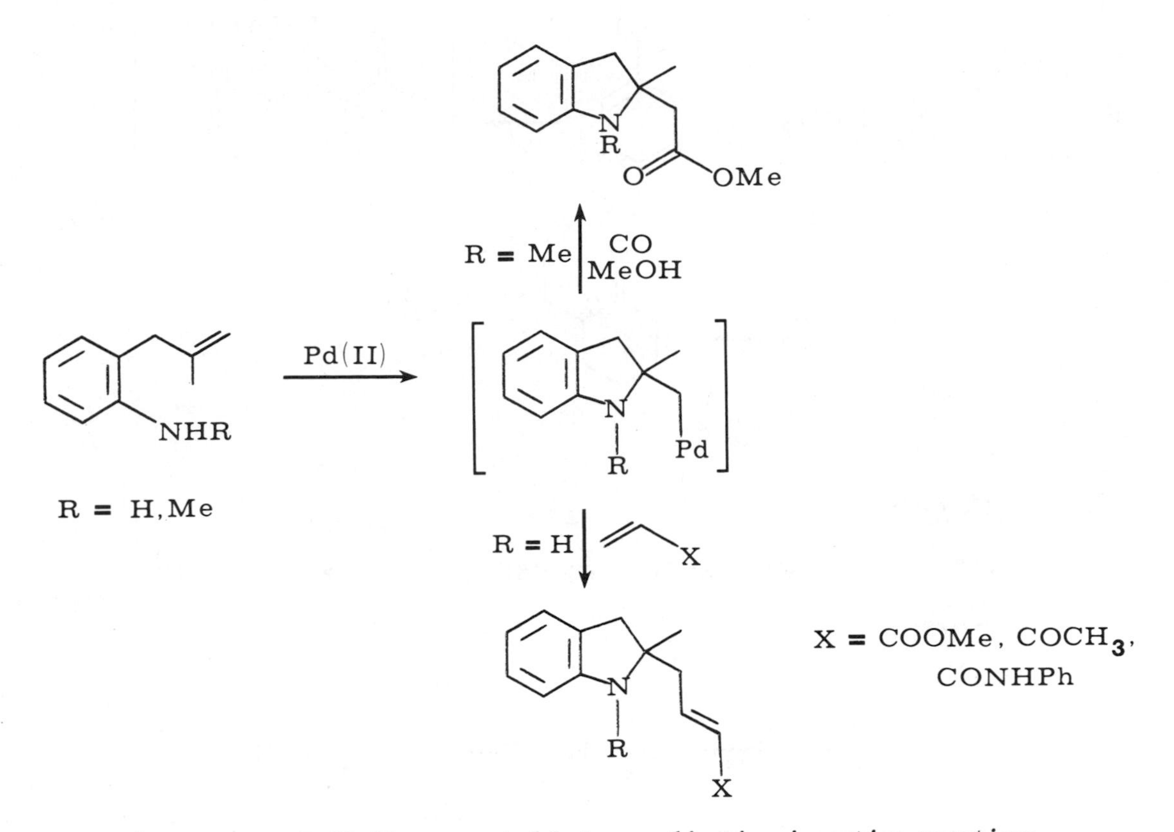


*Figure 11.9. Palladium-promoted heterocyclization-insertion reactions.*

insertion is intramolecular, alkylation occurs in the opposite sense (11.89). This multiple-ring closure sequence is of significant potential utility for the synthesis of natural products.

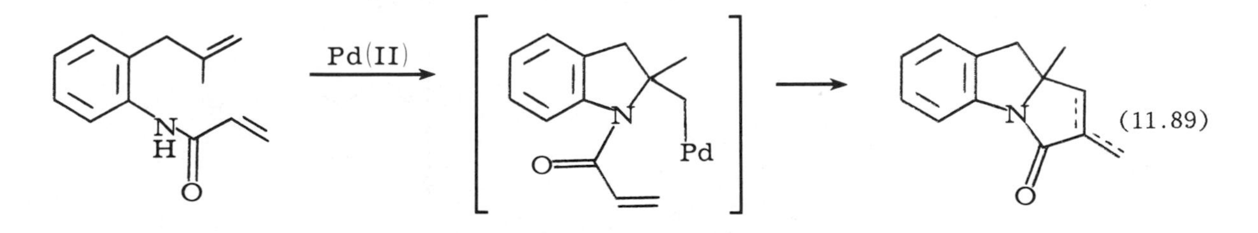


Finally, the alkylmetal complexes resulting from alkylation of olefin-metal complexes undergo a number of these same insertion processes; (Equations 11.90 to 11.92) [115, 116]. The chemistry presented in these equations is attractive synthetically because it

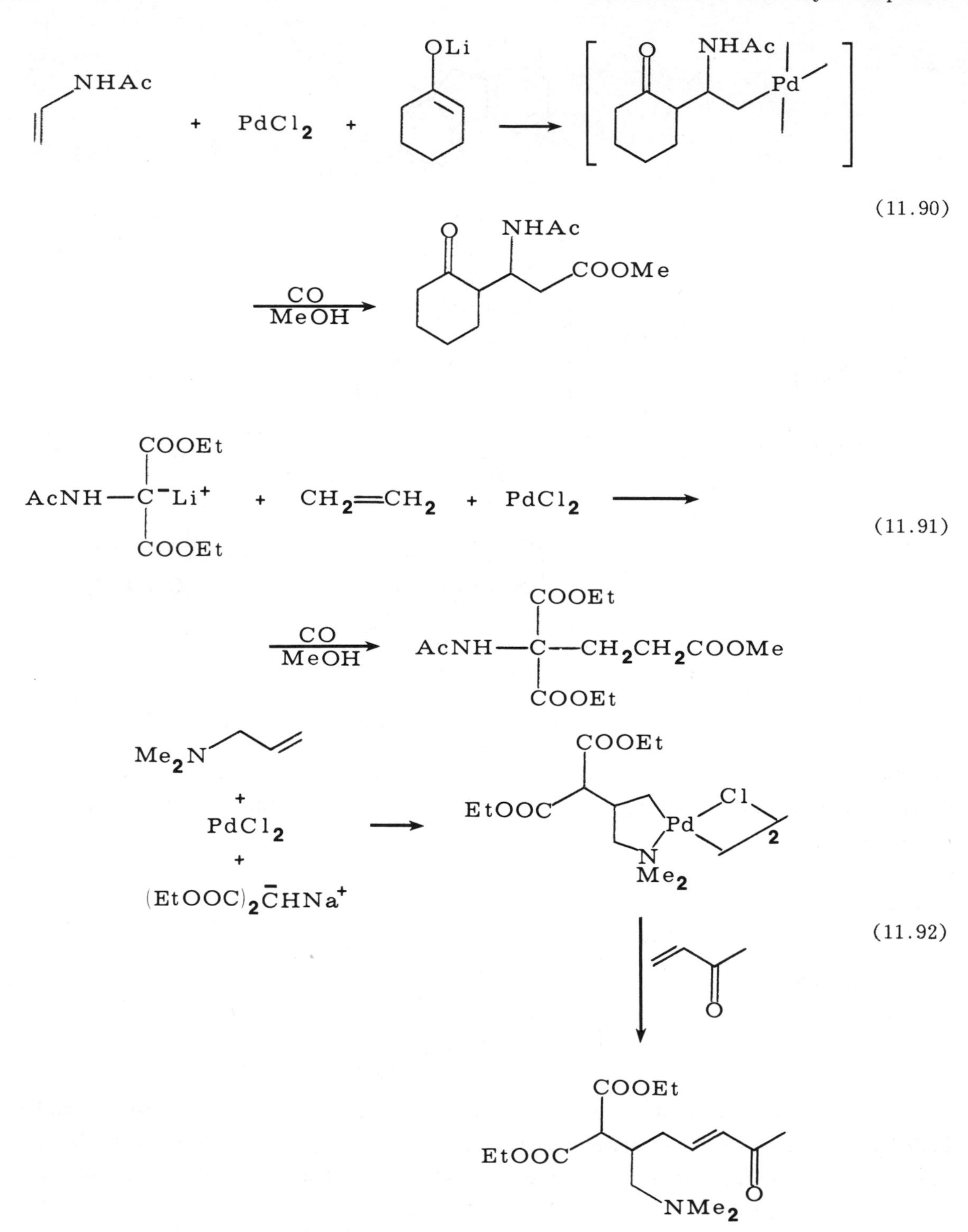
NHAc
+ PdCl2 +
OLi
O
NHAc
Pd
(11.90)
CO
MeOH
COOMe
COOEt
AcNH—C⁻Li⁺
COOEt
+ CH2═CH2 + PdCl2
(11.91)
AcNH—C—CH2CH2COOMe
Me2N
+
PdCl2
+
(EtOOC)2C̄HNa⁺
EtOOC
Cl
2
N
Me2
(11.92)
NMe2

allows the metal to effect two reactions sequentially, resulting in polyfunctionalizations in a minimum number of separate steps. Even more impressive in this regard is the chemistry of dicarbonyl $\eta^5$-cyclopentadienyl-$\eta^1$-allyliron complexes, studied by Rosenblum, in which the iron effects three different reactions [117]. These complexes can be prepared in a variety of ways, the most common of which is the reaction of $CpFe(CO)_2^-$ with allylic halides or tosylates. These complexes react with powerful electrophiles, such as TCNE or isocyanates, to produce cyclic products formally the result of a (3+2) cycloaddition. The reaction is thought to proceed sequentially, however, with the initial step being nucleophilic attack of the σ-allyl group on the electrophile. The resulting adduct now is a cationic iron-olefin complex, which undergoes nucleophilic attack by the negatively charged end of the electrophile to close the ring and generate a neutral alkyliron complex, which can undergo further insertion or cleavage reactions if desired [118] (Equation 11.93). This is a general reaction for a large

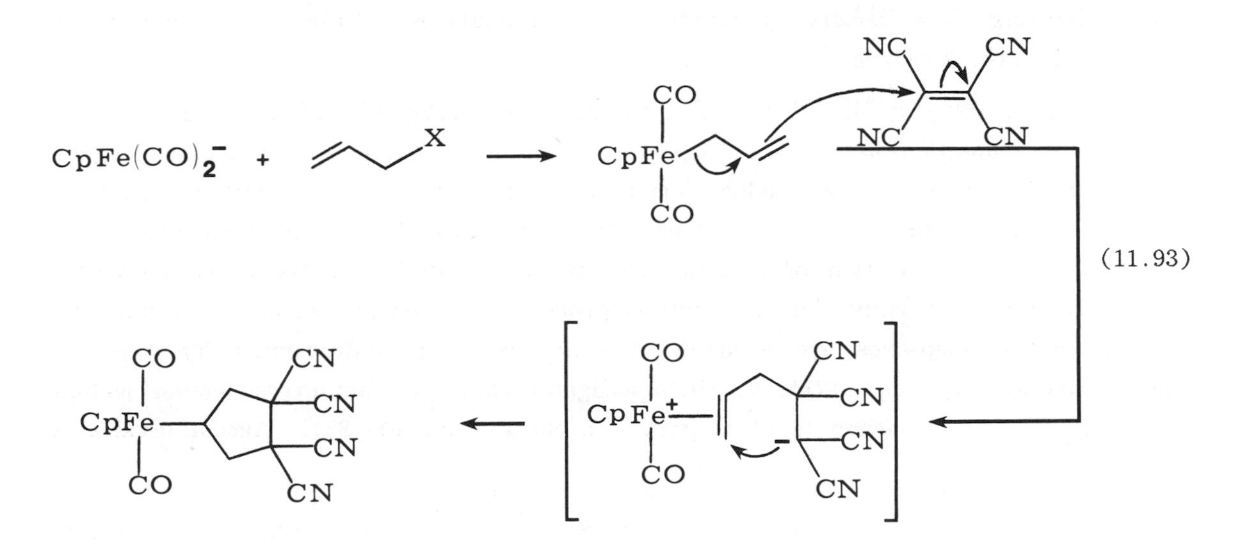


(11.93)

number of differently substituted allyl complexes ($\eta^1$-allenyl and propargyl complexes react in a similar fashion), but is restricted to very strong acceptor components, such as TNCE, trichloroacetyl isocyanate, dichlorodicyanoquinone, and dimethyl methylene malonate. This reaction has been put to use in a very clever synthesis of the hydroazulene ring system [119]. This reaction works because tropyliumiron carbonyl is a sufficiently strong electrophile to undergo attack by the $\eta^1$-allyliron complex (11.94). The resulting iron complex can, in principle, be further functionalized in a number of

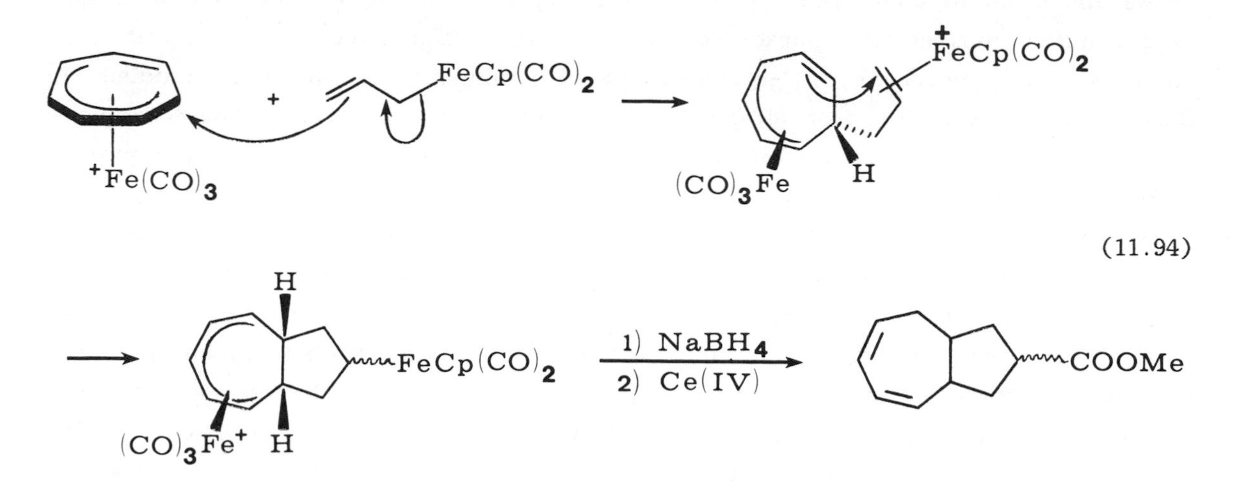


ways, resulting in a variety of substituted hydroazulenes. Only the ester has been reported to date, however.

## 11.9. Reactions of Metal Alkyls from Cyclometallation Reactions.

All the transition-metal alkyl complexes discussed above require the organic group to be functionalized in some fashion prior to the formation of the metal-carbon bond. That is, none are directly available from the hydrocarbon itself. Hydrocarbon activation [120] (direct insertion of a metal into an unactivated C-H bond) is an area of active research, but is not yet a practical process in a general sense. However, certain alkylmetal complexes are available directly from the hydrocarbon by a process termed "cyclometallation" [121], in which a ligand undergoes an intramolecular metallation to form a metal-carbon bond as part of a chelate ring (11.95). Almost invariably,

$$\text{L}\frown\text{CH} + \text{MX} \longrightarrow \text{(L}\rightarrow\text{M}-\text{C)} + \text{HX} \qquad (11.95)$$

five-membered metallacycles are formed, although both four- and six-membered [122] metallacycles have also been formed in this fashion. The ligand is usually nitrogen, phosphorus, or sulfur. There are probably several mechanisms for cyclometallation, strongly dependent on the particular ligand and the particular metal. In some cases, for example, with low-valent metals such as Ir(I) and Rh(I), the reaction proceeds by an intramolecular oxidative addition; in other cases, particularly that of Pd(II), the reaction involves an electrophilic substitution [121].

From the standpoint of utility in organic synthesis, only Pd(II) complexes resulting from o-metallation of aromatic compounds having nitrogen in a benzylic position have received much attention. A wide variety of substrates having nitrogen in this position react with $Li_2PdCl_4$ under mild conditions to form stable, ortho-palladated complexes in high yield. Although the nitrogen almost invariably must be benzylic and somewhat sterically hindered, its chemical nature is less important. This is indicated by the facility with which such diverse substrates as dialkylbenzylamines, α-naphthylamines, azobenzenes, 2-phenylpyridines, and benzaldehyde oximes, imines, and hydrazones readily ortho-palladate (11.96). The thus-formed arylpalladium(II) complexes undergo

$$C_6H_5(H)\text{—Y—(N)} + Li_2PdCl_4 \longrightarrow [C_6H_4\text{—Y—(N)}\rightarrow Pd(\mu\text{-}Cl)]_2 + HCl + 2\,LiCl \qquad (11.96)$$

Y—(N) = $—CH_2NMe_2$, —CH=NOH, CH=NR, CH=N—NHR, 2-pyridyl, —N=NPh

all the reactions presented above, and thus permit the ortho functionalization of these aromatic substrates.

Heck studied the carbonylation of the ortho-palladation products of α-arylnitrogen derivatives, which lead to a number of interesting heterocyclic compounds [123]. The azobenzene complex reacts with carbon monoxide in methanol to produce 2-phenyl-1H-indazolone in 70 percent yield (11.97). Benzaldehyde or acetophenone imine complexes, hydrazone complexes and N,N-dialkylbenzylamine complexes undergo similar carbonylations (11.98 to 11.100). (These reactions are more complex than presented, and more detail is available in the primary reference.) Surprisingly, aromatic oxime complexes do not insert CO. Rather, stable Pd(II) carbonyl complexes form [124] (Equation 11.101). Ortho-palladated benzylamines also add to (insert) conjugated enones [125]; this reaction (11.102) goes in high yield with enones having unsubstituted or α substituted vinyl groups, but fails entirely with enones having β-substituents. Since the o-palladation reaction is often regiospecific with substituted aromatic substrates, selective elaboration of aromatics is possible by this method [126].

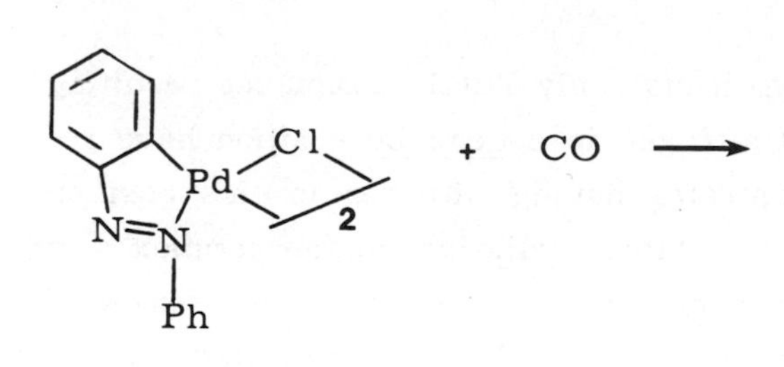

+ CO ⟶

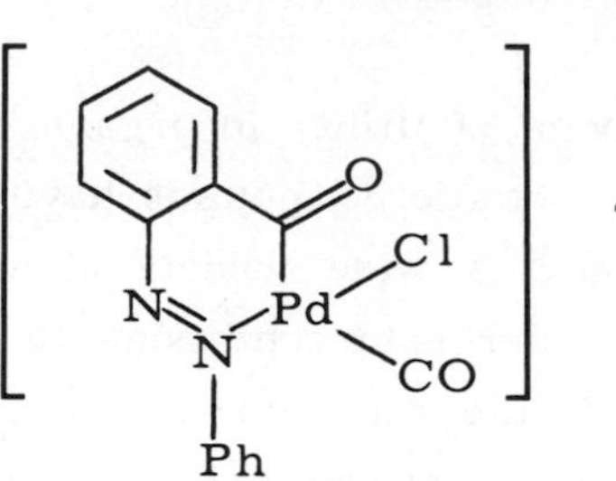

(11.97)

HCl + Pd(O) + $[CH_2O]$ +

MeOH ⟵

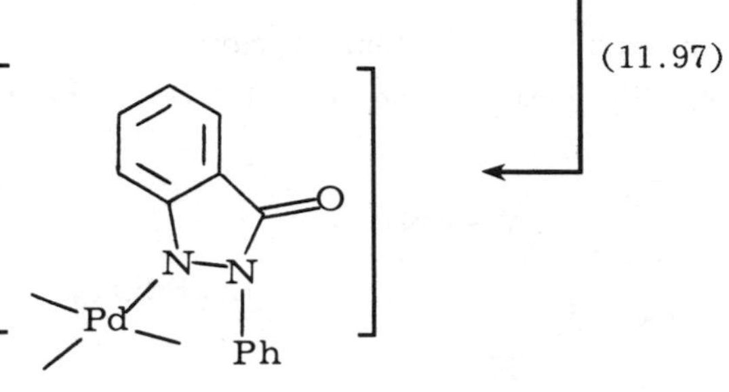

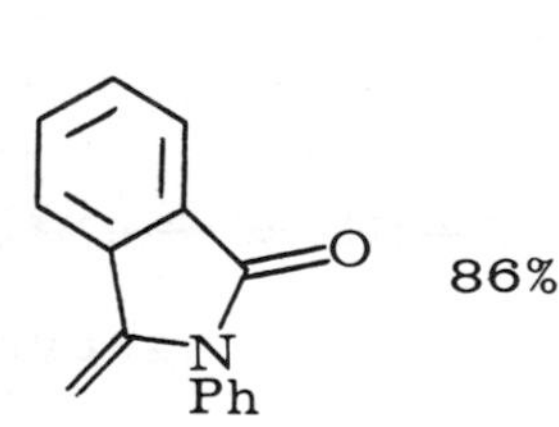

86%

$-$PdH | R = Me

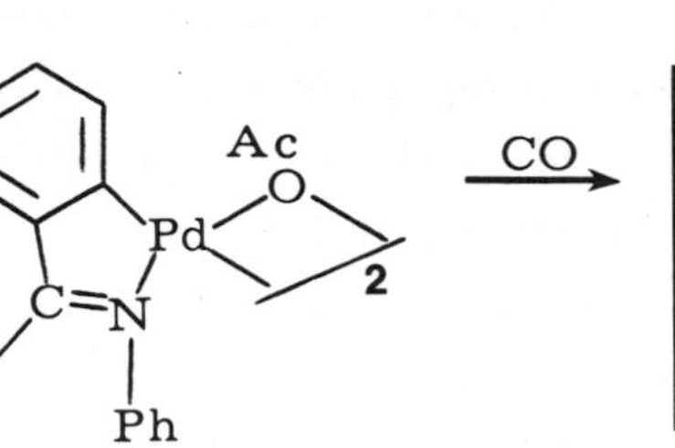

CO ⟶

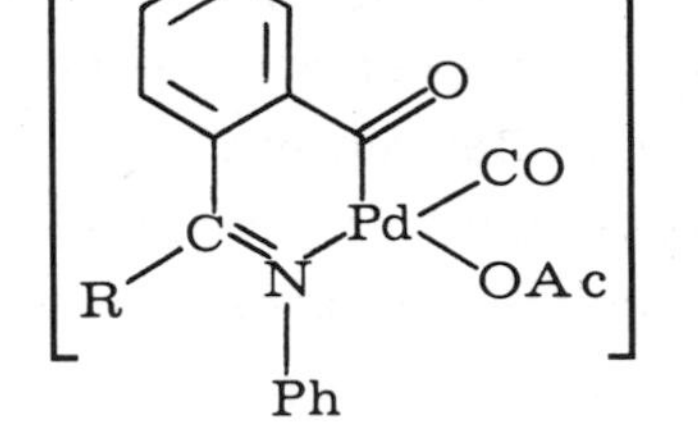

⟶

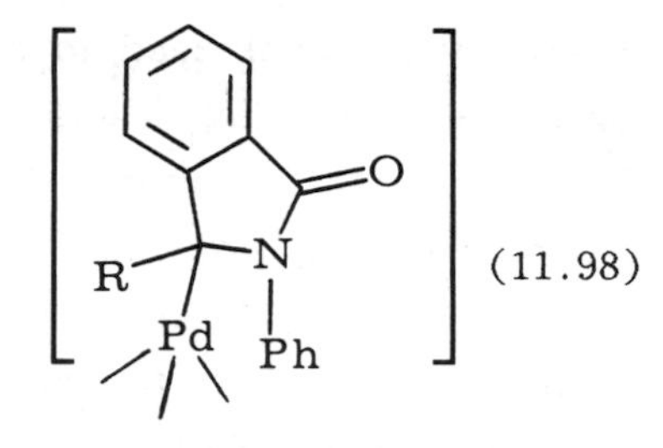

(11.98)

AcO$^-$ | R = H

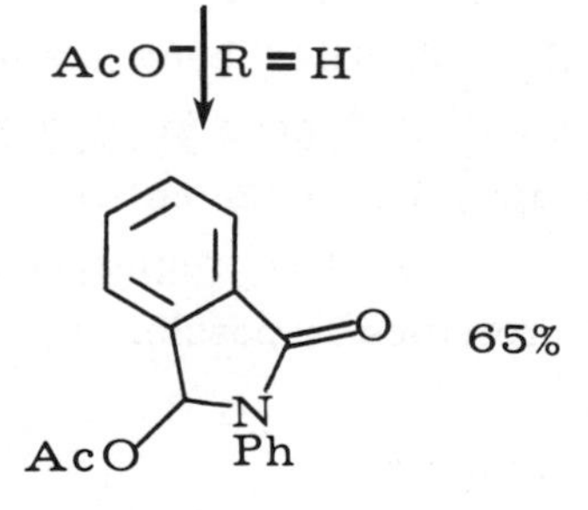

65%

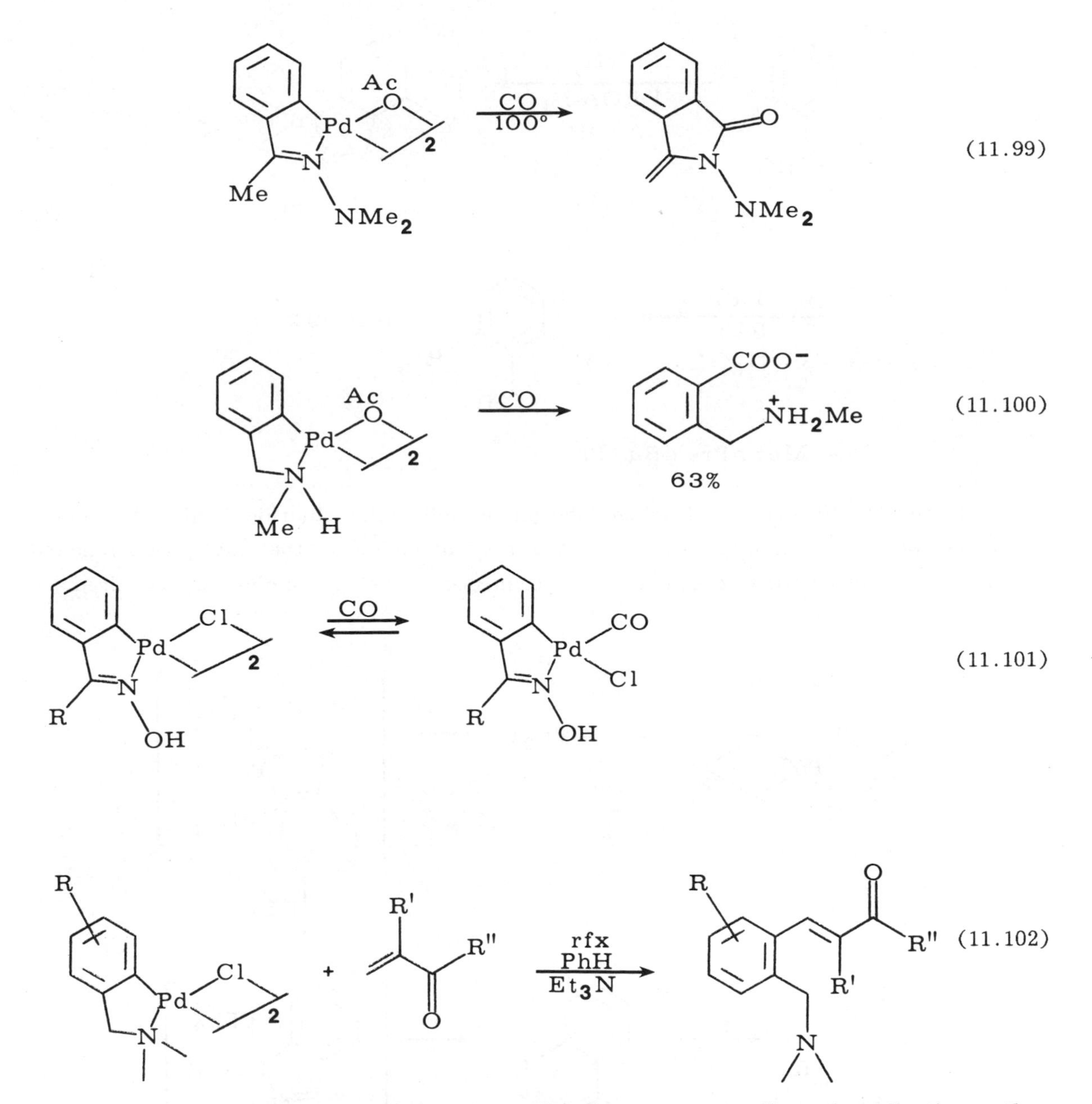


Finally, Murahashi used this ortho-palladation route to effect the selective ortho-alkylation and arylation of benzaldehydes, azobenzenes, and tertiary benzylic amines [127]. Treatment of a variety of ortho-palladated aromatics with triphenylphosphine and either Grignard reagents or organolithium reagents results in selective replacement of the palladium by the anionic R group (11.103). The reaction is thought to proceed

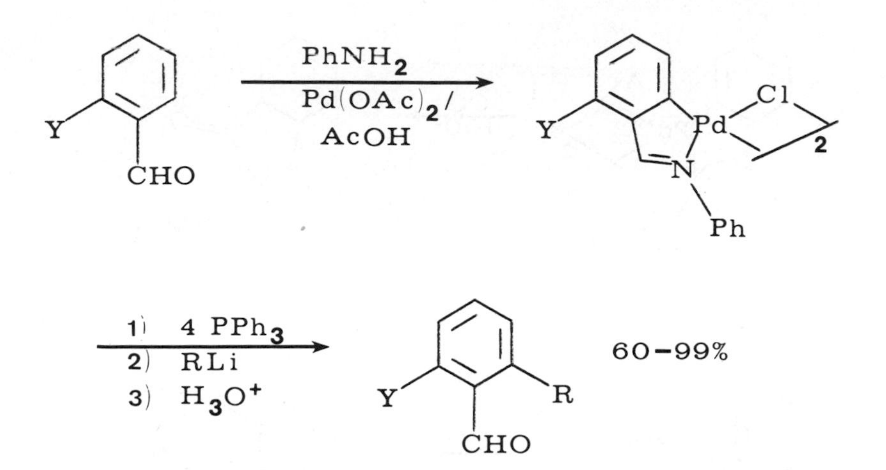


(11.103)

as in Figure 11.10 [114]. Triphenylphosphine splits the chloride bridge to form a monomeric phosphine-chloro complex. Replacement of halide by the alkyl group followed by reductive elimination completes the sequence. The phosphine is required to

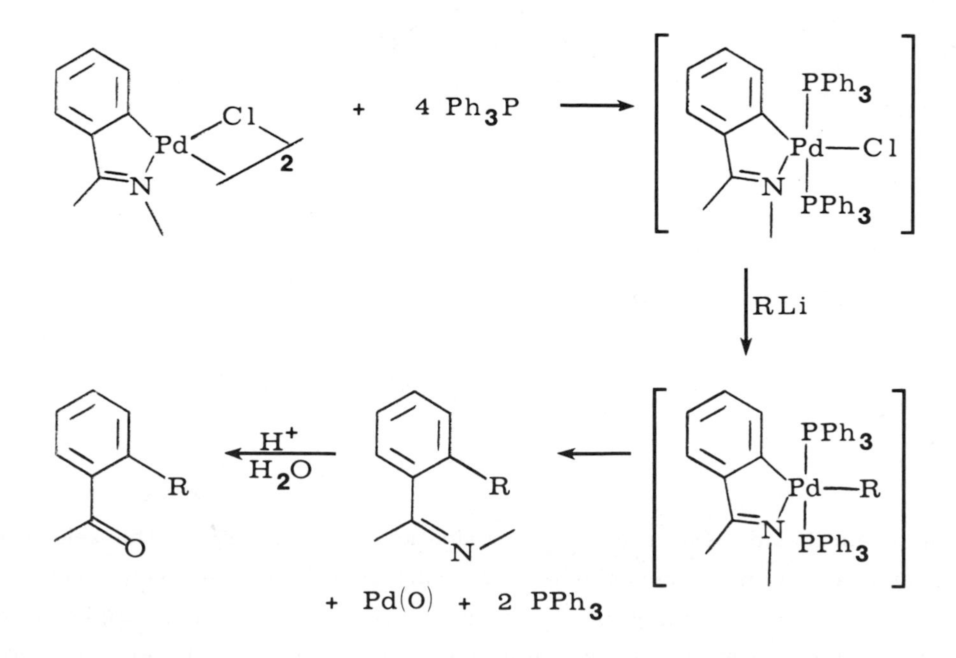


*Figure 11.10. Alkylation of o-palladated aromatics.*

stabilize the intermediate Pd-R complex, preventing both simple reduction of the Pd(II) to metallic Pd, and β-hydride elimination from alkyl groups having β-hydrogens such as n-butyl. The alkylation only proceeds with unstabilized carbanions, which prefer to attack palladium. Stabilized carbanions, which prefer to attack directly on Pd-coordinated carbon, fail. The reaction is also restricted to primary alkyl Grignard or lithium reagents. Secondary and tertiary carbanions lead to low yields of direct alkylation, the major products being the starting aldehyde and minor (15 to 20 percent) amounts of primary alkyl-substituted products. That is, with i-propylmagnesium chloride, the major alkylation product is the n-propylbenzaldehyde. This is reminiscent of the Rh(I)-assisted alkylation of acid chlorides presented above, and is likely to proceed by the same β-hydride elimination-readdition mechanism cited there. Within constraints, however, this is a very useful reaction. Furthermore, alkylation of σ-alkylpalladium complexes formed by other methods should be possible.

## Notes.

1. For reviews on hydrozirconation, see:
   (a) J. Schwartz and J. A. Labinger, Angew. Chem. Int. Ed., 15, 333 (1976);
   (b) D. B. Carr, M. Yoshifuji, L. I. Shoer, K. I. Gall, and J. Schwartz, Annals N.Y. Acad. Sci., 295, 127 (1977).
2. D. W. Hart and J. Schwartz, J. Amer. Chem. Soc., 96, 8115 (1974).
3. D. W. Hart, T. F. Blackburn, and J. Schwartz, J. Amer. Chem. Soc., 97, 679 (1975).
4. P. L. Bock, D. J. Boschetto, J. R. Rasmussen, J. P. Demers, and G. M. Whitesides, J. Amer. Chem. Soc., 96, 2814 (1974).
5. J. A. Labinger, D. W. Hart, W. E. Seibert, III, and J. Schwartz, J. Amer. Chem. Soc., 97, 3851 (1975).
6. C. A. Bertelo and J. Schwartz, J. Amer. Chem. Soc., 97, 228 (1975).
7. C. A. Bertelo and J. Schwartz, J. Amer. Chem. Soc., 98, 262 (1976).
8. T. F. Blackburn, J. A. Labinger, and J. Schwartz, Tetrahedron Lett., 3041 (1975).
9. D. B. Carr and J. Schwartz, J. Amer. Chem. Soc., 99, 638 (1977).
10. M. Yoshifuji, M. J. Loots, and J. Schwartz, Tetrahedron Lett., 1303 (1977).
11. M. J. Loots and J. Schwartz, J. Amer. Chem. Soc., 99, 8045 (1977).
12. M. J. Loots and J. Schwartz, Tetrahedron Lett., 4381 (1978).
13. E.-i. Negishi and D. E. Van Horn, J. Amer. Chem. Soc., 99, 3168 (1977).

14. D. E. Van Horn and E.-i. Negishi, J. Amer. Chem. Soc., 100, 2252 (1978).
15. N. Okukado and E.-i. Negishi, Tetrahedron Lett., 2357 (1978).
16. E.-i. Negishi, N. Okukado, A. O. King, D. E. Van Horn, and B. I. Spiegel, J. Amer. Chem. Soc., 100, 2254 (1978).
17. For a useful review, see: J. R. Normant, Pure and Appl. Chem., 50, 709 (1978).
18. This scheme adapted from a forthcoming book by G. H. Posner entitled An Introduction to Synthesis using Organocopper Reagents.
19. M. S. Kharasch and P. O. Tawney, J. Amer. Chem. Soc., 63, 2308 (1941).
20. This reaction is reviewed in depth by G. H. Posner, Org. React., 22, 253 (1975).
21. This reaction is reviewed in depth by Ibid, 19, 1 (1972).
22. R. G. Pearson and C. D. Gregory, J. Amer. Chem. Soc., 98, 4098 (1976).
23. C. R. Johnson and G. A. Dutra, J. Amer. Chem. Soc., 95, 7783 (1973).
24. K. Kitatani, T. Hiyama, and H. Nozaki, Bull. Chem. Soc. Japan, 50, 1600 (1977).
25. P. W. Raynolds, M. J. Manning, and J. S. Swenton, J.C.S. Chem. Comm., 499 (1977).
26. S. Rauscher, Tetrahedron Lett., 1161 (1976).
27. D. W. Knight and G. Pattenden, J.C.S. Perkin I, 641 (1975).
28. K. Mori, S. Tamada, and M. Matsui, Tetrahedron Lett., 901 (1978).
29. J. P. Marino, Ann. Repts. Med. Chem., 10, 327 (1975).
30. This scheme and summary were taken from a paper by C. P. Casey and M. C. Cesa, J. Amer. Chem. Soc., 101, 4236 (1979).
31. H. O. House, Accts. Chem. Res., 9, 59 (1976); H. O. House and J. M. Wilkins, J. Org. Chem., 43, 2443 (1978).
32. C. Luthy, P. Konstantin, and K. Untch, J. Amer. Chem. Soc., 100, 6211 (1978).
33. H. C. Arndt, W. G. Biddlcom, G. P. Peruzzotti, and W. D. Woessner, Prostaglandins, 7, 387 (1974).
34. J. W. Huffman and P. G. Harris, J. Org. Chem., 42, 2357 (1977).
35. F. E. Ziegler, G. R. Reid, W. L. Studt, and P. A. Wender, J. Org. Chem., 42, 1991 (1977).
36. Although the lithium enolate has been shown to be formed upon conjugated addition of $R_2CuLi$ to the enone (H. O. House and J. M. Wilkins, J. Org. Chem., 43, 2443 (1978)), the presence of less than 1 percent copper can drastically alter the chemistry of this lithium enolate (G. H. Posner and C. M. Lenz, J. Amer. Chem. Soc., 101, 934 (1979)).

37. G. H. Posner, J. J. Sterling, C. E. Whitten, C. M. Lenz, and D. J. Brunelle, J. Amer. Chem. Soc., 97, 107 (1975).
38. R. Boeckman, Jr., J. Amer. Chem. Soc., 95, 6867 (1973).
39. G. H. Posner, M. J. Chapdelaine, and C. M. Lentz, J. Org. Chem., 44, 3661 (1979).
40. M. Bertrand, G. Gil, and J. Viala, Tetrahedron Lett., 1785 (1977).
41. G. H. Posner, C. E. Whitten, and J. J. Sterling, J. Amer. Chem. Soc., 95, 7788 (1973).
42. R. T. Buckler and D. L. Garland, Tetrahedron Lett., 2257 (1978).
43. R. F. Newton, C. C. Howard, D. P. Reynolds, A. H. Wadsworth, N. M. Crossland, and S. M. Roberts, J.C.S. Chem. Comm., 662 (1978).
44. M. F. Semmelhack, A. Yamashita, J. C. Tomesch, and K. Hirotsu, J. Amer. Chem. Soc., 100, 5565 (1978).
45. E. Piers, I. Nagakura, and H. E. Morton, J. Org. Chem., 43, 3630 (1978).
46. P. A. Grieco, C. L. Wang, and G. Majetich, J. Org. Chem., 41, 726 (1976).
47. E. J. Corey and R. H. Wollenberg, J. Amer. Chem. Soc., 96, 5581 (1974).
48. H. Westmijze, J. Meijer, and P. Vermeer, Rec. Trav. Chem., 96, 168 (1977).
49. A. B. Levy, P. Talley, and J. A. Dunford, Tetrahedron Lett. 3545 (1977).
50. H. Westmijze, J. Meijer, H. J. T. Bos, and P. Vermeer, Rec. Trav. Chim., 95, 299 (1976).
51. H. Westmijze, J. Meijer, H. J. T. Bos and P. Vermeer, Rec. Trav. Chim., 95, 304 (1976).
52. P. R. McGuirk, A. Marfat, and P. Helquist, Tetrahedron Lett., 2973 (1978).
53. P. R. McGuirk, A. Marfat, and P. Helquist, Tetrahedron Lett., 1363 (1978).
54. J. F. Normant, J. Organometal. Chem. Libr., 1, 219 (1976).
55. A. Alexakis, G. Cahiez, J. F. Normant, and J. Villieras, Bull Soc. Chim. Fr., 693 (1977).
56. W. E. Truce and M. J. Lusch, J. Org. Chem., 43, 2252 (1978).
57. H. Westmijze, H. Kleijn, and P. Vermeer, Synthesis, 454 (1978).
58. P. Vermeer, H. Westmijze, H. Kleijn, and L. A. van Dijck, Rec. Trav. Chim., 97, 56 (1978).
59. G. Tadema, R. H. Everhardus, H. Westmize and P. Vermeer, Tetrahedron Lett., 3935 (1978).
60. For a review on this topic, see H. Felkin and G. Swierczewski, Tetrahedron, 31, 2735 (1975).

61. K. Tamao, K. Sumitani, Y. Kiso, M. Zembayashi, A. Fijioka, S-i. Kodama, I. Nakajima, A. Minato, and M. Kumada, Bull. Chem. Soc. Japan, 49, 1958 (1976).
62. T. T. Tsou and J. K. Kochi, J. Amer. Chem. Soc., 101, 7547 (1979).
63. E. D. Thorsett and F. R. Stermitz, J. Heterocyclic Chem., 10, 243 (1973).
64. L. N. Pridgen, J. Heterocyclic Chem., 12, 443 (1975).
65. K. Tamao, M. Zembayashi, and M. Kumada, Chem. Lett., 1237 (1976).
66. K. Tamao, S.-i. Kodama, T. Nakatsuka, Y. Kiso, and M. Kumada, J. Amer. Chem. Soc., 97, 4405 (1975).
67. T. Hayashi, M. Tajika, K. Tamao, and M. Kumada, J. Amer. Chem. Soc., 98, 3719 (1976); K. Tamao, H. Matsumoto, H. Yamamoto, and M. Kumada, Tetrahedron Lett., 7155 (1979).
68. L. S. Hegedus, P. M. Kendall, S. M. Lo, and J. R. Sheats, J. Amer. Chem. Soc., 97, 5448 (1975).
69. R. F. Heck, Proc. of the Welch Foundation, XVII, 53 (1973).
70. R. C. Larock, Angew. Chem. Int. Ed., 17, 27 (1978).
71. R. C. Larock, J. Organometal. Chem. Libr., 1, 257 (1976).
72. R. C. Larock and B. Riefling, J. Org. Chem., 43, 1468 (1978).
73. R. C. Larock, J. C. Bernhardt, and R. J. Driggs, J. Organometal. Chem., 156, 45 (1978).
74. D. E. Bergstrom and J. L. Ruth, J. Amer. Chem. Soc., 98, 1587 (1976); C. F. Bigge, P. Kalaritis, and M. P. Mertes, Tetrahedron Lett., 1653 (1979).
75. J. L. Ruth and D. E. Bergstrom, J. Org. Chem., 43, 2870 (1978).
76. I. Arai and G. D. Daves, Jr., J. Org. Chem., 43, 4110 (1978).
77. I. Arai and G. D. Daves, Jr., J. Amer. Chem. Soc., 100, 287 (1978).
78. B. M. Trost and Y. Tanigawa, J. Amer. Chem. Soc., 101, 4743 (1979).
79. R. C. Larock and M. A. Mitchell, J. Amer. Chem. Soc., 100, 180 (1978).
80. R. C. Larock, B. Riefling, and C. A. Fellows, J. Org. Chem., 43, 131 (1978).
81. For a review of R. H. Heck's work in this area, see R. F. Heck, Pure and Appl. Chem., 50, 691 (1978).
82. J. K. Stille and P. Kwan Wong, J. Org. Chem., 40, 532 (1975).
83. A. Schoenberg and R. F. Heck, J. Org. Chem., 39, 3327 (1974).
84. A. Cowell and J. K. Stille, Tetrahedron Lett., 133 (1979).
85. M. Mori, K. Chiba, and Y. Ban, J. Org. Chem., 43, 1684 (1978).
86. M. Mori, K. Chiba, and Y. Ban, Heterocycles, 6, 1841 (1977).
87. J. E. Plevyak and R. F. Heck, J. Org. Chem., 43, 2454 (1978).

88. W. C. Frank, Y. C. Kim, and R. F. Heck, J. Org. Chem., 43, 2947 (1978).
89. I. Arai and G. D. Daves, Jr., J. Heterocyclic Chem., 15, 351 (1978).
90. N. A. Cortese, C. B. Ziegler, Jr., B. J. Hrnjez, and R. F. Heck, J. Org. Chem., 43, 2952 (1978).
91. C. B. Zeigler, Jr., and R. F. Heck, J. Org. Chem., 43, 2941 (1978).
92. C. B. Ziegler, Jr., and R. F. Heck, J. Org. Chem., 43, 2949 (1978).
93. A. J. Chalk and S. A. Magennis, J. Org. Chem., 41, 273 (1976).
94. J. B. Melpolder and R. F. Heck, J. Org. Chem., 41, 265 (1976).
95. Y. Tamaru, Y. Yamada, and Z.-i. Yoshida, Tetrahedron Lett., 919 (1978).
96. Y. Tamaru, Y. Yamada, and Z.-i. Yoshida, Chem. Lett., 529 (1978).
97. Y. Tamaru, Y. Yamada, and Z.-i. Yoshida, Chem. Lett., 975 (1978).
98. Y. Tamaru, Y. Yamada, and Z.-i. Yoshida, J. Org. Chem., 43, 3396 (1978).
99. M. Mori, K. Chiba, and Y. Ban, Tetrahedron Lett., 1037 (1977).
100. M. Mori and Y. Ban, Tetrahedron Lett., 1803 (1976); 1133 (1979).
101. M. Mori and Y. Ban, Tetrahedron Lett., 1807 (1976).
102. M. Mori and Y. Ban, Heterocycles, 9, 391 (1978).
103. D. Milstein and J. K. Stille, J. Amer. Chem. Soc., 101, 4992; 4981 (1979).
104. M. Kosugi, Y. Shimizu, and T. Migata, Chem. Lett., 1423 (1977).
105. H. P. Dang and G. Linstrumelle, Tetrahedron Lett., 191 (1978).
106. J. K. Stille and L. F. Hines, J. Amer. Chem. Soc., 94, 485 (1972).
107. D. Medema, R. Van Helden, and C. F. Kohle, Inorg. Chim. Acta, 3, 255 (1969).
108. J. K. Stille, D. E. James, and L. F. Hines, J. Amer. Chem. Soc., 95, 5062 (1973).
109. R. A. Holton and R. A. Kjonaas, J. Organometal. Chem., 133, C5 (1977).
110. L. S. Hegedus, O. P. Anderson, K. Zetterberg, G. Allen, K. Siirala-Hansen, D. J. Olsen, and A. B. Packard, Inorg. Chem., 16, 1887 (1977).
111. J. E. Backvall, Tetrahedron Lett., 2225 (1975).
112. G. W. Daub, Prog. Inorg. Chem., 22, 409 (1977).
113. J. E. Backvall, J.C.S. Chem. Comm., 413 (1977).
114. L. S. Hegedus, G. Allen, and D. J. Olsen, J. Amer. Chem. Soc., 102, 3583 (1980).
115. L. S. Hegedus and W. H. Darlington, J. Amer. Chem. Soc., in press.
116. R. A. Holton and R. A. Kjonaas, J. Amer. Chem. Soc., 99, 4177 (1977).
117. M. Rosenblum, Accts. Chem. Res., 7, 122 (1974).

118. A. Cutler, D. Ehntholt, W. P. Giering, P. Lennon, S. Raghu, A. Rosan, M. Rosenblum, J. Tancrede, and D. Wells, J. Amer. Chem. Soc., 98, 3495 (1976).
119. N. Genco, D. Marten, S. Raghu, and M. Rosenblum, J. Amer. Chem. Soc., 98, 848 (1976).
120. For reviews on hydrocarbon activation, see G. W. Parshall, Accts. Chem. Res., 8, 113 (1975); D. E. Webster, Adv. Organometal. Chem., 15, 147 (1977).
121. For reveiws on cyclometallation, see: M. I. Bruce, Angew. Chem. Int. Ed., 16, 73 (1977); H.-P. Abicht and K. Issleib, Z. Chem., 17, 1 (1977); J. Dehand and M. Pfeffer, Coord. Chem. Rev., 18, 327 (1976).
122. N. D. Cameron and M. Kilner, J.C.S. Chem. Comm., 687 (1975).
123. J. M. Thompson and R. F. Heck, J. Org. Chem., 40, 2667 (1975).
124. H. Onoue, K. Nakagawa, and I. Moritani, J. Organometal. Chem., 35, 217 (1972).
125. R. A. Holton, Tetrahedron Lett., 355 (1977).
126. R. A. Holton and R. G. Davis, J. Amer. Chem. Soc., 99, 4175 (1977).
127. S.-I. Murahashi, Y. Tamba, M. Yamamura, and N. Yoshimura, J. Org. Chem., 43, 4099 (1978).

# 12

## Transition-Metal Olefin Complexes

As was discussed in Chapter 3, two general types of metal-olefin complexes representing the two extremes of the Dewar-Chatt bonding model are known. Those formed by the reaction of low-valent (usually zero-valent) metals with olefins can be considered to be either trigonal metal(0)-olefin complexes with significant electron transfer from the metal to the olefin, 1a, or metal(II) metallacyclopropane complexes formed by a formal oxidative addition of the olefin to the metal (1b) [1]. These

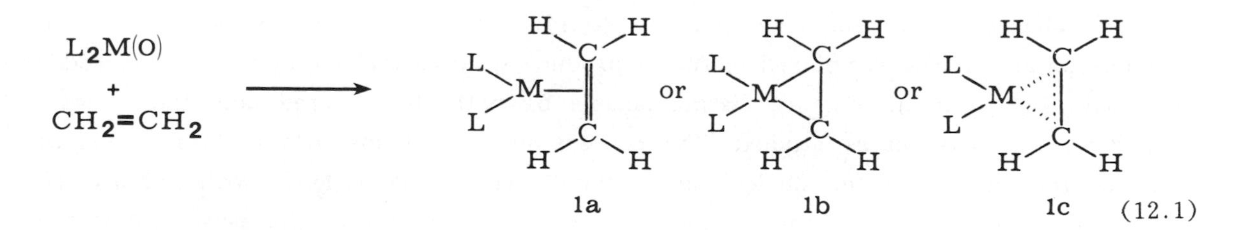


(12.1)

complexes are most common for zero-valent ($d^{10}$) nickel, palladium, and platinum. Although reported to react with acids, and expected to react with electrophiles in general, they have found little use in synthesis.

The more common and more useful type of olefin complex is that involving $d^8$ transition metals, particularly Pd(II), Pt(II), Fe(II), and, to a lesser extent, Rh(I) and Ir(I). In these complexes, often formed by treatment of the metal salt with an olefin, the olefin is bonded perpendicular to the remainder of the metal-ligand plane (12.2). The olefin is relatively labile, and replaceable by a variety of ligands, particularly phosphines and amines. Olefins coordinated to Pd(II), Pt(II), and Fe(II)

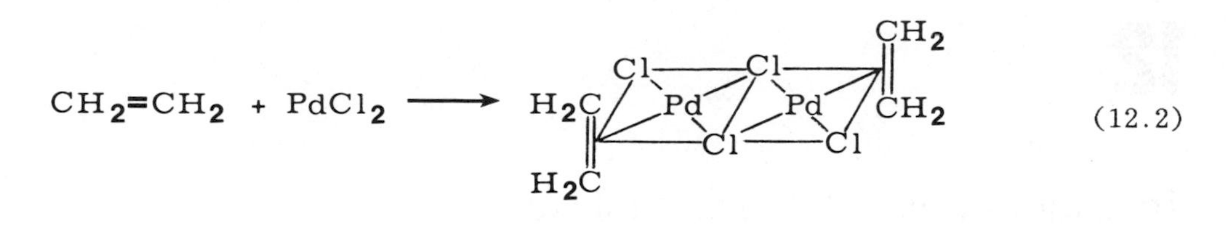


(12.2)

in this fashion react readily with nucleophiles, and are quite resistant to electrophilic attack, perhaps because a substantial reduction of electron density in the $\pi$ (bonding) molecular orbital of the olefin upon complexation to the metal leads to structures of type 2. In contrast, Rh(I) and Ir(I) olefin complexes do not react readily with nucleophiles,

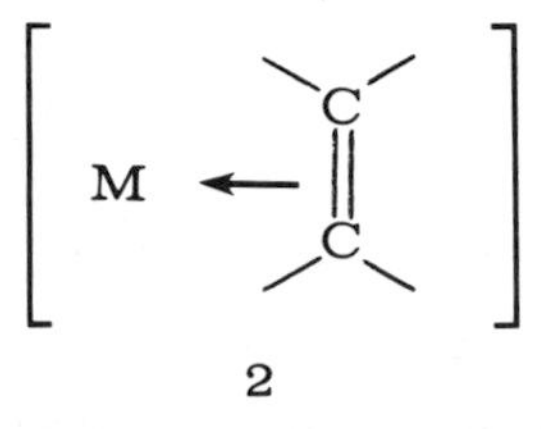


2

but rather suffer electrophilic attack, perhaps because the lower oxidation state of rhodium (I vs. II) makes the metal a poorer acceptor and diminishes the importance of structure 2.

## 12.1. Nucleophilic Attack on Metal-Olefin Complexes.

The characteristic reaction of Pd(II), Pt(II), and Fe(II) olefin complexes is attack of nucleophiles on the complexed olefin to produce σ-alkylmetal complexes. Two modes of attack are possible, and are distinguishable by both their stereo- and their regiochemistry. In most cases studied, the process involves trans attack (from the face opposite the metal) by the nucleophile, without prior coordination. With substituted olefins, this type of nucleophilic attack occurs predominantly at the more-substituted olefin terminus. In contrast, some nucleophiles ($OAc^-$, $Cl^-$, and some carbanions, for example) appear to attack the metal first. This is followed by insertion of the olefin into the metal-nucleophile bond. Processes such as these, involving prior coordination of the nucleophile, result in cis additions. In the absence of unusual electronic effects, attack occurs at the less-substituted olefin terminus (12.3). Reactions of both types are generally restricted to mono- or disubstituted olefins, since tri- and tetrasubstituted olefins coordinate only weakly and are easily displaced. Displacement of olefin by nucleophile is a major competing reaction in all nucleophilic attacks on monoolefin complexes, particularly if the nucleophile is a good ligand for the metal (12.4).

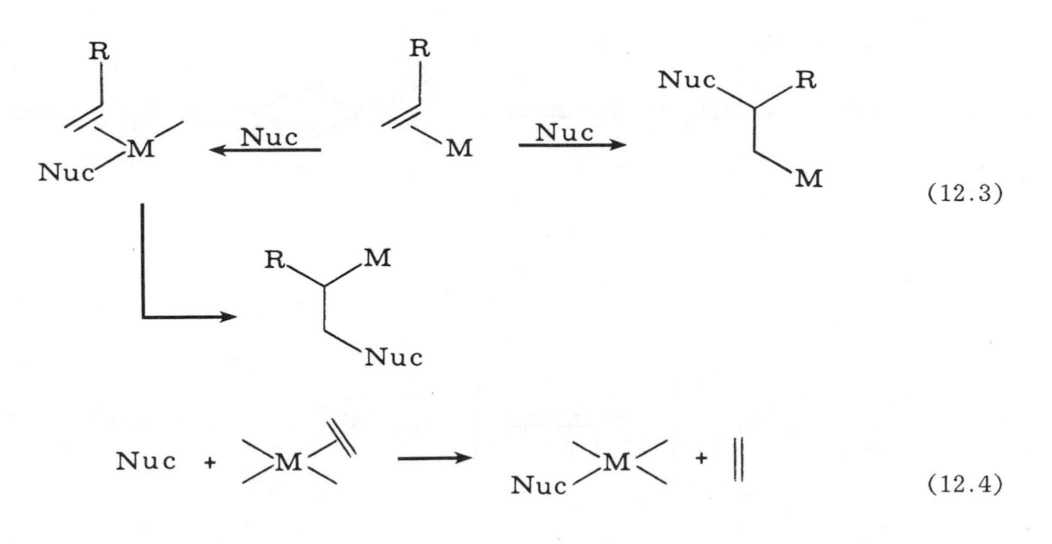


Coordination of an olefin to a metal also activates the allylic hydrogens, and facilitates their removal by bases as weak as carbonate. Thus another possible reaction in competition with nucleophilic attack on the olefin is generation of $\eta^3$-allyl metal complexes by allylic proton abstraction (12.5). Both of these side reactions have been observed, although olefin displacement appears to be the more common.

Nuc + (H–CH₂–CH=CH₂)M ⟶ (η³-allyl)M + NucH⁺ (12.5)

Nucleophilic attack on palladium-olefin complexes has been studied extensively, starting with the industrially important Wacker process for the conversion of ethylene to acetaldehyde (12.6). It is clear that the key step in this process is nucleophilic attack

$$CH_2{=}CH_2 + PdCl_2 + H_2O \longrightarrow CH_3\overset{O}{\overset{\|}{C}}H + Pd(0) + 2\,HCl \quad (12.6)$$

(Pd(0) recycled to $PdCl_2$ by $O_2$, Cu(II))

of $OH^-$ (or $H_2O$) on the Pd-complexed olefin. Kinetic results suggested prior coordination of the water (nucleophile) by the metal, and the long-accepted mechanism for the Wacker process is shown in Figure 12.1, which involves coordination of $H_2O$, loss of a

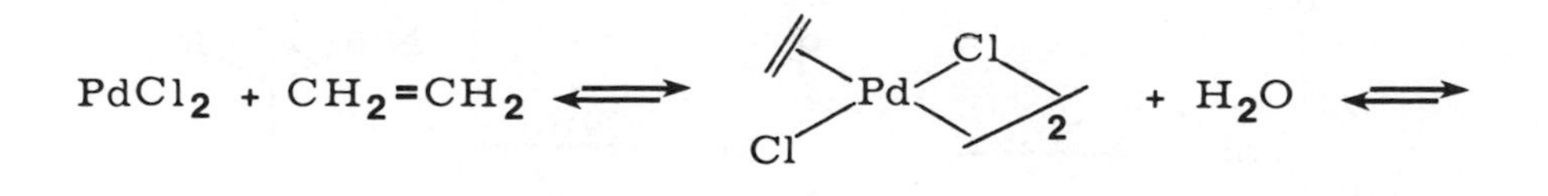


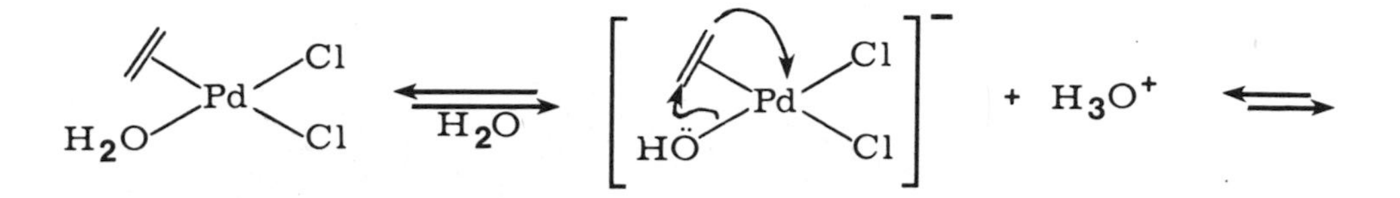


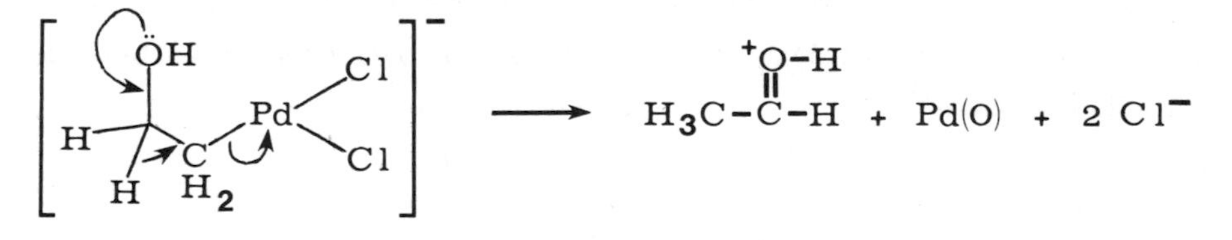


*Figure 12.1. Proposed mechanism of the Wacker oxidation.*

proton, insertion of the olefin into the Pd-OH bond, and collapse to acetaldehyde. Labeling studies show that all the hydrogens in the product acetaldehyde come from the ethylene, not from the solvent. Hence a hydride transfer from $C_1$ to $C_2$ of the initial ethene must occur. This may either be a concerted rearrangement, as shown in Figure 12.1, or a β-Pd-H elimination followed by readdition of Pd-H in the opposite sense, followed by β-Pd-H elimination from the OH group to form the carbonyl group. However, this is one of the few examples in which nucleophilic attack on the coordinated olefin is claimed to occur cis with respect to the metal. Since the stereochemical course of nucleophilic attack cannot be directly probed in this process, cis attack has not been demonstrated, only inferred from kinetic data. However, equally consistent with the kinetic data is a mechanism involving trans attack of water on coordinated olefin (12.7). This process is consistent with the stereochemistry for nucleophilic attack found in

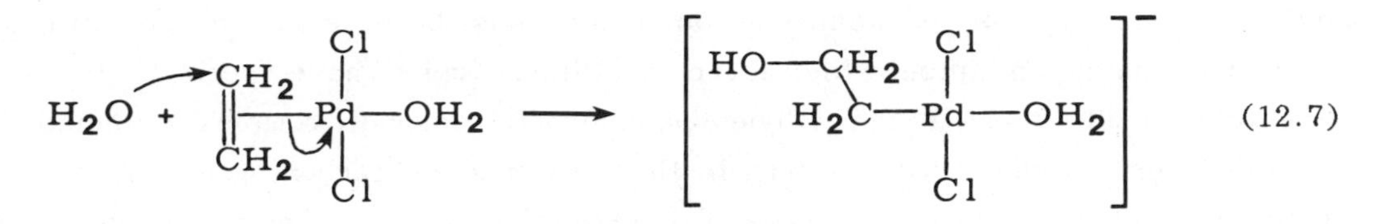


(12.7)

other systems which can be directly observed. Several recent studies also suggest that this is the stereochemistry of the Wacker process.

Åkermark converted trans-1,2-dideuterioethene stereospecifically to threo-1,2-dideuteriochloroethanol by treatment with a catalytic system ($PdCl_2$-$CuCl_2$-LiCl) similar to the Wacker catalyst [2] (Equation 12.8). Since the copper(II) chloride-induced

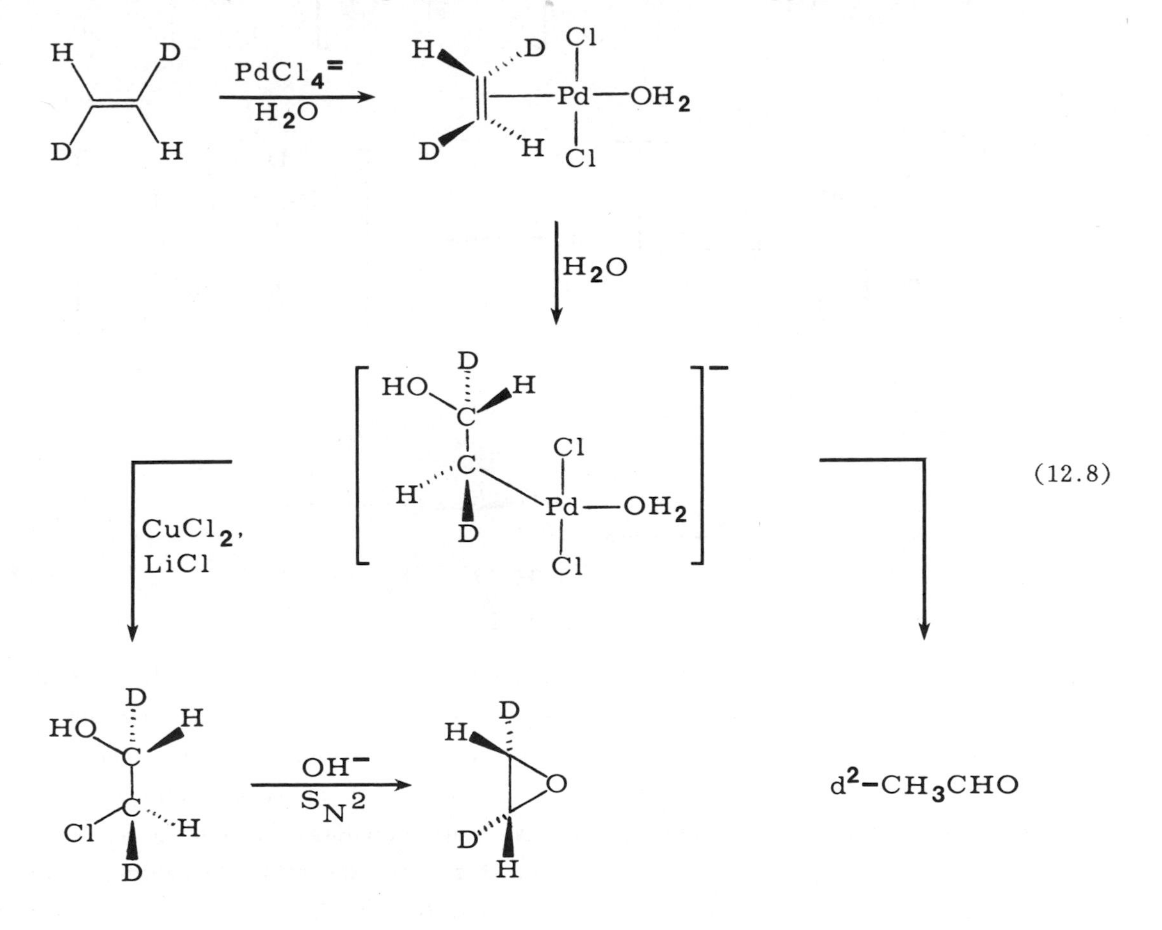


(12.8)

oxidative cleavage of palladium-carbon bonds has been shown to go with inversion, hydroxy-palladation under the above conditions must have been a trans process. Similarly, Stille treated the ethylenepalladium(II)-chloride complex obtained from cis-1,2-dideuterioethylene with water in the presence of carbon monoxide and produced trans-2,3-dideuterio-β-propiolactone [3] (Equation 12.9). Since the insertion of CO

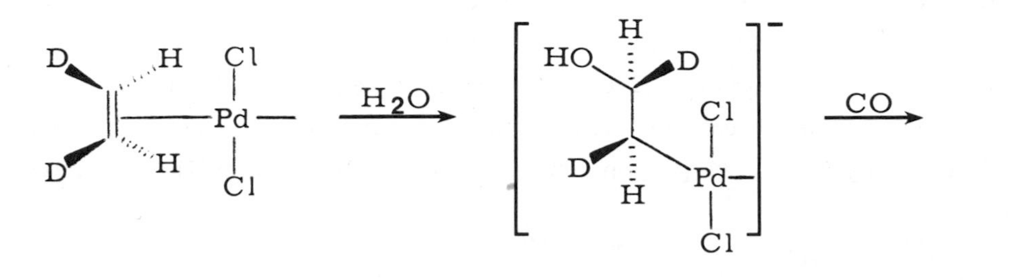


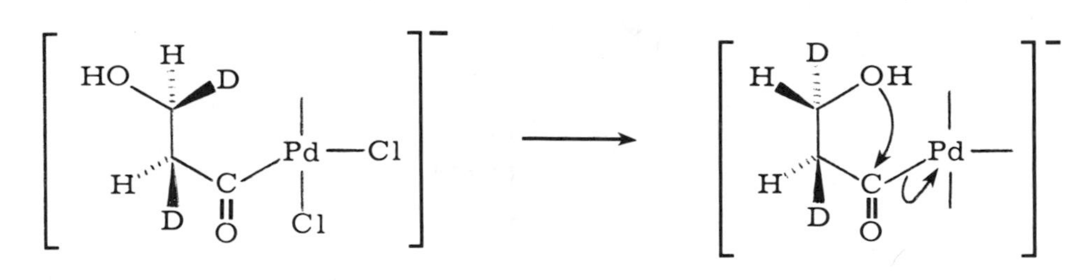


(12.9)

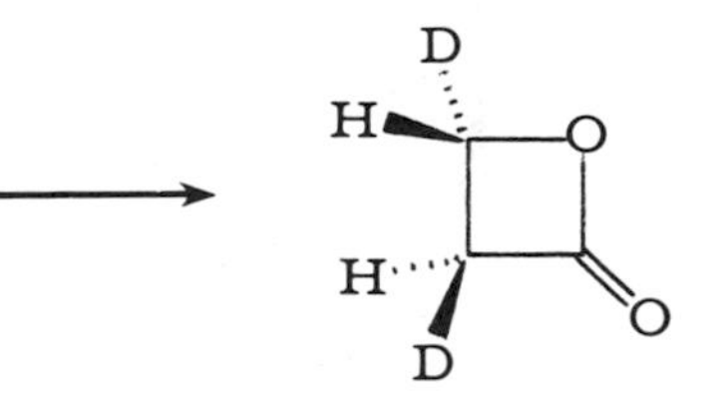


into palladium-carbon bonds is known to proceed with retention, the hydroxy-palladation step must again have been a trans process. Finally, the σ-alkylpalladium complex resulting from methoxide attack on cis-dideuterioethene in the complex $[(\eta^5\text{-}C_5H_5)Pd(Ph_3P)(CHD{=}CHD)]^+ClO_4^-$ has been isolated and characterized by Kurosawa. NMR studies unambiguously established that methoxypalladation had occurred stereospecifically trans [4] (Equation 12.10). (In this case, cis attack involving prior coordination of the nucleophile is impossible, since the Pd is coordinatively saturated.) Thus it seems likely that the Wacker process is also a trans hydroxypalladation.

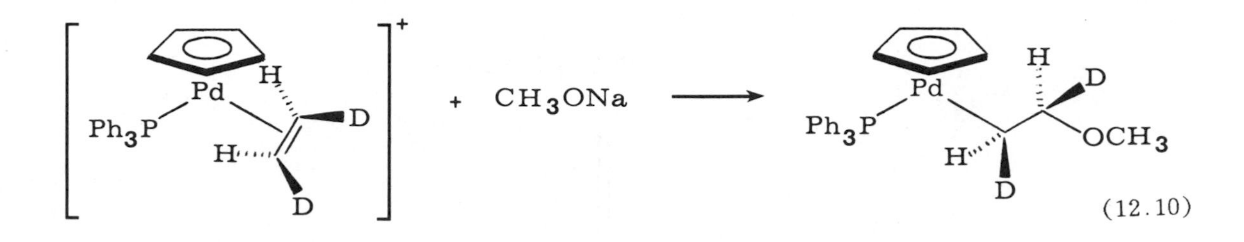


(12.10)

Other oxygen nucleophiles such as acetate and alcohols have also long been known to attack olefin-palladium complexes, and industrial procedures for the production of vinyl acetate and acetals based on chemistry related to the Wacker process are well-known. All of them are thought to involve nucleophilic attack on the palladium-bound olefin, elimination of Pd(0) from the σ-alkylpalladium complex, and reoxidation of Pd(0) to Pd(II), using air and copper salts or some other suitable oxidant.

In contrast, the reaction of palladium-bound olefins with other nucleophiles is much less general. Åkermark showed that olefins can be aminated in a reaction using palladium-olefin complexes provided the reaction is carried out at low temperatures [5]. A reasonable mechanism, as yet unproven experimentally, is shown in Figure 12.2. At temperatures in excess of -20°, olefin displacement, rather than amination, is the predominant reaction. The reaction proceeds best with nonhindered secondary amines and terminal olefins. Primary amines and internal olefins react but in lower (~50%) yield. Trisubstituted olefins react in very low yield, as does ammonia. Three equivalents of amine are required for high conversion, and the reaction may proceed either through a cationic bis-amine complex or a zwitterionic σ-alkyl complex, with the third equivalent of amine acting simply as a base. The reaction is not catalytic, since a reduction step is needed to prevent formation of enamines by β-elimination. The stereochemistry of this amination is cleanly trans [6], and the reaction is regioselective (but not regiospecific) for the most substituted carbon of the olefin. Note that this regiochemistry is consistent with trans attack.

Hegedus showed that olefins are alkylated by stabilized carbanions ($pK_a$ 10-17) under conditions essentially identical to those required for amination [7] (Equation 12.11). Saturated products are obtained by exposure of the reaction mixture to hydrogen. Again, attack occurs primarily (but not exclusively) at the most-substituted position of the olefin, implying trans attack by carbanion without prior coordination. In the absence of added triethylamine, alkylation proceeds in very low yield, if at all,

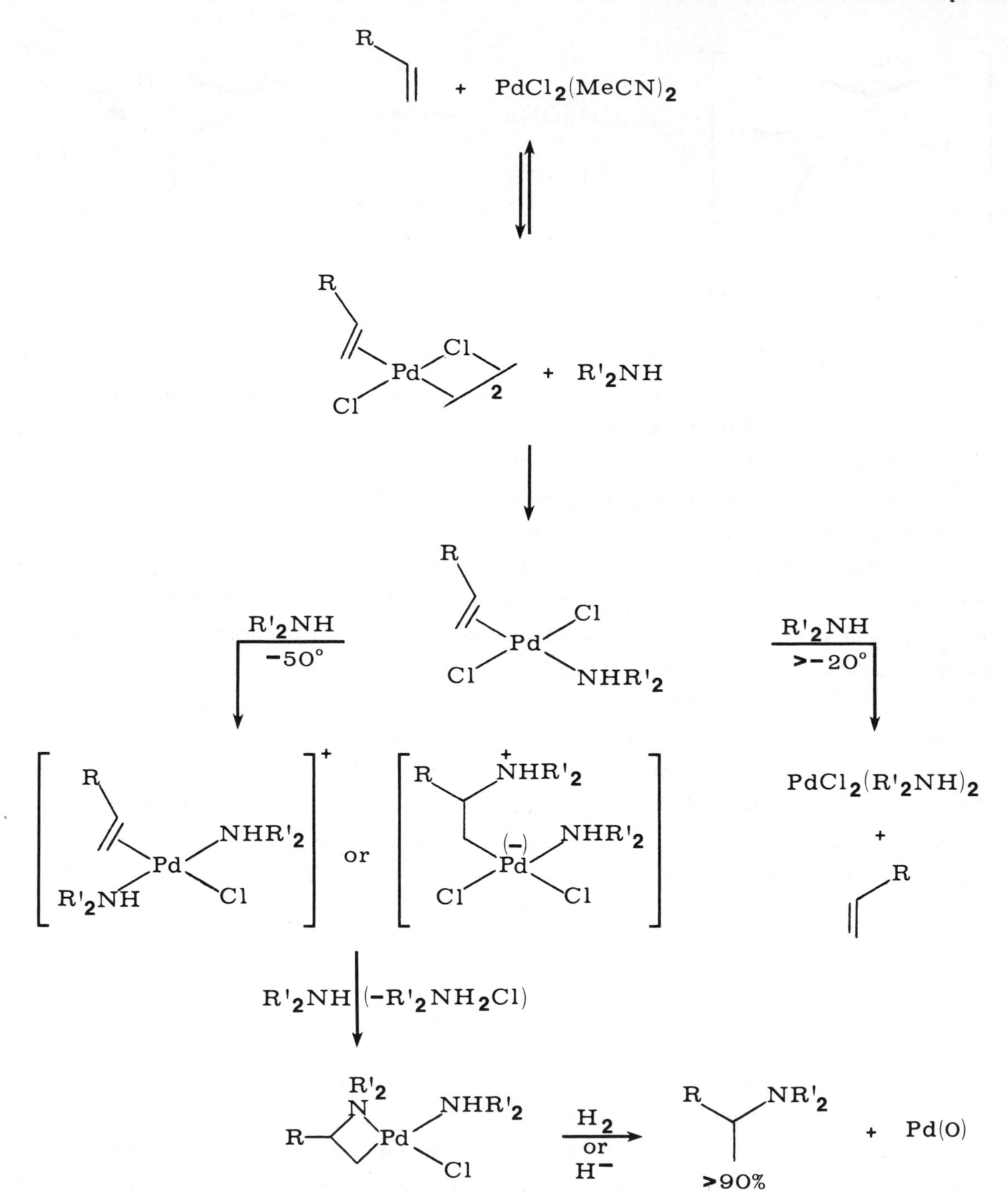


*Figure 12.2. Proposed mechanism for the Pd-assisted amination of olefins.*

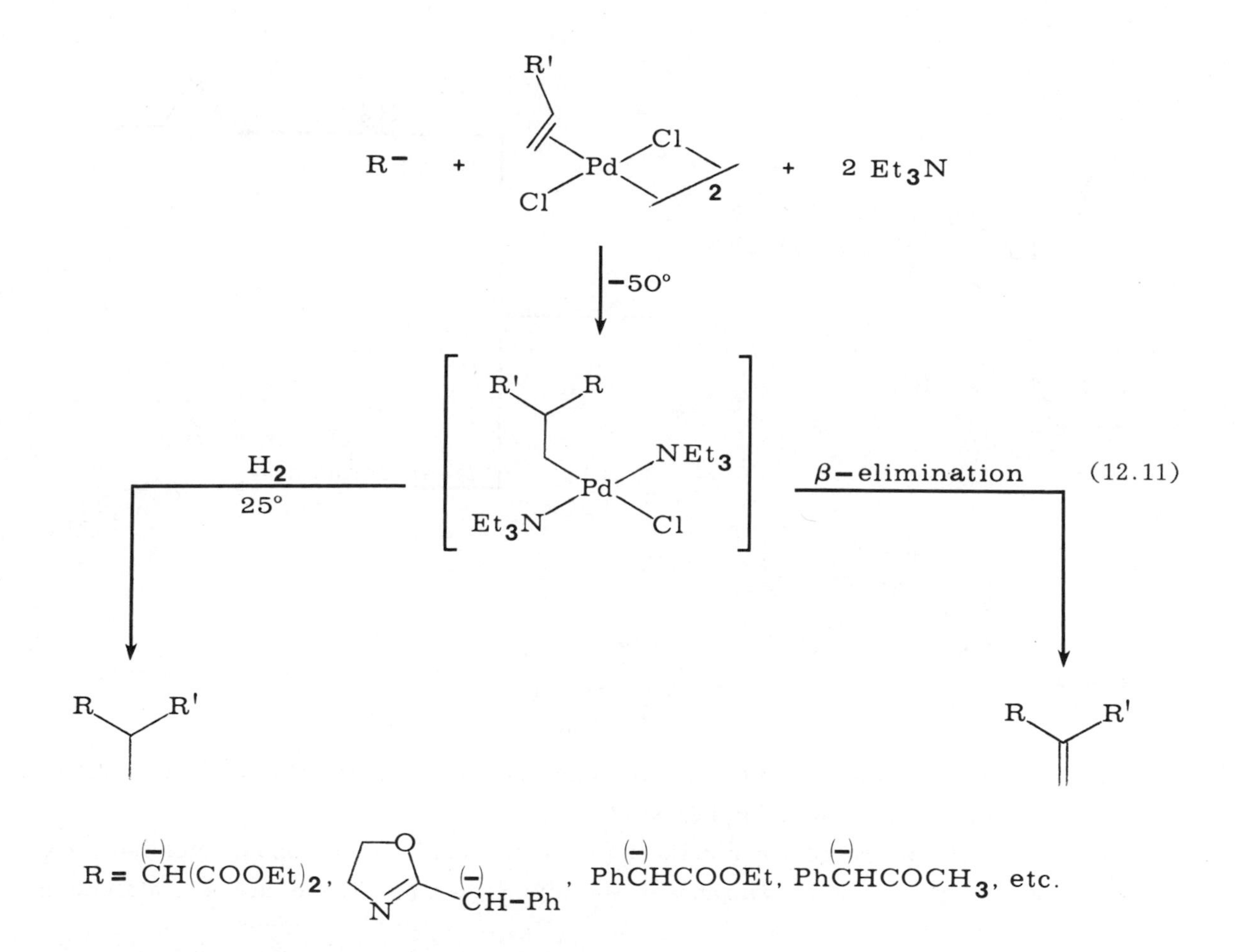


implicating the intermediacy of an olefin-palladium-amine complex, and demonstrating the the sensitivity of the reaction to ligand environment around the metal.

Under the above conditions, less-stabilized carbanions alkylate the olefin only in very low yield. However, addition of HMPA to the olefin-palladium complex prior to addition of triethylamine and carbanion permits this alkylation to proceed with nonstabilized carbanions having $pK_a$s of up to ~36. The role of HMPA is not yet clear, but its presence is required. With these less-stabilized carbanions, attack occurs exclusively at the less-substituted olefin terminus with most olefins (12.12). This regiochemistry is consistent with a *cis* addition, involving prior coordination of the carbanion. However, not all olefins fit this pattern, and there are serious differences between the regiochemistry observed with this system, and that observed in the "Heck" arylation, a

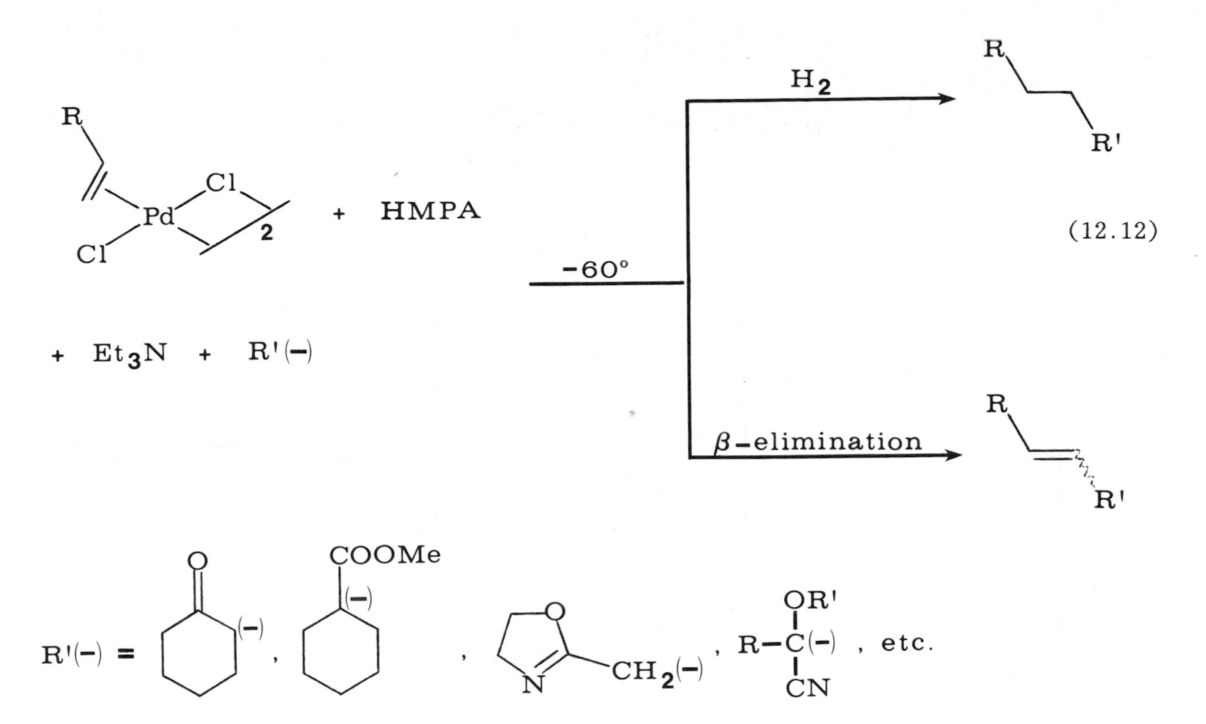


known cis process, with the same olefins (Figure 11.8). Hence the mechanism of the alkylation of olefins must await further study.

A number of intramolecular reactions (12.13 to 12.17) of the above type have also been developed, primarily by Murahashi and by Hegedus. These are all catalytic in

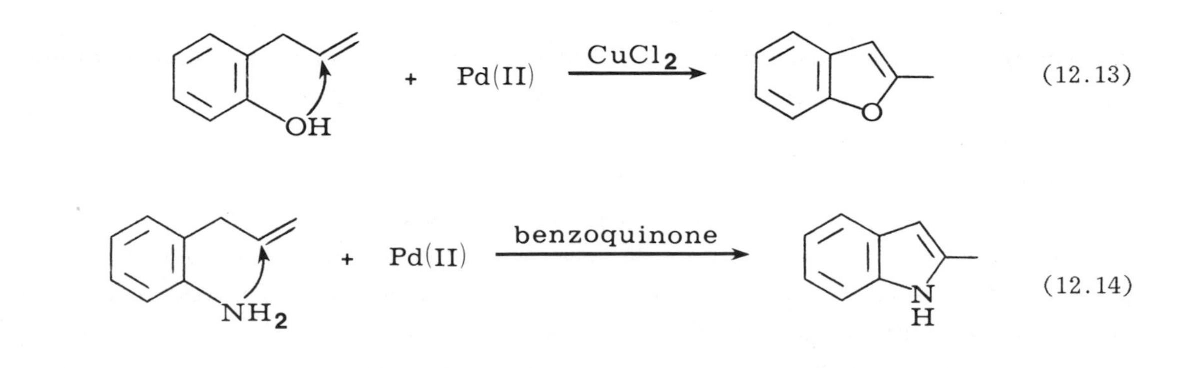

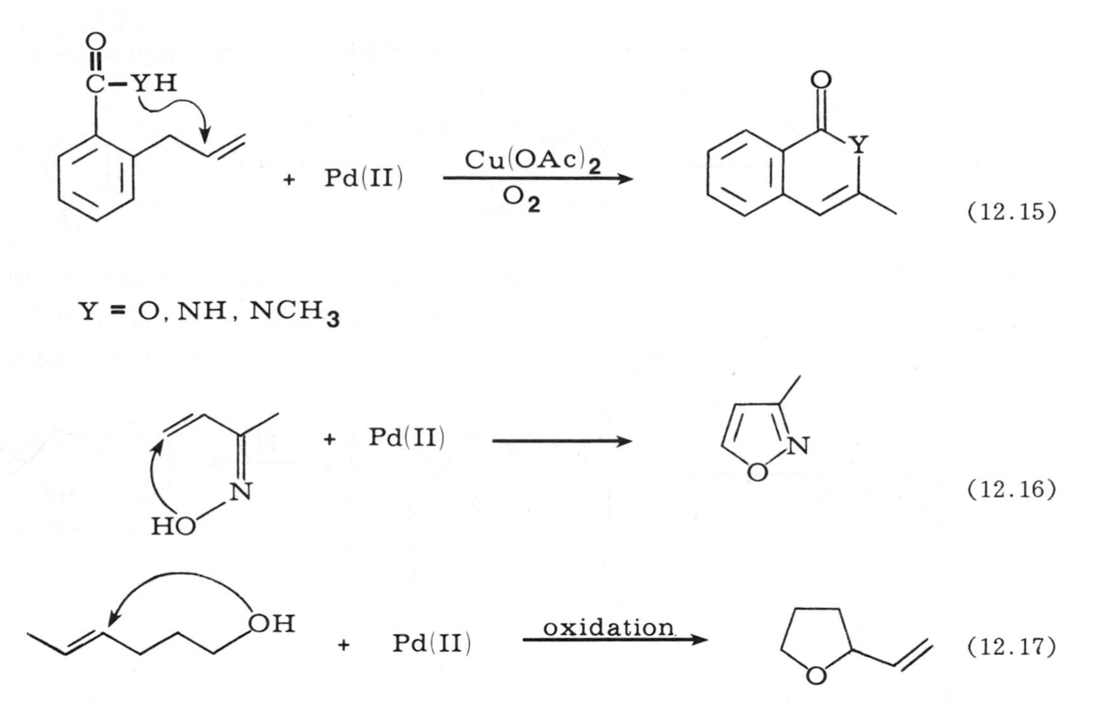


palladium, and lead to the synthesis of a variety of heterocyclic systems. They are generally more facile than corresponding intermolecular reactions. They all appear to involve nucleophilic attack on the metal-complexed olefin, followed by β-hydride elimination. The initially formed olefin often rearranges to the most stable unsaturated system under the conditions of the reaction. They are all catalytic in Pd, and depend on the use of some oxidizing agent to carry the Pd(II)-Pd(0)-Pd(II) redox system [8-12].

The regioselectivity of these intramolecular reactions (i.e., which olefinic carbon is attacked) depends on several factors, including the substitution pattern on the olefin, the ring size formed, and the specific conditions under which the reaction is carried out. For example, with o-allylanilines, the terminally unsubstituted olefin reacts at the most-substituted olefin terminus, producing indoles (Equation 12.14). With terminal disubstitution, the quinoline ring system is formed, again by attack at the most-substituted position (12.18). With terminal monosubstitution, the size of the ring

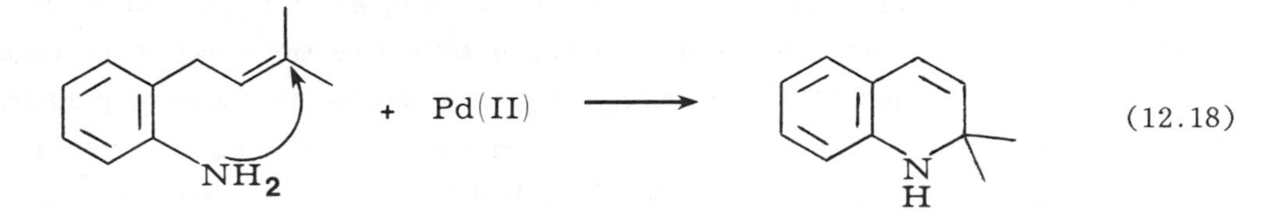

formed depends on reaction conditions (12.19). With o-(2-methallyl)aniline, no ring

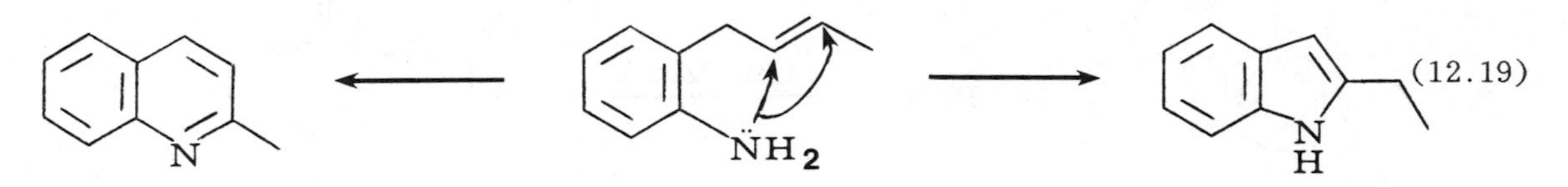


closure is observed under catalytic conditions, whereas using stoichiometric amounts of $PdCl_2$ and a reductive isolation procedure gives low yields of 2,2-dimethylindoline (12.20). Note that closure to give a five-membered ring produces a complex which

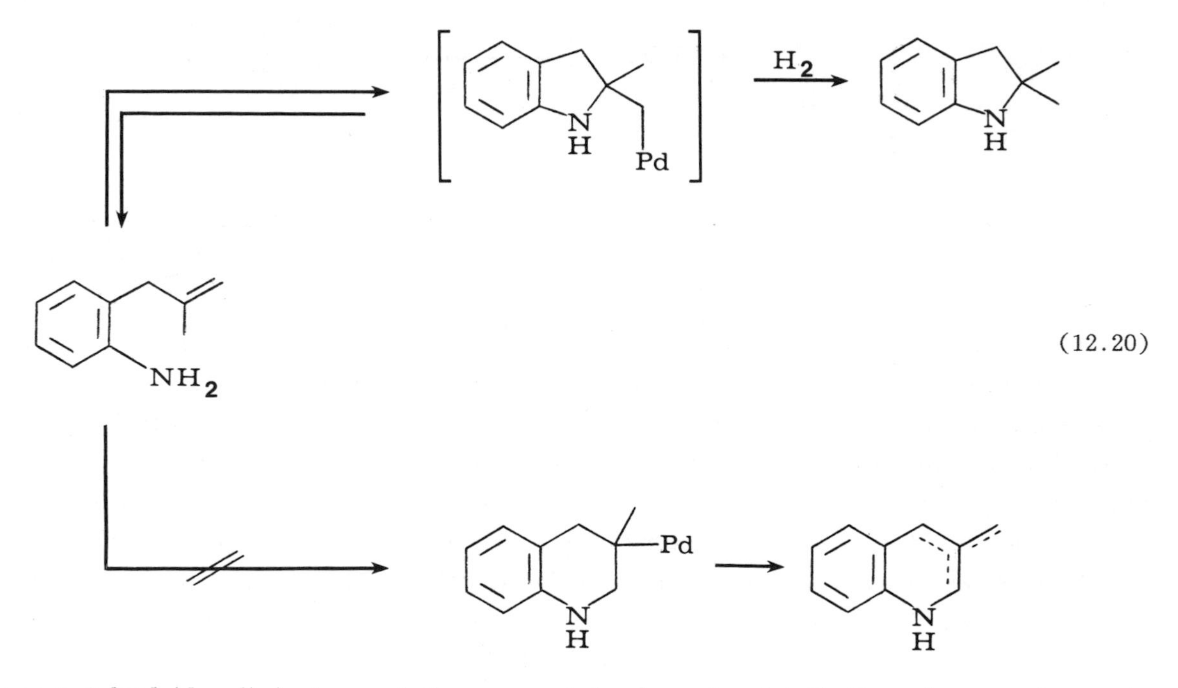


cannot β-hydride eliminate, and thus cannot be formed catalytically, whereas closure to give a six-membered ring is quite possible catalytically, but doesn't occur [9].

In contrast, the analogous o-allylbenzoic acids always close to give six-membered rings when possible [10] (Equation 12.21). With the 2-methallyl side-chain, for which six-membered ring formation would result in a complex incapable of β-elimination, rearrangement of the olefin followed by closure to a five-membered ring results (12.22)

Although platinum(II) forms olefin complexes similar to those of palladium(II), and these complexes also activate the olefin toward nucleophilic attack, useful catalytic or stoichiometric systems based on platinum(II)-olefin complexes are rare indeed. There

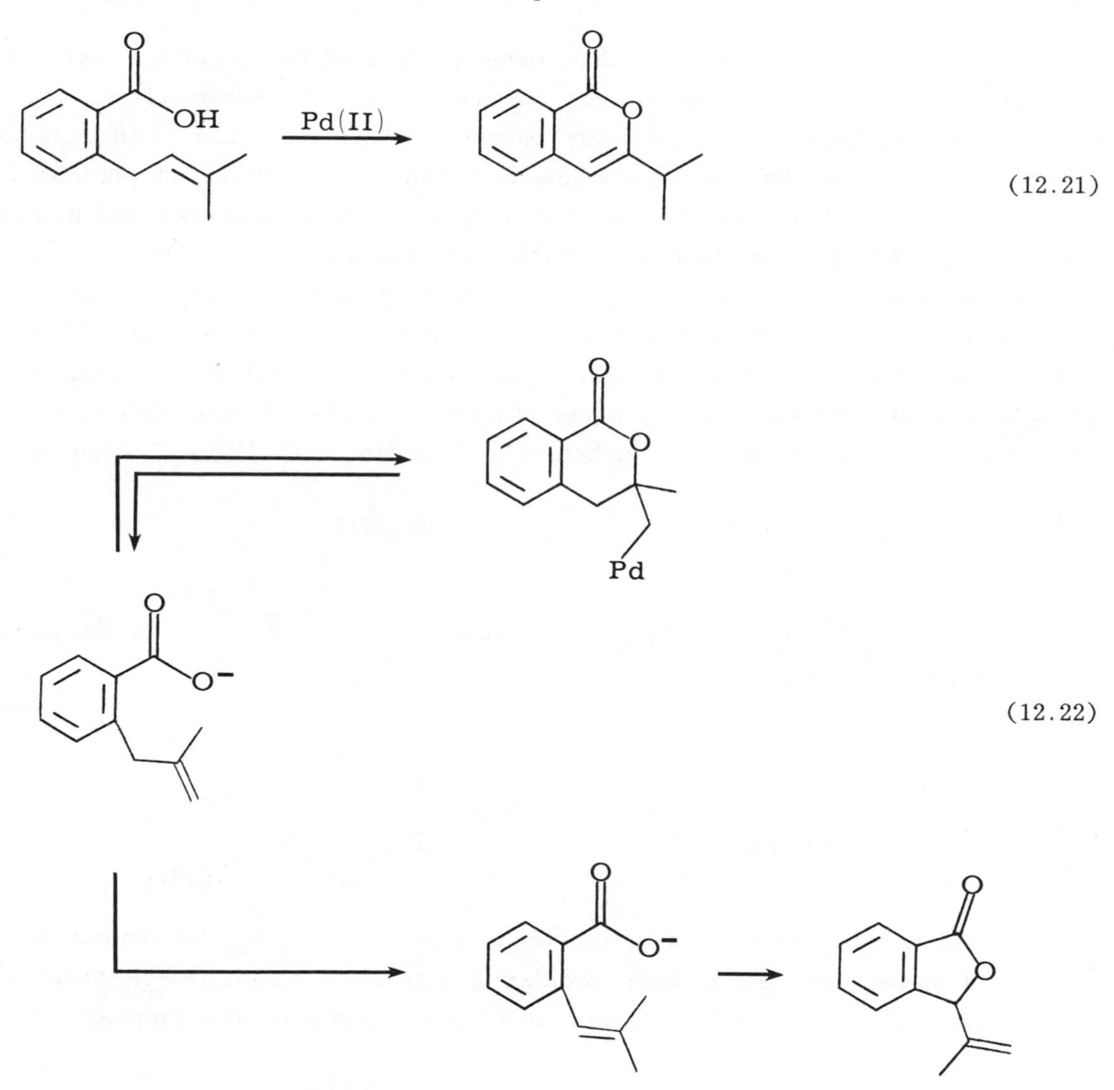


are at least two reasons for this. Equilibria between platinum(II) salts and olefins are established very slowly compared with palladium (days rather than minutes), making catalytic cycles based on platinum impossibly slow. Once formed, platinum-olefin complexes are more stable than the corresponding palladium complexes, and undergo reaction with nucleophiles more slowly than do palladium-olefin complexes. In addition, the resulting σ-alkylplatinum complexes are often quite stable, requiring an additional chemical step for removal of the metal and generation of the desired organic product. Hence, platinum-based systems are less practical than palladium-based systems for use in organic synthesis.

However, the very features that make platinum-olefin complexes synthetically awkward make them very attractive for mechanistic and stereochemical studies, since intermediate complexes are more easily isolated and characterized. An outstanding example of this is a series of experiments concerning the amination of platinum-bound olefins. As noted above, the mechanism of palladium-assisted amination and hydroxylation of olefins (Wacker process) is the subject of some controversy, since the proposed intermediate complexes are not stable enough to be isolated and characterized. However, with similar platinum-olefin complexes, the intermediates are isolable. Thus treatment of cis dichloro(ethylene)triphenylphosphine platinum(II) with diethylamine produces an isolable σ-alkylplatinum complex whose elemental analysis, IR and NMR spectra, and chemistry are consistent with the structure in Equation 12.23 [13]. Furthermore, the

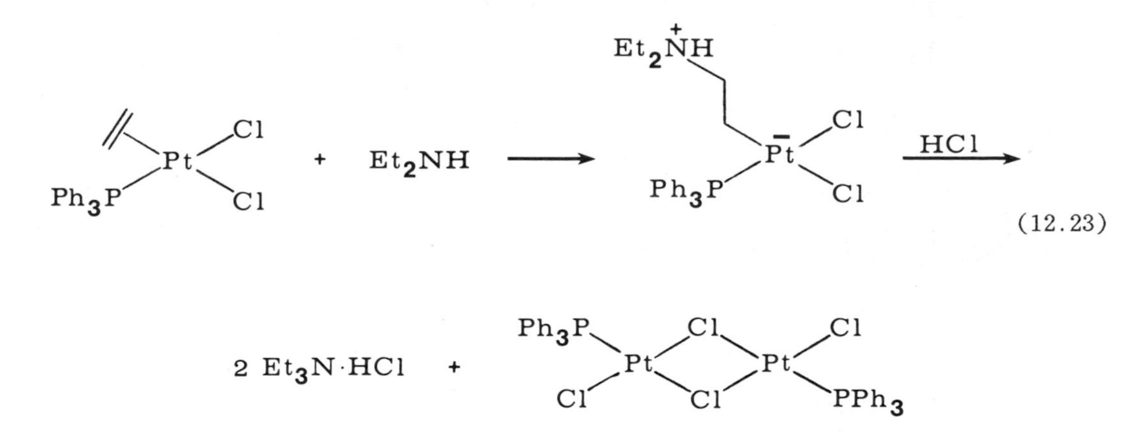


(12.23)

stereochemistry of amination was unequivocally shown to be trans by Panunzi by using a resolved diastereoisomeric platinum complex of a prochiral olefin, (+)-cis-dichloro[(S)-1-butene][(S)-α-methylbenzylamine]platinum(II) and examining the stereochemistry of

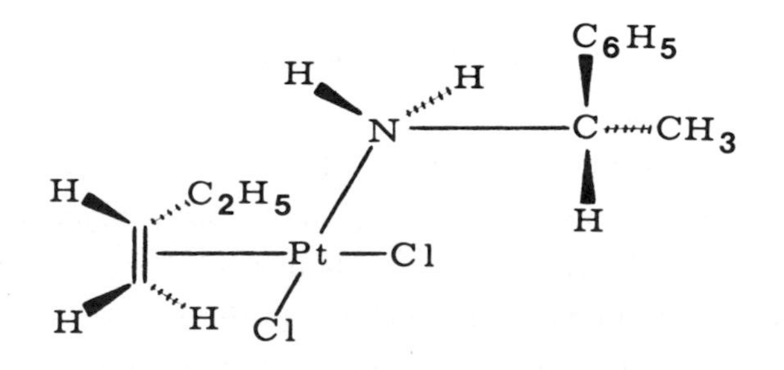

the resulting chiral amine [14]. To the extent that the behavior of platinum can be extended to palladium, these experiments provide both mechanistic and stereochemical insight into palladium-assisted nucleophilic attack on monoolefins.

Rosenblum has developed the chemistry of cationic $\eta^5$-cyclopentadienyl(olefin)iron dicarbonyl complexes, which also undergo nucleophilic attack at the olefin by a wide range of nucleophiles to produce stable, neutral σ-alkyliron complexes (12.24).

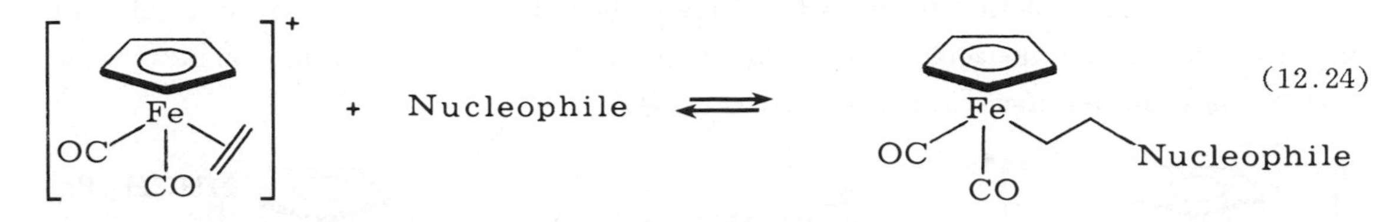


Cationic olefin complexes should be more susceptible to attack by nucleophiles than neutral complexes, since the metal is better able to accommodate the pair of electrons it formally gains as the result of a nucleophilic attack. This is indeed a general observation. With monosubstituted olefin complexes and nucleophiles such as methanol, amines, phosphines, phosphites, and thiols, the reaction is highly regioselective, attack occurring primarily or exclusively at the most-substituted olefinic position [15] in the absence of excessive steric demands (i.e., t-butylmercaptan gives a 1.5-to-1 mixture of regioisomers).

Amines often undergo further reaction to produce bis adducts (12.25). Phosphines and phosphites lead to phosphonium salts in good yield.

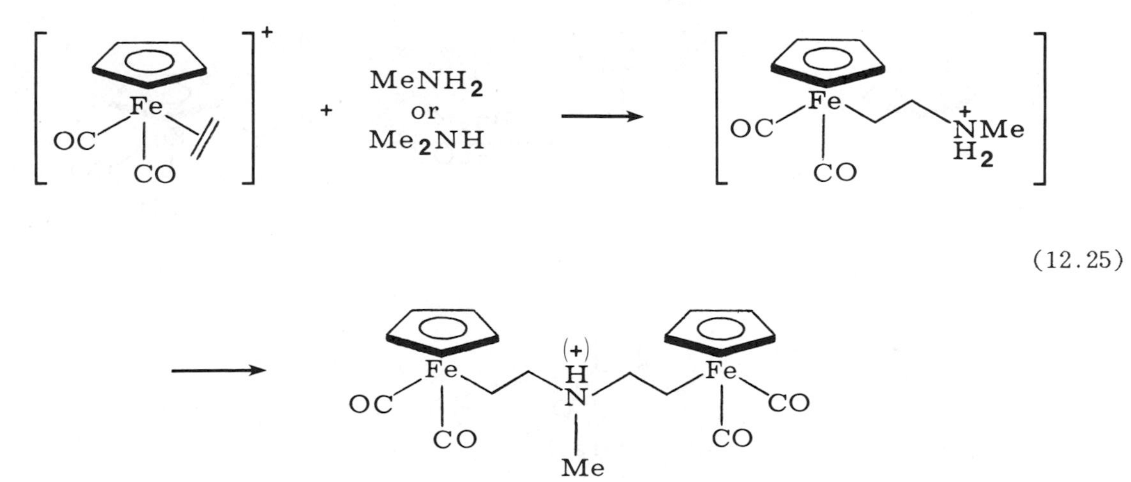

In contrast to palladium-based systems, the σ-alkyliron complexes do not readily decompose to free the organic portion, and further chemically induced decomposition does not lead to the desired substituted olefins. Hence, these complexes have found little use in organic synthesis of complex molecules. A notable exception is the observation that oxidation of an aminated olefin complex leads to carbonyl insertion. This observation led to the development of an elegant approach to β-lactams utilizing amination of an Fe-bound olefin, followed by oxidatively induced CO insertion and subsequent ring closure to the β-lactam [16] (Equation 12.26). With ω-olefinic amines as substrates, bicyclic β-lactams are produced (12.27).

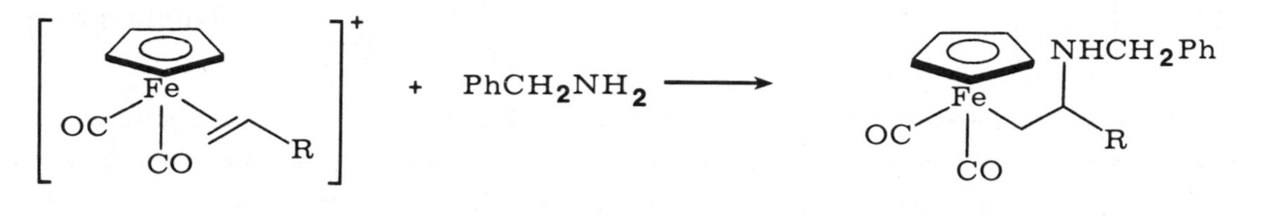


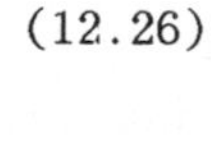


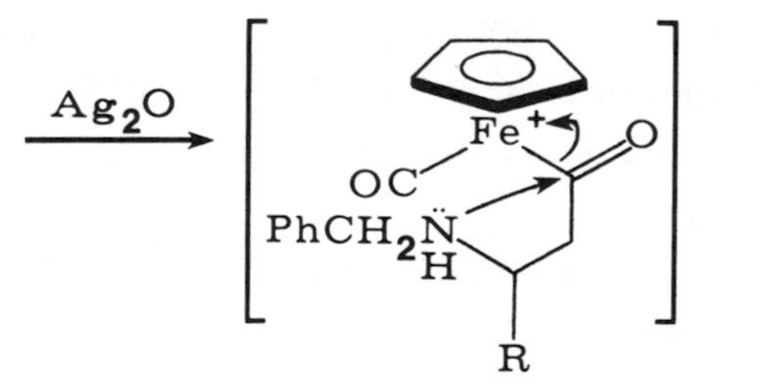


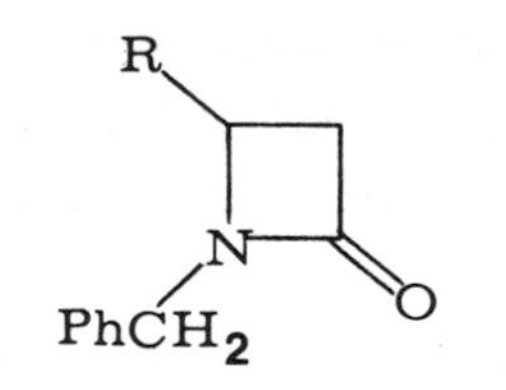


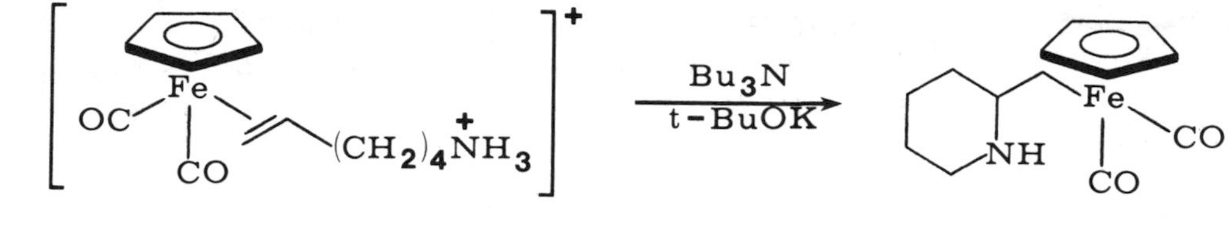


(12.27)

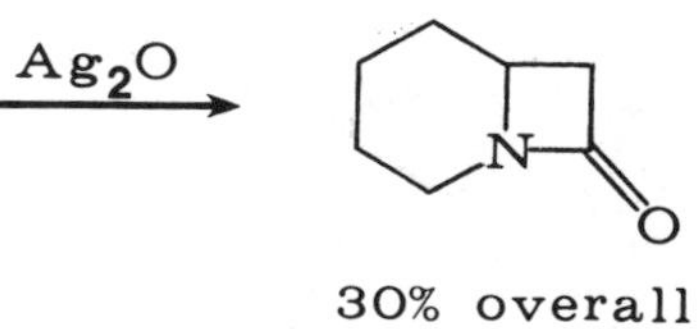

A wide variety of carbanions also attack cationic iron-olefin complexes [17]. Enolates of nitromethane, acetoacetates, malonates, and cyanoacetates, as well as the enamines of isobutyraldehyde, cyclopentanone, and cyclohexanone react with cationic iron-olefin complexes of ethene, propene, styrene, cyclopentene, cyclohexene, allene, and butadiene (12.28). Regioselectivity is low with propene, but high (attack on the

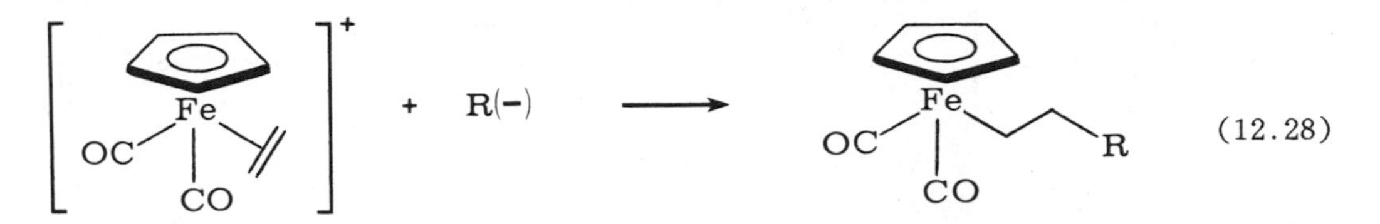


(12.28)

benzyl carbon) for styrene. Organolithium and Grignard reagents reduce the metal complexes and afford no alkylation, but lithium dialkylcuprates do successfully alkylate the olefin. Stabilized ylids also alkylate the complexed olefin, and the resulting product can be further condensed with benzaldehyde to introduce further functionalization (12.29). Again, stable σ-alkyliron complexes result from the alkylation reaction, and

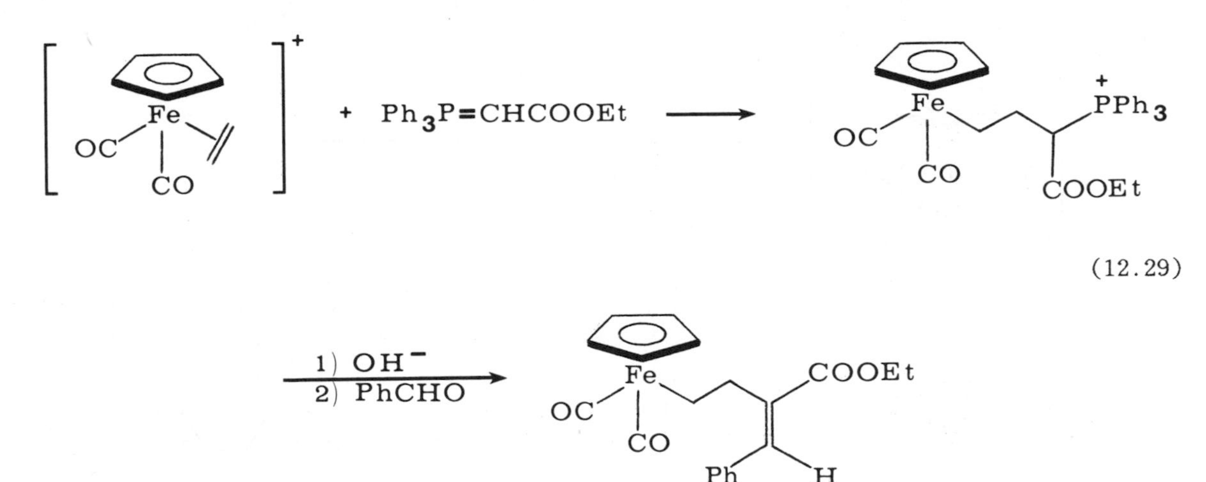


(12.29)

the alkylated olefin has not often been freed from the metal.

The same features of bonding that activate these metal-bound olefins to undergo nucleophilic attack make the olefins more resistant to electrophilic attack. Thus, the cyclopentadienyliron dicarbonyl cation has been used by Nicholas as an olefin-protecting group against electrophiles. The complex $\eta^5$-$C_5H_5Fe(CO)_2$(isobutylene)$^+BF_4^-$ exchanges isobutylene for a variety of other olefins. With dienes, the least substituted or most strained olefin coordinates preferentially [18]. With enynes, the olefin coordinates, leaving the alkyne uncomplexed [19]. Decomplexation of the olefin is readily accomplished with sodium iodide (12.30-12.32).

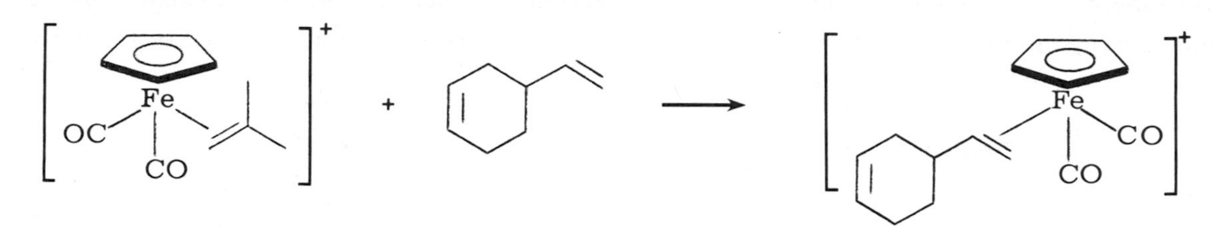


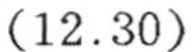


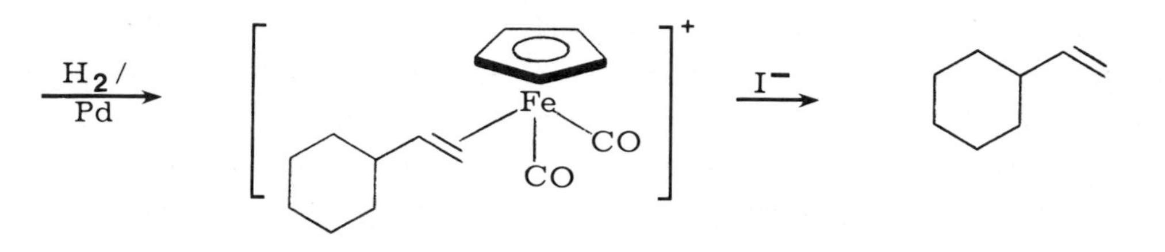


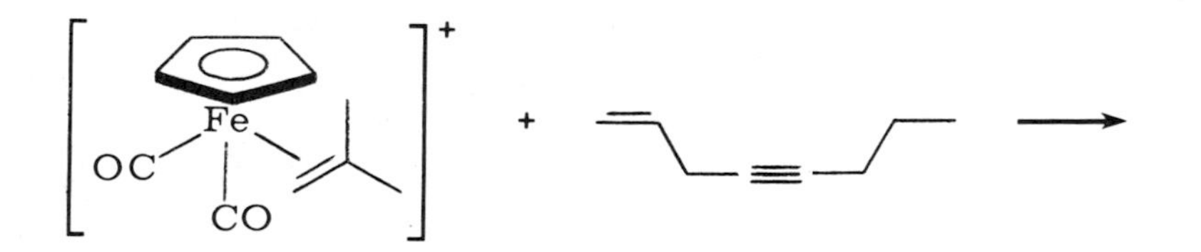


(12.31)

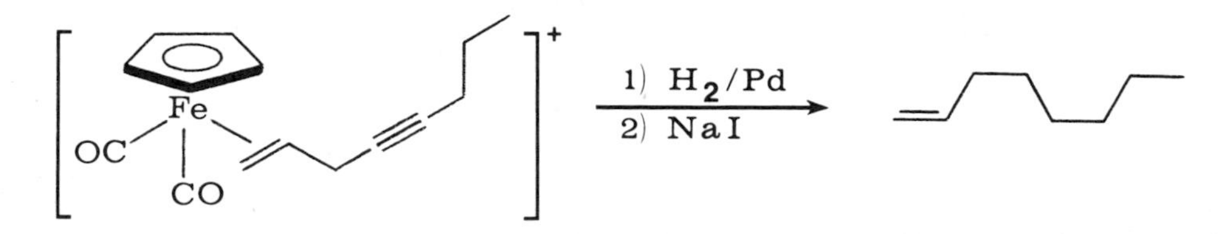


The above discussion has centered on simple monoolefin complexes, which, although they undergo nucleophilic attack in a useful manner, suffer several competing reactions (particularly olefin displacement). In contrast, olefins held in coordination by chelation, such as chelating diolefins and allylamines and sulfides, undergo clean nucleophilic attack by a wide variety of nucleophiles with little complication. Norbornadiene and

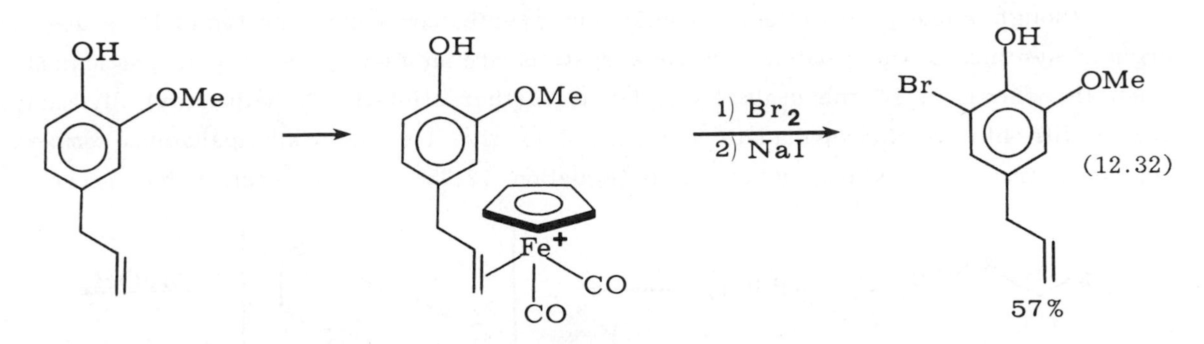


cyclooctadiene complexes of both palladium and platinum react with methoxide [20], amines [21], azide [22], and stabilized carbanions [23a] to generate a σ-alkyl-π-olefin complex which resists further attack at the remaining coordinated olefin (12.33). The

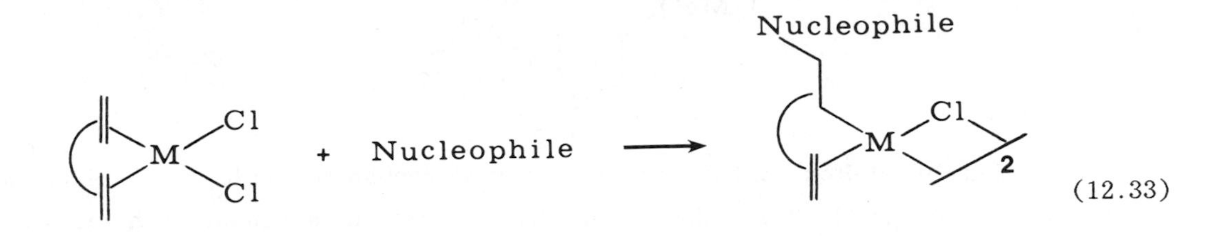


M = Pd, Pt

Nucleophile = $RO^-$, $RNH_2$, $N_3^-$, $\bar{C}H(COOR)_2$, $\bar{C}H(COR)_2$

attack is always <u>trans</u>, from the face opposite the metal. This is a much more facile reaction than with monoolefin complexes, and displacement of the olefin by the nucleophile is not observed. However, even with these systems, unexpected results are obtained [23b] (Equation 12.34).

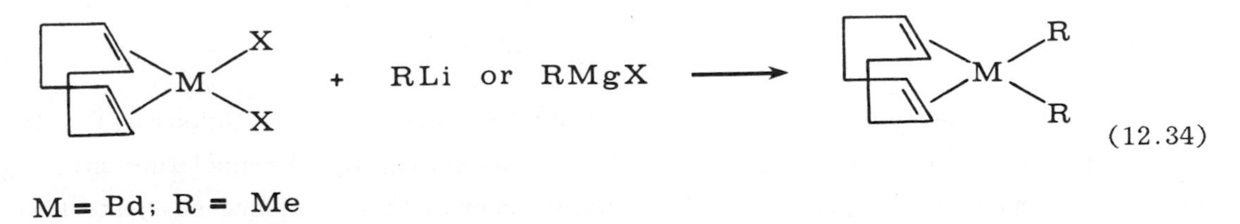


M = Pd; R = Me

M = Pt; R = Me, Et, Ph, o–tolyl

Although nucleophilic attack on chelating diolefin complexes has found little use in organic synthesis, the related allylamine systems are in fact quite useful, particularly when the chemistry of the σ-alkyl complex is further utilized. N,N-dimethyl allylamine reacts with lithium chloropalladate in methanol to give a stable σ-alkylpalladium complex resulting from methoxypalladation [24] (Equation 12.35). This reaction has recently

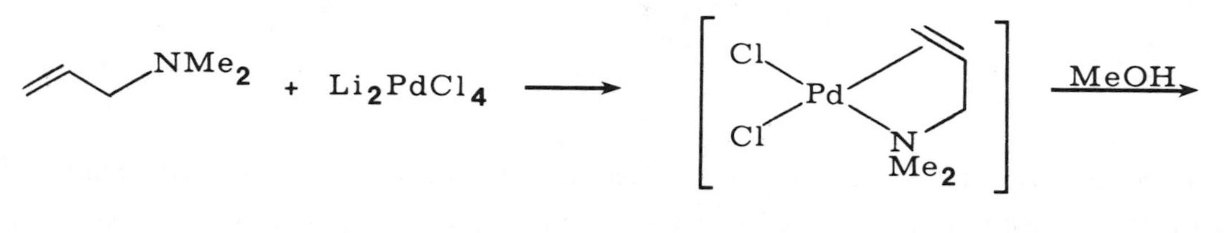


(12.35)

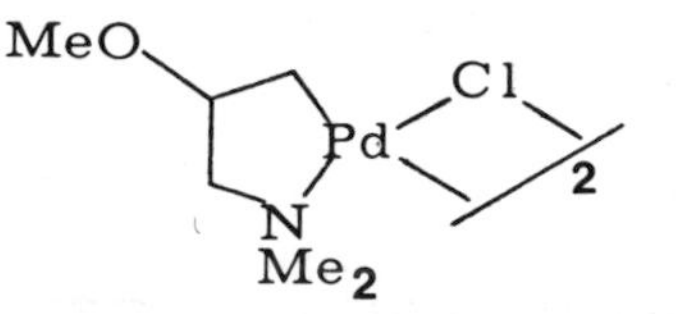


been extended by Holton to the use of stabilized carbanions ($pK_a$ < 17) such as keto esters [25] (Equation 12.36). Homoallylic amines react in a similar fashion, attack occurring at the terminal carbon with these substrates [26] (12.37). Longer-chain ω-aminoolefins do not undergo similar reactions. Both the regiochemistry and stability of

$$\text{CH}_2\!=\!\text{CHCH}_2\text{NMe}_2 + {}^{(-)}\text{CH(COOEt)}_2 + \text{Li}_2\text{PdCl}_4 \longrightarrow$$

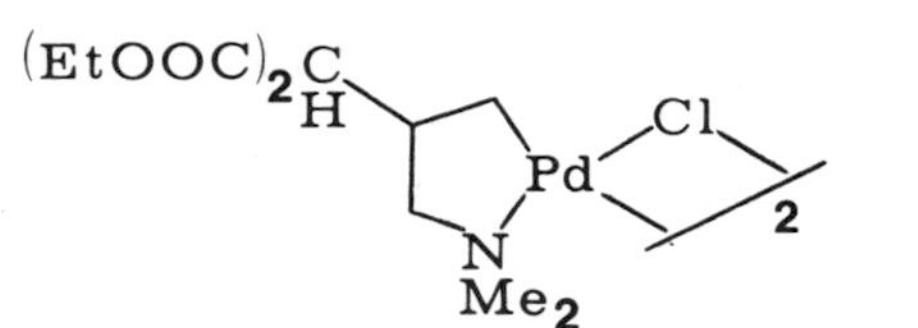


(12.36)

these complexes merit comment. That nucleophilic attack on these substrates is facile is not surprising, since the olefin is held in coordination by the chelating amino group and displacement is suppressed. Continually accumulating evidence suggests that five-membered-ring metallacycles are considerably more stable than either four- or six-membered-ring systems. The complexes in Equations 12.35-12.37 are quite stable at

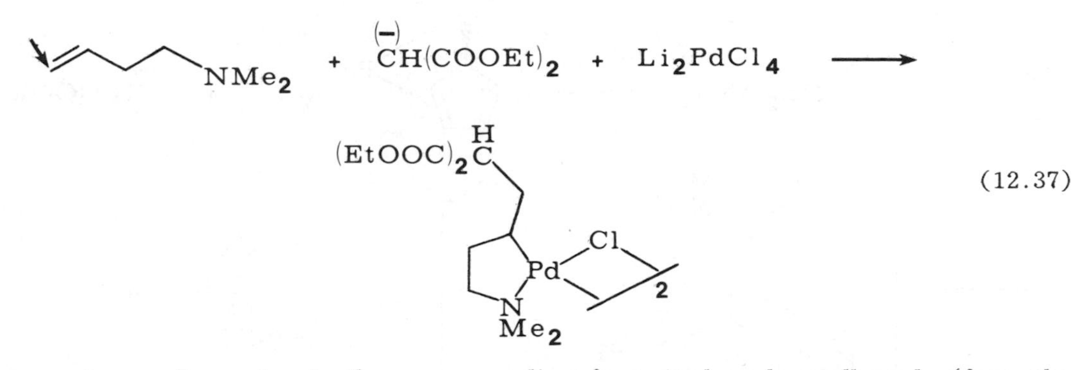


(12.37)

room temperature. In contrast, the corresponding four-membered metallacycle (from the amination of olefins) decomposes above -20° (Figure 12.2). Thus it is likely that ring size determines the site of nucleophilic attack in these cases, both reactions producing the favored five-membered ring. Longer-chain ω-amino olefins fail to react in this manner probably because they can neither chelate nor form five-membered metallacycles.

This inherent stability of five-membered metallacycles is further emphasized by carbonylation studies. A characteristic reaction of σ-alkylpalladium complexes is the insertion of CO into the carbon-metal bond. Treatment of the unstable four-membered aminopalladiacycle produces a very stable five-membered β-aminoacyl complex [27,28] (Equation 12.38). In contrast, similar treatment of the five-membered aminopalladiacycle

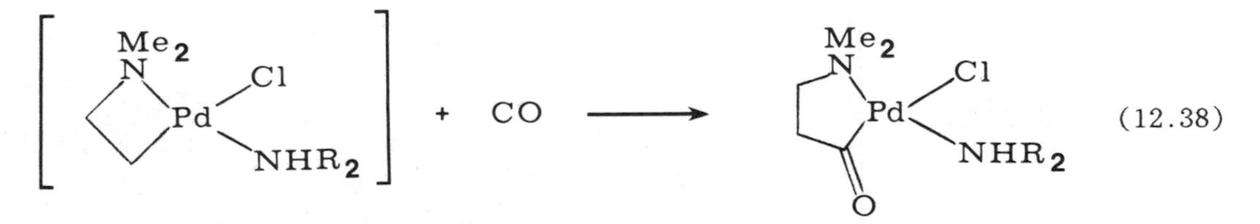


(12.38)

with CO gives a (proposed) six-membered aminoacyl complex which is not sufficiently stable to be isolated, but decomposes into the β-amino ester [29] (Equation 12.39). When chelation is not possible, neither the σ-alkyl nor the σ-acyl complex is stable [30] (Equation 12.40). Reactions 12.38 to 12.40 are useful for the difunctionalization of olefinic substrates.

In addition to carbon monoxide, σ-alkylpalladium complexes also add to (insert) α,β-unsaturated enones [31]. Holton [32] has combined this with the chemistry just discussed to develop the elegant approach to prostaglandins described in Figure 12.3. The first step is carbopalladation of the allylamine as in Equation 12.36. Again, attack occurs regiospecifically to form the five-membered metallacycle, and stereospecifically trans to the palladium. β-Elimination of "PdH" generates the alkylated allylamine.

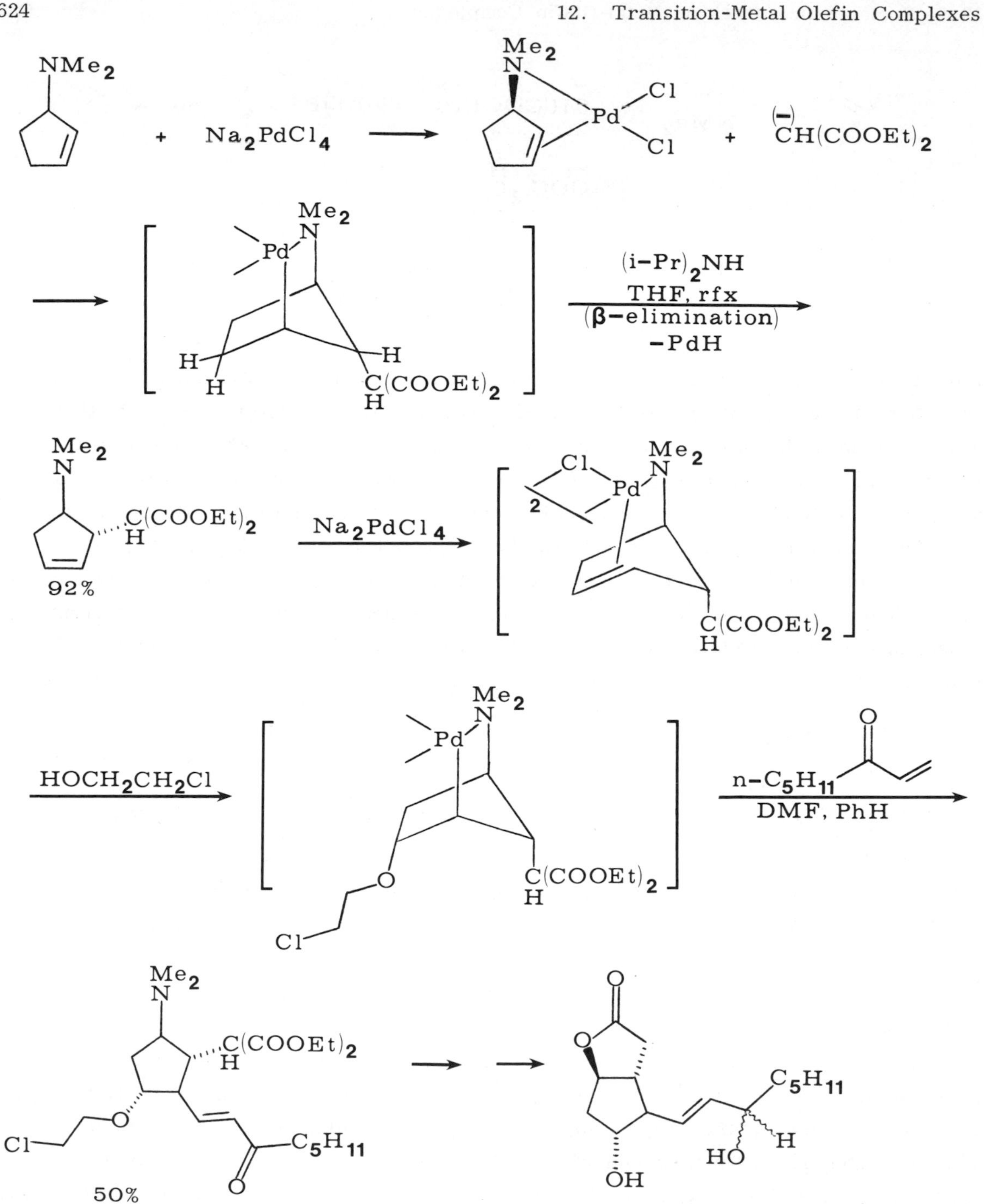


*Figure 12.3. Pd-assisted synthesis of prostaglandins.*

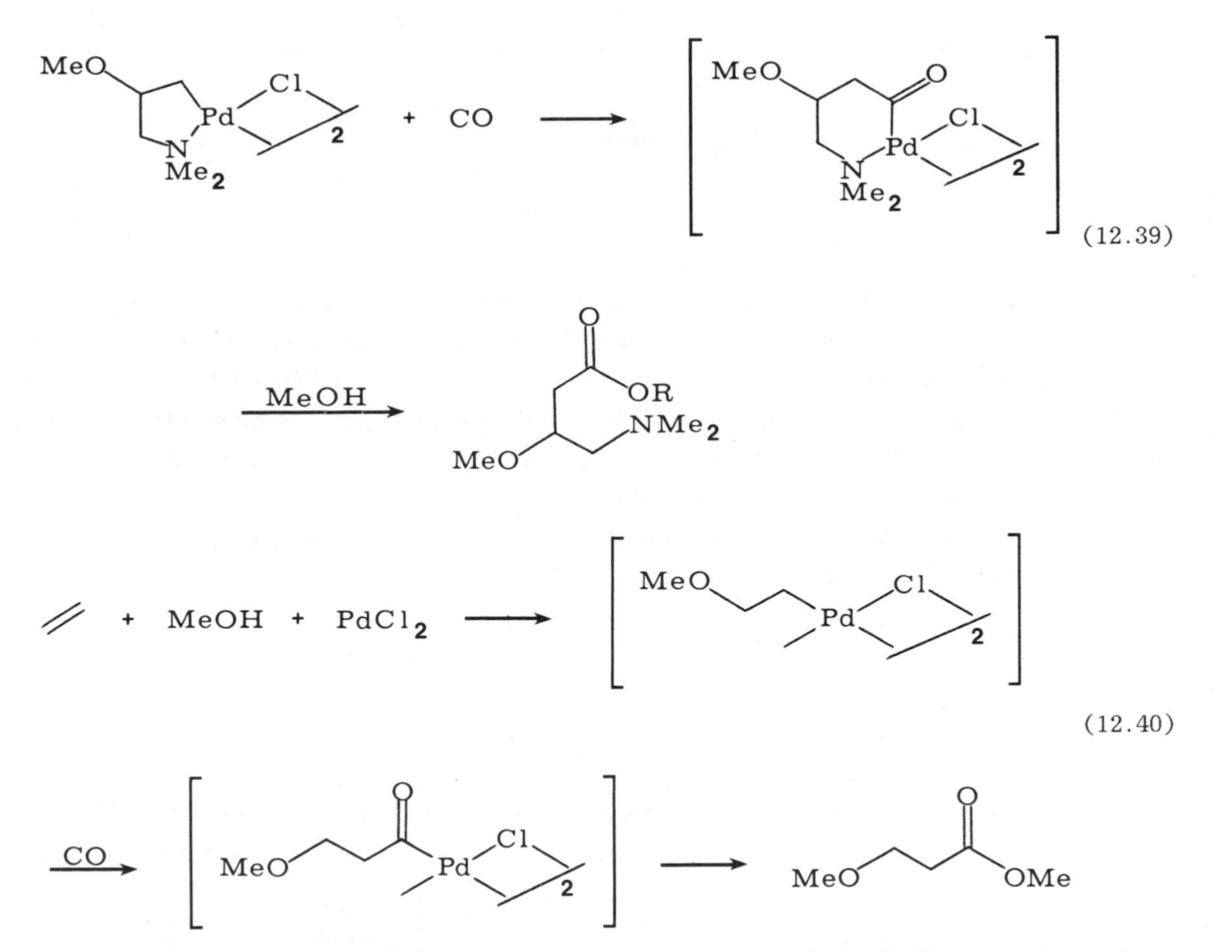


Alkoxypalladation of this allylamine again generates a σ-alkylpalladium complex whose regio- and stereochemistry is controlled by the same factors that affect the first step. This complex inserts pentyl vinyl ketone to produce the trisubstituted allylamine stereospecifically. This material is then converted to "Corey's lactone," which has already been converted to several prostaglandins.

Other useful insertion reactions of σ-alkylmetal complexes that result from nucleophilic attack on olefin complexes are discussed in Chapter 10.

## 12.2. Conjugated Diene-Olefin Complexes.

Although chelating diolefins coordinate best to Pd(II), Pt(II), and Rh(I) metals, conjugated dienes such as butadiene complex most strongly to iron(0). Simply heating iron pentacarbonyl and butadiene at ~140° in a bomb produces η[4]-butadiene iron(0) tricarbonyl (12.41). Such severe conditions are required because $Fe(CO)_5$ is

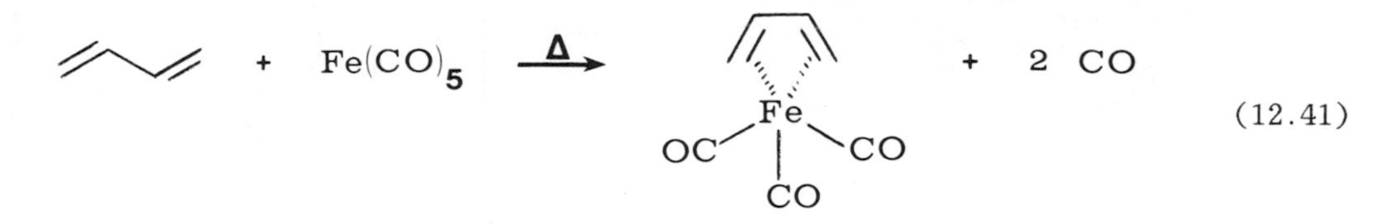


(12.41)

coordinatively saturated and relatively substitution-inert. This complex can be made under considerably milder conditions using $Fe_2(CO)_9$, a source of "$Fe(CO)_4$." This diene complex is so stable that the butadiene ligand undergoes Friedel Crafts acylation without suffering decomplexation from the metal (Chapter 3). Nonconjugated dienes often rearrange in the presence of iron carbonyl to give conjugated diene complexes, again the stability of these being the driving force.

Butadiene complexes are also produced by the reaction of 1,4-dihalo-2-butenes with $Fe_2(CO)_9$ (Equation 12.42). An especially useful example of this reaction is the

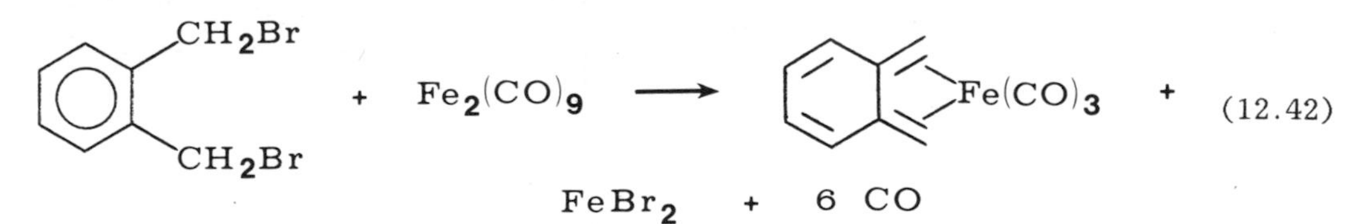


(12.42)

preparation of η[4]-cyclobutadieneiron tricarbonyl from dichlorocyclobutene [33] (Equation 12.43). This reaction illustrates the stabilization of extremely unstable organic

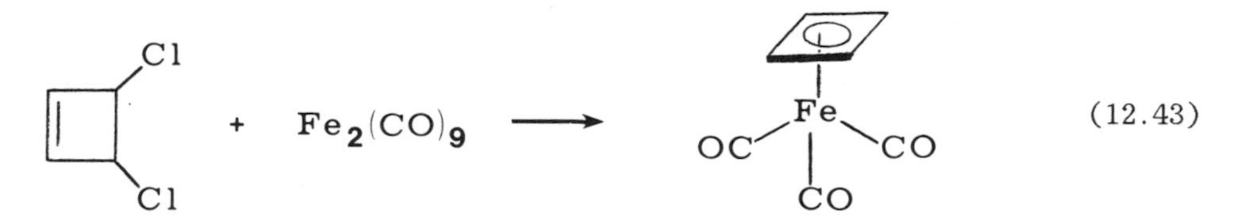


(12.43)

molecules by complexation to appropriate transition metals. Free cyclobutadiene itself has only fleeting existence, since it rapidly dimerizes. However, cyclobutadieneiron tricarbonyl is a very stable complex. The complexed cyclobutadiene undergoes a variety of electrophilic substitution reactions including Friedel Crafts acylation, formylation, chloromethylation, and aminomethylation. The keto group of the acylated cyclobutadieneiron tricarbonyl complex can be reduced by hydride reagents without decomposition [34]. On the other hand, oxidation of the complex with cerium(IV), triethylamine oxide, or pyridine N-oxide frees the cyclobutadiene ligand for use in synthesis. The use of cyclobutadieneiron tricarbonyl by Pettit [35] as a stable source of highly reactive cyclobutadiene in organic synthesis is illustrated in the synthesis of cubane (Figure 12.4). That this type of reaction actually involves free cyclobutadiene, rather

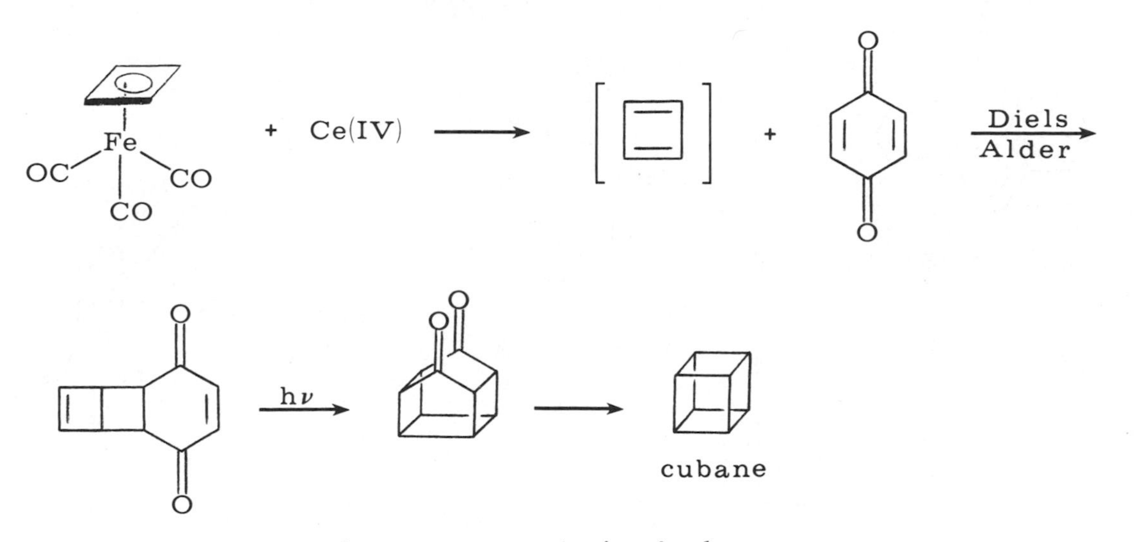


*Figure 12.4. Synthesis of cubane.*

than iron-coordinated intermediates, was demonstrated by Rebec in an elegant fashion using a "three-phase test" of general utility for the detection of reactive intermediates [36]. The test involves the attachment of the cyclobutadieneiron complex to one insoluble, crosslinked polymer phase, and a cyclobutadiene trap (maleimide, in this case) to a different insoluble polymer support. The principle is that substrates in two different crosslinked polymer phases cannot react directly with each other, but each polymer phase can react with reagents in solvents which swell (penetrate) the polymer phases. Pyridine N-oxide in solution frees the butadiene from the complex in polymer phase-1, and this intermediate is trapped by the maleimide in polymer phase-2, confirming the existence of free cyclobutadiene (Figure 12.5).

Although conjugated diene-iron complexes are quite stable and rather inert to attack on the coordinated diene, hydride abstraction generates a dienyl cation ligand which is subject to nucleophilic attack. This chemistry has been most extensively studied by Birch with cyclohexadiene complexes. Birch reduction of aromatics produces 1,4-cyclohexadienes, which react with $Fe_3(CO)_{12}$ to give the corresponding 1,3-diene

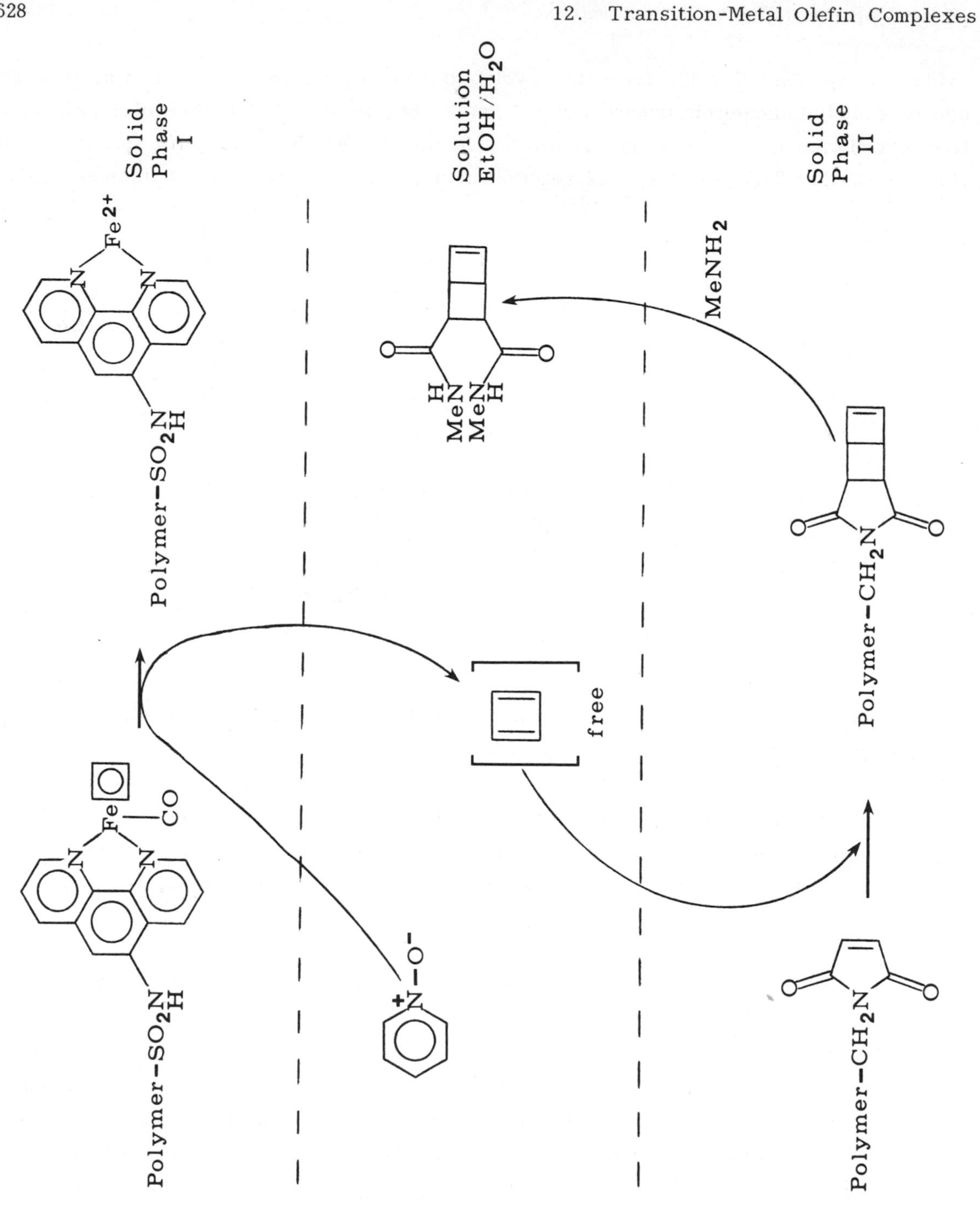


*Figure 12.5. Three-phase test for free cyclobutadiene.*

iron tricarbonyl complex, which in turn undergoes hydride abstraction to give the cationic cyclohexadienyliron complex (12.44).

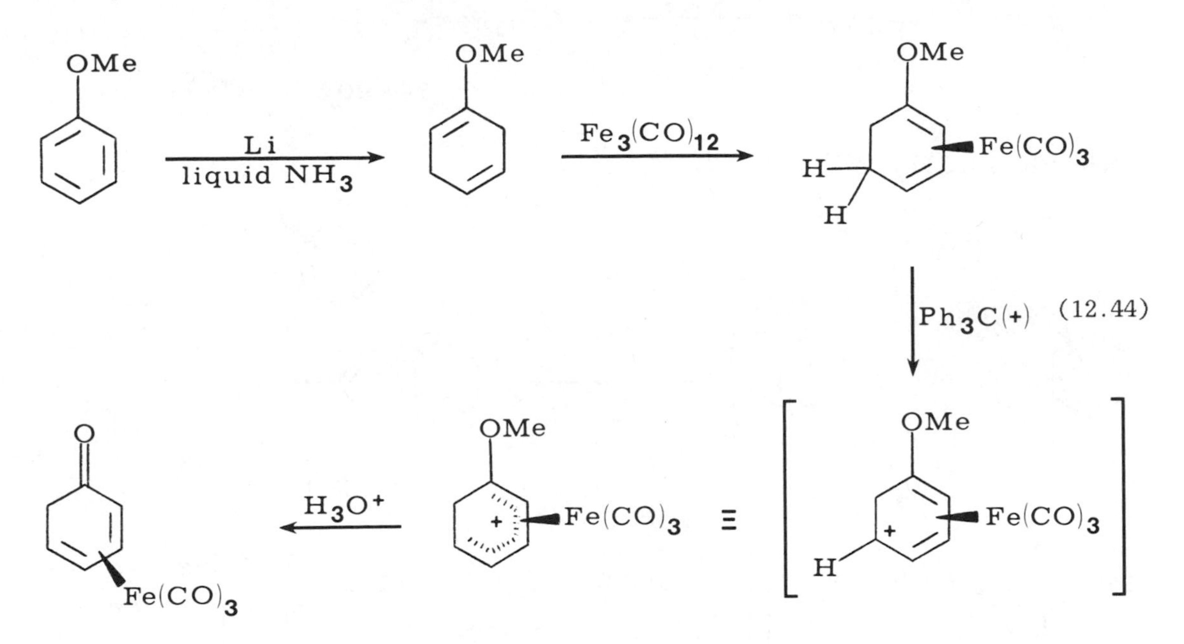


This complex hydrolyzes to give the diene complex corresponding to the keto form of phenol. Again, the great stability of conjugated diene-iron complexes stabilizes a normally very unstable organic compound [37].

Cationic dienyl complexes of iron undergo a number of interesting reactions [37]. Nucleophiles such as cyanide, enamines, enolate anions, and organocadmium reagents attack, regenerating the 1,3-butadieneiron-tricarbonyl system (12.45-12.48).

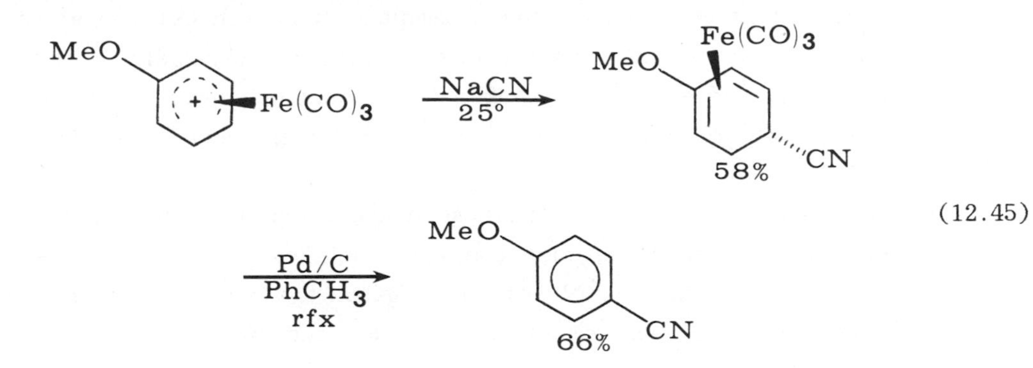

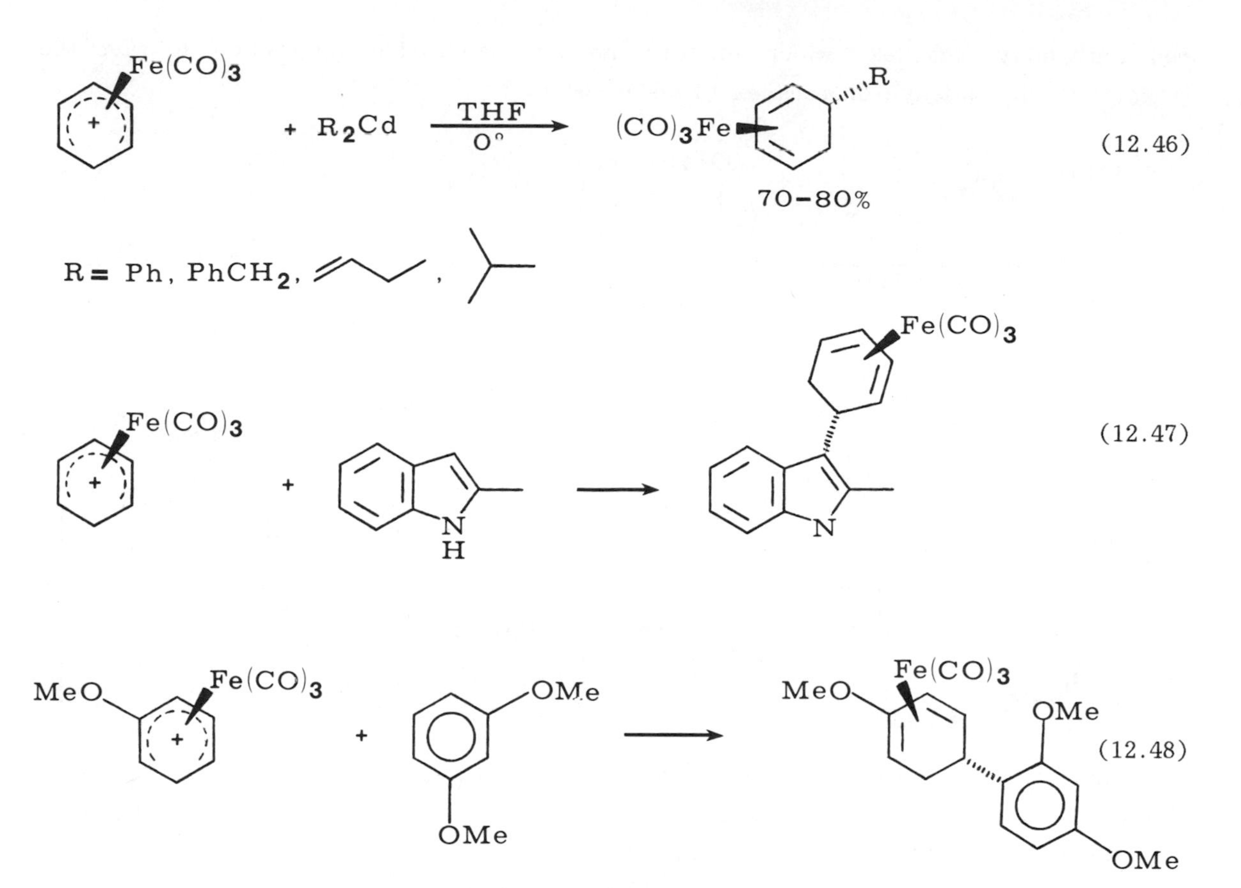


The 1,3-cyclohexadienyliron tricarbonyl complex itself is sufficiently electrophilic to react with both indole and m-dimethoxybenzene. Note that nucleophilic attack always takes place from the face opposite the iron. At the time most of this work was done, the best way of removing iron from these complexes was with Ce(IV), a strong oxidizing agent which frequently destroys the organic product as rapidly as it frees it from iron. More recently, amine oxides, much milder oxidizing agents, have been found to free 1,3-dienes from iron, and offer real utility for the above reactions in organic synthesis.

Many transition metals other than those discussed in this chapter form complexes with olefins, chelating diolefins, and conjugated olefins. Many of the chemical and tructural features of these complexes are similar to those discussed above. The complexes considered in detail are those for which the most generally useful organic chemistry has been developed.

## NOTES.


1a. P. M. Maitlis, The Organic Chemistry of Palladium (Academic Press, 1971), Vol. I, 129.

1b. F. R. Hartley, The Chemistry of Platinum and Palladium, (Wiley, 1973), pp. 400-406.

2. J. E. Backvall, B. Åkermark, and S. O. Ljunggren, J.C.S. Chem. Comm., 264 (1977); J. Amer. Chem. Soc., 101, 2411 (1979).

3. J. K. Stille and R. Divakaruni, J. Amer. Chem. Soc., 100, 1303 (1978); J. Orgnomet. Chem., 169, 239 (1979).

4. T. Majima and H. Kurosawa, J.C.S. Chem. Comm., 610 (1977).

5. B. Akermark, J. E. Backvall, L. S. Hegedus, K. Siirala-Hansen, K. Sjoberg, and K. Zetterberg, J. Organomet. Chem., 72, 127 (1974).

6. B. Akermark, J. E. Backvall, K. Siirala-Hansen, K. Sjoberg, and K. Zetterberg, Tetrahedron Lett., 1363 (1974).

7. T. Hayashi and L. S. Hegedus, J. Amer. Chem. Soc., 99, 7093 (1977); L. S. Hegedus, R. E. Williams, T. Hayashi, and M. A. McGuire, J. Amer. Chem. Soc., in press.

8. T. Hosokawa, H. Ohkata, and I. Moritani, Bull. Chem. Soc. Japan, 48, 1533 (1975); T. Hosokawa, S. Miyagi, S-I. Murahashi and A. Sonada, J. Org. Chem., 43, 2752 (1978).

9. L. S. Hegedus, G. F. Allen, J. J. Bozell, and E. L. Waterman, J. Amer. Chem. Soc., 100, 5800 (1978).

10. D. E. Korte, L. S. Hegedus, and R. K. Wirth, J. Org. Chem., 42, 1329 (1977).

11. K. Maeda, T. Hosokawa, S-I. Murahashi, and I. Moritani, Tetrahedron Lett., 5075 (1973).

12. T. Hosokawa, M. Hirata, S-I. Murahashi, and A. Sonoda, Tetrahedron Lett., 1821 (1976).

13. A. Panunzi, A. DeRenzi, R. Palumbo, and G. Paiaro, J. Amer. Chem. Soc., 91, 3879 (1969).

14. A. Panunzi, A. DeRenzi, and G. Paiaro, J. Amer. Chem. Soc., 92, 3488 (1970).

15. P. Lennon, M. Madhavarao, A. Rosan, and M. Rosenblum, J. Organomet. Chem., 108, 93 (1976).

16. P. K. Wong, M. Madhavarao, D. F. Marten, and M. Rosenblum, J. Amer. Chem. Soc., 99, 2823 (1977).

17. P. Lennon, A. M. Rosan, and M. Rosenblum, J. Amer. Chem. Soc., 99, 8426 (1977).
18. P. F. Boyle and K. M. Nicholas, J. Org. Chem., 40, 2682 (1975).
19. K. M. Nicholas, J. Amer. Chem. Soc., 97, 3254 (1975).
20. J. Chatt, L. M. Vallarino, and L. M. Venanzi, J. Chem. Soc., 2496 (1957).
21. R. Palumbo, A. DeRenzi, A. Panunzi, and G. Paiaro, J. Amer. Chem. Soc., 91, 3874 and 3879 (1969).
22. M. Tada, Y. Kuroda, and T. Sato, Tetrahedron Lett., 2871 (1969).
23a. J. Tsuji and H. Takahashi, J. Amer. Chem. Soc., 87, 3275 (1965).
23b. C. R. Kistner, J. H. Hutchinson, J. R. Doyle and J. C. Storlie, Inorg. Chem., 2, 1255 (1963).
24. A. C. Cope, J. M. Kliegman, and E. C. Friederich, J. Amer. Chem. Soc., 89, 287 (1967).
25. R. A. Holton and R. A. Kjonaas, J. Amer. Chem. Soc., 99, 4177 (1977).
26. R. A. Holton and R. A. Kjonaas, J. Organomet. Chem., 142, C15 (1977).
27. L. S. Hegedus and K. Siirala-Hansen, J. Amer. Chem. Soc., 97, 1184 (1975).
28. L. S. Hegedus, O. P. Anderson, K. Zetterberg, G. Allen, K. Siirala-Hansen, D. J. Olsen, and A. B. Packard, Inorg. Chem., 16, 1887 (1977).
29. D. Medema, R. van Helden, and C. F. Kohll, Inorg. Chim. Acta, 3, 255 (1969).
30. J. K. Stille, D. E. James, and L. F. Hines, J. Amer. Chem. Soc., 95, 5062 (1973).
31. R. A. Holton and R. A. Kjonaas, J. Organomet. Chem., 133, C5 (1977).
32. R. A. Holton, J. Amer. Chem. Soc., 99, 8083 (1977).
33. G. F. Emerson, L. Watts, and R. Pettit, J. Amer. Chem. Soc., 87, 131 (1965).
34. J. D. Fitzpatrick, L. Watts, G. F. Emerson, and R. Pettit, J. Amer. Chem. Soc., 87, 3254 (1965).
35. J. C. Barborak, L. Watts, and R. Pettit, J. Amer. Chem. Soc., 88, 1328 (1966).
36. J. Rebec, Jr., and F. Gavina, J. Amer. Chem. Soc., 97, 3453 (1975).
37. A. J. Birch and I. D. Jenkins, in H. Alper, ed., Transition-Metal Organometallics in Organic Synthesis (Academic Press, 1976), Vol. I, 1-75.

# 13

# Transition-Metal Alkyne Complexes

Although virtually all transition metals react with alkynes, relatively few form simple, stable metal-alkyne complexes analogous to the corresponding metal-olefin complexes. This is because many alkyne complexes are quite reactive toward additional alkyne, and react further to produce more elaborate complexes or organic products. However, some stable metal alkynes are known. Among the best-characterized of these are complexes arising from the reaction (13.1) of platinum(0) complexes with substituted acetylenes.

$$(Ph_3P)_3Pt \quad + \quad RC{\equiv}CR \quad \longrightarrow \quad (Ph_3P)_2Pt{-}\|\!\!\!\|\,(RC{\equiv}CR) \quad \text{or} \quad (Ph_3P)_2Pt\langle\!\!\!\|\,(RC{=}CR) \qquad (13.1)$$

These complexes are similar to the previously discussed "platinum(0)-olefin complexes" (Chapter 12), in that the C≡C is somewhat lengthened upon complexation (from 1.20 Å in free alkynes to ~ 1.28 Å upon complexation). In addition, the alkyl groups are no longer colinear, but rather distort toward "$sp^2$" hydridization (C≡C-R angle ~130-140°) [1]. Formally, complexes of this type may be considered to be either Pt(0)-alkyne complexes or Pt(II)-metallacyclopropene complexes.

This distortion of alkyne geometry upon complexation has been used by Bennett [2] to permit the stabilization of small-ring cycloalkynes. The smallest unsubstituted cycloalkyne known in the free state is cyclooctyne. Because undue strain is needed to incorporate the four linear carbons of an alkyne group (C-C≡C-C) into a smaller ring, cycloheptyne and cyclohexyne have never been synthesized, although their existence as unstable intermediates has been demonstrated by a number of trapping experiments. However, both cycloheptyne and cyclohexyne give fairly stable platinum complexes (13.2). Unfortunately, these cycloalkyne ligands have not yet been successfully

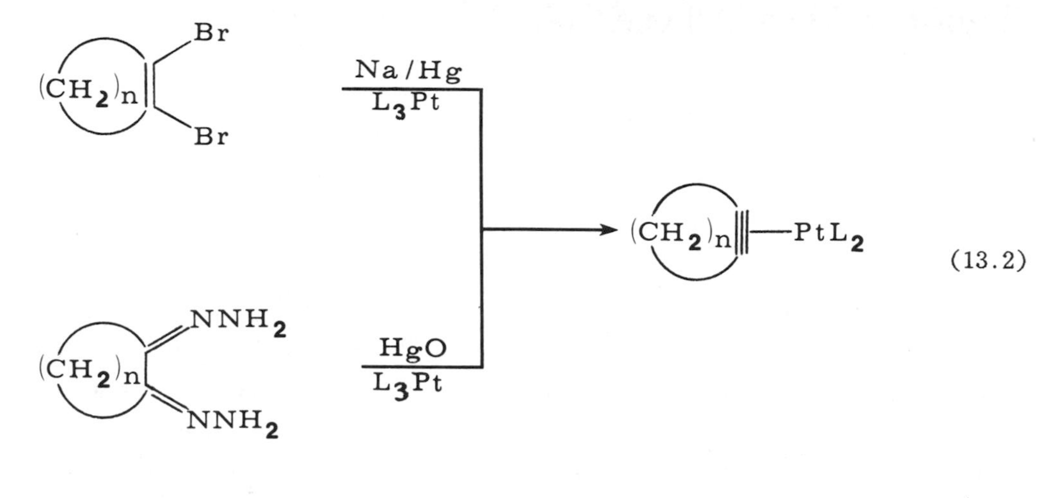


(13.2)

displaced from platinum; hence these complexes are not a stable synthetic source of cycloheptyne or cyclohexyne. However, these complexes are far from inert. Reaction with a number of proton acids produces σ-alkenylplatinum complexes (13.3). Hydrogen

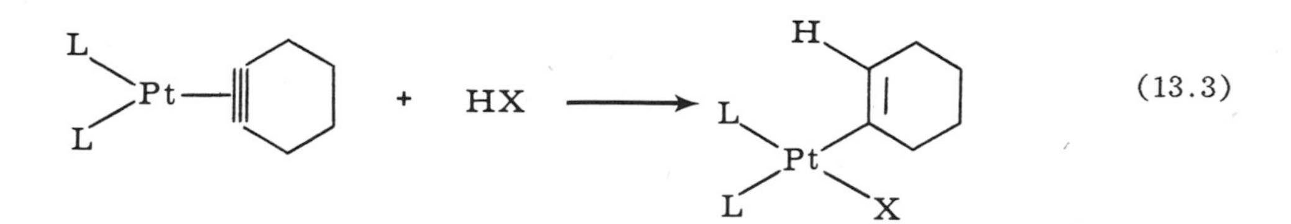


(13.3)

chloride or trifluoroacetic acid are required with the cycloheptyne complex. In contrast, the cyclohexyne complex reacts in the same fashion with much weaker acids, including water, primary alcohols, amides, and even carbon acids such as nitromethane and acetonitrile. This is considered to be a manifestation of the severe strain remaining in the complexed cyclohexyne. Acyclic platinum-alkyne complexes also react with hydrogen chloride or trifluoroacetic acid to give σ-alkenylplatinum complexes. These react further to give the uncomplexed olefin (13.4).

A rather different type of platinum-alkyne complex is formed by the reaction of trans $PtCl(CH_3)(Ph_3P)_2$, $AgPF_6$, and alkynes (13.5). These complexes are structurally

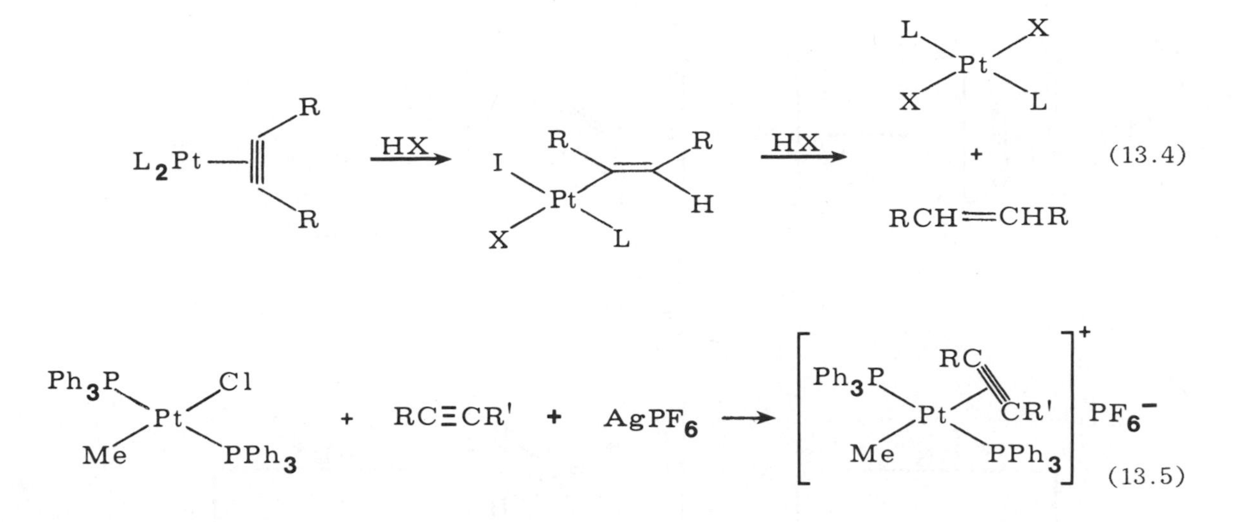

analogous to the platinum and palladium(II)-olefin complexes discussed in Chapter 12. They are only isolable when both R and R' are alkyl or aryl groups. Monosubstituted acetylenes and disubstituted acetylenes containing electron-withdrawing groups react further, and complexes of these cannot be isolated as such. These complexes, studied most extensively by Chisholm and Clark [3], show reactivity characteristic of carbonium ions. Thus, complexes of disubstituted acetylenes undergo attack by nucleophiles such as methanol, much like the corresponding platinum(II)-olefin complexes (13.6). In

$$\left[ Me{-}PtL_2{-}(RC{\equiv}CR') \right]^+ + MeOH \longrightarrow Me{-}PtL_2{-}C(R'){=}C(R){-}OMe + H^+ \quad (13.6)$$

contrast, complexes of monosubstituted acetylenes cannot normally be isolated, but rather undergo a variety of rearrangement and/or insertion reactions (13.7) characteristic of a carbonium-ion intermediate.

Although palladium forms some alkyne complexes analogous to the platinum complexes discussed above, often they are less stable and more reactive than their platinum analogs. Thus palladium(0) complexes of both linear and cyclic alkynes are known, and react like those of platinum (Equations 13.2 to 13.4). Alkyne complexes of palladium(II) are much more rare. Often the reaction of an alkyne (particularly a terminal one) with a palladium(II) salt produces a bewildering number of organic compounds and organometallic complexes containing more than one alkyne unit. Although

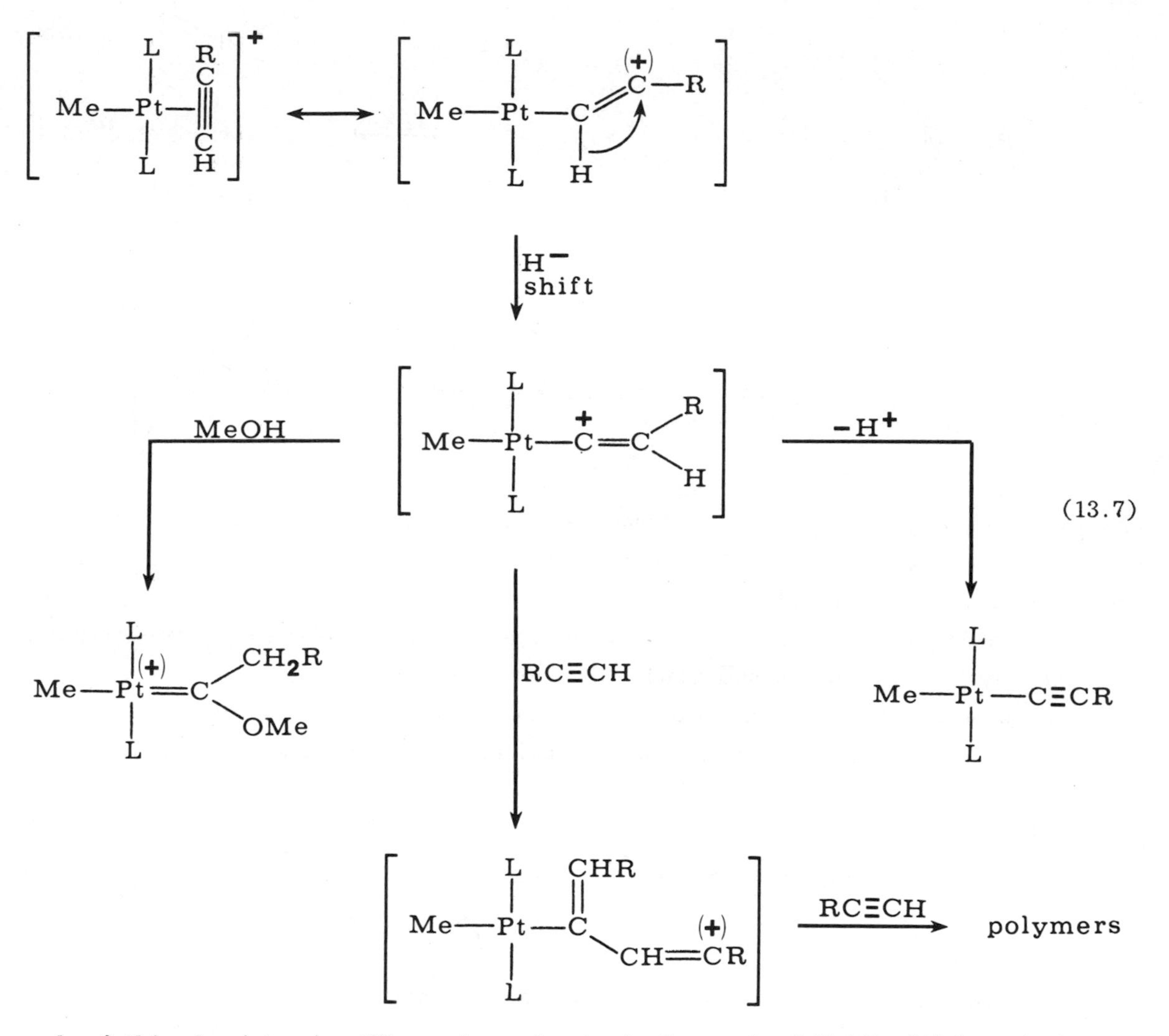


much of this chemistry is still poorly understood, the work of Maitlis [4] has shed some light on this subject.

The reaction which ensues between $PdCl_2$ and an alkyne depends very much on the steric bulk of the alkyne, but is remarkably insensitive to the electronic nature of the alkyne. Thus, the very bulky di-t-butylacetylene forms a simple palladium(II)-alkyne complex, one of the very few known for this metal (13.8). In contrast, phenyl t-butylacetylene forms a cyclobutadiene complex (13.9), whereas diphenylacetylene forms both a cyclobutadiene complex and hexaphenylbenzene (13.10). Finally, dimethyl

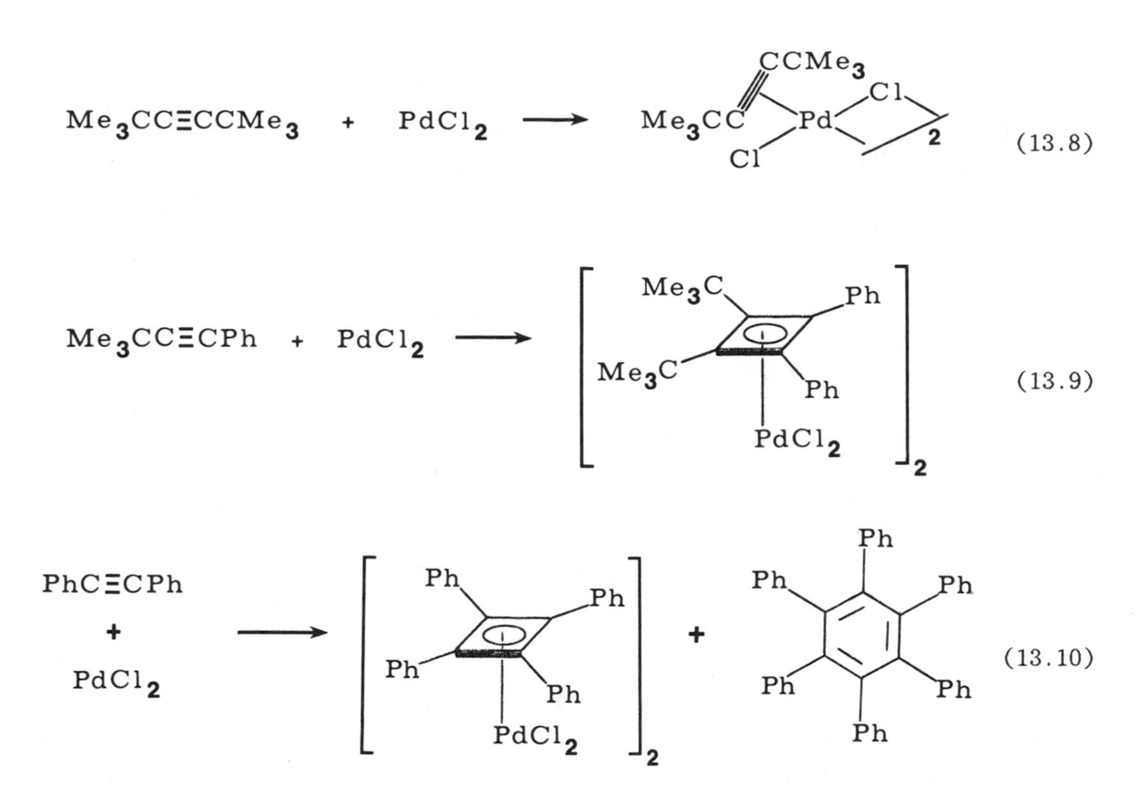


acetylenedicarboxylate gives a complex incorporating three alkyne units which, upon heating or treatment with strong ligands, produces hexamethyl mellitate (13.11).

On the surface this is very complex and confusing chemistry. However, extensive careful study, including the isolation of proposed intermediates and their conversion to products, has produced the coherent scheme, involving a series of stepwise cis insertions of coordinated acetylenes, summarized in Figure 13.1.

The first step involves formation of a simple palladium(II)-alkyne complex analogous to the palladium(II)-olefin complexes discussed in Chapter 12. With the very bulky di-t-butylacetylene, the reaction stops here, and the complex can be isolated and characterized. With virtually all other alkynes, a very rapid cis insertion of the coordinated alkyne into a Pd-Cl bond (chloropalladation) ensues, producing a σ-vinylpalladium(II) complex, which inserts another alkyne in a cis fashion, producing a σ-butadienyl complex. (Complexes of this type have been trapped and isolated from reactions involving t-butylacetylene.) This intermediate can react in at least two ways, depending on the alkyne substituents. With large substituents, coordination and insertion of additional alkyne is inhibited, and the σ-butadienyl complex is proposed to

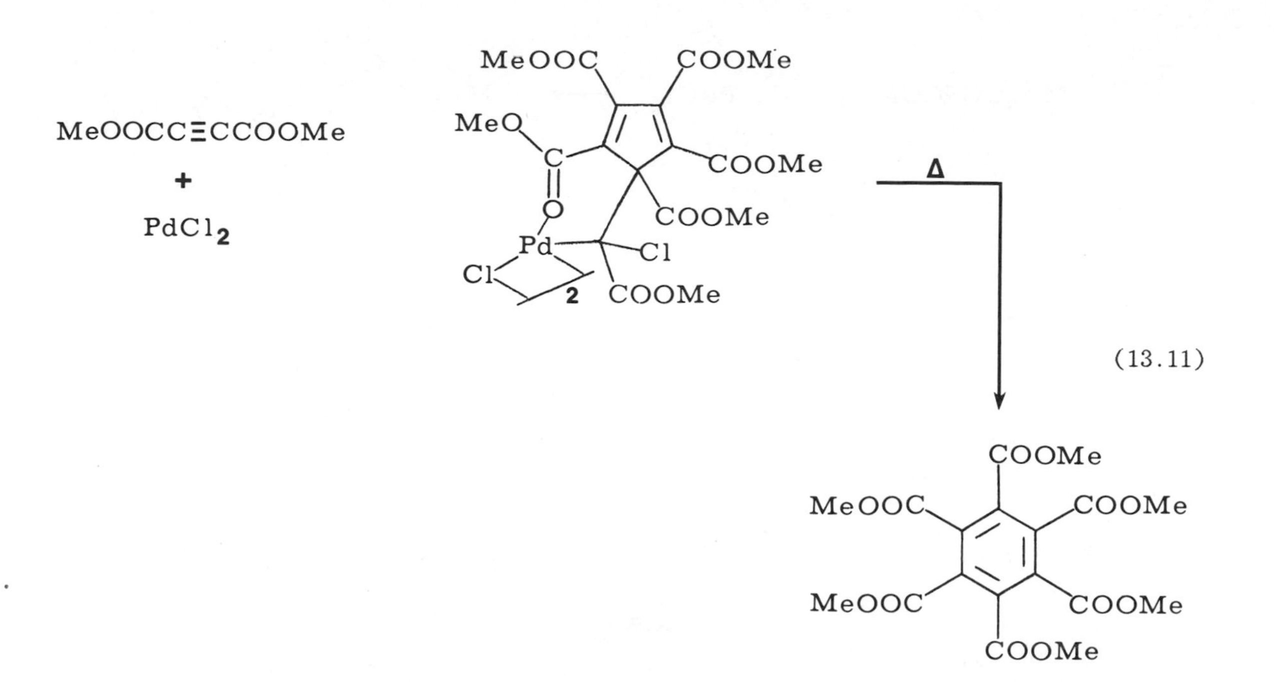


(13.11)

undergo a conrotatory cyclization to produce a π-cyclobutadiene complex. Complexes of this type are well-known, and form in the reaction of palladium chloride with a number of alkynes. They are quite stable and do not react with alkynes to produce benzenes. Instead, cyclotrimers appear to result from insertion of another molecule of alkyne into the σ-butadienyl complex, followed by an internal insertion of the terminal olefinic group into the Pd-carbon bond, forming a σ-alkylpalladium complex. Again, complexes of this type have been isolated and shown to convert to substituted benzenes by an as-yet unspecified mechanism.

It should be noted that Figure 13.1 contains only the broad, somewhat simplified, outlines of the processes involved in the palladium-catalyzed oligomerization of alkynes, and many details remain to be discovered. It is of interest that metallacyclopentadienes do not appear to be involved in these palladium(II)-catalyzed reactions. In contrast, many of the other transition metals that cyclotrimerize alkynes react via metallacyclopentadienes (Chapter 10).

Some of the earliest work on the metal-catalyzed oligomerization of acetylenes was the work of Reppe [5,6], who showed that acetylene itself could be cyclotetramerized to cyclooctatetraene in the presence of nickel catalysts at 80-120° and 10-25 atm (13.12). Monosubstituted alkynes lead to tetrasubstituted cyclooctatetraenes, but disubstituted

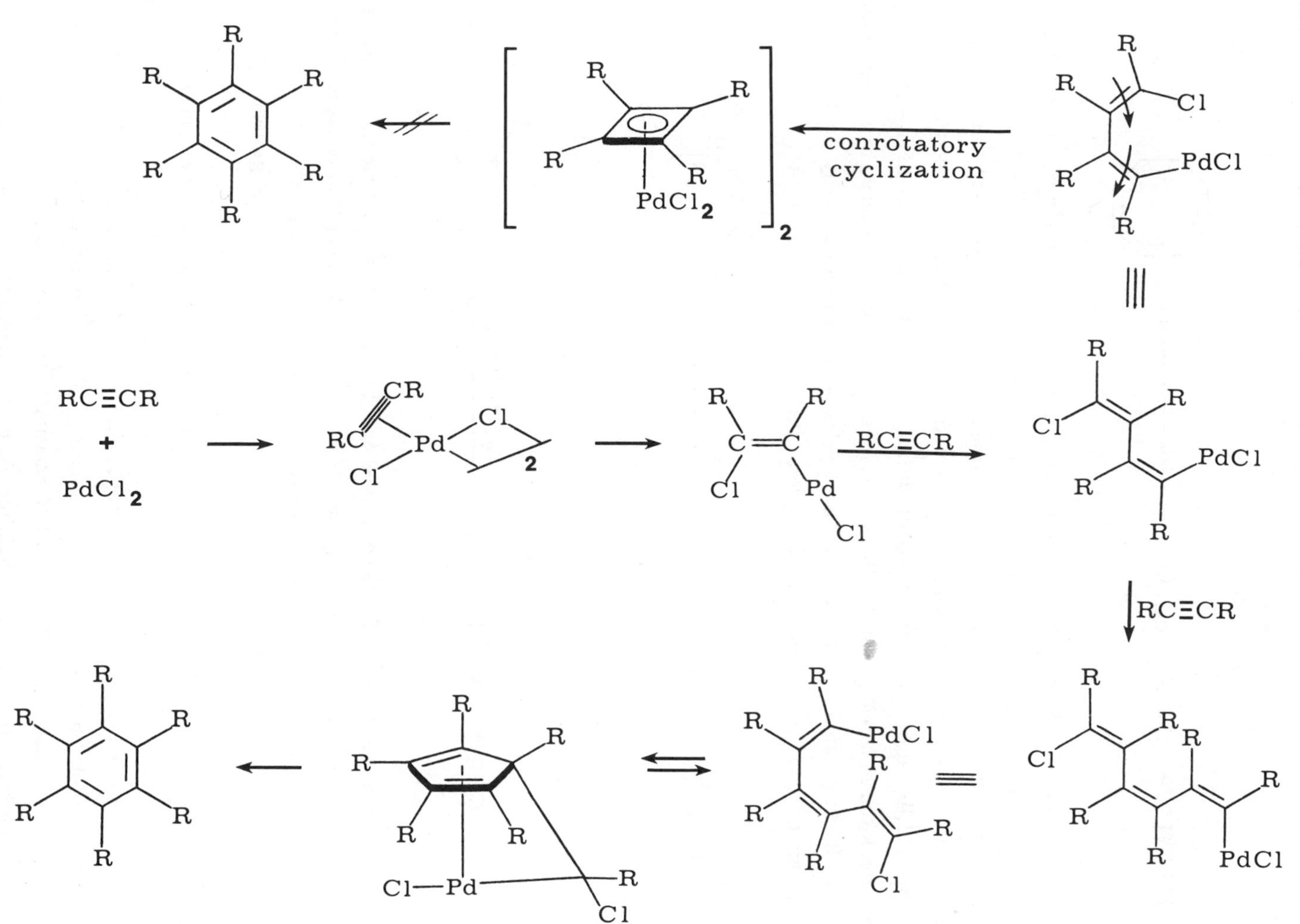


*Figure 13.1. Palladium-catalyzed oligomerization of alkynes.*

$$4\ CH{\equiv}CH\ +\ NiX_2\ (cat.)\ \longrightarrow\ \text{[cyclooctatetraene]}\quad 70\%\qquad (13.12)$$

alkynes do not react. Although this work is more than 30 years old, the mechanism of this reaction is still unknown, although many have been proposed.

By introducing ligands, such as phosphines, which can compete with alkyne for a coordination site, the reaction can be directed toward cyclotrimerization to produce mixtures of 1,2,4- and 1,3,5-trisubstituted benzenes. Linear oligomerization competes with this cyclotrimerization, and, in some cases, linear oligomers are the principle products. Although disubstituted alkynes are not cyclotrimerized by nickel catalysts, they often can be cooligomerized with acetylene itself (13.13 and 13.14).

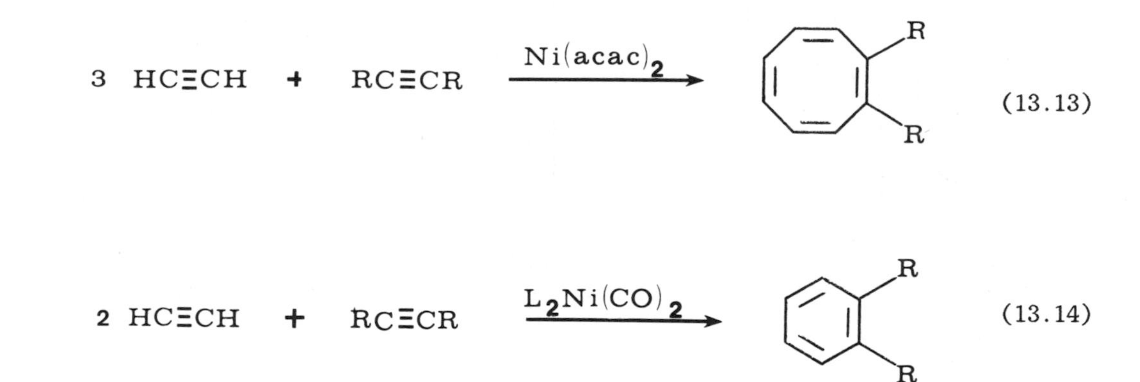


Oligomerization reactions of these types are by no means restricted to the nickel triad, but are a common theme throughout transition-metal-alkyne chemistry. Mechanistically, the most extensively studied system is the cyclotrimerization of alkynes by $\eta^5$-$C_5H_5Co(CO)_2$ [7]. Although some details are not certain, the general process is believed to be that described in Figure 13.2. This process, and related ones, are discussed in detail in Chapter 10, and only a brief outline will be presented here.

The initial steps involve coordination of two alkynes. If the initial complex cannot coordinate two alkynes (i.e., if CO is replaced by $Et_3P$) because its ligands are too firmly bound, cyclotrimerization does not ensue. The bis-alkyne complex then undergoes an oxidative coupling (C-C bond formation) to form the metallacyclopentadiene complex. Complexes of this type, arising from the reaction of metal complexes with

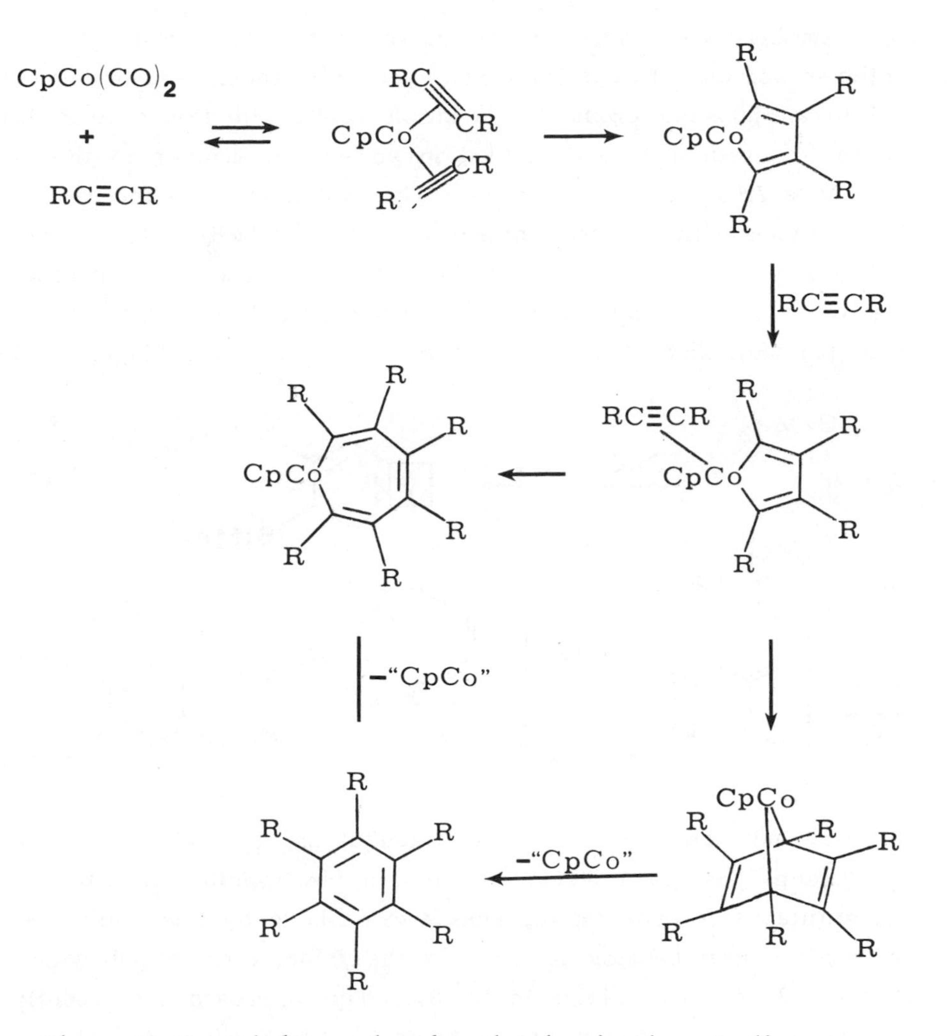


*Figure 13.2. Cobalt-catalyzed cyclotrimerization of alkynes.*

alkynes, are known for a wide variety of transition metals. This metallacyclopentadiene has several options, depending on substrate, ligand, metal, and reaction conditions. For cyclotrimerization to occur, the metallacyclopentadiene complex must coordinate another alkyne. This can either insert to form a metallacycloheptatriene complex or undergo a cycloaddition to produce a bridged bicyclic ($\pi$)complex. Loss of metal from either of these produces the substituted aromatic cyclotrimerization product and regenerates a catalytically active species. It is rather difficult to distinguish experimentally among these pathways, and, indeed, different catalyst systems may produce the same products by different routes.

This cyclotrimerization reaction has been known for many years, but was of little use for synthesis because attempts at mixed cyclotrimerizations, involving two different alkynes, led to all possible products. A simple solution to this problem has recently been found by Vollhardt [7], and applied in an elegant manner to the synthesis of polycyclic products from simple precursors. The system involves the co-cyclotrimerization of 1,5-hexadiynes with bis-(trimethylsilyl)acetylene (BTMSA), an alkyne which does not itself cyclotrimerize. The initial product of the reaction is a benzocyclobutane, which is reactive in its own right, and will undergo Diels-Alder reactions with a variety of dienophiles [8] (Equation 13.15). The success of the system hinges on the inability

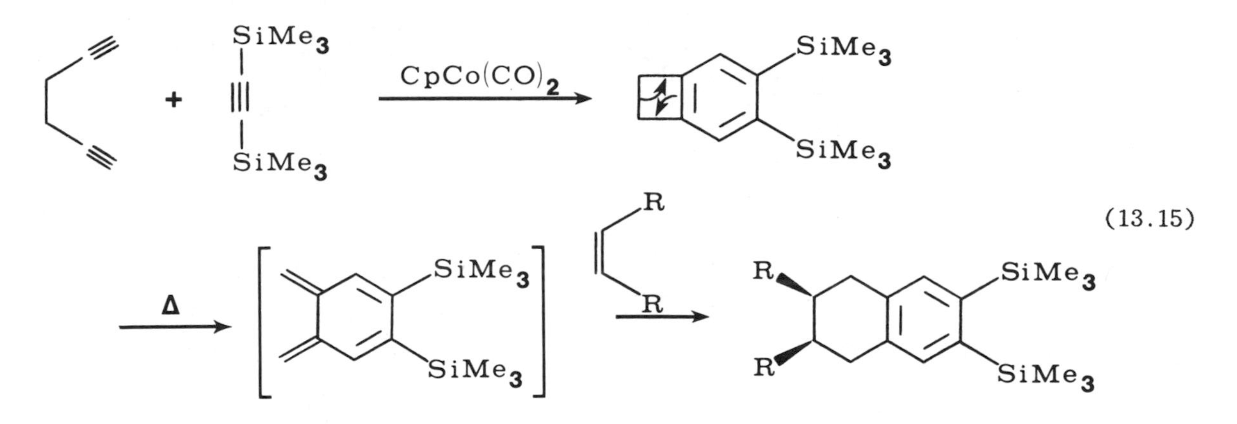


of BTMSA to self-condense, thereby facilitating co-cyclotrimerization. Since trimethylsilyl groups are good leaving groups in electrophilic substitution reactions, variously substituted tetrahydronaphthalenes are available by this route. By incorporating the dienophile into the same molecule as the diyne, complex polycyclic compounds are available in a single step (13.16 to 13.18). This approach has recently been employed in an elegant synthesis of racemic estrone [9,10] (Equation 13.19).

Organic nitriles are another triply bonded species which do not self-trimerize, but can be cotrimerized with alkynes to produce pyridines. With unsymmetrical alkynes, however, all possible isomers are obtained [11,12] (Equation 13.20). On the other hand, when combined with diynes, the isoquinoline nucleus is formed [7] (Equation 13.21). Isocyanides, $CS_2$, and RNCS have also been co-cyclotrimerized with alkynes [1].

The chemistry presented above is not restricted to $CpCo(CO)_2$. Many different transition-metal complexes engage in one or more of the reactions presented in Figure 13.2, and both cyclobutadiene and metallacyclopentadiene complexes are frequently

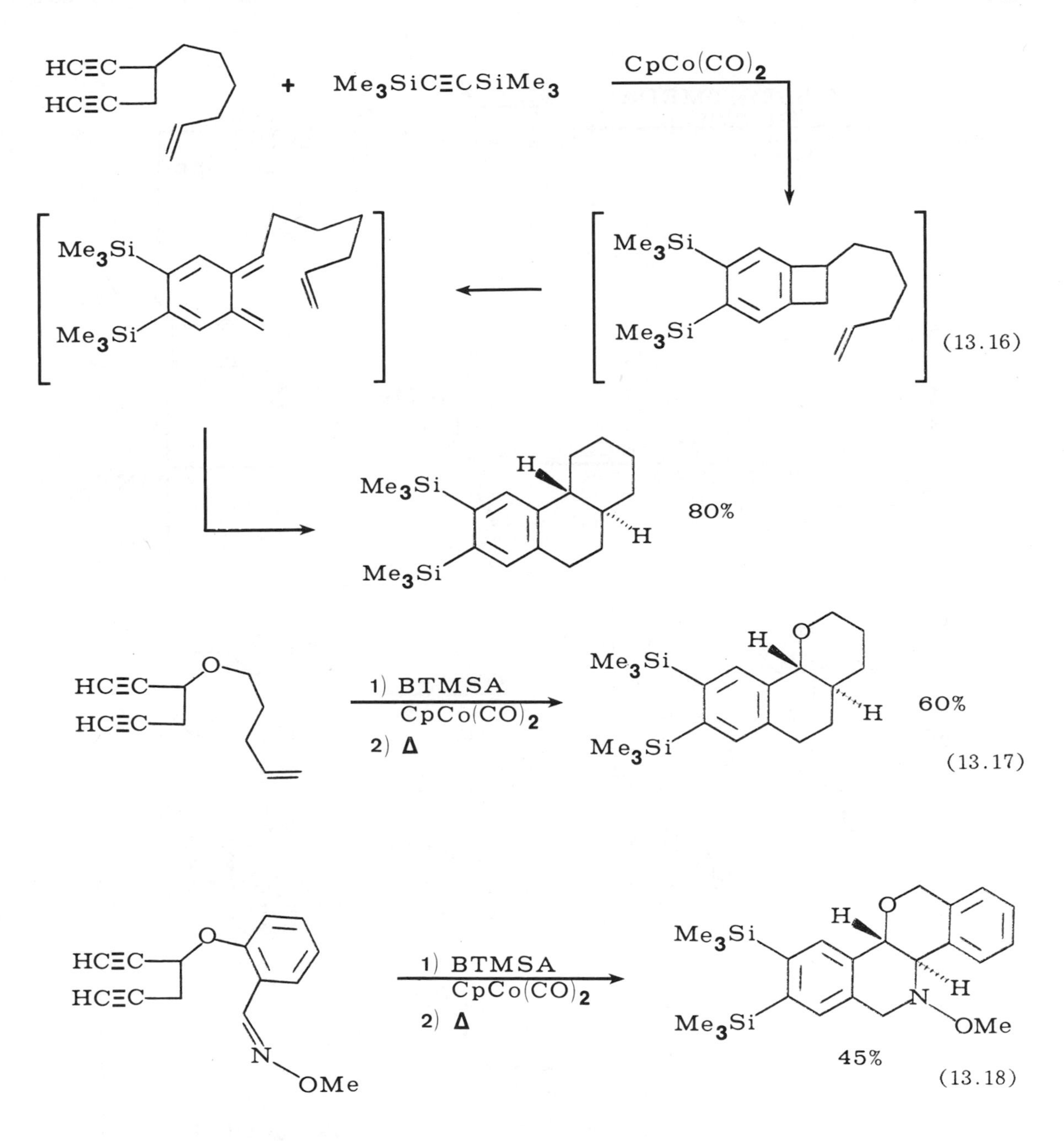
HC≡C
HC≡C
+
Me3SiC≡CSiMe3
CpCo(CO)2
Me3Si
Me3Si
Me3Si
Me3Si
(13.16)
H
Me3Si
Me3Si
H
80%
HC≡C
HC≡C
O
1) BTMSA
CpCo(CO)2
2) Δ
H
O
Me3Si
Me3Si
H
60%
(13.17)
HC≡C
HC≡C
O
N
OMe
1) BTMSA
CpCo(CO)2
2) Δ
H
O
Me3Si
Me3Si
H
N
OMe
45%
(13.18)

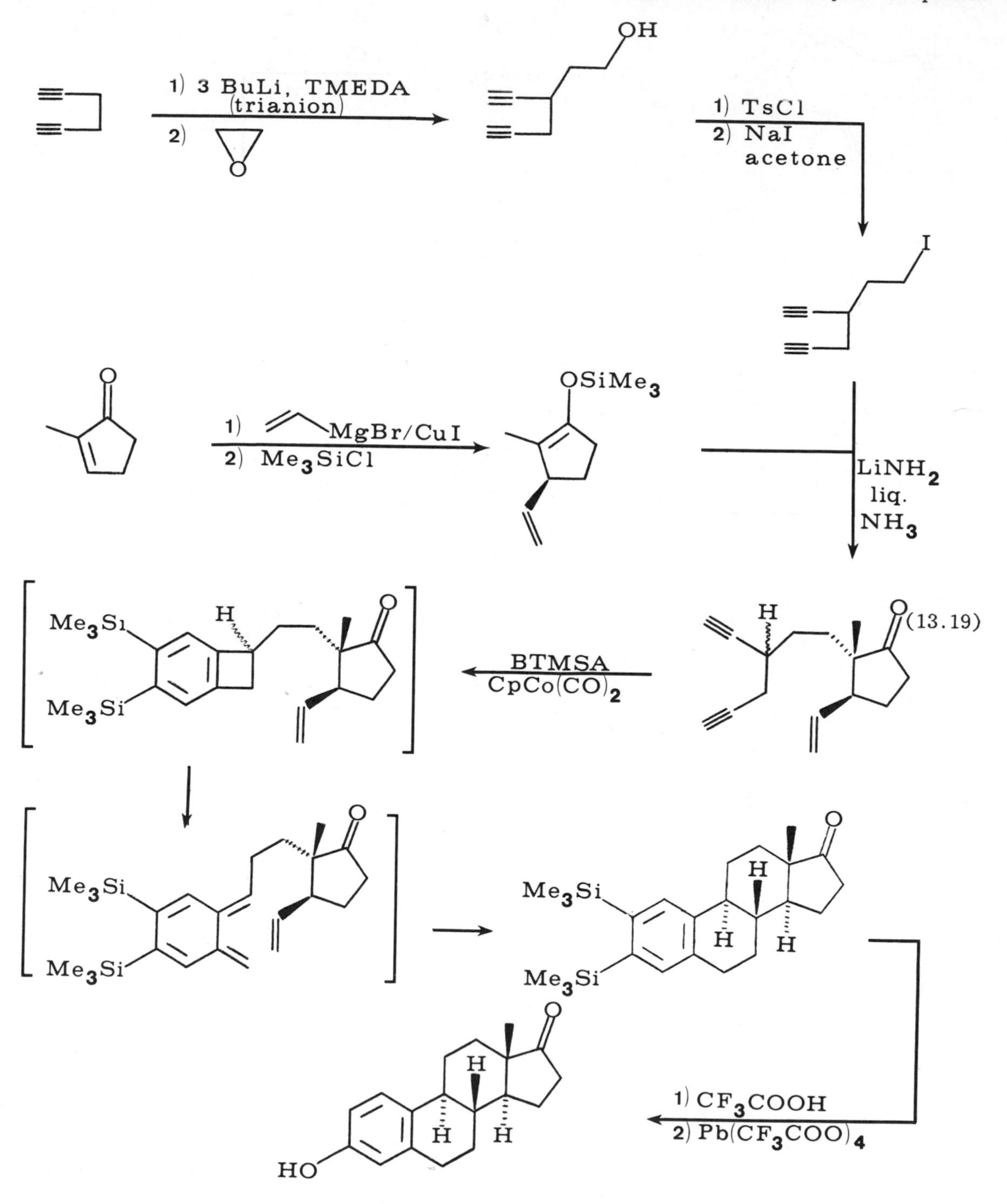
OH
1) 3 BuLi, TMEDA
(trianion)
2)
1) TsCl
2) NaI
acetone
I
OSiMe3
1) MgBr/CuI
2) Me3SiCl
LiNH2
liq.
NH3
Me3Si
Me3Si
H
H
(13.19)
BTMSA
CpCo(CO)2
Me3Si
Me3Si
Me3Si
Me3Si
H
H
H
1) CF3COOH
2) Pb(CF3COO)4
H
H
H
HO

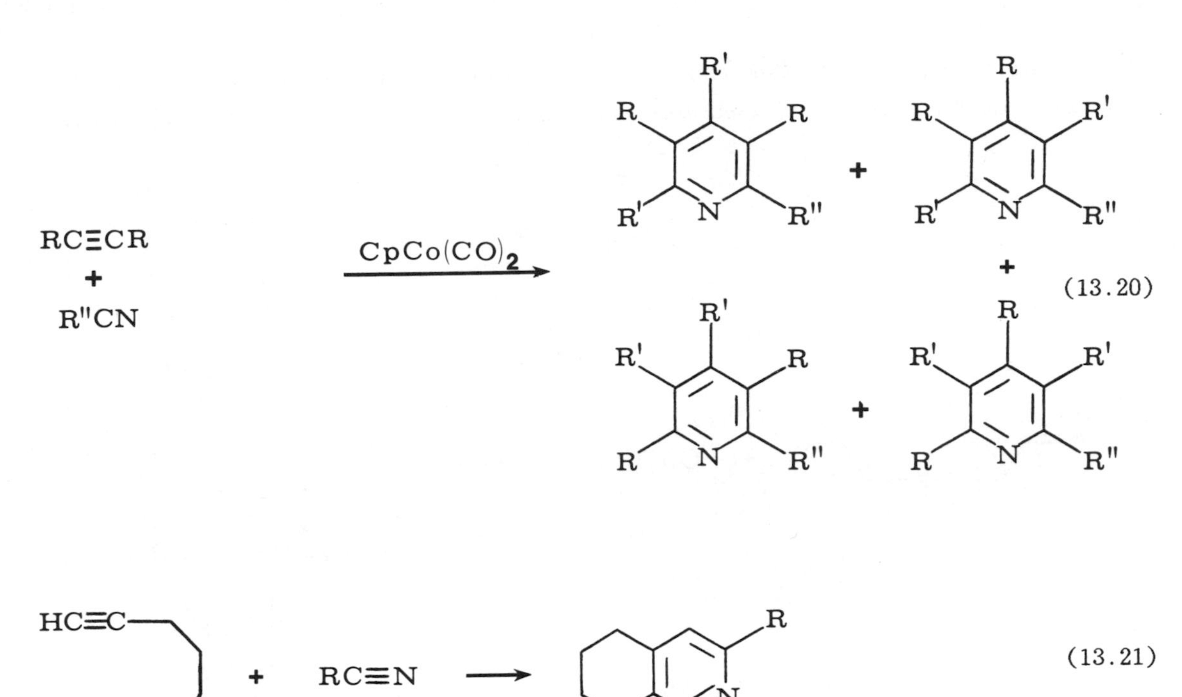


isolated from these reactions. However, $CpCo(CO)_2$ performs the cleanest cyclotrimerization reactions, whereas other systems tend to give varying amounts of linear oligomers, larger-ring cyclic products, or both.

Most of the alkyne complexes of cobalt involved in the above cyclotrimerization reactions are relatively reactive, and difficult to isolate and characterize. In contrast, dicobalt octacarbonyl [$Co_2(CO)_8$] forms very stable complexes with alkynes [13]. In these complexes the alkyne bridges the two metals, using both of its $\pi$-bonding orbitals (all 4 $\pi$-electrons) for complexation (13.22). As a consequence, the reactivity of the

$$RC\equiv CR \quad + \quad Co_2(CO)_8 \quad \longrightarrow \quad (CO)_3Co\text{—}Co(CO)_3 \qquad (13.22)$$

R R C C

alkyne complexed in this fashion is drastically <u>reduced</u>. In fact, alkynes have been protected from electrophilic addition and hydroboration by complexation to $Co_2(CO)_8$. Pettit selectively reduced the olefinic portion of an enyne by complexation of the alkyne by $Co_2(CO)_8$, reduction of the olefin with diimide, and regeneration of the free alkyne by oxidative removal of the cobalt. The olefin could also be hydroborated [14] (Equation 13.23). Nicholas used this same principle to stabilize propargyl cations for

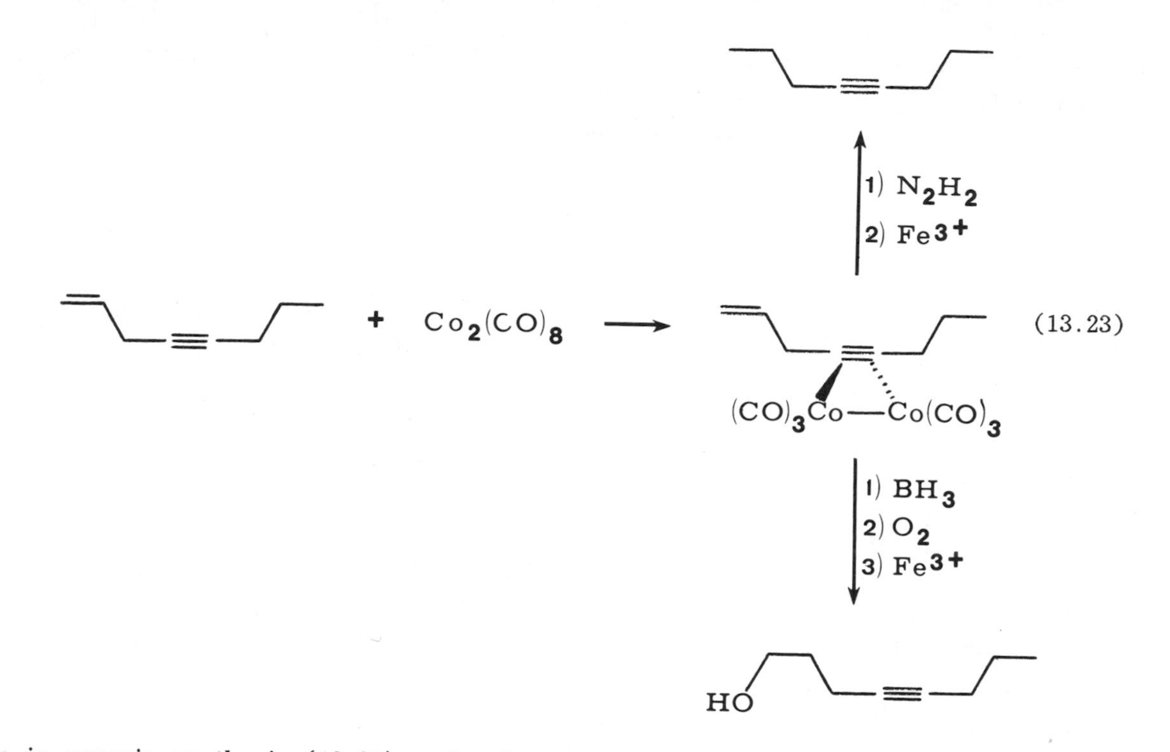


use in organic synthesis (13.24). Complexation of the alkyne group not only protects it from self-reaction, but greatly increases the stability and ease of formation of the propargylic cation. These species react with aromatic compounds by an electrophilic aromatic substitution mechanism [15], and are alkylated by β-diketone anions [16] and trimethylsilylenol ethers [17]. (Cobalt is not unique in its ability to incorporate alkynes in a bridging position between two metals. Many other transition metals coordinate alkynes in this manner. However, the cobalt complexes are the only ones to have been used synthetically, to date.)

Although these cobalt-alkyne complexes are stable to electrophiles, they do react with carbon monoxide to incorporate CO into the alkyne group. Thus exposure of the alkyne complex of $Co_2(CO)_8$ to high CO pressures (75°, 210 atm) results in the formation of a lactone complex, with a net incorporation of two molecules of CO [18] (Equation 13.25). Under even more severe conditions (90°, 950 atm), dimers of this lactone ligand are produced catalytically from CO and alkyne. Under different conditions,

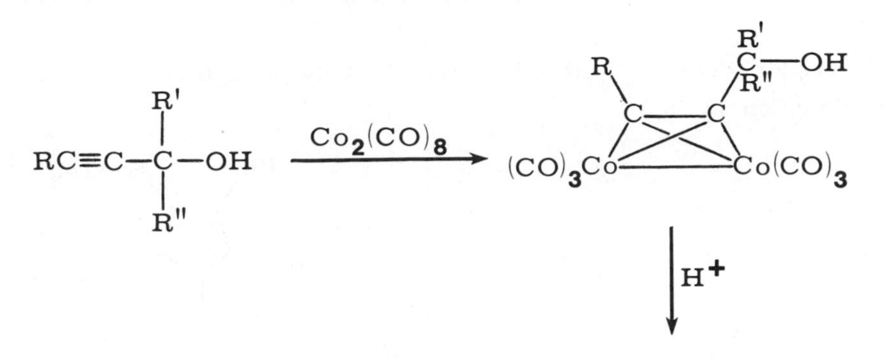
RC≡C—C—OH
R'
R''
Co2(CO)8
(CO)3Co
Co(CO)3
H+

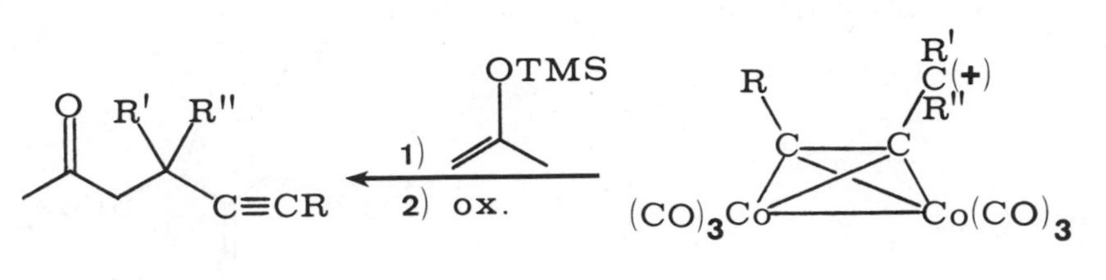
OTMS
1)
2) ox.
C≡CR
(CO)3Co
Co(CO)3

(13.24)

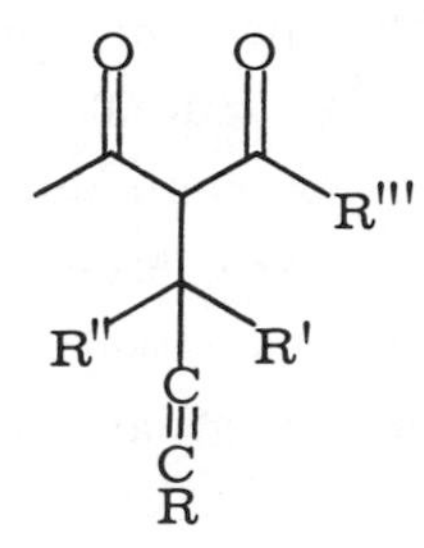

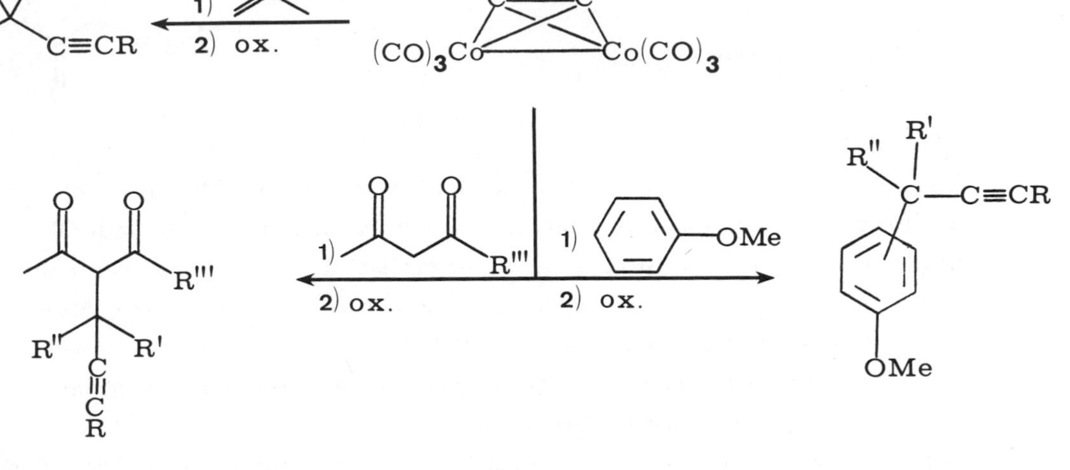
1)
2) ox.
R'''
OMe

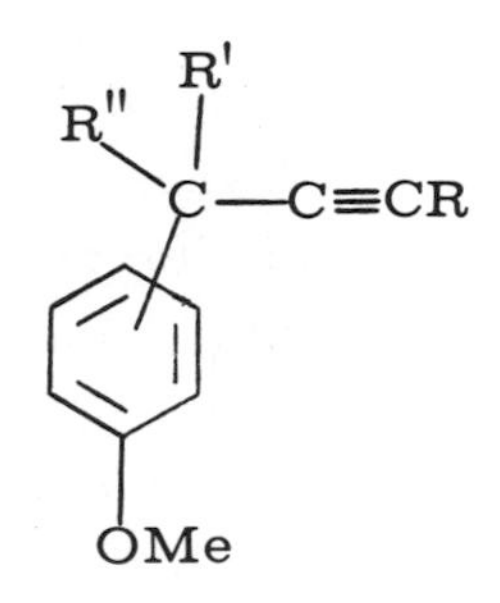
C≡CR
OMe

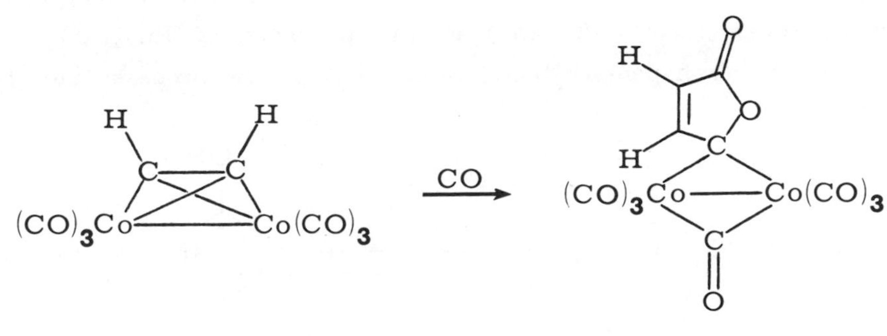
CO
(CO)3Co
Co(CO)3

(13.25)

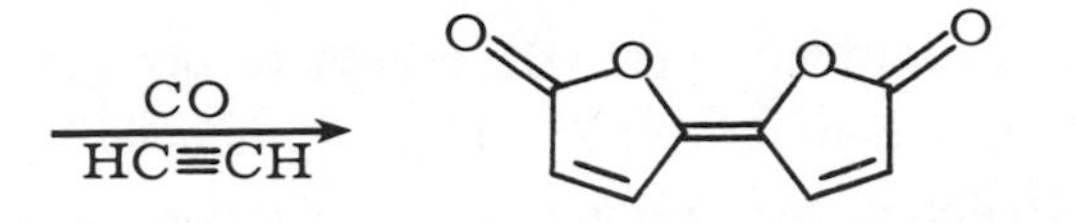
CO
HC≡CH

the diphenylacetylene complex reacts with bis-(trimethylsilyl)acetylene to give cyclopentadienones [19] (Equation 13.26).

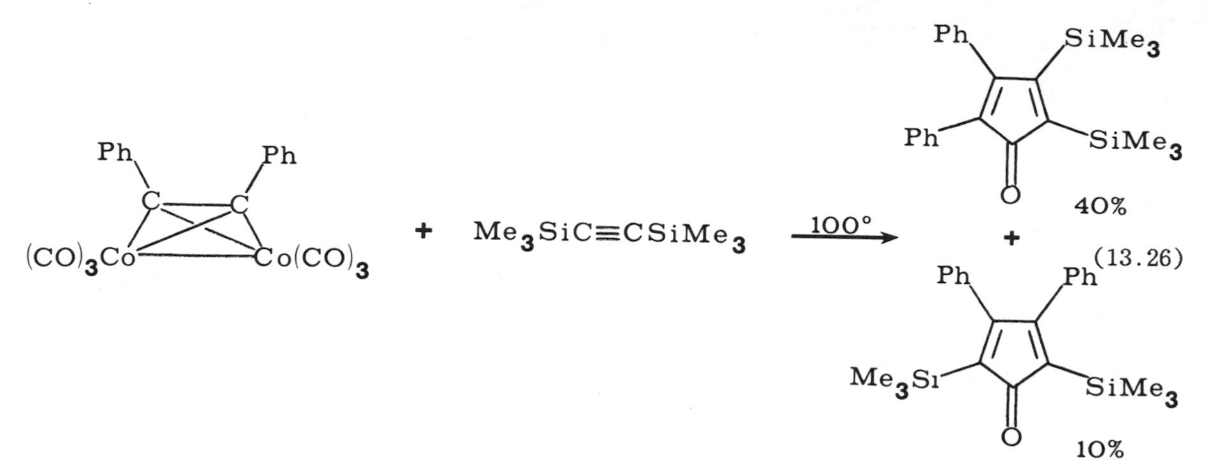


Alkynes react with many other transition metals in the presence of carbon monoxide to produce compounds containing both carbon monoxide and alkyne-derived fragments. In addition, a profusion of organometallic complexes containing ligands constructed of alkyne-derived fragments and carbon monoxide are often obtained. Reactions of this type are usually quite complex, and few have been tamed sufficiently to be of use in the synthesis of organic compounds. However, a great deal of this potentially useful chemistry awaits development.

Hydroquinone is produced in modest (~60%) yield by the reaction of acetylene, carbon monoxide, and hydrogen (200°, 120 atm) in the presence of $Ru_3(CO)_{12}$ as catalyst [18] (Equation 13.27). Iron pentacarbonyl catalyzes the same process, but the

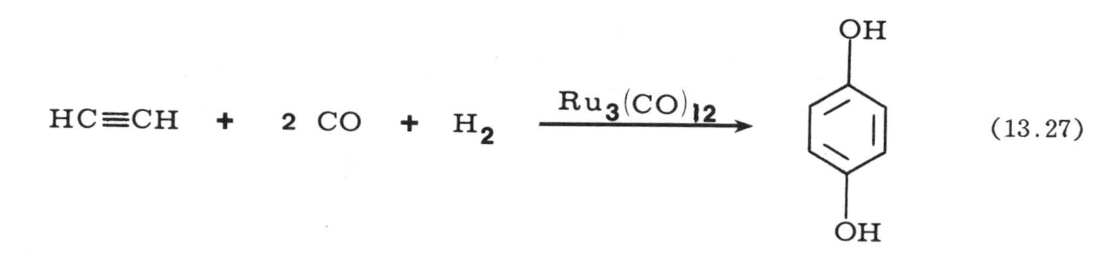


yield is somewhat lower. Under milder conditions, $Fe(CO)_5$ reacts to produce a host of iron complexes containing coordinated cyclopentadienones, benzoquinones, cyclobutadienes, or cycloheptatrienones, all constructed by co-cyclooligomerization of the alkyne and carbon monoxide [19]. Recently, even carbon dioxide has been cyclooligomerized with alkynes [20] (Equation 13.28). The generality of this process is not yet known.

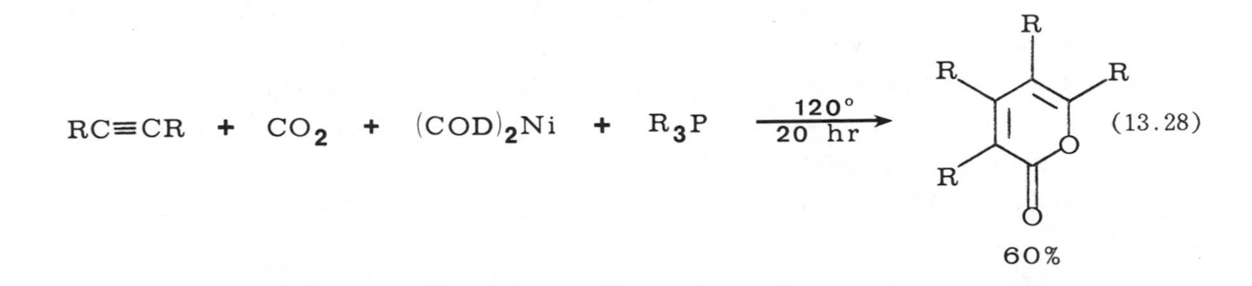


Many more complex reactions between transition metals and alkynes are known, but are beyond the scope of this discussion [for discussion of them, see 1,3-7,13,18,19].

## NOTES.

1. S. Otsuka and A. Nakamura, Adv. Organomet. Chem., 14, 245 (1976).
2. M. A. Bennett and T. Yoshida, J. Amer. Chem. Soc., 100, 1750 (1978).
3. M. H. Chisholm and H. C. Clark, Accts. Chem. Res., 6, 202 (1973).
4. P. M. Maitlis, Accts. Chem. Res., 9, 93 (1976).
5. P. W. Jolley and G. Wilke, The Organic Chemistry of Nickel (Academic Press, 1975), Vol. II, 94-105.
6. C. Hoogzand and W. Hubel, "Cyclic Polymerization of Acetylenes by Metal Carbonyl Compounds," in I. Wender and P. Pino, eds., Organic Synthesis via Metal Carbonyls, (Wiley Interscience, 1968), Vol. I, 343-369.
7. K. P. C. Vollhardt, Accts. Chem. Res., 10, 1 (1977), and references cited therein.
8. W. G. L. Aalbersberg, A. J. Barkovich, R. L. Funk, R. L. Hillard, III, and K. P. C. Vollhardt, J. Amer. Chem. Soc., 97, 5600 (1975).
9. R. L. Funk and K. P. C. Vollhardt, J. Amer. Chem. Soc., 99, 5483 (1977).
10. R. L. Funk and K. P. C. Vollhardt, J. Amer. Chem. Soc., 101, 215 (1979).
11. Y. Wakatsuki and H. Yamazaki, J.C.S. Chem. Comm., 280 (1973).
12. H. Bönnemann, R. Brinkmann, and H. Schenkluhn, Synthesis, 575 (1974).
13. R. S. Dickson and P. J. Fraser, Adv. Organomet. Chem., 12, 323 (1974).
14. K. M. Nicholas and R. Pettit, Tetrahedron Lett., 3475 (1971).
15. R. F. Lockwood and K. M. Nicholas, Tetrahedron Lett., 4163 (1977).
16. H. D. Hodes and K. M. Nicholas, Tetrahedron Lett., 4349 (1978).

17. K. M. Nicholas, M. Mulvaney and M. Bayer, J. Amer. Chem. Soc., 102, 2508 (1980).

18. P. Pino and G. Braca, "Carbon Monoxide Addition to Acetylenic Substrates," in I. Wender and P. Pino, eds., Organic Synthesis via Metal Carbonyls, (Wiley Interscience, 1977), Vol. II, 420-516.

19. W. Hübel, "Organometallic Derivatives from Metal Carbonyls and Acetylene Compounds," in I. Wender and P. Pino, eds., Organic Synthesis via Metal Carbonyls, (Wiley Interscience, 1968), Vol. I, 273-340.

20. Y. Inoue, Y. Itoh, and H. Hashimoto, Chem. Lett., 633 (1978).

# 14

## $\eta^6$-Arene Transition-Metal Complexes

Although several transition metals (e.g., Cr, Mo, W, Fe, V, Mn, Rh) form stable, characterizable complexes with arenes, only those of chromium have found significant application in organic synthesis. There are several practical reasons for this. Arene-chromium tricarbonyl compounds are readily prepared by several convenient methods. Simply boiling $Cr(CO)_6$ in benzene or dioxane produces $(Arene)Cr(CO)_3$ in excellent yield [1]. Another approach consists of heating $Cr(NH_3)_3(CO)_3$ [2] or Cr(2-picoline)$(CO)_5$ [3] with excess arene (14.1). As a consequence of this ease of preparation and the high yields obtained, literally hundreds of arenechromium tricarbonyl

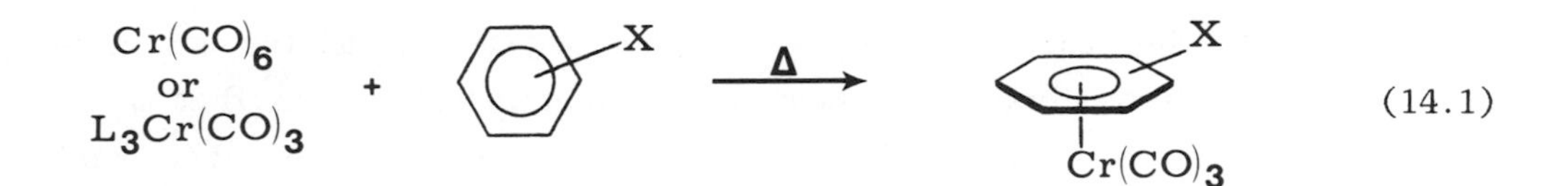


complexes, some with complex aryl groups, are known, and a number are commercially available [4]. Among the more common substituted-arene complexes of chromium are those of toluene, aniline, anisole, chloro- and iodobenzene, t-butylbenzene, acetophenone, and thiophene. Those of benzaldehyde and benzoic acid are not accessible by these direct methods, but are stable when prepared by indirect routes. Nitrobenzene is unknown as a $\eta^6$-ligand. In addition to their ease of preparation, $\eta^6$-arenechromium tricarbonyl complexes are air-stable, soluble, crystalline solids, easily characterized by spectroscopic methods, and easily purified by chromatographic techniques. Finally, the arene group is readily released from complexation by mild oxidation of the metal by aqueous Ce(IV) or iodine [1,5] or by exposure to sunlight and air. This point is very important for use in organic synthesis, since the metal is not normally desired in the final product.

The arene ligand (i.e., benzene) is considered to be a $\eta^6$-bonding, six-electron system occupying three coordination sites. Several lines of evidence suggest that in arenechromium tricarbonyl complexes, the arene group is net electron-donating relative to the chromium; equivalently, the chromium is an electron-withdrawing group toward the arene. For example, the dipole moment of $\eta^6$-benzenechromium tricarbonyl is 5.08 D [6]; the $pK_a$ of $\eta^6$-benzoic acid chromium tricarbonyl is 4.77 compared with 5.75 for benzoic acid itself; and the $pK_b$ of $\eta^6$-anilinechromium tricarbonyl is 13.31 compared to 11.70 for aniline itself. These data suggest that the "electron-withdrawing" power of a $Cr(CO)_3$ group is similar to that of a p-nitro group. The infrared stretching frequencies of the chromium carbonyl groups shift from 2000 $cm^{-1}$ in $Cr(CO)_6$ to 1987 and 1917 $cm^{-1}$ in $(C_6H_6)Cr(CO)_3$, indicating increased electron density on the metal as translated into increased backbonding of this electron density into the $\pi^*$ orbitals of coordinated CO. (Recall that CO's are strong $\pi$-acceptor ligands and are well-suited to accepting increased electron density from the metal.) Although in the proton NMR spectrum of $(C_6H_6)Cr(CO)_3$, the aryl protons are shifted 1.5-2.0 ppm upfield from uncomplexed benzene, correlation of this shift with changes in electron density on the ring is hazardous. The most compelling evidence for the reduction of electron density of an aromatic ring upon complexation is found by examining the chemistry of these complexes.

One of the earliest studies of the organic chemistry of $\eta^6$-arenechromium tricarbonyl complexes was an examination of their reactivity in that most typical of reactions of aromatics, electrophilic aromatic substitution. To no one's surprise, at least in retrospect, $\eta^6$-arenechromium tricarbonyl complexes do not undergo electrophilic aromatic substitution easily. Furthermore, electron donating substituents on the aromatic ring do not enhance this reaction [1]. $\eta^6$-Benzenechromium tricarbonyl does undergo Friedel-Crafts acylation by acetyl chloride and aluminum bromide. Carrying out this reaction on $\eta^6$-toluenechromium tricarbonyl produces a 46:39:15 ratio of p:o:m isomers, in contrast to free toluene (92:8:0 p:o:m). Since the $\eta^6$-arenechromium tricarbonyl complexes are unstable to strong acids, electrophilic nitration and sulfonation cannot be carried out. Because of these features, electrophilic aromatic substitution reactions of these complexes have found little use in organic synthesis. However, complexation of an arene to chromium does alter its chemistry significantly in ways that have potential application to organic synthesis. Figure 14.1 summarizes the effects that have been utilized to date [7].

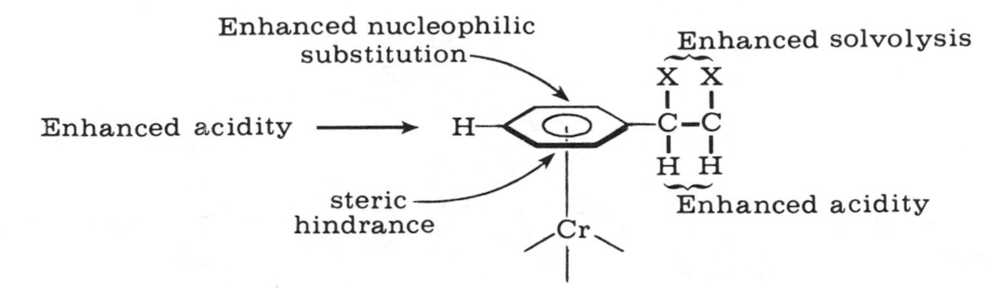


*Figure 14.1. Effects of complexation of aromatic compounds to chromium carbonyl.*

Although the arene ring in these complexes is unreactive toward electrophilic aromatic substitution, it is quite reactive toward nucleophilic aromatic substitution, and it is this reaction which is the most extensively studied and most used synthetically. Reactivity toward nucleophiles is expected because of the demonstrated electron-withdrawing capabilities of the $Cr(CO)_3$ group. However, as is often the case in organometallic reactions, several potential sites for attack exist. These include the metal and the carbonyl groups, in addition to the aromatic ring.

Immediately following the synthesis of $\eta^6$-chlorobenzenechromium tricarbonyl, its reactions with nucleophiles were probed. Reaction with sodium methoxide in methanol leads to production of the $\eta^6$-anisole complex in excellent yield (14.2). The rate of

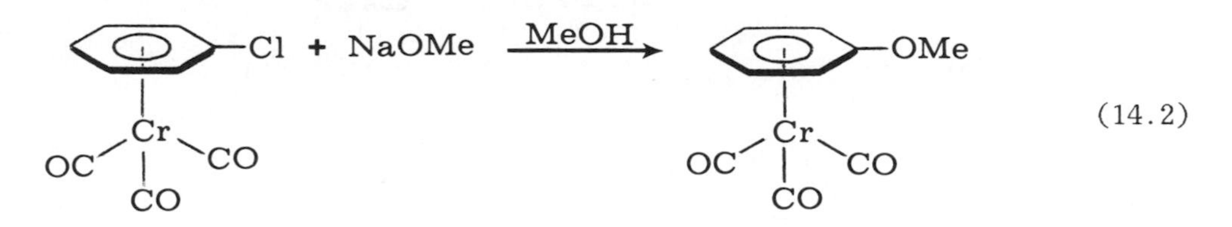


(14.2)

substitution of chloride by methoxide in this reaction approximates that of uncomplexed p-nitrochlorobenzene [8]. Later studies showed that sodium phenoxide and aniline react similarly, producing diphenylether and diphenylamine, respectively [9]. Surprisingly, Grignard reagents do not react with these complexes of chlorobenzene. $\eta^6$-Fluorobenzenechromium tricarbonyl reacts in a similar fashion, undergoing facile displacement of fluoride by a variety of nucleophiles, including alkoxide, amines, and cyanide [10]. Based on kinetic data, the reaction of this complex with amines is thought to proceed by a rapid formation of a $\eta^5$-cyclohexadienyl complex followed by rate-limiting loss of fluoride from the endo side of the ring [11] (Equation 14.3). It is

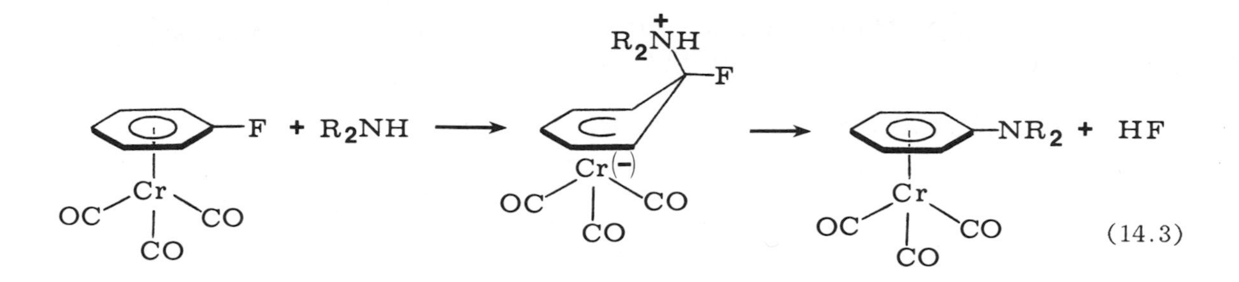


(14.3)

likely that this mechanism is common to all nucleophilic displacements of aromatic halides in these complexes.

Of greater interest for synthesis is the reaction of $\eta^6$-chlorobenzenechromium tricarbonyl complexes with carbanions, leading to direct alkylation of the aromatic ring. This area has been developed primarily by Semmelhack [12]. Indeed the chromium tricarbonyl complexes of chloro -and fluorobenzene react smoothly with highly stabilized carbanions, such as dimethyl malonate anion, and with highly reactive tertiary carbanions, such as isobutyronitrile anion, ethyl isobutyrate anion, protected cyanohydrin anions, and the dianion of isobutyric acid, to produce the substituted arene complexes in excellent yield. Oxidation by $I_2$ produces the free alkylated arene (14.4). Cyanide

$$(C_6H_5X)Cr(CO)_3 + R(-) \xrightarrow[15-20\ hr]{25^\circ} (C_6H_5R)Cr(CO)_3 \xrightarrow{I_2} C_6H_5R + CO + Cr(III) \tag{14.4}$$

$$R(-) = (-)CH(COOMe)_2,\ (-)C(Me)_2CN,\ CN(-),\ (-)C(Me)_2COOEt,\ PhC(-)(CN)(OR),\ (-)C(Me)_2COO(-)$$

ion itself also reacts in high yield, producing the $\eta^6$-benzonitrilechromium tricarbonyl complex, a complex not available by standard methods [7]. As might be expected, the fluorobenzene complex reacts faster than the chlorobenzene complex. That these alkylation reactions proceed by an addition-elimination mechanism (14.5a) rather than an

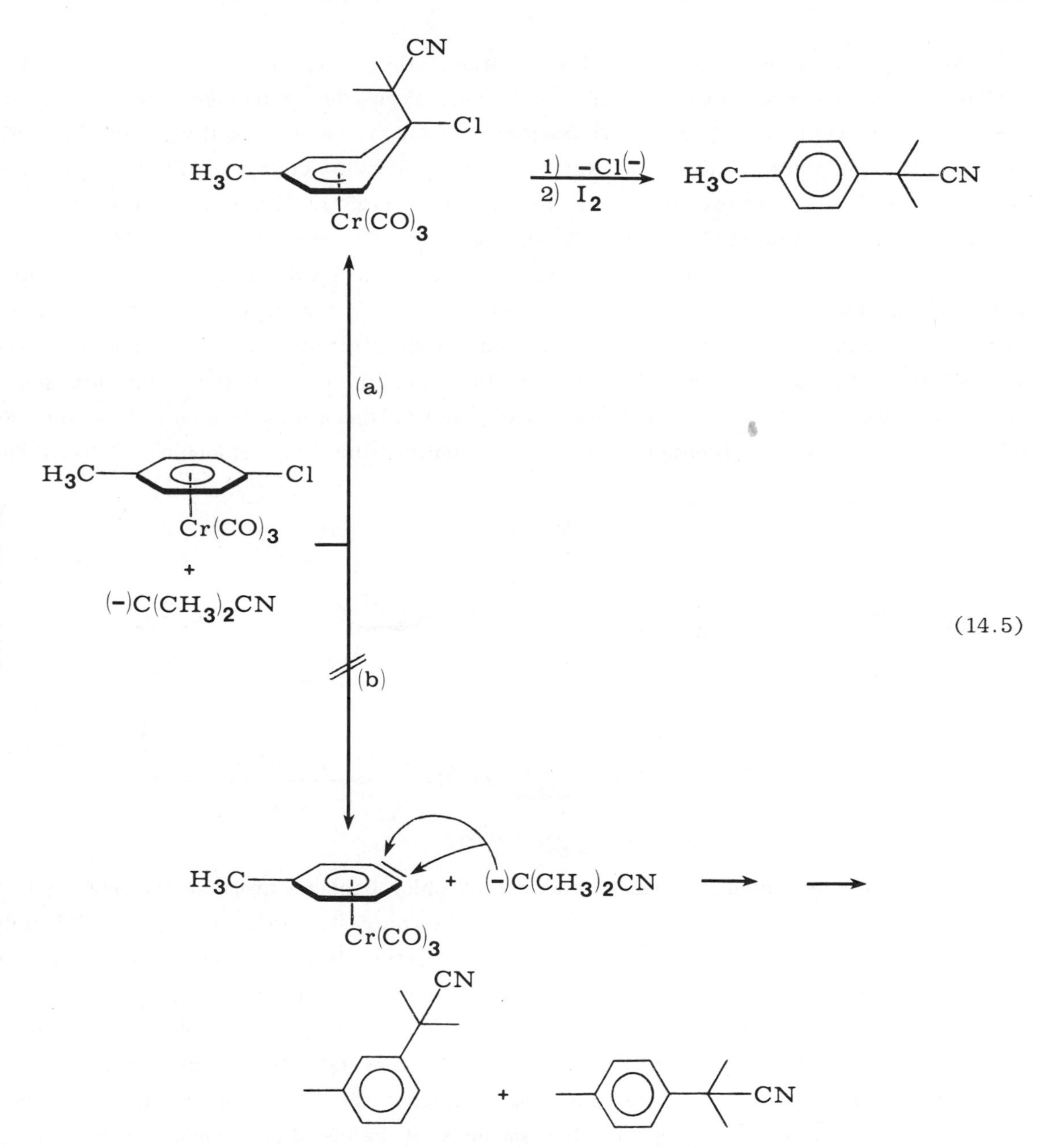


aryne mechanism (14.5b) is demonstrated by the observation that the reaction of $\eta^6$-p-chlorotoluenechromium tricarbonyl with isobutyronitrile anion leads exclusively to p-tolyl isobutyronitrile (after oxidative isolation) with none of the meta isomer detected.

Although this is quite a useful reaction, the small number of well-behaved carbanions is somewhat limiting. Two different types of carbanions fail to alkylate $\eta^6$-chlorobenzenechromium tricarbonyl complexes. Anions such as methyl, allyl, phenyl, and t-butylmagnesium halides, as well as lithium dimethylcuprate and aryl mercuric chlorides, fail to react at all at low temperatures (<25°). At higher temperatures, decomposition to unidentified materials ensues. In contrast, lithium anions such as $LiCH_2CO_2CH_3$, $LiC(CH_3)_2CO_2CH_3$, $LiCH_2CO_2(t\text{-}Bu)$, $LiCH_2CN$, $LiC{\equiv}CH$, and 2-lithio-1,3-dithiane react rapidly, even at -78°, but fail to give significant amounts of substitution for chloride even after extended periods at higher temperatures. Instead, these anions attack the aromatic ring o and m to the chlorine, producing $\eta^5$-cyclohexadienyl complexes, which cannot directly lose chloride, and furthermore are unable to rearrange to the $\eta^5$-cyclohexadienyl complex, which can irreversibly lose chloride (14.6). The

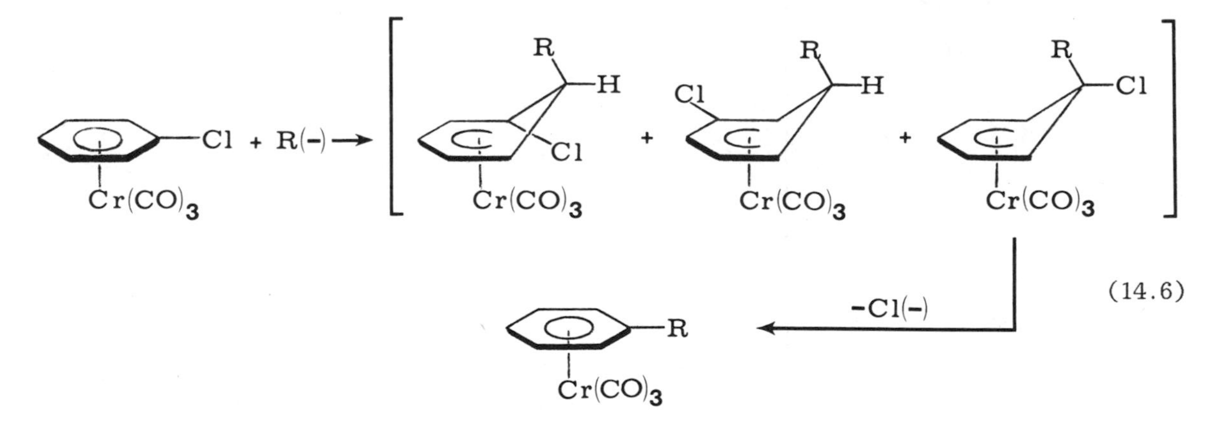


(14.6)

tertiary carbanions which do cleanly alkylate $\eta^6$-chlorobenzenechromium tricarbonyl lead to similar mixtures of initial complexes, which, in contrast, can rearrange to ultimately result in displacement of chloride. Evidence supporting this argument is presented in Equation 14.7. Treatment of $\eta^6$-chlorobenzenechromium tricarbonyl with the lithium anion of t-butyl acetate followed by rapid quenching with $I_2$ produces o and m alkylated chlorobenzenes exclusively. The ratio of products does not change appreciably as the time between reaction and quench increases. In contrast, the same reaction with isobutyronitrile anion produces appreciable amounts of halide displacement product even after a short reaction time, and the amount of this product increases with increased

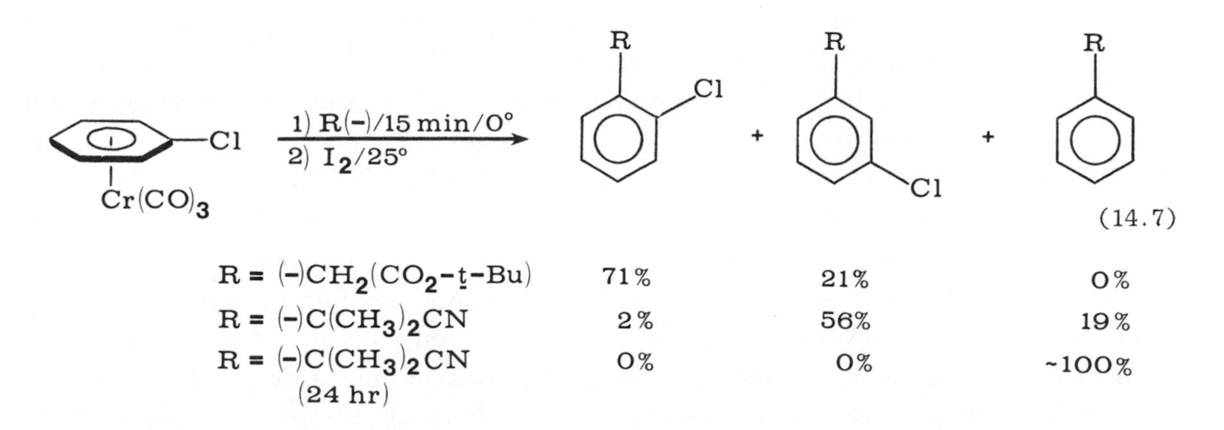


reaction time [12]. The importance of this observation far transcends its use to explain the behavior of various anions toward chlorobenzenechromium tricarbonyl, for it shows that direct alkylation of unfunctionalized aromatic rings should be possible.

Indeed it is. Treatment of $\eta^6$-benzenechromium tricarbonyl itself with a wide variety of reactive carbanions followed by oxidative removal of the metal by treatment with $I_2$ leads to the alkylated benzene in excellent yield. In contrast, reaction of the adduct with electrophiles which are potential hydride acceptors ($CH_3I$, benzophenone, trityl cation, $Et_3B$) regenerates the starting $\eta^6$-arene complex (14.8).

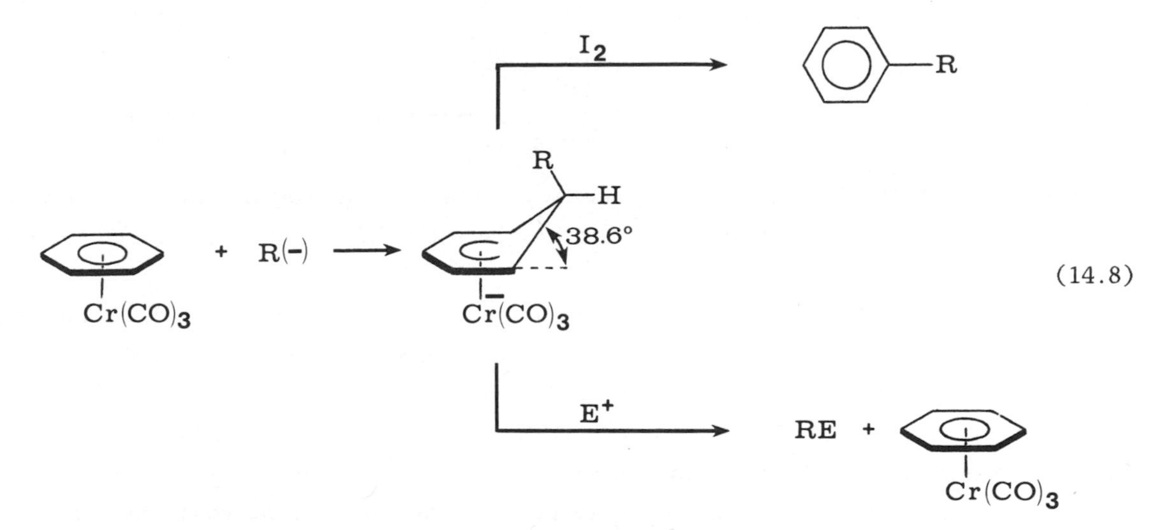


Semmelhack demonstrated that this reaction proceeds through the $\eta^5$-cyclohexadienyl complex by isolation and characterization of the complexes from the anions $^{-}CH_2CN$, $^{-}CH(SPh)_2$, and $^{-}C(CH_3)_2CN$. X-ray diffraction analysis of the crystalline adduct with 2-lithio-1,3-dithiane confirms the exo mode of attack by the carbanion, and shows that C-6 is displaced 38.6° from the plane of the remaining carbons of the cyclohexadienyl group [13,14]. Anions which react in high yield in this process include $LiC(CH_3)_2CN$, $LiCH_2CN$, $LiCH(S(CH_2)_3S)$, LiC(CN)(OR)R, $LiCH(SPh)_2$, $LiC(CH_3)_3$, p-tolyl Li, $LiCH_2CO_2R$, $LiCH(CH_3)CO_2R$, and $LiC(CH_3)_2CO_2R$. Thus direct acylation (via dithiane or protected cyanohydrin-masked acyl anions), arylation, and production of arylacetic acids from complexed benzene is possible by this method. Two types of anions fail to react in a useful fashion. Anions more stable than those of ester enolates (including ketone enolates) lack sufficient reactivity to attack the complexed arene ring at all, and no reaction ensues. Similarly, Grignard reagents, dialkyl cuprates, and alkylmercuric halides are unreactive. In contrast, strongly basic anions such as methyl-or butyllithium abstract an aryl proton to produce the lithiated aromatic, which yields iodobenzene upon oxidative isolation with $I_2$ (Equation 14.9). This is a

$$\text{Cr(CO)}_3 + n\text{-BuLi} \longrightarrow \left[ \text{—Li} \quad \text{Cr(CO)}_3 \right] + I_2 \qquad (14.9)$$

manifestation of the enhancement of acidity of the aryl hydrogens upon complexation, alluded to above. This lithiated complex has potential utility in organic synthesis and will be discussed in greater detail below.

Particularly interesting is the substitution pattern observed in the nucleophilic substitution reactions of substituted $\eta^6$-arenechromium tricarbonyl complexes (Table 14.1) [15,16]. The general reactivity follows that expected for nucleophilic aromatic substitution, namely, PhCl > $PhCH_3$ > $PhOCH_3$. The methoxy group has a very strong meta-directing effect. With $\eta^6$-anisolechromium tricarbonyl, no para alkylation is observed, and less than 5% of the ortho isomer is obtained. Arene complexes containing two methoxy groups, although less reactive, react with exceptionally high regioselectivity. Thus, the complex of 1,3-dimethoxybenzene gives exclusively 1,3,5-trisubstituted aromatics, and 1,4-dimethoxybenzene gives exclusively the 1,2,4-trisubstituted

*Table 14.1. Addition of carbanions to substituted $\eta^6$-arenechromium tricarbonyl complexes.*

A. Monosubstituted arene ligands

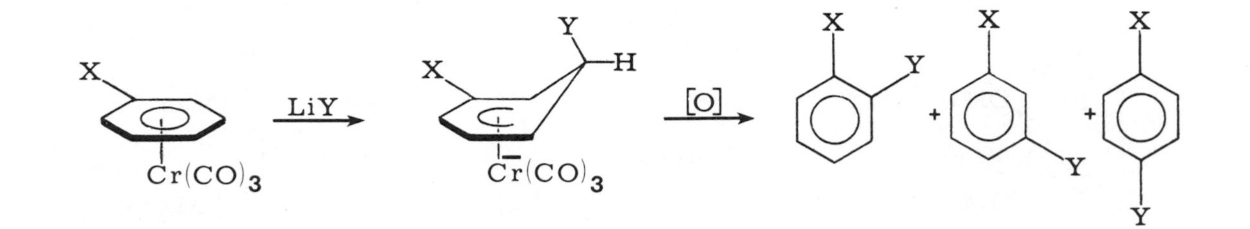


| Substituent (X) | Carbanion (LiY) | Product Ratios o:m:p (combined yield) |
|---|---|---|
| $CH_3$ | $LiCH_2CO_2R$ | 28:72:0 (89%) |
| $OCH_3$ | $LiCH_2CO_2R$ | 4:96:0 (93%) |
| Cl | $LiCH_2CO_2R$ | 54:45:1 (98%) |
| Cl | $LiCH(CH_3)CO_2R$ | 53:46:1 (88%) |
| Cl | $LiC(CH_3)_2CO_2R$ | 5:95:1 (84%) |
| Cl | $LiCH_2CO_2C(CH_3)_3$ | 70:24:0 (87%) |
| Cl | $LiC(CH_3)_2CN$ | 10:89:1 (84%) |
| Cl | Li-(1,3-dithianyl) | 46:53:1 (56%) |
| $Si(CH_3)_3$ | $LiC(CH_3)_2CN$ | 0:2:98 (65%) |
| $CF_3$ | $LiC(CN)(OR_1)CH_3$ | 0:30:70 (33%) |
| $N(CH_3)_2$ | $LiC(CH_3)_2CN$ | 1:99:0 (92%) |
| $C(CH_3)_3$ | $LiC(CN)(OR_1)CH_3$ | 0:35:65 (85%) |
| $CH_2CH_3$ | $LiC(CN)(OR_1)CH_3$ | 0:94:6 (89%) |

*Table 14.1. Continued.*

B. Polysubstituted arene ligands.

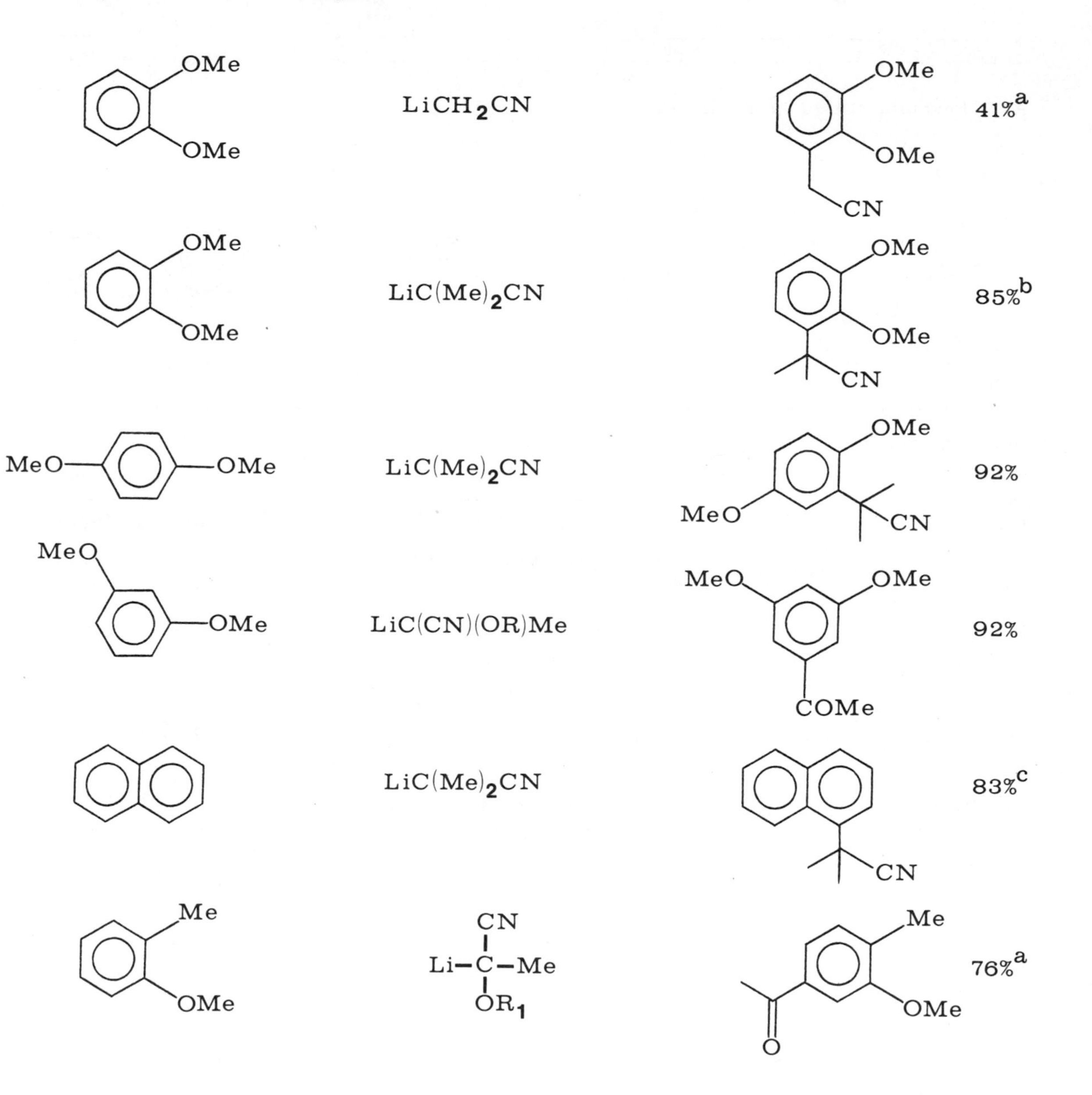

*Table 14.1. Continued.*

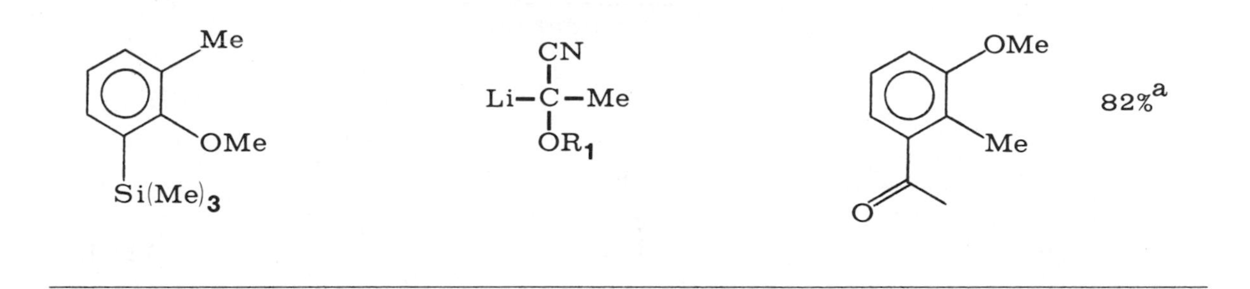


[a]No other isomers were detected.

[b]The 1,2,4-substitution product was obtained in 12% yield.

[c]The β-substituted isomer were detected in 1% yield.

aromatic (the only possible product in this case). Complexes of 1,2-dimethoxybenzene react with small carbanions to give, in addition, minor amounts of the 1,2,4-substituted product. Since these substitution patterns are opposite those obtained from traditional electrophilic aromatic chemistry, these reactions are quite useful for the synthesis of unusually substituted aromatic systems [16]. Similar preference for meta substitution is observed with $\eta^6$-toluenechromium tricarbonyl, although less strongly so. Meta substitution ranges from 43 to 92 percent, depending on carbanion, and ortho substitution makes up the remainder, except for some para substitution, always less than 2 percent. Chloride shows a similar directing effect, except for bulky anions for which meta substitution is very strongly preferred. In contrast, trimethylsilyl and trifluoromethyl favor para substitution.

Although the meta-directing effect of methoxy substituents is consistent with the expected reversal of the usual directing influence in electrophilic substitution, the directing effects of other substituents are less easily explained. As yet, no single theory will encompass all experimental data. However, many of the results are consistent with predictions derived from a consideration of the LUMO coefficients of the free arene ligand [15]. Finally, the indole ring system can be activated to undergo nucleophilic aromatic substitution after complexation to chromium [20] (Equation 14.10).

Intramolecular versions of this aromatic alkylation also lead to some interesting and useful chemistry. Although ester-stabilized carbanions do not react well intramolecularly, cyano-stabilized ones do. The product formed depends on chain length, reaction

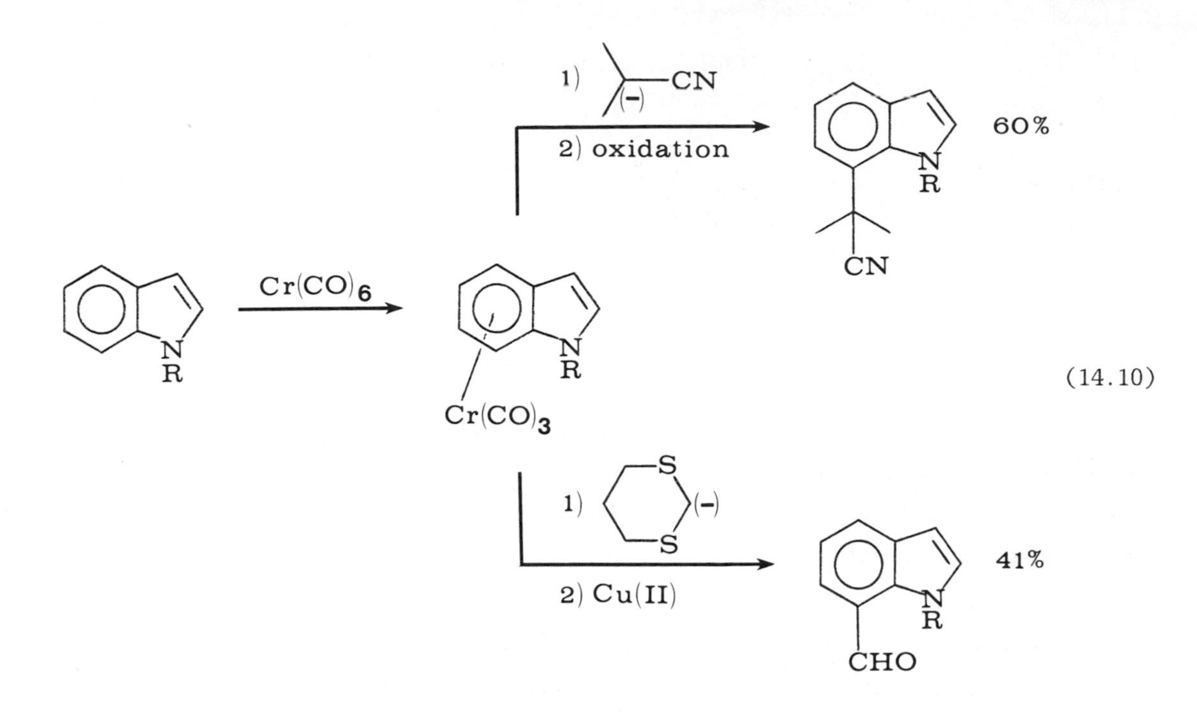


(14.10)

conditions, and other substituents on the aromatic ring. Thus the reaction of the 5-phenylvaleronitrile complex with lithium tetramethylpiperidide (to generate the anion) followed by oxidation with $I_2$ produces the 1-cyanotetrahydronaphthalene derivative in excellent yield (14.11). The next higher homolog, under identical reaction conditions,

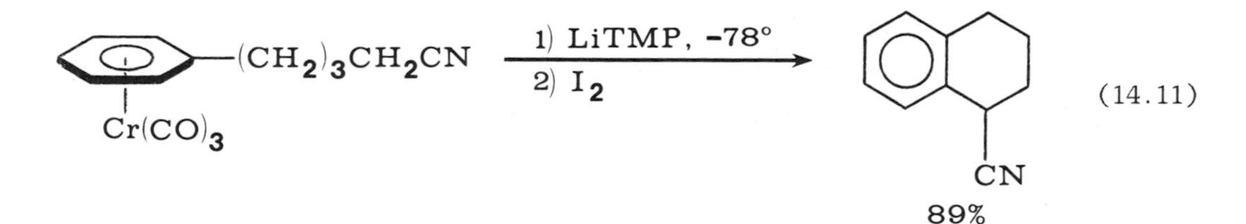


(14.11)

gives a mixture of 1,2 and 1,1 (spiro) fused products. Carrying out the same reaction at higher temperatures (0° rather than -78°) and quenching with trifluoroacetic acid prior to oxidation leads almost exclusively to the spiro-fused compound (14.12). The next lower homolog dimerizes to the metacyclophane rather than cyclizing to form the indane ring system [17] (Equation 14.13). Obviously, the reaction is quite sensitive to structural features of the substrate. This chemistry has been applied by Semmelhack to the total synthesis of acorenone B [18], shown in Figure 14.2.

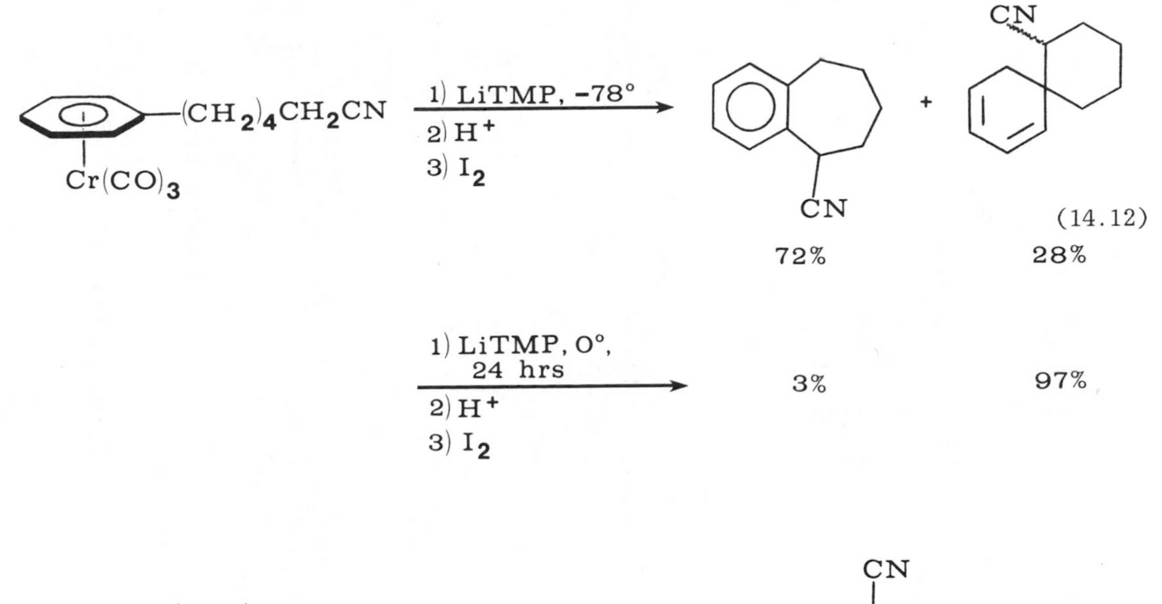


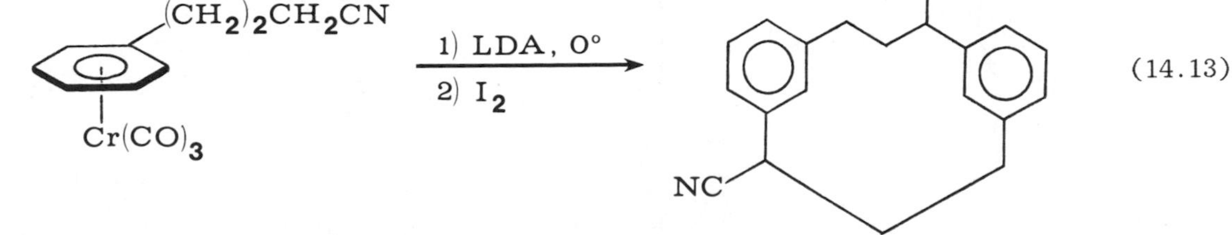


Two features of this synthesis deserve comment. The ring-closure step to form a spiro-ring fusion is at first surprising, since the unsubstituted aromatic with a five-carbon side chain closes exclusively to the fused ring tetralin system rather than to the corresponding spiro compound (Equation 14.11). However, in the acorenone synthesis, the aromatic ring is substituted. In the intermolecular alkylation of o-methyl anisole, the substituents direct the alkylation exclusively para to the methyl group and ortho to the methoxy group (Table 14.1). This same substitution pattern in the intramolecular version leading to acorenone B results in spiro ring formation (that is, alkylation para to the methyl group and meta to the methoxy group). The conversion of A to the cyclohexenone is also of interest. Oxidative cleavage of $\eta^5$-cyclohexadienylchromium complexes regenerates the aromatic compound. In contrast, acid cleavage of these complexes generates free cyclohexadienes. This is a general process, and with $\eta^6$-

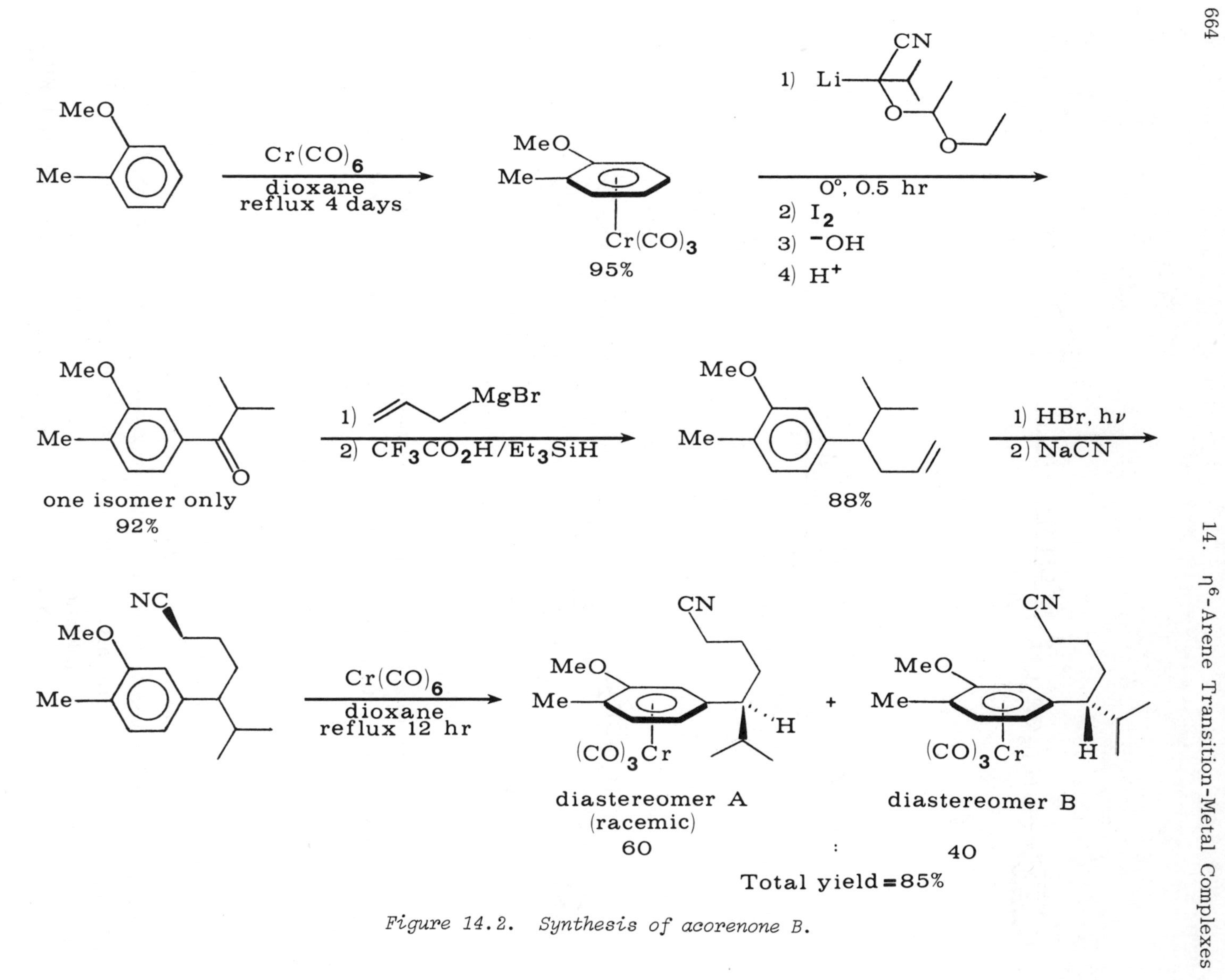


*Figure 14.2. Synthesis of acorenone B.*

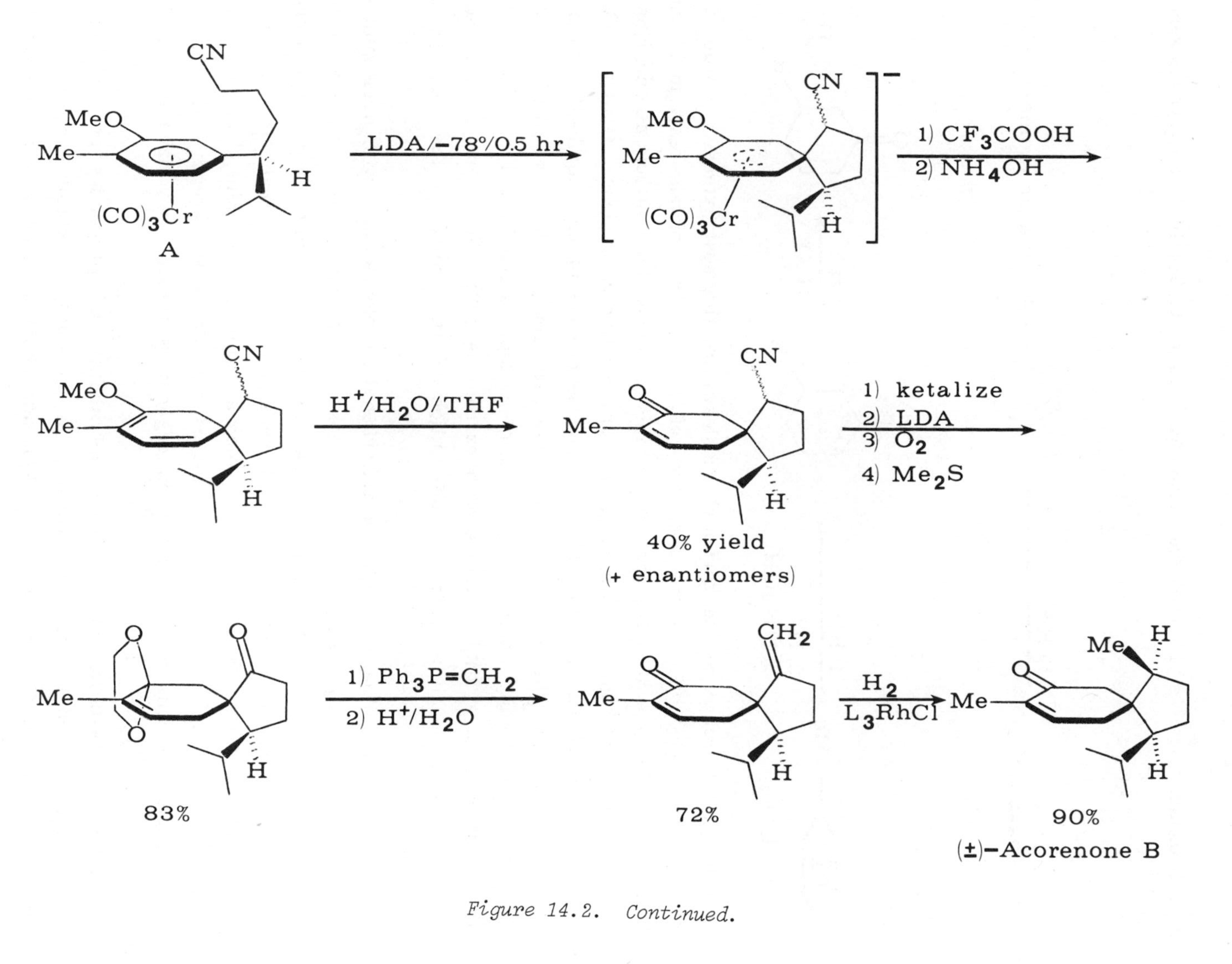


*Figure 14.2. Continued.*

anisolechromium tricarbonyl complexes, provides a route to 3-substituted cyclohexenones [19]:

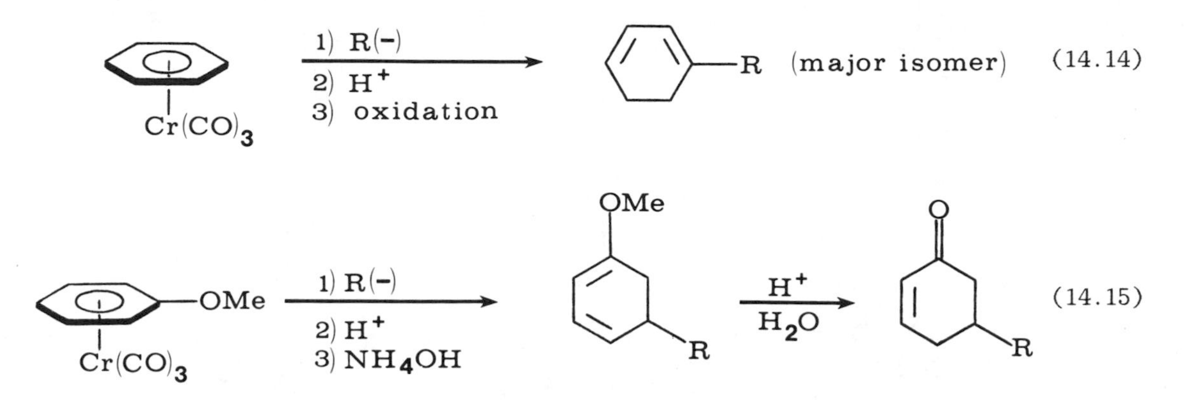


Although $\eta^6$-arenechromium carbonyl complexes are the most extensively studied for reactions with nucleophiles, arene complexes of several other metals react in a similar fashion. Bis-($\eta^6$-arene)iron(II) complexes undergo nucleophilic attack at one ring by stabilized carbanions, and at both rings by more reactive carbanions. Oxidative cleavage frees the alkylated aromatic (14.16). The severe conditions required to form the starting arene complex have prevented this reaction from use in more complex systems [21]. The more readily accessible $\eta^6$-arene-$\eta^5$-cyclopentadienyliron(I) complexes react at the arene ring in a similar fashion with carbanions [22]. A variety of heteroatoms effect classical nucleophilic aromatic substitution of the halide in the corresponding complex of chlorobenzene [23].

Finally, cationic $\eta^6$-arenemanganese tricarbonyl complexes react with nucleophiles in a variety of ways, depending on the nucleophile. Carbanions such as acetoacetate, diethyl malonate, phenyl-, and methyllithium, as well as cyanide, attack the aromatic ring [24,25,26]. In contrast, amines and methoxide attack a coordinated carbonyl to produce acyl complexes [26,27]. Ligands such as triphenylphosphine displace a carbon monoxide [26,28], whereas iodide or acetonitrile displaces the arene [28]. Although many of these reactions may be useful in organic synthesis, they have not yet been used.

Another manifestation of the "electron-withdrawing" properties of the $Cr(CO)_3$ group when complexed to aromatic compounds is its ability to stabilize negative charge at both the C-1 and C-2 positions of alkyl side chains. This was demonstrated by

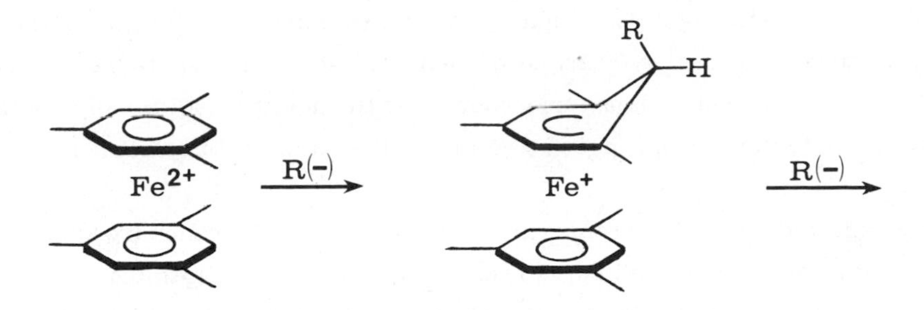


(14.16)

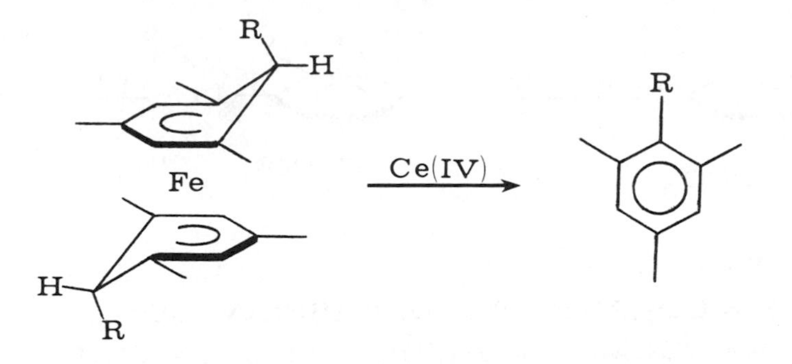


Jaouen by the facile double alkylation of complexed methyl phenyl acetate at the benzyl carbon under conditions where the free ligand does not react (14.17). This reaction

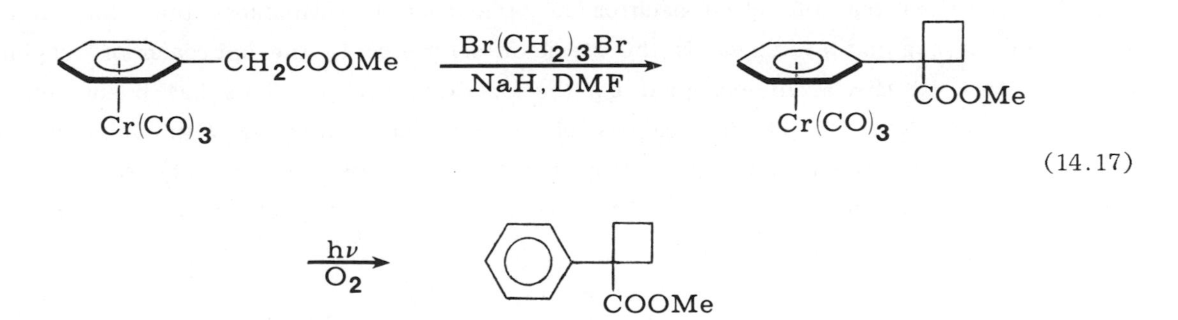


(14.17)

is particularly interesting, since dialkylation predominates even with methyl iodide under mild conditions, whereas replacement of one of the carbon monoxides with a triphenyl phosphite ligand (a better donor than CO) decreases the reactivity of the system and allows moderately specific monoalkylation of the benzylic position [29].

Decomplexation of the arene ligands is achieved by exposure to sunlight in the presence of air. An even more impressive example of benzylic activation in these systems is the reaction of $\eta^6$-ethylbenzenechromium tricarbonyl with methyl iodide and potassium t-butoxide/DMSO to produce isopropyl benzene (71%) and t-butylbenzene (8%) after decomplexation.

Semmelhack showed that the arene ring C-H bonds in $\eta^6$-arenechromium tricarbonyl complexes also have enhanced acidity. Treatment with n-butyllithium results in ring metallation, and the resulting lithio $\eta^6$-arenechromium tricarbonyl complexes react with a variety of electrophiles. With anisole, fluorobenzene, and chlorobenzene complexes, metallation always occurs at an ortho position [30] (Equation 14.18). Since this process

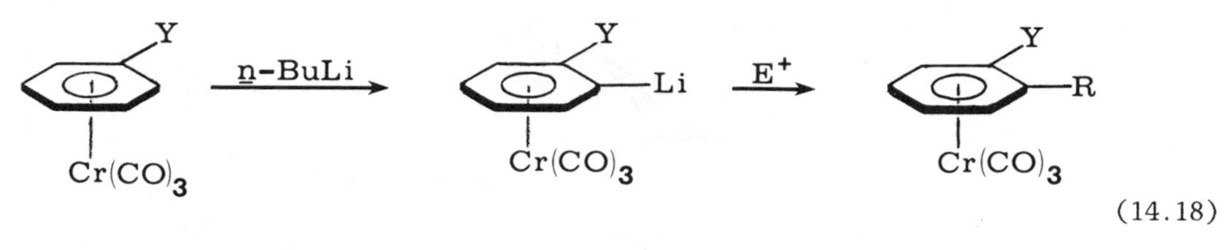


(14.18)

results in a stable $\eta^6$-arenechromium tricarbonyl complex, further functionalization by nucleophilic substitution is possible.

A final feature of $\eta^6$-arenechromium tricarbonyl complexes that has found application in organic synthesis is the steric interference to the approach of reagents from the face of the arene occupied by the $Cr(CO)_3$ group. This has been used in several ways. Reaction of the complex of o-methoxyacetophenone with Grignard reagents results in predominant attack from the face opposite the $Cr(CO)_3$ group. The steric efficiency of the reaction (14.19) depends on the nature of the anion, but specificity is generally rather high [31].

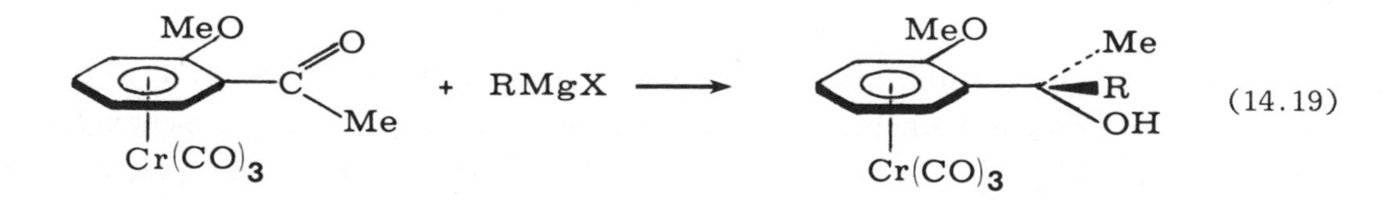


(14.19)

Even higher stereospecificity was obtain by Jaouen in cyclic systems (Figure 14.3). Racemic 1-indanol complexes to chromium hexacarbonyl exclusively on the same face as

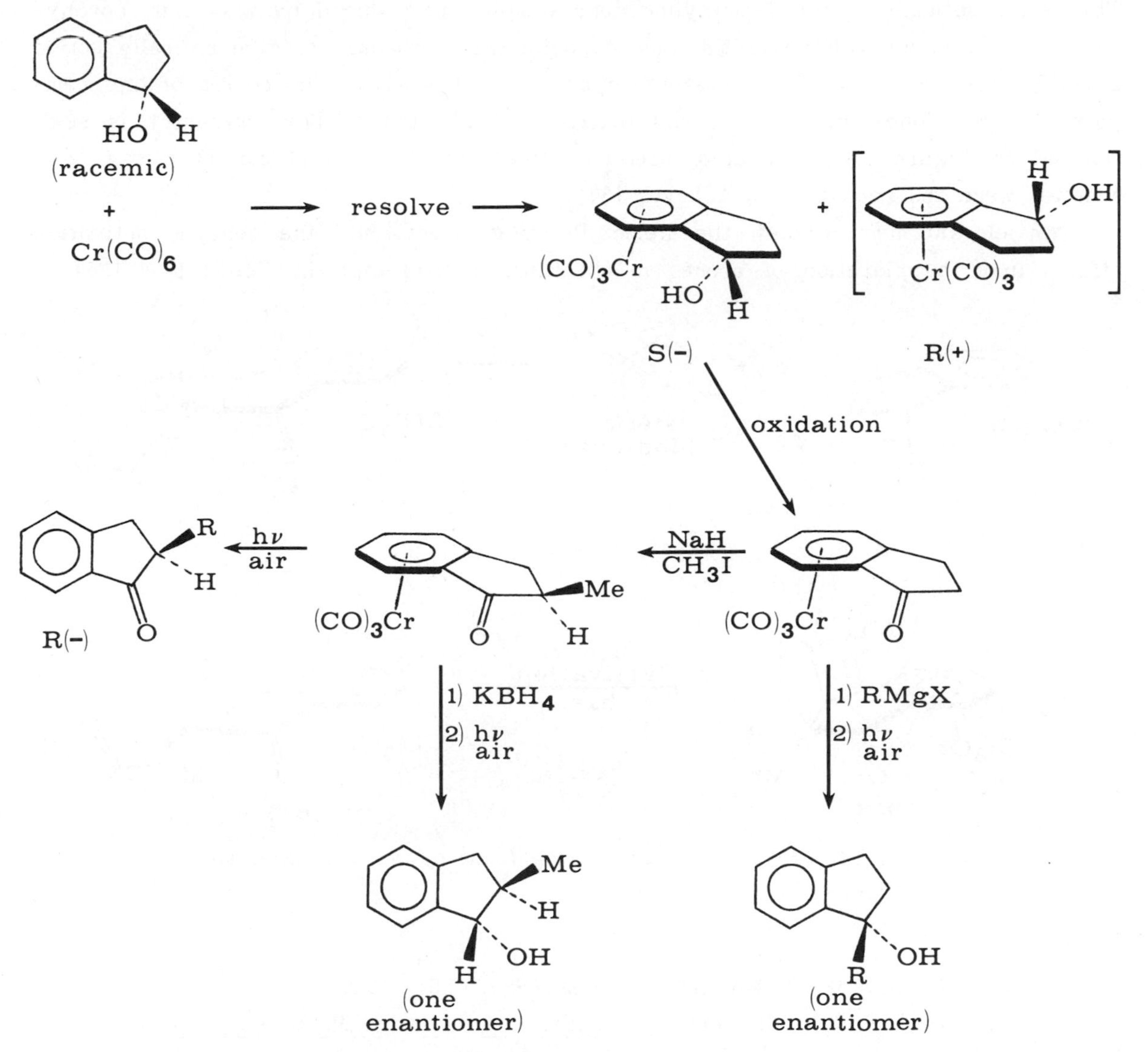


*Figure 14.3. Steric effects of complexation of aromatic compounds to chromium carbonyl.*

the hydroxyl group, and the resulting arenechromium tricarbonyl complexes can be resolved. Oxidation of <u>either</u> enantiomer produces an indanone complex which is optically active solely by virtue of its complexation of a specific face to chromium. Alkylation of this optically active indanone complex with sodium hydride/methyl iodide occurs exclusively from the face opposite the chromium to produce the optically active 2-methylindanone complex, from which the ligand is freed by exposure to sunlight in air. This same optically active 2-methylindanone complex is reduced by potassium borohydride, again exclusively from the face opposite the chromium, to give optically active 2-methyl-1-indanol. Grignard reagents react similarly [32]. The corresponding complexes of tetralone can be made and behave similarly [33]. This chemistry is summarized in Figure 14.3. Related directive effects in the α-alkylation of methyl aryl acetates were observed by des Abbayes [34].

An elegant use of both the steric directing effects and the benzylic activation effects upon complexation of arenes to chromium is presented in Figure 14.4 [35].

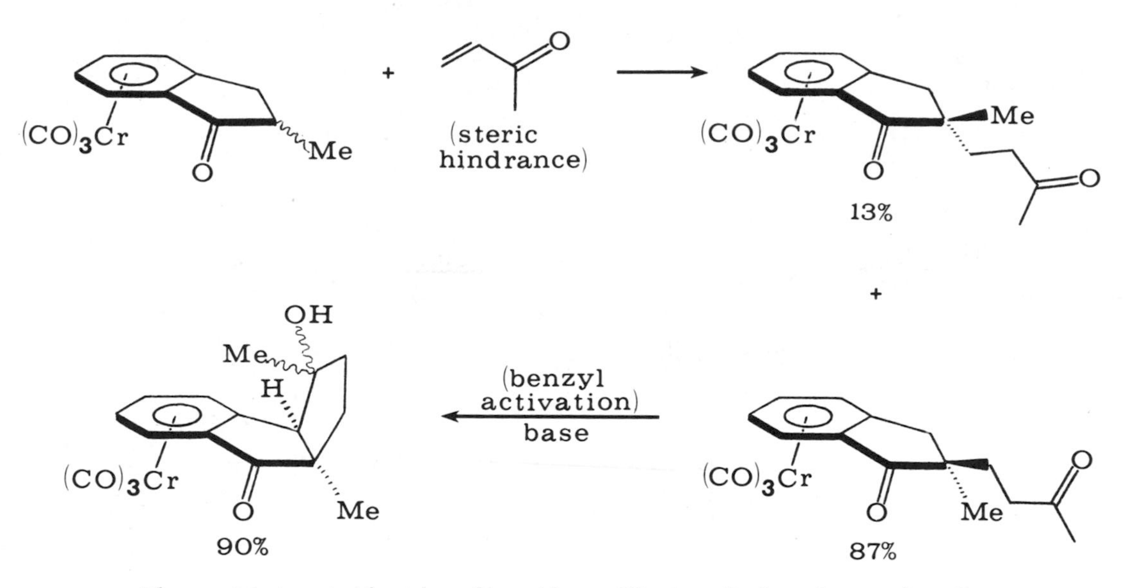


*Figure 14.4. Activation-direction effects of chromium carbonyl.*

## NOTES.

1. B. Nicholls, and M. C. Whiting, J. Chem. Soc., 551 (1959).
2. G. A. Moser, and M. D. Rausch, Synth. React. Inorg. Metal. Org. Chem., 4, 38 (1974).
3. M. D. Rausch, J. Org. Chem., 39, 1787 (1974).
4. W. E. Silverthorn, Adv. Organomet. Chem., 13, 47 (1975).
5. R. J. Card, and W. S. Trayhanovsky, Tetrahedron Lett., 3823 (1973).
6. E. W. Randall, and L. E. Sutton, Proc. Chem. Soc., 93 (1959).

7. This figure is adapted from an excellent review on the chemistry of arenemetal tricarbonyl complexes by M. F. Semmelhack, J. Organomet. Chem. Libr., 1, 361 (1976).
8. D. Brown, and J. R. Raju, J. Chem. Soc. A, 40 and 1617 (1966).
9. S. J. Rosca, and S. Rosca, Rev. Chim., 25, 461 (1974).
10. C. A. L. Mahaffy, and P. L. Pauson, J. Chem. Res., 128 (1979).
11. J. F. Bunnett, and H. Hermann, J. Org. Chem., 36, 4081 (1971).
12. M. F. Semmelhack, and H. T. Hall, Jr., J. Amer. Chem. Soc., 96, 7091 and 7092 (1974).
13. M. F. Semmelhack, H. T. Hall, Jr., R. Farina, M. Yoshifuji, G. Clark, T. Bargar, K. Hirotsu, and J. Clardy, J. Amer. Chem. Soc., 101, 3535 (1979).
14. M. F. Semmelhack, H. T. Hall, Jr., M. Yoshifuji, and G. Clark, J. Amer. Chem. Soc., 97, 1247 (1975).
15. M. F. Semmelhack, G. R. Clark, R. Farina, and M. Saeman, J. Amer. Chem. Soc., 101, 217 (1979).
16. M. F. Semmelhack, and G. Clark, J. Amer. Chem. Soc., 99, 1675 (1977).
17. M. F. Semmelhack, V. Thebtaranonth, and L. Keller, J. Amer. Chem. Soc., 99, 959 (1977).
18. M. F. Semmelhack and A. Yamashita, private communication.
19. M. F. Semmelhack, J. J. Harrison, and Y. Thebtaranoth, J. Org. Chem., 44, 3275 (1979).
20. A. P. Kozikowski and K. Isobe, J.C.S. Chem. Comm., 1076 (1978).
21. J. F. Helling and G. G. Cash, J. Organomet. Chem., 73, C10 (1974).
22. I. U. Khand, P. L. Pauson, and W. E. Watts, J. Chem. Soc. C, 2024 (1969).
23. A. N. Nesmeyanov, N. A. Volkenau, and I. N. Bolesova, Dokl. Akad. Nauk SSSR, 175, 606 (1967); 183, 834 (1968).
24. D. Jones and G. Wilkinson, J. Chem. Soc., 2479 (1964).
25. P. L. Pauson and J. A. Segal, J. Chem. Soc. Dalton, 1683 (1975).
26. P. J. C. Walker and R. J. Mawby, Inorg. Chim. Acta, 7, 621 (1973).
27. R. J. Angelici and L. J. Blacik, Inorg. Chem., 11, 1754 (1972).
28. P. J. C. Walker and R. J. Mawby, J. Chem. Soc. Dalton, 622 (1973).
29. G. Jaouen, A. Meyer, and G. Simonneaux, J.C.S. Chem. Comm., 813 (1975).
30. M. F. Semmelhack, J. Bisaha, and M. Czarny, J. Amer. Chem. Soc., 101, 768 (1979).
31. J. Besancon, J. Tirouflet, A. Card, and Y. Dusausoy, J. Organomet. Chem., 59, 267 (1973).

32. A. Meyer and G. Jaouen, J.C.S. Chem. Comm., 787 (1974).
33. G. Jaouen and A. Meyer, J. Amer. Chem. Soc., 97, 4667 (1975).
34. H. Abbayes and M. A. Boudeville, Tetrahedron Lett., 2137 (1976); J. Org. Chem., 42, 4104 (1977).
35. G. Jaouen and A. Meyer, Tetrahedron Lett., 3547 (1976).

# 15

# $\eta^3$-Allyl Transition-Metal Complexes

The $\eta^3$-allyl ligand, formally a four-electron donor occupying two coordination sites, is among the most common of carbon ligands, forming at least moderately stable complexes with virtually all the transition-metal series. The $\eta^3$-allyl group is normally disposed symmetrically to the transition metal, with equal carbon-to-carbon bond lengths. Proton NMR confirms the symmetrical disposition of $\eta^3$-allyl group, having three sets of signals with an intensity ratio of 1:2:2 ($H_a$:$H_b$:$H_c$). The "internal" ($H_c$) protons appear at highest field because of shielding by the metal (Figure 15.1).

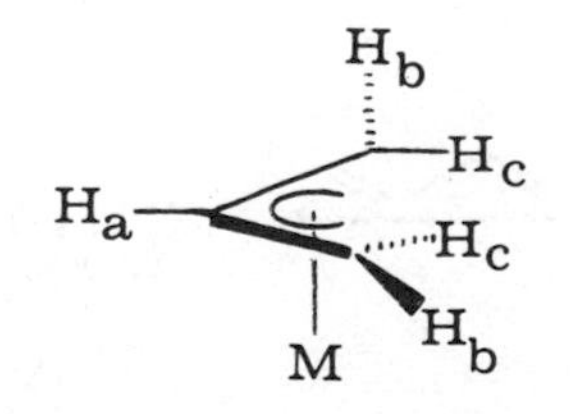


*Figure 15.1. $\eta^3$-Allylmetal complexes.*

"Pure" $\eta^3$-allylmetal complexes (complexes containing no ligands other than $\eta^3$-allyl groups) are known for Ni, Pd, Pt, Co, Fe, Cr, Mo, W, V, Nb, Ta, Ti, Zr, and Th, and are made by the reaction (15.1) of allylmagnesium bromide with the appropriate

$$\text{n allylMgX} + \text{MX}_n \longrightarrow (\eta^3\text{-allyl})_n\text{M} + \text{n MgX}_2 \quad (15.1)$$

transition-metal halide (Wilke). The resulting complexes are purified by sublimation, or by crystallization from hydrocarbon solvents at low temperatures with strict exclusion of air and moisture, since these complexes are often pyrophoric, and easily hydrolyzed [1]. $\eta^3$-Allylmetal complexes containing additional, nonallyl ligands are also very common for most transition metals, and are often formed directly from the corresponding olefin, either by abstraction of an allylic proton by base or by insertion of the transition metal into the allylic C-H bond.

Of the simple $\eta^3$-allylmetal complexes, (bis-$\eta^3$-allyl)nickel is of both practical and historical significance. This complex, as well as a variety of other nickel(0) complexes, catalyzes the cyclotrimerization of butadiene to cyclododecatriene. This reaction was one of the first organometallic reactions subjected to careful mechanistic studies, primarily by Wilke [2]. The reaction is thought to proceed as shown in Figure 15.2.

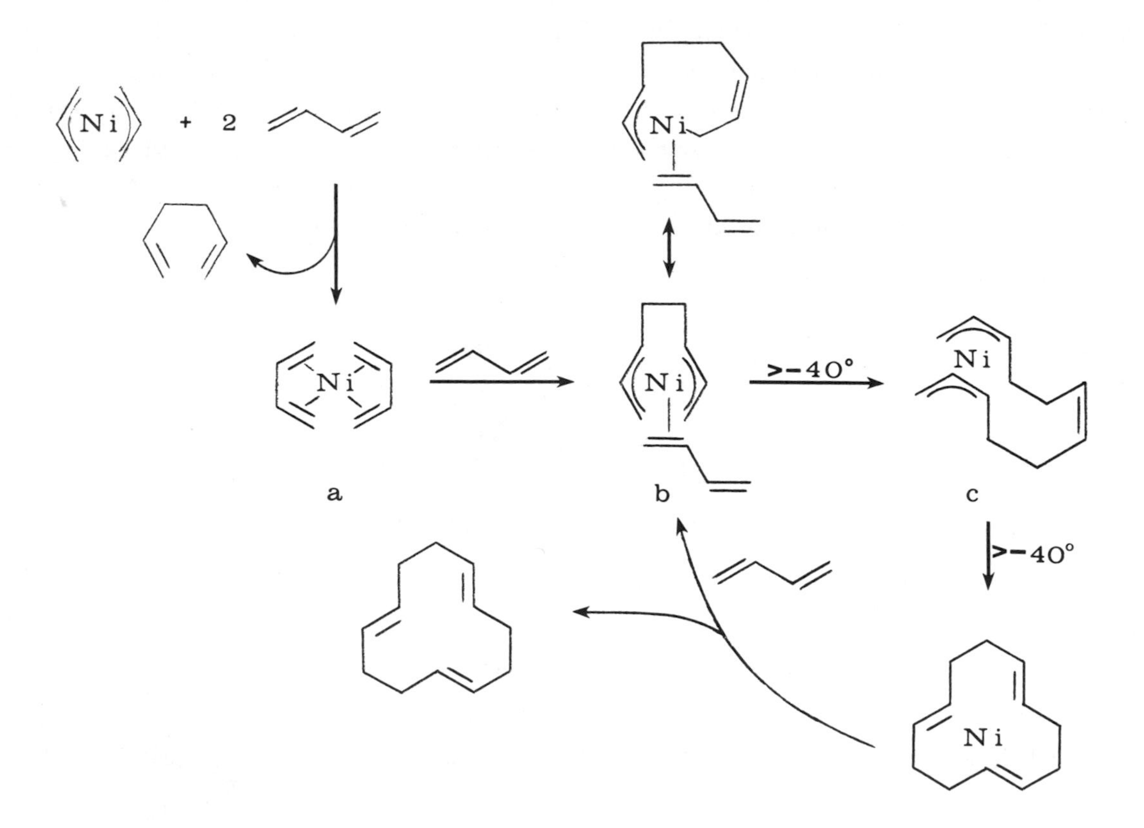


*Figure 15.2. Ni(0) cyclotrimerization of butadiene.*

Complex c has been isolated and characterized, and shown to enter the catalytic cycle as proposed. Phosphine-stabilized analogues of b have also been isolated, and shown to be important in cyclodimerization reactions.

Complex c has a rich chemistry in its own right [3], as shown in Figure 15.3. By carrying out the process shown in Figure 15.3 in the presence of good ligands for

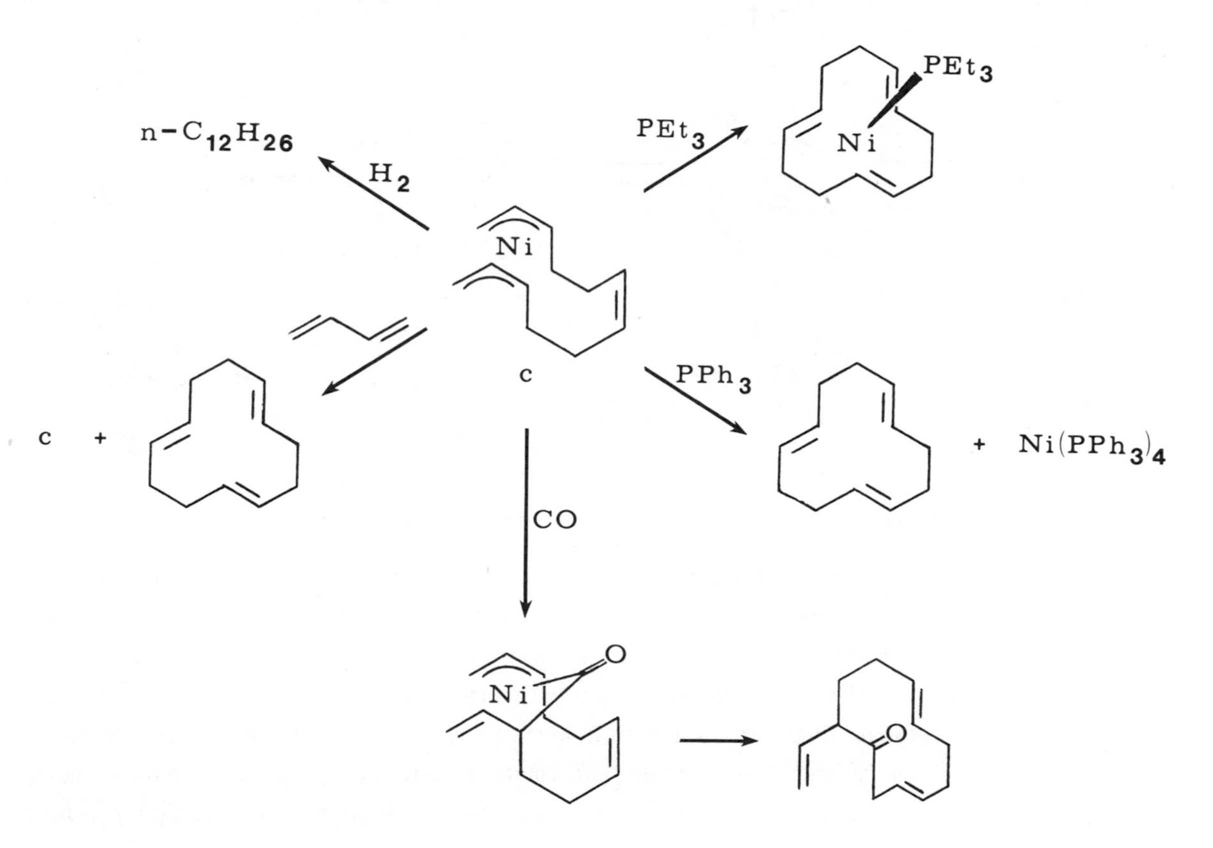


*Figure 15.3. Reactions of the cyclotrimerization intermediate Ni complex.*

nickel(0) complexes, such as phosphines or phosphites, cyclodimerization of butadiene is observed [2] (Figure 15.4). The specific cyclodimer obtained depends upon the ligand and the ligand-to-metal ratio.

In fact, the general reaction of butadiene with nickel(0) catalysts in the presence of various ligands is an exceedingly complex reaction, in which many mobile equilibria are concurrently operating. It is remarkable that it has been possible to control this reaction to give one major product. Recently a new and potentially powerful method for

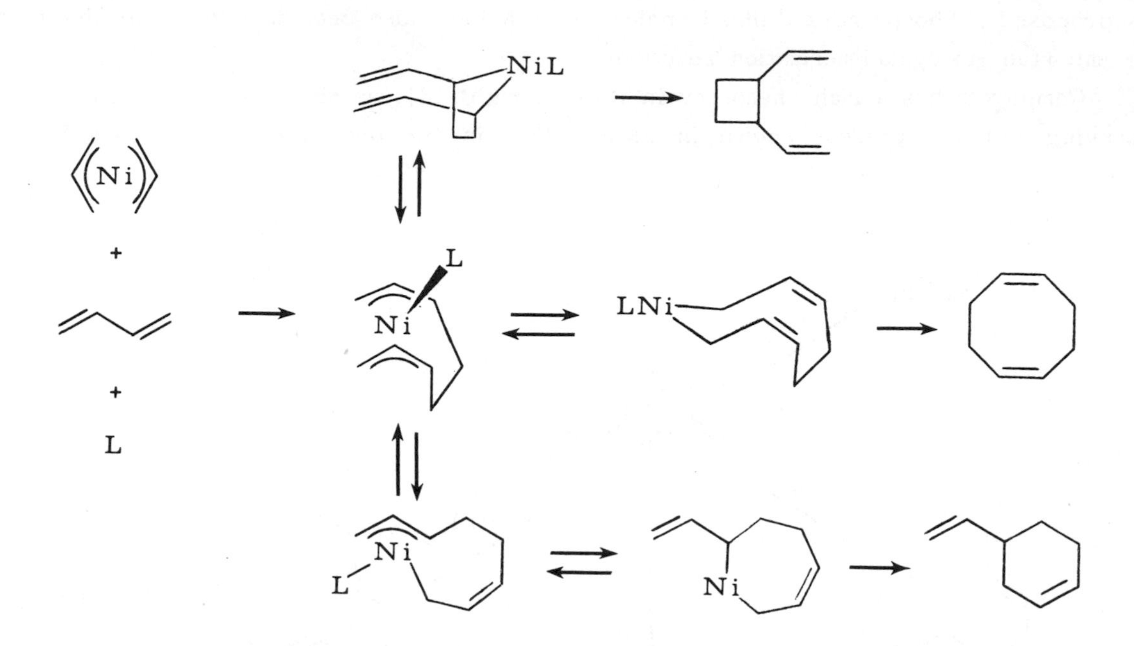


*Figure 15.4. Ni(0) cyclodimerization of butadiene.*

the study and optimization of complex reactions that respond to added ligands with a change in product distribution has been developed by Heimbach [4]. It is called the method of inverse titration. The ratio of the concentrations of the added ligand to the metal complex ($\log[L]_0/[Ni]_0$) is varied logarithmically from -5 to +2, and the product distribution across these changes is plotted, producing "product-ligand concentration-control maps." From these maps, much information can be gleaned; perhaps most importantly, one can optimize the reaction for production of a particular desired product with a minimum number of experiments.

Figure 15.5 shows two such product-ligand concentration-control maps, for the reactions described in Figures 15.3 and 15.4. They dramatically illustrate the sensitivity of the product distribution to both the nature and the concentration of the added ligand. They also allow one to readily choose the optimum conditions for the production of the desired product.

Under different conditions the cooligomerization of butadiene with ethylene to form cyclodecadienes, and with alkynes to produce cyclodecatrienes, proceeds by a similar reaction path. (Bis-η$^3$-allyl)palladium complexes behave somewhat differently, reacting

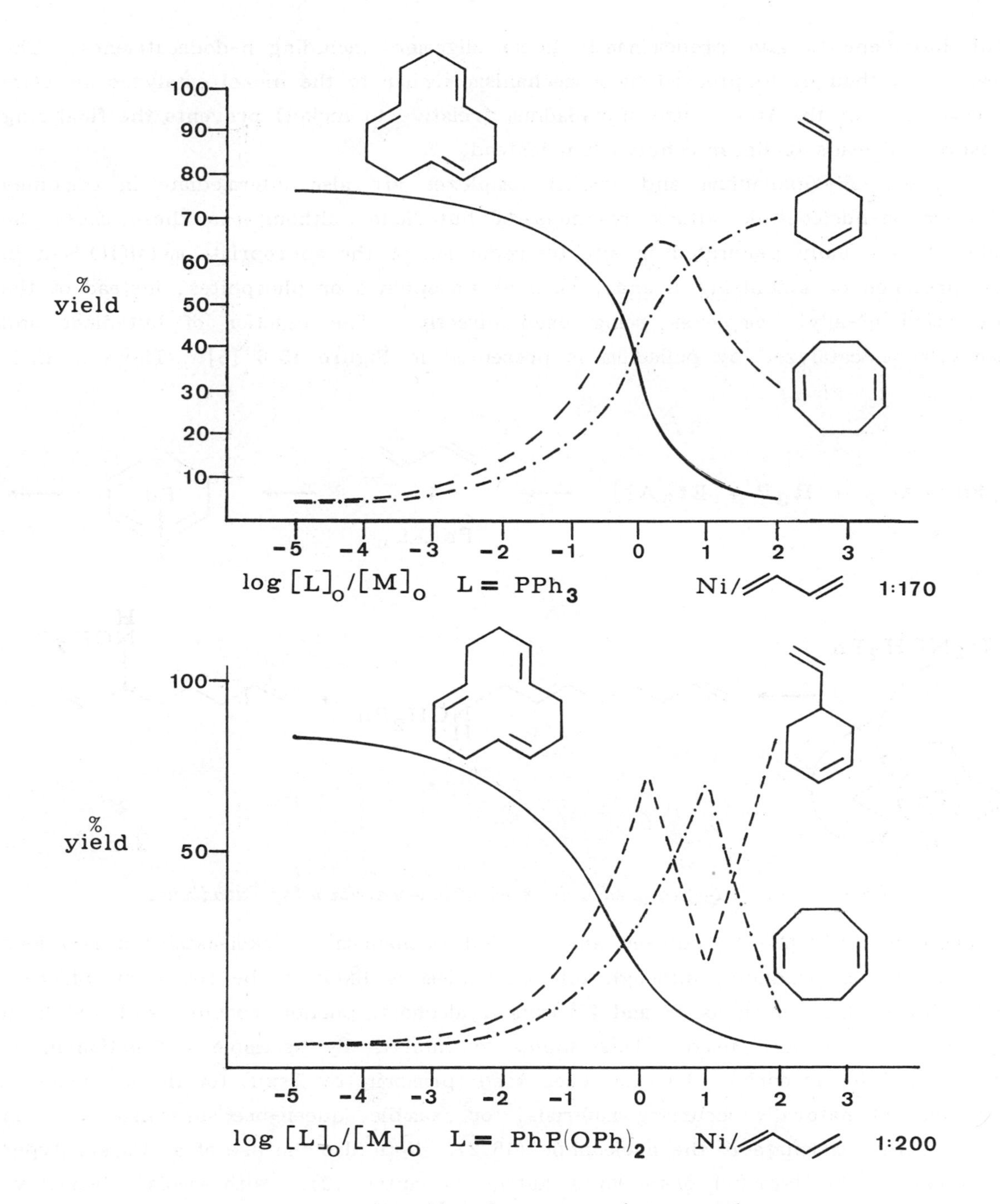


*Figure 15.5. Product-ligand concentration-control maps.*

with butadiene to give predominantly linear oligomers including n-dodecatrienes. The reaction is thought to proceed by a mechanism similar to the nickel catalyzed cyclotrimerization, but the larger size of palladium (relative to nickel) prevents the final ring closure and leads to linear condensation instead.

(Bis-$\eta^3$-allyl)palladium and -nickel complexes are also intermediate in combined dimerization-nucleophilic attack reactions of butadiene, although in these cases the catalyst is usually generated in situ by reduction of the appropriate metal(II) salt in the presence of stabilizing ligands, such as phosphines or phosphites, instead of the preformed $\eta^3$-allyl complexes being used directly. The reaction of butadiene and benzylamine catalyzed by palladium is presented in Figure 15.6 [5]. There is little

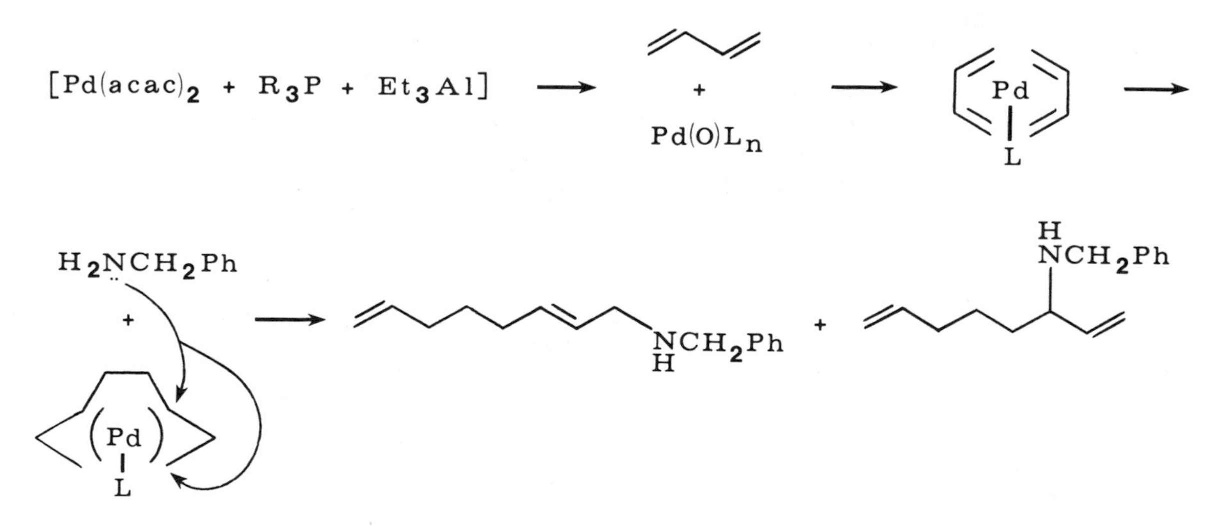


*Figure 15.6. Pd-catalyzed oligomerization-amination of butadiene.*

experimental evidence to support any detailed mechanism. Nickel-catalyzed reactions give analogous products, although this mechanism is likely to be somewhat different [6]. Nucleophiles such as 1° and 2° amines, alcohols, phenol, acetate, and stabilized carbanions have been used. This ability to catalytically assemble a functionalized, unsaturated linear carbon chain has been used (primarily by Tsuji) for the synthesis of a variety of naturally occurring materials, for example, queen-bee substance [7] , in which diethyl malonate is the nucleophile (15.2). Note also the use of a "Wacker type" oxidation of the terminal olefin to a ketone (Chapter 12). With similar chemistry, long-chain $\alpha$-aminoacid derivatives are available in high yield [8].

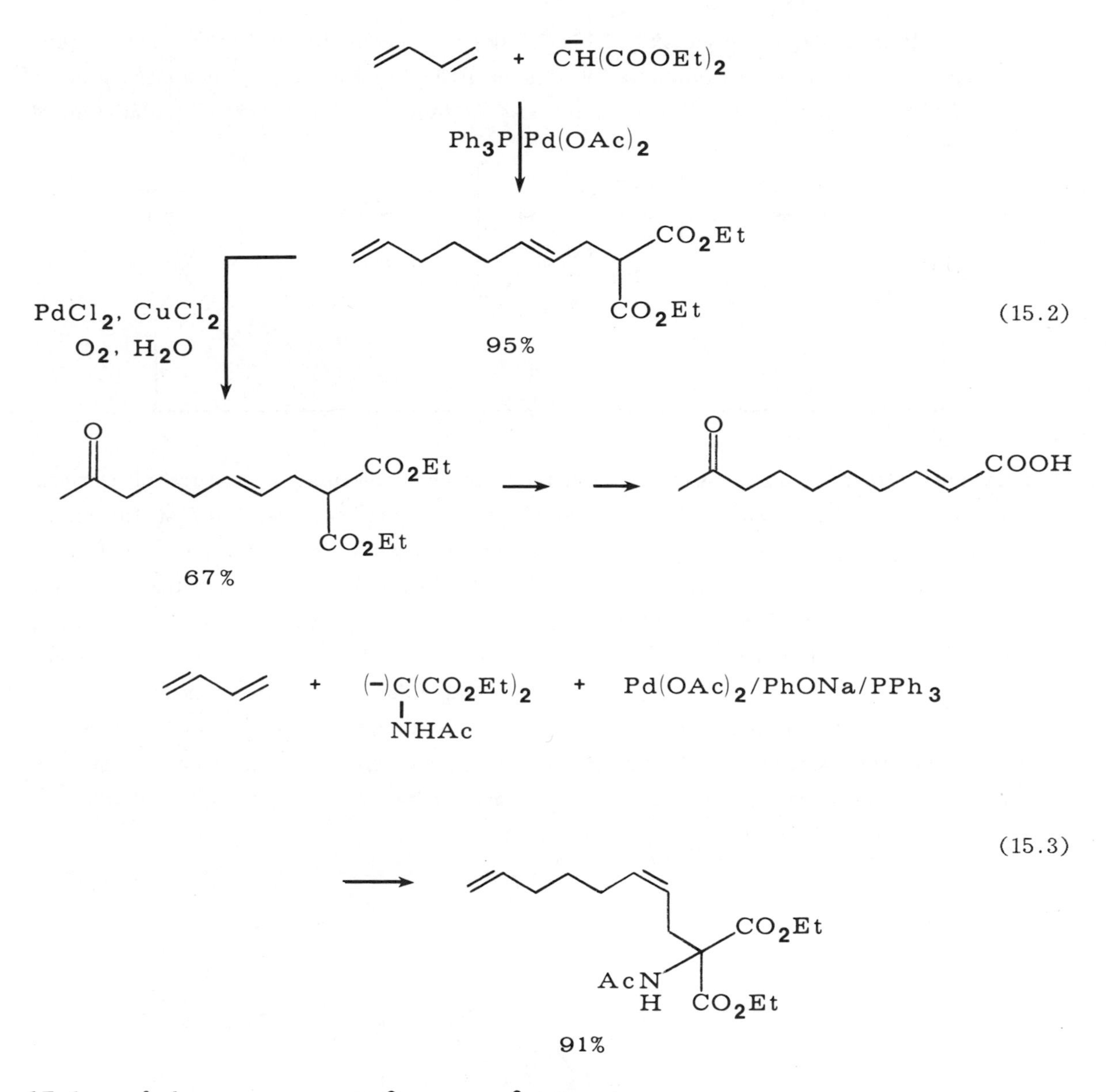


## 15.1. η³-ALLYLPALLADIUM CHLORIDE COMPLEXES.

η³-Allylpalladium chloride complexes are considerably more stable than bis-η³-allylpalladium complexes, and are readily prepared directly from olefins by treatment with palladium chloride and sodium acetate in glacial acetic acid [9]. (Many other conditions work as well.) The reaction involves ultimate abstraction of an allylic proton from the $\pi$-olefin palladium complex. This is likely to occur by insertion of the metal into a syn allylic C-H bond, followed by loss of HCl, but direct proton abstraction by base cannot be ruled out (15.4). This process is quite similar to ortho palladation of

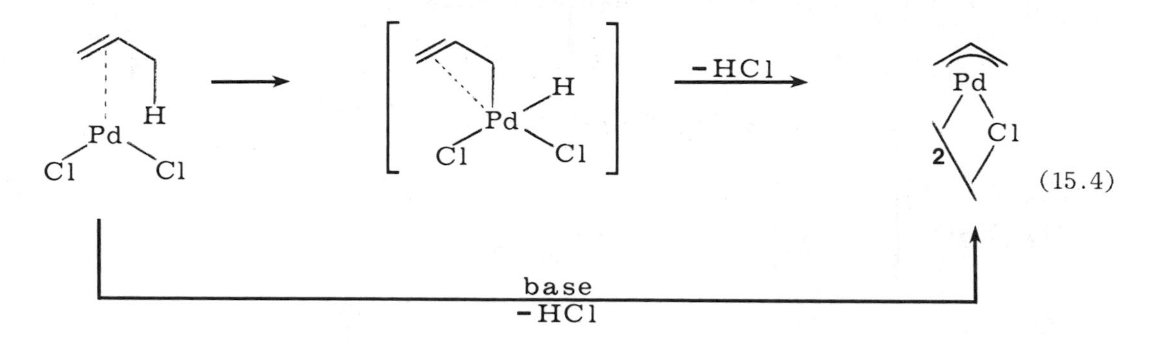


(15.4)

benzylamines (Chapter 11). With unsymmetrical internal olefins, a mixture of isomeric $\eta^3$-allylpalladium halide complexes is obtained, with the general reactivity of the allylic positions being $CH_3 > CH_2 >> CH$, but dependent somewhat on conditions. Olefins in conjugation with carbonyl groups form $\eta^3$-allylpalladium halide complexes less readily, probably because they coordinate only weakly to Pd(II). With alkyl-substituted olefins, the order of ease of formation of $\eta^3$-allyl complexes appears to be tri- > di- > monosubstituted olefins. Again, this order of reactivity depends very much on reaction conditions, and is only a trend, not a rule. The stereochemistry of the olefin does not determine the stereochemistry of the $\eta^3$-allyl complex. For example, both isomers of 4-methyl-3-heptene are converted to the same mixture of (syn)$\eta^3$-allylpalladium complexes (15.5).

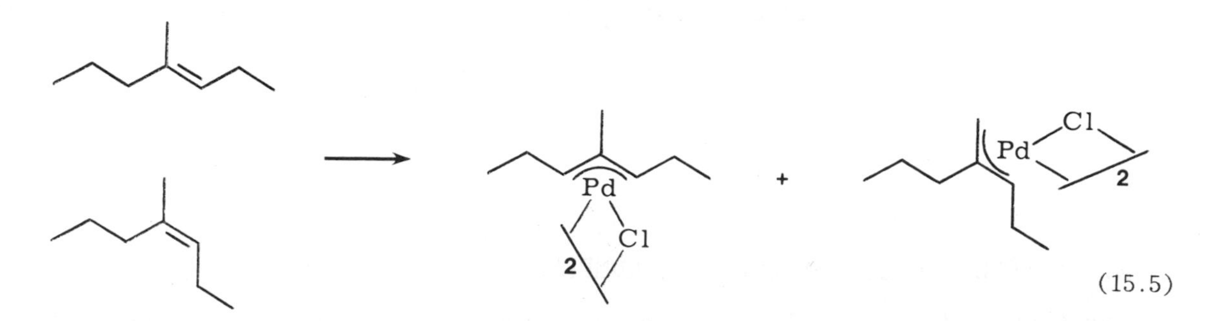


(15.5)

$\eta^3$-Allylpalladium chloride complexes are quite stable, yellow crystalline solids, easily handled and stored in air, and rather inert to a variety of chemical reagents.

Perhaps because of this stability they had found little use in organic synthesis until it was observed that, in the presence of strongly coordinating ligands such as phosphines or DMSO, nucleophiles—particularly stabilized carbanions [10] and enamines [11]—would attack, resulting in allylic alkylation (15.6). (In the absence of excess ligand, no re-

$$\left[(\eta^3\text{-}C_3H_5)Pd(\mu\text{-}Cl)\right]_2 + 2\ Ph_3P + {}^{(-)}CH(COOR)_2 \longrightarrow CH_2{=}CHCH_2CH(COOR)_2 \qquad (15.6)$$

action occurs.) This is a quite general reaction for a wide variety of η$^3$-allylpalladium-halide complexes and stabilized carbanion combinations, and has been studied extensively by Trost. With unsymmetrical η$^3$-allylpalladium complexes, the regioselectivity of attack depends on both the specific structure of the η$^3$-allyl complex and on the carbanions, although alkylation at the less-substituted end of the η$^3$-allyl system usually predominates (15.7).

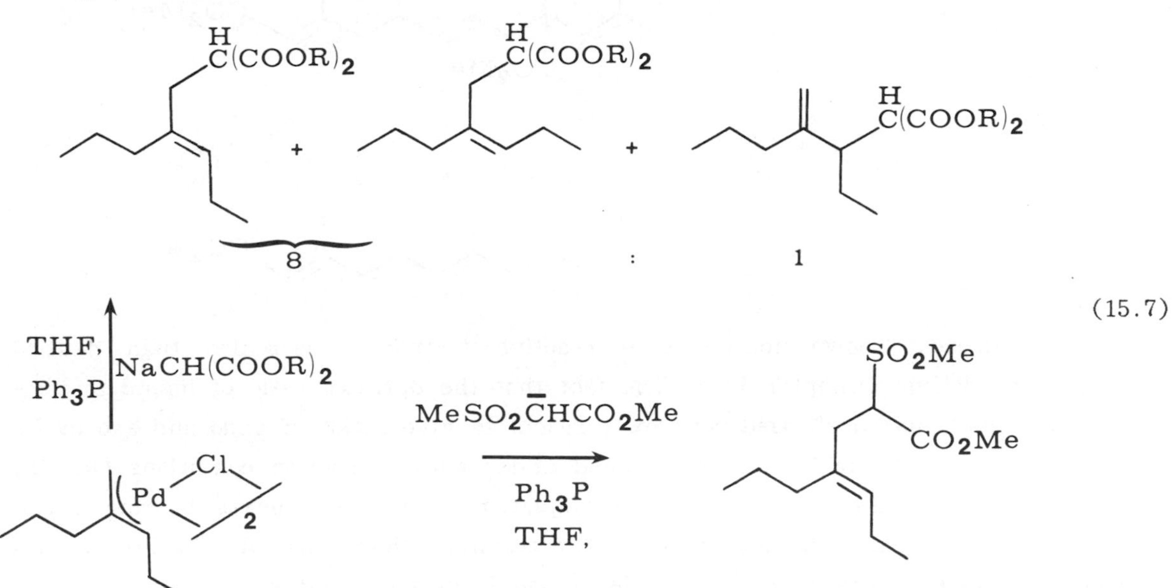


The preference for η$^3$-allylpalladium halide complexes to form with nonconjugated double bonds, and the ability of the sulfur-stabilized carbanions to attack these complexes exclusively at the less-substituted η$^3$-allyl position, provides the basis for a useful "prenylation" procedure [12] (Equation 15.8).

The mechanism of this nucleophilic attack on η$^3$-allylpalladium complexes is not clearly understood. However, several features suggest the involvement of cationic

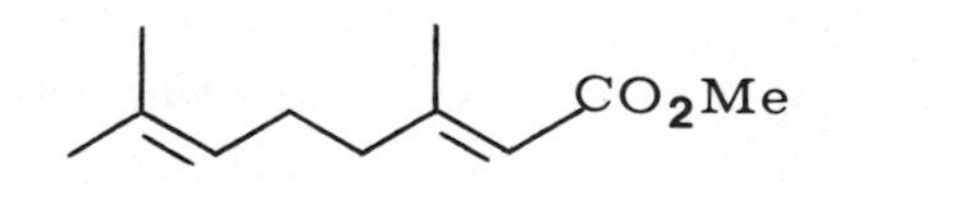


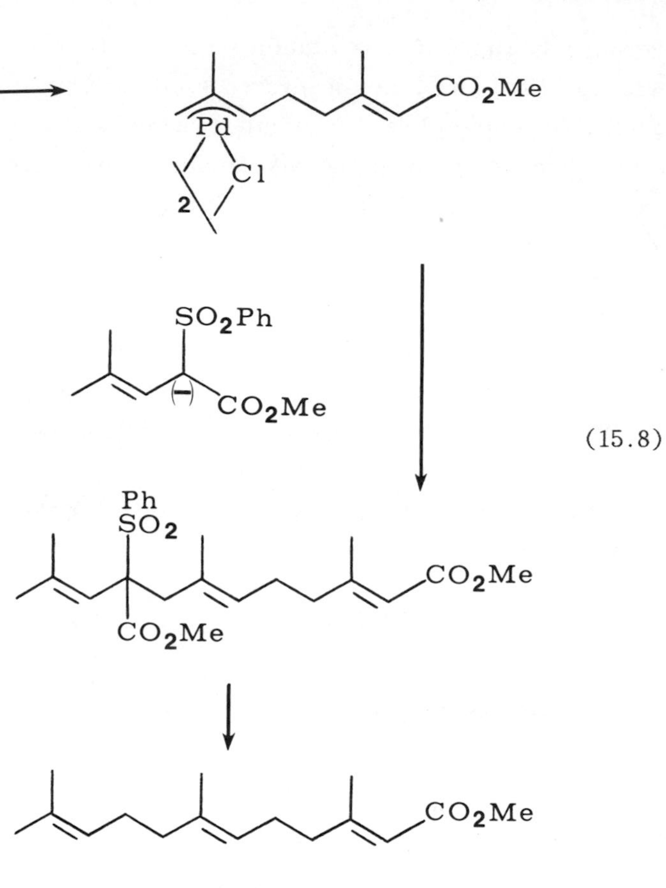


complexes, already shown to be more reactive toward nucleophiles than neutral complexes for olefins (Chapter 12). The fact that the optimum ratio of ligand to complex is 2:1, and that preformed complex 1 reacts to give ratios of endo and exo cyclic alkylation product identical to those obtained under normal reaction conditions (in situ generation) support this supposition [10] (Equation 15.9). Complexes bearing a full formal positive charge should be more electrophilic than neutral complexes, and nucleophilic attack on these cationic complexes should be increased.

The intermediacy of cationic η³-allylpalladium complexes is further supported by the studies of Åkermark. Using amines as the nucleophile [13], he examined the amination of the η³-crotylpalladium chloride complex by dimethylamine (15.10). The best yields and highest rates of reaction are observed when two equivalents of $Ph_3P$ per palladium have been added, and the reaction is run in a polar, ionizing solvent. This again implicates a cationic η³-allyl complex as the reactive species. Even more

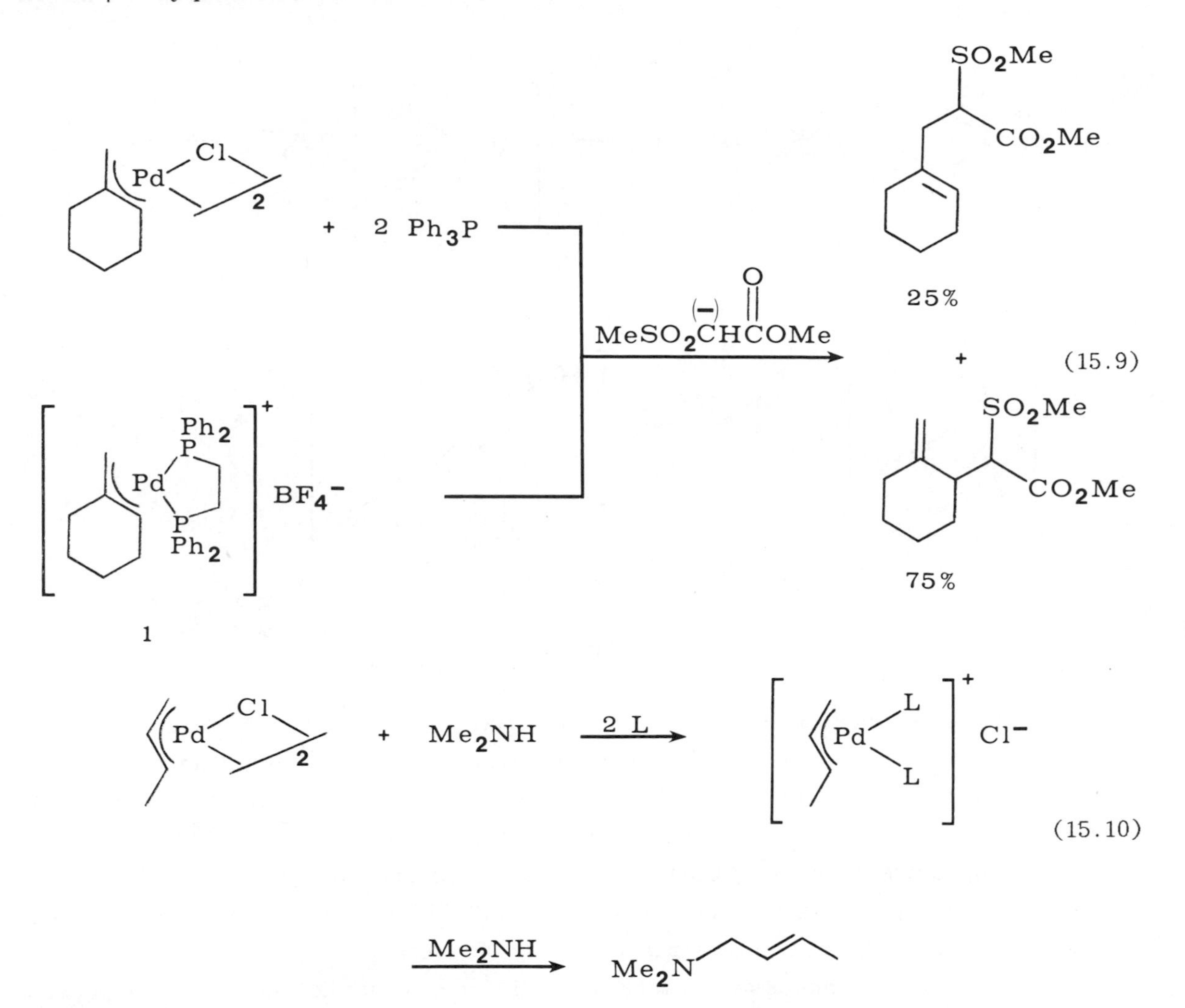


reactive is the complex resulting from treatment of η³-crotylpalladium chloride with $AgBF_4$, generating a cationic complex by removal of halogen. Addition of one equivalent of $Ph_3P$ followed by $Me_2NH$ results in 100% amination in 10 min. Use of two equivalents of phosphine slows the reaction somewhat (97% in 1.5 hr), as does use of no phosphine (33% in 2 hr). The mechanism proposed is outlined in Equation 15.11. Use of Ag(+) permits one equivalent of ligand to give a cationic complex which is highly reactive toward the amine. In the absence of Ag(+), two equivalents of phosphine are required to produce a cationic species.

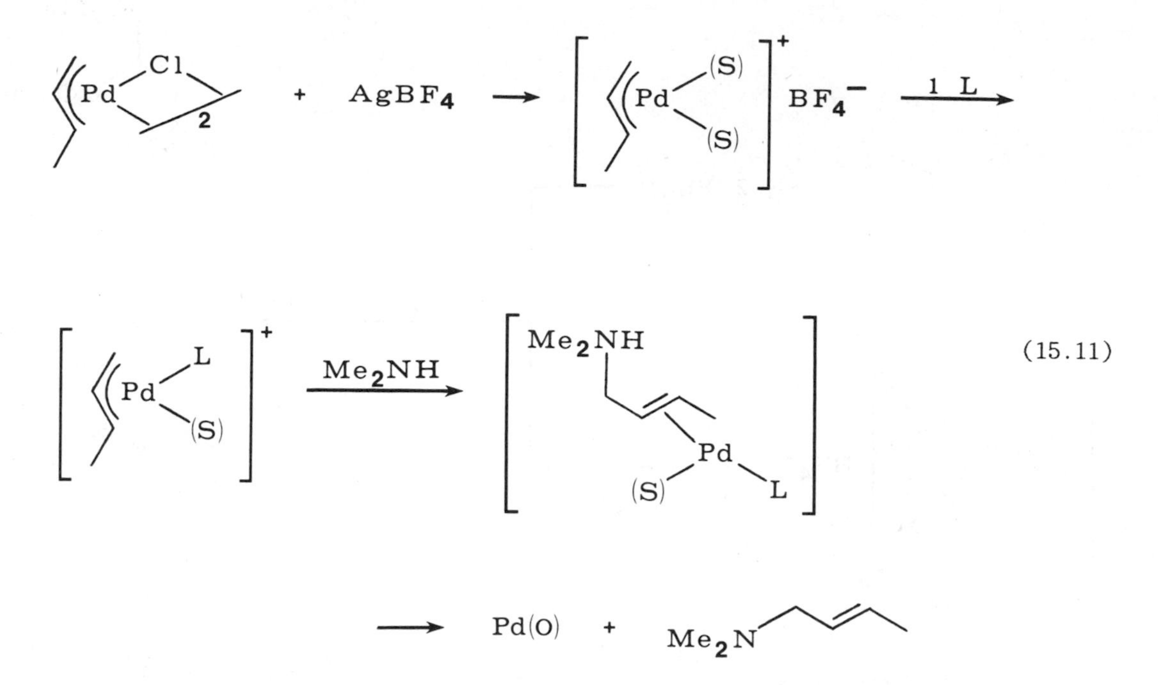


(15.11)

That external nucleophilic attack from the face opposite palladium occurs in these allylic aminations was demonstrated in a closely related system. Butadiene reacts with palladium chloride in the presence of nucleophiles to form substituted $\eta^3$-allylpalladium complexes. Reaction of cyclohexadiene with palladium chloride and dimethylamine under conditions conducive to allylic amination results in the exclusive formation of cis-1,4-(bis-dimethylamino)cyclohexane [14] (Equation 15.12). We shall shortly see evidence that trans attack of the nucleophile may not always be observed.

The phosphine ligands play another important role in allylic substitution reactions by exerting an influence on the stereochemistry of the resulting product. This was best illustrated by Trost in an alkylation using the chiral phosphine ligand (+)DIOP (Equation 15.13). Here up to 22% optical yields of alkylation product are obtained [15]. In other instances, the regiochemistry of this reaction can be effected by changing the steric bulk of the phosphines.

To date, only stabilized carbanions ($pK_a$ 10-20), enamines, and amines react cleanly in the above systems. Less-stabilized carbanions such as methyllithium and lithium dimethylcuprate, and sulfur-stabilized anions such as 2-lithiodithiane, do not result in appreciable yields of allylic alkylation [12].

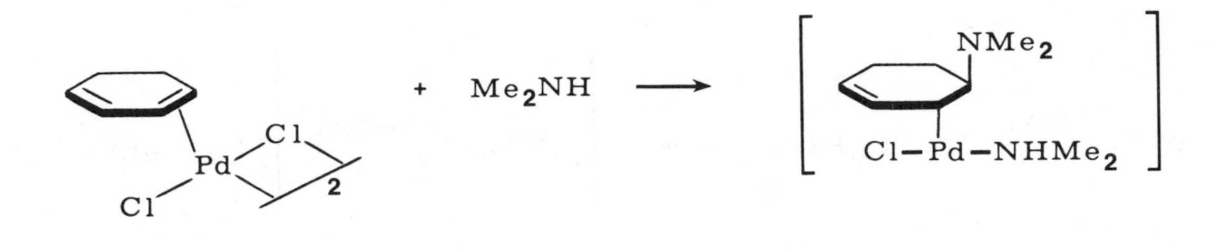


(15.12)

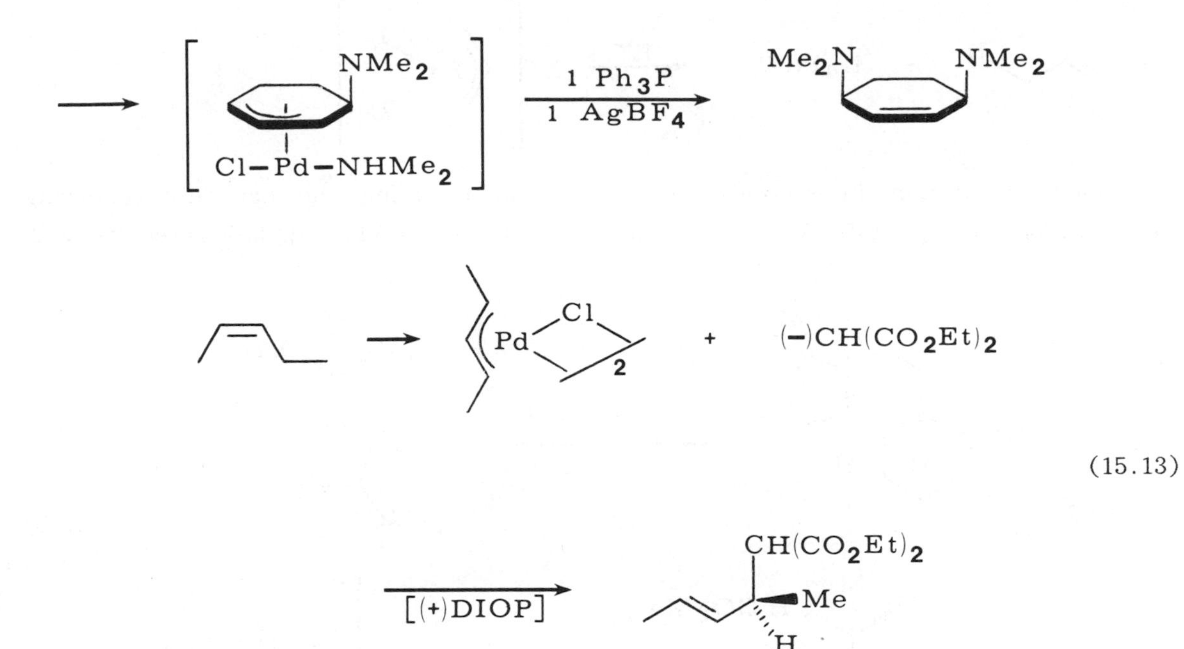


The reactions discussed so far are quite useful, for they allow direct allylic alkylation of olefins. However, they are stoichiometric in palladium, and hence rather expensive to carry out on a large scale. A related and potentially more useful system is the catalytic allylic alkylation of allyl acetates by stabilized carbanions, again developed by Trost [16]. The system is based on the observation that Pd(0) complexes will "oxidatively add" to allylic acetates to generate $\eta^3$-allylpalladium(II) complexes. Since nucleophilic attack on the $\eta^3$-allylpalladium complexes regenerates Pd(0) species, the system is catalytic (15.14). This system was originally developed as a method for

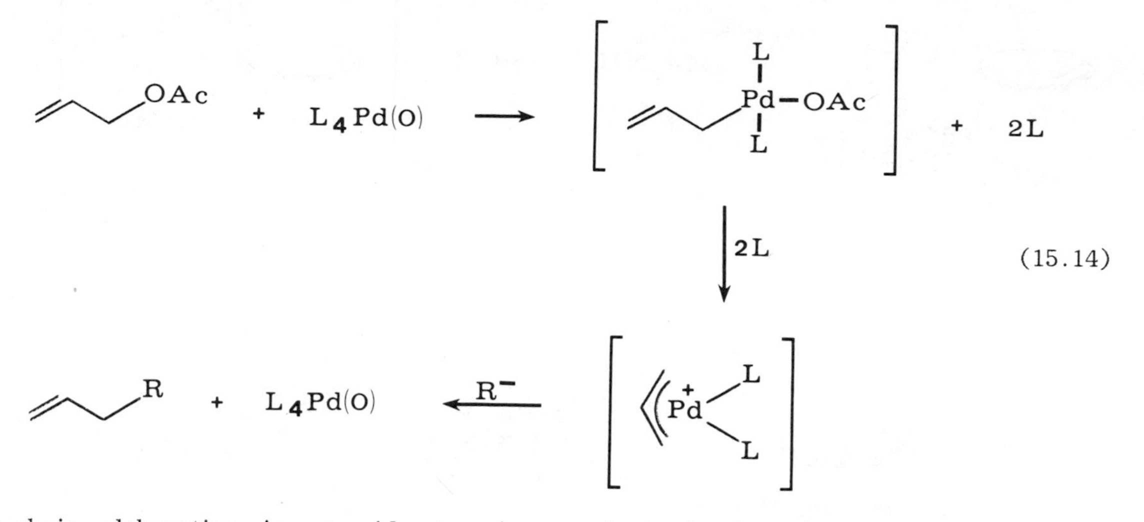


side-chain elaboration in steroids to give products having the naturally occurring configuration at $C_{20}$ (15.15). Note that this catalytic allylic alkylation results with

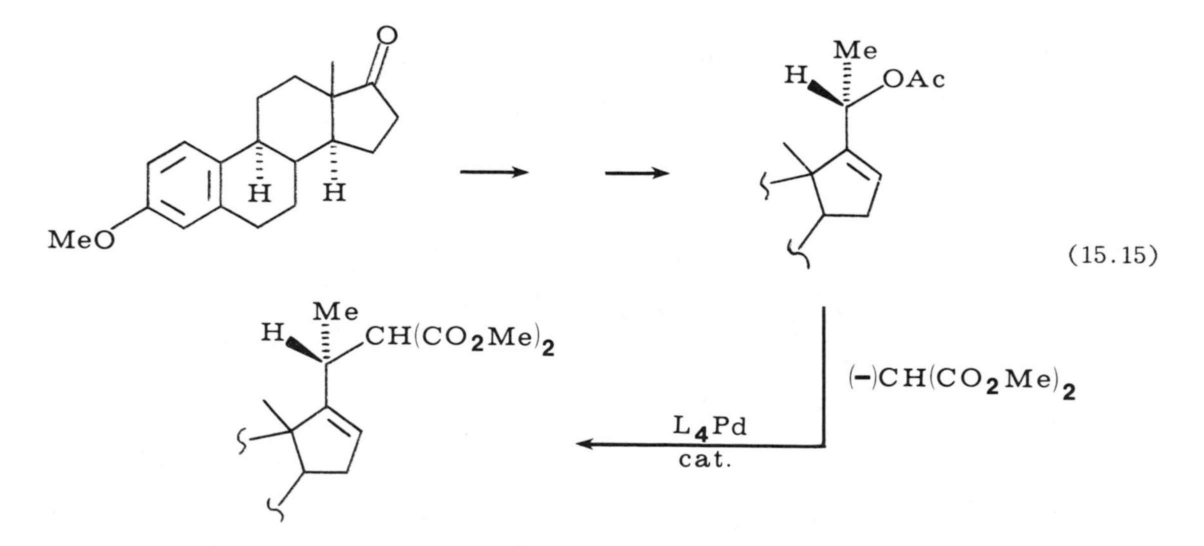


over-all net retention of configuration, since both $\eta^3$-allyl formation (oxidative addition to Pd(0)) and allylic alkylation proceed with net inversion, and two inversions result in net retention [17]. Carrying out this catalytic allylic alkylation using the chiral ligand (+)DIOP results in the production of optically active products with 20 to 37% optical yield (15.16). Note that this is a kinetic resolution of the racemic starting material.

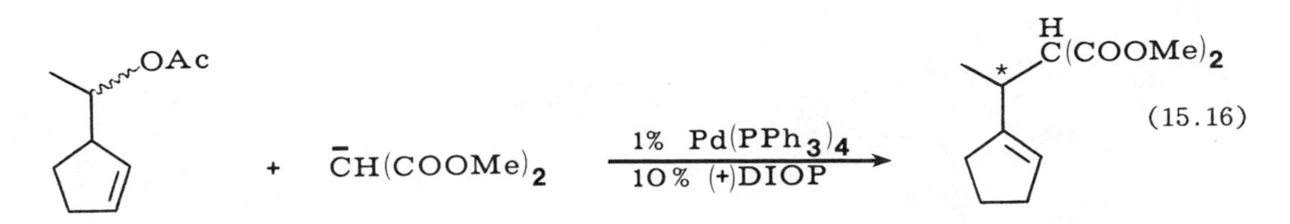


More recently, catalytic allylic alkylation has been carried out intramolecularly under high dilution and used in the synthesis of humulene by Yamamoto [18] (Equation 15.17) and the sixteen-membered macrocyclic lactone exaltolid by Trost [19] (Equation 15.18). Both of these syntheses use the Pd-catalyzed allylic alkylation of allyl acetates

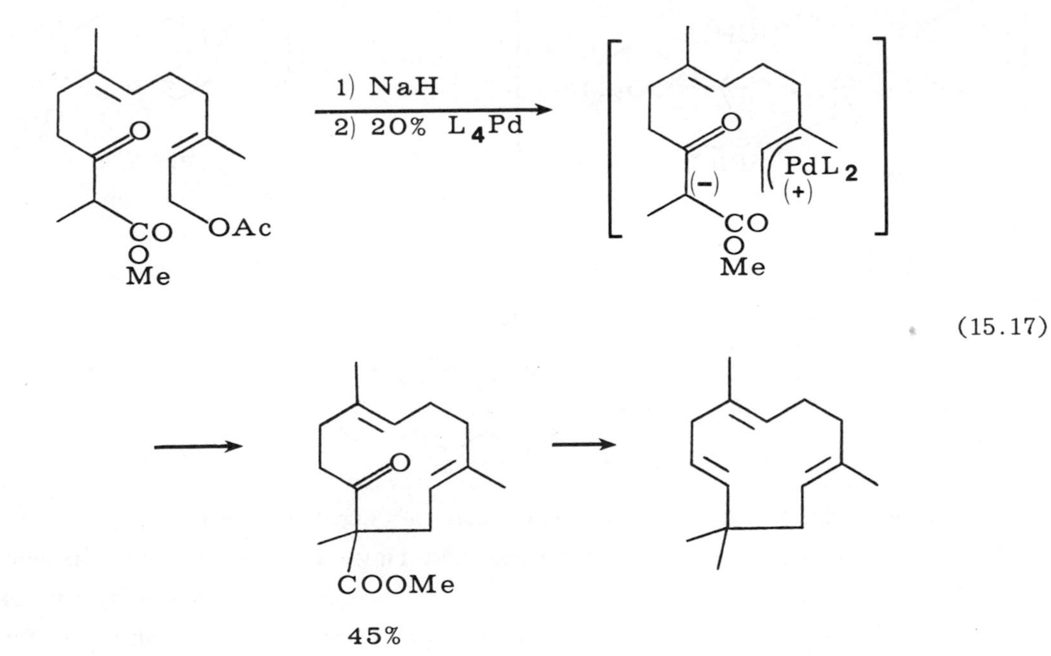


by stabilized carbanions as the key ring-forming step. Note, in the humulene synthesis (15.17), that the olefinic stereochemistry of the allyl acetate is maintained in the product (E,E). Starting with Z,E allyl acetate produces Z,E cyclic material. In both instances, attack by the carbanion occurs at the less-substituted end of the $\eta^3$-allyl complex.

With shorter-chain esters, ring closure always produces the larger ring size, in contrast to other ring closures, which tend to form the most stable ring size. Thus,

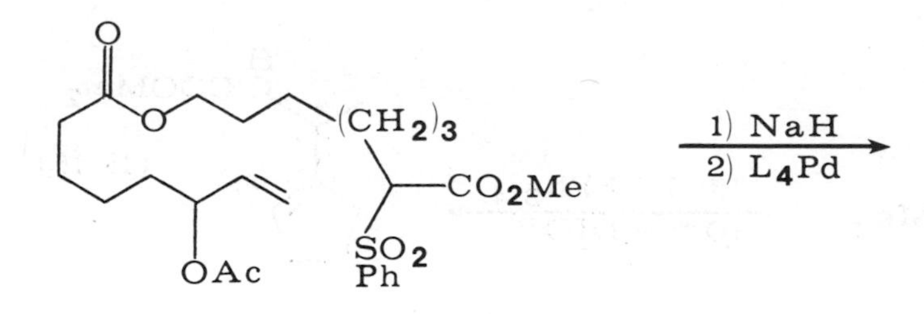


(15.18)

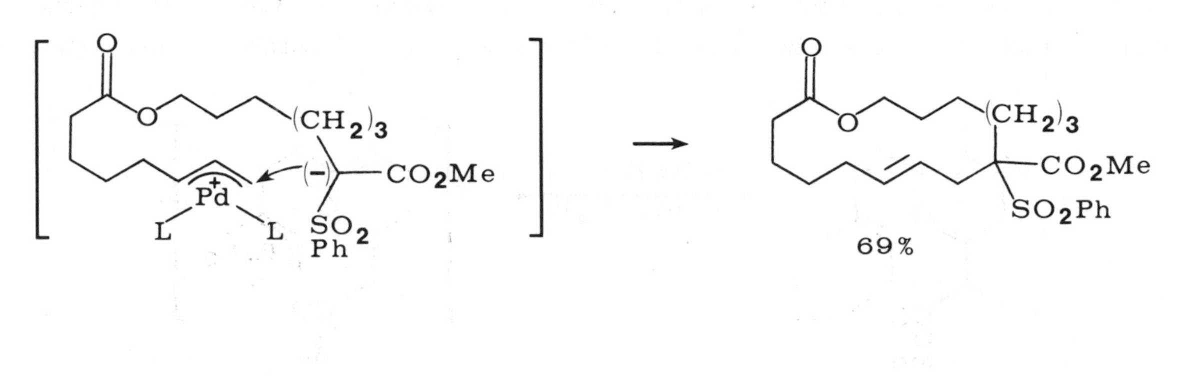


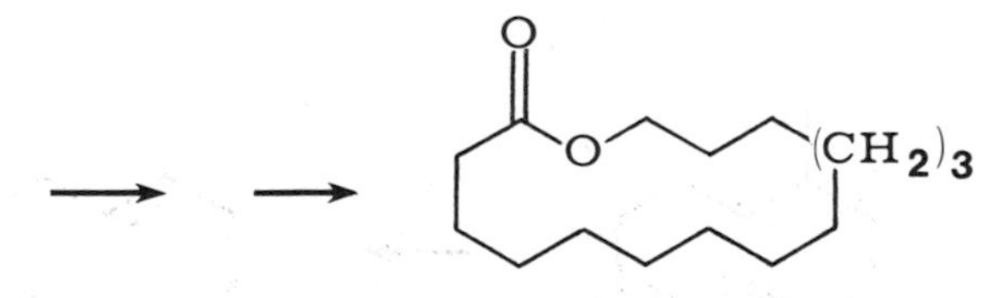


ten-membered rings form to the exclusion of eight-membered rings, nine to the exclusion of seven, and even eight-membered rings form to the virtual exclusion of six-membered rings [20] (Equation 15.19). The reasons for this selectivity are not entirely clear. In all cases studied, large-ring formation corresponds to nucleophilic attack at an unsubstituted allylic terminus, whereas small-ring formation requires the less-favored attack at a substituted allylic terminus. However, with the all-carbon analogues of the 9-7 case, seven-membered-ring product predominates; so other factors, as yet unspecified, are clearly involved.

Using the same approach, allylic acetates were catalytically aminated using Pd(0) catalysts. In this fashion the basic ring systems of the actinabolamine (15.20) ibogamine (15.21) and mesembrine (15.22) alkaloids were synthesized by Trost [21].

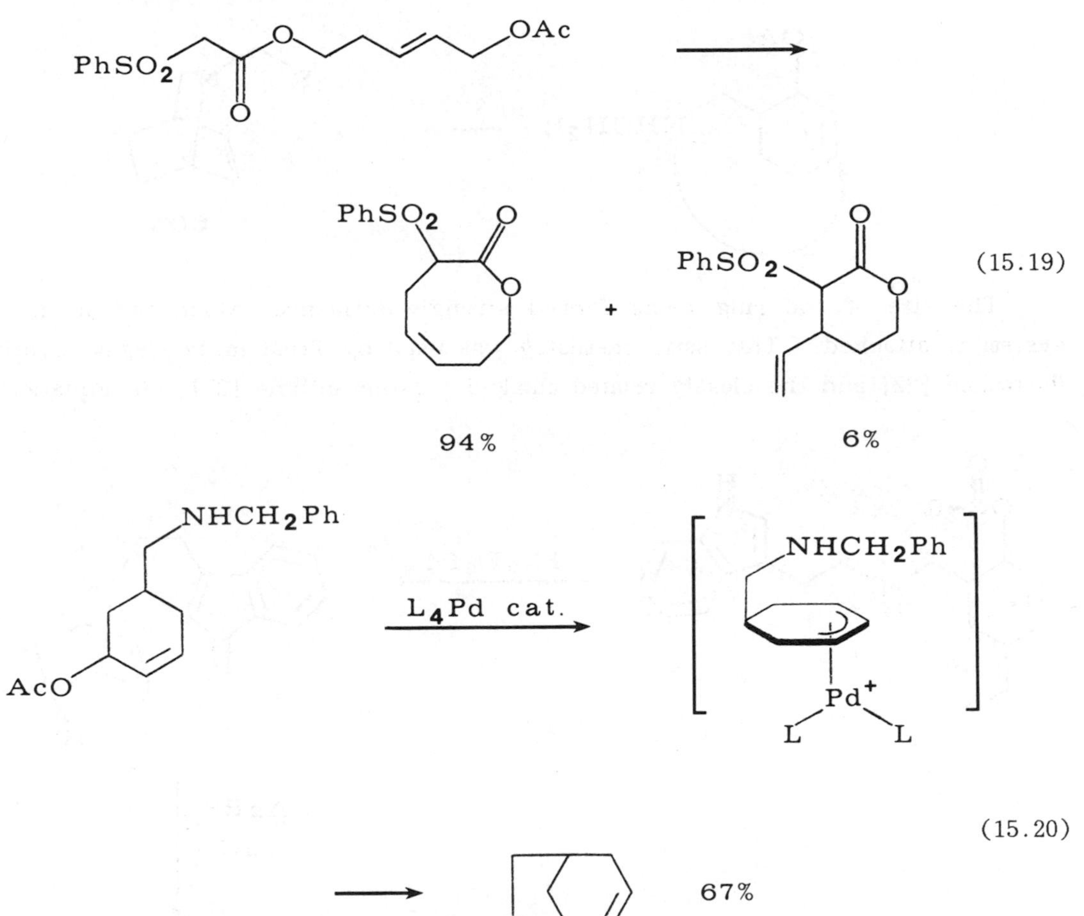
PhSO2
OAc
PhSO2
PhSO2
94%
6%
(15.19)
NHCH2Ph
AcO
L4Pd cat.
NHCH2Ph
Pd+
L
L
(15.20)
67%
N
Ph

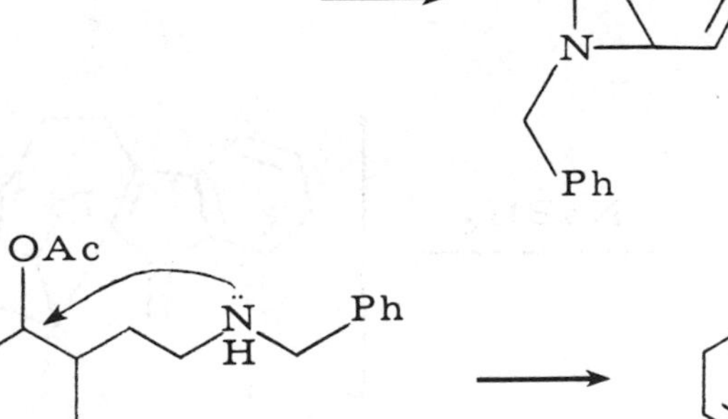
OAc
N
H
Ph

>50%

(15.21)

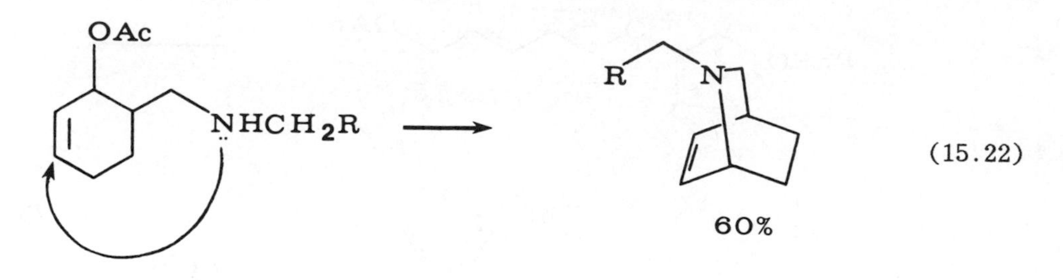


(15.22)

The size of the ring being formed strongly influences which end of the $\eta^3$-allyl system is attacked. This same chemistry was used by Trost in an elegant synthesis of ibogamine [22] and the closely related alkaloid $\pm$-catharanthine [23]. In Equation 15.23,

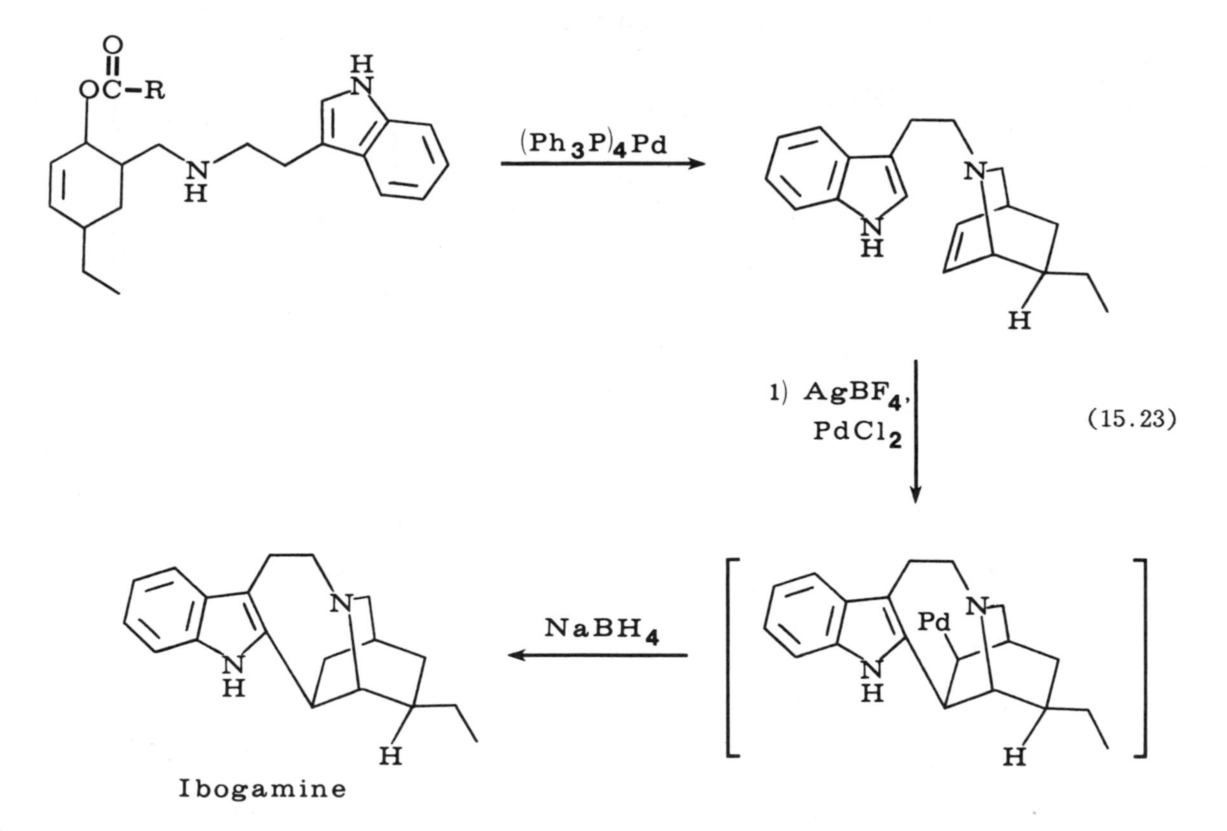


(15.23)

the bridged bicyclic amine ring is formed via catalytic allylic amination of the allyl ester, and the cyclization of the indole 2-position to the bicyclic olefin involves nucleophilic attack on a Pd-complexed olefin (Chapter 12).

One final point deserves attention before leaving this topic. The stereochemistry of nucleophilic attack on η³-allylpalladium complexes has been trans in several studies, but may not always be so. There is accumulating evidence that some nucleophiles, particularly acetate and perhaps amines, can also attack cis, perhaps by prior coordination with the metal. A cis cyclohexenylacetate undergoes a very rapid (one hour) isomerization when exposed to $L_4Pd$ [24] (Equation 15.24). One rational explanation is

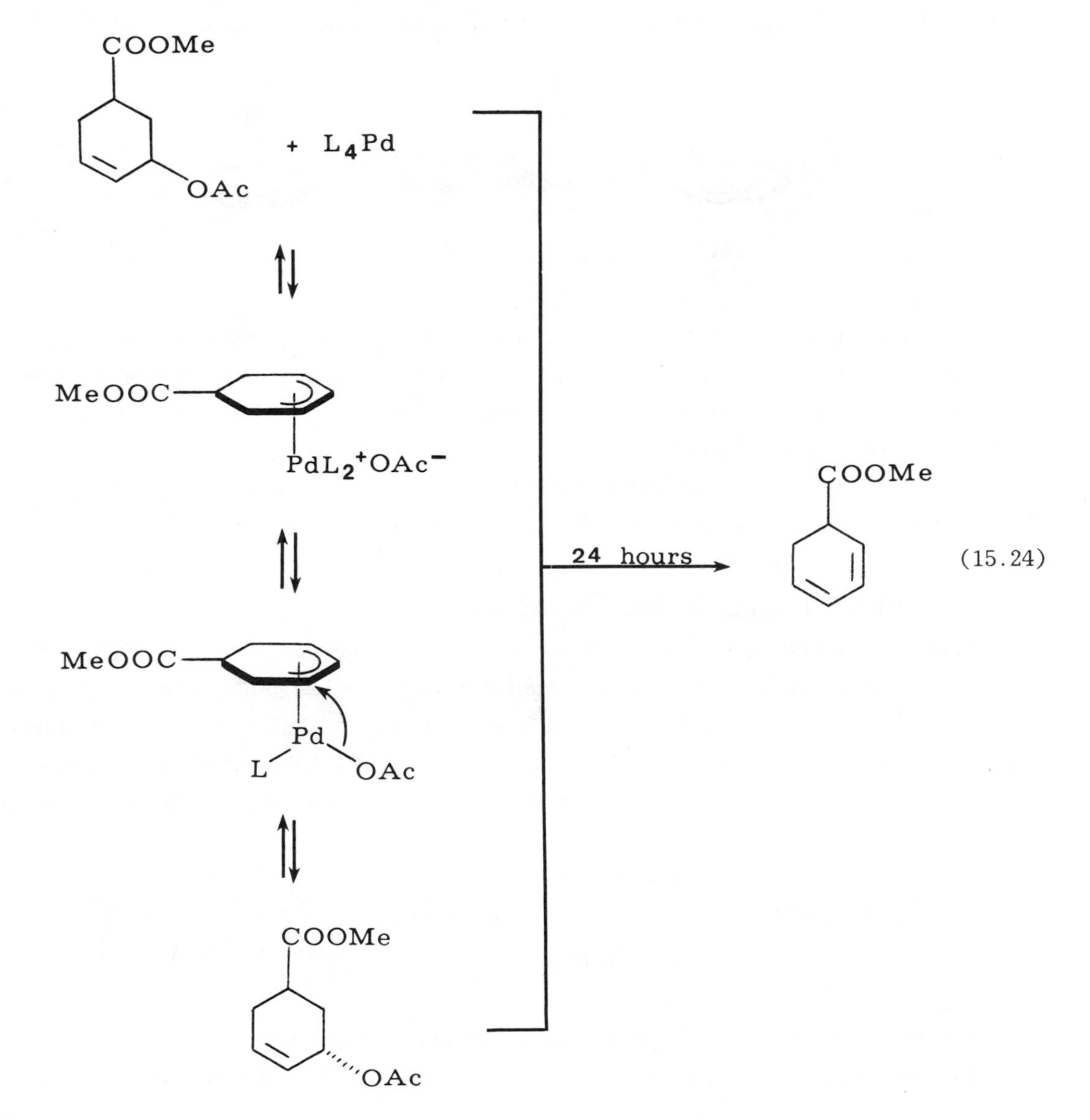

that $\eta^3$-allyl complex formation occurs, and acetate, a good ligand for palladium, coordinates to the metal and is redelivered to the $\eta^3$-allyl system from the same face as the metal. Should this indeed be the case, this duality of mechanisms for nucleophilic attack on $\eta^3$-allylpalladium complexes mirrors the same behavior that is well-known for $\eta^2$-olefinpalladium complexes. However, several other explanations are also consistent with the observations; the most likely one is that Pd is displaced from the $\eta^3$-allyl complex by trans attack by $L_4Pd$, a strong nucleophile in its own right (15.25).

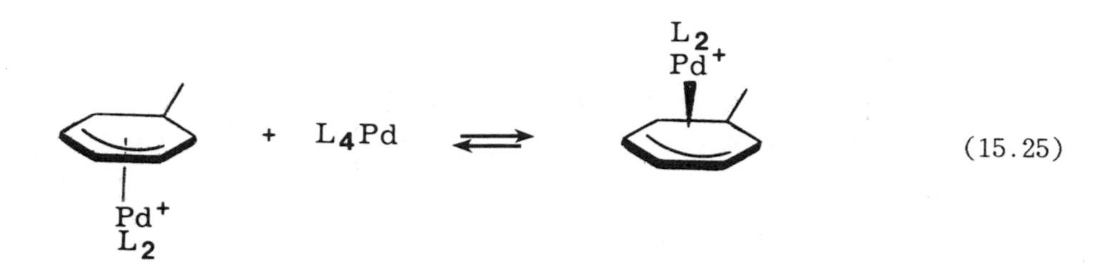


(15.25)

This metal-metal exchange is well-known in other systems involving metal nucleophiles. In this case, the resulting mixture of $\eta^3$-allylpalladium isomers undergoes trans attack by acetate to give the observed isomerization. The answer to this problem awaits experimental elucidation.

It should be noted that when this isomerization reaction is allowed to proceed for extended periods (24 hours), an effective conversion to cyclohexadienes without disproportionation is achieved.

## 15.2. $\eta^3$-Allylnickel Halide Complexes [25].

Another exceedingly useful class of $\eta^3$-allyl complexes is $\eta^3$-allylnickel halide complexes. Although they are isostructural and isoelectronic with $\eta^3$-allylpalladium halides, they have very different chemistry. $\eta^3$-Allylnickel halides are prepared, in high yield, by the reaction of allylic halides with nickel(0) complexes, usually nickel carbonyl or bis(cyclooctadiene)nickel in nonpolar solvents such as benzene (15.26).

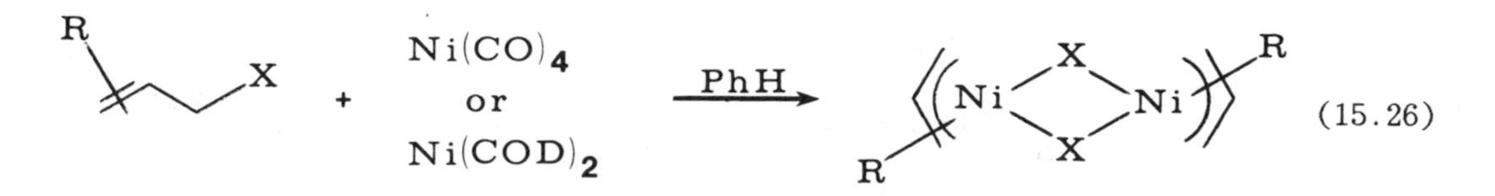


(15.26)

This reaction tolerates a wide variety of functional groups in the allyl chain, and allows the preparation of a number of synthetically useful functionalized complexes.

η³-Allylnickel halides are not directly accessible from olefins, in contrast to the corresponding palladium complexes, at least in part because nickel(II)-olefin complexes are virtually unknown, allowing no pathway for activation of the allylic position for proton removal. η³-Allylnickel halide complexes are deep red to red-brown crystalline solids, quite air-sensitive in solution, but stable in the absence of air for several years.

Although η³-allylpalladium halides are subject to nucleophilic attack, η³-allylnickel halides are not. Instead they behave, at least superficially, as if they were nucleophiles themselves, reacting with organic halides, aldehydes, and ketones, to transfer the allyl group. We shall see presently that these reactions are radical-chain processes rather than nucleophilic reactions. However, the chemistry of η³-allylnickel halides is drastically different from that of the corresponding palladium complexes.

The best-established and most widely used reaction of η³-allylnickel halide complexes is their reaction with organic halides to replace the halogen with the allyl group (15.27). This reaction, developed by Corey, proceeds only in polar, coordinating

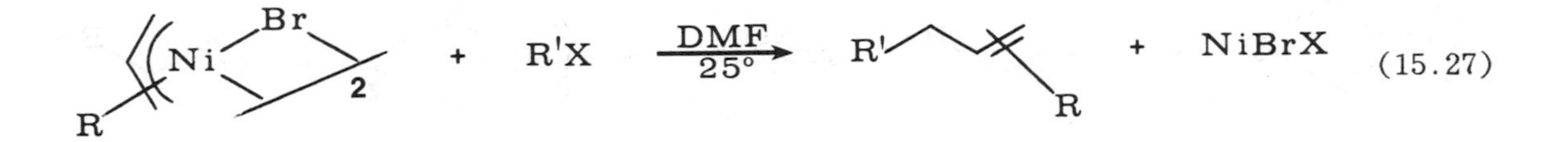

(15.27)

solvents such as DMF, HMPA, or N-methylpyrrolidone. Aryl, vinyl, primary, and secondary alkyl bromides and iodides react in high yields, with aryl and vinyl halides being considerably more reactive than the alkyl halides. Chlorides react much more slowly than bromides or iodides. This reaction tolerates a wide variety of functional groups including hydroxyl, ester, amide, and nitrile. These complexes will react with bromide in preference to chloride in the same molecule, and will tolerate ketones and aldehydes in some instances. With unsymmetrical η³-allyl groups, coupling occurs at the least-substituted terminus exclusively, in contrast to most other allyl organometallics. This property was used to an advantage in the synthesis of α-santalene, and epi-β-santalene [26] (Equation 15.28). Although allylic halides are among the most reactive toward these complexes, coupling normally results to give all possible products because of rapid exchange of η³-allyl ligand with allyl bromide (15.29). If the two allyl groups are somewhat different electronically, selective cross coupling [27,28] is sometimes observed in reasonable yield (15.30 and 15.31). Note that the 2-methoxyallylnickel complex is a convenient source for introduction of the acetonyl group ($CH_3COCH_2$-) into

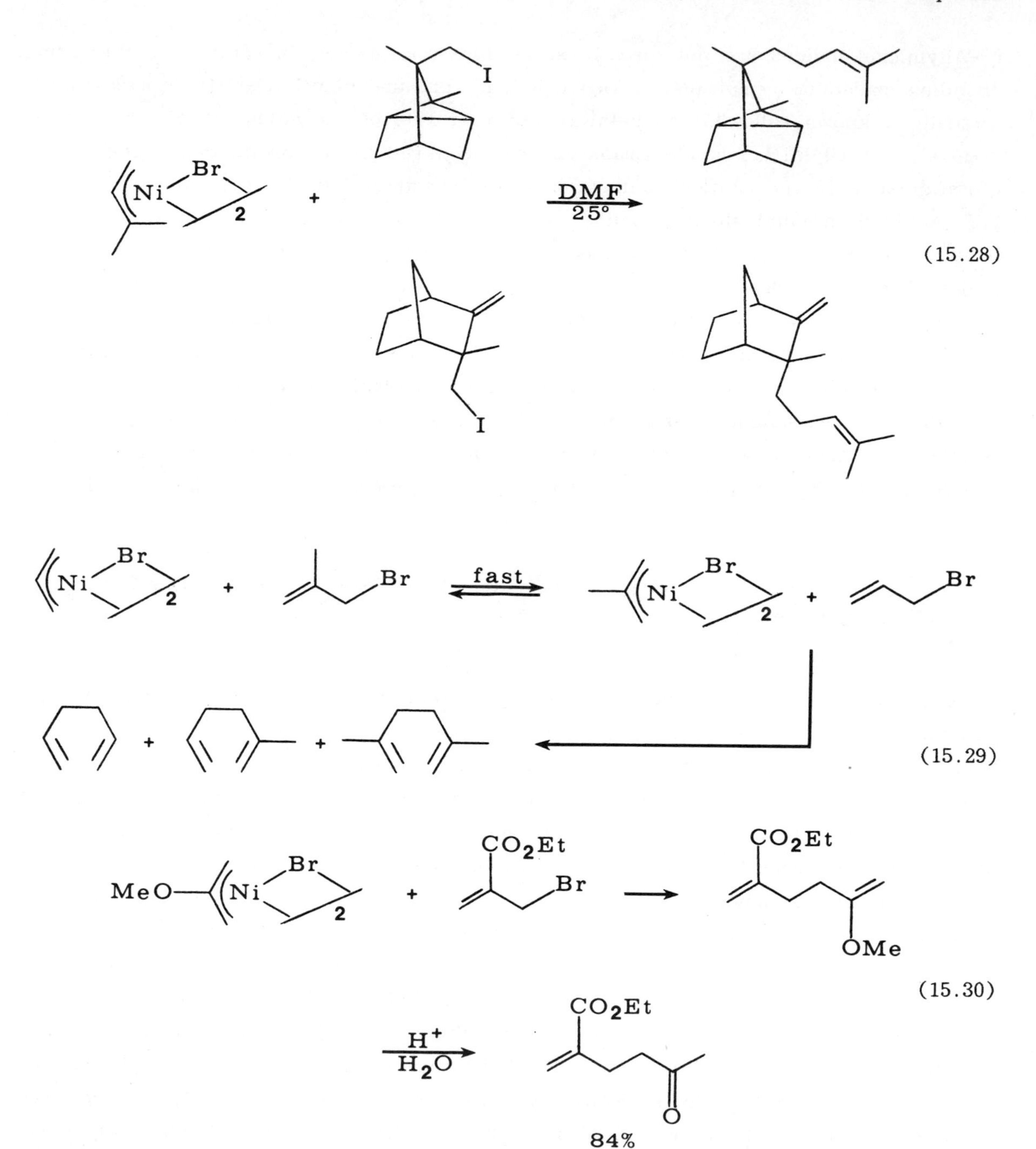
Br
Ni
2
+
I
DMF
25°
I
(15.28)
Br
Ni
2
+
Br
fast
Br
Ni
2
+
Br
+
+
(15.29)
MeO
Br
Ni
2
+
CO2Et
Br
CO2Et
OMe
(15.30)
H+
H2O
CO2Et
O
84%

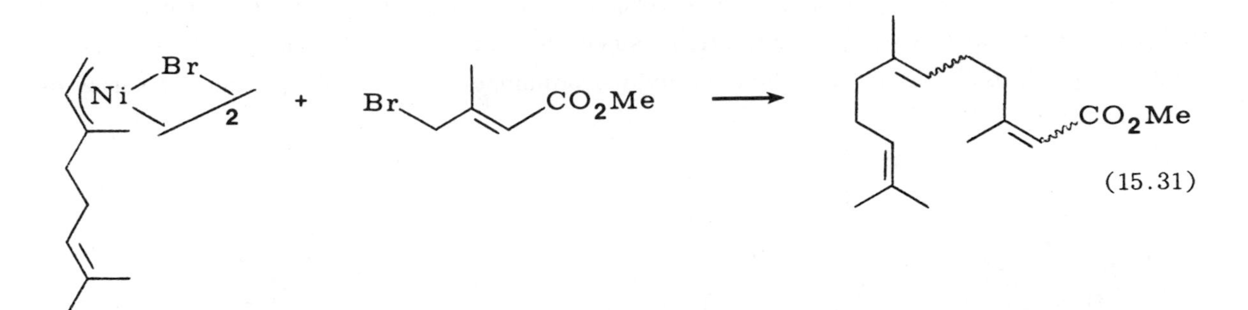


substrates (such as aryl and vinyl halides) normally unreactive to acetone enolates. Note also that the stereochemistry of the olefin in the η$^3$-allyl system is lost in these allyl transfer reactions, but the stereochemistry of the olefin in vinyl halide substrates is normally maintained.

The ability of η$^3$-allylnickel halide complexes to react with aromatic halides under very mild conditions, and to tolerate a wide range of functionality, has been used by Sato in the synthesis of a variety of isoprenoid quinones from the corresponding protected bromohydroquinones. The over-all sequence is shown in Equation 15.32 for the

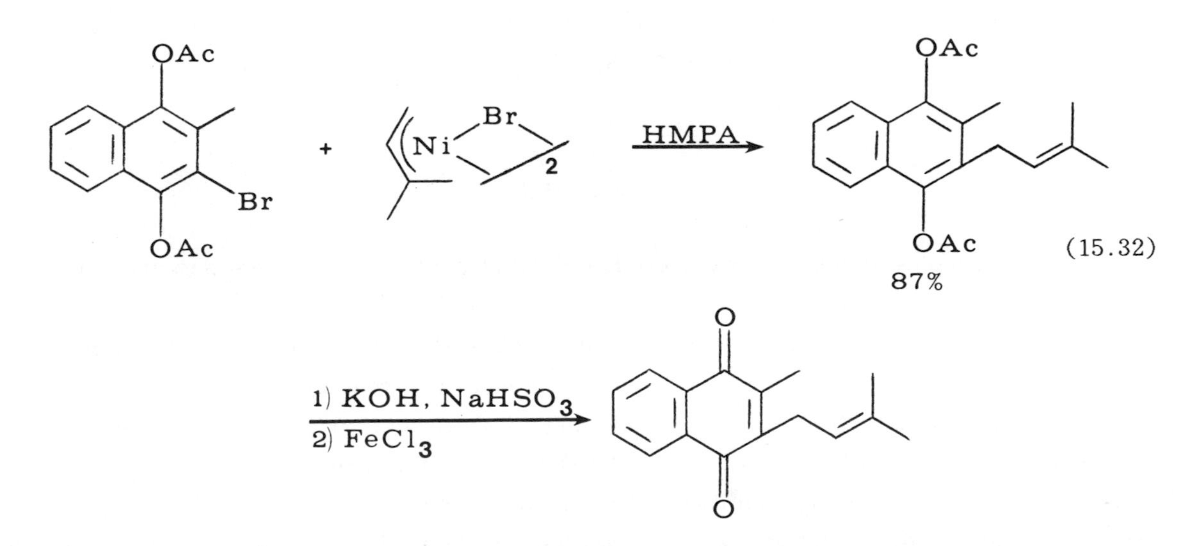


simplest vitamin K analog [29]. The ability of these nickel complexes to react with completely substituted aromatic halides under mild conditions, to couple cleanly at the less-hindered allyl position, and to be inert to the acetate protecting group make this quite an attractive approach. For the longer-chain vitamin Ks' (2 and 3), the requisite π-solanesyl- and π-phytylnickel halide complexes are prepared from the corresponding bromides and used without purification, since they are oils. A similar procedure applied to protected (bromo)dimethoxymethylbenzoquinone was used to produce a series of coenzyme Qs [30].

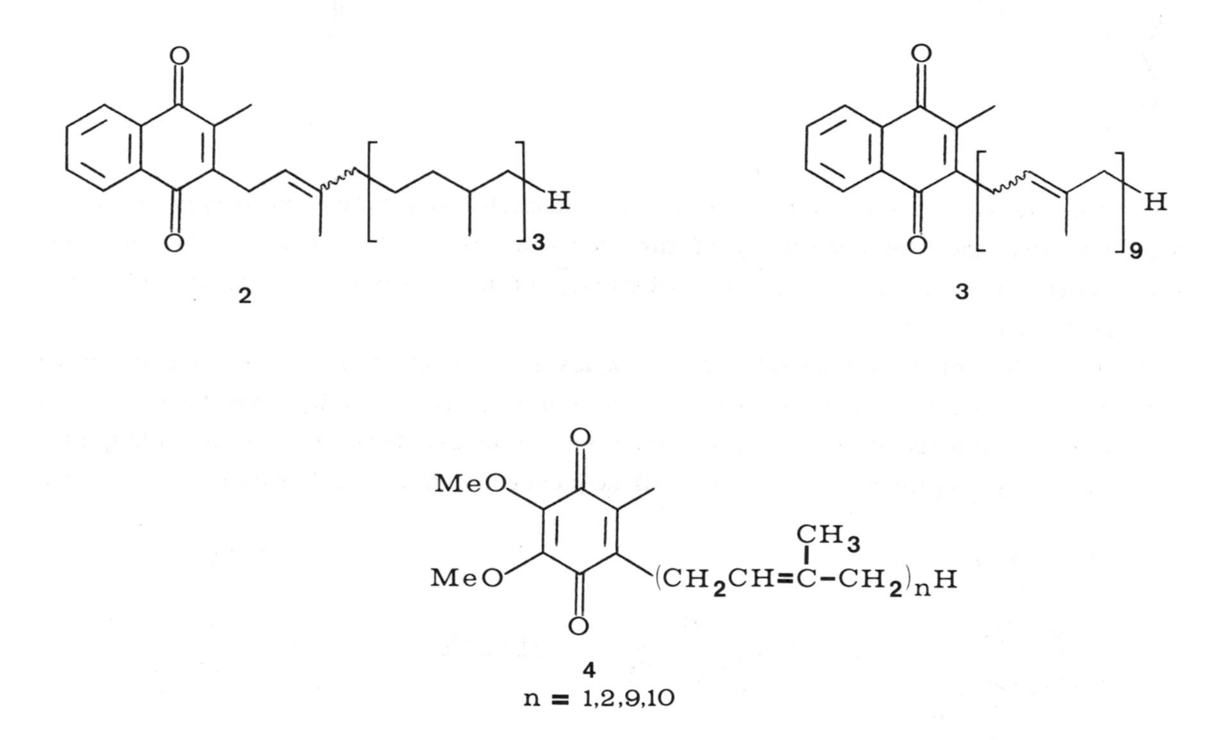


Other examples developed by Hegedus [27,31] involving aryl halides are shown in Equations 15.33 to 15.35.

Since $\eta^3$-allylnickel halides react equally well with aryl, vinyl, and alkyl halides, the reaction is clearly not an $S_N2$ process. Instead, the reaction is thought to proceed by a radical process, based on the following observations by Hegedus. Reaction of $\eta^3$-(2-methoxyallyl)nickel bromide with S-(+)-2-iodooctane produces completely racemic 4-methyl-2-decanone under conditions for which neither starting material nor product are racemized. This reaction, as well as the reaction of $\eta^3$-2-methallylnickel bromide with 2-iodooctane, iodobenzene, and β-bromostyrene, are completely inhibited by addition of 1 mole % of m-dinitrobenzene, a potent radical-anion scavenger. Even with the

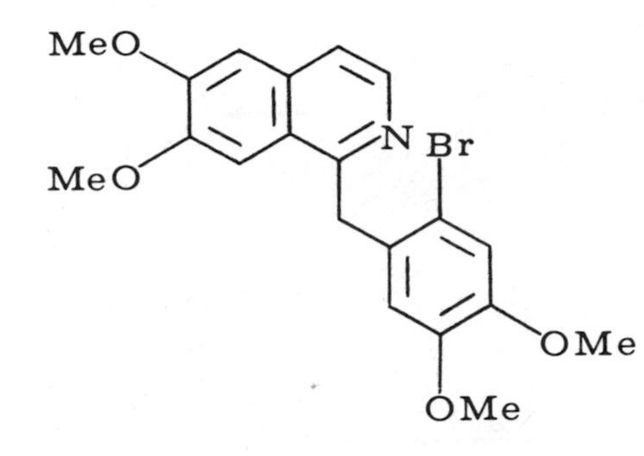


+

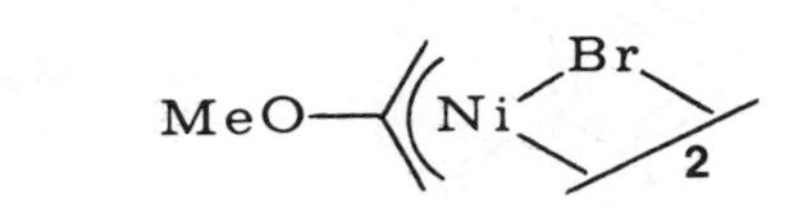


(15.33)

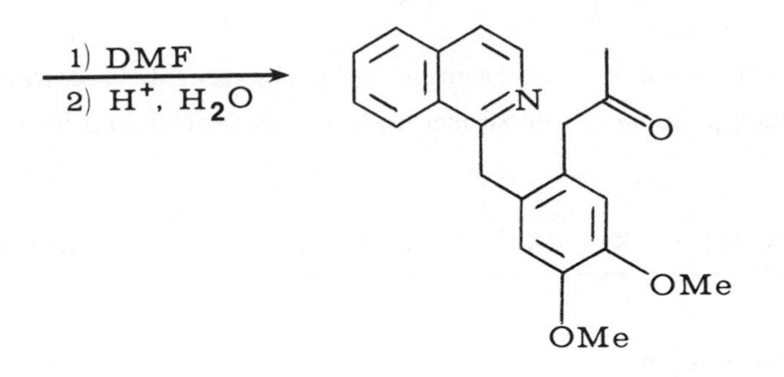


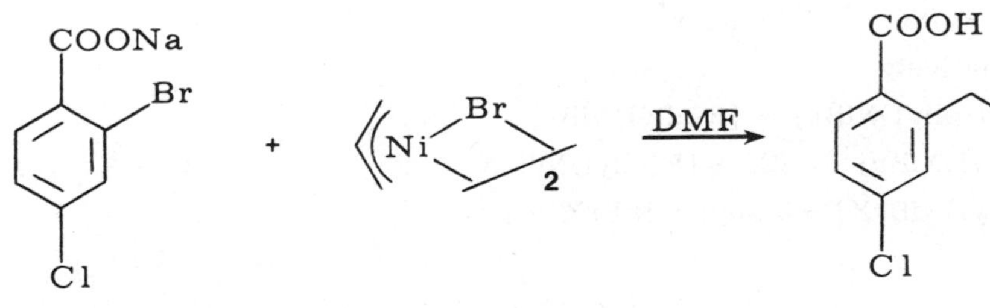


(15.34)

Pd(II)

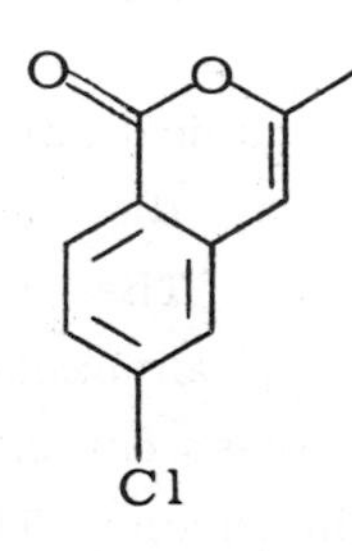

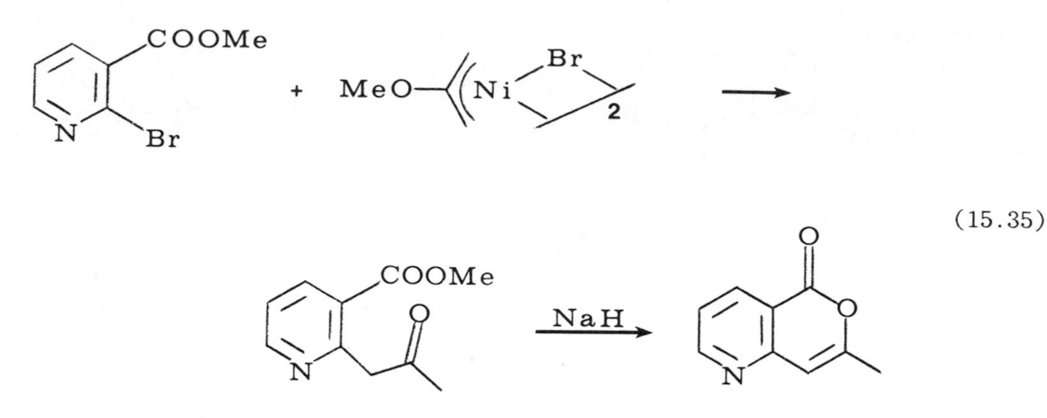


(15.35)

highly reactive allyl bromide, the reaction goes considerably slower in the presence of this inhibitor. The proposed mechanism [32] is presented in Figure 15.7, and involves

$[(\text{allyl})\text{NiBr}] + \text{RX} \rightarrow \text{RX}^{\bullet -} + [(\text{allyl})\text{NiBr}]^{\bullet +}$     a initiation

---

$\text{RX}^{\bullet -} \rightarrow \text{R}\cdot + \text{X}^-$     b

$\text{R}\cdot + [(\text{allyl})\text{NiBr}] \rightarrow \text{Rallyl} + \text{NiBr}\cdot$     c

$\text{NiBr}\cdot + \text{RX} \rightarrow \text{RX}^{\bullet -} + \text{NiBr}^+$     d

Alternatively

$\text{R}\cdot + [(\text{allyl})\text{NiBr}] \rightarrow [\text{R}(\text{allyl})\text{NiBr}]^{\bullet}$     c

$[\text{R}(\text{allyl})\text{NiBr}]^{\bullet} + \text{RX} \rightarrow [\text{R}(\text{allyl})\text{NiBrX}] + \text{R}\cdot$     d

$[\text{R}(\text{allyl})\text{NiBrX}] \rightarrow \text{Rallyl} + \text{NiBrX}$     e

*Figure 15.7. Radical-chain mechanism for $\eta^3$-allylnickel halide reactions.*

an initiation step to generate the radical anion of the organic halide. As shown, a one-electron reduction of the halide by the nickel complex has occurred, although other initiation steps are quite as likely. The organic halide radical anion loses halide, and the resulting radical attacks the $\eta^3$-allylnickel complex to give allylation and a nickel(I) species. This nickel(I) species transfers one electron to substrate to generate the required radical anion to carry the chain. The reaction of the organic radical with the $\eta^3$-allylnickel halide complex (step c) may actually involve a radical chain-type oxidative-addition reductive-elimination sequence as discussed in Chapter 4. Since this is a chain reaction, small amounts of dinitrobenzene are sufficient to completely inhibit it, by interception of a chain carrying intermediate. The intermediacy of alkyl radical species is consistent with the complete racemization of the S-(+)-2-iodooctane upon reaction. Surprisingly, although the reaction with vinyl halides is completely inhibited by dinitrobenzene, the stereochemistry of the double bond is maintained, indicating, at least for vinyl systems, that whatever radicals are involved, they are not entirely free of the influence of the metal. (Vinyl free radicals isomerize very rapidly.) This reaction is one of a growing number of radical chain processes being uncovered in organometallic chemistry.

Although η³-allylnickel halides react with halides in preference to aldehydes or ketones, under slightly more vigorous conditions (50°C vs 20°C) reaction 15.36 to

(15.36)

produce homoallylic alcohols results [26,33]. α-Diketones are the most reactive substrates producing α-ketohomoallylic alcohols. Aldehydes and cyclic ketones, including cholestanone, progesterone, and 5-α-androstane-3,17-dione, react well (the steroids at the most reactive carbonyl group), but simple aliphatic and α,β-unsaturated ketones react only sluggishly. Again, reaction occurs at the less-substituted end of the allyl group, in contrast to other allyl organometallics. The reaction of η³-(2-carboethoxyallyl)nickel bromide with aldehydes and ketones produces spiro-α-methylene-γ-butyrolactones (15.37).

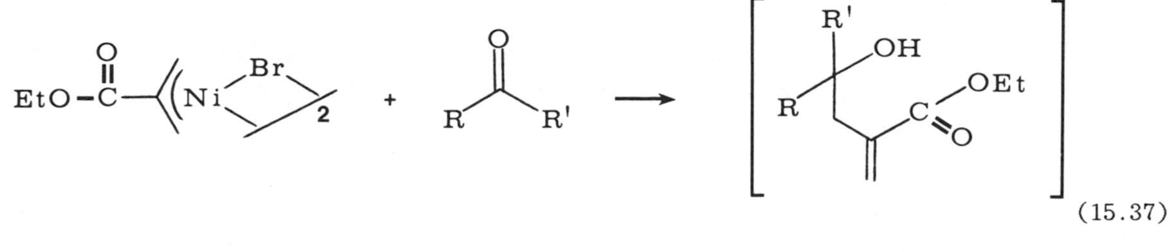


(15.37)

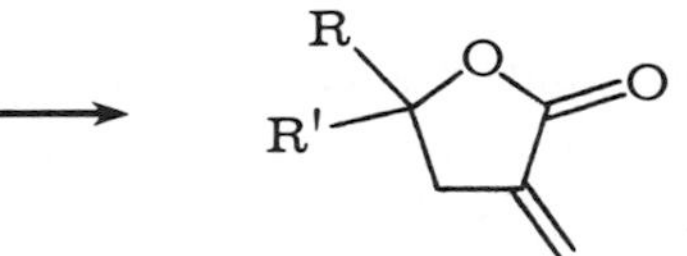

Hegedus [34] showed that quinones also react with $\eta^3$-allylnickel halides to produce allylhydroquinones (15.38). This is a one-step, direct introduction of the allyl side

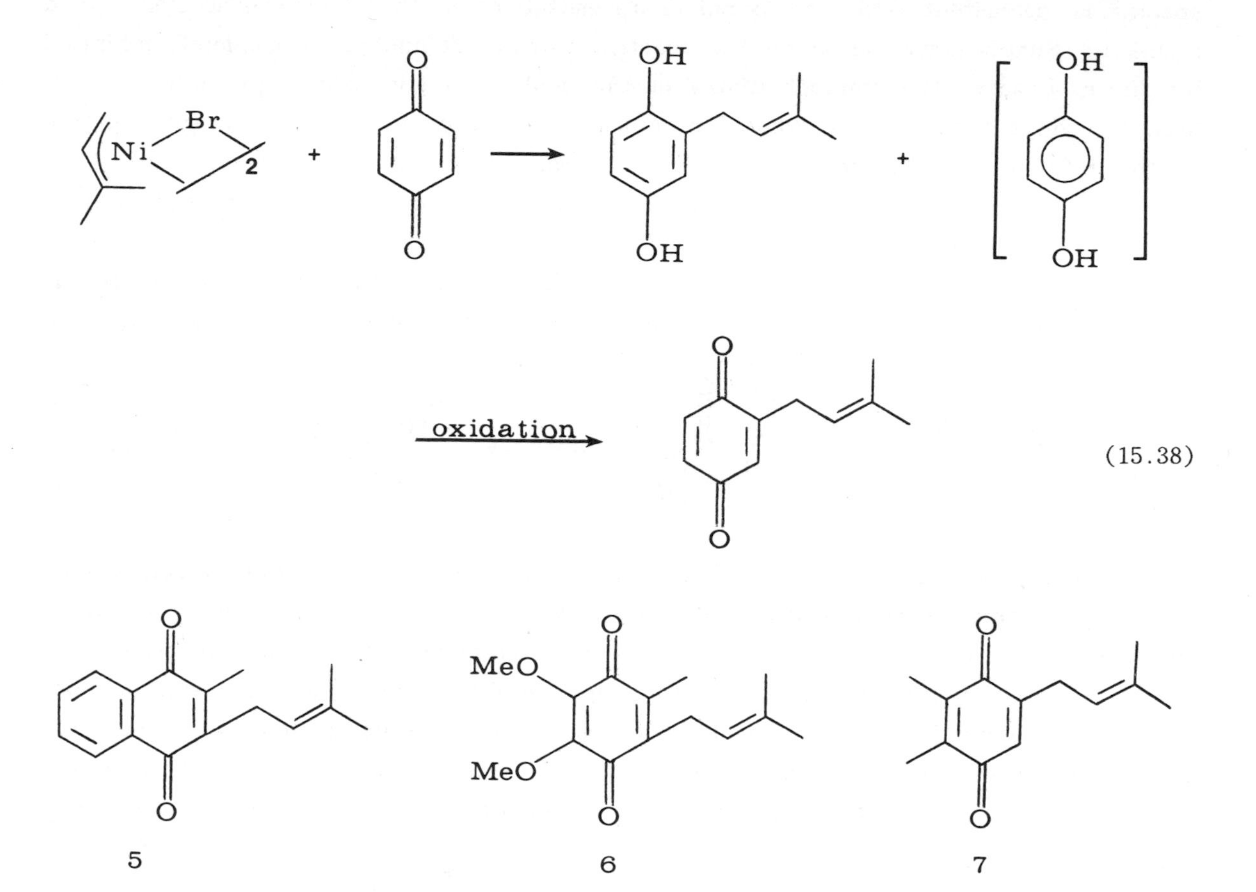


chain into quinones, and allows the synthesis of isoprenoid quinones such as vitamin $K_1$ (5), coenzyme $Q_1$ (6), and plastoquinone (7) directly from the quinone. Although the yields are only moderate (30 to 50%), the major side product is hydroquinone, which can be reused. With methyl-substituted quinones, the major site of allylation corresponds to the ring site of highest spin density in the corresponding quinone radical anion, and an electron-transfer mechanism for allylation was originally proposed. However, subsequent studies showed that the reaction proceeds through quinol intermediates [35]. With unsymmetrical quinones, the particular quinol formed, as well as the mode of rearrangement of the quinol (i.e., [3,3], [1,3], [1,2]) can be controlled, to some extent, by reaction and isolation conditions (15.39).

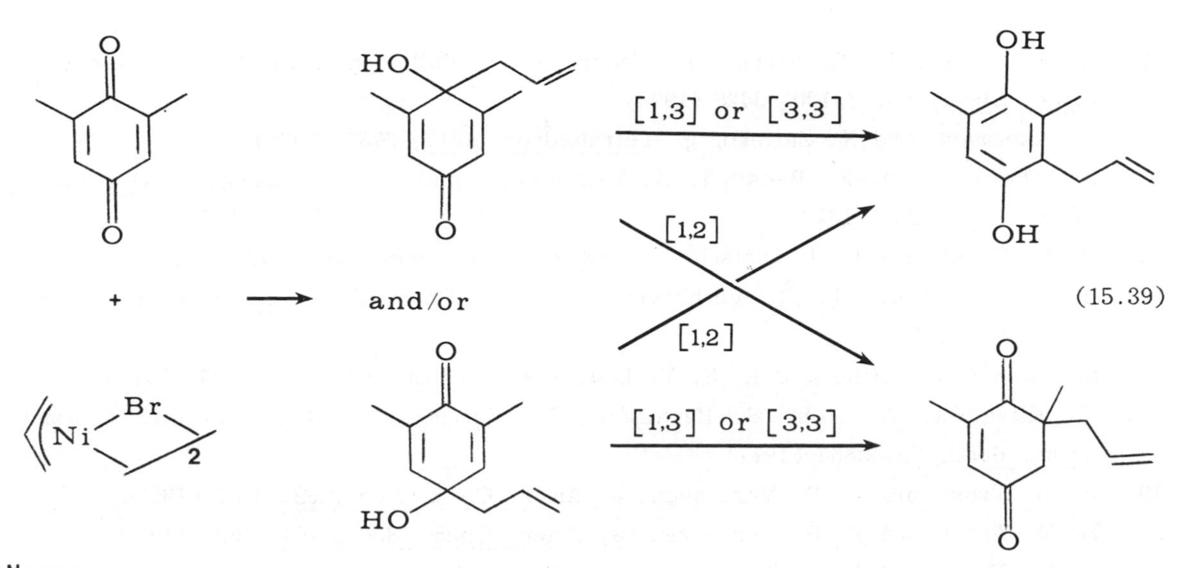


## NOTES.


1. G. Wilke, B. Bogdanovic, P. Hardt, P. Heimbach, W. Keim, M. Kröner, W. Oberkirch, K. Tanaka, E. Steinrucke, D. Walter, and H. Zimmerman, Angew. Chem. Int. Ed. Engl., 5, 151 (1966).
2. For a thorough discussion of this and related reactions, see P. W. Jolly and G. Wilke, The Organic Chemistry of Nickel, (Academic Press, 2 vols., 1974).
3. R. Baker, Chem. Rev., 73, 487 (1973).
4. F. Brille, P. Heimbach, J. Kluth, and H. Schenkluhn, Angew. Chem. Int. Ed. Engl., 18, 400 (1979).
5. S. Takahashi, T. Shibano, and N. Hagihara, Bull. Chem. Soc. Japan, 41, 454 (1968).
6. B. Akermark, G. Akermark, C. Moberg, C. Bjorklund, and K. Siirala-Hansen, J. Organomet. Chem., 164, 97 (1979).
7. J. Tsuji, K. Masaoka, and T. Takahashi, Tetrahedron Lett., 2267 (1977).
8. J.-P. Haudegond, Y. Chauvin, and D. Commereuc, J. Org. Chem., 44, 3063 (1979).
9. B. M. Trost, P. E. Strege, L. Weber, T. J. Fullerton, and T. J. Dietsche, J. Amer. Chem. Soc., 100, 3407 (1978).
10. B. M. Trost, L. Weber, P. E. Strege, T. J. Fullerton, and T. J. Dietsche, J. Amer. Chem. Soc., 100, 3416 (1978).
11. J. Tsuji, H. Takahashi, and M. Morikawa, Tetrahedron Lett., 4387 (1965); J. Tsuji, Bull. Chem. Soc. Japan, 46, 1896 (1973).

12. B. M. Trost, P. E. Strege, L. Weber, T. J. Fullerton, and T. J. Dietsche, J. Amer. Chem. Soc., 100, 3426 (1978).
13. B. Akermark and K. Zetterberg, Tetrahedron Lett., 3733 (1975).
14. B. Akermark, J.-E. Backvall, A. Lowenborg, and K. Zetterberg, J. Organomet. Chem., 166, C33 (1979).
15. B. M. Trost and T. J. Dietsche, J. Amer. Chem. Soc., 95, 8200 (1973).
16. B. M. Trost and T. R. Verhoeven, J. Amer. Chem. Soc., 98, 630 (1976), and 100, 3435 (1978).
17. See also J. K. Stille and K. S. Y. Lau, Accts. Chem. Res., 10, 434 (1977).
18. Y. Kitagawa, A. Itoh, S. Hashimoto, H. Yamamoto, and H. Nozaki, J. Amer. Chem. Soc., 99, 3864 (1977).
19. B. M. Trost and T. R. Verhoeven, J. Amer. Chem. Soc., 99, 3867 (1977).
20. B. M. Trost and T. R. Verhoeven, J. Amer. Chem. Soc., 101, 1595 (1979).
21. B. M. Trost and J. P. Genet, J. Amer. Chem. Soc., 98, 8516 (1976).
22. B. M. Trost, S. A. Godleski, and J. P. Genet, J. Amer. Chem. Soc., 100, 3930 (1978).
23. B. M. Trost, S. A. Godleski, and J. L. Belletire, J. Org. Chem., 44, 2052 (1979).
24. B. M. Trost, T. R. Verhoeven, and J. M. Fortunak, Tetrahedron Lett., 2301 (1979).
25. For reviews, see M. F. Semmelhack, Org. React., 19, 115 (1972); R. Baker, Chem. Rev., 73, 487 (1973); Jolly and Wilke (see Reference 2); and L. S. Hegedus, J. Organomet. Chem. Lib., 1, 329 (1976).
26. E. J. Corey and M. F. Semmelhack, J. Amer. Chem. Soc., 89, 2755 (1967).
27. L. S. Hegedus and R. K. Stiverson, J. Amer. Chem. Soc., 96, 3250 (1974).
28. F. Guerrieri, G. P. Chiusoli, and S. Merzoni, Gazz. Chem. Ital., 104, 557 (1974).
29. K. Saito, S. Inoue, and K. Sato, J.C.S. Chem. Comm., 953 (1972); J.C.S. Perkin I, 2289 (1973).
30. S. Inoue, R. Yamaguchi, K. Saito, and K. Sato, Bull. Chem. Soc. Japan, 47, 3098 (1974); K. Sato, S. Inoue, and R. Yamaguchi, J. Org. Chem., 37, 1889 (1972).
31. L. S. Hegedus, D. E. Korte, and R. K. Wirth, J. Org. Chem., 42, 1329 (1977).
32. L. S. Hegedus and L. L. Miller, J. Amer. Chem. Soc., 97, 459 (1975).
33. L. S. Hegedus, S. D. Wagner, E. L. Waterman, and K. Siirala-Hansen, J. Org. Chem., 40, 593 (1975).

34. L. S. Hegedus, B. R. Evans, D. E. Korte, E. L. Waterman, and K. Sjoberg, J. Amer. Chem. Soc., 98, 3901 (1976).
35. L. S. Hegedus and B. R. Evans, J. Amer. Chem. Soc., 100, 3461 (1978).

# Index

D

I



L

S



T



U



V